FRONTIER PROBLEMS IN
QUANTUM
MECHANICS

FRONTIER PROBLEMS IN
QUANTUM
MECHANICS

Lee Chang
Tsinghua University, China

Molin Ge
Nankai University, China

清華大學出版社
TSINGHUA UNIVERSITY PRESS

World Scientific

NEW JERSEY · LONDON · SINGAPORE · BEIJING · SHANGHAI · HONG KONG · TAIPEI · CHENNAI · TOKYO

Published by

World Scientific Publishing Co. Pte. Ltd.

5 Toh Tuck Link, Singapore 596224

USA office: 27 Warren Street, Suite 401-402, Hackensack, NJ 07601

UK office: 57 Shelton Street, Covent Garden, London WC2H 9HE

British Library Cataloguing-in-Publication Data
A catalogue record for this book is available from the British Library.

FRONTIER PROBLEMS IN QUANTUM MECHANICS

ISBN 978-981-3146-84-6

For any available supplementary material, please visit
https://www.worldscientific.com/worldscibooks/10.1142/10203#t=suppl

Desk Editor: Rhaimie Wahap

Typeset by Stallion Press
Email: enquiries@stallionpress.com

About the Authors

Lee Chang

Lee Chang was born in 1925. He graduated from Fu Jen University, Peking, with a B.Sc. degree, and obtained a Ph.D. degree (physics) from the University of Leningrad in 1956. He joined the faculty of Tsinghua University, Beijing, in 1957. Serving as deputy head of department, he was responsible for the undergraduate and graduate education in the then newly established engineering physics department from 1958 and became the head in 1981. From 1982 to 1984, he served as head of the then restored department of physics. He also served as the director of the Institute of Modern Physics of Tsinghua University from 1982 to 1990.

As a faculty member of Tsinghua University, he has been and still is involved in the teaching of undergraduate and graduate courses in physics and theoretical physics and in the supervision of graduate students working toward their M.Sc. or Ph.D. degrees. His research in theoretical physics includes the theory of annihilation in many electron systems, resulting in a formalism first established for such a problem. The research on spinor helicity technique led to results widely used by colleagues in the high energy physics community for calculating many particle scattering amplitudes, and was later dubbed The Chinese Magic. He has a number of papers published in leading journals in physics. He edited and contributed to a graduate textbook entitled *Developments of Modern Physics* published in Beijing in 1997, which won the title of Priority Textbook awarded by the State Education Commission in 1997. He held visiting appointments abroad, as research associate in the Institute of Physics, University of Leningrad 1956–1957, as visiting scientist, Niels Bohr Institute, Copenhagen 1963–1964, as visiting associate, California Institute of Technology 1980–1981 and as visiting professor of the same Institute in 1987–1988. He is the recipient of the distinguished service certificate, State Education Commission PR China in 1991, First Class Award for Teaching Excellence, Beijing Municipality in 1997, and the Zhou Peiyuan Prize in Physics in 2013.

Molin Ge

Molin Ge was born in 1938. He completed the graduate education (Ph.D. equivalent) in Lanzhou University in 1965. He joined the Chern Institute of Mathematics, Nankai University in 1986 as a professor, and was elected as an academician of the Chinese Academy of Sciences in 2013. He has been responsible for supervising graduate students pursuing doctor's degree and also for teaching various courses in theoretical physics, including electrodynamics, quantum mechanics, mathematical methods in physics, quantum field theory, general theory of relativity and so on. Ge's main research fields included particle physics theory, general theory of relativity, Yang–Mills gauge field theory, soliton, and quantum integrable systems. In recent years he is devoted to the applications of Yang–Baxter equation to quantum information as well as the propagation of electromagnetic waves in complex media. Awards include the State Education Commission Progress of Science Award, State Natural Science Award, He Liang and He Li Progress of Science Award, State Award for Pedagogic Achievements.

Preface

The motivation of writing this book can be traced back to 1983. In that year Professor Chen Ning Yang gave a series of 12 lectures on topics in modern physics on the Stony Brook Campus of the State University of New York. One of us (M.G.) had the fortune to attend, and was deeply enlightened. Professor Yang stressed that a research program in physics is usually concerned with different branches, and the resolution of a problem often depended on combining knowledge from various directions. Graduate students were advised to follow new concepts and developments in various branches of physics, and their mutual relations during their study.

Three years later, Professor C.N. Yang gave another 9 lectures on Phase and Modern Physics in Beijing at the invitation of the Graduate School of the University of Science and Technology of China. The lectures were concerned with the birth, development and acceptance of many important concepts in modern physics. They gave insightful theoretical interpretations of the ideas, and introduced key experiments elucidating the difficult points. After each lecture, Professor Yang had working lunch with representatives of graduate students, and had discussion over wide range of topics. During the lectures, seminar talks and discussions, Professor Yang offered his advice and guidance to the students on their study and work. He re-iterated that the students must follow new ideas and their developments in various branches of physics, read regularly related journals besides carrying out research in their own problems. The other of us (L.C.) attended all lectures and was deeply moved by Professor Yang's rigorous and enlightening lectures and his extensive interests in the development of the students. The two of us met later and shared our experience in attending the lectures. We reached a strong consensus that we should follow Professor Yang's spirit, and do our part by writing a book entitled *Frontier Problems of Quantum Mechanics*.

Many recent advances in physics were connected with conceptual developments in quantum mechanics, and had their origin in its basic formulation. Progress in these frontier problems also promoted the further development of quantum mechanics. For the probabilistic interpretation of quantum mechanics, there was the well known dispute between Einstein and Bohr on the completeness of quantum mechanical description. There has been important progress in research around this problem, both theoretical and experimental. It is difficult to comprehend some basic questions on the basis of intuition acquired in studying classical physics. For example, the diffraction of electrons through a double slit. In textbooks on quantum mechanics, thought experiments were used to explain the result. The experiments which cannot be realized in the laboratory can only teach students how to think according to quantum mechanics, but can hardly convince them. The surprising advances in experimental methods in recent years have changed the thought to reality. A series of new experiments helped to clarify many difficult conceptual points. At present the investigation of quantum mechanics is opening up a challenging direction: the quantum information and quantum computing. Here the information to be processed and transmitted is not classical, it is a superposition of quantum states. This is another example of close association of basic science with an important field of technology. There have been significant progress in understanding the boundary between quantum and classical mechanics, and the quantum coherent properties revealed in meso-scopic as well as macroscopic systems. These have enhanced our comprehension of the nature and fundamentals of quantum mechanics, and will be discussed in Chapter 7.

The research in frontier fields of physics necessitates the analysis based on deep insight into the very nature of matter. In the field of condensed matter physics, the interpretation of fractional quantum Hall effect in terms of wave function of collective states in 2-dimensional electron system revealed the existence of a new incompressible quantum liquid. This example demonstrated vividly that the application of genuine quantum mechanical method can promote the development of various fields and at the same time broaden the range of contributions of quantum mechanics. In Chapters 8–10 we have chosen some representative topics including the cavity quantum electrodynamics, quantum Hall effect and the Bose-Einstein condensation for presentation. We do not intend to give a systematic introduction or an overview on such flourishing directions, but rather emphasize how basic concepts and methods of quantum mechanics are used in some aspects during the development of these directions. No attempts were made to achieve a broad coverage of literature nor to include advanced topics.

Yang-Baxter system is the general theory for treating a large class of non-linear quantum integrable models for many-particle systems. It has become a prosperous branch of mathematical physics after three decades of development. Many problems in theoretical physics, including the symmetry of the simplest system of quantum mechanics — the hydrogen atom, quantization of the phase of wave function and so on are intimately connected with it. From this view point we give an elementary introduction of Yang-Baxter system in Chapters 11~12.

The frontier research directions and research topics are very diverse and extensive. This book can only address a part of all the important aspects. The presentation of this book is based on the knowledge of quantum mechanics (including the second quantization) and statistical mechanics in standard courses. For the necessary theoretical concepts beyond the assumed knowledge, we give a comparatively systematic introduction. Various branches of physics are intimately connected with one another. Concepts, ideas and methods are freely borrowed or transplanted from one field to another. The investigation of a problem usually involves different branches. This appears in many places in this book.

This book is intended for working physicists. The topics covered can frequently be found in monographs, and proceedings of conferences, advanced study institutes, summer schools, as well as in articles in scientific journals. Such resources are often aimed at specialist with a high level of assumed knowledge. They can be hard to follow for people getting into contact of related fields for the first time, especially for graduate students or upper class undergraduates. Terms like it can be shown and it is apparent can hinder reader's progress. This book intends to help them to shorten their process of entering an active research field. We attempt to introduce the aim of the derivation, the key steps, and provide the necessary details. Grasping a new concept in physics can be a difficult and sometime torturous process. A concept can appear in different branches of physics, but the origin is common. We try not to use definition only in introducing concepts, but incorporate necessary explanation about the development of, and the relation between concepts.

Quantum mechanics was founded on the basis of experiments, and its developments depended also on experiments. Highly sophisticated experiments in recent years have been able to clarify some profound concepts and disputes which have lasted for many years, or to achieve deeper understanding in the nature of the objects. Such experiments have opened before us new horizons one after another. We have tried to introduce the ideas of the designs and methods, as well as their significance in deepen our understanding and clarifying questions of controversies with the hope that the readers can understand the essential physics behind the phenomenon. We apologize to the authors for possible omissions of the sophistication of some experimental techniques.

The references given at the ends of each chapter are very far from complete, they represent books and journals we have been consulting frequently. We have reproduced numerous figures from journal articles and also from monographs or conference proceedings. In the figure captions we indicate the sources of the figures. We are grateful to the publishers of the journals and monographs, American

Academy of Sciences, Nature Publishing Group, American Institute of Physics, European Physical Society, Addison-Wesley Publishing Company, Springer Verlag, and World Scientific Publishing Company.

Chapters 1~10 were written by Lee Chang, Chapters 11~12 by Molin-Ge. The two of us discussed in detail the styles of the presentation, the arrangements of chapters and sections. Our knowledge and capabilities are limited in our endeavour to fulfil the primary objective we set for the book. Defects and mistakes are inevitable. We sincerely hope that experts and readers would kindly offer us their criticism and suggestions for improvement.

Lee Chang
Department of Physics and Institute for Advanced Study,
Tsinghua University

Molin-Ge
Institute of Mathematics,
Nankai University and Institute for Advanced Study,
Tsinghua University

Contents

Chapter 1

Wave-Particle Dualism, Complementarity Principle, Bell's Theorem and Related Experiments

Since the founding of quantum mechanics in 1925 disputes over its fundamental principles and its interpretation have been lingering. With the progress in the sophistication of the experimental and theoretical research some of the disputes were resolved, but usually new questions were raised and disputes continued on a higher level. New results continue to emerge from the research frontier.

The wave-particle duality is the foundation of quantum mechanics. Experiments on the diffraction of electrons and neutrons on crystals have become well-known for a long time. C. Jönsson carried out experiments on electron diffraction through double slit in 1961. The double slit experiments are used in textbooks on quantum mechanics to explain characteristic concepts. The main concepts involved are as follows. When electrons fall on the screen, they do so as particles. One has to demonstrate that electrons fall on the screen as randomly distributed stars in the night sky in the beginning, and it takes time for electrons falling on the screen to accumulate and the speckles grouped themselves into the regular interference pattern of maxima and minima. The maxima of interference pattern are the places where the number density of electrons is maximal, and the probability distribution is determined by the wave function. As Dirac pointed out in his "Principles of Quantum Mechanics", an

1

electron interferes with itself. You have to allow an electron to pass through both slits in order to observe the interference pattern. To demonstrate this experimentally, conditions must be created such that there can be only one electron between the slits and the screen at any time. It was difficult to achieve the conditions to observe the accumulation of electrons forming the pattern as well as to ensure the existence of only one electron in the apparatus before the 1980's. The experiment performed by Tonomura in 1989 fulfilled these conditions, which we introduce in Section 1.1.

The double slit experiment for light was initiated by Thomas Young in the beginning of 19th century. The results can be immediately understood in terms of the wave theory of light. But if we consider light as particles on the basis of the wave-particle duality it is equally difficult to comprehend as in the electron double slit experiment that how can one particle pass through two slits at the same time. Moreover the light sources used for this kind of experiments, including laser light, belong to the category of classical light sources. It is impossible to ensure that a photon traverses through the apparatus alone without the simultaneous presence of a second photon. In Section 1.4 we introduce the single photon experiment (1986, Aspect), which was successful to overcome the difficulty with classical light sources.

The multi-photon interferometry which emerged in recent years shows the phenomena of a pair of correlated photons interfering with themselves. It also shows that one photon interference cannot co-exist with double photon interference. The results of experiments deepen our understanding of quantum mechanics. We discuss this topic in Section 1.5.

The center of the debate over the fundamentals of quantum mechanics is the complementarity principle. It involves several interconnected problems. An electron passing through the double slit can interfere because it is provided with a choice of two possible routes, such that its wave nature can be exhibited. If we leave both slits open, and illuminate the slits in order to "see" which slit the electron passes through — to enable it to show its particle nature — then the interference pattern disappears. This is the prediction of the complementarity principle of quantum mechanics. But does this really happen? "Thought experiments" are frequently used in textbooks where practical difficulties prevent the demonstration of the effect in real life. These difficulties mainly arise from the weakness of the interaction of the electron with light. Even if we illuminate the slits with light most electrons would pass through without being "seen". Today the "thought" has become a reality. If we replace the electrons by atoms, tune the illumination to be resonant with an atomic transition, then the interaction between the atom and light turns out to be strong enough such that the electron has almost no chance to escape from being "seen" (1995, Pritchard). The experiment confirmed the prediction of complementarity principle. Another atomic interferometer experiment using standing wave of light as diffraction slits (1998, Rempe) achieved the same. We discuss this development in Section 1.2.

Another relevant question is: Why is the wave nature of the electron destroyed when it is illuminated? The standard answer to this question used to be: If you want to observe the electron, say by illuminating it with light, the scattered photon will necessarily give it a kick, changing its momentum. Such an interaction is *uncontrollable*, because the scattering of light is a probabilistic process and the momentum transmitted has a distribution. In some instances this is the main reason. But a different case may have a different mechanism. For instance in Pritchard's experiment the reason for the loss of interference is the effective phase shift produced in the photon scattering. The effective phase shift is *controllable* in the experiment. The increase of the effective phase shift leads to a decrease of the contrast of the interference pattern, and the process is reversible. Rempe's experiment shows that the entanglement of the route with an observable property of the atom (in this case its internal state) is the reason for the loss of the interference. The traditional explanation of the loss of interference is based on a physical analysis of Heisenberg's uncertainty principle. If the uncertainty is introduced by a measurement, then what happens if a measurement is not carried out? The micromaser experiment can avoid the "uncontrollable" interaction during a measurement. The interference is lost anyway, but this time again due to the correlation (entanglement) between electron and photon degrees of freedom. At the same time coherence can be recovered. This is discussed in Section 1.6.

R. P. Feynman said that interference could only happen between two states physically indistinguishable in an equipment. The experiment of Zou, Wang and Mandel showed that interference happens when ways of distinguishing the states are not provided. The interference disappears when a possibility is provided in the experimental arrangement, even if no detectors are in place to carry out the distinction. We discuss in Section 1.5.4.

There has been a kind of assertion that how does a quantum object behave (exhibits wave or particle nature) depends on the information concerning the experimental set-up it receives. J. A. Wheeler suggested in 1978 an ingenious method to test this assertion. The idea of a "delayed choice experiment" consists of the following. Fix the condition first. Wait until the object has already passed the equipment (i.e., how does it behave is already determined), and then change the condition all at once before the object reaches the detector. What result will turn out? Several examples concerning the delayed choice are discussed in Section 1.9.

The most famous dispute of all is that between A. Einstein and N. Bohr. This is sometimes referred to as Einstein-Podolsky-Rosen (EPR) paradox. There is a general consensus on the correctness of the prediction of quantum mechanics on the properties of microscopic systems. Quantum Mechanics has been applied extensively to scientific and technological problems with great success. The challenge of Einstein is that quantum mechanical description is incomplete, i.e., properties of an object are more than quantum mechanics can describe. Could we possibly find a more complete description? For several decades many physicists have devoted themselves to excavate this precious mine. Some failed because of self-inconsistencies, while the arguments of some others appeared to be unreasonable. A class of theories called "hidden variables theories" or "local realistic theories" set an aim of achieving a more complete description while keeping the same predictions as those of quantum mechanics. Debates have been raging fiercely. J. S. Bell put forth a theorem that if a local realistic theory is to reproduce the predictions of quantum mechanics, then it must satisfy an inequality. This theorem sets a clear criterion for the resolution of the dispute. If the inequality is violated, then the local realistic theory is unable to reproduce the predictions of quantum mechanics. Since the late 1960s a large number of experiments were performed to verify this inequality. With increasing accuracy the experiments showed that the inequality was violated. More than three decades passed since then, the debate continues. The reason is that any experiment is almost not free from loopholes, and people can sound disagreement. The experimental technique of modern physics can convince you that the Bell inequality is shown to be violated by more than 100 standard deviations. This is not the end of the story. Bell's theorem without inequalities appeared such that the local realistic theories can be invalidated by experiments directly. The developments are discussed in Sections 1.7~1.10.

1.1   The Buildup of the Electron Interference Pattern

In textbooks of quantum mechanics electron double slit experiment is used to illustrate the dual nature of electrons. In the experiment electrons fall on the screen after they have passed the double slit and are recorded one by one. The accumulation of the recorded electrons forms eventually the interference pattern. The pattern is formed because the waves ψ_1 and ψ_2 passing through slits 1 and 2 respectively superimpose and interfere. The intensity on the screen is proportional to $|\psi_1+\psi_2|^2$. The conditions for an interference pattern to be formed are that the coherence length of the de Broglie wave of the electron must be larger than the path difference between the routes through the slits and that no attempts are made to determine through which slit the electron passes. If such attempts are made (e.g., by putting a light source close to a slit or by suspending the slit plane freely), what you would get is the sum of patterns obtained by electron diffraction at the respective single slits, i.e., $|\psi_1|^2 + |\psi_2|^2$. Feynman [1] characterized this phenomenon as "which is impossible, *absolutely* impossible, to explain in any classical way, and which has in it the heart of quantum mechanics. In reality, it contains the *only* mystery." He added: "This experiment has never been done in just this way. The trouble is that the apparatus would have to be made on an impossibly small scale to show the effects we are interested in." The problem is that the energy of the electrons should be sufficiently homogeneous. To obtain such a beam electrons have to be accelerated to an energy which turns out to be too large such that the de Broglie wave length becomes too small. Such kinds of experiments were called "thought experiments". They are described in textbooks to "illustrate" (not actually demonstrate) the basic characteristics of quantum mechanics.

The first experiment demonstrating the buildup of interference pattern was carried out by A. Zeilinger's group [2] by using very cold neutrons with a wave length of 2 nm corresponding to a velocity of 200 ms^{-1}. Slits with the width of 22 μm and 23 μm respectively separated by a distance of 104 μm were used. The detecting plane was 5 m downstream from the plane of slits in order to resolve the pattern shown in Fig. 1.1. The solid diffraction line represents the theoretical prediction taking into account of specific features of the apparatus. Moreover, the observed intensity of neutrons was as low as one in every two seconds. The experiment demonstrated that the pattern

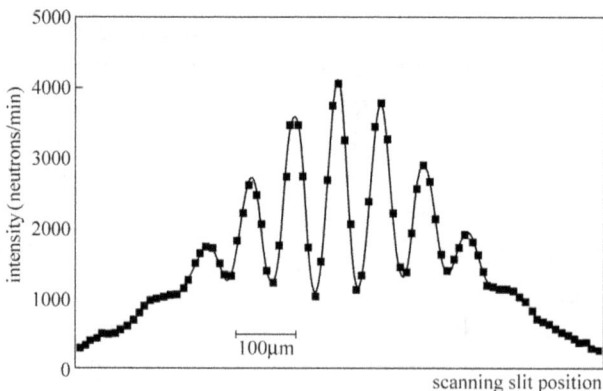

Fig. 1.1   A double slit diffraction pattern measured with very cold neutrons. From Ref. [2]. Reprinted with permission from Anton Zeilinger, Roland Gähler, C. G. Shull, Wolfgang Treimer, and Walter Mampe, *Rev. Mod. Phys.* 60, 1067, 1988. Copyright (2017) by the American Physical Society.

showing wave-like nature was actually formed by neutrons arriving one by one and that a neutron interferes with itself while passing through the apparatus.

Demonstration of the same phenomenon with electrons leads more experimental sophistication. The group led by A. Tonomura [3] overcame the difficulties pointed out by Feynman by using two projector lenses (electron microscopic technique) behind the interference pattern which was able to magnify the pattern 2000 times and by using position-sensitive detection technique to resolve the pattern. In the experiment a biprism interferometer was used. It consists of two parallel grounded plates separated at a distance b with a fine filament of radius a in between, the latter having a positive potential relative the former. In the geometry defined in Fig. 1.2(a) the incident wave $e^{ik_z z}$ enters an electrostatic field with potential $V(x, z)$. In general ψ, the wave function of the electron in an electromagnetic field with 4-potential A_μ is expressed in terms of ψ_0, the wave function in absence of the field in the following way[1]

$$\psi(\xi) = \exp\left(-\frac{ie}{\hbar} \int^\xi A_\mu(\eta) d\eta^\mu\right) \psi_0(\xi). \qquad (1.1.1)$$

The integral on the exponent is taken from an arbitrary reference point to ξ, where ξ and η are 4-dimensional space-time coordinates. In the present case the 4-potential has only a scalar component $A_0 = V(x, z)$ and hence the integral becomes

$$\int V(x, z)dt = \int V(x, z)\frac{ds}{v_z} = \frac{m}{\hbar k_z} \int V(x, z)ds,$$

where $v_z = \hbar k_z/m$ is the velocity of the electron, ds is the line element. The potential is symmetric with respect to x. When the reference point is taken to be $z = -\infty$ the phase factor in (1.1.1) becomes

$$\exp\left(-\frac{iem}{\hbar^2 k_z} \int_{-\infty}^z V(x, z')dz'\right).$$

---

[1]Please refer to Section 3.1.

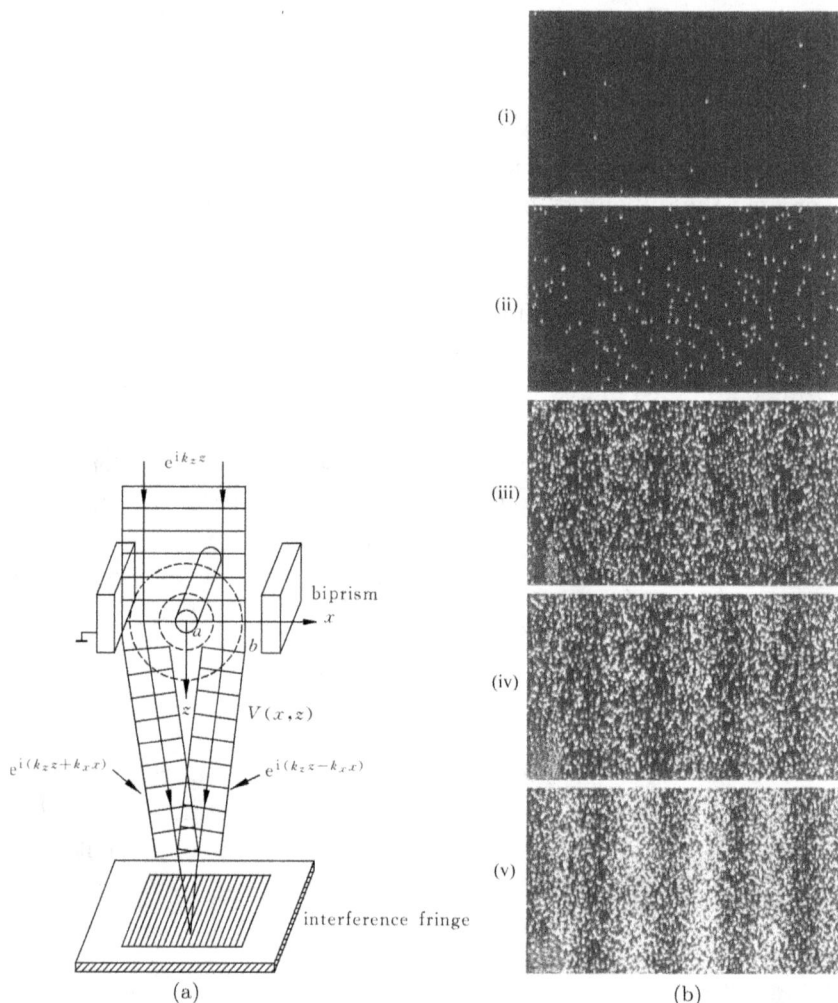

Fig. 1.2   (a) Interference pattern formed by incident plane wave passing through an electron biprism; (b) Buildup of the electron interference pattern. The number of electrons are from (i) to (v) 10, 100, 3000, 20000 and 70000 respectively. From Ref. [3]. Original photo picture kindly provided by Professor Akira Tonomura reprinted with permission.

The wave function of the electron entering the biprism is

$$\psi(x, z) = \exp \mathrm{i}\left(k_z z - \frac{em}{\hbar^2 k_z} \int_{-\infty}^{z} V(x, z')\mathrm{d}z' \right). \qquad (1.1.2)$$

The electron experiences a force in the x-direction with a magnitude $-e\frac{\partial V}{\partial x}$. Expanding V around $x = a$ ($|x| \geq a$) we obtain

$$V(x, z') = V(a, z') + \left.\frac{\partial V(x, z')}{\partial x}\right|_{x=a} x,$$

and for $x \leq -a$ we have $\frac{\partial V(-x,z')}{\partial x} = -\frac{\partial V(x,z')}{\partial x}$. For the time being we restrict ourselves to consider the case $x \geq a$. The wave function (1.1.2) becomes

$$\psi(x, z) = \exp i\left(k_z z - \frac{em}{\hbar^2 k_z}\int_{-\infty}^{z} V(a, z')\mathrm{d}z' - x\frac{em}{\hbar^2 k_z}\int \left.\frac{\partial V}{\partial x}\right|_{x=a}\mathrm{d}z'\right).$$

The second term in the bracket is independent of x leading to a constant phase factor $\phi(z)$ depending on z only. The x component of the momentum acquired by the electron is denoted by $-\hbar k_x$,

$$\int \mathrm{d}t\left(-e\frac{\partial V}{\partial x}\right) = \int \mathrm{d}z'\frac{m}{\hbar k_z}\left(e\frac{\partial V}{\partial x}\right) = -\hbar k_x.$$

Consequently in the expression for ψ, the third term in the bracket is actually $-k_x x$.

Finally we obtain

$$\psi(x, z) = \exp i(k_z z \mp k_x x + \phi(z)). \qquad (1.1.3)$$

The upper sign applies for $x \geq a$ and the lower sign applies to $x \leq -a$. The wave function after the merge of the two diffracted waves is

$$\psi(x, z) = e^{i(k_z z + \phi(z))}\left(e^{-ik_x x} + e^{ik_x x}\right).$$

The interference pattern is given by

$$|\psi(x, z)|^2 = 4\cos^2 k_x x. \qquad (1.1.4)$$

Around a cylindrical fine wire the electrostatic potential is

$$V(x, z) = V_a\frac{\ln(\sqrt{x^2 + z^2}/b)}{\ln(a/b)},$$

where V_a is the potential on the wire. From the expression defining k_x we have

$$k_x = \frac{\pi e V_a}{\hbar v_z} \ln \frac{b}{a}. \tag{1.1.5}$$

The distance between fringes, $d = \pi/k_x$, determined by parameters in the experiment is only 7000 Å. The magnifying projector lenses were able to enlarge it to 1.4 mm. This can be resolved by position-sensitive electron counting device. The buildup of the pattern is shown in Fig. 1.2(b). The number of electrons arriving at the detecting plane is 10^3/s, the distance between the electron source (the tip of electron emission) and the screen is 1.5 m, the velocity of electrons is 1.5×10^8 m/s. For uniformly emitted electrons the average distance between successive electrons is 150 km. The length of a wave packet for the electron is 1 µm. The possibility of overlap of two wave packets is negligible. One can be sure that the probability of having two electrons appearing in the apparatus at a given time is extremely small. A recording time of 20 min was enough to get interference fringes when the number of recorded electrons reached 3000. When the number reached 70000, five fringes could be seen.

This experiment demonstrated clearly the wave-particle dualism of electrons. The interference pattern given by the probability distribution (1.1.4) originates from electrons arriving on the detecting plane one by one, each single electron is responsible for the occurrence of interference.

Feynman gives a further analysis that when the electron is watched by illuminating a given slit and thus revealing whether the electron does pass through this slit or not, the interference will disappear. This can be understood qualitatively by considering the wave function of the electron downstream the slit plane

$$\psi = \frac{1}{\sqrt{2}}(\psi_a + \psi_b),$$

where ψ_a and ψ_b are the waves passing through the slits a and b respectively. The diffraction pattern depends on the relative phase of ψ_a and ψ_b on different points of the screen. The probing photon has to be scattered by either of the waves, otherwise no information can be acquired. This scattering can alter the phase of either ψ_a or ψ_b, albeit in a probabilistic fashion. Consequently the diffraction pattern is to be smeared such that the visibility is lowered or even extinguished. Can this be observed experimentally? We turn to the next section for a discussion.

1.2   Experimental Verification of Complementarity Principle by Atomic Interferometry

The picture of the microscopic world provided by the double slit experiment is entirely different from that of the macroscopic world with which we are familiar from the study of classical physics. If the interference pattern is required (to have wave nature manifest), then no attempt is allowed to find out which slit the electron passes through. If such a judgement is made (providing information of the electron path, i.e., particle nature is made manifest), the interference pattern disappears. It turns out that the electron possesses wave-particle dual nature, but these aspects are mutually exclusive, only one aspect can be manifest at a time, depending on conditions. Furthermore they together offer a complete description of the nature of the electron, i.e., each aspect complements the other. Such a comprehension was summarized by N. Bohr in 1927 as the complementary principle.[2] Dirac characterized such an understanding as "led to a drastic change in the physicist's view of the world, perhaps the biggest that has yet taken place." N. Bohr had been presenting and elaborating this principle in lectures, conference talks, discussion sessions, popular science writings and journal articles throughout his life. The beginning was a difficult one, though. There were numerous skeptics and opponents, among whom the most authoritative scientist was A. Einstein. A direct experimental demonstration of this principle would have a great significance. The difficulty lies in the

---

[2]In the same sense, a distribution-in-position and a distribution-in-momentum are complementary characterizations of a quantum state. In quantum mechanics, coordinate and momentum cannot be used simultaneously to produce a trajectory of motion as in classical mechanics.

fact that electrons and neutrons interact very weakly with light. The experiment became realized only in 1995 by a group at MIT led by D. Pritchard [4] by using an atomic interferometer. Using resonant light to illuminate the atoms in the interferometer, they were able to demonstrate the complementarity. The wave nature of light is suppressed and the particle nature becomes manifest if the scattering of light can provide the information of "which path"[3] the atom chooses in the equipment. The wave nature is kept intact if the path difference induced by photon scattering is less than one half of the de Broglie wave length of the atom such that the scattered photon does not provide the information of the choice of the atom. Furthermore, if only the photons scattered in a narrow range of the phase space are detected, the lost interference can be regained, the price being that the scattered photons do not provide the information of "which path" the atom passed. The result of the experiment clearly shows that whether the light scattering demolishes the wave nature of the atom depends on whether the former can provide the "which path" information of the atom. The loss of coherence does not originate from the momentum transferred from the photon to the atom, but from the random phase shift produced by the photon scattering on the atom. This is a very direct experimental demonstration of the principle of complementarity.

In the experiment a beam of atomic Na with a narrow velocity distribution (<4% rms) is produced by a seeded inert gas supersonic source. The beam is pumped by a laser (wavy line) to the $F = 2, m_F = 2$ state the pumping rate is 95% as confirmed by using a Stern-Gerlach analysis magnet. After being collimated by 2 slits the beam enters a Mach-Zehnder interferometer consisting of three nanofabricated gratings (dotted lines). The paths of the beam splitting and recombination at $z = 0$ and $z = 2L$ respectively and reflection at $z = L$ are achieved by transmission and Bragg reflection at the gratings with the beam incident on the first grating at Bragg angle. The atomic fringe pattern after the third grating is

---

[3] "which path" or " which way" or "welcher Weg" in German has become a proper name for characterizing such kind of experiments.

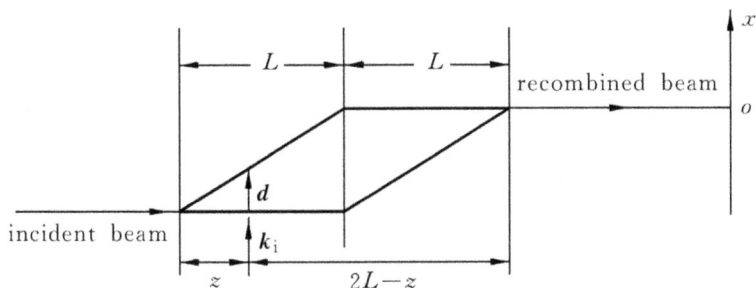

Fig. 1.3 Path of an atom in the interferometer.

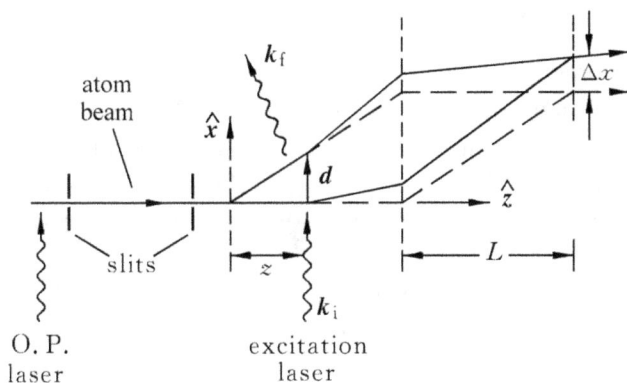

Fig. 1.4 Effect of photon scattering on the interference of an atom. Reprinted with permission from Michael S. Chapman, Troy D. Hammond, Alan Lenef, Jörg Schmiedmayer, Richard A. Rubenstein, Edward Smith, and David E. Pritchard, *Rev. Mod. Phys.* 75, 3783, 1995. Copyright (2017) by the American Physical Society.

$\overline{N}[1 + C\cos(k_g x)]$, where $k_g = 2\pi/\lambda_g$ with λ_g the period of gratings, $\overline{N}$ is the average counting rate, C is the contrast (or visibility) of the pattern. The path of an atom wave in the interferometer is shown in Fig. 1.3. The effect of photon scattering is shown in Fig. 1.4. In this figure the paths of the atom without photon scattering is represented by dashed lines. The atom is resonantly excited to the state $F' = 3, m_F = 3$ by the σ^+ polarized excitation laser with momentum of the photon $\boldsymbol{k}_i$. The resonant excitation with correctly adjusted laser power ensures the occurrence of photon scattering. The excited atom emits a photon with momentum $\boldsymbol{k}_f$ spontaneously and returns to the ground state, here $|\boldsymbol{k}_i| = |\boldsymbol{k}_f| \equiv k$. The altered paths of the atom due to photon scattering are represented by solid lines, $\boldsymbol{d}$ represents the relative distance between the two arms of the interferometer where the photon is scattered. The loss of contrast will be expressed as a function of $\boldsymbol{d}$.

Without photon scattering the wave function of the atom at the third grating is

$$\psi(x) \propto u_1(x) + u_2(x)e^{ik_g x},$$

where u_1 and u_2 are the (real) amplitudes of the upper and lower beams, $k_g = 2\pi/\lambda_g$. The effects of scattering consist of the following:

(1) The momentum change of the photon is $\Delta\boldsymbol{k} = \boldsymbol{k}_f - \boldsymbol{k}_i$, the x component of which is denoted by Δk_x. Consequently the envelope of the interference pattern is shifted in the x direction by $\Delta x = \frac{\Delta k_x}{k_A}(2L - z)$, where k_A is the momentum of the atom, and is related to the de Broglie wave length λ_A of the atom by $k_A = 2\pi/\lambda_A$.

(2) The relative phase shift of beams along the arms of the interferometer changes by $\Delta\phi = \Delta\boldsymbol{k} \cdot \boldsymbol{d} = \Delta k_x d$.

It is important to notice that the quantum process of scattering leads to $\Delta\boldsymbol{k}$ determined by a distribution.

Taking into account these two effects, we obtain the wave function of the atom at the third grating

$$\psi(x, \Delta k_x) \propto u_1(x - \Delta x) + u_2(x - \Delta x)e^{i(k_g x + \Delta\phi)}. \qquad (1.2.1)$$

If all the atoms are observed irrespective of the direction of the scattered photon $\boldsymbol{k}_f$, then the interference pattern corresponds to different values of Δk_x:

$$C' \cos(k_g x + \phi') = \int d(\Delta k_x) P(\Delta k_x) C \cos(k_g x + \Delta k_x d), \qquad (1.2.2)$$

where $P(\Delta k_x)$ is the probability distribution of the transverse momentum transfer given by the radiation pattern of a radiating dipole shown in the inset of Fig. 1.5. The 0 momentum transfer corresponds to forward scattering and momentum transfer $2\hbar k_x$

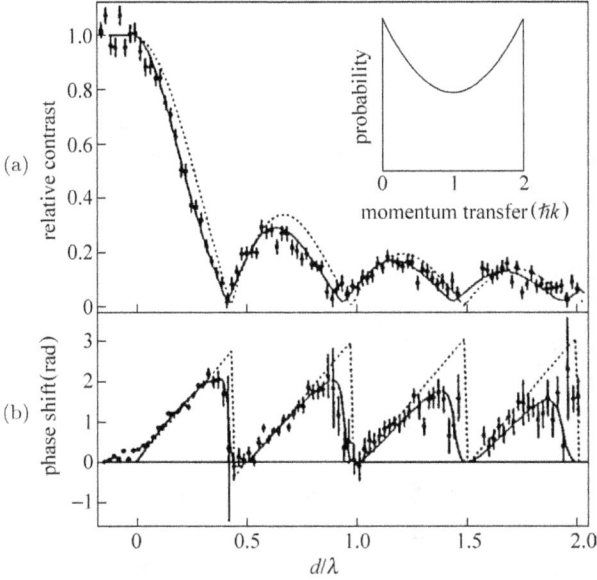

Fig. 1.5   Relative contrast and phase shift as functions of d, λ is the wave length of the laser. Reprinted with permission from Michael S. Chapman, Troy D. Hammond, Alan Lenef, Jörg Schmiedmayer, Richard A. Rubenstein, Edward Smith, and David E. Pritchard, *Rev. Mod. Phys.* 75, 3783, 1995. Copyright (2017) by the American Physical Society.

corresponds to backward scattering. The average value of momentum transfer is $\hbar\overline{\Delta k_x} = \hbar k$. C represents the contrast of the interference pattern when there is no photon scattering. From Eq. (1.2.2) we see that the contrast of interference pattern C' and the phase ϕ' as functions of d can be obtained by taking the Fourier transform of the probability distribution $P(\Delta k)$ of the momentum transfer. The ratio of contrasts C' to C is called the relative contrast. Figure 1.5 gives the relative contrast and phase shift as functions of d. In the experiment the excitation laser beam moves along the z direction, giving different values for d: $d = z\frac{\lambda_A}{\lambda_g}$. For $d < 0$ corresponding to $z < 0$ the laser beam is situated in front of the first grating. The result shows that for $|d|$ small the contrast of the pattern and the phase shift are not much influenced by the photon scattering. For $\overline{\Delta k_x}d \ll \pi$ the phase shift increases with d linearly. For isotropic scattering $\overline{\Delta k_x} = k = \frac{2\pi}{\lambda}$. The phase shift $\overline{\Delta k_x}d = 2\pi\frac{d}{\lambda}$ increases linearly with $\frac{d}{\lambda}$, with slope 2π.

Correspondingly the contrast drops sharply. The solid curves in the figure represent the best fit taking into account contributions from atoms that scatter 0 photons (5%) and 2 photons (18%) respectively, the dashed curves represent theoretical calculations corresponding to single photon scattering. The contrast decreases rapidly with the increase of d, and drops to 0 for $\Delta k_x d \approx \pi$, i.e., $d \approx \lambda/2$. It begins to rise again when d is further increased. The contrast is first partially recovered and then decreases again. Such tendency repeats itself periodically.

In the experiment we see that the relative contrast begins at 1 (one extreme of the complementarity) and decreases toward 0 (the other extreme) depending on the effective phase shift. Between the extremes there is a continuum of the intermediate status.

It is now appropriate to discuss an important issue based on the results obtained. What is the origin of the loss of coherence (or decoherence)? To put the question more general: what mechanism makes the principle of complementarity to hold? In examples most often discussed, e.g., the freely suspended slit plane (Einstein) or the slits illuminated by light (Feynman), the coherence is destroyed by Heisenberg's uncertainty principle of the position and momentum. In the present case it is natural to think about the momentum transferred from the photon to the atom. It could cause the pattern to shift in the x direction, which eventually smears the maxima and minima in the pattern. The shift of the pattern envelope represents this effect (Δx is proportional to Δk_x). The order of magnitude of that really shows up in the experiment corresponds to 100~200 fringes. Its variation can hardly explain the change of contrast of the pattern. On the other hand the phase shift $\Delta\phi$ corresponds at most to several fringes. Actually when the excitation laser beam moves toward small values of z, Δx increases for a given k_i while the loss of coherence and the phase shift decrease hand in hand when $z \to 0$. It is clear that the effective phase shift caused by photon scattering is directly linked to the loss of contrast.

A variant of this experiment deals with the partial revival of the relative contrast, and reveals the role played by the entanglement of final states in the scattering process in this respect. In this variant the atoms observed are correlated with photons scattered into a restricted range of final directions. The coherence lost as d increases may be partially regained. This can be realized by restricting the deflection of the atom Δx which in turn is related to k_x, and hence Δk_x. Experimentally this is achieved by using $10\,\mu m$ collimation slit (compared with $40\,\mu m$ slits for a full scale of measurable momentum transfers) and by covering the third grating with a $10\,\mu m$ wide slit which can be translated along the x direction to preferentially admit atoms with specific momentum transfers (compared with the $50\,\mu m$ full width of the third grating used in full measurement). The resulting distribution $P'(\Delta k_x)$ is now narrower, shown in the lower inset of Fig. 1.6, where the acceptance of detector for each case is shown, I and III refer to predominantly forward and predominantly backward scattering respectively, and II favors intermediate scattering angles. The dotted line represents the original distribution. The solid curves are calculated using the geometry of the equipment and experimental parameters, and are compared with the uncorrelated case (dotted curves). In the upper inset atomic beam profiles at the third grating are shown when the laser is off (thin line) and when the laser is on (thick line). The arrows show the position of the third grating for cases I and II. In the relative contrast diagram result for case II is not presented, it is very close to that of case I. The slopes of phase shift curves correspond to the restricted range of momentum transfer in cases of I, II and III. The slope of case III is nearly 4π, indicating a phase of the pattern determined by predominantly backward scattering. The slope of case I approaches 0 due to predominantly forward scattering. Case II is intermediate between I and III, and the phase is around 3π, corresponding to an average momentum transfer of $1.5\hbar k$. We see that the relative contrast falls off much slowly than the case of full acceptance. Actually the contrast is regained up to 60% at $d \approx \lambda/2$.

It turns out that measuring atoms uncorrelated or correlated with scattered photons does lead to very different results, while the process of scattering of photons on atoms is the same. Photons scattered by atoms have many possible final momenta k_f with atoms recoiling with corresponding final momenta (the third grating is

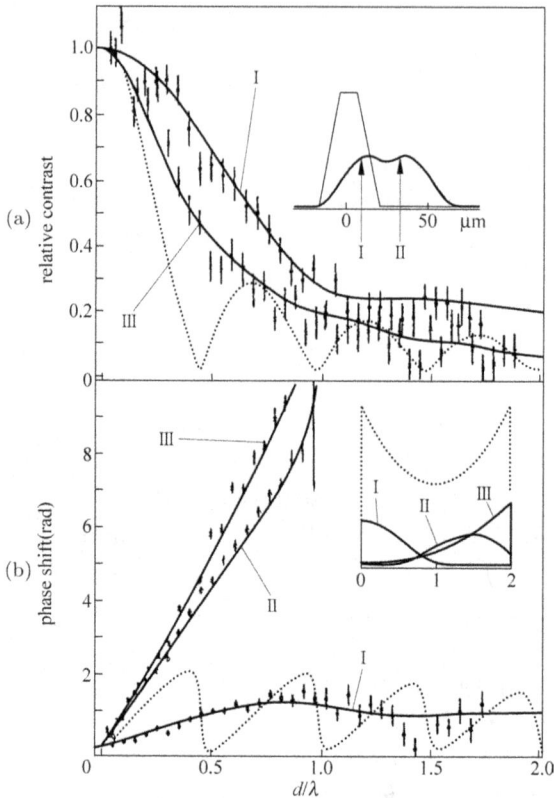

Fig. 1.6   Relative contrast and phase shift as functions of d with atoms correlated with photons scattered into a limited range of directions. Reprinted with permission from Michael S. Chapman, Troy D. Hammond, Alan Lenef, Jörg Schmiedmayer, Richard A. Rubenstein, Edward Smith, and David E. Pritchard, *Rev. Mod. Phys.* 75, 3783, 1995. Copyright (2017) by the American Physical Society.

$50\,\mu$m wide for full scale measurement). The wave function of the atom-photon system is actually described by Schrödinger's entangled state.[4] There is no dissipation in such an interaction, and the

---

[4]Let the atom and radiation field (A and F) interact, leading to the state $\sum_i A_i F_i$. This state cannot be expressed as a simple direct product. Such linear superposition of direct products is called an entangled state (after Schrödinger). Schrödinger said:"entanglement is not only one of the peculiarities of quantum mechanics, it is the only peculiarity."

state evolves according to Schrödinger's equation. The coherence between the atomic waves is not extinguished, but distributed (entangled) in a number of possible final states. The resulting interference pattern is then given by an expression in which there are a number of interference terms, thus the randomness of the relative phase between any two components of the superposition leads to cancellations of the interference. Once the final states are significantly limited, the coherence can be partially regained. One can ask the question: Does the experiment provide information of "which way"? In principle measuring k_f of the scattered photon provides the information of the possible path of the atom on which the photon scatters, but actually k_f is *not measured* in the experiment. Consequently it is sufficient to arrange the experimental equipment such that provision of the "which way" information is possible, and the wave nature retires and particle nature goes on stage. "The importance of this work is that it provides for quantitative tests of basic principles, in a regime simple enough for clean comparison with theory but complex enough to explore a rich nontrivial phenomenology" [54].

This point that a potential possibility for answering the "which way" question is sufficient to suppress the wave aspect was further elaborated in another atomic interferometry experiment carried out by G. Rempe's group [5], the scheme of the interferometer is shown in Fig. 1.7. In the experiment a beam of ^{85}Rb atoms falls on a "diffraction grating" made of a standing wave of light. The frequency ω_0 of light is detuned from the energy level difference ΔE of the atom, with detuning $\Delta = \omega_0 - \frac{\Delta E}{\hbar}$. The intensity of light I is different at nodes and antinodes of the standing wave, and the light shift potential U acting on the atom is $U \propto \frac{I}{\Delta}$ [6]. When the angle of incidence of the atom beam is set at the Bragg angle, the atom beam undergoes the Bragg reflection [7] from the periodic light shift potential. By adjusting the light intensity the standing light wave becomes a 50% beam splitter, and the beam A is split into a Bragg-reflected beam B and a transmitted beam C. When these beams arrive at the second standing wave they are separated by a distance d, corresponding to the distance between the two slits. Beam B is split

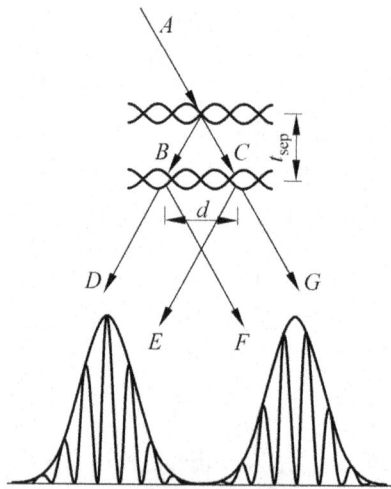

Fig. 1.7 Scheme of the atom interferometer. Reprinted with permission from Macmillan Publishers Ltd: Nature (395, 33), 1998.

into D (transmitted) and F (reflected), beam C into G (transmitted) and E (reflected). Beams in the same direction meet in the far field and interfere. Since the incident beam is not perpendicular to the beam splitter, the interference fringes under the left and right envelopes are not symmetrical. The maxima under the left envelope corresponds to the minima under the right envelope. The distance between fringes is determined by the time duration of the interaction between the atom and the standing wave, adjusted by the switch on and switch off of the light source. The theoretical calculation fits the experimental result.

The "which way" information can be stored in the internal state of the atom, as shown in Fig. 1.8. The excited state $5\,^2P_{3/2}$ of ^{85}Rb atom is denoted by $|e\rangle$, the ground state $5\,^2S_{1/2}$ has two hyperfine states with hyperfine spin $F = 2$ and $F = 3$, denoted by $|2\rangle$ and $|3\rangle$ respectively. The frequency of standing wave is set in the middle of transitions $|2\rangle \rightarrow |e\rangle$ and $|3\rangle \rightarrow |e\rangle$, with detuning parameters $\Delta_{2e} < 0$, $\Delta_{3e} > 0$ and $\Delta_{3e} = -\Delta_{2e}$. Suppose that the incident beam is in the state $|2\rangle$. It passes through a microwave field with frequency ω_{mw} equal to the energy difference between $|2\rangle$ and $|3\rangle$ divided by $\hbar$. The time duration of the microwave pulse is adjusted

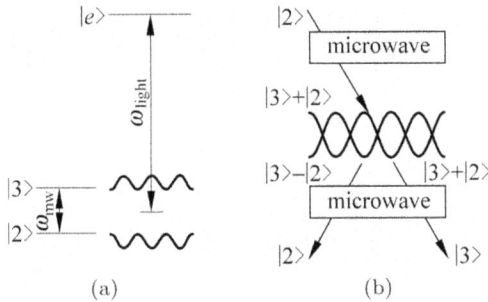

Fig. 1.8 "Which way" information stored in the internal state of the atom (a) simplified energy level of ^{85}Rb; (b) Two $\pi/2$ microwave pulses change the internal state of the atom. Reprinted with permission from Macmillan Publishers Ltd: Nature (395, 33), 1998.

such that $|2\rangle$ and $|3\rangle$ are mixed to become $|2\rangle + |3\rangle$.[5] There is no phase shift for the transmitted beam at the first beam splitter. In the Bragg-reflected beam the light shift potential acting on the state $|2\rangle$ is negative, and this corresponds to a reflection at an optically dense medium, leading to a phase shift π, the light shift potential acting on the state $|3\rangle$ is positive, and there is no phase shift. To summarize, the transmitted beam is $|3\rangle + |2\rangle$ and the Bragg-reflected beam is $|3\rangle - |2\rangle$. These beams separate after the beam splitter, and they pass through another microwave field with the same frequency and pulse duration as the previous one. As the result the transmitted beam becomes $|3\rangle$ and the Bragg-reflected beam becomes $|2\rangle$. In this way, the information of "which way" is stored in the internal state of the atom. When the beams pass through the second beam splitter and arrive at the detecting plane, no interference pattern can be seen (Fig. 1.9). It is interesting to notice that no attempt of distinguishing $|2\rangle$ and $|3\rangle$ on the detecting plane is made, i.e., actually no attempt is made to acquire the "which way" information. Interference pattern disappears whenever there exists the *potential* possibility to acquire such information. If detectors distinguishing $|2\rangle$ and $|3\rangle$ are really equipped on the detecting plane, the counting rate curve has the same

---

[5]We neglect the normalization factor $1/\sqrt{2}$. The mixing of two energy states due to the interaction with an electromagnetic field will be discussed in Section 8.2.

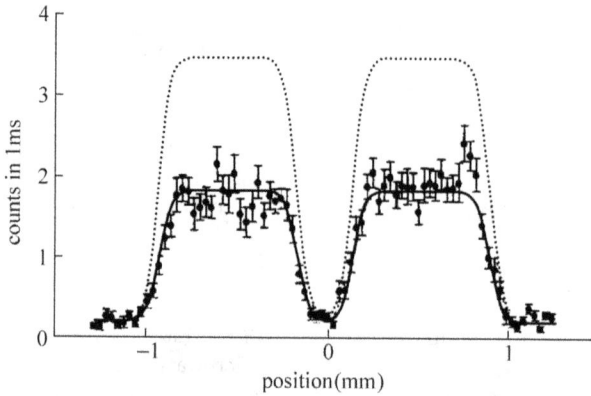

Fig. 1.9 Interference pattern disappears under the envelope due to the storage of the "which way" information in the internal state of the atom. Reprinted with permission from Macmillan Publishers Ltd: Nature (395, 33), 1998.

shape as that in Fig. 1.9, only the absolute magnitude is decreased to $1/2$ of the original value. The reason for the disappearance of the interference pattern is the entanglement of the internal degree of freedom and that of the center of mass motion of the atom. The wave function of the atom after it has passed the second beam splitter is

$$|\psi\rangle \propto -|\psi_D\rangle \otimes |2\rangle + |\psi_E\rangle \otimes |3\rangle + |\psi_F\rangle \otimes |2\rangle + |\psi_G\rangle \otimes |3\rangle,$$

where the entanglement is explicit. Upon reflection once $|2\rangle$ acquires a minus sign. This is the case with the term containing $|\psi_D\rangle$. The term containing $|\psi_F\rangle$ has a plus sign because it is reflected twice. In the far field and under the left envelope (where ψ_F and ψ_G are both 0) the distribution in the position z of the atom is

$$P_{\rm L}(z) \propto |\psi_D(z)|^2 + |\psi_E(z)|^2 - \psi_D^*(z)\psi_E(z)\langle 2|3\rangle - \psi_E^*(z)\psi_D(z)\langle 3|2\rangle.$$

The third and fourth term could have given the effect of interference, but they vanish because $|2\rangle$ and $|3\rangle$ are orthogonal. The role of the "which way" indicator, the internal state, is now clear. It is interesting enough to notice [5] that the microwave field does lead to a momentum change Δp_z of the atom when it prepares the "which way" information. Calculation gives the shift of the position of the atom $\Delta z = \pm 10$ nm on the detecting plane. Actually the scale of the distance between the fringes is of the order of mm. Compared with this Δz is unobservable. The entanglement exists whenever a potential "which way" judgement is arranged, including the illuminated slit experiment (Feynman) and the freely suspending slit plane experiment (Einstein) as well. In these two examples the uncertainty of momentum introduced by the "which way" measurement is just of the right magnitude to explain the disappearance of interference pattern, so that in some textbooks the only reason for the disappearance of interference is ascribed to the momentum uncertainty, and the entanglement is not mentioned at all. From the Pritchard's and Rempe's experiments the decisive role of entanglement becomes transparent.

While more experiments demonstrating the complementarity principle will be discussed later, it is proper to comment here on the apparent disparity between the electromagnetic wave (wave of the photon) and the matter wave (wave of the electron, the neutron or the atom). Conceptual problem does not arise with the electromagnetic wave which represents a space-time variation of the field E and B, carriers of energy and momentum. The matter wave of an electron is interpreted as the probability amplitude, something mathematical rather than realistic. Actually the electromagnetic wave of a photon is also probabilistic. It is the macroscopic number of photons which turn the wave into an energy/momentum distribution varying with time. The same cannot happen with the electron. Only one electron can populate a quantum state because of Pauli's principle, while a macroscopic number of photons can populate a state (e.g., a given mode of electromagnetic field). The wave properties, e.g., interference of an electron, can only be observed by the accumulation of the effect due to a large number of electrons in the same condition (e.g., momentum) repeated many times. This apparent disparity disappears when we have the Bose-Einstein condensates of gaseous atoms. They are bosons "condensed" in the same quantum state. The matter wave of such atoms exhibits very similar properties as laser waves (e.g., the phase coherence). We have an exact parallelism between the wave of photons and the matter wave of the bosonic atoms.

1.3   Verification of Complementarity by Quantum Optical Experiments

As discussed in many textbooks the loss of coherence is ascribed to Heisenberg's uncertainty principle of position and momentum. We have seen that this is actually unnecessary. In fact M. O. Scully, B.-G. Englert and H. Walther [8] thought of aiming at providing the "which way" information while making a detour to avoid the uncertainty principle. Their idea is based on the new achievement of the micromaser in quantum optics. An atom in a metastable excited state will emit a photon while making a transition to a lower state when it passes through a high Q resonant micromaser cavity. The enhancement of emission rate of an atom in a resonant cavity is a problem studied in the cavity quantum electrodynamics (cf. Chapter 8). Scully, Englert and Walther suggested an equipment shown in Fig. 1.10. Rb atoms excited by a resonant laser to a long-live Rydberg state 63 $p_{3/2}$ enter the equipment. If the micromaser cavity is absent, the wave function of the atom which has passed

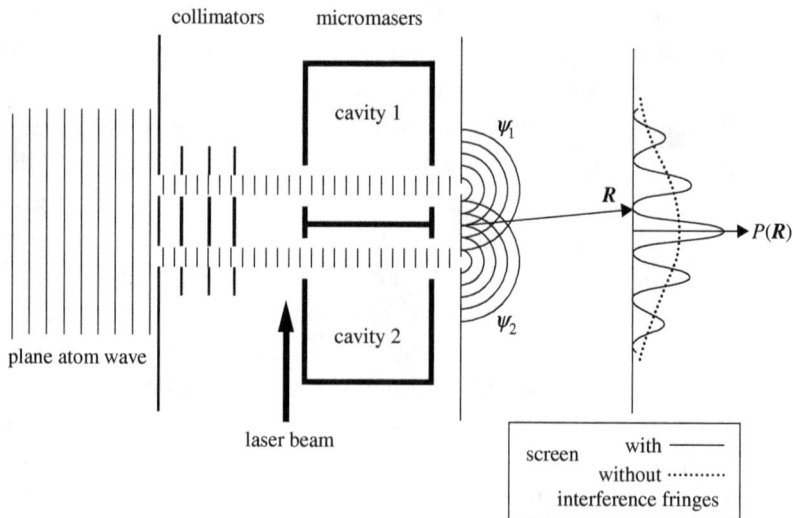

Fig. 1.10   Double slit experiment with beams passing through a micromaser cavity. Redrawn from Ref. [8].

through the double slit is

$$\psi(\boldsymbol{r}) = \frac{1}{\sqrt{2}}(\psi_1(\boldsymbol{r}) + \psi_2(\boldsymbol{r}))|a\rangle, \qquad (1.3.1)$$

where $\boldsymbol{r}$ is the coordinates of the center of mass of the atom, $|a\rangle$ represents the internal state $63p_{3/2}$ of the atom. The probability density distribution at $\boldsymbol{R}$ on the screen is given by $|\psi(\boldsymbol{R})|^2$.

$$P(\boldsymbol{R}) = |\psi|^2 = \frac{1}{2}[|\psi_1|^2 + |\psi_2|^2 + (\psi_1^*\psi_2 + \psi_2^*\psi_1)]\langle a|a\rangle$$

$$= \frac{1}{2}[|\psi_1|^2 + |\psi_2|^2 + (\psi_1^*\psi_2 + \psi_2^*\psi_1)]. \qquad (1.3.2)$$

The third term in the bracket represents the interference effect. If the micromaser is installed, it then induces the transition of the atom from the Rydberg state $63p_{3/2}$ to $61d_{5/2}$(denoted by b) or $61d_{3/2}$(denoted by c) with the emission of a photon (-21GHz). Let the cavity be tuned to be resonant to the transition a→b, then the atom emits a photon when it passes through the cavity. We obtain information on the path of the atom by observing in which cavity a photon is left. After the emission of a photon the internal state of the atom changes while the center of mass motion does not change.[6] After the emission of the photon the internal state of the atom is now entangled with the state of the micromaser cavity. The state of the atom/cavity system is

$$\psi(\boldsymbol{r}) = \frac{1}{\sqrt{2}}[\psi_1(\boldsymbol{r})|1_1 0_2\rangle + \psi_2(\boldsymbol{r})|0_1 1_2\rangle]|b\rangle, \qquad (1.3.3)$$

where $|1_1 0_2\rangle$ represents a state with 1 photon in cavity 1 and 0 photon in cavity 2, and $|0_1 1_2\rangle$ represents 0 photon in cavity 1 and 1 photon in cavity 2. The difference between (1.3.1) and (1.3.3) lies on the fact that (1.3.3) is no more a simple direct product,[7] but an entangled

---

[6]Disregarding the change of momentum of the atom due to the emission of the microwave photon.

[7](1.3.1) actually corresponds to $\frac{1}{\sqrt{2}}[\psi_1(r) + \psi_2(r)]|0_1 0_2\rangle|a\rangle$.

state. The probability distribution on the screen is

$$P(\boldsymbol{R}) = \frac{1}{2}[|\psi_1|^2 + |\psi_2|^2 + \psi_1^*\psi_2\langle 1_1 0_2|0_1 1_2\rangle + \psi_2^*\psi_1\langle 0_1 1_2|1_1 0_2\rangle]\langle b|b\rangle$$

$$= \frac{1}{2}[|\psi_1|^2 + |\psi_2|^2]. \tag{1.3.4}$$

There is no more interference effect in the distribution because $|1_1 0_2\rangle$ and $|0_1 1_2\rangle$ are orthogonal to each other. Consequently the interference disappears once there is a possibility of judging the path, while the actual measurement is not mandatory. Here Heisenberg's uncertainty principle does not play any role, and the disappearance of interference is ascribed to the entanglement of the center-of-mass motion of the atom and the state of the photon.

It is natural to ask a further question, since the center-of-mass motion has not suffered from any "uncontrollable interaction", whether the interference pattern can be recovered when the correlation between the atom and the photon is erased. This is the idea of the quantum eraser shown in Fig. 1.11. In the atom beam/micromaser system the two cavities are separated by a shutter-detector system. When the shutters are closed, the photon remains in the upper or the

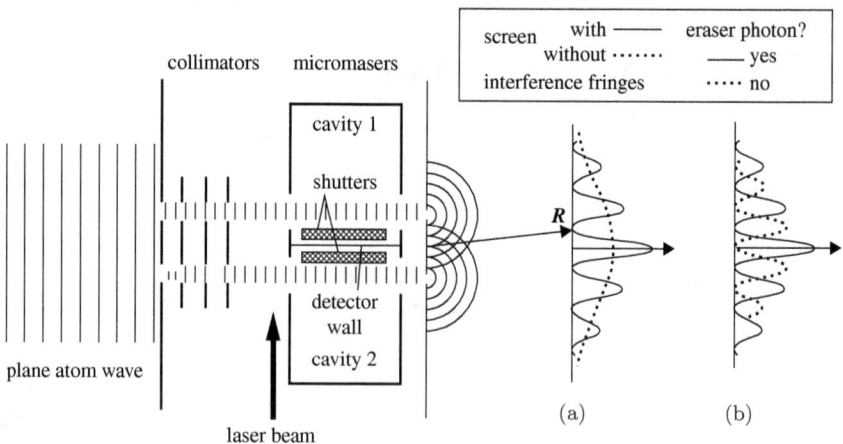

Fig. 1.11   (a) Quantum eraser equipped on the micromaser; (b) Density distribution of atoms on the screen depending on whether a photon is absorbed(yes) or not(no). Redrawn from Ref. [8].

lower cavity. When the atom has already passed through the slits, the shutters are open and the photon can interact with the thin-film semiconductor covering the wall separating the cavities. Upon the absorption of the photon, the detector undergoes a transition from its ground state $|g\rangle$ to an excited state $|e\rangle$. With the photon absorbed, the "which way" information is erased. Can the interference pattern be recovered? The answer is positive, provided the detector possesses a special prescribed property. The important point is, at the time of information erasure, the atom is already about to reach the screen. How can a decision made by an experimentalist to keep the shutters closed or to open them can influence the behavior of the atom (as a particle or a wave) immediately? Is this an action-at-a-distance? This is the question raised by the Einstein-Podolsky-Rosen paradox and answered by the delayed-choice experiments suggested by J. Wheeler.[8] Actually the following discussion on the quantum eraser is also a kind of delayed-choice experiment. Before the shutters are opened, the wave function of the atom/cavity/detector system is

$$\psi(\boldsymbol{r}) = \frac{1}{\sqrt{2}}[\psi_1(\boldsymbol{r})|1_1 0_2\rangle + \psi_2(\boldsymbol{r})|0_1 1_2\rangle]|b\rangle|g\rangle. \tag{1.3.5}$$

Define the symmetric wave function ψ_+ and anti-symmetric wave function ψ_- as

$$\psi_\pm(\boldsymbol{r}) = \frac{1}{\sqrt{2}}(\psi_1(\boldsymbol{r}) \pm \psi_2(\boldsymbol{r})), \tag{1.3.6}$$

and the symmetric state $|+\rangle$ and the anti-symmetric state $|-\rangle$ of the cavity field as

$$|\pm\rangle = \frac{1}{\sqrt{2}}(|1_1 0_2\rangle \pm |0_1 1_2\rangle). \tag{1.3.7}$$

The wave function (1.3.5) can be re-written as

$$\psi(\boldsymbol{r}) = \frac{1}{\sqrt{2}}(\psi_+(\boldsymbol{r})|+\rangle + \psi_-(\boldsymbol{r})|-\rangle)|b\rangle|g\rangle. \tag{1.3.8}$$

---

[8]This issue will be discussed in Sections 1.7 and 1.9.

If the detector is only sensitive to the photon in the state $|+\rangle$ and insensitive to the photon in the state $|-\rangle$,[9] the wave function of the entire system becomes

$$\psi(\boldsymbol{r}) = \frac{1}{\sqrt{2}}[\psi_+(\boldsymbol{r})|0_1 0_2\rangle|e\rangle + \psi_-(\boldsymbol{r})|-\rangle|g\rangle]|b\rangle, \qquad (1.3.9)$$

where the first term originates from the first term of (1.3.8), with the photon $|+\rangle$ absorbed leaving a state $|0_1 0_2\rangle$ for the cavity field and the state $|e\rangle$ for the detector. The density distribution of the atom on the screen is then

$$P(\boldsymbol{R}) = \frac{1}{2}(|\psi_+(\boldsymbol{R})|^2 + |\psi_-(\boldsymbol{R})|^2) = \frac{1}{2}(|\psi_1(\boldsymbol{R})|^2 + |\psi_2(\boldsymbol{R})|^2), \quad (1.3.10)$$

where no interference term exists as long as the final state of the detector is unknown. Under the condition that the detector is found to be in the excited state, the distribution is then

$$P_e(\boldsymbol{R}) = \frac{1}{2}|\psi_+(\boldsymbol{R})|^2 = \frac{1}{4}[|\psi_1(\boldsymbol{R})|^2 + |\psi_2(\boldsymbol{R})|^2 + \mathrm{Re}(\psi_1^*(\boldsymbol{R})\psi_2(\boldsymbol{R}))], \tag{1.3.11}$$

where the interference fringe is recovered (solid line Fig. 1.11(a)). When the detector is in the ground state, the probability density distribution of the atom is

$$P_g(\boldsymbol{R}) = \frac{1}{4}[|\psi_1(\boldsymbol{R})|^2 + |\psi_2(\boldsymbol{R})|^2 - \mathrm{Re}(\psi_1^*(\boldsymbol{R})\psi_2(\boldsymbol{R}))], \qquad (1.3.12)$$

where the interference pattern also persists, giving rise to the "antifringes"(dotted line, Fig. 1.11(b)). The fringes and antifringes add up to wash out the interference pattern, corresponding to the fact that when the state of the detector is irrelevant (entanglement not made use of), the interference disappears. The foregoing discussion demonstrates the principle of the quantum eraser and represents a thought delayed-choice experiment, clarifying that the behavior of the atom depends on the condition of the equipment, and no message

---

[9]This is just the special prescribed property of the detector, which enables both cavities to be in a state with no photons and thus erasing the information on path.

is to be sent from the experimentalist to the atom. The equipment suggested is very difficult to realize and the first quantum eraser was realized in the laboratory by R. Chiao's group using the two-photon interferometry (cf. Section 1.6).

A question can be raised why the radiation field states $|+\rangle$ and $|-\rangle$ have to be chosen? This is necessary because in (1.3.9) the two terms have to be orthogonal in the states of the detector. Then a correlation between the atom and the detector states will give rise to fringes and antifringes which sum up to the absence of interference, clarifying that the correlation/entanglement is the decisive factor governing the behavior of the atom.

There was a dispute when the idea on quantum eraser was presented on a conference [9, 10]. Yanhua Shih *et al.* constructed a "delayed choice" quantum eraser in 2000, we postpone the discussion until Section 1.6.

1.4 Single Photon Interference Experiment

In the first chapter of his classical textbook "Principles of Quantum Mechanics" Dirac discussed the quantum mechanical description of the interference of photons as, each photon interferes only with itself. Several decades of years elapsed when many experiments attempted to verify this conclusion. Different interferometers with very faint light sources, sometimes even with "debunching device" were used. The aim was to ensure that only one photon exists in the equipment in a reasonably long duration. This was to demonstrate that a photon interferes with itself. In those experiments interference pattern was found and people thought that Dirac's conclusion was proved to be true. Research in the 1980s on the non-classical effect in the statistical properties of light launched a challenge to this seemingly flawless assertion.

The Orsay group (P. Grangier, G. Roger and A. Aspect) [11] carried out the study on a beam splitter shown in Fig. 1.12. Light pulses from the source S reach the beam splitter BS. A trigger synchronized with the light pulse opens a window of width w for the counting device. Photomultipliers PM_t and PM_r record the counting

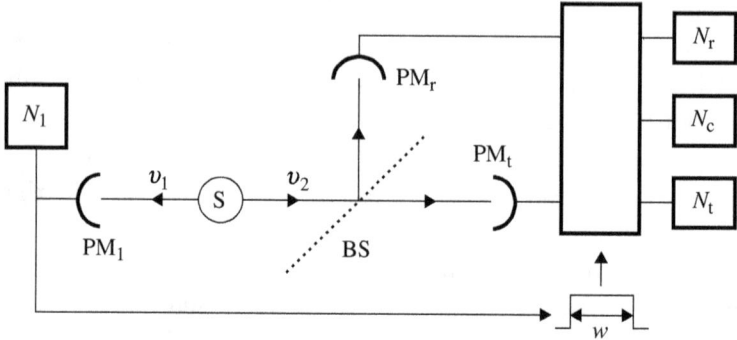

Fig. 1.12 Equipment for checking single photon emission of a light source. Reprinted with permission from P. Grangier, G. Roger and A. Aspect, *Europhys. Lett.* 1, 173, 1986.

rates N_t and N_r in the transmitted and reflected channel respectively. The coincidence counting rate is denoted by N_c. Let N_w be the rate of the gates. The probabilities for reflection, transmission and coincidence counts are respectively

$$P_r = \frac{N_r}{N_w}, \quad P_t = \frac{N_t}{N_w}, \quad P_c = \frac{N_c}{N_w}. \tag{1.4.1}$$

We begin from the classical wave description of light. Let i be the average intensity of light during a time window and α_t be the product of the transmission efficiency and the detection efficiency of the transmission channel. Then

$$P_t = \alpha_t w i.$$

Similarly, we have

$$P_r = \alpha_r w i,$$
$$P_c = \alpha_r \alpha_t w^2 i^2.$$

Carrying out ensemble average over many pulses/windows we obtain

$$P_t = \alpha_t w \langle i \rangle,$$
$$P_r = \alpha_r w \langle i \rangle,$$
$$P_c = \alpha_t \alpha_r w^2 \langle i^2 \rangle.$$

According to the Cauchy-Schwartz inequality

$$\langle i^2 \rangle \geq \langle i \rangle^2$$

we have

$$P_{\rm c} \geq P_{\rm t} P_{\rm r}. \qquad (1.4.2)$$

Define

$$\alpha = \frac{N_{\rm c}/N_{\rm w}}{\frac{N_{\rm t}}{N_{\rm w}} \frac{N_{\rm r}}{N_{\rm w}}}, \qquad (1.4.3)$$

then

$$\alpha \geq 1. \qquad (1.4.4)$$

These inequalities mean that the classical coincidence probability $P_{\rm c}$ is greater than the "accidental coincidence" probability $P_{\rm r} P_{\rm t}$. α is called the anticoincidence parameter. The violation of the inequality (1.4.4) gives an anticorrelation criterion characterizing a non-classical behavior. If we consider that photons are emitted by the source one by one, then $P_{\rm c} = 0$, since a photon can only enter one of the channels, the transmission or the reflection channel. Therefore $\alpha = 0$. The Orsay group carried out measurement using photodiode. Weak light source has an intensity of 1 counts per 1 000 pulses. The detection efficiency is about 10%. Then there are 0.01 photons per pulse. Surprisingly the result of the measurement was $\alpha \approx 1$. This shows that usual light sources, however faint they may be, exhibit classical behavior. In a classical light source macroscopic number of atoms are in the excited states. Many atoms emit almost simultaneously within a very short interval. A classical light source has a statistical property that the probability of emitting a second photon after a time delay τ from the first is finite, no matter how small τ is. The laser also belongs to the category of classical light sources. This means that to demonstrate directly Dirac's assertion non-classical light source with $\alpha \ll 1$ must be used. For a finite width of the window accidental coincidence cannot be avoided, so that the value of α cannot be strictly zero. The Orsay group constructed a single photon light source. A Ca atom is excited via a two-photon

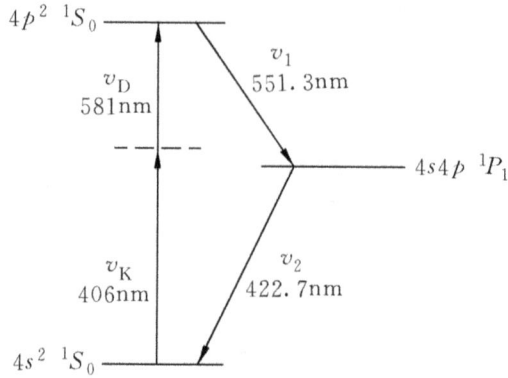

Fig. 1.13   Emission cascade of Ca atom, used for single photon light source.

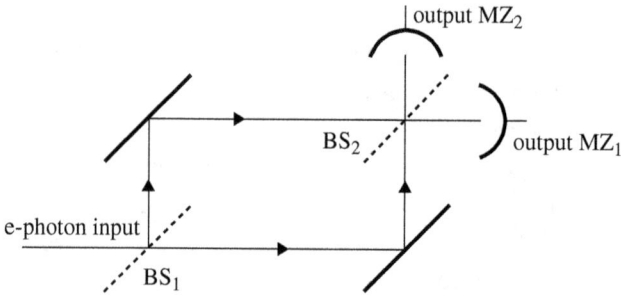

Fig. 1.14   Scheme of the Mach-Zehnder interferometer. Reprinted with permission from P. Grangier, G. Roger and A. Aspect, *Europhys. Lett.* 1, 173, 1986.

resonance to a state $4p^2\ {}^1S_0$. It emits two photons in a cascade separated by 4.7 ns. The energy level diagram is shown in Fig. 1.13. The first photon emitted can be used as the trigger for the photon detector and the second photon enters the beam splitter. Under experimental conditions the possibility that a photon emitted by another Ca atom also enters the beam splitter is highly improbable. Under a triggering rate $8800s^{-1}$ a trial of 5 hrs leads to a value $\alpha = 0.18 \pm 0.06$, which differs from the classical value $\alpha = 1$ by 13 standard deviations.

The Orsay group carried out interference experiments using this single photon source. The scheme of the Mach-Zehnder interferometer is given in Fig. 1.14. Outputs MZ_1 and MZ_2 are directed to photomultipliers. Dashed lines represent beam splitters BS_1 and BS_2. The optical path difference of the two arms can be adjusted by moving the mirrors. At the beam splitters the reflected beam has a phase difference $\pi/2$ behind the transmitted beam. If the two arms have the same length, then the two beams entering MZ_1 have phase difference 0 and the beams entering MZ_2 have phase difference π. There should be no counts in output MZ_2. When the length difference of the two arms is continuously varied the counts in the two detectors vary periodically and have opposite phases. The result is shown in Fig. 1.15. The path difference is expressed in terms of the number of channels, each channel being equivalent to $\lambda/50$. This experiment

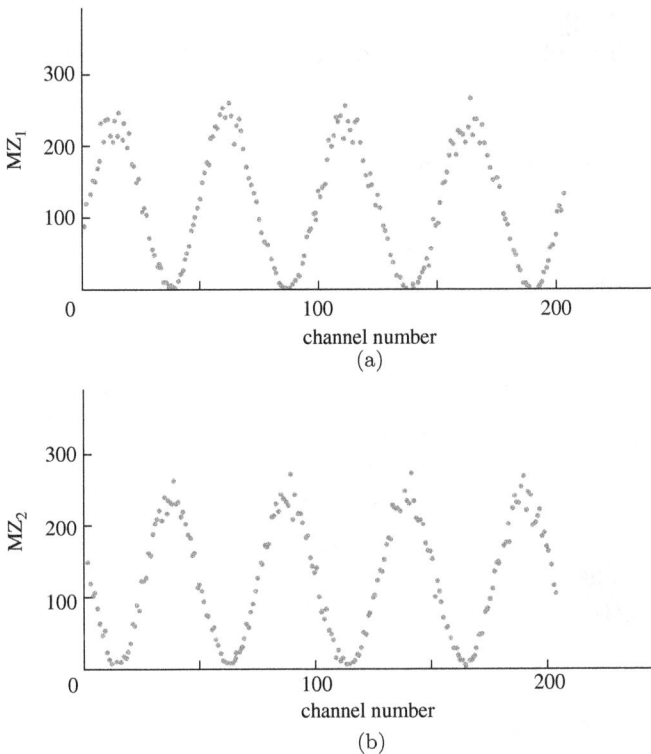

Fig. 1.15 Output counts MZ_1 and MZ_2 versus the optical path difference. Reprinted with permission from P. Grangier, G. Roger and A. Aspect, *Europhys. Lett* . 1, 173, 1986.

succeeded in demonstrating the interference of a photon with itself using a single photon source with parameter $\alpha = 0.18$.

1.4.1 Interference between independent photon beams

A beam of light is split into two components on a beam splitter and the two components are subsequently made to meet and interfere. Dirac [59] emphasized that each photon interferes only with itself, and interferences between two different photons never occurs. Each photon in the original beam splits into two components, all of them. Each photon eventually interferes with itself in the sense that interference occurs between its its two components, not between two separate photons. Dirac's statement does not refer to or deny the possibility of observing interference in the superposition of two incoherent beams.

A.T. Forrester, R.A. Gudmundsen and P.O. Johnson [60] carried out an experiment in which they obtained beats between two incoherent wave trains. They used two Zeeman components of the 546.1 nm line of ^{202}Hg from a thermal source. These components are of course emitted by different atoms of the same source. The light was focused onto a photocathode and the ejected electrons were directed into a microwave cavity and recorded. Because of the incoherence between the spectral lines and incoherence between the beats in different photocathode areas, the effect was a very weak one: the signal-to-noise ratio at the cavity was only 3×10^{-5}. By using optical modulation of the beats while maintaining the total intensity constant, they were able to bring the signal-to-noise ratio to 2 at the indicator. The beat phenomenon can be understood as interference, because it is an effect indicating superposition of electromagnetic fields, and the field strengths add rather than intensities. R. Hanbury Brown and R.Q. Twiss [61] studied the intensity fluctuation in thermal fields using stellar sources. They measured correlations in photocurrents of two detectors, i.e., the correlation function

$$G^{(2)}(\boldsymbol{r}_1, t; \boldsymbol{r}_2, t) = \langle I(\boldsymbol{r}_1, t) I(\boldsymbol{r}_2, t) \rangle, \qquad (1.4.5)$$

where $\langle\ \rangle$ means the ensemble average, as a function of the distance between two detectors $|r_1 - r_2|$. They found the interesting result that $G^{(2)}$ exhibits a maximum at $|r_1 - r_2| = 0$, and decreases with increasing distance until a definite constant level is attained. The critical value of the distance at which the correlation function approaches the constant value is the transverse coherence length of the stellar light on the earth surface. It is a consequence of interference between incoherent sources which the authors used to estimate the diameter of the star. From the quantum mechanical point of view, this phenomenon is the interference between the choice that photon 1 enters detector I while photon 2 enters detector II and the choice of interchanging the outcome of photons.

Studies using laser beams were carried out by many researchers. Interference was observed even if the beams were strongly attenuated. R.L. Pfleeger and L. Mandel [62] used two independently operated single-mode lasers and found the following. Interference takes place even under conditions in which the light intensities are so low that one photon is absorbed before the next one is emitted by one or the other souce. Photon correlation techniques were used to demonstrate the interference because the average number of registered photons per trial was only about 10. The correlation coefficient was obtained. For the theoretical interpretation to the experiments discussed above, the classical theory of electromagnetic fields gives results in full agreement with experiments. This is the case even when the beams are strongly attenuated. The reason is, when Glauber states $|\alpha\rangle \equiv |\alpha|e^{i\theta}$ are used, the attenuation decreases the amplitude $|\alpha|$ but keeps θ intact. The phase relation remains. H. Paul [63] calculated the intensity correlation function (1.4.5) using classical fields of two Hertzian oscillating dipoles located at r_I and r_{II} emitting photons k_1 and k_2 These photons are to be absorbed by detectors situated at r_1 and r_2, and the result is

$$G^{(2)}(r_1, t; r_2, t) = 4f^4\tilde{a}^4 \left\{ 1 + \frac{1}{2}\cos[(k_2 - k_1)(r_{II} - r_I)] \right\}, \quad (1.4.6)$$

where f is the amplitude of the electric field $E^{(+)}$ (associated with $e^{-i\omega t}$), and $\tilde{a}(t) = \exp(-\frac{\Gamma}{2}t)$ is the damping factor of the dipole moment operator. The dependence of $G^{(2)}(r_1, t; r_2, t)$ on $|r_1 - r_2|$ is realized through $k_2 - k_1$ assuming that the distance between the two dipoles is much smaller than the source-detector distance.

Interesting situation happens when we have two excited atoms emitting photons. The resulting field can no more be described by a density operator using Glauber states. The electromagnetic field has to be quantized, and the photons represent the quanta of the field. Let $E^{(\pm)}(r, t)$ be the positive and negative frequency part (annihilation and creation operators respectively) of the electric field at r and t. The interference is described by the intensity correlation function defined as

$$G_{\mathrm{qu}}^{(2)}(r_1, t; r_2, t) = \sum_{i,j=1}^{3} \langle E_i^{(-)}(r_1, t) E_j^{(-)}(r_2, t) E_j^{(+)}(r_2, t) E_i^{(+)}(r_1, t) \rangle.$$

(1.4.7)

The result turns out to be

$$G_{\mathrm{qu}}^{(2)}(r_1, t; r_2, t) = 2f^4 \widetilde{a}^4 \{1 + \cos[(k_2 - k_1)(r_{\mathrm{II}} - r_{\mathrm{I}})]\}. \qquad (1.4.8)$$

Comparing (1.4.8) with (1.4.6) we see that the difference between the quantum and the classical theories is remarkable. While the possible minimum value of $G^{(2)}$ is $1/2$ of the average value in the classical theory (1.4.6), it is 0 in the quantum theory (1.4.8). The quantum theory gives a much stronger correlation. As pointed by H. Paul [63], the equivalence between quantum and classical description in the Glauber representation rests upon the fact that the number of photons is uncertain, without an upper bound. However, when we are dealing with fields produced by two atoms, the number of photons is bound from above, and in this case quantum and classical descriptions differ.

Finally it might be asked whether the effects discussed above in anyway contradict Dirac's statement. L. Mandel [64] pointed out that the answer is that they clearly do not. Actually, there are two photons recorded, and the resulting correlation depends on a relative phase, i.e., interference phenomenon has been observed. Does this mean the interference between two photons emitted by two atoms? The crucial point is that whether one can assign each recorded photon to one or the other atom. If the answer is affirmative, then interference does occur between two independent photons. But it turns out that the opposite happens. The photon recorded in photoelectric measurement automatically rules out the possibility of knowing its momentum, and with it the possibility of assigning the photon to one or the other beam. More specifically Pfleeger and Mandel [62] demonstrated this point using the uncertainty relation. Just as in conventional interferometry, each photon is to be considered as being partly in both beams, and interferes only with itself. Each recorded photon is just a quantum of the quantized electromagnetic field established by two beams (atoms), it is not associated with any specific beam (atom), or it is associated with both beams (atoms). Interference phenomena arising from two photons will be further discussed in Section 1.5. The assertion made in Ref. [63] that Dirac's statement that "interference between two different photons never occurs" is proved to be false is inappropriate.

1.5 Multiparticle Interferometry

Beginning from 1980s the down conversion of photons from a laser beam in a non-linear crystal has been used in many laboratories to produce pairs of correlated photons. By using this process "two-photon entangled state" and two-photon interferometers can be constructed. The first idea of such an application was due to M. A. Horne and A. Zeilinger [12, 13]. A number of groups have established their equipments. L. Mandel's group pioneered in studying the fundamental issues of quantum mechanics, revealing the non-classical nature of light and providing more accurate experimental verification of the violation of Bell's inequality. New results have been emerging in this direction.

1.5.1 Two-particle double-slit interferometry

Let us consider the two-particle interferometer shown schematically in Fig. 1.16, where a particle source of length d is situated at O. A particle at S decays into two daughter particles. Since the decaying particle is at rest, the momenta of the daughters are basically equal

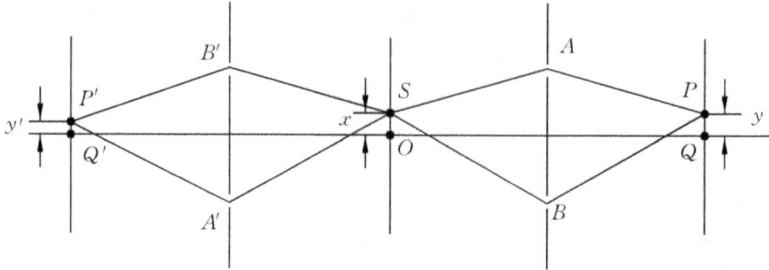

Fig. 1.16   Scheme of a two-particle interferometer.

and opposite. A pair of daughter particles can pass through holes at
A and A' and fall onto P and P' on the screens, but the same pair
can pass through B and B' and fall onto the same points P and P'.
The wave function of such a pair is

$$|\psi\rangle = \frac{1}{\sqrt{2}}(|A\rangle_1|A'\rangle_2 + |B\rangle_1|B'\rangle_2), \tag{1.5.1}$$

where $|A\rangle$ represents a state with momentum pointing toward A.
Others are defined similarly. One of the daughter particles has two
choices of its path, viz. SAP or SBP. The other can then choose
$SA'P'$ or $SB'P'$. Let the distance of the decaying particles from
O be x, and the distances PQ and $P'Q'$ denoted by y and y', the
distance $OAQ = OBQ = L$, the angle $\angle AOQ$ be θ. Then we have

$$SAP = L - x\theta - y\theta,$$
$$SBP = L + x\theta + y\theta.$$

The probability amplitude at P is

$$\psi(P) \propto e^{ikL}(e^{-ik(x+y)\theta} + e^{ik(x+y)\theta})$$
$$\propto \cos\frac{2\pi}{\lambda}(x+y)\theta, \tag{1.5.2}$$

where λ is the de Broglie wave length and the wave vector is $k = 2\pi/\lambda$. The probability amplitude for coincidence counts of a pair of daughter particles at P and P' is

$$\psi(y, y') \propto \frac{1}{d} \int_{-d/2}^{d/2} dx \cos \frac{2\pi}{\lambda}(x + y)\theta \cos \frac{2\pi}{\lambda}(x + y')\theta. \qquad (1.5.3)$$

Rewrite the integrand as

$$\cos \frac{2\pi}{\lambda}(x + y)\theta \cos \frac{2\pi}{\lambda}(x + y')\theta$$

$$= \frac{1}{2} \cos \frac{2\pi}{\lambda}(y - y')\theta + \frac{1}{2} \cos \frac{2\pi}{\lambda}(2x + y + y')\theta,$$

consequently

$$\psi(y, y') \propto \frac{1}{2} \cos \frac{2\pi}{\lambda}(y - y')\theta + \frac{1}{2d} \int_{-d/2}^{d/2} dx \cos \frac{2\pi\theta}{\lambda}(2x + y + y').$$

$$(1.5.4)$$

We discuss two limiting cases

(1) $d \gg \lambda/\theta$. The second term in (1.5.4) is small and can be neglected, and the first term represents the "conditional fringe", i.e., the periodic variation of the coincidence counts under a fixed P' and a variable P. For any pair of particles there is a choice between the two alternatives, and consequently there is the interference between the two alternatives.

(2) $d \ll \lambda/\theta$. Then x in the integrand of (1.5.3) can be neglected and it follows that

$$\psi(y, y') \propto \cos \frac{2\pi}{\lambda}y\theta \cos \frac{2\pi}{\lambda}y'\theta. \qquad (1.5.5)$$

This is the product of the interference amplitudes of two independent particles. Each particle interferes with itself independently. We can understand the above limits from an analysis in the momentum space.

(1) The uncertainty in transverse momentum is $\delta k_\perp \propto 1/d$ according to the uncertainty principle. If $d \gg \lambda/\theta$ then it follows that

$$\delta k_\perp \ll \frac{\theta}{\lambda}, \quad \frac{\delta k_\perp}{k_\perp} \ll \frac{\theta}{2\pi}. \tag{1.5.6}$$

The uncertainty of the transverse momentum is too small for both of the particles that the holes A and B (or A' and B') cannot be illuminated simultaneously by the waves corresponding to each of the particles. Thus each of the particles cannot interfere with itself, and the single particle interference fringes are absent. The conditional fringes of two-particle interference remain.

(2) If $d \ll \lambda/\theta$, then it implies that $\frac{\delta k_\perp}{k_\perp} \gg \frac{\theta}{2\pi}$. Each particle can illuminate the holes on its own side, and single particle interference ensues. The uncertainty in transverse momenta is, however, too large such that when one daughter particle passes through A, there is no guarantee that her twin sister particle can pass through A'. The same applies to the holes B and B'. The two-particle entangled state (1.5.1) does not hold. There will be no two-particle interference.

From the foregoing discussions we see an important difference between the two-particle interference and the single particle interference. The single particle interference of a photon can be described by the classical electromagnetic wave theory, but the two-particle interference begins with the two-particle entangled state, which does not have a classical counterpart, and is based on quantum mechanical principles. It is a pure non-classical phenomenon.

1.5.2   Interference experiment of a pair of down-converted photons

The down-conversion phenomenon happens when a laser beam passes through a non-linear crystal, where the original photon with wave number $\boldsymbol{k}$ is transformed into two correlated photons with wave vectors $\boldsymbol{k}_A$ and $\boldsymbol{k}_C$ respectively, with $\boldsymbol{k} = \boldsymbol{k}_A + \boldsymbol{k}_C$. Theory [14] and experiment [15] show that the momenta of the down-converted pair are asymmetrically oriented with respect to the momentum of the original photon in the case $|\boldsymbol{k}_A| \neq |\boldsymbol{k}_C|$, as shown in Fig. 1.17. On an aperture downstream of the non-linear crystals S four pinholes are

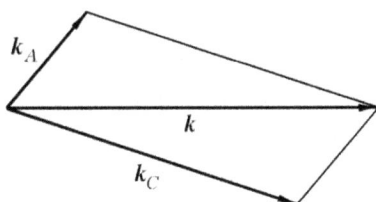

Fig. 1.17   Asymmetric down-converted pair.

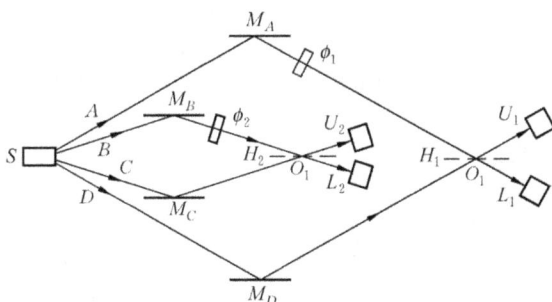

Fig. 1.18   Scheme of a two-photon interferometer. Reprinted with permission from Michael A. Horne, Abner Shimony, and Anton Zeilinger, *Phys. Rev. Lett.* 62, 2209, 1989. Copyright (2017) by the American Physical Society.

drilled such that the down-converted pair can have a choice, either the photons pass through A and C, or they pass through B and D, with

$$|\mathbf{k}_A| = |\mathbf{k}_D|, \quad |\mathbf{k}_B| = |\mathbf{k}_C|. \tag{1.5.7}$$

In this way a pair of correlated photons is obtained in an entangled state:

$$|\psi\rangle = \frac{1}{\sqrt{2}}[|\mathbf{k}_A\rangle_1|\mathbf{k}_C\rangle_2 + |\mathbf{k}_D\rangle_1|\mathbf{k}_B\rangle_2]. \tag{1.5.8}$$

In the two-photon interferometer shown in Fig. 1.18 the beam A is reflected by the mirror M_A, passes through the phase shifter ϕ_1, arrives at the beam splitter H_1 and ends up in the detector U_1 or L_1. The partner beam C is reflected by the mirror M_C, arrives at the beam splitter H_2 and ends up in the detector U_2 or L_2. The above story represents one of the two alternatives that the pair can choose. The other alternative is that the first photon passes along D to M_D,

H_1 and enters U_1 or L_1, the partner photon passes along B to M_B, ϕ_2, H_2 and enters U_2 or L_2. The detectors U_1 and L_1 record the interference between the paths that the first photon can choose (A or D), U_2 and L_2 record the interference between the paths that the second photon can choose (B or C). What can be recorded are the coincidence counts for two photons as functions of ϕ_1 and ϕ_2, the results are then compared with the calculations using quantum mechanics.

Let the quantum efficiency of the detector be η. Then the coincidence counting rate of U_1 and U_2 is $\eta^2 |A(U_1 U_2 \mid \phi_1 \phi_2)|^2$, where $|A(U_1 U_2 \mid \phi_1 \phi_2)|$ is the coincidence probability amplitude of U_1 and U_2 as a function of ϕ_1 and ϕ_2. It is the superposition of probability amplitudes corresponding to the two correlated pairs of paths AC and DB:

$$A(U_1 U_2 | \phi_1 \phi_2) = \frac{1}{\sqrt{2}} \left[\left(\frac{1}{\sqrt{2}} i e^{i\phi_1} \right) \frac{1}{\sqrt{2}} + e^{i\theta} \left(\frac{1}{\sqrt{2}} \right) \left(\frac{1}{\sqrt{2}} i e^{i\phi_2} \right) \right],$$
$$(1.5.9)$$

where $e^{i\phi_1}, e^{i\phi_2}$ are the phase factors acquired by beams A and B through phase shifters, $\frac{1}{\sqrt{2}}$ and $\frac{1}{\sqrt{2}} i$ correspond to the transmission and reflection respectively at the beam splitters. Each term in the bracket has two factors corresponding to the correlated paths of the first and the second photon respectively. The factor $e^{i\theta}$ depends on the arrangement of reflecting mirrors and beam splitters, and is independent of ϕ_1 and ϕ_2. Similarly we obtain

$$A(U_1 L_2 | \phi_1 \phi_2) = \frac{1}{\sqrt{2}} \left[\left(\frac{1}{\sqrt{2}} i e^{i\phi_1} \right) \left(\frac{1}{\sqrt{2}} i \right) + e^{i\theta} \left(\frac{1}{\sqrt{2}} e^{i\phi_2} \right) \left(\frac{1}{\sqrt{2}} \right) \right],$$

and $A(L_1 U_2 \mid \phi_1 \phi_2)$, $A(L_1 L_2 \mid \phi_1 \phi_2)$. The coincidence counting rate between two detectors is the product of η^2 and the modulus squared of the coincidence amplitude:

$$P(U_1 U_2 | \phi_1 \phi_2) = P(L_1 L_2 | \phi_1 \phi_2) = \eta^2 \left[\frac{1}{4} + \frac{1}{4} \cos(\phi_2 - \phi_1 + \theta) \right],$$
$$(1.5.10)$$

$$P(U_1 L_2 | \phi_1 \phi_2) = P(L_1 U_2 | \phi_1 \phi_2) = \eta^2 \left[\frac{1}{4} - \frac{1}{4} \cos(\phi_2 - \phi_1 + \theta) \right].$$
$$(1.5.11)$$

The coincidence counting rates show the two-photon interference. The counting rate of any one detector, for instance U_1, is

$$P(U_1|\phi_1\phi_2) = P(U_1U_2|\phi_1\phi_2) + P(U_1L_2|\phi_1\phi_2) = \frac{\eta^2}{2}, \qquad (1.5.12)$$

and similarly

$$P(U_2|\phi_1\phi_2) = P(L_1|\phi_1\phi_2) = P(L_2|\phi_1\phi_2) = \frac{\eta^2}{2}. \qquad (1.5.13)$$

The two-photon interference generalizes the well-known teachings of P. Dirac, "A pair of correlated photons interfere with the pair itself."

Does this contradict the other half of Dirac's teachings that interference between different photons never occurs? It does not. The two down-converted photons are in an entangled state. The pair has an entangled choice of paths: either A and C or D and B. The interference occurs between these choices, not between the two photons. Actually they never meet, one ends up in detectors U_1 or L_1, and the other ends up in U_2 or L_2. The situation becomes more clear in an experiment by Hong, Ou and Mandel [65], in which a pair of down-converted photons mix at a beam splitter such that each can enter detector A or detector B. The counts in either of the detectors do not show any interference (therefore there is no interference between the two photons), while the coincidence counts of A and B does show interference. It is apparent that the interference occurs between the two components of each photon produced at the beam splitter. The research on quantum beat of two single photons by Legero *et al.* [66] serves as another interesting example. Two long pulse independent photons arrive on a beam splitter (there is a time interval between the arrivals in general, of course the interval can also be zero), and both can therefore enter detector A or detector B. One of the detectors clicks first. This act entangles the originally independent photons, and the coincidence between the first and the second detector measures the coherence of the entangled pair. Both photons can enter either A or B, but the counts of A and B do not show any interference between the two photons. Only the coincidence counts of the detectors reveal the existence of interference between the modes (components) of each photon.

1.5.3   Interference between different emission times

L. Mandel [16] pointed out that two independent and spatially separated single atom light sources can interfere in the sense that two detectors on the receiving plane have coincidence counts as a function of the distance separating them, provided that it is impossible to trace back from which source does each photon originate. A similar idea is due to J. Franson [17] that two-photon interference occurs when it is impossible to judge the time at which the photons are emitted. Based on Franson's idea R. Chiao's group [18] constructed an interferometer of high visibility,[10] shown schematically in Fig. 1.19(a). In this figure M_1 and M_2 are reflecting mirrors, $B1_1$, $B1_2$, $B2_1$, $B2_2$ are beam splitters, F_1 and F_2 are filters, D_1 and D_2 are detectors. A simplified version of the equipment is shown in Fig. 1.19(b), where a pair of down-converted photons arrive simultaneously at detectors D_1 and D_2. They have the choice of both taking the long route or the short route. The traversing time difference between long and short routes in the experiment is 4ns. The simultaneity for the detectors really means to be within a window of 1ns. Since the photons are produced in a down-conversion they have to take jointly the long or the short routes. There is no arrangement to judge the time of emission in the equipment, therefore the state of the photon pair is an entangled one,

$$|\psi\rangle = \frac{1}{\sqrt{2}}(|s\rangle_1|s\rangle_2 + |l\rangle_1|l\rangle_2), \qquad (1.5.14)$$

where s and l represent the short and long routes respectively. The relative phase between the two terms in (1.5.14) can be varied by changing the length of the long route of one of the photons. This in turn can be achieved by moving the trombone prism in the equipment. Let $\Delta L_i = L_i - S_i$, where $i = 1, 2$, with L_i and S_i representing the length of the long and short routes for photon i.

---

[10]This equipment was also used to give a high precision demonstration of the violation of Bell's inequality. (cf. Section 1.8).

(a)

——— mirror

---- beam splitter

(b)

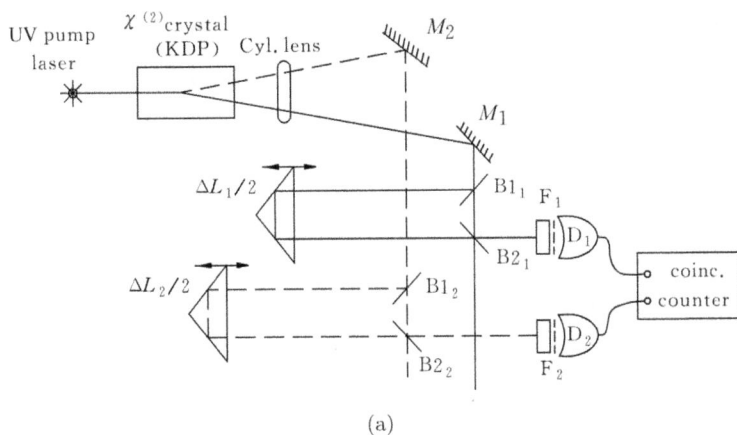

Fig. 1.19 (a) Schematic of the emission time interferometer. Reprinted with permission from P. G. Kwiat, A. M. Steinberg, and R. Y. Chiao, *Phys. Rev. A* 47, R2472, 1993. Copyright (2017) by the American Physical Society. (b) The simplified diagram.

The relative phase between the two terms in $|\psi\rangle$ is

$$\Delta\phi = \omega_1 \frac{\Delta L_1}{c} + \omega_2 \frac{\Delta L_2}{c}$$

$$= \frac{\omega_1 + \omega_2}{2c}(\Delta L_1 + \Delta L_2) + \frac{\omega_1 - \omega_2}{2c}(\Delta L_1 - \Delta L_2),$$

where ω_1 and ω_2 are the angular frequencies of the two photons. In the experiment it is chosen that $\omega_1 \approx \omega_2$, i.e., $\omega_1 - \omega_2 \approx 0$. Denote the angular frequency of the parent photon by ω_p, $\omega_p = \omega_1 + \omega_2$. Then

$$\Delta\phi = \frac{\omega_p}{2c}(\Delta L_1 + \Delta L_2). \qquad (1.5.15)$$

The coincidence counting rate R satisfies

$$R \propto |1 + e^{i\Delta\phi}|^2 = 2 + 2\cos\left[\frac{\omega_p}{2c}(\Delta L_1 + \Delta L_2)\right]. \qquad (1.5.16)$$

Theoretically the visibility can reach 100%, while the experimental value is 80.4%.

In a later version of the experiment by W. Tittel *et al.* [19], the down-converted photons of the same energy are sent through optical fibers to two interferometers separated by a distance of 10.5 km. The two-photon interference observed has a visibility of 81.6%. That an entangled state can be kept through such a large distance is important theoretically for demonstrating that the quantum mechanical correlation (not the spooky interaction) is at work, and also practically for being applicable to the quantum cryptography and quantum teleportation.

It is now appropriate to stay on a point we discussed in Section 1.4. Usually a light source with a large number of radiating atoms is considered a classical source, while a source consisting of a very small number of atoms can hardly be considered a classical one. For a description of such a source, classical electromagnetic theory is no more appropriate and quantized field theory has to be used instead. In Ref. [16] L. Mandel found that such an approach gives results very different from those of the classical electromagnetic theory. These two approaches coincide in the limit of large number of atoms. This is to be expected. Surprisingly, when the number is not large but subject to Poisson fluctuations, the quantum prediction is also reduced to that of the classical theory. Therefore a thermal source, no matter how faint it is, is still classical with an anti-coincidence parameter α close to 1. Here the fluctuation plays a decisive role.

1.5.4 Coherence and the distinguishability of paths

A Feynman's dictum says that states interfere with each other only when. they cannot be physically distinguished in a particular experimental setup. Only physically distinguishable states in an equipment can interfere. A direct demonstration to this principle was given by X. Y. Zou, L. J. Wang and L. Mandel [20]. Their interferometer is shown schematically in Fig. 1.20(a), and its simplified version in Fig. 1.20(b). In the latter dotted lines A and C represent beam splitters. The incident ultraviolet light beam is split at A into two beams,

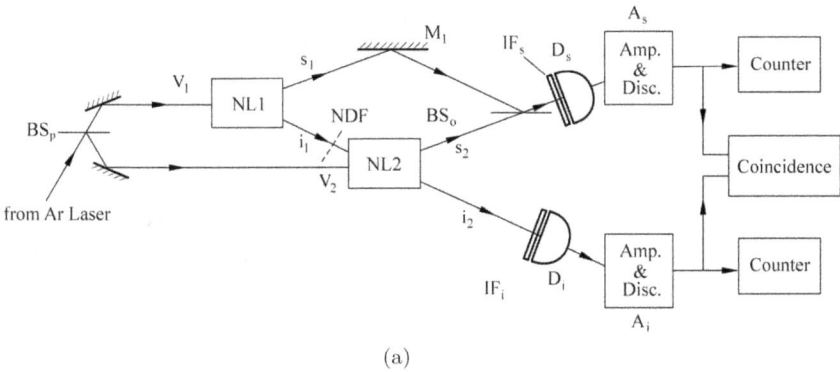

(a)

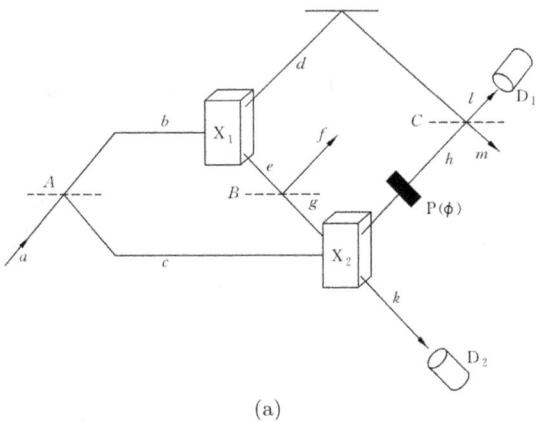

(a)

Fig. 1.20 (a) Scheme of the Zou-Wang-Mandel interferometer, Reprinted with permission from X. Y. Zou, L. J. Wang, and L. Mandel, *Phys. Rev. Lett.* 67, 318, 1991. Copyright (2017) by the American Physical Society. (b) A simplified diagram.

each enters a non-linear crystal X_1 and X_2 respectively. Each incident photon can produce the down-conversion only in one of the non-linear crystals, and the result is the formation of an entangled state

$$\frac{1}{\sqrt{2}}(|d\rangle_1|e\rangle_2 + |h\rangle_1|k\rangle_2).$$

Beams $|d\rangle$ and $|h\rangle$ can both enter the detector D_1. Can they interfere? If beams e and k are monitored, then it is possible to predict from which crystal the down-converted photon recorded by D_1 comes. If e gives a count, then it is certainly d which arrives at D_1. If k gives a count, then it is h which arrives at D_1. Then there will be no interference. Such a judgement becomes impossible if the paths e and k are aligned, and consequently the interference ensues. In this case the state is

$$\frac{1}{\sqrt{2}}(|d\rangle + |h\rangle)_1|k\rangle_2,$$

which is a direct product but not an entangled state. Adjusting the phase shifter P, interference pattern will appear in D_1. Specifically, the beam a becomes a linear superposition of beams b and c at the 50% beam splitter A, and

$$|a\rangle \rightarrow \frac{|b\rangle + i|c\rangle}{\sqrt{2}},$$

where the factor i represents a phase difference $\pi/2$ for the reflected beam with respect to the transmitted beam. For convenience of the further discussions we denote by T and R (both real) the transmissivity and reflectivity of the beam splitter B. We will see that this beam splitter serves as a monitor of beam e. Then

$$|e\rangle \rightarrow T|g\rangle + iR|f\rangle.$$

At the 50% beam splitter C,

$$|h\rangle \rightarrow \frac{|l\rangle + i|m\rangle}{\sqrt{2}}, \quad |d\rangle \rightarrow \frac{|m\rangle + i|l\rangle}{\sqrt{2}}.$$

At the phase shifter

$$|h\rangle \rightarrow e^{i\phi}|h\rangle,$$

and at the non-linear crystals

$$|b\rangle \rightarrow \eta|d\rangle_1|e\rangle_2,$$
$$|c\rangle \rightarrow \eta|h\rangle_1|k\rangle_2,$$

where η is the coefficient of down-conversion, typically of the order of 10^{-6}. Aligning g and h, we have

$$|g\rangle \rightarrow |k\rangle.$$

Summarizing all the processes above, we obtain

$$|a\rangle \rightarrow \frac{1}{\sqrt{2}}(|b\rangle + i|c\rangle) \rightarrow \frac{\eta}{\sqrt{2}}(|d\rangle_1|e\rangle_2 + i|h\rangle_1|k\rangle_2)$$

$$\rightarrow \frac{\eta}{2}[(T - e^{i\phi})|m\rangle + i(T + e^{i\phi})|l\rangle]_1|k\rangle_2 + i\frac{\eta}{2}R(|m\rangle + i|l\rangle)_1|f\rangle_2.$$
$$(1.5.17)$$

To find the coincidence counting rate of D$_1$ and D$_2$, we take the modulus squared of the coefficient of $|l\rangle_1|k\rangle_2$ in (1.5.17) to obtain

$$\frac{\eta^2}{4}[(T + \cos\phi)^2 + \sin^2\phi] = \frac{\eta^2}{4}(1 + T^2 + 2T\cos\phi).$$

The contrast of the interference pattern V is

$$V = \frac{2T}{1 + T^2}. \qquad (1.5.18)$$

It increases with increasing T, beginning from $V = 0$ for $T = 0$ until $V = 1$ for $T = 1$. If only the counting rate in D$_1$ is required, it can be obtained by taking the sum of the moduli squared of the coefficients of $|l\rangle_1|b\rangle_2$ and $|l\rangle_1|f\rangle_2$:

$$\frac{\eta^2}{4}[(T + \cos\phi)^2 + \sin^2\phi] + \frac{\eta^2}{4}R^2 = \frac{\eta^2}{4}[1 + T^2 + 2T\cos\phi + R^2]$$

$$= \frac{\eta^2}{2}(1 + T\cos\phi).$$

The last equality is obtained by considering $T^2 + R^2 = 1$. The contrast of the interference fringe is

$$V = T. \qquad (1.5.19)$$

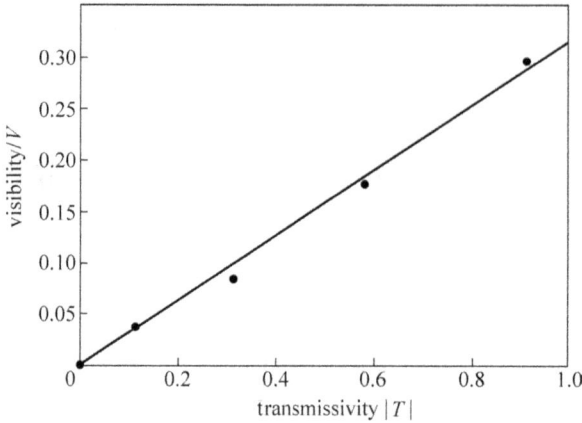

Fig. 1.21 Visibility versus transmissivity. Reprinted with permission from X. Y. Zou, L. J. Wang, and L. Mandel, *Phys. Rev. Lett.* 67, 318, 1991. Copyright (2017) by the American Physical Society.

Taking into account the actual experimental parameters viz. fractions of the incident light energy which are down-converted at the crystals, the intensities and cross-correlation of the pump waves, V is proportional to T, with the constant of proportionality equal to 1 only in extreme cases [20]. The experimental result is plotted in Fig. 1.21. It is important to notice that notwithstanding the beams toward D_2 are not included in the coherent paths, they can influence the visibility of the interference pattern observed in D_1. The role of the paths toward D_2 is to render the "which way" (i.e., from X_1 or X_2) judgement impossible. A pair of down-converted photons are correlated. Manipulating one of the pair can influence the other. That $T = 0$ leads to $V = 0$ is due to the fact that by stopping the beam e the judgement whether the photon in D_1 comes from the crystal X_1 (if there is no photon in k) or from X_2 (if there is a photon in k) becomes possible. In such a situation no interference fringes appears in the data of detector D_1. In Fig. 1.22 the counting rate R_s in detector D_1 is shown by curve A as a function of beam splitter BS_O displacement (Fig. 1.20(a), or equivalently of phase ϕ in the phase shifter, Fig. 1.20(b)). When the idle beam e is blocked ($T = 0$) the rate is shown by curve B. It is demonstrated once again in this experiment that it is the potential possibility of acquiring the

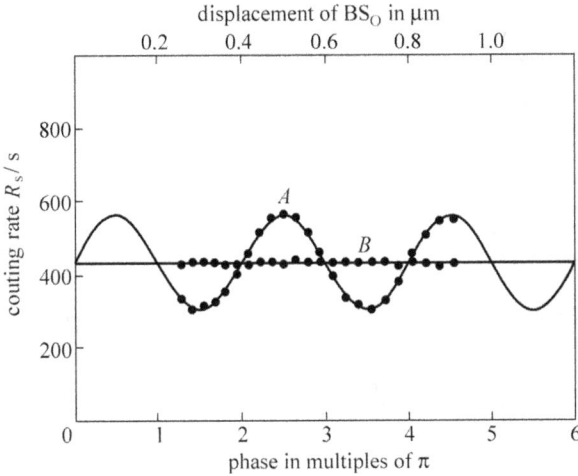

Fig. 1.22 Counting rate in detector D_1. Curve A neutral-density filter between NL1 and NL2, curve B beam stop inserted between NL1 and NL2, $T = 0$. Reprinted with permission from X. Y. Zou, L. J. Wang, and L. Mandel, *Phys. Rev. Lett.* 67, 318, 1991. Copyright (2017) by the American Physical Society.

information, not necessarily the actual information acquired, which destroys the coherence. Adjusting the transmissivity T of the beam splitter B from 1 to 0, we change the degree of certainty of monitoring the beam e and correspondingly the visibility of the interference fringes. Another way of monitoring the beam is to misalign $|g\rangle$ and $|k\rangle$. If we do not align $|g\rangle$ and $|k\rangle$, the factor $\frac{\eta}{2}i(T + e^{i\phi})|l\rangle_1|k\rangle_2$ in (1.5.17) becomes $\frac{\eta}{2}(ie^{i\phi}|l\rangle_1|k\rangle_2 + iT|l\rangle_1|g\rangle_2)$. Looking for coincidence in detectors D_1 and D_2 from the new expression, we find that the coherence is lost without actually measuring $|g\rangle$ and $|k\rangle$.

1.6 Double Photon Interferometer as a Quantum Eraser

We have discussed the concept of a quantum eraser in Section 1.3. It was realized for the first time by Raymond R. Chiao's group in a double photon interferometer [21]. The experimental setup is shown schematically in Fig. 1.23. The non-linear KDP crystal is pumped by the Ar ion laser of wave length 351.1 nm. The incident photon is down-converted to a pair of conjugate photons of average wave length

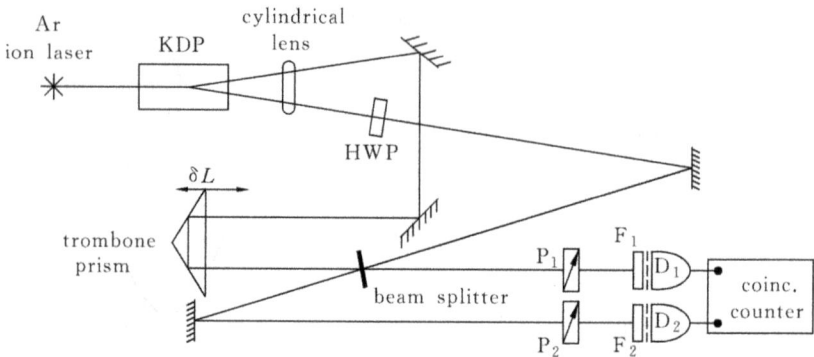

Fig. 1.23   Schematic of experiment to observe quantum erasure. Reprinted with permission from P. G. Kwiat, Aephraim M. Steinberg, and Raymond Y. Chiao, *Phys. Rev. A* 45, 7729, 1992. Copyright (2017) by the American Physical Society.

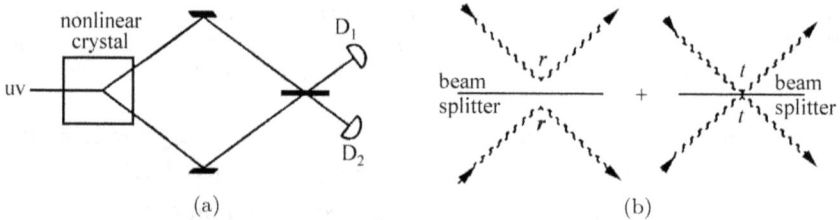

Fig. 1.24   (a) Simplified setup; (b) Paths for coincidence detection. Reprinted with permission from Paul G. Kwiat, Aephraim M. Steinberg, and Raymond Y. Chiao, *Phys. Rev. A* 45, 7729, 1992. Copyright (2017) by the American Physical Society.

702.2 nm. The band width is restricted to 10nm by the filters. The correlated photons reach the beam splitter simultaneously after being reflected by mirrors. The avalanche photodiodes D_1 and D_2 record outputs of two channels and also their coincidence. The simplified setup and path for coincidence detection is shown in Fig. 1.24. In Fig. 1.24(b) we see paths where the pair of photons are both reflected ($r \times r$) or both transmitted ($t \times t$). When the path difference is 0, the coincidence counting rate is

$$P_c = |r \times r + t \times t|^2 = \left| \frac{i}{\sqrt{2}} \times \frac{i}{\sqrt{2}} + \frac{1}{\sqrt{2}} \times \frac{1}{\sqrt{2}} \right|^2 = 0. \qquad (1.6.1)$$

Factors i represent phase difference $\pi/2$ of the reflected beam at beam splitter. When the path difference is much larger than the coherent length, the paths $(r \times r)$ and $(t \times t)$ do not interfere any more, and the counting rate is the sum of moduli squared of these paths

$$P_c = \frac{1}{2}. \qquad (1.6.2)$$

Since the probabilities of reflection and of transmission are both 50%, the coincidence counts amount to 1/2 of the total. The other half represents the case where the two photons enter the same detector. Actually the wave function behind the beam splitter is

$$|\psi\rangle_{\Delta x=0} = \frac{1}{2}(|1_1 1_2\rangle + i^2|1_1 1_2\rangle + i|2_1 0_2\rangle + i|0_1 2_2\rangle)$$

$$= \frac{i}{2}(|2_1 0_2\rangle + |0_1 2_2\rangle). \qquad (1.6.3)$$

In Eq. (1.6.3) we see that for zero path difference the two entangled photons leave the beam splitter in the same direction (to D_1 or D_2). This phenomenon was first discovered by Hong, Ou and Mandel [65], and was later referred to as the photon coalescence. It is interesting to find that for two independent photons the same happens, for instance, in a study of quantum beats by T. Legero *et al.* [66] mentioned in Section 1.5. The two photon pulses are recorded successively. The first detection entangles the two photons, which then determines the probability of detecting the second photon with one or the other detector. For photons of the same frequency the photon coalescence is given explicitly.

The Fock state $|nm\rangle$ represents a state with n and m photons heading toward detectors D_1 and D_2 respectively. $\Delta x = 0$ means vanishing path difference. If the trombone prism in Fig. 1.23 is moved to adjust the path difference δL of the two paths, the coincidence will increase with the increasing of path difference until the latter approaches the coherent length. Then the coincidence counts approach 1/2 and becomes independent of the path difference, and the interference disappears. In equipment with longer coherent lengths periodic interference pattern can be observed, for instance in Ref. [61].

The down-converted photons are horizontally polarized. If a half wave plate (HWP in Fig. 1.23) with its optical axis making angle $\phi/2$ with the horizontal direction is put in the path of one of the down-converted photons, the polarization of this photon will make an angle ϕ with the horizontal direction. If $\phi/2 = 45°$ then the polarization of the photon becomes vertical. In this way the polarization of the photon becomes a new parameter which can be used to distinguish along "which way" the photon arrives. As can be expected, the interference will appear. Let H and V represent the horizontal and vertical polarization respectively. The polarization of the photon passing through the half wave plate becomes $H + \phi$, i.e., making an angle ϕ with the horizontal direction.

$$|1^{H+\phi}\rangle = |1^H\rangle \cos \phi + |1^V\rangle \sin \phi. \qquad (1.6.4)$$

The wave function behind the beam splitter after the half wave plate is installed is

$$|\psi\rangle_{\Delta x=0} = \frac{1}{2}[1_1^H 1_2^{H+\phi}\rangle + i^2 |1_1^{H+\phi} 1_2^H\rangle + i|1_1^{H+\phi}$$

$$+ 1_1^H, 0_2\rangle + i|0_1, 1_2^H + 1_2^{H+\phi}\rangle]. \qquad (1.6.5)$$

The terms proportional to $|1_1^H 1_2^H\rangle \cos \phi$ cancel in (1.6.5) because of (1.6.4). The remaining terms can be written as

$$|\psi\rangle_{\Delta x=0} = \frac{1}{2}[1_1^H 1_2^V\rangle \sin \phi - |1_1^V 1_2^H\rangle \sin \phi + i|2_1^H 0_2\rangle \cos \phi + i|1_1^V$$

$$+ 1_1^H, 0_2\rangle \sin \phi + i|0_1 2_2^H\rangle \cos \phi + i|0_1, 1_2^H + 1_2^V\rangle \sin \phi]. \qquad (1.6.6)$$

Only the first two terms are concerned with coincidence counts, since the Fock states of all remaining terms are either $|02\rangle$ or $|20\rangle$. The result of coincidence measurement is given by

$$P_c = \frac{1}{4}(2 \sin^2 \phi) = \frac{1}{2}\sin^2 \phi. \qquad (1.6.7)$$

The result is obtained by summing the moduli squared of the coefficients of $|1_1^V 1_2^H\rangle$ and $|1_1^H 1_2^V\rangle$ in (1.6.6). In these states there is one photon in each of the detectors. Why there is no interference term? This is because H and V polarization states are orthogonal to each other. The use of a half wave plate in the path of one of the photons makes the terms which would have interfered to become orthogonal. This situation is similar to the potential possibility of distinguishing "which way", the same result happens no matter an actual measurement is made or not. In Fig. 1.25(a) different half wave plate orientations lead to different effects on interference. In Fig. 1.25(b) the visibility as a function of half wave plate angle is given.

Making the quantum erasure is going the backward way. Each detecting channel receives the photon from two alternative paths, but with the photon earmarked to reveal the "which way" information. The quantum eraser deletes this information in the very last moment before the photons enter the detectors. The polarizers P_1 and P_2 installed in front of the detectors are set at angles θ_1 and θ_2 respectively. Assume that the half wave plate has already been set at $\phi/2 = 45°$, then (1.6.6) becomes

$$|\psi\rangle_{\Delta x=0} = \frac{1}{2}[-|1_1^V 1_2^H\rangle + |1_1^H 1_2^V\rangle + \cdots]. \qquad (1.6.8)$$

The action of polarizers consists of projecting the state (1.6.8) onto

$$\langle\theta_1| = \langle 1_1^H|\cos\theta_1 + \langle 1_1^V|\sin\theta_1$$

and

$$\langle\theta_2| = \langle 1_2^H|\cos\theta_2 + \langle 1_2^V|\sin\theta_2.$$

The very projection mixes the polarizations and consequently deletes the "which way" information. The result is

$$\langle\theta_1\theta_2|\psi\rangle_{\Delta x=0} = \frac{1}{2}(\cos\theta_1\sin\theta_2 - \sin\theta_1\cos\theta_2)$$

$$= \frac{1}{2}\sin(\theta_2 - \theta_1),$$

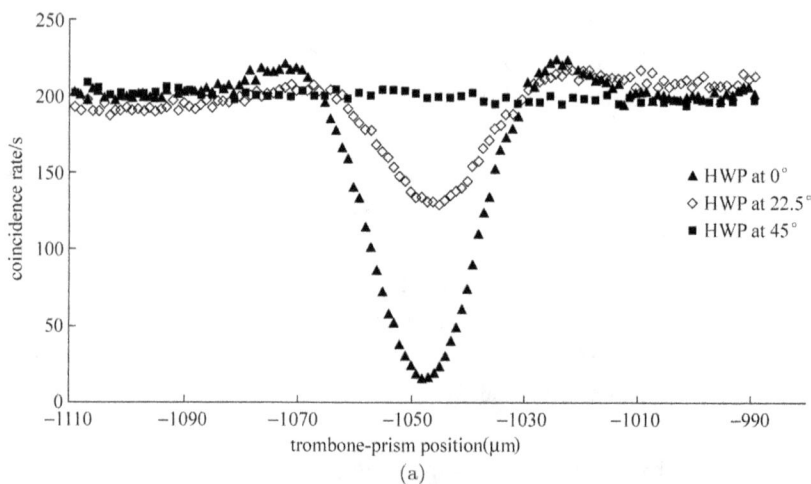

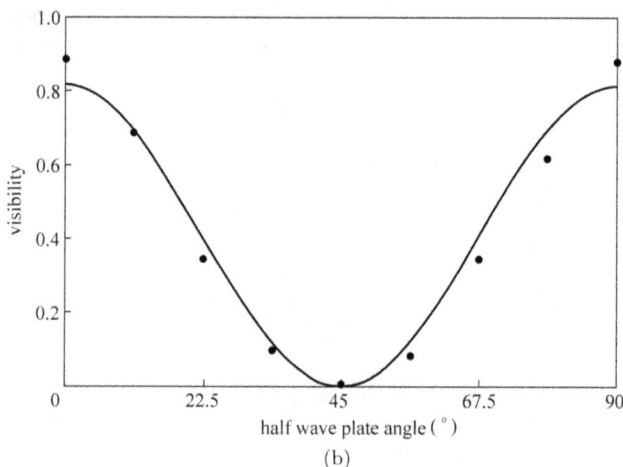

Fig. 1.25   (a)Profile of interference dip in coincidence rate for different half wave plate orientations; (b)Visibility as a function of half wave plate angle. Reprinted with permission from Paul G. Kwiat, Aephraim M. Steinberg, and Raymond Y. Chiao, *Phys. Rev. A* 45, 7729, 1992. Copyright (2017) by the American Physical Society.

and

$$P_c(0) = |\langle\theta_1\theta_2|\psi\rangle_{\Delta x=0}|^2 = \frac{1}{4}\sin^2(\theta_2 - \theta_1).$$

If $\theta_1 = \theta_2$, then $P_c(0) = 0$ is fully restored to the value of quantum interference. If the path difference is larger than the coherent length the two terms of (1.6.8) do not interfere any more. P_c is then determined by the sum of moduli squared of their separate projections on $\langle\theta_1\theta_2|$:

$$P_c(x > c\tau) \approx \left|\frac{1}{2}\langle\theta_1\theta_2|1_1^H 1_2^V\rangle\right|^2 + \left|\frac{1}{2}\langle\theta_1\theta_2|1_1^V H_2^H\rangle\right|^2$$

$$= \frac{1}{8}[\sin^2(\theta_2 - \theta_1) + \sin^2(\theta_2 + \theta_1)]. \qquad (1.6.9)$$

Due to the action of the half wave plate, there are information on "which way" in both channels. Polarizers are to be installed in front of both detectors to erase them. It is to be noticed that the "which way" information is erased behind the output channels of the interferometer, interference is restored just at the coincidence measurement in the two detectors. The behavior of the pair of conjugate photons depend solely on the angles set at the polarizers. The comparison between theory and experiment is shown in Fig. 1.26. The theoretical curves for the effect of erasure for different angles set on the polarizers are presented in Fig. 1.26(a). In Fig. 1.26(b) the points represent the experimental data while the curves represent theoretical calculations corrected to the actual visibility of 91%. Far from the dip, there is no interference and the angle is irrelevant. At the dip the nonlocal collapse of the polarization of photon 2 leads to the sinusoidal variation. The translation of the prism is effected by an inchworm piezoelectric motor system with a position resolution of $0.1\,\mu\text{m}$.

Yanhua Shih *et al.* [67] reported a delayed choice quantum eraser experiment, where the choice was made randomly by a photon at a beam splitter. The principle is illustrated in Fig. 1.27. An atom labeled by A or B is excited by a weak laser pulse. A pair of entangled quanta, the photon 1 and photon 2, are emitted from either A or B (the excited atom) by cascade decay. Photon 1, propagating to the right, is registered by detector D_0, which can be scanned along the x axis for the observation of interference fringes. Photon 2, propagating to the left, will meet a 50/50 beam splitter. If it is emitted by the atom A, it will follow path A and meet the beam splitter BSA, with 50% chance of being

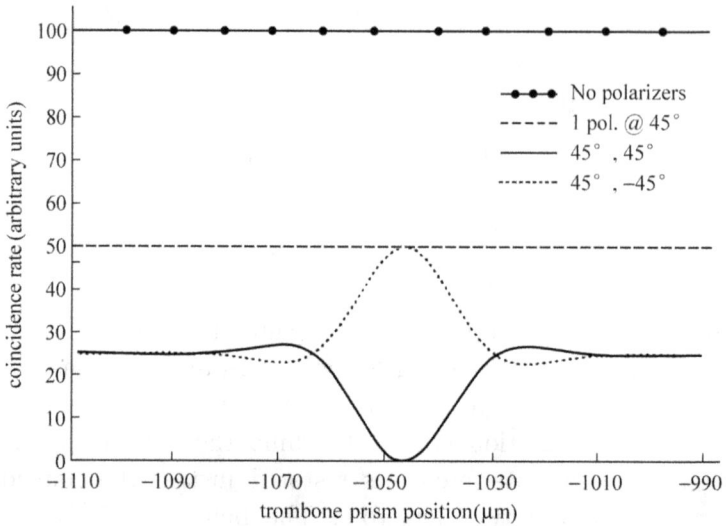

(a)

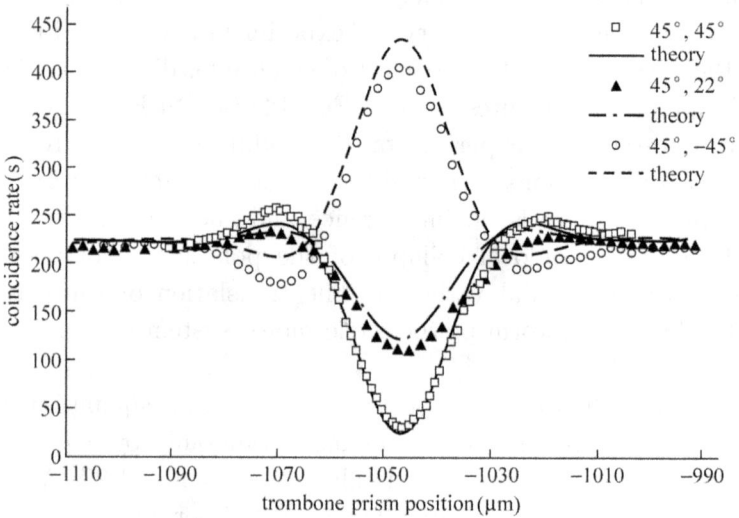

(b)

Fig. 1.26   (a) Theoretical curves of coincidence rate as a function of trombone prism position at different angles of the polarizers; (b) Comparison between theory and experiment. Reprinted with permission from Paul G. Kwiat, Aephraim M. Steinberg, and Raymond Y. Chiao, *Phys. Rev. A* 45, 7729, 1992. Copyright (2017) by the American Physical Society.

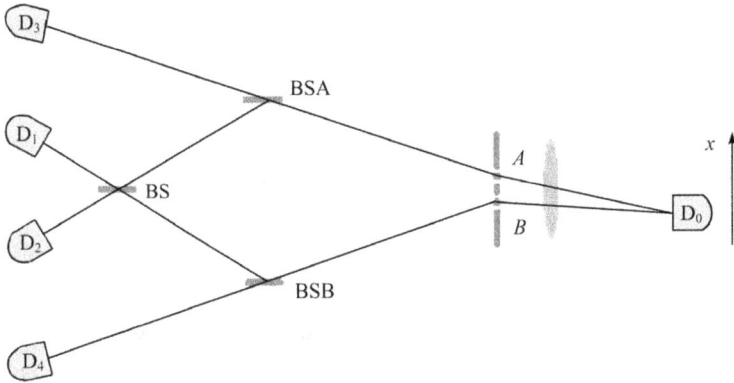

Fig. 1.27   Proposed quantum eraser experiment. Reprinted with permission from Yoon-Ho Kim, Rong Yu, Sergei P. Kulik, and Yanhua Shih, *Phys. Rev. Lett.* 84, 1, 2000. Copyright (2017) by the American Physical Society.

reflected or transmitted. The photon 2 emitted by atom B will meet BSB, again with 50% chance of being reflected or transmitted. In the case of transmission at either of the 2 beam splitters, photon 2 is collected by the detector D_3 or D_4. The registration of D_3 or D_4 provides the "which path" information for photon 2, and consequently the information for photon 1 because of the entangled state of two photons generated by cascade decay is provided by the wave function

$$|\psi\rangle = \frac{1}{\sqrt{2}}[|1\rangle_A|2\rangle_A + |1\rangle_B|2\rangle_B].$$

In case of reflection at either of the beam splitters photon 2 will meet another 50/50 beam splitter BS and then registered by either of D_1 or D_2. The triggering of detectors D_1 or D_2 erases the which-path information of photon 1 owing to the beam splitter BS. Therefore either the absence of interference or its restoration can be arranged via an appropriately contrived photon correlation arrangement. The state of two photons can be re-expressed by

$$|\psi\rangle = \frac{1}{\sqrt{2}}[|1\rangle_+|2\rangle_+ + |1\rangle_-|2\rangle_-]$$

where

$$|1\rangle_\pm = \frac{1}{\sqrt{2}}[|1\rangle_A \pm |1\rangle_B], \quad |2\rangle_\pm = \frac{1}{\sqrt{2}}[|2\rangle_A \pm |2\rangle_B].$$

These states represent coherent superpositions of two states of the same photon, i.e., the typical double-slit wave functions. Registration of D_1 or D_2 for photon 2 leads to the collapse of wave function, revealing that the state of photon 1 is given by $|1\rangle_+$ or $|1\rangle_-$, with interference fringe preserved.

The experiment is designed in such a way that the optical distance between atoms A, B and the detector D_0 is much shorter than the optical distance between A, B and the beam splitters BSA and BSB, where the which-path or both-path choice is made randomly by photon 2. Thus after the triggering of detector D_0 by photon 1, photon 2 is still on its way to the beam splitters. This delayed choice is a stronger variation of Wheeler's main theme described above. The delayed coincidence rate between the countings of D_0 and one of the detectors D_i ($i = 1, 2, 3, 4$) denoted by R_{0i} guarantees that the events must have resulted from the same photon pair. The joint counting rate R_{01} and R_{02} will show an interference pattern as a function of D_0's position x. This reflects the wave nature (both-path) of photon 1. For the joint counting rate R_{03} and R_{04} no interference pattern will be observed, because which-path information for photon 1 is provided.

In the actual experiment, the pump laser beam is divided by a double slit forming two regions A and B inside the BBO crystal. A pair of down-converted photons is generated from either the A or B region. The delay of photon 2 arrival relative to the signal photon 1 is 7.7 ns. The schematic is shown in Fig. 1.28.

In Fig. 1.29 we show R_{01} and R_{02} as functions of x. Standard Young's double-slit patterns are observed. These are the conjugate fringes with a phase shift of π. Their sum is of course featureless (Fig. 1.30). Fig. 1.31 shows the absence of interference for R_{03}.

The delayed choice eraser idea stirred up considerable controversy since the idea was first proposed by M. O. Scully and K. Druhl. It appears that the registration of photon 1 made in the past would

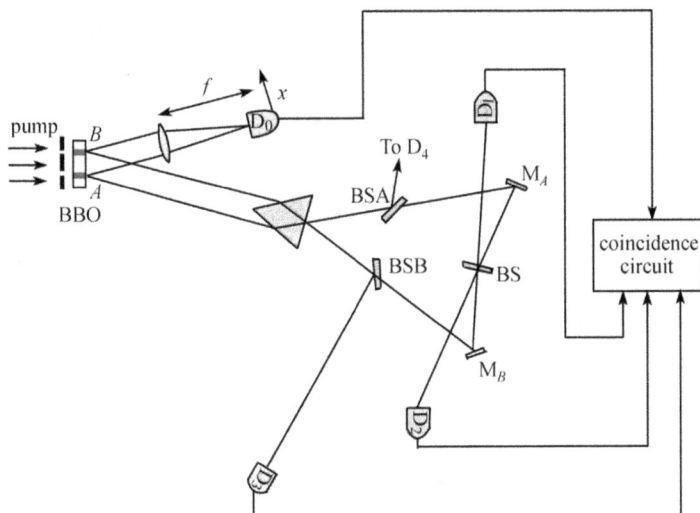

Fig. 1.28 Schematic of the actual experimental setup. Reprinted with permission from Yoon-Ho Kim, Rong Yu, Sergei P. Kulik, and Yanhua Shih, *Phys. Rev. Lett.* 84, 1, 2000. Copyright (2017) by the American Physical Society.

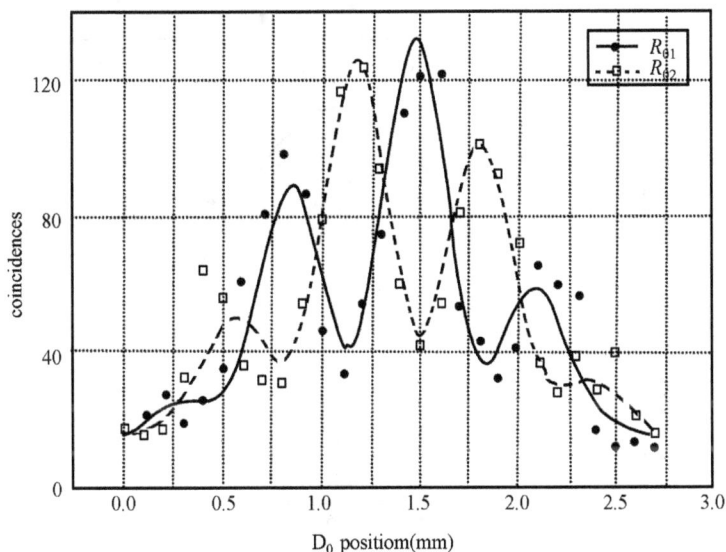

Fig. 1.29 Conjugate fringes of R_{01} and R_{02}. Reprinted with permission from Yoon-Ho Kim, Rong Yu, Sergei P. Kulik, and Yanhua Shih, *Phys. Rev. Lett.* 84, 1, 2000. Copyright (2017) by the American Physical Society.

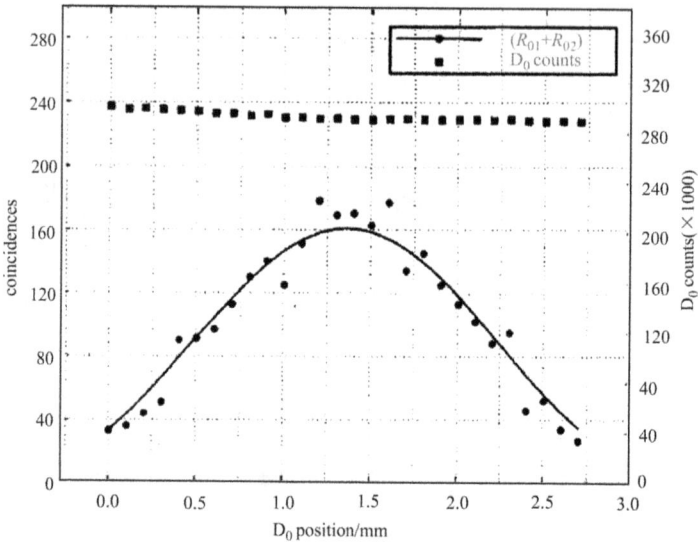

Fig. 1.30   The sum of R_{01} and R_{02} shows no interference. Reprinted with permission from Yoon-Ho Kim, Rong Yu, Sergei P. Kulik, and Yanhua Shih, *Phys. Rev. Lett.* 84, 1, 2000. Copyright (2017) by the American Physical Society.

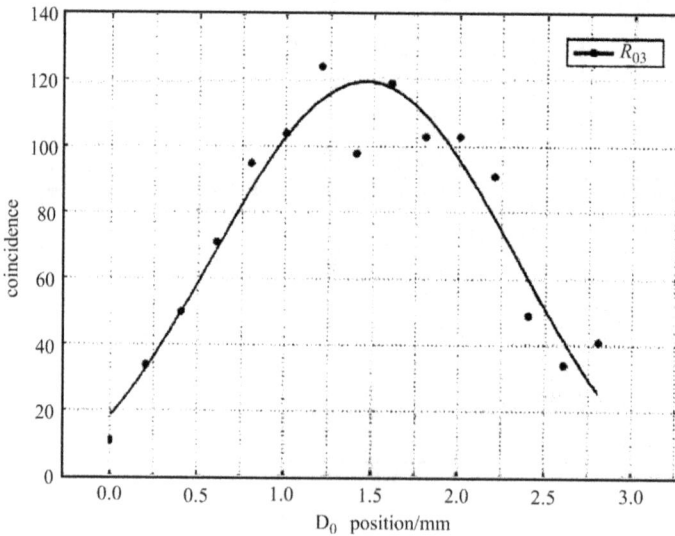

Fig. 1.31   Absence of interference for R_{03}. Reprinted with permission from Yoon-Ho Kim, Rong Yu, Sergei P. Kulik, and Yanhua Shih, *Phys. Rev. Lett.* 84, 1, 2000. Copyright (2017) by the American Physical Society.

be influenced by the behavior of photon 2 in the future. Actually the physics is determined by the entanglement of $|\psi\rangle$. Whether the first variant (which-path) or the second variant (both-path) is used depends randomly on the behavior of photon 2. Practically a sorting to separate the events detected by each of D_1, D_2, D_3, and D_4 is made only after the experiment has been completed. This discloses the full content of the physics. A mere registration of the photon 1 by detector D_0 tells us very little (Fig. 1.30, squares).

1.7 Dispute between Einstein and Bohr on Quantum Mechanics, Bell's Theorem

Disputes on the nature of quantum mechanics between the two greatest physicists of the 20th century A. Einstein and N. Bohr broke out twice in 1930 and 1935. The dispute in 1935 was referred to as the EPR paradox. The discussions then were limited to theoretical arguments or thought experiments, and were not conclusive. In 1965 discussions on this issue was revitalized. The origin was a paper by J. S. Bell. He pointed out that the "hidden variable theory" based on Einstein's viewpoint can be checked by experiment, the criterion being the "Bell's inequality". Since the late 1980s the experiments have already reached high precision to give convincing judgements on the violation of Bell's inequality.

1.7.1 Criticism by Einstein in 1930: "quantum mechanics is not self-consistent"

At the Solvay Congress in Brussels in 1930 Einstein gave an example of thought experiment. There are radiations in a closed box, the mass of which is accurately measured. The aperture of the box is opened for a time duration T and then closed by an exact clock mechanism. During this period a photon has escaped. After the closure the mass of the box is measured again, the difference turns out to be the energy of the escaped photon E. Since the measurements of time T (by the clock mechanism) and the energy E (by measuring the mass difference) are independent and they do not disturb each other, consequently the respective exactness of measurement do not

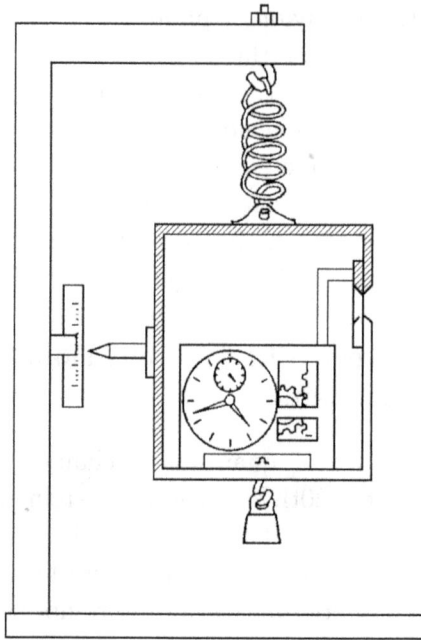

Fig. 1.32   Einstein's box.

impose constraint on each other. Thus the uncertainty $\Delta E \cdot \Delta T \geq \hbar$ is violated. Einstein concluded:"quantum mechanics is not self-consistent". For some time N. Bohr was puzzled. He believed firmly in the correctness of the uncertainty principle but was unable to find any flaw in Einstein's argument. An overnight meditation produced the illustration shown in Fig. 1.32, which Bohr drew on the blackboard the next morning. Bohr pointed out that it is possible to find out the relation between the uncertainties of energy and time measurements upon serious consideration of the measuring process. At the time the photon escapes the box receives an upward momentum

$$p \leq T \frac{E}{c^2} g, \qquad (1.7.1)$$

where g is the acceleration of gravity. The uncertainty of momentum is

$$\Delta p \leq T \frac{\Delta E}{c^2} g. \qquad (1.7.2)$$

The difference in positions Δx of the box is connected with Δp:

$$\Delta p \geq \frac{\hbar}{\Delta x}. \tag{1.7.3}$$

Consequently

$$\hbar \leq T\frac{\Delta E}{c^2}\Delta xg. \tag{1.7.4}$$

The uncertainty in T measurement is conditioned by the change in the tempo of the clock due to the difference in gravitational potential at different altitude Δx. According to the formula for gravitational red-shift

$$\frac{\Delta T}{T} = \frac{g\Delta x}{c^2}. \tag{1.7.5}$$

Substituting (1.7.5) into (1.7.4) we obtain

$$\Delta E\Delta T \geq \hbar. \tag{1.7.6}$$

Bohr replied using the gravitational red-shift discovered by Einstein himself in demonstrating the self-consistency of quantum mechanics. Since then Einstein has not returned to this issue any more.

1.7.2 The Einstein-Podolsky-Rosen paradox: "quantum mechanical description is incomplete"

A. Einstein, B. Podolsky and N. Rosen voiced a new criticism in 1935 [22], which is usually referred to as the EPR paradox. EPR intended to judge whether quantum mechanical description is complete, beginning with an example of a 1-dimensional system of two particles. They put forward three premises: (1) Some of the quantum mechanical descriptions concerning observations on a certain type of systems, consisting of two spatially separated particles, are correct. (2) A criterion for the existence of an element of physical reality is proposed: "If, without in any way disturbing a system, we can predict with certainty (i.e., a probability equal to unity) the value of a physical quantity, then there exists an element of physical reality corresponding to this physical quantity." (3) There is no action-at-a-distance in nature. Equivalently this is also described as the "separability of distant systems". In the 1-dimensional system of particles considered, the operators x_1 and p_1 pertaining to particle 1 do not commute, and so do x_2 and p_2 of particle 2. But $x_1 - x_2$ does commute with $p_1 + p_2$. There exists therefore a state with eigenvalue a of $x_1 - x_2$ and eigenvalue 0 of $p_1 + p_2$ with the wave function $\delta(x_1 - x_2 - a)$. Now we measure the position of particle 1 and find it to be x, so we can determine that the position of particle 2 must be $x - a$. If we measure the momentum of particle 1 and find it to be p, we can determine that the momentum of particle 2 must be $-p$. Let a be sufficiently large, then any physical manipulation on particle 1 does not in any way disturb particle 2 (premise (3)). According to premise (2) there exist 4 elements of independent physical reality corresponding to x_1, x_2, p_1 and p_2. But according to the principle of quantum mechanics x_1 and p_1 do not commute and x_2 and p_2 do not commute either. Therefore corresponding to any pair of quantities there exists one element of physical reality, and corresponding to the system, only two elements of physical reality. EPR hence conclude:"quantum mechanical description is incomplete". In order to have the discussion closer to experiments we introduce another version of the paradox, viz. D. Bohm's version [23] and carry out discussion on it.

In Bohm's version three components of spin pertaining to two particles are considered instead of position and momentum. Consider a system consisting of two spin $1/2$ particles with a total spin 0 (the singlet state) described by the wave function

$$\psi = \frac{1}{\sqrt{2}}[u_\uparrow(1) \times u_\downarrow(2) - u_\downarrow(1) \times u_\uparrow(2)] \qquad (1.7.7)$$

where $u_\uparrow$, $u_\downarrow$ are spinor wave functions with spin component $1/2$ and $-1/2$ respectively. The direction $\hat{n}$ of the axis of quantization for spin is arbitrary. Since the system is in the singlet state, the two particles have their spins antiparallel. Let the particles be separated by a large distance. We measure the x component of spin of particle 1. From (1.7.7) we know that the probabilities of finding the results $1/2$ and $-1/2$ are both equal to $1/2$. Furthermore if we find that the measurement of x component of spin of particle 1 gives $1/2$, then correspondingly the x component of spin of particle 2 must be $-1/2$. An observer can predict with certainty the x component of spin of particle 2 without disturbing it at all. Similarly an observer can predict with certainty the y component of spin of particle 2 without disturbing it by measuring the y component of spin of particle 1. Now we can carry out an analysis according to three premises of EPR. By measuring $S_x(1)$, $S_y(1)$, or $S_z(1)$ we can predict with certainty $S_x(2)$, $S_y(2)$ or $S_z(2)$ without in any way disturbing particle 2 (premise (3) no action-at-a-distance). By premise (2) we infer that corresponding to $S_x(2)$, $S_y(2)$ and $S_z(2)$ there exist three independent elements of physical reality. According to quantum mechanics $S_x(2)$, $S_y(2)$ and $S_z(2)$ do not commute with one another. Therefore there can be only one element of physical reality. The conclusion inferred from the three premises would have been : the quantum mechanical description is incomplete.

The use of premise (2) is, however, problematic. In premise (2) there is a phrase "we can predict with certainty". We can predict $S_x(2)$ with certainty only *when the experimental arrangement is made for measuring $S_x(1)$. In this situation we cannot predict anything about $S_y(2)$ and $S_z(2)$.* In Bohr's reply to EPR [24] he pointed out that "there is essentially the question of *an influence on the very condition which defines the possible type of predictions regarding the future behavior of the system*". By "the very condition" he actually means the relevant experimental arrangement. Since the arrangements for the three components of spin are mutually exclusive, therefore only one component can be chosen and correspondingly only one physical quantity can be predicted with certainty. The conclusion is that it is out of question whether the quantum mechanical description is incomplete.

It is worthwhile to notice the phrase in premise (2) "without in any way disturbing a system". In early literatures and textbooks on quantum mechanics the reason that two non-commuting observables A and B cannot have definite values at the same time is ascribed to the disturbance suffered by B during a measurement of A. Therefore to EPR once particle 2 does not experience any disturbance from the measurement on A, all observables of particle 2 must have definite values. In a discussion with Einstein on the matrix formulation of quantum mechanics Heisenberg said that only observable quantities can enter the theory. To this Einstein replied [25]: "What quantities are observables should not be our choice, but should be given, should be indicated to us by the theory." Whether the three components of spin can be observables simultaneously, cannot be decided by transplanting classical mechanics to the microscopic world, but should be decided by the experimentally verified theory of quantum mechanics. In the question, whether there exist three elements of physical reality corresponding to the three components of spin of a particle, Einstein did not follow his own correct judgement. As L. Rosenfeld commented in this connection [25], "that was a very wise remark which Einstein ought perhaps to have remembered later on."

When two particles are separated spatially by a large distance, from the result of a measurement on the spin component of one of the particles we can immediately predict the spin component of the other particle. This happens because of the correlation of spin components of the particles built in the state of the system. Remember that in the spin singlet state the spin components of any of the two particles are undetermined before the measurement. Once a measurement is carried out on one particle, the relevant term in the entangled wave function is singled out (collapse of wave function) in which information on the spin component of the other particle is contained. The measurement does not send any message or exert any interaction to the other particle.[11] Should it be so, that would really be an action-at-a-distance. The delayed choice experiment discussed in Section 1.9 will give experimental proof of the absence of "communication" between particles. In Section 2.2.2 on the Schmidt decomposition of an entangled state a specific demonstration will be given that there is no information exchange possible between the two particles when a measurement is carried out on one of them.

---

[11]In a version of hidden variable theory, causal relations between the particles to save locality is worked out explicitly, but in a rather strange way which Einstein would have liked least [25].

Einstein could not accept the quantum mechanical description of the entanglement, he characterized the correlation between the spin components of the particles as being realized by a "spooklike" action-at-a-distance. The premise (3) EPR put forward was aimed at ruling out the correlation/entanglement, which is just an important characteristic of quantum mechanics. The question EPR raised was actually not a paradox, but a suggestion to create a new theory on the basis of the premises. Their challenge led to an spurt of the study of "hidden variable theories". According to such theories all dynamical variables have definite values. Some of them do not appear in the quantum mechanical description, they are called "the hidden variables" [26]. For example a state of spin angular momentum is characterized in quantum mechanics by S^2, S_z, here S_y and S_z are the hidden variables. Hidden variable theories can be classified into two categories. One category consists of theories attempting to reproduce all observable conclusions of quantum mechanics. The consequence is that some dynamical variables may show strange properties or some strange interaction has to be introduced. The other category consists of theories based strictly on the three premises, in some cases quantum mechanical predictions are reproduced, but other cases different results are obtained. Such theories attract attention of physicists, because crucial experiments can be designed and carried out to judge the correctness of such theories. In the literature such theories are sometimes called local (premise (3)) and realistic (premise (2)) theories, or local deterministic theories. Concerning the disputes between Einstein and Bohr a majority of physicists (at least from the 1940s) have been on Bohr's side. For about two decades this has been an issue in discussions among physicists, but researches in this direction have not been among the mainstream problems. The situation changed in 1965 due to the research of J. S. Bell, which played the decisive role.

1.7.3 Bell's theorem

Is a hidden variable theory compatible with the commutation relations and the correlations of quantum mechanics? There have been many discussions. J. S. Bell [27][12] proved a theorem[1] in 1965 that it is impossible to construct a local, deterministic hidden variable theory which could lead to results in agreement with *all* quantum mechanical predictions. Beginning from Bohm's version of EPR paradox, Bell proved the following. The hidden variable theory leads to an inequality which is unable to cover quantum mechanical predictions. Let $A_{\hat{a}}$ and $B_{\hat{b}}$ represent results of measurement of spin components of particle 1 in the direction $\hat{a}$ and of particle 2 in the direction $\hat{b}$ in units of $\hbar/2$ respectively, where $\hat{a}$ and $\hat{b}$ are arbitrary unit vectors. Consider the product of $A_{\hat{a}}$ and $B_{\hat{b}}$. According to quantum mechanics it is the average value of the Hermitian operator $E(\hat{a}, \hat{b}) = \sigma_1 \cdot \hat{a} \, \sigma_2 \cdot \hat{b}$ in the state (1.7.7), i.e.,

$$A_{\hat{a}} B_{\hat{b}} = [E(\hat{a}, \hat{b})]_\psi = \langle \psi | \sigma_1 \cdot \hat{a} \sigma_2 \cdot \hat{b} | \psi \rangle$$

$$= -\hat{a} \cdot \hat{b}. \tag{1.7.8}$$

A special case $\hat{a} = \hat{b}$ gives

$$[E(\hat{a}, \hat{a})]_\psi = -1. \tag{1.7.9}$$

This is the example discussed in Bohm's version. Can the average value (1.7.8) be obtained from a local deterministic theory by taking the statistical average? From the viewpoint of the hidden variable theories, all observables have definite values (determinism), while probability distributions for the results of measurement in quantum mechanics is due to the fact that the measured system is in different states distinguished from one another by the values of hidden variables, and that these states form a statistical ensemble. Quantum mechanics is ignorant about this and is content with the probability distribution. In the hidden variable theories there exists a complete specification of the state in which the determinism is manifest. Let such a state be characterized by means of a set of parameters denoted by a simple symbol λ. Let $\rho(\lambda)$ be the probability distribution with

---

[12]The theoretical and experimental developments concerning Bell's theorem before February 1978 were reviewed by J. F. Clauser and A. Shimony [28].

the normalization condition

$$\int_\Lambda \rho(\lambda)\mathrm{d}\lambda = 1. \tag{1.7.10}$$

The theory is local, and there is no action-at-a-distance, therefore the measurement of the $\hat{a}$ component of spin of particle 1 can only depend on λ and $\hat{a}$, and is independent of $\hat{b}$. Similarly, the measurement of the $\hat{b}$ component of spin of particle 2 is independent of $\hat{a}$. For arbitrary $\hat{a}$, $\hat{b}$ the following relation follows

$$(A_{\hat{a}} B_{\hat{b}})(\lambda) = A_{\hat{a}}(\lambda) B_{\hat{b}}(\lambda). \tag{1.7.11}$$

For any λ, the dynamical variable $E(\hat{a}, \hat{b})$ has definite values. Carrying out the ensemble average, we obtain

$$E(\hat{a}, \hat{b}) = \int_\Lambda A_{\hat{a}}(\lambda) B_{\hat{b}}(\lambda) \rho(\lambda) \mathrm{d}\lambda. \tag{1.7.12}$$

In the following we denote by $E(\hat{a}, \hat{b})$ its value in the hidden variable theory, and by $[E(\hat{a}, \hat{b})]_\psi$ its quantum mechanical expectation value. Bell was going to prove the following: Under the condition of locality (1.7.11) reproducing the quantum mechanical prediction (1.7.9) will lead to an inequality to be satisfied by $E(\hat{a}, \hat{b})$. Since $E(\hat{a}, \hat{a}) = -1$, Eq. (1.7.11) gives

$$A_{\hat{a}}(\lambda) = -B_{\hat{a}}(\lambda). \tag{1.7.13}$$

Let $\hat{c}$ be another unit vector, then we have

$$E(\hat{a}, \hat{b}) - E(\hat{a}, \hat{c}) = \int_\Lambda (A_{\hat{a}}(\lambda) B_{\hat{b}}(\lambda) - A_{\hat{a}}(\lambda) B_{\hat{c}}(\lambda)) \rho(\lambda) \mathrm{d}\lambda$$

$$= \int_\Lambda A_{\hat{a}}(\lambda) B_{\hat{b}}(\lambda)(1 - A_{\hat{b}}(\lambda) A_{\hat{c}}(\lambda)) \rho(\lambda) \mathrm{d}\lambda, \tag{1.7.14}$$

where (1.7.13) and the relation $A_{\hat{b}}(\lambda) A_{\hat{b}}(\lambda) = 1$ (which follows from $A_{\hat{b}}(\lambda) = \pm 1$) are used. In (1.7.14) the factor $A_{\hat{a}}(\lambda) B_{\hat{b}}(\lambda)$ can be +1 or -1 for different λ. This factor leads to partial cancellation in the

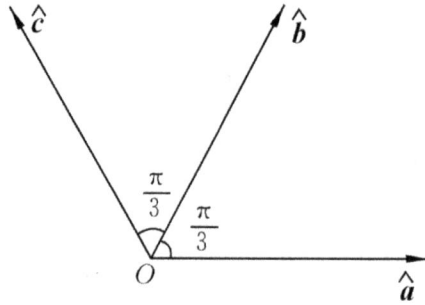

Fig. 1.33   A choice of three coplanar unit vectors $\hat{a}, \hat{b}$ and $\hat{c}$.

integral. Consequently

$$|E(\hat{a}, \hat{b}) - E(\hat{a}, \hat{c})| \leq \int_{\Lambda} (1 - A_{\hat{b}}(\lambda) A_{\hat{c}}(\lambda)) \rho(\lambda) d\lambda.$$

Using $A_{\hat{c}}(\lambda) = -B_{\hat{c}}(\lambda)$, we obtain

$$|E(\hat{a}, \hat{b}) - E(\hat{a}, \hat{c})| \leq 1 + E(\hat{b}, \hat{c}). \qquad (1.7.15)$$

This is one of the Bell inequalities. By making ingenious choice of $\hat{a}, \hat{b}$ and $\hat{c}$ the inequality can be made to contradict quantum mechanical predictions. For example, if the unit vectors are chosen as shown in Fig. 1.33, the quantum mechanical prediction is

$$|[E(\hat{a}, \hat{b})]_{\psi} - [E(\hat{a}, \hat{c})]_{\psi}| = |-\hat{a} \cdot \hat{b} + \hat{a} \cdot \hat{c}| = 1,$$

but $1 + [E(\hat{b}, \hat{c})]_{\psi} = 1/2$, violating Bell's inequality (1.7.15).

From such an antithesis Bell's theorem is proved. The power of Bell's theorem lies on its universality. It does not aim at studying a specific form of the hidden variable theory to prove its invalidity. It proves that the EPR premises on locality, realism etc. can be led in *certain circumstances* to contradict quantum mechanical predictions.

1.7.4 Bell's inequality generalized to real systems

In the foregoing discussions Bell's inequality is shown for ideal systems. In order to confront the theory with experiment it is necessary to deal with real systems, for which some conditions used in the proof are no more valid, for instance the efficiency of detectors is always smaller than unity, analyzers have attenuations and leakage to the orthogonal channel etc. In general one of the four following possibilities can happen: (1) both particles are detected, (2) particle 1 is detected and particle 2 is lost, (3) particle 2 is detected and particle 1 is lost, (4) both particles are lost. If only the first possibility is used to be compared with the experiment we are making the tacit assumption that the relative weight of sub-ensemble 1 in the total ensemble is a constant, independent of $\hat{a}$ and $\hat{b}$. But the experimental setup and the measuring process cannot guarantee this. In 1971 Bell generalized his proof [29] to include situations in a real system. For each measurement of spin components there are three possibilities: $+1$ represents spin upward, -1 represents spin downward, and 0 represents particle loss. We have then

$$
A_{\hat{a}}(\lambda) = \begin{cases} +1 \\ 0 \\ -1 \end{cases}, \quad B_{\hat{b}}(\lambda) = \begin{cases} +1 \\ 0 \\ -1 \end{cases}. \tag{1.7.16}
$$

Denote by $\overline{A_{\hat{a}}}(\lambda)$ and $\overline{B_{\hat{b}}}(\lambda)$ expectation values of $A_{\hat{a}}(\lambda)$ and $B_{\hat{b}}(\lambda)$ in the state λ. Since particle loss is possible, we have

$$
|\overline{A_{\hat{a}}}(\lambda)| \le 1, \ |\overline{B_{\hat{b}}}(\lambda)| \le 1. \tag{1.7.17}
$$

The locality premise leads to

$$
E(\hat{a}, \hat{b}) = \int \overline{A_{\hat{a}}}(\lambda)\overline{B_{\hat{b}}}(\lambda)\rho(\lambda)\mathrm{d}\lambda, \tag{1.7.18}
$$

where $\rho(\lambda)$ is independent of $\hat{a}$ and $\hat{b}$. Let $\hat{a}'$ and $\hat{b}'$ be other orientations of the spin measurements for particle 1 and 2 respectively. From Eq. (1.7.18) we have[13]

$$
E(\hat{a}, \hat{b}) - E(\hat{a}, \hat{b}') = \int_{\Lambda} \overline{A_{\hat{a}}}(\lambda)\overline{B_{\hat{b}}}(\lambda)[1 \pm \overline{A'_{\hat{a}}}(\lambda)\overline{B'_{\hat{b}}}(\lambda)]\rho(\lambda)\mathrm{d}\lambda
$$

$$
- \int_{\Lambda} \overline{A_{\hat{a}}}(\lambda)\overline{B_{\hat{b}}}(\lambda)[1 \pm \overline{A_{\hat{a}'}}(\lambda)\overline{B_{\hat{b}'}}(\lambda)]\rho(\lambda)\mathrm{d}\lambda.
$$

---

[13]The $\pm$ terms in the bracket of the right hand side combine to cancel each other. They are added for further convenience to obtain (1.7.19).

Using (1.7.17), we obtain

$$|E(\hat{a},\hat{b}) - E(\hat{a},\hat{b}')| \leq \int_{\Lambda} [1 \pm \overline{A_{\hat{a}'}(\lambda)B_{\hat{b}'}(\lambda)}]\rho(\lambda)\mathrm{d}\lambda$$

$$+ \int_{\Lambda} [1 \pm \overline{A_{\hat{a}'}(\lambda)B_{\hat{b}}(\lambda)}]\rho(\lambda)\mathrm{d}\lambda$$

$$= 2 \pm (E(\hat{a}',\hat{b}') + E(\hat{a}',\hat{b})),$$

and consequently[14]

$$-2 \leq E(\hat{a},\hat{b}) - E(\hat{a},\hat{b}') + E(\hat{a}',\hat{b}) + E(\hat{a}',\hat{b}') \leq 2. \qquad (1.7.19)$$

The minus sign can be made to appear in front of any of the 4 terms by redefining $\hat{a},\hat{b},\hat{a}'$ and $\hat{b}'$. (1.7.19) represents another class of Bell's inequalities.

Quantum mechanics gives

$$[E(\hat{a},\hat{b})]_{\psi} = -C\hat{a}\cdot\hat{b}. \qquad (1.7.20)$$

A real positive number $C \leq 1$ is introduced to take into account the situation that in real systems the correlation is not so complete. $C = 1$ only in ideal systems. Choose $\hat{a},\hat{b},\hat{a}'$ and $\hat{b}'$ as shown in Fig. 1.34, the quantum mechanical predictions can violate Bell's inequality. The two kinds of geometry in the figure give

$$[E(\hat{a},\hat{b}) - E(\hat{a},\hat{b}') + E(\hat{a}',\hat{b}) + E(\hat{a}',\hat{b}')]_{\psi} = \pm 2\sqrt{2}C. \qquad (1.7.21)$$

Comparing with inequality (1.7.19) we see that in a wide range of the parameter C Bell's inequality is violated. From Fig. 1.35 we see that values of the quantum mechanical predictions fall within the range between 2 and $2\sqrt{2}$, and also between -2 and $-2\sqrt{2}$ (shown by the wavy line area), where the hidden variable theory cannot reproduce the quantum mechanical predictions. For C too small due to the deficiencies of the equipment, the values given by quantum mechanics can satisfy Bell's inequality, and therefore no definite judgement can be made. The verification of Bell's inequality

---

[14]This inequality is often referred to as the CHSH inequality, after Clauser,Horne, Shimony and Holt.

is designated to answer the question whether quantum mechanical predictions can be reproduced by hidden variable theories. If a series of inequalities under a number of ingeniously chosen circumstances are found to be violated, we conclude that the hidden variable theories cannot describe the microscopic world correctly.

1.8 Experimental Verifications of Bell's Inequality

From the analysis mentioned above we see that it is not always possible to distinguish between a hidden variable theory and the quantum mechanics in *any* experiment. It is necessary to find sensitive regions of the parameters in the experiment, for example the geometry shown in Fig. 1.34, and to use sensitive detectors and instruments.

Early experiments tried to use the spin correlation between two photons created during the annihilation of a positronium with 0 angular momentum. The difficulty was that it is impossible to find an efficient polarizer for photons of such high energy (0.51 MeV). Indirect measurements were carried out via the Compton effect cross-section produced by the photons. First measurements did not bring consistent results. The situation improved somewhat and the results pointed to the violation of the Bell's inequality. There were disputes because of the indirect implications.

Convincing experiments made use of the cascade radiations of total angular momentum 0 state of an atom. High efficiency of

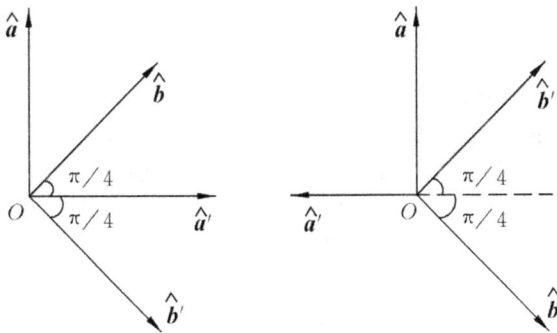

Fig. 1.34 Two choices of four coplanar vectors $\widehat{a}, \widehat{b}, \widehat{a}'$ and $\widehat{b}'$.

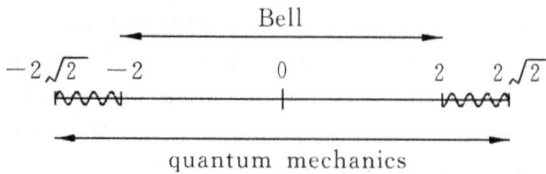

Fig. 1.35   Comparison between quantum mechanics and hidden variable theories using Bell's inequality.

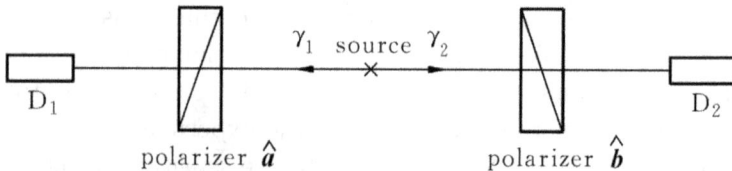

Fig. 1.36   Scheme of the equipment measuring the spin correlation of photons.

polarizers can be achieved for photons of such low energies. Research in this direction was pioneered by J. F. Clauser, M. A. Horne, A. Shimony and R. A. Holt [30]. A number of similar experiments followed. Figure 1.35 gives the scheme of the experimental setup, where two photons γ_1, γ_2 are emitted in a cascade from the source. Two polarizers set at directions of polarization $\hat{a}$ and $\hat{b}$. The coincidence counts of two detectors as a function of the angle θ between $\hat{a}$ and $\hat{b}$ are measured, which is a measure of the angular momentum correlation of the photons. For this the hidden variable theory and quantum mechanics give different results. Choose the cascade transitions

$$J = 0 \xrightarrow{\gamma_1} J = 1 \xrightarrow{\gamma_2} J = 0,$$

where J is the angular momentum of atomic energy states. Since the initial and the final states have the same angular momentum, the sum of the angular momenta of the emitted photons must be 0. Let $R(\theta)$ be the coincidence counting rate of the two detectors when the angle between $\hat{a}$ and $\hat{b}$ is θ . The result given by Clauser, Shimony *et al.* [30] for the hidden variable theory is

$$\left| \frac{R(\pi/8)}{R_0} - \frac{R(3\pi/8)}{R_0} \right|_{\text{HVT}} \leq \frac{1}{4},$$

where HVT is the abbreviation for the hidden variable theory, R_0 is the coincidence counting rate without the polarizers. The quantum mechanical result is

$$\left| \frac{R(\pi/8)}{R_0} - \frac{R(3\pi/8)}{R_0} \right|_\psi = \frac{1}{4}\sqrt{2},$$

which does not satisfy the Bell's inequality. Let $R \equiv \frac{R(\pi/8)}{R_0} - \frac{R(3\pi/8)}{R_0}$, the results of measurements were

(1) The cascade radiation of Ca

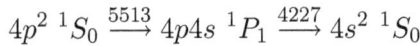

$$4p^2 \, {}^1S_0 \xrightarrow{5513} 4p4s \, {}^1P_1 \xrightarrow{4227} 4s^2 \, {}^1S_0$$

was used by Freeman and Clauser [31] in 1972. The result was $R = 0.300 \pm 0.008$, violating Bell's inequality.

(2) The cascade radiations from ^{198}Hg

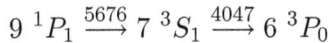

$$9 \, {}^1P_1 \xrightarrow{5676} 7 \, {}^3S_1 \xrightarrow{4047} 6 \, {}^3P_0$$

was used by Holt and Pipkin [32]. This is a cascade transition $J = 1 \to J = 1 \to J = 0$, for which the quantum mechanical formula has to be changed [26]. Their result was $R=0.216\pm0.013$, compatible with Bell's inequality, while the quantum mechanical result needed a value of C (Eq. (1.7.20)) too small to be understood. It was found that the stress in the walls of the Pyrex bulb used to contain the electron gun and mercury vapor was an origin of systematic error.

(3) Clauser [33] repeated the experiment by Holt and Pipkin. The result was $R = 0.2885 \pm 0.0093$, violating Bell's inequality.

(4) The cascade radiations

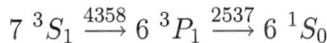

$$7 \, {}^3S_1 \xrightarrow{4358} 6 \, {}^3P_1 \xrightarrow{2537} 6 \, {}^1S_0$$

from ^{200}Hg was used by Fry and Thompson [34]. The result was $R = 0.296 \pm 0.014$, violating Bell's inequality. Many experiments before February 1978 were reviewed by J. F. Clauser and A. Shimony [28]. The results pointed rather convincingly to the violation of Bell's inequality. Most of the experiments used "single channel" recording, i.e., parallel polarization was recorded as 1, while perpendicular

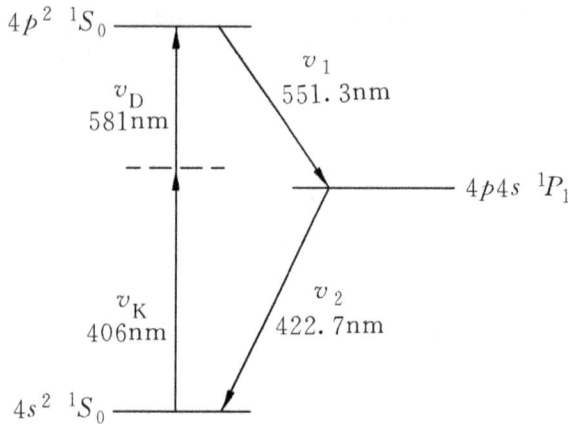

Fig. 1.37   Cascade radiations from Ca atom. Reprinted with permission from Alain Aspect, Philippe Grangier, Gérard Roger, *Phys. Rev. Lett.* 47, 460, 1981. Copyright (2017) by the American Physical Society.

polarization was not recorded. Critics considered this as "incomplete measurement" and did not accept the conclusion.

The experimental conditions have been improved since the early 1980s, and results have become more and more convincing. Some are already close to the ideal case. A. Aspect [35] used cascade radiations from Ca (Fig. 1.37). Ca atom is excited to the state $4p^2$ $^1S_0$ by a two-photon process (Kr ion laser and tunable dye laser). Two photons with total angular momentum 0 are emitted during de-excitation. The spin correlation $R(\theta)/R_0$ is measured. Since laser excitation is used, the light beam has very small dimensions ($0.5\,\text{mm} \times 0.05\,\text{mm} \times 0.05\,\text{mm}$). This is very favorable for the detector channel geometry. Through careful control of the experimental conditions the drift and fluctuation of the data for several hours of experiment was kept under 1%. The single channel result is shown in Fig. 1.38. The quantum mechanical result (solid line) with the solid angle allowance of polarizers and the detection efficiency taken into account agrees well with the experiment. Bell's inequality is violated, deviating from the data by 9 standard deviations. Aspect's group carried out a double channel experiment, schematically shown in Fig. 1.39(a). In the experiment polarimeters are used, enabling photons polarized parallel to the direction set by the polarimeter to

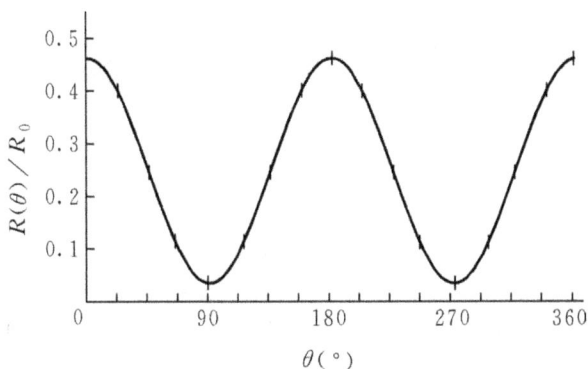

Fig. 1.38 Normalized coincidence rate as a function of the relative polarizer orientation. Reprinted with permission from Alain Aspect, Philippe Grangier, Gérard Roger, *Phys. Rev. Lett.* 47, 460, 1981. Copyright (2017) by the American Physical Society.

pass through directly, while photons polarized perpendicular to this direction are deflected by 90°. Photons with either polarizations are recorded by photomultipliers (PM). A circuit of coincidence counting gives $N_{++}(\hat{a}, \hat{b})$, $N_{--}(\hat{a}, \hat{b})$, $N_{+-}(\hat{a}, \hat{b})$, $N_{-+}(\hat{a}, \hat{b})$, where, for example, $N_{+-}(\hat{a}, \hat{b})$ represents the counts for one photon polarized parallel to $\hat{a}$ (denoted by +) and the other polarized perpendicular to $\hat{b}$ (denoted by −). Using these quantities a function $E(\hat{a}, \hat{b})$ can be calculated according to

$$E(\hat{a}, \hat{b}) = \frac{N_{++}(\hat{a}, \hat{b}) + N_{--}(\hat{a}, \hat{b}) - N_{+-}(\hat{a}, \hat{b}) - N_{-+}(\hat{a}, \hat{b})}{N_{++}(\hat{a}, \hat{b}) + N_{--}(\hat{a}, \hat{b}) + N_{+-}(\hat{a}, \hat{b}) + N_{-+}(\hat{a}, \hat{b})}.$$

(1.8.1)

The experimental data are shown in Fig. 1.39(b), where $E(\theta)$ represents the value of $E(\hat{a}, \hat{b})$ when the angle between $\hat{a}$ and $\hat{b}$ is θ. We see that the experimental equipment is close to ideal that $E(0)$ is very close to 1 for $\theta = 0$. Let

$$S = E(\hat{a}, \hat{b}) - E(\hat{a}, \hat{b}') + E(\hat{a}', \hat{b}) + E(\hat{a}', \hat{b}').$$

(1.8.2)

According to the arrangements shown in Fig. 1.40, the quantum mechanical result for ideal case is

$$S_{\text{QM}} = 2\sqrt{2},$$

(1.8.3)

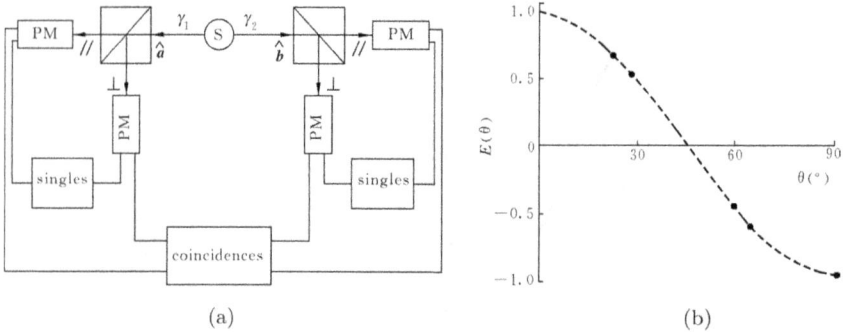

Fig. 1.39   (a) Scheme of the double channel experiment setup; (b) Correlation of polarimeters $E(\theta)$. Reprinted with permission from Alain Aspect, Philippe Grangier, Gérard Roger, *Phys. Rev. Lett.* 47, 460, 1981. Copyright (2017) by the American Physical Society.

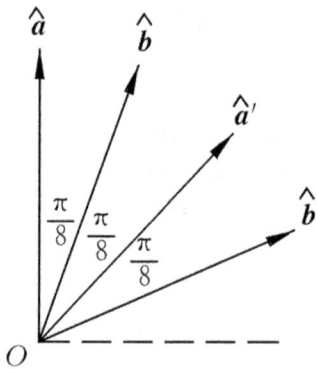

Fig. 1.40   A choice of $\widehat{a}$, $\widehat{b}$, $\widehat{a}'$, $\widehat{b}'$. Reprinted with permission from Alain Aspect, Philippe Grangier, Gérard Roger, *Phys. Rev. Lett.* 47, 460, 1981. Copyright (2017) by the American Physical Society.

while the hidden variable theory result is

$$-2 \leq S_{\text{HVT}} \leq 2. \tag{1.8.4}$$

The experimental result is 2.697 ± 0.015. Taking into account the realistic situation (solid angle allowance of the polarimeters and detection efficiency) the quantum mechanical result is $S_{\text{QM}} = 2.70 \pm 0.05$. The experimental result agrees excellently with the quantum mechanical prediction, violating Bell's inequality by 40 standard deviations.

Ever since Bell's theorem was established, large numbers of theoretical and experimental papers have been published. Disputes kept going on for about two decades as to whether the hidden variable theories can describe the microscopic world correctly and whether is the quantum mechanical description incomplete. When an experimental result is published contradicting the Bell's inequality, critics follow to voice discontent, that there are loopholes in the experiment. Such criticism often led to more experiments attempting at the removal of the loopholes, and were criticized again for other loopholes.

The down-conversion of a photon in a non-linear crystal provides a real example for discussing the EPR paradox. The sum of the energies of a pair of conjugate photons is equal to the energy of the pump photon, viz. $k_1 + k_2 = k$, a constant. The time of emission of the photons is the same, i.e., $t_1 - t_2 = 0$. This situation is similar to the EPR case of $p_1 + p_2 = $ const. and $x_1 - x_2 = $ const. The high visibility interferometer mentioned at the end of Section 1.5.3 (Ref. [18]) was constructed for the purpose of testing the Bell's inequality. R. Chiao analyzed the interference between the pair of down-converted photons in the interferometers. Since ΔL is much longer than the coherent length of a single photon, there is no single photon interference in both interferometers. According to the local hidden variable theories the visibility of two-photon interference can at most be 50%. The experimental result is $80.4\% \pm 0.6\%$, violating Bell's inequality by 16 standard deviations.

On the basis of this experiment, R. Chiao's group suggested a "loophole free test" of Bell's inequality [36]. The method involves the use of two non-linear crystals, and is complicated in its realization. P. G. Kwiat *et al.* [37] provided a simple method relying on non-collinear type II phase matching. Non-linear BBO (beta barium borate) crystal is used to produce a polarized entangled state. With type II phase matching, the down-converted photons are emitted into two cones, one ordinary polarized and the other extraordinary polarized. By appropriately adjusting the angle between the pump

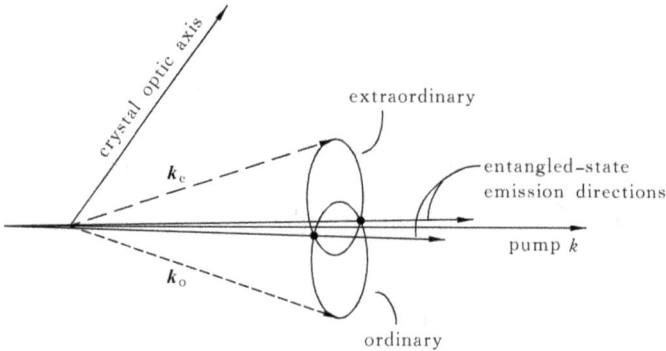

Fig. 1.41 Polarized entangled states of down-converted photons with type II phase matching. Reprinted with permission from Paul G. Kwiat, Klaus Mattie, Harald Weinfurter, and Anton Zeilinger, *Phys. Rev. Lett.* 75, 4337, 1995. Copyright (2017) by the American Physical Society.

beam and the optic axis of the crystal the two cones can be made to intersect along two lines, as shown in Fig. 1.41. The resulting entangled state is described by

$$|\psi\rangle = \frac{1}{\sqrt{2}}(|H_1 V_2\rangle + e^{i\alpha}|V_1 H_2\rangle), \qquad (1.8.5)$$

where H and V indicate horizontal (extraordinary) and vertical (ordinary) polarization, respectively. The relative phase α arises from the crystal birefringence. Using an additional birefringence phase shifter the value of α can be set to 0 or π. Similarly, a half-wave plate in one path can be used to change horizontal polarization to vertical polarization and vice versa. One can thus produce very easily one of the four EPR-Bell states[15]

$$\left.\begin{array}{l} |\psi^{\pm}\rangle = \dfrac{1}{\sqrt{2}}(|H_1 V_2\rangle \pm |V_1 H_2\rangle) \\[2ex] |\varphi^{\pm}\rangle = \dfrac{1}{\sqrt{2}}(|H_1 H_2\rangle \pm |V_1 V_2\rangle) \end{array}\right\}. \qquad (1.8.6)$$

---

[15]These states form the complete maximally entangled basis of the two-particle Hilbert space, and are important in their applications to quantum information (Section 2.2).

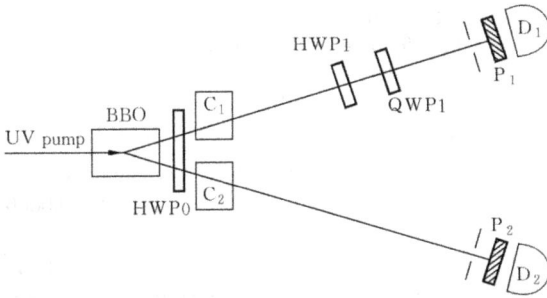

Fig. 1.42 Schematic of the equipment for producing polarization entangled states by down-conversion. Reprinted with permission from Paul G. Kwiat, Klaus Mattie, Harald Weinfurter, and Anton Zeilinger, *Phys. Rev. Lett.* 75, 4337, 1995. Copyright (2017) by the American Physical Society.

For example starting from $|\psi^\pm\rangle$ by putting a half wave plate in the second beam one obtains $|\varphi^\pm\rangle$. The birefringent nature of the down-conversion crystal complicates the situation. The group velocities of light in the two beams are different, leading to time dilation and longitudinal and transverse walk-offs of the paths. These deviations must be constrained within the coherent time and the coherent length, in order that interference can still be observed. The experimental setup is shown in Fig. 1.42. The extra birefringent crystals C_1 and C_2 and the half wave plate HWP0 are used to compensate the birefringent walk-off effects from the BBO crystal. By appropriately setting the half wave plate HWP1 and the quarter wave plate QWP1 one can produce any one of the four EPR-Bell states. P_1 and P_2 are polarizers, D_1 and D_2 are cooled silicon avalanche photodiode detectors. Coincidence rates $C(\theta_1, \theta_2)$ were recorded as a function of the polarizer settings θ_1 and θ_2. The maximum visibility obtained is 97.8%±1.0%. Formula (1.8.1)∼(1.8.4) are used for the test of Bell's inequality. Define

$$E(\theta_1, \theta_2) = \frac{C(\theta_1, \theta_2) + C(\theta_1^\perp, \theta_2^\perp) - C(\theta_1, \theta_2^\perp) - C(\theta_1^\perp, \theta_2)}{C(\theta_1, \theta_2) + C(\theta_1^\perp, \theta_2^\perp) + C(\theta_1, \theta_2^\perp) + C(\theta_1^\perp, \theta_2)},$$

$$(1.8.7)$$

where

$$\theta_1^\perp = \frac{\pi}{2} + \theta_1, \quad \theta_2^\perp = \frac{\pi}{2} + \theta_2.$$

Table 1.8.1   Coincidence rate predictions and measured values for the four different EPR-Bell states.

EPR-Bell states	$C(\theta_1, \theta_2)$	S
$|\psi^+\rangle$	$\sin^2(\theta_1 + \theta_2)$	-2.6489 ± 0.0064
$|\psi^-\rangle$	$\sin^2(\theta_1 - \theta_2)$	-2.6900 ± 0.0066
$|\varphi^+\rangle$	$\cos^2(\theta_1 - \theta_2)$	2.557 ± 0.014
$|\varphi^-\rangle$	$\cos^2(\theta_1 + \theta_2)$	2.529 ± 0.013

The observed correlation is expressed in terms of the parameter S, where

$$S = E(\theta_1, \theta_2) + E(\theta_1', \theta_2) + E(\theta_1, \theta_2') - E(\theta_1', \theta_2'). \qquad (1.8.8)$$

The local realistic theory gives $|S| \leq 2$. Under the experimental setting $\theta_1 = -22.5°, \theta_1' = 22.5°, \theta_2 = -45°, \theta_2' = 0°$ the data are tabulated in Table 1.8.1.

From the table we see that the CHSH inequality (1.7.19) is strongly violated. The source of the entangled photons is more than an order of magnitude brighter than previous sources since the entanglement is produced directly from the crystal. In less than 5 min. the statistics accumulated was enough for a 102 standard deviation violation of the Bell's inequality for $|\psi^+\rangle$. This is a large step toward the "loophole free" experiments.

1.9   Wheeler's Delayed Choice Experiment

The question whether there exists an action-at-a-distance was raised in the discussion on EPR paradox. When S_x of one particle of a correlated pair is measured, the S_x of its distant partner can be predicted with certainty. When S_y of one of the particle is measured, then the S_y of its distant partner is known immediately. The hidden variable theory does not accept the quantum mechanical correlation between the particles and attempts at obtaining the same prediction as that from quantum mechanics by an ensemble average over deterministically specified states. From the violation of Bell's inequality we can see that such attempts fail. It is interesting anyway to try to find out experimentally whether some kind of communication between the particles does happen when one of them is measured or at least when an experimental arrangement dedicated for such a measurement is set. J. A. Wheeler [38] raised a question that if the choice between the path (particle nature) and the interference (wave nature) is made only when the photon has already passed through the slits (the delayed choice), will the result be the same as that for the choice made beforehand? One could argue in the following way: If the experimental arrangement (choice) is ready beforehand, then the arrangement could send a message to the photon such that it could be ready to respond: be a wave or be a particle. But when the photon has already passed through the slits it would be impossible to change its mind even if a message is received about the change in detection choice. A thought experiment based on Wheeler's idea is shown in Fig. 1.43. A single photon pulse enters the interferometer through the beam splitter BS_1. Without the splitter BS_2 the detectors D_1 and D_2 ascertain on "which way" (path 1 or 2) the photon traveled. Then only one detector receives the photon and there is no interference. When BS_2 is installed, the "which way " information is erased, and the detectors show interference signature, implying that the arriving photon has travelled both routes. According to Wheeler, in the delayed choice version of the

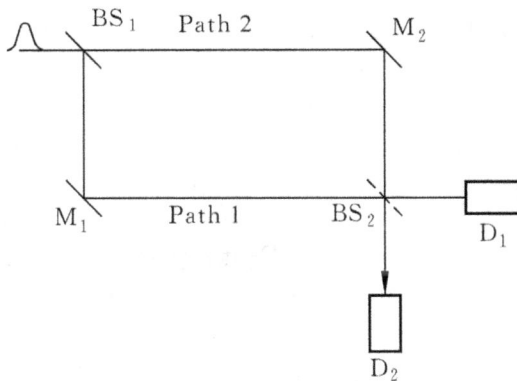

Fig. 1.43   Experimental setup for photon interference. Reprinted with permission from T. Hellmuth, H. Walther, A. Zajonc, and W. Schleich, *Phys. Rev. A* 35, 2532, 1987. Copyright (2017) by the American Physical Society.

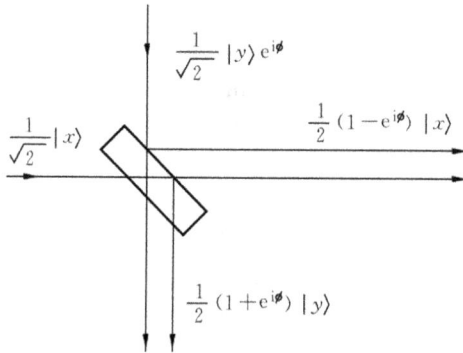

Fig. 1.44   The phase relations of coherent beams formed at the beam splitter. Reprinted with permission from T. Hellmuth, H. Walther, A. Zajonc, and W. Schleich, *Phys. Rev. A* 35, 2532, 1987. Copyright (2017) by the American Physical Society.

Table 1.9.1   Counting rates of detectors for different phase differences.

Phase difference ϕ	N_1	N_2
0	N	0
$\pi/2$	$N/2$	$N/2$
π	0	N

experiment one decides whether to put in the second beam splitter or not at the very last minute. Fig. 1.44 shows the change of the relative phase of the beams during transmission and reflection. Let ϕ be the phase difference between the beams when they arrived at BS$_2$. The phase of the partial wave $\frac{1}{\sqrt{2}}|y\rangle e^{i\phi}$ is shifted by π since it is reflected at the optically denser medium. From the phases of coherent beams entering detectors D$_1$ and D$_2$ it is possible to find the counting rates of detectors N_1 and N_2 as functions of initial phase difference ϕ. The counting rates for $\phi = 0, \pi/2$ and π are shown in Table 1.9.1. For any given ϕ, $N_1 + N_2 = N$, the total counting rate. The central part of the delayed-choice experiment carried out by the Garching group (T. Hellmuth, H. Walther, A. Zajonc, and W. Schleich) [39] is the Pockel cell, which can be activated and de-activated within a

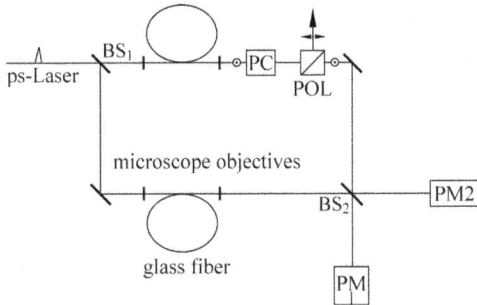

Fig. 1.45   Pockel cell and the polarizing prism for realizing the delayed choice. Reprinted with permission from T. Hellmuth, H. Walther, A. Zajonc, and W. Schleich, *Phys. Rev. A* 35, 2532, 1987. Copyright (2017) by the American Physical Society.

time of a few nanoseconds. It can rotate the plane of polarization of light by 90° when activated. It is followed by a Glan polarizing prism (POL) which deflects the light when the polarization direction is rotated. The Pockel cell and the polarizing prism are inserted in the path between BS_1 and M_1, as shown in Fig. 1.45. The photon from BS_1 will be reflected and enter the detector D_3 if the cell is activated. It will enter M_2 if the cell is not activated. The choice made is: photon path measured (particle nature) when the Pockel cell is activated or not measured (wave nature ensured) when the cell is not activated. In the experiment a krypton ion laser emits pulses of 150 ps pulse duration. Pulse repetition rate is lowered by an acousto-optical switch to ensure the operation of the Pockel cell and also to guarantee that the time between two pulses is much lower than the transit time of the light through the interferometer. An optical attenuator ensures that the average number of photons per pulse is less than 0.2. The incident light passes through the beam splitter BS_1 and the two split beams are directed and focused into the separate single-mode optical fibers of 5 m in length such that the Pockel cell can be operated properly. The polarization of light leaving the fiber is linear. The interference of the recombined beams is detected by the photomultipliers D_1 and D_2. In the normal mode of the experiment the Pockel cell is not activated during the whole transit time of the pulse through the setup.

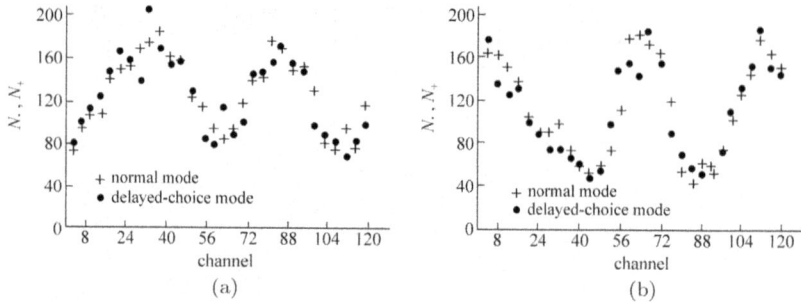

Fig. 1.46   Comparison of interference patterns for normal and delayed-choice modes. (a) Detector D_1; (b) Detector D_2. points: normal mode, crosses: delayed-choice mode. Reprinted with permission from T. Hellmuth, H. Walther, A. Zajonc, and W. Schleich, *Phys. Rev. A* 35, 2532, 1987. Copyright (2017) by the American Physical Society.

In the delayed-choice mode the Pockel is activated at first, but de-activated 5ns after the pulse has passed through the first beam splitter. Therefore the light pulse is well inside the optical fiber when the Pockel cell is opened. The pulse will find the cell opened when it arrives. The sufficient length of the optical fiber ensures the operation. In the experiment the laser pulses enter alternatively into the normal and the delayed-choice modes, with data denoted by points $(\cdot)$ and crosses $(+)$ respectively, as shown in Fig. 1.46. Counts in D_1 and D_2 in 30s are recorded in Fig. 1.46(a) and (b) respectively. The counts are recorded in a multichannel analyzer and the abscissa denotes the time in terms of the number of channels. Since the path difference of the two arms is strongly influenced by temperature-induced refractive index variations in the fibers the interference pattern changed steadily in time. In any condition the signals in detectors D_1 and D_2 are phase inverted. The data shows that these two modes give the same interference pattern. To make this more definite, the ratio of the counts in two modes is plotted in Fig. 1.47 (a) and (b) for detectors D_1 and D_2 respectively. The average values are $N_{\cdot}/N_{+} = 1.00 \pm 0.02$ for detector D_1 and $N_{\cdot}/N_{+} = 0.99 \pm 0.02$ for detector D_2 respectively, in full agreement with the quantum mechanical prediction.

Is there any real problem for the quantum mechanical description? Here we quote R. Feynman[40]: "We always have a great deal

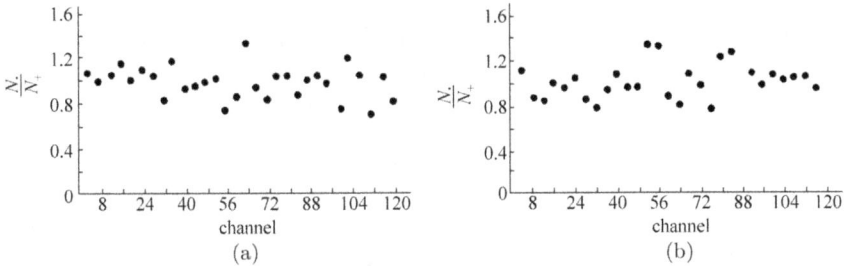

Fig. 1.47   Ratio $N./N_+$ (a) In detector D_1; (b) In detector D_2. Reprinted with permission from T. Hellmuth, H. Walther, A. Zajonc, and W. Schleich, *Phys. Rev. A* 35, 2532, 1987. Copyright (2017) by the American Physical Society.

of difficulty in understanding the world view that the quantum mechanics represents. Every new idea, it takes a generation or two until it becomes obvious that there is no real problem. I cannot define the real problem, therefore I suspect there is no real problem, but I'm not sure there is no real problem."

1.10   Bell's Theorem without Inequalities

1.10.1   Complete correlation between three particles

The hidden variable theory comes inevitably to contradict quantum mechanics because it does not accept the correlation in quantum mechanics and insists on the locality demand, as was made clear in the proof of Bell's theorem. If a property of particle 2 can be predicted with certainty when a measurement on particle 1 is carried out, the correlation is called *complete*. If the property of particle 2 can be predicted with a probability distribution only when a measurement on particle 1 is made, the correlation is called *statistical*. In the Bohm version of the EPR paradox considered in Section 1.7.3, a prediction of the spin component of particle 2 along $\widehat{b}$ is made after the spin component along $\widehat{a}$ of particle 1 is measured. In the case of $\widehat{a} = \widehat{b}$ the correlation is complete, while in the case of $\widehat{a} \neq \widehat{b}$ the correlation is statistical. Bell's inequality is derived in the case of $\widehat{a} \neq \widehat{b}$. The contradiction between EPR and quantum mechanics is not manifest in the case of complete correlation between *two* particles. D. M. Greenberger, M. A. Horne and A. Zeilinger proved Bell's theorem in

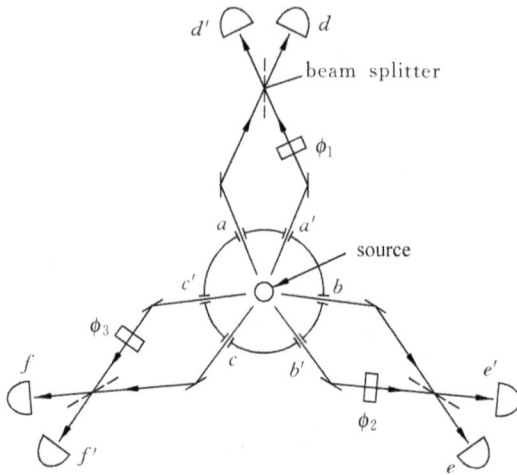

Fig. 1.48   A thought experiment of 3-particle interferometry. From Ref. [42].

a new way [41, 42] by analyzing a system consisting of *three or more* correlated spin 1/2 particles. The incompatibility of EPR premises with quantum mechanics can be demonstrated referring solely to *complete correlations* without resorting to inequalities. N. D. Mermin [43] had similar considerations.

We begin with a thought experiment using a three-particle interferometer shown in Fig. 1.48. A particle at rest decays into three particles of the same mass. If the decay products have the same energy, then they fly apart at an angle 120° with one another. Practically this can be realized by putting energy filters in front of the detectors such that only particles with the same energy are recorded. The central source is surrounded by an array of six apertures a, b and c at 120° angle and a', b', and c' also at 120° angle with one another. The decay products must emerge either through a, b and c or through a', b' and c'. The state of the decay products beyond the aperture is described by

$$|\Psi\rangle = \frac{1}{\sqrt{2}}(|a\rangle_1|b\rangle_2|c\rangle_3 + |a'\rangle_1|b'\rangle_2|c'\rangle_3), \qquad (1.10.1)$$

where $|a\rangle_1$ denotes the particle 1 in beam a, and so on. Such entangled states are usually referred to as the GHZ states. The relative phase between the two terms is set equal to 0, it has no influence on the subsequent discussions. Beyond the apertures beams a and a' are reflected and overlap at a 50%-50% beam splitter, with the outgoing beams monitored by detectors d and d'. The phase shifter *en route* beam a' produces an adjustable phase shift ϕ_1. The time evolution of beams $|a\rangle_1$ and $|a'\rangle_1$ are given by

$$|a\rangle_1 \to \frac{1}{\sqrt{2}}(|d\rangle_1 + \mathrm{i}|d'\rangle_1), \tag{1.10.2}$$

$$|a'\rangle_1 \to \frac{1}{\sqrt{2}}e^{\mathrm{i}\phi_1}(|d'\rangle_1 + \mathrm{i}|d\rangle_1). \tag{1.10.3}$$

There is a phase shift $\pi/2$ for the reflected beam at the beam splitter. The evolution of particle 2 beams and particle 3 beams are similar. The evolution of these beams leads to a change of the entangled state (1.10.1), which results in a linear superposition of 8 terms:

$$
\begin{aligned}
|\Psi\rangle \to \frac{1}{4}[&(1 - \mathrm{i}e^{\mathrm{i}(\phi_1+\phi_2+\phi_3)})|d\rangle_1|e\rangle_2|f\rangle_3 \\
&+ (\mathrm{i} - e^{\mathrm{i}(\phi_1+\phi_2+\phi_3)})|d\rangle_1|e\rangle_2|f'\rangle_3 \\
&+ (\mathrm{i} - e^{\mathrm{i}(\phi_1+\phi_2+\phi_3)})|d\rangle_1|e'\rangle_2|f\rangle_3 \\
&+ (\mathrm{i} - e^{\mathrm{i}(\phi_1+\phi_2+\phi_3)})|d'\rangle_1|e\rangle_2|f\rangle_3 \\
&+ (-1 + \mathrm{i}e^{\mathrm{i}(\phi_1+\phi_2+\phi_3)})|d\rangle_1|e'\rangle_2|f'\rangle_3 \\
&+ (-1 + \mathrm{i}e^{\mathrm{i}(\phi_1+\phi_2+\phi_3)})|d'\rangle_1|e\rangle_2|f'\rangle_3 \\
&+ (-1 + \mathrm{i}e^{\mathrm{i}(\phi_1+\phi_2+\phi_3)})|d'\rangle_1|e'\rangle_2|f\rangle_3 \\
&+ (-\mathrm{i} + e^{\mathrm{i}(\phi_1+\phi_2+\phi_3)})|d'\rangle_1|e'\rangle_2|f'\rangle_3].
\end{aligned} \tag{1.10.4}
$$

In the first term the detectors d, e, f receive the transmitted beams from a, b, c (with coefficient 1) and also the reflected beams (with phase shifts) from a', b', c' (with coefficient $\mathrm{i}^3 = -\mathrm{i}$). In the second, third and fourth terms there is one primed detector with the interchange of a transmitted beam and a reflected beam. The coefficients become i and $\mathrm{i}^2 = -1$. In the fifth, sixth and seventh terms there are two primed detectors, the coefficients become $\mathrm{i}^2 = -1$ and i. The last term has three primed detectors, and the coefficients becomes $i^3 = -i$ and 1. Provided that the detectors are complete, each set of 3 decay products can make each of d and d', each of e and e', each of f and f' to click. From (1.10.4) we obtain the detection probabilities

$$P_{def}(\phi_1, \phi_2, \phi_3) = \frac{1}{8}[1 + \sin(\phi_1 + \phi_2 + \phi_3)], \qquad (1.10.5)$$

$$P_{d'ef}(\phi_1, \phi_2, \phi_3) = \frac{1}{8}[1 - \sin(\phi_1 + \phi_2 + \phi_3)]. \qquad (1.10.6)$$

The sign of the second term in the bracket is $+(-)$ if there are even (odd) number of primed detectors. Let the result be $+1$ when a particle enters an unprimed detector and be -1 when a particle enters a primed detector. Then the expectation value of the measurement is

$$
\begin{aligned}
E(\phi_1, \phi_2, \phi_3) &= P_{def}(\phi_1, \phi_2, \phi_3) + P_{de'f'}(\phi_1, \phi_2, \phi_3) \\
&\quad + P_{d'ef'}(\phi_1, \phi_2, \phi_3) + P_{d'e'f}(\phi_1, \phi_2, \phi_3) \\
&\quad - P_{d'ef}(\phi_1, \phi_2, \phi_3) - P_{de'f}(\phi_1, \phi_2, \phi_3) \\
&\quad - P_{def'}(\phi_1, \phi_2, \phi_3) - P_{d'e'f'}(\phi_1, \phi_2, \phi_3) \\
&= \sin(\phi_1 + \phi_2 + \phi_3). \qquad (1.10.7)
\end{aligned}
$$

If we choose

$$\phi_1 + \phi_2 + \phi_3 = \pi/2,$$

then

$$E(\phi_1, \phi_2, \phi_3) = 1; \qquad (1.10.8)$$

If we choose instead

$$\phi_1 + \phi_2 + \phi_3 = 3\pi/2,$$

then

$$E(\phi_1, \phi_2, \phi_3) = -1. \qquad (1.10.9)$$

The foregoing are the quantum mechanical predictions. In the following we demonstrate that the EPR premises lead to inconsistencies if they are required to reproduce the quantum mechanical predictions. Because of the locality, separate measurements on the three particles are independent of one another. Let measurement be carried on particle 1 (phase shift ϕ_1) in a state λ, and the result be $A_\lambda(\phi_1)$. Similarly measurements on particles 2 and 3 give $B_\lambda(\phi_2)$ and $C_\lambda(\phi_3)$. These are independent, and therefore corresponding to the case (1.10.8) we must have

$$A_\lambda(\phi_1)B_\lambda(\phi_2)C_\lambda(\phi_3) = 1,$$
$$\phi_1 + \phi_2 + \phi_3 = \pi/2; \tag{1.10.10a}$$

and corresponding to the case (1.10.9) we have

$$A_\lambda(\phi_1)B_\lambda(\phi_2)C_\lambda(\phi_3) = -1,$$
$$\phi_1 + \phi_2 + \phi_3 = 3\pi/2. \tag{1.10.10b}$$

From Eq. (1.10.10a) we have

$$A_\lambda(0)B_\lambda(0)C_\lambda\left(\frac{\pi}{2}\right) = 1, \tag{1.10.11a}$$

$$A_\lambda\left(\frac{\pi}{2}\right)B_\lambda(0)C_\lambda(0) = 1, \tag{1.10.11b}$$

$$A_\lambda(0)B_\lambda\left(\frac{\pi}{2}\right)C_\lambda(0) = 1. \tag{1.10.11c}$$

Being 1 and -1, each number is the reciprocal of itself. Therefore from Eqs. (1.10.11a) and (1.10.11b) we have

$$A_\lambda(0)C_\lambda(0)A_\lambda\left(\frac{\pi}{2}\right)C_\lambda\left(\frac{\pi}{2}\right) = 1, \tag{1.10.11d}$$

and from (1.10.11c) we have

$$A_\lambda(0)C_\lambda(0) = \frac{1}{B_\lambda\left(\frac{\pi}{2}\right)} = B_\lambda\left(\frac{\pi}{2}\right). \tag{1.10.11e}$$

Equations (1.10.11d) and (1.10.11e) give

$$A_\lambda\left(\frac{\pi}{2}\right)B_\lambda\left(\frac{\pi}{2}\right)C_\lambda\left(\frac{\pi}{2}\right) = 1. \tag{1.10.11f}$$

Equation (1.10.11f) belongs to the category $\phi_1 + \phi_2 + \phi_3 = 3\pi/2$ and therefore contradicts (1.10.10b). This demonstrates that the quantum mechanical result (1.10.7) cannot be reproduced by the local realistic theory. The experimental setup shown in Fig. 1.48 is called a three-particle interferometer. When the phases of any two beams are fixed, the coincidence counting rates in (1.10.5) and (1.10.6) as functions of the phase of the remaining beam oscillates sinusoidally, with minimum 0 and maximum equal to 1/4 times the emission rate. This can be understood as the fringe of 3-particle interference. If only the coincidence counts between two detectors are measured, then from (1.10.5) and (1.10.6) it follows that

$$P_{ef}(\phi_1, \phi_2) = P_{def}(\phi_1, \phi_2, \phi_3) + P_{d'ef}(\phi_1, \phi_2, \phi_3) = \frac{1}{4}$$

This means that a 3-particle interferometer does not give 2-particle interference pattern. Similarly by summing $P_{def}, P_{d'ef}, P_{de'f}$ and $P_{d'e'f}$ we obtain $P_f = 1/2$, also independent of ϕ, i.e., there is no 1-particle interference. It is in general difficult to construct a 3-particle interferometer in the laboratory. Theoretical research has led to the proof of Bell's theorem without inequalities using 2-particle systems.

1.10.2 Bell's theorem without inequalities: Two-particle correlation

The following is the proof given by L. Hardy [44]. Consider the entangled state of two particles $i = 1, 2$ composed of the orthonormal basis $|\pm\rangle_i$,

$$|\Psi\rangle = \alpha|+\rangle_1|+\rangle_2 - \beta|-\rangle_1|-\rangle_2, \tag{1.10.12}$$

where α and β are real and satisfy

$$\alpha^2 + \beta^2 = 1. \tag{1.10.13}$$

Introduce another set of basis $|u_i\rangle, |v_i\rangle$ by

$$|+\rangle_i = b|u_i\rangle + ia^*|v_i\rangle, \tag{1.10.14a}$$
$$|-\rangle_i = ia|u_i\rangle + b^*|v_i\rangle, \tag{1.10.14b}$$

where

$$|a|^2 + |b|^2 = 1. \tag{1.10.15}$$

The orthonormality of the old basis guarantees the new basis being also orthonormal. The inverse transformation of (1.10.14) is

$$|u_i\rangle = b^*|+\rangle_i - ia^*|-\rangle_i, \qquad (1.10.16\text{a})$$
$$|v_i\rangle = -ia|+\rangle_i + b|-\rangle_i. \qquad (1.10.16\text{b})$$

The entangled state expressed in terms of the new basis is

$$|\Psi\rangle = (\alpha b^2 + \beta a^2)|u_1\rangle|u_2\rangle + i(\alpha a^* b - \beta a b^*)|u_1\rangle|v_2\rangle$$
$$+ i(\alpha a^* b - \beta a b^*)|v_1\rangle|u_2\rangle - (\alpha a^{*2} + \beta b^{*2})|v_1\rangle v_2\rangle. \quad (1.10.17)$$

We demand that the coefficients of $|u_1\rangle|u_2\rangle$ vanishes,[16] i.e.,

$$\frac{a^2}{\alpha} = -\frac{b^2}{\beta} \equiv k^2,$$

or

$$a = k\sqrt{\alpha}, \quad b = ik\sqrt{\beta}, \qquad (1.10.18)$$

where the plus sign is chosen for the square root and k is made real by appropriately choosing the phases of a and b. Equations (1.10.15) and (1.10.18) lead to

$$k^2 = \frac{1}{|\alpha| + |\beta|}. \qquad (1.10.19)$$

By substituting (1.10.18) back into (1.10.17) and using (1.10.19) we obtain

$$|\Psi\rangle = -\left[-\frac{\alpha\beta}{|\alpha| - |\beta|}|u_1\rangle|u_2\rangle + \sqrt{\alpha\beta}|u_1\rangle|v_2\rangle \right.$$
$$\left. + \sqrt{\alpha\beta}|v_1\rangle|u_2\rangle + (|\alpha| - |\beta|)|v_1\rangle|v_2\rangle \right]$$

---

[16]This is a crucial step, which is required to lead to an important result, viz. Eq. (1.10.28a).

$$= \left[\frac{\sqrt{\alpha\beta}}{\sqrt{|\alpha| - |\beta|}} |u_1\rangle + \sqrt{|\alpha| - |\beta|} |v_1\rangle \right]$$

$$\times \left[\frac{\sqrt{\alpha\beta}}{\sqrt{|\alpha| - |\beta|}} |u_2\rangle + \sqrt{|\alpha| - |\beta|} |v_2\rangle \right]. \tag{1.10.20}$$

Choose the third set of basis $|c_i\rangle, |d_i\rangle$ as

$$|c_i\rangle = A|u_i\rangle + B|v_i\rangle, \tag{1.10.21a}$$
$$|d_i\rangle = -B^*|u_i\rangle + A^*|v_i\rangle. \tag{1.10.21b}$$

The inverse transformation of (1.10.21) is

$$|u_i\rangle = A^*|c_i\rangle - B|d_i\rangle, \tag{1.10.22a}$$
$$|v_i\rangle = B^*|c_i\rangle + A|d_i\rangle. \tag{1.10.22b}$$

The coefficients A and B are given by

$$A = \frac{\sqrt{\alpha\beta}}{\sqrt{1 - |\alpha\beta|}}, \quad B = \frac{|\alpha| - |\beta|}{\sqrt{1 - |\alpha\beta|}}. \tag{1.10.23}$$

From the normalization condition for α and β (1.10.13) the normalization condition for A and B follows:

$$|A|^2 + |B|^2 = 1. \tag{1.10.24}$$

The entangled state $|\Psi\rangle$ can be expressed in terms of the basis $|c_i\rangle$ and $|u_i\rangle$.

$$|\Psi\rangle = N \left(|c_1\rangle|c_2\rangle - A^2|u_1\rangle|u_2\rangle \right), \tag{1.10.25}$$

where

$$N = \frac{1 - |\alpha\beta|}{|\alpha| - |\beta|}.$$

Starting from (1.10.25) $|\Psi\rangle$ can be expressed in four equivalent forms using different basis.

(1) using $u_1, v_1; u_2, v_2$ (the original form)

$$|\Psi\rangle = N[AB|u_1\rangle|v_2\rangle + AB|v_1\rangle|u_2\rangle + B^2|v_1\rangle|v_2\rangle], \qquad (1.10.26a)$$

(2) using $c_1, d_1; u_2, v_2$ (c_2 is replaced by (1.10.21a) and u_1 is replaced by using (1.10.22a))

$$|\Psi\rangle = N[|c_1\rangle (A|u_2\rangle + B|v_2\rangle) - A^2(A^*|c_1\rangle - B|d_1\rangle)|u_2\rangle], \qquad (1.10.26b)$$

(3) using $u_1, v_1; c_2, d_2$ (c_1 is replaced by (1.10.21a) and u_2 is replaced by using (1.10.22a))

$$|\Psi\rangle = N\left[(A|u_1\rangle + B|v_1\rangle)|c_2\rangle - A^2|u_1\rangle (A^*|c_2\rangle - B|d_2\rangle)\right], \qquad (1.10.26c)$$

(4) using $c_1, d_1; c_2, d_2$ (u_1 and u_2 are replaced by using (1.10.22a))

$$|\Psi\rangle = N[|c_1\rangle|c_2\rangle - A^2(A^*|c_1\rangle - B|d_1\rangle)(A^*|c_2\rangle - B|d_2\rangle)]. \quad (1.10.26d)$$

Define observables U_i and D_i by the operators

$$\hat{U}_i = |u_i\rangle\langle u_i|, \quad \hat{D}_i = |d_i\rangle\langle d_i|. \qquad (1.10.27)$$

Their eigenvalues are 0 and 1, for instance $\hat{U}_i|u_i\rangle = |u_i\rangle$, $\hat{U}_i|v_i\rangle = 0$. In general $\hat{U}_i$ and $\hat{D}_i$ do not commute, i.e., it is impossible to determine U_i and D_i precisely for any particle at the same time. From Eq. (1.10.26a) we see that simultaneous measurement of U_1 and U_2 leads to

$$U_1 U_2 = 0. \qquad (1.10.28a)$$

This is the consequence of the demand that the coefficient of $|u_1\rangle|u_2\rangle$ in (1.10.17) vanishes. In the following a pitfall is dug for the hidden variable theory that $U_1(\lambda)U_2(\lambda) = 1$, contradicting (1.10.28a). It follows from Eq. (1.10.26b) that if D_1 is measured for particle 1 and U_2 is measured for particle 2, the following statement holds.

$$\text{If } D_1 = 1, \quad \text{then } U_2 = 1, \qquad (1.10.28b)$$

since the only term in (1.10.26b) containing $|d_1\rangle$ is $|d_1\rangle|u_2\rangle$. Similarly from Eq. (1.10.26c) if U_1 is measured for particle 1 and D_2 measured for particle 2, the following statement holds.

$$\text{If } D_2 = 1, \quad \text{then } U_1 = 1. \qquad (1.10.28c)$$

Finally from Eq. (1.10.26d) we know that if D_1 is measured for particle 1 and D_2 is measured for particle 2, then

the probability to obtain $D_1 = 1$ and $D_2 = 1$ is $|NA^2B^2|^2$.

$$(1.10.28d)$$

All the foregoing predictions are due to quantum mechanics.

In the following a proof will be given that the local realistic theory is incompatible with quantum mechanics. The state of a system of two particles is described by an ensemble with the deterministic dynamical variable λ, as in Section 1.7.3. We begin the measurement process when the particles are separated by a large distance. Let a measurement give $D_1 = 1$ and $D_2 = 1$. This is allowed by quantum mechanics (1.10.28d). Now $D_1 = 1$, then according to (1.10.28b) any measurement of U_2 must yield $U_2 = 1$. According to locality, this result is completely independent of any measurement on particle 1, i.e., $U_2(\lambda) = 1$. Similarly, according to (1.10.28c) $D_2 = 1$ implies $U_1 = 1$. By locality $U_1(\lambda) = 1$. To summarize, $U_1(\lambda)U_2(\lambda) = 1$, i.e., if instead of D_1 and D_2 we measure U_1 and U_2, locality implies $U_1U_2 = 1$, contradicting $U_1U_2 = 0$ by (1.10.28a). By locality Eqs. (1.10.28b) and (1.10.28c) predict independently $U_2 = 1$ and $U_1 = 1$ (with probability 1), and furthermore the measurement of D_1 does not disturb particle 2 and the measurement of D_2 does not disturb particle 1. It follows that there exist elements of physical reality corresponding to U_1 and U_2. The foregoing proof depends on the value of the probability of obtaining $D_1 = 1$, $D_2 = 1$ in the measurement on the state $|\Psi\rangle$,

$$|NA^2B^2|^2 = \left[\frac{(|\alpha| - |\beta|)\,|\alpha\beta|}{1 - |\alpha\beta|}\right]^2.$$

If any one of α and β vanishes, the probability also vanishes. This corresponds to a direct product of states, not an entangled state. The probability vanishes also for $|\alpha| = |\beta|$, corresponding to the maximally entangled state. Hardy's proof states that Bell's theorem is valid for any two-particle entangled state except for the maximally entangled one. T. F. Jordan [45] and S. Goldstein [46] also gave independent proof of Bell's theorem for two-particle systems.

1.10.3 Experimental verification of Bell's theorem without inequalities for two-particle systems

The scheme of the experiment carried out by L. Mandel's group [47, 48] is shown in Fig. 1.49. Parametric down-conversion in the non-linear crystal (PDC) generates two photons. By using a rotator (denoted by $\frac{\pi}{2}$ Rot.) their polarizations are made orthogonal, denoted

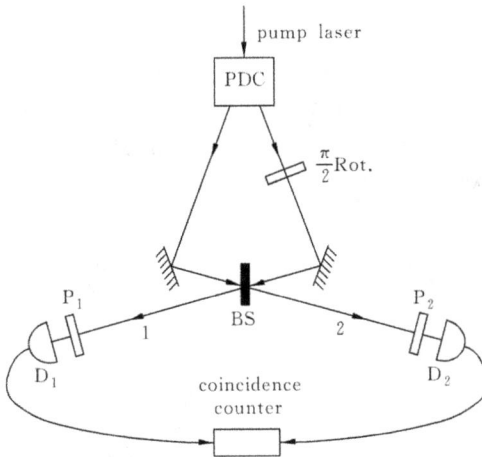

Fig. 1.49 Scheme of the experimental setup. Reprinted with permission from J. Torgerson, D. Branning, L. Mandel, *Applied Physics B*, 60, 1995, 267. Copyright (1995) Springer.

by x and y respectively. The photons are sent in nearly opposite directions through the beam splitter BS. The beam splitter is not symmetric, with T and T' representing the coefficients of transmission in directions 1 and 2, R and R' representing the coefficients of reflection in directions 1 and 2 respectively. Because the photons arrive close to normal incidence, these coefficients can be assumed to be independent of polarization. The state of photons leaving BS is described by

$$
\begin{aligned}
|\psi\rangle = {} & TT'|1\rangle_{1x}|1\rangle_{2y} \\
& + RR'|1\rangle_{1y}|1\rangle_{2x} + TR'|1\rangle_{1x}|1\rangle_{1y} + T'R|1\rangle_{2x}|1\rangle_{2y},
\end{aligned}
$$

$$(1.10.29)$$

where $|1\rangle_{2y}$ represents the state of one photon propagating in the direction 2 with polarization y, etc. The third and fourth terms on the right hand side do not contribute to the coincidence counts. After passing through the polarizers P_1 and P_2 the photons enter detectors D_1 and D_2. Denote by $P_j(\theta)$ the probability of detecting a photon in the detector D_j $(j = 1, 2)$, where θ is the angle set for the polarizer, $\bar{\theta} = \theta + \frac{\pi}{2}$ is the direction of polarization orthogonal to θ. Complete polarizer and detector combination ensures $P_j(\theta) + P_j(\bar{\theta}) = 1$. Let $P_{12}(\theta_1, \theta_2)$ be the probability of coincident detection of a photon in each of the detectors when the polarizers are set at θ_1 and θ_2, and $P_{12}(\theta_1, -)$ be the probability of coincident detection of a photon in each of the detectors when P_1 is set at θ_1 and the polarizer in the direction 2 is absent. They satisfy the following relation

$$P_{12}(\theta_1, -) = P_{12}(\theta_1, \theta_2) + P_{12}(\theta_1, \bar{\theta}_2), \qquad (1.10.30)$$

where θ_2 is arbitrary. Similarly we have

$$P_{12}(-, \theta_2) = P_{12}(\theta_1, \theta_2) + P_{12}(\bar{\theta}_1, \theta_2). \qquad (1.10.31)$$

It is possible to find specific values of $\theta_1, \theta_2, \theta_1'$ and θ_2' under given parameters T, T', R and R', such that the following relations hold [47].

$$P_{12}(\theta_1, \bar{\theta}_2') = 0, \qquad (1.10.32a)$$
$$P_{12}(\bar{\theta}_1', \theta_2) = 0, \qquad (1.10.32b)$$

$$P_{12}(\theta'_1, \theta'_2) = 0, \qquad\qquad (1.10.32c)$$

$$P_{12}(\theta_1, \theta_2) > 0. \qquad\qquad (1.10.32d)$$

The foregoing relations are quantum mechanical predictions. From Eqs. (1.10.30) and (1.10.32a) we obtain

$$P_{12}(\theta_1, -) = P_{12}(\theta_1, \theta'_2). \qquad\qquad (1.10.33)$$

This equality implies that the probability that the photon in direction 2 has a polarization θ'_2 is 1 when the photon in direction 1 has a polarization θ_1. According to the EPR premise (2), corresponding to the polarization of the photon in direction 2 there exists an element of physical reality because it can be predicted with certainty and is not disturbed when the measurement of polarization is carried out on the photon in direction 1. From (1.10.31) and (1.10.32b) we obtain

$$P_{12}(-, \theta_2) = P_{12}(\theta'_1, \theta_2). \qquad\qquad (1.10.34)$$

Following the same argument, we can judge that corresponding to θ'_1 there is also an element of physical reality. This situation is similar to (1.10.28b) and (1.10.28c). Next we consider (1.10.32d) $P_{12}(\theta_1, \theta_2) > 0$. Since θ_1 appears, it implies that $\Pi^\lambda(2) = \theta'_2$, where $\Pi^\lambda(2)$ is the polarization of the photon in direction 2 in the state specified by the parameter λ. From EPR premise (2), it is θ'_2. Similarly $\Pi^\lambda(1) = \theta'_1$. Then $P_{12}(\theta'_1, \theta'_2) > 0$ must follow, in contradiction with (1.10.32c). The argument goes similar to that in Section 1.10.2. In the experiments [46, 47] $|T|^2 = 0.70, |R|^2 = 0.30$, and the following values for the angles are computed: $\theta_1 = 74.3°, \theta_2 = 15.7°, \theta'_1 = -56.8°, \theta'_2 = -33.2°$. The probabilities determined by measurements are $P_{12}(\theta'_1, \theta_2) = P_{12}(\theta_1, \theta'_2) = 0.98$, $P_{12}(\theta_1, \theta_2) = 0.099$. According to EPR premise on locality, $P_{12}(\theta'_1, \theta'_2)$ should be

$$P_{12}(\theta'_1, \theta'_2) = 0.099 \times 0.98 \times 0.98 = 0.095$$

while its experimental value is

$$P_{12}(\theta'_1, \theta'_2) = 0.0070 \pm 0.0005$$

deviating from the EPR prediction by 45 standard deviations.

1.10.4   Experimental realization of 3-photon entangled states

The superiority of the 3-photon entangled states lies in the possibility of providing a "non-statistical" experimental verification of the contradiction between the hidden variable theories and quantum mechanics, while a statistical verification is based on the buildup of data by measuring many pairs of entangled particles. The Innsbruck group (A. Zeilinger [49]) suggested to realize a 3-photon entangled GHZ state (Section 1.10.1) by using two pairs of polarization entangled photons. This idea has been realized experimentally by the Innsbruck group [50]. In the scheme of experimental setup shown in Fig. 1.50 a short ultraviolet pulse generates *two* pairs of polarization entangled photons in a BBO crystal. The state of such a pair is described as

$$\frac{1}{\sqrt{2}}\left(|H\rangle_a|V\rangle_b - |V\rangle_a|H\rangle_b\right),$$

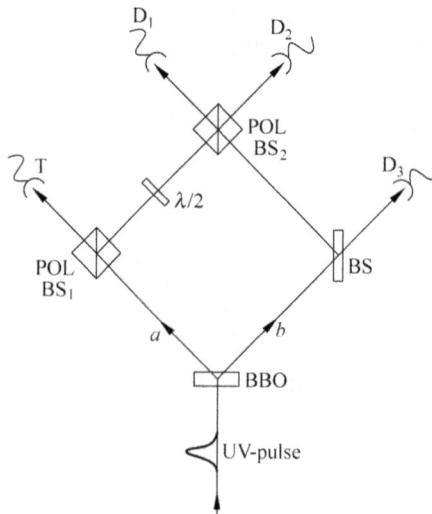

Fig. 1.50   Scheme showing the experimental setup for producing 3-photon entangled states. Reprinted with permission from Dik Bouwmeester, Jian-Wei Pan, Matthew Daniell, Harald Weinfurter, and Anton Zeilinger, *Phys. Rev. Lett.* 82, 1345, 1999. Copyright (2017) by the American Physical Society.

which gives the two possibilities of polarizations of the photons in arm a and arm b. H denotes the horizontal polarization and V denotes the vertical polarization. These possibilities are coherently superimposed with a phase difference π. This phase difference is allowed to have any value, as long as it is fixed for all pairs created. The arm a continues towards a polarizing beam splitter (POL BS$_1$), where V polarized photons are reflected and H polarized photons are transmitted towards the trigger detector T. Arm b continues towards a 50%–50% polarization independent beam splitter BS. From each beam splitter one output is directed to the final polarizing beam splitter (POL BS$_2$). The vertical polarization of the photon reflected from POL BS$_1$ is rotated to 45° polarization by using a $\lambda/2$ plate. Besides the photon hitting the detector T, the remaining three output arms continue through interference filters ($\delta\lambda = 3.6$ nm) and single mode fibers towards the single photon detectors D$_1$,D$_2$ and D$_3$.

Consider the two pairs generated by a single short pulse, and that the four photons are all detected one by each detector T, D$_1$,D$_2$ and D$_3$. In the following it can be shown that under the fourfold coincident detection and the brief duration of the pulse and the narrowness of the filters, the photons recorded by detectors D$_1$,D$_2$ and D$_3$ form a GHZ entangled state. When the coincident detection occurs, the photon detected by detector T is H polarized, and its partner (arm b) is V polarized, which is either (50% chance) transmitted at the beam splitter BS towards detector D$_3$, or reflected (50% chance) towards the final polarizing beam splitter POL BS$_2$, where it will be reflected towards D$_2$. In the first possibility the photons entering D$_1$ and D$_2$ must be originated from the other pair of down-converted photons. The photon along arm a is V polarized and is reflected at POL BS$_1$, but after passing through the $\lambda/2$ plate it becomes polarized at 45° (equal weight superposition of H and V polarizations). Its companion is H polarized and is to be detected by D$_1$. What D$_2$ detects must be the photon with the H component of the 45° polarization. This scenario corresponds to the detection of the state

$$|H\rangle_1|H\rangle_2|V\rangle_3. \tag{1.10.35}$$

In the second possibility, the triggering photon (H polarized) and its companion (V polarized) are detected by T and D_2 respectively. The second pair has a V polarized photon in arm a which is reflected at POL BS$_1$ towards POL BS$_2$. It V polarized component after the $\lambda/2$ plate is reflected at POL BS$_2$ towards detector D_1. The H polarized companion is in arm b towards detector D_3. The combination is

$$|V\rangle_1|V\rangle_2|H\rangle_3. \tag{1.10.36}$$

In general the two possible states (1.10.35) and (1.10.36) do not form a coherent superposition because they originate from two independent pairs of photons which are in principle distinguishable, for instance by the difference in emission time. The state (1.10.35) is characterized by the synchronization of counts in T and D_3, as well as synchronization of counts in D_1 and D_2, while state (1.10.36) is characterized by the synchronization of counts in T and D_2, as well as in D_1 and D_3. Such information can be erased by the brief duration of the ultraviolet pulse (200 fs) and the narrowness of the interference filters leading to a long coherent time (500fs). Under such conditions the resulting state is a GHZ entangled state

$$\frac{1}{\sqrt{2}}\left(|H\rangle_1|H\rangle_2|V\rangle_3 + |V\rangle_1|V\rangle_2|H\rangle_3\right). \tag{1.10.37}$$

The sign + needs some elaborations. The state of the initial two pairs is

$$\frac{1}{2}(|H\rangle_a|V\rangle_b - |V\rangle_a|H\rangle_b)(|H\rangle'_a|V\rangle'_b - |V\rangle'_a|H\rangle'_b), \tag{1.10.38}$$

where the unprimed and the primed polarizations refer to the first pair and second pair respectively. The evolution of individual components of (1.10.38) through the equipment is given by

$$|H\rangle_a \rightarrow |H\rangle_T,$$

$$|H\rangle_b \rightarrow \frac{1}{\sqrt{2}}(|H\rangle_1 + |H\rangle_3),$$

$$|V\rangle_a \rightarrow \frac{1}{\sqrt{2}}(|V\rangle_1 + |V\rangle_2),$$

$$|V\rangle_b \rightarrow \frac{1}{\sqrt{2}}(|V\rangle_2 + |V\rangle_3) \tag{1.10.39}$$

with the same expressions for the second pair. Restricting ourselves to those terms contributing to the fourfold coincidence, we obtain

$$\frac{1}{2}[|H\rangle_T(|H\rangle'_1|H\rangle'_2|V\rangle_3 + |V\rangle'_1|V\rangle_2|H\rangle'_3)$$

$$+ |H\rangle'_T((|H\rangle_1|H\rangle_2|V\rangle'_3 + |V\rangle_1|V\rangle'_2|H\rangle_3)]. \tag{1.10.40}$$

For primed and unprimed states indistinguishable, we obtain finally

$$\frac{1}{\sqrt{2}}|H\rangle_T((|H\rangle_1|H\rangle_2|V\rangle_3 + |V\rangle_1|V\rangle_2|H\rangle_3) \tag{1.10.41}$$

confirming (1.10.37).

The experimental demonstration of the GHZ entanglement consists of two steps. First it is verified that under the detection of a photon by the trigger detector, both $H_1H_2V_3$ and $V_1V_2H_3$ are observed, but no others. The second step is the demonstration of the *coherence* of the two states in (1.10.37). The polarizer in front of D_1 is set at 45°, i.e.,

$$|45°\rangle_1 = \frac{1}{\sqrt{2}}(|H\rangle_1 + |V\rangle_1).$$

The state (1.10.37) is then projected onto

$$\frac{1}{\sqrt{2}}|45°\rangle_1(|H\rangle_2|V\rangle_3 + |V\rangle_2|H\rangle_3).$$

Expressing the states of photons 2 and 3 in terms of the basis $|45°\rangle$, $|-45°\rangle$, we have

$$\frac{1}{\sqrt{2}}(|45°\rangle_2|45°\rangle_3 - |-45°\rangle_2|-45°\rangle_3)$$

i.e., photons 2 and 3 have identical polarizations. The absence of states $|45°\rangle_2|-45°\rangle_3$ and $|-45°\rangle_2|45°\rangle_3$ is a clear demonstration of the GHZ entangled state. The data shown in Fig. 1.51(a) are obtained from polarization analysis of the photon at D_3, conditioned by the detection of the trigger, $|45°\rangle_1$ and $|-45°\rangle_2$. The two curves show the fourfold coincidences for a polarizer oriented at $-45°$ (squares) and $45°$ (circles) in front of D_3 as a function of delay in path a. With zero delay the polarization of the photon at D_3 is oriented along $-45°$, in accordance with the GHZ state. The visibility is as high as 75%. For increasing delay the two pairs become distinguishable and the coherence is destroyed. One can alternatively conclude that at zero delay the photons at D_1 and D_3 have been projected onto a two-particle entangled state by the projection of the photon at D_2 onto $-45°$. To obtain additional information the polarizer at D_1 is set at $0°$ (V polarization). For GHZ state the remaining two photons should be in the state $|V\rangle_2|H\rangle_3$, which cannot give rise to any correlation between these photons in the $45°$ basis. In Fig. 1.51(b) the two angles set for D_3 make no difference at any delay in path a. The data clearly indicates the absence of two-photon correlations.

The same group of authors continued to test the quantum non-locality by using another GHZ state [68]:

$$|\Psi\rangle = \frac{1}{\sqrt{2}}(|H\rangle_1|H\rangle_2|H\rangle_3 + |V\rangle_1|V\rangle_2|V\rangle_3).$$

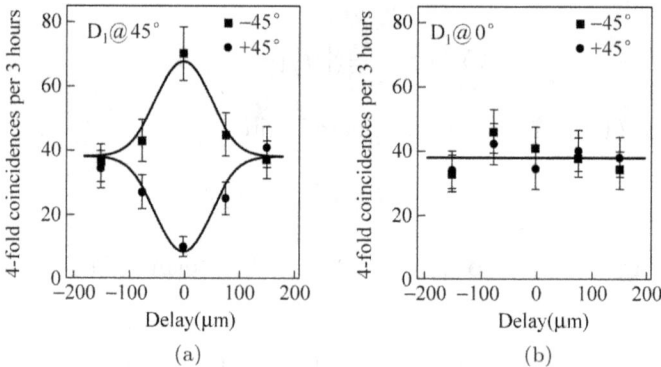

Fig. 1.51 Experimental confirmation of GHZ entanglement. (a) Fourfold coincidence for D_1 at $45°$, D_2 at $-45°$ and D_3 at $45°$ (circles) or $-45°$(squares) as a function of delay in path a (b) D_1 at $0°$. Reprinted with permission from Dik Bouwmeester, Jian-Wei Pan, Matthew Daniell, Harald Weinfurter, and Anton Zeilinger, *Phys. Rev. Lett.* 82, 1345, 1999. Copyright (2017) by the American Physical Society.

Consider alternative measurements of linear polarization along directions H'/V' rotated by $45°$ with respect to the original H/V directions, or of circular polarization L/R. The new polarizations are related to the original ones by:

$$|H'\rangle = \frac{1}{\sqrt{2}}(|H\rangle + |V\rangle), \quad |V'\rangle = \frac{1}{\sqrt{2}}(|H\rangle - |V\rangle),$$

$$|R\rangle = \frac{1}{\sqrt{2}}(|H\rangle + i|V\rangle), \quad |L\rangle = \frac{1}{\sqrt{2}}(|H\rangle - i|V\rangle).$$

For convenience, a measurement of H'/V' is referred to as a x measurement and one of L/R as a y measurement. Depending on the measurements carried out, the GHZ state can be expressed in terms of the new basis, for instance for a yyx measurement the GHZ state can be described by

$$|\Psi\rangle = \frac{1}{2}(|R\rangle_1|L\rangle_2|H'\rangle_3 + |L\rangle_1|R\rangle_2|H'\rangle_3 + |R\rangle_1|R\rangle_2|V'\rangle_3$$

$$+ |L\rangle_1|L\rangle_2|V'\rangle_3).$$

By cyclic permutation, we obtain alternative expressions for the GHZ state for yxy and xyy measurements. Such states imply that given measurement results for two of the photons, say 1 and 2, the corresponding polarization of the remaining photon can be predicted with certainty. Since 3-photon entangled states are firmly established by experiments, local realists have to give their own interpretation of these states. As the predictions are independent of the spatial separation of the photons being measured, as well as the time order of the measurements, let us assume that measurements are carried out simultaneously (in the frame of reference of the equipment say) on three photons. EPR locality demands that any specific measurement result obtained for any photon never depends on which specific measurements are performed simultaneously on the other two nor on their outcome. The only way to confront the reality of GHZ states is to assume that each photon carries elements of reality for both x and y measurements that determine the specific individual measurement results. Let us call these elements of reality X_i with values $+1(-1)$ for $H'(V')$, and Y_i with values $+1(-1)$ for $R(L)$, the yyx measurement requires $Y_1 Y_2 X_3 = -1$, $Y_1 X_2 Y_3 = -1$, and $X_1 Y_2 Y_3 = -1$ in order to be able to reproduce the quantum predictions of the GHZ state $|\Psi\rangle$ and similarly for its permutations. According to the local realists any specific measurement for x must be independent of whether x or y measurement is performed on the other photons. As $Y_i Y_i = +1$, we can write $X_1 X_2 X_3 = (X_1 Y_2 Y_3)(Y_1 X_2 Y_3)(Y_1 Y_2 X_3)$ and obtain $X_1 X_2 X_3 = -1$. The conclusion is that according to local realism the only possible results for an xxx measurement are $V'V'V'$, $H'H'V'$, $H'V'H'$, and $V'H'H'$. Quantum mechanics gives unambiguous prediction for an xxx measurement of the state $|\Psi\rangle$ when it is re-expressed in terms of the new basis H' and V':

$$|\Psi\rangle = \frac{1}{2}(|H'\rangle_1|H'\rangle_2|H'\rangle_3 + |H'\rangle_1|V'\rangle_2|V'\rangle_3 + |V'\rangle_1|H'\rangle_2|V'\rangle_3$$

$$+ |V'\rangle_1|V'\rangle_2|H'\rangle_3).$$

This is in direct contradiction with the local realistic theory. Schematically the situation is represented in Fig. 1.52.

1.10.5 EPR paradox in the context of entanglement and non-locality

The intent of EPR was to show that the quantum mechanics is incomplete in the sense that for the one-dimensional motion of a 2-particle system the non-commuting observables x_2 and p_2 could be known with certainty such that the uncertainty relation $(\Delta x_2)^2 (\Delta p_2)^2 \geq \hbar^2/4$ is violated.

Consider the EPR operators $\hat{x}_1 + \hat{x}_2$ and $\hat{p}_1 - \hat{p}_2$. The value of the joint uncertainty product $(\Delta(\hat{x}_1 + \hat{x}_2))^2 (\Delta(\hat{p}_1 - \hat{p}_2))^2$ depends critically on the nature of the quantum mechanical state, whether it is separable or entangled. Beginning from a definition of separability, L.-M. Duan *et al.* [54] proved a theorem that for a separable continuous variable state the product is bound from below by a value resulting from the uncertainty relation. Therefore a violation of this lower bound is a sufficient criterion for inseparability (entanglement). Furthermore, for all Gaussian states the existence of a lower bound

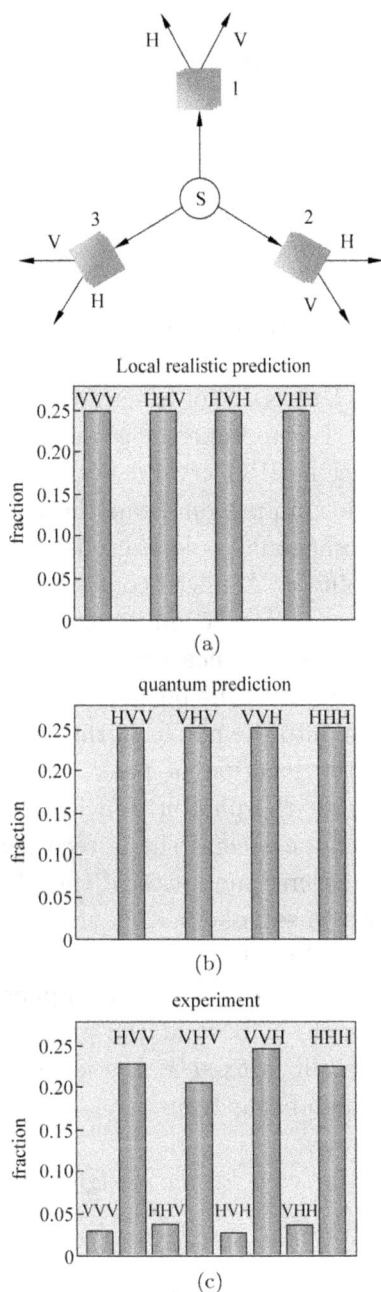

Fig. 1.52   Quantum nonlocality of 3-photon GHZ state.

turns out to be a necessary and sufficient criterion for separability. S. Manchini *et al.* [55] proved a theorem that for any separable quantum state the operators $u = q_1 + q_2$ and $v = p_1 - p_2$ and $[q_j, p_j]$ (being a *c*-number, $j = 1, 2$), the following inequality follows $\langle(\Delta u)^2\rangle\langle(\Delta v)^2\rangle \geq \langle[q_1, p_1]\rangle|^2$. For the present case, we have simply $(\Delta(\hat{x}_1 + \hat{x}_2))^2(\Delta(\hat{p}_1 - \hat{p}_2))^2 \geq \hbar^2$.

The maximally entangled EPR state cannot be realized in the laboratory, but a pair of parametric down converted photons can be made to approximate it under suitable conditions. The position and momentum correlations of such a photon pair at large distances were measured by the University of Rochester group [56]. The source of entangled photons is spontaneous parametric down conversion generated by pumping a BBO crystal with a 390 nm laser beam. A prism separates the pump light from the down-converted light. The photons have orthogonal polarizations and are separated by a polarizing beam splitter. A microscope objective after the filter focusses the transmitted light to an avalanche photodiode single-photon counting module. To measure position correlation of the photons, a lens placed prior to the beam splitter is used to image the exit face of the crystal onto the planes of the two slits (Fig. 1.53(a)). One slit is fixed at the location of peak signal intensity and the other is moved such that the photon coincidence rate as a function of the displacement of the second slit is recorded. To measure the correlation of the transverse momenta of the photons, the imaging lens is replaced by two lenses, one in each arm, a distance f from the planes of the two slits (Fig. 1.53(b)). These lenses map transverse momenta to transverse positions, such that a photon with transverse momentum $\hbar k_\perp$ comes to a focus at the point $fk_\perp/k$ in the plane of the slit. Again, one slit is fixed at the location of the maximum counting rate while the other is translated to obtain the coincidence distribution.

By normalizing the coincidence distributions, we obtain the conditional probability density functions $P(x_2|x_1)$ and $P(p_2|p_1)$ (Fig. 1.54).

These *conditional* probability density functions are then used to calculate the uncertainty in the inferred position and momentum of

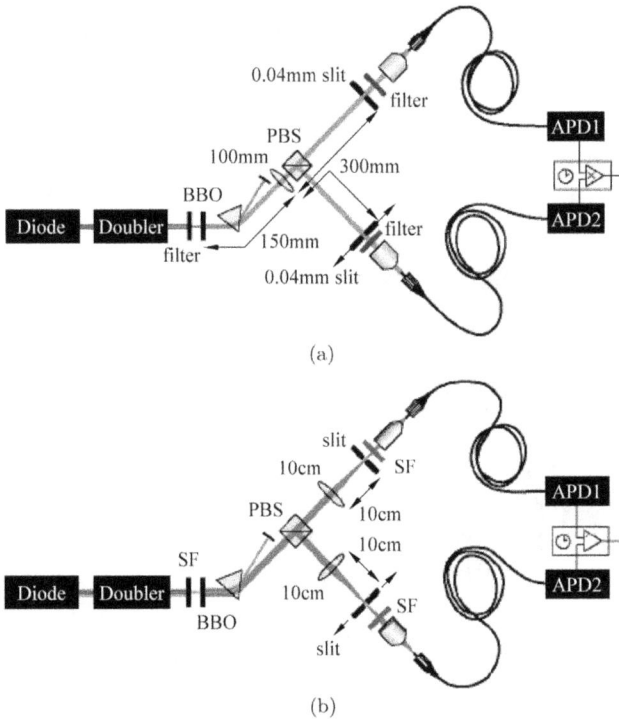

Fig. 1.53 Experimental setup for measuring photon correlations. Reprinted with permission from John C. Howell, Ryan S. Bennink, Sean J. Bentley, and R. W. Boyd, *Phys. Rev. Lett.* 92, 210403, 2004. Copyright (2017) by the American Physical Society.

photon 2 given the position *or* momentum of photon 1:

$$(\Delta x_2|_{x_1})^2 = \int x_2^2 P(x_2|x_1)\mathrm{d}x_2 - \left(\int x_2 P(x_2|x_1)\mathrm{d}x_2\right)^2,$$

$$(\Delta p_2|_{x_1})^2 = \int p_2^2 P(x_2|x_1)\mathrm{d}x_2 - \left(\int p_2 P(x_2|x_1)\mathrm{d}x_2\right)^2.$$

$$(1.10.42)$$

The data finally gives

$$(\Delta x_2|_{x_1})^2(\Delta p_2|_{p_1})^2 = 0.01\hbar^2.$$

$$(1.10.43)$$

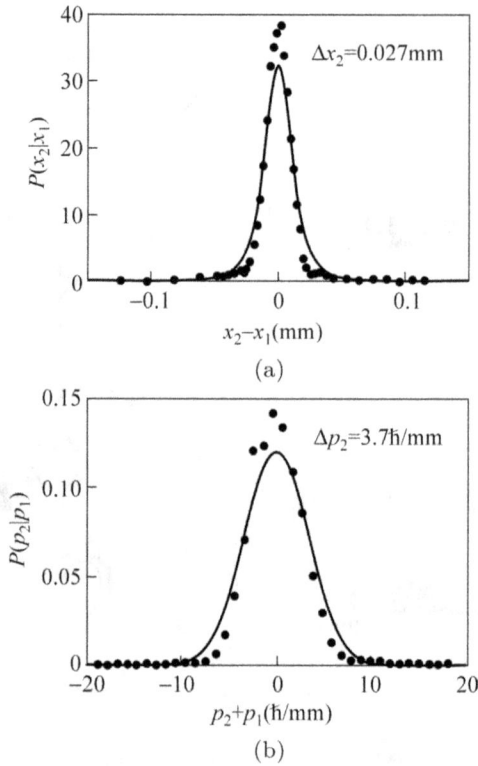

Fig. 1.54 The conditional probability distribution functions. Reprinted with permission from John C. Howell, Ryan S. Bennink, Sean J. Bentley, and R.W. Boyd, *Phys. Rev. Lett.* 92, 210403, 2004. Copyright (2017) by the American Physical Society.

This relation shows how good the correlation between photons is and how badly the EPR locality requirement is violated. This relation is not to be mixed up with the uncertainty relation, because the variances are taken at different conditions. The uncertainty relation $(\Delta x_2)^2 (\Delta p_2)^2 \geq \hbar^2/4$ is actually *not* violated, because the simultaneous vanishing of the variances cannot be realized by any quantum mechanical state. For a separable state, we calculate the variances using the *joint* probabilities $P(x_1, x_2)$ and $P(p_1, p_2)$, and the result would be $(\Delta(\hat{x}_1 + \hat{x}_2))^2 (\Delta(\hat{p}_1 - \hat{p}_2))^2 \geq \hbar^2$.

1.10.6 Non-locality of a single photon

The experimental test of Bell's inequalities discussed in Section 1.8 involves the correlation between two entangled particles. Consider a single photon which impinges on a beam splitter. It has equal probability amplitudes for reflection and transmission. The state is described by

$$\frac{1}{\sqrt{2}}[|1\rangle_T |0\rangle_R + |0\rangle_T |1\rangle_R], \qquad (1.10.44)$$

where T(R) refers to the transmitted (reflected) arm of the beam splitter. In the photon number basis, the correlation between T and R is mutually exclusive: if detector in the T arm clicks, the detector in the R arm should be idle, and vice versa. But this is not a quantum mechanical correlation involving a phase complementary to the particle number. Comparing (1.10.44) with the Bell states of (1.8.6), we find that they are mathematically isomorphic, this means that a single photon state can also be used to test the Bell inequalities, if a phase can be appropriately arranged into the scheme. Then the detectors in the T and the R arms can also show correlations between the two "versions" of the same photon, even if the detectors are very far away. The experiment was suggested by S.M. Tan, D.F. Walls and M.J. Collett [57]. The correlation between the photon pairs in (1.8.6) is detected by adjusting the polarizer settings. Here the single photon is mixed with a local oscillator on a beam splitter in such a way that the detector is unable to tell if the detected photon comes from the signal or from the local oscillator which produces a coherent state of photon $\alpha \exp(i\theta)$ (Fig. 1.55). The phase θ plays the same role as the polarizer setting in experiments discussed in Section 1.8. The experimental demonstration was achieved by B. Hessmo *et al.* [58] with some variations of the original proposal [57]. The experimental setup is shown in Fig. 1.56.

The source laser produces femtosecond pulses at 390nm, which in turn pump a BBO down-converter. The idler photon is sent to the detector D_T. Detection of D_T indicates that the other (signal) photon

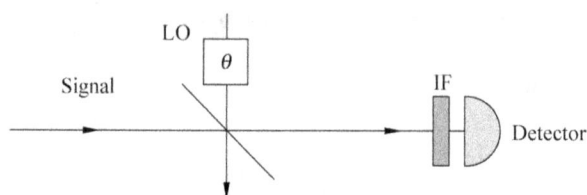

Fig. 1.55   Mixing of the signal photon with the photon produced by the local oscillator (LO). Reprinted with permission from Björn Hessmo, Pavel Usachev, Hoshang Heydari, and Gunnar Björk, *Phys. Rev. Lett.* 92, 180401, 2004. Copyright (2017) by the American Physical Society.

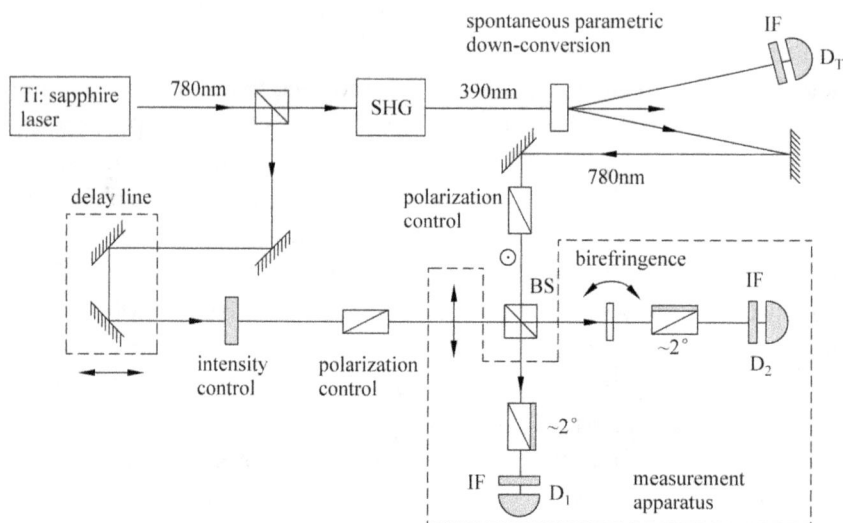

Fig. 1.56   Experimental setup: a single photon produced in a parametric down-conversion is overlapped with a local oscillator on a beam splitter. In each output arm of the beam splitter a detector is arranged. Reprinted with permission from Björn Hessmo, Pavel Usachev, Hoshang Heydari, and Gunnar Björk, *Phys. Rev. Lett.* 92, 180401, 2004. Copyright (2017) by the American Physical Society.

of the converted pair is to enter the beam splitter of the measurement apparatus. To generate the local oscillator, some light is picked off the main beam of the source laser. A delay line is adjusted such that the oscillator photon arrives the beam splitter simultaneously with the signal photon. The intensity control matches the intensity of

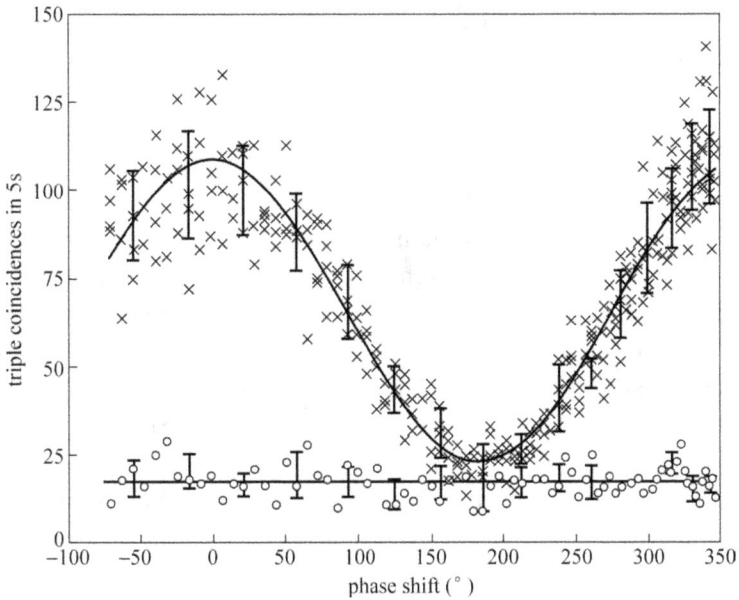

Fig. 1.57 Experimental data and curve fits. The oscillating curve shows the non-classical correlation of the two "versions" of a single photon as a function of phase shift. Reprinted with permission from Björn Hessmo, Pavel Usachev, Hoshang Heydari, and Gunnar Björk, *Phys. Rev. Lett.* 92, 180401, 2004. Copyright (2017) by the American Physical Society.

the oscillator with the signal ensuring high visibility in the correlation measurements. The polarization of the local oscillator is adjusted to be orthogonal to the signal photon polarization. The relative phase between the local oscillators is adjusted by tilting a birefringent quartz around its optic axis in one of the arms. The triple coincidence between D_T, D_1 and D_2 is recorded as a function of the phase shift between the two arms as shown in Fig. 1.57. If either the signal photon or the local oscillator is missing, a flat, phase independent correlation curve is obtained. This background correlation is used as a correction to the correlation vs phase shift curve, making its visibility to become $(91\pm3)\%$. The limit for violation of the Bell inequality is 71%. As pointed out in Ref. [57], this correlation is purely quantum mechanical, because the result differs from that of a classical wave theory, let aside from that of a classical particle theory.

1.11   A Brief Introduction to Quantum Non-Demolition Experiment

Higher and higher precision is demanded in the process of the development of research in basic science, for example in the detection of gravitational waves, trap of a single atom etc. Quantum measurements are very important in the processing and transmission of quantum information. Although being constrained by the principle of uncertainty in general, quantum non-demolition (QND) experiments and back action evading (BAE) experiments can help to obtain optimum results [51]. We begin with an example. The Weber bar of a gravitational wave detector is made of Al (Si, Nb) with tons of weight. It oscillates under the driving force of a gravitational wave with a very small amplitude (of the order of $\delta x \approx 10^{-19}$ cm). When such a high precision is demanded, the massive metal bar is to be treated by using quantum mechanics. The time interval between successive measurements is $\tau \approx 10^{-3}$s. If the amplitude of oscillation is measured with an uncertainty Δx, a corresponding uncertainty in the momentum $\Delta p \geq \frac{\hbar}{2\Delta x}$ is introduced, this leads to an uncertainty of velocity $\Delta v \geq \frac{\hbar}{2m\Delta x}$. This uncertainty in turn leads to an uncertainty in position $(\Delta x)' \geq \frac{\hbar\tau}{2m\Delta x}$ in a time interval τ. Substituting $\Delta x = 10^{-19}$ cm, $m = 10t = 10^7$g, we obtain $(\Delta x)' \geq 5 \times 10^{-19}$ cm, larger than the value allowed by the precision required for position measurement. This means that the next amplitude measurement will not meet the precision required. To reduce the uncertainty $(\Delta x)'$ one can use a still heavier bar, which is not practical. To further decrease the time interval between measurements will decrease the signal of gravitational waves. Let us consider more carefully about the cause of the trouble. The position measurement is a back action of the apparatus on the system, bringing a disturbance on the momentum, the conjugate variable of the position. This "pollution" is fed back to the system, destroying the possibility of next precision position measurement. Is the feedback of pollution inevitable? A smarter experimentalist could try to measure the momentum of the bar as a measurement of the intensity of the gravitational wave, which has driven the bar to acquire the momentum. Let the precision required by the momentum measurement be $\Delta p \approx 10^{-9}$gcm/s. The corresponding uncertainty in position measurement is $\Delta x \geq \frac{\hbar}{2\Delta p} \approx 5 \times 10^{-19}$ cm. Can this uncertainty in position induce a momentum uncertainty during the evolution of the system? It turns out that the evolution of the system is such that the momentum of the bar is conserved after it is set into motion driven by the gravitational wave during an interval of time much shorter than the period of oscillation, because the bar is freely suspended. The uncertainty in position cannot induce a new uncertainty in momentum $(\Delta p)'$ during the evolution of the system, and the precision of the next momentum measurement is not influenced.[17] Successive momentum measurements do induce position uncertainties, which can build up during the process, but they do not cause any unfavorable influence on the successive momentum measurements. We conclude that the measurement of momentum of a freely moving body can be QND, but a position measurement cannot be. In other words the momentum of a free motion is a QND observable, and the position is not.

1.11.1 The standard quantum limit and the Back Action Evading (BAE) experiments

Consider a free particle. Let its position be measured at $t = 0$ with an uncertainty $(\Delta x)_1$. The back action of the apparatus introduces a momentum uncertainty $(\Delta p)_{\text{ba}} = \frac{\hbar}{2(\Delta x)_1}$. This pollution fed back to the system causes an uncertainty in position $(\Delta x)'$ after time τ, where

$$(\Delta x)' = \frac{(\Delta p)_{\text{ba}}\tau}{m} = \frac{\hbar\tau}{2m(\Delta x)_1}.$$

The position is again measured at time τ with uncertainty $(\Delta x)_2$. The successive measurements in position determine the momentum

---

[17]Just contrast the case of position measurement, the evolution of the system in time τ gives a change of the position $\delta x = p\tau/m$, any uncertainty in p can be fed back to yield new uncertainty in x. In the case of momentum measurement, the evolution gives $p=$const, independent of x. Therefore the uncertainty in x cannot be fed back through this relation to yield a new uncertainty in momentum.

of the particle

$$p = m\frac{x_2 - x_1}{\tau}. \tag{1.11.1}$$

The uncertainty in momentum is

$$\Delta p = \frac{m}{\tau}\left[(\Delta x)_1^2 + \frac{\hbar^2\tau^2}{4m^2(\Delta x)_1^2} + (\Delta x)_2^2\right]^{1/2}. \tag{1.11.2}$$

What is the value of $(\Delta x)_1$ that makes Δp minimum? A simple calculation gives

$$(\Delta x)_{1\,\text{min}} = \sqrt{\frac{\hbar\tau}{2m}} \equiv \Delta x_{\text{SQL}}, \tag{1.11.3}$$

which leads to the momentum uncertainty Δp_{SQL} conditioned by $(\Delta x)_1$ (1.11.3) equal to

$$\Delta p_{\text{SQL}} = \sqrt{\frac{\hbar m}{2\tau}}. \tag{1.11.4}$$

Equations (1.11.3) and (1.11.4) define the standard quantum limit (SQL) of the uncertainties of x and p. The definition of SQL is to some extent arbitrary, it satisfies

$$\Delta x_{\text{SQL}}\Delta p_{\text{SQL}} = \frac{\hbar}{2}.$$

The Δp in (1.11.2) is apparently larger than Δp_{SQL}, but is of the same order of magnitude:

$$\Delta p \gtrsim \Delta p_{\text{SQL}}. \tag{1.11.5}$$

Consider the simple harmonic oscillator

$$x(t) = x(0)\cos\omega t + \frac{p(0)}{m\omega}\sin\omega t$$

$$\equiv x_1\cos\omega t + x_2\sin\omega t. \tag{1.11.6}$$

x_1 and x_2 are called the quadrature amplitudes of the harmonic oscillator. They are in fact constants of integration. From the uncertainty relation between x and p the precision of defining the quadrature amplitudes must satisfy

$$\Delta x_1 \Delta x_2 \geq \frac{\hbar}{2m\omega}. \qquad (1.11.7)$$

If they have the same precision, then

$$\Delta x_1 = \Delta x_2 \geq \sqrt{\frac{\hbar}{2m\omega}} \qquad (1.11.8)$$

which is the standard quantum limit

$$\Delta x_{\text{SQL}} = \sqrt{\frac{\hbar}{2m\omega}}. \qquad (1.11.9)$$

The energy of the harmonic oscillator is

$$E = \frac{1}{2}m\omega^2 A^2 = \frac{1}{2}m\omega^2(x_1^2 + x_2^2),$$

where A is the amplitude. If the amplitude is small, then

$$\Delta E = m\omega^2 A \Delta A_{\text{SQL}} = m\omega^2 \sqrt{\frac{2E}{m\omega^2}} \sqrt{\frac{\hbar}{2m\omega}} = \sqrt{\hbar \omega E}.$$

$$(1.11.10)$$

Can the SQL be superseded? The answer is positive. The uncertainty principle is used in deriving the SQL, but they are not identical. The SQL depends on how the measurement is made. Again we consider the example of the harmonic oscillator. Its energy depends on the amplitude A, and is independent of the phase ωt. To determine the quadrature amplitude, both amplitude and phase need to be measured. The information on phase is irrelevant to the determination of energy, but the back action of apparatus for measuring phase can be fed back to the system leading to an additional uncertainty of the energy. The best way of determining the energy is to avoid any information of the phase. This was just the method P. N. Lebedev used in determining the energy density of the electromagnetic radiation by measuring the radiation pressure. Since the information on the phase is irrelevant, the measurement can be carried out in as long a duration of time as required. Then the uncertainty in energy introduced is as small as required. In a cavity of dimension d a movable wall can be used (momentum uncertainty Δp) to measure the electromagnetic energy E, the uncertainty turns out to be

$$\Delta E \approx \frac{d}{\tau}\Delta p, \qquad (1.11.11)$$

where τ is the time duration of the measurement. If a long time period is used to make a precise momentum measurement ΔE can be made sufficiently small, even smaller than the SQL. Such kind of measurement is called the back action evading (BAE) experiment. The condition for a BAE experiment can be obtained by considering the interaction between the system and the apparatus. Denote by $\widehat{H}_1$ the Hamiltonian of the coupling between the observable $\widehat{A}$ of the system and the observable $\widehat{M}$, i.e., the readout of the apparatus.

$$\hat{H}_1 = f(\hat{A}, \hat{M}). \qquad (1.11.12)$$

This interaction is just the back action of the apparatus on the system. The time development of $\widehat{A}$ is determined by the quantum equation of motion

$$\frac{d\hat{A}}{dt} = \frac{i}{\hbar}[\hat{H}_1, \hat{A}]. \qquad (1.11.13)$$

Consequently, if $\widehat{A}$ and $\widehat{H}_1$ commute, then there will be no back action to $\widehat{A}$ during the measurement. If $\widehat{H}_1$ does not contain any observable not commuting with $\widehat{A}$, then $\widehat{A}$ and $\widehat{H}_1$ commute, i.e., $\widehat{H}_1$ does not induce any transition between eigenstates of $\widehat{A}$. A rigorous proof of the foregoing heuristic argument is given in Ref. [51].

1.11.2   The quantum non-demolition experiment

The QND experiment of an observable $\widehat{A}$ is defined to be a succession of precise measurements on $\widehat{A}$, between which in the evolution of the system no pollution of observables non-commuting with $\widehat{A}$ is fed back to the system. Only very special observables are immune to the feedback of pollutions. They are called the QND variables. The observable $\widehat{A}$ is a QND observable if, and only if the operators $\widehat{A}(t)$ in the Heisenberg picture at different times commute with one another, i.e.,

$$[\hat{A}(t_i), \hat{A}(t_j)] = 0, \quad t_i \neq t_j. \tag{1.11.14}$$

We illustrate the importance of QND experiments by using the example of gravitational wave detection again. The experiment needs the repetition of measurements on the same observable in different time. If we have a large number of such equipments in identical conditions we could carry out measurements of the observable in different time on different equipment, without worrying about the possibility of influencing the system in subsequent evolution due to the measurements. But for the gravitational wave detection the equipment is unique. Measurements on the same bar have to be repeated for many times. The demand that these measurements do not influence the later ones becomes absolutely necessary. Let $\hat{A}$ be a QND observable. In the Heisenberg picture a measurement at time t_0 gives $A(t_0)$. This measurement prepares the state $|\psi_0\rangle$, which is an eigenstate of $\hat{A}(t_0)$ corresponding to the eigenvalue $A(t_0)$. Since $\hat{A}(t_0), \hat{A}(t_1), \hat{A}(t_2) \cdots$ all commute with one another, $|\psi_0\rangle$ is also an eigenstate of $\hat{A}(t_1), \hat{A}(t_2) \cdots$. From the result of the first measurement it is possible to calculate the eigenvalues of the observable in any subsequent time. The measurements carried out in subsequent times $t_1, t_2, \cdots$ should give the known eigenvalues. The following two relations are of crucial importance.

$$[\hat{A}(t), \hat{H}_1] = 0, \tag{1.11.15}$$

$$[\hat{A}(t_i), \hat{A}(t_j)] = 0, \quad t_i \neq t_j. \tag{1.11.16}$$

The first relation guarantees the absence of back action of the apparatus during the measurement which would pollute the observable $\hat{A}$. Other observables, e.g., $\hat{C}$ which do not commute with $\hat{H}_1$ will suffer from the back action from the apparatus and get polluted. The second relation guarantees that the polluted observable $\hat{C}$ does not transmit the pollution to $\hat{A}$ during the free evolution of the system, provided that the interaction $\hat{H}_1$ has been switched off. The LIGO group of Caltech [52] considered the realistic situation of the gravitational wave detection viz. the persistence of interaction $\widehat{H}_1$ etc. and gave the condition for $\widehat{A}$ being a QND observable. The gravitational wave detection aims at measuring the force $F(t)$ experienced by the bar, and its momentum is actually changing in a given period of time. There are a number of theoretical and experimental discussions concerning the subtle situations [53]. From the fundamental principles of quantum mechanics the relations (1.11.13) and (1.11.14) are necessary for all cases.

References

[1] R. P. Feynman, R. B. Leighton and M. Sands. *The Feynman Lectures on Physics*, Vol. III. Reading, Addison-Wesley, 1965, 1-1~1-9.

[2] A. Zeilinger, R. Gaehler, C. G. Shull and W. Treimer in *Proceedings of the Conference on Neutron Scattering*, edited by J. Faber Jr., New York, AIP, 1982; *Rev. Mod. Phys.* 60,1067, 1988.

[3] A. Tonomura, J. Endo, T. Matsuda, T. Kawasaki and H. Ezawa, *Am. J. Phys.*, 57, 117, 1989.

[4] M. S. Chapman, T. D. Hammond, A. Lenef, J. Schmiedmeyer, R. A. Rubinstein, E. Smith and D. E. Pritchard, *Phys. Rev. Lett.* 75, 3783, 1995.

[5] S. Dürr, T. Nonn and G. Rempe, *Nature* 395, 33, 1998.

[6] C. Cohen-Tannoudji, *Atoms in Electromagnetic Fields*, Singapore, World Scientific, 1994, pp. 343–348.

[7] S. Kunze, S. Dürr and G. Rempe, *Europhys. Lett.* 34, 343, 1996.

[8] M. O. Scully, B.-G. Englert and H. Walther, *Nature* 351, 111, 1991.

[9] B.-G. Englert, H. Fearn, M. O. Scully and H. Walther, in *Quantum Interferometry*, edited by F. De Martini, G. Denardo and A. Zeilinger, Singapore, World Scientific, 1994.

[10] P. Storey, S. Tan, M. Collett and D. Walls, in *Quantum Interferometry*, edited by F. De Martini, G. Denardo and A. Zeilinger, Singapore, World Scientific, 1994.

[11] P. Grangier, G. Roger and A. Aspect, *Europhys. Lett.* 1, 173, 1986.

[12] M. A. Horne, A. Shimony and A. Zeilinger, *Phys. Rev. Lett.* 19, 2209, 1989.

[13] D. M. Greenberger, M. A. Horne and A. Zeilinger, *Phys. Today* 46(8), 22, 1993.

[14] D. C. Burnham and D. L. Weinberg, *Phys. Rev. Lett.* 25, 84, 1970.

[15] C. K. Hong and L. Mandel, *Phys. Rev. A* 31, 2409, 1985.

[16] L. Mandel, *Phys. Rev. A* 28, 929, 1983.

[17] J. D. Franson, *Phys. Rev. Lett.* 62, 2205,1989.

[18] P. G. Kwiat, A. M. Steinberg and R. Y. Chiao, *Phys. Rev. A* 47, R2472, 1993.

[19] W. Tittel, J. Brendel, B. Gisin, T. Herzog, H. Zbinden and N. Gisin, *Phys. Rev. A* 57, 3229, 1998.

[20] X. Y. Zou, L. J. Wang and L. Mandel, *Phys. Rev. Lett.* 67, 318, 1991.

[21] P. G. Kwiat, A. M. Steinberg and R. Y. Chiao, *Phys. Rev. A* 45, 7729, 1992.

[22] A. Einstein, B. Podolsky and N. Rosen, *Phys. Rev.* 47, 777,1935.

[23] D. Bohm, *Quantum Theory*, Englewood Cliffs, Printice Hall, 1951.

[24] N. Bohr, *Phys. Rev.* 48, 696, 1935.

[25] L. Rosenfeld, The Wave-Particle Dilemma, in *The Physicist's Conception of Nature*, edited by J. Mehra, Dordrecht, D. Reidel, 1973.

[26] J. S. Bell, *Rev. Mod. Phys.* 38, 447, 1966.

[27] J. S. Bell, *Physics* 1, 195, 1965.

[28] J. F. Clauser and A. Shimony, *Rep. Prog. Phys.* 41, 1881, 1978.

[29] J. S. Bell, in *Foundations of Quantum Mechanics*, edited by B. d'Espagnat, New York and London, Academic, 1971.

[30] J. F. Clauser, M. A. Horne, A. Shimony and R. A. Holt, *Phys. Rev. Lett.* 23, 880, 1969.

[31] S. J. Freeman and J. F. Clauser, *Phys. Rev. Lett.* 28, 938, 1972.

[32] R. A. Holt and F. M. Pipkin, Harvard Preprint, 1973.

[33] J. F. Clauser, *Phys. Rev. Lett.* 36, 1223, 1976.

[34] E. S. Fry and R. C. Thompson, *Phys. Rev. Lett.* 37, 465, 1976.

[35] A. Aspect, P. Grangier and G. Roger, *Phys. Rev. Lett.* 47, 460, 1981; 49, 91, 1982.

[36] P. G. Kwiat, P. H. Eberhard, A. M. Steinberg and R. Y. Chiao, *Phys. Rev. A* 49,3209,1994.

[37] P. G. Kwiat, K. Mattle H. Weinfurter, A. Zeilinger, A. V. Sergienko and Y. Shih, *Phys. Rev. Lett.* 75, 4337, 1995.

[38] J. A. Wheeler, in *Mathematical Foundations of Quantum Theory*, edited by A. R. Marlow, New York and London, Academic, 1978.

[39] T. Hellmuth, H. Walther, A. Zajonc and W. Schleich, *Phys. Rev. A* 35, 2532, 1987.

[40] R. P. Feynman, *Int. J. Theor. Phys.* 21, 471, 1982.

[41] D. M. Greenberger, M. A. Horne and A. Zeilinger, in *Bell's Theorem, Quantum Theory and Conception of the Universe*, edited by M. Kafatos, Dordrecht, Kluwer Academic, 1989 pp. 73–76.

[42] D. M. Greenberger, M. A. Horne, A. Shimony and A. Zeilinger, *Am. J. Phys.* 58, 1131, 1990.

[43] N. D. Mermin, *Am. J. Phys.* 58, 731, 1990; *Phys. Today* 43(6), 9, 1990.

[44] L. Hardy, *Phys. Rev. Lett.* 68, 2981, 1992; 71, 1665, 1993.

[45] F. F. Jordan, *Phys. Rev. A* 50, 62, 1994.

[46] S. Goldstein, *Phys. Rev. Lett.* 72, 1951, 1994.

[47] J. R. Torgerson, D. Branning, C. H. Monken and L. Mandel, *Phys. Lett. A* 204, 323, 1995.

[48] J. R. Torgerson, D. Branning and L. Mandel, *App. Phys. B* 60, 267, 1995.

[49] A. Zeilinger, M. A. Horne, H. Weinfurter and M. Zukowski, *Phys. Rev. Lett.* 78, 3031, 1997.

[50] D. Bouwmeester, J.-W. Pan, M. Daniell, H. Weinfurter and A. Zeilinger, *Phys. Rev. Lett.* 82, 1345, 1999.

[51] V. B. Braginsky and F. Yu. Khalili, *Quantum Measurement*, Cambridge, Cambridge University Press 1992.

[52] C. M. Caves, K. S. Thorne, R. M. P. Drever, V. D. Sandberg and M. Zimmermann, *Rev. Mod. Phys.* 52, 341, 1980.

[53] V. B. Braginsky, Yu. I. Vorontsov and K. S. Thorne, *Science* 209, 547, 1980.

[54] L.-M. Duan, G. Giedke, J. I. Cirac and P. Zoller, *Phys. Rev. Lett.* 84, 2722, 2000.

[55] S. Manchini, V. Giovannetti, D. Vitali and P. Tombesi, *Phys. Rev. Lett.* 88, 12041, 2002.

[56] J. C. Howell, R. S. Bennick, S. J. Bentley and R. W. Boyd, *Phys. Rev. Lett.* 92, 210403, 2004.

[57] S. M. Tan, D. F. Walls and M. J. Collett, *Phys. Rev. Lett.* 66, 252, 1991.

[58] B. Hessno, P. Usachev, H. Heydari and G. Bjork, *Phys. Rev Lett.* 92, 18041, 2004.

[59] P. A. M. Dirac, *Quantum Mechanics* (4th edition), Wxford, Clarendon Press, 1958. Chapter 1, p. 9.

[60] A. T. Forrester, R. A. Gudmundsen and P. Q. Johnson, *Phys. Rev.* 99, 1691, 1955.

[61] R. Handbury Brown and R. Q. Twiss, *Nature* 178, 1046, 1956.

[62] R. L. Pfleeger and L. Mandel, *Phys. Rev.* 159, 1084, 1967.

[63] H. Paul, *Rev. Mod. Phys.* 58, 209, 1986.

[64] L. Mandel, *Phys. Rev.* 134, A10, 1964.

[65] C. K. Hong. Z. Y. Ou and L. Mandel, *Phys. Rev. Lett.* 59, 2044, 1987.

[66] T. Legero, T. Wilk, M. Hennrich, G. Rempe and A. Kuhn, *Phys. Rev. Lett.* 93, 070503, 2004.

[67] Y.-H. Kim, R. Yu, S. P. Kulik, Y. Shih and M. O. Scully, *Phys. Rev. Lett.* 84, 1, 2000.

[68] J.-W. Pan, D. Bouwmeester, M. Daniell, H. Wein furtur and A. Zeilinger, *Nature* 403, 515, 2008.

Chapter 2

Quantum Entanglement and Its Applications to Quantum Information and Quantum Computing

The concept of quantum computing began in the early 1980s. In 1982 P. Benioff [1] proposed that a computer could in principle function in a pure quantum fashion. A quantum computer may be prone to errors and may have troubles in correcting them. Then how can it probably be superior to a classical computer? A classical computer can store vectors, rotate them, project them onto orthogonal axes, etc. In fact it can simulate a quantum process, but the simulation becomes more and more inefficient when the dimension of the Hilbert space becomes large. In 1982 R. Feynman [2] pointed out that quantum computers can simulate a quantum process much more easily and can therefore perform certain tasks that are beyond the reach of any realistic classical computers. If a quantum computer had parallel architecture it could mimic any quantum system in real time independent of its size while a classical computer generally requires a number of steps that rises exponentially both with the size and with the time period of its evolution. Quantum information and quantum computing have undergone a spurt of new developments especially since 1990s. As the aim of the present book is to introduce basic principles of quantum mechanics which lie on the foundation of new emerging disciplines, we elaborate further properties of the quantum entanglement most relevant to quantum information and quantum

125

computing. For this purpose some necessary fundamental concepts on quantum information are also briefly introduced. There are extensive review articles and also monographs on this subject [3, 45].

2.1   A Brief Introduction to Quantum Computing

2.1.1   Quantum data and data processing

A classical bit is represented by two distinct states 0 and 1. A classical n-bit is represented by a string of n bits, each can be 0 or 1, altogether 2^n distinct states, from $000 \cdots 0$ to $111 \cdots 1$. A quantum bit (qubit) is typically realized in a two level microscopic system such as an atom or nuclear spin or photon polarization. We can, for example, denote by $|0\rangle$ and $|1\rangle$ the states spined up and down respectively. The most characteristic feature that distinguishes a qubit from a classical bit is that it can be in a state $\alpha |0\rangle + \beta |1\rangle$, where α and β are complex numbers with $|\alpha|^2 + |\beta|^2 = 1$. A pair of qubits can be in one of the Boolean bits $|00\rangle$, $|01\rangle$, $|10\rangle$ and $|11\rangle$, and it can also be in all possible superpositions of them, e.g.,

$$\frac{1}{\sqrt{2}}(|00\rangle + |01\rangle) = |0\rangle \otimes \frac{1}{\sqrt{2}}(|0\rangle + |1\rangle). \qquad (2.1.1)$$

This is a direct product state (separable for the two registers). The superposition

$$\frac{1}{\sqrt{2}}(|00\rangle + |11\rangle) \qquad (2.1.2)$$

cannot be expressed as a direct product (not separable) and is called an entangled state.

A string of n qubits can exist in any state of the form

$$|\psi\rangle = \sum_{x=(00\cdots0)}^{(11\cdots1)} c_x |x\rangle, \qquad (2.1.3)$$

where c_x are complex numbers satisfying

$$\sum_x |c_x|^2 = 1. \qquad (2.1.4)$$

A quantum state of n qubits is represented by a vector in a Hilbert space of 2^n dimensions, one for each possible classical state.

Operations in classical computation can be expressed as a sequence of 1- and 2-bit operations, e.g. the NOT and AND gates. In the 1-bit operation NOT flips the bit, i.e., $0 \to 1, 1 \to 0$. In the 2-bit operation AND on a and b the result $x = a \wedge b$ gives 1 if and only if a and b are both 1, otherwise x is 0. Quantum computation is likewise expressed as a sequence of 1- and 2-qubit quantum gates. A 1-qubit state can be expressed as

$$\begin{pmatrix} |0\rangle \\ |1\rangle \end{pmatrix},$$

and the quantum gate acting on it is expressed as a unitary matrix

$$\begin{pmatrix} \alpha & \beta \\ \gamma & \delta \end{pmatrix},$$

which maps $|0\rangle$ to $\alpha |0\rangle + \beta |1\rangle$ and $|1\rangle$ to $\gamma |0\rangle + \delta |1\rangle$. While the Boolean states $|0\rangle$ and $|1\rangle$ have their classical counterparts 0 and 1, the states $\alpha |0\rangle + \beta |1\rangle$ have no classical correspondence. For the 2-qubit Boolean states we have $|00\rangle, |01\rangle, |10\rangle$ and $|11\rangle$. The operation CNOT (controlled NOT) flips the second (target) input if the first (controlled) input is 1 and does nothing if the first input is 0, i.e., it interchanges $|10\rangle$ and $|11\rangle$ and keeps $|00\rangle$ and $|01\rangle$ unchanged, or for a 2-qubit we have

$$\text{CNOT} : |a, b\rangle \to |a, a \oplus b\rangle, \tag{2.1.5}$$

where $\oplus$ is the addition modulo 2. The quantum gate XOR (exclusive OR) is defined as

$$U_{\text{XOR}} |x, 0\rangle = |x, x\rangle. \tag{2.1.6}$$

It leaves the first qubit unchanged and the second qubit becoming a copy of the first. One is tempted to try the cloning of a typical qubit

$$|\psi\rangle = \alpha |0\rangle + \beta |1\rangle. \tag{2.1.7}$$

The result is

$$U_{\text{XOR}} |\psi, 0\rangle = \alpha U_{\text{XOR}} |0, 0\rangle + \beta U_{\text{XOR}} |1, 0\rangle$$
$$= \alpha |0, 0\rangle + \beta |1, 1\rangle, \tag{2.1.8}$$

which is not the expected $|\psi, \psi\rangle$, but an entangled state instead, in which neither the output qubit alone has a definite state. This is the consequence of no-cloning theorem for a general quantum state, which will be elaborated in Section 2.3.3.

Any unitary operation U on quantum data can be constructed from the 2-qubit XOR or CNOT and the 1-qubit unitary operation $\begin{pmatrix} \alpha & \beta \\ \gamma & \delta \end{pmatrix}$.

2.1.2   Quantum parallelism and efficient quantum algorithms

Feynman's idea mentioned above was put in a more concrete form by D. Deutsch [6]. We begin with an example. The function $f(x)$ takes a single bit x to a single bit. We want to decide whether $f(0) = f(1)$ or $f(0) \neq f(1)$. A classical computer has to calculate $f(0)$ and $f(1)$ separately (making computation twice) and then make the judgement by comparing them. Can a quantum computer do a better job? Let us define a unitary transformation U_f.

$$U_f : |x\rangle |y\rangle \rightarrow |x\rangle |y \oplus f(x)\rangle, \tag{2.1.9}$$

which flips the second qubit if $f(x)$ is 1 and does nothing if $f(x)$ is 0. For $|y\rangle = \frac{1}{\sqrt{2}}(|0\rangle - |1\rangle)$ we have

$$U_f : |x\rangle \frac{1}{\sqrt{2}} (|0\rangle - |1\rangle) \rightarrow |x\rangle \frac{1}{\sqrt{2}} (|f(x)\rangle - |1 \oplus f(x)\rangle)$$

$$= |x\rangle (-1)^{f(x)} \frac{1}{\sqrt{2}} (|0\rangle - |1\rangle). \tag{2.1.10}$$

Further we set $|x\rangle = \frac{1}{\sqrt{2}}(|0\rangle + |1\rangle)$, the result is

$$U_f : \frac{1}{\sqrt{2}} (|0\rangle + |1\rangle) \frac{1}{\sqrt{2}} (|0\rangle - |1\rangle)$$

$$\rightarrow \frac{1}{\sqrt{2}} \left[(-1)^{f(0)} |0\rangle + (-1)^{f(1)} |1\rangle \right] \frac{1}{\sqrt{2}} (|0\rangle - |1\rangle). \tag{2.1.11}$$

Finally we perform a measurement that projects the first qubit onto the basis

$$|\pm\rangle = \frac{1}{\sqrt{2}} (|0\rangle \pm |1\rangle), \qquad (2.1.12)$$

and we obtain $|+\rangle$ if $f(0) = f(1)$ and $|-\rangle$ if $f(0) \neq f(1)$. We can imagine the quantum computer as a black box. The input is a qubit and the operator. The result of computing is projected on the basis $\{|0\rangle, |1\rangle\}$. If the projection gives 1 and 0 respectively, then the result is 0, on the other hand, if the projection gives 0 and 1 respectively, the result is 1. In the present example, the input qubit is $\frac{1}{\sqrt{2}}(|0\rangle + |1\rangle) \frac{1}{\sqrt{2}}(|0\rangle - |1\rangle)$, the operator is U_f. Somewhat different from the usual computation, the final result is projected on the basis $\{|+\rangle, |-\rangle\}$, the superpositions of $|0\rangle$ and $|1\rangle$. If the result turns out to be $\{1, 0\}$, then $f(0) = f(1)$. If it is $\{0, 1\}$, then $f(0) \neq f(1)$.

The quantum computer does the job in one go. This algorithm was implemented experimentally by I. Chuang *et al.* [7] This is an example of quantum parallelism, which can extract global information depending on both $f(0)$ and $f(1)$. Let $f(x)$ be a function of N-bit, i.e., of 2^N possible arguments. With a quantum computer acting according to

$$U_f : |x\rangle |0\rangle \rightarrow |x\rangle |f(x)\rangle, \qquad (2.1.13)$$

we choose the input

$$\left[\frac{1}{\sqrt{2}} (|0\rangle + |1\rangle) \right]^N = \frac{1}{2^{N/2}} \sum_{x=0}^{2^N-1} |x\rangle, \qquad (2.1.14)$$

i.e., each qubit is a linear superposition $\frac{1}{\sqrt{2}}(|0\rangle + |1\rangle)$. By implementing U_f once, we generate a state

$$\frac{1}{2^{N/2}} \sum_{x=0}^{2^N-1} |x\rangle |f(x)\rangle, \qquad (2.1.15)$$

in which global properties of f depending on 2^N functions $f(x)$ are encoded in this state. Better ways to make efficient use of it must be found to design powerful algorithms. Well known examples of these algorithms are

(1) factorization of a large number (P. Shor 1994) [8]. The time required is a polynomial in the logarithm of best classical time. This exponential speed-up depends on using the destructive interference among a number of parallel computation paths.

(2) searching an object in an unsorted database ("searching a needle in a haystack", L.K. Grover 1997) [9]. To find a specific object among N objects of identical appearance the classical algorithm needs on the average to search $\frac{N}{2}$ times with a probability $\frac{1}{2}$ of success. Using entangled qubits the result of a search will influence the subsequent operation. Therefore the probability of success will be approximately $\frac{1}{2}$ after $\sqrt{N}$ searches. This algorithm was implemented experimentally by I. L. Chuang *et al.* using nuclear magnetic resonance (NMR) technique [10]. However it is worth mentioning that the entanglement property of NMR experiment is still under deliberations and the power of quantum computing needs further investigation [11].

(3) quantum simulation. To solve a many-body problem a classical computer needs a time rising exponentially with the number of particles. S. Wiesner [12] and D. Abrams and S. Lloyd [17] followed Feynman's idea that a quantum computer has no difficulties to simulate a quantum system, because the laws governing them are the same. In a quantum computer the unitary operations (designed in accordance with the Hamiltonian) on the initial state lead to the final state without resorting to solving the Schrödinger equation. Such algorithms are by far superior compared to numerical methods for solving Schrödinger equation and Monte Carlo simulation on conventional computers.

2.1.3   Quantum information

In discussing basic questions such as the measure of entanglement as well as practical questions such as quantum data compression we need the basic notions from information theory, classical and quantum. In the classical information one is concerned with the questions (1) how much can a message be compressed (to get rid of redundancy) and (2) how much redundancy is required to protect a message against errors brought by noise. C. Shannon (1948) characterized the redundancy by using entropy. Consider a message with n letters, $n \gg 1$, each letter is chosen from binary letters 0 and 1 with *a priori* probability of occurrence of $1 - p$ and p respectively, $0 \leqslant p \leqslant 1$. The question is, is it possible to use a shorter string of letters that can convey essentially the same information. For very large n, the string contains about $n(1 - p)$ letters 0 and about np letters 1. The number of distinct strings of this form is $\binom{n}{np}$. From Stirling's formula for large n

$$\log n! = n \log n - n + O(\log n), \tag{2.1.16}$$

we have

$$\log \binom{n}{np} = \log \frac{n!}{(np)![n(1-p)]!}$$

$$\simeq n \log n - n - [np \log np - np$$

$$+ n(1-p) \log n(1-p) - n(1-p)]$$

$$= nH(p), \tag{2.1.17}$$

where

$$H(p) = -p \log p - (1 - p) \log(1 - p) \tag{2.1.18}$$

is called the entropy function. The base of the logarithm is taken to be 2, and the number of strings of this form (the typical strings) is $2^{nH(p)}$. Strings with different numbers of letters 0 and 1 (the atypical ones) occur with vanishing probability in the asymptotic $n \to \infty$ limit. Therefore, to convey the information carried by a string of n bits, it is sufficient to choose a block code which assigns a positive integer to each of the typical strings. This block code has $2^{nH(p)}$ letters (equally probable), so we may specify each of the letters with a binary string of length $nH(p)$. Since $0 \leqslant H(p) \leqslant 1$ for $0 \leqslant p \leqslant 1$, and $H(p) = 1$ only for $p = \frac{1}{2}$. The block code shortens the message for any $p \neq \frac{1}{2}$. So one obtains a significant compression for p very different from $\frac{1}{2}$. The argument can be generalized to an alphabet of k letters instead of 2, with probability $p(x)$ of occurrence for letter x.

In a string of n letters x occurs about $np(x)$ times, and the number of typical strings is of the order

$$\frac{n!}{\prod_{x}(np(x))!} \simeq 2^{nH(X)}, \tag{2.1.19}$$

where

$$H(X) = \sum_{x} -p(x)\log p(x) \tag{2.1.20}$$

is the Shannon entropy of the ensemble of letters $X = \{x, p(x)\}$. A letter x from this ensemble carries on the average $H(x)$ bits of information. The above argument can be made more precise by using the (ε, δ) statement. A particular n letter message $x_1 x_2 \cdots x_n$ occurs with a priori probability

$$\left.\begin{array}{l} P(x_1 \cdots x_n) = p(x_1) \cdots p(x_n) \\[2mm] \log P(x_1 \cdots x_n) = \sum_{i=1}^{n} \log p(x_i) \end{array}\right\}. \tag{2.1.21}$$

The central limit theorem gives

$$-\frac{1}{n}\log P(x_1 \cdots x_n) - \langle -\log p(x)\rangle \equiv \frac{\sum_x -p(x)\log p(x)}{\sum_x p(x)} = H(X) \tag{2.1.22}$$

For any $\varepsilon, \delta > 0$ and for x sufficiently large, each typical string has a probability P satisfying

$$H(X) - \delta < -\frac{1}{n}\log P(x_1 \cdots x_n) < H(X) + \delta, \tag{2.1.23}$$

and the total probability of all typical strings exceeds $1 - \varepsilon$. Shannon's noiseless coding theorem asserts that the optimal code compresses each letter to $H(x)$ bits asymptotically.

When the message is sent through a noisy channel, the message y received may differ from the message x sent. The noisy channel is characterized by the conditional probabilities $p(y \mid x)$, the probability that y is received for x sent. The probability of receiving y for all possible x is

$$p(y) = \sum_x p(y \mid x) p(x). \tag{2.1.24}$$

After y is received, the probability distribution for x is updated by Bayes' rule

$$p(x \mid y) = \frac{p(y \mid x) p(x)}{p(y)}. \tag{2.1.25}$$

This updated distribution for x is no more a priori, but with some knowledge of x after y is received. Instead of the entropy function $H(X)$ we have now the conditional entropy

$$H(X \mid Y) = \langle -\log p(x \mid y) \rangle, \tag{2.1.26}$$

which quantifies the bits of information carried per letter using an optimal code. The joint probability $p(x, y)$ is given by

$$p(x, y) = p(x \mid y) p(y) = p(y \mid x) p(x). \tag{2.1.27}$$

$H(X \mid Y)$ becomes then

$$H(X \mid Y) = \langle -\log p(x \mid y) \rangle = \langle -\log p(x, y) \rangle - H(Y). \tag{2.1.28}$$

On the other hand

$$H(Y \mid X) = \langle -\log p(y \mid x) \rangle = \langle -\log p(x, y) \rangle - H(X). \tag{2.1.29}$$

We *gain* information about X when we learn about Y. This can be quantified by how much the number of bits per letter needed to specify X is reduced when Y is known. It is

$$I(X;Y) \equiv H(X) - H(X \mid Y) = H(Y) - H(Y \mid X)$$
$$= H(X) + H(Y) - H(X, Y), \tag{2.1.30}$$

which is called the mutual information, and is symmetrical with respect to X and Y. It can be shown that $H(X) \geqslant H(X \mid Y) \geqslant 0$, so that I is non-negative. $I(X;Y)$ is the number of bits of information per letter about X that we can acquire by reading Y or vice versa. If $p(y \mid x)$'s characterize a noisy channel, the $I(X;Y)$ is the amount of information per letter that can be transmitted through the channel, given the a priori probability distribution $p(x)$. This is Shannon's noisy channel coding theorem. When there is no correlation between X and Y, we have $p(x, y) = p(x)p(y)$, then $I(X; Y) = 0$, i.e., we know nothing about X by learning the uncorrelated Y.

The above consideration can now be generalized to quantum information. The ensemble of quantum states is now the source that prepares messages of n letters. Each letter is characterized by its density matrix ρ_x with a priori probability p_x. The density matrix for the ensemble is

$$\rho = \sum_x p_x \rho_x, \qquad (2.1.31)$$

for which the von Neumann entropy is defined by

$$S(\rho) = -\text{tr}(\rho \log \rho). \qquad (2.1.32)$$

If an orthonormal basis $\{|a\rangle\}$ that diagonalizes ρ is chosen, then

$$\rho = \sum_a \lambda_a |a\rangle \langle a|, \qquad (2.1.33)$$

and

$$S(\rho) = \sum_a -\lambda_a \log \lambda_a = H(A), \qquad (2.1.34)$$

where $H(A)$ is the entropy function of the ensemble $A = \{a, \lambda_a\}$.

When x consists of mutually orthogonal pure states, the quantum source reduces to a classical one, and $S(\rho) = H(X)$. The quantum source is non-trivial when the states ρ_x are not mutually commuting. The von Neumann entropy plays a dual role. It quantifies the quantum information content, i.e., the minimum number of qubits per letter needed to reliably encode the information, as well as its classical information content, i.e., the maximum amount of information per letter (in bits) that we can gain about the preparation by making the best possible measurement. Von Neumann entropy of the reduced density matrix also characterizes the entanglement of a bipartite pure state.

2.2 Quantum Entanglement

We have discussed the EPR-Bell states in connection with Bell's theorem. We elaborate here further some points especially in context of its application to the quantum information and computing.

2.2.1 Density matrix characterization of an entangled state

Consider a system composed of two subsystems A and B, described by the correlated wave function

$$|\psi\rangle = a\,|0\rangle_A \otimes |0\rangle_B + b\,|1\rangle_A \otimes |1\rangle_B, \qquad (2.2.1)$$

where $|0\rangle$ and $|1\rangle$ are orthogonal. Measurements of an observable M_A are carried out on A. The relevant operator for the entire system is $M_A \otimes 1_B$, and the expectation value of this operator in the state $|\psi\rangle$ is

$$
\begin{aligned}
\langle\psi|M_A \otimes 1_B|\psi\rangle &= (a_A^*\,\langle 0| \otimes_B \langle 0| + b_A^*\,\langle 1| \otimes_B \langle 1|) \\
&\quad \cdot (M_A \otimes 1_B)(a\,|0\rangle_A \otimes |0\rangle_B + b\,|1\rangle_A\,|1\rangle_B) \\
&= |a|^2\,_A\langle 0|M_A|0\rangle_A + |b|^2\,_A\langle 1|M_A|1\rangle_A \quad (2.2.2)
\end{aligned}
$$

where we have used the orthogonality

$$_B\langle 0|1_B|1\rangle_B = 0. \qquad (2.2.3)$$

This relation can be expressed in terms of the density matrix ρ_A

$$\langle M_A\rangle = \mathrm{tr}(M_A\rho_A), \qquad (2.2.4)$$

$$\rho_A = |a|^2\,|0\rangle_A\,_A\langle 0| + |b|^2\,|1\rangle_A\,_A\langle 1|. \qquad (2.2.5)$$

The above example can be generalized to an arbitrary state of any bipartite system, the Hilbert space of which is $\mathcal{H}_A \otimes \mathcal{H}_B$. Any pure state of the entire system can be expanded as

$$|\psi\rangle_{AB} = \sum_{i,\mu} a_{i\mu}\,|i\rangle_A \otimes |\mu\rangle_B, \qquad (2.2.6)$$

with

$$\sum_{i,\mu} |a_{i\mu}|^2 = 1. \tag{2.2.7}$$

The expectation value of M_A is

$$
\begin{aligned}
\langle M_A \rangle &= {}_{AB}\langle \psi | M_A \otimes 1_B | \psi \rangle_{AB} \\
&= \sum_{j,\nu} a_{j\nu}^* \left({}_A\langle j | \otimes {}_B\langle \nu | \right) (M_A \otimes 1_B) \sum_{i,\mu} a_{i\mu} \left(|i\rangle_A \otimes |\mu\rangle_B \right) \\
&= \sum_{i,j,\nu} a_{j\nu}^* a_{i\nu} \, {}_A\langle j | M_A | i \rangle_A \\
&= \operatorname{tr}(M_A \rho_A),
\end{aligned} \tag{2.2.8}
$$

where[1]

$$\rho_A = \operatorname{tr}_B |\psi\rangle_{AB}\,{}_{AB}\langle \psi| = \sum_{i,j,\mu} a_{i\mu} a_{j\mu}^* |i\rangle_A\,{}_A\langle j|. \tag{2.2.9}$$

ρ_A can be diagonalized, and its eigenvalues are non-negative and the sum of all eigenvalues is one.

$$\operatorname{tr}\rho_A = \sum_{i,\mu} |a_{i\mu}|^2 = 1. \tag{2.2.10}$$

Let the basis which diagonalizes ρ_A be denoted by $\{\varphi_\alpha\}$, then

$$\rho_A = \sum_\alpha p_\alpha |\varphi_\alpha\rangle \langle \varphi_\alpha|, \tag{2.2.11}$$

where $0 < p_\alpha \leqslant 1$ and $\sum p_\alpha = 1$. If there is only one non-zero eigenvalue of ρ_A, then $\rho_A^2 = \rho_A$, and the state represented by ρ_A is pure. Otherwise $\operatorname{tr} \rho_A^2 < \operatorname{tr} \rho_A$, the state is mixed, consisting of incoherent superposition of states φ_α with $p_\alpha \neq 0$. A pure state $|\psi\rangle_{AB}$ does not necessarily imply that the subsystem A is also a pure state. Such is the case only for a direct product $|\psi\rangle_{AB} = |\varphi\rangle_A \otimes |\chi\rangle_B$ which results from non-interacting subsystems. When a system A interacts

---

[1] $\operatorname{tr} |i\rangle \langle j| = \sum_k \langle k|i\rangle \langle j|k\rangle = \sum_k \langle j|k\rangle \langle k|i\rangle = \langle j \mid i\rangle.$

with a system B, they become correlated, or entangled, such is the case considered in the example (2.2.6). Even if the interaction exists for a brief period of time, the entanglement destroys the coherence of a superposition of states of A such that some of the relative phases become inaccessible experimentally if we carry out measurement on subsystem A. For a 2-state $(|0\rangle, |1\rangle)$ system the 2×2 density matrix can be expressed in terms of the Pauli matrices and the unit matrix.

$$\rho(\boldsymbol{P}) = \frac{1}{2}(1 + \boldsymbol{P} \cdot \boldsymbol{\sigma})$$

$$= \frac{1}{2}\begin{pmatrix} 1 + P_3 & P_1 - iP_2 \\ P_1 + iP_2 & 1 - P_3 \end{pmatrix}. \tag{2.2.12}$$

This form follows the requirement $\mathrm{tr}\rho = 1$ since the Pauli matrices are traceless. From (2.2.12) we obtain

$$\det \rho = \frac{1}{4}\left(1 - \boldsymbol{P}^2\right). \tag{2.2.13}$$

Let the eigenvalues of ρ be p_1 and p_2, then $\det \rho = p_1 p_2$. The demand that ρ has non-negative eigenvalues implies that $\boldsymbol{P}^2 \leqslant 1$. The condition is also sufficient, because $\mathrm{tr}\rho = p_1 + p_2 = 1$, p_1 and p_2 cannot be both negative. The vector $\boldsymbol{P}$ with $0 \leqslant |\boldsymbol{P}| \leqslant 1$ characterizing the state ρ lies within or on the surface of a sphere of radius 1. This sphere is called the Bloch sphere. Points on the surface $|\boldsymbol{P}| = 1$ belong to a state $\det \rho = 0$, i.e., ρ has eigenvalues 0 and 1, characterizing a pure state. In such cases $\boldsymbol{P}$ is given by the direction $\hat{n}\,(\sin\theta\cos\varphi, \sin\theta\sin\varphi, \cos\theta)$,

$$\rho(\hat{n}) = \frac{1}{2}(1 + \hat{n} \cdot \boldsymbol{\sigma}). \tag{2.2.14}$$

$\rho(\hat{n})$ satisfies the relation

$$(\hat{n} \cdot \boldsymbol{\sigma})\,\rho(\hat{n}) = \rho(\hat{n})\,(\hat{n} \cdot \boldsymbol{\sigma}) = \rho(\hat{n}), \tag{2.2.15}$$

i.e., $\rho(\hat{n})$ is an eigenstate of $\hat{n} \cdot \boldsymbol{\sigma}$ with eigenvalue 1, which can be interpreted as a spin pointing in the direction $\hat{n}\,(\theta, \varphi)$. Such a state is represented by the eigenspinor

$$|\psi(\theta,\varphi)\rangle = \begin{pmatrix} e^{-i\varphi/2}\cos\dfrac{\theta}{2} \\ e^{i\varphi/2}\sin\dfrac{\theta}{2} \end{pmatrix}. \tag{2.2.16}$$

This can be easily checked by a direct calculation

$$
\begin{aligned}
\rho(\hat{n}) &= |\psi(\theta,\varphi)\rangle\langle\psi(\theta,\varphi)| \\
&= \begin{pmatrix} \cos^2\dfrac{\theta}{2} & \cos\dfrac{\theta}{2}\sin\dfrac{\theta}{2}e^{-i\varphi} \\ \cos\dfrac{\theta}{2}\sin\dfrac{\theta}{2}e^{i\varphi} & \sin^2\dfrac{\theta}{2} \end{pmatrix} \\
&= \dfrac{1}{2} + \dfrac{1}{2}\begin{pmatrix} \cos\theta & \sin\theta e^{-i\varphi} \\ \sin\theta e^{i\varphi} & -\cos\theta \end{pmatrix} \\
&= \dfrac{1}{2}(1 + \hat{n}\cdot\boldsymbol{\sigma}).
\end{aligned}
\tag{2.2.17}
$$

2.2.2   Schmidt decomposition

A bipartite pure state can be expressed in a standard form, the Schmidt decomposition, as in (2.2.6),

$$|\psi\rangle_{AB} = \sum_{i,\mu} a_{i\mu}|i\rangle_A|\mu\rangle_B. \tag{2.2.18}$$

Define another set of basis

$$|\tilde{i}\rangle_B = \sum_{\mu} a_{i\mu}|\mu\rangle_B, \tag{2.2.19}$$

which needs not be orthonormal. Then

$$|\psi\rangle_{AB} = \sum_i |i\rangle_A|\tilde{i}\rangle_B. \tag{2.2.20}$$

Here $\tilde{i}$ bears a one-to-one correspondence with i, over which the sum is to be taken. Suppose that the basis $\{|i\rangle_A\}$ is chosen to make ρ_A

diagonal,

$$\rho_A = \sum_i p_i \, |i\rangle_A \,\, _A\langle i| \,. \tag{2.2.21}$$

On the other hand we can obtain ρ_A by taking trace with respect to B

$$\rho_A = \mathrm{tr}_B \left(|\psi\rangle_{AB} \,\, _{AB}\langle \psi| \right)$$

$$= \mathrm{tr}_B \left(\sum_{i,j} |i\rangle_A \,\, _A\langle j| \otimes |\tilde{i}\rangle_B \,\, _B\langle \tilde{j}| \right)$$

$$= \sum_{i,j} \,\, _B\langle \tilde{j} \mid \tilde{i} \rangle_B \left(|i\rangle_A \,\, _A\langle i| \right) \,. \tag{2.2.22}$$

Comparing (2.2.22) with (2.2.21) we see that

$$_B\langle \tilde{j} \mid \tilde{i} \rangle_B = p_i \delta_{ij}. \tag{2.2.23}$$

The basis $\{|\tilde{i}\rangle_B\}$ is actually orthogonal, though not normalized. Defining

$$|i'\rangle_B = \frac{|\tilde{i}\rangle_B}{\sqrt{p_i}} \tag{2.2.24}$$

to make the set $\{|i'\rangle_B\}$ orthonormal, we obtain the standard form of Schmidt decomposition

$$|\psi\rangle_{AB} = \sum_i \sqrt{p_i} \, |i\rangle_A \, |i'\rangle_B \,. \tag{2.2.25}$$

The coefficients $\{\sqrt{p_i}\}$ are called the Schmidt coefficients and the number of nonzero eigenvalues p_i of ρ_A and ρ_B, the Schmidt number.

In general two different pure states $|\psi\rangle_{AB}$ and $|\varphi\rangle_{AB}$ cannot be expanded simultaneously using the same orthonormal basis for $\mathcal{H}_A$ and $\mathcal{H}_B$. Starting from (2.2.23) we can also take the trace over A to obtain

$$\rho_B = \mathrm{tr}_A \, |\psi\rangle_{AB} \,\, _{AB}\langle \psi| = \sum_i p_i \, |i'\rangle_B \,\, _B\langle i'| \,. \tag{2.2.26}$$

We see that ρ_A and ρ_B share the same set of nonzero eigenvalues. Since $\mathscr{H}_A$ and $\mathscr{H}_B$ can have different dimensions, so the number of zero eigenvalues of ρ_A and ρ_B can differ. If ρ_A and hence ρ_B, have no degenerate eigenvalues other than 0, then the Schmidt decomposition of $|\psi\rangle_{AB}$ is essentially uniquely determined. We can diagonalize ρ_A and ρ_B to find $|i\rangle_A$ and $|i'\rangle_B$ and pair them up with $\sqrt{p_i}$ to obtain $|\psi\rangle_{AB}$. The remaining freedom for $|\psi\rangle_{AB}$ is the phase.

We return briefly to the EPR paradox. One of EPR's objection is that the collapse of state caused by a measurement of s_z at B giving $|\uparrow_z\rangle_B$ sends instantaneously a signal to A such that a measurement of s_z there reveals $|\downarrow_z\rangle_B$ unambiguously. EPR interpreted this as an action at a distance. Consider a state of total spin 1,

$$|\psi\rangle_{AB} = \frac{1}{\sqrt{2}} \left(|\uparrow_z\rangle_A |\uparrow_z\rangle_B + |\downarrow_z\rangle_A |\downarrow_z\rangle_B \right). \qquad (2.2.27)$$

A measurement at B leads to the preparation of spin at A in the ensemble

$$\rho_A = \frac{1}{2} \left(|\uparrow_z\rangle_A \, _A\langle \uparrow_z | + \frac{1}{2} | \downarrow_z \rangle_{A \ A\langle\downarrow_z|} \right), \qquad (2.2.28)$$

which has nonzero degenerate eigenvalues, and therefore the Schmidt decomposition (2.2.27) is not unique. Another decomposition is

$$|\psi\rangle_{AB} = \frac{1}{\sqrt{2}} \left(|\uparrow_x\rangle_A |\uparrow_x\rangle_B + |\downarrow_x\rangle_A |\downarrow_x\rangle_B \right), \qquad (2.2.29)$$

which differs from (2.2.27) by a different choice of basis $\mathscr{H}_A$ and $\mathscr{H}_B$. Both (2.2.27) and (2.2.29) describe the same physical state, a spin $s = 1$ bipartite entangled state with the only difference of quantization axis for spin and can be transformed into each other by simultaneously carrying out a unitary transformation on $|\cdot\rangle_A$ and on $|\cdot\rangle_B$. On carrying out measurement of s_x at B prepares the spins in an ensemble $\{|\uparrow_x\rangle_A, |\downarrow_x\rangle_B\}$. Measurement at A of s_z leads to the same density matrix (2.2.28)! Therefore, measuring s_z at A does not reveal the message sent by the measurement at B, namely the quantization axis at B can be any direction. This property will be further discussed in connection with the quantum key distribution.

The Schmidt number is used to characterize the entanglement for a bipartite pure state. A direct product of pure states in $\mathscr{H}_A$ and $\mathscr{H}_B$ is a separable bipartite pure state

$$|\psi\rangle_{AB} = |\varphi\rangle_A \otimes |\chi\rangle_B. \qquad (2.2.30)$$

For this state the Schmidt number is 1. Any state for which the Schmidt number is larger than 1 cannot be expressed in this form and is entangled.

Finally we mention briefly that the entanglement of a system of two distinguishable particles is well defined and its properties sufficiently explored. For indistinguishable 2-particle systems the problem is non-trivial. J. Schliemann *et al.* [14] discussed the entanglement in two-fermion systems and found that the results for distinguishable 2-particle system can be transplanted to this case. Concerning two-boson systems, there have been reports with different approaches [15, 16].

2.2.3 More about EPR-Bell states

In Section 1.3 we discussed EPR-Bell states, in qubit language they are

$$\left.\begin{aligned}
|\phi^{\pm}\rangle &= \frac{1}{\sqrt{2}}\left(|00\rangle \pm |11\rangle\right) \\
|\psi^{\pm}\rangle &= \frac{1}{\sqrt{2}}\left(|01\rangle \pm |10\rangle\right)
\end{aligned}\right\} \qquad (2.2.31)$$

They are "maximally entangled states" in the sense that a trace over qubit B leads to

$$\rho_A = \mathrm{tr}_B\left(|\phi^{+}\rangle_{AB}\,_{AB}\langle\phi^{+}|\right) = \frac{1}{2}I_A \qquad (2.2.32)$$

and similarly a trace over A leads to $\rho_B = \frac{1}{2}I_B$. This means that if we measure spin A along *any* axis, the resulting spin B is completely random–spin up and down with equal probability $\frac{1}{2}$. The information of two qubits contained in such entangled states cannot be acquired by measuring the spin A or B.

Table 2.1 Phase and parity bits of EPR-Bell states.

States	Parity bit	Phase bit
ϕ^+	$+$	$+$
ϕ^-	$+$	$-$
ψ^+	$-$	$+$
ψ^-	$-$	$-$

The entangled basis states (2.2.31) are simultaneous eigenstates of two commuting observables $\sigma_1^{(A)}\sigma_1^{(B)}$ and $\sigma_3^{(A)}\sigma_3^{(B)}$. An eigenvalue of $\sigma_3^{(A)}\sigma_3^{(B)}$ is called the "parity bit"(telling whether the spins are aligned or anti-aligned), and an eigenvalue of $\sigma_1^{(A)}\sigma_1^{(B)}$ is called the "phase bit"(giving $+$ or $-$ sign in the superposition). Each state carries two bits of information as given in Table 2.1.

The sender and receiver of information are kept at a large distance, they share the qubits of an entangled state, and each can manipulate the information realized by carrying out unitary transformations locally at A and B. For example, Alice can apply σ_3 to her qubit at A, flipping the relative phase of $|0\rangle_A$ and $|1\rangle_A$, resulting in the change

$$|\phi^+\rangle \leftrightarrow |\phi^-\rangle$$
$$|\psi^+\rangle \leftrightarrow |\psi^-\rangle. \qquad (2.2.33)$$

She can also apply σ_1 to her qubit, flipping the spin $|0\rangle_A \leftrightarrow |1\rangle_A$, resulting in the change

$$|\phi^+\rangle \leftrightarrow |\psi^+\rangle$$
$$|\phi^-\rangle \leftrightarrow |\psi^-\rangle. \qquad (2.2.34)$$

Such local unitary transformations change one of the maximally entangled basis state to another. Bob can carry out similar operations at B. During such transformations $\rho_A = \rho_B = \frac{1}{2}I$ does not change, i.e., information stored in any one of the states cannot be acquired by either of them using local operations, even if Alice and Bob can exchange classical messages about outcomes of their measurements, so that they can learn about the correlation of their measurements. Let both of them measure σ_3. Then by combining their results the parity bit is known, since both $\sigma_3^{(A)}$ and $\sigma_3^{(B)}$ commute with the operator $\sigma_3^{(A)}\sigma_3^{(B)}$, their measurements do not disturb the parity bit. But each of them does not commute with $\sigma_1^{(A)}\sigma_1^{(B)}$, their measurements do disturb the phase bit. They can alternatively measure $\sigma_1^{(A)}$ and $\sigma_1^{(B)}$. Then they acquire the information of the phase bit at the cost of disturbing the parity bit. If $\sigma_3^{(A)}\sigma_3^{(B)}$ can be measured without acquiring any information about $\sigma_3^{(A)}$ and $\sigma_3^{(B)}$ separately, $\sigma_1^{(A)}\sigma_1^{(B)}$ can be left intact. This cannot be done locally, and must be carried out by their joint effort. For this purpose we introduce the single qubit unitary transformation H, the Hadamard transform

$$H = \frac{1}{\sqrt{2}}\begin{pmatrix} 1 & 1 \\ 1 & -1 \end{pmatrix} = \frac{1}{\sqrt{2}}(\sigma_1 + \sigma_3), \tag{2.2.35}$$

which has the following properties

$$\left.\begin{array}{l} H^2 = 1 \\ H\sigma_1 H = \sigma_3 \\ H\sigma_3 H = \sigma_1. \end{array}\right\} \tag{2.2.36}$$

In quantum circuit notation, the 1-qubit unitary operation denoted by H can be envisioned as a $\theta = \pi$ rotation about the axis $\hat{n} = \frac{1}{2}(\hat{n}_1 + \hat{n}_3)$ which rotates $\hat{x}$ to $\hat{z}$ and vice versa.

$$R(\hat{n}, \theta) = \cos\frac{\theta}{2} + i\hat{n}\cdot\sigma\sin\frac{\theta}{2} \xrightarrow{\theta=\pi} i\frac{1}{\sqrt{2}}(\sigma_1 + \sigma_3) = iH. \tag{2.2.37}$$

We still need the two-qubit transformation CNOT introduced in Section 2.1.1

$$\text{CNOT} : |a, b\rangle \rightarrow |a, a \oplus b\rangle, \tag{2.2.38}$$

with

$$(\text{CNOT})^2 = 1. \tag{2.2.39}$$

In quantum circuit this is shown schematically in Fig. 2.1.

By composing these transformations we have the quantum circuit shown in Fig. 2.2.

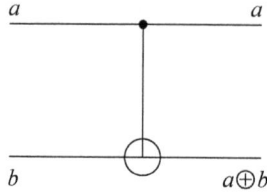

Fig. 2.1   CNOT.

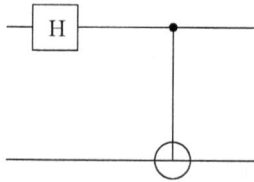

Fig. 2.2   Successive application of H and CNOT.

This means applying H to the qubit a and then applying CNOT to the result. Acting on the un-entangled basis (the Boolean basis) states this combined transformation leads to the EPR-Bell states[2]

$$
\left.
\begin{aligned}
|00\rangle \xrightarrow{H} \frac{1}{\sqrt{2}}(|0\rangle + |1\rangle)|0\rangle \xrightarrow{\text{CNOT}} |\phi^+\rangle \\
|01\rangle \xrightarrow{H} \frac{1}{\sqrt{2}}(|0\rangle + |1\rangle)|1\rangle \xrightarrow{\text{CNOT}} |\psi^+\rangle \\
|10\rangle \xrightarrow{H} \frac{1}{\sqrt{2}}(|0\rangle - |1\rangle)|0\rangle \xrightarrow{\text{CNOT}} |\phi^-\rangle \\
|11\rangle \xrightarrow{H} \frac{1}{\sqrt{2}}(|0\rangle - |1\rangle)|1\rangle \xrightarrow{\text{CNOT}} |\psi^-\rangle
\end{aligned}
\right\}.
\tag{2.2.40}
$$

The inverse of the above operation brings the EPR-Bell states back to the Boolean basis states. It is important to notice that CNOT

---

[2]It is straightforward to see that

$$H|0\rangle = \frac{1}{\sqrt{2}}\begin{pmatrix}1&1\\1&-1\end{pmatrix}\begin{pmatrix}1\\0\end{pmatrix} = \frac{1}{\sqrt{2}}\begin{pmatrix}1\\1\end{pmatrix} = \frac{1}{\sqrt{2}}(|0\rangle + |1\rangle)$$

$$H|1\rangle = \frac{1}{\sqrt{2}}\begin{pmatrix}1&1\\1&-1\end{pmatrix}\begin{pmatrix}0\\1\end{pmatrix} = \frac{1}{\sqrt{2}}\begin{pmatrix}1\\-1\end{pmatrix} = \frac{1}{\sqrt{2}}(|0\rangle - |1\rangle).$$

is a non-local operation, because its action (bit flip) on the target bit depends on the control bit. The Boolean states can now be measured locally and the outcome reveals the 2 qubits contained in the original EPR-Bell states.

2.3 Entanglement in Applications to Quantum Information

In Section 2.1.3, examples are given in which by exploiting the entangled states we can perform tasks that are otherwise very difficult or even impossible. We elaborate here some other uses of this basic concept of quantum mechanics.

2.3.1 Dense coding

There is an advantage to sending qubits instead of classical bits, as we shall demonstrate in an example. Alice can prepare $|\uparrow_z\rangle$ and $|\downarrow_z\rangle$, and Bob can measure to infer the choice. The transmission of a classical bit (0 or 1) via a qubit leads to the superiority of quantum coding. Alice can actually make use of entangled pairs of qubits, e.g., the state $|\phi^+\rangle$. She keeps one qubit for herself and sends the other qubit to Bob. They keep this entangled pairs for emergency needs of sending a brief message. Alice can perform one of the following unitary transformations on her qubit:

(1) identity transformation 1.
(2) rotation about $\hat{x}$ by $\pi : e^{i\frac{\sigma_1}{2}\pi} = i\sigma_1$.
(3) rotation about $\hat{y}$ by $\pi : e^{i\frac{\sigma_2}{2}\pi} = i\sigma_2$.
(4) rotation about $\hat{z}$ by $\pi : e^{i\frac{\sigma_3}{2}\pi} = i\sigma_3$.

These transformations acting on $|\phi^+\rangle_{AB}$ lead to the states $|\phi^+\rangle_{AB}, |\psi^+\rangle_{AB}, |\psi^-\rangle_{AB}$ and $|\phi^-\rangle_{AB}$ respectively. Each of these states carries two qubits of information (its parity and phase bits). Alice can choose her operation according to the parity and phase qubits of the resulting EPR-Bell states (this is the message that she wants to send to Bob), and then operate on her qubit before sending the result to Bob. On receiving the qubit A, Bob can carry out the orthogonal collective measurement described in Section 2.2.3 on the pair and thus determine the qubits carried by the EPR-Bell state. In this action Alice sends one qubit to Bob and Bob receives instead 2 qubits of information. This is out of reach of the classical information. Hence the name dense coding. Of course the pre-existing entanglement was exploited for the more efficient communication. Dense coding was implemented by the Innsbruck group [17].

2.3.2   Quantum cryptography: EPR quantum key distribution

The communication between Alice and Bob is highly confidential. In order to be free from eavesdropping they use a key which is known exclusively to the two of them. They can therefore encode and decode the message. But how can this key be kept in absolute security? The best way is to establish the key during their communication — yet no third party can find out what the key is — and use the key only once. The first realization of cryptography was accomplished by C. H. Bennett and G. Brassard [18] using photon polarizations without invoking entanglement. Let Alice and Bob share a supply of entangled pairs, each prepared in the state $|\psi^-\rangle$. They agree to a protocol in order to establish a private key. For each qubit in each party's possession, he/she decides to measure either σ_1 or σ_3. The decision is arbitrary for each qubit. After the measurements are performed, they both publicly announce what observables they measured in the string of measurements but do not reveal the outcomes. For those cases in which they measured different observables the results are discarded, because those outcomes are uncorrelated. When they measured the same observable, their results are perfectly anti-correlated, though random. A key is consequently established, of which the eavesdropper is ignorant. The protocol is secure even if the attacker attempted to tamper with the pairs in possession of Alice and Bob, and some of the pairs may no more be $|\psi^-\rangle$'s, but entangled with the attacker's qubits. The attacker can wait until Alice and Bob make their announcements and then measure his own qubits in order to acquire information about the results of Alice and Bob. Can the attacker succeed? The most general possible state of AB pair and a set of eavesdropper(E)'s qubits is

$$|\gamma\rangle_{ABE} = |00\rangle_{AB} |e_{00}\rangle_E + |01\rangle_{AB} |e_{01}\rangle_E$$
$$+ |10\rangle_{AB} |e_{10}\rangle_E + |11\rangle_{AB} |e_{11}\rangle_E. \qquad (2.3.1)$$

Since $|\psi^-\rangle$ is an eigenstate of both $\sigma_1^{(A)}\sigma_1^{(B)}$ and $\sigma_3^{(A)}\sigma_3^{(B)}$, Alice and Bob are able to verify whether the pairs in their possession do have this property. They do so by sacrificing a portion of the key and publicly comparing their measurement outcomes for this portion. If the result of verification is confirmative, viz. the pairs do have both parity and phase qubits negative, then (2.3.1) must satisfy the following requirements. To have $\sigma_1^{(A)}\sigma_1^{(B)} = -1$, (2.3.1) must be

$$|\gamma\rangle_{ABE} = |01\rangle_{AB} |e_{01}\rangle_E + |10\rangle_{AB} |e_{10}\rangle_E, \qquad (2.3.2)$$

and for $\sigma_3^{(A)}\sigma_3^{(B)} = -1$, (2.3.1) must be

$$|\gamma\rangle_{ABE} = \frac{1}{\sqrt{2}} (|01\rangle - |10\rangle) |e\rangle_E, \qquad (2.3.3)$$

where

$$\frac{1}{\sqrt{2}} |e\rangle_E \equiv |e_{10}\rangle_E = -|e_{01}\rangle_E. \qquad (2.3.4)$$

Because of (2.3.4) the requirement of $\sigma_3^{(A)}\sigma_3^{(B)} = -1$ is also reduced to (2.3.3), i.e., the AB pair is completely un-entangled with the qubit at E. Therefore, the eavesdropper is unable to learn anything about the outcomes of Alice's and Bob's measurements by measuring his own qubit, and Alice and Bob can be sure that the key is secure. If the attacker did succeed to entangle his qubits with those of Alice and Bob so that the verification results fail for some of the pairs, then Alice and Bob have detected the presence of the eavesdrop activity. They can abandon the present key and establish a new one. The above protocol is the theme described in [19], and there are variations to it.[3]

---

[3]The protocol was realized by the Innsbruck group [20].

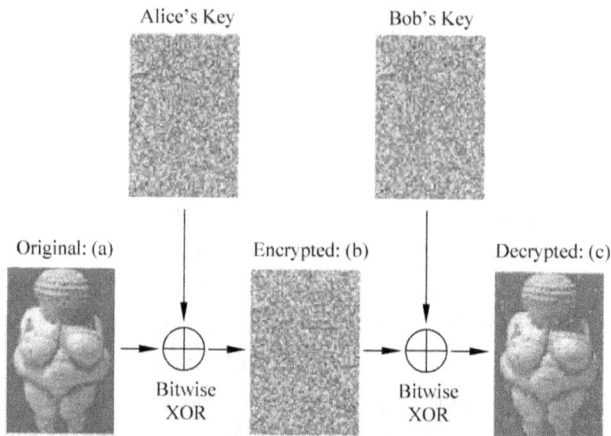

Fig. 2.3   Secure transmission of an image "Venus von Willendorf" effigy (Naturhistorische Museum, Vienna). Reprinted with permission from Thomas Jennewein, Christoph Simon, Gregor Weihs, Harald Weinfurter, and Anton Zeilinger, *Phys. Rev. Lett.* 84, 4729, 2000. Copyright (2017) by the American Physical Society.

In this realization Alice and Bob use a 49984 bit key generated by the scheme [18]. Alice encrypts the image via bitwise XOR operation with her key and transmits the encrypted image to Bob via the computer network. Bob decrypts the image with his key, resulting in an image with a few errors due to the remaining bit errors in the keys. The result is shown in Fig. 2.3.

2.3.3   The quantum no-cloning theorem

A smart attacker could have tried to copy the key qubits in Alice's and Bob's possession without disturbing them, and then obtain the key by measuring the copies. The quantum key is secured by the fact that it is impossible to acquire information that distinguishes between *non-orthogonal* quantum states without *disturbing* them. Let $|\psi\rangle$ and $|\varphi\rangle$ be two non-orthogonal states in the AB Hilbert space, $\langle \psi \mid \varphi \rangle \neq 0$. The attacker tries to copy them by performing a unitary transformation U:

$$U:$$
$$|\psi\rangle \otimes |0\rangle_E \rightarrow |\psi\rangle \otimes |i\rangle_E$$
$$|\varphi\rangle \otimes |0\rangle_E \rightarrow |\varphi\rangle \otimes |j\rangle_E . \tag{2.3.5}$$

Unitarity implies that

$$\langle \psi \mid \varphi \rangle = ({}_E \langle 0| \otimes \langle \psi|) \, (|\varphi\rangle \otimes |0\rangle_E)$$

$$= ({}_E \langle i| \otimes \langle \psi|) \, (|\varphi\rangle \otimes |j\rangle_E)$$

$$= \langle \psi \mid \varphi \rangle \,\, {}_E \langle i \mid j \rangle_E . \qquad (2.3.6)$$

For $\langle \psi \mid \varphi \rangle \neq 0$ we have $_E \langle i \mid j \rangle_E = 1$, i.e., $|i\rangle = |j\rangle$. The attacker obtains identical copies for different states $|\psi\rangle$ and $|\varphi\rangle$ and therefore is unable to distinguish non-orthogonal states. The situation is different if $|\psi\rangle$ and $|\varphi\rangle$ are orthogonal. The orthogonal states $|0\rangle$ and $|1\rangle$ are in principle not immune to copying without being disturbed, just as with classical bits. The unitary transformation U acts as

$$U : |0\rangle_A |0\rangle_E \rightarrow |0\rangle_A |0\rangle_E \qquad (2.3.7)$$

$$|1\rangle_A |0\rangle_E \rightarrow |1\rangle_A |1\rangle_E$$

i.e., it copies the state in A onto E. But if instead the state in A is a general qubit

$$|\varphi_A\rangle = a \, |0\rangle_A + b \, |1\rangle_A , \qquad (2.3.8)$$

then

$$U : (a \, |0\rangle_A + b \, |1\rangle_A) \, |0\rangle_E \qquad (2.3.9)$$

$$\rightarrow a \, |0\rangle_A |0\rangle_E + b \, |1\rangle_A |1\rangle_E ,$$

which is different from $|\psi\rangle_A |\psi\rangle_E$, a direct product of the original and the copy. Therefore, a general qubit which is not orthogonal to $|0\rangle$ or $|1\rangle$, cannot be copied. We could relax the situation by enlarging the Hilbert space beyond $\mathscr{H}_A$ (the original) $\otimes$ $\mathscr{H}_E$ (the copy) and include a supplementary $\mathscr{H}_F$ and define the most general copying unitary transformation

$$U : |\varphi\rangle_A |0\rangle_E |0\rangle_F \rightarrow |\psi\rangle_A |\psi\rangle_E |i\rangle_F \qquad (2.3.10)$$

$$|\varphi\rangle_A |0\rangle_E |0\rangle_F \rightarrow |\varphi\rangle_A |\varphi\rangle_E |j\rangle_F ,$$

where $|\psi\rangle$ and $|\varphi\rangle$ are distinct non-orthogonal states. Then unitarity implies

$$_A \langle \psi \mid \varphi \rangle_A = {}_A \langle \psi \mid \varphi \rangle_A \,\, {}_E \langle \psi \mid \varphi \rangle_E \,\, {}_F \langle i \mid j \rangle_F ,$$

i.e.,

$$1 = {}_E\langle \psi \mid \varphi \rangle_E \, {}_E\langle i \mid j \rangle_F.$$

With normalized states we conclude that

$$\langle \psi \mid \varphi \rangle = 1, \tag{2.3.11}$$

so that $|\psi\rangle$ and $|\varphi\rangle$ actually represent the same ray (wave function differ only by a phase factor). No unitary transformation can make a copy of both $|\psi\rangle$ and $|\varphi\rangle$, if they are *distinct*, non-orthogonal states. This is the quantum no-cloning theorem.

2.3.4   Quantum teleportation

Imagine a strange situation: Alice has a qubit, but she does not know its state. Bob needs to create this state in his disposal. Between them there exists only a classical channel. In addition, they share the entangled pair $|\phi^+\rangle_{AB}$. It is this entangled state that helps them to realize their wish according to the following protocol. Alice carries out a Bell measurement described in Section 2.2.3 on the unknown to her qubit denoted by $|\chi\rangle_C$ and her member of the shared entangled pair, projecting onto one of the four EPR-Bell states $|\phi^\pm\rangle_{CA}, |\psi^\pm\rangle_{CA}$, and she then sends the outcome of measurement (one parity bit, one phase bit) through the classical channel to Bob. On receiving the message Bob carries out a unitary transformation on his qubit $|\cdot\rangle_B$

Table 2.2   Protocol for quantum teleportation.

Unitary operation	If the Bell state is
1_B	$|\phi^+\rangle_{CA}$
$\sigma_1^{(B)}$	$|\psi^+\rangle_{CA}$
$\sigma_2^{(B)}$	$|\psi^-\rangle_{CA}$
$\sigma_3^{(B)}$	$|\phi^-\rangle_{CA}$

in accordance with the EPR-Bell state described in the following protocol.[4]

Then Bob can be sure that in his disposal is the state $|\chi\rangle_B$. The protocol is well-founded quantum mechanically. Let $|\chi\rangle_C = a\,|0\rangle_C + b\,|1\rangle_C$, then we have

$$
\begin{aligned}
|\chi\rangle_C\,|\phi^+\rangle_{AB} & \\
&= (a\,|0\rangle_C + b\,|1\rangle_C)\frac{1}{\sqrt{2}}\,(|00\rangle_{AB} + |11\rangle_{AB}) \\
&= \frac{1}{\sqrt{2}}\,(a\,|000\rangle_{CAB} + a\,|011\rangle_{CAB})\,(b\,|100\rangle_{CAB} + b\,|111\rangle_{CAB}) \\
&= \frac{1}{2}a\,(|\phi^+\rangle_{CA} + |\phi^-\rangle_{CA})\,|0\rangle_B + \frac{1}{2}a\,(|\psi^+\rangle_{CA} + |\psi^-\rangle_{CA})\,|1\rangle_B \\
&\quad + \frac{1}{2}b\,(|\psi^+\rangle_{CA} - |\psi^-\rangle_{CA})\,|0\rangle_B + \frac{1}{2}b\,(|\phi^+\rangle_{CA} - |\phi^-\rangle_{CA})\,|1\rangle_B \\
&= \frac{1}{2}|\phi^+\rangle_{CA}\,(a\,|0\rangle_B + b\,|1\rangle_B) + \frac{1}{2}|\psi^+\rangle_{CA}\,(a\,|1\rangle_B + b\,|0\rangle_B) \\
&\quad + \frac{1}{2}|\psi^-\rangle_{CA}\,(a\,|1\rangle_B - b\,|0\rangle_B) + \frac{1}{2}|\phi^-\rangle_{CA}\,(a\,|0\rangle_B - b\,|1\rangle_B) \\
&= \frac{1}{2}|\phi^+\rangle_{CA}\,|\chi\rangle_B + \frac{1}{2}|\psi^+\rangle_{CA}\,\sigma_1\,|\chi\rangle_B \\
&\quad + \frac{1}{2}|\psi^-\rangle_{CA}\,(-i\sigma_2)\,|\chi\rangle_B + \frac{1}{2}|\phi^-\rangle_{CA}\,\sigma_3\,|\chi\rangle_B .
\end{aligned} \tag{2.3.12}
$$

The above equation shows that the qubit $|\chi\rangle_C$ and Alice's member $|\cdot\rangle_A$ of entangled pair $|\phi^+\rangle_{AB}$ undergoing the Bell measurement yield one of the Bell states with equal probability with the corresponding qubit states at Bob's disposal. It is the measurement process which projects the CA pair to one of the Bell states and at the same time changes the state of Bob's qubit. The prescription in Table 2.2 is designed to recover $|\chi\rangle$ when the appropriate unitary

---

[4]As given in Section 2.3.1 operator $i\sigma_1, i\sigma_2, i\sigma_3$ are unitary operators (rotation by π about $\hat{x}, \hat{y}, \hat{z}$ respectively).

transformation is operated on the corresponding qubit at B. The state $|\chi\rangle$ is in effect tele-transported from A to B. It is puzzling how the classical channel could transmit information about a qubit. Actually it is the shared entangled pair which plays the role of quantum channel. It is also important to note that the teleportation is completely consistent with the no-cloning theorem. The qubit $|\chi\rangle_C$ is destroyed in the measurement process before the copy $|\chi\rangle_B$ is created. This trick is called quantum teleportation [21].

Quantum teleportation using photon polarization was experimentally implemented by the Innsbruck group [22] with success probability 25% and later demonstrated by Y. H. Shih's group with success probability 100% [23]. Teleportation with continuous variables was suggested by L. Vaidman [24] and achieved experimentally by the Caltech group [25]. A high fidelity teleportation of photons over a distance of 600 meters across the River Danube in Vienna was realized by A. Zeilinger's group [42]. The quantum channel (an optical fiber installed in a tunnel underneath the river) is exposed to temperature fluctuations and other environmental factors. A subtle question can be raised that Bob has to perform a unitary transformation on his qubit according to Alice's result of Bell measurement, in other words, he has to prepare his apparatus before his share of the entangled photon arrives. In this practical case [42] the result of Alice's Bell measurement is communicated to Bob via the classical microwave channel, which arrives before the shared qubit by $1.5\,\mu$s, since the latter is communicated through the optical fiber with a velocity of $2c/3$. Bob can manage to activate his electro-optical modulator in this time interval.

2.4 Six-photon Singlet — An Example of Decoherence Free States

A quantum superposition state can experience decoherence from various interactions with the environment. But usually there is one specific class of interactions that is most significant. Therefore it is desirable to encode informations into states that are immune against this most significant class of interactions. If the interaction, no matter how strong, possesses some kind of symmetry, there exist states that are invariant against this interaction. If two photons are entangled into a singlet state, then a "collective noise" (i.e. a noise that can change the polarization of a photon, but the influence is simultaneous and identical for the two photons) will not produce decoherence. H. Weinfurter *et al.* [43] succeeded to obtain two- and four-photon entangled states that are invariant with respect to the collective noise. M. Bourennane *et al.* [44] synthesized a six-photon singlet that is invariant against identical rotations of polarizations of each constituent. The authors used parametric down conversion of ultra-violet pulses on non-linear crystals. They focused on events where three indistinguishable pairs are produced, shown in Fig. 2.4. The photons of each pair are emitted in different directions so that the six photons are separated into two sets of three photons, shown as solid dots in the figure. The photons are then separated using beam splitters and enter the polarization analyzers each containing a polarization beam splitter (PBS in the figure) and two photon detectors.

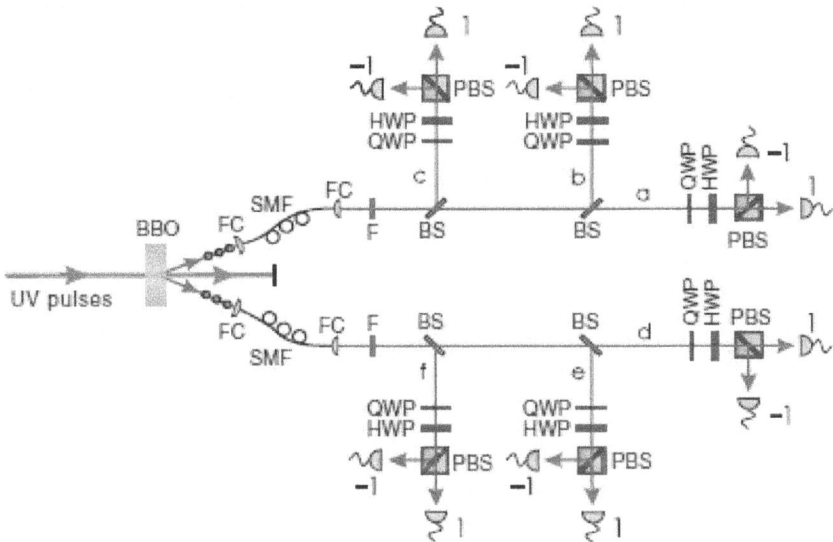

Fig. 2.4 Experimental setup for generating and analyzing the 6-photon polarization entangled state. Reprinted with permission from Magnus Radmark, Marek Żukowski, and Mohamed Bourennane, *Phys. Rev. Lett.* 103, 150501, 2009. Copyright (2017) by the American Physical Society.

This state is a superpositon of a 6-photon Greenberger-Horne-Zeilinger (GHZ) state and two products of 3-photon W-states, viz.

$$|\Psi_6^-\rangle = \frac{1}{\sqrt{2}}|GHZ_6^-\rangle + \frac{1}{2}\left(|\tilde{W}_3\rangle|W_3\rangle - |W_3\rangle|\tilde{W}_3\rangle\right) \qquad (2.4.1)$$

where

$$|GHZ_6^-\rangle = \frac{1}{\sqrt{2}}\left(|HHHVVV\rangle - |VVVHHH\rangle\right)$$

$$|W_3\rangle = \frac{1}{\sqrt{3}}\left(|HHV\rangle + |HVH\rangle + |VHH\rangle\right) \qquad (2.4.2)$$

$$|\tilde{W}_3\rangle = \frac{1}{\sqrt{3}}\left(|VVH\rangle + |VHV\rangle + |HVV\rangle\right)$$

The authors demonstrated the "rotational invariance" by performing a full set of measurements in three complementary polarization bases, viz. (H,V), (L,R) and (D,A), where

$$|L(R)\rangle = [|H\rangle \pm i\,|V\rangle]/\sqrt{2},$$
$$|D(R)\rangle = [|H\rangle \pm |V\rangle]/\sqrt{2}. \qquad (2.4.3)$$

The patterns obtained are almost indistinguishable. The overlap between experimentally observed and theoretically expected results was 88%. This is a significant progress toward the quantum information manipulation because 6 photons is the minimum needed to encode an arbitrary logical state of two quantum bits immune to the collective noise. The difference in the orientations of polarization in the definition of V or H between the sender and the receiver of the information on two continents amounts only to a unitary transformation that acts on all the constituents identically, and is therefore irrelevant.

We know that with each pump pulse the parametric down conversion process a random number of pairs of photons is produced. The problem of choosing a 6-photon entangled singlet (and nondestructively!) for practical use is still far from being solved.

2.5 Epilogue

The applications of quantum mechanics to quantum information and quantum computing have already created a new discipline with a spurt of theoretical and experimental activities. As a feedback, the problems posed in the new field have also advanced the understanding of some basic concepts in quantum mechanics, especially on the entangled states. In the above discussions, only bipartite pure entangled states are concerned. In the following, we would like to mention some of the new developments pushed forward by the attempts to realize quantum computers.

One of the main problems in this direction is the vulnerability of quantum computing to all kinds of disturbance leading to decoherence of a superimposed state, to deterioration of entanglement, to errors during transmission and operations of qubits, etc. The classical computation is also prone to making mistakes, and the cure is the error correcting code: making back-up copies of the information bits which can be used as the control. The same cannot be adopted in quantum computing because of the quantum no-cloning theorem. The quantum correction code [26] requires redundancy to enable data to be transmitted through a noisy channel. The code encodes the input qubit into an entangled state of 5 qubits,[5] in such a way that if any is corrupted in the channel, the decoder can restore the first qubit to its original state while funneling the effect of the error into the remaining four qubits which are subsequently discarded. The quantum fault tolerating computation (reviewed in Ref. [27]) makes use of the quantum error-correcting code to spread to logical state $|\psi_L\rangle$ being stored or processed over a number of qubits carried by a bundle of parallel wires. The bundle passes periodically restorative gate arrays in which clean ancillas qubits are provided, correcting the errors by funneling through into the

---

[5]Let $\psi = a|0\rangle + b|1\rangle$. The cloning $\psi \to \psi \otimes \psi \otimes \psi$ is forbidden by quantum mechanics, but $\psi = a|0\rangle + b|1\rangle \to a|000\rangle + b|111\rangle \neq \psi \otimes \psi \otimes \psi$ is allowed. This is the generalization of the classical "triple repetition". The quantum error correction needs 5 qubits instead of 3 because superposition of Boolean states is to be protected.

ancillas which are subsequently discarded. Errors can happen in the restorative operation too, but they can be corrected in the next restorative array. Multiparticle entangled states are involved in the error correction. In Section 1.10.4 the 3-photon entangled states [28] are used to test the "local realistic theory" which are examples of one of the 8 kinds of maximally entangled Greenberger-Horne-Zeilinger states [29]. Four-photon entanglement was obtained experimentally by the same group [30]. An experimental demonstration of five photon entanglement was achieved by the group led by J.-W. Pan at the University of Science and Technology of China in Hefei [41]. The manipulation of five or more entangled particles is required for the universal error correction and the 'open-destination' teleportation in which the teleported state can be read out at any of the N particles, by a projection measurement on the remaining particles. Four entangled trapped ions was realized by the NIST group [31]. Decoherence during processing and transmission leads to mixed sate entanglement [32]. To recover maximal entanglement a purification (distillation) process is necessary [33]. This was accomplished experimentally by the Innsbruck group [34]. The physics of many-particle entanglement and mixed-state entanglement is extremely rich and challenging, and there remain many open questions pending further research.

In order to preserve coherence a scheme can be adopted to pairing each qubit with an ancilla qubit and encoding them into a coherence preserving state [35], and more generally by working in decoherence-free subspaces [36], which was realized experimentally [37].

Experimental research in various disciplines of physics is stimulated by the enormous progress in theory. In 1995 C.H. Bennett and D.P. DiVincenzo published a paper entitled "Quantum Computing–Towards an Engineering Era" [38]. It was a time when skeptics bet on that the sun would burn up before a 500-digit number is factorized by a quantum computer. The odds in favor of the quantum computer has been increased, but toward realizing a quantum computer there is a long way to go. D.P. DiVincenzo and D. Loss [39] gave a list of milestones which must be achieved to reach the final goal. For the latest experimental results the readers can refer to the monograph [40].

References

[1] P. Benioff, *J. Stat. Phys.* 22, 563, 1980; 29, 515, 1882; *Phys. Rev. Lett.* 48, 1581, 1982.

[2] R. Feynman, *Int. J. Theor. Phys.* 21, 467, 1982; *Optics News* 11, 11, 1985.

[3] For example R. Landauer, *Phys. Today* 44(5), 23, 1991; C. H. Bennett and D. P. DiVincenzo, *Nature* 404 247, 2000.

[4] R. Feynman, *Feynman Lectures on Computation*, ed. by A. J. G. Hey and R. W. Allen, Addison-Wesley, Reading, Massachusetts, 1996; M. A. Nielsen and I. L. Chuang *Quantum Computation and Quantum Information*, Cambridge University Press, 2000.

[5] J. Preskill, Lecture Notes on Quantum Information and Computation 1998. http://www.theory.caltech.edu/~preskill/ph229.

[6] W. Deutsch, *Proc. Roy. Soc. London Ser. A* 400, 97, 1985.

[7] I. L. Chuang *et al.*, *Nature* 393, 143, 1998.

[8] P. W. Shor in *Proceedings of the 35th Annual Symposium on the Foundations of Computer Science* 124–133 IEEE Computer Society Press, Los Alamitos, California 1994.

[9] L. K. Grover, *Phys. Rev. Lett.* 79, 325, 1997.

[10] I. L. Chuang, N. Gershenfeld and M. Kubinec, *Phys. Rev. Lett.* 80, 3408, 1998.

[11] C. Seife, *Science* 293, 2026, 2001.

[12] S. Wiesner, Los Alamos Preprint quant-ph/9603028, 1996.

[13] K. Mattle, H. Weinfurter, P. G. Kwiat and A. Zeilinger, *Phys. Rev. Lett.* 76, 4656, 1996.

[14] J. Schliemann, D. Loss, A. H. MacDonald, *Phys. Rev. B* 63, 085311, 2001.
J. Schliemann, J. I. Cirac, M. Kus, M. Lewenstein and D. Loss quant-ph/0012094.

[15] Y. S. Li, B. Zeng, X. S. Liu and G. L. Long, *Phys. Rev. A* 64, 016112, 2001.

[16] R. Paskauskas and L. You, quant-ph/0106117.

[17] D. S. Abrams and S. Lloyd, *Phys. Rev. Lett.* 79, 2586, 1997.
D. Bouwmeester, J.-W. Pan, K. Mattle, M. Eibl, H. Weinfurter, *Nature* 390, 575, 1997.

[18] C. H. Bennett and G. Brassard in: *Proc. International Conf. on Computer Systems and Signal Processing*, Bangalore 1984.

[19] A. K. Ekert, *Phys. Rev. Lett.*, 67, 661, 1991; C. H. Bennett, G. Brassard, N. D. Mermin, *Phys. Rev. Lett.* 68, 557, 1992.

[20] T. Jennewein, C. Simon, G. Weihs, H. Weinfurter and A. Zeililnger, *Phys. Rev. Lett.* 84, 4729, 2000.

[21] C. H. Bennett, G. Brassard, C. Crepean, R. Josza, A. Peres and W. K. Wooters, *Phys. Rev. Lett.* 70, 1895, 1993.

[22] D. Bouwmeester, J.-W. Pan, K. Mattle, M. Eibl, H. Weinfurter, *Nature* 390, 575, 1997.

[23] Y. H. Kim, S. P. Kulik and Y. Shih, *Phys. Rev. Lett.* 86, 1370, 2001.

[24] L. Vaidman, *Phys. Rev. A* 49, 1473, 1994.

[25] A. Furusawa, J. L. Sφrenson, S. L. Braunstein, C. A. Fuchs, H. J. Kimble, E. S. Polzik, *Science* 282, 706, 1998.

[26] P. Shor, *Phys. Rev. A* 52 2493, 1995.

[27] J. Preskill, *Proc. Roy. Soc. London A* 454, 385, 1998.

[28] D. Bouwmeester, J. W. Pan, M. Daniell, H. Weinfurter and A. Zeilinger, *Phys. Rev. Lett.* 82, 1375, 1999.

[29] D. M. Greenberger, M. Horne and A. Zeilinger in *Bell's Theorem, Quantum Theory and Conceptions of the Universe*, ed. by M. Kafatos, Kluwer, Dordrecht, 1989.

[30] J. W. Pan, M. Daniell, G. Gasparoni, G. Weihs, A. Zeilinger, *Phys. Rev. Lett.* 86, 4435, 2001.

[31] C. A. Sackett *et al.*, *Nature* 404, 256, 2000.

[32] C. H. Bennett, D. P. DiVincenzo, J. Smolin, W. K. Wooters, *Phys. Rev. A* 54, 3824, 1996.

[33] C. H. Bennett *et al.*, *Phys. Rev. Lett.* 76, 722, 1996; D. Deutsch *et al.*, *Phys. Rev. Lett.*, 77, 2818, 1996.

[34] J. W. Pan, C. Simon, C. Bruckner and A. Zeilinger, *Nature* 410, 1067, 2001.

[35] L.-M. Duan and G.-C. Guo, *Phys. Rev. Lett.* 79, 1953, 1997.

[36] D. A. Lidar, I. L. Chuang and K. B. Whaley, *Phys. Rev. Lett.* 81, 2594, 1998.

[37] P. G. Kwiat, A. J. Berlund, J. b. Altepeter and A. G. White, *Science* 290, 498, 2000.

[38] C. H. Bennett and D. P. DiVincenzo, *Nature* 377, 389, 1995.

[39] C. H. Bennett and D. P. DiVincenzo, *Superlatt. Microstruct.* 23, 419, 1998.

[40] D. Bouwmeester, A. Ekert, A. Zeilinger (eds.) *The Physics of Quantum Information*, Berlin, Springer, 2000.

[41] Z. Zhao *et al.*, *Nature*, 430, 54, 2004.

[42] R. Vrsin *et al.*, *Nature* 430, 849, 2004.

[43] M. Bourennane *et al.*, *Phys. Rev. Lett.* 92, 087902, 2004.

[44] M. Radmark *et al.*, *Phys. Rev. A* 80, 040302, 2009; *Phys. Rev. Lett.* 103, 150501, 2009.

Chapter 3

Geometrical Phases in Quantum Mechanics

Consider an infinitely long tube of flux, with all the flux contained within the tube. The magnetic field strength is zero outside. Do the electrons moving outside feel the existence of the tube of flux? The first reaction might be the answer no, because in the space in which the electrons move there is no magnetic field. A second thought may arise when one becomes aware that it is the 4-potential that enters the Schrödinger equation, rather than the field strengths $\boldsymbol{B}$ and $\boldsymbol{E}$. Although $\boldsymbol{B} = 0$ outside the flux tube, the vector potential is non-vanishing there. The study of this problem led to the well-known Aharonov-Bohm effect. It turns out that a localized electron state does not feel the existence of the flux tube because the field vanishes *locally*, while an extended electron state with a wave function which is finite around the flux tube does feel its *global* effect. This conclusion was a surprise to many physicists, some of them could not accept it even after long debate. This study also led to the well-known conclusion that in classical electrodynamics of charged particles the physics is determined by the field strengths, but for charged particles abiding the quantum mechanics their motion in electromagnetic field is determined by the *non-integrable phase factor* introduced by C. N. Yang and T. T. Wu. In this case the field strengths are insufficient, they underdetermine physics, while the potentials overdetermine physics. The above developments are introduced in Sections 3.1–3.3.

The Aharonov-Bohm effect can be interpreted in terms of concepts in differential geometry such as parallel transport, connexion and so on, which we introduce in Section 3.4, preparing the reader for yet another surprise — the Berry phase. The phase of a wave function in quantum mechanics is a subtle concept. Consider the solution of time-dependent Schrödinger equation, with a potential changing slowly with time. When the potential evolves according to a prescribed periodic function of time and returns to initial form, the problems become non-trivial. Does the wave function return to its initial form? No. A time dependent phase appears. Can the wave function be re-defined, absorbing the phase factor into it? A careful study should be made before answering this question. The outcome — the Berry phase and related topics is the subject of Sections 3.5–3.7. Berry's phase has far-reaching significance not only in quantum mechanics, but also in many other branches of physics.

3.1 Aharonov-Bohm Effect

The problem of the motion of a charged particle, for instance an electron, in given electromagnetic field was solved in the early days of the development of quantum mechanics. Classical examples include the motion of an electron in the Coulomb field — the problem of hydrogen-like atoms, and the motion of an electron in the uniform magnetic field — the Landau level problem. The Hamiltonian of an electron in the electromagnetic field is

$$H = \frac{1}{2m} \left[-i\hbar\nabla - \frac{e}{c} \boldsymbol{A}\left(x\right) \right]^2 + eA_0\left(x\right), \qquad (3.1.1)$$

where $(A_0, \boldsymbol{A})$ is the 4-potential (scalar and vector potentials) of the electromagnetic field. The relations between the field strengths and the potentials are

$$\boldsymbol{B} = \nabla \times \boldsymbol{A}, \quad \boldsymbol{E} = -\nabla A_0 - \frac{1}{c}\frac{\partial \boldsymbol{A}}{\partial t} \qquad (3.1.2)$$

Under the gauge transformation of the potential $A_\mu(A_0, \boldsymbol{A})$,

$$A_\mu \to A_\mu + \partial_\mu \Lambda \qquad (3.1.3)$$

i.e.,

$$A_0 \rightarrow A_0 - \frac{1}{c}\frac{\partial \Lambda}{\partial t},$$

$$\boldsymbol{A} \rightarrow \boldsymbol{A} + \nabla \Lambda,$$

the field strengths $\boldsymbol{B}$ and $\boldsymbol{E}$ are invariants. The function Λ characterizing the transformation is an arbitrary function of the space-time coordinates. Because of the gauge freedom the correspondence between potentials and field strengths is not one-to-one, but many-to-one. Since in classical electrodynamics the field strengths determine the physics of the fields, including the motion of a charged particle in it, the potentials are considered as useful concepts, but they belong to the category of "derived concepts". In quantum mechanics the potentials, not the field strengths, appear in Schrödinger equation. Under the gauge transformation of the potential (3.1.3) the wave function must transform correspondingly

$$\psi(x) \rightarrow \psi'(x) = \psi(x)\,\mathrm{e}^{(\mathrm{i}e/\hbar c)\Lambda(x)} \tag{3.1.4}$$

in order that the Schrödinger equation

$$\mathrm{i}\hbar\frac{\partial \psi}{\partial t} = H\psi$$

remains invariant. In the gauge transformation (3.1.3), (3.1.4), Λ is an arbitrary function of space-time coordinates. Such transformation is called a *local* gauge transformation.

Let the eigenfunction of the operator H be $\psi(\boldsymbol{x})$, and the eigenfunction of the operator $H_0 = \frac{1}{2m}(-\mathrm{i}\hbar\nabla)^2 + eA_0$ be $\psi_0(\boldsymbol{x})$, then $\psi(\boldsymbol{x})$ is related to $\psi_0(\boldsymbol{x})$ by the following equation:

$$\psi(\boldsymbol{x}) = \psi_0(\boldsymbol{x})\exp\left[(\mathrm{i}e/\hbar c)\int^{\boldsymbol{x}}\boldsymbol{A}(\boldsymbol{x}')\cdot\mathrm{d}\boldsymbol{x}'\right]. \tag{3.1.5}$$

This relation can be verified by substituting it into the equation $H\psi = E\psi$. The upper limit in the integral in (3.1.5) is $\boldsymbol{x}$ while any point $\boldsymbol{x}_0$ in space can be chosen as the lower limit, which is not explicitly written. In general the line integral from $\boldsymbol{x}_0$ to $\boldsymbol{x}$ is *path*

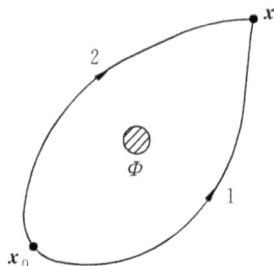

Fig. 3.1   The path dependence of the non-integrable phase factor.

dependent. Figure 3.1 shows two paths 1 and 2 from $\boldsymbol{x}_0$ to $\boldsymbol{x}$ on a plane through which a magnetic flux Φ passes. Taking the difference of the line integral along different paths 1 and 2, we obtain

$$\int_{\boldsymbol{x}_0(1)}^{\boldsymbol{x}} \boldsymbol{A}(\boldsymbol{x}') \cdot \mathrm{d}\boldsymbol{x}' - \int_{\boldsymbol{x}_0(2)}^{\boldsymbol{x}} \boldsymbol{A}(\boldsymbol{x}') \cdot \mathrm{d}\boldsymbol{x}' = \left(\int_{\boldsymbol{x}_0(1)}^{\boldsymbol{x}} + \int_{\boldsymbol{x}(2)}^{\boldsymbol{x}_0} \right) \boldsymbol{A}(\boldsymbol{x}') \cdot \mathrm{d}\boldsymbol{x}'$$

$$= \oint \boldsymbol{A}(\boldsymbol{x}') \cdot \mathrm{d}\boldsymbol{x}'.$$

By using Stokes theorem it is transformed into

$$\oint \boldsymbol{A}(\boldsymbol{x}') \cdot \mathrm{d}\boldsymbol{x}' = \int_S \nabla \times \boldsymbol{A} \cdot \mathrm{d}\boldsymbol{S} = \int_S \boldsymbol{B} \cdot \mathrm{d}\boldsymbol{S} = \Phi, \qquad (3.1.6)$$

where S is the area bounded by the closed paths with the direction of the surface element $\mathrm{d}\boldsymbol{S}$ perpendicular to the plane; $\boldsymbol{B}$ is the field strength determined by the vector potential $\boldsymbol{A}$, namely, $\boldsymbol{B} = \nabla \times \boldsymbol{A}$. The presence of the flux Φ makes the space multiply connected. Since the integral depends on path, the phase factor in (3.1.5) cannot be written as a single-valued function. It is called a non-integrable phase factor.

In quantum mechanics the overall phase of a wave function does not enter the expression for any observable, and hence the phase factor is arbitrary and can be set equal to 1. But if the wave function is a superposition of two parts, viz.

$$\psi = \psi_1 + \psi_2,$$

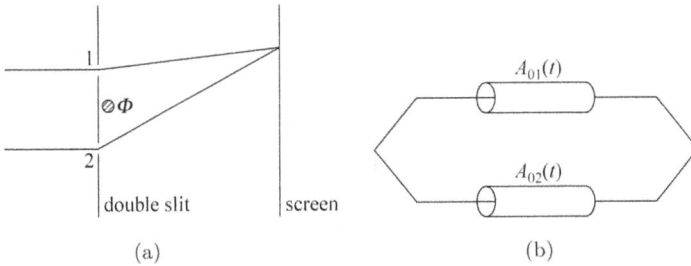

Fig. 3.2 Aharonov-Bohm effect (a) Electron wave passes through the space with vector potential; (b) Electron wave passes through conducting cylinders with different scalar potential.

then the relative phase between ψ_1 and ψ_2 is a very important quantity. It determines the interference between ψ_1 and ψ_2.

Y. Aharonov and D. Bohm [1, 2] published in 1959 a paper entitled "Significance of electromagnetic potentials in quantum theory". Consider the double slit experiment of an electron shown in Fig. 3.2(a). Immediately behind the plane of slits there is a very fine flux tube, the flux in it is Φ. This can be realized with a very thin and long solenoid, with the magnetic field strength $\boldsymbol{B}$ completely confined within it. Outside the solenoid $\boldsymbol{B} = 0$ everywhere. Can a change in the flux in the tube influence the interference pattern on the screen? Since the electron does not experience any force from the magnetic field $\boldsymbol{B}$, it seems that the interference pattern should not be influenced by the change of Φ. We try to analyze the situation strictly according to quantum mechanics. The wave function of the electron behind the plane of slits under the condition $\Phi = 0$ is

$$\psi^{(0)}(\boldsymbol{x}) = \psi_1^{(0)}(\boldsymbol{x}) + \psi_2^{(0)}(\boldsymbol{x}), \qquad (3.1.7)$$

and in the presence of the flux Φ is

$$\psi(\boldsymbol{x}) = \exp\left[(\mathrm{i}e/\hbar c)\int_{(1)}^{\boldsymbol{x}} \boldsymbol{A}(\boldsymbol{x}')\right]\psi_1^0(\boldsymbol{x})$$

$$+ \exp\left[(\mathrm{i}e/\hbar c)\int_{(2)}^{\boldsymbol{x}} \boldsymbol{A}(\boldsymbol{x}')\cdot \mathrm{d}\boldsymbol{x}'\right]\psi_2^{(0)}(\boldsymbol{x}). \qquad (3.1.8)$$

In this equation the path of the line integrals passing through slits 1 and 2 are denoted by 1 and 2 respectively. Neglecting an overall phase factor in Eq. (3.1.8) we get

$$\psi(\boldsymbol{x}) = \psi_1^{(0)}(\boldsymbol{x}) + \exp\left[(ie/\hbar c)\oint A(\boldsymbol{x}')\cdot d\boldsymbol{x}'\right]\psi_2^{(0)}(\boldsymbol{x})$$

$$= \psi_1^0(\boldsymbol{x}) + e^{(ie/\hbar c)\,\Phi}\psi_2^{(0)}(\boldsymbol{x}).$$

Compared with the relative phase in (3.1.7) the relative phase of the two coherent waves in the above equation has changed by $\frac{e}{\hbar c}\Phi$, which is called the Aharonov-Bohm phase S_{AB}

$$S_{AB} = \frac{e}{\hbar c}\Phi. \tag{3.1.9}$$

When Φ changes the interference fringes will shift, and the fringes change by one period when $\Delta\Phi = \frac{\hbar c}{e}2\pi = \frac{hc}{e}$, and the original pattern is restored.

In an accompanying situation, consider the separated electron wave beams enter the cylinders made of ideal conductors with scalar potentials $A_{01}(t)$ and $A_{02}(t)$ shown in Fig. 3.2(b). The electron does not experience any electric field strength, while the Hamiltonian with (without) the scalar potential H (H_0) are given respectively by

$$\left.\begin{array}{l} H = -\dfrac{\hbar^2}{2m}\nabla^2 + eA_0\left(\boldsymbol{x},t\right) \\[3mm] H_0 = -\dfrac{\hbar^2}{2m}\nabla^2 \end{array}\right\}. \tag{3.1.10}$$

The solutions to the respective Schrödinger equations are related by

$$\psi(\boldsymbol{x},t) = \psi(\boldsymbol{x},t)\exp\left[(-ie/\hbar c)\int^t A_0(\boldsymbol{x},t')dt'\right]. \tag{3.1.11}$$

The wave function behind the cylinders in Fig. 3.1.2(b) has the coherently superimposed parts

$$\psi(\boldsymbol{x},t) = \psi_1^0(\boldsymbol{x},t)$$

$$+ \exp\left[(-ie/\hbar c)\int^t [A_{02}(\boldsymbol{x},t') - A_{01}(\boldsymbol{x},t')]dt'\right]\psi_2^{(0)}(\boldsymbol{x},t). \tag{3.1.12}$$

It follows that the shift of the interference pattern should be observed depending on the change of A_{01} and A_{02}. The shift of interference pattern due to the change of the electromagnetic potential $(A_0, \boldsymbol{A})$ in a field free space through which the electron passes is called the Aharonov-Bohm effect.

The relative phase factor in (3.1.9) is gauge invariant, since Φ is determined by $\boldsymbol{B}$ (within the flux tube), and $\boldsymbol{B}$ is gauge invariant. The relative phase factor in (3.1.12) is also gauge invariant, since both A_{01} and A_{02} change by $\frac{1}{c}\frac{\partial \Lambda}{\partial t}$, and their difference does not change. The two cases considered above can be united to one non-integrable phase factor

$$\exp\left[(ie/\hbar c)\int^x A_\mu dx^\mu\right]$$

$$= \exp\left[(ie/\hbar c)\left(\int^x \boldsymbol{A}(\boldsymbol{x}')\cdot d\boldsymbol{x}' - \int^x A_0\left(t'\right)dt'\right)\right]. \quad (3.1.13)$$

The significance of Aharonov-Bohm effect is far-reaching. It demonstrates that in the quantum theory, for the electromagnetic phenomena, the field strengths underdetermine the physics, since the interference pattern shifts when Φ is varying, while the field strength $\boldsymbol{B}$ in the space through which the electron passed remains vanishing: they are not sufficient. Furthermore, for the electromagnetic phenomena, the potentials overdetermine, because two sets of potentials connected with each other by a gauge transformation give rise to the same phase factor: they are more than sufficient. The physics of electromagnetic phenomena with charged particles governed by the quantum mechanics is *completely determined* by the *non-integrable phase factor* $\exp[\frac{ie}{\hbar c}\int^x A_\mu dx^\mu]$ which contains the necessary and sufficient description of the phenomena. This important principle is formulated in 1975 by C. N. Yang and T. T. Wu [3].

Aharonov-Bohm effect is based on the fundamental principles of quantum mechanics without new principle or assumption introduced, but it came beyond the expectation of many physicists. R. Feynman wrote [4]: "It is interesting that something like this can be around for thirty years but, because of certain prejudices of what is and is not significant, continues to be ignored." In the Lagrangian and Hamiltonian of classical electrodynamics of a charged particle the potentials $(A_0, \boldsymbol{A})$ appear. When the equation of motion for the particle is derived from the Lagrangian or Hamiltonian, the potentials are replaced by the field strengths. Since the Schrödinger equation is determined by the Hamiltonian, the potentials remain there. There were attempts to replace them by the field strengths. Such attempts failed, and this failure turns out to be deeply rooted in the difference between classical and quantum mechanics.

The flux quantization is in a certain sense similar to the Aharonov-Bohm effect. It was discovered by Deaver and Fairbank, and also by Doll and Näbauer that the flux through a hollow superconducting cylinder is quantized to be an integral multiple of the flux quantum $\frac{hc}{2e}$. N. Byers and C. N. Yang [5] pointed out that this is a consequence of the formation of Cooper pairs in the superconductor. Let the wave function of the Cooper pair be ψ[1]

$$\psi = \sqrt{\rho}e^{iS}, \tag{3.1.14}$$

where S is a real function. Assumed that ρ is constant in the bulk superconductor shown in Fig. 3.3. The wave function ψ is an eigenfunction of the canonical momentum operator $\hat{\boldsymbol{p}} = -i\hbar\nabla$:

$$\hat{\boldsymbol{p}}\psi = \hbar\sqrt{\rho}\nabla Se^{iS} = \hbar\nabla S\psi, \tag{3.1.15}$$

where the eigenvalue $\hbar\nabla S$ is the value of the conjugate momentum of the Cooper pair. The kinetic momentum of the Cooper pair $2m\boldsymbol{v}$

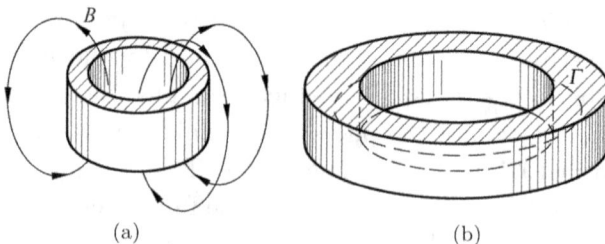

(a)   (b)

Fig. 3.3   Quantization of flux through a superconducting cylinder (a) Flux through a superconducting cylinder; (b) The closed contour for integration.

---

[1]Please refer to Sections 6.1 and 9.3 for details.

(m is the electronic mass, $\boldsymbol{v}$ is the velocity, and the mass of the Cooper pair is $2m$) is then

$$2m\boldsymbol{v} = \hbar\nabla S + 2\frac{e}{c}\boldsymbol{A}, \qquad (3.1.16)$$

where e is the absolute value of the electronic charge. Choose a closed contour Γ well within the bulk superconductor (Fig. 3.3(b)) for the integral of (3.1.16), we get

$$2m\oint \boldsymbol{v}\cdot\mathrm{d}\boldsymbol{s} = \hbar\oint_\Gamma \nabla S\cdot\mathrm{d}\boldsymbol{s} + 2\frac{e}{c}\oint_\Gamma \boldsymbol{A}\cdot\mathrm{d}\boldsymbol{s}. \qquad (3.1.17)$$

The superconducting current exists only on the surfaces (inner and outer) of the cylinder, and the penetration length is much smaller than the thickness of the cylinder. Hence $\boldsymbol{v} = 0$ on Γ. The wave function of the Cooper pair is single-valued, therefore

$$\oint_\Gamma \nabla S\cdot\mathrm{d}\boldsymbol{s} = 2\pi n, \qquad (3.1.18)$$

i.e., we begin with a point on Γ and make a complete winding along Γ back to its original position, and the phase can only change by $2\pi n$, with n being an integer or 0. Equation (3.1.17) becomes

$$\oint_\Gamma \boldsymbol{A}\cdot\mathrm{d}\boldsymbol{s} = \Phi = \frac{\hbar c}{2e}2n\pi = \frac{hc}{2e}n. \qquad (3.1.19)$$

where Φ is the magnetic flux through the hollow cylinder. It is an integral multiple of the flux quantum $\frac{hc}{2e}$. Due to the Meissner effect there is no magnetic field within the superconductor. The Cooper pairs in the superconductor do not experience the action of the magnetic field strength, but the single-valuedness of the wave function does have an effect on the global properties of the electromagnetic field, viz. the amount of magnetic flux passing through the hollow cylinder is quantized.

3.2 Experimental Verifications of Aharonov-Bohm Effect

To many physicists Aharonov-Bohm effect was a surprise. Some of them did not want to accept it. The first experimental test

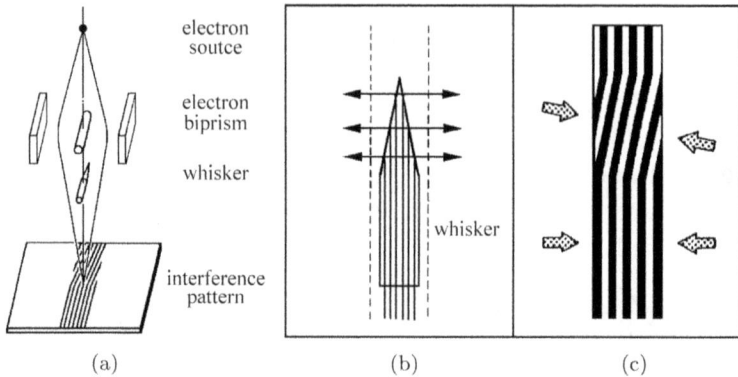

Fig. 3.4   Chambers experiment (a) Electron-optical system; (b) Magnetic lines of force around the whisker; (c) Interference fringes. Reprinted with permission from R. G. Chambers, *Phys. Rev. Lett.* 5, 3, 1960. Copyright (2017) by the American Physical Society.

was carried out by R. G. Chambers [6] in 1960, a scheme of this experiment is shown in Fig. 3.4(a). Electron beam from a point source is incident on an electron biprism consisting of a central fine filament with positive potential and a pair of grounded electrodes on either side. Electron waves pass by both sides of the filament are attracted towards the filament. They converge, overlap on the lower plane to form an interference pattern, with fringes parallel to the filament. When a thin magnetized iron whisker of diameter $\sim$1 μm is put beneath the filament, the fringes are tilted (Fig. 3.4(c)) where the whisker tapers, elsewhere the fringes remain parallel to the filament. The magnetic lines of force around the whisker are shown in Fig. 3.4(b). Without the whisker the electron beams converge in a plane perpendicular to the filament of the biprism. Because of the magnetic flux outside the conical tip of the whisker the electron waves are deflected slightly in the opposite axial directions (arrows in Fig. 3.4(c)) so that the plane on which the beams merge is rotated by a small angle. The tilted fringes connect the two parallel sets of fringes, revealing the shift of fringes due to the flux within the whisker. This experiment was considered not very convincing, because there are magnetic flux coming out of the whisker.

The experiment by Möllenstedt and Bayh in 1962 [7] used also the electron biprism, and a long solenoid was placed in the midpoint

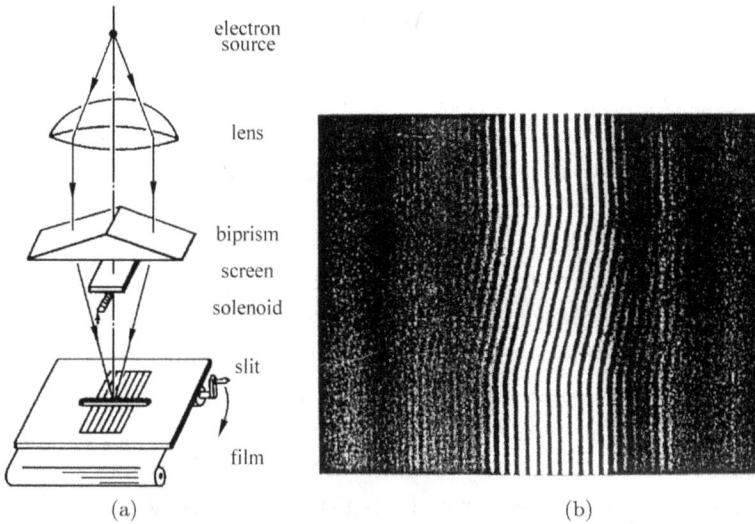

Fig. 3.5 Möllenstedt and Bayh experiment (a) Electron-optical system; (b) Interference pattern. From Ref. [7].

between the two beams of electron waves, as shown in Fig. 3.5(a). By increasing the current in the solenoid the fringes were made to move laterally. In order to record this behavior, only a part of the pattern was recorded on the film through a slit perpendicular to the fringe direction. The film was made to move while the current was increased continuously, giving a pattern shown in Fig. 3.5(b). While fringe tilting resulted from the electric field induced by the increasing magnetic flux, a fringe shift persisted even after the increase in magnetic flux stopped. This demonstrated the existence of the Aharonov-Bohm effect.

Ever since the prediction of Aharonov-Bohm effect various controversies have been going on.[2] There were attempts to re-formulate quantum electrodynamics without referring to the potentials. Others tried to interpret the shift of the interference pattern by a different physical mechanism. Some considered Aharonov-Bohm effect

---

[2]A historical approach to outlining the development is given by A. Tonomura in *The Aharonov-Bohm Effect* [1] p. 51. *et seq*

as mathematical concoction. As time progressed the experimental evidence of the Aharonov-Bohm effect was also questioned. The shift of the interference pattern was interpreted in various ways due to the leakage of magnetic flux from the solenoid, the interaction between the electron and the solenoid, and the possibility of the electron passing through the solenoid, etc.

By 1975 there were developments in theoretical physics (the electroweak unification using the non-Abelian gauge theory, the non-integrable phase factor) which enhanced the significance of potentials. A more precise and unambiguous experimental demonstration of the Aharonov-Bohm effect was called for. A. Tonomura began the experiments using small toroidal ferromagnet covered with superconducting layer to completely confine the flux by Meissner effect.[3] A new technique called electron holography [8] was adopted. Except for the use of electron wave instead of optical wave for producing the hologram, the principle is the same as the optical holography of D. Gabor. The "object" producing the hologram is nothing but the toroid magnet. The electron wave passing through the hole of the torus and that passing outside it develop a phase difference. All information concerning the overlapping waves are recorded on the hologram ready to be reconstructed in the next step. Then the optical image of the original object is reconstructed by illuminating the hologram by a light wave. Though the wave lengths of the two kinds of waves differ by a factor of 200,000, the method works because both the imaging (interference between the object and reference wave producing the hologram) and the reconstruction (diffraction on the hologram) are processes taking place in units of the wave lengths, i.e., they work for any wave length. The use of optical techniques helps to obtain new information inaccessible by means of electron microscopy, because of the convenience of optical components.

In 1986 Tonomura *et al.* [9] carried out an experiment which fully satisfied the required conditions for the experimental verification of the Aharonov-Bohm effect. The preparation of the toroidal magnets

---

[3]Proposed by C. N. Yang in 1983 on the International Conference on the Foundations of Quantum Mechanics, Tokyo.

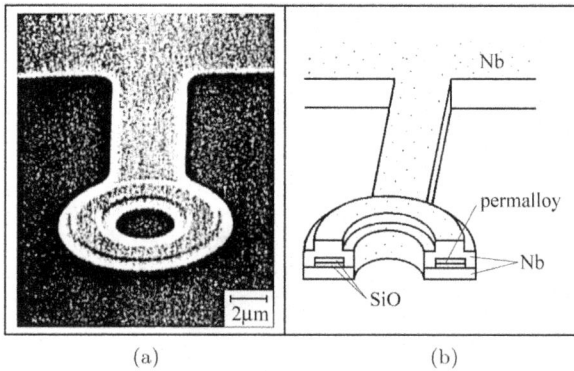

(a)   (b)

Fig. 3.6   Toroidal ferromagnet covered with niobium layer: (a) Scanning electron micrograph; (b) Cross-section. Reprinted with permission from Akira Tonomura, Nobuyuki Osakabe, Tsuyoshi Matsuda, Takeshi Kawasaki, and Junji Endo, *Phys. Rev. Lett.* 56, 792, 1986. Copyright (2017) by the American Physical Society.

was greatly facilitated by the new technique of microlithography and vacuum evaporation. A Permalloy film of thickness 200 Å is prepared by vacuum evaporation on a silicon wafer covered with a niobium film 2500 Å thick. After the toroid shape was cut out from the Permalloy film a 3000 Å niobium film was sputtered on its surface. Finally a copper film of 500 Å–2000 Å thick was evaporated on all of its surfaces, which prevented the penetration of any electron into the magnet. An estimate gave a value of 10^{-6} for the probability of penetration of an electron. In the subsequent experiments the thickness of copper layer was varied from 500 Å to 2000 Å without leading to any change of the interference pattern of electron waves. A scanning electron micrograph of the sample is shown in Fig. 3.6(a) and the cross-section of the magnet shown in Fig. 3.6(b). The leakage of flux from toroidal magnet was checked for each sample by holographic interference microscopy at the room temperature. Only toroids with a leakage $< hc/20e$ were selected for the experiments. Under the experimental conditions (Nb in superconducting state) the leakage should be much smaller. The thickness of the Nb film is 2500 Å, while the penetration depth of the magnetic flux is only 1100 Å.

The electron holograms were taken by using experimental setup shown in Fig. 3.7(a). The highly coherent electron source was

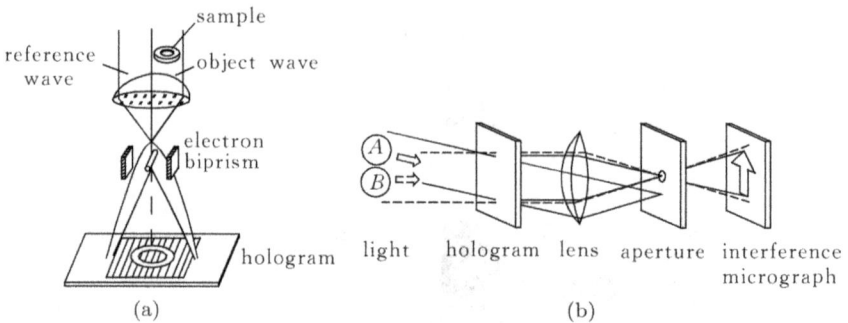

Fig. 3.7   (a) Schematic diagram for electron hologram formation; (b) Scheme of the optical reconstruction system. Reprinted with permission from Akira Tonomura, Nobuyuki Osakabe, Tsuyoshi Matsuda, Takeshi Kawasaki, and Junji Endo, *Phys. Rev. Lett.* 56, 792, 1986. Copyright (2017) by the American Physical Society.

provided by a 150 kV field emission electron microscope (de Broglie wave length 0.030 Å) with excellent collimation. Compared with thermal electron source the number of fringes increased to 3000 from 300. The toroidal sample is situated in one half of the specimen plane, the other half is for the reference wave. The object wave and the reference wave are brought together by the electron biprism to form an interference pattern. The pattern is enlarged 1000 times by electron lenses and recorded on film as a hologram. The phase shift caused by the toroid magnet is reconstructed by a He-Ne laser, shown schematically in Fig. 3.7(b). Interferograms obtained at the temperature 4.5 K are shown in Fig. 3.8. Although samples having various values of magnetic flux (i.e., values when prepared at a temperature higher than T_c) were measured, only two values of phase shifts were obtained, viz. 0 and π. The phase shift produced by a flux Φ is $\frac{e}{\hbar c}\Phi = \pi\frac{\Phi}{hc/2\pi}$. The flux in the toroid covered by a layer of superconductor is quantized (Section 3.1), $\Phi = n\frac{hc}{2\pi}$, and therefore the phase shift is $n\pi$, i.e., 0 or π.

At temperatures higher than T_c for Nb (9.2 K) the phase shift can vary continuously. Figure 3.9 shows a phase shift of 0.8π at $T = 15$ K. The interferogram of a sample taken at $T > T_c$ and $T < T_c$ is completely reversible. This experiment demonstrates Aharonov-Bohm effect and the quantization of flux very convincingly.

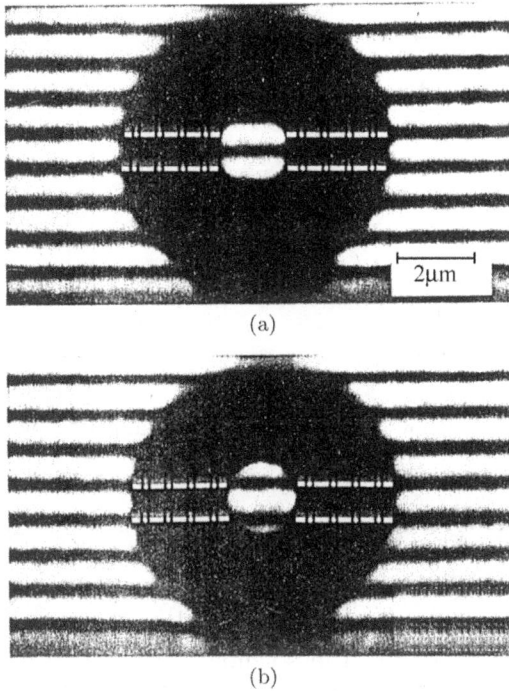

Fig. 3.8 Interferograms of toroidal ferromagnets covered with superconducting layers at 4.5 K (a) Phase shift = 0; (b) Phase shift = π. The dashed lines serve as vision guide. Reprinted with permission from Akira Tonomura, Nobuyuki Osakabe, Tsuyoshi Matsuda, Takeshi Kawasaki, and Junji Endo, *Phys. Rev. Lett.* 56, 792, 1986. Copyright (2017) by the American Physical Society.

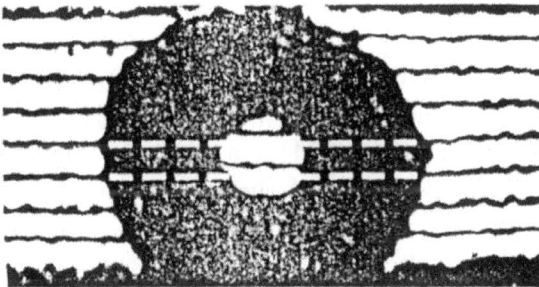

Fig. 3.9 Interferogram of a sample at 15 K showing a phase shift of 0.8π. Reprinted with permission from Akira Tonomura, Nobuyuki Osakabe, Tsuyoshi Matsuda, Takeshi Kawasaki, and Junji Endo, *Phys. Rev. Lett.* 56, 792, 1986. Copyright (2017) by the American Physical Society.

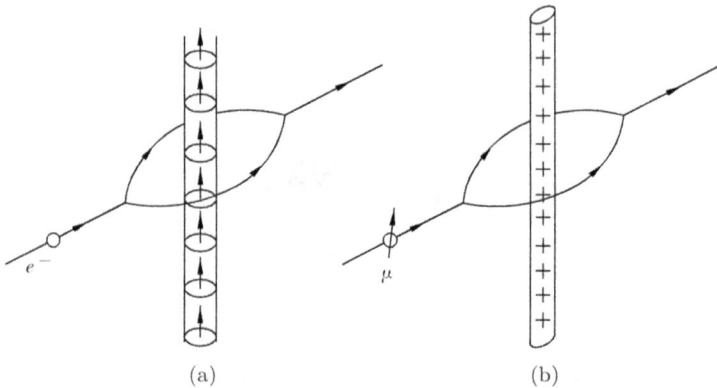

(a)   (b)

Fig. 3.10   (a) Aharonov-Bohm effect; (b) Its dual Aharonov-Casher effect.

3.3 Aharonov-Casher Effect

The long solenoid producing Aharonov-Bohm phase can alternatively be considered as many magnetic dipoles aligned along the axis. Electron waves encircling the solenoid from different sides acquire a phase difference (Fig. 3.10(a)). According to the duality of electromagnetism the stack of magnetic dipoles can be replaced by a uniform linear distribution of electric charge parallel to the axis, while the electron can be correspondingly replaced by a charge-neutral magnetic dipole (e.g., a neutron) with a magnetic moment parallel to the axis. Waves of the charge-neutral dipole is then to develop a phase difference by encircling the linear charge distribution from different sides (Fig. 3.10(b)). This was predicted by Y. Aharonov and A. Casher [10] in 1984. There is, however, a subtlety involved. A magnetic dipole moving with a velocity v in an electric field E experiences a magnetic field in a direction parallel to $v \times E$. In the present case the field of the linear charge is in the radial direction in planes perpendicular to the axis. Since v also lies in the plane, the magnetic field is parallel to the axis. A magnetic dipole lying in the same direction therefore experiences 0 net force. Therefore, the parallelism is intact. Consider a solenoid with mass M, position vector R, velocity V interacting with a charged particle with mass m, position vector r, and velocity v. The Lagrangian of the system is

$$L = \frac{1}{2}mv^2 + \frac{1}{2}MV^2 + \frac{e}{c}A\left(r - R\right) \cdot \left(v - V\right). \tag{3.3.1}$$

The equation of motion of the particle can be derived from this Lagrangian

$$m\dot{v}_j + \frac{e}{c}\frac{\partial}{\partial r_i}A_j\left(v_i - V_i\right) - \frac{e}{c}\frac{\partial}{\partial r_j}A_i(v_j - V_j) = 0,$$

i.e.,

$$m\dot{v} = \frac{e}{c}(v - V) \times (\nabla_r \times A(r - R)). \tag{3.3.2}$$

The equation of motion of the solenoid is similar, with $-M\dot{V}$ equal to the right hand side of (3.3.2), i.e.,

$$M\dot{V} + m\dot{v} = 0, \tag{3.3.3}$$

It follows that

$$mV + mv = \text{const.} \tag{3.3.4}$$

The Lagrangian and the equations of motion are Galilei invariant. From (3.3.1) the canonical momenta can be obtained

$$\left.\begin{aligned} p &= \frac{\partial L}{\partial v} = mv + \frac{e}{c}A \\ P &= \frac{\partial L}{\partial V} = MV - \frac{e}{c}A \end{aligned}\right\} \tag{3.3.5}$$

Equations (3.3.4), (3.3.5) give the conservation of momentum

$$p + P = mv + MV = \text{const.} \tag{3.3.6}$$

In the Lagrangian of the system (3.3.1) the interchange symmetry is manifest: either the set (m, r, v) or the set (M, R, V) can be ascribed to the charge (or the dipole), since $\boldsymbol{A}$ depends only on the relative position vector $\boldsymbol{r} - \boldsymbol{R}$. In both situations the direction of the axis remains the same. It implies that a magnetic moment moving in the field of a linear uniformly distributed charge does not experience a force, while the waves of the dipole passing by the different sides of the linear charge distribution develop a phase difference, viz. the Aharonov-Casher phase

$$S_{\text{AC}} = -\oint \frac{e}{\hbar c} \boldsymbol{A}(\boldsymbol{r} - \boldsymbol{R}) \cdot \mathrm{d}\boldsymbol{R} = \frac{e}{\hbar c} \Phi. \qquad (3.3.7)$$

Since the flux Φ is equal to the dipole moment μ divided by the length of the solenoid ξ, it follows that

$$S_{\text{AC}} = \frac{1}{\hbar c} \frac{e}{\xi} \mu = \frac{1}{\hbar c} \lambda \mu,$$

where λ is the linear density of charge. The magnetic moment of the neutron is $g\frac{e\hbar}{2Mc}$, consequently

$$S_{\text{AC}} = \frac{\lambda}{e} g 2\pi\alpha \frac{\hbar}{Mc}, \qquad (3.3.8)$$

where $\alpha = e^2/4\pi\hbar c$ is the fine structure constant. Making an order of magnitude estimate, we set $g = O(1)$, the neutron Compton wave length $\frac{\hbar}{Mc} = 2 \times 10^{-14}$ cm. In order to have a phase shift easily observable, we set $S_{\text{AC}} = \pi/2$. Thus the linear charge density required is extremely large:

$$\lambda \approx e/10^{-15} \ (\text{cm}^{-1}). \qquad (3.3.9)$$

In laboratory conditions the phase observed is rather small. The University of Melbourne and University of Missouri team [11] reported in 1989 their results. As shown in Fig. 3.11 thermal neutrons from the reactor enter a single crystal silicon neutron interferometer. The central electrode between the two arms is charged to 45 kV. For this setup the expected phase difference is 1.5 mrad. To achieve an acceptable statistics the data taking required several months. The final result was (2.19 ± 0.52) mrad.

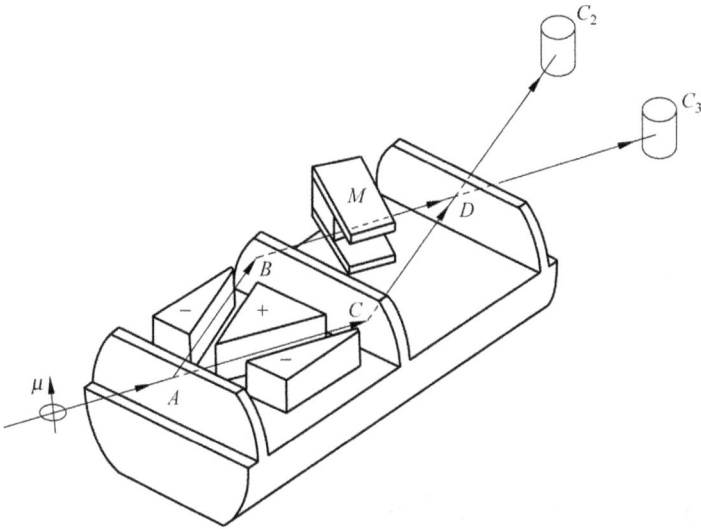

Fig. 3.11   A perspective view of the neutron beams in the interferometer.

3.4   Parallel Transport, Connexion, Curvature and Anholonomy

Concepts of differential geometry were first applied to physical problems by A. Einstein in his general theory of relativity. Encouraged by the intimate connection between gravity and geometry in the general theory of relativity, H. Weyl attempted to ascribe geometrical significance also to the electromagnetic field. He assumed that space-time has a scale which is different at different points. From a point x^μ to a neighboring point $x^\mu + dx^\mu$ the scale changes by a factor $1 + S_\mu dx^\mu$. A space-time function $f(x)$ changes from x^μ to $x^\mu + dx^\mu$ by

$$f(x) \rightarrow (f + \partial_\mu f dx^\mu)(1 + S_\mu dx^\mu) \approx f + (\partial_\mu + S_\mu) f dx^\mu. \quad (3.4.1)$$

Weyl attempted to interrelate the scale function S_μ to the electromagnetic potential A_μ. His attempt did not succeed. From quantum mechanics we know that the operator of kinetic momentum is $\left(-i\partial_\mu - \frac{e}{c} A_\mu\right)$. Comparing with (3.4.1) we see that S_μ corresponds to iA_μ. The electromagnetic potential does not provide a real scale

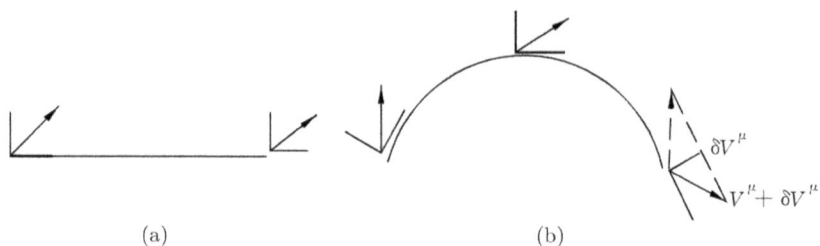

Fig. 3.12   Parallel transport (a) In a flat space; (b) In a curved space.

function but is connected to the non-integrable phase factor (3.1.13) through an imaginary factor i. On the basis of this comprehension Weyl established the *gauge* invariance of electromagnetism. He did not change the old terminology (gauge — a standard of *scale*) though. The term *gauge invariance* has been accepted ever since.

It is necessary to introduce the concept of parallel transport[4] when we deal with a curved space. In order to compare the vector field $V_\mu(x)$ with $V_\mu(x')$ at two different points, we have to transport V_μ from x to x' while keeping the transported vector parallel to the original direction. For such a transport no special definition is required in a flat space (Fig. 3.12(a)). Next we consider the case of transporting a vector along a plane curve on a curved surface. We can define the parallel transport of a vector by demanding the vector at any point on the curve to remain at a fixed angle to the tangent at that point (Fig. 3.12(b)). The vector transported parallel from x to $x' = x + dx$ becomes $V_\mu + \delta V_\mu$. δV_μ depends linearly on V^ν and dx^λ:

$$\delta V_\mu = \Gamma^\upsilon_{\nu\lambda} V_\nu dx^\lambda \tag{3.4.2}$$

Equation (3.4.2a) is the definition for $\Gamma^\nu_{\mu\lambda}$, the affine connexion, which is a function of coordinates. For flat surfaces $\Gamma^\nu_{\mu\lambda} = 0$. Since $V_\mu V^\mu$ is an invariant, we have $\delta(V_\mu V^\mu) = 0$, and consequently

$$\delta V^\mu = -\Gamma^\mu_{\nu\lambda} V^\nu dx^\lambda \tag{3.4.3}$$

---

[4]Parallel transport is sometimes referred to as parallel displacement in the literature.

Define the covariant differential of the vector field (denoted by DV_μ) between two points x and $x + dx$ by

$$DV_\mu = V_\mu\,(x + dx) - [V_\mu\,(x) + \delta V_\mu]$$
$$= (\partial_\lambda V_\mu - \Gamma^\nu_{\mu\lambda} V_\nu)dx_\lambda \qquad (3.4.4)$$

The quantity in parenthesis is called the covariant derivative. Equivalently we have

$$DV^\mu = V^\mu(x') - [V^\mu\,(x) + \delta V^\mu]$$
$$= (\partial_\lambda V^\mu + \Gamma^\mu_{\nu\lambda} V^\nu)dx^\lambda. \qquad (3.4.5)$$

Parallel transport along curves in a curved manifold[5] is significantly different from that along curves in a flat manifold in that it is *path dependent*. The vector obtained by transporting it from a point P to a point Q depends on the route taken between P and Q. In transporting a vector around a closed loop, the direction of the transported vector is in general different from the initial direction of the vector. By parallel transport of a vector along a closed curve the curvature can be defined. In Fig. 3.13 parallel transport of a vector along a closed contour $1 \to 2 \to 3 \to 4$ in flat space leads to the same vector to begin with without any change. On a spherical surface the

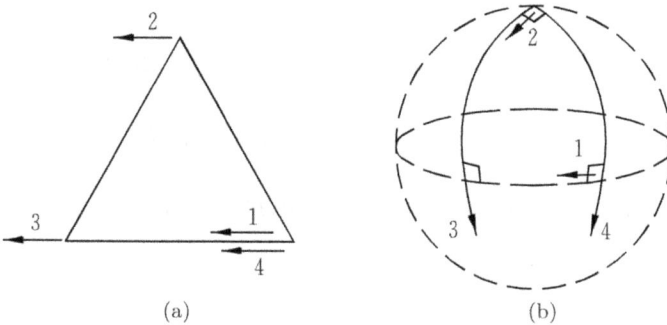

(a)   (b)

Fig. 3.13 The parallel transport of a vector field around a closed contour (a) In flat space; (b) On a spherical surface.

---

[5] Actually (3.4.3) and (3.4.4) are the general definitions of the parallel transport in a curved space in terms of the affine connexion Γ," the Christoffel symbol of the second kind" in differential geometry defined in connection with geodesics.

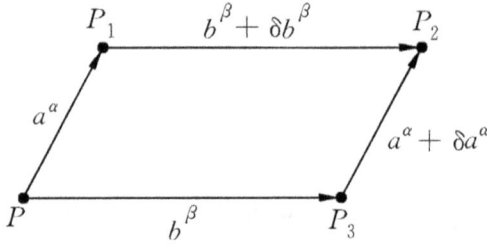

Fig. 3.14   The parallel transport of a vector field along different paths in a curved manifold.

parallel transport along closed contour $1 \to 2 \to 3 \to 4$ leads to a rotation $\pi/2$ of the vector. In general, in a curved manifold, the parallel transports of a vector V along PP_1P_2 and PP_3P_2 (as shown in Fig. 3.14) give different results. PP_1 is denoted by the vector a^α, PP_3 by b^β. P_1P_2 is the parallel transport of b, i.e., $b + \delta b$, where δb is

$$\delta b^\beta = -\Gamma^\beta_{\xi\eta} b^\xi dx^\eta. \tag{3.4.6}$$

P_3P_2 is the parallel transport of a, i.e., $a + \delta a$, where δa is

$$\delta a^\alpha = -\Gamma^\alpha_{\xi\eta} a^\xi dx^\eta. \tag{3.4.7}$$

The vector transported along PP_1P_2 acquires a change

$$\delta V_\mu = \left(\Gamma^\nu_{\mu\alpha} V_\nu\right)_P a^\alpha + \left(\Gamma^\nu_{\mu\beta} V_\nu\right)_{P_1} \left(b^\beta + \delta b^\beta\right), \tag{3.4.8}$$

while the vector transported along PP_3P_2 acquires a change

$$\delta V'_\mu = \left(\Gamma^\nu_{\mu\beta} V_\nu\right)_P b^\beta + \left(\Gamma^\nu_{\mu\alpha} V_\nu\right)_{P_3} \left(a^\alpha + \delta a^\alpha\right). \tag{3.4.9}$$

The function $\Gamma^\nu_{\mu\beta} V_\nu$ with values taken at P_1 and P_3 can be expressed in terms of the value taken at P by

$$\left.\begin{aligned}
\left(\Gamma^\nu_{\mu\beta} V_\nu\right)_{P_1} &= \left(\Gamma^\nu_{\mu\beta} + \partial_\alpha \Gamma^\nu_{\mu\beta} a^\alpha\right)\left(V_\nu + \Gamma^\sigma_{\nu\alpha} V_\sigma a^\alpha\right) \\
\left(\Gamma^\nu_{\mu\alpha} V_\nu\right)_{P_3} &= \left(\Gamma^\nu_{\mu\alpha} + \partial_\beta \Gamma^\nu_{\mu\alpha} b^\beta\right)\left(V_\nu + \Gamma^\sigma_{\nu\beta} V_\sigma b^\beta\right)
\end{aligned}\right\}. \tag{3.4.10}$$

Substituting (3.4.6), (3.4.7) and (3.4.10) in (3.4.8) and (3.4.9) and taking the difference, we obtain

$$\Delta V_\mu = \delta V_\mu - \delta V_\mu' = R^\nu_{\mu\alpha\beta} V_\nu a^\alpha b^\beta \qquad (3.4.11)$$

where

$$R^\nu_{\mu\alpha\beta} = \partial_\alpha \Gamma^\nu_{\mu\beta} - \partial_\beta \Gamma^\nu_{\mu\alpha} + \Gamma^\lambda_{\mu\beta}\Gamma^\nu_{\lambda\alpha} - \Gamma^\lambda_{\mu\alpha}\Gamma^\nu_{\lambda\beta}. \qquad (3.4.12)$$

Equation (3.4.11) demonstrates that the difference ΔV_μ of the changes of a vector transported parallel along different paths is proportional to the area enclosed by the paths $\sigma^{\alpha\beta} = a^\alpha b^\beta$ and to the transported vector V_μ , and the constant of proportionality is the curvature tensor $R^\nu_{\mu\alpha\beta}$ (3.4.12). In the case of Abelian gauge field (e.g., electromagnetism) the covariant derivative of the wave function of the fermion is

$$D_\mu \psi = \left(\partial_\mu + i\frac{e}{c}A_\mu\right)\Psi. \qquad (3.4.13)$$

For the non-Abelian gauge fields,[6] we have

$$D_\mu \psi = (\partial_\mu + igA_\mu)\psi, \qquad (3.4.14)$$

where ψ is a n-component wave function with internal symmetry. For $SU(2)$ symmetry it is a 2-component wave function. A_μ is correspondingly a $n \times n$ matrix. For $SU(2)$ symmetry it is $A_\mu = \frac{\tau_a}{2}A^a_\mu$, $(a = 1, 2, 3)$, where repeated indices need to be summed, and τ^a are the Pauli Matrices. The change due to parallel transport from x to $x + dx$ is

$$\delta\psi = igA_\mu\psi dx^\mu. \qquad (3.4.15)$$

Comparing it with (3.4.2), we see that the gauge potential is equivalent to the affine connexion and the change due to the parallel

---

[6]In Section 7.6 a brief introduction to non-Abelian gauge field will be given. It is sufficient to follow mathematical relations here.

transport from x to x' is

$$P\left(x',x\right)\psi = \exp\left[ig\int_x^{x'} A_\mu(y)dy^\mu\right]\psi. \qquad (3.4.16)$$

Since $A_\mu = \frac{\tau_a}{2} A_\mu^a$, and $\frac{\tau_a}{2}$ are the generators of the $SU(2)$ group, any path $x \to x'$ is a group element of $SU(2)$. $P(x'x)$ is the non-integrable phase factor, which is equivalent to the parallel transport. In non-Abelian gauge field theory, the gauge field strength and the gauge potential are related by

$$F_{\mu\nu} = \partial_\mu A_\nu - \partial_\nu A_\mu - [A_\mu, A_\nu]. \qquad (3.4.17)$$

Like A_μ, the field strengths $F_{\mu\nu}$ are also $n \times n$ matrices. Comparing (3.4.17) and (3.4.12), we see that the gauge field strengths is equivalent to the curvature. The discussion on the geometrical significance of the gauge fields concerns the fiber bundle theory. T. T. Wu and C. N. Yang [3] gave the correspondence between gauge field concepts with concepts in the fiber bundle theory.

The importance of introducing the covariant derivative can be understood from another viewpoint. The free fermion field has a Lagrangian density

$$\mathscr{L} = i\gamma^\mu \partial_\mu \psi - m\,\bar{\psi}\psi. \qquad (3.4.18)$$

While a global phase transformation $\psi \to e^{i\alpha}\psi$ leaves this Lagrangian density invariant, a local gauge transformation $\psi \to e^{i\alpha(x)}\psi$ does not keep it invariant. The way to keep the local gauge invariance of the fermion field is to introduce the gauge field A_μ interacting with the fermion field in a *unique* way.

$$\mathscr{L} = i\gamma^\mu \left(\partial_\mu + i\frac{e}{c}A_\mu\right)\psi - m\,\bar{\psi}\psi$$

$$= i\gamma^\mu D_\mu \psi - m\,\bar{\psi}\psi. \qquad 3.4.19$$

The way the gauge field is introduced provides the affine connexion for the spacetime. In the generalization to non-Abelian gauge theories for electroweak and strong interactions, the demand of local gauge invariance as a first principle unambiguously fix the interactions between the fermion fields (leptons and quarks) and the

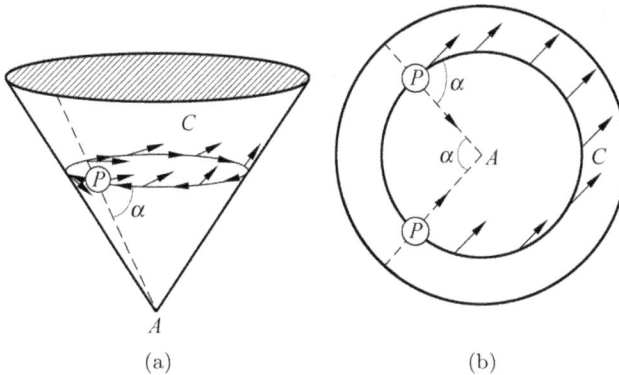

Fig. 3.15 Parallel transport of a vector on a cone (a) 3-dimensional view of the cone; (b) 2-dimensional view of the cut cone.[7]

corresponding gauge fields (the $W^{\pm}$ and Z bosons, the photon and the gluons). The gauge invariance plays the crucial role to guarantee the renormalizability of the theories. This solid framework has been verified with the highest precision of experimental physics.

It turns out that Aharonov-Bohm phase is an example of parallel transport. The electron is situated in a space of vanishing field strengths, while the gauge potential is different from zero. The space is flat, since its curvature (gauge field strengths) is zero. The space around the flux tube is equivalent to a conical surface shown in Fig. 3.15(a). The curvature is everywhere zero except at the apex, where it is ∞. It is the point where all the flux is concentrated (a singular point). Aharonov-Bohm phase factor is the parallel transport of a vector. The surface of a cone can be cut along a dashed line which passes through the apex and be unfolded and laid flat onto a plane (Fig. 3.15(b)). The resulting flat surface is pie-shaped, with a wedge angle α. The parallel transport around a circuit C on the surface of the cone which encloses the apex can be carried out by beginning with a point marked P on the dashed line. On the unfolded plane the direction of the vector remains unchanged, but after traversing through the circuit, a single particle on the cone

---

[7]Adopted from R. Chiao in *Quantum Coherence* edited by J. S. Anandan, Singapore, World Scientific, 1990.

becomes, after the cutting and unfolding, two points on the wedge (both denoted by P). We see that the vector after the circuit makes an angle α with its initial direction. This angle is called the angle of anholonomy and is equal to the apex angle of the cone, depending on the amount of flux enclosed in the flux tube. In case the circuit does not enclose the apex, the anholonomy angle is zero. The circuit can be deformed arbitrarily with the condition that it does not touch the singularity A without changing the value of the anholonomy angle. Hence the effect is topological.

Another anholonomy phenomenon is the Foucault pendulum.[8] Its direction of swing described by the unit vector $\hat{e}$ is always perpendicular to the local vertical $\hat{r}$ (the unit vector connecting the center of the earth to the local point). The condition for the parallel transport is the constraint that $\hat{e}$ does not rotate about $\hat{r}$. After a day $\hat{r}$ has completed a circuit C (a circle of latitude), but the final $\hat{e}_f$ does not return to its original direction $\hat{e}_i$, as shown in Fig. 3.16. This is characterized by "global change without local change". Absence of local change is the constraint that $\hat{e}$ remains perpendicular to $\hat{r}$

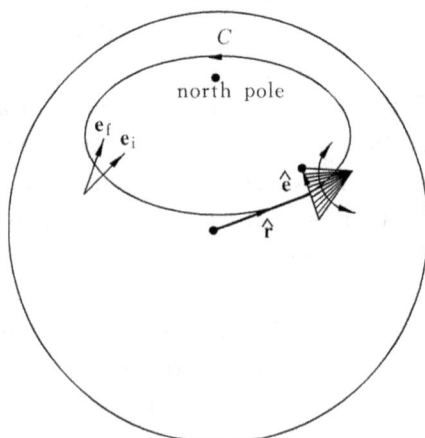

Fig. 3.16   Foucault pendulum: The parallel transport of vector $\hat{e}$ over the diurnal circuit C leads to the angle of anholonomy.

---

[8]For various aspects of the geometrical phases the readers can refer to an article written by M. V. Berry [12].

without rotating about it; the global change is the angle between $\hat{e}_i$ and $\hat{e}_f$, the angle of anholonomy. This angle is equal to the solid angle subtended by the circuit C at the center of the earth.

$$\Omega = 2\pi(1 - \cos\theta) = 2\pi(1 - \sin\phi),$$

where θ is the polar angle of the latitude circle, ϕ is the longitude. This angle of anholonomy is independent of the speed with which $\hat{r}$ traverses the circuit, and is consequently geometrical. But it depends on the magnitude of the solid angle, and is consequently not topological.

3.5 Berry's Phase

In a certain class of problems in quantum mechanics the observables can be divided into two sets. Observables in one set vary with time fast, while observables in the other set vary with time slowly. In solving such problems one can temporarily treat the slow variables as time dependent parameters of the system and solve the quantum mechanical problem of the fast variables. Then one releases the slow variables and treats them as operators and finds the solution to the entire problem using the solution to the fast observables. The quantum mechanical problem of the molecule is solved in this way by Born-Oppenheimer approximation. Let $\boldsymbol{P}, \boldsymbol{R}$ be the slow observables (e.g., the momentum and position of the nucleus), $\boldsymbol{p}, \boldsymbol{r}$ be the fast variables (e.g., the momentum and position of the electron). The Hamiltonian of the system is

$$H = \frac{\boldsymbol{P}^2}{2M} + \frac{\boldsymbol{p}^2}{2m} + V(\boldsymbol{R}, \boldsymbol{r}). \tag{3.5.1}$$

First we treat the slow observables $\boldsymbol{P}$ and $\boldsymbol{R}$ as time dependent c-number parameters. The Hamiltonian of the fast problem is

$$h = \frac{\boldsymbol{p}^2}{2m} + V(\boldsymbol{R}, \boldsymbol{r}), \tag{3.5.2}$$

assumed that the eigenproblem of the operator h has been solved,

$$h(\boldsymbol{p}, \boldsymbol{r}, \boldsymbol{R}) \, |m; \boldsymbol{R}\rangle = \varepsilon_m(\boldsymbol{R}) \, |m; \boldsymbol{R}\rangle, \tag{3.5.3}$$

where $|m; \boldsymbol{R}\rangle$ is an eigenstate of the fast problem, and m is a quantum number characterizing the state. We assume that the eigenvalue spectrum is discrete and non-degenerate, and states with different m form a complete set of orthonormal eigenfunctions. We begin to solve the time evolution problem of the wave function with time, with $\boldsymbol{R}$ treated as a slowly varying c-number parameter. The Schrödinger equation is

$$i\hbar\frac{\partial \psi}{\partial t} = h\psi. \tag{3.5.4}$$

We expand ψ in terms of $|m; \boldsymbol{R}\rangle$:

$$\psi = \sum_m a_m(t) \exp\left[-\frac{i}{\hbar}\int_0^t \varepsilon_m\left(t'\right) dt'\right] |m; \boldsymbol{R}\rangle, \tag{3.5.5}$$

where the exponential factor is the dynamical phase factor, and the change of ε_m with time is caused by the slow variation of $\boldsymbol{R}$ with time. Substituting (3.5.5) in (3.5.4) and using (3.5.3), multiplying the result from the left by $\langle k; \boldsymbol{R}|$, we obtain the time derivative of the expansion coefficient $a_k(t)$[9]:

$$\dot{a}_k(t) = -\sum_m a_m \left\langle k; \boldsymbol{R}\left|\frac{\partial}{\partial t}m; \boldsymbol{R}\right.\right\rangle \exp\left\{-\frac{i}{\hbar}\int_0^t \left[\varepsilon_m\left(t'\right) - \varepsilon_k\left(t'\right)\right] dt'\right\}$$

$$\tag{3.5.6}$$

Actually we can express $\frac{\partial}{\partial t}|m; \boldsymbol{R}\rangle$ in terms of $\frac{\partial h}{\partial t}$ by using (3.5.3), namely, by differentiating (3.5.3) with respect to t, multiplying from

---

[9]For a stationary state the solution to the Schrödinger equation is

$$\psi = \sum_m a_m e^{-\frac{i}{\hbar} E_m t}|m\rangle$$

where a_m is independent of t. In the present problem $|m; \boldsymbol{R}\rangle$ is time dependent due to $\boldsymbol{R}$, and $\exp[-\frac{i}{\hbar}\int_0^t \varepsilon_m(t')dt']|m; \boldsymbol{R}\rangle$ alone does not satisfy the Schrödinger equation. Consequently the coefficients a_m in the expansion (3.5.5) are functions of time. In (3.5.6) we see that $\dot{a}$ is different from 0 because of $\frac{\partial}{\partial t}|m; \boldsymbol{R}\rangle$.

the left by $\langle k; \boldsymbol{R}|$, we obtain

$$\left\langle k; \boldsymbol{R} \left| \frac{\partial}{\partial t} m; \boldsymbol{R} \right.\right\rangle = \frac{1}{\varepsilon_m - \varepsilon_k} \left\langle k; \boldsymbol{R} \left| \frac{\partial h}{\partial t} \right| m; \boldsymbol{R} \right\rangle, \quad k \neq m. \quad (3.5.7)$$

For $k = m$, by using the normalization condition $\langle k; \boldsymbol{R} \,|k; \boldsymbol{R}\rangle = 1$, we obtain

$$\left\langle k; \boldsymbol{R} \left| \frac{\partial}{\partial t} k; \boldsymbol{R} \right.\right\rangle + \left\langle \frac{\partial}{\partial t} k; \boldsymbol{R} | k; \boldsymbol{R} \right\rangle = 0,$$

i.e.,

$$\left\langle k; \boldsymbol{R} \left| \frac{\partial}{\partial t} k; \boldsymbol{R} \right.\right\rangle = i\alpha_k(t). \quad (3.5.8)$$

The right hand side of the above equation is pure imaginary. Let the system be in a state $|n; \boldsymbol{R}(0)\rangle$ at $t = 0$, i.e., $a_m(0) = \delta_{mn}$, we have

$$\dot{a}_k = \frac{1}{\varepsilon_k - \varepsilon_n} \left\langle k, \boldsymbol{R} \left| \frac{\partial h}{\partial t} \right| n; \boldsymbol{R} \right\rangle \exp\left[\frac{i}{\hbar}(\varepsilon_k - \varepsilon_n)t\right], \quad k \neq n.$$

After integration we have

$$a_k(t) \approx \frac{1}{i\hbar(\varepsilon_k - \varepsilon_n)^2} \left\langle k; \boldsymbol{R} \left| \frac{\partial h}{\partial t} \right| n; \boldsymbol{R} \right\rangle$$

$$\times \exp\left[\frac{i}{\hbar}(\varepsilon_k - \varepsilon_n)t\right], \quad k \neq n. \quad (3.5.9)$$

For states $k \neq n$, the probability amplitudes oscillate with time and do not show the behavior of a stable growth. We reach the conclusion that from $t = 0$ to a finite time t the state $|n; \boldsymbol{R}(0)\rangle$ and its energy eigenvalue $\varepsilon_n(\boldsymbol{R})$ may have changed appreciably, but the system is most likely still be in the state $|n; \boldsymbol{R}(t)\rangle$. Historically Einstein and Ehrenfest came to this conclusion which was then referred to as the adiabatic hypothesis in 1911 before the birth of quantum mechanics. It was proved by M. Born and V. Fock in 1928. The foregoing argument shows the persistence of a state $|n; \boldsymbol{R}(0)\rangle$ to remain adiabatically in the same state $|n; \boldsymbol{R}(t)\rangle$. The condition that the Hamiltonian $H(t)$ evolves slowly enough is encoded in the equality [13]

$$\hbar \left| \frac{\left\langle n; \boldsymbol{R} \left| \frac{\partial}{\partial t} m; \boldsymbol{R} \right\rangle \right.}{\varepsilon_n(t) - \varepsilon_m(t)} \right| \ll 1, \qquad n \neq m, \quad t \in [0, T],$$

where T is the time of evolution. K.-P. Marzlin and B.C. Sanders [14] pointed out that a perfunctory application of the theorem to systems in which the change in the adiabatic eigenstate is significant may lead to inconsistency. D.M. Tong *et al.* [15] pointed out that the above condition is sufficient for the validity of the quantum adiabatic thoerem as long as the time dependence of both $\langle n; \boldsymbol{R} | \frac{\partial}{\partial t} m; \boldsymbol{R} \rangle$ and $\varepsilon_n(t) - \varepsilon_m(t)$ are negligible. With this demand satisfied, the transition between adiabatic states can be neglected. Otherwise the change of adiabatic states may become significant after a sufficiently long time of evolution such that the transition between some of them can be dynamically driven. When more general quantum system is considered, additional conditions are necessary to guarantee the validity of the theorem [15]. The central question is how does a wave function change with time when the Hamiltonian changes adiabatically. The motivation is the following. Consider, for example, a magnetic field changes slowly its direction while keeping its magnitude unchanged. The spin of a charged particle follows precession around the slowly changing magnetic field. When the magnetic field return to its original direction after a certain period of time, will the spin of the particle return to its original direction? To answer this question we have to be careful in analyzing the time evolution of the state of the spin following the slowly changing direction of the magnetic field. For an exact stationary state $|n\rangle$ the time evolution is given by the dynamical phase factor $e^{-i\varepsilon_n t/\hbar}$. For a time varying $\boldsymbol{R}(t)$, this factor can be generalized to be $\exp[-\frac{i}{\hbar} \int_0^t \varepsilon_n(t') dt']$.

The time evolution could be given by

$$\Psi(t) = \exp\left[-\frac{i}{\hbar} \int_0^t \varepsilon_n(t') dt' \right] |n; \boldsymbol{R}(t)\rangle,$$

where $|n; \boldsymbol{R}(t)\rangle$ is the adiabatic eigenstate. This is the quantum mechanical adiabatic theorem. Substituting this expression back to the Schrodinger equation with a time-dependent Hamiltonian, one obtains

$$i\hbar \frac{\partial}{\partial t}\psi(t) = H(t)\psi(t) + \exp\left[-\frac{i}{\hbar}\varepsilon_n(t')dt'\right]i\hbar\frac{\partial}{\partial t}|n; \boldsymbol{R}(t)\rangle.$$

The second term vanishes in the adiabatic limit.

$$\frac{\partial}{\partial t}|n; \boldsymbol{R}(t)\rangle = \frac{\partial \boldsymbol{R}}{\partial t} \cdot \partial_{\boldsymbol{R}}|n; \boldsymbol{R}(t)\rangle = 0.$$

Problem arises when one deals with adiabatic *cyclic* processes for which $\boldsymbol{R}(0) = \boldsymbol{R}(T)$. To handle such processes, M. V. Berry [16] added a phase factor $e^{i\gamma_n(t)}$

$$\psi(t) = \exp\left[-\frac{i}{\hbar}\int_0^t \varepsilon_n(t')dt'\right]e^{i\gamma_n(t)}|n; \boldsymbol{R}(t)\rangle. \tag{3.5.10}$$

Questions can be raised that Eq. (3.5.3) does not determine the phase factor for the eigenstates, why the phase factor $e^{i\gamma_n(t)}$ cannot be absorbed into $|n; \boldsymbol{R}(t)\rangle$? The answer will come a little later. Substituting (3.5.10) into (3.5.4) we get the equation for $\dot{\gamma}_n(t)$.

$$\dot{\gamma}_n(t) = i\left\langle n; \boldsymbol{R}|\frac{\partial}{\partial t}n; \boldsymbol{R}\right\rangle = i\dot{\boldsymbol{R}}(t) \cdot \langle n; \boldsymbol{R}|\nabla_{\boldsymbol{R}}n; \boldsymbol{R}\rangle. \tag{3.5.11}$$

The last equality is based on the observation that $|n; \boldsymbol{R}\rangle$ depends on time only through $\boldsymbol{R}$. $\nabla_{\boldsymbol{R}}$ denotes the gradient with respect to the coordinates $\boldsymbol{R}$. As Eq. (3.5.8) shows, $\langle n; \boldsymbol{R} |\frac{\partial}{\partial t}n; \boldsymbol{R}\rangle$ is pure imaginary, hence $\dot{\gamma}_n$ is real, implying that if γ_n is real initially, it will remain real in subsequent times. Let

$$\boldsymbol{A}\left(\boldsymbol{R}(t)\right) \equiv i\langle n; \boldsymbol{R}|\nabla_{\boldsymbol{R}}n; \boldsymbol{R}\rangle, \tag{3.5.12}$$

Equation (3.5.11) then becomes

$$\dot{\gamma}_n(t) = \dot{\boldsymbol{R}}(t) \cdot \boldsymbol{A}\left(\boldsymbol{R}\right). \tag{3.5.13}$$

Let $\boldsymbol{R}(t)$ change slowly with time from $\boldsymbol{R}(0)$ to $\boldsymbol{R}(T)$, where T is the period of the adiabatic change, $\boldsymbol{R}(T) = \boldsymbol{R}(0)$. Will $\gamma_n(T)$ return to its initial value $\gamma_n(0$) ? We can easily calculate

$$\gamma_n\,(T) - \gamma_n(0) = \int_0^T \mathrm{d}t \dot{\gamma}_n(t) = \int_0^T \mathrm{d}t \dot{\boldsymbol{R}}(t) \cdot \boldsymbol{A}\,(\boldsymbol{R})$$

$$= \oint_C \mathrm{d}\boldsymbol{R} \cdot \boldsymbol{A}(\boldsymbol{R}),$$

where C is the closed circuit described by $\boldsymbol{R}(t)$ extending from $t = 0$ to $t = T$. Using Stokes theorem and denoting the closed integral around C by $\gamma_n(C)$, we have Berry's phase

$$\gamma_n(C) = \int_S \mathrm{d}\boldsymbol{S} \cdot \nabla_R \times \boldsymbol{A} = \mathrm{i} \int_S \mathrm{d}\boldsymbol{S} \cdot \nabla_R \times \langle n; \boldsymbol{R} | \nabla_R n; \boldsymbol{R} \rangle.$$

$$(3.5.14)$$

Here S is the surface enclosed by C. In general, $\boldsymbol{A}$ is not irrotational, hence the closed integral does not vanish, i.e., $\gamma_n(T) \neq \gamma_n(0$). This means that $\int_{\boldsymbol{R}_1}^{\boldsymbol{R}_2} \mathrm{d}\boldsymbol{R} \cdot \boldsymbol{A}(\boldsymbol{R})$ is path dependent, and $\gamma_n(t)$ is non-integrable which cannot be expressed as a single-valued function of $\boldsymbol{R}$. Since $|n; \boldsymbol{R}\rangle$ depends on t only through $\boldsymbol{R}$, it cannot absorb the phase factor $\mathrm{e}^{\mathrm{i}\gamma_n(t)}$. This explains the necessity of including the phase factor in Eq. (3.5.10). Since $\gamma_n(C)$ (3.5.14) does not depend on the time required for $\boldsymbol{R}$ to complete the closed circuit, Berry characterized the phase as a "geometrical phase", in contrast to the dynamical phase $-\frac{\mathrm{i}}{\hbar} \int_0^T \varepsilon_m(t) \mathrm{d}t$. In general this "geometry" is one of the parameter space. $\boldsymbol{R}$ can be any slow observable of the system. The closed integral is taken along a circuit in the abstract parameter space. In contrast, Aharonov-Bohm phase is expressed in terms of the closed integral $-\frac{\mathrm{i}e}{\hbar c} \oint \boldsymbol{A} \cdot \mathrm{d}\boldsymbol{s}$, and is therefore a geometrical phase in the real space.

Equation (3.5.12) defines a vector function. The notation is not accidentally chosen. Consider a change of the phase of $|n; \boldsymbol{R}\rangle$,

$$|n; \boldsymbol{R}\rangle \rightarrow \mathrm{e}^{\mathrm{i}\Theta(\boldsymbol{R})} |n; \boldsymbol{R}\rangle, \qquad (3.5.15)$$

then the corresponding changes are

$$
\left.
\begin{aligned}
|\nabla_R n; \boldsymbol{R}\rangle &\to (\mathrm{i}\nabla_R \Theta)\mathrm{e}^{\mathrm{i}\Theta(\boldsymbol{R})}|n; \boldsymbol{R}\rangle + \mathrm{e}^{\mathrm{i}\Theta(\boldsymbol{R})}|\nabla_R n; \boldsymbol{R}\rangle \\
\boldsymbol{A} &\to -\nabla_R \Theta + \mathrm{i}\langle n; \boldsymbol{R}|\nabla_R n; \boldsymbol{R}\rangle = \boldsymbol{A} - \nabla_R \Theta
\end{aligned}
\right\}. \qquad (3.5.16)
$$

Equations (3.5.15) and (3.5.16) specify a gauge transformation. Since $\gamma_n(C)$ is related to $\nabla \times \boldsymbol{A}$ (3.5.14), it is invariant under this transformation which is a necessary condition for the phase to be an observalle. Correspondingly $\boldsymbol{A}$ and $\nabla \times \boldsymbol{A}$ are the gauge affine connexion and the curvature respectively. From (3.5.11) we know that $\langle n; \boldsymbol{R} \mid \nabla_R n; \boldsymbol{R}\rangle$ is pure imaginary. Equations (3.5.12) and (3.5.14) give

$$
\gamma_n(C) = \mathrm{i}\int_S \nabla_R \times \langle n; \boldsymbol{R}|\,\nabla_R n; \boldsymbol{R}\rangle \cdot \mathrm{d}\boldsymbol{S}
$$

$$
= -\mathrm{Im}\left[\int \nabla_R \times \langle n; \boldsymbol{R}|\,\nabla_R n; \boldsymbol{R}\rangle \cdot \mathrm{d}\boldsymbol{S}\right]
$$

$$
= -\mathrm{Im}\left[\int \langle \nabla_R n; \boldsymbol{R}| \times |\nabla_R n; \boldsymbol{R}\rangle \cdot \mathrm{d}\boldsymbol{S}\right]
$$

$$
= -\mathrm{Im}\left[\int \sum_{m \neq n} \langle \nabla_R n; \boldsymbol{R}|m; \boldsymbol{R}\rangle \times \langle m; \boldsymbol{R}|\nabla_R n; \boldsymbol{R}\rangle \cdot \mathrm{d}\boldsymbol{S}\right].
$$

$$
(3.5.17)
$$

To obtain the last equality the completeness condition $\sum_m |m; \boldsymbol{R}\rangle\langle m; \boldsymbol{R}| = 1$ has been inserted. The sum in (3.5.17) is taken for $m \neq n$, since $\langle n; \boldsymbol{R} \mid \nabla_R n; \boldsymbol{R}\rangle$ is pure imaginary, and therefore $\mathrm{Im}[\langle \nabla_R n; \boldsymbol{R} \mid n; \boldsymbol{R}\rangle \times \langle n; \boldsymbol{R} \mid \nabla_R n; \boldsymbol{R}\rangle] = 0$.

Taking ∇_R of both side of the eigenequation $h|n; \boldsymbol{R}\rangle = \varepsilon_n(\boldsymbol{R})|n; \boldsymbol{R}\rangle$, we have

$$
\nabla_R h|n; \boldsymbol{R}\rangle + h|\nabla_R n; \boldsymbol{R}\rangle = \nabla_R \varepsilon_n|n; \boldsymbol{R}\rangle + \varepsilon_n|\nabla_R n; \boldsymbol{R}\rangle.
$$

Multiplying from the left by $\langle m; \boldsymbol{R}|$ and using the orthogonality of the wave functions, we obtain

$$
\langle m, \boldsymbol{R}|\nabla_R h|n; \boldsymbol{R}\rangle + \langle m; \boldsymbol{R}|h|\nabla_R n; \boldsymbol{R}\rangle = \varepsilon_n\langle m; \boldsymbol{R}|\nabla_R n; \boldsymbol{R}\rangle,
$$

i.e.,

$$\langle m; \boldsymbol{R} | \nabla_R n; \boldsymbol{R} \rangle = \frac{\langle m; \boldsymbol{R} | \nabla_R h | n; \boldsymbol{R} \rangle}{\varepsilon_n(\boldsymbol{R}) - \varepsilon_m(\boldsymbol{R})}. \tag{3.5.18}$$

Substituting (3.5.18) in (3.5.17) we obtain the expression of $\gamma_n(C)$ in terms of the matrix elements of $\nabla_R h$.

$$\gamma_n(C) = - \int d\boldsymbol{S} \cdot$$

$$\mathrm{Im} \left[\sum_{m \neq n} \frac{\langle n; \boldsymbol{R} | \nabla_R h(\boldsymbol{R}) | m; \boldsymbol{R} \rangle \times \langle m; \boldsymbol{R} | \nabla_R h(\boldsymbol{R}) | n; \boldsymbol{R} \rangle}{[\varepsilon_n(\boldsymbol{R}) - \varepsilon_n(\boldsymbol{R})]^2} \right]$$

$$\equiv - \int d\boldsymbol{S} \cdot \mathrm{Im}[\boldsymbol{V}(\boldsymbol{R})]. \tag{3.5.19}$$

The remarkable contribution of M. Berry is to prove the existence of a non-integrable phase factor in the case of adiabatic cyclic processes and reveal its geometrical significance. We consider an example given by Berry of 2-level system described by a $2\times$ matrix

$$h = \frac{1}{2} \begin{pmatrix} R_2 & R_1 - iR_2 \\ R_1 + iR_2 & -R_3 \end{pmatrix} = \frac{1}{2} \boldsymbol{\sigma} \cdot \boldsymbol{R}.$$

This Hamiltonian can be diagonalized to give the eigenvalues

$$E_+(\boldsymbol{R}) = -E_-(\boldsymbol{R}) = \frac{1}{2}(R_1^2 + R_2^2 + R_3^2)^{1/2} \equiv \frac{1}{2}R.$$

It is easy to see that the accidental degeneracy occurs at $R = 0$ where $E_+(0) = E_-(0) = 0$. We have $\nabla h = \frac{1}{2}\boldsymbol{\sigma}$ and $\boldsymbol{V}_\pm(\boldsymbol{R}) = \frac{\boldsymbol{R}}{2R^3}$. In this case $\boldsymbol{V}$ is the field produced by a monopole at $R = 0$ in the parameter space where the accidental degeneracy occurs. $\gamma(C)$ is the flux of this monopole through the area encircled by the closed circuit in the parameter space, and is proportional to the solid angle subtended by the circuit ant the point $\boldsymbol{R} = 0$.

Finally we consider the correction to the Born-Oppenheimer approximation based on Berry's point of view. We do not prescribe for $\boldsymbol{R}$ its variation with time and consider $\boldsymbol{R}$ as an observable. The wave function of a complete system is

$$\Psi(\boldsymbol{R}, \boldsymbol{r}) = \psi(\boldsymbol{R})|n; \boldsymbol{R}\rangle. \tag{3.5.20}$$

Substituting (3.5.20) into the Schrödinger equation

$$H\Psi = \left[\frac{\boldsymbol{P}}{2M} + \frac{\boldsymbol{p}^2}{2m} + V(\boldsymbol{R}, \boldsymbol{r})\right]\Psi = E\Psi, \tag{3.5.21}$$

and using (3.5.3) in the form

$$\left[\frac{\boldsymbol{p}^2}{2m} + V(\boldsymbol{R}, \boldsymbol{r})\right]|n; \boldsymbol{R}\rangle = \varepsilon_n(\boldsymbol{R})|n; \boldsymbol{R}\rangle,$$

we obtain

$$-\frac{\hbar^2}{2M}\nabla_R^2\psi(\boldsymbol{R})|n; \boldsymbol{R}\rangle + \psi(\boldsymbol{R})\varepsilon_n(\boldsymbol{R})|n; \boldsymbol{R}\rangle = E\psi(\boldsymbol{R})|n; \boldsymbol{R}\rangle. \tag{3.5.22}$$

Calculating explicitly the first term, multiplying both sides of the equation from the left by $\langle n; \boldsymbol{R}|$, we obtain

$$-\frac{\hbar^2}{2M}\nabla_R^2\psi(\boldsymbol{R})$$

$$-\frac{\hbar^2}{2M}2\langle n; \boldsymbol{R}|\nabla_R n; \boldsymbol{R}\rangle \cdot \nabla_R\psi(\boldsymbol{R}) - \frac{\hbar^2}{2M}\psi(\boldsymbol{R})\langle n; \boldsymbol{R}|\nabla_R^2 n; \boldsymbol{R}\rangle$$

$$+\varepsilon_n(\boldsymbol{R})\psi(\boldsymbol{R}) = E\psi(\boldsymbol{R}).$$

This equation can be rewritten in the form

$$\left[\frac{1}{2M}(\boldsymbol{P} - \hbar\boldsymbol{A})^2 + \mathcal{V}(\boldsymbol{R})\right]\psi(\boldsymbol{R}) = E\psi(\boldsymbol{R}), \tag{3.5.23}$$

where

$$\mathcal{V}(\boldsymbol{R}) = \varepsilon_n(\boldsymbol{R}) + \frac{\hbar^2}{2M}\left(\langle\nabla_R n; \boldsymbol{R}|\nabla_R n; \boldsymbol{R}\rangle - \boldsymbol{A}^2\right). \tag{3.5.24}$$

Equation (3.5.23) is the Schrödinger energy eigenequation for the slow variables. The effective potential energy $\mathcal{V}(\boldsymbol{R})$ for the slow variables is provided mainly by the eigenvalue of the fast motion $\varepsilon_n(\boldsymbol{R})$. The second term of (3.5.24) is the correction term. The appearance of $\boldsymbol{A}^2$ distinguishes the present treatment from the previous ones.[10] If the phase γ_n can be "absorbed" into $|n; \boldsymbol{R}\rangle$ to make it real at any time, $\boldsymbol{A}$ vanishes. As we have indicated, this is in general impossible.

It is easy to verify that $\mathcal{V}(\boldsymbol{R})$ (3.5.24) is invariant under the transformation (3.5.15), (3.5.16). Therefore by demanding $\psi(\boldsymbol{R}) \rightarrow e^{-i\Theta}$ $\psi(\boldsymbol{R})$ Eq. (3.5.23) can be made invariant. The phase transformation of $\psi(\boldsymbol{R})$ and of $|n; \boldsymbol{R}\rangle$ (3.5.15) have phases of opposite sign. They make the total wave function of the system $\Psi = \psi(\boldsymbol{R})|n; \boldsymbol{R}\rangle$ invariant under the transformation.

If there exists degeneracy in the eigenstates, then corresponding to one ε_n there are states labelled by $a, b, c, \ldots$. This results in the connexion in the form of matrix elements.

$$\boldsymbol{A}_{ab} = i\langle n, a; \boldsymbol{R}|\nabla n, b; \boldsymbol{R}\rangle$$

The corresponding field strength is

$$\boldsymbol{B}_{ab} = \nabla \times \boldsymbol{A}_{ab} + i\left(\boldsymbol{A} \times \boldsymbol{A}\right)_{ab}. \qquad (3.5.25)$$

This corresponds to the non-Abelian gauge fields [17–19].

Berry's phase appears 25 years after Aharonov-Bohm phase. Situations are in a certain sense similar. The manifestation of Berry's phase in many branches of physics has been widely studied. It is very interesting to notice that precursors of the idea on geometrical phase appeared several times before Berry's discovery: S. Pancharatnam on the polarization of light in 1956, G. Herzberg, H. C. Longuet-Higgins on the molecular structure in 1963 and 1976, C. A. Mead and D. G. Truhlar on Born-Oppenheimer approximation in 1979. The book "Geometric Phases in Physics" [20] contains survey articles and collection of relevant journal articles on geometrical phases in a variety of fields. When the cyclic evolution is fast enough, the adiabatic approximation is no more valid, and higher corrections can be introduced [21].

---

[10]C. A. Mead and D. G. Truhlar (J. Chem. Phys. 70, 2284, 1979) first pointed out that the anholonomy of the electron wave function in Born-Oppenheimer approximation can be described by including the gauge potential in the Hamiltonian kof the nucleus.

3.6 Aharonov-Anandan Phase

Y. Aharonov and J. S. Anandan [22] studied in 1987 the geometrical phase connected with a quantum state in a cyclic evolution in general, in which the state of a system returns to its original situation. Let the period of the cyclic evolution be τ. The initial and final states of the system $\psi(0)$ and $\psi(\tau)$ are connected by

$$|\psi(\tau)\rangle = e^{i\phi}|\psi(0)\rangle. \tag{3.6.1}$$

The phase factor $e^{i\phi}$ can have observational effect. In the Hilbert space $\mathscr{H}$ the states $\psi(0)$ and $\psi(\tau)$ are different vectors for $\phi \neq 0$. Then the path C described by $\psi(t)$ in $\mathscr{H}$ from $t = 0$ to $t = \tau$ is not a closed one. Let us consider the projective Hilbert space $\mathscr{P}$ of rays in $\mathscr{H}$, which does not distinguish $|\psi\rangle$ and $e^{if}|\psi\rangle$ for real f. Hence $|\psi(0)\rangle$ and $|\psi(\tau)\rangle$ are the same vector in $\mathscr{P}$, and the path $\widehat{C}$ traced in $\mathscr{P}$ by $\psi(t)$ from $t = 0$ to $t = \tau$ is a closed circuit. It is more convenient to describe cyclic evolution in $\mathscr{P}$. Let $|\psi\rangle \in \mathscr{H}$ evolves in time according to Schrödinger equation

$$H(t)|\psi(t)\rangle = i\hbar\frac{d}{dt}|\psi(t)\rangle, \tag{3.6.2}$$

and completes a cycle in time τ. The relation between $\psi(0)$ and $\psi(\tau)$ is given by (3.6.1). Define

$$|\tilde{\psi}(t)\rangle = e^{-if(t)}|\psi(t)\rangle. \tag{3.6.3}$$

Hence we have

$$|\tilde{\psi}(\tau)\rangle = e^{-if(\tau)}|\psi(\tau)\rangle = e^{-i[f(\tau)-\phi]}|\psi(0)\rangle$$
$$= e^{-i[f(\tau)-f(0)-\phi]}|\tilde{\psi}(0)\rangle. \tag{3.6.4}$$

If we demand

$$f(\tau) - f(0) = \phi \tag{3.6.5}$$

then it follows

$$|\tilde{\psi}(\tau)\rangle = |\tilde{\psi}(0)\rangle. \tag{3.6.6}$$

$|\widetilde{\psi}(t)\rangle$ is a vector in $\mathscr{P}$, it describes a closed circuit in a cycle of evolution. The substitution of (3.6.3) into (3.6.2) gives the equation satisfied by $f(t)$:

$$H|\psi\rangle = i\hbar \frac{d}{dt} e^{if} |\widetilde{\psi}\rangle = -\hbar \frac{df}{dt} e^{if} |\widetilde{\psi}\rangle + i\hbar e^{if} \frac{d}{dt} |\widetilde{\psi}\rangle.$$

By multiplying this equation from the left by $\frac{1}{\hbar}\langle\psi| = \frac{1}{\hbar}e^{-if}\langle\widetilde{\psi}|$ and reorganizing the result we get

$$-\frac{df}{dt} = \frac{1}{\hbar}\langle\psi|H|\psi\rangle - \left\langle \widetilde{\psi} \left| i\frac{d}{dt} \right| \widetilde{\psi} \right\rangle, \tag{3.6.7}$$

To obtain the phase due to cyclic evolution, we need to subtract the dynamical phase[11]

$$\phi = f(\tau) - f(0) = -\frac{1}{\hbar} \int_0^\tau \langle\psi| H |\psi\rangle \, dt + \int_0^\tau \left\langle \widetilde{\psi} \left| i\frac{d}{dt} \right| \widetilde{\psi} \right\rangle dt. \tag{3.6.8}$$

The first term in the right hand side of (3.6.8) is the dynamical phase $\phi_d(\tau)$ and consequently the second term is the geometrical phase β of the cyclic evolution

$$\phi_d(t) = -\frac{1}{\hbar} \int_0^t \langle\psi(t')|H|\psi(t')\rangle dt'. \tag{3.6.9}$$

The first term in the right-hand side of (3.6.8) is just $\phi_d(\tau)$, and consequently the second term is the geometrical phase β for the cyclic evolution:

$$\beta = \int_0^\tau \langle\widetilde{\varphi}|i\frac{d}{dt}|\widetilde{\varphi}\rangle dt. \tag{3.6.10}$$

The important fact is that β is expressed solely in terms of states in the projective Hilbert space $\mathscr{P}$. This implies that this geometrical phase is universal, in the sense that a closed circuit $\widehat{C}$ in $\mathscr{P}$ corresponds to infinitely many paths C in $\mathscr{H}$, and states undergoing

---

[11]Compare with the adiabatic case $\phi_d = -\frac{1}{\hbar}\int_0^t E(t')dt'$, where $E(t)$ is the adiabatic energy. In the general case we replace $E(t')$ by $\langle\psi(t') | H | \psi(t')\rangle$.

cyclic evolution along these paths driven by the corresponding $H(t)$ all lead to the same, unique phase β, the Aharonov-Anandan phase.

3.7 Experimental Revelations of Berry's Phase

There are two classes of situations for the appearance of Berry's phase. If the observable $\boldsymbol{R}$ is under control in the laboratory, the states with different phases can be prepared, and the interference between them reveals the Berry's phase. When $\boldsymbol{R}$ is an observable of a larger system, e.g., in the Born-Oppenheimer approximation of molecular structure, it is necessary to measure the eigenvalues of the system and compare the results with theoretical predictions based on (3.5.23) and (3.5.24). From the comparison Berry's phase can be determined. In the following, we discuss the cases in which $\boldsymbol{R}$ can be manipulated.

Berry [16] gave the geometrical phase which the spin of a particle acquires after an adiabatic cycle. The Hamiltonian of a particle of spin S interacting with a slowly changing magnetic field $\boldsymbol{B}$ is

$$h(\boldsymbol{B}) = \kappa \hbar \boldsymbol{B} \cdot \hat{\boldsymbol{S}}, \qquad (3.7.1)$$

where κ is a constant related to the gyromagnetic ratio, and $\hat{\boldsymbol{S}}$ is the spin operator. The energy eigenvalue is

$$E_n(\boldsymbol{B}) = \kappa \hbar B n, \qquad (3.7.2)$$

where n is the component of the spin in the direction of $\boldsymbol{B}$, with n taking the value $-S \leqslant n \leqslant S$. Equation (3.5.18) gives the Berry's phase

$$\gamma_n(C) = -\int d\boldsymbol{S} \cdot \boldsymbol{V}_n, \qquad (3.7.3)$$

where

$$V_n = \text{Im} \left[\frac{1}{B^2} \sum_{m \neq n} \frac{\langle n; \boldsymbol{B}| \hat{\boldsymbol{S}} |m; \boldsymbol{B}\rangle \times \langle m; \boldsymbol{B}| \hat{\boldsymbol{S}} |n; \boldsymbol{B}\rangle}{(m-n)^2} \right]. \qquad (3.7.4)$$

In order to calculate the matrix elements in (3.7.4), we fix the magnetic field $\boldsymbol{B}$ in the z-direction and use

$$\left. \begin{aligned} (\hat{S}_x \pm i\hat{S}_y)\,|n\rangle &= [s\,(s+1) - n\,(n \pm 1)]^{\frac{1}{2}}\,|n \pm 1\rangle \\ \hat{S}_z\,|n\rangle &= n\,|n\rangle \end{aligned} \right\}. \qquad (3.7.5)$$

In calculating V_{nx} and V_{ny} we observe that all vector products contributing to them involve the matrix element of S_z, and the non-diagonal matrix elements $(m \neq n)$ vanish. Therefore, $V_{nx} = V_{ny} = 0$. To V_{nz}, only matrix elements with $m = n \pm 1$ contribute. Finally we have

$$\begin{aligned} V_{nz}(\boldsymbol{B}) = \operatorname{Im} \Big\{ &\frac{1}{B^2}[\langle n|\hat{S}_x|n+1\rangle\langle n+1|\hat{S}_y|n\rangle \\ &- \langle n|\hat{S}_y|n+1\rangle\langle n+1|\hat{S}_x|n\rangle + \langle n|\hat{S}_x|n-1\rangle\langle n-1|\hat{S}_y|n\rangle \\ &- \langle n|\hat{S}_y|n-1\rangle\langle n-1|\hat{S}_x|n\rangle] \Big\} \\ = &\frac{n}{B^2}. \end{aligned} \qquad (3.7.6)$$

For an arbitrary orientation of the magnetic field, we have

$$\boldsymbol{V}_n\,(\boldsymbol{B}) = n\frac{\boldsymbol{B}}{B^3}. \qquad (3.7.7)$$

Finally the Berry's phase is

$$\gamma_n(C) = -\int d\boldsymbol{S} \cdot \boldsymbol{V}_n = -n\Omega(C). \qquad (3.7.8)$$

In the parameter space of $\boldsymbol{B}(B_x, B_y, B_z)$, $\int d\boldsymbol{S} \cdot \frac{\boldsymbol{B}}{B^3}$ is just the solid angle $\Omega(C)$ subtended at the origin of the parameter space by the closed circuit C which $\boldsymbol{B}$ traces. Berry's phase factor is $e^{i\gamma_n(C)} = e^{-in\Omega(C)}$. This is the geometrical phase acquired by the spin when the latter completes a cycle of rotation around the magnetic field. The phase is proportional to n. If a state of the particle can be prepared as a superposition of two states with different values of n (therefore with different Berry's phases) interference between these terms will reveal the existence of Berry's phase.

3.7.1 Manifestation of Berry's phase in the rotation of polarization of a photon

R. Y. Chiao and Y.-S. Wu [23] suggested using helically wound optical fiber in which the wave vector k of a photon varies continuously. When the optical fiber completes a whole winding such that the initial and final propagation directions coincide, the vector k traces a complete path in the parameter k space. It is a circle on a spherical surface. The circle subtends a cone at the origin with apex angle θ, and the solid angle of the cone is

$$\Omega(C) = 2\pi(1 - \cos\theta). \tag{3.7.9}$$

Since the photon is massless the component of spin in the direction k can only be ± 1. The spin follows the vector k in its adiabatic variation, and the respective Berry's phase after one cycle is

$$\gamma(C) = -2\pi s_k(1 - \cos\theta). \tag{3.7.10}$$

On the basis of this suggestion, A. Tomita and R. Y. Chiao [24] carried out an experiment, the scheme of the setup is shown in Fig. 3.17 (a). A He-Ne laser and a pair of linear polarizers, one at the input and the other at the output end of the optical fiber were used to measure the rotation of the plane of polarization in a 180-cm-long single mode fiber. The two polarization states $s_k = +1$ and $s_k = -1$ have equal and opposite Berry's phases such that the plane of linear polarization composed of these states rotates. A uniformly wound

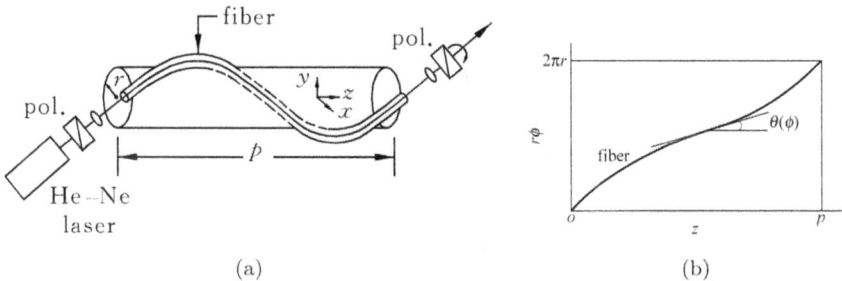

Fig. 3.17   (a) Experimental setup; (b) Optical fiber on the unfolded cylinder. Reprinted with permission from Akira Tomita and Raymond Y. Chiao, *Phys. Rev. Lett.* 57, 937, 1986. Copyright (2017) by the American Physical Society.

fiber is shown in Fig. 3.17 (b) with the cylinder unwrapped. The length of the fiber is l, length of the cylinder is p, and θ is the apex angle, i.e., the angle between the fiber and the helix axis. They are related by

$$\cos\theta = \frac{p}{l}. \tag{3.7.11}$$

Consequently

$$\gamma(C) = -2\pi s_k \left(1 - \frac{p}{l}\right). \tag{3.7.12}$$

In the experiment, different winding geometries were used for comparison. The result is shown in Fig. 3.18, indicating that the angle of rotation of the plane of polarization is proportional to the solid angle in the parametric space. Different point shapes represent different winding geometries, showing that deformation of the paths in parameter space does not influence the relation between $\gamma(C)$ and $\Omega(C)$: all points lie on the same straight line. This is the first experimental demonstration of Berry's phase.

Using photons is a smart choice, since a large number of photons can be in the same state such that experimenting on these photons

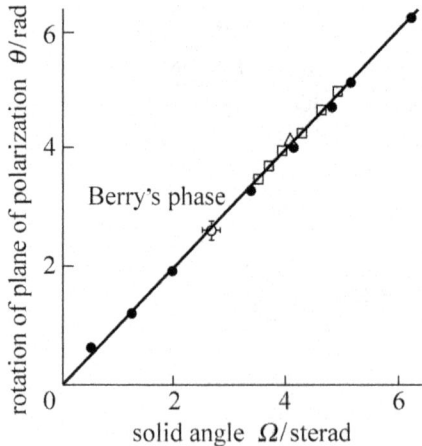

Fig. 3.18   Relation between the angle of rotation of the plane of polarization and the solid angle $\Omega(C)$. Reprinted with permission from Akira Tomita and Raymond Y. Chiao, *Phys. Rev. Lett.* **57**, 937, 1986. Copyright (2017) by the American Physical Society.

can give information on the behavior of a single photon in that state. Interesting enough, the appearance of geometrical phase for the polarization of photons following the adiabatic change of the direction of propagation can be interpreted also by classical electromagnetism. The rotation of the plane of polarization can be interpreted as the angle of holonomy produced by the parallel transport of the electric vector $\boldsymbol{E}$ of the light wave, which can be obtained by using Maxwell equations [25]. This is correct, but the classical theory fails for low photon numbers when fluctuation sets in, whereas the quantum theory still holds. This experiment actually demonstrates the existence of Berry's phase and its topological properties *at the classical level*. These effects are topological features of Maxwell theory which originate at the quantum level, but survives the correspondence limit $\hbar \to 0$ into the classical level.

3.7.2 Manifestation of Berry's phase in neutron spin rotation

T. Bitter and D. Dubbers [26] measured the Berry's phase acquired by the spin of neutron in a slowly rotating magnetic field. In a helical static magnetic field a neutron moving with a slow uniform speed along the axis of the helix experiences an adiabatically rotating magnetic field. Through its interaction with the magnetic field the neutron spin follows the adiabatic change of the field and acquires a Berry's phase after a complete cycle. Figure 3.19(a) shows the adiabatic transport of the tip of vector $\boldsymbol{B}$ around the closed loop C. Figure 3.19(b) shows the helical coil producing the field $\boldsymbol{B}_1$, the projection of $\boldsymbol{B}$ in a plane perpendicular to the axis of the helix. The solenoid producing the axial component of $\boldsymbol{B}$ is not shown. Let the neutron be in a state $|m\rangle$ at $t = 0$ where m is the z-component of the spin. The time for $\boldsymbol{B}_1$ to complete a cycle (i.e., the time required for the neutron in traversing a helix distance) is denoted by T. During this time the neutron picks up a dynamical phase

$$\phi_d = \kappa \int_0^T B(t)\mathrm{d}t = \kappa BT.$$

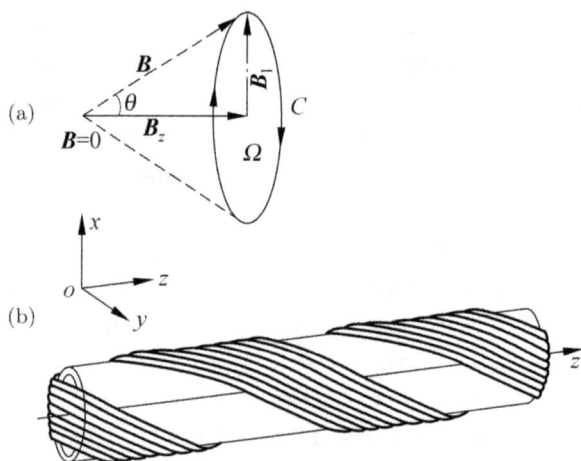

Fig. 3.19   (a) Adiabatic transport of B around a closed path C; (b) Helical coils producing the field component perpendicular to the axis of the helix. Reprinted with permission from T. Bitter and D. Dubbers, *Phys. Rev. Lett.* 59, 251, 1987. Copyright (2017) by the American Physical Society.

The Berry's phase is according to (3.7.8) $-2\pi(1 - \cos\theta)$ and the total phase Φ_T is

$$\Phi_T = \kappa BT - 2\pi(1 - \cos\theta). \qquad (3.7.13)$$

In the experiment monochromatic beam of polarized neutrons (velocity $v \lesssim 500$ ms^{-1}, polarization $P \simeq 97\%$) from a neutron guide of the Institut Laue-Langevin high flux reactor was used. Let $P_\alpha(0)$ and $P_\beta(T)$ be the components of the polarization of the neutron at $t = 0$ and $t = T$ respectively, $\alpha, \beta = x, y, z$. The initial and final polarizations are measured experimentally. Using the relation

$$P_\beta(T) = G_{\beta\alpha}P_\alpha(0),$$

the coefficients $G_{\beta\alpha}$ are determined and compared with the theoretical predictions [27].

For the case $B_z = 0$ (then $\theta = \pi/2, \gamma = -2\pi$), measured values of G_{yy} is shown in Fig. 3.20 (a),[12] the comparison between the values of

---

[12]Let $P_y(0) = 1$, then $P_y(T) = G_{yy}$. $G_{yy} = 1$ means that the spin does not rotate (modulo 2π); $G_{yy} = 0$ means that the spin rotates by $\frac{\pi}{2}$ and $G_{yy} = -1$ means that the spin rotates by π.

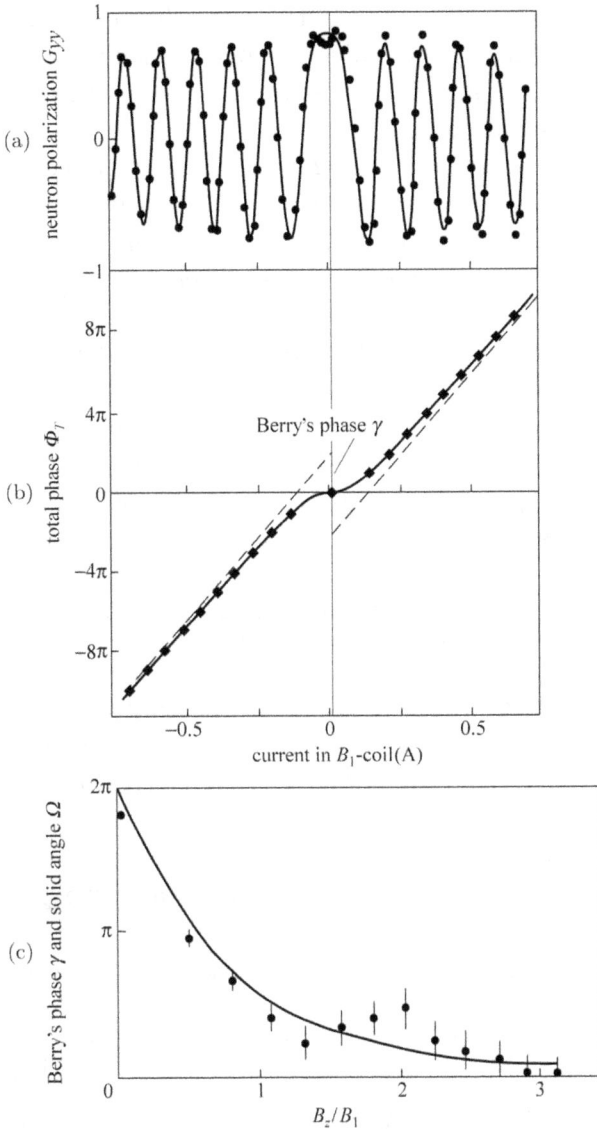

Fig. 3.20 (a) Neutron spin rotation of the transverse neutron spin component in the helical B_1 field. Without Berry's phase the maxima of this curve should be equidistant; (b) Observed and calculated phase shifts Φ_T; (c) Berry's phase γ (measured) at different solid angles Ω of the B field. Reprinted with permission from T. Bitter and D. Dubbers, *Phys. Rev. Lett.* **59**, 251, 1987. Copyright (2017) by the American Physical Society.

Φ obtained from the measurement and the theoretical predictions is shown in Fig. 3.20 (b). The value of γ determined is 2π. The relation between Berry's phase and the solid angle is given as a function of B_z/B_1 as shown in Fig. 3.20 (c).

D. J. Richardson *et al.* [24] carried out a complementary experiment. Spin polarized ultracold neutrons from Institut Laue-Langevin reactor are subjected to a temporally varying magnetic field. Ultracold neutrons are those having velocities $\lesssim 5$ m/s, and can be stored in a "bottle". Since the Fermi potentials of Be and BeO surface is sufficiently high, the ultracold neutrons will be reflected from the surface with negligible probability of loss for all incident angles. The time of containment is only limited by the neutron lifetime. The relation between Berry's phase and the solid angle in the parameter space, as well as the additivity of Berry's angles for repeated cycles are verified.

3.7.3 Adiabatic rotational splitting and Berry's phase in nuclear quadrupole resonance

R. Tycho [28] reported a research on the splitting of the nuclear quadrupole resonance spectrum due to the adiabatic rotation of the nuclear spin. The splitting is interpreted as a manifestation of Berry's phase. A simple crystal of $NaClO_3$ contains nuclear spin of ^{35}Cl ($S=3/2$) with an axially symmetric quadrupole coupling with coupling constant ω_Q, and the direction of the principal axis

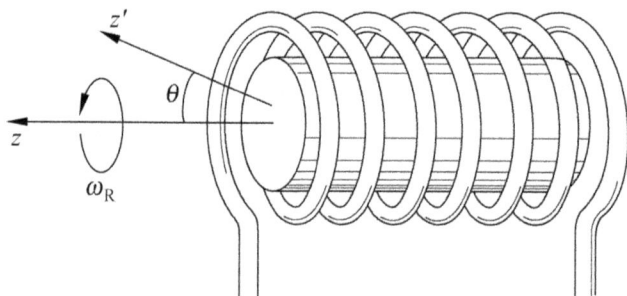

Fig. 3.21 Experimental arrangement for observations of adiabatic rotational splitting in nuclear quadrupole resonance. Reprinted with permission from Robert Tycko, *Phys. Rev. Lett.* 58, 2281, 1987. Copyright (2017) by the American Physical Society.

is denoted by z'. The Hamiltonian is (Fig. 3.21)

$$h = \omega_Q S_{z'}^2. \tag{3.7.14}$$

By applying radio frequency pulses the ^{35}Cl nucleus can be in a state of linear superposition of different spin component values. A solenoid wound along z is used to detect the free-induction decay signals. States with different $|S_{z'}|$ values give two energy levels, and the resonance occurs at 29.94 MHz. The principal axis makes an angle θ with z, and the crystal is slowly rotated about z axis with a frequency ω_R ($\omega_R \ll \omega_Q$). The operator $S_{z'}$ and the Hamiltonian H become time dependent.

$$S_{z'}(t) = S_z \cos\theta + S_x \sin\theta \cos\omega_R t + S_y \sin\theta \sin\omega_R t, \tag{3.7.15}$$

$$H(t) = \omega_Q S_{z'}^2(t). \tag{3.7.16}$$

A basis of time-dependent eigenstates written in terms of eigenstates of $S_{z'}$ is chosen.

$$
\begin{aligned}
|a\rangle &= \left|\frac{3}{2}\right\rangle \\
|b\rangle &= \cos\frac{1}{2}\xi \left|\frac{1}{2}\right\rangle - \sin\frac{1}{2}\xi \left|-\frac{1}{2}\right\rangle \\
|c\rangle &= \sin\frac{1}{2}\xi \left|\frac{1}{2}\right\rangle + \cos\frac{1}{2}\xi \left|-\frac{1}{2}\right\rangle \\
|d\rangle &= \left|-\frac{3}{2}\right\rangle,
\end{aligned}
\tag{3.7.17}
$$

where

$$\tan\xi = 2\tan\theta.$$

Taking into account that the operator $S_{z'}$ and some of the eigenstates are θ dependent, the Berry's phase is written as

$$\beta = 2\pi \int_0^\theta d\theta' \sin\theta' \langle\psi| S_{z'} |\psi\rangle. \tag{3.7.18}$$

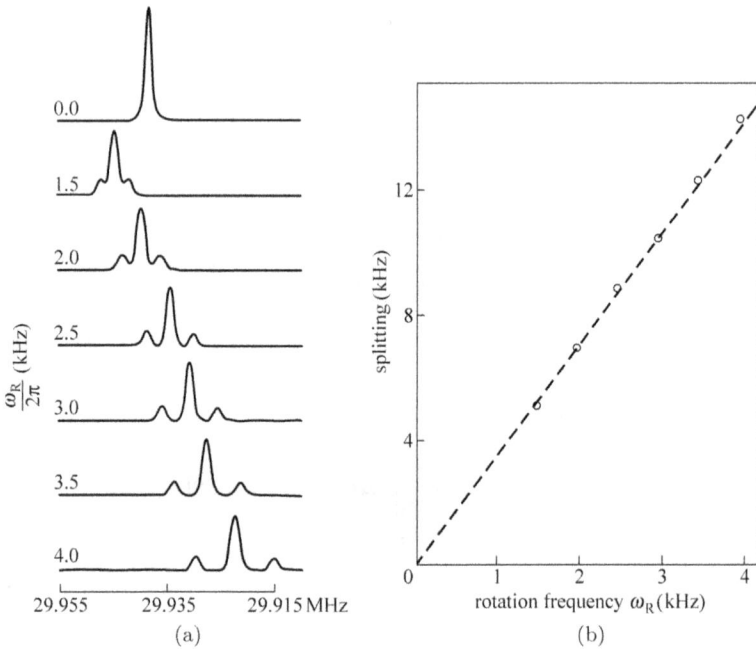

Fig. 3.22   (a) ^{35}Cl nuclear quadrupole resonance spectra of NaClO$_3$ vs rotation frequencies; (b) Splitting between the countermost lines vs ω_R. Reprinted with permission from Robert Tycko, *Phys. Rev. Lett.* 58, 2281, 1987. Copyright (2017) by the American Physical Society.

For the states (3.7.17) the corresponding Berry's phases are

$$
\left.
\begin{aligned}
\beta_a &= 3\pi(\cos\theta - 1) \\
\beta_b &= -\pi[(4 - 3\cos^2\theta)^{1/2} - 1] \\
\beta_c &= \pi[(4 - 3\cos^2\theta)^{1/2} - 1] \\
\beta_d &= -3\pi(\cos\theta - 1).
\end{aligned}
\right\}
\qquad (3.7.19)
$$

The experiment was carried out at an angle $\cos^2\theta = 1/3$. For this value the possible phase differences are $-2\sqrt{3}\pi, 0, 2\sqrt{3}\pi$ (modulo 2π.) Consider any two eigenstates $|\psi_1\rangle$ and $|\psi_2\rangle$ prepared at $t = 0$. At $t = T = 2\pi/\omega_R$ the difference of Berry's phase is $\gamma_1 - \gamma_2$. This phase difference is built up as time goes on. The time rate of change of the phase difference corresponds to a frequency shift $\Delta\omega = \frac{\gamma_1 - \gamma_2}{2\pi}\omega_R$. The original resonance line is now split into three, with $\Delta\omega = -\sqrt{3}\,\omega_R, 0, \sqrt{3}\omega_R$. Experiments confirm this splitting

(Fig. 3.22). The overall shifts of resonance lines are artifacts due to changes in the crystal temperature. The splitting between the two side lines versus ω_R is a straight line with slope $2\sqrt{3}$.

References

[1] M. Peshkin and A. Tonomura *The Aharonov-Bohm Effects*, Berlin, Springer Verlag, 1989.

[2] Y. Aharonnov and D. Bohm, *Phys. Rev.* 115, 485, 1959.

[3] T. T. Wu and C. N. Yang, *Phys. Rev. D* 12, 3845, 1975.

[4] R. P. Feynman, R. B. Leighton and M. Sands, *The Feynman Lectures on Physics* vol. 2, Reading, Addison-Wesley, 1964, pp. 15–12.

[5] N. Byers and C. N. Yang, *Phys. Rev. Lett.* 7, 46, 1961.

[6] R. G. Chambers, *Phys. Rev. Lett.* 5, 3, 1960.

[7] G. Möllenstedt and W. Bayh, *Phys. Bl.* 18, 299, 1962; *Naturwissen.* 4, 81, 1962; W. Bayh, *Z. Phys.* 169, 492, 1962.

[8] A. Tonomura, *Rev. Mod. Phys.* 59, 639, 1987.

[9] A. Tonomura, N. Osakabe, T. Matsuda, T. Kawasaki, J. Endo, S. Yano and H. Yamada, *Phys. Rev. Lett.* 56, 792, 1986; A. Tonomura, T. Matsuda, J. Endo, H. Todokoro, T. Komoda, *J. Electron Microsc.* 28, 1, 1979.

[10] Y. Aharonov and A. Casher, *Phys. Rev. Lett.* 53, 319, 1984.

[11] A. Cimmino, G. I. Opat, A. G. Klein, H. Kaiser, S. A. Werner, M. Arif and R. Clothier, *Phys. Rev. Lett.* 63, 380, 1989.

[12] M. V. Berry, *Phys. Today* 43(12), 34, 1990.

[13] L. Schiff *Quantum Mechanics* 3rd edition McGraw-Hill Inc. New York, 1968.

[14] K.-P. Marzlin and B.C. Sanders, *Phys. Rev. Lett.* 93, 160408, 2004.

[15] D.M. Tong, K Singh, L.C. Kwek and C.H. Oh, *Phys. Rev. Lett.* 95, 110407, 2005.

[16] M. V. Berry, *Proc. Roy. Soc. London* A392, 45, 1984.

[17] F. Wilczek and A. Zee, *Phys. Rev. Lett.* 52, 2111, 1984.

[18] J. Moody, A. Shapere and F. Wilczek, *Phys. Rev. Lett.* 56, 893, 1986.

[19] R. Jackiw, *Phys. Rev. Lett.* 56, 2779, 1986.

[20] A. Shapere and W. Wilczek, eds. *Geometric Phases in Physics*, Singapore, World Scientific, 1989.

[21] C. P. Sun, *J. Phys. A* 21, 1595, 1988.

[22] Y. Aharonov and J. S. Anandan, *Phys. Rev. Lett.* 58, 1593, 1987.

[23] R. Y. Chiao and Y. S. Wu, *Phys. Rev. Lett.* 57, 933, 1986.

[24] A. Tomita and R. Y. Chiao, *Phys. Rev. Lett.* 57, 937, 1986.

[25] M. V. Berry, *Nature* 326, 277, 1987.

[26] T. Bitter and D. Dubbers, *Phys. Rev. Lett.* 59, 251, 1987.

[27] E. Muskat, D. Dubbers and O. Schärpf, *Phys. Rev. Lett.* 58, 2047, 1987.

[28] D. J. Richardson, A. I. Kilvington K. Green and S. K. Lamoreaux, *Phys. Rev. Lett.* 61, 2030, 1988.

[29] R. Tycho, *Phys. Rev. Lett.* 58, 2281, 1987.

Chapter 4

Boundary between Quantum and Classical Mechanics, Entanglement and Decoherence

In a course on quantum mechanics we learn that quantum mechanics is the theory of microscopic world, beginning with atoms and molecules. Being a more refined, more general theory it should be valid also for the macroscopic world as well, and classical mechanics is only the limiting case of quantum mechanics when the action involved in the relevant problem is much larger than $\hbar$. The task of demonstrating the above statement has been the subject of hot discussion and dispute since the inception of quantum mechanics three quarters of a century ago without reaching a consensus, let alone giving a rigorous proof.

We begin with a simple problem in quantum mechanics. The eigenstates of a harmonic oscillator behave drastically different from states of a classical oscillator. Does a quantum mechanical oscillator possess a state which looks somewhat similar to a classical oscillator? Schrödinger succeeded in 1926 to form a wave packet from the eigenstates which mimics the to-and-fro motion of a classical oscillator. Encouraged by the success he devoted to forming a wave packet of the hydrogen atom which could mimic the Kepler motion. He failed to do so because the oscillator problem the energy levels are equidistant so that the wave packet formed does not diffuse, and for the hydrogen atom the energy levels are not equidistant. Based on the observation that the energy levels of the H atom get more and more close to uniform distribution when the quantum number n gets larger and larger, it might be possible to form a more or less stable wave packet by superimposing eigenstates of large n. The highly excited circular orbit wave packet was obtained in the mid-1970s. It does assimilate the Kepler circular motion, but as a quantum mechanical object, the wave function broadens and diffuses. To form an elliptic orbit wave packet it is necessary to find a conserved quantity in quantum mechanics corresponding to the classical Runge-Lenz vector which characterizes the shape of the orbit. This task was completed and a highly excited elliptic orbit wave packet was found in the late 1980s. These developments form the content of Sections 4.1, 4.2 and 4.4.

Is it possible to solve a quantum mechanical energy eigenvalue problem without solving the Schrödinger equation? This becomes possible when the dynamical symmetry of the problem is exactly known. For the H atom the dynamical symmetry is $SO(4)$, the generators of which are composed of the angular momentum operators and the quantum mechanical Runge-Lenz operators under some restrictions. Then the energy spectrum of the bound state of the H atom can be obtained. This is discussed in Section 4.3. The superposition principle plays a very important role in quantum mechanics. Its predictions work extremely good for the microscopic world, but in terms of everyday classical reality a prediction of this principle may seem awesome. It allows states which can never coexist in everyday life, e.g., a live cat and its own corpse superimposed, not only exist simultaneously, but interfere with each other! This is the well-known Schrödinger cat. There must be some mechanism which can quench the interference between various terms in a superposition when they represent macroscopic states, i.e., to make them incoherent. As a result, one has various outcomes with corresponding probabilities, quite acceptable according to the classical wisdom. The mechanism of decoherence has been the subject of intensive study and disputes for decades. Physicists work on models for demonstrating the phenomena. A growing consensus was reached that the origin of decoherence is the influence of the environment. In Section 4.5–4.8 we discuss the mechanism of decoherence and a dynamical model for it, and the question how the classical limit of quantum mechanics is reached. Finally section 4.9 concerns the realization of Schrödinger cat in the laboratory, i.e., two macroscopically distinguishable states superimposed coherently.

The equation of motion for observables in quantum mechanics leads to the result that the expectation value of the position variable $\langle x \rangle = \int \psi^* x \psi dx$ of a particle of mass m in a state described by a wave function $\psi(x, t)$ under the action of in a 1-dimensional potential $V(x)$ satisfies the following equation

$$m \frac{d^2}{dt^2} \langle x \rangle = - \left\langle \frac{\partial V}{\partial x} \right\rangle.$$

This is the Ehrenfest theorem. Formally it looks very much like the classical equation of motion. As a matter of fact, the variation with time of $\langle x \rangle$ in quantum mechanics is similar to the motion of a particle under the force $-\frac{\partial V}{\partial x}$ in classical mechanics only under restricted conditions. The wave packet in quantum mechanics has a wave function significantly different from zero only in the vicinity of its center $\bar{x} = \langle x \rangle$, and is therefore the counterpart of a particle in classical mechanics. The question is whether the motion of $\bar{x}$ in quantum mechanics is given by the classical equation of motion

$$m \frac{d^2}{dt^2} \bar{x} = - \frac{\partial V(\bar{x})}{\partial x}.$$

Notice the difference between the right hand side of this equation with that of the previous one, which is a quantum mechanical equation. In general $\langle \frac{\partial V}{\partial x} \rangle |_{\bar{x}} \neq \frac{\partial V(\bar{x})}{\partial \bar{x}}$. Assume that $V(x)$ is a slowly varying function, we expand $\frac{\partial V}{\partial x}$ around $\bar{x}$, up to the second order of $\langle \Delta x^2 \rangle$:

$$\frac{\partial V}{\partial x} = \frac{\partial V(\bar{x})}{\partial \bar{x}} + \frac{1}{2} \frac{\partial^3 V(\bar{x})}{\partial \bar{x}^3} \langle \Delta x^2 \rangle.$$

Consequently only under the condition

$$\left| \frac{\partial V(\bar{x})}{\partial \bar{x}} \right| \gg \frac{1}{2} \left| \frac{\partial^3 V(\bar{x})}{\partial \bar{x}^3} \right| \langle \Delta x^2 \rangle$$

the motion of the center of the wave packet is approximately equivalent to a classical particle. The important point is that a wave packet is to diffuse in general. Even if the above condition is fulfilled initially it will break down as time goes on, and the quantum mechanical motion of the center of a wave packet deviates more and more from the motion of a classical particle.

In the following we begin with the wave packet of a harmonic oscillator given in 1926 by E. Schrödinger. The correspondence between quantum and classical mechanics in this case is direct and exact. Next we discuss the wave packet of the H atom, and we will find its correspondence with the Kepler motion is incomplete: besides the similarity the wave nature of a quantum object is manifest.

4.1 Schrödinger's Harmonic Oscillator Wave Packets, Coherent States

In his pioneering works on quantum mechanics W. Heisenberg expressed strongly his point of view that concepts like the orbit of electron which is not observable should not appear in the theory. In contrast, E. Schrödinger paid more attention to the connection between quantum and classical mechanics. The 1-dimensional harmonic oscillator in quantum mechanics is a good example. In the low excited states the probability distributions of the position of the particle look much different from the classical distribution. When the quantum number becomes sufficiently large, the probability distribution becomes closer to the classical one. This is a revelation of Bohr's correspondence principle. The energy eigenstates are stationary states, i.e., the probability distribution of the position variable is independent of time. Is it possible to find a time dependent wave function in quantum mechanics which describes a motion mimicking the to-and-fro classical motion of the oscillator? In other words, is it possible to find a wave function $\Psi(x,t)$ which satisfies

$$\langle \Psi(t)| \, x \, | \Psi(t) \rangle = A \cos(\omega t + \alpha). \qquad (4.1.1)$$

This problem was solved by Schrödinger in 1926 [1]. To construct the required solution to the Schrödinger equation

$$i\hbar \frac{\partial \Psi(x,t)}{\partial t} = -\frac{\hbar^2}{2m} \frac{\partial^2 \Psi(x,t)}{\partial x^2} + \frac{1}{2}m\omega^2 x^2 \Psi(x,t). \qquad (4.1.2)$$

we can expand $\Psi(x,0)$ using the eigenstates of the harmonic oscillator, and let each term of the expansion evolve in time according to the respective energy eigenvalues, i.e., each term is to be multiplied by the dynamical phase factor $e^{-iE_n t/\hbar}$. The eigenfunctions and eigenvalues of the 1-dimensional harmonic oscillator are

$$\psi_n = C_n e^{-(m\omega/2\hbar)x^2} H_n\left[\sqrt{\frac{m\omega}{\hbar}}x\right], \qquad (4.1.3)$$

$$E_n = \hbar\omega\left(n + \frac{1}{2}\right), \qquad (4.1.4)$$

where C_n are the normalization constants,

$$C_n = 2^{-n/2}(n!)^{-1/2}\left(\frac{m\omega}{\hbar\pi}\right)^{1/4}, \qquad (4.1.5)$$

H_n is the Hermite polynomial. Expanding $\Psi(x,0)$ using ψ_n, we get

$$\Psi(x,0) = \sum_{n=0}^{\infty} c_n C_n e^{-(m\omega/2\hbar)x^2} H_n\left[\sqrt{\frac{m\omega}{\hbar}}x\right]. \qquad (4.1.6)$$

The expansion coefficients c_n are determined by

$$c_n = \int_{-\infty}^{\infty} \psi_n^*(x)\,\Psi(x,0)\,\mathrm{d}x. \qquad (4.1.7)$$

The wave packet at time t is then

$$\Psi(x,t) = \sum_{n=0}^{\infty} c_n \psi_n(x)\, e^{-i\left(n+\frac{1}{2}\right)\omega t}. \qquad (4.1.8)$$

The center of the wave packet is

$$\langle x \rangle = \int_{-\infty}^{\infty} x \, | \Psi(x,t) |^2 \, \mathrm{d}x$$

$$= \sum_{n=0}^{\infty} \sum_{k=0}^{\infty} c_n^* c_k \mathrm{e}^{-\mathrm{i}(n-k)\omega t} \int_{-\infty}^{\infty} \psi_n^*(x) \, x \psi_k(x) \, \mathrm{d}x.$$

Denoting the integral of the last equality by x_{nk}, the foregoing equation becomes

$$\langle x \rangle = \sum_{n=0}^{\infty} \sum_{k=0}^{\infty} c_n^* c_k x_{nk} \mathrm{e}^{-\mathrm{i}(n-k)\omega t}. \tag{4.1.9}$$

Using the eigenfunctions. (4.1.3) we can calculate x_{nk}.

$$x_{nk} = \sqrt{\frac{\hbar}{m\omega}} \left[\sqrt{\frac{n}{2}} \delta_{k,n-1} + \sqrt{\frac{n+1}{2}} \delta_{k,n+1} \right] \tag{4.1.10}$$

Substituting (4.1.10) into (4.1.9) we get

$$\langle x \rangle = \sqrt{\frac{\hbar}{2m\omega}} \sum_{n=1}^{\infty} \sqrt{n} \left(c_n^* c_{n-1} \mathrm{e}^{\mathrm{i}\omega t} + c_{n-1}^* c_n \mathrm{e}^{-\mathrm{i}\omega t} \right). \tag{4.1.11}$$

In order to obtain a function $\Psi(x,t)$ mimicking (4.1.2) we need to put some constraint on the initial wave function $\Psi(x,0)$. For the expansion coefficients $c_n = | c_n | \, \mathrm{e}^{\mathrm{i}\phi_n}$ of $\Psi(x,0)$, we demand that $| c_n |$ vary slowly with n, $| c_n | \approx | c_{n-1} |$, $| \phi_n - \phi_{n-1} | = a$, a constant independent of n, and let c_n be different from 0 appreciably for n large (such that $E_n \approx n\hbar\omega$). Under these conditions equation (4.1.11) becomes

$$\langle x \rangle = \sqrt{\frac{2}{m\omega^2}} \left(\sum_{n=0}^{\infty} \sqrt{E_n} \, |c_n|^2 \right) \cos(\omega t + \alpha)$$

$$= \sqrt{\frac{2}{m\omega^2}} \left\langle \sqrt{E} \right\rangle \cos(\omega t + \alpha). \tag{4.1.12}$$

since

$$\left\langle \sqrt{E} \right\rangle = \sum_{n=0}^{\infty} \sqrt{E_n} \, |c_n|^2 \,. \qquad (4.1.13)$$

Equation (4.1.12) gives the behavior of a classical oscillator varying with time.

We go back to equation (4.1.8). This is a periodic function of period $T = 2\pi/\omega$, i.e., $\Psi(x.t) = \Psi(x,0)$ for $t = T$ or any integral multiple of it. The wave packet does not diffuse with time increasing. The state constructed above is a *coherent state*. Such states can find important applications in many branches of physics [2].

4.1.1 Basic properties of coherent states

A coherent state is a vector $|l\rangle$ in Hilbert space, where l is in general a multicomponent label. There are different types of coherent states, they share two properties in common, namely continuity and completeness (resolution of unity). Define the norm of vector $|\psi\rangle$ as $\| \, |\psi\rangle \, \| \equiv \langle \psi \mid \psi \rangle^{\frac{1}{2}}$. For $|l\rangle \neq 0, \| \, |l\rangle \, \|$ is real and positive.

(1) continuity property

When the label l varies continuously and approaches l', $l \rightarrow l'$, continuity implies $\| \, |l'\rangle - |l\rangle \, \| \rightarrow 0$.

The continuity property rules out as coherent states two familiar sets of vectors. A set of discrete orthonormal vectors $\{|n\rangle : n = 0, 1, 2, \ldots\}$ with $\langle n \mid m \rangle = \delta_{mn}$ are not coherent states, because they are not continuous. Coherent states $|l\rangle$ are actually not orthogonal $\langle l' \mid l \rangle \neq 0$ for $l \neq l'$. A set of δ-normalized, continuum orthogonal vectors $\{|x\rangle : -\infty < x < \infty\}$ with $\langle x \mid x' \rangle = \delta(x - x')$ are not coherent states either, because although the label x is continuous, but the states is not continuous, in the sense that $|x\rangle$ and $|x\rangle'$ remain orthogonal as x approaches x' continuously while remaining in the open neighborhood of x'. The conclusion is that the eigenvectors in the normal or generalized sense of any self-adjoint (Hermitian) operator *never* constitute a set of coherent states.

(2) completeness property

There exists a positive measure δl on the label space $\mathscr{L}$ ($l \in \mathscr{L}$) such that the unit operator I admits the "resolution of unity",

$$I = \int |l\rangle \langle l| \, \delta l, \qquad (4.1.14)$$

when integrated over $\mathscr{L}$. This completeness property is different from that of eigenvectors of a self-adjoint operator in that for the latter the completeness condition in terms of orthonormal states $|n\rangle$,

$$\sum_n |n\rangle \langle n| = I$$

follows automatically, but for the coherent states this property is to be verified, thus imposing severe restrictions on the choice of the set of states. If (4.1.14) is satisfied, then it induces in the Hilbert space $\mathscr{H}$ a *continuous representation*, such that any vector $|\varphi\rangle$ or any operator B can be expressed as

$$|\varphi\rangle = \int |l\rangle \langle l| \varphi\rangle \delta l, \qquad (4.1.15)$$

$$\langle \varphi'| B |\varphi\rangle = \int \langle \varphi'| l'\rangle \langle l'| B |l\rangle \langle l| \varphi\rangle \delta l \delta l'. \qquad (4.1.16)$$

Here $\langle l \mid \varphi\rangle$ and $\langle l' \mid B \mid l\rangle$ are respectively the representative of the vector $|\varphi\rangle$ and the operator B in the continuous representation. For a vector with finite norm, its representative $\langle l \mid \varphi\rangle$ is square integrable, viz. $\langle \varphi \mid \varphi\rangle = \int \langle \varphi \mid l\rangle\langle l \mid \varphi\rangle \delta l = \int \mid \langle l \mid \varphi\rangle \mid^2 \delta l$ is finite. The functional representation $\langle l \mid \varphi\rangle$ of the state $|\varphi\rangle$ satisfies the integral equation

$$\langle l| \varphi\rangle = \int \langle l| l'\rangle \langle l'| \varphi\rangle \delta l', \qquad (4.1.17)$$

the kernel of which

$$\mathscr{K}(l; l') = \langle l| l'\rangle \qquad (4.1.18)$$

is called the reproducing kernel. It is a jointly continuous function, non-zero for $l' \neq l$ and therefore non-zero for all l' in an open neighborhood of l. This implies that the integral equation (4.1.17) is a real restriction on admissible functions. In contrast, a similar integral equation exists also for the conventional bases, but because of the orthonormality it is a trivial identity, for example

$$\langle x \mid \varphi \rangle = \int \delta(x - x')\langle x' \mid \varphi \rangle \mathrm{d}x'.$$

Also in distinction from ordinary bases, coherent states are linearly dependent.

$$|l'\rangle = \int |l\rangle \langle l| l'\rangle \delta l', \tag{4.1.19}$$

i.e., any coherent state can be expressed as a linear sum or integral of all other coherent states. Therefore, coherent states are "overcomplete".

Lastly, we note that since

$$\| |l\rangle - |l'\rangle \|^2 = \langle l' \mid l'\rangle + \langle l \mid l\rangle - 2\mathrm{Re}\langle l' \mid l\rangle$$

the continuity of the vectors is a simple consequence of the joint continuity of the reproducing kernel.

The foregoing discussions apply to any specific types of coherent states. In addition, each type has specific properties of its own. In the following, we discuss the canonical coherent states, and the spin coherent states will be discussed in Chapter 6.

4.1.2 Canonical coherent states

Canonical coherent states are constructed by using the annihilation operator a and the creation operator $a^\dagger$. The commutation relation satisfied by the operators is

$$[a, a^\dagger] = I. \tag{4.1.20}$$

The normalized vacuum state $|0\rangle$ has the property

$$a |0\rangle = 0. \tag{4.1.21}$$

The states obtained by repeatedly applying $a^\dagger$ to $|0\rangle$ span the *state space*. The orthonormal basis states are

$$|n\rangle = \frac{1}{\sqrt{n!}} \left(a^\dagger\right)^n |0\rangle. \tag{4.1.22}$$

From the commutation relation (4.1.20) we know that $|n\rangle$ is the eigenstate of the *number operator*

$$N = a^\dagger a \tag{4.1.23}$$

with eigenvalue n,

$$N|n\rangle = n|n\rangle. \tag{4.1.24}$$

It is easy to verify that a and $a^\dagger$ lowers and raises the number eigenvalue of the state respectively:

$$\left.\begin{array}{l} Na|n\rangle = (n-1)|n\rangle \\ Na^\dagger|n\rangle = (n+1)|n\rangle \end{array}\right\} \tag{4.1.25}$$

The canonical coherent state is obtained by applying a unitary transformation on $|0\rangle$:

$$|z\rangle = e^{za^\dagger - z^* a} |0\rangle. \tag{4.1.26}$$

One of the most useful formulas in quantum mechanics[1] is the Baker-Campbell-Hausdorff formula

$$e^{A+B} = e^A e^{-\frac{1}{2}[A,B]} e^B = e^B e^{\frac{1}{2}[A,B]} e^A. \tag{4.1.27}$$

An alternative form of this formula is

$$e^A B e^{-A} = B + [A, B] + \frac{1}{2!}[A, [A, B]] + \cdots$$

$$+ \frac{1}{n!}[A, [A, [\cdots [A, B]]\cdots]] + \cdots. \tag{4.1.28}$$

---

[1] A chapter on operator algebra in the book "Quantum Statistical Properties of Radiation" by W. H. Louisell (John Wiley, 1973) contains very useful theorems and formulas.

If $[A, B]$ commutes with A and B, then

$$e^{A+B} = e^A e^B e^{-\frac{1}{2}[A,B]} = e^B e^A e^{\frac{1}{2}[A,B]}. \tag{4.1.29}$$

Since $[za^\dagger, z^* a] = -zz^*$ is a c-number, equation (4.1.26) can be rewritten by using (4.1.29).

$$|z\rangle = e^{-\frac{1}{2}|z|^2} e^{za^\dagger} e^{-z^* a} |0\rangle. \tag{4.1.30}$$

Using (4.1.21) and (4.1.22), we have

$$|z\rangle = e^{-\frac{1}{2}|z|^2} e^{za^\dagger} |0\rangle = e^{-\frac{1}{2}|z|^2} \sum_{n=0}^{\infty} \frac{1}{\sqrt{n!}} z^n |n\rangle. \tag{4.1.31}$$

In defining (4.1.26) z^* appears at first, but because of (4.1.21) it does not enter (4.1.31). Hence it is sufficient to label the canonical coherent state by using z only. From (4.1.31) we obtain

$$\langle z_2 | z_1 \rangle = e^{-\frac{1}{2}|z_2|^2 + z_2^* z_1 - \frac{1}{2}|z_1|^2}, \tag{4.1.32}$$

which is a continuous function of z_1 and z_2, and does not vanish for any two complex numbers z_1 and z_2. We now verify that the resolution of unity holds. The integration measure for the complex variable z is, using polar coordinates,

$$d^2 z = d\,(\mathrm{Re} z)\, d\,(\mathrm{Im} z) = |z|\, d\,|z|\, d\theta. \tag{4.1.33}$$

Consequently

$$\frac{1}{\pi} \int |z\rangle \langle z|\, d^2 z = \frac{1}{\pi} \sum_{n,m} \frac{1}{\sqrt{n!m!}} \int e^{-|z|^2} z^{*n} z^m |m\rangle \langle n|\, d^2 z$$

$$= \sum_n \frac{1}{n!} \int e^{-|z|^2} |z|^{2n} |n\rangle \langle n|\, d\,|z|^2$$

$$= \sum_n |n\rangle \langle n| = I. \tag{4.1.34}$$

In the foregoing derivation the following relations

$$z^{*n}z^m = |z|^{m+n} e^{i(m-n)\theta}$$

$$\int_0^{2\pi} d\theta e^{-i(m-n)\theta} = 2\pi\delta_{mn}$$

are used, and the final equality follows, because $|n\rangle$ (4.1.22) form a complete orthonormal set.

We often encounter with the evaluation of a commutator like $[e^{-za^\dagger}, a]$. Observing that if a function $f(a^\dagger)$ can be expressed in terms of $a^\dagger$, then we have

$$[f(a^\dagger), a] = \frac{\partial f}{\partial a^\dagger}[a^\dagger, a] = -\frac{\partial f}{\partial a^\dagger}, \tag{4.1.35}$$

$$[f(a), a^\dagger] = \frac{\partial f}{\partial a}[a, a^\dagger] = \frac{\partial f}{\partial a}. \tag{4.1.36}$$

From the above, we have

$$[e^{-za^\dagger}, a] = z e^{-za^\dagger},$$

and hence

$$e^{-za^\dagger} a e^{za^\dagger} = a + z. \tag{4.1.37}$$

Thence an important property of the canonical coherent state follows. Operating the left hand side of (4.1.37) on $|0\rangle$, and using (4.1.31) we obtain

$$e^{-za^\dagger} a e^{za^\dagger} |0\rangle = e^{-za^\dagger} a e^{\frac{1}{2}|z|^2} |z\rangle$$
$$= e^{-za^\dagger} e^{\frac{1}{2}|z|^2} a |z\rangle,$$

while operating the right hand side on $|0\rangle$, we obtain

$$(a + z)|0\rangle = z|0\rangle.$$

Equating the right hand sides of the two above equations, we get

$$a|z\rangle = e^{za^\dagger} e^{-\frac{1}{2}|z|^2} z|0\rangle = z|z\rangle. \tag{4.1.38}$$

This implies that the canonical coherent state $|z\rangle$ is an eigenstate of the annihilation operator a corresponding to the eigenvalue z. It is labelled by the eigenvalue of a. From (4.1.38) we obtain

$$\langle z|a|z \rangle = z.$$

The uncertainty relation for the operators a and $a^\dagger$ is

$$\left\langle a^\dagger a \right\rangle \geqslant \left\langle a^\dagger \right\rangle \langle a \rangle. \tag{4.1.39}$$

For the canonical coherent state we have[2]

$$\left\langle z \left| a^\dagger a \right| z \right\rangle = z^* z = \left\langle z \left| a^\dagger \right| z \right\rangle \langle z |a| z \rangle, \tag{4.1.40}$$

i.e., the equality sign of (4.1.39) holds, implying that the canonical coherent state is a minimum uncertainty state.

Normal ordering of a product of the operators is an ordering to put all creation operators to the left of annihilation operators. The notation for a normal product is putting a colon ":" on each side of the operator product. For example : $aa^\dagger := a^\dagger a$. The matrix element of a normal ordered product of operators between two canonical coherent states is, because of (4.1.38),

$$\left\langle z \left| : F\left(a^\dagger a\right) : \right| z' \right\rangle = F\left(z^*, z'\right) \langle z| z' \rangle. \tag{4.1.41}$$

Equipped with methods dealing with coherent sates we come back to the harmonic oscillator problem. A harmonic oscillator can be described in terms of the creation and annihilation operators by using

$$a = \sqrt{\frac{m\omega}{2\hbar}} \left(q + i\frac{p}{m\omega} \right), \tag{4.1.42a}$$

$$a^\dagger = \sqrt{\frac{m\omega}{2\hbar}} \left(q - i\frac{p}{m\omega} \right). \tag{4.1.42b}$$

where q and p are the coordinate and momentum operators of the oscillator. The Hamiltonian becomes

$$H = \frac{1}{2m}p^2 + \frac{1}{2}m\omega^2 q^2 = \hbar\omega \left(a^\dagger a + \frac{1}{2} \right). \tag{4.1.43}$$

---

[2] Note that $\langle z \mid a \mid z \rangle = z$ and that $\langle z \mid a^\dagger = \langle z \mid z^*$, therefore $\langle z \mid a^\dagger \mid z \rangle = z^*$.

The number operator $N = a^\dagger a$ has eigenvalues $n(n = 0, 1, 2 \cdots)$, and the state with quantum number n of the harmonic oscillator is now a state with particle number n in the number representation. The canonical coherent state is now expressed in terms of the harmonic oscillator states $|n\rangle$ with energies $E_n = (n + \frac{1}{2})\hbar\omega$.

$$|z\rangle = e^{-\frac{1}{2}|z|^2} \sum_{n=0}^{\infty} \frac{1}{\sqrt{n!}} z^n |n\rangle. \tag{4.1.44}$$

How does this state evolve with time? We get the answer by operating from the left on $|z\rangle$ by the time development operator $e^{-iHt/\hbar} = e^{-i\omega a^\dagger a t}$, where we have neglected the zero point energy, which contributes a constant phase factor and does not influence the result:

$$e^{-i\omega a^\dagger a t}|z\rangle = e^{-\frac{1}{2}|z|^2} \sum_{n=0}^{\infty} \frac{1}{\sqrt{n!}} z^n e^{-i\omega n t} |n\rangle$$

$$= e^{-\frac{1}{2}|z|^2} \sum_{n=0}^{\infty} \frac{1}{\sqrt{n!}} \left(z e^{-i\omega t}\right)^n |n\rangle = \left|e^{-i\omega t} z\right\rangle. \tag{4.1.45}$$

The result is a new coherent state whose label is $z e^{-i\omega t}$. The time evolution of the canonical coherent *state* of a harmonic oscillator is reduced to the evolution of its *label*. The evolution is stationary, i.e., the state remains coherent, a minimum uncertainty state. The harmonic oscillator wave packet (4.1.6) given by Schrödinger and its time evolution (4.1.8) is just such a coherent state. Expressing x (i.e., q in (4.1.42)) in terms of a and $a^\dagger$, we have

$$x = \sqrt{\frac{\hbar}{2m\omega}}(a^\dagger + a). \tag{4.1.46}$$

Hence

$$\langle z(t) \mid x \mid z(t)\rangle = \sqrt{\frac{\hbar}{2m\omega}} \langle z \mid e^{i\omega t a^\dagger a}(a^\dagger + a)e^{-i\omega t a^\dagger a} \mid z\rangle$$

$$= \sqrt{\frac{\hbar}{2m\omega}}[\langle z \mid a^\dagger \mid z\rangle e^{i\omega t} + \langle z \mid a \mid z\rangle e^{-i\omega t}]$$

$$= \sqrt{\frac{\hbar}{2m\omega}}[z^*e^{i\omega t} + ze^{-i\omega t}]$$

$$= \sqrt{\frac{2\hbar}{m\omega}}|z|\cos(\omega t - \phi).$$

This completes the demonstration.

For a forced harmonic oscillator, in case the external force depends linearly on the coordinate q and momentum p, or equivalently on a and $a^\dagger$, the properties of a coherent state still remains.[3]

4.2 Circular Orbit Wave Packet and the Radial Wave Packet of the Hydrogen Atom

Schrödinger mentioned in his 1926 paper [1] that the electron orbit in the H atom can be treated in a similar way as in the harmonic oscillator problem, viz. mimicking the classical orbit by superimposing energy eigenstates to obtain a quantum mechanical wave packet. He wrote in 1929 to Lorentz, informing that he had met great mathematical difficulties. He did not mention about this anymore. Examining the construction of the harmonic oscillator wave packet in more detail we can understand the difficulty in constructing the wave packet for the Kepler motion. The wave function (3.1.8) is a strictly periodic function, it returns to the original value after a period $T = 2\pi/\omega$ or an integral multiple of it. This is conditioned by the fact that the dynamical phase factor is

$$e^{-i(E_n/h)t} = e^{-i(n+1/2)\omega t}$$

and by performing the sum over n does not change the period of $\Psi(x,t)$. The energy eigenvalue of the harmonic oscillator depends on n linearly, i.e., the energy levels are equidistant. The H atom energy level system is far from equidistant, especially for the low-lying states. In order to assimilate the Kepler motion of an electron in H atom, we

---

[3]See *Quantum Mechanics* by E. Merzbacher, 2nd edition, Wiley, 1970, Chapter 15, Section 9.

have to limit ourselves to the highly excited states — the Rydberg states. The energy eigenvalue of H atom is

$$E_n = -\frac{(Ze^2)\,m}{2\hbar^2}\frac{1}{n^2}, \tag{4.2.1}$$

for which we have

$$\frac{\partial E_n}{\partial n} = \frac{(Ze^2)^2\,m}{\hbar^2}\frac{1}{n^3} \equiv \hbar\omega_{cl}, \tag{4.2.2}$$

where ω_{cl} is the classical angular frequency of the electron orbit n.[4] For two neighboring levels of very large n the energy difference is $O(n^{-3})$. Hopefully we can get wave packets mimicking the Kepler motion by superimposing eigenfunctions of very large n.

L. S. Brown [3] succeeded in constructing wave packet of high circular orbit of the H atom. The quantum states of the H atom most close to the classical circular orbits are the $n, l = n-1, m = \pm l$ states. The wave function for $l = n - 1$ do not have radial nodes. The spherical harmonics $(l, m = \pm l)$ has a factor $(\sin\theta)^l$, such that ψ has maximum at $\theta = \pi/2$ (the xy plane). Since $m = +l$ and $m = -l$ states have factors $e^{\pm il\phi}$, which can combine with the dynamical phase factor to give wave packets moving in the opposite sense. The radial wave function for a $l = n - 1$ state is

$$u_{nl}(r) = \text{const } r^n e^{-\kappa_n r}, \quad \kappa_n = \frac{Ze^2 m}{\hbar^2 n}. \tag{4.2.3}$$

Let $\kappa_n r \equiv x$, then the wave function becomes

$$f(x) = x^n e^{-x}$$

a Poisson distribution. It has a maximum at $x = n$. Let $x = n + \xi$, where $\xi \ll n$. For large n we have

$$x = n\left(1 + \frac{\xi}{n}\right) \approx n\exp\left[\frac{\xi}{n} - \frac{1}{2}\left(\frac{\xi}{n}\right)^2\right],$$

---

[4]From Bohr's theory we have $\frac{Ze^2}{r^2} = mr\omega_{cl}^2$ and $r_n = \frac{n^2\hbar^2}{Ze^2 m}$. Therefore (4.2.2) follows.

and

$$f(x) \approx n^n e^{-n} \exp\left[-\frac{1}{2}n\left(\frac{\xi}{n}\right)^2\right], \quad n \gg 1. \tag{4.2.4}$$

Back to (4.2.3),[5] using $l = n - 1$ as the label, we write down the wave function

$$\frac{1}{r}u_{l+1,l}(r) = \text{const} \, \exp\left[-\frac{1}{2l}\left(l - \frac{r}{la}\right)^2\right], \quad l \gg 1. \tag{4.2.5}$$

It exhibits the Gaussian form, with r reaching its maximum at $r_p = l^2 a$, and the rms value is $\Delta r = (2l)^{-\frac{1}{2}} r_p$. The angular part is the spherical harmonics

$$Y_l^l(\theta, \phi) = \text{const} \, e^{il\phi} (\sin\theta)^l.$$

For large l we can use the approximation

$$\sin\theta = \cos\left(\theta - \frac{1}{2}\pi\right) \approx 1 - \frac{1}{2}\left(\theta - \frac{1}{2}\pi\right)^2 \approx \exp\left[-\frac{1}{2}\left(\theta - \frac{1}{2}\pi\right)^2\right].$$

The complete wave function is finally

$$\psi_{l+1,l,l}(r) = \text{const} \, e^{-(r-l^2a)^2/2l^3a^2} e^{il\phi} \exp\left[-\frac{1}{2}l(\theta - \pi/2)^2\right]. \tag{4.2.6}$$

This is a ring-shaped stationary wave packet with the radius $l^2 a$. In order to get wave packets of the circular motion with angular frequency ω_{cl}, we need to superimpose such wave functions for l

---

[5] Writing the last factor in (4.2.4) as $e^{-\frac{1}{2l}(\kappa r - l)^2}$, $\kappa = \frac{1}{na}$, where a is the radius of the ground state of the Bohr atom $a = \hbar^2/Ze^2 m$, we keep in (4.2.5) terms involving r on the exponential, and include all others in the constant in front.

values around a large $\bar{l}$:

$$\psi(r,t) = \sum A_l e^{-i(E_l/\hbar)t} \psi_{l+1,l,l}(r), \qquad (4.2.7)$$

where the coefficient A_l is peaked at large $\bar{l}$. The sum is realized over the index s:

$$l = \bar{l} + s, \quad s \ll \bar{l}. \qquad (4.2.8)$$

$$E_l = E_{\bar{l}} + \frac{\partial E_l}{\partial l}\bigg|_{\bar{l}} s + \frac{1}{2}\frac{\partial^2 E_l}{\partial l^2}\bigg|_{\bar{l}} s^2 + \cdots . \qquad (4.2.9)$$

From $E_n = -(Ze^2)^2 m/2\hbar^2 n^2$, it follows that

$$\frac{\partial E_l}{\partial l} = \frac{(Ze^2)^2 m}{\hbar^2 (l+1)^2} = \hbar\omega_{c1}, \qquad (4.2.10)$$

where ω_{c_1} is the angular frequency of Bohr classical orbit.

$$\omega_{c1} = \frac{(Ze^2)^2 m}{n^3 \hbar^3}. \qquad (4.2.11)$$

In addition we need

$$\frac{1}{2}\frac{\partial^2 E_l}{\partial l^2} = -\frac{3(Ze^2)^2 m}{2\hbar(l+1)^4} = -\frac{3}{2}\frac{\hbar\omega_{c1}}{l+1}. \qquad (4.2.12)$$

Substituting (4.2.8)–(4.2.12) back into (4.2.7), we obtain

$$\Psi(r,t) = e^{-i(E_{\bar{l}}/\hbar)t} \psi_{\bar{l}+1,\bar{l},\bar{l}}(r) F(\phi - \omega_{c1}t, t), \qquad (4.2.13)$$

where

$$F(\phi - \omega_{c1}t, t) = \sum_s A_{\bar{l}+s} e^{is(\phi-\omega_{c1}t)} \exp\left(i\frac{3}{2}\frac{\omega_{c1}t}{\bar{l}+1}s^2\right). \qquad (4.2.14)$$

The second exponential factor is close to 1 for small t ($\omega_{cl} t \ll \bar{l}$). Putting it temporarily equal to 1, then

$$F(\phi - \omega_{cl} t, t) \approx f(\phi - \omega_{cl} t) = \sum_s A_{\bar{l}+s} e^{is(\phi - \omega_{cl} t)} \tag{4.2.15}$$

where F depends on t only through the combination $\phi - \omega_{cl} t$. If we choose $A_{\bar{l}+s}$ such that $f(\phi)$ is peaked at $\phi = \phi_p + \omega_{cl} t$ at time t. This is equivalent to a uniform motion on a circular orbit, and is strictly periodic. But for sufficiently long time $\omega_{cl} t \approx \bar{l}$, the second exponential factor begins to exert its influence: it makes the wave packet to diffuse. This factor originates from $\frac{\partial^2 E_l}{\partial l^2}|_{\bar{l}}$, exhibiting the deviation of the energy levels from equidistant even $\bar{l}$ is very large. The variation of the phase factor with l causes the diffusion. Consider a Gaussian model

$$A_{\bar{l}+s} = \text{const } e^{-s^2 [\Delta\phi(0)]^2}. \tag{4.2.16}$$

Equations (4.2.14) gives

$$F(\phi - \omega_{cl} t, t) = \text{const} \int_{-\infty}^{\infty} ds\, e^{-s^2 [\Delta\phi(0)]^2} e^{is(\phi - \omega_{cl} t)} \exp\left(i\frac{3}{2}\frac{\omega_{cl} t}{\bar{l}} s^2\right)$$

$$= \text{const } \exp\left\{-\frac{1}{4}\frac{(\phi - \omega_{cl} t)^2}{[\Delta\phi(0)]^2 - \frac{3}{2} i\omega_{cl}/\bar{l}}\right\}. \tag{4.2.17}$$

Consequently

$$|F(\phi - \omega_{cl} t, t)|^2 = \text{const } \exp\left\{-\frac{(\phi - \omega_{cl} t)^2}{2[\Delta\phi(t)]^2}\right\}, \tag{4.2.18}$$

where

$$[\Delta\phi(t)]^2 = [\Delta\phi(0)]^2 + \left(\frac{3\omega_{cl} t}{2\bar{l}\Delta\phi(0)}\right)^2. \tag{4.2.19}$$

The distribution given by (4.2.14) keeps the Gaussian form, but its width given by (4.2.19) increases with time.

In the late 1980's the discussion on the wave packets formed by using highly excited atomic states revived, thanks to the development of short pulse lasers, which makes possible the construction of wave packets in the laboratory. The calculation of J. Parker and C. R. Stroud [4] shows that the frequency spectrum of a ps short pulse laser is wide enough to excite the atom coherently to a certain number of Rydberg states. After the pulse the wave packet formed begins to oscillate and decay by emitting radiations, similar to the electron moving in a classical atom. Besides the classical particle features, the wave packet broadens, diffuses and revives, exhibiting non-classical wave properties. The formation of wave packets can be described by using a semi-classical model. Let the amplitudes of the ground and the excited states in the interaction picture be $a_g(t)$ and $a_n(t)$ satisfying equations

$$\dot{a}_g = -\frac{1}{2}i \sum_n \Omega_n a_n\left(t\right) f\left(t\right) e^{-i\Delta_n t}, \qquad (4.2.20)$$

$$\dot{a}_n = -\frac{1}{2}i\Omega_n a_g\left(t\right) f\left(t\right) e^{i\Delta_n t}. \qquad (4.2.21)$$

where $f(t)$ is the envelope of the laser pulse, Δ_n is the detuning of the center of the pulse frequencies from the $g \to n$ transition frequency and Ω_n is the Rabi frequency[6] of the transition. Equations (4.2.20) and (4.2.21) are integrated numerically for various cases, with 6ps–10ps pulses (FWHM) and the central frequency set for exciting the atom to the Rydberg state $\bar{n}$=85. Numerical computation shows that 5–10 states around $\bar{n}$ can be significantly excited. It is therefore safe to construct the wave packet by including Rydberg states $60 \leqslant n \leqslant 110$. In the computation of the decay by spontaneous emission of the excited atom all states from the ground state up to $n = 59$ are considered as possible final states. The wave function of the wave packet formed by superimposing Rydberg states is

$$\Psi_R\left(r,t\right) = \sum_n a_n\left(t\right) e^{-i\omega_n t} u_n\left(r\right), \qquad (4.2.22)$$

---

[6]Let the system be in the state i at $t = 0$. The probability that the system remains in state i at time t is $P = \cos^2 \Omega_{ij} t$, where Ω_{ij} is the Rabi frequency.

where ω_n is the transition frequency from the ground state to the state n, and $u_n(r)$ is the radial part of the wave function. Wave packets formed by superimposing radial wave functions only are called the radial wave packets. It oscillates between $r_{\min}$ to $r_{\max}$. Because of the angular momentum selection rules, the laser cannot

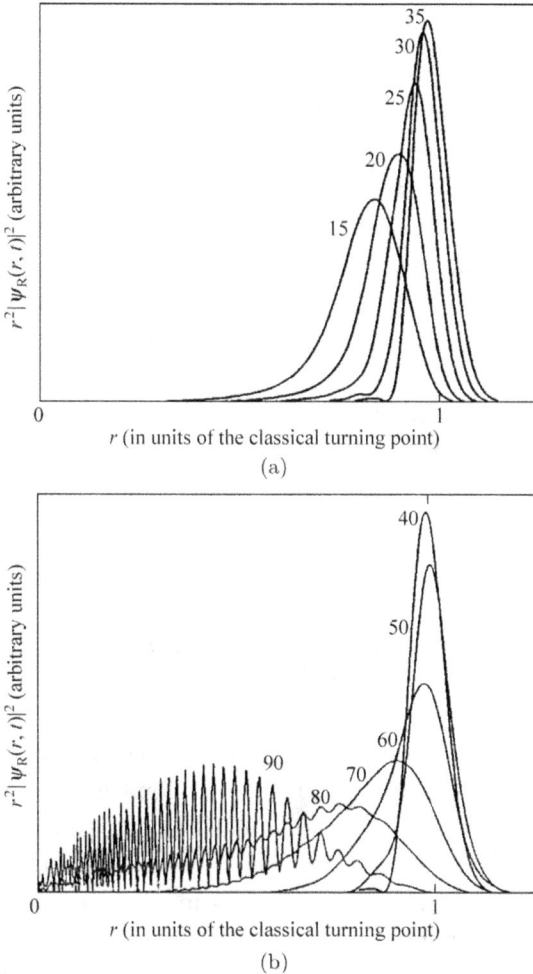

Fig. 4.1 Motion of the radial wave packet (a) Wave packet moves towards the classical turning point; (b) Wave packet returns. Reprinted with permission from Johathan Parker and C. R. Stroud, Jr., *Phys. Rev. Lett.* 56, 716, 1986. Copyright (2017) by the American Physical Society.

excite enough angular momentum states. Hence the wave packet is not localized angle-wise. Figure 4.1(a) shows the development of the wave packet $r^2 \mid \psi_R(r,t) \mid^2$ after the pulse. The numbers represent the time (ps) after the pulse. The wave packet is formed, and subsequently moves towards the classical turning point and contracts. Using Gaussian envelope of the pulse, the uncertainty of the wave packet at the turning point is $\Delta p \Delta r = 0.53\hbar$, close to the minimum uncertainty value. Since the turning point is equivalent to the aphelion point of the orbit, the electron speed is minimum at that point. Therefore, the wave packet becomes steeper and narrower. Figure 4.1(b) describes the return journey of the wave packet. When it approaches the nucleus, it is scattered by the Coulomb potential. The period of the wave packet is equal to the Kepler period of the orbit corresponding to $\bar{n}=85$, and is 93.4 ps in this case. After several periods the wave packet diffuses. It takes a very long time before it revives. If we increase the frequency of the laser pulse ($\bar{n}$ increased) while keeping Δn unchanged (then we have a longer pulse with narrow frequency spectrum) the wave packet can be kept for a longer time before it diffuses.

The power of the spontaneous dipole radiation of the atom is

$$
\begin{aligned}
P(t) &= \frac{4e^2}{3c^3} \langle \Psi_R \mid \ddot{\boldsymbol{r}}(t) \cdot \ddot{\boldsymbol{r}}(t) \mid \Psi_R \rangle \\
&= \frac{4e^2}{3c^3} \left\langle \Psi_R \left| \frac{e^2 \boldsymbol{r}}{mr^3} \cdot \frac{e^2 \boldsymbol{r}}{mr^3} \right| \Psi_R \right\rangle .
\end{aligned} \tag{4.2.23}
$$

The expectation value of the square of acceleration for the wave packet Ψ_R is maximum at the perihelion, and is minimum at aphelion. Figure 4.2 shows the intensity of spontaneous radiation for a time duration of 3 ns, which is approximately equal to 32 periods. In the beginning the radiation intensity exhibits clearly the peak value at the perihelion, and the time interval is just 93.4 ps. The regularity repeats itself at 2.6 ns, signaling a revival of the wave packet. In between there are complicated patterns, but the maxima of intensity repeat themselves at frequencies 2, 3, 4 times the fundamental frequency. The time variation of radiation intensity can be found by substituting the integral of (4.2.21) into (4.2.22) and

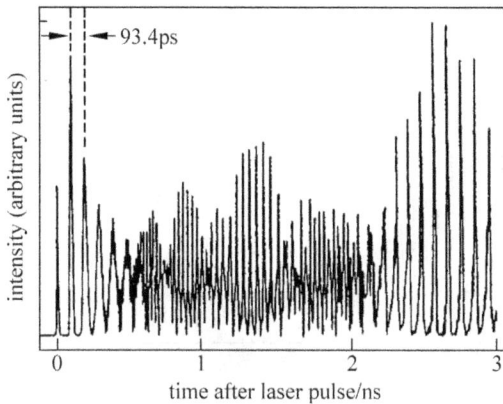

Fig. 4.2 Intensity of spontaneous emission. Reprinted with permission from Johathan Parker and C. R. Stroud, Jr., *Phys. Rev. Lett.* 56, 716, 1986. Copyright (2017) by the American Physical Society.

(4.2.23), and the result is

$$P(t) \propto |\Phi(t)|^2,$$

where

$$\Phi(t) \propto \sum_n e^{i(n-\bar{n})\omega_{c1}t} \exp\left[i\frac{3}{2}\frac{\omega_{c1}}{\bar{n}}(n-\bar{n})^2 t\right]. \qquad (4.2.24)$$

The sum can be performed by including 5~10 energy levels around $\bar{n}$. The first exponential factor in (4.2.24) is periodic, while the second exponential factor represents the effect of deviation of the energy level interval from a constant value. We have already seen this factor in (4.2.14), which causes the wave packet to diffuse. After a long time $t = \frac{\bar{n}}{3}\frac{2\pi}{\omega_{c1}} = \frac{\bar{n}}{3}T_0$ (T_0 is the period of classical motion) the second factor becomes $e^{i\pi(n-\bar{n})^2}$, the wave packet revives, and $P(t)$ also returns to the initial value. The period of revival is

$$T_{\text{rev}} = \frac{\bar{n}}{3}T_0. \qquad (4.2.25)$$

In the calculations of Parker and Stroud, $\bar{n} = 85$, $T_{\text{rev}} = 2.636$ ns. Between the onset and revival there are fractional revivals. There are still higher order terms of $(n-\bar{n})$ which we did not include in (4.2.24). The corresponding revival times of these terms are proportional to $\bar{n}^2 T_0, \bar{n}^3 T_0, \ldots$ respectively. For example for $t = \bar{n}^3 T_0 = 57\,352$ ns all

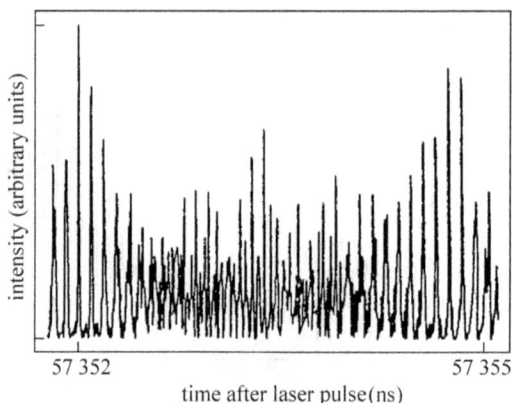

Fig. 4.3 Higher order revivals of the wave packet. Reprinted with permission from Johathan Parker and C. R. Stroud, Jr., *Phys. Rev. Lett.* 56, 716, 1986. Copyright (2017) by the American Physical Society.

terms up to $(n - \bar{n})^4$ give rise to periodic variation of $P(t)$. Figure 4.3 shows such a higher order revival.

4.3 *SO*(4) Symmetry of H atom, the Runge-Lenz Vector and the Kepler Elliptic Orbit Wave Packet

In constructing the wave packet of elliptic orbits a conserved quantity characterizing the orbit shape plays the crucial role. It is the quantum mechanical generalization of the Runge-Lenz vector. Furthermore, this vector leads to a deeper understanding of the symmetry of the H atom. It turns out that the H atom possesses a dynamical symmetry $SO(4)$. This was found by W. Pauli in 1926 [5] and later by V. A. Fock [6]. A complete knowledge of this symmetry helps to find all the properties of the system including its energy eigenvalues. The quantum mechanical Runge-Lenz vector is the generator necessary to enlarge $SO(3)$, the symmetry of the 3-dimensional rotation symmetry of the central field problem, to $SO(4)$, the dynamical symmetry for the Coulomb potential.[7]

---

[7]A more general and extensive algebraic structure, the Yangian, can be constructed using the Runge-Lenz vector operator and the angular momentum operators. See Chapter 11.

4.3.1 The Runge-Lenz Vector of the Classical Kepler motion

Classical Kepler motion is the motion of a particle of mass m in a central field potential[8]

$$V(r) = -\frac{k}{r}. \tag{4.3.1}$$

For hydrogen like atoms $k = Ze^2$. The trajectory of Kepler motion is the conic section, which is described in polar coordinates (r, θ) as

$$\frac{1}{r} = \frac{mk}{l^2}(1 + e\cos\theta), \tag{4.3.2}$$

where l is the magnitude of the angular momentum $\boldsymbol{L} = \boldsymbol{r} \times \boldsymbol{p}$, and e is the eccentricity of the orbit. The relation between e, l and the energy E is given by

$$e = \left(1 + \frac{2El^2}{mk^2}\right)^{1/2}. \tag{4.3.3}$$

E and $\boldsymbol{L}$ are constants of motion. $E > 0$ implies $e > 1$, the orbit is a hyperbola. $E < 0$ implies $e < 1$, the orbit is an ellipse. A special case of the ellipse is the circle, for which $e = 0$, the energy for a circular orbit is $E = -\frac{mk^2}{2l^2}$. An important property of the elliptic orbit is that its semi-major axis is determined solely by its energy. This is simply proved in the following. The *apsidal distances* r_2 and r_1 are defined as the distances between the origin (a focus of the conic section) to the farthermost and the nearest points on the orbit respectively. At these two points the radial velocity of the particle vanishes, and the azimuthal velocity is $\frac{l}{mr}$. The conservation of energy gives a relation valid at the apsidal distances:

$$E = -\frac{k}{r} + \frac{1}{2}\frac{l^2}{mr^2}, \tag{4.3.4}$$

---

[8]For the Kepler motion of a classical particle, see H. Goldstein "*Classical Mechanics*" 2nd edition, Reading, Addison-Wesley 1980. For the Runge-Lenz vector and its subsequent quantum mechanical generalization, see L. Schiff "*Quantum Mechanics*", 3rd edition, New York, McGraw Hill, 1968.

where r takes values r_2 or r_1. Rearranging this into an equation for r, we get

$$r^2 + \frac{k}{E}r - \frac{l^2}{2mE} = 0. \tag{4.3.5}$$

The sum of the two roots of Eq. (4.3.5) is

$$r_1 + r_2 = -\frac{k}{E},$$

which leads to

$$a = \frac{r_1 + r_2}{2} = -\frac{k}{2E}, \tag{4.3.6}$$

and this completes the proof. The eccentricity of the ellipse is related to the semi-major axis and the semi-minor axis in the following way:

$$e = \left(a^2 - b^2\right)^{1/2}/a. \tag{4.3.7}$$

Equations (4.3.3) and (4.3.6) give

$$e = \left(1 - \frac{l^2}{mka}\right)^{1/2}. \tag{4.3.8}$$

When E and $\boldsymbol{L}$ are fixed, the shape of the orbit and the plane of the orbit are determined. But the orientation of the orbit on the plane remains undetermined. There exists yet another constant of motion. Starting from the equation of motion

$$\dot{\boldsymbol{p}} = -\frac{k}{r^3}\boldsymbol{r}, \tag{4.3.9}$$

we take the vector product of $\boldsymbol{L}$ with both sides and obtain

$$\dot{\boldsymbol{p}} \times \boldsymbol{L} = -\frac{k}{r^3}\boldsymbol{r} \times (\boldsymbol{r} \times m\dot{\boldsymbol{r}}) = mk\left(\frac{\dot{\boldsymbol{r}}}{r} - \frac{\dot{r}\boldsymbol{r}}{r^2}\right).$$

Since $\frac{d\boldsymbol{L}}{dt} = 0$, the left-hand side can be written as $\frac{d}{dt}(\boldsymbol{p} \times \boldsymbol{L})$ and the right-hand side can be written as $mk\frac{d}{dt}\left(\frac{\boldsymbol{r}}{r}\right)$. Consequently

$$\frac{d}{dt}\left(\boldsymbol{p} \times \boldsymbol{L} - mk\frac{\boldsymbol{r}}{r}\right) = 0. \tag{4.3.10}$$

Define the Runge-Lenz vector $\boldsymbol{M}$ as

$$\boldsymbol{M} = \frac{1}{m}\boldsymbol{p} \times \boldsymbol{L} - k\frac{\boldsymbol{r}}{r}, \tag{4.3.11}$$

then (4.3.10) gives

$$\frac{d\boldsymbol{M}}{dt} = 0, \tag{4.3.12}$$

implying that $\boldsymbol{M}$ is a conserved quantity. Since $\boldsymbol{L}$ is perpendicular to the plane of the orbit, we have

$$\boldsymbol{M} \cdot \boldsymbol{L} = 0, \tag{4.3.13}$$

i.e., $\boldsymbol{M}$ lies in the orbit plane. Denoting by θ the angle between $\boldsymbol{M}$ and $\boldsymbol{r}$, we calculate $\boldsymbol{M} \cdot \boldsymbol{r}$ to get

$$\boldsymbol{M} \cdot \boldsymbol{r} = \frac{1}{m}\boldsymbol{r} \cdot \boldsymbol{p} \times \boldsymbol{L} - kr = \frac{1}{m}l^2 - kr.$$

i.e.,

$$Mr\cos\theta = \frac{1}{m}l^2 - kr,$$

or

$$\frac{1}{r} = \frac{mk}{l^2}\left(1 + \frac{M}{k}\cos\theta\right). \tag{4.3.14}$$

Comparing (4.3.14) with the equation of Kepler orbit (4.3.2) we see that the angle θ defined above is just the azimuthal angle, which implies that $\boldsymbol{M}$ is in the direction of the major axis, and furthermore

$$M = ke. \tag{4.3.15}$$

From the definition of $\boldsymbol{M}$ (4.3.9), we obtain

$$M^2 = \frac{2H}{m}L^2 + k^2, \tag{4.3.16}$$

where H is the Hamiltonian

$$H = \frac{1}{2m}p^2 - \frac{k}{r}. \tag{4.3.17}$$

4.3.2   The quantum mechanical Runge-Lenz vector, the dynamical symmetry and the energy spectrum of the hydrogen atom

To generalize the Runge-Lenz vector defined in (4.3.11) to a quantum mechanical operator, we encounter the problem of ordering the operators $\boldsymbol{p}$ and $\boldsymbol{L}$. W. Pauli [5] introduced the symmetric ordering operator,[9]

$$\boldsymbol{M} = \frac{1}{2m} (\boldsymbol{p} \times \boldsymbol{L} - \boldsymbol{L} \times \boldsymbol{p}) - k\frac{\boldsymbol{r}}{r}. \tag{4.3.18}$$

From this definition we still have

$$\boldsymbol{L} \cdot \boldsymbol{M} = \boldsymbol{M} \cdot \boldsymbol{L} = 0 \tag{4.3.19}$$

From (4.3.18) we obtain

$$\boldsymbol{M}^2 = \frac{2H}{m} \left(\boldsymbol{L}^2 + \hbar^2\right) + k^2. \tag{4.3.20}$$

which differs from the classical relation (4.3.16) by a term $\frac{2H\hbar^2}{m}$.

In physics, the conserved quantities are related to the symmetry. In this context we need to know all the commutation relations between the operators associated with the conserved observables, i.e., the algebra. In the Kepler problem L^2, L_z are associated with the 3-dimensional rotation $SO(3)$. This is a symmetry of the space. In defining $\boldsymbol{M}$, the Coulomb potential is involved. This concerns the Hamiltonian and we are dealing with the dynamical symmetry. Altogether we have 6 operators $L_x, L_y, L_z; M_x, M_y, M_z$. Known commutation relations are

$$[L_x, L_y] = i\hbar L_z, \tag{4.3.21}$$

and two other relations obtained by cyclic permutations. The commutation relations between components of $\boldsymbol{M}$ and those

---

[9]For this subsection, also see Refs. [6, 7].

of $\boldsymbol{L}$ are

$$[M_x, L_x] = 0, \tag{4.3.22}$$

$$[M_x, L_y] = ihM_z, \tag{4.3.23}$$

$$[M_x, L_z] = -ihM_y, \tag{4.3.24}$$

and 6 other relations obtained by cyclic permutations. The commutation relations between components of $\boldsymbol{M}$ are more complicated, because the Hamiltonian H is involved. One of them is

$$[M_x, M_y] = -\frac{2i\hbar}{m}HL_z. \tag{4.3.25}$$

The other two can be obtained by cyclic permutation. Since H appears in the commutation relations, the algebra of 6 operators $\boldsymbol{L}, \boldsymbol{M}$ is not closed. But if we confine the problem within a subspace of the Hilbert space $\mathcal{H}$ defined by a special eigenvalue E, then H can be simply replaced by E. We rescale the operator by defining the vector $\boldsymbol{M'}$:

$$\boldsymbol{M'} = \left(-\frac{m}{2E}\right)^{\frac{1}{2}} \boldsymbol{M}, \tag{4.3.26}$$

Then we have

$$[M_x', M_y'] = ihL_z \tag{4.3.27}$$

and other two relations obtained by cyclic permutations. Since (4.3.22)~(4.3.24) are linear relations with respect to the components of $\boldsymbol{M}$, they remain linear with respect to the components of $\boldsymbol{M'}$ after the rescaling. Consequently by confining to a subspace of $\mathcal{H}$ and rescaling the generators, a closed Lie algebra is obtained. The crucial step is the dependence of the rescaling on E, which leads eventually to the *energy spectrum determined by the Lie algebra*.

Consider a 4-dimensional rotation group $SO(4)$ with generators

$$L_{ij} = r_i p_j - r_j p_i, \quad i, j = 1, 2, 3, 4 \tag{4.3.28}$$

Since the diagonal elements $i = j$ vanish, and $L_{ij} = -L_{ji}$, there are only 6 independent generators. Let

$$\left. \begin{array}{l} \boldsymbol{L} = (L_{23}, L_{31}, L_{12}) \\ \boldsymbol{M'} = (L_{14}, L_{24}, L_{34}) \end{array} \right\} \tag{4.3.29}$$

Then we can obtain the commutation relations between the components of $\boldsymbol{L}$ and $\boldsymbol{M'}$ on the basis of the commutation relations of x_i and p_j.

$$[r_i, p_j] = i\hbar\delta_{ij}, \quad i, j = 1, 2, 3, 4. \tag{4.3.30}$$

The $so(4)$ algebra[10] consists of two $su(2)$ subalgebra:

$$so(4) \supset su(2) \oplus su(2).$$

Define

$$\left. \begin{array}{l} \boldsymbol{I} = \dfrac{1}{2}\left(\boldsymbol{L} + \boldsymbol{M'}\right) \\[2mm] \boldsymbol{K} = \dfrac{1}{2}\left(\boldsymbol{L} - \boldsymbol{M'}\right) \end{array} \right\}. \tag{4.3.31}$$

It is easy to verify that any component of $\boldsymbol{I}$ commutes with any component of $\boldsymbol{K}$, and $\boldsymbol{I}$ and $\boldsymbol{K}$ constitute two independent $su(2)$ algebra, i.e.,

$$[I_x, I_y] = i\hbar I_z, \tag{4.3.32a}$$

$$[K_x, K_y] = i\hbar K_z, \tag{4.3.32b}$$

and other relations obtained by cyclic permutations of the indices x, y, and z. The two algebras are independent, i.e.,

$$[\boldsymbol{I}, \boldsymbol{K}] = 0. \tag{4.3.33}$$

Furthermore,

$$[\boldsymbol{I}, H] = 0, \tag{4.3.34}$$
$$[\boldsymbol{K}, H] = 0.$$

---

[10]We denote the Lie algebra by small letters and the corresponding Lie group by capital letters.

Commutation relations (4.3.29)~(4.3.31) lead to

$$\left.\begin{array}{l} \boldsymbol{I}^2 = i(i+1)\hbar^2 \\ \boldsymbol{K}^2 = \kappa(\kappa+1)\hbar^2 \end{array}\right\} \tag{4.3.35}$$

$$i, \kappa = 0, \frac{1}{2}, 1\cdots,$$

$so(4)$ is an algebra of rank 2, and there are two Casimir operators. They are composed of the generators of the algebra, and they commute with all of them. Of course $\boldsymbol{I}^2$ and $\boldsymbol{K}^2$ are candidates, but we choose their linear combinations

$$C = \boldsymbol{I}^2 + \boldsymbol{K}^2 = \frac{1}{2}\left(\boldsymbol{L}^2 + \boldsymbol{M}'^2\right)$$

$$\tag{4.3.36}$$

$$C' = \boldsymbol{I}^2 - \boldsymbol{K}^2 = \boldsymbol{L} \cdot \boldsymbol{M}'.$$

Because of the peculiarity of the Kepler problem, we have $\boldsymbol{L} \cdot \boldsymbol{M}' = 0$. The corresponding $so(4)$ is a special case $\boldsymbol{I}^2 = \boldsymbol{K}^2$, i.e., $i = \kappa$. Consequently

$$C = 2\kappa\left(\kappa+1\right)\hbar^2, \quad \kappa = 0, \frac{1}{2}, 1, \ldots \tag{4.3.37}$$

From Eqs. (4.3.20) and (4.3.26), we obtain

$$C = \frac{1}{2}\left(\boldsymbol{L}^2 - \frac{M}{2E}\boldsymbol{M}^2\right) = -\frac{mk^2}{4E} - \frac{1}{2}\hbar^2.$$

Comparing this with (4.3.37) and solving for E, we obtain

$$E = -\frac{mk^2}{2\hbar^2\left(2\kappa+1\right)^2}. \tag{4.3.38}$$

Let $n = 2\kappa+1$, which takes the value $n = 1, 2, 3, \ldots$. Replacing k by Ze^2 we obtain

$$E_n = -\frac{mZ^2e^4}{2\hbar^2 n^2}, \tag{4.3.39}$$

which is just the energy eigenvalues of the H atom. We obtain this without resorting to Schrödinger equation, only relying on the algebraic properties of the dynamical symmetry.

The dynamical symmetry plays important roles in the study of nuclear collective modes. This is the well-known "interacting boson model", which is a powerful demonstration how the group representation theory is applied to physics.

4.3.3 Construction of the Kepler Elliptic Orbit

M. Nauenberg [8] constructed the quantum mechanical wave packet of the Kepler elliptic orbit using the quantum Runge-Lenz vector. The classical orbit with eccentricity e lies on the xy-plane. The corresponding operators M_x, M_y and L_z constitute the generators for the $SO(3)$ symmetry of the Coulomb potential[11]:

$$[M_x, M_y] = -2\mathrm{i}HL_z, \quad [L_z, M_x] = \mathrm{i}M_y, \quad [L_z, M_y] = -\mathrm{i}M_x, \tag{4.3.40}$$

where

$$H = \frac{p^2}{2} - \frac{1}{r}. \tag{4.3.41}$$

The problem is actually one for 2-dimensional H atom.[12] The energy eigenvalues for the bound states are

$$E_n = -\frac{m\left(Ze^2\right)^2}{2\hbar^2 n^2}, \quad n = \frac{1}{2}, \frac{3}{2}, \dots \tag{4.3.42}$$

with the corresponding eigenfunctions.

$$\psi(\rho, \phi) \propto \mathrm{e}^{\mathrm{i}l\phi} \rho^{|l|} \mathrm{e}^{-\rho/n} F(-n_\rho, \, 2\,|l| + 1, \, 2\rho/n), \tag{4.3.43}$$

where (ρ, ϕ) are the polar coordinates, F are the confluent hypergeometric functions, and among the three quantum numbers only two

---

[11] We follow Ref. [8] in using the atomic units $Ze^2 = \hbar = m = 1$.

[12] See H. Mavromatis "*Exercises in Quantum Mechanics*" 2nd edition, Dordrecht, Kluwer, 1992, example 8.3. In (4.3.42) we restored m, Ze^2, and $\hbar$.

are independent.

$$l = 0, \pm 1, \pm 2, \ldots$$
$$n_\rho = 0, 1, 2, \ldots$$
$$n = n_\rho + |l| + \frac{1}{2} = \frac{1}{2}, \frac{3}{2}, \frac{5}{2}, \ldots$$

$$\left.\begin{array}{l}\\\\\\\end{array}\right\} \qquad (4.3.44)$$

From the commutation relations between M_x and M_y (4.3.40) the corresponding uncertainty relations can be obtained

$$\Delta M_x \Delta M_y \geqslant \frac{1}{2} |\langle -2HL_z \rangle| . \qquad (4.3.45)$$

We demand that the basis functions from which the wave packets are constructed to be eigenfunctions of H and have minimum uncertainty in $\Delta M_x \Delta M_y$. This can be achieved by demanding these states to satisfy the eigenequation

$$(M_x + i\delta M_y)\psi = \eta\psi, \qquad (4.3.46)$$

where δ is a real parameter (to be connected with the eccentricity), η is the eigenvalue of the non-hermitian operator $M_x + i\delta M_y$, and hence can be complex. Since M_x and M_y do not commute, ψ is not the simultaneous eigenfunction of them. ψ is at the same time an eigenfunction of H:

$$H\psi = E_n\psi. \qquad (4.3.47)$$

The eigenfunctions for constructing the wave packet are chosen from solutions to Eqs. (4.3.46) and (4.3.47). They should be explicitly expressed in terms of the eccentricity e. Equation (4.3.46) gives

$$\langle M_x \rangle + i\delta \langle M_y \rangle = \eta. \qquad (4.3.48)$$

Multiplying (4.3.46) from the left by $M_x - i\delta M_y$, we obtain

$$(M_x^2 + i\delta[M_x, M_y] + \delta^2 M_y^2)\psi = (M_x^2 + 2\delta HL_z + \delta^2 M_y^2)\psi$$
$$= \eta(M_x - i\delta M_y)\psi,$$

which further leads to

$$\langle M_x^2 \rangle + 2\delta E_n \langle L_z \rangle + \delta^2 \langle M_y^2 \rangle = (\eta \langle M_x \rangle - i\delta \langle M_y \rangle)$$
$$= \langle M_x \rangle^2 + \delta^2 \langle M_y \rangle^2,$$

where the last equality is obtained by using (4.3.48). Moving the right-hand side to the left, we get

$$(\Delta M_x)^2 + \delta^2 (\Delta M_y)^2 + 2\delta E_n \langle L_z \rangle = 0. \tag{4.3.49}$$

We denote $\Delta M_x \Delta M_y$ by K, Eq. (4.3.49) gives the condition for minimizing K:

$$(\Delta M_x)^2 = \delta K, \quad (\Delta M_y)^2 = \frac{K}{\delta}. \tag{4.3.50}$$

Substituting back into (4.3.49), we get

$$K = -E_n \langle L_z \rangle.$$

Consequently the minimum uncertainty can be reached under the following conditions.

$$\left. \begin{array}{l} (\Delta M_x)^2 = -E_n \delta \langle L_z \rangle \\[2mm] (\Delta M_y)^2 = -\dfrac{E_n}{\delta} \langle L_z \rangle \end{array} \right\} \tag{4.3.51}$$

This completes the proof that the simultaneous solutions of (4.3.46) and (4.3.47) are the minimum uncertainty states, i.e., the equality sign of (4.3.45) holds for these solutions. Properties of these eigenfunctions concerning the eccentricity can be revealed by introducing the ladder operators

$$A_\pm = \pm \frac{1}{(-2H)^{1/2}} (\delta M_x + iM_y) - (1 - \delta^2)^{1/2} L_z, \tag{4.3.52}$$

where the parameter δ is restricted by $0 \leqslant \delta^2 \leqslant 1$. Consider $A_+\psi$:

$$A_+\psi = \left[\frac{1}{(-2H)^{1/2}} (\delta M_x + iM_y) - (1 - \delta^2)^{1/2} L_z \right] \psi. \tag{4.3.53}$$

Multiplying this equation from the left by $M_x + i\delta M_y$, and commuting A_+ through to the left using commutation relations (4.3.40), we get

$$(M_x + i\delta M_y)A_+\psi = A_+(M_x + i\delta M_y)\psi + A_+(-2H)^{1/2}(1-\delta^2)^{1/2}\psi$$

$$= \eta A_+\psi + (-2E_n)^{1/2}\left(1-\delta^2\right)^{1/2}A_+\psi$$

$$= \left\{\eta + \left[-2E_n\left(1-\delta^2\right)\right]^{1/2}\right\}A_+\psi.$$

This process can be repeated for $A_+^2\psi, A_+^3\psi, \ldots$

$$(M_x + i\delta M_y)A_+^m\psi = \left\{\eta + m\left[-2E_n\left(1-\delta^2\right)\right]^{1/2}\right\}A_+^m\psi. \quad (4.3.54)$$

We see that $A_+^m\psi$ remains to be an eigenstate of $M_x + i\delta M_y$, corresponding to the eigenvalue $\eta + m[-2E_n(1-\delta^2)]^{\frac{1}{2}}$. Similar derivation shows that $A_-^m\psi$ remains to be an eigenstate of $M_x + i\delta M_y$, corresponding to the eigenvalue $\eta - m[-2E_n(1-\delta^2)]^{\frac{1}{2}}$. The Runge-Lenz vector is directed along the major axis of the ellipse, and its magnitude is e (in atomic units). If we choose in (4.3.48) a real η, then

$$\langle M_x \rangle = \eta, \quad \langle M_y \rangle = 0. \quad (4.3.55)$$

We begin with the eigenfunction for $\eta = 0$. Applying m times to it by A_+ we reach a state with

$$e = \langle M_x \rangle = m\left[-2E_n\left(1-\delta^2\right)\right]^{1/2}, \quad \langle M_y \rangle = 0. \quad (4.3.56)$$

Apparently the successive application of the raising and lowering operators will lead eventually to the maximum (minimum) eccentricity. Consider a special simultaneous eigenstate ψ_n^δ, the *energy coherent state*, of H and $M_x + i\delta M_y$, which satisfies

$$A_+\psi_n^\delta = \left[\frac{1}{(-2H)^{1/2}}\left(\delta M_x + iM_y\right) - \left(1-\delta^2\right)^{1/2}L_z\right]\psi_n^\delta = 0. \quad (4.3.57)$$

From (4.3.57) we see that the expectation value $\langle M_x \rangle$ of ψ_n^δ satisfies the following relation:

$$\delta \langle M_x \rangle = (-2E_n)^{1/2} \left(1 - \delta^2\right)^{1/2} \langle L_z \rangle, \qquad (4.3.58)$$

since we have set $\langle M_y \rangle = 0$. For fixed δ and L_z, the expectation value $\langle M_x \rangle$ has reached its maximum value for the state ψ_n^δ, i.e., the value of m in Eqn. (4.3.56) has reached its maximum possible value denoted by l_n, i.e., $\langle M_x \rangle = l_n [-2E_n(1-\delta^2)]^{\frac{1}{2}}$. Equation (4.3.58) then gives $\langle L_z \rangle = l_n \delta$. Eventually we have a complete understanding of the energy coherent state. It is an eigenstate of H, with eigenvalue $-\frac{1}{2n^2}$; it is an eigenstate of $M_x + i\delta M_y$, with eigenvalue $\frac{l_n}{n}(1-\delta^2)^{\frac{1}{2}}$ and expectation values $\langle M_x \rangle = \frac{l_n}{n}(1-\delta^2)^{\frac{1}{2}}$ and $\langle M_y \rangle = 0$; it is a minimum uncertainty state of $\Delta M_x \Delta M_y$. The quantum number l_n is connected with the maximum eccentricity. From Sommerfeld's theory of elliptic orbit we know that the angular momentum quantum number of the orbit with maximum eccentricity is[13]

$$l_n = n - 1 \text{ (3-dimensional H atom)},$$

$$l_n = n - \frac{1}{2}\text{(2-dimensional H atom).} \qquad (4.3.59)$$

The important point is, when n becomes large, $e = \langle M_x \rangle \rightarrow (1-\delta^2)^{\frac{1}{2}}$, *independent of l_n*. To assimilate a classical orbit (e definite), eigenfunctions with different n, (i.e., different l_n) are required to form the wave packet. The independence of e on l_n makes the formation of wave packet with a fixed e clear-cut and exact. From the relation $\langle L_z \rangle = l_n \delta$ we have, for n large, $\delta = \frac{\langle L_z \rangle}{l_n} \rightarrow \frac{\langle L_z \rangle}{n} = (-2E_n)^{\frac{1}{2}} \langle L_z \rangle$. Consequently

$$e = \left(1 - \delta^2\right)^{1/2} = \left(1 + 2E_n \langle L_z \rangle^2\right)^{1/2}. \qquad (4.3.60)$$

---

[13]See (4.3.44). The l_n here is the $n_\rho + |l|$ of (4.3.44).

Comparing with the eccentricity of classical orbit (4.3.3) (in atomic units)

$$e = (1 + 2El^2)^{\frac{1}{2}}$$

we find a complete agreement.

What we have observed are the remarkable properties of the energy coherent states, but not yet the specific form of them. By expanding $\psi_n^\delta(\boldsymbol{r})$ in terms of the eigenfunctions of 2-dimensional H atom $\psi_{n,l}(\boldsymbol{r})$ (4.3.43):

$$\psi_n^\delta(\boldsymbol{r}) = \sum_{l=-l_n}^{l_n} C_{n,l}^\delta \psi_{n,l}(\boldsymbol{r}) \tag{4.3.61}$$

and demanding ψ_n^δ to satisfy

$$(M_x + i\delta M_y)\,\psi_n^\delta = \frac{l_n}{n}\left(1 - \delta^2\right)^{1/2}\psi_n^\delta, \tag{4.3.62}$$

the coefficients $C_{n,l}^\delta$ in (4.3.61) are determined:

$$C_{n,l}^\delta = \frac{1}{2^{l_n}}\left[\frac{(2l_n)!}{(l_n + l)!\,(l_n - l)!}\right]^{1/2}\left(1 - \delta^2\right)^{l_n/2}\left(\frac{1+\delta}{1-\delta}\right)^{1/2}. \tag{4.3.63}$$

When l_n gets large, $C_{n,l}^\delta$ can be approximated by a Gaussian distribution in l:

$$C_{n,l}^\delta \approx \left[\frac{\pi}{2}l_n\left(1 - \delta^2\right)\right]^{-1/4}\exp\left[-\frac{(l - \delta l_n)^2}{l_n\left(1 - \delta^2\right)}\right]. \tag{4.3.64}$$

The energy coherent state $\psi_n^\delta(\boldsymbol{r})$ (4.3.61) has a definite energy value $-\frac{1}{2n^2}$, its parameter δ results from superimposing eigenfunctions of different l with coefficients (4.3.63). Its spatial distribution is profoundly peaked on the classical orbit with eccentricity e. Figure 4.4 shows the probability distribution of $\psi_n^\delta(\boldsymbol{r})$, $l_n = 40$, $\delta = 0.8$ corresponding to $e = 0.6$, $\langle L_z \rangle = 32$. The peak at aphelion coincides with the classical distribution. The orbital velocity is minimum at this point. The wave packet described above is stationary. To obtain

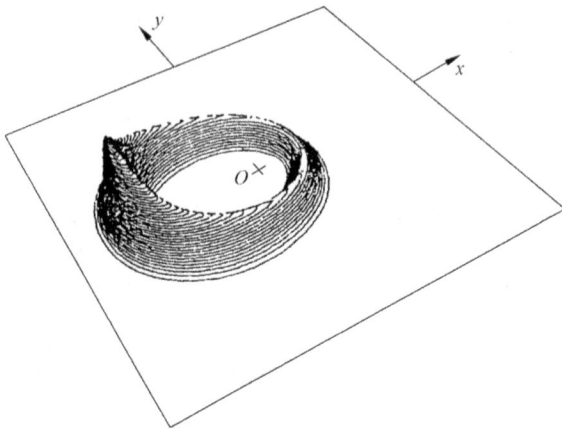

Fig. 4.4　The spatial probability distribution of an energy coherent state. Reprinted with permission from Michael Nauenberg, *Phys. Rev.* A40, 1133, 1989. Copyright (2017) by the American Physical Society.

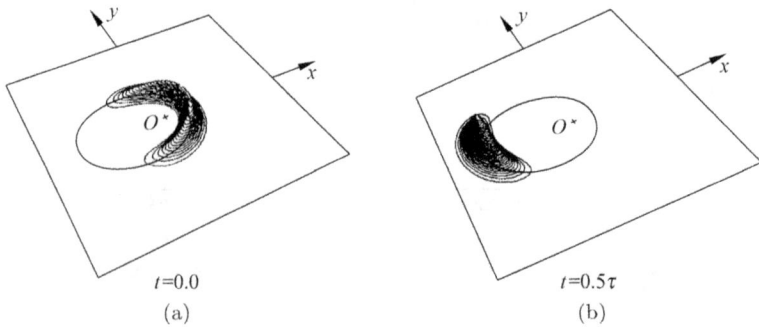

Fig. 4.5　Motion of the wave packet (a) $t = 0$; (b) $t = 0.5T_0$, T_0 is the Kepler period. Reprinted with permission from Michael Nauenberg, *Phys. Rev.* A40, 1133, 1989. Copyright (2017) by the American Physical Society.

a moving, localized wave packet, we need to superimpose the energy eigenstates

$$\Psi^\delta(\boldsymbol{r}, t) = \sum_n a_n \psi_n^\delta(\boldsymbol{r}) \, e^{-iE_n t}. \qquad (4.3.65)$$

We can use Gaussian distribution for the coefficients a_n,

$$a_n = \left(2\pi\sigma^2\right)^{-1/4} e^{-(l_n - l_0)^2/4\sigma^2}. \qquad (4.3.66)$$

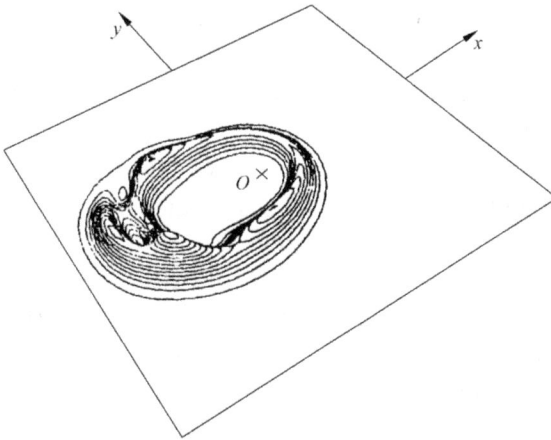

Fig. 4.6   The diffusion of wave packet at $t = 2T_0$. Reprinted with permission from Michael Nauenberg, *Phys. Rev.* A40, 1133, 1989. Copyright (2017) by the American Physical Society.

In Fig. 4.5 the 2-dimensional motion of the wave packet is presented: $e = 0.6, \langle L_z \rangle = 32, \delta = 0.8, l_0 = 40, \sigma^2 = 3.0$. We observe that starting from the perihelion, the wave packet gets slower, contracts and becomes rather steep at the aphelion. When the packet returns to the perihelion its width is larger than that at $t = 0$. After two Kepler periods (Fig. 4.6) the head of the packet catches up with the tail and quantum interference effects occur, causing non-uniform variations of the amplitude of the wave packet along the ellipse. The wave packet diffuses, fractional revivals intermit in between, and revival occurs after long time. In fact these quantum effects mark a breakdown of the semiclassical approximation at long times.

4.3.4   Rutherford atom in quantum mechanics

By Rutherford atom we mean the atomic model in which the electrons perform planetary motion around the nucleus with orbits unquantized. Is it possible to construct a non-spreading quantum mechanical wave packet mimicking closely the Kepler motion of the electron in an atom? This turns out to be possible when some subsidiary role is present. It can be an external field, or another electron as the third body. The story began with the Trojan asteroids orbiting the Sun at the stable Lagrange points of the Sun-Jupiter system. Lagrange points are points of equilibrium in the restricted three-body problem of the Sun, a planet and a third test body with mass m. The planet rotates about the Sun (more exactly about the center of mass of the Sun-planet system) with angular frequency ω. In the reference frame co-rotating with the planet there exist point of equilibrium at which the resultant of gravitational attractions of the Sun and the planet just balances the centrifugal force. The test mass situated at such points can remain at rest in the rotating frame, or rotates with the same angular frequency ω about the Sun in the inertial reference frame of the Sun-planet system. If the Lagrange points are stable equilibrium points, the test mass can oscillate around these points. In the rotating frame, the gravitational force of the planet acting on the test mass is a constant vector, or a vector rotating about the center of mass of the Sun-planet system with angular frequency ω in the inertial frame. I. Bialynicki-Birula, M.Kalinski and J.H. Eberly [37, 38] suggested that such a system can be realized in a hydrogenic atom in a circularly polarized electromagnetic field of angular frequency ω. There exist stable, though non-stationary quantum states of an electron moving on circular orbits in a Coulomb field and the field of a circularly polarized electromagnetic field. The Hamiltonian of such a system in atomic units $(e = m = \hbar = 1)$ is given by Ref. [39][14]

$$H = \frac{\vec{p}^{\,2}}{2} - \frac{1}{4} + \varepsilon x - \omega L_z, \qquad (4.3.67)$$

where ε is the electric field strength, and the field is along the x axis of the frame of reference rotating with angular frequency ω. The eigenfunction of this Hamiltonian with eigenvalue $E^j(\varepsilon)$ is expanded

---

[14]In the atomic unit system, the unit of length is the Bohr orbit radius $a_0 = \frac{\hbar^2}{e^2 m_e} = 5.29 \times 10^{-11}$ m, the unit of energy is the binding energy of hydrogen atom $\frac{m_e e^4}{2\hbar^2} = 13.6$ eV. All other units can be derived from these fundamental ones. For instance, the unit of time is that required for the light to traverse a distance equal to 1 Bohr orbit radius, viz. 1.765×10^{-19} s.

in terms of hydrogenic wave functions

$$\Psi_\varepsilon(r,\theta,\phi) = \sum c_{nlm}(\varepsilon) R_{nl}(r) Y_{lm}(\theta,\phi). \qquad (4.3.68)$$

The stationary Schrödinger equation leads to equations for the expansion coefficients

$$\varepsilon \sum x_{nlm}^{n'l'm'} c_{n'l'm'}^j = [E^j(\varepsilon) - E_n + m\omega]c_{nlm}^j,$$

where j labels the energy eigenvalue of (4.3.67), $E_n = -1/2n^2$ is the eigenvalue of the Hydrogen atom, and $x_{nlm}^{n'l'm'}$ is the dipole matrix element. As we have discussed in section 4.2, the wave packet is to be constructed by using Rydberg circular states $|n, n-1, n-1\rangle$ for n large. We consider manifolds $|n, n-1-k, n-1-k-s\rangle$ labeled by k and s, where $k \ll n, s \ll n$. Specific expressions for dipole matrix elements are known [40], and under conditions $k \ll n, s \ll n$ for n large the dipole matrix elements between different manifolds are small and can be neglected compared with matrix elements in the same manifold. Additionally we concentrate on rotating frame eigenfunctions well localized in angular momentum space, namely, those with only a limited number of terms in the expansion (4.3.68) centered around some value n_0. Under the above constraints we can take $x_{nlm}^{n'l'm'} = n_0^2/2 = r_0/2$ for all matrix elements assumed to be non-zero, where r_0 is the radius of the Bohr orbit with quantum number n_0. The localization in n allows us to expand the hydrogen energy eigenvalue around n_0 up to second order, namely

$$E_n = -\frac{1}{2n_0^2} + \omega_c \delta n - \frac{3}{2}\frac{\delta n^2}{r_0^2}, \qquad (4.3.69)$$

where $\omega_c = 1/n_0^3$ is the Kepler frequency of the orbit n_0 and $\delta n = n - n_0$. The eigenequation (4.3.69) is simplified and is found to be the Mathieu equation in Fourier space. The equation in configuration space written in the $e^{i\delta n\phi}$ basis has the form

$$\left[\frac{3}{2}\frac{1}{r_0^2}\frac{\partial}{\partial\phi^2} + \varepsilon r_0 \cos\phi\right] f = [E^j(\varepsilon) - E_{n0} + (n_0 - k - s)\omega]f.$$

$$(4.3.70)$$

It is the Schrödinger equation for a quantum pendulum of mass $-\frac{1}{3}$. The eigenvalue $E_{ks}^{j}(\varepsilon)$ can be expressed analytically for weak and strong fields [39] where the quantum numbers k and s are associated with the angular momentum and the quantum number j is associated with the excitations of the quantum pendulum. Without going into details we state briefly the results obtained. In Ref. [38] the authors confirmed numerically the presence of shape invariant Rydbery wave packets. Figure 4.7 shows the contour plot of the time evolution of the wave packet in the laboratory frame. Consecutive snapshots are plotted every 1/3 of the period for the 1st, 2nd, 3rd, 5th, 7th and 10th periods, for the following parameters

$$\omega = 209.049 \text{ GHz}$$
$$\varepsilon = -2243.36 \text{ V/m}$$
$$r_0 = 3394.71(\text{unit : Bohr radius})$$

The figure gives a strong impression of shape perseverance of the wave packets. This results from the suppression of the non-linearity of the Coulomb spectrum. A remarkable result of Ref. [39] is the observation that the Trojan wave packet can be obtained from

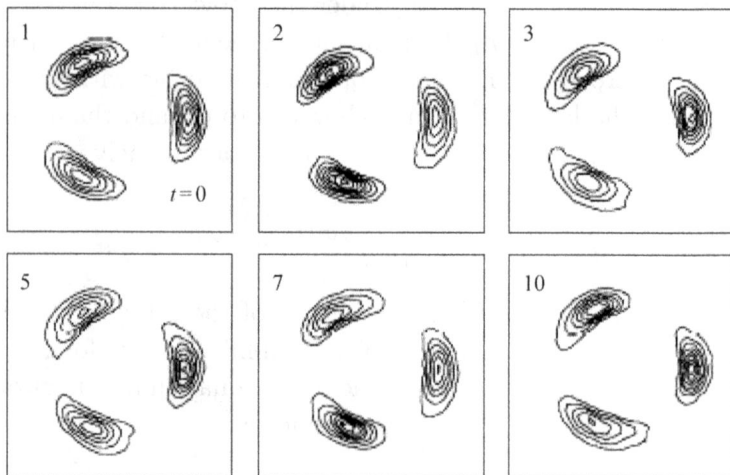

Fig. 4.7   Time evolution of the wave packet in laboratory frame. Reprinted with permission from Maciej Kalinski, J. H. Eberly, and Iwo Bialynicki-Birula, *Phys. Rev.* A52, 2460, 1995. Copyright (2017) by the American Physical Society.

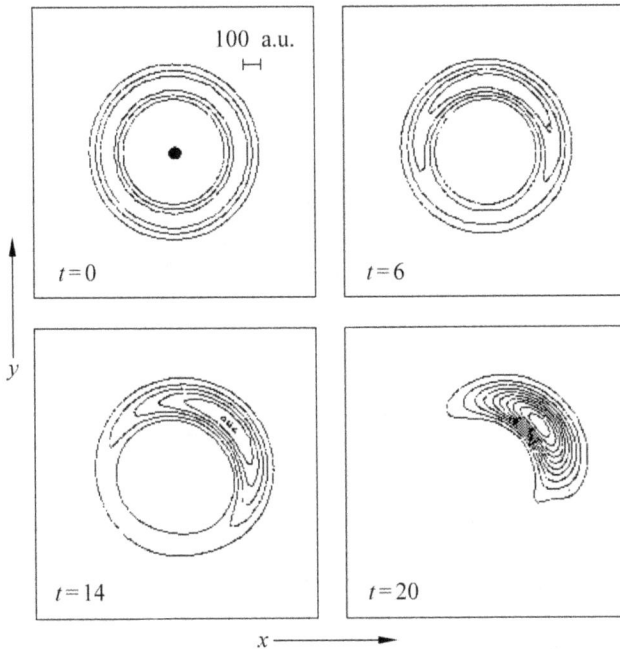

Fig. 4.8 Adiabatic angular localization of the electron probability density. Reprinted with permission from Maciej Kalinski and J. H. Eberly, *Phys. Rev.* A53, 1715, 1996. Copyright (2017) by the American Physical Society.

its parent circular state by switching on the electromagnetic field adiabatically. Figure 4.8 shows the angular localization of electron probability density. Snapshots of the density at $t = 0$ and after 6, 14, abd 20 periods show increasing angular bunching during the exponential switching of the field. The initial state is the circular state of $n_0 = 20$.

Non-spreading wave packets of electron states in atoms without the aid of external fields can also be realized as a direct analogy to Trojan asteroids in the Sun-Jupiter system [41, 42]. In a two-electron atom the electrons are excited to asymmetric Rydberg states with $\langle r_1 \rangle \gg \langle r_2 \rangle$. In the classical adiabatic model the inner electron completes many revolutions per single revolution of the outer electron. The difference in time scale allows the use of adiabatic approximation. The outer electron experiences the screening effect of the inner electron such that Z_{eff} is significantly less than 1. The

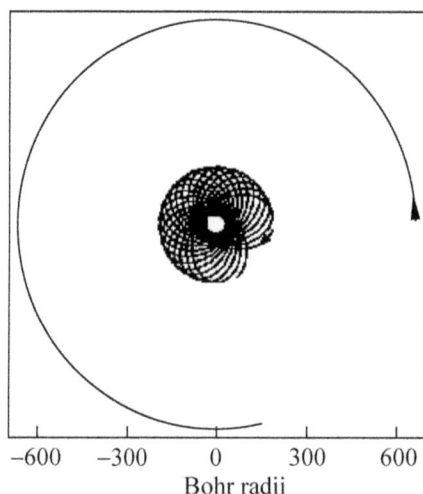

Fig. 4.9   Numerical simulation of classical orbits of the electrons. Reprinted with permission from J. A. West, Z. D. Gaeta, and C. R. Stroud, Jr., *Phys. Rev. A58*, 186, 1998.] Copyright (2017) by the American Physical Society.

outer electron provides a rotating electric field under which the core is polarized and the elliptical orbit of the inner electron precesses. The two electrons revolve in opposite sense such that the major axjs of the elliptic orbit of the inner electron follows adiabatically the outer electron, as shown in Fig. 4.9. The time dependent Hartree equation for the two-electron atom is

$$H_i \Psi_i = i\frac{\partial \Psi_i}{\partial t}, \qquad i = 1, 2 \tag{4.3.71}$$

where

$$H_i = -\frac{\nabla^2}{2} - \frac{2}{|\boldsymbol{r}|} + \int \frac{\rho_j(\boldsymbol{r}', t)}{|\boldsymbol{r} - \boldsymbol{r}'|} \mathrm{d}\boldsymbol{r}', \qquad i \neq j \tag{4.3.72}$$

and

$$\rho(\boldsymbol{r}, t) = \rho_1(\boldsymbol{r}, t) + \rho_2(\boldsymbol{r}, t) = |\Psi_1(\boldsymbol{r}, t)|^2 + |\Psi_2(\boldsymbol{r}, t)|^2. \tag{4.3.73}$$

The exchange potential (Fock term) can be safely neglected since the inner and outer electrons are described by wave packets very distant from each other such that $\Psi(\boldsymbol{r})\Psi_2(\boldsymbol{r})$ is effectively zero for all values of $\boldsymbol{r}$. This situation allows us to retain the dipole term of the multipole expansion of the interaction term in H_1 for the outer electron and the monopole term in H_2 for the inner electron. The Hartree equations then look like Schrödinger equations for two separate electrons

$$H_1 = -\frac{\nabla_1^2}{2} - \frac{1}{|r_1|} + \frac{d(t) \cdot r_1}{|r_1|^3}, \tag{4.3.74}$$

$$H_2 = -\frac{\nabla_2^2}{2} - \frac{2}{|r_2|} + \frac{1}{|r_2 - R(t)|}, \tag{4.3.75}$$

where

$$R(t) = \int \Psi_1^*(r_1, t) r_1 \Psi(r_1, t) dr_1,$$

$$d(t) = \int \Psi_2^*(r_2, t) r_2 \Psi_2(r_2, t) dr_2. \tag{4.3.76}$$

For R and d both rotating with the same angular frequency ω and same phase we have

$$\left.\begin{array}{l} R(t) = R(\hat{x}\cos\omega t + \hat{y}\sin\omega t) \\ d(t) = d(\hat{x}\cos\omega t + \hat{y}\sin\omega t) \end{array}\right\}. \tag{4.3.77}$$

With the time dependent terms in (4.3.74) and (4.3.75) linearized, the coupled stationary Hartree equations become

$$\left[-\frac{\nabla^2}{2} - \frac{1}{|r|} - E_1 x - \omega L_z\right]\phi_1 = \epsilon_1\phi_1,$$

$$\left[-\frac{\nabla^2}{2} - \frac{2}{|r|} - E_2 x - \omega L_z\right]\phi_2 = \epsilon_2\phi_2. \tag{4.3.78}$$

The coupling is realized through E_1 and E_2. where

$$E_1 = \frac{2d}{R^3}, \qquad E_2 = \frac{1}{R^2}. \tag{4.3.79}$$

We see the equivalence of two pictures, one with external field and the other with a third particle. We shall skip the construction of wave packets, giving only the results [42]. The stroboscopic snapshots of the time dependent probability density of the inner and outer electrons are given in Fig. 4.10. The outer electron is in a Trojan packet state with $n_1 = 60$ and the dimensionless parameter

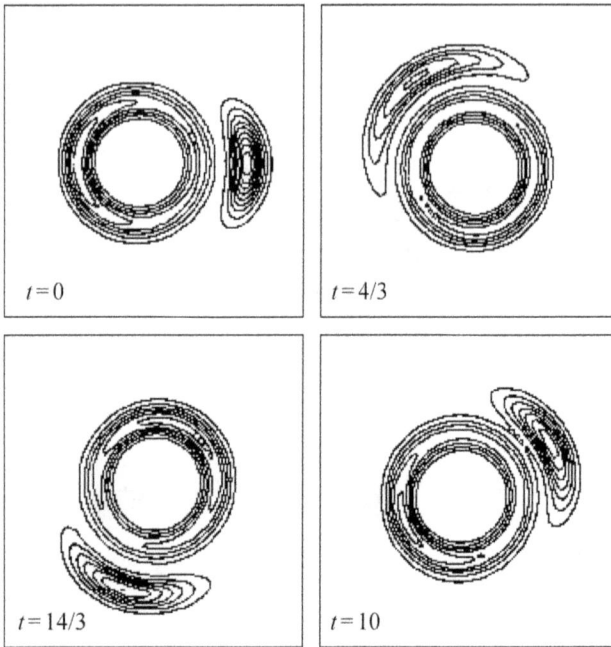

Fig. 4.10   The time evolution of probability density of the inner and outer electrons. Reprinted with permission from J. A. West, Z. D. Gaeta, and C. R. Stroud, Jr., *Phys. Rev.* A58, 186, 1998.] Copyright (2017) by the American Physical Society.

$q = 1/R^3 \omega^2 = 0.9562$, the inner electron is in the elliptic orbit with eccentricity $\epsilon = 0.25$ and $n_2 = 21$. The outer electron square covers the space region of $10\,800 \times 10\,800$ a.u., while the inner is magnified. The time is in units of periods of rotation.

Experimentally, non-dispersing wave packets from atomic Li Rydberg states under microwave field were found by H. Maeda and T.F. Gallagher [43]. The experiment was upgraded to one showing the microwave manipulation of the wave packet [44]. In the above discussions we emphasized the shape perseverance of the wave packet. Actually the packet moves in a *classical* orbit, in the sense that the orbit is not quantized. The Bohr radius r_0 and the angular frequency of revolution ω are both unquantized, i.e., they can be varied continuously. In the experiment [44] by increasing or decreasing the applied microwave frequency between 13 GHz and

19 GHz the binding energy and the orbital size of the electron change continuously.

The energy of the hydrogen atom is $E(n) = \frac{Ry}{n^2}$, where Ry is the Rydberg constant. The energy spacing between adjacent levels is $\Delta E = -\frac{2Ry}{n^3}$. The wave packet consists of Δn states. The classical Kepler frequency is

$$f_K = \frac{\Delta E}{h} = \frac{2Ry}{hn^3} = 6.6 \times 10^{15} \frac{1}{n^3} \text{Hz}.$$

For $n = 70$, $f_K = 19.2$ GHz, in the microwave region. In a microwave field linearly polarized in the x direction, the most localized wave packet is one in which the electrons moves synchronously with field on a highly eccentric orbit aligned along the x axis, as shown in Fig. 4.11. Alkali atom Li is used. Atoms in which the valence electron has been excited are exposed to a microwave pulse tuned

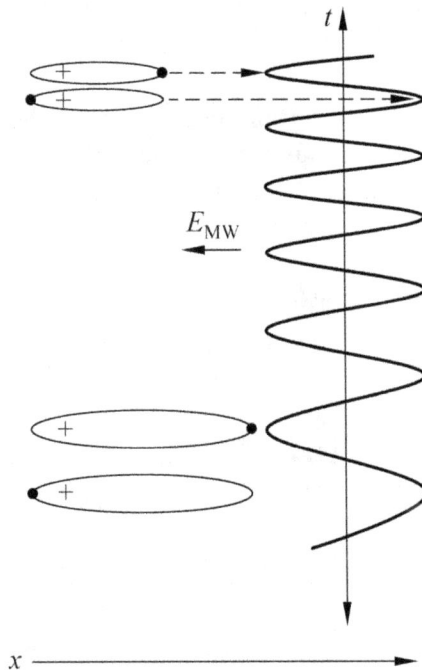

Fig. 4.11   Electron orbit changes with the microwave frequency chirp. Reprinted with permission from H. Maeda, D. V. L. Norum and T. F. Gallagher, *Science*, 307, 1757, 2005. Copyright (2017) by AAAS.

initially to the Kepler frequency, and are then chirped to higher or lower frequency. The microwave field first phase-locks the electron motion and then increases or decreases the Kepler frequency during the chirp leading simultaneously to a change of orbit size and binding energy. When the frequency is lowered, the orbit size increases and binding energy decreases. In the experiment the change in n and the persistent phase locking of the electron to the microwave field throughout the chirp are verified. When the microwave frequency is chirped from 19 to 13 Ghz, an atom initially in $n = 70$ state is converted to $n = 79$ state.

The manipulation of electron orbit is carried out in the following way. Valence electron of Li atoms is excited to $65 < n < 90$ by laser pulses, and the microwave field is tuned initially to the Kepler frequency. The change of electron orbit and the binding energy with microwave frequency is shown in Fig. 4.11. The change in n and the persistent phase-locking during the continuous chirp is verified by analyzing the final states after the microwave pulse. The atoms are exposed to a dc ionization field from 0 to 40V/cm in 1µs, as shown in Fig. 4.12. Ionization occurs when the field lowers the potential on one side of the atom sufficiently that the electron can escape. Higher n states lose their electrons before lower n states

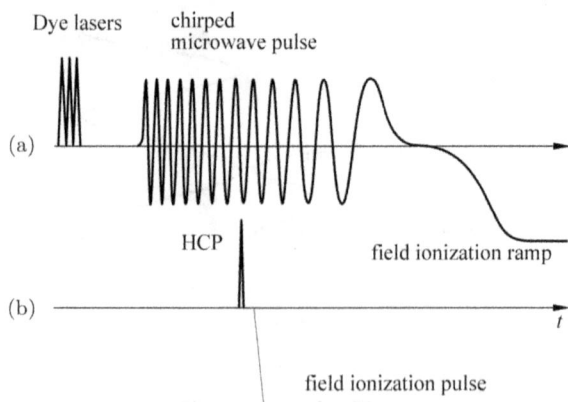

Fig. 4.12   Timing for the experiment. Reprinted with permission from H. Maeda, D. V. L. Norum and T. F. Gallagher, *Science*, 307, 1757, 2005. Copyright (2017) by AAAS.

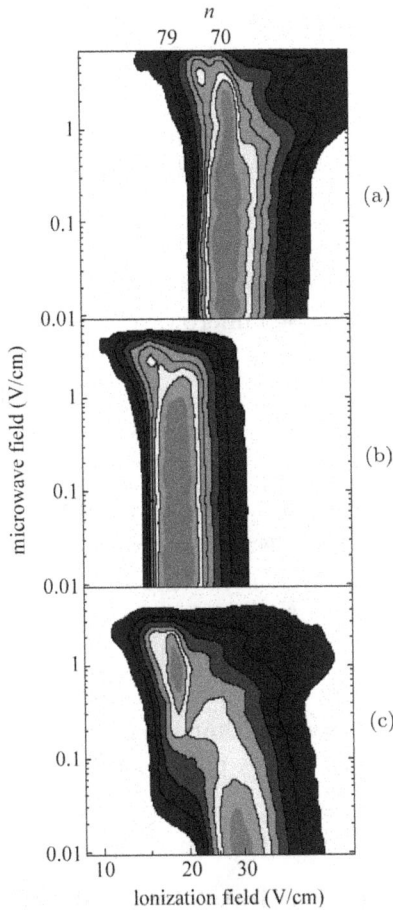

Fig. 4.13 Final state distribution after exposure to the microwave pulses. Reprinted with permission from H. Maeda, D. V. L. Norum and T. F. Gallagher, *Science*, 307, 1757, 2005. Copyright (2017) by AAAS.

do. The electrons resulting from field ionization are detected and the time resolved field ionization signal is used to infer the final state distribution. Figure 4.13 shows the final state distribution. The ionization field strength is indicated on the bottom horizontal axis, the corresponding n on the top horizontal axis, and the microwave field amplitude on the vertical axis. (a) When $n = 70$ atoms are exposed to a 19 GHz pulse, they remain predominantly there. This state is ionized at ionization field $F = 27\text{V/cm}$. When the microwave field becomes too strong $(-4\text{V}/\text{cm})$, it ionizes the atoms completely. (b) When $n = 79$ atoms are exposed to a 19 GHz pulse, they remain in that state because the microwave frequency is too far from the $n = 79$ Kepler frequency to substantially redistribute the atoms. The atoms are ionized at the ionization field $18\text{V}/\text{cm}$. (c) When $n = 70$ atoms are exposed to a pulse chirped from 19 to 13GHz continuously with microwave field between 0.5 and $1.4\text{V}/\text{cm}$, they are largely transferred from $n = 70$ to $n = 79$. The above results demonstrate that the chirped microwave field phase locks the motion of the electron in reducing its orbital frequency from 19 to 13GHz, increasing n from 70 to 79. The phase locking can be further checked by using momentum-selective ionization, with a subpicosecond half cycle pulse (HCP), shown in Fig. 4.12. The electron orbit is highly eccentric, aligned along the x-axis, i.e., the microwave polarization axis. The HCP gives the electron a momentum transfer Δp in the $+x$ direction. If the electron is moving in the $+x$ direction, it gains enough energy to escape from the atom, if it is moving in the $-x$ direction, it loses energy and ionization does not occur. The atoms are exposed to HCP at several times during the chirping of the microwave frequency to verify that the electron momentum remains synchronized to the microwave field. Detailed analysis confirms the phase locking.

4.4   Revivals and Fractional Revivals of the Wave Packets

Z. D. Gaeta and C. R. Stroud, Jr. [9] calculated the motion, spreading, fractional revivals and revivals of the circular Kepler orbit of the H atom. A clear picture of the evolution of the wave packet is obtained. Following L. Brown [3], the circular eigenstates $\psi_{n,n-1,n-1}(\boldsymbol{r})$ for different n are superimposed with a Gaussian weight function which is centered at a large principal quantum number $\bar{n}$ (in atomic units):

$$\Psi\left(\boldsymbol{r},t\right) = \frac{1}{(2\pi\sigma_n^2)^{1/4}} \sum_{n=1}^{\infty} e^{-(n-\bar{n})^2/4\sigma_n^2} \psi_{n,n-1,n-1}\left(\boldsymbol{r}\right) e^{\mathrm{i}t/2n^2}, \quad (4.4.1)$$

where $\bar{n}$ and σ_n are the mean and standard deviation of the Gaussian distribution. In order to get a narrow and slowly spreading wave packet, a large $\bar{n}$ is adopted ($\bar{n} = 320, \sigma_n = 2.5$). The motion, spreading, fractional revivals and revivals all originate from the dynamical phase factor. We have seen similar situations in Section 4.2, Eqs. (4.2.14), (4.2.24). Expanding $\frac{t}{2n^2}$ in a series of $\Delta n = n - \bar{n}$, we obtain

$$\frac{t}{2n^2} = \frac{t}{2\bar{n}^2} \left[1 - 2\frac{\Delta n}{\bar{n}} + 3\left(\frac{\Delta n}{\bar{n}}\right)^2 - 4\left(\frac{\Delta n}{\bar{n}}\right)^3 + \cdots \right], \qquad (4.4.2)$$

If only the linear term is retained, (4.4.1) becomes

$$\Psi_{cl}(t) = \sum_{n=1}^{\infty} w_n \psi_{n,n-1,n-1} e^{-2\pi i \Delta n t / T_0}, \qquad (4.4.3)$$

where only the variable t is explicitly shown; w_n is the Gaussian distribution; T_0 is the classical Kepler period corresponding to the orbit $\bar{n}$, $T_0 = 2\pi \bar{n}^3$. This is the exactly periodic "classical" wave function. The motion of the wave packet in time $0 \leqslant t \leqslant T_0$ is shown in Fig. 4.14. The spreading of wave packet due to the second order term in (4.4.2) is given by eqn. (4.2.19). Using (4.2.25) for the revival time $T_{rev} = \frac{n}{3} T_0$ we rewrite (4.2.19) as

$$[\Delta\phi(t)]^2 = [\Delta\phi(0)]^2 + \frac{\pi^2}{[\Delta\phi(0)]^2} \left(\frac{t}{T_{rev}}\right)^2. \qquad (4.4.4)$$

When the right hand side of (4.4.4) is equal to $\frac{\pi^2}{3}$ (the variance of a uniformly distributed random variable around the circle), the wave packet spreads completely. The time required to reach this stage is denoted by T_{sp} (the spreading time),

$$T_{sp} = T_{rev}/8.713 = 12.2T_0.$$

Figure 4.15 shows the spreading of the wave packet to the entire circular orbit. From now on the typical quantum interference sets in, and fractional revivals occur, as described in Section 4.2. I. Sh. Averbuch and N.F. Perelman [10] gave the condition that for time $t = \frac{k_1}{k_2} T_{rev}$ fractional revivals occur, where k_1 and k_2 are integers

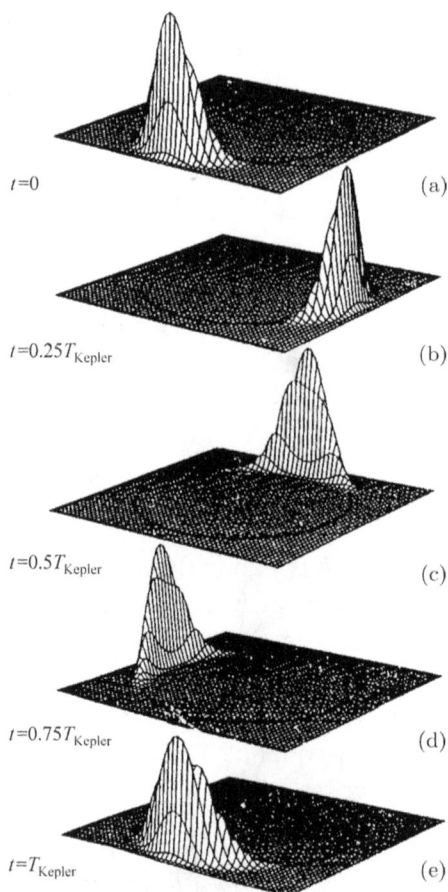

Fig. 4.14   The initial stage of the evolution of the circular orbit wave packet. Reprinted with permission from Z. Dačić Gaeta and C. R. Stroud, Jr., *Phys. Rev. A42*, 6308, 1990. Copyright (2017) by the American Physical Society.

prime to each other, and $k_1 < k_2$. At fractional revivals the wave packet is split into K almost identical small packets:

$$\Psi\left(t\right) = \sum_{k=0}^{K-1} a_k \, \Psi_{c1}\left(t + \frac{kT_0}{K}\right), \qquad (4.4.5)$$

where a_k is the phase factor, Ψ_{cl} is defined in (4.4.3). Figure 4.16 describes the fractional revivals corresponding to $K = 2, 3, 4, 5, 6, 7$

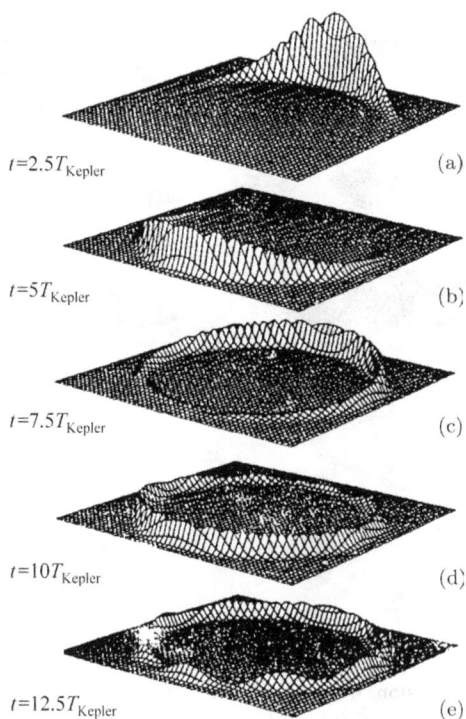

Fig. 4.15   Spreading of the wave packet. Reprinted with permission from Z. Dačić Gaeta and C. R. Stroud, Jr., *Phys. Rev.* A42, 6308, 1990. Copyright (2017) by the American Physical Society.

with times of fractional revivals indicated. For larger K the fractional revivals become less noticeable.

In the experimental side, J. A. Yeazell and C. R. Stroud Jr. [11] first observed the motion of localized wave packet and the fractional revivals. Fractional revivals for radial wave packet were observed earlier by G. Alber, H. Ritsch and P. Zoller [12] using time-delayed 2-photon method (one for exciting the atom and the other for probing).

The Kepler orbit wave packet exhibits classical properties, but it is quantum mechanical in essence. Classical properties are manifest before the spreading of the wave packet ($t < T_{sp}$), and also at revivals ($t = T_{rev}$ and higher order revivals), while the quantum feature dominates over the remaining time. This is somewhat similar to the

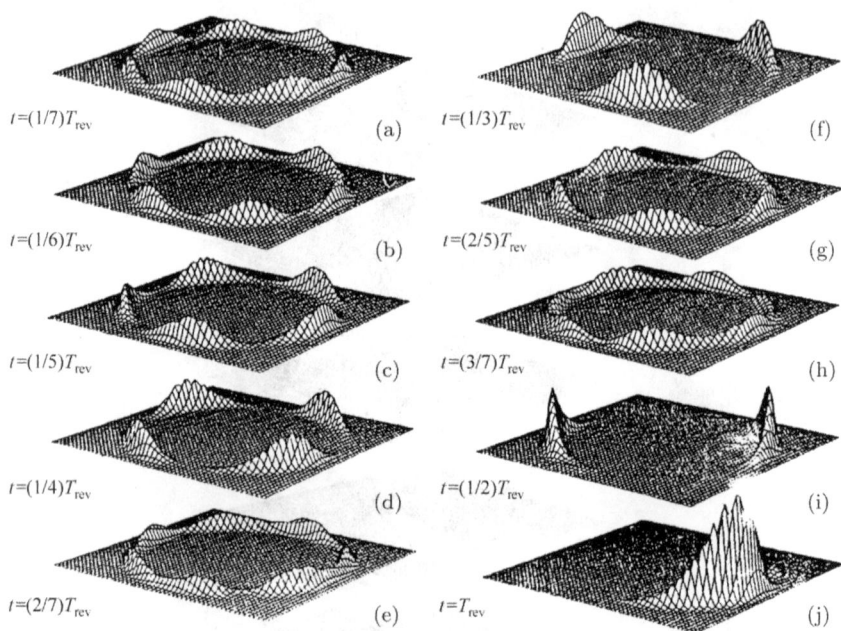

Fig. 4.16   Fractional revivals and revival of the circular orbit wave packet. Reprinted with permission from Z. Dačić Gaeta and C. R. Stroud, Jr., *Phys. Rev.* A42, 6308, 1990. Copyright (2017) by the American Physical Society.

situation that rational numbers are imbedded in the continuum of real numbers.

The foregoing discussion helps to understand the manifestation of classical properties in quantum mechanics and its limitations.

4.5   Principle of Superposition of States and the Quantum Decoherence

The principle of superposition of states serves to illustrate the contrast between quantum and classical physics most significantly. Because of the linearity of the Schrödinger equation, under an arbitrary initial condition, the wave function evolves into a linear superposition of the energy eigenfunctions of the Hamiltonian. Take the simplest example of a two-state system, the wave function is

$$\psi = c_1\psi_1 + c_2\psi_2, \tag{4.5.1}$$

with

$$|c_1|^2 + |c_2|^2 = 1. \qquad (4.5.2)$$

When a measurement to decide the state of the system is carried out, the outcome is either ψ_1 or ψ_2 with probabilities $|c_1|^2$ and $|c_2|^2$ respectively. The pure state (4.5.1) can be expressed in terms of its density matrix ρ:

$$\rho = |\psi\rangle \langle\psi|$$
$$= |c_1|^2 |\psi_1\rangle \langle\psi_1| + c_1 c_2^* |\psi_1\rangle \langle\psi_2| + c_1 c_2^* |\psi_2\rangle \langle\psi_1| + |c_2|^2 |\psi_2\rangle \langle\psi_2|. \qquad (4.5.3)$$

The measurement changes the state of the system from the pure state (4.5.1) to a mixed state ψ_1 with probability $|c_1|^2$ and ψ_2 with probability $|c_2|^2$. A mixed state is most conveniently expressed in terms of its density matrix ρ_m:

$$\rho_m = |c_1|^2 |\psi_1\rangle \langle\psi_1| + |c_2|^2 |\psi_2\rangle \langle\psi_2|. \qquad (4.5.4)$$

In matrix form ρ and ρ_m are expressed respectively as

$$\rho = \begin{pmatrix} |c_2|^2 & c_1 c_2^* \\ c_1^* c_2 & |c_2|^2 \end{pmatrix}, \qquad (4.5.5)$$

and

$$\rho_m = \begin{pmatrix} |c_1|^2 & 0 \\ 0 & |c_2|^2 \end{pmatrix}. \qquad (4.5.6)$$

The off-diagonal elements disappear in the process of measurement. This is called the reduction (or collapse) of wave function. The main difference between the pure and the mixed states is the phase relation kept in the off-diagonal elements, i.e., the relative phase between ψ_1 and ψ_2. In classical physics, the evolution of a system never leads to a coherent superposition. If for some reasons, e.g., the initial condition is not precisely determined, the prediction about the state of the system can be probabilistic. But that is a kind of prediction similar to the mixed state as given in (4.5.6), but not of a coherent superposition of two states as given by (4.5.5). The question is, if quantum mechanics is considered as a correct theory, it should be applicable to macroscopic and microscopic systems as well. The principle of superposition should also be applicable to macroscopic system. What makes the off-diagonal elements of the density matrix disappear? A good example to illustrate this awkward situation is the well-known Schrödinger cat. A cat is confined in a "diabolical device": a closed box containing a small amount of radioactive substance. This substance has a 50% chance of decay after one hour. If it does decay, then the radiation emitted will trigger a sensitive mechanism which will release a deadly poison, and the cat will be killed. The radioactive nucleus abides laws of quantum mechanics, hence its state after one hour will be $\frac{1}{\sqrt{2}}(|\uparrow\rangle + |\downarrow\rangle)$, i.e., a coherent superposition, where $|\uparrow\rangle$ denotes the original state of the nucleus, and $|\downarrow\rangle$ denotes the decayed state. Let $|\smile\rangle$ and $|\frown\rangle$ denote the state of a live and dead cat respectively. The state of the total system (cat plus the radioactive nucleus) will be after one hour

$$\frac{1}{\sqrt{2}}\left(|\smile\rangle\,|\uparrow\rangle + |\frown\rangle\,|\downarrow\rangle\right). \tag{4.5.7}$$

Now the superposition of states for the nucleus is entangled with the life-or-death of the cat in a subtle way. Equation (4.5.7) predicts that the probability of finding a live cat and a dead cat after one hour is 50% each. This is acceptable in classical physics. But actually (4.5.7) offers more: the states of live and dead cat coexist and are superimposed coherently, this is unacceptable in classical physics. To begin with quantum mechanics and end up with a picture acceptable by a common wisdom, some mechanism must be in work to erase the phase coherence in the superposition. The transition of quantum mechanics (superposition of states) to classical mechanics (different outcomes with probabilities allowed, but the outcomes are mutually exclusive) has been the topic of hot debate since the inception of quantum mechanics. The first explanation was given by N. Bohr, who insisted that a borderline must be drawn between the classical and quantum realms. When a classical apparatus is used to carry out measurements, the classical laws, not the quantum laws, are to be applied. Bohr himself was aware the borderline must be mobile, since an apparatus can in its turn be analyzed as a quantum system by a classical device appropriate to carry out this task. Such an approach is not satisfactory, since first of all quantum theory is universal. Firstly the constituents of any macroscopic body (considered as to obey classical laws)—molecules, atoms, electrons, nuclei all obey the quantum laws. Classical laws have to emerge as the limiting case of quantum laws under some circumstances. This is just the main problem that physicists devote themselves to disentangle. Secondly, the identity of "classical" with "macroscopic" has broken down, e.g., a squeezed state in quantum electronics involves macroscopic number of photons; the superconducting state of a metal involves a macroscopic number of Cooper pairs. All these are quantum mechanical systems. The problem is reduced to the following. The quantum theory should be able to explain how a superposition of alternatives of a quantum system turns into the realization of one of the alternatives in a measurement. In the recent years there has been a growing consensus as to the role of environment in leading to a "decoherence" which transforms a pure state into a mixed one. According to quantum mechanics, the state of a macroscopic system in general is also a linear superposition of possible macroscopic states. Due to the interaction of the system with the environment part of the information of the system encoded in the non-diagonal elements of the density matrix leaks to the environment. Since the environment consists of an enormous number of degrees of freedom, the flow of the information is irreversible. The speed of leakage depends on the size of the system and the degree of coupling of the system with the environment. For a macroscopic system the speed can be so large that the decoherence can be considered instantaneous. After the decoherence the density matrix has only diagonal elements, i.e., the system is in a mixed state. As to the specific details there are different theoretical approaches and models along with much controversies. In the following we give an outline of the theory of decoherence due to W. Zurek [13–15].

In contrast to N. Bohr, J. von Neumann begins with a quantum (rather than classical) apparatus [16], the behavior of which is also governed by quantum mechanics. Consider a two-state system S with a Hilbert space spanned by orthonormal states $|\uparrow\rangle$ and $|\downarrow\rangle$.

Alternative states of the systems are

$$
\left.\begin{aligned}
|\oplus\rangle &= \frac{1}{\sqrt{2}}[|\uparrow\rangle + |\downarrow\rangle] \\
|\otimes\rangle &= \frac{1}{\sqrt{2}}[|\uparrow\rangle - |\downarrow\rangle]
\end{aligned}\right\} \tag{4.5.8}
$$

and

$$
\left.\begin{aligned}
|\rightarrow\rangle &= \frac{1}{\sqrt{2}}[|\uparrow\rangle + i|\downarrow\rangle] \\
|\leftarrow\rangle &= \frac{1}{\sqrt{2}}[|\uparrow\rangle - i|\downarrow\rangle]
\end{aligned}\right\} \tag{4.5.9}
$$

If $|\uparrow\rangle$ and $|\downarrow\rangle$ are eigenstates of σ_z, then $|\oplus\rangle$ and $|\otimes\rangle$ are eigenstates of σ_x and $|\rightarrow\rangle$ and $|\leftarrow\rangle$ are eigenstates of σ_y. A detector D has a Hilbert space spanned by $|\smile\rangle$ and $|\frown\rangle$. Alternative states are

$$
\left.\begin{aligned}
|+\rangle &= \frac{1}{\sqrt{2}}[|\smile\rangle + |\frown\rangle] \\
|-\rangle &= \frac{1}{\sqrt{2}}[|\smile\rangle - |\frown\rangle]
\end{aligned}\right\} \tag{4.5.10}
$$

and

$$
\left.\begin{aligned}
|\wedge\rangle &= \frac{1}{\sqrt{2}}[|\smile\rangle + i|\frown\rangle] \\
|\vee\rangle &= \frac{1}{\sqrt{2}}[|\smile\rangle - i|\frown\rangle]
\end{aligned}\right\}. \tag{4.5.11}
$$

The system and the detector have no self-Hamiltonian but are coupled through an interaction

$$
H^{\mathrm{SD}} = g[|\wedge\rangle\langle\wedge| - |\vee\rangle > \vee|] \otimes [|\uparrow\rangle > \uparrow| - |\downarrow\rangle\langle\downarrow|] \equiv gH, \tag{4.5.12}
$$

where g is the coupling constant. The operator in the parenthesis before the direct product sign acts on the apparatus states and that after the sign acts on the system states. We notice that

$$(H^{SD})^2 = g^2 H^2 = g^2 [|\wedge\rangle\langle\wedge| + |\vee\rangle\langle\vee|] \otimes [|\uparrow\rangle\langle\uparrow| + |\downarrow\rangle\langle\downarrow|]$$

$$= g^2 I_D \otimes I_S \equiv g^2 I, \tag{4.5.13}$$

where I is the identity operator. The evolution operator of the system-detector complex is, by using (4.5.13),

$$\exp\left(-i\frac{H^{SD}}{\hbar}t\right) = \exp\left(-i\frac{g}{\hbar}tH\right) = \left(\cos\frac{g}{\hbar}t\right) I - i\left(\sin\frac{g}{\hbar}t\right) H. \tag{4.5.14}$$

Let the complex begin to evolve from $t = 0$:

$$|\Psi(t = 0)\rangle = [a|\uparrow\rangle + b|\downarrow\rangle] \otimes |\smile\rangle. \tag{4.5.15}$$

It follows from (4.5.12) that

$$H = -i[|\smile\rangle\langle\frown| + |\frown\rangle\langle\smile|] \otimes [|\uparrow\rangle\langle\uparrow| - |\downarrow\rangle\langle\downarrow|]$$

which leads to

$$|\Psi(t)\rangle = \exp\left(-i\frac{g}{\hbar}tH\right)|\Psi(0)\rangle = \cos\frac{g}{\hbar}t[a|\uparrow\rangle + b|\downarrow\rangle] \otimes |\smile\rangle$$

$$- \sin\frac{g}{\hbar}t[a|\uparrow\rangle - b|\downarrow\rangle] \otimes |\frown\rangle. \tag{4.5.16}$$

To see the result of evolution, put in $t = \pi\hbar/4g$ and obtain

$$\left|\Psi\left(t = \frac{\pi\hbar}{4g}\right)\right\rangle = a|\uparrow\rangle \otimes |-\rangle + b|\downarrow\rangle \otimes |+\rangle. \tag{4.5.17}$$

It is remarkable to notice that the interaction between system and detector has led to correlation between their states as the result of unitary evolution due to the interaction H^{SD}. But unitary evolution is not enough to solve the measurement problem, because we readily see as time goes on,

$$\left|\Psi\left(t = \frac{\pi\hbar}{2g}\right)\right\rangle = [a|\uparrow\rangle - b|\downarrow\rangle] \otimes |\frown\rangle, \tag{4.5.18}$$

the correlation disappears! Furthermore the system and detector are correlated again at $t = 3\pi\hbar/4g$ in a way opposite to that at $t = \pi\hbar/4g$:

$$\left| \Psi\left(t = \frac{3\pi\hbar}{4g}\right) \right\rangle = -[a|\uparrow\rangle \otimes |+\rangle + b|\downarrow\rangle \otimes |-\rangle]. \qquad (4.5.19)$$

It is probably not surprising, since we see from (4.5.16) that the evolution is periodic. Still another problem poses itself. Does the detector know what it is going to measure? From (4.5.17) it seems that it measures states $|\uparrow\rangle$ and $|\downarrow\rangle$. But we can rewrite (4.5.17) by substituting $|\smile\rangle, |\frown\rangle$ for $|+\rangle, |-\rangle$ to obtain

$$\left| \Psi\left(t = \frac{\pi\hbar}{4g}\right) \right\rangle = \frac{1}{\sqrt{2}}[a|\uparrow\rangle + b|\downarrow\rangle] \otimes |\smile\rangle + \frac{1}{\sqrt{2}}[a|\uparrow\rangle - b|\downarrow\rangle] \otimes |\frown\rangle,$$

$$(4.5.20)$$

and for $a = b = \frac{1}{\sqrt{2}}$ this is

$$\left| \Psi\left(t = \frac{\pi\hbar}{4g}\right) \right\rangle = \frac{1}{\sqrt{2}}[|\oplus\rangle \otimes |\smile\rangle + |\otimes\rangle \oplus |\frown\rangle], \qquad (4.5.21)$$

the detector is now correlated with the states $|\oplus\rangle$ and $|\otimes\rangle$ of the system. To sharpen the confrontation between classical and quantum correlations, we follow J. A. Wheeler [17]. When a coin is split into two halves (one with the head and the other with the tail) and these are put into envelopes and sent to two observers separated at an arbitrary large distance. Then firstly the alternative is certain, i.e., either the head or the tail is in a selected envelope. Secondly, the results of observation of the envelopes are correlated. Such a situation is represented by a density matrix of a mixed state,

$$\rho = \frac{1}{\sqrt{2}}[|h_1\rangle\langle h_1||t_2\rangle\langle t_2| + |t_1\rangle\langle t_1||h_2\rangle\langle h_2|]. \qquad (4.5.22)$$

The quantum correlation is really non-classical in the sense that even the alternatives are uncertain. Von Neumann was aware of the difficulties. He added to the unitary evolution the *ad hoc* process of a non-unitary reduction of the state vector (collapse of the wave function). Let the density matrix of the pure state (4.5.17) be represented by ρ^c:

$$\rho^c = |a|^2|\uparrow\rangle\langle\uparrow||-\rangle\langle-| + ab^*|\uparrow\rangle\langle\downarrow||-\rangle\langle+|$$
$$+ a^*b|\downarrow\rangle\langle\uparrow||+\rangle\langle-| + |b|^2|\downarrow\rangle\langle\downarrow||+\rangle\langle+|. \qquad (4.5.23)$$

As a result of reduction of the state vector, the non-diagonal elements disappear and the reduced density matrix ρ^r characterizing a mixed state is established

$$\rho^c \to \rho^r = |a|^2|\uparrow\rangle\langle\uparrow||-\rangle\langle-| + |b|^2|\downarrow\rangle\langle\downarrow||+\rangle\langle+|. \qquad (4.5.24)$$

where the coefficients of diagonal elements can be interpreted as probabilities of occurrence of $|\uparrow\rangle$ and $|\downarrow\rangle$ respectively. The gain of giving up information stored in the non-diagonal elements (the phase coherence between $|\uparrow\rangle$ and $|\downarrow\rangle$) is that the alternative of measurement is now certain. It is $|\uparrow\rangle$ and $|\downarrow\rangle$ instead of $|\oplus\rangle$ and $|\otimes\rangle$. If the non-diagonal elements of the density matrix corresponding to (4.5.20) are removed, the choice is the other way round. The non-unitary process of von Neumann is *"added in by hands"*, not the consequence of the quantum theory itself. Zurek attempted to include the decoherence as an integral part of the theory, viz. a decoherence caused by the interaction between the system and the environment.

4.6 Decoherence due to Interaction with Environment

Macroscopic quantum systems are never isolated from their environments. Consider the environment E, also a quantum object, interacting with the system S and the detector D. It can be a part of the apparatus, possessing very many degrees of freedom, and its function is to absorb informations from the non-diagonal elements of the density matrix ρ^c and thereby causing the collapse of states and consequently defining the observable to be measured (called the "pointer observable") with its eigenstates (called the "pointer basis"). Specifically the full SDE complex evolves in the following way. The state of the complex at $t = 0$ is

$$|\Phi (t = 0)\rangle = |\psi\rangle \otimes |D_0\rangle \otimes |E (t = 0)\rangle, \qquad (4.6.1)$$

The states in the right-hand side separated by the direct product signs are that of the system, the detector and the environment respectively. $|\Phi\rangle$ evolves unitarily to time t_1 due to interaction between S and D, which produces the correlation between the system and the detector. Up to this time the interaction with the environment has not yet been invoked.

$$|\Phi (t = t_1)\rangle = \left\{ \sum_n c_n |n\rangle \otimes |D_n\rangle \right\} \otimes |E (t_1)\rangle. \qquad (4.6.2)$$

At time t_1 the pointer basis has not yet been fixed. In (4.6.2) the SD complex can be represented in terms of other bases as well. The state evolves from t_1 to t_2 unitarily due to the interaction between D and E.

$$|\Phi (t = t_2)\rangle = \sum_n c_n |n\rangle \otimes |D_n\rangle \otimes |E (t_2)\rangle, \qquad (4.6.3)$$

The system-detector-environment correlation is established, and consequently the pointer basis $|n\rangle$ is fixed. The SD complex has lost its freedom to choose correlations other than $|n\rangle \otimes |D_n\rangle$. This is the important role of the environment. In the future development the environment is to transform the SD complex from a pure state to a mixed one, i.e., to diminish gradually the off-diagonal elements of the density matrix. But before that it is necessary to determine beforehand what basis states is to be put on the diagonal elements. Since there is no measurement carried out on the environment, we look for the reduced density matrix of the SD complex at time $t > t_2$ by taking trace over the environment states:

$$\rho = \mathrm{tr}_E |\Phi (t > t_2)\rangle \langle \Phi (t > t_2)|$$

$$= \sum_{m,n} c_n c_m^* |D_n\rangle \langle D_m| \otimes |n\rangle \langle m| \otimes \mathrm{tr}_E |E_n (t)\rangle \langle E_m (t)|. \qquad (4.6.4)$$

We shall demonstrate in the following using a solvable model that for t large enough, the partial trace $\mathrm{tr}_E|E_n(t)\rangle\langle E_m(t)|$

$=< E_m(t) \mid E_n(t) >$ approaches exponentially δ_{mn}, i.e., the off-diagonal elements of ρ decays exponentially:

$$\rho \approx \sum_n |c_n|^2 \, |D_n\rangle \, \langle D_n| \otimes |n\rangle \, \langle n| \,. \qquad (4.6.5)$$

The SD complex is in a mixed state. It can be in *any state* $|n\rangle \otimes |D_n\rangle$ with probability $|c_n|^2$, but states with different n are now *independent*, and superposition disappears. This is how decoherence takes place.

The environment consists of a large number N of two state systems, the k-th can occupy either $| \smile)_k$ or $| \frown)_k$. Here round bracket and subscript k are used to specify environment states. Again there is no self-Hamiltonian for the environment atoms, and there is no interaction between them either. the environment interacts with the detector with the Hamiltonian

$$H^{\mathrm{DE}} = \sum_k H_k^{\mathrm{DE}}, \qquad (4.6.6)$$

where

$$H_k^{\mathrm{DE}} = g_k \, [|\smile\rangle \, \langle\smile| - |\frown\rangle \, \langle\frown|] \otimes [|\smile) \, (\smile| - |\frown) \, (\frown|]_k \otimes \prod_{j \neq k} 1_j . \qquad (4.6.7)$$

We observe that eigenstates of the Hamiltonian H_k^{DE} are direct products of any of the detector states $| \smile \rangle$ or $| \frown \rangle$ and any of the atomic states $| \smile)_k$ or $| \frown)_k$. The eigenvalues are easily determined to be $\pm g_k$. Let the full complex at $t = 0$ be denoted by $|\Phi(0)\rangle$, when DE interaction begins to act and S and D are already correlated[15]:

$$|\Phi(0)\rangle = [a \, |{\uparrow}\rangle \otimes |\smile\rangle + b \, |{\downarrow}\rangle \otimes |\frown\rangle] \prod_{k=1}^{n} \otimes [\alpha_k \, |\smile)_k + \beta_k \, |\frown)_k]. \qquad (4.6.8)$$

This state evolves under H^{DE} straightforwardly to time t, since direct products of detector and atomic states are eigenstates of H^{DE} and

---

[15] In (4.6.2) this time was denoted by t_1.

the state at time t is represented by

$$|\Phi(t)\rangle = a\,|\!\uparrow\rangle \otimes |\smile\rangle \prod_{k=1}^{N} \otimes \left[\alpha_k e^{-ig_k t/\hbar}\,|\smile)_k + \beta_k e^{ig_k t/\hbar}\,|\frown)_k\right]$$

$$+\, b\,|\!\downarrow\rangle \otimes |\frown\rangle \prod_{k=1}^{N} \otimes \left[\alpha_k e^{ig_k t/\hbar}\,|\smile)_k + \beta_k e^{-ig_k t/\hbar}\,|\frown)_k\right].$$

$$(4.6.9)$$

Abbreviating the environment states by

$$|E_{|\smile\rangle}(t)\rangle \equiv \prod_{k=1}^{N} \otimes [\alpha_k e^{-ig_k t/\hbar}|\,\smile)_k + \beta_k e^{ig_k t/\hbar}|\,\frown)_k],$$

$$|E_{|\frown\rangle}(t)\rangle \equiv \prod_{k=1}^{N} \otimes [\alpha_k e^{ig_k t/\hbar}|\,\smile)_k + \beta_k e^{-ig_k t/\hbar}|\,\frown)_k],$$

$$(4.6.10)$$

where the subscript of E denotes the detector state to which the environment is correlated, we rewrite (4.6.9) in the following form:

$$|\Phi(t)\rangle = a\,|\!\uparrow\rangle \otimes |\smile\rangle \otimes |E_{|\smile\rangle}(t)\rangle + b\,|\!\downarrow\rangle \otimes |\frown\rangle \otimes |E_{|\frown\rangle}(t)\rangle. \qquad (4.6.11)$$

Evolution from $|\Phi(0)\rangle$ to $|\Phi(t)\rangle$ establishes the pointer observable

$$\hat{\Lambda} = \lambda_1\,|\smile\rangle \langle\smile| + \lambda_2\,|\frown\rangle \langle\frown|, \qquad (4.6.12)$$

where λ_1 and λ_2 are real. Actually the pointer basis $|\smile\rangle$ and $|\frown\rangle$ are those states of the detector which enter the Hamiltonian H^{DE}, in other words, the states entering the interaction Hamiltonian will become the pointer basis. Although the SD complex at $t = 0$ can be expressed in terms of alternative detector basis, it is the pointer basis $|\smile\rangle$ and $|\frown\rangle$ which is picked up by the environment. The reduced density matrix of the system-detector complex is

$$\rho^{\mathrm{SD}} = \mathrm{tr}_E\,|\Phi(t)\rangle \langle\Phi(t)|$$

$$= |a|^2\,|\!\uparrow\rangle \langle\uparrow| \otimes |\smile\rangle \langle\smile| + z(t)\,ab^*\,|\!\uparrow\rangle \langle\downarrow| \otimes |\smile\rangle \langle\frown|$$

$$+\, z^*(t)\,a^*b\,|\!\downarrow\rangle \langle\uparrow| \otimes |\frown\rangle \langle\smile| + |b|^2\,|\!\downarrow\rangle \langle\downarrow| \otimes |\frown\rangle \langle\frown|, \qquad (4.6.13)$$

where $z(t)$ is called the *correlation amplitude:*

$$z\left(t\right) = \left\langle E_{|\smile\rangle}\left(t\right)\middle| E_{|\frown\rangle}\left(t\right)\right\rangle$$

$$= \prod_{k=1}^{N}\left[\cos\left(2g_k t/\hbar\right) - \mathrm{i}\left(|\alpha_k|^2 - |\beta_k|^2\right)\sin\left(2g_k t/\hbar\right)\right]. \quad (4.6.14)$$

It is therefore the correlation amplitude $z(t)$ which plays a decisive role in decoherence. It has the following properties:

$$z\left(0\right) = 1, \quad (4.6.15)$$

$$|z\left(t\right)|^2 \leqslant 1, \quad (4.6.16)$$

$$\left\langle z\left(t\right)\right\rangle = \lim_{T\to\infty} T^{-1}\int_0^T z\left(t\right)\mathrm{d}t = 0, \quad (4.6.17)$$

$$\left\langle |z\left(t\right)|^2\right\rangle = 2^{-N}\prod\left[1 + \left(|\alpha_k|^2 - |\beta_k|^2\right)^2\right]. \quad (4.6.18)$$

Equation (4.6.18) implies that the expected value of $|z(t)|^2$ is much less than 1 unless the initial state of the environment coincides with one of the eigenstates of the Hamiltonian. Figure 4.19 gives the correlation amplitude as a function of t, obtained from (4.6.14) for $|\alpha_k| = |\beta_k|$, i.e., $z(t) = \prod \cos 2g_k t/\hbar$. In Fig. 4.19 (a), (b), (c) $N = 5, 10, 15$ respectively, and the coupling constant g_k are chosen randomly from the interval $(0, 1)$. It is obvious that such a small number as 15 is quite effective to bring about decoherence, and the effect of increasing N is also obvious. In the thermodynamic limit $N \to \infty$ the interaction H^{DE} becomes irreversible. The interaction establishes the detector-environment correlation at the price of erasing the most part of the system-detector correlation, i.e., the SD correlation is now fixed, and does a job of fixing the pointer basis and collapsing the state vector at the same time. Thus the dynamics of the complex prevents the stable existence of states corresponding to a superposition of the pointer states. In this model, the system, the detector and the environment all abide the quantum

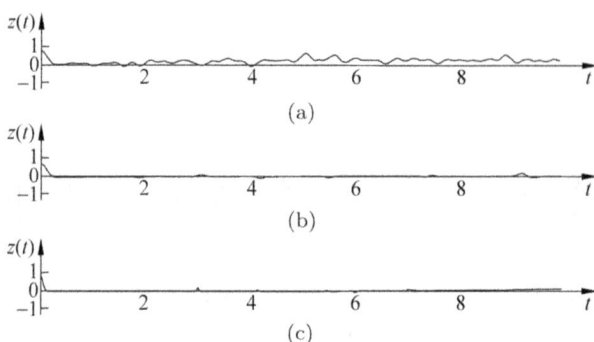

Fig. 4.17 The evolution in time of the correlation amplitude. Reprinted figure with permission from [W. H. Zurek, *Phys. Rev.* D26, 1862, 1982.] Copyright (2017) by the American Physical Society.

mechanics. The complex of its constituent parts evolve according to respective Hamiltonians. The appearance of the pointer basis is the result of dynamic evolution. There is no non-unitary process added in by hand. The coherence between pointer states in any linear superposition decays because of the interaction of the system-detector complex with the environment with an enormous number of degrees of freedom. Like any physical quantity in statistical physics, the correlation amplitude has fluctuations and also Poincaré recurrence. In Ref. [13] (1982) a discussion on the attenuation of coherence, fluctuations and recurrences in their dependence on the size of the environment can be found.

In writing down the Hamiltonian (4.5.11) the self-Hamiltonian is neglected. The convenience of this neglect is that the pointer states are stationary states, consequently the evolution of the complex can be obtained explicitly (4.6.9). The neglect is justifiable in idealized models, and is not applicable to realistic macroscopic systems. But the basic ideas regarding the openness of macroscopic quantum systems and the function of environment in the erasure of coherence and therefore in the violation of the superposition principle on the classical level can still serve as guiding principles. When each of the pointer states is no more stationary, it is necessary to find the generalization of them, viz. a set of states which are least prone to deteriorate into a mixture of states. W. Zurek, S. Habib and J. Paz

[18] studied a harmonic oscillator undergoing quantum Brownian motion. From the master equation for the density matrix,[16] the entropy is defined and the condition for minimum increase of entropy is found. It turns out that the minimum entropy producing initial states are the coherent states.

4.7 A Dynamical Model of Decoherence

The quantum theory of open systems has been intensively pursued since the 1980s. The motivation was that many of the experimentally accessible systems fall in this category. Many of such systems are macroscopic quantum systems, for which the environment problem is of utmost importance. The treatment (modelling) of the environment can be directly applied to the problem of decoherence as well. A tractable model of the environment is afforded by a collection of harmonic oscillators or, equivalently, by a massless scalar quantum field. The former is the approach of A. Caldeira and A. Leggett [19], based on the theory of a quantum system interacting with a linear dissipating system by R. Feynman and F. Vernon [20]. This approach is described in detail in Section 6.5. The latter approach is that of W. Zurek and W. Unruh [21]. We give a brief introduction to the underlying ideas of this approach without going into technical details. A harmonic oscillator of unit mass with position coordinate x interacts with a massless scalar field $\varphi(q,t)$ propagating in the direction q through the Hamiltonian

$$H_{\text{int}} = \varepsilon x \frac{\partial \varphi}{\partial t}, \qquad (4.7.1)$$

where ε is the coupling constant. Starting from the action of the system $\int \mathscr{L} dt dq$, where $\mathscr{L}$ is the Lagrangian density, equations of motion of the harmonic oscillator and of the field can be derived. The equation of motion of the particle is

$$\ddot{x} + \frac{\varepsilon^2}{2} \dot{x} + \Omega_0^2 x = -\varepsilon \dot{\varphi}_0, \qquad (4.7.2)$$

---

[16] See Section 4.7.

where Ω_0 is the angular frequency of the oscillator and φ_0 is the solution of the free propagating field. Equation (4.7.2) can be considered as a generalized Langévin equation. The effect of system-environment interaction is two-fold. It leads to a damping force $-\eta\dot{x}$, where $\eta = \frac{\varepsilon^2}{2}$ is the viscosity, and also a random fluctuating force $-\varepsilon\varphi_0^2$. In the usual Langévin equation the fluctuating force would have the time dependence of the white noise: $\langle\dot{\varphi}_0(t)\,\dot{\varphi}_0(0)\rangle \sim \delta(t)$. For the present case, this will be true only in the high temperature limit

$$\langle\dot{\varphi}_0(t)\,\dot{\varphi}_0(0)\rangle \approx \frac{k_B T}{2}\delta(t), \qquad (4.7.3)$$

in which the terms independent of T have been neglected. The density matrix $\rho(x, x')$ of the particle in the position representation evolves according to the master equation

$$\frac{d\rho}{dt} = -\frac{i}{\hbar}[H, \rho] - \gamma(x - x')\left(\frac{d\rho}{dx} - \frac{d\rho}{dx'}\right) - \frac{2m\gamma kT}{\hbar^2}(x - x')^2\rho,$$
$$(4.7.4)$$

where H is the Hamiltonian with the potential $V(x)$ adjusted for H_{int}. In (4.7.4), T is the temperature of the field (environment), and γ is the relaxation rate depending on the viscosity η[17]:

$$\gamma = \frac{\eta}{2m}, \quad \eta = \frac{\varepsilon^2}{2}. \qquad (4.7.5)$$

The first term in the right hand side of (4.7.4), is the dynamical evolution, the consequence of the Schrödinger equation. In the classical limit it gives the equation of motion of the averages of observables (Ehrenfest theorem). Because the particle is coupled to the environment, its evolution is also governed by γ. The second term is the dissipation corresponding to the second term in (4.7.2). The last term is responsible for the fluctuation leading to the Brownian motion corresponding to the right hand side of (4.7.2). For our

---

[17]The mass m has been restored.

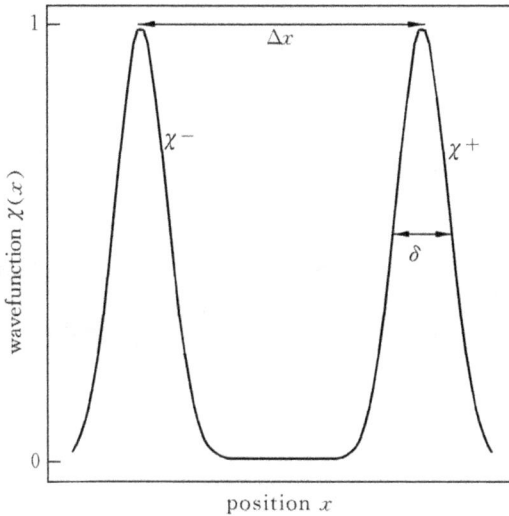

Fig. 4.18 The coherent superposition of two Gaussian Wave packets. From Ref. [15].

case it leads to decoherence. Zurek illustrates this with an example. Consider a coherent superposition of two Gaussian wave packets with width δ, and their centers separated by a distance Δx, with $\Delta x \gg \delta$, as shown in Fig. 4.18. This state is represented by the wave function

$$\chi(x) = \chi^+(x) + \chi^-(x). \tag{4.7.6}$$

The density matrix

$$\rho(x, x') = \chi(x)\chi^*(x'), \tag{4.7.7}$$

has four peaks on the (x, x') plane shown in Fig. 4.19, two from diagonal elements and two from the off-diagonal elements. The peaks from the off-diagonal elements occur at very different x and x'. The effect of the last term in the right hand side of (4.7.4) on the diagonal peaks are very small due to the factor $(x - x')$. In contrast, its effect on the off-diagonal peaks is large, which leads to the decay of the

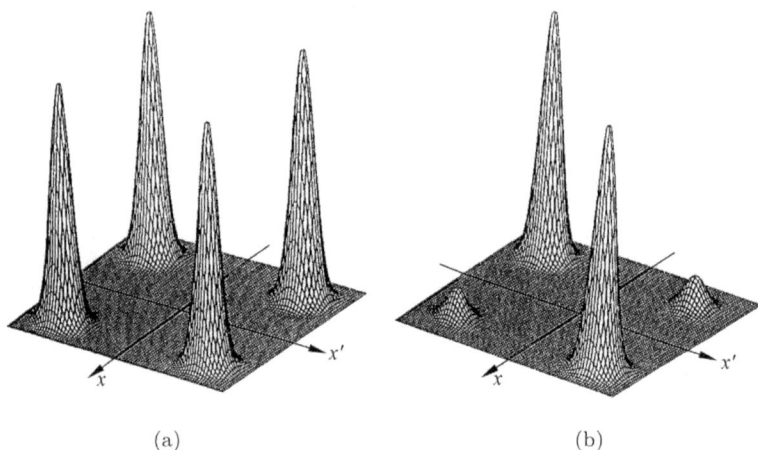

(a)   (b)

Fig. 4.19 Density matrix of the wave packet χ (a) at $t = 0$; (b) partially decohered. From Ref. [15].

off-diagonal peaks at the rate

$$\tau_D^{-1} \approx 2\gamma \frac{mkT \left(\Delta x\right)^2}{\hbar^2},$$

and the decoherence time τ_D is

$$\tau_D \approx \tau_R \frac{\hbar^2}{2mkT \left(\Delta x\right)^2} = \tau_R \left(\frac{\lambda_T}{\Delta x}\right)^2. \qquad (4.7.8)$$

where $\tau_R = \frac{1}{\gamma}$ is the relaxation time and $\lambda_T = \frac{\hbar}{\sqrt{2mk_BT}}$ is the thermal De Broglie wave length. For macroscopic bodies $\tau_D \ll \tau_R$, e.g., for $T = 300\text{K}, m = 1\text{g}, \Delta x = 1\text{cm}, \tau_D/\tau_R = 10^{-40}$. The decoherence between macroscopically distinguishable positions can be nearly instantaneous even for rather well isolated systems. In Fig. 4.12 the off-diagonal peaks die away and then the diagonal peaks can be considered as the classical probability distribution. On the other hand, a gravitational wave detector (cryogenic Weber bar) serves as an opposite example. A bar with $m = 100\text{kg}$ operating at $T = 10^{-3}\text{K}$ and $\Delta x = 10^{-19}\text{cm}$, τ_D/τ_R can be as large as 10^{-2}, i.e., decoherence is suppressed by low temperature and high precision of position measurement. Such a giant bar has to be treated as a quantum mechanical object.

4.8   Classical Limit of Quantum Dynamics

The quantum mechanics is formulated in the Hilbert space while the classical mechanics is formulated in the phase space. The most direct relationship between quantum dynamics and its classical limit can be made evident by using the Wigner transformation [22] of a wave function defined by

$$W(x,p) = \frac{1}{2\pi\hbar} \int_{-\infty}^{\infty} e^{ipy/\hbar} \psi^* \left(x + \frac{y}{2}\right) \psi \left(x - \frac{y}{2}\right) dy. \qquad (4.8.1)$$

The Wigner distribution $W(x,p)$ is real, but it can be negative. As such, it cannot be interpreted as distribution of observables, e.g., a probability distribution in coordinates. Nevertheless when integrated over p it gives the distribution in x and vice versa. This is demonstrated in the following. Integrating the distribution $W(x,p)$ with respect to p, we get

$$\int_{-\infty}^{\infty} W(x,p)dp = \int_{-\infty}^{\infty} \delta(y)\psi^* \left(x + \frac{y}{2}\right) \psi \left(x - \frac{y}{2}\right) dy = \psi^*(x)\psi(x),$$
$$(4.8.2)$$

This is the distribution in x. Conversely, consider the Fourier transform of ψ,

$$\varphi(p) = \frac{1}{\sqrt{2\pi\hbar}} \int_{-\infty}^{\infty} e^{-ip\xi/\hbar} \psi(\xi)d\xi. \qquad (4.8.3)$$

Hence, the distribution in p is

$$\varphi^*(p)\varphi(p) = \frac{1}{2\pi\hbar} \int_{-\infty}^{\infty} e^{ip(\eta-\xi)/\hbar} \psi^*(\eta)\psi(\xi)d\xi d\eta. \qquad (4.8.4)$$

Let

$$\eta = x + \frac{y}{2}, \ \xi = x - \frac{y}{2}. \qquad (4.8.5)$$

The distribution in p becomes

$$\varphi^*(p)\varphi(p) = \frac{1}{2\pi h} \int_{-\infty}^{\infty} e^{ipy/\hbar}\psi^* \left(x + \frac{y}{2}\right) \psi \left(x - \frac{y}{2}\right) dx dy$$

$$= \int_{-\infty}^{\infty} W(x,p) dx, \qquad (4.8.6)$$

Hence, integration of $W(x,p)$ with respect to x gives the distribution in p. For a minimum uncertainty wave packet

$$\psi(x) \approx \exp \left[-\frac{(x-x_0)^2}{\delta^2} + i\frac{p_0 x}{\hbar} \right], \qquad (4.8.7)$$

the Wigner distribution is

$$W(x,p) = \frac{1}{\pi \hbar} \exp \left[-\frac{(x-x_0)^2}{\delta^2} - \frac{(p-p_0)^2 \delta^2}{\hbar^2} \right]. \qquad (4.8.8)$$

Here both the distribution in x and the distribution in p are Gaussian, giving a minimum uncertainty. This is the closest analogy to a point in phase space in which a vector in Hilbert space can give. Wigner distribution can be generalized to the density matrix

$$W(x,p) = \frac{1}{2\pi \hbar} \int_{-\infty}^{\infty} e^{ipy/\hbar} \rho \left(x - \frac{y}{2}, x + \frac{y}{2}\right) dy. \qquad (4.8.9)$$

For a non-classical object as the superposition of two wave packets (4.7.6) and (4.7.7) the Wigner distribution is[18]

$$W \approx \frac{W^+ + W^-}{2} + \frac{1}{\pi \hbar} \exp \left[-\frac{p^2 \delta^2}{\hbar^2} - \frac{x^2}{\delta^2} \right] \cos \left(\frac{\Delta x}{\hbar} p\right) \qquad (4.8.10)$$

where W^+ and W^- are the Wigner distributions of χ^+ and χ^- respectively. The resulting W (4.8.10) is shown in Fig. 4.20, in which two wave packets having Gaussian distribution in x and p are

---

[18] See Zurek [14].

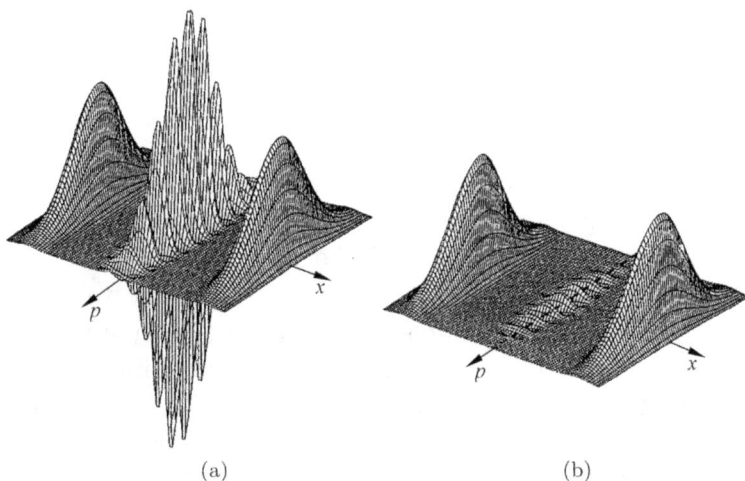

(a)   (b)

Fig. 4.20 Decoherence caused by damping (a) Wigner distribution of the superposition of coherent wave packets; (b) Decoherence caused by the diffusion of the distribution in momentum space. From Ref. [15].

shown alongside with an oscillatory distribution due to the second term. W cannot be interpreted as a classical distribution in the phase space. The equation of motion for W can be obtained from that for $\rho(x, x')$ (4.7.4). For harmonic oscillator the equation turns out to be

$$\frac{dW}{dt} = \left(-\frac{p}{m} \frac{\partial W}{\partial x} + \frac{\partial V}{\partial x} \frac{\partial W}{\partial p} \right) + 2\gamma \frac{\partial(pW)}{\partial p} + D \frac{\partial^2 W}{\partial p^2}, \quad (4.8.11)$$

where V is the potential and

$$D = 2m\gamma kT = \eta kT. \quad (4.8.12)$$

The three terms on the right hand side correspond to those in Eq. (4.7.4). The first term is identified with the Poisson bracket

$$[H, W]_{\text{Poisson}} = \frac{\partial H}{\partial x} \frac{\partial W}{\partial p} - \frac{\partial H}{\partial p} \frac{\partial W}{\partial x}, \quad (4.8.13)$$

since

$$\frac{\partial H}{\partial x} = \frac{\partial V}{\partial x},$$

and

$$\frac{\partial H}{\partial p} = \dot{x} = \frac{p}{m}.$$

In the case of harmonic oscillator the classical dynamics is given directly by the quantum dynamics, while for the more general case there are quantum corrections to the order $\hbar$ on the right hand side of (4.8.11). The second term in (4.8.11) is due to damping, and the third term is the diffusion of W in momentum space with diffusion coefficient D (4.8.12). The effect of the diffusion term on the Wigner distribution (4.8.10) is fairly simple. Since the oscillatory term $\cos(\frac{\Delta x}{\hbar}p)$ is an eigenfunction of the diffusion operator $\frac{\partial^2}{\partial p^2}$, the diffusion term tends to wash out this oscillation with a rate of $\tau_D^{-1} = 2m\gamma k_B T(\frac{\Delta x}{\hbar})^2$. The negative valley of W in Fig. 4.13(a) will be filled in a time of order τ_D, and now the two separate Gaussian packets in x and p can be interpreted as two separate points in the phase space (two equally probable classical alternative). In the above example damping and decoherence are intimately connected.

Now we turn to the other side of the coin. We have seen that the macroscopic number of environment atoms helps to erase the phase coherence between the states of quantum system in superposition. The inverse question is: whether states of a quantum system in superposition can still keep the coherence in the presence of large number of environment atoms. The discussion of this question forms a part of chapter 6. Both sides of the coin show their support of a common view that quantum mechanical laws are universal and can be applied to macroscopic and microscopic systems as well. Under certain conditions, a macroscopic system exhibits quantum behavior, e.g., a superconducting ring with a Josephson junction or a gravitational wave detector weighing tons, and in other cases exhibits classical behavior as a consequence of quantum mechanical laws.

There is a division of opinion among physicists upon. Whether the wave function collapse (or decoherence) is a postulate from outside quantum mechanics, or a consequence of it. In the 1970s those holding the second view belonged to the minority. Their number has been growing since then. There are many approaches in the direction of enhancing the second view.[19] There have been hot discussions on the papers by Zurek, as can be found in numerous letters to the editor of Physics Today [26].

4.9 Schrödinger Cat Realized in the Laboratory

4.9.1 A Schrödinger cat at single atom level

A quantum system which can be microscopic or macroscopic is in a state of coherent superposition of two *macroscopically distinguishable* states is defined as a Schrödinger's cat. D. J. Wineland and C. Monroe *et al.* [27–29] generated in a NIST laboratory a Schrödinger cat like state at the single atom level. Macroscopic distinguishability in this case means two *spatially separated* coherent harmonic states of *one single atom*. $^9\text{Be}^+$ ion (one electron in the $n = 2$ state) is trapped in an ion trap after laser cooling. The trap is equivalent to a harmonic oscillator potential well with frequencies $\frac{1}{2\pi}(\omega_x, \omega_y, \omega_z) = (11.2, 18.2, 29.8)$ MHz. The electronic (internal) energy levels of the ion are $^2S_{\frac{1}{2}}(F = 2, m_F = -2)$ and $^2S_{\frac{1}{2}}(F = 1, m_F = -1)$, denoted by $|\uparrow\rangle_i$ and $|\downarrow\rangle_i$ respectively. The nuclear spin $\boldsymbol{I}$ of ^9Be is $\frac{3}{2}$, the hyperfine spin $\boldsymbol{F} = \boldsymbol{s} + \boldsymbol{I}$ composed of the electron spin $\boldsymbol{s}$ and the nuclear spin $\boldsymbol{I}$ can be $F = 2$ or 1. the projection of $\boldsymbol{F}$ in the direction of an external magnetic field is m_F. The center of mass motion (external) levels are those in the potential of harmonic oscillators. The internal and external levels of the ion are manipulated by laser beams. Figure 4.21(a) shows the energy levels, the splitting of the hyperfine ground states $|\uparrow\rangle_i$ and $|\downarrow\rangle_i$ is $\omega_{\text{HF}} = 1.250$ GHz. Quantum numbers 0, 1, 2 refer to the harmonic oscillator energy levels (the external levels). The two-photon coupling between $|\uparrow\rangle_i$ and $|\downarrow\rangle_i$ is realized by the laser beams a and b with a frequency difference ω_{HF}, which can excite the states $|\uparrow\rangle_i$ and $|\downarrow\rangle_i$ respectively to a virtual level close to $^2P_{\frac{1}{2}}(2, -2)$ with detuning $\Delta \approx -12$ GHz. This kind of two-photon coupling is called a Raman

---

[19]See Refs. [23–25].

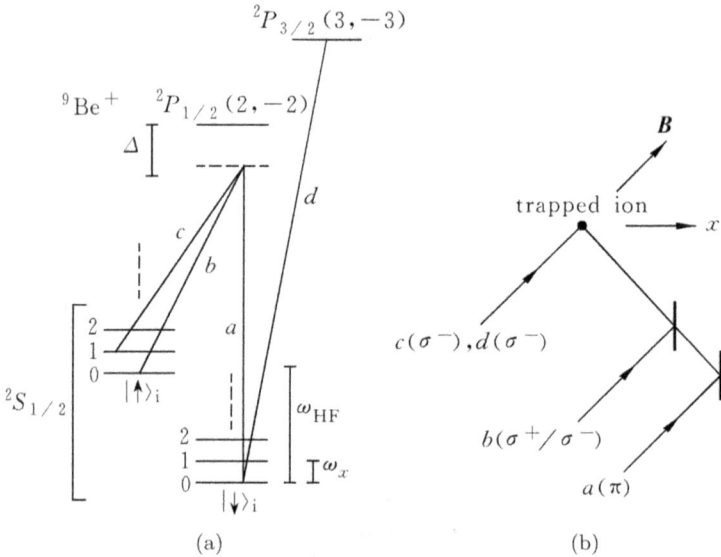

Fig. 4.21   (a) Electronic and center-of-mass motion energy levels of Be^+ ion, coupled by laser beams; (b) The geometry of the laser beams with polarizations indicated. Reprinted with permission from C. Monroe, D. M. Meekhof, B. E. King, D. J. Wineland, *Science*, 272, 1131, 1996. Copyright (2017) by the AAAS.

coupling, which induces a *Rabi oscillation* of the ion between the levels $|\uparrow\rangle_i$ and $|\downarrow\rangle_i$. The geometry of the laser beams is shown in Fig. 4.21 (b). Then the internal state of the ion is a coherent superposition of $|\uparrow\rangle_i$ and $|\downarrow\rangle_i$. The experimental steps are:

(1) By adjusting the time of illumination by the *carrier beams a* and *b*, the coefficients of superposition can be varied. If the ion is initially in the state $|\uparrow\rangle_i$, then by using a $\pi/2$ pulse (time equal to $1/4$ period of Rabi oscillation) the superimposed state of the ion is

$$\Psi_1 = \frac{1}{\sqrt{2}} \left[|\downarrow\rangle_i |0\rangle_e - ie^{i\mu} |\uparrow\rangle_i |0\rangle_e \right], \qquad (4.9.1)$$

where $|0\rangle_e$ is the ground state of the external motion. The laser beams used are shown in Fig. 4.21(b) and the evolution of the external motion is shown in Fig. 4.22(a) and (b), the parabola in the figure represents the harmonic oscillator potential. A phase difference is produced when the carrier beams are used. Such beams are used 3

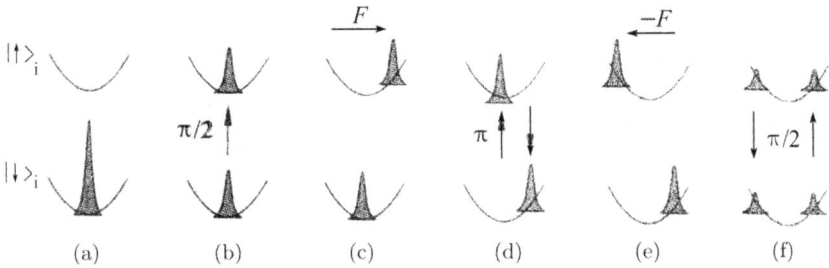

Fig. 4.22 Evolution of the wave packet entangled with the internal states $|\uparrow\rangle_i$ and $|\downarrow\rangle_i$ (a) initial position of the wave packet; (b) the first carrier $\pi/2$ pulse, 0.5 µs; (c) displacement -10 µs; (d) the second carrier π pulse, 1.0 µs; (e) displacement -10 µs; (f) the third carrier $\pi/2$ pulse, 0.5 µs. Reprinted with permission from C. Monroe, D. M. Meekhof, B. E. King, D. J. Wineland, *Science*, 272, 1131, 1996. Copyright (2017) by the AAAS.

times during the experiment. The phase is relative, and we define the phase of the last carrying to be 0. The phase of the first carrying is $-\mu$.

(2) By using the *displacement beams b* and *c* with a frequency difference $\omega_x/2\pi = 11.2$ MHz which couple the two neighboring states of the harmonic oscillator, we can manipulate the motion of the center of mass of the ion, such that the two states can be separated by a macroscopically distinguishable distance. When the time of illumination is sufficiently long, successive transitions can occur, such that a coherent state of the external motion can be formed.

$$|\beta\rangle_e = e^{-|\beta|^2/2} \sum_n \frac{\beta^n}{(n!)^{1/2}} |n\rangle_e \,,$$

where $\beta = \alpha e^{i\theta}$, α and θ are real. The result is shown in Fig. 4.22(c), where the oscillating wave packet in the first row is shown at its maximum displacement. The average quantum number of the coherent state is $\langle n \rangle = \alpha^2$, the corresponding amplitude of harmonic oscillator is $2\alpha\sqrt{\frac{\hbar}{2m\omega_x}}$. By arranging the polarizations of the laser beams $a(\pi), b(\sigma^+/\sigma^-), c(\sigma^-),$[20] it is possible that only the state $|\uparrow\rangle_i$

---

[20]By σ and π we denote polarizations parallel and perpendicular to the direction of external magnetic field (same as that of the laser beams) respectively.

is influenced by the displacement beams, because the σ^- polarization of beam c cannot couple $|\downarrow\rangle_i$ to any of the virtual $^2P_{\frac{1}{2}}$ states. The displacement beams achieve the *entanglement* of the external and internal states. From the value of w_x it follows that the root mean square value of the width of the coherent state wave packet $x_0 = 7.1$ nm. The resulting state after the operation is

$$\Psi_2 = \frac{1}{\sqrt{2}}\left(|\downarrow\rangle_i\,|0\rangle_e - \mathrm{ie}^{\mathrm{i}\mu}\,|\uparrow\rangle_i\,\Big|\alpha e^{-\mathrm{i}\phi/2}\Big\rangle_e\right). \qquad (4.9.2)$$

The displacement in the positive $\hat{x}$ direction yields a phase $-\phi/2$.

(3) In order to enlarge the distance between the two states of the superposition, their internal states are interchanged first. The transitions $|\uparrow\rangle_i \leftrightarrow |\downarrow\rangle_i$ are induced by applying a π pulse (time equal to $1/2$ of the Rabi period) carrier beams, with a phase factor $-\mathrm{ie}^{\mathrm{i}\nu}$. The state after the operation (shown in Fig. 4.22 (d)) is

$$\Psi_3 = \frac{1}{\sqrt{2}}\left(\mathrm{ie}^{-\mathrm{i}\nu}\,|\uparrow\rangle_i\,|0\rangle_e + \mathrm{ie}^{\mathrm{i}(\nu-\mu)}\,|\downarrow\rangle_i\,\Big|\alpha e^{-\mathrm{i}\phi/2}\Big\rangle_e\right), \qquad (4.9.3)$$

where a common minus sign has been neglected.

(4) The state $|\uparrow\rangle_i|0\rangle_e$ is displaced in the $-\hat{x}$ direction with phase $\phi/2$, while $|\downarrow\rangle_i$ is not influenced. The state after this operation (shown in Fig. 4.22 (e)) is

$$\Psi_4 = \frac{1}{\sqrt{2}}\left(\mathrm{ie}^{-\mathrm{i}\nu}\,|\uparrow\rangle_i\,\Big|\alpha e^{\mathrm{i}\phi/2}\Big\rangle_e + \mathrm{ie}^{\mathrm{i}(\nu-\mu)}\,|\downarrow\rangle_i\,\Big|\alpha e^{-\mathrm{i}\phi/2}\Big\rangle_e\right) \qquad (4.9.4)$$

Now the distance between the two states reaches the maximum. This state is one most close to the Schrödinger cat. There is but *one* ion, whose state is the coherent superposition of *two* states. The external motion of these two states is Gaussian wave packets separated at a distance. So far as wave packets are concerned, the distance is really "macroscopic": in the experiment the maximum distance reached is 80 nm, while the width of a single packet is 7 nm, and the radius of the atom is 0.1 nm.

(5) Finally a $\pi/2$ pulse of the carrier beams splits each of $|\uparrow\rangle_i$ and $|\downarrow\rangle_i$ into two. The state after the operation is (Fig. 4.22 (f))

$$\Psi_5 = \frac{1}{2}|\downarrow\rangle_i \left(\left|ae^{-i\phi/2}\right\rangle_e - e^{i\delta}\left|ae^{i\phi/2}\right\rangle_e\right)$$

$$- \frac{i}{2}|\uparrow\rangle_i \left(\left|ae^{-i\phi/2}\right\rangle_e + e^{i\delta}\left|ae^{i\phi/2}\right\rangle_e\right), \qquad (4.9.5)$$

where the phase of the carrier beams is 0 (the standard value). In the above equation a common phase factor $e^{i(\nu-\mu)}$ is neglected, and $\delta = \mu - 2\nu + \pi$. Ψ_5 can further be rewritten as

$$\Psi_5 = |\downarrow\rangle_i |S_-\rangle_e - i|\uparrow\rangle_i |S_+\rangle_e, \qquad (4.9.6)$$

where

$$|S_\pm\rangle_e = \frac{1}{2}\left(\left|ae^{-i\phi/2}\right\rangle_e \pm e^{i\delta}\left|ae^{i\phi/2}\right\rangle_e\right). \qquad (4.9.7)$$

For $\phi = \pi, \delta = 0$, $|S_\pm\rangle$ is called the "even cat" and "odd cat" respectively.

In the experiment the probability $P_\downarrow(\phi)$ that the internal state of the ion is $|\downarrow\rangle_i$ as a function of the coherent state phase separation ϕ is measured. Detection of the internal state is accomplished by first illuminating the ion with σ^--polarized detection beam d which drives the transition $^2S_{\frac{1}{2}}(F = 2, m_F = -2) \rightarrow ^2P_{\frac{3}{2}}(F = 3, m_F = -3)$, and then observing the scattered fluorescence. The state correlated with $|\downarrow\rangle_i$ is $|S_-\rangle_e$. The two terms constituting $|S_-\rangle_e$ (4.9.7) represent wave packets separated at a distance depending on ϕ. When the wave packets are closer, the interference effect is more remarkable. The probability distribution of the wave packets is $|\langle x | S_-\rangle_e|^2$. Since the time evolution of a coherent state $|z\rangle$ is stationary in the parameter z (4.1.45), ϕ is linear in time t. For the case $\delta = 0$, the two wave packets are the most distant from each other when $\phi = \pm\pi$, and they coincide when $\phi = 0$. But for $\delta = 0$, S_- is the odd cat, the two packets cancel each other. $P_\downarrow(\phi)$ is the integral over x of the distribution function:

$$P_{\downarrow}(\phi) = \int_{-\infty}^{\infty} |\langle x | S_{-} \rangle|^2 \, dx$$

$$= \frac{1}{2} \left[1 - e^{-\alpha^2(1-\cos\phi)} \cos\left(\delta + \alpha^2 \sin\phi\right) \right]. \qquad (4.9.8)$$

$P_{\downarrow}(\phi)$ shows oscillatory behavior close to $\phi = 0$ for sufficiently large α, this is the interference between the two wave packets. In the cat analogy, this is the interference between the live and dead cat. The measured and fit interference signal $P_{\downarrow}(\phi)$ versus the phase difference ϕ of two coherent states for $\delta = 0$ is shown in Fig. 4.23.

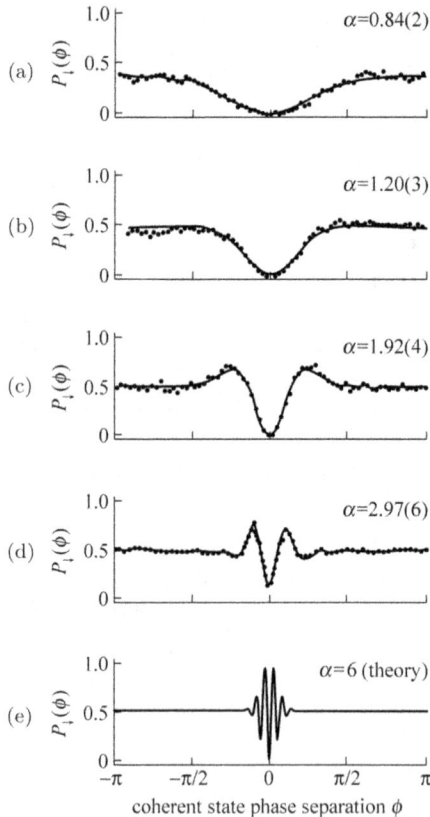

Fig. 4.23   Comparison between theory and experiment for $P_{\downarrow}(\phi)$, for $\delta = 0$. Reprinted with permission from C. Monroe, D. M. Meekhof, B. E. King, D. J. Wineland, *Science*, 272, 1131, 1996. Copyright (2017) by the AAAS.

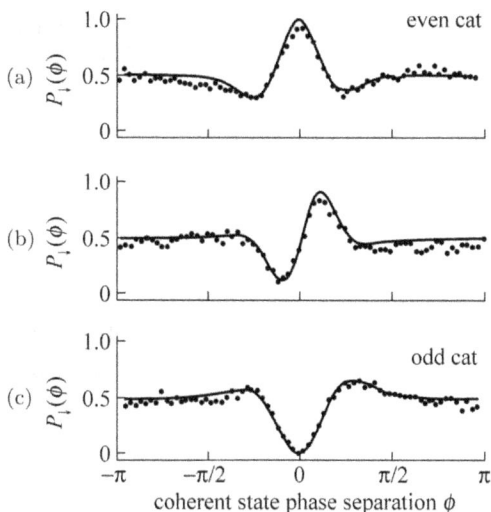

Fig. 4.24 $P_\downarrow(\phi)$ for different values of δ. Reprinted with permission from C. Monroe, D. M. Meekhof, B. E. King, D. J. Wineland, *Science*, 272, 1131, 1996. Copyright (2017) by the AAAS.

Figure 4.23(a)∼(d) give the comparison of experiment (points) with the theory (curves), while (e) is the theory for a larger value of α. Measurements are repeatedly made for each value of δ: cooling, preparation of states, detection. If decoherence occurs, no phase coherence exists between the two wave packets, then $P_\downarrow(\phi) = \frac{1}{2}$. The value of δ can be measured by blocking the displacement beams $(\alpha = 0)$, then $P_\downarrow(\phi) = \sin^2 \frac{\delta}{2}$ gives δ directly. As a function of ϕ, $P_\downarrow(\phi)$ is different for different δ. Figure 4.24 gives $P_\downarrow(\phi)$ for different values of δ, remembering that ϕ is inversely proportional to δ, for $\alpha \approx 1.5$. Figure 4.27 (a) $\delta = 1.03\pi$, close to the even cat. A constructive interference occurs at $\phi = 0$. In Fig. 4.24 (b) $\delta = 0.48$. In Fig. 4.24 (c) $\delta = 0.06\pi$, close to the odd cat, destructive interference occurs at $\phi = 0$.

The study of Schrödinger cat in the laboratory is important not only for the sake of the fundamental principles of quantum mechanics, but also for a potential application — the quantum computing. Since the quantum bits must keep the state of coherent superposition before the entire manipulation on it is completed, the time of decoherence is a quality of utmost importance. The NIST group [30] carried out a research on the rate of decoherence based on the experiment described above. In this experiment the decoherence is induced by coupling the trapped atom to an engineered reservoir, in which the coupling between the system and the environment is controllable. The system is an oscillator in a superposition of coherent states

$$|\psi\rangle = N(|\alpha_1\rangle + |\alpha_2\rangle) \qquad (4.9.9)$$

coupled to the reservoir through an interaction proportional to the product of the system oscillation and the amplitude of the fluctuation of the reservoir atom. A scaling law[21] states that the remaining coherence $C(t)$ is

$$C(t) = e^{-|\alpha_1 - \alpha_2|^2 \xi t}, \qquad (4.9.10)$$

where ξ is the coupling between the system and the reservoir, $|\alpha_1 - \alpha_2|$ can be considered as the "distance in the Hilbert space", or the "size of the superposition". The reservoir consists of pseudo-random uniform electric fields applied along the axis of the trap, oscillating near the axial motion frequency of the ion. The coherence of superposition is measured by single-atom interferometry. From the initial spin state of the ion $F = 1, m_F = -1$ (denoted by $|\downarrow\rangle$) and the ground state of external motion the cat state is produced by the method described above [29]. The cat state is then coupled to the reservoir, and the perturbed superposition is recombined by reversing the steps which initially created it, and the final state is tested for coherence. Detailed follow-up of the process [30] gives the probability of finding the ion in the $|\downarrow\rangle$ state

$$P_\downarrow = \frac{1}{2}(1 - e^{-2|\Delta\alpha|^2\sigma^2} \cos\delta), \qquad (4.9.11)$$

where $\sigma^2 \propto \langle V^2 \rangle$, V is the applied random voltage, and δ is the phase of the last carrier beams relative to the first. The result is shown in the Fig. 4.25, where the fringe contrast is scaled to unity at $\langle V \rangle^2 = 0$. The size of the superposition $|\Delta\alpha|$ varies linearly with the pulse time

---

[21] See "*Quantum Optics*" by D. F. Walls and G. J. Milburn, Berlin, Springer, 1994.

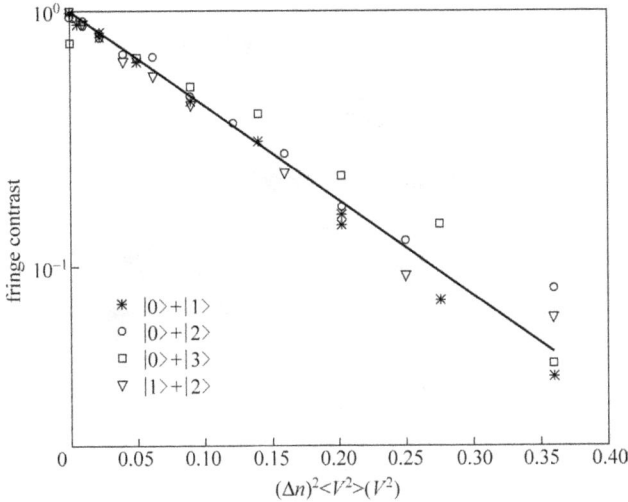

Fig. 4.25 Decoherence of superpositions of Fock states coupled to the phase reservoir. The solid line is a fit to an exponential. Reprinted with permission from Macmillan Publishers Ltd: *Nature* (403, 269), 2000.

for the displacement beams. We see that the experimental data agree well with the theoretical predictions.

Can a single atom be considered as a cat? Views of physicists differ. W. Zurek preferred to call it a Schrödinger kitten.

4.9.2 A phase Schrödinger cat

The experiments of the ENS(l'Ecole Normale Supérieure) group of C. Haroche [31] prepared a superposition of the coherent states of electromagnetic oscillations in a cavity.[22] Figure 4.26 shows the scheme of the experimental setup. C is a high Q superconducting Nb cavity. Rb atoms effusing from the oven O are excited by prescribed laser beams, and emerge from the box B in "circular Rydberg states" $n = 51, l = 50$. The interval between the arrivals of the atoms is made sufficiently large (-1.5 ms), such that a single atom passes through the cavity at a time. The low Q cavity R_1 is tuned in resonance with the transition from $n = 51, l = 50$ (denoted by e) to $n = 50, l = 49$

---

[22]See Section 8.2 for the physics of formation of such states.

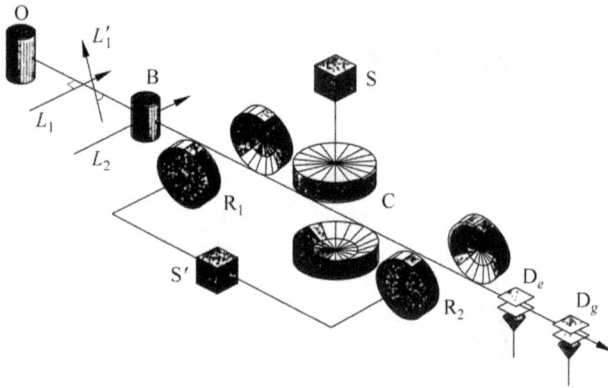

Fig. 4.26   Scheme of the experimental setup. Reprinted with permission from
M. Brune, E. Hagley, J. Dreyer, X. Maître, A. Maali, C. Wunderlich, J. M.
Raimond, and S. Haroche, *Phys. Rev. Lett.* **77**, 4887, 1996. Copyright (2017)
by the American Physical Society.

(denoted by g) at $\nu_0 = 51.099$ GHz. Resonant $\pi/2$ pulse is applied,
such that the atom passing through R_1 is in a state $\frac{1}{\sqrt{2}}(|e\rangle + |g\rangle)$. The
cavity C is fed by the microwave source S and tuned with a detuning
δ from the transition frequency ω_{eg}. The electromagnetic field in the
cavity is in a Glauber coherent state $|\alpha\rangle$.[23]

$$|\alpha\rangle = e^{-|\alpha|^2/2} \sum_n \frac{\alpha^n}{(n!)^{1/2}} |n\rangle .$$

Because of the detuning, the atom cannot undergo transitions
between e and g, but the cavity field undergoes a phase change
depending on the state of the atom passing through the cavity
because of the interaction between the atom and the cavity field.[24]
Therefore the coherence between the states e and g of the atom
induces a coherence between the states of the field. The state of the
atom-cavity system is

$$|\psi_1\rangle = \frac{1}{\sqrt{2}} \left(\left| e, \alpha e^{i\phi} \right\rangle + \left| g, \alpha e^{-i\phi} \right\rangle \right). \tag{4.9.12}$$

---

[23]This is the canonical coherent state in section 4.1, and n is the number of
photons here.
[24]See Section 8.6 on the atom-caviy dispersive phase shift effect.

The phase shift ϕ depends on the detuning δ, and is inversely proportional to it. Since the superimposed states are distinguished by the phase ϕ and their "distance" D is

$$D = \mid \alpha e^{i\phi} - \alpha e^{-i\phi} \mid = 2\alpha \sin \phi,$$

the superposition can be called a phase cat if D is, for instance, $0(1)$. Through an adjustment of the detuning different phase shift can be obtained. R_2 is a cavity of the same structure as R_1 and both are fed by the source S' ($\pi/2$ pulse). The atoms passing through the cavities C and R_2 now enter the field ionization detectors D_e and D_g successively. Different voltages are applied to them, such that a lower voltage for D_e is just enough to ionize an atom in state e, and a higher voltage for D_g can ionize an atom in state g. Varying the frequency ν of source S' such that it sweeps through ν_0, the probability $P_g^{(1)}(\nu)$ that the atom passing through the setup is in the state g is measured. Altogether 50000 events were recorded in 10min., and result is shown in Fig. 4.20. In (a) the result when there is no field in the cavity C is shown. The atom experiences the action of $\pi/2$ pulse twice, in R_1 and R_2 respectively in a time interval of 230 μs. The $e \rightarrow g$ transition can happen in either of R_1 and R_2, and consequently interference pattern occurs, the *Ramsey fringes* are observed.[25] In (b)–(d) results are shown for cavity fields with $\mid \alpha \mid = \sqrt{9.5} = 3.1$, (i.e., the average photon number is 9.5, and this is sometimes referred to as a mesoscopic cat), and with detuning $\delta/2\pi = 712, 347$ and 104 kHz respectively. The insets on the right give the relative phase between the fields in the superposition, with the angle between the lines given by ϕ. From the figure we can see the shift of fringes, and the decrease of visibility of the fringes with increasing angle ϕ. The atom is in the state given by (4.9.12) when it leaves the cavity C. When ϕ is small, the information about the states of the atom acquired in measuring the phase of the field is not so much, therefore the interference between the two paths remains significant. With ϕ increasing, the information about the state of the

---

[25]See Section 8.6 for the Ramsey interferometer.

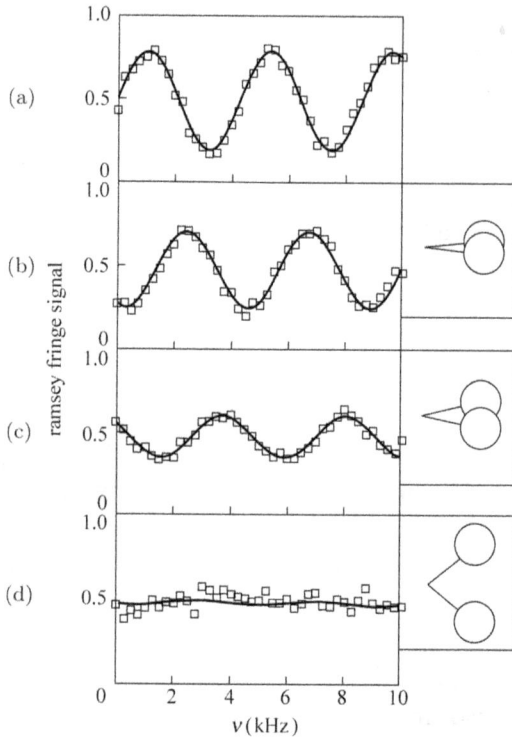

Fig. 4.27   The Ramsey fringes of $P_g^{(1)}(\nu)$. Reprinted with permission from M. Brune, E. Hagley, J. Dreyer, X. Maître, A. Maali, C. Wunderlich, J. M. Raimond, and S. Haroche, *Phys. Rev. Lett.* 77, 4887, 1996. Copyright (2017) by the American Physical Society.

atom acquired in measuring the phase of the field is increased, and the interference is inevitably weakened.

From Fig. 4.27 we can see that the closer to the macroscopic distinguishability the state of the field is (i.e., a larger $|\phi|$), the weaker is the interference. Through the measurement of the state of the atom, we obtain information about the state of the field in a cavity separated at a large distance from the detectors. Furthermore, if R_2 is absent, then the atom remains in the state given by (4.9.12), the measurement of the state of the atom (in e or g) gives the phase of the field ϕ and $-\phi$ respectively in the far-away cavity. If R_2 is in place, then it mixes once again the two Rydberg states e and g. By adjusting the phase of the S' pulse, it is possible to make the atom after passing through R_2 to be in the states $|e\rangle \rightarrow \frac{1}{\sqrt{2}}(|g\rangle + |e\rangle)$ and $|g\rangle \rightarrow \frac{1}{\sqrt{2}}(|g\rangle - |e\rangle)$, and the field remains in its original state. The state of the atom after passing through R_2 is

$$\psi_2 = |e\rangle \frac{|\alpha e^{i\phi}\rangle - |\alpha e^{-i\phi}\rangle}{\sqrt{2}} + |g\rangle \frac{|\alpha e^{i\phi}\rangle + |\alpha e^{-i\phi}\rangle}{\sqrt{2}}. \qquad (4.9.13)$$

The state of the field is now $\frac{1}{\sqrt{2}}(|\alpha e^{i\phi}\rangle - |\alpha e^{-i\phi}\rangle)$ and $\frac{1}{\sqrt{2}}(|\alpha e^{i\phi}\rangle + |\alpha e^{-i\phi}\rangle)$ respectively for the state of the atom measured $|e\rangle$ and $|g\rangle$. This shows not only the "non-locality", i.e., the quantum correlation of two objects at a large distance, and also gives a "delayed choice", i.e., by deciding whether to put R_2 in place *after* the atom has passed through the cavity C. Different choices give different state for the field.

This experiment also enables a study of decoherence. The first atom produces two macroscopically distinguishable states of the field (phases $-\phi$ and ϕ). A second atom then passes through the cavity and shifts the phase again, and the result is three different states of the field with phases $-2\phi, 0, 2\phi$. There are two possibilities that the field with phase 0 to be formed, either by the first atom passing in state e and the second passing in state g (denoted by (e, g)), or the other way round, i.e., (g, e). These two possibilities are indistinguishable, and consequently interference occurs when joint probabilities $P_{ee}^{(2)}, P_{eg}^{(2)}, P_{ge}^{(2)}$ and $P_{gg}^{(2)}$ are measured. Denote the two atom correlation signal by η:

$$\eta = \frac{P_{ee}^{(2)}}{P_{ee}^{(2)} + P_{eg}^{(2)}} - \frac{P_{ge}^{(2)}}{P_{ge}^{(2)} + P_{gg}^{(2)}}. \qquad (4.9.14)$$

If the field is in the coherent superposition state $\frac{1}{\sqrt{2}}(\alpha e^{i\phi}\rangle \pm |\alpha e^{-i\phi}\rangle)$, then η is a constant. If the field has deteriorated to an incoherent mixture, then η vanishes. Fig. 4.28 gives the relation between η and $\tau/T_{\rm r}$, where τ is the time interval between the passage of two atoms through the cavity, and $T_{\rm r}$ is the lifetime of the photon determined by the quality factor Q of the cavity. In this experiment $T_{\rm r} = 160$ μs. From the figure we can see that η approaches 0 when $\tau/T_{\rm r}$ approaches 1, i.e., the interference does not occur any more.

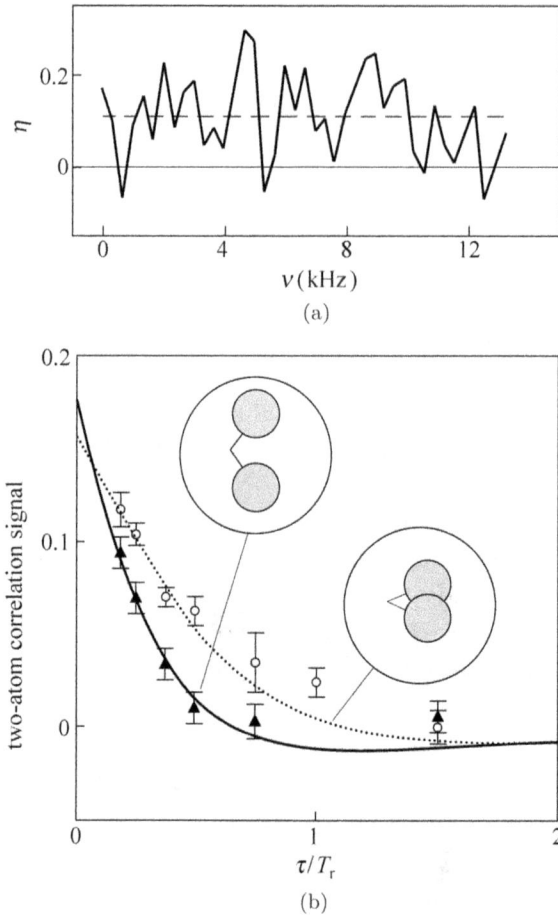

Fig. 4.28   Two-atom correlation signal versus τ/T_r. Reprinted with permission from M. Brune, E. Hagley, J. Dreyer, X. Maître, A. Maali, C. Wunderlich, J. M. Raimond, and S. Haroche, *Phys. Rev. Lett.* 77, 4887, 1996. Copyright (2017) by the American Physical Society.

Besides, the time of decoherence depends on the magnitude of ϕ, for large ϕ the time of decoherence is much shorter. A less technical paper by C. Haroche [32] covers the research described above.

The irreversibility of the decoherence has its origin in the enormous number of degrees of freedom of the reservoir. Another paper by the ENS group describes the blueprint of an experiment in which the decoherence of a mesoscopic superposition of radiation states becomes reversible. By controlling the size of the reservoir it is possible to design the situations in which the coherence of the cat collapses and revives periodically. The experimental setup is the same as described in Fig. 4.26 except an additional resonance cavity C_1 playing the role of "single mode reservoir" is coupled to the cavity C containing the cat.

4.9.3 A macroscopic cat

A SUNY Stony Brook group (C. J. Friedman) [33] succeeded in experimentally verifying the macroscopic quantum coherence in a SQUID (superconducting quantum interference device) by measuring directly the tunnel splitting of the superimposed state. Macroscopic quantum systems have been known since 1950s when N. Bogoliubov introduced the wave function of Bose-Einstein condensates, which applies to superconductors (condensates of Cooper pairs). Josephson systems have long been the object of the study of macroscopic quantum phenomena.[26] The SQUID is a superconducting loop of inductance L interrupted by a thin Josephson tunnel junction with capacitance C and critical current I_C. In equilibrium, a supercurrent I can flow around the loop producing a flux LI. When an external magnetic flux Φ_x is present, a resulting flux $\Phi = \Phi_x - LI$ threads the loop. The minus sign applies when the supercurrent flows in a direction such that the flux produced is opposite to Φ_x. When the supercurrent flows in the other direction, the flux produced tends to augment Φ_x. The wave function of Cooper pairs is single-valued, hence the phase of the wave function must vary continuously around the loop (except at the Josephson junction), increasing by an integer f times 2π when one winds once around the loop. The quantum number f defines the *fluxoid state* of the SQUID. In the case of $f = 0$ (or 1), the supercurrent tends to cancel (or augment) the flux Φ_x. In this experiment the supercurrent is $> 1\mu A$, and the local magnetic moment can be $10^7 \mu_B$. The system can be considered as macroscopic. The dynamics of the SQUID is described

---

[26]See Sections 6.1–6.3.

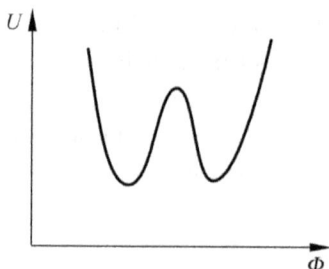

Fig. 4.29   Symmetric double well of the SQUID.

by the flux Φ threading the loop,[27] and the Hamiltonian of the system is

$$H = \frac{1}{2}C\,\dot{\Phi}^2 + V(\Phi). \qquad (4.9.15)$$

The first term in the right hand side is the analog of the kinetic energy, and the second term is the potential energy

$$V(\Phi) = \frac{(\Phi_x - \Phi)^2}{2L} - I_c \frac{\Phi_0}{2\pi}\cos 2\pi\frac{\Phi}{\Phi_0}, \qquad (4.9.16)$$

where Φ_0 is the flux quantum. For $\Phi_x = \frac{\Phi_0}{2}$, $V(\Phi)$ is a symmetric double well shown in Fig. 4.29. When Φ_x deviates from this value, the well is asymmetric, as shown in Fig. 4.30 (a). The two fluxoid state $f = 0$ and 1 correspond to the states in the left and right well respectively. The double well is an interesting topic in quantum mechanics. Consider a single particle in a symmetric well first.

The low lying levels $|L\rangle$ and $|R\rangle$ are mainly localized in the left and right wells respectively. When the levels get higher and higher in energy, the probability of tunneling through the barrier between the wells becomes non-negligible. They get mixed in the sense that the eigenstates are $\frac{1}{\sqrt{2}}(|L\rangle \pm |R\rangle)$ with a difference in energy Δ called the *tunnel splitting*, because the tunneling lifts the

---

[27]Although Φ is in general macroscopic, it should display the quantum behavior. The flux is directly related to the supercurrent, which is connected with the phase of the wave function. In the Hamiltonian (4.9.15) the capacitive energy is much smaller, and hence Φ can be described with sufficient precision with a tolerable uncertainty in the number of particles.

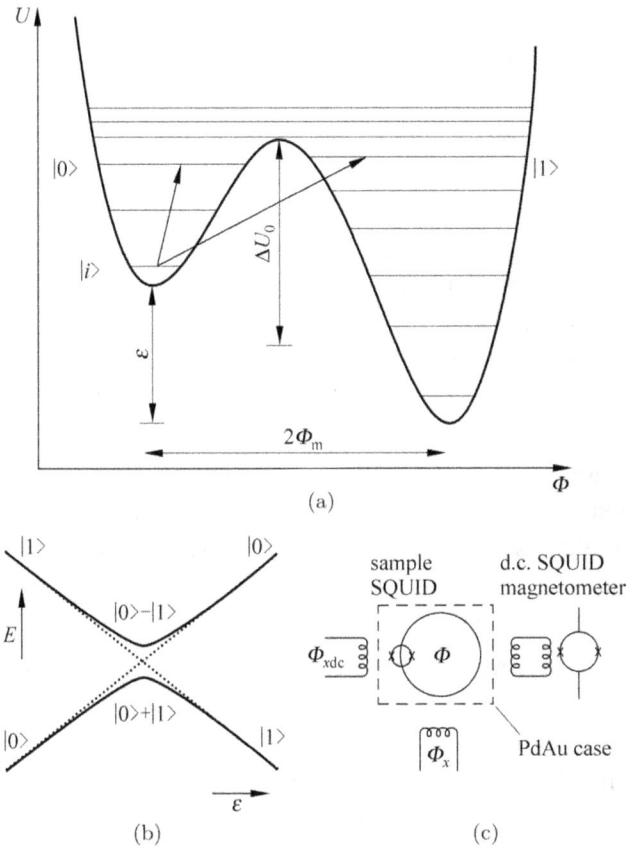

Fig. 4.30 (a) SQUID potential; (b)Energy level anticrossing; (c) Experimental set-up. Reprinted with permission from Macmillan Publishers Ltd: *Nature* (406, 43), 2000.

degeneracy of $|L\rangle$ and $|R\rangle$, with the symmetric superposition being the ground state. If the state $|L\rangle$ is prepared initially (it is not an eigenstate any more), then the wave function of the system will oscillate between $|L\rangle$ and $|R\rangle$ with a frequency $\Gamma = \Delta/\hbar$. This is the quantum coherence problem, and the mechanism is extremely important for the chemical bond, e.g., the NH$_3$ molecule.[28] The question is, in the macroscopic system as a SQUID, whether the

---

[28] Read Section 8.6 of The Feynman Lectures on Physics vol. III.

superposition of the fluxoid states $|0\rangle$ and $|1\rangle$, i.e., $\frac{1}{\sqrt{2}}(|0\rangle \pm |1\rangle)$ still makes sense. The quantum coherence for a single particle occurs on the basis of tunneling. But for a macroscopic system which is always coupled to the environment, the situation is much more complicated. A. Caldeira and A. Leggett gave a systematic answer to this question in 1983.[29] If the coupling is strong enough, then the system is trapped in one of the wells. If the coupling is weak enough, then the system can tunnel regularly, and the linear superposition $\frac{1}{\sqrt{2}}(|0\rangle \pm |1\rangle)$ are the eigenstates of the system with a tunnel splitting. These are the macroscopic Schrödinger cat states. In the intermediate coupling case, tunneling is possible, but the system hops through the barrier *incoherently*. This is called the *sequential hopping*. The cat states do not make sense in this case. The resonance tunneling of a SQUID state has been previously verified in experiment [34]. The experimental verification of the linear superposition lies in the direct measurement of the tunnel splitting. For this purpose we consider in more detail the energy variations of $|g\rangle = \frac{1}{\sqrt{2}}(|0\rangle + |1\rangle)$ and $|e\rangle = \frac{1}{\sqrt{2}}(|0\rangle - |1\rangle)$ with parameters of the potential (4.9.16), the tilt ε and the energy barrier at zero tilt ΔU_0 (Fig. 4.30(a)). With increasing tilt ε the energy of $|0\rangle$ and $|1\rangle$ increases and decreases respectively, as shown in Fig. 4.30 (b). They would cross at $\varepsilon = 0$ (the symmetric well), but this will not happen because the tunneling removes the degeneracy. This is called the *anticrossing*, or the *avoided crossing*. The tunnel splitting is very small, and its measurement is difficult because a necessary condition for resolving this small splitting is that the experimental line widths of the states must be smaller than Δ. The SQUID is extremely sensitive to external noise and dissipation, both tend to broaden the line width. The experimental challenges include the weak coupling of the system to the measuring apparatus, the sufficient strength of the signal to resolve the closed levels, the shielding of the system from the external noise. These were what have frustrated the previous attempts to observe coherence in SQUID. The sketch of the experimental setup is illustrated in

---

[29]See Section 6.6.

Fig. 4.30(c). The SQUID is encased in a PdAu box to reduce the noise. Applied flux Φ_{xdc} is used to control ΔU_0 through its action on the Josephson junction which regulates I_c (Eq. (4.9.16)). The flux Φ_x is used to control the tilt ε. The tunneling of a SQUID state accompanied by a reversal of flux Φ is measured by the dc SQUID magnetometer. In the experiment anticrossing of two excited levels in the potential is probed by using microwaves to produce photon-assisted tunneling. The tunneling through a much narrower barrier significantly enhances the signal. The system is initially prepared in the lowest state $|i\rangle$ of the left well with the height of the barrier large enough such that the tunneling is negligible. Microwave radiation is then applied. When the difference in energy between the initial state and an excited state matches the frequency of the microwave radiation, the system has an appreciable probability to be excited into this state and subsequently tunnel to the right well. From the SQUID potential the energy levels of the SQUID are solved numerically in the zero damping limit, viz. the energy eigenvalue problem of the following Hamiltonian is solved,

$$H = p_\Phi^2/2C + V(\Phi, \Phi_x, \Phi_{x,dc}),$$

where $p_\Phi = -i\partial/\partial\Phi$, and V is the potential (4.9.16). In Fig. 4.31 calculated levels for $\Delta U_0 = 9.117\text{K}$ are plotted as a function of Φ_x (thin solid lines). The calculated top of the barrier is indicated by a thick solid line. The dot-dashed line represents level $|i\rangle$ shifted upward by the microwave radiation. At values of Φ_x for which this line intersects one of the excited levels indicated by arrows, the system absorbs a photon and tunnels. For the reduced barrier $\Delta U_0 = 8.956$ K, the corresponding levels are represented by dotted lines and the photon absorption occurs at different values of Φ_x. By progressively reducing the barrier and thus moving the levels downward, the anticrossing can be mapped. The probability of the photon-assisted tunneling is plotted as a function of Φ_x. Two peaks correspond to the absorption of photons by the lower and upper branches of the anticrossing respectively. The peaks move closer together and then separate without crossing. Numerically calculated energy values agree well with the measured data. Predictions made

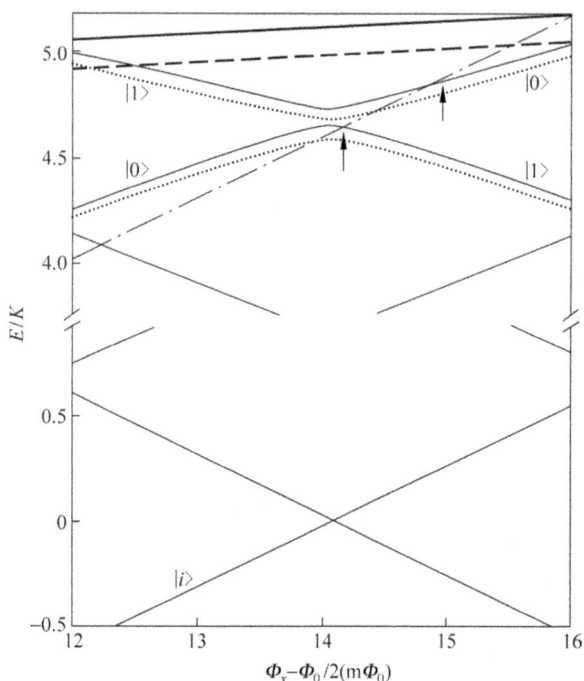

Fig. 4.31   Calculated energy levels versus Φ, and photon assisted tunneling. Reprinted with permission from Macmillan Publishers Ltd: *Nature* (406, 43), 2000.

by Caldeira and Leggett are eventually verified experimentally after 17 years.

4.9.4   Decoherence by the emission of thermal radiations

The interaction of a quantum system with environment entangles the two and distributes the quantum coherence over the degree of freedom of the environment as to render it unobservable. Large molecules are particularly suitable for the investigation of decoherence because they possess numerous internal modes, the excitation energy can be released by radiation, which induces decoherence by disclosing the "which way" information An example is the C_{70} molecule, which has 204 vibrational modes. The University of Vienna group led by A. Zeilinger [35] investigated the decoherence of molecular matter waves by changing the internal temperature of the molecules in a controlled way before they enter the interferometer and observing the corresponding visibility of interference pattern. The result is that at temperature below 1000 K the molecules exhibit perfect quantum interference and they gradually lose the quantum behavior when the internal temperature is increased to 3000 K. The quantum-to-classical transition is traced in a controlled and quantitative way.

A beam of C_{70} molecules pass a heating stage where they cross a focused laser beam up to 16 times. The temperature is monitored, and it is found that the hottest molecules have a temperature about 300 K where they enter the interferometer. The Talbot-vonLau interferometer consists of 3 identical gratings separated by equal distance. The first grating serves as a periodic array of narrow slit sources, the second as the diffraction element, and the third as a scanning detection mask. The transmitted molecules are ionized by a laser beam and their intensity is recorded as a function of the lateral displacement of the third grating. The fringe visibility $V = (I_{max} - I_{min})/(I_{max} + I_{min})$ characterizes the coherence of the molecular evolution. The variation of V with the heating power is shown in Fig. 4.32.

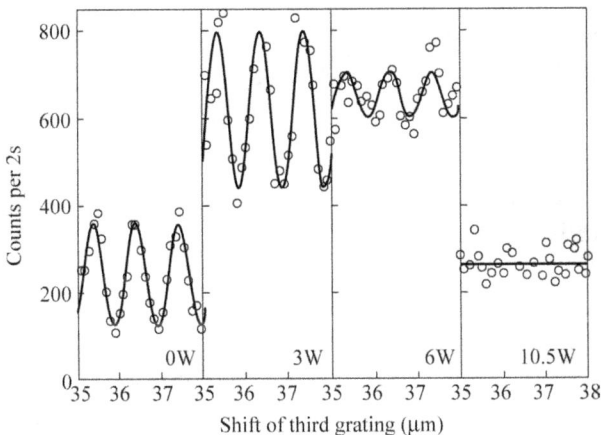

Fig. 4.32   Molecule interferograms for C_{70} at a speed 190 ms^{-1} for increasing laser heating powers P. Reprinted with permission from Macmillan Publishers Ltd: *Nature* (427, 711), copyright (2004).

The fringe visibility decreases with heating power owing to the rising emission probability of photons: 47% for 0 W, 29% for 3 W, 7% for 6 W, and 0% for 10.5 W. The absolute counting rate grows initially with P, and falls again because of ionization and fragmentation in the heating stage.

The decoherence depends also on the velocity of the molecules: the normalized visibility drops more quickly with the heating power for molecules of lower velocity. The decoherence-heating power relation agrees well with the decoherence theory [36].

4.10   The Collapse of Wave Function and the Quantum Zeno Effect

The concept of the collapse of wave function due to a measurement process introduced by J. von Neumann works successively in the standard interpretation of quantum mechanics. Is there any experimental confirmation of this concept? In 1977 Misra and Sudarshan [45] proposed what they called the quantum Zeno effect as a demonstration of the collapse of wave function. An excited atom in the state Ψ_2 of natural lifetime τ is subjected to repeated measurements. Each measurement collapses the wave function, resetting the clock, and the expected transition to the ground state Ψ_1 can be delayed indefinitely if the measurements are made frequently enough. Actually, for times t much shorter than τ, the probability of transition is

$$P_{2\to1} = \frac{t}{\tau}. \tag{4.10.1}$$

If a measurement is made at time t then the probability that the atom is still in excited state is

$$P_2(t) = 1 - \frac{t}{\tau}. \tag{4.10.2}$$

Suppose this is the case, then the wave function collapses back to Ψ_2. If a second measurement is made at time $2t$, the probability that the atom is still in the excited state is

$$\left(1 - \frac{t}{\tau}\right)^2 \approx 1 - \frac{2t}{\tau}, \tag{4.10.3}$$

which is the same as it would have been if no measurement was made at time t. This is what one can naturally expect.

However, for *extremely* short times, the probability $P_{2 \to 1}$ is proportional to t^2 instead of t:

$$P_{2 \to 1} = \alpha t^2. \tag{4.10.4}$$

Some explanation is in order. In any standard text on quantum mechanics one finds in the theory of quantum transitions the formula for transition probability $P_{2 \to 1} \propto \frac{\sin^2(\omega t/2)}{\omega^2}$, where ω is the angular frequency of transition. This expression as a function of ω is a sharp spike with height $t^2/4$ and width $4\pi/t$. The width is in accordance with the uncertainty relation between energy and time. Integration over ω gives the area under the spike proportional to t, and the probability $P_{2 \to 1}$ is therefore proportional to t which leads to a finite probability per unit time and Fermi's Golden rule. For extremely small t the picture becomes very different. We have a series of low bumps of large widths and instead of integrating over the spike the integral over all ω must be taken. We have $\frac{1}{4}t^2 \int \rho(\omega)d\omega$, and the probability $P_{2 \to 1}$ is proportional to t^2. In this case the probability that the atom is still in the excited state after two measurements is

$$(1 - \alpha t^2)^2 \approx 1 - 2\alpha t^2, \tag{4.10.5}$$

whereas if we have not carried out the first measurement it would have been

$$1 - \alpha(2t)^2 \approx 1 - 4\alpha t^2. \tag{4.10.6}$$

Evidently measurement made in the interval of time of the "quadratic dependence regime" suppresses the transition. Indeed if n (a large number) equally spaced measurements are made in the interval from 0 to T such that T/n is the "quadratic dependence range", the probability that the atom is still in the excited state is

$$\left(1 - \alpha\left(\frac{T}{n}\right)^2\right)^n = \left(1 - \alpha\frac{T^2}{n}\frac{1}{n}\right)^n \xrightarrow{n \to \infty} e^{-\alpha T^2/n}, \tag{4.10.7}$$

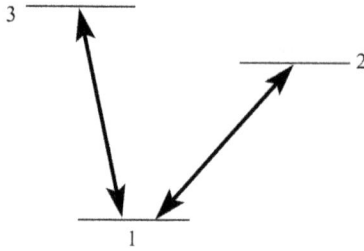

Fig. 4.33   Energy level scheme.

which approaches 1 for $n \to \infty$, a continuously monitored excited atom never decays at all. The dependence of lifetime on the frequency of measurement is very difficult to observe for spontaneous emission of atom, but it can be observed for induced transitions. D.J. Wineland *et al.* [46] realized the original proposal by R.J. Cook to observe this dependence for a single trapped ion. Figure 4.33 shows the level scheme. Levels 2 and 1 are coupled by resonance radio-frequency perturbation to form a coherent superposition state. Measurements are carried out by continuous dye laser beams tuned to resonance to excite the ion from level 1 to level 3. In principle scattered photons observed confirms that the ion is in state 1. Suppose the ion is in level 1 at time $\tau = 0$. The resonant rf π pulse (for an interval $T = \pi/\Omega$, where Ω is the Rabi frequency) is applied. Without the measurement laser pulses the probability of the ion being in state 2 $P_2(T)$ at time $\tau = T$ is 1. Let the measurement laser pulses be applied at times $\tau = kT/n = k\pi/(n\Omega)$, where $k = 1, 2, \ldots, n$. By using the vector representation of a two-level system [47] $P_2(T)$ is calculated to be [46]

$$P_2(T) \approx \frac{1}{2}\left[1 - \exp\left(-\frac{1}{2}\pi^2/n\right)\right]. \qquad (4.10.8)$$

Therefore $P_2(T)$ approaches 0 for $n \to \infty$. Figures 4.34 and 4.35 show the experimental and calculated $1 \to 2$ and $2 \to 1$ transition probabilities as a function of the number of measurement pulses n given in [46] using $^9\mathrm{Be}^+$ ions. The agreement between theory and experiment is remarkable.

A comment made by L.E. Ballentine [48] illustrates an unusual situation in quantum mechanics that, while there is general

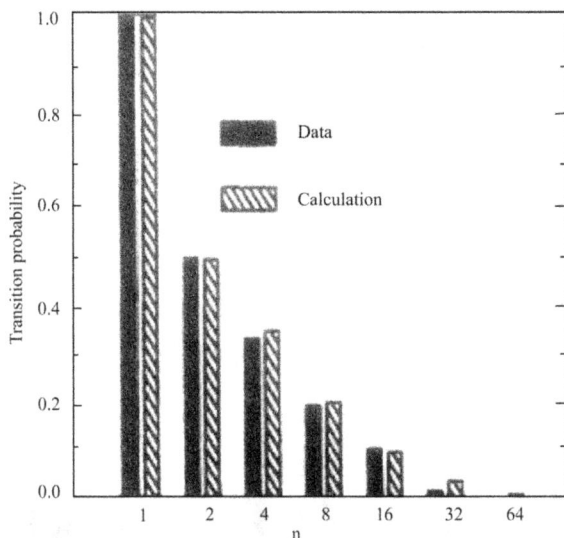

Fig. 4.34   Experimental and calculated 1→2 transition probabilities as a function of the number of measurement pulses. Reprinted with permission from Wayne M. Itano, D. J. Heinzen, J. J. Bollinger, and D. J. Wineland, *Phys. Rev.* A41, 2295, 1990. Copyright (2017) by the American Physical Society.

agreement on the experimental consequences, there is endless debate on matters of interpretation. Ballentine explains the same experimental result without using the collapse of wave function, which he considers not necessary to quantum mechanics in his "statistical interpretation" [49]. In his interpretation the effect is a result of strong perturbation of the laser pulse on the ion. The authors of [46] answered in a reply [50] that quantum mechanics can be interpreted in different ways. In this case, interpretations with and without collapse postulate correctly predict the experimental data. Therefore the experiment [46] neither verifies nor falsifies the notion of "wave function collapse".

4.11   The Squeeze Operator and the Squeezed Coherent State

Consider a quantized single-mode electric field of angular frequency ω

$$\mathbf{E}(t) = E\,\hat{\epsilon}\left(a\,e^{-i\omega t} + a^\dagger e^{i\omega t}\right), \qquad (4.11.1)$$

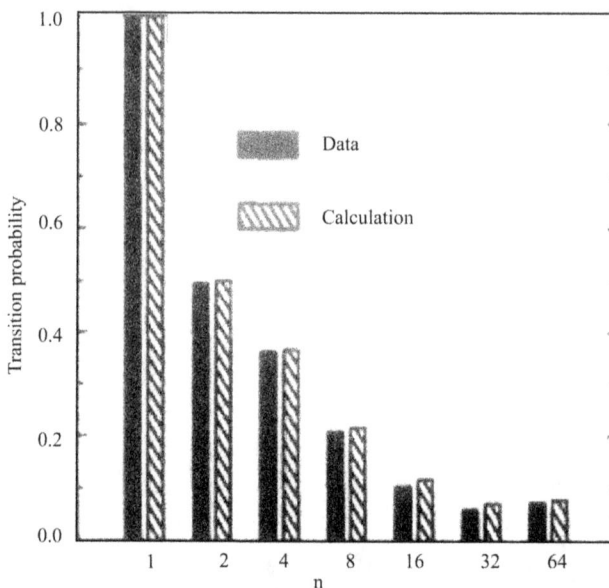

Fig. 4.35   Experimental and calculated 2→1 transition probabilities as a function of the number of measurement pulses. Reprinted with permission from Wayne M. Itano, D. J. Heinzen, J. J. Bollinger, and D. J. Wineland, *Phys. Rev.* A41, 2295, 1990. Copyright (2017) by the American Physical Society.

with the annihilation and creation operators a and $a^\dagger$ satisfying the commutation relations

$$\left[a, a^\dagger\right] = 1. \tag{4.11.2}$$

We introduce Hermitian operators X_1 and X_2 in terms of operators a and $a^\dagger$:

$$X_1 = \frac{1}{2}\left(a + a^\dagger\right), \tag{4.11.3}$$

$$X_2 = \frac{1}{2i}\left(a - a^\dagger\right). \tag{4.11.4}$$

Comparing with the quantum mechanical problem of a harmonic oscillator of mass m and angular frequency ω we see that X_1 and X_2

are the dimensionless operators x and p:

$$x = \sqrt{\frac{\hbar}{m\omega}} \frac{a + a^\dagger}{\sqrt{2}}, \qquad (4.11.5)$$

$$p = \sqrt{m\hbar\omega} \frac{a - a^\dagger}{\sqrt{2}i}. \qquad (4.11.6)$$

Operators X_1 and X_2 obey the commutation relation

$$[X_1, X_2] = \frac{i}{2}. \qquad (4.11.7)$$

In terms of X_1 and X_2 the quantized field can be expressed as

$$\mathbf{E}(t) = E\,\hat{\epsilon}\,(X_1 \cos \omega t + X_2 \cos \omega t). \qquad (4.11.8)$$

Now X_1 and X_2 are seen to be the amplitudes of the two quadratures of the field having a phase difference of $\pi/2$. On a complex plane the complex amplitude $X = X_1 + iX_2$ is represented as a point. The commutation relation (4.11.7) leads to the uncertainty relation

$$\Delta X_1 \Delta X_2 \geqslant \frac{1}{4}. \qquad (4.11.9)$$

Consider the coherent state $|\alpha\rangle$. In Section 4.1 we learned that it is a state of minimum uncertainty with respect to x and p. We now confirm directly that it is a minimum uncertainty state with respect to X_1 and X_2:

$$(\Delta X_1)^2 = \langle\alpha|\,X_1^2\,|\alpha\rangle - \langle\alpha|\,X_1\,|\alpha\rangle^2$$

$$= \frac{1}{4}\langle\alpha|\left(a^2 + aa^\dagger + a^\dagger a + a^{\dagger 2}\right)|\alpha\rangle - \frac{1}{4}\left[\langle\alpha|\left(a + a^\dagger\right)|\alpha\rangle\right]^2$$

$$= \frac{1}{4},$$

$$(4.11.10)$$

where we have used $a|\alpha\rangle = \alpha|\alpha\rangle$. Similarly $(\Delta X_2)^2 = \frac{1}{4}$, and eventually

$$\Delta X_1 \Delta X_2 = \frac{1}{4}. \qquad (4.11.11)$$

The uncertainty product is a minimum with errors of the two quadratures balanced $\Delta X_1 = \Delta X_2$. In Fig. 4.36 the coherent state and the error circle of the quadrature amplitudes are represented on the complex plane (X_1, X_2).

We now define a new set of quadrature amplitudes by rotating the complex amplitudes $X = X_1 + \mathrm{i} X_2$ by an angle $\theta/2$:

$$Y_1 + \mathrm{i} Y_2 = (X_1 + \mathrm{i} X_2)\ \mathrm{e}^{-\mathrm{i}\theta/2}. \qquad (4.11.12)$$

Correspondingly we define the unitary squeeze operator

$$S(\xi) = \exp\left(\frac{1}{2}\xi^* a^2 - \frac{1}{2}\xi a^{\dagger 2}\right), \qquad (4.11.13)$$

where

$$\xi = r\ \mathrm{e}^{\mathrm{i}\theta} \qquad (4.11.14)$$

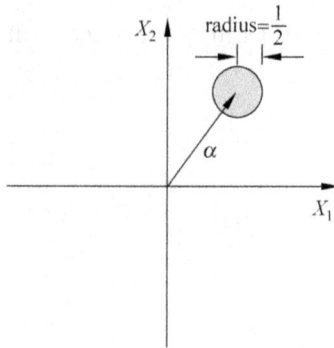

Fig. 4.36   Error circle for coherent state $|\alpha\rangle$.

is a complex number. It is easy to see that

$$S^\dagger(\xi) = S^{-1}(\xi) = S(-\xi). \tag{4.11.15}$$

From Baker-Campbell-Hausdorff formula

$$e^A B e^{-A} = B + [A, B] + \frac{1}{2!}[A[A, B]] + \cdots \tag{4.11.16}$$

it follows that

$$S^\dagger(\xi)\, a\, S(\xi) = a \cosh r - a^\dagger\, e^{i\theta} \sinh r, \tag{4.11.17}$$

$$S^\dagger(\xi)\, a^\dagger\, S(\xi) = a^\dagger \cosh r - a\, e^{-i\theta} \sinh r. \tag{4.11.18}$$

A squeezed coherent state $|\alpha, \xi\rangle$ is obtained by operating the squeeze operator $S(\xi)$ on the coherent state $|\alpha\rangle$

$$|\alpha, \xi\rangle = S(\xi)\,|\alpha\rangle. \tag{4.11.19}$$

We will see that the squeezed state has the variances of the rotated quadratures Y_1 and Y_2 unequal (one squeezed and the other relaxed). The variances of the new set Y_1 and Y_2 for the squeezed state $|\alpha, \xi\rangle$ can be determined by using (4.11.12), I(4.11.17) and (4.11.18). It follows from (4.11.12) that

$$Y_1 = \frac{ae^{-i\theta/2} + a^\dagger e^{i\theta/2}}{2}, \tag{4.11.20}$$

$$Y_2 = \frac{ae^{-i\theta/2} - a^\dagger e^{i\theta/2}}{2i}. \tag{4.11.21}$$

Consequently using (4.11.17) and (4.11.18) we obtain

$$\begin{aligned}
S^\dagger Y_1 S &= \frac{1}{2}(ae^{-i\theta/2} \cosh r - a^\dagger e^{i\theta/2} \sinh r \\
&\quad + a^\dagger e^{i\theta/2} \cosh r - ae^{-i\theta/2} \sinh r) \\
&= \frac{1}{2}(ae^{-i\theta/2} + a^\dagger e^{i\theta/2})e^{-r}.
\end{aligned} \tag{4.11.22}$$

By using (4.11.22) we obtain

$$(\Delta Y_1)^2 = \langle \alpha, \xi | Y_1^2 | \alpha, \xi \rangle - \langle \alpha, \xi | Y_1 | \alpha, \xi \rangle^2$$

$$= \langle \alpha | S^\dagger Y_1 S S^\dagger Y_1 S | \alpha \rangle - \langle \alpha | S^\dagger Y_1 S | \alpha \rangle^2 \qquad (4.11.23)$$

$$= \frac{1}{4} e^{-2r},$$

and similarly

$$(\Delta Y_2)^2 = \frac{1}{4} e^{2r}, \qquad (4.11.24)$$

so that

$$\Delta Y_1 \Delta Y_2 = \frac{1}{4}. \qquad (4.11.25)$$

We see that the variances of the two quadratures are no more equal, one squeezed and the other relaxed, depending on the value of r larger or smaller than 1. The uncertainty product remains unaltered. In the complex amplitude plane the error contour of $|\alpha, \xi\rangle$ becomes an ellipse (Fig. 4.37) of the same area of the error circle for $|\alpha\rangle$. The principal axes of the ellipse lie along directions of Y_1 and Y_2 rotated at an angle $\theta/2$ from X_1 and X_2 respectively.

Squeezed state are extensively used in quantum optics and quantum information studies.

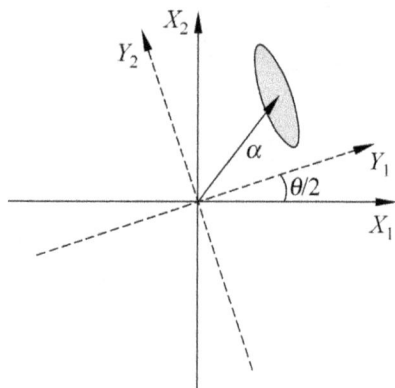

Fig. 4.37   Error ellipse for squeezed coherent state $|\alpha, \xi\rangle$.

References

[1] E. Schrödinger, *Naturwissenschaften*, 14, 137, 1926.

[2] J. R. Klauder and B.-S. Skagerstam, *Coherent States*, Singapore, World Scientific, 1985.

[3] L. M. Brown, *Am. J. Phys.* 41, 525, 1973.

[4] J. Parker and C. R. Stroud, Jr., *Phys. Rev. Lett.* 56, 716, 1986.

[5] W. Pauli, *Z. Phys.* 36, 336, 1926.

[6] V. A. Fock, *Z. Phys.* 98, 145, 1935.

[7] M. Bander and C. Itzykson, *Rev. Mod. Phys.* 38, 330, 1966; 38, 346, 1966.

[8] M. Nauenberg, *Phys. Rev. A* 40, 1133, 1989.

[9] Z. D. Gaeta and C. R. Stroud, Jr., *Phys. Rev. A* 42, 6308, 1990.

[10] I. Sh. Averbuch and N. F. Perelman, *Phys. Lett. A* 19, 449, 1989.

[11] J. A. Yeazell and C. R. Stroud, Jr., *Phys. Rev. A* 35, 2806, 1987.

[12] G. Alber, H. Ritsch and P. Zoller, *Phys. Rev. A* 34, 1058, 1986.

[13] W. H. Zurek, *Phys. Rev. D* 24, 1516, 1980; 26, 1862, 1982.

[14] W. H. Zurek, *Prog. Theor. Phys.* 89, 281, 1993.

[15] W. H. Zurek, *Phys. Today* 44(10), 36, 1991.

[16] J. von Neumann, *Mathematical Foundations of Quantum Mechanics*, Princeton, Princeton University Press, 1955.

[17] J. A. Wheeler, in *Problems in Foundation of Physics,* edited by Toraldo di Francia, Amsterdam, North Holland, 1979.

[18] W. Zurek, S. Habib, and J. P. Paz, *Phys. Rev. Lett.* 70, 1187, 1993.

[19] A. O. Caldeira and A. J. Leggett, *Ann. Phys. (N.Y.)* 149, 374, 1983.

[20] R. P. Feynman and F. L. Vernon, Jr., *Ann. Phys. (N.Y.)* 24, 118, 1963.

[21] W. G. Unruh, W. H. Zurek, *Phys. Rev. D* 40, 1071, 1989.

[22] E. P. Wigner, *Phys. Rev.* 40, 749, 1932.

[23] J. A. Wheeler and W. H. Zurek, editors, *Quantum Theory and Measurement*, Princeton, Princeton University Press, 1983.

[24] V. B. Braginsky and F. Ya. Khalili, *Quantum Measurement,* Cambridge, Cambridge University Press, 1992.

[25] M. Namiki and S. Pascazio, *Phys. Repts* 232, 301, 1993.

[26] letters to editors, *Phys. Today* 46(4), 13, 82, 1993.

[27] C. Monroe, D. M. Meekhof, B. E. King, D. J. Wineland, *Science* 272, 1131, 1996.

[28] C. Monroe, D. M. Meekhof, B. E. King, S. R. Jefferts, W. M. Itano, D. J. Wieland and P. Gould, *Phys. Rev. Lett.* 75, 4011, 1995.

[29] D. M. Meekhof, C. Monroe, B. E. King, W. M. Itano, D. J. Wieland, *Phys. Rev. Lett.* 76, 1765, 1996.

[30] C. J. Myatt, B. E. King, Q. A. Turchette, C. A. Sackett, D. Kilpinski, W. M. Itano, C. Monroe and D. J. Wineland, *Nature* 403, 269, 2000.

[31] M. Brune, E. Hagley, J. Dreyer, X. Maitre, A. Maali, C. Wunderlich, J. M. Raimond and C. Haroche, *Phys. Rev. Lett.* 76, 1796, 1996.

[32] C. Haroche, *Phys. Today* 51(7), 36, 1998.

[33] J. R. Friedmann, V. Patel, W. Chen, S. K. Tolpygo and J. E. Lukens, *Nature* 406, 43, 2000.

[34] R. Rouse, S. Han and J. Lukens, *Phys. Rev. Lett.* 75, 1614, 1995.

[35] L. Hackermüller, K. Hornberger, B. Brezger, A. Zeilinger and M. Arndt, *Nature* 427, 711, 2004.

[36] E. Joos and H. D. Zeh, *Z. Phys.* B 59, 223, 1985.

[37] I. Bialynicki-Birula, M. Kalinski and J. H. Eberly, *Phys. Rev. Lett.* 73, 1777, 1994.

[38] M. Kalinski, J. H. Eberly and I. Bialynicki-Birula, *Phys. Rev. A* 52, 2460, 1995.

[39] M. Kalinski and J. H. Eberly, *Phys. Rev. A* 53, 1715, 1996.

[40] H. A. Bethe and E. E. Salpeter, *Quantum Mechanics of One and Two-Electron Atoms*, Springer-Verlag, Berlin, 1957.

[41] J. A. West, Z. D. Gaeta and C. R. Stroud Jr., *Phys. Rev. A* 58, 186, 1998.

[42] M. Kalinski, J. H. Eberly, J. A. West and C. R. Stroud Jr., *Phys. Rev. A* 67, 032503, 2003.

[43] H. Maeda and T. F. Gallagher, *Phys. Rev. Lett.* 92, 133004, 2004.

[44] H. Maeda, D. V. L. Notum, T. F. Gallagher, *Science* 307, 1757, 2005.

[45] B. Misra and E. C. G. Sudarshan, *J. Math. Phys.* 18, 756, 1977.

[46] W. M. ltano, D. J. Heinzen, J. J. Bollinger, and D. J. Wineland, *Phys. Rev. A* 41, 2295, 1990.

[47] I. I. Rabi, N. F. Ramsey, and J. Schwinger, *Rev. Mod. Phys.* 26, 167, 1954.

[48] L. E. Ballentine, *Phys. Rev. A* 43, 5165, 1991.

[49] L. E. Ballentine, *Rev. Mod. Phys.* 42, 358, 1970.

[50] W. M. Itano, D. J. Heinzen, J. J. Bollinger, and D. J. Wineland, *Phys. Rev. A* 43, 5168, 1991.

Chapter 5

Path Integral Formulation of Quantum Mechanics

Working on his Ph.D. thesis under the supervision of J. A. Wheeler in Princeton University, R. P. Feynman tried to rid the quantum electrodynamics from infinite self-mass of the electron by adopting an approach to discard fields altogether and considering only charged particles with the retarded interaction. He did not succeed in his primary aim, but did create an equivalent way of formulating quantum mechanics alongside with those of Schrödinger, Heisenberg and Dirac — the path integral formalism. Many equivalent formalisms of a given theory can be different in the ways of handling specific problems. The method of instantons based on path integral formalism is one most efficient to deal with tunneling through barrier and the related decay rate problem. The macroscopic quantum phenomena (Chapter 6) and the theta vacuum of non-Abelian gauge field (Section 7.6) are treated using this method. Path integral provides an ideal framework for quantizing non-Abelian gauge field theory (Faddeev and Popov). A method of dealing with the quantum mechanics of dissipative systems is also based on this formalism (Section 6.5). We devote this chapter to the path integral formalism.

5.1   The Path Integral in Quantum Mechanics

The path integral method is one of the formulation of quantum mechanics created by R. P. Feynman [1, 2]. Instead of using the state vectors and operators of the Hilbert space, this method expresses the *transition amplitude,* a fundamental quantity in quantum mechanics, in terms of an integral as the *sum over possible histories.* This integral is called the path integral. In this formulation, all quantities are c-numbers. All physical information of a quantum system can be obtained from the path integral, with the *action functional* of the classical dynamics playing the key role. Path integral is directly connected with the partition function of the statistical mechanics. Therefore, this method finds numerous applications in statistical mechanics [3]. The generalization to the quantum field theory is direct. In the quantization of gauge field the path integral provides a direct and clear-cut alternative, promoting the advance of the theory. Path integral has been applied in many branches of physics. A number of monographs on this subject has been published [4, 5].

Let $|q\rangle$ be the eigenstate of the position operator $\widehat{Q}$ in the Schrödinger picture, $\widehat{Q}|q\rangle = q|q\rangle$, and $|q;t\rangle$ the corresponding state in the Heisenberg picture $|q;t\rangle = e^{iHt/\hbar}|q\rangle$.[1] The transition amplitude is the overlap between the initial and final state vectors

$$\langle q';t' \mid q;t\rangle = \Big\langle q' \mid e^{-\frac{i}{\hbar}\widehat{H}(t'-t)} \mid q\Big\rangle. \tag{5.1.1}$$

Dividing the interval $t' - t$ into n equal parts $\delta t = \frac{t'-t}{n}$, the transition amplitude can be rewritten as

$$\Big\langle q' \Big| \exp\left[-\frac{i}{\hbar}\widehat{H}\,(t'-t)\right] \Big| q\Big\rangle$$

$$= \int dq_1 \cdots dq_{n-1} \Big\langle q' \Big| \exp\left(-\frac{i}{\hbar}\widehat{H}\delta t\right) \Big| q_{n-1}\Big\rangle$$

---

[1] The Heisenberg states do not carry time dependence. The parameter t in $|q;t\rangle$ reminds that it coincide with the Schrödinger vector $|q(t)\rangle$.

$$\cdot \left\langle q_{n-1} \left| \exp\left(-\frac{i}{\hbar}\hat{H}\delta t\right) \right| q_2 \right\rangle \cdots \left\langle q_1 \left| \exp\left(-\frac{i}{\hbar}\hat{H}\delta t\right) \right| q \right\rangle,$$

$$(5.1.2)$$

where the identity $1 = \int dq |q\rangle\langle q|$ for $q_1, q_2, \cdots, q_{n-1}$ (the complete-ness of the set of eigenstates) has been inserted in the derivation. We concentrate now on one of the matrix elements of (5.1.2) and transform it in the following way:

$$\left\langle q_2 \left| \exp\left(-\frac{i}{\hbar}\hat{H}\delta t\right) \right| q_1 \right\rangle = \left\langle q_2 \left| \left(1 - \frac{i}{\hbar}\hat{H}\delta t\right) \right| q_1 \right\rangle + O(\delta t^2).$$

$$(5.1.3)$$

The operator form of the Hamiltonian is

$$\hat{H}(\hat{P};\hat{Q}) = \frac{\hat{P}^2}{2m} + V(\hat{Q}).$$

Its matrix element is

$$\left\langle q_2 \left| \hat{H}(\hat{P};\hat{Q}) \right| q_1 \right\rangle = \left\langle q_2 \left| \frac{\hat{P}^2}{2m} \right| q_1 \right\rangle + V\left(\frac{q_2 + q_1}{2}\right) \delta(q_2 - q_1)$$

$$= \int \frac{dp}{2\pi} \langle q_2 | p \rangle \left\langle p \left| \frac{\hat{P}^2}{2m} \right| q_1 \right\rangle$$

$$+ V\left(\frac{q_2 + q_1}{2}\right) \int \frac{dp}{2\pi} e^{ip(q_2 - q_1)}$$

$$= \int \frac{dp}{2\pi} e^{ip(q_2 - q_1)} \left[\frac{p^2}{2m} + V\left(\frac{q_2 + q_1}{2}\right)\right].$$

In the above $1 = \int \frac{dp}{2\pi} |p\rangle\langle p|$ has been used, where $|p\rangle$ is the eigenstate of operator $\hat{P}$ with eigenvalue p, and normalized according to $\langle q \mid p \rangle = e^{ipq/\hbar}$. The matrix element (5.1.3) can be written by using the foregoing equation as

$$\left\langle q_2 \left| \exp\left(-\frac{i}{\hbar}\hat{H}\delta t\right) \right| q_1 \right\rangle$$

$$\approx \int \frac{dp}{2\pi} \exp\left[i\frac{p}{\hbar}(q_2 - q_1)\right] \left\{ 1 - \frac{i}{\hbar}\delta t \left[\frac{p^2}{2m} + V\left(\frac{q_2 + q_1}{2}\right) \right] \right\}$$

$$\approx \int \frac{dp}{2\pi} \exp\left[i\frac{p}{\hbar}(q_2 - q_1)\right] \exp\left[-\frac{i}{\hbar}H\left(p, \frac{q_2 + q_1}{2}\right)\delta t\right].$$

$$(5.1.4)$$

It is to be emphasized that in (5.1.4) the Hamiltonian H is already a c-number function. The transition amplitude (5.1.2) now becomes

$$\left\langle q' \left| \exp\left[-\frac{i}{\hbar}\hat{H}(t'-t)\right] \right| q \right\rangle$$

$$= \int \frac{dp_1}{2\pi} \cdots \frac{dp_n}{2\pi} \int dq_1 \cdots dq_{n-1}$$

$$\times \exp\left\{ \frac{i}{\hbar}\sum_{i=1}^{n} \delta t \left[p_i \left(\frac{q_i - q_{i-1}}{\delta t} \right) - H\left(p_i, \frac{q_i + q_{i-1}}{2}\right) \right] \right\}.$$

$$(5.1.5)$$

There are n variables of integration for p and $n-1$ variables of integration for q. Adopting a formal notation for the integration measure, we rewrite (5.1.5) as

$$\left\langle q' \left| \exp\left[-\frac{i}{\hbar}\hat{H}(t'-t)\right] \right| q \right\rangle$$

$$= \int \left[\frac{dqdp}{2\pi} \right] \exp\left\{ \frac{i}{\hbar} \int_t^{t'} dt \left[p\dot{q} - H(p,q) \right] \right\}. \qquad (5.1.6)$$

By using the Gaussian integration formula

$$\int_{-\infty}^{\infty} \frac{dx}{2\pi} e^{-ax^2 + bx} = \frac{1}{\sqrt{4\pi a}} e^{b^2/4a}$$

the integration with respect to p can be taken

$$\int \frac{dp_i}{2\pi} \exp\left[-\frac{i}{2m\hbar}\delta t \, p_i^2 + i\frac{p_i}{\hbar}(q_i - q_{i-1}) \right]$$

$$= \left(\frac{m\hbar}{2\pi i\hbar\delta t} \right)^{1/2} \exp\left[\frac{im(q_i - q_{i-1})^2}{2\hbar\delta t} \right].$$

In the above the integrand is oscillating, hence $i\delta t$ is treated formally as real, which is equivalent to an analytical continuation to the Euclidean time. Combining the above exponential factor with others in (5.1.5) to obtain

$$\frac{i}{\hbar}\delta t \left[\frac{m}{2}\left(\frac{q_i - q_{i-1}}{\delta t} \right)^2 - V\left(\frac{q_i + q_{i-1}}{2} \right) \right]$$

such that (5.1.5) finally becomes

$$\left\langle q' \left| \exp\left[-\frac{i}{\hbar}\hat{H}(t' - t) \right] \right| q \right\rangle$$

$$= \lim_{n\to\infty} \left(\frac{m\hbar}{2\pi i\delta t} \right)^{\frac{n}{2}} \int \prod_{i}^{n-1} dq_i$$

$$\times \exp\left\{ \frac{i}{\hbar}\sum_{i=1}^{n}\delta t \left[\frac{m}{2}\left(\frac{q_i - q_{i-1}}{\delta t} \right)^2 - V(q_i) \right] \right\}$$

$$\equiv N\int [Dq]\exp\left(\frac{i}{\hbar}\int_{t}^{t'} d\tau \left[\frac{m}{2}\dot{q}^2 - V(q) \right] \right)$$

$$= N\int [Dq]\exp\left[\frac{i}{\hbar}\int_{t}^{t'} d\tau L(q, \dot{q}) \right]$$

$$= N\int [Dq]\exp\left(\frac{i}{\hbar}S \right). \tag{5.1.7}$$

This is an infinite-dimensional integral. In (5.1.7) the product of the normalization constant and the integration measure $[Dq]$ is defined in the limit after the first equality sign. $L(q, \dot{q})$ is the

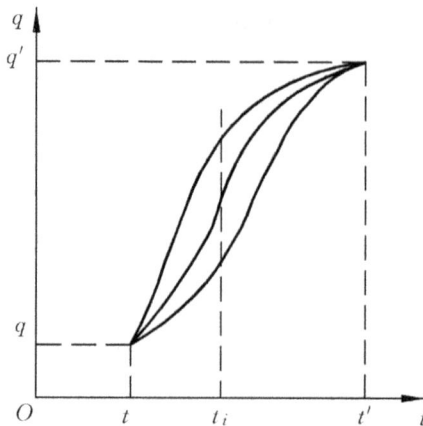

Fig. 5.1   Different histories from (q, t) to (q', t').

Lagrangian, and S is the action of the path $q(t)$. The transition amplitude can be interpreted as the sum of contributions over all paths that connect (q, t) and (q', t') weighted by a phase factor with phase $S/\hbar$, the action of the path in units of $\hbar$. Some possible paths are shown in Fig. 5.1. Each integration variable q_i is the possible value of the particle coordinate at the time t_i, i.e., the intersections of the straight line at t_i with different possible paths. The correspondence of (5.1.7) with the classical dynamics is clear. Among all the possible paths, one history is special. It is the classical path which minimizes the action. The difference of (5.1.7) from classical dynamics is that all possible paths contribute with corresponding weight factors. This will be demonstrated further in the developments following.

How is the problem of energy eigenvalues and eigenfunctions solved in the path integral? Denote the eigenstates of $\hat{H}$ by $|n\rangle$, with corresponding eigenvalues E_n, i.e.,

$$\hat{H} |n\rangle = E_n |n\rangle. \qquad (5.1.8)$$

The transition amplitude $\langle x_f, t_f \mid x_i, t_i \rangle$ is

$$\langle x_f t_f | x_i t_i \rangle = \sum_n \exp\left(-i\frac{E_n}{\hbar}t\right) \langle x_f|n\rangle \langle n|x_i\rangle, \qquad (5.1.9)$$

where $t = t_f - t_i$. When the transition amplitude of a specific problem has been found, the eigenvalue E_n and $\psi_n = \langle x \mid n \rangle$ can be found by comparing the transition amplitude with the right hand side of (5.1.9). We give an example of 1-dimensional harmonic oscillator. If only the low lying states is looked for, we can use the imaginary time

$$t \equiv -i\tau, \tag{5.1.10}$$

and write the transition amplitude as

$$\left\langle x_f \left| \exp\left(-\frac{\hat{H}}{\hbar}\tau \right) \right| x_i \right\rangle = \sum_n \exp\left(-\frac{E_n}{\hbar}\tau \right) \langle x_f | n \rangle \langle n | x_i \rangle. \tag{5.1.11}$$

When the path integral is calculated, the result can be expanded for large τ. Comparing the expansion with the right hand side of (5.1.11), the low-lying energy eigenvalues and the corresponding eigenfunctions can be found. The Lagrangian of a harmonic oscillator of mass unity is

$$L = \frac{1}{2}\left(\frac{dx}{dt} \right)^2 - V(x), \tag{5.1.12}$$

and the action is

$$S = \int_{t_i}^{t_f} L\left(x, \frac{dx}{dt} \right) dt. \tag{5.1.13}$$

The phase factor in the path integral in Euclidean space-time $(x, \tau = it)$ is

$$\exp\left(\frac{i}{\hbar}S\left[x(t) \right] \right) = \exp\left\{ -\frac{1}{\hbar}\int_{\tau_i}^{\tau_f} \left[\frac{1}{2}\left(\frac{dx}{d\tau} \right)^2 + V(x) \right] d\tau \right\}$$

$$\equiv \exp\left(-\frac{1}{\hbar}S_E \right), \tag{5.1.14}$$

where the Euclidean action S_E is

$$S_E = \int_{\tau_i}^{\tau_f} \left[\frac{1}{2}\dot{x}^2 + V(x) \right] d\tau, \qquad (5.1.15)$$

and $\dot{x}$ is the abbreviation for $\frac{dx}{d\tau}$. We neglect the subscript E in the following. The transition amplitude to be calculated is

$$\left\langle x_f \left| \exp\left(-\frac{\hat{H}}{\hbar}\tau \right) \right| x_i \right\rangle = N \int [Dx]\, e^{-S/\hbar}. \qquad (5.1.16)$$

The initial and final position coordinates are

$$x_i = x(\tau_i), \quad x_f = x(\tau_f). \qquad (5.1.17)$$

The main contribution to the path integral (5.1.16) comes from the path with minimum action S_0, i.e., the classical path denoted by $X(\tau)$. Paths far away from the classical one contributes negligibly. The deviation of a neighboring arbitrary path from the classical path is denoted by $\delta x(\tau)$:

$$x(\tau) = X(\tau) + \delta x(\tau). \qquad (5.1.18)$$

Expanding $\delta x(\tau)$ in terms of a complete set of orthonormal functions $x_n(\tau)$:

$$x(\tau) = X(\tau) + \sum_n c_n x_n(\tau). \qquad (5.1.19)$$

The measure $[Dx]$ can now be written as

$$[Dx] = \prod_n \frac{dc_n}{\sqrt{2\pi\hbar}}. \qquad (5.1.20)$$

Numerical factors are irrelevant before the normalization factor assumes a definite value. In the following, we are going to prove that the path integral (5.1.16) is proportional to $e^{-S_0/\hbar}$, where $S_0 = S[X(\tau)]$ is the action of the classical path. The constant of proportionality is called the prefactor, determined by contributions from all possible paths deviating from the classical one. Expressing the action functional $S[x(\tau)]$ of an arbitrary path $x(\tau)$ in terms of S_0 and all orders of the variations of S:

$$S\left[x\left(\tau\right)\right] = \int_{\tau_i}^{\tau_f} \left[\frac{1}{2}\left(\frac{\mathrm{d}x}{\mathrm{d}\tau}\right)^2 + V\left(x\right)\right] \mathrm{d}\tau$$

$$= S\left[X\left(\tau\right) + \delta x\left(\tau\right)\right] = S\left[X\left(\tau\right)\right] + \delta S + \delta^2 S + \cdots .$$

$$(5.1.21)$$

Here $S[x(\tau)] - S[X(\tau)]$ is expanded in a series of δx, the coefficients of the first order of δx is the first variation δS, and so on for higher orders. Carrying out integration by parts in (5.1.21), we can make the following replacements

$$2\frac{\mathrm{d}X}{\mathrm{d}\tau}\frac{\mathrm{d}}{\mathrm{d}\tau}\delta x \rightarrow -\frac{\mathrm{d}^2 X}{\mathrm{d}\tau^2}\delta x,$$

$$\left(\frac{\mathrm{d}}{\mathrm{d}\tau}\delta x\right)^2 \rightarrow -\left(\frac{\mathrm{d}^2}{\mathrm{d}\tau^2}\delta x\right)\delta x,$$

under the integral because of the boundary conditions $\delta x(\tau_i) = \delta x(\tau_j) = 0$. Using also the relations

$$V\left(x\right) - V\left(X\right) = V'\left(X\right)\delta x + \frac{1}{2}V''\left(X\right)\left(\delta x\right)^2,$$

$$\left(\frac{\mathrm{d}X}{\mathrm{d}\tau} + \frac{\mathrm{d}}{\mathrm{d}\tau}\delta x\right)^2 - \left(\frac{\mathrm{d}X}{\mathrm{d}\tau}\right)^2 = 2\frac{\mathrm{d}X}{\mathrm{d}\tau}\frac{\mathrm{d}}{\mathrm{d}\tau}\delta x + \left(\frac{\mathrm{d}}{\mathrm{d}\tau}\delta x\right)^2,$$

we eventually obtain

$$\delta S = \int_{\tau_i}^{\tau_f} \mathrm{d}\tau \left[-\frac{\mathrm{d}^2 X}{\mathrm{d}\tau^2} + V'\left(X\right)\right]\delta x. \qquad (5.1.22)$$

The demand of least action $\delta S = 0$ leads to

$$-\frac{\mathrm{d}^2 X}{\mathrm{d}\tau^2} + V'(X) = 0. \qquad (5.1.23)$$

This is the Euler-Lagrange equation for the solution with optimum action (cf. (5.1.21)), i.e., the equation satisfied by the classical path.

Since $\delta S = 0$ it follows that

$$S = S_0 + \delta^2 S, \tag{5.1.24}$$

$$\delta^2 S = \int_{\tau_i}^{\tau_f} \left[-\frac{1}{2} \frac{d^2}{d\tau^2} \delta x + \frac{1}{2} V''(X) \delta x \right] \delta x d\tau. \tag{5.1.25}$$

The second variation $\delta^2 S$ is the contribution from the paths deviating from the classical one. To get $\delta^2 S$ we impose additional conditions on x_n of (5.1.19). Let the eigenequation for x_n be

$$\left[-\frac{d^2}{d\tau^2} + V''(X) \right] x_n = \lambda_n x_n, \tag{5.1.26}$$

and the boundary condition be

$$x_n(\tau_i) = x_n(\tau_f) = 0,$$

The functions x_n are orthonormal:

$$\int_{\tau_i}^{\tau_f} x_m(\tau) \, x_n(\tau) \, d\tau = \delta_{mn}.$$

Substituting $\delta x = \sum c_n x_n$ into (5.1.25), using the orthonormality and (5.1.26), we obtain

$$\delta^2 S = \int d\tau \sum_{m,n} c_m x_m \left[-\frac{1}{2} \frac{d^2}{d\tau^2} + \frac{1}{2} V''(X) \right] c_n x_n$$

$$= \sum_{mn} \frac{1}{2} \lambda_n c_m c_n \delta_{mn} = \frac{1}{2} \sum_n c_n^2 \lambda_n. \tag{5.1.27}$$

Eventually we obtain the path integral

$$\left\langle x_f \left| \exp\left(-\frac{\hat{H}}{h} \tau \right) \right| x_i \right\rangle$$

$$= N \int [Dx] \, e^{-S/\hbar} = N e^{-S_0/\hbar} \int \prod_n \frac{dc_n}{\sqrt{2\pi\hbar}} e^{-\delta^2 S/\hbar}$$

$$= N e^{-S_0/\hbar} \int \prod_n \frac{\mathrm{d}c_n}{\sqrt{2\pi\hbar}} \exp\left(-\frac{1}{2\hbar}\sum c_n^2 \lambda_n\right)$$

$$= N \left(\prod_n \lambda_n^{-\frac{1}{2}}\right) e^{-S_0/\hbar}. \tag{5.1.28}$$

Since λ_n is the eigenvalue of the operator $-\frac{\mathrm{d}^2}{\mathrm{d}\tau^2} + V''(X)$, we have

$$\prod \lambda_n^{-\frac{1}{2}} = [\det(-\partial_\tau^2 + V''(X))]^{-\frac{1}{2}}, \tag{5.1.29}$$

and Eq. (5.1.28) finally becomes

$$\left\langle x_{\mathrm{f}} \left| \exp\left(-\frac{\hat{H}}{\hbar}\tau\right) \right| x_{\mathrm{i}} \right\rangle = N \left[\det\left(-\partial_\tau^2 + V''(X)\right)\right]^{-1/2} e^{-S_0/\hbar}. \tag{5.1.30}$$

This is the result up to the second variation $\delta^2 S$. It is in fact a semiclassical approximation. Higher order approximations lead to $O(\hbar)$ corrections, namely

$$\left\langle x_{\mathrm{f}} \left| \exp\left(-\frac{\hat{H}}{\hbar}\tau\right) \right| x_{\mathrm{i}} \right\rangle$$
$$= N \left[\det\left(-\partial_\tau^2 + V''(X)\right)\right]^{-1/2} e^{-S_0/\hbar}(1 + O(\hbar)).$$

As a simple example, consider an inverted bell potential shown in Fig. 5.2. We choose the boundary conditions to be $x_{\mathrm{i}} = x_{\mathrm{f}} = 0$. Then the classical path is $X(\tau) = 0$ with classical action $S_0 = 0$. Denote $V''(X(\tau)) = V''(x = 0)$ by ω^2, we have

$$\left\langle x_{\mathrm{f}} \left| \exp\left(-\frac{\hat{H}}{\hbar}\tau\right) \right| x_{\mathrm{i}} \right\rangle = N \left[\det\left(-\partial_\tau^2 + \omega^2\right)\right]^{-1/2} e^{-S_0/\hbar}(1 + O(\hbar)).$$

We are going to prove that for $\tau_{\mathrm{i}} = -\frac{\tau_0}{2}$, $\tau_{\mathrm{f}} = \frac{\tau_0}{2}$ and τ_0 large, we have

$$\left\langle 0 \left| \exp\left(-\frac{\hat{H}}{\hbar}\tau\right) \right| 0 \right\rangle = N \left[\det\left(-\partial_\tau^2 + \omega^2\right)\right]^{-1/2} = \left(\frac{\omega}{\hbar\pi}\right)^{\frac{1}{2}} e^{-\omega\frac{\tau_0}{2}}. \tag{5.1.31}$$

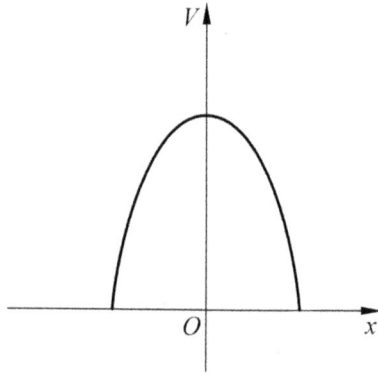

Fig. 5.2   An inverted bell potential.

Then comparing with (5.1.11) we obtain for the ground state $|0\rangle$ the following eigenstate and the wave function at the origin.

$$E_0 = \frac{1}{2}\hbar\omega(1 + O(\hbar)),$$

$$\langle x = 0 \mid 0 \rangle = \left(\frac{\omega}{\pi\hbar}\right)^{\frac{1}{2}}(1 + O(\hbar)).$$

In the following, we give the proof originally given in the appendix A of Ref. [6]. Consider the differential equation

$$(-\partial_\tau^2 + W(\tau))\psi = \mu\psi, \tag{5.1.32}$$

where $W(\tau)$ is some bounded function of τ, and μ is a parameter. Let $\psi_\mu(\tau)$ be the solution of (5.1.32) satisfying the boundary conditions[2]

$$\psi_\mu\left(-\frac{\tau_0}{2}\right) = 0, \quad \partial_\tau\psi_\mu\left(-\frac{\tau_0}{2}\right) = 1. \tag{5.1.33}$$

The operator $(-\partial_\tau^2 + W(\tau))$ has *eigenvalues* λ_n corresponding to *eigenfunctions* ψ_n, satisfying boundary conditions

$$\psi_{\lambda_n}\left(-\frac{\tau_0}{2}\right) = \psi_{\lambda_n}\left(\frac{\tau_0}{2}\right) = 0.$$

---

[2]Notice that this is a problem different from (5.1.26), it has a solution for any value of the parameter μ while (5.1.26) has solutions for usually discrete λ_n (the eigenvalues).

Rewrite (5.1.29) as

$$\det(-\partial_\tau^2 + W(\tau)) = \prod_n \lambda_n. \qquad (5.1.34)$$

Now let $W^{(1)}$ and $W^{(2)}$ be two different functions of τ, and let $\psi_\mu^{(1)}$ and $\psi_\mu^{(2)}$ be the corresponding solutions of Eq. (5.1.32). Consider

$$\frac{\det(-\partial_\tau^2 + W^{(1)} - \mu)}{\det(-\partial_\tau^2 + W^{(2)} - \mu)} = \frac{\psi_\mu^{(1)}\left(\frac{\tau_0}{2}\right)}{\psi_\mu^{(2)}\left(\frac{\tau_0}{2}\right)}. \qquad (5.1.35)$$

The left-hand side is a meromorphic function[3] of the complex variable μ, with a simple zero at each $\mu = \lambda_n^{(1)}$, and a simple pole at each $\mu = \lambda_n^{(2)}$, the λ's are real and positive. This function is analytic everywhere except on the positive real axis. Now we take the limit $\mu \to \infty$ in any direction on the complex plane except along the real axis. This function then approaches 1 in the limit. For a function analytic in a given region and its value along a continuous curve in this region is a constant, then it is *equal* to this constant in this entire region. Therefore, the right-hand side is also equal to 1 except on the positive real axis. We define a *constant* quantity N by

$$\frac{\det(-\partial_\tau^2 + W)}{\psi_0\left(\frac{\tau_0}{2}\right)} = \pi\hbar N^2, \qquad (5.1.36)$$

since we know that for $\mu = 0$ the left-hand side should be independent of W by (5.1.35). Consequently we have

$$N[\det(-\partial_\tau^2 + W)]^{-\frac{1}{2}} = \left[\pi\hbar\psi_0\left(\frac{\tau_0}{2}\right)\right]^{-\frac{1}{2}}. \qquad (5.1.37)$$

This relation is valid for any bounded function $W(\tau)$ and the corresponding solution ψ_μ for $\mu = 0$. For the harmonic oscillator, $W = \omega^2$, a constant, and ψ_0 satisfying (5.1.32) and (5.1.33) is

$$\psi_0(\tau) = \frac{1}{\omega}\sinh\omega\left(\tau + \frac{\tau_0}{2}\right). \qquad (5.1.38)$$

---

[3] A function of complex variable having only poles as singularities is called a meromorphic function.

Now $\psi_0\left(\frac{T_0}{2}\right) = \frac{1}{\omega}\sinh\omega T_0$, (5.1.36) for a harmonic oscillator is simply

$$N[\det(-\partial_\tau^2 + \omega^2)]^{-\frac{1}{2}} = \left(\frac{\omega}{\hbar\pi}\right)^{\frac{1}{2}} e^{-\omega\frac{T_0}{2}},$$

and (5.1.31) is proved.

In the following, we give an outline of the calculation of a general path integral for the 1-dimensional harmonic oscillator[4] using Minkowski space-time. The path integral to be calculated is

$$\langle x_b t_b | x_a t_a\rangle = \int [Dx]\exp\left(\frac{i}{\hbar}S\right) \equiv \exp\left(\frac{i}{\hbar}S_0\right)F_\omega\left(t_b - t_a\right),$$

$$(5.1.39)$$

where

$$S = \int_{t_a}^{t_b}\frac{m}{2}\left(\dot{x}^2 - \omega^2 x^2\right)dt.$$

$$(5.1.40)$$

The equation for the classical path is

$$\ddot{x} = -\omega^2 x.$$

$$(5.1.41)$$

The solution satisfying the boundary conditions is

$$X(t) = \frac{x_b\sin\omega\left(t - t_a\right) + x_a\sin\omega\left(t_b - t\right)}{\sin\omega\left(t_b - t_a\right)}.$$

$$(5.1.42)$$

It is easy to obtain S_0, but the calculation of F_ω is relatively complicated. The result turns out to be

$$\langle x_b t_b | x_a t_a\rangle = \frac{1}{\sqrt{2\pi i\hbar/m}}\sqrt{\frac{\omega}{\sin\omega\left(t_b - t_a\right)}}$$

$$\times \exp\left\{\frac{i}{\hbar}\frac{m\omega}{2\sin\omega\left(t_b - t_a\right)}\right.$$

$$\left.\times\left[\left(x_b^2 + x_a^2\right)\cos\omega\left(t_b - t_a\right) - 2x_b x_a\right]\right\}.$$

$$(5.1.43)$$

---

[4]See Sections 2.4 and 9.3 of Ref. [7].

Similar to Eq. (5.1.11), path integral can be expressed in terms of the eigenfunctions and eigenvalues of H.

$$\langle x_b t_b | x_a t_a \rangle = \sum_n \psi_n(x_b) \, \psi_n^*(x_a) \, e^{-iE_n(t_b - t_a)/\hbar}, \qquad (5.1.44)$$

We need the Mehler's formula[5] to transform the right hand side of (5.1.43) before we can compare it with the right hand side of (5.1.44),

$$\frac{1}{\sqrt{1-a^2}} \exp\left[-\frac{1}{1-a^2}\left(x^2 + x'^2 - 2xx'a \right) \right]$$

$$= e^{-x^2 - x'^2} \sum_{n=0}^{\infty} \frac{a^n}{2^n n!} H_n(x) \, H_n(x'),$$

where $H_n(x)$ is the Hermite polynomial. Substitute $a = e^{-i\omega(t_b - t_a)}$, $x = \frac{x_b}{\lambda}$, $x' = \frac{x_a}{\lambda}$, $\lambda = \sqrt{\frac{\hbar}{m\omega}}$ in the Mehler formula, the comparison gives

$$\left.\begin{aligned}
\psi_n(x) &= N_n \lambda^{-1/2} e^{-x^2/2\lambda^2} H_n(x/\lambda) \\[2mm]
N_n &= \left(2^n n! \sqrt{\pi} \right)^{-1/2} \\[2mm]
E_n &= \hbar\omega \left(n + \frac{1}{2} \right)
\end{aligned}\right\} \qquad (5.1.45)$$

The result is identical to that obtained by solving Schrödinger equation.

5.2 The Instanton and the Coherence Splitting of Energy Levels in a Double Well

Consider the double well potential shown in Fig. 5.3(a). The potential is symmetric with respect to the coordinate x,

$$V(-x) = V(x). \qquad (5.2.1)$$

---

[5]See P. M. Morse and H. Feshbach, *Methods of Theoretical Physics* vol. 1, p. 781, New York, McGraw Hill, 1953.

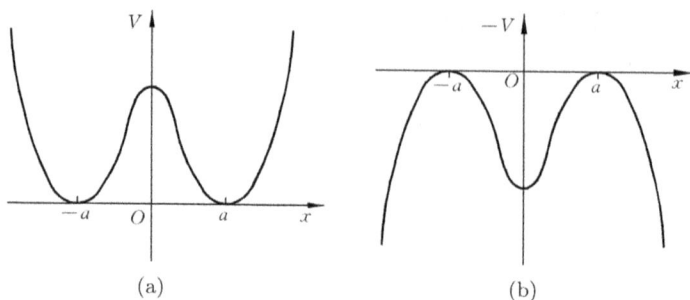

Fig. 5.3   (a) The double well potential; (b) Its mechanical analog.

and has minima at $x = \pm a$, and $V''(\pm a) = w^2$. This means that the potential around $x = \pm a$ can be approximated by a harmonic potential with frequency w. The equation of motion for the classical path in Euclidean space-time is given by (5.1.23)

$$\frac{d^2 X}{d\tau^2} = V'(X).$$

Comparison with the equation of motion in Minkowski space-time $(m = 1)$,

$$\frac{d^2 x}{dt^2} = -V'(x)$$

leads to an analog between the path $X(\tau)$ in imaginary time and the path of a *classical* particle in the *inverted potential* $-V$ in real time, shown in Fig. 5.3(b). This enables an intuitive way of contemplating the classical path using the well-known examples in *classical* mechanics. This method is called the mechanical analog. From such an analog for the double well potential we see immediately that there are two kinds of simple classical paths: (1) $X(\tau) = a$ or $X(\tau) = -a$ for all τ, (2) $X\left(-\frac{\tau}{2}\right) = -a, X\left(\frac{\tau}{2}\right) = a$ for $\tau \to \infty$; or $X\left(-\frac{\tau}{2}\right) = a, X\left(\frac{\tau}{2}\right) = -a$ for $\tau \to \infty$. The transition amplitudes to be evaluated are:

$$\left\langle -a \left| \exp\left(-\frac{\hat{H}}{\hbar}\tau\right) \right| -a \right\rangle = \left\langle a \left| \exp\left(-\frac{\hat{H}}{\hbar}\tau\right) \right| a \right\rangle, \qquad (5.2.2)$$

and

$$\left\langle a \left| \exp\left(-\frac{\hat{H}}{\hbar}\tau\right) \right| -a \right\rangle = \left\langle -a \left| \exp\left(-\frac{\hat{H}}{\hbar}\tau\right) \right| a \right\rangle. \qquad (5.2.3)$$

Consider the second kind of solution. For $\tau \to \infty$, the velocity of the particle is 0 at $x = \pm a$. Hence, both the kinetic and the potential energies are 0 there. From the conservation of energy we know that this is a *zero energy solution*. The first integral of the equation of motion is therefore

$$\frac{1}{2}\left(\frac{dx}{d\tau}\right)^2 - V(x) = 0.$$

Hence, the classical path satisfies

$$\frac{dX}{d\tau} = \sqrt{2V}. \qquad (5.2.4)$$

Its solution is

$$\tau = \tau_c + \int_0^X dx'\, (2V\,(x'))^{-1/2}, \qquad (5.2.5)$$

where the arbitrary constant τ_c is the imaginary time corresponding to $X = 0$. Figure 5.4 shows the solution (5.2.5). It is an *instanton* with the center at τ_c. It is also referred to as a *kink solution*. The suffix "on" originates from the similarity in mathematical structure of (5.2.5) to the particle-like *soliton* in the field theory.

For a very long period of imaginary time the particle is situated at $x = -a$. Then it sets out to move and almost instantaneously it lashes through the point $x = 0$. Thence the headword *instant*. This instant can actually be calculated. For $\tau \to \infty$, $X \approx a$, V can be approximately written as $\frac{1}{2}\omega^2(a - X)^2$. From Eq. (5.2.4) we obtain $\frac{dX}{d\tau} \approx \omega(a - X)$, which is solved by $a - X = e^{-\omega\tau}$. Consequently only for an brief instant $1/\omega$ the position X deviates appreciably from a. The solution for $X(-\infty) = a$, $X(\infty) = -a$ is called an anti-instanton. From Eq. (5.2.4), the classical action for an instanton or

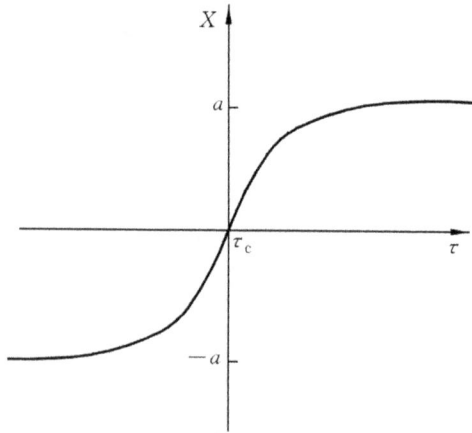

Fig. 5.4   The instanton.

an anti-instanton is

$$S_0 = \int d\tau \left[\frac{1}{2} \left(\frac{dX}{d\tau} \right)^2 + V \right] = \int d\tau \left(\frac{dX}{d\tau} \right)^2 = \int_{-a}^{a} dx \sqrt{2V}.$$

$$(5.2.6)$$

Since the instant $1/\omega$ is completely negligible in comparison with the entire period of evolution $(-\infty, \infty)$, the instanton (anti-instanton) cannot be the only approximate classical path. By connecting a series of instanton-anti-instantons one obtains another approximate classical path. Figure 5.5 shows an example of such a series, with $\tau_1, \tau_2, \cdots$ as the centers of instantons/anti-instantons. In the figure $1/\omega$ is taken to be zero. For a series of n instanton-anti-instantons, the classical action is nS_0, since by (5.2.6) the action of an instanton comes mainly from the instant for which $X \approx 0$. The calculation of the prefactor $[\det(-\partial_\tau^2 + V''(X))]^{-\frac{1}{2}}$ is more involved. If there were no such instants $\tau_1, \tau_2, \cdots$, the particle remains at $x = a$ or $x = -a$, where $V'' = \omega^2$. This is the situation of a harmonic oscillator and we can rely on Eq. (5.1.31) which gives $\left(\frac{\omega}{\pi \hbar} \right)^{\frac{1}{2}} e^{-\omega T_0/2}$. Let the correction due to a single instanton-anti-instanton pair be a factor K which will be given later. Then a series of n instanton-anti-instantons

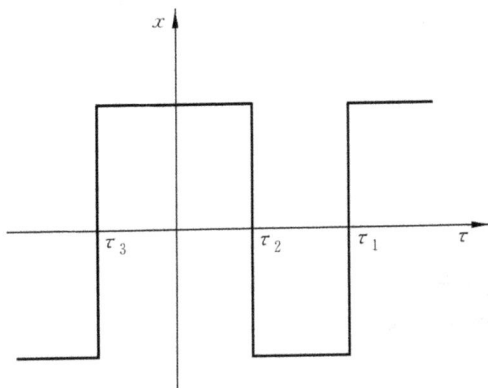

Fig. 5.5   A series of instanton-anti-instantons.

gives the prefactor

$$\left(\frac{\omega}{\pi\hbar}\right)^{1/2} e^{-\omega\tau_0/2} K^n. \tag{5.2.7}$$

Since the positions of $\tau_1, \tau_2, \cdots$ are arbitrary, the sum over all possible paths involves the integrations over the τ's. We notice that the integrand is independent of the positions of $\tau_1, \tau_2, \cdots$, so that the integration leads to a factor

$$\int_{-\tau_0/2}^{\tau_0/2} d\tau_1 \int_{-\tau_0/2}^{\tau_1} d\tau_2 \cdots \int_{-\tau_0/2}^{\tau_{n-1}} d\tau_n = \frac{\tau_0^n}{n!}. \tag{5.2.8}$$

There is yet another delicate point. If $X(-\frac{\tau_0}{2}) = X(\frac{\tau_0}{2})$, then an even number of instanton-anti-instantons has to be taken, while if $X(-\frac{\tau_0}{2}) = -X(\frac{\tau_0}{2})$, an odd number has to be taken. We have, for instance,

$$\left\langle -a \left| \exp\left(-\frac{\hat{H}}{\hbar}\tau\right) \right| a \right\rangle = \left(\frac{\omega}{\pi\hbar}\right)^{1/2} e^{-\omega\tau_0/2} \sum_{n(\text{odd})} \frac{K^n e^{-nS_0/\hbar}\tau_0^n}{n!}$$

$$= \left(\frac{\omega}{\pi\hbar}\right)^{1/2} e^{-\omega\tau_0/2} \frac{1}{2}\left[\exp\left(Ke^{-S_0/\hbar}\tau_0\right) \right.$$

$$\left. - \exp\left(-Ke^{-S_0/\hbar}\tau_0\right) \right], \tag{5.2.9}$$

$$\left\langle -a \left| \exp\left(-\frac{\hat{H}}{\hbar}\tau \right) \right| -a \right\rangle = \left(\frac{\omega}{\pi\hbar} \right)^{1/2} e^{-\omega\tau_0/2} \sum_{n(\text{even})} \frac{K^n e^{-nS_0/\hbar}\tau_0}{n!}$$

$$= \left(\frac{\omega}{\pi\hbar} \right)^{1/2} e^{-\omega\tau_0/2} \frac{1}{2} \left[\exp\left(Ke^{-S_0/\hbar}\tau_0 \right) \right.$$

$$\left. - \exp\left(-Ke^{-S_0/\hbar}\tau_0 \right) \right]. \tag{5.2.10}$$

Comparing with (5.1.32) we see that the two low-lying energy states have eigenvalues

$$E_{\pm} = \frac{1}{2}\hbar\omega \pm \hbar Ke^{-S_0/\hbar}. \tag{5.2.11}$$

Denoting the corresponding eigenstates by $|+\rangle$ and $|-\rangle$, we find their scalar products with $|a\rangle$ and $|-a\rangle$ to be

$$|\langle +| \pm a\rangle|^2 = |\langle -| \pm a\rangle|^2 = \langle a|-\rangle\langle -|-a\rangle = -\langle a|+\rangle\langle +|-a\rangle$$

$$= \frac{1}{2}\left(\frac{\omega}{\pi\hbar} \right)^{\frac{1}{2}}. \tag{5.2.12}$$

We see that $|\pm\rangle$ are proportional to linear combinations of $|a\rangle$ and $|-a\rangle$. More exactly, from (5.2.10) we infer that

$$\left\langle +\left| \exp\left(-\frac{\hat{H}}{\hbar}\tau \right) \right| + \right\rangle = \exp\left(-\frac{E_+}{\hbar}\tau_0 \right)$$

$$= e^{-\omega_0\tau_0/2} \exp\left(-Ke^{-S_0/\hbar}\tau_0 \right), \tag{5.2.13}$$

$$\left\langle -\left| \exp\left(-\frac{\hat{H}}{\hbar}\tau \right) \right| - \right\rangle = \exp\left(-\frac{E_-}{\hbar}\tau_0 \right)$$

$$= e^{-\omega_0\tau_0/2} \exp\left(Ke^{-S_0/\hbar}\tau_0 \right) \tag{5.2.14}$$

and we conclude that the ground state $|-\rangle$ is the even combination of $|a\rangle$ and $|-a\rangle$, and the excited state $|+\rangle$ is the odd combination. The splitting of energy given in (5.2.11) is proportional to the tunneling factor $e^{-S_0/\hbar}$. Its appearance is natural. If the barrier between the wells is indefinitely high such that no tunneling can occur, the levels localized in the wells are degenerate. Only finite tunneling can lead to the removal of the degeneracy and $|\pm\rangle$ becomes the energy eigenstates with a difference of energy proportional to the tunneling factor. This splitting is called the tunnel splitting. Since the energy eigenstates are the coherent superposition of the states localized in the two wells, the splitting is also called the coherence splitting.

Finally we turn to the calculation of the prefactor K of a single instanton. Let $x(\tau, \tau_c)$ be an instanton with center at τ_c. When τ_c is displaced by a small amount $\delta \tau_c$ the instanton solution changes by

$$X(\tau, \tau_c) - X(\tau, \tau_c + \delta \tau_c) = -\frac{\partial}{\partial \tau_c} X(\tau, \tau_c) \delta \tau_c = \frac{\partial}{\partial \tau} X(\tau, \tau_c) \delta \tau_c.$$

The last step of the above equation holds because displacing τ_c is equivalent to displacing the origin of τ in the opposite direction by the same amount. Since the action is not changed when the instanton is displaced as a whole due to time translation invariance, it follows from (5.1.29) that

$$\left[-\frac{d^2}{d\tau^2} + V''(X) \right] \delta X = 0,$$

i.e.,

$$\left[-\frac{d^2}{d\tau^2} + V''(X) \right] \frac{\partial}{\partial \tau} X(\tau, \tau_c) = 0. \qquad (5.2.15)$$

From this we conclude that $\frac{\partial X}{\partial \tau}$ is an eigenfunction of the operator $[-\frac{d^2}{d\tau^2} + V''(X)]$ corresponding to the eigenvalue 0, or it is the zero mode solution. Denoting the normalized zero mode solution by x_0,

$$x_0 = C \frac{dX}{d\tau},$$

By using Eq. (5.2.6) we see that the normalization constant C satisfies

$$1 = C^2 \int d\tau \left(\frac{dX}{d\tau} \right)^2 = C^2 S_0. \qquad (5.2.16)$$

Hence

$$x_0 = (S_0)^{-1/2} \frac{dX}{d\tau}. \qquad (5.2.17)$$

When an arbitrary solution x is expanded around X we have

$$x(\tau) = X(\tau) + c_0 x_0 + \sum_{n=1} c_n x_n. \qquad (5.2.18)$$

The integration with respect to x_0 gives $\lambda_0^{-\frac{1}{2}}$ with $\lambda_0 = 0$. But this integration is equivalent to the integration with respect to τ_c, and the latter has already been performed in (5.2.8). It is only necessary to find the factor of proportionality of this equivalence. The change of $x(\tau)$ due to a change of $d\tau_c$ is

$$dx(\tau) = -\frac{dX}{d\tau} d\tau_c = -S_0^{1/2} x_0 d\tau_c,$$

while the change due to varying c_0 is

$$dx(\tau) = x_0 dc_0.$$

Equating these changes gives

$$\frac{1}{\sqrt{2\pi\hbar}} dc_0 = \sqrt{\frac{S_0}{2\pi\hbar}} d\tau_c. \qquad (5.2.19)$$

Up to this point, the contribution to the path integral from a single-instanton has been calculated in full:

$$\left\langle a \left| \exp\left(-\frac{\hat{H}}{\hbar}\tau\right) \right| -a \right\rangle_1$$

$$= N\tau_0 \left(\frac{S_0}{2\pi\hbar}\right)^{\frac{1}{2}} e^{-S_0/\hbar} \det{}' \left[-\frac{\partial^2}{\partial\tau^2} + V''(X)\right]^{-\frac{1}{2}}. \qquad (5.2.20)$$

The prime above the det sign means that the zero eigenvalue is not included, and the integral over $\frac{dc_0}{\sqrt{2\pi}}$ has been replaced by $\tau_0(\frac{S_0}{2\pi\hbar})^{\frac{1}{2}}$. Extracting the single instanton contribution from (5.2.9)

$$\left\langle a \left| \exp\left(-\frac{\hat{H}}{\hbar}\tau\right) \right| -a \right\rangle_1 = \left(\frac{\omega}{\pi\hbar}\right)^{\frac{1}{2}} e^{-\omega\tau_0/2} K e^{-S_0/\hbar} \tau_0,$$

and using the prefactor for the harmonic oscillator (5.1.35)

$$
N \det \left[-\frac{\partial^2}{\partial \tau^2} + \omega^2 \right]^{-1/2} = \left(\frac{\omega}{\pi \hbar} \right)^{\frac{1}{2}} e^{-\omega \tau_0 / 2},
$$

we obtain

$$
K = \left(\frac{S_0}{2\pi\hbar} \right)^{\frac{1}{2}} \left[\frac{\det \left(-\frac{\partial^2}{\partial \tau^2} + \omega^2 \right)}{\det{}' \left(-\frac{\partial^2}{\partial \tau^2} + V''(X) \right)} \right]^{\frac{1}{2}}. \tag{5.2.21}
$$

In the Schrödinger formulation of quantum mechanics the problem of coherence splitting is handled by solving the Schrödinger equation.[6] The solution makes use of the WKB approximation, but it is not quite trivial, because the usual connection formula cannot be directly applied. A substitute way is devised and the result agrees with that of the instanton method.

We have mentioned that the treatment in this section is equivalent to the semiclassical approximation. We obtained in (5.2.10) as the correction to the energy $\frac{1}{2}\hbar\omega$ a term proportional to the exponential factor $e^{-S_0/\hbar}$. The semiclassical approximation makes sense when the result is $O(\hbar)$. For $S_0 \gg \hbar$ the correction can be much smaller than $O(\hbar^2)$. Does it make sense to keep such a correction? The answer is, as a correction term it is irrelevant; but as the splitting between $|+\rangle$ and $|-\rangle$ it is the leading term, and hence makes complete sense.

The presentation of this section follows Ref. [6].

5.3 The Density Matrix and the Path Integral

Consider a quantum operator $\widehat{K}$. Its discrete eigenstates $\{|k_n\rangle\}$ form a complete set of orthonormal functions. A wave function $|\psi\rangle$ of a quantum system is a superposition of this set of functions.

$$
|\psi\rangle = \sum c_n |k_n\rangle. \tag{5.3.1}
$$

---

[6]See appendix B of Ref. [6].

The complex coefficients c_n satisfies the condition $\sum |c_n|^2 = 1$. $|c_n|^2$ is the probability that $|\psi\rangle$ is found in the eigenstate $|k_n\rangle$. The superposition (5.3.1) offers more information than the statement above, because the complex $\{c_n\}$ contains the relative phase between various terms in the *coherent* superposition. In this case, $|\psi\rangle$ represents a *pure state*. It can happen that the state of the quantum system is known to the extent of probabilities that it is in the state $|k_n\rangle$, i.e., only $\{w_n = |c_n|^2\}$ is known, where $w_n \leqslant 1$ are real numbers and $\sum w_n = 1$. When the number of non-vanishing w_n is larger than 1, the state is a *statistical mixture*. The density matrix

$$\rho = \sum w_n |k_n\rangle \langle k_n| \tag{5.3.2}$$

is an efficient way of describing a statistical mixture. ρ can be considered as a diagonal matrix with elements $\rho_{nm} = w_n \delta_{nm}$.[7] The condition $\sum |c_n|^2 = 1$ leads to

$$\mathrm{tr}\rho = 1. \tag{5.3.3}$$

The density matrix is Hermitian

$$\rho^\dagger = \rho. \tag{5.3.4}$$

Let us calculate ρ^2:

$$\rho^2 = \sum_{m,n} w_n w_m |k_n\rangle \langle k_n| k_m\rangle \langle k_m|$$

$$= \sum_n w_n^2 |k_n\rangle \langle k_n|, \tag{5.3.5}$$

due to the fact that $\langle k_n \mid k_m\rangle = \delta_{mn}$. In general, $\sum w_n = 1$ implies $w_n^2 < w_n$ except for the case that only one of the coefficients $w_l = 1$

---

[7]ρ is diagonal because of the representation. A change of representation makes it non-diagonal.

and all others vanish. In general case, we have

$$\mathrm{tr}\rho^2 < \mathrm{tr}\rho, \qquad (5.3.6a)$$

and in special case

$$\rho^2 = \rho. \qquad (5.3.6b)$$

The special case is $\rho = |k_l\rangle\langle k_l|$, which corresponds to $|\psi\rangle = |k_l\rangle$, a pure state. In this case

$$\rho = |\psi\rangle\langle\psi|. \qquad (5.3.7)$$

Therefore, (5.3.6a) is the condition for a statistical mixture, while (5.3.6b) is the condition for a pure state. The density matrix is a versatile way of describing the state of a quantum system for both pure states or statistical mixtures. If the pure state $|\psi\rangle$ in (5.3.7) is a superposition (5.3.1), then we have

$$\rho = \sum_{m,n} c_m c_n^* |k_m\rangle\langle k_n|, \qquad (5.3.8)$$

here the matrix has off-diagonal elements, because (5.3.1) can be considered as a change of the representation. The form (5.3.7) is a diagonal matrix itself, with $w_\psi = 1$ and all other w'_φs vanish, where the φ's are functions orthogonal to ψ. To distinguish whether the density matrix represents a pure state or a statistical mixture by its matrix form, it is necessary to diagonalize the matrix first, and then make the judgement by counting the number n of the non-vanishing elements. If $n = 1$, then it is a pure state, and if $n > 1$, it is a statistical mixture. One can also judge by examining whether ρ satisfies (5.3.6a) or (5.3.6b). For a non-diagonal density matrix there is no guarantee that it represents a pure state. The non-diagonal matrix elements do carry phase information, but not necessarily the *complete* phase information.

In the x-representation,

$$\rho\left(x'x\right) \equiv \langle x' |\rho| x\rangle = \sum_k w_k \langle x'|k\rangle \langle k|x\rangle = \sum_k w_k \psi_k\left(x'\right) \psi_k^*(x).$$

$$(5.3.9)$$

The expectation value of any observable $\hat{A}$ in the state represented by ρ is, by using the diagonal form (5.3.2),

$$
\begin{aligned}
\left\langle \hat{A} \right\rangle &= \sum_k w_k \left\langle k \mid \hat{A} \mid k \right\rangle \\
&= \int \mathrm{d}x \mathrm{d}x' w_k \left\langle k \mid x \right\rangle \left\langle x' \mid A \mid x \right\rangle \left\langle x' \mid k \right\rangle \\
&= \int \mathrm{d}x \mathrm{d}x' \left\langle x \mid \rho \mid x' \right\rangle \left\langle x' \mid A \mid x \right\rangle \\
&= \int \mathrm{d}x \mathrm{d}x' \rho\left(x, x'\right) A\left(x', x\right) = \mathrm{tr}\rho A,
\end{aligned}
\tag{5.3.10}
$$

which is actually independent of representations.

We now go over to the statistical mechanics and consider a system in equilibrium at temperature T. The set of energy eigenstates $\{|\varphi\rangle\}$ satisfies the eigenequation

$$
\hat{H}\left|\varphi_i\right\rangle = E_i\left|\varphi_i\right\rangle.
\tag{5.3.11}
$$

The probability that the system is in the state $|\varphi_i\rangle$ is

$$
w_i = \frac{1}{Q}\mathrm{e}^{-\beta E_i},
\tag{5.3.12}
$$

where

$$
\beta = \frac{1}{kT},
\tag{5.3.13}
$$

$$
Q = \sum_i \mathrm{e}^{-\beta E_i}.
\tag{5.3.14}
$$

Q is the partition function. The density matrix can be written as

$$
\rho = \frac{1}{Q}\sum_i \mathrm{e}^{-\beta E_i}\left|\varphi_i\right\rangle\left\langle\varphi_i\right| = \frac{1}{Q}\mathrm{e}^{-\beta\hat{H}}.
\tag{5.3.15}
$$

The last inequality is obtained by writing $\mathrm{e}^{-\beta E_i}|\varphi_i\rangle$ as $\mathrm{e}^{-\beta\hat{H}}|\varphi_i\rangle$ and inserting the identity $\sum_i |\varphi_i\rangle\langle\varphi_i| = 1$. For convenience the un-normalized density matrix $\rho(\beta)$ is sometimes used

$$\rho(\beta) = e^{-\beta\hat{H}}. \tag{5.3.16}$$

Care must be exercised when expectation values are calculated using such density matrices. The equation satisfied by $\rho(\beta)$ is

$$-\frac{\partial\rho}{\partial\beta} = \hat{H}\rho, \tag{5.3.17}$$

$$\rho(0) = 1. \tag{5.3.18}$$

In $x-$representation equations (5.3.16)–(5.3.18) have the following form:

$$\rho(x, x'; \beta) = \left\langle x \left| e^{-\beta\hat{H}} \right| x' \right\rangle, \tag{5.3.19}$$

$$-\frac{\partial}{\partial\beta}\rho(x, x'; \beta) = \hat{H}_x\rho(x, x'; \beta), \tag{5.3.20a}$$

$$\rho(x, x'; 0) = \delta\left(x - x'\right). \tag{5.3.20b}$$

Comparing (5.3.19) with the path integral in Euclidean space $\langle x|e^{-\frac{H}{\hbar}\tau}|x'\rangle$, we find that they are closely resemble to each other. It is sufficient to set

$$\tau = \hbar\beta = \frac{\hbar}{kT}, \tag{5.3.21}$$

so that the density matrix can be expressed as a path integral

$$\rho(x, x'; \beta) = N \int [Dx\,(u)] \exp\left[-\frac{1}{\hbar}\int_0^U \left\{\frac{1}{2}m\dot{x}^2\,(u) + V\left[x\,(u)\right]\right\} du\right], \tag{5.3.22}$$

where $U = \hbar\beta$. All properties of the system in equilibrium can be obtained from the density matrix.

We calculate the density matrix of a harmonic oscillator in Euclidean space-time. The equation of motion for the classical path is

$$\ddot{X} - \omega^2 X = 0, \quad X(0) = x, \quad X(U) = x'. \tag{5.3.23}$$

which is solved by

$$X = \frac{\left(x' - xe^{-\omega U}\right)e^{\omega u} + \left(xe^{\omega U} - x'\right)e^{-\omega u}}{2\sinh \omega U}. \tag{5.3.24}$$

The classical action can be calculated directly

$$S_0 = \frac{m\omega}{2\sinh \omega U}\left[\left(x'^2 + x^2\right)\cosh \omega U - 2xx'\right]. \tag{5.3.25}$$

An arbitrary path satisfying the boundary conditions can be expressed as

$$x(u) = X(u) + y(u), \quad y(0) = y(U) = 0. \tag{5.3.26}$$

The second order variation is

$$\delta^2 S = \frac{m}{2}\int_0^U \left(\dot{y}^2 + \omega^2 y^2\right)\mathrm{d}u. \tag{5.3.27}$$

The prefactor and the path integral are respectively

$$F(U) \equiv \int [\mathrm{D}y]\exp\left[-\frac{1}{\hbar}\int_0^U \left(\frac{m}{2}\dot{y}^2 + \frac{m\omega^2}{2}y^2\right)\mathrm{d}u\right], \tag{5.3.28}$$

$$\rho(x, x'; U) = F(U)e^{-S_0/\hbar}. \tag{5.3.29}$$

As a path integral, the density matrix possesses a remarkable transition property:

$$\rho(x, x'; U_1 + U_2) = \int \rho(x, x''; U_2)\,\rho(x'', x; U_1)\,\mathrm{d}x''. \tag{5.3.30}$$

A history can be divided by an intermediate position x'' into two successive histories extending over periods U_1 and U_2 respectively.

To sum over histories, the intermediate position x'' is arbitrary, and to be integrated over. Substituting (5.3.29) into (5.3.30), we have

$$F(U_1 + U_2) \exp\left[-\frac{S_0(x, x'; U_1 + U_2)}{\hbar}\right]$$

$$= \int \exp\left\{-\frac{1}{\hbar}[S_0(x, x''; U_2) + S_0(x'', x'; U_1)]\right\}$$

$$\times F(U_1) F(U_2) \, dx''.$$

With the classical action S_0 in Eq. (5.3.26) substituted in and integration over x'' performed, we obtain the equation for $F(U)$:

$$F(U_1 + U_2) \left[\frac{2\hbar\pi \sinh \omega (U_1 + U_2)}{m\omega}\right]^{\frac{1}{2}}$$

$$= F(U_1) \left[\frac{2\hbar\pi \sinh \omega U_1}{m\omega}\right]^{\frac{1}{2}} F(U_2) \left[\frac{2\hbar\pi \sinh \omega U_2}{m\omega}\right]^{\frac{1}{2}} \quad (5.3.31)$$

The solution is

$$\left[\frac{2\hbar\pi \sinh U}{m\omega}\right]^{\frac{1}{2}} F(U) = e^{\alpha U},$$

or

$$F(U) = \left[\frac{m\omega}{2\hbar\pi \sinh U}\right]^{\frac{1}{2}} e^{\alpha U}. \quad (5.3.32)$$

Here α is determined by normalization. The result agrees with (5.1.43). We made use of the transition property (5.3.31) and did not perform the path integral (5.3.29). This is an enlightening approach [3].

We turn now to the density matrix of coupled systems, in which we will find numerous applications in the quantum mechanics of open systems. A particle with mass M and coordinate q in a potential $V(q)$ interacts with a harmonic oscillator of frequency ω, mass m and coordinate x through a coupling $-\gamma xq$. The Hamiltonian of the coupled system is

$$H = \frac{m\dot{x}^2}{2} + \frac{m\omega^2}{2}x^2 + \frac{M\dot{q}^2}{2} + V(q) - \gamma x q. \tag{5.3.33}$$

The density matrix of the system is

$$\rho(x, x'; U) = N \int [Dq][Dx] \exp\left[-\frac{1}{\hbar}\int_0^U \left(\frac{m\dot{x}^2}{2} + \frac{m\omega^2}{2}x^2 - \gamma qx\right)du\right]$$

$$\times \exp\left[-\frac{1}{\hbar}\int_0^U \left(\frac{M\dot{q}^2}{2} + V(q)\right)du\right], \tag{5.3.34}$$

with boundary conditions

$$x(0) = x, \quad x(U) = x'.$$

In general, $V(q)$ can have various forms and it is not possible to evaluate the path integral exactly, so the strategy is to take the functional integral in x, thus obtaining an effective action for the system q. We slightly change the notation, and concentrate on the following path integral:

$$F[f; x, x'] = \int [Dx] \exp\left[-\frac{1}{\hbar}\int_0^U \left(\frac{m\dot{x}^2}{2} + \frac{m\omega^2}{2}x^2 + if(u)x\right)du\right], \tag{5.3.35}$$

$$x(0) = x, \quad x(U) = x'.$$

The Euler-Lagrange equation for the oscillator is

$$\ddot{X} - \omega^2 X = \frac{i}{m}f(u), \tag{5.3.36}$$

and this is the equation of motion for a forced oscillator. Using $x = X + y$, we rewrite the action as

$$S = \int_0^U \left[\frac{m}{2}\left(\dot{X} + \dot{y}\right)^2 + \frac{m\omega^2}{2}(X + y)^2 + if(X + y)\right]du$$

$$= \int \frac{m}{2}\left[\dot{X}^2 + \omega^2 X^2 + \frac{2}{m}ifX\right]du$$

$$+ \int m \left[\dot{X} \dot{y} + \omega^2 X y + \frac{2}{m} i f y \right] du$$

$$+ \int \frac{m}{2} \left(\dot{y}^2 + \omega^2 y \right) du.$$

The first term of the second integral can be transformed as $\int \dot{X} y |_0^U - \int_0^U \ddot{X} y du$, so the second integral becomes $\int m[-\ddot{X} + \omega^2 X + \frac{i}{m} f(u)] y du$, it vanishes because of the equation of motion (5.3.38). The third integral gives the prefactor for the harmonic oscillator, i.e., $\sqrt{m\omega/2\pi\hbar \sinh \omega U}$ (Eq. (5.1.35)). Equation (5.3.38) can be solved by using the method of variation of parameters,

$$X(u) = (c_1 + y_1)e^{\omega u} + (c_2 + y_2)e^{-\omega u}, \qquad (5.3.37)$$

where

$$y_1 = \frac{i}{2m\omega} \int_0^u f(u) e^{-\omega u} du,$$

$$y_2 = -\frac{i}{2m\omega} \int_0^u f(u) e^{\omega u} du \qquad (5.3.38)$$

Using $X(0) = x$, $X(U) = x'$, the constants c_1 and c_2 can be determined, they are functions of x, x', U and the constants A and B, where

$$A = \frac{1}{2m\omega} \int_0^U e^{-\omega u} f(u) du, \qquad (5.3.39)$$

$$B = \frac{1}{2m\omega} \int_0^U e^{-\omega(U-u)} f(u) du.$$

The classical action can be calculated as follows:

$$S_0 = \frac{m}{2} \int_0^U \left(\dot{X}^2 + \omega^2 X^2 + \frac{2}{m} i f(u) X \right) du$$

$$= \frac{m}{2} \left[\dot{X} X \Big|_0^U + \int_0^U \left(-\ddot{X} + \omega^2 X^2 + \frac{2}{m} i f(u) X \right) X du \right]$$

$$= \frac{m}{2} \left[\dot{X} X \Big|_0^U + \int_0^U \frac{i}{m} f(u) X du \right].$$

Substituting X of (5.3.37), (5.3.38) in (5.3.39) and (5.3.40) into S_0, we obtain

$$
\begin{aligned}
S_0 &= \frac{1}{4m\omega} \int_0^U \int_0^U e^{-\omega|u-u'|} f(u) f(u') \, du \, du' \\
&+ \frac{m\omega}{2\sinh\omega U}[(x^2 + x'^2)\cosh\omega U - 2xx' + 2A(xe^{\omega U} - x') \\
&+ 2B(x'e^{\omega U} - x) + (A^2 + B^2)e^{\omega U} - 2AB]
\end{aligned}
\tag{5.3.40}
$$

Up to this point, we have calculated F (5.3.35):

$$
F[f, x, x'] = \sqrt{\frac{m\omega}{2\pi\hbar\sinh\omega U}} e^{-S_0/\hbar}.
\tag{5.3.41}
$$

In practical applications (Section 6.5.5), we need the following functional of f:

$$
\mathcal{E}[f] \equiv \frac{\displaystyle\int F[f, x, x] \, dx}{\displaystyle\int F[0, x, x] \, dx}.
\tag{5.3.42}
$$

The prefactors of F can be taken out of the integral and they cancel in the ratio. The integral in the denominator is

$$
\int \exp\left[-\frac{m\omega}{\hbar\sinh\omega U}(\cosh\omega U - 1)x^2\right] dx = \left(\frac{\pi\hbar\sinh\omega U}{m\omega(\cosh\omega U - 1)}\right)^{1/2}.
\tag{5.3.43}
$$

$S_0 \big|_{x=x'}$ in the numerator $F[f, x, x]$ is

$$
\begin{aligned}
S_0 \big|_{x'=x} &= \frac{1}{4m\omega} \int_0^U \int_0^U e^{-\omega|u-u'|} f(u) f(u') \, du \, du' \\
&+ \frac{m\omega}{\sinh\omega U}\Big[x^2(\cosh\omega U - 1) + (A+B)(e^{\omega U} - 1)x \\
&+ \left(\frac{A^2 + B^2}{2}\right)e^{\omega U} - AB \Big].
\end{aligned}
\tag{5.3.44}
$$

The second term becomes after some rearrangement

$$\frac{m\omega}{\sinh \omega U}$$

$$\times \left\{ (\cosh \omega U - 1) \left[x + \frac{1}{2} \frac{(A+B)(e^{\omega U} - 1)}{\cosh \omega U - 1} \right]^2 - AB(1 + e^{\omega U}) \right\}.$$

$$(5.3.45)$$

The first term in the curly bracket above is x dependent which enters the integral $\int e^{-\frac{1}{\hbar} S_0|_{x=x'}} dx$ in the numerator of (5.3.42), and the result is $[\frac{\pi \hbar \sinh \omega U}{m\omega(\cosh \omega U - 1)}]^{\frac{1}{2}}$, which cancels exactly the denominator (5.3.43). Up to now terms, contributing to $\mathcal{E}[f]$ are the first term of (5.3.44) and the second term in the curly bracket (5.3.45).

$$\mathcal{E}[f] = \exp \left[-\frac{1}{4m\omega\hbar} \int_0^U \int_0^U e^{-\omega|u-u'|} f(u) f(u') \, du du' \right.$$

$$\left. + \frac{m\omega}{\hbar \sinh \omega U} (1 + e^{\omega U}) AB \right]. \qquad (5.3.46)$$

Rewriting (5.3.39) as

$$AB = -\frac{1}{4m^2\omega^2} \int_0^U \int_0^U e^{-\omega U} e^{-\omega(u-u')} f(u) f(u') \, du du' \quad (5.3.47)$$

and substituting (5.3.49) into (5.3.48), we obtain after some rearrangement

$$\mathcal{E}[f] = \exp \left[-\frac{1}{4m\omega\hbar} \int_0^U \int_0^U \frac{\cosh \left(\omega|u-u'| - \frac{\omega U}{2} \right)}{\sinh \frac{\omega U}{2}} f(u) f(u') du du' \right].$$

$$(5.3.48)$$

5.4 Instanton Method for a Decaying State

Consider the potential shown in Fig. 5.5(a) and its mechanical analog (b). If the tunnelling through the barrier is neglected, this potential has an energy eigenstate at the bottom of the well. Is it possible to calculate the correction to the energy using the path

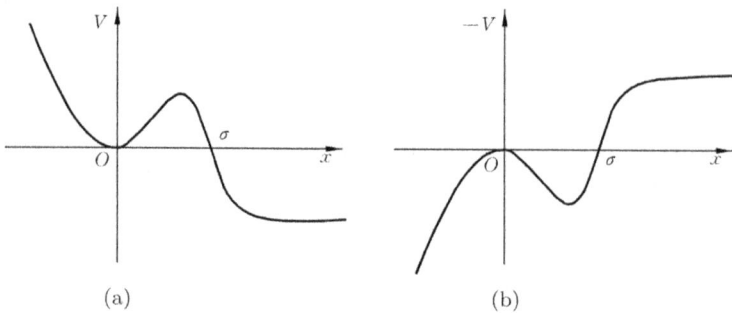

(a)   (b)

Fig. 5.6   (a) The potential; (b) The inverted potential.

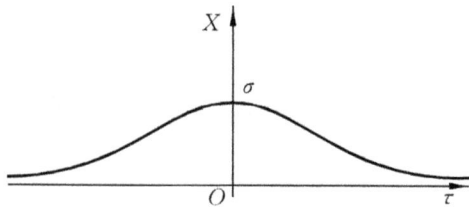

Fig. 5.7   The bounce solution of the classical path.

integral when tunneling is taken into account? From the mechanical analog we see that in the classical path the particle starts from the top of the hill at $x = 0$ in the infinite past and *bounces off* at the classical turning point σ, and return to the top of the hill, as shown in Fig. 5.6. This classical path is called *the bounce*. If we begin to calculate $\langle 0 \mid e^{-\hat{H}\tau_0/\hbar} \mid 0 \rangle$, it is probable that the result will be similar to that in the double well, viz. the sum over many bounce paths gives something like $\left(\frac{\omega}{\pi\hbar}\right)^{\frac{1}{2}} e^{-\omega\tau_0/2} e^{K\tau_0} e^{-S_0/\hbar}$ (cf. (5.2.13)), although ω and S_0 would be different from the double well case.

Actually the situation is not that simple. First, just as in the case we discussed in Section 5.2, the correction itself is smaller than what is neglected in the approximation, but the difference is that here we do not have a splitting to calculate for which the correction itself is the leading term. Second, the bounce path has a maximum where $\frac{dX}{d\tau} = 0$, i.e., $\frac{dX}{d\tau}$ has a node. From the general properties of the bound levels we see that since the zero mode solution has a node, there must be another solution without a node of still lower energy, i.e., a negative energy solution. Then the factor K proportional to $\prod \lambda_i^{-\frac{1}{2}}$ would be imaginary. Furthermore, from the potential in Fig. 5.5(a) we observe that the barrier penetration would render the states in the well unstable.

In fact, the state in the well is unstable, so that its energy has an imaginary part. This imaginary part determines the decay rate of the unstable state. The decay rate should be calculable using the path integral. If the energy correction is real, then it is not reasonable to keep it, because it is in fact smaller than what is neglected in the approximation. But as long as this correction is imaginary, that is just what we want: it gives the decay rate, and it is the leading term, because the zero order term, the energy without barrier penetration, is real. The result will turn out to be

$$\mathrm{Im} E_0 = \frac{\Gamma}{2} = \hbar |K| e^{-S_0/\hbar}, \qquad (5.4.1)$$

where Γ is the width of the energy level of the decaying state. In the calculation of the prefactor, let λ_{-1} be the negative eigenvalue of the eigenproblem and c_{-1} the coefficient of the negative eigenvalue state in the expansion of $x(\tau)$. The integral with respect to c_{-1} will be something like

$$\int \frac{dc_{-1}}{\sqrt{2\pi\hbar}} \exp\left(-\frac{1}{2\hbar} c_{-1}^2 \lambda_{-1}\right) = \int \frac{dc_{-1}}{\sqrt{2\pi\hbar}} \exp\left(\frac{1}{2\hbar} |\lambda_{-1}| c_{-1}^2\right), \qquad (5.4.2)$$

and it is divergent. From another point of view, the factor in $\prod \lambda_i^{-\frac{1}{2}}$ corresponding to the negative eigenvalue is $\frac{1}{\sqrt{\lambda_{-1}}} = \frac{i}{\sqrt{|\lambda_{-1}|}}$. Can we obtain this from (5.4.2)? Let us try analytic continuation. Assume first that λ_{-1} in the left hand side is positive, then the integral gives $\frac{1}{\sqrt{\lambda_{-1}}}$. Analytic continuation to negative values of λ_{-1} gives $\frac{i}{\sqrt{|\lambda_{-1}|}}$. Then we would have $\mathrm{Im} E_0 = \frac{\Gamma}{2} = \hbar |K| e^{-S_0/\hbar}$, only a factor of $\frac{1}{2}$ short from the correct result. This gives us a hint that the analytic continuation works, but has to be taken with more care. The well-known Callan-Coleman method [8] shows how to carry out the calculation. In order to have a closer look on the analytic

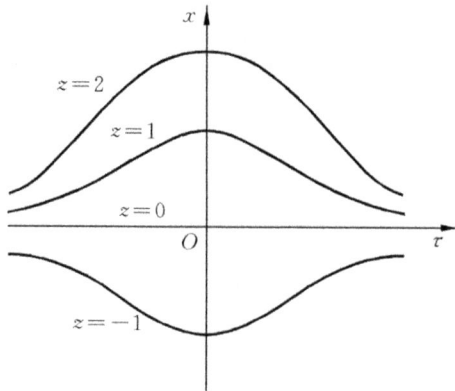

Fig. 5.8   A family of paths characterized by the parameter z.

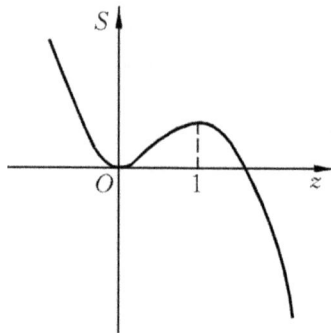

Fig. 5.9   The action of the paths for the potential in Fig. 5.6(a).

continuation, we consider only a family of histories characterized by a real parameter z in part of the function space.

$$J(z) = \int \frac{dz}{\sqrt{2\pi\hbar}} e^{-S(z)/\hbar}. \tag{5.4.3}$$

Figure 5.8 shows these paths. Path with $z = 1$ corresponds to the bounce path, and its action is a maximum. Paths with still larger values of z have smaller action because the particle spends more (imaginary) time in places with larger x values, where the potential V is more negative. The relation between the action and the parameter z of the paths is shown in Fig. 5.9. Due to the fact that $S(z)$ becomes more and more negative with increasing z the

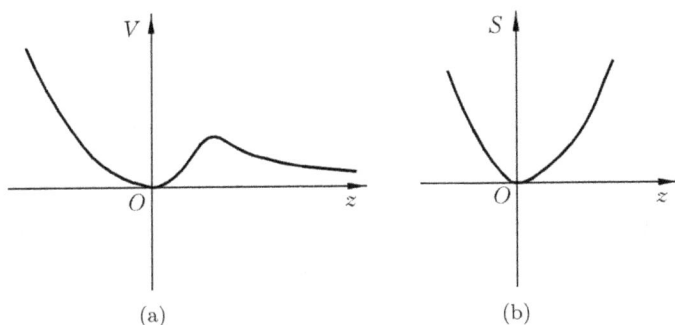

Fig. 5.10 (a) A potential giving rise to bound states; (b) The action of the paths.

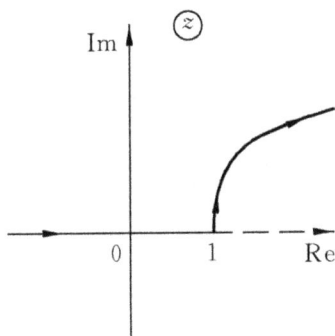

Fig. 5.11 The deformed contour of integration.

integral (5.4.3) diverges. Since such calculations are carried out for stationary states, not decaying states, the direct adaptation to a potential in Fig. 5.6(a) is problematic. A better way of calculating the integral is to begin with a potential shown in Fig. 5.10(a) and deform it gradually to the form in Fig. 5.6(a). After the action reaches its maximum for $z = 1$, the action of the real potential Fig. 5.6(a) deviates from the behavior shown in Fig. 5.10(b). As the action curve begins to bend down quickly from $z = 1$, we cannot continue to integrate along the real z axis. To obtain a finite integral with an imaginary part, we have to deform the path of integration at $z = 1$ to the complex plane. Let us assume that the deformation of the potential dictates a deformation of the path of integration upward to the upper complex plane as shown in the Fig. 5.11. The integral from $z = 1$ upward together with the integral along the real axis up to $z = 1$ gives

$$J = \left[\int_{-\infty}^{1} + \int_{1}^{1+i\infty} \right] \frac{dz}{\sqrt{2\pi\hbar}} e^{-S(z)/\hbar}.$$

The first integral gives a real number. Let $z = 1 + iy$, the second integral calculated by using the method of steepest descent gives

$$J' = i \int_{0}^{\infty} dy \frac{1}{\sqrt{2\pi\hbar}} \exp\left(- \left[S\left(1 \right) + \frac{1}{2} S''\left(1 \right) (iy)^2 \right] \Big/ \hbar \right).$$

Here $S''(y)$ is negative at $z = 1$. Hence the imaginary part of J is

$$\mathrm{Im} J = \int_{0}^{\infty} \frac{dy}{\sqrt{2\pi\hbar}} \exp\left(- \left[S\left(1 \right) - \frac{1}{2} S''\left(1 \right) y^2 \right] \Big/ \hbar \right)$$

$$= \frac{1}{2} e^{-S(1)/\hbar} \left[S''\left(1 \right) \right]^{-1/2}. \tag{5.4.4}$$

The origin of the important factor $\frac{1}{2}$ is: the Gaussian integral here is taken from 0 to ∞, which gives a value of one-half of that from $-\infty$ to ∞. One can generalize the above discussion finally to the imaginary part of K as well as to the decay rate.

$$\mathrm{Im} K = \frac{1}{2} \left(\frac{S_0}{2\pi\hbar} \right)^{1/2} \left[\frac{\det'\left(-\frac{\partial^2}{\partial \tau^2} + V''\left(X \right) \right)}{\det\left(-\frac{\partial^2}{\partial \tau^2} + \omega^2 \right)} \right]^{-1/2},$$

$$\Gamma = 2\hbar |K| e^{-S_0/\hbar} = \hbar \left(\frac{S_0}{2\pi\hbar} \right)^{1/2} e^{-S_0/\hbar} \left[\frac{\det'\left(-\frac{\partial^2}{\partial \tau^2} + V''(X) \right)}{\det\left(-\frac{\partial^2}{\partial \tau^2} + \omega^2 \right)} \right]^{-1/2}. \tag{5.4.5}$$

The Callan-Coleman method is a field theoretic adaptation of the method introduced by J. S. Langer [9] in treating a phase transition problem in statistical mechanics. A specific example on the tunnelling through the "quadratic plus cubic" potential will be given in 5.4.1, in which the imaginary part of the energy is directly computed. In 5.4.2

a method of directly computing the path integral of the second variation without resorting to the expansion in the eigenfunctions of the operator $-\frac{\partial^2}{\partial\tau^2} + V''(X)$ is given.

In the foregoing discussions the instanton method is applied to the coherence splitting and the decay by barrier penetration. This method is an approximation scheme in which the main contribution comes from the classical path. The vanishing of the first order variation of the action gives the classical path, for which the action is an extremum. The second order variation $\delta^2 S$ is the contribution from the quantum fluctuations. The use of the imaginary time facilitates the sum over histories deviating from the classical path. For imaginary time, the weight factor for the paths becomes $e^{-S/\hbar}$, which is an exponentially decreasing function when the path deviates further from the classical path. This method is sometimes referred to as the method of background fields. For more complicated problems, where the classical equation is non-linear, this method is a convenient framework for the semiclassical approximation. The quantization is carried out on the basis of the classical path, viz. the classical soliton.

It is appropriate to emphasize that the path integral method is a formulation of quantum mechanics, it does not involve any approximation, the case of the harmonic oscillator serves as an example. But it *provides* a convenient framework for making approximations. Also the quantization scheme is by no means unique. The quantization on the basis of the soliton is just an example. The transition of a quantum theory to its limiting case of classical theory is realized in a unique way. Given a classical theory, the way to construct a corresponding quantum theory is by no means unique. As L. Faddeev puts it, this involves a "deformation" of the Planck's constant. He characterized the transition from a quantum theory to its classical limit as a "contraction". The transition from the classical Poisson bracket to the quantum Poisson bracket (the commutator) is a generalization. This generalization does not work for fermions. The anticommutator has to be introduced. How can one judge the correctness of a scheme of quantization? A quantum theory must contain the classical theory as the limiting case. The theory must have internal consistency. The most important of all, the theoretical predictions of the theory using the given quantization scheme must be checked by experiments. Canonical quantization is the most familiar quantization scheme. The *correspondence principle quantization* is a variation on the theme of the canonical quantization. Let ψ be a representation of a symmetry group of the Hamiltonian. Let $\widehat{Q}$ be the generator corresponding to a conserved observable which generates a symmetry transformation. Under the symmetry transformation ψ undergoes a change $\delta\psi$ (a c-number). $\widehat{Q}$ is to be expressed in terms of the second quantized operators $\widehat{\psi}$ and $\widehat{\psi}^\dagger$ satisfying certain commutation relations. The correspondence principle quantization demands that $\delta\psi \propto [\widehat{\psi}, \widehat{Q}]$, this demand leads to the expression of $\widehat{Q}$ in terms of $\widehat{\psi}$ and $\widehat{\psi}^\dagger$. In the left hand side of the equation $\delta\psi \propto [\widehat{\psi}, \widehat{Q}]$ the quantity $\delta\psi$ is a c-number, while the $\widehat{\psi}$ in the right hand side is an operator, hence the name correspondence principle quantization. In the usual linear problems, the canonical commutation relations for $\widehat{\psi}$ and $\widehat{\psi}^\dagger$ lead to the well-known expressions for the generator $\widehat{Q}$. But in non-linear problems the usual form may be incorrect. In such cases use of the classical variation $\delta\psi$ can determine the form of the corresponding operator $\widehat{Q}$. The path integral quantization can be generalized to the field theory. The path integral quantization of the non-Abelian gauge fields by L. D. Faddeev and V. N. Popov [10] provides a natural framework unifying unitarity and renormalizability in one theoretical structure.

Imaginary time is used in many calculations of path integral. The use of $t = i\tau$ is called a Wick rotation. This is a method frequently used in theoretical physics. Some mathematicians voice their discontent, considering this process as not rigorous.

5.4.1 Penetration of the "quadratic plus cubic" potential

The decay rate of penetration of the quadratic plus cubic potential was first obtained by A. O. Caldeira (Ph.D. Thesis, University of Sussex, 1980). The result was cited in Ref. [11]. We give here details of the calculation.

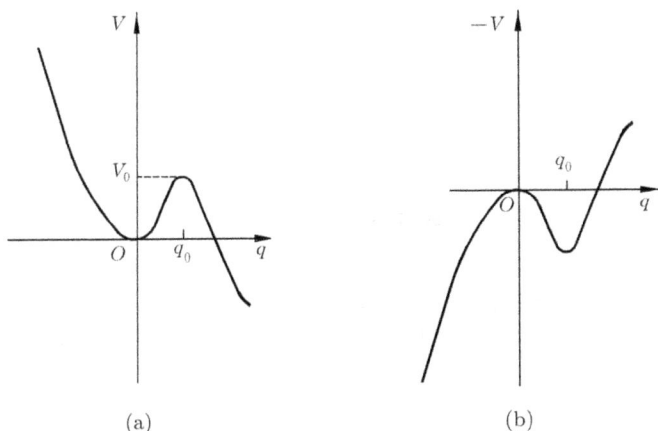

Fig. 5.12 (a) The quadratic plus cubic potential; (b) The inverted potential.

Consider the potential

$$V(q) = \frac{1}{2}m\omega^2 q^2 - \beta q^3$$

$$= \frac{27}{4}V_0\left[\left(\frac{q}{q_0}\right)^2 - \left(\frac{q}{q_0}\right)^3\right], \qquad (5.4.6)$$

shown in Fig. 5.12(a), where

$$q_0 = \frac{m\omega^2}{2\beta}, \quad V_0 = \frac{1}{54}\frac{\left(m\omega^2\right)^3}{\beta^2}.$$

The calculation of the tunneling rate consists of the usual steps dealing with the classical action, the classical path and the prefactor. From the mechanical analogy the classical path (the bounce) is $q(-\frac{\tau_0}{2}) = 0$, $q(0) = q_0$, $q(\frac{\tau_0}{2}) = 0$ for $\tau_0 \to \infty$. The classical path is the zero energy solution and therefore the classical action is

$$S_0 = \int \left(\frac{1}{2}m\dot{q}^2 + V(q)\right) d\tau$$

$$= \int m\dot{q}^2 d\tau = \int m\dot{q}dq = 2\int_0^{q_0} \sqrt{2mV(q)}dq.$$

Substituting (5.4.6) in, we get

$$S_0 = 2q_0 \left(\frac{27}{2}mV_0\right)^{1/2} \int_0^{q_0} \left[\left(\frac{q}{q_0}\right)^2 - \left(\frac{q}{q_0}\right)^3\right]^{1/2} d\left(\frac{q}{q_0}\right) = \frac{36V_0}{5\omega}.$$

$$(5.4.7)$$

The equation for the classical path is

$$\frac{1}{2}m\dot{q}^2 = \frac{27}{4}V_0\left\{\left(\frac{q}{q_0}\right)^2 - \left(\frac{q}{q_0}\right)^3\right\}.$$

The solution satisfying the boundary conditions $q(-\infty) = q(\infty) = 0$, $q(0) = q_0$ is

$$X(\tau) = q_0^2\operatorname{sech}^2\frac{1}{2q_0}\sqrt{\frac{27V_0}{2m}}\tau.$$

$$(5.4.8)$$

To calculate the prefactor we need to solve the eigenequation

$$-m\frac{d^2x_n}{d\tau^2} + V''(X)x_n = \lambda_n x_n,$$

$$(5.4.9)$$

where V'' is the second derivative of V

$$V''(q) = \frac{27V_0}{2q_0^2}\left(1 - 3\frac{q}{q_0}\right),$$

which, by using (5.4.8), becomes

$$V''(X) = \frac{27V_0}{2q_0^2}\left[1 - \frac{3}{\cosh^2\frac{1}{2q_0}\sqrt{\frac{27V_0}{2m}}\tau}\right].$$

$$(5.4.10)$$

Defining parameters

$$E_n = \frac{\lambda_n - \frac{27}{2}V_0\frac{1}{q_0^2}}{2m} \equiv \frac{\lambda_n - \omega^2}{2m},$$

$$V = \frac{81V_0}{4mq_0^2}, \quad a = 2q_0\sqrt{\frac{2m}{27V_0}}$$

$$(5.4.11)$$

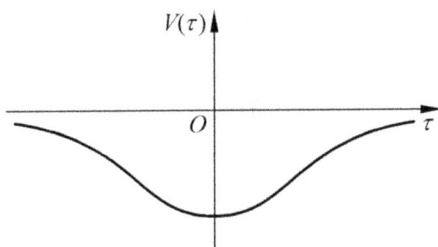

Fig. 5.13 The potential for the eigenequation.

and rearranging Eq. (5.4.9), we obtain[8]

$$\frac{d^2 x_n}{d\tau^2} + 2\left[E_n + \frac{V}{\cosh^2 \frac{\tau}{a}}\right] x_n = 0. \tag{5.4.12}$$

The "potential energy" of Eq. (5.4.12) is

$$V(\tau) = \frac{V}{\cosh^2 \frac{\tau}{a}},$$

which is shown in Fig. 5.13. The positive eigenvalues of Eq. (5.4.12) are continuous; negative eigenvalues are discrete. The negative eigenvalues are:

$$E_n = -\frac{1}{2a^2}\left[\frac{1}{2}\sqrt{8V_a^2 + 1} - \left(n + \frac{1}{2}\right)\right]^2$$

$$= -\frac{\omega^2}{8m}(3-n)^2, \quad n = 0, 1, 2, \tag{5.4.13}$$

which give the corresponding λ_n by using (5.4.11).

$$\lambda_n = \omega^2\left(1 - \frac{1}{4}(3-n)^2\right). \tag{5.4.14}$$

The three discrete eigenvalues are

$$\lambda_0 = -\frac{5}{4}\omega^2, \ \lambda_1 = 0, \ \lambda_2 = \frac{3}{4}\omega^2. \tag{5.4.15}$$

---

[8]See [12], p. 73.

For the continuous eigenvalues, we define $k = \sqrt{2mE}$ and use (5.4.11) to get

$$k = \sqrt{2mE} = \sqrt{\lambda_n - \omega^2}. \tag{5.4.16}$$

Remember the eigenequation for the harmonic oscillator problem Eq. (5.1.33)

$$-m\frac{d^2x_n}{d\tau^2} + \omega^2 x_n = \lambda_n x_n$$

with eigenvalues λ_n (5.1.34)

$$\lambda_n = \omega^2 + \frac{n^2\pi^2}{\tau_0^2}.$$

They are obtained by solving the equation with the boundary conditions $x_n\left(\pm\frac{\tau_0}{2}\right) = 0$. The corresponding values for k is

$$k = \sqrt{\lambda_n - \omega^2} = \frac{n\pi}{\tau_0}. \tag{5.4.17}$$

From this, we infer that the spectrum of k is quasi-continuous for τ_0 large but finite, the interval between levels is π/τ_0. The eigenequation (5.4.12) is to be solved by imposing boundary conditions $x_n(\pm\frac{\tau_0}{2}) = 0$. We define

$$\xi = \tanh\frac{\omega\tau}{2}. \tag{5.4.18}$$

The solution to Eq. (5.4.12) is the hypergeometric function

$$x = \left(1 - \xi^2\right)^{-i\frac{k}{\omega}} F\left[-i\frac{2k}{\omega} - 3, -i\frac{2k}{\omega} + 4, -i\frac{2k}{\omega} + 1, \frac{1}{2}(1 - \xi)\right], \tag{5.4.19}$$

where ξ is the independent variable involving τ; k is related to λ_n through (5.4.16) and is to be determined by boundary conditions; and ω is defined in (5.4.11)

$$\omega = \frac{27}{2}V_0\frac{1}{q_0^2}, \tag{5.4.20}$$

where V_0 and q_0 are parameters of the quadratic plus cubic potential. The equation (5.4.12) has two linearly independent solutions, of which only one is adopted (5.4.19), because the other is singular at the origin. It is sufficient to know the asymptotic behavior of x to apply boundary conditions for τ_0 large. For $\tau \to \pm\infty$ the limits for the independent variables are

$$\left.\begin{array}{l} \tau \to \infty, \tanh \dfrac{\omega\tau}{2} \to 1 - 2\mathrm{e}^{-\omega\tau}, \ \dfrac{1}{2}(1+\xi) \to 1, \ \dfrac{1}{2}(1-\xi) \to \mathrm{e}^{-\omega\tau} \\[2mm] \left(1 - \xi^2\right)^{-\mathrm{i}\frac{k}{\omega}} \to \mathrm{e}^{\mathrm{i}k\tau} \\[2mm] \tau \to -\infty, \tanh \dfrac{\omega\tau}{2} \to -1 + 2\mathrm{e}^{\omega\tau}, \ \dfrac{1}{2}(1-\xi) \to 1, \ \dfrac{1}{2}(1+\xi) \to \mathrm{e}^{\omega\tau} \\[2mm] \left(1 - \xi^2\right)^{-\mathrm{i}\frac{k}{\omega}} \to \mathrm{e}^{-\mathrm{i}k\tau} \end{array}\right\}.$$

$$(5.4.21)$$

The asymptotic behavior of the solutions x (5.4.19) is

$$\left.\begin{array}{l} \tau \to +\infty, \ x \to \mathrm{e}^{\mathrm{i}k\tau} F(\,,\,,\,,\,0) = \mathrm{e}^{\mathrm{i}k\tau} \\[2mm] \tau \to -\infty, \ x \to \mathrm{e}^{\mathrm{i}k\tau} F(\,,\,,\,,\,1) \end{array}\right\}.$$

$$(5.4.22)$$

Using formula[9]

$$F\left(\alpha, \beta, \gamma, \frac{1}{2}(1-\xi)\right) = \frac{\Gamma(\gamma)\,\Gamma(\gamma - \alpha - \beta)}{\Gamma(\gamma - \alpha)\,\Gamma(\gamma - \beta)}$$

$$\times F\left[\alpha, \beta, \alpha + \beta + 1 - \gamma; \frac{1}{2}(1+\xi)\right]$$

$$+ \frac{\Gamma(\gamma)\,\Gamma(\alpha + \beta - \gamma)}{\Gamma(\alpha)\,\Gamma(\beta)} \left[\frac{1}{2}(1+\xi)\right]^{\gamma - \alpha - \beta}$$

$$\times F\left[\gamma - \alpha, \gamma - \beta, \gamma + 1 - \alpha - \beta, \frac{1}{2}(1+\xi)\right],$$

$$(5.4.23)$$

it is possible to connect the solutions at $\tau \to \infty$ and $\tau \to -\infty$. In the solutions (5.4.19) $\gamma - \beta = 3$ and $\Gamma(-3) = \infty$. Hence, the first term

---

[9]See [12], Appendix E, p. 659.

in (5.4.23) vanishes. For $\tau \to -\infty$,

$$F\left(-i\frac{2k}{\omega} - 3, -i\frac{2k}{\omega} + 4, -i\frac{2k}{\omega} + 1, 1\right)$$

$$= -\frac{\left(1 + i\frac{2k}{\omega}\right)\left(1 + i\frac{k}{\omega}\right)\left(1 + i\frac{2k}{3\omega}\right)}{\left(1 - i\frac{2k}{\omega}\right)\left(1 - i\frac{k}{\omega}\right)\left(1 - i\frac{2k}{3\omega}\right)} e^{i2k\tau}, \tag{5.4.24}$$

i.e., for $\tau \to \infty$, we have

$$x = -e^{ik\tau}\frac{\left(1 + i\frac{2k}{\omega}\right)\left(1 + i\frac{k}{\omega}\right)\left(1 + i\frac{2k}{3\omega}\right)}{\left(1 - i\frac{2k}{\omega}\right)\left(1 - i\frac{k}{\omega}\right)\left(1 - i\frac{2k}{3\omega}\right)}. \tag{5.4.25}$$

The first solution of (5.4.22) gives $x = e^{ik\tau}$ for $\tau \to \infty$, therefore x should be $x = e^{ik\tau}e^{i\delta_k}$ for $\tau \to -\infty$, where δ_k is the scattering phase shift. Equation (5.4.25) gives

$$e^{i\delta_k} = -\frac{\left(1 + i\frac{2k}{\omega}\right)\left(1 + i\frac{k}{\omega}\right)\left(1 + i\frac{2k}{3\omega}\right)}{\left(1 - i\frac{2k}{\omega}\right)\left(1 - i\frac{k}{\omega}\right)\left(1 - i\frac{2k}{3\omega}\right)}. \tag{5.4.26}$$

Equation (5.4.12) is symmetric with respect to $\tau \to -\tau$, i.e., the propagation in $-\tau$ is the same as propagation in τ. From the above discussions we infer that there are two linearly independent solutions for the continuous states of the equation

$$x_k(\tau) \to \begin{cases} e^{ik\tau}, & \tau \to +\infty \\ e^{ik\tau + i\delta_k}, & \tau \to -\infty \end{cases} \tag{5.4.27}$$

$$x_k(\tau) \to \begin{cases} e^{-ik\tau}, & \tau \to +\infty \\ e^{-ik\tau + i\delta_k}, & \tau \to -\infty \end{cases} \tag{5.4.28}$$

and the general solution is

$$x_k = Ax_k(\tau) + Bx_k(-\tau). \tag{5.4.29}$$

Boundary conditions give

$$Ax_k\left(\frac{\tau_0}{2}\right) + Bx_k\left(-\frac{\tau_0}{2}\right) = 0,$$

$$Ax_k\left(-\frac{\tau_0}{2}\right) + Bx_k\left(\frac{\tau_0}{2}\right) = 0. \tag{5.4.30}$$

The condition for the existence of nonvanishing A and B is

$$\frac{x_k\,(\tau_0/2)}{x_k\,(-\tau_0/2)} = \pm 1. \tag{5.4.31}$$

For sufficiently large τ_0, equations (5.4.31), (5.4.27) and (5.4.28) give

$$\exp\left(ik\frac{\tau_0}{2}\right) = \pm \exp\left(-ik\frac{\tau_0}{2} + i\delta_k\right),$$

i.e.,

$$e^{ik\tau_0 - i\delta_k} = \pm 1,$$

$$k\tau_0 - \delta_k = n\pi, \quad n = 0,1,2,\cdots. \tag{5.4.32}$$

The n-th eigenvalue $\tilde{k}_n{}^{10}$ is

$$\tilde{k}_n = \frac{n\pi + \delta_k}{\tau_0}. \tag{5.4.33}$$

The corresponding λ_n is

$$\lambda_n = \tilde{k}_n^2 + \omega^2. \tag{5.4.34}$$

In the following, we calculate for the discrete eigenvalues the ratio $\frac{\det'(-\partial_\tau^2 + V'')}{\det(-\partial_\tau^2 + \omega^2)}$:

$$\frac{-\frac{5}{4}\omega^2 \cdot \frac{3}{4}\omega^2}{\omega^2 \cdot \omega^2 \cdot \omega^2} = -\frac{15}{16\omega^2}. \tag{5.4.35}$$

The zero eigenvalue is to be excluded in $\det'$. For the quasi-continuous spectrum, we have

$$\prod_{n=1}^{\infty} \frac{\omega^2 + \tilde{k}_n^2}{\omega^2 + k_n^2} = \exp\left[\sum_n \ln\left(1 + \frac{\tilde{k}_n^2 - k_n^2}{\omega^2 + k_n^2}\right)\right]. \tag{5.4.36}$$

The sum on the exponential can be calculated as follows. For τ_0 sufficiently large we have $\tilde{k}_n^2 - k_n^2 \approx 2k_n(\tilde{k}_n - k_n)$, and also

---

[10] The tilde serves to distinguish from the eigenvalues for the harmonic oscillator.

$\tilde{k}_n - k_n = \frac{\delta_k}{\tau_0}$, consequently

$$\ln\left(1 + \frac{\tilde{k}_n^2 - k_n^2}{\omega^2 + k_n^2}\right) \approx \frac{2k_n\delta_k}{\tau_0\left(\omega^2 + k_n^2\right)}.$$

From this we have

$$\sum_n \ln\left(1 + \frac{\tilde{k}_n^2 - k_n^2}{\omega^2 + k_n^2}\right) \approx \frac{\tau_0}{\pi}\int_0^\infty dk\frac{2k\delta_k}{\tau_0\left(\omega^2 + k_n^2\right)}$$

$$= \frac{1}{\pi}\int_{k=0}^{k=\infty} \delta_k d\ln\left(\omega^2 + k^2\right). \qquad (5.4.37)$$

The above quantity is independent of τ_0. After integrating by parts and using (5.4.26), it becomes $\ln\frac{4}{225}$. Therefore (5.4.36) is just $\frac{4}{225}$. Now include the contribution of (5.4.35), the value for $\frac{\det'}{\det}$ is $-\frac{15}{16\omega^2}\frac{4}{225} = -\frac{1}{60\omega^2}$. Therefore,

$$\left[\frac{\det'\left(-\partial_\tau^2 + V''\left(X\right)\right)}{\det\left(-\partial_\tau^2 + \omega^2\right)}\right]^{-\frac{1}{2}} = i\sqrt{60}\omega,$$

The width Γ is given by (5.4.5):

$$\Gamma = \sqrt{60}\omega\left(\frac{S}{2\pi\hbar}\right)^{\frac{1}{2}}e^{-S_0/\hbar}$$

$$= \sqrt{60}\omega\left(\frac{18V_0}{5\pi\hbar\omega}\right)^{\frac{1}{2}}e^{-36V_0/5\hbar\omega}. \qquad (5.4.38)$$

5.4.2 Shifting method for evaluating the second order variation in path integrals

This method makes use of a mapping to evaluate the contribution of the second order variation to the path integral. The method was first used by R. Dashan, B. Hasslacher and A. Neveu [13] in field theory in Minkowski space, and was adapted by J.-Q. Liang and H. J. W. Müller-Kirsten [14] to the quantum tunneling problem using imaginary time. The following is a brief introduction to the method. An arbitrary path x is written as $x = x_c + y$, where x_c is the classical path. Then the contribution of the second order variation to the path integral is

$$I = \int [Dy]\, e^{-\delta^2 S_E} = \int [Dy] \exp\left(-\int_{-T}^{T} y \hat{M} y \mathrm{d}\tau\right), \qquad (5.4.39)$$

where

$$\hat{M} = -\frac{1}{2}\frac{\mathrm{d}^2}{\mathrm{d}\tau^2} + V''(x_c). \qquad (5.4.40)$$

The boundary condition for y is

$$y(\pm T) = 0.$$

A mapping $y(\tau) \to z(\tau)$ is introduced,

$$z(\tau) = y(\tau) - \int_{-T}^{T} \frac{\dot{N}(\tau')}{N(\tau')} y(\tau')\, \mathrm{d}\tau', \qquad (5.4.41)$$

where $N(\tau)$ satisfies the equation

$$\left[\frac{\mathrm{d}^2}{\mathrm{d}\tau^2} - 2V''(x_c)\right] N(\tau) = 0, \qquad (5.4.42)$$

i.e., $N(\tau) = \frac{\mathrm{d}x_c}{\mathrm{d}\tau}$ is the zero mode solution of $\hat{M}$. From (5.4.41) we obtain

$$z(-T) = y(-T) = 0.$$

The other boundary condition for z has to be found with some effort. Equation. (5.4.41) together with the conditions $y(-T) = 0$, $z(-T) = 0$ leads to an *inverse* mapping $z \to y$ [13]:

$$y(\tau) = z(\tau) + N(\tau) \int_{-T}^{T} \frac{\dot{N}(\tau')}{N^2(\tau')} z(\tau')\, \mathrm{d}\tau', \qquad (5.4.43)$$

which leads to

$$z(T) + N(T) \int_{-T}^{T} \frac{\dot{N}(\tau')}{N^2(\tau')} z(\tau')\, \mathrm{d}\tau' = 0. \qquad (5.4.44)$$

This is a constraint for $z(T)$, which depends on $z(\tau)$ from $\tau = -T$ to $\tau = T$. The simplification that the mapping brings us is the second order variation of the action becomes[11]

$$\delta^2 S_{\mathrm{E}} = \frac{1}{2} \int_{-T}^{T} z^2(\tau) \mathrm{d}\tau. \tag{5.4.45}$$

As the price to pay, $z(T)$ is under constraint. The constraint can be taken care of by the Lagrange multiplier. Denoting $N(\tau) = \dot{x}_{\mathrm{c}}(\tau)$, the integral I (5.4.39) can be rewritten as

$$I = \int [\mathrm{D}z] \, \mathrm{d}\alpha \left| \frac{\partial y}{\partial z} \right|$$

$$\times \exp\left[-\int_{-T}^{T} \mathrm{d}\tau \left\{ \frac{1}{2}\dot{z}^2 + \alpha\{z(T) + \dot{x}_{\mathrm{c}}(T) \int_{-T}^{T} \frac{\ddot{x}_{\mathrm{c}}}{\dot{x}_{\mathrm{c}}^2(\tau')} z(\tau') \mathrm{d}\tau'\} \right\} \right], \tag{5.4.46}$$

where $\left| \frac{\partial y}{\partial z} \right|$ is the Jacobian of the mapping. The exponential contains the Lagrange multiplier α, exhibiting the constraint on $N(T)$. After a much involved integration, I can be calculated directly

$$I = \frac{1}{\sqrt{2\pi}} [\dot{x}_{\mathrm{c}}(-T)\, \dot{x}_{\mathrm{c}}(T)]^{-1/2} \left[\int_{-T}^{T} \frac{\mathrm{d}\tau}{\dot{x}_{\mathrm{c}}^2(T)} \right]^{-1/2}. \tag{5.4.47}$$

An important implication is ready in (5.4.47): for the kink solution $\dot{x}_{\mathrm{c}}(-T) = \dot{x}(T)$, hence the factor in the first bracket is positive, and for the bounce solution $\dot{x}_{\mathrm{c}}(-T) = -\dot{x}_{\mathrm{c}}(T)$, this factor is negative, and I is imaginary.

5.4.3   Alternative Evaluation of the Prefactor for the Harmonic Oscillator

In the following we follow Ref. [7].

The Lagrangian of a harmonic oscillator in the Euclidean 1+1 space time is

$$L = \frac{1}{2} \left(\frac{\mathrm{d}x}{\mathrm{d}\tau} \right)^2 + \frac{1}{2}\omega^2 x^2. \tag{5.4.48}$$

---

[11]See Appendix B of Ref. [15].

From the potential energy we get

$$V'' = \omega^2, \tag{5.4.49}$$

which is a constant. The equation for the classical path is

$$-\frac{d^2 X}{d\tau^2} + \omega^2 X = 0.$$

For the purpose of illustration we begin with a simple case: $X(\tau) = 0$ for any τ. The classical oscillator is stationary at the origin. The classical action is

$$S_0 = 0.$$

The prefactor can be calculated by considering the equation for x_n

$$\left(-\frac{d^2}{d\tau^2} + \omega^2\right) x_n = \lambda_n x_n. \tag{5.4.50}$$

The boundary conditions are

$$x_n\left(-\frac{\tau_0}{2}\right) = 0, \ x_n\left(\frac{\tau_0}{2}\right) = 0.$$

The boundary conditions lead to the eigenvalues of this equation[12]

$$\lambda_n = \omega^2 + \frac{n^2\pi^2}{\tau_0^2} = \frac{n^2\pi^2}{\tau_0^2}\left(1 + \frac{\omega^2\tau_0^2}{n^2\pi^2}\right), \ n = 1,2,\cdots \tag{5.4.51}$$

and

$$\prod_n \lambda_n^{-1/2} = \left(\prod_n \frac{n^2\pi^2}{\tau_0^2}\right)^{-1/2}\left[\prod_n\left(1 + \frac{\omega^2\tau_0^2}{n^2\pi^2}\right)\right]^{-1/2}.$$

---

[12]The solutions of (5.A.45) and the corresponding eigenvalues are:

$$x_n(\tau) = \cos\sqrt{\lambda_n - \omega^2}\tau, (\lambda_n - \omega^2)^{\frac{1}{2}} = (2n+1)\frac{\pi}{\tau_0}, \ n = 0,1,2,3\cdots.$$

$$x_n(\tau) = \sin\sqrt{\lambda_n - \omega^2}\tau, (\lambda_n - \omega^2)^{\frac{1}{2}} = 2n\frac{\pi}{\tau_0}, \ n = 1,2,3\cdots.$$

The spectrum of eigenvalues is given by $\lambda_n = \omega^2 + \frac{n^2\pi^2}{\tau_0^2}, n = 1,2,3\cdots.$

The two factors of the right hand side of this equation can be separately evaluated. Consider the special case of a free particle, $\omega = 0$. From (5.1.34) we have $\lambda_n = \frac{n^2\pi^2}{\tau_0^2}$, hence

$$\left\langle 0 \left| \exp\left(-\frac{p^2}{2\hbar}\tau_0\right) \right| 0 \right\rangle = N \left(\prod_n \frac{n^2\pi^2}{\tau_0^2} \right)^{-1/2}.$$

The left-hand side can be evaluated directly. By using (5.1.11) we have

$$\left\langle 0 \left| \exp\left(-\frac{p^2}{2\hbar}\tau_0\right) \right| 0 \right\rangle = \sum_n \exp\left(-\frac{p_n^2}{2\hbar}\tau_0\right) \psi_n(0)\, \psi_n^*(0)$$

$$= \sum_n \exp\left(-\frac{p_n^2}{2\hbar}\tau_0\right)$$

$$= \int_{-\infty}^{\infty} \frac{dp}{2\pi} e^{-p^2\tau_0/2\hbar} = \frac{1}{\sqrt{2\pi\tau_0/\hbar}}.$$

since $\psi_n(0) = 1$ for a free particle. Consequently,

$$N \left(\prod_n \frac{n^2\pi^2}{\tau_0^2} \right)^{-1/2} = \frac{1}{\sqrt{2\pi\tau_0/\hbar}}.$$

The second factor can be calculated by using the formula[13]

$$\pi y \prod_n \left(1 + \frac{y^2}{n^2}\right) = \sinh \pi y.$$

The two factors are brought together to give

$$\left\langle 0 \left| e^{-\hat{H}\tau_0/\hbar} \right| 0 \right\rangle = \left(\frac{\omega}{\pi\hbar}\right)^{1/2} (2\sinh \omega\tau_0)^{-1/2}. \qquad (5.4.52)$$

---

[13]See I. S. Gradshtein and I. M. Ryzhik, *"Table of Integrals, Series and Products,"* New York, Academic, 1968. Just for fun, we can consider the left-hand side as a polynomial of infinite degree with zeros at $\pm in/\pi$, $n = 0, 1, 2, \ldots$. The right-hand side has the same zeros. Being an analytic function on the finite complex plane, both sides must be equal.

For large values of τ_0, using the expansion

$$(2\sinh\omega\tau_0)^{-1/2} = e^{\omega\tau_0/2}\left(1 - e^{-2\omega\tau_0}\right)^{1/2}$$

$$\approx e^{-\omega\tau_0/2}\left(1 + \frac{1}{2}e^{-2\omega\tau_0} + \cdots\right),$$

we have

$$\lim \sum_n \exp\left(-\frac{E_n}{\hbar}\tau_0\right)|\langle 0|n\rangle|^2$$

$$= \left(\frac{\omega}{\pi\hbar}\right)^{1/2} e^{-\omega\tau_0/2}\left(1 + \frac{1}{2}e^{-2\omega\tau_0} + \cdots\right). \tag{5.4.53}$$

Comparing the right hand side of the Eq. (5.1.36) with that of (5.1.11), we obtain the energy eigenvalue of the lowest even states $(n = 0, 2)$ and the corresponding eigenfunctions:

$$E_0 = \frac{\hbar\omega_0}{2}, \quad \psi_0^2(0) = \left(\frac{\omega}{\hbar\pi}\right)^{1/2}, \tag{5.4.54}$$

$$E_0 = \frac{5\hbar\omega_0}{2}, \quad \psi_2^2(0) = \frac{1}{2}\left(\frac{\omega}{\hbar\pi}\right)^{1/2}. \tag{5.4.55}$$

The odd parity states do not appear, because we have calculated the path integral in a special case $x_i = x_j = 0$.

References

[1] R. P. Feynman, *Rev. Mod. Phys.* 20, 367, 1948.
[2] R. P. Feynman and A. R. Hibbs, *Quantum Mechanics and Path Integrals*, New York, McGraw Hill, 1965.
[3] R. P. Feynman, *Statistical Mechanics*, Reading, W. A. Benjamin, 1972.
[4] M. S. Swanson, *Path Integrals and Quantum Processes*, Boston, Academic, 1992.
[5] H. Kleinert, *Path Integrals in Quantum Mechanics, Statistics and Polymer Science*, Singapore, World Scientific, 1990.
[6] S. Coleman, The Use of Instantons, in *The Whys of Subnuclear Physics*, edited by A. Zichichi, New York, Plenum Press, 1977.
[7] A. I. Vainshtein, V. I. Zakharov, V. A. Novikov and M. A. Shifman, *Sov. Phys. Uspekhi* 25, 195, 1982.
[8] C. Callen and S. Coleman, *Phys. Rev. D* 16, 1762, 1977.

[9] J. S. Langer, *Ann. Phys. (N.Y.)* 108, 41, 1967.

[10] L. D. Faddeev and V. N. Popov, *Phys. Lett.* 25B, 29, 1967.

[11] A. O. Caldeira and A. J. Leggett, *Ann. Phys. (N.Y.)* 149, 374, 1983.

[12] L. D. Landau and E. M. Lifshitz, *Quantum Mechanics*, 3rd edition, Oxford, Pergamon Press, 1977.

[13] R. F. Dashan, B. Hasslacher and A. Neveu, *Phys. Rev. D* 16, 1762, 1974.

[14] J. Q. Liang and H. J. W. Müller-Kirsten, *Phys. Rev. D* 46, 4685, 1992.

[15] J. Q. Liang and H. J. W. Müller-Kirsten, *Phys. Rev. D* 50, 6519, 1994.

Chapter 6

Quantum Mechanics on the Macroscopic Level

In Chapter 4 we concentrated on the relation between quantum and classical mechanics, the main idea being the applicability of quantum mechanics to macroscopic systems as the more general theory. As long as the quantum mechanical principle of superposition is not observed in macroscopic systems, we must begin with quantum mechanics and find a mechanism invalidating the superposition principle. This is the decoherence due to interaction between the system and the environment. But on the other hand under special conditions phase coherence can also survive in macroscopic systems.

In quantum mechanics a many-particle system is described by a wave function of coordinates of all the particles. There exist very exceptional cases in which system of macroscopic number of particles is described by a single particle wave function. There is only one position vector as the argument of the wave function. Not only is the number of particles macroscopic, but also the spatial region in which the wave function is appreciably non-vanishing has a macroscopic scale. That means the particles are in a high degree of coherence and consequently the system possesses very characteristic properties. The system can be the Cooper pairs of superconductivity or a superfluid. For the former the electrical resistance vanishes and the superconductor expels magnetic flux in the Meissner effect, for the latter the viscosity vanishes and the thermal conductivity becomes infinite. These are "super" properties. The phenonmena are macroscopic. There are macroscopic systems which exhibit pure quantum mechanical properties, viz. the tunneling of barrier and phase coherence. One example of such systems is the Josephson junction exhibiting the quantum tunneling phenomena. In order to show quantum mechanical properties, a macroscopic system must overcome the influence of environment, because of the tendency of the latter to destroy phase coherence. Both superconductivity and superfluidity are associated with a quantum statistical phenomena — the Bose-Einstein condensation. This will be discussed in Chapter 10. Another example is the macroscopic quantum phenomena of magnetism. The giant spin of single domain ferromagnetic and antiferromagnetic particles can tunnel through the barrier formed from the magnetic anisotropy, and it is possible to have macroscopic quantum coherence on the basis of this tunneling. These phenomena concern the basics of quantum mechanics and are closely connected with problems of practical significance, the information storage.

There have been significant developments in the quantum mechanics of dissipative systems in recent years. An important method in this direction (Feynman and Vernon, Leggett) treats the environment as an assembly of large number of harmonic oscillators, interacting with the system. Writing down the path integral for system-environment complex and integrating over the coordinates of the environment, one gets an effective theory of the system interacting with the environment including the dissipative effect.

6.1   Wave Function with a Macroscopic Significance

In quantum mechanics the wave function of a many-particle system $\psi(\boldsymbol{r}_1, \boldsymbol{r}_2, \ldots, \boldsymbol{r}_N, t)$ depends on the coordinates of the particles. When the number of particles begins to grow, the problem gets more and more complicated. In the early years, quantum mechanics is considered as a study of microscopic systems. Since the discovery of superfluidity F. London was the first to point out that the essence of such a phenomenon is the population of a macroscopic number of bosonic atoms in the ground state of the system, thus forming a Bose-Einstein condensate. In his theory of weakly interacting Bose gas, N. Bogoliubov first introduced the concept of the wave function of the condensate[1]

$$\psi\left(\boldsymbol{r}, t\right) = \sqrt{\rho\left(\boldsymbol{r}, t\right)} e^{i\theta(\boldsymbol{r},t)}, \tag{6.1.1}$$

where ρ is the density of the condensate atoms. The wave function is written as a product of the amplitude and the phase factor, where θ is the phase. Notice that ψ (6.1.1) is a single particle wave function, to be distinguished from the many-particle wave function $\psi(\boldsymbol{r}_1, \boldsymbol{r}_2, \ldots, \boldsymbol{r}_N,\ t)$. Since ρ is a macroscopic quantity, $|\psi|^2 = \rho$ has already a macroscopic significance. The current density calculated from the wave function (6.1.1) is

$$\boldsymbol{j} = \frac{i\hbar}{2m}\left(\psi\nabla\psi^* - \psi^*\nabla\psi\right) = \frac{\hbar}{m}\rho\nabla\theta. \tag{6.1.2}$$

Here ρ and $\boldsymbol{j}$ are measurable macroscopic quantities. The gradient of the phase can be measured. With

$$\boldsymbol{j} = \rho\boldsymbol{v}_{\mathrm{s}}, \tag{6.1.3}$$

where $\boldsymbol{v}_{\mathrm{s}}$ is the superfluid velocity, we have

$$\boldsymbol{v}_{\mathrm{s}} = \frac{\hbar}{m}\nabla\theta. \tag{6.1.4}$$

The wave function ψ describes the properties of the entire condensate, which is a coherent aggregate of macroscopic number of atoms.

L. D. Landau published his theory of phase transition in 1937. Consider the ferromagnetic phase transition as an example. Let $\boldsymbol{S}_i$ be the spin at the lattice site i of a cubic lattice. The Hamiltonian of the system is[2]

$$H = -J\sum_{\langle i,j \rangle} \boldsymbol{S}_i \cdot \boldsymbol{S}_j, \tag{6.1.5}$$

---

[1]See Section 10.4.
[2]See Section 7.1.

where J is the exchange energy, $\langle i, j \rangle$ specifies that the sum is to be taken over nearest neighbors. The operator of magnetization of the system is defined as[3]

$$M = \frac{1}{V} \sum_i S_i. \tag{6.1.6}$$

where V is the volume of the sample. Phase transition occurs at a temperature T_c which is of the order $O(\frac{J}{k})$. For $T < T_c$ the spontaneous magnetization $\langle M \rangle \neq 0$, and for $T > T_c$ the thermal fluctuation introduces disorder, and $\langle M \rangle = 0$. $\langle M \rangle$ is the order parameter of the transition. To explain the phenomenon, Landau gave the density of the free energy of the system

$$f(\langle M \rangle, T) = f(0, T) + \alpha(T) \langle M \rangle^2 + \frac{\beta(T)}{2} \langle M \rangle^4. \tag{6.1.7}$$

and prescribed the behavior of α and β at temperatures close to the critical temperature T_c,

$$\alpha(T) = \alpha_0 \left(\frac{T}{T_c} - 1 \right),$$
$$\beta(T) = \beta_0. \tag{6.1.8}$$

Above T_c, both α and β are positive, and the free energy acquires a minimum for $\langle M \rangle = 0$. This is the paramagnetic phase (no spontaneous magnetization). When $T < T_c$, $\beta > 0$, $\alpha < 0$. The free energy as a function of $\langle M \rangle$ behaves as shown in Fig. 6.1. The minima of the free energy occurs at $|\langle M \rangle| = \sqrt{-\alpha(T)/\beta(T)}$, giving rise to a non-vanishing spontaneous magnetization

$$|\langle M \rangle| = \sqrt{\frac{\alpha_0}{\beta_0} \left(1 - \frac{T}{T_c} \right)}. \tag{6.1.9}$$

---

[3]The magnetic moment is put equal to 1.

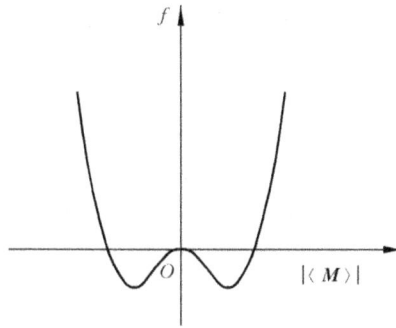

Fig. 6.1   Relation between the free energy and $|\langle \boldsymbol{M} \rangle|$ for $T < T_{\mathrm{c}}$.

In determining the minima for the free energy, we can only determine the magnitude of $\langle \boldsymbol{M} \rangle$, not its direction. This degeneracy is due to the rotational invariance of the Hamiltonian (6.1.5). But when the system adopts a definite direction of $\langle \boldsymbol{M} \rangle$ in equilibrium, this state does not respect the rotational invariance of the system. The breakdown of symmetry of the system by its ground state is termed a *spontaneous breakdown of symmetry*.

V. L. Ginzburg and L. D. Landau applied the theory of phase transition to superconductivity and put forward a phenomenological theory in 1950. This was at a time when the microscopic mechanism of superconductivity has not yet been discovered.[4] Let ψ be the order of parameter of the superconductor. The Ginzburg-Landau free energy is written as

$$F = F_n + \int \left\{ \frac{\hbar^2}{2m^*} |\nabla\psi|^2 + a\,|\psi|^2 + \frac{1}{2}b\,|\psi|^4 \right\} \mathrm{d}^3x, \qquad (6.1.10)$$

where F_n is the free energy in the normal state (i.e., a state with $\psi = 0$). As in the Landau theory of phase transition, $b > 0$ and a is written as

$$a = \alpha\,(T - T_{\mathrm{c}}), \qquad \alpha > 0. \qquad (6.1.11)$$

---

[4]The concept of Cooper pairs came into being in 1956.

The first term in the bracket of (6.1.10) depends on the spatial variation of the order parameter, and its gradient gives rise to the kinetic energy, and m^* is a parameter with the meaning of effective mass. In the study of superconductivity, external magnetic field is applied, with $\boldsymbol{B} = \nabla \times \boldsymbol{A}$, the free energy is subjected to modifications. First, the energy density of the magnetic field $\boldsymbol{B}^2/8\pi$ is to be added. Second, the gauge invariance of the theory demands the substitution

$$\nabla \to \nabla - \frac{ie^*}{\hbar c}\boldsymbol{A}, \tag{6.1.12}$$

where e^* is the effective charge parameter. Consequently we have

$$F = F_{n0} + \int \left\{ \frac{\boldsymbol{B}^2}{8\pi} + \frac{\hbar^2}{2m^*}\left|\left(\nabla - \frac{ie^*}{\hbar c}\boldsymbol{A}\right)\psi\right|^2 + a\,|\psi|^2 + \frac{1}{2}b\,|\psi|^4 \right\} \mathrm{d}^3x, \tag{6.1.13}$$

where F_{n0} is the free energy of the normal state without the magnetic field. In determining the minimum of the free energy $\delta F = 0$ the variation of ψ and ψ^* must be taken independently since ψ is a complex function. Taking variations of ψ and ψ^* leads to the same result, namely

$$\frac{1}{2m^*}\left(-i\hbar\nabla - \frac{e^*}{c}\boldsymbol{A}\right)^2\psi + a\psi + b\,|\psi|^2\,\psi = 0. \tag{6.1.14}$$

The variation of $\boldsymbol{A}$ gives

$$\nabla \times \boldsymbol{B} = \frac{4\pi}{c}\boldsymbol{j}, \tag{6.1.15}$$

where

$$\boldsymbol{j} = -\frac{ie^*\hbar}{2m^*}\left(\psi^*\nabla\psi - \psi\nabla\psi^*\right) - \frac{e^{*2}}{m^*c}\,|\psi|^2\,\boldsymbol{A}. \tag{6.1.16}$$

From (6.1.15) it follows that $\nabla \cdot \boldsymbol{j} = 0$. Equations (6.1.14)~ (6.1.16) are the Ginzburg-Landau equations. This system of equations possess simple solutions corresponding to the symmetrical and spontaneously broken symmetry cases discussed in the beginning of this paragraph:

(1) $\psi = 0$ and $\boldsymbol{B} = \nabla \times \boldsymbol{A}$. This is the normal state solution.

(2) $\psi = \psi_0 \equiv [-a(T)/b(T)]^{1/2}$, $\boldsymbol{A} = 0$. This is the solution with spontaneously broken symmetry, and ψ_0 is the constant order of parameter in the bulk for $T < T_c$. Surface effect is neglected here. Close to the surface, we consider the spatial variation of ψ with z, the coordinate perpendicular to the surface. Neglecting $\boldsymbol{A}$ and putting $\psi(z) = \psi_0 \phi(z)$, we have the equation for ϕ:

$$-\frac{\hbar^2}{2m^*a(T)}\frac{\partial^2 \phi}{\partial z^2} + \phi - \phi^3 = 0. \tag{6.1.17}$$

The parameter characterizing the spatial variation of ψ is called the coherent length ξ, defined by

$$\xi^2 = -\frac{\hbar^2}{2m^*a(T)}. \tag{6.1.18}$$

The equation for ϕ is solved with the boundary condition $\phi = 0$ at $z = 0$ by

$$\phi(z) = \tanh\frac{z}{\sqrt{2}\xi(T)}, \quad z \geq 0. \tag{6.1.19}$$

It is apparent that ψ reaches the bulk value ψ_0 within a distance of order ξ. This parameter also characterizes the distance for which ψ can deviate appreciably from the bulk value ψ_0. For this reason it is also called the healing length. Write ψ as

$$\psi(\boldsymbol{r}) = \sqrt{\rho(\boldsymbol{r})}e^{i\theta(\boldsymbol{r})} \tag{6.1.20}$$

Substituting (6.1.20) into (6.1.16) we get

$$\boldsymbol{j} = \frac{\hbar\rho}{m^*}\left[\nabla\theta - \frac{e^*}{\hbar c}\boldsymbol{A}\right]. \tag{6.1.21}$$

In the application of the theory to specific problems of superconductivity, physicists found that the parameters in the theory assume certain values if agreement with experiments is required: the effective mass and charge are respectively $m^* = 2m$ and $e^* = 2e$, where m and e are the mass and charge of the electron respectively, and ρ is one-half of the density of superconducting electrons. For some time the microscopic significance of the order parameter ψ was not known. Only after the Bardeen-Cooper-Schrieffer theory of superconductivity was accepted by physicists and L. P. Gorkov [3] gave a microscopic derivation of Ginzburg-Landau equations, the significance of the order parameter ψ becomes known: it is the wave function of the Bose-Einstein condensates of the Cooper pairs. The macroscopic quantum quantities ρ, $\boldsymbol{j}$, $\nabla\theta$ are physical observables. Similar to (6.1.3), we have here

$$\boldsymbol{j} = \rho \boldsymbol{v} \qquad (6.1.22)$$

$$m^* \boldsymbol{v} = \hbar \nabla \theta - \frac{e^*}{c} \boldsymbol{A} \qquad (6.1.23)$$

$\hbar \nabla \theta$ is equivalent to the canonical momentum. Equation (6.1.20) is invariant under the gauge transformation

$$\boldsymbol{A} \to \boldsymbol{A} + \nabla \chi$$
$$\theta \to \theta + \frac{e^*}{\hbar c} \chi \qquad (6.1.24)$$

Equation (6.1.23) has important implications. In bulk superconductors $\rho = \text{const}$, hence the equation of continuity is $\nabla \cdot \boldsymbol{j} = 0$. Using the Coulomb gauge $\nabla \cdot \boldsymbol{A} = 0$, and taking divergence of (6.1.23), we have

$$\nabla^2 \theta = 0. \qquad (6.1.25)$$

In a bulk material (6.1.25) can only be satisfied by

$$\theta = \text{const} \qquad (6.1.26)$$

Then (6.1.21) becomes

$$\boldsymbol{j} = -\frac{e^*}{m^* c} \rho \boldsymbol{A}. \qquad (6.1.27)$$

In order to explain superconducting phenomena, H. London and F. London suggested in 1935 that $\boldsymbol{j}$ should be proportional to $\boldsymbol{A}$. Now this assertion is a consequence of Ginzburg-Landau phenomenological theory. Equation (6.1.27) serves to explain the

Meissner effect. Together with the equation $\nabla^2 \boldsymbol{A} = -\frac{4\pi}{c}\boldsymbol{j}$, which in the gauge $\nabla \cdot \boldsymbol{A} = 0$ follows from the Maxwell equation $\nabla \times \boldsymbol{B} = \frac{4\pi}{c}\boldsymbol{j}$, Eq. (6.1.27) gives

$$\nabla^2 \boldsymbol{A} = \lambda^2 \boldsymbol{A},$$

where

$$\lambda^2 = \rho\frac{4\pi e}{mc^2}. \tag{6.1.28}$$

λ is just London's penetration length. λ and ξ are the two characteristic parameters of Ginzburg-Landau theory. This theory has been very successful [1, 2] e.g., in predicting the critical magnetic field, the critical current density and the surface energy between the superconducting and normal phases. It turns out that the surface energy is positive for $\kappa \equiv \lambda/\xi < 1/\sqrt{2}$, and negative for $\kappa > 1/\sqrt{2}$. The latter case is critical for the existence of "type II superconductors", predicted by A. Abrikosov in 1952. V. Ginzburg and A. Abrikosov shared the Nobel Prize in Physics 2003 with A. Leggett (^{3}He superfluidity).

Consider a superconducting ring,[5] with magnetic flux Φ threading through it. In the bulk of superconductor there is no supercurrent $\boldsymbol{j} = 0$. The flux originates from current flowing on the surface of the superconducting ring. Performing a closed line integral along a contour winding the ring with the contour well within the bulk, we have

$$\oint \nabla\theta \cdot \mathrm{d}\boldsymbol{s} = \frac{e^*}{\hbar c}\oint \boldsymbol{A} \cdot \mathrm{d}\boldsymbol{s} = \frac{e^*}{\hbar c}\Phi. \tag{6.1.29}$$

Because of the single valuedness of the wave function, the left-hand side is the change of the phase θ for one winding, and can only be $2\pi n$. From this the quantization of flux through a superconducting ring follows

$$\Phi = \frac{\hbar c}{e^*}2\pi n = n\frac{hc}{2e}. \tag{6.1.30}$$

---

[5] We follow here Sections 21-5 to 21-9 of the Feynman lectures [4].

We have discussed this point in connection with the Aharonov-Bohm effect.[6]

The equations for ρ and θ can be obtained by substituting $\psi = \sqrt{\rho}\, e^{i\theta}$ into the Schrödinger equation

$$i\hbar\frac{\partial\psi}{\partial t} = \frac{1}{2m^*}\left(-i\hbar\nabla - \frac{e^*}{c}\boldsymbol{A}\right)^2\psi + e^*\varphi\psi,$$

where $(\varphi, \boldsymbol{A})$ is the 4-potential. The result is

$$\frac{\partial\rho}{\partial t} = -\boldsymbol{v}\cdot\nabla\rho - \rho\nabla\cdot\boldsymbol{v} = -\nabla\cdot(\rho\boldsymbol{v}), \qquad (6.1.31)$$

$$\hbar\frac{\partial\theta}{\partial t} = -\frac{m^*}{2}v^2 - e^*\varphi + \frac{\hbar^2}{2m^*}\frac{1}{\sqrt{\rho}}\nabla^2\sqrt{\rho}. \qquad (6.1.32)$$

Equation (6.1.31) is the equation of continuity accounting for the conservation of current, and except for the last term Eq. (6.1.32) is identical to the equation of motion for irrotational flow in hydrodynamics. The last term is the quantum correction. From Eq. (6.1.29) the following relation follows:

$$\nabla\times\left(\boldsymbol{v} + \frac{e^*}{m^*c}\boldsymbol{A}\right) = 0,$$

i.e.,

$$\nabla\times\boldsymbol{v} = -\frac{e^*}{m^*c}\boldsymbol{B}. \qquad (6.1.33)$$

The flow of ideal fluid is irrotational $\nabla\times\boldsymbol{v} = 0$, but for a charged fluid in magnetic field Eq. (6.1.33) holds.

6.2 Coupled Superconductors, Josephson Effect

Two superconductors connected by a thin layer of insulator form a Josephson junction as shown in Fig. 6.2. When the thickness of the insulator is ~ 30Å, and a voltage of U ($U > 2\Delta$, Δ is the energy gap of the superconductor) is applied between these two superconductors, single electron (more precisely, quasiparticle) can

---

[6]In (6.1.26) the charge e of the electron is the absolute value.

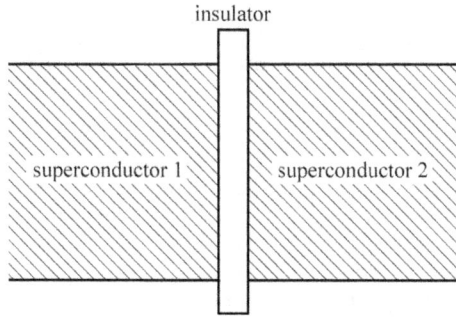

Fig. 6.2   Coupled superconductors, Josephson junction.

tunnel through the insulating layer. This is the Giaever tunneling. When the thickness of the insulating layer is further reduced to -10Å, Cooper pairs can tunnel through. This phenomenon is called the Josephson effect, predicted in 1962 by B. D. Josephson [5, 6] and experimentally confirmed by J. M. Rowell [7] in 1963. In this case the wave functions of the condensates of Cooper pairs in the two pieces of superconductors are coupled. Because the off-diagonal long range order is transmitted from one side to the other, the phases of the wave functions on both sides begin to establish certain relations. Let ψ_1 and ψ_2 be the wave functions of the condensates of Cooper pairs on both sides. They satisfy the following coupled equations[7]:

$$\left.\begin{aligned} i\hbar\frac{\partial\psi_1}{\partial t} &= U_1\psi_1 + K\psi_2 \\ i\hbar\frac{\partial\psi_2}{\partial t} &= U_2\psi_2 + K\psi_1 \end{aligned}\right\}, \qquad (6.2.1)$$

where K represents the coupling between the two pieces of superconductors, $U_1 - U_2 = e^*U$. For convenience, the zero of energy is set at $\frac{U_1+U_2}{2}$, and hence $U_1 = \frac{e^*U}{2}$, $U_2 = -\frac{e^*U}{2}$, and then Eqs. (6.2.1) become

$$\left.\begin{aligned} i\hbar\frac{\partial\psi_1}{\partial t} &= \frac{e^*U}{2}\psi_1 + K\psi_2 \\ i\hbar\frac{\partial\psi_2}{\partial t} &= -\frac{e^*U}{2}\psi_2 + K\psi_1 \end{aligned}\right\} \qquad (6.2.2)$$

---

[7]See [4] Section 21-9.

Substituting $\psi_1 = \sqrt{\rho_1}e^{i\theta_1}$, and $\psi_2 = \sqrt{\rho_2}e^{i\theta_2}$ in (6.2.2), and separating the real and imaginary parts, and denoting

$$\phi = \theta_2 - \theta_1, \tag{6.2.3}$$

we obtain the equations satisfied by the density and phase on both sides.

$$\left.\begin{aligned} \dot{\rho}_1 &= \frac{2}{\hbar}K\sqrt{\rho_1\rho_2}\sin\phi \\ \dot{\rho}_2 &= -\frac{2}{\hbar}K\sqrt{\rho_1\rho_2}\sin\phi \end{aligned}\right\}, \tag{6.2.4}$$

$$\left.\begin{aligned} \dot{\theta}_1 &= -\frac{K}{\hbar}\sqrt{\frac{\rho_2}{\rho_1}}\cos\phi - \frac{e^*U}{2\hbar} \\ \dot{\theta}_2 &= -\frac{K}{\hbar}\sqrt{\frac{\rho_1}{\rho_2}}\cos\phi + \frac{e^*U}{2\hbar} \end{aligned}\right\}. \tag{6.2.5}$$

From (6.2.4) we get $\dot{\rho}_1 = -\dot{\rho}_2$, and this is the particle current J from 2 to 1. Equations (6.2.1) give only the coupling between two sides, $\dot{\rho}_1$ and $\dot{\rho}_2$ represent the would-be rate of change of the density on both sides. Actually the two superconductors are connected in a circuit, the particle current does not lead to particle accumulation. Consequently we can set in the right-hand side of (6.2.4) $\rho_1 = \rho_2 = \rho_0$. Define the critical current $J_c = \frac{2K}{\hbar}\rho_0$, the first equation of (6.2.4) gives on using $\dot{\rho}_1 = J$

$$J = J_c \sin\phi. \tag{6.2.6}$$

The direction of the tunneling supercurrent is opposite to that of the particle current. Denoting the corresponding electric currents by I and I_c, we have

$$I = I_c \sin\phi. \tag{6.2.7}$$

The critical current I_c depends on the density of Cooper pairs and the coupling between the two superconductors. Equation (6.2.5) give the *Josephson relation*

$$\dot{\phi} = \dot{\theta}_2 - \dot{\theta}_1 = \frac{e^*U}{\hbar}. \tag{6.2.8}$$

Integrating with respect to t gives

$$\phi(t) = \phi_0 + \frac{e^*}{\hbar} \int U \, dt. \tag{6.2.9}$$

If U is a constant (a D C voltage), then ϕ increases linearly with time, and I is an alternating current:

$$\phi(t) = \phi_0 + \frac{2eU}{\hbar} t, \tag{6.2.10}$$

$$J = J_c \sin(\phi_0 + \omega t), \tag{6.2.11}$$

$$\omega = \frac{2|e|U}{\hbar}. \tag{6.2.12}$$

Equation (6.2.12) gives $\frac{\nu}{U} = 48.36$ MHz/μV which is a very high frequency. If $U = 0$, then ϕ does not change with time $I = I_c \sin \phi_0$, the tunneling current persists. Fig. 6.3 shows the V–A characteristics curve of a Sn–Sn_xO_y–Sn Josephson junction at $T = 1.52$K. The ordinate represents the current, and the absissa — the voltage. We see from Fig. 6.3 that $U = 0$ when $I < I_c$. When I exceeds I_c the regime becomes the Giaever tunneling with a non-vanishing voltage. Hence the name critical current. The left and right branches of the curve represent the Giaever tunneling (with dissipation) and the central branch is the Josephson tunneling (without dissipation).

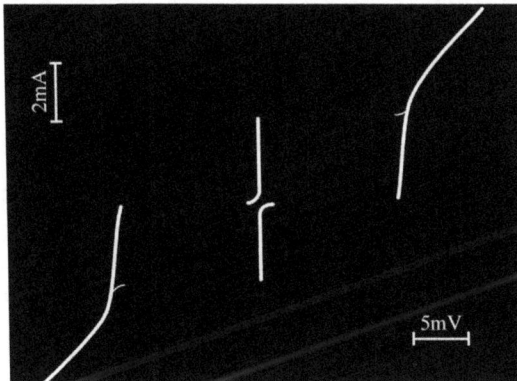

Fig. 6.3 V–A characteristics of a Sn–Sn_xO_y–Sn Josephson junction at 1.52K.

Remember that $\boldsymbol{j}$ is proportional to $\nabla\theta$ in a bulk superconductor (Eq. 6.1.21). Here the current is proportional to the difference of phases on both sides. Consequently the physics of Josephson effect is similar to the phenomenon in the bulk superconductor.

The change of energy of the system is equal to the work done by the voltage. This work is reversible (dissipationless). Denoting by $\mathrm{d}V$ the differential of work done, we have

$$\mathrm{d}V\left(\phi\right) = -IU\mathrm{d}t. \tag{6.2.13}$$

The Josephson relation (6.2.8) gives

$$\dot{\phi} = -\frac{2\left|e\right|U}{\hbar} = -\frac{2\pi c}{\Phi_0}U, \tag{6.2.14}$$

where Φ_0 is the flux quantum

$$\Phi_0 = \frac{hc}{2\left|e\right|}. \tag{6.2.15}$$

Equations (6.2.13) and (6.2.14) infer that

$$\mathrm{d}V = \frac{I}{c}\frac{\Phi_0}{2\pi}\mathrm{d}\phi = \frac{I_c}{c}\sin\phi\frac{\Phi_0}{2\pi}\mathrm{d}\phi,$$

$$\tag{6.2.16}$$

$$V(\phi) = -\frac{I_c}{c}\frac{\Phi_0}{2\pi}\cos\phi,$$

where the integration constant has been set equal to zero. $V(\phi)$ is called the *Josephson coupling energy*, which is a minimum at $\phi = 0$. The supercurrent and the Josephson coupling energy are related by

$$I = \frac{2\pi c}{\Phi_0}\frac{\partial V(\phi)}{\partial \phi}. \tag{6.2.17}$$

Under the equilibrium condition when the state of the system is time independent, the supercurrent passing through the junction is equal to the external current of a *current-biased Josephson junction* (abbreviated CBJ). This equality can be obtained by adding to (6.2.16) a term due to the external current $-I_{ext}\phi\frac{\Phi_0}{2\pi}$,[8] and

---

[8]The work done is $\int I_{ext}U\mathrm{d}t = -I_{ext}\frac{\Phi_0}{2\pi}\int\dot{\phi}\mathrm{d}t = -I_{ext}\frac{\Phi_0}{2\pi}\phi$.

demanding the total energy to be minimum in equilibrium. The Josephson coupling energy for a CBJ is now

$$V(\phi) = -\frac{I_c}{c}\frac{\Phi_0}{2\pi}\cos\phi - \frac{I_{ext}}{c}\phi\frac{\Phi_0}{2\pi}. \tag{6.2.18}$$

The minimization of V at equilibrium

$$\left.\frac{\partial V(\phi)}{\partial \phi}\right|_{eq} = 0 \tag{6.2.19}$$

leads to

$$I_{ext} = I_c \sin\phi, \tag{6.2.20}$$

i.e., the supercurrent and the external current are equal at equilibrium.

The Aharonov-Bohm effect can be demonstrated using two Josephson junctions connected in parallel [4], as shown in Fig. 6.4. The magnetic flux is confined in a tube through the loop. Although the magnetic field $\boldsymbol{B}$ vanishes everywhere outside the flux tube, but the vector potential exists. Let J_a and J_b be the currents passing through the junctions a and b, and the total current be $J_{tot} = J_a + J_b$. Along the route PaQ and PbQ the changes of phase of wave function

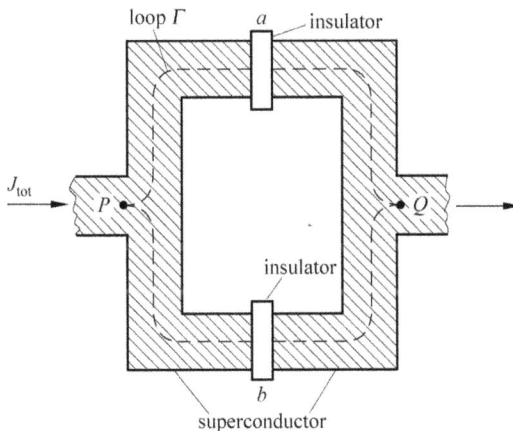

Fig. 6.4   Two Josephson junctions connected in parallel.

are respectively

$$
\left.\begin{aligned}
\Delta\theta\,(PaQ) &= \phi_a + \frac{e^*}{\hbar c}\int_{P(a)}^{Q} \boldsymbol{A}\cdot\mathrm{d}\boldsymbol{s}, \\[2mm]
\Delta\theta\,(PbQ) &= \phi_b + \frac{e^*}{\hbar c}\int_{P(b)}^{Q} \boldsymbol{A}\cdot\mathrm{d}\boldsymbol{s},
\end{aligned}\right\}, \qquad (6.2.21)
$$

where ϕ_a and ϕ_b are the changes of phase across the junctions a and b and the line integrals are path dependent with paths indicated in parentheses. The phase difference between P and Q is unique, and hence the two expressions for the phase difference must be equal, which leads to

$$
\begin{aligned}
\phi_b - \phi_a &= \frac{e^*}{\hbar c}\left(\int_{P(a)}^{Q} - \int_{P(b)}^{Q}\right)\boldsymbol{A}\cdot\mathrm{d}\boldsymbol{s} \\[2mm]
&= \frac{e^*}{\hbar c}\oint \boldsymbol{A}\cdot\mathrm{d}\boldsymbol{s} = -\frac{e^*}{\hbar c}\int \boldsymbol{B}\cdot\mathrm{d}\boldsymbol{s} = -\frac{e^*}{\hbar c}\Phi. \qquad (6.2.22)
\end{aligned}
$$

The convention used here is that $\boldsymbol{A}$ due to Φ is counter clockwise, with the contour of line integral specified. Since $e^* = 2e$, we can write for convenience

$$
\left.\begin{aligned}
\phi_a &= \phi_0 + \frac{e}{\hbar c}\Phi \\[2mm]
\phi_b &= \phi_0 - \frac{e}{\hbar c}\Phi
\end{aligned}\right\}. \qquad (6.2.23)
$$

The relation between the Josephson current and the phase difference is now

$$
\left.\begin{aligned}
J_a &= J_c \sin\left(\phi_0 + \frac{e}{\hbar c}\Phi\right) \\[2mm]
J_b &= J_c \sin\left(\phi_0 - \frac{e}{\hbar c}\Phi\right)
\end{aligned}\right\} \qquad (6.2.24)
$$

and the total current is

$$
J_{\mathrm{tot}} = J_a + J_b = 2J_c \sin\phi_0 \cos\frac{e\Phi}{\hbar c}. \qquad (6.2.25)
$$

J_{tot} is a periodic function of Φ, the condition for maximum is

$$\Phi = n\frac{\pi\hbar c}{e} = n\Phi_0. \qquad (6.2.26)$$

The phenomenon shown above is the interference between the Josephson current; the phase difference between them is regulated by the flux Φ (vector potential $\boldsymbol{A}$) while the supercurrents flow in space of zero magnetic field. The total Josephson current through a double junction was measured by R. C. Jaklevic *et al.* [8] as a function of the magnetic field in the area of the flux tube, as shown in Fig. 6.5. There is a general background of current from various effects, but the rapid oscillations of the current with changes in $\frac{e\Phi}{\hbar c}$ show clearly the interference.

Remember the relation between the wave function in the presence of a vector potential and the wave function without it given in Section 3.1:

$$\psi(\boldsymbol{x}) = \psi_0(\boldsymbol{x}) \exp\left(-i\frac{e^*}{\hbar c}\int^x \boldsymbol{A}\cdot d\boldsymbol{s}\right). \qquad (6.2.27)$$

Then the scalar product of the wave functions across the junction is

$$\langle 2\mid 1\rangle = \langle 2\mid 1\rangle_0 \exp\left(-\frac{ie^*}{\hbar c}\int_1^2 \boldsymbol{A}\cdot d\boldsymbol{s}\right). \qquad (6.2.28)$$

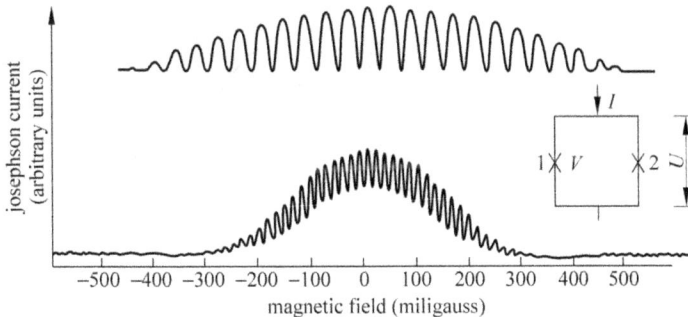

Fig. 6.5   Josephson current through two junctions in parallel.

Hence, in the presence of a vector potential $\boldsymbol{A}$ the Eqns. (6.2.2) become

$$
\left.
\begin{aligned}
i\hbar \frac{\partial \psi_1}{\partial t} &= \frac{e^* U}{2} \psi_1 + K \left[\exp\left(-i \frac{e^*}{\hbar c} \int_1^2 \boldsymbol{A} \cdot \mathrm{d}\boldsymbol{s} \right) \right] \psi_2 \\
i\hbar \frac{\partial \psi_2}{\partial t} &= -\frac{e^* U}{2} \psi_2 + K \left[\exp\left(-i \frac{e^*}{\hbar c} \int_1^2 \boldsymbol{A} \cdot \mathrm{d}\boldsymbol{s} \right) \right] \psi_1
\end{aligned}
\right\}.
\tag{6.2.29}
$$

The integral on the exponential is carried out through the junction. The tunneling current and Josephson relation are respectively

$$
J = J_c \sin\left(\phi + \frac{e^*}{\hbar c} \int_1^2 \boldsymbol{A} \cdot \mathrm{d}\boldsymbol{s} \right),
\tag{6.2.30}
$$

$$
\frac{\partial}{\partial t} \left(\phi + \frac{e^*}{\hbar c} \int_1^2 \boldsymbol{A} \cdot \mathrm{d}\boldsymbol{s} \right) \equiv \frac{\partial}{\partial t} \phi^* = \frac{e^*}{\hbar} U.
\tag{6.2.31}
$$

The phase difference ϕ of wave function across the junction is now replaced by ϕ^*, the gauge invariant phase difference.

6.3 Superconducting Loop with a Josephson Junction, the SQUID

The main part of a superconducting quantum interference device (SQUID) is a superconducting loop interrupted by a single Josephson junction, as shown in Fig. 6.6. As the CBJ, the SQUID is also widely used in physics research and applications. The Josephson current in

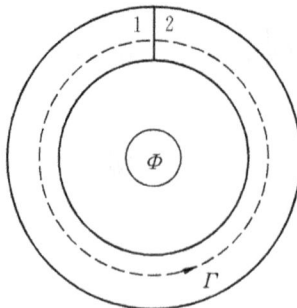

Fig. 6.6   Scheme of a SQUID.

the loop produces a flux Φ threading it. The Josephson current is related to the particle current of the Cooper pairs by

$$\boldsymbol{J}_{\mathrm{s}} = -2e\boldsymbol{J}. \tag{6.3.1}$$

According to (6.1.18) we write

$$\boldsymbol{J}_{\mathrm{s}} = -\frac{e\hbar\rho}{m}\left(\nabla\theta + \frac{2\,|e|}{\hbar c}\boldsymbol{A}\right). \tag{6.3.2}$$

In the bulk of the loop $\boldsymbol{J}_{\mathrm{s}} = 0$ and hence

$$\nabla\theta = -\frac{2\,|e|}{\hbar c}\boldsymbol{A} = -\frac{2\pi}{\Phi_0}\boldsymbol{A}. \tag{6.3.3}$$

Assuming that the supercurrent flows counter clockwise, we integrate both sides of (6.3.3) along the contour Γ. Because of the single-valuedness of the wave function, the left-hand side gives $2\pi n$,

$$2\pi n = -\frac{2\pi}{\Phi_0}\int_{1\Gamma}^{2}\boldsymbol{A}\cdot d\boldsymbol{s} + \theta_1 - \theta_2. \tag{6.3.4}$$

The convention here is to define $\phi = \theta_1 - \theta_2$ for supercurrent flowing from 2 to 1. When a vector potential is present, ϕ is to be replaced by ϕ^* (6.2.31)

$$\phi^* = \phi + \frac{2\pi}{\Phi_0}\int_{2J}^{1}\boldsymbol{A}\cdot d\boldsymbol{s}. \tag{6.3.5}$$

In (6.3.4) and (6.3.5), the subscripts Γ and J of the integrals denote the path being taken along the loop Γ and across the junction respectively. These equations lead to

$$\phi^* = 2\pi n + \frac{2\pi}{\Phi_0}\left[\int_{1\Gamma}^{2}\boldsymbol{A}\cdot d\boldsymbol{s} + \int_{2J}^{1}\boldsymbol{A}\cdot d\boldsymbol{s}\right]$$

$$= 2\pi n + \frac{2\pi}{\Phi_0}\oint\boldsymbol{A}\cdot d\boldsymbol{s}$$

$$= 2\pi n + \frac{2\pi}{\Phi_0}\Phi. \tag{6.3.6}$$

Comparing (6.3.6) with (6.1.30) we find their essential difference. Equation (6.1.30) exhibits the quantization of flux due to the supercurrent on the surface of a superconducting loop; while (6.3.6) shows that, when a Josephson junction is present, the flux threading the loop is no more quantized.

Consider an external magnetic field with flux Φ_{ext} is imposed on the SQUID loop. Let L be the self- inductance of the loop. Φ_{ext} induces supercurrent on the loop, which tends to screen a part of the external flux. The resultant flux Φ through the loop is[9]

$$\Phi = \Phi_{\text{ext}} - LI. \tag{6.3.7}$$

The supercurrent is determined by the Josephson junction

$$I = I_{\text{c}} \sin \phi^* = I_{\text{c}} \sin \left(2\pi \frac{\Phi}{\Phi_0} \right). \tag{6.3.8}$$

The energy of the loop consists of two parts,

$$V(\Phi) = -I_{\text{c}} \frac{\Phi_0}{2\pi} \cos 2\pi \frac{\Phi}{\Phi_0} + \frac{(\Phi_{\text{ext}} - \Phi)^2}{2L}. \tag{6.3.9}$$

The first term is the coupling energy, the second term is the energy of the self-inductance $\frac{1}{2}LI^2$. Equation (6.3.9) is equivalent to the energy of CBJ (6.2.18). The minimization of (6.3.9) with respect to Φ gives (6.3.7), similar to the CBJ case. From the equation for the loop energy (6.3.9) we obtain the stable and metastable states of the loop.[10] First consider the case $\Phi_{\text{ext}} = 0$. Equation (6.3.9) can be rewritten as

$$\frac{V(\Phi)}{\Phi_0^2/2L} = \left(\frac{\Phi}{\Phi_0} \right)^2 - \frac{LI_{\text{c}}}{\pi\Phi_0} \cos 2\pi \frac{\Phi}{\Phi_0}. \tag{6.3.10}$$

Figure 6.7 shows $\frac{V(\Phi)}{\Phi_0^2/2L}$ as functions of $\frac{\Phi}{\Phi_0}$ for different values of $\frac{LI_{\text{c}}}{\pi\Phi_0}$. In Fig. 6.7(a), $\frac{LI_{\text{c}}}{\pi\Phi_0} = 2$. There are 5 minima, including the stable state at $\Phi = 0$ and other 4 metastable states. In Fig. 6.7(b), $\frac{LI_{\text{c}}}{\pi\Phi_0} = 3$

---

[9]The use of Φ_{ext} is to adjust the potential energy of the system to set a stage for the macroscopic quantum tunneling or macroscopic quantum coherence of the flux Φ.

[10]From this point on to the end of Section 6.6 we adopt the SI system of units in order to keep the same convention as most of the relevant literature.

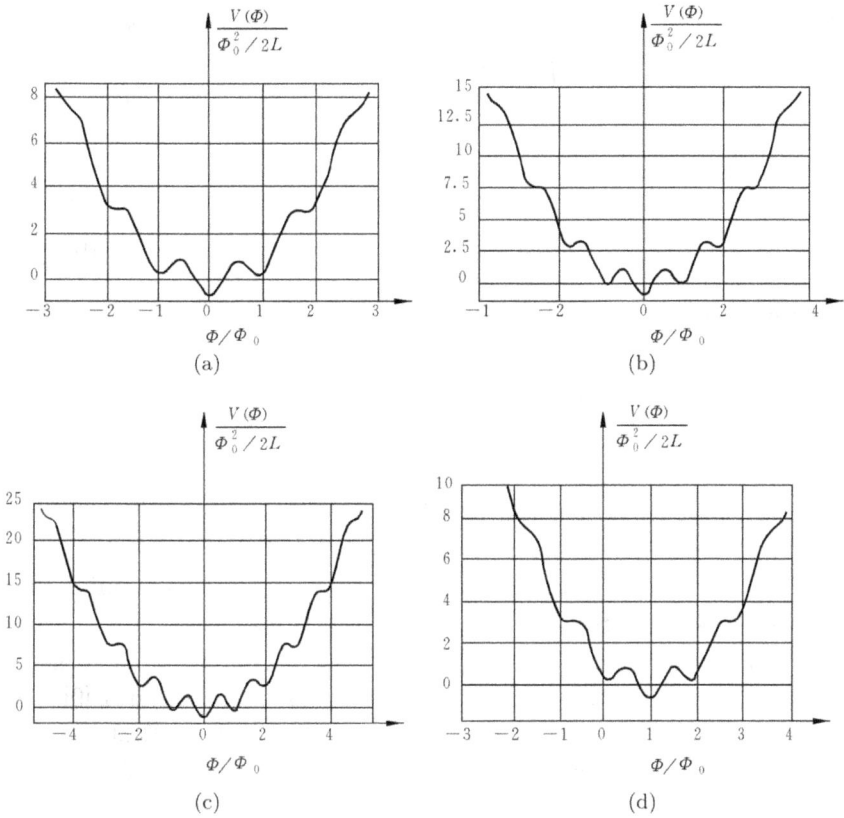

Fig. 6.7   Stable and metastable states of the SQUID.

and In Fig. 6.7(c) $\frac{LI_c}{\pi\Phi_0} = 4$. There are 7 and 9 minima respectively. Next consider the case for finite Φ_{ext}. It can vary continuously. For convenience, we take $\Phi_{\text{ext}} = n\Phi_0$ in order to shift the stable state to $n\Phi_0$. Equation (6.3.9) can also be written as

$$V\left(\Phi\right) = \frac{\left(\Phi - n\Phi_0\right)^2}{2L} - \frac{I_c\Phi_0}{2\pi}\cos\left(2\pi\frac{\Phi - n\Phi_0}{\Phi_0}\right). \qquad (6.3.11)$$

In Fig. 6.7(d), $\Phi_{\text{ext}} = \Phi_0$, $\frac{I_c\Phi_0}{2\pi} = 2$. We see that the well of the metastable states becomes more and more shallow when they get further away from the stable state. Φ_{ext} can be used to adjust the depth of the well at the metastable state at $\Phi = 0$. For example, for

$\Phi_{\text{ext}} = \Phi_0$, at $\Phi = 0$ we find the first metastable state. For $\Phi_{\text{ext}} = 2\Phi_0$ at $\Phi = 0$ we find the second metastable state. For $\Phi_{\text{ext}} = 3\Phi_0$ no well can be found at $\Phi = 0$ and the system is unstable there — it slides down along the $V(\Phi)$ curve.

The case for $\Phi_{\text{ext}} = \frac{1}{2}\Phi_0$ is interesting. Let

$$\tilde{\Phi} = \Phi - \frac{1}{2}\Phi_0, \qquad (6.3.12)$$

then we have

$$\frac{V(\tilde{\Phi})}{\Phi_0^2/2L} = \left(\frac{\tilde{\Phi}}{\Phi_0}\right)^2 + \frac{LI_c}{\pi\Phi_0}\cos\left(2\pi\frac{\tilde{\Phi}}{\Phi_0}\right). \qquad (6.3.13)$$

Figure 6.8 gives $\frac{V(\tilde{\Phi})}{\Phi_0^2/2L}$ as a function of $\frac{\tilde{\Phi}}{\Phi_0}$. It is a double well, and provides the stage for macroscopic quantum coherence.

If a SQUID or CBJ is prepared at a certain local minimum on the $V(\Phi)$ (for SQUID) or $V(\phi)$ (for CBJ) curve, how does it evolve in time? We need a study of the dynamics of flux Φ or phase ϕ. A frequently used model is given by J. Kurkijärvi [9], the "resistively shunted junction model" (abbreviated an RSJ model). Let the current through the Josephson junction be I_J. Connected

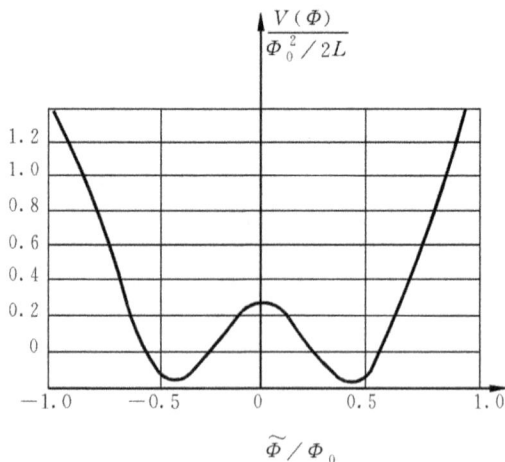

Fig. 6.8   A double well.

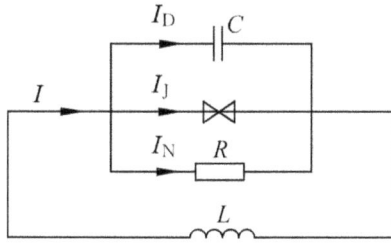

Fig. 6.9   The RSJ model.

in parallel with the junction is a resistance, through which normal current I_N flows. The junction has a capacitance, and there exists the displacement current I_D. The CBJ can now be put in the RSJ model as shown in Fig. 6.9. Eqns. (6.2.30) and (6.3.6) give

$$I_J = I_c \sin \phi^* = I_c \sin \left(2\pi n + \frac{2\pi}{\Phi_0} \Phi \right) = I_c \sin \frac{2\pi}{\Phi_0} \Phi. \qquad (6.3.14)$$

Using $(6.2.14)^{11}$ and (6.3.6), we have

$$I_N = \frac{U}{R} = \frac{1}{R} \left(\frac{\Phi_0}{2\pi} \dot{\phi}^* \right) = \frac{1}{R} \dot{\Phi}, \qquad (6.3.15)$$

$$I_D = C\dot{U} = C \left(\frac{\Phi_0}{2\pi} \ddot{\phi}^* \right) = C\ddot{\Phi}. \qquad (6.3.16)$$

To get the equation of motion for Φ we begin with the energy of capacitance $\frac{1}{2}CU^2 = \frac{1}{2}C\dot{\Phi}^2$ which can be considered as the kinetic energy term of mass C. The equation of motion would have been

$$C\ddot{\Phi} = -\frac{\partial V(\Phi)}{\partial \Phi}. \qquad (6.3.17)$$

But the system is dissipative (current I_N flows through R). The dissipative term is $\frac{\dot{\Phi}}{R}$,[12] proportional to the velocity. The equation of

---

[11]The convention then was $\phi = \theta_1 - \theta_2$, different by a sign with Section 6.3, hence the sign of the right-hand side of (6.2.14) has to be reversed.
[12]See (6.5.16).

motion is

$$C\ddot{\Phi} + \frac{\dot{\Phi}}{R} = -\frac{\partial V}{\partial \Phi} = -I_c \sin 2\pi \frac{\Phi}{\Phi_0} + \frac{(\Phi_{\text{ext}} - \Phi)}{L}. \tag{6.3.18}$$

From Eqs. (6.3.14) to (6.3.16) we find that (6.3.18) is actually

$$I_D + I_N + I_J = I. \tag{6.3.19}$$

Under the condition $\Phi_{\text{ext}} = 0$, when Φ is close to the stable state $\Phi = 0$, we get from (6.3.18)

$$\ddot{\Phi} = -\frac{1}{C}\left(\frac{1}{L} + I_c\frac{2\pi}{\Phi_0}\right)\Phi. \tag{6.3.20}$$

The angular frequency of oscillation around the stable state is

$$\omega = \frac{1}{\sqrt{LC}}\left(1 + \frac{2\pi L I_c}{\Phi_0}\right)^{1/2} \equiv \frac{1}{\sqrt{LC}}(1 + \beta_L)^{1/2}. \tag{6.3.21}$$

In the foregoing equation an important parameter β_L is defined. For the following discussions we need to define some more parameters of the SQUID [10]. A metastable state exists in a well shown in Fig. 6.10(a), the depth of the well is V_0, the angular frequency of oscillation at the bottom of the well is ω_0. By increasing Φ_{ext} up to Φ_{x0}, then the well disappears as shown in Fig. 6.10(b), and no more

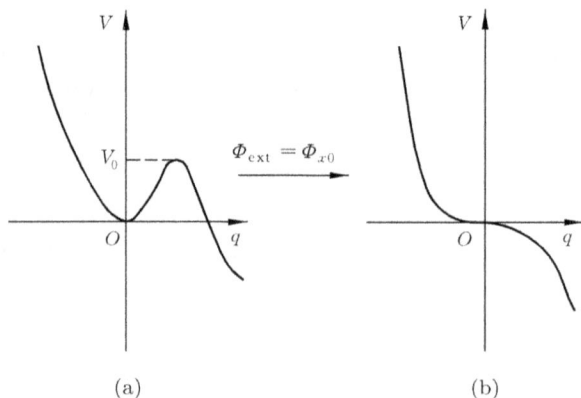

(a)   (b)

Fig. 6.10   The potential energy curve changes with the external magnetic flux.

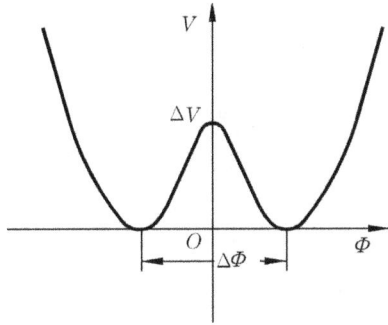

Fig. 6.11   Parameters of the double well.

metastable state can exist. Let $\delta \Phi_x = \Phi_{x0} - \Phi_{\text{ext}}$, $\omega_{LC} = \frac{1}{\sqrt{LC}}$, we have

$$
\left.
\begin{aligned}
\omega_0 &= \omega_{LC}(2\pi\sqrt{2})^{1/2} \left(\frac{LI_c}{\Phi_0}\right)^{1/4} \left(\frac{\delta \Phi_x}{\Phi_0}\right)^{1/4} \\
V_0 &= \frac{2\sqrt{2}}{3\pi} \frac{\Phi_0^2}{L} \left(\frac{\Phi_0}{LI_c}\right)^{1/4} \left(\frac{\delta \Phi_x}{\Phi_0}\right)^{3/2}
\end{aligned}
\right\}
\tag{6.3.22}
$$

For the double well shown in Fig. 6.11, the spacing between the wells is $\Delta \Phi$, the angular frequency at the bottom of the well is ω_0, we have

$$
\left.
\begin{aligned}
\Delta \Phi &= \frac{\Phi_0}{\pi} \left[6 \left(\beta_L - 1\right)/\beta_L\right]^{1/2} \\
\Delta V &= \frac{3}{8\pi^2} \frac{\Phi_0^2}{L} \frac{(\beta_L - 1)^2}{\beta_L} \\
\omega_0 &= \left[2 \left(\beta_L - 1\right)\right]^{1/2} \omega_{LC}
\end{aligned}
\right\}
\tag{6.3.23}
$$

6.4   Macroscopic Quantum Tunneling and Macroscopic Quantum Coherence in Josephson Systems

The flux Φ in a SQUID is a macroscopic quantity. Applying external flux to SQUID helps to arrange potential wells. The crucial question is, Eq. (6.3.16) is a classical equation of motion, and Φ is a c-number.

What is the condition for treating Φ as a quantum observable?[13] The c-number Φ must be the expectation value of this observable. Does the quantum fluctuation around the expectation value make sense? We neglect dissipation, and consider Φ as the generalized coordinate. The Lagrangian of the system is

$$L = \frac{1}{2}C\dot{\Phi}^2 - V(\Phi). \tag{6.4.1}$$

The conjugate momentum is

$$p_\Phi = \frac{\partial L}{\partial \dot{\Phi}} = C\dot{\Phi}. \tag{6.4.2}$$

Equations (6.2.31) and (6.3.5) give

$$p_\Phi = C\frac{\Phi_0}{2\pi}\dot{\phi}^* = -CU = -Q, \tag{6.4.3}$$

where Q is the charge stored in the capacitance. Consider the case $\Phi_{\text{ext}} = 0$. The SQUID is in the stable state $\Phi = 0$. The quantum fluctuation can be estimated as follows

$$\frac{1}{2}C\omega^2\left\langle(\Delta\Phi)^2\right\rangle = \frac{1}{2}\cdot\frac{1}{2}\hbar\omega. \tag{6.4.4}$$

The left-hand side is the analog of the harmonic oscillator potential energy $\frac{1}{2}m\omega^2x^2$, the right-hand side is half of the zero-point energy. Consequently

$$\left\langle(\Delta\Phi)^2\right\rangle = \frac{\hbar}{2C\omega}. \tag{6.4.5}$$

The Heisenberg uncertainty principle gives

$$\Delta\Phi\cdot(e\Delta N) \geq \frac{\hbar}{2}, \tag{6.4.6}$$

Let the imbalance of the number of electrons across the junction be denoted by N, then $eN = Q$. Equation (6.4.6) gives

$$\left\langle(\Delta N)^2\right\rangle \gtrsim \frac{\hbar^2}{4e^2}\frac{1}{\left\langle(\Delta\Phi)^2\right\rangle} = \frac{\hbar C\omega}{2e^2}. \tag{6.4.7}$$

---

[13]In this and the following section we follow closely Refs. [9, 10].

Define a parameter for characterizing the importance of quantum effect λ by the following relation

$$\lambda \equiv \left(8CI_c\,\Phi_0^3/\pi^3\hbar^2\right)^{1/2} = \left(8CI_c\hbar/e^3\right)^{1/2}. \qquad (6.4.8)$$

From Eq. (6.3.19), for $\beta_L \gg 1$, we have

$$\omega \approx \left(\frac{2\pi I_c}{C\,\Phi_0}\right)^{\frac{1}{2}},$$

Both $\langle(\Delta N)^2\rangle$ and $\langle(\Delta\Phi)^2\rangle$ can be expressed in terms of λ :

$$\langle(\Delta N)^2\rangle = \frac{\lambda}{4},$$

$$\langle(\Delta\Phi)^2\rangle = \frac{h^2}{4e^2}\frac{4}{\lambda} = \frac{\Phi_0^2}{\pi^2\lambda}. \qquad (6.4.9)$$

If $\lambda \gg 1$, $\Delta N \gg 1$, $\Delta\Phi$ can be very small and the quantum fluctuation is unimportant. This is the correspondence principle limit. For $\lambda \approx 1$, N can be large and ΔN negligible. The values of Φ are scattered. This is the quantum case. In order that the quantum effect be significant, it should be in a favorable position compared with competitors, besides that the condition $\lambda \approx 1$ should hold. From (6.3.11) we see that the local variation of the potential curve which produces a well to accommodate a metastable state is $\frac{I_c\Phi_0}{\pi}$. If the system is in an environment with a temperature sufficiently high such that $kT \gtrsim \frac{I_c\Phi_0}{\pi}$, the thermal activated transition occurs. Therefore we demand in addition that $kT < \frac{I_c\Phi_0}{\pi}$. Combining the demands, we have

$$1 \approx \left(\frac{8CI_c\hbar}{e^3}\right)^{1/2} > \left(\frac{8C}{e^2}kT\right)^{1/2},$$

which leads to

$$C \lesssim \frac{e^2}{8kT}. \qquad (6.4.10)$$

This relation can be interpreted as that the energy of an electron in the capacitance $\frac{e^2}{2C}$ should be smaller than $4kT$. For $T = 1\text{mK}$, $C \lesssim 0.1\text{pF}$. This is easily achievable in technology.

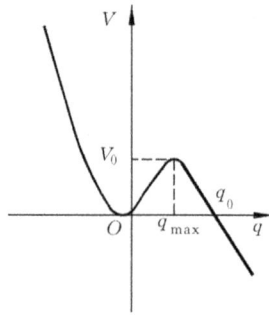

Fig. 6.12   The potential curve close to a metastable state.

A typical quantum behavior is the barrier penetration. Around a metastable state the potential curve behaves as shown in Fig. 6.12, where q is used to denote the flux Φ. The tunnelling rate at $T = 0$ can be calculated by using the WKB approximation in quantum mechanics, i.e.,

$$P = \text{const } \omega_0 \left(\frac{V_0}{\hbar\omega_0}\right)^{\frac{1}{2}} \exp\left[-\frac{2}{\hbar}\int_0^{q_0} \sqrt{2mV(q)}\,dq\right]. \qquad (6.4.11)$$

The Callan-Coleman method can give more precise results.[14] The potential curve in Fig. 6.12 can be approximated by quadratic+cubic potential.

$$V(q) = \frac{1}{2}M\omega_0^2 q^2 - \beta q^3$$

$$\equiv \frac{27}{4}V_0\left[\left(\frac{q}{q_0}\right)^2 - \left(\frac{q}{q_0}\right)^3\right]. \qquad (6.4.12)$$

The parameters in (6.4.12) are related to the SQUID parameters by

$$q_0 = \frac{M\omega_0^2}{2\beta}, \quad V_0 = \frac{1}{54}\frac{\left(M\omega_0^2\right)^3}{\beta^2}.$$

---

[14]See Section 5.4 and Appendix 1 to Chapter 5.

For this potential the instanton method gives the tunneling rate

$$P = \omega_0 \left(\frac{60V_0}{\hbar\omega_0} \right)^{1/2} \left(\frac{18}{5\pi} \right)^{1/2} \exp\left(-\frac{36}{5} \frac{V_0}{\hbar\omega_0} \right). \qquad (6.4.13)$$

Strictly speaking, the quadratic+cubic potential is pathological, because it is not bound from below. The system can progressively go to lower and lower states, releasing more and more energy without a limit. But in reality (Fig. 6.8) this does not happen. Remember that in Section 5.4 the integral contributing to the classical action from the bounce solution is taken up to the bouncing point only. The potential beyond this point has influence on the result of tunneling rate only through the prefactor, and the influence is small. Thus we conclude that the result of our calculation is not very different from a more reasonable potential.

At finite temperature the decay rate due to thermal activation is [13]

$$P_{c1} = \text{const } \omega_0 \exp\left(-\frac{V_0}{kT} \right). \qquad (6.4.14)$$

In order that the thermal transition does not disturb the quantum tunneling it is necessary that

$$kT \ll a_0 \hbar\omega_0,$$

where a_0 is a constant. For quadratic+cubic potential $a_0 = \frac{5}{36}$. Other factors, e.g., ω_0 and V_0, need also to be considered. A. J. Leggett [10] pointed out that for $T < 100\,\text{mK}$ quantum tunneling begins to show up.

The double well potential shown in Fig. 6.12 provides a possibility to realize the macroscopic quantum coherence. This is the macroscopic version of the two-level problem in quantum mechanics (Section 5.2). If the barrier between the wells is very high, then the state ψ_L and ψ_R in the left and right well are degenerate in energy. When the barrier penetration is appreciable these states get mixed and the energy eigenstates turn out to be

$$\left.\begin{array}{c} \psi_0 = \dfrac{1}{\sqrt{2}}\left(\psi_L + \psi_R\right) \\[2mm] \psi_e = \dfrac{1}{\sqrt{2}}\left(\psi_L - \psi_R\right) \end{array}\right\}, \tag{6.4.15}$$

where ψ_0 and ψ_e are the ground and excited states respectively. For convenience we use the WKB formula for the tunneling splitting

$$\Gamma = \text{const } \omega_0 \exp\left[-\frac{1}{\hbar}\int_0^R \sqrt{2CV\left(\Phi\right)}\mathrm{d}\,\Phi\right], \tag{6.4.16}$$

$$E_e - E_0 = \hbar\Gamma.$$

If at time $t = t_0$ the system is in the state ψ_L, then at time t the wave function is

$$\begin{aligned} \psi\left(t\right) &= \frac{1}{\sqrt{2}}\left[\psi_0 \exp\left(-\mathrm{i}\frac{E_0}{h}t\right) + \psi_e \exp\left(-\mathrm{i}\frac{E_e}{h}t\right)\right] \\[2mm] &= \frac{1}{\sqrt{2}}\left\{\psi_L\left[\exp\left(-\mathrm{i}\frac{E_0}{h}t\right) + \exp\left(-\mathrm{i}\frac{E_e}{h}t\right)\right]\right. \\[2mm] &\quad \left. + \psi_R\left[\exp\left(-\mathrm{i}\frac{E_0}{h}t\right) - \exp\left(-\mathrm{i}\frac{E_e}{h}t\right)\right]\right\}. \tag{6.4.17} \end{aligned}$$

Let $W_L(t)$ and $W_R(t)$ be the probabilities that the system is in the left and right wells respectively, i.e.,

$$\left.\begin{array}{c} W_L\left(t\right) = \left|\langle\psi_L \mid \psi\left(t\right)\rangle\right|^2 \\[2mm] W_R\left(t\right) = \left|\langle\psi_R \mid \psi\left(t\right)\rangle\right|^2 \end{array}\right\} \tag{6.4.18}$$

Equation (6.4.17) gives

$$\frac{W_L\left(t\right)}{W_R\left(t\right)} = \frac{\left|1 + e^{-\mathrm{i}\Gamma t}\right|^2}{\left|1 - e^{-\mathrm{i}\Gamma t}\right|^2} = \frac{1 + \cos\Gamma t}{1 - \cos\Gamma t}.$$

Using $W_L(t) + W_R(t) = 1$, we obtain

$$\left. \begin{array}{l} W_L(t) = \dfrac{1}{2}\left(1 + \cos \Gamma t\right) \\[2mm] W_R(t) = \dfrac{1}{2}\left(1 - \cos \Gamma t\right) \end{array} \right\}. \tag{6.4.19}$$

For the SQUID, L and R means $\tilde{\Phi} = -\frac{\Phi_0}{2}$ and $\tilde{\Phi} = \frac{\Phi_0}{2}$ respectively. These are macroscopically distinguishable states, and ψ_0 and ψ_e are Schrödinger cats in the laboratory. To discuss the plausibility of realizing cat states in the laboratory, the most important factor is the parameter β_L (6.3.19). For $\beta_L \gg 1$, the integral on the exponential of (6.4.16) is $e^{-\lambda/\sqrt{2}}$ [10]. For $\beta_L \gg 1$, the corresponding λ is also large, this is far from the quantum regime. We need to study the case $\beta_L \lesssim 1$. Let $\beta_L - 1 \equiv K$, Eq. (6.3.21) gives

$$\Delta V \approx \frac{3}{8\pi^2} \frac{\Phi_0^2}{L} K^2, \tag{6.4.20}$$

$$\omega_0 = 2K^{\frac{1}{2}}\omega_{LC}.$$

The integral on the exponential can be taken as that for the quadratic+cubic potential, with a factor $\frac{1}{2}$ added,

$$\Gamma \approx \omega_0 \exp\left(-\frac{18}{5}\frac{\Delta V}{\hbar\omega_0}\right).$$

Using (6.4.20), we rewrite this as

$$\Gamma \approx \omega_0 \exp\left(-\frac{1}{2}K^{3/2}\lambda\right). \tag{6.4.21}$$

Since the thermal transition rate is

$$\Gamma_{\text{cl}} \approx \omega_0 e^{-\Delta V/kT}.$$

it is necessary that $\Delta V \gg kT$ for the classical effect giving way to quantum coherence. Simplifying (6.4.20) we have $\Delta V \approx \frac{3}{4}K^2 I_c \Phi_0$. The condition $\Delta V \gg kT$ becomes

$$\frac{3}{4}K^2 I_c \Phi_0 \gg kT. \tag{6.4.22}$$

For quantum coherence to occur, the value of $K^{\frac{3}{2}}\lambda$ on the exponential of (6.4.21) cannot be large. If we demand it be $O(1)$, then

$$(K^{\frac{3}{2}}\lambda)^{\frac{4}{3}} = K^2\lambda^{\frac{4}{3}} \approx 1$$

Combining with (6.4.22), i.e.,

$$\frac{kT}{I_c\Phi_0} \ll \frac{3}{4}K^2$$

we must have

$$\frac{kT}{I_c\Phi_0} \ll \lambda^{-4/3}. \tag{6.4.23}$$

Since $\lambda \propto \sqrt{C}$, and $\lambda \approx 1$, we must have $C \approx 10^{-14}\text{F} = 10^{-2}\text{pF}$. This is rather demanding. Furthermore, the main role has not yet appeared on the stage. The characteristic property of a macroscopic system is that it cannot be isolated from the environment. The interaction with environment will have its share of demand.

6.5   The Influence of Environment on the Macroscopic Quantum Phenomena

One of the basic characteristics of a macroscopic quantum system is its interaction with the environment. In general such interaction is irreversible, and tends to erase the coherence in the system. In the study of macroscopic quantum phenomena a discussion concerning the system-environment interaction is indispensable. Take the example of SQUID or CBJ. Even under the condition $T \ll T_c$ the number of normal electrons is finite, and they lead to dissipation. The RSJ model is successful, demonstrating the necessity of including dissipation in the theory from the beginning. The quantum mechanics of open systems is a direction which has been flourishing in recent years [14]. In 1963 R. P. Feynman and F. L. Vernon formulated a theory of the interaction between a quantum system and a linear dissipative system [15].

On the basis of this approach, A. J. Leggett and collaborators [10–12] studied systematically the macroscopic quantum phenomena in relation to dissipation. The resistance involved in a SQUID or Josephson junction is measurable. This is the only known macroscopic parameter. The question is that a study of macroscopic quantum phenomena has to begin with the quantum theory, the macroscopic dissipation parameter is only the *result* of the interaction between the system and the environment. How can this macroscopic parameter be incorporated into a quantum theory? A. O. Caldeira and A. J. Leggett [11] gave a specific program. How can the environment be described? Feynman and Vernon used a "harmonic oscillator bath", which was followed by Caldeira and Leggett. The environment consists of a large number of harmonic oscillators with mass, frequency and coordinate $m_i, \omega_i, x_i(t)$ respectively, $i = 1, 2, \ldots$. They interact with the "system" with mass M, coordinate $q(t)$, under the action of a potential $V(q)$ and a time dependent external force $F_{\text{ext}}(t)$. The system-environment interaction can be assumed as a postulate, but as we shall see demand is to be imposed on it.

Then the Hamiltonian of the system-environment complex can be written down. Eventually it is the evolution of the *system* that we are interested in. An ideal framework is provided by the path integral formalism. If the integral over all possible histories of the environmental harmonic oscillators is carried out, the result will be an effective Hamiltonian of the system, the influence of the environment is manifested in it as an effective interaction, in which the only macroscopic dissipation parameter is to be incorporated. For the purpose of applying instanton or bounce Caldeira and Leggett used imaginary time. With the effective Hamiltonian the quantum properties of the system can be studied.

Two important issues emerge from this program. First, is the integration over all possible histories of the huge number of harmonic oscillators possible? How can the large numbers of parameters m_i, ω_i and coupling constants c_i, d_i be incorporated into a single macroscopic parameter in the effective Hamiltonian? Second, is the Hamiltonian general enough? A specific effective Hamiltonian is a model. There can be competing models. Caldeira and Leggett made a great effort to argue that this Hamiltonian is almost unique. In the following we enter the program step by step.

6.5.1 On the Canonical Transformations and the Adiabatic Approximation

Consider the Lagrangian of a system

$$L = L_0 + \Delta L_1,$$

$$L_0 = \frac{1}{2}m \left(\dot{x}^2 + \dot{y}^2 \right) - V(x, y),$$

$$\Delta L_1 = \varepsilon \dot{x} y. \tag{6.5.1}$$

where ΔL_1 is the velocity dependent coupling, ε is a constant. Adding a complete differential of a function of time to the Lagrangian does not change the equation of motion, because its contribution to the action is a constant. We can use alternative coupling

$$\left.\begin{aligned}
\Delta L_2 &= -\varepsilon x \dot{y} \\
\Delta L_3 &= \frac{\varepsilon}{2} \left(\dot{x} y - x \dot{y} \right)
\end{aligned}\right\}. \tag{6.5.2}$$

They differ from the coupling ΔL_1 only by a differential of a function of time, and therefore lead to the same equation of motion. The Hamiltonians corresponding to the three diffeerent couplings are respectively

$$\left.\begin{aligned}
H_1 &= \frac{(p_x - \varepsilon y)^2}{2m} + \frac{p_y^2}{2m} + V(x, y) \\
H_2 &= \frac{p_x^2}{2m} + \frac{(p_y + \varepsilon x)^2}{2m} + V(x, y) \\
H_3 &= \frac{(p_x - \varepsilon y/2)^2}{2m} + \frac{(p_y + \varepsilon x/2)^2}{2m} + V(x, y)
\end{aligned}\right\}. \tag{6.5.3}$$

From the form of the Hamiltonians we know that they describe the motion of a particle in the field of potential $V(x, y)$ and a magnetic field in z-direction. The vector potential of the magnetic field is

$$\left.\begin{array}{lll} A_x^{(1)} = -\varepsilon y, & A_y^{(1)} = A_z^{(1)} = 0 \\[2mm] A_y^{(2)} = \varepsilon x, & A_x^{(2)} = A_z^{(2)} = 0 \\[2mm] A_x^{(3)} = -\dfrac{\varepsilon}{2} y, & A_y^{(3)} = \dfrac{\varepsilon}{2} x, & A_z^{(3)} = 0 \end{array}\right\}. \qquad (6.5.4)$$

The correspond to different choices of the gauge.

Consider the Hamiltonian for a 2-dimensional harmonic oscillator in a magnetic field in the z-direction,

$$H = \frac{p_x^2}{2m} + \frac{(p_y + \varepsilon x)^2}{2m} + \frac{1}{2}\left(m\omega_1^2 x^2 + m\omega_2^2 y^2\right). \qquad (6.5.5)$$

We carry out a canonical transformation with the generating function

$$F(y, z) = -m\omega_2 y z, \qquad (6.5.6)$$

with transformation relations

$$\begin{aligned} p_y &= \frac{\partial F}{\partial y} = -m\omega_2 z, \\[2mm] p_z &= -\frac{\partial F}{\partial z} = m\omega_2 y. \end{aligned} \qquad (6.5.7)$$

The transformation replaces the old momentum p_y by the new coordinate z and the old coordinate y by the new momentum p_z. The Hamiltonian becomes

$$H' = \frac{1}{2m}\left(p_x^2 + p_z^2\right) + \frac{1}{2}m\omega_1^2 x^2 + \frac{1}{2}m\left(\omega_2 z - \frac{\varepsilon}{m}x\right)^2. \qquad (6.5.8)$$

We notice that the $p_y x$ coupling in (6.5.5) becomes the xz coupling in (6.5.8).

In studying the problem of molecular structure, the Born-Oppenheimer approximation is often used in quantum mechanics [16]. The idea is to treat the nuclear observables (coordinate q, momentum p_q) as the slow variables and the electron observables (coordinate ξ_i, momentum p_i, $i = 1, 2, \ldots, N$) as the fast variables. In the adiabatic approximation, q is considered as a parameter at first. Then the Schrödinger equation for the electrons is solved

$$\left[-\frac{\hbar^2}{2m} \nabla_i^2 + V(q) \right] \chi_k(\xi_i, q) = U_k(q) \chi_k(\xi_i, q). \tag{6.5.9}$$

In the right hand side of this equation, $U_k(q)$ is the eigenvalue and $\chi_k(\xi_i, q)$ is the eigenfunction of state k, with q entering as a parameter. The electron energy eigenvalue in its turn provide potential for the nuclear motion

$$\left[-\frac{\hbar^2}{2m} \nabla_q^2 + U_k(q) \right] \varphi_{ik}(q) = E_i \varphi_{ik}(q). \tag{6.5.10}$$

In this way the motion of the electrons is "separated" effectively from the motion of the nucleus. The wave function of the molecule is

$$\psi_{ik}(q, \{\xi\}) = \varphi_{ik}(q) \chi_k(\{\xi\}, q), \tag{6.5.11}$$

where i, j denote the state of the nucleus; h, l denote the states of the many-electron wave function; and $\{\xi\}$ denotes the set of coordinates of the electrons. The coupling in (6.5.11) is explicitly expressed: the nuclear wave function $\varphi_{ik}(q)$ depends on the state k of the electrons and the electron wave function $\chi_k(\{\xi\}, q)$ depends on the nuclear coordinate q. The condition for validity of the adiabatic approximation is that the following matrix element can be neglected:

$$\left\langle ik \left| \Delta \hat{H} \right| jl \right\rangle$$

$$= -\frac{\hbar^2}{2m} \int dq \int d\xi \left\{ 2\varphi_{ik}^*(q) \frac{\partial}{\partial q} \varphi_{jl}(q) \chi_k^*(\xi, q) \frac{\partial}{\partial q} \chi(\xi, q) \right.$$

$$+ \left. \varphi_{ik}^*(q) \varphi_{jl}(q) \chi_k^*(\xi, q) \frac{\partial^2}{\partial q^2} \chi_l(\xi, q) \right\}. \tag{6.5.12}$$

Under the adiabatic approximation the motion of the nucleus and of the electrons are conservative. There is no energy exchange between them.

6.5.2   The Hamiltonian of a Dissipative Electromagnetic System

We begin with a simple system of a superconducting loop with a flux Φ threading it. The Lagrangian of the system is

$$L = \frac{1}{2}C\dot{\Phi}^2 - \frac{1}{2L}\Phi^2. \qquad (6.5.13)$$

The corresponding equation of motion is

$$C\ddot{\Phi} + \frac{\Phi}{L} = 0. \qquad (6.5.14)$$

The environment consists of the electrons in normal state. The interaction Lagrangian is

$$\Delta L = \int \boldsymbol{j}\left(\boldsymbol{r}\right) \cdot \boldsymbol{A}\left(\boldsymbol{r}\right) \mathrm{d}^3 x,$$

$$\boldsymbol{j}\left(\boldsymbol{r}\right) = \sum_i e v_i \delta^3 \left(\boldsymbol{r} - \boldsymbol{r}_i\right), \qquad (6.5.15)$$

where $\boldsymbol{j}$ is the current density of the normal electrons, $\boldsymbol{v}$ is the electron velocity, and the integral is taken over the volume of the loop. Let $\boldsymbol{l}$ be an arc along the loop, its direction coincides with that of $\boldsymbol{j}$. Let the cross section area of the loop be S. Then we have $\boldsymbol{j}\,\mathrm{d}^3 x = S\boldsymbol{j}\,|\mathrm{d}\boldsymbol{l}| = I_n \mathrm{d}\boldsymbol{l}$, where I_n is the normal current. Hence

$$\Delta L = I_n \oint \boldsymbol{A} \cdot \mathrm{d}\boldsymbol{l},$$

where the integral is taken around the loop. Using Stokes' theorem, we obtain $\Delta L = I_n \Phi$. We use the Coulomb gauge $\boldsymbol{E} = -\dot{\boldsymbol{A}}$, then $I_n = jS = \sigma E S = -\sigma|\dot{\boldsymbol{A}}|S = -\sigma\dot{\Phi}\frac{S}{l}$, where σ is the electric conductivity, and l is the length of the loop. Since the electrical resistance is $R_n = \left(\frac{\sigma S}{l}\right)^{-1}$, the above equation becomes $I_n = -\frac{\dot{\Phi}}{R_n}$. The equation of motion can be written as

$$C\ddot{\Phi} + \frac{\dot{\Phi}}{R_n} + \frac{\Phi}{L} = 0. \qquad (6.5.16)$$

Our aim is to express the dissipation term in terms of the *microscopic* observables of the environment oscillators, then write down the

Lagrangian of the system-environment complex, which will lead to the equation of motion (6.5.16) involving only one macroscopic parameter R_n. This will witness the correctness of the choice of the Lagrangian, and then the Lagrangian acquires the right to be used in calculations of any macroscopic quantum phenomenon. Define the charge $\mathcal{Q}_n$ of a flow of normal current in time t by

$$\mathcal{Q}_n\left(t\right) = \int_0^t I_n\left(t'\right) dt',$$

i.e.,

$$I_n\left(t\right) = \frac{d\mathcal{Q}_n\left(t\right)}{dt}. \tag{6.5.17}$$

The interaction Lagrangian ΔL can be rewritten as

$$\Delta L = -\mathcal{Q}_n \dot{\Phi}, \tag{6.5.18}$$

since it differs from the original expression only by a time derivative of a function. $\mathcal{Q}_n$ is related to the resistance, and th latter is the effect of large number of environment oscillators, whose normal modes are denoted by Q_α, where α is the mode index. In general we have

$$\mathcal{Q}_n = \sum_\alpha \tilde{c}_\alpha Q_\alpha. \tag{6.5.19}$$

We apply once again the "difference in a time derivative" trick, obtaining

$$\Delta L = \Phi \sum_\alpha \tilde{c}_\alpha \dot{Q}_\alpha.$$

The environment is a set of oscillator modes Q_α with mass m_a, frequency w_a, and has its own Lagrangian. The Lagrangian of the total complex is

$$L = \frac{1}{2}C\dot{\Phi}^2 - \frac{1}{2L}\Phi^2 + \Phi \sum_\alpha \tilde{c}_\alpha \dot{Q}_\alpha$$

$$- \sum_\alpha \frac{1}{2}m_a w_\alpha^2 Q_\alpha^2 + \sum_\alpha \frac{1}{2}m_a \dot{Q}_\alpha^2. \tag{6.5.20}$$

The first two terms in the right-hand side refer to the system, the last two terms refer to the environment, and the middle term is the system-environment interaction. The conjugate momenta of Φ and Q_α are

$$P_\Phi = \frac{\partial L}{\partial \dot{\Phi}} = C\dot{\Phi} \left. \begin{array}{l} \\ \\ \\ \end{array} \right\}$$
$$P_{Q_\alpha} = \frac{\partial L}{\partial \dot{Q}} = m_\alpha \dot{Q}_\alpha + \Phi \sum_\alpha \tilde{c}_\alpha$$

$$(6.5.21)$$

The Hamiltonian of the total complex is

$$H = \frac{1}{2C}P_\Phi^2 + \frac{1}{2L}\Phi^2 + \sum_\alpha \left(\frac{1}{2m_\alpha}P_{Q_\alpha}^2 + \frac{1}{2}m_\alpha\omega_\alpha^2 Q_\alpha^2 \right)$$
$$- \Phi \sum_\alpha \frac{1}{m_\alpha}\tilde{c}_\alpha P_{Q_\alpha} + \Phi^2 \sum_\alpha \frac{\tilde{c}_\alpha^2}{2m_\alpha}. \qquad (6.5.22)$$

Remember that from Eq. (6.5.5) to Eq. (6.5.8) the coupling changes from the momentum-coordinate type to the coordinate-coordinate type. We repeat the process here, using the generating function

$$F(x_\alpha, Q_\alpha) = \sum_\alpha m_\alpha \omega_\alpha Q_\alpha x_\alpha \qquad (6.5.23)$$

with the transformation relations

$$P_{x_\alpha} = -m_\alpha \omega_\alpha Q_\alpha,$$
$$P_{Q_\alpha} = m_\alpha \omega_\alpha x_\alpha. \qquad (6.5.24)$$

Replacing the old coordinate Q_α and momentum P_{Q_α} by new coordinate x_α and new momentum P_{x_α}, we get the Hamiltonian

$$H' = \frac{1}{2C}\dot{\Phi}^2 + \frac{1}{2L}\Phi^2 + \sum_\alpha \left(\frac{1}{2}m_\alpha\omega_\alpha^2 x_\alpha^2 + \frac{1}{2m_\alpha}P_{x_\alpha}^2 \right)$$
$$- \Phi \sum_\alpha \tilde{c}_\alpha \omega_\alpha x_\alpha + \Phi^2 \sum_\alpha \frac{\tilde{c}_\alpha^2}{2m_\alpha}. \qquad (6.5.25)$$

Let $\tilde{c}_\alpha \omega_\alpha = c_\alpha$, the Lagrangian of the total complex is

$$L' = \frac{1}{2}C\dot{\Phi}^2 - \frac{1}{2L}\Phi^2 + \sum_\alpha \left(\frac{1}{2}m_\alpha \dot{x}_\alpha^2 - \frac{1}{2}m_\alpha \omega_\alpha^2 x_\alpha^2\right)$$

$$+ \Phi \sum_\alpha c_\alpha x_\alpha - \Phi^2 \sum_\alpha \frac{c_\alpha^2}{2m_\alpha \omega_\alpha^2}. \tag{6.5.26}$$

The interaction term exhibits the environment coordinates x_α coupled to the system coordinate Φ with corresponding strength c_α. The last term of (6.5.26) is a renormalization term, which can be combined to $-\frac{1}{2L}\Phi^2$ leading to a change of the system parameter. When calculations are made using a Lagrangian, for example (6.5.26), the interaction term can lead in general to changes of the parameters of the system (renormalization of mass, coupling constants, etc.). The last term of (6.5.26) can serve as a "counter term", which can cancel in part or in full the change caused by the interaction. In a phenomenological theory it is the physics judgement which decides whether such a counter term is desirable or not. The Eq. (6.5.26) is the standard form of system-environment Lagrangian. The coordinate-coordinate coupling facilitates the evaluation of path integral over environment coordinates.

6.5.3   The Non-adiabatic Correction, The Lagrangian of a Dissipative System

If the condition for adiabatic approximation cannot be fulfilled, then (6.5.12) is the non-adiabatic correction. There is a special feature of this correction when the adiabatic approximation is applied to macroscopic systems. The microscopic states $\chi_k(\xi)$ which are coupled to each other by $\Delta\hat{H}$ (6.5.12) are only different in the behavior of a small fraction $\left(-\frac{1}{N}\right)$ of the macroscopically many particles involved. Consequently the effective potential $U_k(q)$ for the slow degree of freedom q depends on the state k very weakly, and so the states $\varphi_{ik}(q)$ are effectively independent of k. In this situation Eq. (6.5.10) becomes

$$\left[-\frac{\hbar^2}{2M}\nabla_q^2 + U(q)\right]\varphi_i(q) = E_i \varphi_i(q), \tag{6.5.27}$$

and the non-adiabatic correction becomes

$$\langle ik|\Delta\hat{H}|ij\rangle$$

$$= -\frac{\hbar^2}{2M}\int dq \int d\xi \left\{ 2\varphi_i^*(q)\frac{\partial}{\partial q}\varphi_j(q)\chi_k^*(\xi,q)\frac{\partial}{\partial q}\chi_l(\xi,q) \right.$$

$$\left. + \varphi_i^*(q)\varphi_j(q)\chi_k^*(\xi,q)\frac{\partial^2}{\partial q^2}\chi_l(\xi,q) \right\}. \tag{6.5.28}$$

Further simplilfication of (6.5.28) can be achieved by introducing [11] a linear operator $\hat{K}$ in the space spanned by the microscopic wave functions $\chi_k(\xi,q)$.

$$\hat{K}_{kl}(q) \equiv \langle k|\hat{K}(q)|l\rangle = \int \chi_k^*(\xi,q)\left(-i\hbar\frac{\partial}{\partial q}\right)\chi_l(\xi,q)d\xi, \tag{6.5.29}$$

After some algebra (6.5.28) becomes

$$\langle ik|\Delta\hat{H}|ij\rangle = \frac{1}{2M}\langle i|p\hat{K}_{kl}(q) + \hat{K}_{kl}(q)p + (\hat{K}^2)_{kl}(q)|j\rangle, \tag{6.5.30}$$

where

$$\langle i|p|j\rangle = \int \varphi_i^*(q)\left(-i\hbar\frac{\partial}{\partial q}\right)\varphi_j(q)dq. \tag{6.5.31}$$

The Hamiltonian of the system-environment complex is

$$\hat{H} = \frac{1}{2M}(p + \hat{K}(q))^2 + U(q) + \hat{H}_{\text{env}}, \tag{6.5.32}$$

where $\hat{H}_{\text{env}}$ is the Hamiltonian of the environment, depending on $\{\xi\}$ only, $U(q)$ is the effective potential for q under adiabatic approximation, and $\hat{K}(q)$ represents the non-adiabatic correction. As an operator in the space spanned by the environment wave functions, $\hat{K}(q)$ depends on $\hat{x}_j$ and $\hat{p}_j$, the generalized coordinate and momentum operators of the normal mode j of the environment. In Fock space representation, the state of the environment is

$|n_1, n_2, \ldots, n_j, \ldots\rangle$, n_j being the occupation number of the mode j. The non-zero matrix elements of $\hat{x}_j$ and $\hat{p}_j$ are

$$\left.\begin{aligned}
\langle n_j + 1 | \hat{x}_j | n_j \rangle &= \langle n_j | \hat{x}_j | n_j + 1 \rangle = (n_j + 1)^{1/2} \left(\frac{\hbar}{2m_j\omega_j}\right)^{1/2} \\
\langle n_j + 1 | \hat{p}_j | n_j \rangle &= - \langle n_j | \hat{p}_j | n_j + 1 \rangle \\
&= im_j\omega_j (n_j + 1)^{1/2} \left(\frac{\hbar}{2m_j\omega_j}\right)^{1/2}
\end{aligned}\right\}.$$

$$(6.5.33)$$

We assume that $\hat{K}(q)$ depends on $\hat{x}_j$ and $\hat{p}_j$ linearly. This does not mean that the interaction between the system and environment is weak. The interaction is mainly represented by the adiabatic potential $U(q)$, here $\hat{K}(q)$ is only the non-adiabatic correction. $\hat{K}$ can change the number of particles in any mode by 1, but it can induce changes in many modes simultaneously. From (6.5.32) we see that the time reversal invariance of $\hat{H}$ demands that the time reversal behavior of $\hat{K}$ must be the same as that of $\hat{p}$. Since $\hat{x}$ and q remain invariant under time reversal, the form of $\hat{K}$ under the linear dependence assumption can only be

$$\hat{K}(q) = \sum_j K_j(q)\hat{p}_j. \qquad (6.5.34)$$

The Hamiltonian (6.5.32) becomes

$$\hat{H} = \frac{1}{2M} \left(\hat{p} + \sum_j K_j(q)\hat{p}_j^2\right) + U(q) + \hat{H}_{\text{env}}. \qquad (6.5.35)$$

The system-environment interaction is of the momentum-momentum type, with the system coordinate q appearing at the same time. To obtain the result we look for (coordinate-coordinate type), we use the method introduced in subsection 6.5.2 in three steps:

(1) introducing new co-ordinates y_i by canonical transformation with the generating function

$$F = \sum_j m_j\omega_j x_j y_j, \qquad (6.5.36)$$

with the transformation relations

$$
\left.\begin{array}{l}
p_j = \dfrac{\partial F}{\partial x_j} = m_j\omega_j y_j \\[3mm]
p_{y_j} = -\dfrac{\partial F}{\partial y_j} = -m_j\omega_j y_j
\end{array}\right\}. \tag{6.5.37}
$$

Substituting (6.5.37) into (6.5.35) we obtain $\hat{H}'(q,p,y_i,p_{y_i})$. Returning to the Lagrangian we have

$$
L'\left(q,\dot{q};y_j,\dot{y}_j\right) = \frac{1}{2}m\dot{q}^2 - U(q) + \sum_j \frac{1}{2}m_i\dot{y}_i^2 - \sum_j \frac{1}{2}m_j\omega_j^2 y_j^2
$$

$$
- \sum_j m_j\omega_j K_j(q)\dot{q}y_j. \tag{6.5.38}
$$

The coupling is now of the velocity-coordinate type. Before we carry out the second canonical transformation, it is necessary to move the time derivative from q to y_i. This is accomplished by step (2).

(2) the "difference by a time derivative trick"

Rewrite first $K_j(q)\dot{q}$:

$$
K_j(q)\dot{q} = \frac{\mathrm{d}}{\mathrm{d}t}\int_0^q K_j\left(q'\right)\mathrm{d}q'.
$$

Then the coupling becomes $-\sum_j m_j\omega_j y_j \frac{\mathrm{d}}{\mathrm{d}t}\int_0^q K_j(q')\mathrm{d}q'$. Adding a time derivative term $\sum_j m_j\omega_j \frac{\mathrm{d}}{\mathrm{d}t}\left[y_i \int_0^q K_j(q')\mathrm{d}q'\right]$, we obtain a new Lagrangian

$$
L'' = \frac{1}{2}m\dot{q}^2 - U(q) + \sum_j \frac{1}{2}m_j\dot{y}_j^2 - \sum_j \frac{1}{2}m_j\omega_j^2 y_j^2
$$

$$
+ \sum_j m_j\omega_j \dot{y}_j \int_0^q K_j\left(q'\right)\mathrm{d}q'. \tag{6.5.39}
$$

The time derivative now appears on y_i. Write down the Hamiltonian

$$
H'' = \frac{p^2}{2M} + U(q) + \sum_j \left[\frac{1}{2m}\left(p_{y_j} - \frac{F_j(q)}{\omega_j}\right)^2_{\text{sym}} + \frac{1}{2}m_j\omega_j y_j^2\right], \tag{6.5.40}
$$

where

$$F_j(q) = m_j \omega_j^2 \int_0^q K_j(q') \, dq'. \tag{6.5.41}$$

We are ready to carry out the second canonical transformation now.

(3) a second canonical transformation has the generating function and transformation relations given below:

$$F' = \sum_j m_j \omega_j z_j y_j, \tag{6.5.42}$$

$$\left. \begin{array}{l} p_{y_j} = \dfrac{\partial F'}{\partial y_j} = m_j \omega_j z_j \\[3mm] p_{z_j} = \dfrac{\partial F'}{\partial z_j} = -m_j \omega_j y_j \end{array} \right\}. \tag{6.5.43}$$

The new Hamiltonian and Lagrangian are respectively

$$H''' = \frac{p^2}{2M} + U(q) + \sum_j \left[\frac{1}{2m_j} p_{z_j}^2 + \frac{1}{2} m_j \omega_j^2 \left(z_j - \frac{F_j(q)}{m_j \omega_j^2} \right)^2 \right], \tag{6.5.44}$$

$$L''' = \frac{1}{2} m \dot{q}^2 - U(q) + \sum_j \left(\frac{1}{2} m_j \dot{z}_j^2 - \frac{1}{2} m_j \omega_j^2 z_j^2 \right)$$
$$+ \sum_j F_j(q) z_j - \sum_j \frac{F_j^2(q)}{2 m_j \omega_j^2}. \tag{6.5.45}$$

Including the external force $F_{\text{ext}}(t)$, changing z to x, U to V, we obtain the final standard form

$$L = \frac{1}{2} m \dot{q}^2 - V(q) + \sum_j \left(\frac{1}{2} m_j \dot{x}_j^2 - \frac{1}{2} m_j \omega_j^2 x_j^2 \right) + \sum_j F_j(q) x_j$$
$$- \sum_j \frac{F_j^2(q)}{2 m_j \omega_j^2} + q F_{\text{ext}}(t). \tag{6.5.46}$$

The Lagrangian (6.5.26) of the dissipative electromagnetic system is a special example of this Lagrangian. Introducing the linear dependence assumption[15]

$$F_j(q) = C_j q. \tag{6.5.47}$$

We find that the coupling term becomes $q \sum_j C_j x_j$, the renormalization term becomes

$$-\sum_j \frac{C_j^2}{2 m_j \omega_j} q^2 \equiv \frac{1}{2} M (\Delta \omega)^2 q^2. \tag{6.5.48}$$

Lagrangian (6.5.46) leads to the equation of motion

$$\left. \begin{aligned} M\ddot{q} &= -\frac{\partial V}{\partial q} + M (\Delta \omega)^2 q + F_{\text{ext}}(t) - \sum_j C_j x_j \\ m_j \ddot{x}_j &= -m_j \omega_j^2 x_j - C_j q \end{aligned} \right\}. \tag{6.5.49}$$

Carrying out Fourier transform to $\tilde{q}(\omega)$ and $\tilde{x}_j(\omega)$, and eliminating $\tilde{x}_j$ from the two equations, we obtain the equation of motion for the system

$$-M\omega^2 \tilde{q}(\omega) = -\left(\frac{\partial V}{\partial q} \right)(\omega) + \tilde{F}_{\text{ext}}(\omega) + K(\omega)\tilde{q}(\omega), \tag{6.5.50}$$

where

$$K(\omega) = \sum_j \frac{C_j^2}{m_j^2 \left(\omega_j^2 - \omega^2 \right)}. \tag{6.5.51}$$

The frequency renormalization $(\Delta \omega)^2$ is included in V, thus defining a renormalized potential. The remarkable result is that in the equation for the macroscopic system only one macroscopic parameter $K(\omega)$ is involved, in which all relevant microscopic environment parameters C_j, m_j, ω_j are incorporated.

---

[15]This will be relaxed in subsequent discussions.

6.5.4   Relation between Macroscopic Dissipation Parameter and the Microscopic Parameters

The macroscopic equation of motion of the system is

$$M\ddot{q} + \eta\dot{q} + \frac{\partial V}{\partial q} = F_{\text{ext}}, \tag{6.5.52}$$

its Fourier transform is

$$-M\omega^2\tilde{q}(\omega) - i\omega\eta\tilde{q}(\omega) + \left(\frac{\partial V}{\partial q}\right)(\omega) = \tilde{F}_{\text{ext}}(\omega). \tag{6.5.53}$$

Comparing this with Eq. (6.5.50) we obtain

$$K(\omega) = i\omega\eta. \tag{6.5.54}$$

Before identifying this with (6.5.51), we need to handle the pole of (6.5.51) at ω_j. in an integral with respect to ω the detour of the integration path around the pole at ω_j is

$$\frac{1}{\omega_j^2 - \omega^2} = \frac{1}{(\omega_j + \omega)(\omega_j - \omega - i\varepsilon)},$$

i.e.,

$$\text{Im}\frac{1}{\omega_j^2 - \omega^2} = \frac{\pi}{2\omega_j}\delta(\omega_j - \omega).$$

Equation (6.5.51) gives

$$\left.\begin{array}{l} \text{Im}K(\omega) = \dfrac{\pi}{2}\displaystyle\sum_j \dfrac{C_j^2}{m_j\omega_j}\delta(\omega_j - \omega) \equiv J(\omega) \\[2ex] \text{Re}K(\omega) = 0. \end{array}\right\} \tag{6.5.55}$$

The first equation defines the *spectral density* $J(\omega)$. Equation (6.5.54) gives therefore

$$J(\omega) = \eta\omega \tag{6.5.56}$$

The phenomenological Eq. (6.2.52) is valid up to a characteristic frequency ω_c. Equation (6.5.56) is assumed to hold for ω much less than the cut-off frequency ω_c of the environment. The assumption of linear dependence (6.5.47) can actually be relaxed, then (6.5.55) is to be modified as

$$\frac{\pi}{2} \sum_j \frac{1}{m_j \omega_j^2} \left(\frac{\partial F_j}{\partial q} \right)^2 \delta(\omega - \omega_j) = \eta(q). \tag{6.5.57}$$

Up to this point the Lagrangian (6.5.46) is seen to lead to the phenomenological equation of motion for the system with the macroscopic parameter of dissipation η expressed in terms of the microscopic environment parameters exclusively. This Lagrangian is the starting point for making predictions on the macroscopic quantum properties of the system.

Besides the approach of Caldeira and Leggett, there is an alternative method due to V. Ambegaokar, U. Eckern and G. Schön [17, 18], starting from the microscopic model of a superconducting junction, treating the quasiparticle degrees of freedom as the cause of dissipation and noise. C. P. Sun and L. H. Yu [19–21] studied a system under harmonic oscillator potential and a constant external force interacting with the environment. For this specific system the authors used Laplace transform to express the system and the environment coordinates $q(t)$ and $x_j(t)$ explicitly in terms of the initial conditions, so that more detailed discussions can be carried out.

6.5.5 Dissipation and the Macroscopic Quantum Tunneling

When the external force is absent, the Hamiltonian corresponding to the Lagrangian (6.5.46), (6.5.47) is

$$H = \frac{p^2}{2M} + V(q) + \sum_i \left(\frac{p_i^2}{2m_i} + \frac{1}{2} m_i \omega_i^2 x_i^2 \right)$$

$$- q \sum_i C_i x_i + q^2 \sum_i \frac{C_i^2}{2m_i} \omega_i^2. \tag{6.5.58}$$

We begin with a qualitative discussion [10]. The potential contours in a multi-dimensional space (x_i, q) are shown in Fig. 6.13(a), while

(a)

(b)

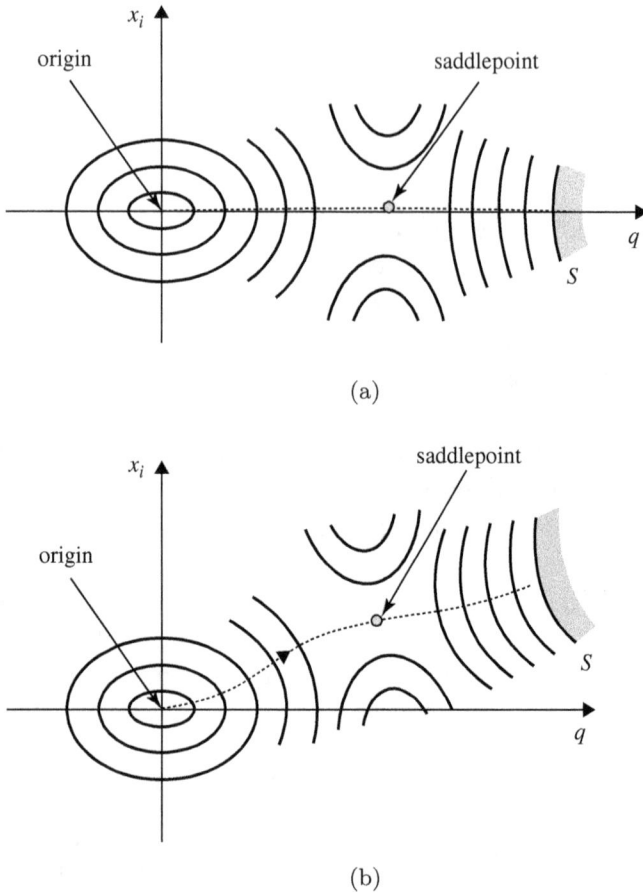

Fig. 6.13 Potential contours in the multi-dimensional space (x_i, q) (a) No dissipation; (b) With dissipation. Redrawn from Ref. [10].

the potential energy curve $V(q)$ is shown in Fig. 6.12. The potential energy is negative in the region $q > q_0$. This region is shaded and denoted by S in Fig. 6.13. The maximum V_0 of the potential energy $V(q)$ is reached at q_{max} as indicated in Fig. 6.12. The point $(q_{max}, x_i = 0)$ in Fig. 6.13(a) is the saddle point of the potential energy surface. It is a maximum for the potential energy of the system, and a minimum for the potential energy of the environment. The classical path of tunneling corresponds to the path from the origin

$(q = 0, x_i = 0)$ along the q-axis and ending at S through the saddle point. When dissipation exists the potential energy becomes

$$V(q) + \frac{1}{2}m\omega^2 x^2 - Cqx + q^2\frac{C^2}{2m\omega^2}$$

$$= V(q) + \frac{1}{2}m\omega^2 \left(x - \frac{C}{m\omega}q \right)^2 ,$$

where only one environment coordinate is explicitly given. The potential energy contours are shown in Fig. 6.13(b). The saddle point now moves to $\left(q_{\max}, \frac{C}{m\omega}q_{\max} \right)$. The potential energy at this point remains to be $V(q_{\max})$. The potential energy along the path of integration remains the same, but the path becomes longer. The WKB exponent is proportional to $\int [V(q, x_i)]^{\frac{1}{2}}ds$, and we see that dissipation decreases the tunneling rate.

In quantitative discussions [11] we use the density matrix of the system at temperature $T = \frac{1}{k\beta}$,

$$\rho(q_i \{x_i\}, q_f\{x_f\}; \beta) = \sum_n e^{-\beta E_n} \psi_n^*(q_i \{x_i\})\psi_n(q_f, \{x_f, \}),$$

$$(6.5.59)$$

where i and f denote the initial and final states respectively. The integral over all possible histories of the environment coordinates is to be taken. Such histories are represented by periodic motions of the oscillator modes, i.e., $\{x_i\} = \{x_f\} = \{\bar{x}\}$, with $\bar{x}$ taking all possible values. Integration over all possible histories includes the integration over all possible $\{\bar{x}_\alpha\}$'s, where α is the mode index. The reduced density matrix can be written as

$$\rho(q_i, q_f; \beta) = \int \prod_\alpha d\bar{x}_\alpha \int_{q(0)=q_i}^{q(\beta)=q_f} [Dq(\tau)] \int_{x_\alpha(0)=\bar{x}_\alpha}^{x_\alpha(\beta)=\bar{x}_\alpha} [Dx_\alpha(\tau)]$$

$$\times \exp\left[-\frac{1}{\hbar} \int_0^\beta L_E (q, \dot{q}; \{x_\alpha, \dot{x}_\alpha\}) d\tau \right], \qquad (6.5.60)$$

where L_E is the Euclidean Lagrangian $L_E = T + V$:

$$L_E(q, \dot{q}; \{x_\alpha, \dot{x}_\alpha\}) = \frac{1}{2}M\dot{q}^2 + V(q) + \frac{1}{2}\sum_\alpha m_\alpha \left(\dot{x}_\alpha^2 + \omega_\alpha^2 x_\alpha^2\right)$$

$$+ q\sum_\alpha C_\alpha x_\alpha + \frac{1}{2}M\left|\Delta\omega\right|^2 q^2. \qquad (6.5.61)$$

The Gaussian integral with respect to $[Dx_\alpha(\tau)]$ is given by (5.3.50)

$$\int d\bar{x}_a \int_{x_\alpha(0)=\bar{x}_a}^{x_\alpha(\beta)=\bar{x}_a}$$

$$\times [Dx_\alpha(\tau)]\exp\left\{-\frac{1}{\hbar}\int_0^\beta \left[\frac{1}{2}m\left(\dot{x}_\alpha^2 + \omega^2 x_\alpha^2\right) + x_\alpha C q(\tau)\right]d\tau\right\}$$

$$= I(0)\exp\left\{\frac{C^2}{4m\hbar\omega}\int_0^\beta d\tau \int_0^\beta d\tau'\right.$$

$$\left.\times\frac{q(\tau)q(\tau')\cosh\omega\left(|\tau-\tau'|-\frac{\beta}{2}\right)}{\sinh\omega\beta/2}\right\}, \qquad (6.5.62)$$

where $I(0)$ is the value of the integral for $C = 0$, and is just the integral of the density matrix of the free harmonic oscillator $\rho_{\text{osc}}(\bar{x}, \bar{x}, \beta)$ with respect to $\bar{x}$.

$$I(0) = \int d\bar{x}\int_{x(0)=\bar{x}}^{x(\beta)=\bar{x}} [Dx(\tau)]\exp\left[-\frac{1}{\hbar}\int_0^\beta \frac{1}{2}\left(m\dot{x}^2 + \omega^2 x^2\right)d\tau\right]$$

$$\equiv \int d\bar{x}\rho_{\text{osc}}(\bar{x}, \bar{x}; \beta). \qquad (6.5.63)$$

With the classical action given by

$$S_0 = \frac{m\omega}{\cosh\frac{\hbar\omega\beta}{2}}\bar{x}^2,$$

the density matrix of a harmonic oscillator is given by $(5.3.26)$,[16] $(5.3.30)$ and $(5.3.33)$

$$\rho(\bar{x}, \bar{x}; \beta) = \exp\left(-\frac{S_0}{\hbar}\right) F(\beta)$$

$$= \left(\frac{m\omega}{2\hbar\pi \sinh(\hbar\omega\beta)}\right)^{\frac{1}{2}} \exp\left[-\frac{m\omega}{\hbar \cosh\frac{\hbar\omega\beta}{2}}\bar{x}^2\right]$$

and consequently $(6.5.63)$ becomes

$$I\left(0\right) = \int_{-\infty}^{\infty} d\bar{x}\rho(\bar{x}, \bar{x}; \beta) = \frac{1}{2}\mathrm{csch}\frac{\hbar\omega\beta}{2}. \qquad (6.5.64)$$

The integral on the exponential of $(6.5.62)$ can be further transformed. Extend $q(\tau')$ beyond the region of definition $0 \leq \tau' < \beta$ by defining $q(\tau' + \beta) = q(\tau')$[17], $(6.5.62)$ can be written as

$$\frac{1}{2}\mathrm{csch}\frac{\hbar\omega\beta}{2}\exp\left\{\frac{C^2}{4m\hbar\omega}\int_{-\infty}^{\infty} d\tau' \int_{0}^{\beta} d\tau e^{-\omega|\tau-\tau'|}q\left(\tau\right)q\left(\tau'\right)\right\},$$

$$(6.5.65)$$

Substituting $(6.5.65)$ back to $(6.5.60)$, we obtain

$$\rho(q_i, q_f; \beta) = \rho_0(\beta) \int_{q(0)=q_i}^{q(\beta)=q_f} [Dq\left(\tau\right)]$$

$$\times \exp\left\{\int_{0}^{\beta}\left[\frac{1}{2}M\dot{q}^2 + V(q)\right] d\tau\right\}\exp\left\{\Lambda[q\left(\tau\right)]/\hbar\right\},$$

where

$$\rho_0(\beta) = \prod_{\alpha}\frac{1}{2}\mathrm{csch}\frac{\hbar\omega_{\alpha}\beta}{2}$$

---

[16]Put $x = x' = \bar{x}$.
[17]This enables an extension of the limits of integration to $\pm\infty$, since when τ is limited to the range $0 \leq \tau < \beta$, the integrand in $(6.5.65)$ $e^{-\omega|\tau-\tau'|}$ tends to 0 exponentially.

is the reduced density matrix of the environment oscillators, and

$$\Lambda[q\,(\tau)] = \int_0^\beta \frac{1}{2} M\,(\Delta\omega)^2\,q^2\,(\tau)\,d\tau$$

$$+ \sum_\alpha \left\{ \frac{C_\alpha^2}{4m_\alpha\omega_\alpha} \int_{-\infty}^\infty d\tau' e^{-\omega|\tau-\tau'|} q\,(\tau)\,q(\tau') \right\} \qquad (6.5.66)$$

represents the influence of the system-environment interaction. Its form can be further simplified. Substituting the identity

$$q(\tau)q(\tau') \equiv \frac{1}{2}\{q^2(\tau) + q^2(\tau') - (q(\tau) - q(\tau'))^2\}$$

into (6.5.66) and integrating $q^2(\tau)$ and $q^2(\tau')$ with respect to τ and τ', we see that the result cancels exactly the first term of (6.5.66) due to (6.5.48). Consequently

$$\Lambda[q(\tau)] = -\frac{1}{2} \int_{-\infty}^\infty d\tau' \int_0^\beta d\tau \alpha\,(\tau - \tau')\,\{q(\tau) - q\,(\tau')\}^2, \quad (6.5.67)$$

where

$$\alpha\,(\tau - \tau') = \sum_\alpha \frac{C_\alpha^2}{4m_\alpha\omega_\alpha} e^{-\omega_\alpha|\tau-\tau'|}$$

$$= \frac{1}{2\pi} \int_0^\infty J\,(\omega)\,e^{-\omega|\tau-\tau'|} d\omega \geq 0. \qquad (6.5.68)$$

The spectral density $J(\omega)$ is defined by (6.5.55). If we adopt (6.5.56), i.e., $J = \eta\omega$, then we have

$$\alpha\,(\tau - \tau') = \frac{\eta}{2\pi} \frac{1}{(\tau - \tau')^2}. \qquad (6.5.69)$$

It is valid under the condition $|\tau - \tau'| \gtrsim \omega_c^{-1}$. The final form of the reduced matrix element is

$$\rho(q_i, q_f; \beta) = \int_{q_i}^{q_f} [Dq\,(\tau)]\exp\left(-\frac{1}{\hbar}S_{\text{eff}}[q(\tau)]\right). \qquad (6.5.70)$$

The effective action of the system is then

$$S_{\text{eff}}[q(\tau)] = \int_0^\beta d\tau \left\{ \frac{1}{2} M \dot{q}^2 + V(q) \right\}$$

$$+ \frac{1}{2} \int_0^\beta d\tau \int_0^\beta d\tau' \alpha \left(\tau - \tau'\right) \left\{ q(\tau) - q(\tau') \right\}^2. \quad (6.5.71)$$

Actually the contribution of $\alpha(\tau - \tau')$ to (6.5.71) is very small for $|\tau - \tau'|$ small. The reduced matrix element gives all the equilibrium state properties of the system. The tunneling rate can now be calculated by using Callan-Coleman method. In order to demonstrate the correctness of (6.5.71), the authors of Ref. [11] include in Appendix B the case for $V(q) = \frac{1}{2} m\omega^2 q^2$, i.e., the case of a damped oscillator. The result agrees with the exact expression of the reduced density matrix of a damped oscillator. The equation of motion given by the effective action (6.5.71) is

$$M\ddot{q} = \frac{\partial V}{\partial q} + \frac{\eta}{\pi} \int_{-\infty}^\infty d\tau' \frac{q(\tau) - q(\tau')}{(\tau - \tau')^2}, \quad (6.5.72)$$

where the second term in the right-hand side is the effective force exerted on the system by the environment. The principal value is to be taken at $\tau = \tau'$ for the integral. Solutions of the equation is the classical path. Assumed that $V(q)$ has a minimum 0 at $q = 0$, a maximum at $q = q_{\max}$ decreases to zero at $q = q_0$, and becomes negative for $q > q_0$. A metastable state at $q = 0$ can tunnel through the barrier to q_0 and will not come back. Substituting the classical path into (6.5.71) we get the classical action B.

$$B = \int_{-\infty}^\infty d\tau \left[\frac{1}{2} M \dot{q}^2 + V(q) \right] + \frac{\eta}{4\pi} \int_{-\infty}^\infty d\tau \int_{-\infty}^\infty d\tau' \left[\frac{q(\tau) - q(\tau')}{(\tau - \tau')^2} \right]^2.$$
$$(6.5.73)$$

The tunneling rate Γ is

$$\Gamma = A \exp\left(-\frac{B}{\hbar} \right), \quad (6.5.74)$$

where q in the integrand is the bounce solution. The bounce action is the decisive quantity determining the decay rate. The prefactor A is the contribution of paths deviating from the classical path. Referring to Eq. (5.2.21) we have

$$A = \left(\frac{B^*}{2\pi\hbar}\right)^{1/2} \left|\frac{\det \hat{\mathcal{D}}_0}{\det' \hat{\mathcal{D}}_1}\right|^{1/2}, \qquad (65.75)$$

in which the operators $\hat{\mathcal{D}}_0$ and $\hat{\mathcal{D}}_1$ are defined respectively as

$$\hat{\mathcal{D}}_0 q(\tau) \equiv \left(-\frac{d^2}{d\tau^2} + \omega_0^2\right) q(\tau) + \frac{\eta}{\pi M} \int_{-\infty}^{\infty} \frac{q(\tau) - q(\tau')}{(\tau - \tau')^2} d\tau', \quad (6.5.76)$$

$$\hat{\mathcal{D}}_1 q(\tau) \equiv \left(-\frac{d^2}{d\tau^2} + \frac{1}{M} V''\left[q_{cl}(\tau)\right]\right) q(\tau) + \frac{\eta}{\pi M} \int_{-\infty}^{\infty} \frac{q(\tau) - q(\tau')}{(\tau - \tau')^2} d\tau'. \tag{6.5.77}$$

In (6.5.77), $q_{cl}(\tau)$ is the bounce solution of (6.5.72). In (6.5.75), B^* is[18]

$$B^* = \int_{-\infty}^{\infty} M \dot{q}_{cl}^2 d\tau. \tag{6.5.78}$$

For the general case, we can begin with finding a numerical solution to Eqn. (6.5.72), then substitute this solution to (6.5.73) to find the bounce action. One can also get some feeling on the magnitudes of important quantities by using the analytic results from the quadratic+cubic potential. Analytic results are available for the weak damping limit $\alpha = \frac{\eta}{2m\omega} \to 0$ and the strong damping limit $\alpha \to \infty$. Here we give only results of Ref. [11]. Without dissipation the bounce action and the tunneling rate are $B_0 = \frac{36V_0}{5\omega_0}$ (5.A.2) and $\Gamma = \sqrt{60}\omega_0 \left(\frac{B_0}{2\pi\hbar}\right)^{\frac{1}{2}} e^{-B_0/\hbar}$ (5.A.31) respectively. One of the effects of dissipation is to increase the bounce action

$$B = B_0 + \Delta B \tag{6.5.79}$$

---

[18] B^* originates from the handling of the zero mode solution, and for a zero energy solution the action is twice the time integral of the kinetic energy.

where

$$\Delta B = \Phi(\alpha)\, \eta q_0^2 = \Phi(\alpha)\, 2M\omega_0 \alpha q_0^2. \tag{6.5.80}$$

Here Φ is a number of $O(1)$ depending on α. For weak damping $\Phi = 12\zeta(3)/\pi^3 \approx 0.47$; for strong damping $\Phi = 2\pi/9 \approx 0.70$. The other effect of dissipation is to change the prefactor. A general analysis shows that the prefactor has a stronger dependence on α, but prefactor affects the tunneling rate much less than the classical action on the exponential. By (5.A.1) the q_0^2 in (6.5.80) can be written as

$$q_0^2 = \frac{27 V_0}{2M\omega_0^2}. \tag{6.5.81}$$

Substituting (6.5.81) into (6.5.80), we finally obtain

$$\Gamma(\alpha) = f(\alpha)\, \omega_0 \left(\frac{60 V_0}{\hbar\omega_0}\right)^{1/2} \left(\frac{18}{5\pi}\right)^{1/2} \exp\left\{-\frac{36}{5}\frac{V_0}{\hbar\omega_0}\left(1 + \frac{15}{4}\Phi\alpha\right)\right\}, \tag{6.5.82}$$

where $f(\alpha)$ is a correction to the prefactor due to dissipation. For $\Delta B > 0$ dissipation always tends to lower the tunneling rate. This is the inevitable conclusion in the instanton method. The instanton method is a semi-classical approximation, but is non-perturbative. When the instanton method is not applicable, for instance, in a shallow double well the quantum effect is very important, dissipation can lead to enhancing the tunneling rate [22].

6.5.6 Dissipation and the Macroscopic Quantum Coherence

The influence of dissipation on macroscopic quantum coherence is much more complicated than that on macroscopic quantum tunneling, and is much less understood. A macroscopic quantum coherence can either be characterized by a coherent superposition of ψ_L and ψ_R leading to the removal of degeneracy or by an oscillation between the wells of an initially prepared state in one of the wells.

The coherent superposition can happen because a state starting in one of the wells can oscillate regularly and indefinitely between the wells. A macroscopic quantum tunneling is a tunneling once for all. In the following, we give a semi-quantitative discussion based on Ref. [10].[19] Consider the double well potential shown in Fig. 6.9. The potential contour for the double well plus the environment potential is shown in Fig. 6.14. Consider the two level problem, i.e. two bound states ψ_L and ψ_R in the left and right well respectively when the barrier between the wells is very high. They are degenerate in energy. When the barrier is lowered, we can expect a tunnel splitting, with the even combination of ψ_L and ψ_R being the ground state, and the odd combination being the excited state. The two-level problem can be conveniently expressed in terms of the Pauli matrices. In the representation in which σ_z is diagonal, $\sigma_z = +1(-1)$ corresponds to the right (left) well $+q_0$ $(-q_0)$. In other words, the eigenstates of σ_z represent ψ_R and ψ_L. The Hamiltonian of the system can be written as

$$H_{\mathrm{s}} = -\frac{\hbar\Gamma}{2}\sigma_x, \qquad (6.5.83)$$

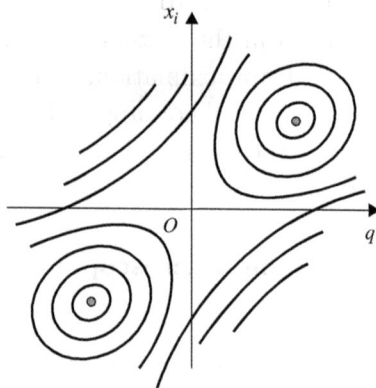

Fig. 6.14   Potential contours of double well plus environment in the multi-dimensional space (q, x_i). Redrawn from Ref. [10].

---

[19]More detailed discussions can be found in Ref. [12].

which leads to a transition between ψ_R and ψ_L, and is usually called the hopping term. The eigenstates and the corresponding eigenvalues of H_s are

$$
\left.
\begin{aligned}
\psi_0 &= \frac{1}{2}\begin{pmatrix} 1 \\ 1 \end{pmatrix} = \frac{1}{\sqrt{2}}(\psi_R + \psi_L), & \varepsilon_0 &= -\frac{\hbar\Gamma}{2} \\
\psi_e &= \frac{1}{\sqrt{2}}\begin{pmatrix} 1 \\ -1 \end{pmatrix} = \frac{1}{2}(\psi_R - \psi_L), & \varepsilon_e &= -\frac{\hbar\Gamma}{2}
\end{aligned}
\right\}.
\tag{6.5.84}
$$

The energy splitting is $\hbar\Gamma$. According to (6.5.58), the Hamiltonian of the system-environment complex is

$$
H = -\frac{\hbar\Gamma}{2}\sigma_x - q_0\sigma_z \sum_i C_i x_i + \frac{1}{2}\sum_i\left(\frac{p^2}{2m_i} + m_i\omega_i^2 x_i^2\right)
$$
$$
+ q_0^2 \sum_i \frac{C_i^2}{2m_i\omega_i^2}.
\tag{6.5.85}
$$

The interaction term in (6.5.58) $-q\sum_i C_i x_i$ is now replaced by $-q_0\sigma_z \sum_i C_i x_i$. In Section 6.4 we learned that the coherent splitting Γ is the resonance frequency, i.e., the frequency for the system to oscillate between the two wells. For the environment frequencies $\omega_i \gg \Gamma$ the system-environment interaction can be described by the adiabatic approximation. The wave function of the complex is

$$
\psi(\sigma_z, \{x_i\}, t) \approx \sum_{\alpha=\pm 1} C_\alpha(t)\chi_\alpha(\sigma_z)\prod_i \psi_\alpha^0(x_i),
\tag{6.5.86}
$$

where $\chi_\alpha(\alpha = +1, -1)$ is the eigenstate (R, L) of σ_z. We define h_i as the following

$$
h_i = -q_0\sigma_z \sum_i C_i x_i + \frac{1}{2}\sum_i\left(\frac{p^2}{2m_i} + m_i\omega_i^2 x_i^2\right) + q_0^2 \sum_i \frac{C_i^2}{2m_i\omega_i^2},
$$

such that we have

$$
H = -\frac{\hbar\Gamma}{2}\sigma_x + \sum_i h_i.
$$

When h_i is operated on ψ_R and ψ_L, σ_z can be replaced by $+1$ and -1 respectively. Therefore,

$$h_i = -q_0\alpha C_i x_i + \frac{p_i^2}{2m} + \frac{1}{2}m_i\omega_i^2 x_i^2 \tag{6.5.87a}$$

$$h_i = \frac{p_i^2}{2m} + \frac{1}{2}m_i\omega_i^2\left(x_i - \frac{\alpha C_i q_0}{m_i\omega_i^2}\right)^2 + \text{const.} \tag{6.5.87b}$$

We denote the eigenstates of h_i by $\psi_\alpha^0(x_i)$. The first two terms of (6.5.87b) is the Hamiltonian of a harmonic oscillator with position of equilibrium at $\frac{\alpha C_i q_0}{m_i\omega_i^2}$. Let the wave function of a harmonic oscillator with the position of equilibrium at the origin be $\psi_0(x)$, then

$$\psi_\alpha^0(x_i) = \psi_0\left(x_i - \frac{\alpha C_i q_0}{m_i\omega_i^2}\right). \tag{6.5.88}$$

When the system is in the left well, the wave function of the complex is $\chi_{-1}\prod_i \psi_{-1}^0(x_i)$, while when the system is in the right well, the wave function is $\chi_{+1}\prod_i \psi_{+1}^0(x_i)$. When the environment is taken into account, the resonance frequency becomes

$$\tilde{\Gamma} = \Gamma\left\langle \chi_+ \prod_i \psi_{+1}^0(x_i) \mid \sigma_x \mid \chi_- \prod_i \psi_{-1}^0(x_i) \right\rangle,$$

$$\Gamma = \prod_i (\psi_{+1}^0(x_i), \psi_{-1}^0(x_i)). \tag{6.5.89}$$

Simple integration gives[20]

$$\tilde{\Gamma} = \Gamma \exp\left[-\sum_i \frac{C_i^2 q_0^2}{\hbar m_i\omega_i^3}\right]. \tag{6.5.90}$$

Equation (6.5.90) shows that the influence of environment tends to lower the tunneling rate. This can also be seen from Fig. 6.14. Because of the coupling, the position of the bottoms of the wells

---

[20] From $\psi_0(x) = \left(\frac{m\omega}{\hbar\pi}\right)^{\frac{1}{4}} e^{-m\omega x^2/2\hbar}$, we have

$$(\psi_{+1}^0(x_i), \psi_{-1}^0(x_i)) = \int_{-\infty}^{+\infty} dx \left(\frac{m\omega}{\hbar\pi}\right)^{\frac{1}{2}} e^{-\frac{C_i q_0^2}{\hbar m_i\omega_i^2}} e^{-\frac{m\omega}{\hbar}x^2} = \exp\left(-\frac{C_i^2 q_0^2}{\hbar m_i\omega_i^2}\right)$$

move from $\pm q_0$ on the q-axis towards $\pm x$-direction. This enhances the difficulty of the tunneling. Consider the right-hand side of Eq. (6.5.90). The exponential function depends on the spectral density $J(\omega)$. From Eq. (6.5.55) it follows that

$$\int \frac{J(\omega)}{\omega^2} d\omega = \frac{\pi}{2} \sum_i \int \frac{C_i^2}{m_i \omega_i \omega^2} \delta(\omega_i - \omega) d\omega = \frac{\pi}{2} \sum_i \frac{C_i^2}{m_i \omega_i^3}.$$

Consequently

$$\exp\left[-\frac{q_0^2}{\hbar} \sum_i \frac{C_i^2}{m_i \omega_i^3}\right] = \exp\left[-\frac{2q_0^2}{\pi\hbar} \int \frac{J(\omega)}{\omega^2} d\omega\right].$$

The upper limit of the integral is the cut-off frequency $\bar{\omega}$, the lower limit is ω_Γ, the boundary of the applicability of adiabatic approximation. Equation (6.5.90) gives

$$\tilde{\Gamma} = \Gamma \exp\left[-\frac{2q_0^2}{\pi\hbar} \int_{\omega_\Gamma}^{\bar{\omega}} \frac{J(\omega)}{\omega^2} d\omega\right]. \tag{6.5.91}$$

When $J = \eta\omega$, it follows that

$$\tilde{\Gamma} = \Gamma \left(\frac{\omega_\Gamma}{\bar{\omega}}\right)^{\eta/\eta_c} = \bar{\omega} \left(\frac{\Gamma}{\bar{\omega}}\right) \left(\frac{\omega_\Gamma}{\bar{\omega}}\right)^{\eta/\eta_c}, \tag{6.5.92}$$

where

$$\eta_c = \frac{\pi\hbar}{2q_0^2}.$$

The above discussion is unsatisfactory in the sense that a lower limit ω_Γ is taken. Then what is the contribution of those environment oscillators whose frequencies are lower than ω_Γ? We choose another lower limit Γ, such that the environment oscillators with frequencies lower than Γ cannot follow the slow motion of the system and consequently can be safely neglected. Equation (6.5.92) gives

$$\tilde{\Gamma} = \bar{\omega} \frac{\Gamma}{\bar{\omega}} \left(\frac{\Gamma}{\bar{\omega}}\right)^{\eta/\eta_c} = \bar{\omega} \left(\frac{\Gamma}{\bar{\omega}}\right)^{1+\eta/\eta_c}. \tag{6.5.93}$$

This result is encouraging: the real frequency $\tilde{\Gamma} = \Gamma \left(\frac{\Gamma}{\bar{\omega}}\right)^{\frac{\eta}{\eta_c}} < \Gamma$. Take a step further. Why not choose $\tilde{\Gamma}$ as the lower limit? It is sufficient that $\omega_i \gg \tilde{\Gamma}$, such that the adiabatic approximation is valid. Then we have

$$\tilde{\Gamma}' = \bar{\omega}\frac{\Gamma}{\bar{\omega}}\left(\frac{\tilde{\Gamma}}{\bar{\omega}}\right)^{\eta/\eta_c} = \bar{\omega}\frac{\Gamma}{\bar{\omega}}\left[\left(\frac{\Gamma}{\bar{\omega}}\right)^{1+\eta/\eta_c}\right]^{\eta/\eta_c}$$

$$= \bar{\omega}\left(\frac{\Gamma}{\bar{\omega}}\right)^{1+\frac{\eta}{\eta_c}+\left(\frac{\eta}{\eta_c}\right)^2}.$$

The frequency is further lowered. We continue this process, and two possible outcomes can emerge.

$$\left.\begin{array}{ll} \tilde{\Gamma} = \bar{\omega}\left(\dfrac{\Gamma}{\bar{\omega}}\right)^{\frac{1}{1-\eta/\eta_c}}, & \eta < \eta_c \\[2mm] \tilde{\Gamma} = 0, & \eta > \eta_c \end{array}\right\}. \qquad (6.5.94)$$

Such renormalization-group-like argument leads finally to a very interesting result. If η is lower than the critical value η_c, the tunneling frequency is lowered. For η sufficiently small, the quantum coherence can be maintained, i.e., the coherence splitting is still at work. This is rather demanding. Only quite recently the coherence splitting has been experimentally established in a SQUID under condition of strictly controlled coupling between system and environment. For η larger but still smaller than η_c, tunneling can still be detected, but the coherence splitting evades measurement. This situation is called the sequential tunneling, i.e., tunneling without coherence. If $\eta > \eta_c$, the system sits in one of the wells, with vanishing probability of tunneling: the state is *localized*.

6.6   Experiments on the Macroscopic Quantum Tunneling in Josephson Systems

Two experiments reported in 1981 [23, 24] both used current biased junctions. Equation (6.2.18) gives the potential energy

$$V(\phi) = -\frac{\Phi_0}{2\pi}\left(I_c \cos\phi + I_{\text{ext}}\phi\right). \qquad (6.6.1)$$

When $I_{\text{ext}}/I_{\text{c}} \equiv x < 1$, $V(\phi)$ has a series of local minima (wells), for which $\sin \phi_{\text{min}} = x$, between them are barriers. The difference in energy between the metastable state in a well (with zero point energy $\hbar\omega_0/2$ above the bottom of the well) and the top of the barrier is

$$H\left(x\right) = \frac{I_{\text{c}}\Phi_0}{\pi}\left[\left(1-x^2\right)^{1/2} - x\cos^{-1}x\right] - \frac{\hbar\omega_0}{2}. \qquad (6.6.2)$$

The barrier height decreases when I approaches the critical value. At sufficiently high temperature the metastable state can be thermally activated to escape, the transition rate being

$$\tau_{\text{th}}^{-1} = \frac{\omega_0}{2\pi}e^{-H(x)/kT}. \qquad (6.6.3)$$

The transition rate decreases with lowering temperature. The thermal activation rate approaches 0 when $T \to 0$. The quantum mechanical tunneling rate is independent of T. In a CBJ the macroscopic observable is the phase ϕ of the junction. The tunneling rate is

$$\tau_{\text{MQT}}^{-1} = \frac{\omega_0}{2\pi}\left(\frac{b}{2\pi}\right)^{1/2}e^{-b}. \qquad (6.6.4)$$

The quantity b for a CBJ (shunt resistance R, capacitance C) and for $x \lesssim 1$ is

$$b = \frac{H\left(x\right)}{\hbar\omega_0}\left(7.2 + \frac{8A}{\omega_0 RC}\right). \qquad (6.6.5)$$

The second term in the parenthesis in the right-hand side is the dissipation effect of the environment, where $A = O(1)$. When ϕ is in the well, it oscillates with frequency ω_0 and $\langle\phi\rangle$ is at the bottom of the well. When the tunneling through the barrier takes place, ϕ will roll down the potential curve, a finite $\dot{\phi}$ produces a voltage determined by the Josephson relation. In the experiment of Ref. [24] the junction is kept at a given temperature ranging from $5\,\text{mK}$ to $1.7\,\text{K}$, and the current is increased in each case. When the current reaches a certain value the junction switches from a superconducting state (voltage across the junction $= 0$) to the normal state (finite voltage across the junction). The switching probability distribution $P(I)$ versus applied current is recorded. With decreasing temperature the distribution

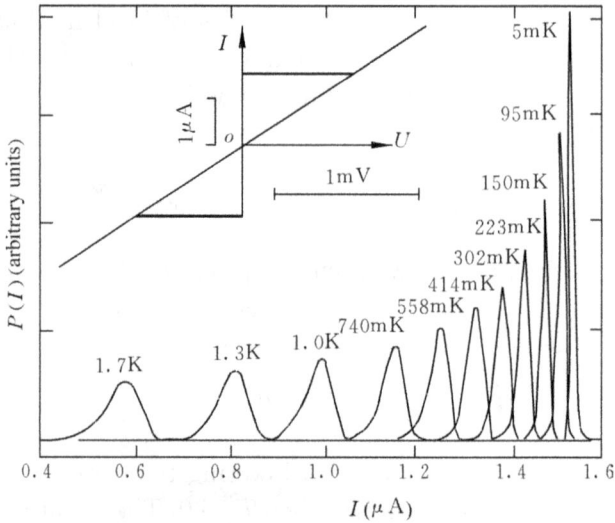

Fig. 6.15   The distribution in transition current $P(I)$ from superconducting to normal states of a Josephson junction. Reprinted with permission from Richard F. Voss and Richard A. Webb, *Phys. Rev. Lett.* 47, 265, 1981. Copyright (2017) by the American Physical Society.

$P(I)$ narrows and moves to higher values of I as shown in Fig. 6.15. The inset shows the current-voltage characteristics of the junction at 95 mK. The two vertical segments represent the superconducting state of the junction. The junction switches to normal state (inclined straight lines) through the horizontal segments. The voltage is produced momentarily, signaling the penetration of barrier by ϕ. The transition rate $\tau^{-1}(I)$ can be calculated from the distribution $P(I)$. Figure 6.16 gives the $\tau^{-1}(I)$ curve for two junctions. The $\ln \tau^{-1}(I)$ values are approximately proportional to I. When the temperature is lowered $\tau^{-1}(I)$ shifts to higher current values and inclinations of curves increase. For $T \lesssim 100$ mK the inclination almost does not change. The solid lines represent theoretical result of the thermal activation (6.6.3), and dashed lines represent the quantum tunneling rate without dissipation and the dotted lines represent the rate with dissipation (6.6.4), (6.6.5). They are independent of temperature. A fit of experimental results with theoretical prediction determines the value of parameter A characterizing the dissipation: $A \approx 4.5$.

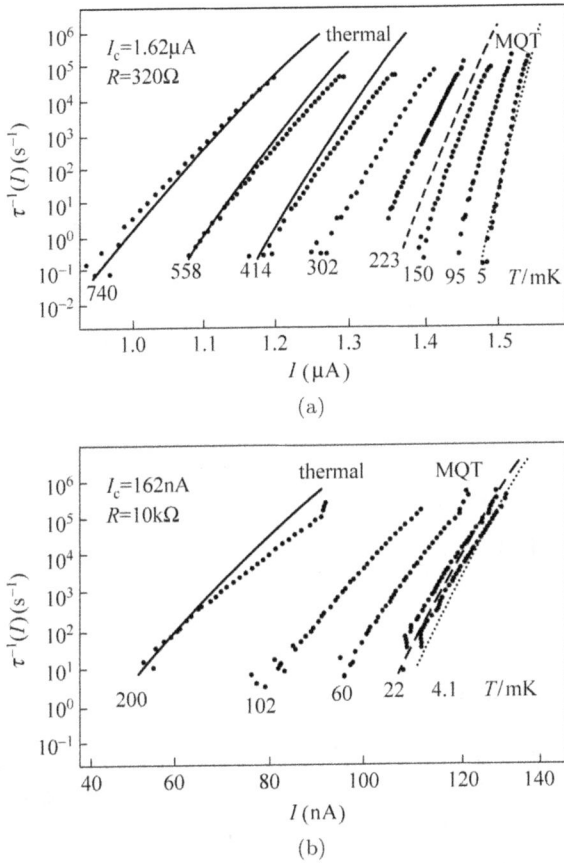

Fig. 6.16   Transition rate $\tau^{-1}(I)$ for two junctions. Reprinted with permission from Richard F. Voss and Richard A. Webb, *Phys. Rev. Lett.* 47, 265, 1981. Copyright (2017) by the American Physical Society.

Figure 6.17 gives the widths of $P(I)$ for two junctions, $\Delta I = \langle (I - \langle I \rangle^2 \rangle^{\frac{1}{2}}$ as a function of T. Solid, dashed and dotted lines represent thermal activation, quantum tunneling without and with dissipation ($A = 4.5$) respectively. Another group [25] studied the relation between tunneling rate and dissipation. The results agree with theory. The macroscopic observable in the well is the phase ϕ. It can penetrate potential barrier, and have different quantum states in the well [26]. From the late 1980s through 1990s there have been a significant number of precise experiments.

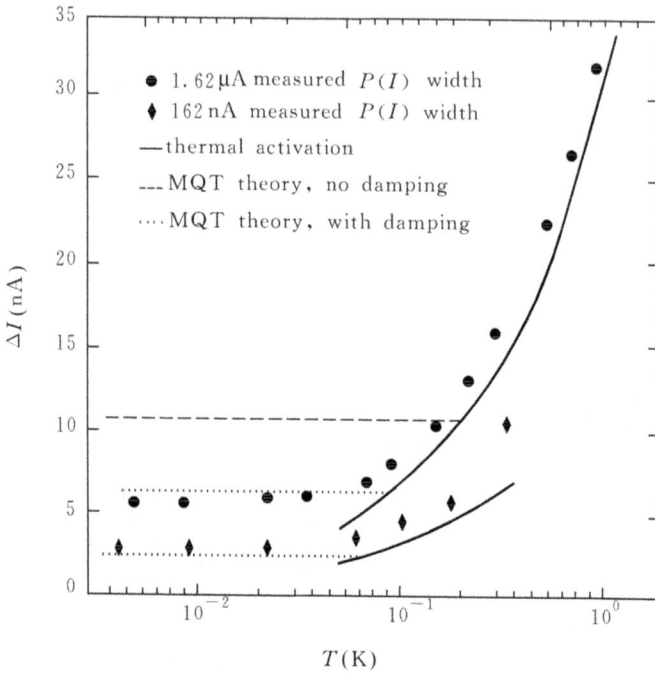

Fig. 6.17   The width of transition current distribution $P(I)$ versus temperature. Reprinted with permission from Richard F. Voss and Richard A. Webb, *Phys. Rev. Lett.* 47, 265, 1981. Copyright (2017) by the American Physical Society.

The experimental discovery of macroscopic quantum coherence has been discussed in section 4.9 in connection with the Schrödinger cat. It was really a significant breakthrough.

6.7   Macroscopic Quantum Tunneling of Magnetism, the Spin Coherent State

Before the late 1980s discussions on macroscopic quantum phenomena were mainly limited to the Josephson systems. In 1988 discussions on the theory of macroscopic quantum tunneling of magnetism began [27, 28]. Experimental research followed quickly. Large number of published papers appeared in the early 1990s. Due to the strong exchange interaction between atoms, ferromagnetic and anti-ferromagnetic single domain particles can have magnetic moment $10^3 \sim 10^6$ Bohr magnetons, which exhibit macroscopic quantum properties in external magnetic field as a united whole. The domain walls can involve some 10^{10} elementary spins. These are called giant spin phenomena.

The method of spin coherent states is used in the theory of the tunneling of magnetism. The following is a brief introduction of the subject [29, 30]. The construction of spin coherent states is based on the spin operators. The components of a spin operator S are (S_1, S_2, S_3). S^2 is the Casimir operator with eigenvalue $s(s+1)$, s can be integers or half-integers. Take the eigenstate $|m\rangle$ of S_3 as the fiducial state, $-s \leq m \leq s$. Let the polar and azimuthal angle of a unit vector $\hat{\Omega}$ be θ and ϕ respectively. The rotation operator $U(\theta, \phi)$ rotates the z-axis onto the direction of $\hat{\Omega}(\theta, \phi)$. The expression of $U(\theta, \phi)$ is

$$U(\theta, \phi) = e^{-i\phi S_3} e^{-i\theta S_2}. \tag{6.7.1}$$

Applying $U(\theta, \phi)$ to $|m\rangle$ leads to a state of spin s with component m in the direction of $\hat{\Omega}(\theta, \phi)$, called the spin coherent state, denoted by $|\Omega\rangle$ or $|\theta, \phi\rangle$.

$$|\Omega\rangle \equiv |\theta, \phi\rangle = U(\theta, \phi)|m\rangle = e^{-i\phi S_3} e^{-i\theta S_2}|m\rangle. \tag{6.7.2}$$

Inserting the identity $I = \sum_{n=-s}^{s}|n\rangle\langle n|$ between the two exponential factors, we obtain

$$|\theta, \phi\rangle = \sum_{n=-s}^{s} e^{-in\phi}\langle n|e^{-i\theta S_2}|m\rangle|n\rangle, \tag{6.7.3}$$

where the matrix element in (6.7.3) is called the reduced Wigner coefficient in the representation of angular momentum s.

$$d_{nm}^{s}(\theta) \equiv \langle n|e^{-i\theta S_2}|m\rangle. \tag{6.7.4}$$

The expression for d is simple when $m = s$ (the highest weight):

$$d_{ns}^{s}(\theta) = \left(\frac{2s}{s-n}\right)^{1/2} \cos^{s+n}\frac{\theta}{2} \sin^{s-n}\frac{\theta}{2}. \tag{6.7.5}$$

The resolution of unity is[21]

$$\frac{2s+1}{4\pi} \int |\theta, \phi\rangle \langle \theta, \phi| \sin\theta d\phi d\theta = I. \tag{6.7.6}$$

The overlapping of spin coherent state is

$$\langle \theta, \phi \mid \theta' \phi' \rangle = \langle m | e^{i\theta S_2} e^{i(\phi - \phi')S_3} e^{-i\theta S_2} | m \rangle$$

$$= \sum_{n=-s}^{s} d_{mn}^s(\theta) d_{nm}^s(-\theta') e^{i(\phi - \phi')n}. \tag{6.7.7}$$

For the highest weight $m = s$ the above equality becomes

$$\langle \theta, \phi \mid \theta', \phi' \rangle = \left[\cos\frac{\theta}{2} \cos\frac{\theta'}{2} e^{i\frac{1}{2}(\phi-\phi')} + \sin\frac{\theta}{2} \sin\frac{\theta'}{2} e^{-i\frac{1}{2}(\phi-\phi')} \right]^{2s}$$

$$= \left[\cos\frac{\theta}{2} \cos\frac{\theta'}{2} e^{i\frac{1}{2}(\phi-\phi')} \right]^{2s} \left[1 + \tan\frac{\theta}{2} \tan\frac{\theta'}{2} e^{i(\phi-\phi')} \right]^{2s}. \tag{6.7.8}$$

The above relation can also be written in the following form

$$\langle \Omega \mid \Omega' \rangle = |\langle \Omega \mid \Omega' \rangle| e^{-i\Phi},$$

where

$$|\langle \Omega \mid \Omega' \rangle| = \left[\left(\cos\frac{\theta}{2} \cos\frac{\theta'}{2} + \sin\frac{\theta}{2} \sin\frac{\theta'}{2} \cos\left(\phi' - \phi\right) \right)^2 \right.$$

$$\left. + \left(\sin\frac{\theta}{2} \sin\frac{\theta'}{2} \sin\left(\phi' - \phi\right) \right)^2 \right]^s$$

$$= \left[\frac{1}{2} \left(1 + \hat{\Omega} \cdot \hat{\Omega}' \right) \right]^s.$$

The angle Φ between $\hat{\Omega}$ and $\hat{\Omega}'$ is given by

$$\tan\Phi = \frac{\sin\frac{\theta}{2} \sin\frac{\theta'}{2} \sin\left(\phi' - \phi\right)}{\cos\frac{\theta}{2} \cos\frac{\theta'}{2} + \sin\frac{\theta}{2} \sin\frac{\theta'}{2} \sin\left(\phi' - \phi\right)} 2s. \tag{6.7.9}$$

---

[21] The proof of (6.7.6) and (6.7.7) is straightforward. See Refs. [29, 30].

In the foregoing discussions, many properties common to the coherent states are explicit. The parameters characterizing a coherent state is continuous, coherent states with different parameters are not orthogonal. They constitute an overcomplete set, and through the resolution of unity an arbitrary wave function can be expressed in term of coherent states.

For a quantum spin system, the spin coherent states provide a natural framework to apply the semiclassical approximation. Using imaginary time, the path integral for the transition amplitude can be written as

$$\langle \Omega_f | e^{-\hat{H}(\tau_f - \tau_i)/\hbar} | \Omega_i \rangle$$

$$= \prod_{n=1}^{N} \left[\int \left(\frac{2s+1}{4\pi} d\Omega_n \right) \right] \prod_{n=1}^{N+1} \langle \Omega_n | e^{-\varepsilon \hat{H}/\hbar} | \Omega_{n-1} \rangle. \quad (6.7.10)$$

In the above the imaginary time interval $\tau_f - \tau_i$ is divided into N equal small segments ε, and N resolutions of unity are inserted between the exponential factors. A typical factor in (6.7.10) is

$$\langle \Omega_n | e^{-\varepsilon \hat{H}/\hbar} | \Omega_{n-1} \rangle = \langle \Omega_n | 1 - \varepsilon \hat{H}/\hbar \, | \, \Omega_{n-1} \rangle$$

$$= \langle \Omega_n \, | \, \Omega_{n-1} \rangle \left[1 - \frac{\varepsilon}{\hbar} \frac{\langle \Omega_n | \hat{H} | \Omega_{n-1} \rangle}{\langle \Omega_n \, | \, \Omega_{n-1} \rangle} \right]. \quad (6.7.11)$$

In the following, we calculate each factor of this amplitude. Factoring the amplitude and phase factor of $\langle \Omega_n \, | \, \Omega_{n-1} \rangle$ we have

$$\langle \Omega_n \, | \, \Omega_{n-1} \rangle = |\langle \Omega_n \, | \, \Omega_{n-1} \rangle| e^{-i\Phi}.$$

According to (6.7.9a) the amplitude is $\left[\frac{1}{2}(1 + \hat{\Omega}_n \cdot \hat{\Omega}_{n-1}) \right]^s$. In the limit $N \to \infty$, $\varepsilon \to 0$ it approaches 1. The phase angle in the limit is, according to (6.7.9b)

$$\Phi \approx \tan \Phi \approx 2 \sin^2 \frac{\theta}{2} \Delta\phi = s \, (1 - \cos \theta) \, \Delta\phi,$$

where $\Delta\phi = \phi_n - \phi_{n-1}$. Since in the following a sum is to be taken over many $\Delta\phi$'s, this leading term in $\Delta\phi$ should be kept. Therefore

$$\langle \Omega_n \mid \Omega_{n-1} \rangle = e^{-is(1-\cos\theta)\Delta\phi} = e^{-is(1-\cos\theta)\dot\phi_n\varepsilon}.$$

Since in (6.7.11) the second term in the bracket already contains a factor ε, the $O(\varepsilon)$ term in the quotient of the matrix elements can be neglected, then

$$\frac{\langle \Omega_n|\hat H|\Omega_{n-1}\rangle}{\langle \Omega_n \mid \Omega_{n-1}\rangle} = \langle \Omega_n|\hat H|\Omega_n\rangle + O\left(\varepsilon\right) \approx H\left(\Omega_n\right).$$

$H(\Omega_n)$ is already a c-number, and it is the expected value of the Hamiltonian in the coherent state $|\Omega_n\rangle$. The matrix element (6.7.11) has been fully calculated:

$$\langle \Omega_n|e^{-\varepsilon\hat H/\hbar}|\Omega_{n-1}\rangle = \exp\left[-i\varepsilon s\left(1-\cos\theta\right)\dot\phi_n - \frac{\varepsilon}{\hbar}H\left(\Omega_n\right)\right].$$

The transition amplitude (Eq. (6.7.10)) can now be written as a path integral

$$\langle \Omega_f|\exp\left[-H\left(\tau_f - \tau_i\right)/\hbar\right]|\Omega_i\rangle$$

$$= \int_{\Omega(\tau_i)}^{\Omega(\tau_f)} [D\Omega(\tau)]$$

$$\times \exp\left[-is\int_{\tau_i}^{\tau_f}\left(1-\cos\theta\right)\dot\phi d\tau - \frac{1}{\hbar}\int_{\tau_i}^{\tau_f}H(\Omega(\tau))\right]. \quad (6.7.12)$$

The quantity on the exponential of the path integral is a function of $\Omega(\tau)$ and is proportional to the action $S[\Omega(\tau)]$.

$$\frac{1}{\hbar}S\left[\Omega(\tau)\right] = \int_{\tau_i}^{\tau_f}\left[-is\left(1-\cos\theta\right)\dot\phi(\tau) + \frac{1}{\hbar}H\left(\Omega(\tau)\right)\right]d\tau. \quad (6.7.13)$$

We discuss the significance of the first term of (6.7.13). In Fig. 6.18, on the surface of a unit sphere the shaded area of a spherical triangle

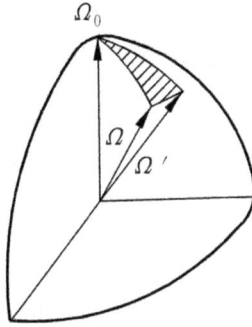

Fig. 6.18 The geometrical significance of Wess-Zumino term.

with vertices at $\hat{\Omega}(\theta, \phi)$, $\hat{\Omega}'(\theta + d\theta, \phi + d\phi)$ and $\hat{\Omega}_0(\theta = 0)$ is

$$d\omega\left[\Omega(\tau)\right] = \int_0^\theta \sin\theta' d\theta' d\phi = (1 - \cos\theta)\, d\phi. \qquad (6.7.14)$$

If in the path integral we have $\Omega(\tau_i) = \Omega(\tau_f)$, i.e., $\Omega(\tau)$ describes a closed contour, then

$$\omega\left[\Omega\right] = \int_{\tau_i}^{\tau_f} (1 - \cos\theta)\, \dot{\phi}(\tau)\, d\tau = \int_{\tau_i}^{\tau_f} (1 - \cos\theta)\, d\phi \qquad (6.7.15)$$

is the area enclosed by the closed contour on the spherical surfaces. The interesting fact is, the spherical surface does not have a boundary. The inside and outside of a closed curve on it cannot be distinguished from each other. Let the area on both sides of a closed contour be S_1 and S_2, and their sign is determined by a convention between the sense of winding the contour and the outside normal of the area (for example, a right hand screw). Then $S_1 - S_2 = 4\pi$. In (6.7.12), the factor $e^{-is\int(1-\cos\theta)\dot{\phi}d\tau} = e^{-isS}$, where s is an integer or a half-integer. Since $e^{-is4\pi} = 1$, no matter s is integral or half-integral, we have $e^{-isS_1} = e^{-is(4\pi+S_2)} = e^{-isS_2}$, i.e., each of S_1 or S_2 gives the same result for the phase factor. This factor has a geometrical origin, because it does not depend on the speed with which Ω traces a closed contour. The term $-is(1-\cos\theta)\dot{\phi}$ in the Lagrangian is called the Wess-Zumino term. Its significance will be further discussed in Sections 6.8 and 7.2.

6.8   Macroscopic Quantum Phenomena of Single Domain Ferromagnetic Particles

Consider a single domain ferromagnetic particle. Its magnetic moment M exhibits macroscopic quantum behavior and the magnitude of M remains constant during the quantum process. Let the unit vector in the direction of M be denoted by $\hat{n}(\theta, \phi)$. The crystalline magnetic anisotropy is described by the anisotropy energy $E_K(\theta, \phi)$. The direction in which a magnetic moment acquires the lowest energy is called the easy direction. The magnetic moment of the particle M is proportional to the total spin s of the particle $M = \gamma s$, where $\gamma = g\mu_B$, g is the gyromagnetic ratio and μ_B is the Bohr magneton. In the presence of an external magnetic field H, the energy of the magnetic moment is

$$E = -M \cdot H + E_K(\theta, \phi). \qquad (6.8.1)$$

The equation of motion of the magnetic moment is called the Landau-Lifshitz equation or the Bloch equation.

$$\frac{dM}{dt} = -\gamma M \times \frac{\delta E}{\delta M}. \qquad (6.8.2)$$

The effective magnetic field experienced by the magnetic moment is $-\frac{\delta E}{\delta M}$ which consists of the external magnetic field plus the crystalline anisotropy field. The magnetic moment presesses under this effective magnetic field. Since M remains a constant, the variable during the motion of the magnetic moment is $\hat{n}$. Equation (6.8.2) expressed in terms of $\hat{n}$ is

$$s\frac{d\hat{n}}{dt} = -\hat{n} \times \frac{\delta E}{\delta \hat{n}}, \qquad (6.8.3)$$

or in terms of θ and ϕ

$$\dot{\theta}\sin\theta = -\frac{\gamma}{M}\frac{\partial E}{\partial \phi}, \qquad (6.8.4)$$

$$\dot{\phi}\sin\theta = \frac{\gamma}{M}\frac{\partial E}{\partial \theta}. \qquad (6.8.5)$$

This pair of equations can be derived from the action

$$I = \int \mathrm{d}tL = \int \mathrm{d}t\{s\dot{\phi}\cos\theta - E\left(\theta, \phi\right)\}. \qquad (6.8.6)$$

Since adding a time derivative to L does not change the equation of motion, we can rewrite the action as

$$I = \int \mathrm{d}t\{-s(1 - \cos\theta)\dot{\phi} - E(\theta, \phi)\}. \qquad (6.8.7)$$

The Euclildean action is

$$S \equiv -\mathrm{i}I = \int \{\mathrm{i}s(1 - \cos\theta)\dot{\phi}(\tau) + E(\theta, \phi)\}\mathrm{d}\tau$$

$$= \mathrm{i}s\omega\left[\hat{\boldsymbol{n}}\right] + \int \mathrm{d}\tau E\left(\theta, \phi\right). \qquad (6.8.8)$$

in which the first term is the Wess-Zumino action. From comparison with (6.7.12) we see that (6.8.8) is just the factor on the exponential of the spin coherent state path integral. Strictly speaking the Wess-Zumino term is the consequence of the spin coherent states, which cannot be unambiguously obtained from the equation of motion.[22]

Macroscopic quantum tunneling of ferromagnetic particles can be treated by using the Callan-Coleman method. Research in this direction was pioneered by E. Chudnovsky and L. Gunther [27, 28].

6.8.1 Quantum Coherence: Tunneling Splitting of Energy States

Consider the crystalline anisotropy energy

$$E\left(\theta, \phi\right) = K_1 \cos^2\theta + K_2 \sin^2\theta \sin^2\phi, \quad K_1 > K_2 > 0. \qquad (6.8.9)$$

The x-direction ($\theta = \pi/2, \phi = 0$) is the easy axis, y-direction is the medium axis, and z-direction is the hard axis. The energy of the magnetic moment is the same when it is aligned along the positive or the negative x-axis. Since $\boldsymbol{M}$ is a pseudo-vector, an initially x-oriented magnetic moment will become $-x$-oriented under space

---

[22]To obtain the Wess-Zumino term in (6.8.8) the time derivative term is added ad hoc.

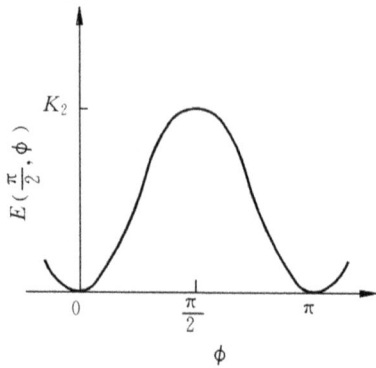

Fig. 6.19   Anisotropy energy in xy-plane $E\left(\frac{\pi}{2},\phi\right)$.

inversion. Consequently its properties must be invariant with respect to the exchange of $+x$ and $-x$-directions. Figure 6.19 shows the anisotropy energy on the xy-plane. Substituting (6.8.9) into the equations of motion (6.8.4) and (6.8.5) we obtain

$$\left.\begin{array}{l} \dot{\theta} = -\dfrac{2\gamma}{M}K_2\sin\theta\sin\phi\cos\phi \\[3mm] \dot{\phi} = -\dfrac{2\gamma}{M}K_1\left(1-\lambda\sin^2\phi\right)\cos\theta \end{array}\right\}, \qquad (6.8.10)$$

where $\lambda = K_2/K_1$. From (6.8.10) we can derive the equation satisfied by the instanton kink solution. Actually Eq. (6.8.10) gives

$$\frac{d\theta}{d\phi} = \frac{K_2\sin\theta\sin\phi\cos\phi}{K_1\left(1-\lambda\sin^2\phi\right)\cos\theta}.$$

Integrating once, we obtain

$$\ln\sin\theta = -\frac{1}{2}\ln\left(1-\lambda\sin^2\phi\right) + \text{const}, \qquad (6.8.11)$$

or

$$\sin^2\theta\left(1-\lambda\sin^2\phi\right) = C. \qquad (6.8.12)$$

Using this equation to eliminate the $\cos\theta$ term in (6.8.10), we obtain

$$\left(\frac{d\phi}{d\tau}\right)^2 = -\left(\frac{2\gamma}{M}\right)^2 K_1^2\left(1-\lambda\sin^2\phi\right)\left[-C+\left(1-\lambda\sin^2\phi\right)\right].$$

In the following, we look for the kink solution, i.e., $\phi \to 0, \pi$ and $\dot{\phi} \to 0$ for $\tau \to \pm\infty$. These conditions determine the constant $C = 1$ in the above equation. Using $\omega_0 = \frac{2\gamma}{M}(K_1 K_2)^{\frac{1}{2}}$, (6.8.12) becomes

$$\left(\frac{d\phi}{d\tau}\right)^2 = \omega_0^2 \left(1 - \lambda \sin^2 \phi\right) \sin^2 \phi. \qquad (6.8.13)$$

This is the equation for kink solution, and it is solved by

$$\phi = \arccos \frac{(1-\lambda)^{1/2} \tanh \omega_0 \tau}{\left(1 - \lambda \tanh^2 \omega_0 \tau\right)^{1/2}}. \qquad (6.8.14)$$

Equation (6.8.8) gives the Euclidean action by the help of (6.8.13)

$$S_{\text{inst}} = K_2 \int_{-\infty}^{\infty} d\tau\, 2 \sin^2 \phi.$$

Using the kink solution (6.8.14) we obtain the action for the kink

$$S_{\text{inst}} = -\frac{M}{\gamma} \ln \frac{1 - \sqrt{\lambda}}{1 + \sqrt{\lambda}}. \qquad (6.8.15)$$

The corresponding tunneling rate is

$$\Gamma = A \exp\left(-\frac{S_{\text{inst}}}{\hbar}\right) = A \exp\left(\frac{M}{\hbar\gamma} \ln \frac{1 - \sqrt{\lambda}}{1 + \sqrt{\lambda}}\right) = A \left(\frac{1 - \sqrt{\lambda}}{1 + \sqrt{\lambda}}\right)^{\frac{M}{\hbar\gamma}}, \qquad (6.8.16)$$

where A is the prefactor. Its evaluation is more involved. A. Garg and G.-H. Kim [31] gave a general method for calculating the prefactor of magnetic tunneling. The tunnel splitting of the energy level is $\Delta E = 2\hbar\Gamma$. Equation (6.8.16) shows that $\Gamma \to 0$ when $\lambda \to 1$. The reason is that $E \to \text{const}(M_x^2 + M_y^2) = \text{const}(M^2 - M_z^2)$ for $\lambda \to 1$. E_K then commutes with M_z which makes it a constant of motion. The value of λ determines the barrier height. The barrier is lower for smaller λ, and the tunneling becomes easier.

6.8.2 The Quantum Tunneling

Let z-direction be the easy axis, y-direction be the hard axis. An external magnetic field H is applied in the $-z$-direction. Then the

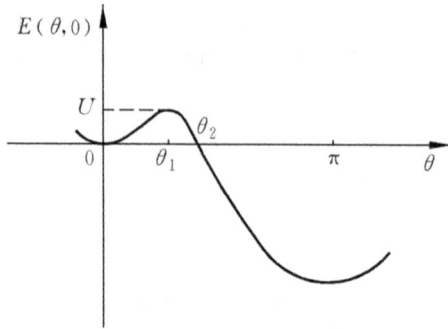

Fig. 6.20   Anisotropy energy curve, external magnetic field along the easy axis, and non-vanishing transverse anisotropy.

anisotropy energy is

$$E\left(\theta, \phi\right) = \left(K_1 + K_2 \sin^2 \phi\right) \sin^2 \theta - MH\left(1 - \cos \theta\right). \qquad (6.8.17)$$

Figure 6.20 shows the anisotropy energy curve. The effect of the magnetic field is to make $\theta = 0$ the metastable state and $\theta = \pi$ the stable state, and the height and position of the barrier can be adjusted by H. The position of the barrier is at θ_1, $\cos \theta_1 = H/H_c$, $H_c = 2K_1/M$. Let $\varepsilon = 1 - \frac{H}{H_c}$, then the barrier height is $U = K_1 \varepsilon^2$. H_c is the critical field strength. The potential barrier disappears when H reaches this value. The quantum tunneling is facilitated by the transverse anisotropy parameter K_2. M_z ceases to be a constant of motion due to a non-vanishing K_2. In this problem analytic solution is available only when an approximation is made. Because of a non-vanishing K_2 the classical path will not reach the region of large ϕ. We assume in addition that ε is small in order to avoid a high barrier.

Under this condition $\theta_1 \approx \sqrt{2\varepsilon}$, $\theta_2 \approx 2\sqrt{\varepsilon}$. From $E(\theta, \phi)$ the equation of motion is obtained

$$\dot{\theta} = \frac{\gamma}{M} 2K_2 \sin \theta \sin \phi \cos \phi, \qquad (6.8.18)$$

$$\dot{\phi} = -\frac{\gamma}{M}\left(2K_1 \cos \theta - MH\right). \qquad (6.8.19)$$

In (6.8.19) a term $K_2 \sin^2 \phi$ is neglected compared with K_1. Equation (6.8.19) gives the relation between $\dot{\phi}$ and $\cos \theta$. To obtain the equation of motion for θ, we need to eliminate ϕ in the equation

(6.8.18). For this purpose we differentiate it with respect to τ. The result gives $\ddot{\theta}$ proportional to $\cos^2\phi\dot{\phi}$, where small terms proportional to $\sin\phi$ have been neglected. Finally $\dot{\phi}$ is eliminated by using (6.8.19), leading to an equation for θ. Since both θ_1 and θ_2 are small, and the bounce solution is limited to small θ, we have finally

$$\ddot{\theta} \approx \omega_0^2 \left(\varepsilon\theta - \frac{\theta^3}{2}\right), \qquad (6.8.20)$$

where $\omega_0 = \frac{2\gamma}{M}(K_1 K_2)^{\frac{1}{2}}$. Notice that both θ_1 and θ_2 are proportional to $\sqrt{\varepsilon}$, $\varepsilon\theta$ and θ^3 are both proportional to $\varepsilon^{\frac{3}{2}}$.

The boundary conditions for the bounce solution are: $\theta = 0$ for $\tau \to \pm\infty$, $\theta = \theta_2$ and $\dot{\theta} = 0$ for $\tau = 0$. Equation (6.8.20) is solved for the boundary conditions to yield

$$\theta = \frac{\theta_2}{\cosh(\omega_0\sqrt{\varepsilon}\tau)}. \qquad (6.8.21)$$

The action for the bounce solution is

$$S_{\mathrm{E}} = \frac{8M}{3\gamma}(K_1 K_2)^{1/2}\,\varepsilon^{3/2}, \qquad (6.8.22)$$

and the tunneling rate is

$$\Gamma = A\exp\left[-\frac{8M}{3\hbar\gamma}\left(\frac{K_1}{K_2}\right)^{1/2}\varepsilon^{3/2}\right], \qquad (6.8.23)$$

where A is the prefactor. At higher temperature, the transition across the barrier is mainly due to thermal activation, and the rate of relaxation is proportional to $e^{-U/kT}$, where U is the barrier height, $U = K_1\varepsilon^2$. When the temperature is lowered, the quantum tunneling becomes dominant. The tunneling rate is independent of temperature. The crossover temperature T_{c} between these regimes is determined by

$$\frac{K_1\varepsilon^2}{kT_{\mathrm{c}}} = \frac{8M}{3\hbar\gamma}\left(\frac{K_1}{K_2}\right)^{1/2}\varepsilon^{3/2},$$

i.e.,

$$T_{\mathrm{c}} = 3\hbar\gamma\,(K_1 K_2)^{1/2}\,\sqrt{\varepsilon}/8kM. \qquad (6.8.24)$$

6.8.3 Quantum Interference Phenomena, the Topological Quenching

In the Lagrangian for magnetic tunneling there is a Wess-Zumino term which does not in any way affect the equation of motion. Two papers [32, 33] in 1992 pointed out a special effect of this factor in the path integral, viz. that the phases from symmetry related paths connecting two classically degenerate minima can cancel each other exactly, yielding a vanishing tunneling splitting. This cancellation depends on the spin, i.e., cancellation takes place for half-integral spin, just as in the phenomenon of the Haldane gap in one-dimensional antiferromagnetic chain.[23]

Consider the case of macroscopic quantum coherence shown in Fig. 6.20. The Hamiltonian of the system is

$$\hat{\mathscr{H}} = k_1 J_x^2 + k_2 J_y^2. \tag{6.8.25}$$

The classical energy is given by (6.8.9)

$$E\left(\theta, \phi\right) = \langle \hat{n}|\hat{H}|\hat{n}\rangle = K_1 \cos^2\theta + K_2 \sin^2\theta \sin^2\phi,$$

where $K_1 = k_1 J^2$, $K_2 = k_2 J^2$. The Hamiltonian (6.8.25) is time reversal invariant. According to Kramers theorem [34], *all* energy eigenstates with half-integral spin are doubly degenerate, when time reversal invariance holds. This implies that the ground state is degenerate, or the tunneling splitting vanishes for half-integral spin. How does this happen when the tunneling splitting in our case is calculated by the path integral? We consider the term $\int$ is $\dot{\phi}$ dτ from the Wess-Zumino term (6.8.8). It contributes a phase angle to the transition amplitude

$$\text{is} \int \dot{\phi} \mathrm{d}\tau = \text{is} \left\{\phi(\tau_{\mathrm{f}}) - \phi\left(\tau_{\mathrm{i}}\right)\right\}. \tag{6.8.26}$$

Consider the kink solution as the classical path. The kink with boundary conditions $\phi(\tau_{\mathrm{i}}) = 0, \phi(\tau_{\mathrm{f}}) = \pi$ is a solution, while the kink with boundary conditions $\phi(\tau_{\mathrm{i}}) = 0, \phi(\tau_{\mathrm{f}}) = -\pi$ is another distinct

---

[23]See section 7.2.

solution; they have to be summed in the path integral as different histories. But physically the initial and final states are identical. From the symmetry of these solutions, their classical actions and prefactors are also identical. The total contribution from the twin paths to the transition amplitude is

$$A \left(e^{is\pi} + e^{-is\pi} \right) e^{-S_E/\hbar} = 2A \cos s\pi e^{-S_E/\hbar}. \tag{6.8.27}$$

For half integral spin it gives zero. The total time differential part of the Wess-Zumino term guarantees the validity of the Kramers theorem.[24]

A. Garg [35] added a term $-\gamma H J_z$ to the Hamiltonian (6.8.25)

$$\hat{\mathscr{H}} = k_1 J_x^2 + k_2 J_y^2 - \gamma H J_z. \tag{6.8.28}$$

This can be realized by applying a magnetic field H along the z-direction. The minima of the potential now move to $\theta = \theta_0$, $\phi = 0, \pi$; where $\cos \theta_0 = \frac{H}{H_c}$, $H_c = \frac{2k_1 s}{\gamma}$. The quantum coherence situation remains intact. Only the time reversal invariance is broken and Kramers theorem no more applies. How does the Wess-Zumino term act now? Equation of motion (6.8.2) expressed in imaginary time is

$$is\frac{d\hat{n}}{d\tau} = -\hat{n} \times \frac{\partial E(\theta, \phi)}{\partial \hat{n}}. \tag{6.8.29}$$

It leads to energy conservation

$$\frac{dE}{d\tau} = 0.$$

Let $u = \cos \theta$, $u_0 = \cos \theta_0$, $\lambda = K_2/K_1$, and rewrite the Hamiltonian by adding a constant to it,

$$E(u, \phi) = K_1 \{ (u - u_0)^2 + \lambda(1 - u^2) \sin^2 \phi \}. \tag{6.8.30}$$

Energy conservation gives

$$E(u, \phi) = E(u_0, 0) = 0. \tag{6.8.31}$$

---

[24]See Section 6.12.

Solving (6.8.30), we have

$$u = \frac{u_0 + i\lambda^{1/2} \sin\phi \left(1 - u_0^2 - \lambda \sin^2\phi\right)^{1/2}}{\left(1 - \lambda \sin^2\phi\right)}. \qquad (6.8.32)$$

Since τ is imaginary in the equation of motion (6.8.28), θ and ϕ can be complex. We adjust the magnetic field such that $u_0^2 < 1 - \lambda$ and choose $\phi(\tau)$ to be real, then the square root of (6.8.31) is also real. Because of symmetry, these are two instanton (kink) paths, viz. $\theta_\pm(\tau), \phi_\pm(\tau)$. They start from the same point $(\theta_0, 0)$ and wind around the hard axis in opposite directions and arrive at the same point $(\theta_0, \pm\pi)$ as shown in Fig. 6.21. We have $\phi_\pm(-\infty) = 0$, $\phi_\pm(+\infty) = \pm\pi$, $\phi_+(\tau) = -\phi_-(\tau)$. From (6.8.32) we can calculate the real part ω_τ of $\omega[\hat{n}]$:

$$\omega_{r,\pm} = \int_0^{\pm\pi} \left(1 - \frac{u_0}{1 - \lambda \sin^2\phi}\right) d\phi = \pm\pi[1 - u_0(1 - \lambda)^{\frac{1}{2}}]. \quad (6.8.33)$$

It will give the phase factor part of $e^{-S_E/\hbar}$. By adding the contributions from both paths the tunneling splitting is found to be

$$\Gamma = De^{-S_r/\hbar}\left(e^{is\omega_{r+}} + e^{is\omega_{r-}}\right) = 2De^{-S_r/\hbar}\cos\phi\,(H), \qquad (6.8.34)$$

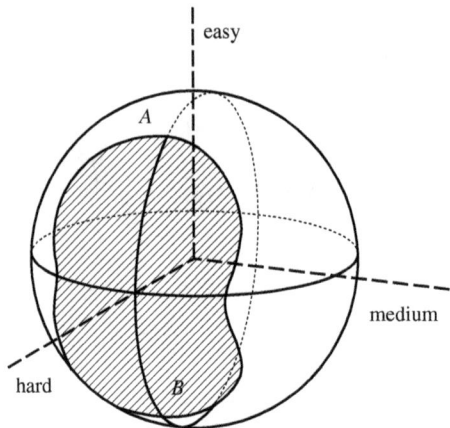

Fig. 6.21   Unit sphere showing different instanton paths between degenerate minima A and B of the potential. Redrawn from Ref. [35].

where the subscript r means the real part, D is a constant, and

$$\phi(H) = \frac{\omega_{r+} - \omega_{r-}}{2}. \tag{6.8.35}$$

From (6.8.32) we have

$$(\omega_{r+} - \omega_{r-})/2 = \pi[1 - u_0(1-\lambda)^{-1/2}],$$

which is just the shaded area in Fig. 6.22. When

$$s\pi\left[1 - u_0(1-\lambda)^{-1/2}\right] = \left(n + \frac{1}{2}\right)\pi,$$

i.e.

$$\frac{H}{H_c} = u_0 = (1-\lambda)^{1/2}\frac{1}{s}\left(s - n - \frac{1}{2}\right), \tag{6.8.36}$$

The tunneling splitting vanishes. Figure 6.22 shows the tunneling splitting as a function of H/H_c, for $K_1 = 1$, $\lambda = 0.1$. The spin for the solid line is $s = 10$, and for dotted line $s = 19/2$. The tunnel splitting oscillates periodically as a function of H, no matter the spin is integral or half-integral. The important special case is $H = 0$, where the tunnel splitting is quenched for half-integral spin. This

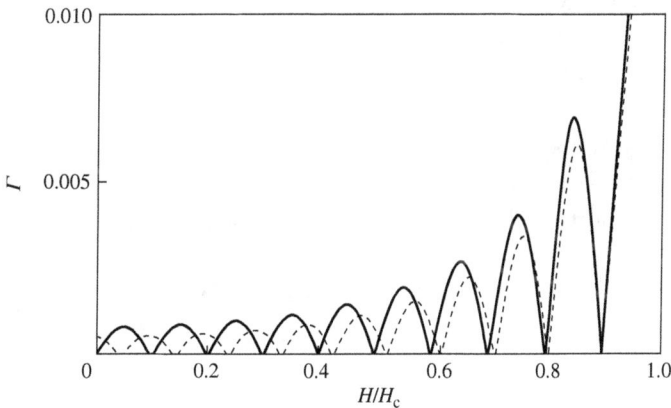

Fig. 6.22 Tunneling splitting as a function of H/H_c. Redrawn from Ref. [35].

is the only point where Kramers theorem applies. The oscillation in tunneling rate originates from the interference of Wess-Zumino terms from two symmetric instanton solutions. Such a phenomenon is called the *quantum interference* or *topological quenching*. This effect has no analog in the Josephson systems.

The foregoing discussion concerns the macroscopic quantum coherence. Does the quantum tunneling also show topological interference? In the macroscopic quantum tunneling the bounce solution dominates the process. Since the initial and final positions are the same $\hat{n}(\tau_i) = \hat{n}(\tau_f)$, the classical path is unique, and for each bounce $\omega_r = 0$, consequently there is no quantum interference. Chudnovsky and DiVincenzo [36] considered the case of the asymmetric double-well, with the asymmetry regulated by an external magnetic field. The metastable state at the bottom of the shallow well penetrates through the barrier to a state of approximately equal energy in the deep well. The tunneling rate is also oscillating with the varying magnetic field. Garg [37] gave a thorough analysis of the oscillation phenomena. It turns out that the situation depends on the width of the levels in the deep well and the distance between levels. If these levels are narrow and the interlevel distances are large, then the tunneling rate oscillates with the varying magnetic field, and this is analogous to the case of quantum interference where the system oscillates between the wells (resonance). When the levels in the deep well are densely populated and have large widths such that the level density is quasi-continuous, the situation is close to the case of quantum tunneling and the tunneling rate does not oscillate with the varying field.

The systems discussed above have Hamiltonians which are invariant under a rotation about z-axis by π, i.e., they possess a bi-axial rotational symmetry. The topological quenching can be generalized to the case of M-fold symmetry, i.e., of the invariance by a rotation about z-axis by an angle $\frac{2\pi}{M}$. When the spin quantum number s is not equal to an integral times $\frac{M}{2}$, the transition between $|s\rangle$ and $|-s\rangle$ is frozen.

$$s \neq 0 \,(\mathrm{mod}M/2) \Rightarrow \langle -s|\mathrm{e}^{-\mathrm{i}\hat{H}t}|s\rangle = 0.$$

This is the *spin-parity effect*.[25] There are a number of published works in this direction [39–41]. The spin-parity effect discussed above is based on the Wess-Zumino term in the Lagrangian. Does it have a deeper root in the basic principles of quantum mechanics? We return to this question in Section 6.12.

6.9   Macroscopic Quantum Phenomena of Single Domain Antiferromagnetic Particles

The magnetic properties of antiferromagetic particles can be considered as due to bipartite lattices of ferromagnetic atoms. One sublattice has magnetization m_1 and the other m_2. Due to the strong exchange interaction each sublattice behaves as a united whole in quantum processes. m_1 and m_2 have about the same magnitude $m_1 \approx m_2$, and are oppositely directed. $m_1 + m_2 = m$ is a small remnant from cancellation between them: $m \ll m_1, m_2$. The order parameter of anti-ferromagnetism is the Néel vector, defined as

$$l = \frac{m_1 - m_2}{2m_1}. \tag{6.9.1}$$

To derive the Lagrangian of a single domain ferromagnetic particle we follow Chudnovsky [42] using the bipartite lattice. The result can be easily generalized to ferri-magnetism. There is a strong exchange coupling $\chi_\perp^{-1} m_1 \cdot m_2$ between the magnetizations m_1 and m_2, with $\chi_\perp \ll 1$. Let $(\theta_1 \phi_1)$ and $(\theta_2 \phi_2)$ be the directions of m_1 and m_2 respectively. For the time being we neglect the crystalline anisotropy energy. The Lagrangian of the bipartite lattice is

$$\mathcal{L}_0 = V \Big\{ i \frac{m_1}{\gamma} \dot{\phi}_1 \left(1 - \cos \theta_1 \right) + i \frac{m_2}{\gamma} \dot{\phi}_2 \left(1 - \cos \theta_2 \right)$$

$$+ \frac{1}{\chi_\perp} m_1 m_2 \left[\sin \theta_1 \sin \theta_2 \cos \left(\phi_1 - \phi_2 \right) + \cos \theta_1 \cos \theta_2 + 1 \right] \Big\},$$
$$\tag{6.9.2}$$

---

[25] Here parity refers to the even or odd multiple of $\hbar/2$ for the spin, see Ref. [38].

where $\dot{\phi}_1$ and $\dot{\phi}_2$ represent derivatives with respect to the imaginary time τ. A constant 1 has been added in the curly bracket. Here the volume V is explicitly written, such that $M_1 = m_1 V$, $M_2 = m_2 V$. In subsequent derivations the anisotropy parameter K also refers to unit volume.

In the following, we consider only low action paths between initial and final states. Since $\chi_\perp^{-1} \gg 1$ and m_1 and m_2 are approximately oppositely directed, it is convenient to define

$$\theta_2 = \pi - \theta_1 - \varepsilon_\theta, \quad \phi_2 = \pi + \phi_1 + \varepsilon_\phi, \tag{6.9.3}$$

where

$$|\varepsilon_\theta|, |\varepsilon_\phi| \ll 1.$$

We have $\varepsilon_\theta = \varepsilon_\phi = 0$ at $\tau = \tau_i$ and $\tau = \tau_f$. Substituting (6.9.3) into (6.9.2) and keeping quantities up to second order in ε_θ and ε_ϕ when first order quantities are absent, we obtain

$$\mathscr{L}_0 = V \left\{ i \frac{m_1 + m_2}{\gamma} \dot{\phi}_1 - i \frac{m}{\gamma} \dot{\phi}_1 \cos\theta - i \frac{m_2}{\gamma} \varepsilon_\theta \dot{\phi}_1 \sin\theta_1 + i \frac{m_2}{\gamma} \varepsilon_\phi \sin\theta_1 \dot{\theta}_1 \right.$$
$$\left. + \frac{1}{2\chi_\perp} m_1 m_2 \varepsilon_\theta^2 + \frac{1}{2\chi_\perp} m_1 m_2 (\sin\theta \varepsilon_\phi)^2 \right\}. \tag{6.9.4}$$

The path integral for transition amplitude is

$$\int [D\theta_1] [D\phi_1] [D\varepsilon_\theta] [D\varepsilon_\phi] \exp\left[-\frac{1}{\hbar} \int \mathscr{L}_0 d\tau \right].$$

Substituting (6.9.4) for $\mathscr{L}_0$, performing the Gaussian integrals with respect to ε_θ and $\varepsilon_\phi \sin\theta_1$, and rewriting (θ, ϕ) for (θ_1, ϕ_1) after integration, we obtain finally

$$\int [D\theta] [D\phi] \exp\left[-\frac{1}{\hbar} \int \mathscr{L}_0^{\mathrm{eff}} d\tau \right],$$

where

$$\mathscr{L}_0^{\mathrm{eff}} = V \left\{ i \frac{m_1 + m_2}{\gamma} \dot{\phi} - i \frac{m}{\gamma} \dot{\phi} \cos\theta + \frac{\chi_\perp}{2\gamma^2} \left(\dot{\theta}^2 + \dot{\phi}^2 \sin^2\theta \right) \right\}. \tag{6.9.5}$$

Notice that (θ_1, ϕ_1) and (θ_2, ϕ_2) are independent histories at the outset. But after introducing (6.9.3) and performing integrations over ε_θ and ε_ϕ, only one pair (θ, ϕ) is left as the generalized coordinates describing the antiferromagnetic particle. The deviation from anti-parallelism of m_1 and m_2 is also incorporated in $\mathscr{L}_{\mathrm{eff}}$. When the magnetic anisotropy energy per unit volume $E_K(\theta, \phi)$ is added, the transition amplitude path integral is finally

$$\int [\mathrm{D}\theta(\tau)]\,[\mathrm{D}\phi(\tau)]\exp\left[-\frac{1}{\hbar}\int \mathscr{L}\mathrm{d}\tau\right],$$

where

$$\mathscr{L} = V\int \left\{ i\frac{m_1+m_2}{\gamma}\dot{\phi} - i\frac{m}{\gamma}\dot{\phi}\cos\theta \right.$$
$$\left. +\frac{\chi_\perp}{2\gamma^2}\left(\dot{\theta}^2 + \dot{\phi}^2\sin^2\theta\right) + E_K(\theta, \phi) \right\}\mathrm{d}\tau. \qquad (6.9.6)$$

For the complete anti-ferromagnetic case $m = 0$ the Euclidean Lagrangian is

$$\mathscr{L} = V\int \left\{ i\frac{m_1+m_2}{\gamma}\dot{\phi} + \frac{\chi_\perp}{2\gamma^2}\left(\dot{\theta}^2 + \dot{\phi}^2\sin^2\theta\right) + E_K(\theta, \phi) \right\}\mathrm{d}\tau. \qquad (6.9.7)$$

It should be emphasized that the early research in 1990 devoted to antiferromagnetic tunneling [43, 44] was based on the Lagrangian derived from the equation of motion, where the Wess-Zumino term is not included. In the following we start from (6.9.7) to calculate the tunneling rate. Let the x-direction be the easy axis, z-direction be the hard axis, and the anisotropy energy is

$$E_K = K_\perp\cos^2\theta + K_{\|}\sin^2\theta\sin^2\phi, \quad K_\perp \gg K_{\|}. \qquad (6.9.8)$$

The initial and final states in the tunneling is shown in Fig. 6.23. Since $K_\perp \gg K_1$, the magnetic moment is mainly confined in the vicinity of xy-plane such that we can neglect the $\dot{\theta}^2$ term in (6.9.7) and then Gaussian integration over $\cos\theta$ can be performed. The

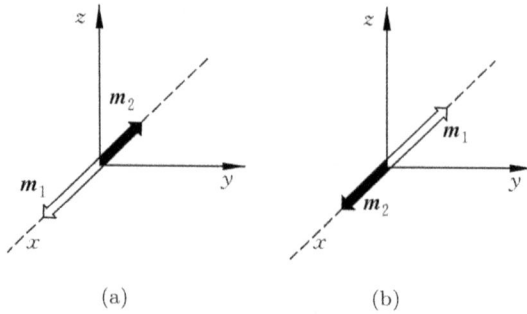

(a)   (b)

Fig. 6.23   (a) The initial; (b) The final states of the tunneling of an anti-ferromagnetic particle.

transition amplitude has now the form

$$\int [D\phi(\tau)] \exp\left\{-\frac{V}{\hbar} \int d\tau \left[i\frac{m_1 + m_2}{\gamma}\dot{\phi}\right.\right.$$

$$\left.\left. +\frac{1}{2}(I_f + I_a)\dot{\phi}^2 + K_{\parallel}\sin^2\phi\right]\right\}, \tag{6.9.9}$$

where

$$I_f = \frac{m^2}{2\gamma^2 K_\perp}, \quad I_a = \frac{\chi_\perp}{\gamma^2}$$

are the ferromagnetic and anti-ferromagnetic moments of inertia respectively. The path integral is dominated by the classical instanton solution and paths in the immediate vicinity of it. The equation of motion is

$$\ddot{\phi} = \frac{K_{\parallel}}{I_f + I_a}\sin 2\phi = \omega_0^2 \sin 2\phi, \tag{6.9.10}$$

where ω_0 is defined as

$$\omega_0 = \left(\frac{K_{\parallel}}{I_f + I_a}\right)^{1/2} = \gamma\left(\frac{2K_{\parallel}K_\perp}{m^2 + 2\chi_\perp K_\perp}\right)^{1/2}. \tag{6.9.11}$$

The solution to (6.9.10) is the instanton kink solution

$$\phi = \pm 2\arctan e^{\sqrt{2}\omega_0 r}. \tag{6.9.12}$$

There are two different paths with the same initial and final states corresponding to $\phi(-\infty) = 0$ and $\phi(\infty) = \pm\pi$, and they will add to give the topological interference. The first term of action (6.9.8) gives

$$V \int i \frac{m_1 + m_2}{\gamma} d\phi.$$

Its contribution to the phase factor is

$$\exp\left[-\frac{V}{\hbar\gamma}i\left(m_1 + m_2\right)\Delta\phi\right] = e^{-i(S_1+S_2)\Delta\phi},$$

where S_1 and S_2 are the total spin of the sublattice 1 and 2 respectively, and $\Delta\phi$ is

$$\Delta\phi = \begin{cases} \pi, & \text{for tunneling with counterclockwise rotation} \\ -\pi, & \text{for tunneling with clockwise rotation} \end{cases}.$$

Sum over histories gives $2\cos(S_1 + S_2)\pi = \cos(S\pi + 2S_2\pi)$, where $S = S_1 - S_2$ is the uncompensated spin. Since $2S_2$ is an integer, the foregoing factor is $\pm 2\cos S_1\pi$. If S_1 is an integer, it gives ± 1, if S_1 is a half-integer, it gives 0. The classical action is given by the classical path:

$$S_{\mathrm{E}} = V \int d\tau \left\{ \frac{1}{2}\left(I_{\mathrm{f}} + I_{\mathrm{a}}\right)\dot{\phi}^2 + K_{||}\sin^2\phi \right\} = V \int d\tau 2K_{||}\sin^2\phi.$$

The integral is taken directly to yield

$$\int_{-\infty}^{\infty} \sin^2\phi \, d\tau = \frac{1}{\sqrt{2\omega_0}} \tanh\sqrt{2}\omega_0\tau \,\big|_{-\infty}^{\infty} = \frac{\sqrt{2}}{\omega_0}.$$

Substituting it back into the previous equation for S_{E}, we obtain

$$S_{\mathrm{E}} = V2K_{||}\frac{\sqrt{2}}{\omega_0} = V\frac{2}{\gamma}\left(2\chi_\perp K_{||} + m^2\frac{K_{||}}{K_\perp}\right)^{1/2}. \tag{6.9.13}$$

Without a prefactor the tunneling rate is

$$\Gamma \approx \cos\left(S\pi\right)\omega_0\exp\left[-\frac{2V}{\hbar\gamma}\left(2\chi_\perp K_{||} + m^2\frac{K_{||}}{K_\perp}\right)^{1/2}\right]. \tag{6.9.14}$$

In the ferromagnetic limit $(m_1 \gg m_2, \; m \gg \sqrt{\chi_\perp K_\perp})$ the above result agrees with that of Ref. [27]. In the anti-ferromagnetic limit ($m_1 \approx m_2$, $m \ll \sqrt{\chi_\perp K_\perp}$) the above result agrees with that of Refs. [43, 44] except for the phase factor from the topological interference. From (6.9.14) we see that for $m \neq 0$ (no matter how small it is) $K_\perp \neq 0$ is necessary for a non-vanishing tunneling rate. Since for $K_\perp = 0$, m is conserved and cannot tunnel to change its sign. For ferromagnetic particles, $S_E \propto \sqrt{\frac{K_\parallel}{K_\perp}}$, the tunneling rate can be different from vanishingly small only for $K_\perp \gg K_\parallel$; while for anti-ferromagnetic particles, $S_E \propto \sqrt{K_\parallel \chi_\perp}$. Since $\chi_\perp \ll 1$, the tunneling rate is in general much larger than that of ferromagnetic particles.

In the foregoing discussions we did not touch the problem of influence of environment on quantum tunneling. Actually this is an important issue. The agencies of dissipation is different in different solids, for instance free electrons, vortices, magnons, phonons, etc. Garg [45] has made systematic studies on this problem.

6.10   Experiments on the Macroscopic Quantum Tunneling of Magnetic Systems

In most experiments of the quantum tunneling magnetic relaxation was used. Barbara *et al.* [46] measured ferromagnetic particles $Tb_{0.2}Ce_{0.5}Fe_2$ with the average size of 15 nm. The sample is placed in a high magnetic field of 8T to reach magnetic saturation. The field is decreased to 0, and then reversed to a field strength $\lesssim H_c$. The magnetization is then in a metastable state as in Fig. 6.20. Then the magnetization of the sample is measured as a function of time. For a single domain particle without interaction the magnetic relaxation follows the following law:

$$M(t) = M(0) e^{-\Gamma t}. \qquad (6.10.1)$$

At high temperatures the thermal activation is the dominant mechanism for transition across the barrier, and the tunneling rate is

$$\Gamma = \omega e^{-U/kT^*}, \qquad (6.10.2)$$

where T^* is called the escape temperature. For high temperature T of the sample, $T^* = T$. When T is decreased to a certain value,

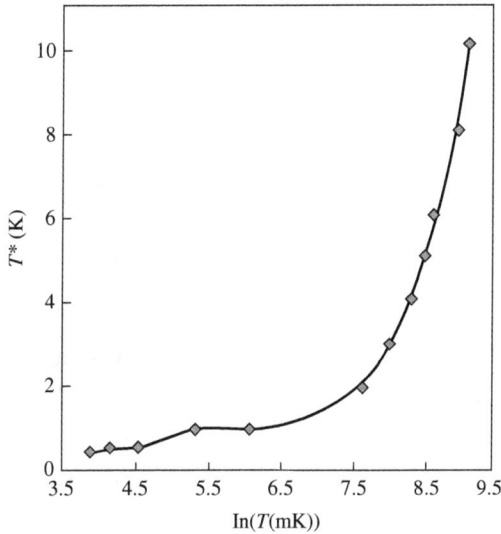

Fig. 6.24   Change of the escape temperature T^* with temperature T of the sample. Reprinted with permission from E. M. Chudnovsky, *J. Magn. Magn. Mater.* 140-144, 1821, 1995. Copyright (2017) by Elsevier.

Γ ceases to fall exponentially with $1/T$. Instead it approaches a value independent of temperature, i.e. T^* approaches a constant. This shows that the quantum tunneling begins to take part in the relaxation process, and finally becomes dominant. Figure 6.24 shows this process. Most magnetic systems consist of many single domain particles, and the potential barrier has a fairly wide distribution. They do not follow the exponential law in the relaxation process. The evolution of the magnetization is [47]

$$M\left(t\right) = M\left(t_0\right)\left[1 - S\left(T, H\right)\ln\frac{t}{t_0}\right], \qquad (6.10.3)$$

where $S(T, H)$ is the *magnetic viscosity*, its dependence on the temperature and the magnetic field strength characterizes the relaxation behavior of the system. J. Tejada and X.-X. Zhang [47, 48] carried out measurements on magnetic viscosities of many systems. The magnetic viscosity decreases with decreasing temperature when T is high and tends to a constant at the cross-over temperature between the thermal activation and the quantum tunneling regimes. Figure 6.25

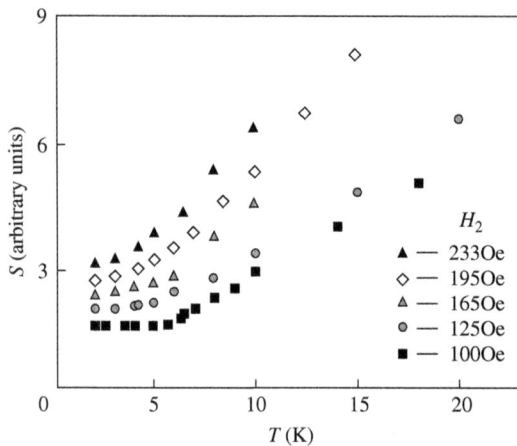

Fig. 6.25   Variation of magnetic viscosity of random TbFe₃ film with temperature. From Ref. [48].

shows the variation of magnetic viscosity of random TbFe$_3$ films with temperature. Since the average barrier decreases with increasing H, the quantum tunneling rate increases correspondingly, giving a higher escape temperature T^*. Reference [48] also gives the exponential relaxation of the TbFeO$_3$ anti-ferromagnetic single crystals as shown in Fig. 6.26. From the experimental data we get the barrier height as a function of H and the attempt frequency ω, which makes possible a comparison with the theory.

A quite different type of experiments was carried out by D. Awschalom and collaborators [49–51] using small anti-ferromagnetic particles. Consider a double well separated by an energy barrier that arises from the magnetic anisotropy (crystalline, shape or surface). This is a stage set for the macroscopic quantum coherence, in which the Néel vector l can oscillate between the two wells. If the magnetic moments of the two sublattices do not cancel exactly, the remaining moment m oscillates together with l.[27] The

---

[26]In the electromagnetic units (emu), the unit of magnetization M is erg/G.cm^2. In SI, the unit of M is the same as H, i.e., $A \cdot m^{-1}$. The conversion is 1emu $= 10^3 A \cdot m^{-1}$.

[27]Since the Néel vector l is not coupled to the apparatus, and is hence not observable, the oscillation of l can be inferred from the oscillation of m, which is observable.

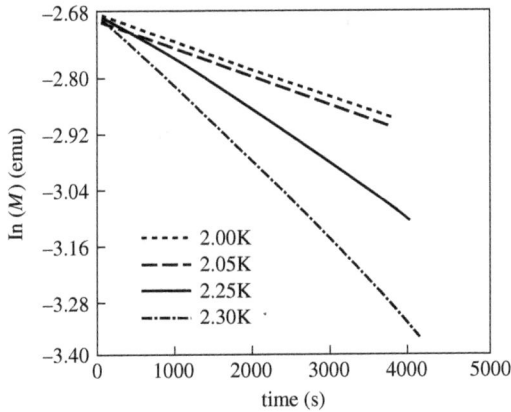

Fig. 6.26 Exponential relaxation of the magnetization of the antiferromagnetic single crystal TbFeO$_3$[26] From Ref. [48].

nanometer magnetic particles are the ferritin protein from horse spleen, each particle contains about 4 500 Fe^{3+} spin $\frac{5}{2}$ ions. They are single domain anti-ferromagnetic below the Néel temperature 240K. Owing to the large surface-to-volume ratio at this reduced length scales, they have a relatively large number of uncompensated surface spins. The excess spins serve as tracers for the anti-ferromagnetic particle dynamics. The magnetic anisotropy has the single uniaxial form. The magnetic interaction between different particles can be continuously controlled by mixing the solution of a ferritin with a solution of apoferritin, a protein whose structure is almost the same as that of a ferritin, but is non-magnetic because the ion core is absent. The reduction of interaction is evidenced by the experimental result. In the experiments the frequency dependent magnetic susceptibility $\chi(\omega)$ and the magnetic noise $S(\omega)$ of ferritin protein are measured by a fully integrated thin film SQUID susceptometer. When the uncompensated magnetic moment oscillates with l, the correlation function of the magnetization is

$$S(\tau) = \langle M(t) M(t + \tau) \rangle = M_0^2 \cos(\omega_{res}\tau), \qquad (6.10.4)$$

where M_0 is the magnitude of the uncompensated magnetic moment, ω_{res} is the resonance angular frequency. The Fourier transform of

$S(\tau)$ is just the measured magnetic noise.

$$S(\omega) = \pi M_0^2 \delta (\omega - \omega_{\mathrm{res}}). \qquad (6.10.5)$$

The measured $S(\omega)$ and $\chi(\omega)$ are represented in Fig. 6.27(a) and (b). In the experiment $1\,000 : 1$ diluted solutions at $T = 29.7$mK. The resonance frequency is 9.4×10^5Hz. The slight difference in resonance frequency is due to the effect of stray fields. The frequency is very sensitive to the magnetic field. If the dilution is not sufficient, the

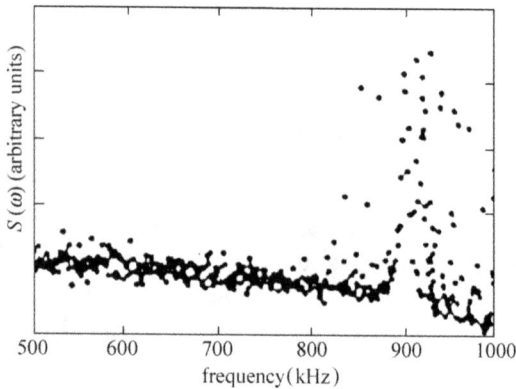

(a)

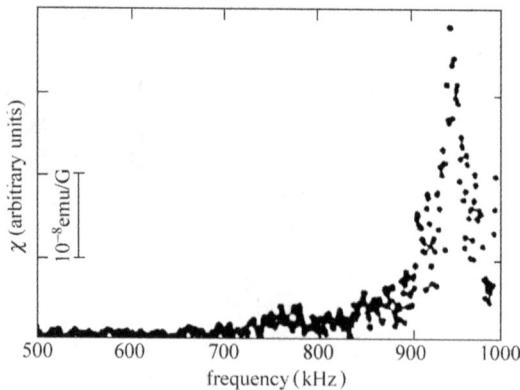

(b)

Fig. 6.27   (a) Magnetic noise spectrum ($B = 10^{-5}$G); (b) Magnetic susceptibility ($B = 10^{-6}$G) as functions of the frequency. Reprinted with permission from A. Garg and G.-H. Kim, *Phys. Rev. Lett.* **63**, 2512, 1989. Copyright (2017) by the American Physical Society.

resonance frequency will increase so much that it escapes from the range of measurement. The coincidence of resonance frequency for $S(\omega)$ and the imaginary part of the susceptibility $\chi''(\omega)$ is an important test. According to the fluctuation-dissipation theorem we have

$$\chi''(\omega) = (1 - e^{-\hbar\omega/kT})S(\omega)\frac{1}{2\hbar}$$

$$\approx \frac{\omega}{2kT}S(\omega) = \frac{\pi N\omega M_0^2}{2kT}\delta(\omega - \omega_{res}) \equiv \chi_{res}\delta(\omega - \omega_{res}),$$

$$(6.10.6)$$

where N is the number of ferritin particles and χ_{res} is defined to be

$$\chi_{res} = \frac{\pi N\omega M_0^2}{2kT}. \qquad (6.10.7)$$

According to this relation, χ_{res} and the temperature are related such that $T\chi_{res}/S(\omega_{res})$ is a constant, which is also experimentally verified Ref. [50] reported the relation between the resonance frequency and the volume of the ferritin particle, using artificial ferritin which has a variable volume of the protein shell. The result confirms the exponential fall of the frequency versus the volume ratio V/V_0 where V_0 is the volume of the natural ferritin. as shown in Fig. 6.28. This agrees with the quantum mechanical prediction.

Different opinions are expressed by various authors: Prokofiev and Stamp [52], Garg [53], Braun and Loss [54].

6.11 Macroscopic Quantum Phenomena of Macromolecules

In 1980, a magnetic crystalline compound manganese acetate ($Mn_{12}Ac$) was synthesized. The chemical formula of the crystal is $[Mn_{12}O_{12}(CH_3COO)_{16}(H_2O)_4] \cdot 2CH_3COOH \cdot 4H_2O$. The scheme of its structure is shown in Fig. 6.29. The large circles represent Mn ions. Eight of the Mn^{3+} ions (spin 2) form the outer octahedron with total spin 16, and four of the Mn^{4+} ions (spin 3/2) form the inner tetrahedron with total spin 6. The total spin 10 of the molecule

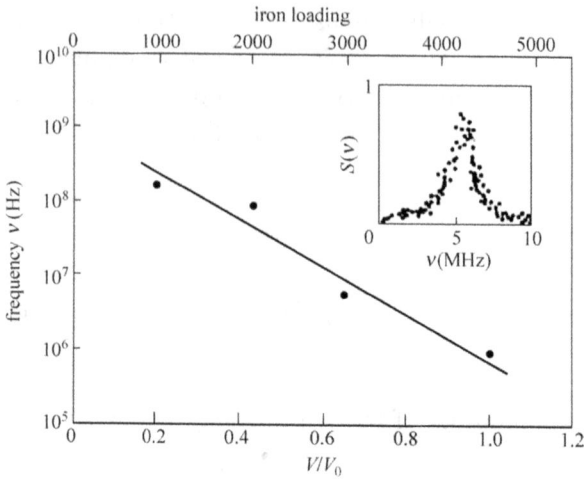

Fig. 6.28   Dependence of resonance frequency on particle size. Reprinted with permission from A. Garg and G.-H. Kim, *Phys. Rev. B* 43, 712, 1991. Copyright (2017) by the American Physical Society.

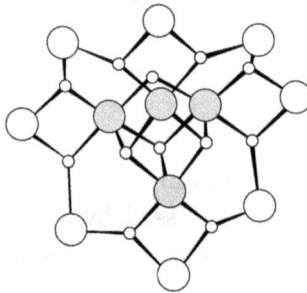

Fig. 6.29   Structure of the $Mn_{12}Ac$ molecule.

results from the antiferromagnetic interaction between the two clusters. The small circles are the bridging O ions. The acetate bridging ligands and the water molecules are not shown. In the crystal lattice all the molecules have the S_4 symmetry axis parallel to the crystallographic c axis, and the shortest inter-cluster contacts between Mn ions are larger than 7 Å. The exchange interaction between Mn ions and the molecule is strong so that the Mn_{12} molecule is effectively a single spin 10 object, and the interaction between ions of different molecules are very weak. Experiments established a strong magnetic anisotropy. Magnetic relaxation experiment verified a single relaxation law (thermal activation) $\tau = \tau_0 e^{-\Delta E/kT}$ above 2.1K. For lower temperatures quantum tunnelling may show up.

This is an ideal system for studying the macroscopic quantum phenomena. In Section 6.10 we mentioned that the single domain particles have varying sizes such that the barrier height has a distribution. Information on quantum tunneling can be obtained only after an average process which smears out some of the characteristics. The interaction between particles also causes problems. Such problems do not exist for Mn_{12}. Experimentally this molecule has the advantage that almost perfect samples can be prepared with identical molecules in a crystalline array. The investigations of quantum tunneling began in 1996, the groups of J. Friedman [55], J. Tejada [56] and B. Barbara [57] discovered the staircase structure in the hysteresis curve of Mn_{12} shown in Fig. 6.30, where jumps and plateaus appear alternatively. Since all molecules are identical, and are well aligned in the crystal, quantum tunneling occurs in all molecules when the magnetic field reaches a resonance value, and a unique macroscopic

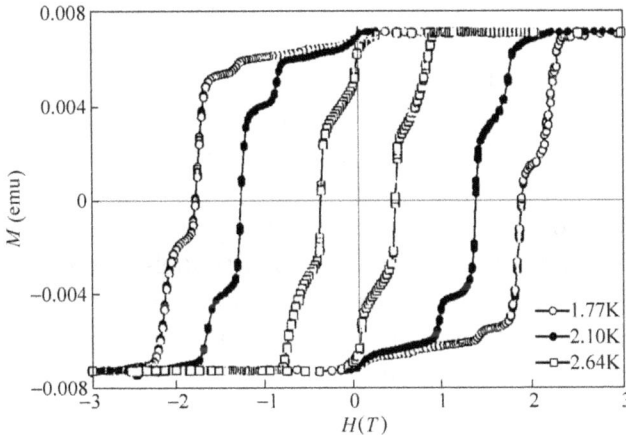

Fig. 6.30 The magnetization hysteresis curve of Mn_{12} measured at different temperatures. Reprinted with permission from D. D. Awschalom, J. F. Smyth, G. Grinstein, D. P. DiVincenzo and D. Loss, *Phys. Rev. Lett.* **68**, 3092, 1992. Copyright (2017) by the American Physical Society.

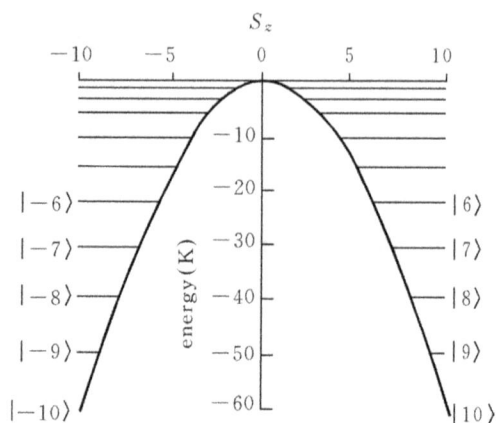

Fig. 6.31   Spin states of Mn$_{12}$.

signal is formed. Let the z-direction be the crystallographic c-axis, the spin dependence of the energy is

$$E = -DS_z^2 - g\mu_z H S_z. \qquad (6.11.1)$$

The spin of the molecule is 10, therefore there are altogether 21 states, $s_z = -10, -9, \ldots, 9, 10$. In the absence of external magnetic field the 10 pairs of levels are degenerate, as shown in Fig. 6.31. For the lowest states $|-10\rangle$ and $|10\rangle$ the barrier penetration accompanied by the spin reversal is difficult at low temperatures. The relaxation time was estimated to be $\sim$ 3 years [58, 59]. The penetration from an excited state is much easier, and thermally assisted tunneling is much more probable, i.e., the molecule is first thermally excited to a higher state and then the magnetization tunnels through the barrier from the excited state. For the theoretical interpretation, the group of J. Villain [59, 60] considered the mechanism of tunneling. In order that the tunneling may occur, there must be an anisotropy energy. The S_4 symmetry demands the invariance of the Hamiltonian with respect to the change $S_x \rightarrow S_y$, $S_y \rightarrow -S_x$. We have to include in the Hamiltonian an operator which does not commute with S_z and satisfies the invariance indicated above. A candidate is $A(S_+^2 + S_-^2)$, but it does not work, because it is just $A(S^2 - S_z^2)$, commuting with S_z. The next choice is

$$H = -DS_z^2 - C\left(S_+^4 + S_-^4\right). \tag{6.11.2}$$

Other possibilities include the interaction between spin and the phonons.

The structure of the hysteresis curve and the time of relaxation, especially their correlation need a thorough interpretation. When external magnetic field exists, levels with S_z positive and those with S_z negative shift relative to each other. When a state $|-m\rangle$ coincides with a state $|m - n\rangle$, the resonance tunneling occurs. Without taking into account the S_+^4, S_-^4 terms, (6.11.1) gives

$$-Dm^2 + g\mu_B m H = -D\left(m - n\right)^2 - g\mu_B\left(m - n\right)H,$$

i.e.,

$$g\mu_B H_n = nD, \tag{6.11.3}$$

independent of m, and consequently the field strength at which the resonance tunneling takes place is determined by n only:

$$H_n = \frac{D}{g\mu_B}n, \tag{6.11.4}$$

where $n = 0, 1, \ldots$ Whenever one pair of levels coincide, there are some other pairs also coincide in energy. When the magnetic field reaches these resonant values the magnetization changes abruptly. Because of the resonant tunneling the relaxation time drops to a minimum. Figure 6.32 [57] gives the relaxation time versus H from $0\,T$ to $2.64\,T$. The quantum jump of magnetization occurs with evenly spaced values of H with $\Delta H = 0.44\,T$, and the jumps occur only when the direction of H is opposite to the residual magnetization. The value of ΔH implies $D = 0.60\,K$.

A new experimental method was developed by Wernsdorfer and Sessoli [61] to measure very small tunnel splitting in molecular clusters of eight iron atoms, Fe$_8$, which behave like a nanomagnet with a spin ground state of $S = 10$ at low temperature. They observed a clear oscillation of the tunnel splitting as a function of the external

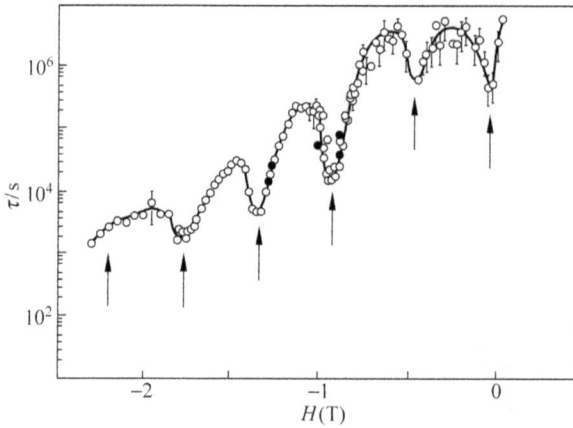

Fig. 6.32   Relaxation time as a function of H.

magnetic field along the hard axis, which are due to topological quantum interference of two tunnel paths of opposite windings. This observation is a direct evidence of the role of the topological spin phase which we discussed in Section 6.8.3.

To deal with a system with spin 10, the method of spin coherent states path integral may not be appropriate. In this formalism the spin is considered as a classical variable. Actually $[\phi, S_z] = i$ is replaced by $s[\phi, \cos\theta] = i$, here s is a c-number. If we attempt to calculate $[S_x, S_y]$ where $S_x = s\sin\theta\cos\phi$, $S_y = s\sin\theta\sin\phi$, we find that

$$[S_x, S_y] = iS_z + O\left(\frac{1}{s^3}\right). \qquad (6.11.5)$$

This means that the formalism is valid for large s. The potential description of the spin system developed by Zaslavskii and collaborators [62] begins with the Schrödinger equation of the spin particle, and the spin-coordinate correspondence is exact; its validity does not depend on the magnitude of the spin. This method was adapted to the problem of magnetic tunneling by E. M. Chudnovsky and D. A. Garanin [63] and by J.-Q. Liang *et al.* [64]. Subsequent study led to a result in agreement with the experiment of Wernsdorfer and Sessoli.

6.12   Quantum Mechanical Foundation of the Spin-parity Effect

The spin parity-effect is discussed in Section 6.8 using the spin coherent state path integral method, which is valid for large s. Has this effect a deeper root in the fundamental principles of quantum mechanics? As B.-Z. Li and F. C. Pu and collaborators [65, 66] pointed out, it is directly related to selection rules due to symmetries.

Consider a M-fold rotational symmetry around the z-axis. This implies that the Hamiltonian of the system commutes with the rotation operator $e^{i\frac{2\pi}{M}\hat{S}_z}$. Consequently, for $\hbar = 1$,

$$\exp(-i\hat{H}t) = \exp\left(-i\frac{2\pi}{M}\hat{S}_z\right)\exp(-i\hat{H}t)\exp\left(i\frac{2\pi}{M}\hat{S}_z\right). \quad (6.12.1)$$

The transition matrix element between eigenstates $|m\rangle$ and $|m'\rangle$ of S_z is

$$\langle m'|\exp(-i\hat{H}t)|m\rangle$$

$$= \left\langle m'\left|\exp\left(-i\frac{2\pi}{M}\hat{S}_z\right)\exp(-i\hat{H}t)\exp\left(i\frac{2\pi}{M}\hat{S}_z\right)\right|m\right\rangle$$

$$= \exp\left[i\frac{2\pi}{M}(m'-m)\right]\langle m'|\exp(-i\hat{H}t)|m\rangle, \quad (6.12.2)$$

where $-S \le m, m' \le S'$. From (6.12.2), it follows that

$$m' - m \ne 0\,(\text{mod}M) \Rightarrow \langle m'|e^{-i\hat{H}t}|m\rangle = 0. \quad (6.12.3)$$

Let $m' = -S$, $m = S$, we have

$$S \ne 0\left(\text{mod}\frac{M}{2}\right) \Rightarrow \langle -S|e^{-i\hat{H}t}|S\rangle = 0, \quad (6.12.4)$$

This is the spin-parity effect.[28]

The above discussion can be generalized to a system of N spins. Let $\hat{\mathbf{S}}_\alpha = (\hat{S}_\alpha^x, \hat{S}_\alpha^y, \hat{S}_\alpha^z)$ be the spin operator of the α-th spin, and S_α

---

[28]Notice that for example the integer 1 does not satisfy the criterion of the left-hand side of (6.12.4) while the half-integer 1/2 does.

the spin quantum number, $|m_\alpha\rangle$ the eigenstate of $\hat{S}_\alpha^z$ with eigenvalue m_α, $-S_\alpha \leq m_\alpha \leq S_\alpha$. The eigenstate of the z-component of the *total* spin of the system $\sum_\alpha \hat{S}_\alpha^z$ is the tensor product

$$|\{m_\alpha\}\rangle = |m_1\rangle |m_2\rangle \cdots |m_N\rangle, \qquad (6.12.5)$$

with the eigenvalue $\sum_\alpha m_\alpha$. Equation (6.12.3) can be directly generalized to

$$\sum_\alpha (m_\alpha - m_\alpha') \neq 0 \,(\mathrm{mod}M) \Rightarrow \langle\{m_\alpha'\}|e^{-i\hat{H}t}|\{m_\alpha\}\rangle = 0.$$

$$(6.12.6)$$

Equations (6.12.3) and (6.12.6) are the spin-parity effect originating from M-fold rotational invariance about the z-axis in the form of a selection rule of quantum mechanics. It applies to spin of any value, large or small, to a single spin or a system of N spins.

Consider the selection rule originating from the time reversal invariance. Let $\hat{T}$ be the time reversal operator. It is an anti-linear, anti-unitary operator[29] satisfying the following relations:

$$\hat{T}i\hat{T}^{-1} = -i, \qquad (6.12.7a)$$

$$\hat{T}\hat{S}T^{-1} = -\hat{S}, \qquad (6.12.7b)$$

$$\hat{T}^2 = (-1)^{2S}. \qquad (6.12.7c)$$

Let $|\psi^T\rangle$ be the time-reversed state of $|\psi\rangle$:

$$|\psi^T\rangle = \hat{T}|\psi\rangle, \qquad (6.12.8)$$

Then we have

$$\langle\psi^T|\varphi^T\rangle = \langle\varphi|\psi\rangle. \qquad (6.12.9)$$

The system possesses time reversal invariance, i.e., $\hat{H}$ commutes with $\hat{T}$. By using (6.12.7a) we have

$$e^{-i\hat{H}t} = \hat{T}e^{i\hat{H}t}\hat{T}^{-1}, \qquad (6.12.10)$$

---

[29] See Ref. [16].

Hence

$$\langle m'|e^{-i\hat{H}t}|m\rangle = \langle m'|\hat{T}e^{i\hat{H}t}T^{-1}|m\rangle. \tag{6.12.11}$$

Defining $|\mu\rangle$ by

$$|\mu\rangle = e^{i\hat{H}t}\hat{T}^{-1}|m\rangle, \tag{6.12.12}$$

Using (6.12.7c) and (6.12.9), (6.12.11) can be written as

$$\langle m'|e^{-i\hat{H}t}|m\rangle = \langle m'|\hat{T}|\mu\rangle = \langle \mu \mid T^{-1}m'\rangle. \tag{6.12.13}$$

For further deduction we need to know the result of applying $\hat{T}$ and $\hat{T}^{-1}$ to the eigenstate $|m\rangle$ of $\hat{S}_z$. The resulting phases have to be carefully determined. Let us suppose the orthonormal $|m\rangle$ to be the *standard basis*[30] of the state space satisfying the following relations:

$$\hat{S}_z|m\rangle = m|m\rangle, \tag{6.12.14}$$

$$\hat{S}_\pm|m\rangle = [(S \mp m)(S \pm m + 1)]^{1/2}|m \pm 1\rangle. \tag{6.14.15}$$

Since

$$\hat{S}_\pm = \hat{S}_x \pm i\hat{S}_y \tag{6.12.16}$$

we have

$$\hat{T}\hat{S}_\pm\hat{T}^{-1} = -\hat{S}_\mp, \quad \hat{T}\hat{S}_z\hat{T}^{-1} = -\hat{S}_z; \tag{6.12.17}$$

where (6.12.7a,b) have been used. Then using (6.12.17) and (6.12.14), (6.12.15) we obtain

$$\hat{S}_z\hat{T}|m\rangle = -m\hat{T}|m\rangle, \tag{6.12.18}$$

$$\hat{S}_\pm\hat{T}|m\rangle = -[(S \pm m)(S \mp m + 1)]^{1/2}\hat{T}|m \mp 1\rangle. \tag{6.12.19}$$

---

[30]This refers to the phase convention of (6.12.14) and (6.12.15). Otherwise we can fix (6.12.14) and put a phase factor in (6.12.15). In the following, we see that this will not influence the result.

Since the eigenvalues of $\hat{S}_z$ are all non-degenerate , by comparing (6.12.14) and (6.12.18), we have

$$\hat{T}\,|m\rangle = \theta_m\,|-m\rangle\,, \qquad (6.12.20)$$

where θ_m is a phase factor depending on m. Substituting (6.12.20) into (6.12.19) we get

$$\theta_m = (-1)^{S-m}\theta, \qquad (6.12.21)$$

where

$$\theta \equiv \theta_{-S}.$$

Equation (6.12.21) gives the relation between phase factors with different values of m. Then (6.12.20) can be written as

$$\hat{T}\,|m\rangle = (-1)^{S-m}\theta\,|m\rangle\,. \qquad (6.12.22)$$

Applying T to both sides of (6.12.22), we get

$$\theta^2 = 1, \qquad (6.12.23)$$

i.e., $\theta = \pm 1$.

It can be proved that $\theta = 1$ [66], but for deriving the selection rule, (6.12.23) is sufficient. (6.12.22) gives

$$\hat{T}\,|-m\rangle = (-1)^{S+m}\theta\,|m\rangle\,. \qquad (6.12.24)$$

Applying $\hat{T}^{-1}$ to both sides of (6.12.24), then multiplying both sides by $(-1)^{S+m}\theta$, and considering $(-1)^{2S+2m} = 1$, we obtain

$$\hat{T}^{-1}\,|m\rangle = (-1)^{S+m}\theta\,|-m\rangle\,. \qquad (6.12.25)$$

Using (6.2.25), we get from (6.2.12)

$$|\mu\rangle = \mathrm{e}^{\mathrm{i}\hat{H}t}T^{-1}\,|m\rangle = \mathrm{e}^{\mathrm{i}\hat{H}t}(-1)^{S+m}\theta\,|-m\rangle\,,$$

i.e.,

$$|\mu\rangle = \langle -m|\,\mathrm{e}^{-\mathrm{i}\hat{H}t}(-1)^{S+m}\theta. \qquad (6.12.26)$$

Now we return to (6.2.13). Using (6.12.25), (6.12.26), we get

$$\left\langle m' \middle| e^{-i\hat{H}t} \middle| m \right\rangle = \left\langle \mu \middle| (-1)^{S+m'} \theta \middle| -m' \right\rangle$$

$$= \left\langle -m \middle| e^{-i\hat{H}t}(-1)^{2S+m+m'} \middle| -m' \right\rangle.$$

Let $m' = -m$, the above relation becomes

$$\left\langle -m \middle| e^{-i\hat{H}t} \middle| m \right\rangle = (-1)^{2S}\left\langle -m \middle| e^{-i\hat{H}t} \middle| m \right\rangle. \tag{6.12.27}$$

Consequently

$$S = \text{half integer} \implies \left\langle -m \middle| e^{-i\hat{H}t} \middle| m \right\rangle = 0. \tag{6.12.28}$$

The result can easily be generalized to a system of spins. Since the implication (6.12.28) remains valid if we multiply it by a phase factor, so the assumption of $|m\rangle$, being the standard basis, is unnecessary.

We discussed in Section 6.8 that in the absence of an external magnetic field the topological quenching is equivalent to Kramers theorem. The foregoing discussion shows that the transition from $|S\rangle$ to $|-S\rangle$, being forbidden for half-integral spin, is the rigorous consequence of the time reversal invariance. Generally speaking the states $|S\rangle$ and $|-S\rangle$ need not be the *eigenstates* of the Hamiltonian. Therefore, the forbidden transition from $|S\rangle$ to $|-S\rangle$ has no direct relationship with the double degeneracy of the eigenstates of H. Both are the consequence of time reversal invariance, but they do not imply each other. In the discussion of Section 6.8 in a double well the ground state of half-integral spin is doubly degenerate. The expectation value of the spin of these two states is just $-S$ and S. In semi-classical approximation we do not distinguish between the expected value and the eigenvalue.

The spin-parity effect of the macroscopic quantum coherence of magnetization originates from the *Wess-Zumino term*. Its topological effect does not only manifests itself when time reversal invariance is valid. When the existence of an external magnetic field breaks the time reversal invariance, the topological effect manifests itself in the oscillation of the tunneling rate with the magnetic field strength. That the *time reversal invariance* distinguishes the integral and self-integral spin is based on the property $\hat{T}^2 = (-1)^{2S}$. No matter whether the Wess-Zumino term is present, the selection rule does exist. The two origins are independent of each other, but sometimes both come across each other.

References

[1] E. M. Lifshitz and L. P. Pitaevskii, *Statistical Physics* part 2, Oxford, Pergamon Press, 1980.

[2] J. R. Schrieffer, *Theory of Superconductivity* Reading, Benjamin/ Cummings, 1964.

[3] L. P. Gorkov, *Sov. Phys. JETP* 9, 1364, 1959.

[4] R. P. Feynman, R. B. Leighton and M. Sands, *Feynman Lectures on Physics*, vol III, Reading, Addison-Wesley, 1965.

[5] B. D. Josephson, *Phys. Lett.* 1, 251, 1962.

[6] P. W. Anderson, *Phys. Today* 23(11), 23, 1970.

[7] J. W. Rowell, *Phys. Rev. Lett.* 11, 200, 1963.

[8] R. C. Jaklevic, J. Lambe, A. H. Silver and J. E. Mercereau, *Phys. Rev. Lett.* 12, 159, 1964; 12, 274, 1964.

[9] J. Kurkijärvi, *Phys. Rev. B* 6, 832, 1972.

[10] A. J. Leggett, Macroscopic Quantum Tunneling and Related Effects in Josephson Systems, in *Percolation, Localization and Superconductivity*, edited by M. Goldman and S. A. Wolf, New York, Plenum, 1984.

[11] A. O. Caldeira and A. J. Leggett, *Ann. Phys. (N.Y.)* 149, 374, 1983.

[12] A. J. Leggett, S. Chakravarty, A. T. Dorsey, M. Fisher, A. Garg and W. Zwerger, *Rev. Mod. Phys.* 59, 1,1987.

[13] H. A. Kramers, Physica, 7, 284, 1940; H. A. Kramers, *Physica* 7, 284, 1940.

[14] U. Weiss, *Quantum Dissipative Systems*. Singapore, World Scientific, 1993.

[15] R. P. Feynman and F. L. Vernon Jr., *Ann. Phys. (N.Y.)* 24, 118, 1963.

[16] L. I. Schiff, *Quantum Mechanics*, 3rd edition, New York, McGraw Hill, 1968.

[17] V. Ambegaokar, U. Eckern and G. Schön, *Phys. Rev. Lett.* 48, 1745, 1982.

[18] V. Ambegaokar, U. Eckern and G. Schön, *Phys. Rev. B* 30, 6419, 1984.

[19] L. H. Yu and C. P. Sun, *Phys. Rev. A* 49, 592, 1994.

[20] C. P. Sun and L. H. Yu, *Phys. Rev. A* 51, 1845, 1995.

[21] C. P. Sun, Y. B Gao, H. F. Dong and S. R. Zhao, *Phys. Rev. E* 57, 3900, 1998.

[22] K. Fujikawa, S. Iso, M. Sasaki and H. Suzuki, *Phys. Rev. Lett.* 68, 1093, 1992.

[23] L. D. Jackel *et al.*, *Phys. Rev. Lett.* 47, 697, 1981.

[24] R. F. Voss and R. A. Webb, *Phys. Rev. Lett.* 47, 265, 1981.

[25] S. Washburn, R. A. Webb, R. F. Voss and S. M. Farris, *Phys. Rev. Lett.* 54, 2712, 1985.

[26] J. M. Martinis, M. H. Devoret and J. Clarke, *Phys. Rev. Lett.* 55, 1543, 1985.

[27] E. M. Chudnovsky and L. Gunther, *Phys. Rev. Lett.* 60, 661, 1988.

[28] E. M. Chudnovsky and L. Gunther, *Phys. Rev. B* 37, 9455, 1988.

[29] A. Auerbach, *Interacting Electrons and Quantum Magnetism*, New York, Springer, 1994.

[30] J. R. Klauder and B.-S. Skagerstam, *Coherent States*, Singapore, World Scientific, 1985.

[31] A. Garg and G.-H. Kim, *Phys. Rev. B* 45, 12921, 1992.

[32] D. Loss, D. P. DiVincenzo and G. Grinstein, *Phys. Rev. Lett.* 69, 3232, 1992.

[33] J. von Delft and C. L. Henley, *Phys. Rev. Lett.* 69, 3236, 1992.

[34] A. Messiah, *Quantum Mechanics*, vol. 2, Amsterdam, North Holland, 1962.

[35] A. Garg, *Europhys. Lett.* 22, 205, 1993.

[36] E. M. Chudnovsky and D. P. DiVincenzo, *Phys. Rev. B* 48, 10548, 1993.

[37] A. Garg, *Phys. Rev. B* 51, 15161, 1995.

[38] L. Gunther and B. Barbara, eds. *Quantum Tunneling of Magnetism*, Dordrecht, Kluwer 1995.

[39] X. B. Wang and F. C. Pu, *J. Phys. Cond. Matter* 9, 693, 1997.

[40] R. Lü, P. Zhang, J.-L. Zhu and L. Chang, *Phys. Rev. B* 56, 10993, 1997.

[41] R. Lü, J.-L. Zhu, X. Chen and L. Chang, *Eur. Phys. J. B* 4, 223, 1998.

[42] J.-L. Zhu, R. Lü, X. B. Wang, X. Chen, L. Chang and F. C. Pu, *Eur. Phys. J. B* 4, 223, 1998.

[43] R. Lü, J.-L. Zhu, X. B. Wang and L. Chang, *Phys. Rev. B* 58, 8542, 1998.

[44] R. Lü, J.-L. Zhu, J. Wu, X. B. Wang and L. Chang, *Phys. Rev. B* 60, 3435, 1999.

[45] R. Lu, J.-L. Zhu, Y. Zhou and B.-L. Gu, *Phys. Rev. B* 62, 11661 2000. Y. Zhou, R. Lu, J.-L. Zhu and L. Chang, *Phys. Rev. B* 63, 054429 2001.

[46] E. M. Chudnovsky, *J. Magn. Magn. Matter.* 140–144, 1821, 1995.

[47] B. Barbara and E. M. Chudnovsky, *Phys. Lett. A* 145, 205, 1990.

[48] I. V. Krive and D. B. Zaslavskii, *J. Phys. Cond. Matter* 2, 9457, 1990.

[49] A. Garg and G.-H. Kim, *Phys. Rev. Lett.* 63, 2512, 1989.

[50] A. Garg and G.-H. Kim, *Phys. Rev. B* 43, 712, 1991.

[51] A. Garg, *Phys. Rev. Lett.* 70, 1541, 1993.

[52] B. Barbara *et al.*, *J. Appl. Phys.* 73, 6703, 1993.

[53] J. Tejada, X. X. Zhang and E. M. Chudnovsky, *Phys. Rev. B* 47, 14977, 1995.

[54] J. Tejada and X. X. Zhang, *J. Magn. Magn. Matter*, 140–144, 1815, 1995.

[55] D. D. Awschalom, D. P. DiVincenzo and J. F. Smyth, *Science* 258, 414, 1992.

[56] S. Gider, D. D. Awschalom, T. Douglas, S. Mann and M. Chaparala, *Science* 268, 77, 1995.

[57] D. D. Awschalom, J. F. Smyth, G. Grinstein, D. P. DiVincenzo and D. Loss, *Phys. Rev. Lett.* 68, 3092, 1992.

[58] N. V. Prokofiev and P. C. E. Stamp, *J. Phys. Cond. Matter* 5, L663, 1993.

[59] A. Garg, *Phys. Rev. Lett.*, 70, 1541, 1993; 70, 2193, 1993; 71, 4249, 1993.

[60] H. B. Braun and D. J. Loss, *J. Appl. Phys.* 76, 6177, 1994.

[61] J. R. Friedmann, M. P. Sarachik, J. Tejada and R. Ziolo, *Phys. Rev. Lett.* 75, 537, 1996.

[62] J. M. Hernandez, X. X. Zhang, F. Luiz, J. Bartelomé, J. Tejada and R. Ziolo, *Europhys. Lett.* 35, 301, 1996.

[63] L. Thomas, F. Lionti, R. Ballou, D. Gatteschi, R. Sessoli and B. Barbara, *Nature* 383, 145, 1996.

[64] C. Paulsen and J. G. Park, in *Quantum Tunneling of Magnetism*, ed. L. Gunther and B. Barbara, Dordrecht, Kluwer, 1995.

[65] P. Politi, A. Rittori, F. Hartmann-Boutron and J. Villain, *Phys. Rev. Lett.* 75, 537, 1995.

[66] F. Hartmann-Boutron, P. Politi and J. Villain, *Int. J. Mod. Phys. B* 10, 2577, 1996.

[67] W. Wernsdorfer and R. Sessoli, *Science*, 284, 133, 1999.

[68] O. B. Zaslavskii, *Phys. Lett. A* 145, 471, 1990; *Phys. Rev. B* 42, 992, 1990.

[69] V. V. Ulyanov and O. B. Zaslavskii, *Phys. Rep.* 216, 179, 1992.

[70] E. M. Chudnovsky and D. A. Garanin, *Phys. Rev. Lett.* 79, 4469, 1997.

[71] J. Q. Liang, H. J. W. Müller-Kirsten, D. K. Park and F. C. Pu *Phys. Rev. B* 61, 8856, 2000.

[72] Y.-H. Jin, Y.-H. Nie, Z.-J. Li, J.-Q. Liang and F.-Q. Pu, *Mod. Phys. Lett.* B14, 809, 2000.

[73] B. Li, J. Wu, W. Zhong and F. Pu, *Science in China (series A)*, 41, 301, 1998.

[74] B. Li and F. Pu, *Science in China (series A)*, 41, 983, 1998.

Chapter 7

Topological Phase Factors in Quantum Systems

We have seen that the tunneling of spin distinguishes the integral and half-integral values of spin, arising from the Wess-Zumino term in the Lagrangian. This phenomenon was given the name of topological quenching. In the study of spin wave in the Heisenberg model, it was found that the excitations of ferromagnetic spin wave have an energy gap, while the excitations of antiferromagnetic spin wave are gapless, i.e. there are massless quasiparticles. When spin waves are quantized, it is found that the one-dimensional antiferromagnetic chain is unstable with respect to quantum fluctuations, infrared divergence arises when the wave vector k approaches zero. This means that the *quantum* theory leads to complications. The spin wave theory is discussed in Section 7.1.

The Wess-Zumino term appears also in the study of 1-dimensional antiferromagnetic chain. Haldane guessed that the excitation has a gap for integral spin, and it is gapless for half-integral spin. This is called Haldane's conjecture. Observing from a different angle, the large spin limit of the 1-dimensional antiferromagnetic chain is the non-linear sigma model. This model possesses spontaneous symmetry breaking in which massless particles called Goldstone bosons appear. Can spontaneous symmetry breaking happen in the 1-dimensional antiferromagnetic chain? Different approaches give the same conclusion that the chain does not possess a long range

473

order, which one expects if spontaneous symmetry breaking happens. This agrees with Coleman theorem which asserts that spontaneous symmetry breaking cannot happen, i.e., the excitation must be massive, if infrared divergence arises. Then why does the gapless excitation in the half-integer spin case arise? The issue one has to address is the difference between the half-integer spin and the integer spin. It turns out that the difference lies in the fact that the large spin limit of the chain maps onto different field theories with or without a topological term. It is the non-perturbative effect of this topological term which is present in the half-integer spin case suppresses the mass generation. Therefore, the gapless excitation of the half-integer spin is not connected with the spontaneous symmetry breaking, i.e., it is not a Goldstone boson. This is discussed in Sections 7.2 and 7.3.

The Lieb-Schultz-Mattis theorem aims at distinguishing the half-integer spin from the integer spin. It states that for half-integer spin respecting translational symmetry and rotational symmetry in the spin space either has a zero gap or has degenerate ground states of opposite parities. This is discussed in Section 7.4. A simple discussion of the significance of the topological term is given in Section 7.5.

Another important appearance of the topological term is associated with the Θ-vacuum of non-Abelian gauge theory. This is introduced in Section 7.6.

In the study of spontaneous symmetry breaking in chiral $SU(2)$ theory of particle physics, the non-linear sigma model can ascribe the π-mesons to Goldstone bosons. The situation is that pion mass is much smaller than the mass of all others in the family of strongly interacting mesons, and it is natural to interpret the pions as pseudo-Goldstone bosons. There is yet another symmetry in the theory, viz. the chiral phase transformation, which actually does not exist in reality. If we attempt to break this symmetry spontaneously, another Goldstone boson would appear, but as such it has not been found in experiments either. The Adler-Bell-Jackiw anomaly discovered in 1969 helped to overcome this difficulty: the chiral phase symmetry is broken by the anomaly appearing in the quantum correction to the theory. The anomaly is represented by the Wess-Zumino term again.

We give an elementary introduction to the anomalies with examples in a series of occasions in Section 7.7.

Some basic features of physics appear in many research problems in its different branches, thus connecting together many different problems which look very different. The understanding of a basic feature in one problem can often give insight into a different problem. This chapter concerns many problems, some of which surpass the usual border of quantum mechanics. We choose to introduce them to the reader with the aim of demonstrating the unity of physics.

7.1 Spin Wave Theory in the Heisenberg Model

The interaction between spins on lattice sites leads to collective excitations of the spin system called the spin wave. As the ground state of a ferromagnetic system, all spins take the same direction. In the case of an antiferromagnetic system the ions form two sublattices with opposite spin directions resulting in a mutually compensating magnetic moment. In ferrimagnets the sublattices are formed with different kinds of atoms and possess different total spin, and the result is a state with a net magnetic moment. The physics of a system of spins on a lattice is described by the Heisenberg model. The Hamiltonian of the system is

$$H = - \sum_{i,j(i\neq j)} J_{ij} \boldsymbol{S}_i \cdot \boldsymbol{S}_j, \qquad (7.1.1)$$

where $\boldsymbol{S}_i$ is the spin operator of the ion i, J_{ij} is the exchange integral of the atoms i and j, and the sum is taken over all ions of the system. The Hamiltonian (7.1.1) can be considered phenomenologically, but for a system of spin $\frac{1}{2}$ carried by electrons, the interaction energy of electrons is *reproduced* by the Hamiltonian. Firstly, we explain the meaning of the exchange integral. In the Hartree-Fock theory of many-electron systems,[1] the wave function of many-electron system is expressed in terms of single electron wave functions $\varphi_i(\boldsymbol{r}_\alpha)$, where

---

[1] Readers not familiar with this theory can read Ref. [1].

i is the state index and α the electron index. A system of electrons with the same spin is described by a Slater determinant

$$\psi_{ij\cdots}(\boldsymbol{r}_1, \boldsymbol{r}_2, \cdots) = \begin{vmatrix} \varphi_1(\boldsymbol{r}_1) & \varphi_1(\boldsymbol{r}_2) & \cdots \\ \varphi_2(\boldsymbol{r}_1) & \varphi_2(\boldsymbol{r}_2) & \cdots \\ \cdots & \cdots & \cdots \end{vmatrix}, \tag{7.1.2}$$

in which the anti-symmetry with respect to the interchange of any two electrons is manifest. In calculating the interaction energy between electrons, there is a group of terms called the exchange energy

$$E_0 = -\frac{e^2}{2} {\sum_{i,j}}' \int \frac{\varphi_j^*(\boldsymbol{r}_1)\varphi_j(\boldsymbol{r}_2)\varphi_i^*(\boldsymbol{r}_2)\varphi_i(\boldsymbol{r}_1)}{|\boldsymbol{r}_1 - \boldsymbol{r}_2|} \mathrm{d}^3 r_1 \mathrm{d}^3 r_2 + \cdots, \tag{7.1.3}$$

where the prime above the summation reminds that $i \neq j$. Exchange energy exists only between electrons of the same spin. Let us reverse the spin of one of the electrons, then the system loses the exchange energy of this electron with all the remaining $N-1$ electrons. The energy difference is

$$E_1 - E_0 = \frac{e^2}{2} {\sum_j}' \int \frac{\varphi_j^*(\boldsymbol{r}_1)\varphi_j(\boldsymbol{r}_2)\varphi_i^*(\boldsymbol{r}_2)\varphi_i(\boldsymbol{r}_1)}{|\boldsymbol{r}_1 - \boldsymbol{r}_2|} \mathrm{d}^3 r_1 \mathrm{d}^3 r_2 \equiv \frac{1}{2} {\sum_j}' J_{ij}. \tag{7.1.4}$$

This defines J_{ij}, the exchange integral in the Heisenberg Hamiltonian. Now we calculate the interaction energy from the Hamiltonian (7.1.1). Let $\uparrow$ and $\downarrow$ represent the state $s_z = \frac{1}{2}$ and $s_z = -\frac{1}{2}$ respectively. From (7.1.1) the interaction energy between two electrons of the same spin is

$$E_{\uparrow\uparrow} = -J_{ij} \langle \uparrow_i \uparrow_j | \boldsymbol{S}_i \cdot \boldsymbol{S}_j | \uparrow_i \uparrow_j \rangle.$$

Using the spin raising and lowering operators $S^+ = S^x + \mathrm{i} S^y$ and $S^- = S^x - \mathrm{i} S^y$, we have

$$S^+ |\downarrow\rangle = |\uparrow\rangle, \quad S^- |\uparrow\rangle = |\downarrow\rangle,$$

$$S^+ |\uparrow\rangle = S^- |\downarrow\rangle = 0,$$

$$\boldsymbol{S}_i \cdot \boldsymbol{S}_j = \frac{1}{2} \left(S_i^+ S_j^- + S_i^- S_j^+ \right) + S_i^z \cdot S_j^z.$$

The subscripts of a spin operator refer to the lattice sites, while the superscripts refer to the spin components. From these relations we have

$$E_{\uparrow\uparrow} = -\frac{1}{4}J_{ij},$$

$$E_{\uparrow\downarrow} = \frac{1}{4}J_{ij}.$$

The energy difference is $E_{\uparrow\downarrow} - E_{\uparrow\uparrow} = \frac{1}{2}J_{ij}$. If a system has N electrons with all spins parallel, and we reverse the spin of one of the electrons, the energy required for this change is $\frac{1}{2}\sum_{j\neq i} J_{ij}$, which is the same as that given by the Hartree-Fock theory. So we have demonstrated that the Heisenberg model reproduces the energy of interaction of electrons on lattice sites. The sign of the exchange integral is of crucial importance. If $J > 0$, then parallel spins have the lower energy, and $J < 0$ favors antiparallel spins. From the definition of J_{ij} (7.1.4), we see that the exchange between neighboring atoms is most important, because they have wave functions with a larger overlap. In the following, we consider only the exchange between the nearest neighbors, and denote them by $\langle i, j \rangle$ in the sum. Suppose the atom i has ν nearest neighbors, denoted by $i + \delta$, $\delta = 1, \ldots, \nu$. For simple lattices, $J_{i,i+\delta}$ is independent of δ, and is simply denoted by J. Then we have from (7.1.1)

$$H = -J\sum_{\langle ij \rangle} \boldsymbol{S}_i \cdot \boldsymbol{S}_j = -J\sum_{i,\delta} \boldsymbol{S}_i \cdot \boldsymbol{S}_{i+\delta}. \qquad (7.1.5)$$

Let the spin of the ions be s (integral or half-integral, in units of $\hbar$). Consider a state Φ_0 in which all the ions have the z-component of spin equal to s. Denote the state of the n-th ion by $|s\rangle_n$, then we have $\Phi_0 = \prod_n |s\rangle_n$. Expressing (7.1.5) in terms of the spin components, we have

$$H = -J\sum_{\langle ij \rangle} \left[S_i^z \cdot S_j^z + \frac{1}{2}\left(S_i^+ S_j^- + S_i^- S_j^+ \right) \right]. \qquad (7.1.6)$$

Operating H on Φ_0 gives

$$H\Phi_0 = -J \sum_{i,j=i+\delta} s^2\,\Phi_0 = -J\nu N s^2\,\Phi_0 \equiv E_0\,\Phi_0, \qquad (7.1.7)$$

where E_0 is the eigenvalue of H for the state Φ_0, ν is the number of nearest neighbors of an ion, and N is the total number of ions. We have used $S^+|s\rangle = 0$ and the fact that the components of S_i and S_j commute for $i \neq j$. From (7.1.7) we see that Φ_0 is an eigenstate of H, and for $J > 0$ it is the ground state, i.e., the ferromagnetic ground state. Let us reduce the z-component of the m-th ion by one unit, and denote the state obtained by Φ_m,

$$\Phi_m = S_m^- \prod_n |s\rangle_n. \qquad (7.1.8)$$

Operating H on Φ_m, we obtain

$$H\Phi_m = -J \sum_{\langle i,j\rangle} \left[S_i^z S_j^z S_m^- + \frac{1}{2}\left(S_i^+ S_j^- S_m^- + S_i^- S_j^+ S_m^- \right) \right] \Phi_0. \quad (7.1.9)$$

Using the commutation relations

$$\left.\begin{array}{l}
[S_m^+, S_n^-] = 2S_m^z \delta_{mn} \\[4pt]
[S_m^-, S_n^z] = S_m^- \delta_{mn} \\[4pt]
[S_m^+, S_n^z] = -S_m^+ \delta_{mn}
\end{array}\right\} \qquad (7.1.10)$$

we can simplify (7.1.9) to the following form:

$$H\Phi_m = E_0\Phi_m + 2Js\sum_\delta (\Phi_m - \Phi_{m+\delta}). \qquad (7.1.11)$$

Hence, Φ_m is not an eigenstate of H. An inspection of (7.1.11) reveals that H induces a linear transformation in the space $(\Phi_m, \Phi_{m+1}, \ldots, \Phi_{m+\nu})$, so that we can expect to get an eigenstate by taking a linear superposition of Φ_m. The linear superposition has to satisfy the requirement of translational invariance of the lattice,

i.e.,

$$\Phi = \sum_m e^{i\boldsymbol{k}\cdot\boldsymbol{R}_m}\,\Phi_m, \qquad (7.1.12)$$

where $\boldsymbol{k}$ is one of the N vectors in the Brillouin zone, $\boldsymbol{R}_m$ is the radius vector of the atom m. Operating H on Φ gives

$$H\Phi = E_0\Phi + 2Js\nu\,\Phi_m - 2Js\sum_{m,\delta} e^{i\boldsymbol{k}\cdot\boldsymbol{R}_m}\,\Phi_{m+\delta}$$

$$= \left[E_0 + 2J\nu s(1 - \gamma_k)\right]\Phi, \qquad (7.1.13)$$

where

$$\gamma_k = \frac{1}{\nu}\sum_\delta e^{i\boldsymbol{k}\cdot\boldsymbol{R}_\delta}. \qquad (7.1.14)$$

Then Φ constructed in (7.1.12) is an eigenstate of H with an eigenvalue

$$E_k = E_0 + 2J\nu s(1 - \gamma_k). \qquad (7.1.15)$$

The excited state Φ is in a collective excitation mode. When a single atom is excited, it can pass the energy to other atoms through interactions, such that the excitation can be propagated in the form of a wave through the lattice. This kind of wave is called a spin wave. Figure 7.1 shows the spin wave with wave vector $\boldsymbol{k}$ in a 1-dimensional lattice. In order to obtain the dispersion relation for low excitations, we expand (7.1.15) in small k, then obtain

$$E_k = E_0 + Js\sum_\delta (\boldsymbol{k}\cdot\boldsymbol{R}_\delta)^2. \qquad (7.1.16)$$

The spin wave excitations are quantized to be magnons. In the foregoing discussion we considered a simple way of exciting the spin chain. To obtain the general state of excitation we consider the ground state Φ_0 for which the z-component of spin at all sites is $S_z = s$. For a general state, let $n_i = s - s_i^z$ be the deviation of the spin[2] at site i from the maximum value s. We use the occupation

---

[2]We call the z-component of spin simply "spin" in most occasions.

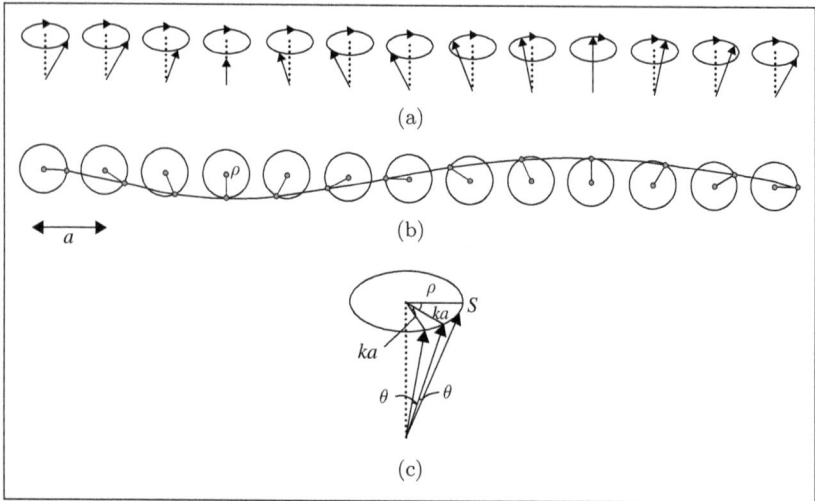

Fig. 7.1   Spin wave on a 1-dimensional chain (a) Perspective; (b) View from above; (c) Spins on 3 neighboring sites. Redrawn from O. Madelung, Introduction to Solid State Theory, translated by B. C. Taylor, Berlin, Springer, 1978.

number representation $|n_1, n_2, \ldots, n_i, \ldots, n_N\rangle$ to characterize the state, identifying the "deviation from the maximum spin" to the "occupation number". The ground state is therefore $|0, 0, \ldots, 0\rangle$. Then by introducing the creation and the annihilation operators for particle number $a_j^\dagger$ and a_j, and the particle number operator $n_j = a_j^\dagger a_j$, we can relate the spin operators S_j^- to $a_j^\dagger$, S_j^+ to a_j, since increasing the spin by one unit is equivalent to decreasing the "number" (deviation) by one unit. From the action of spin operators on state $|n\rangle$ (where the site index is temporarily suppressed)

$$\left. \begin{array}{l} S_+ |n\rangle = \sqrt{2s + 1 - n}\sqrt{n}\,|n - 1\rangle \\ S_- |n\rangle = \sqrt{2s - n}\sqrt{n + 1}\,|n + 1\rangle \\ S_z |n\rangle = (s - n)\,|n\rangle \end{array} \right\} \qquad (7.1.17)$$

we intend to obtain the action of the a and $a^\dagger$ operators

$$\left. \begin{array}{l} a^\dagger |n\rangle = \sqrt{n + 1}\,|n + 1\rangle \\ a |n\rangle = \sqrt{n}\,|n - 1\rangle \end{array} \right\}. \qquad (7.1.18)$$

On comparing (7.1.17) with (7.1.18) we obtain the connection between these two sets of operators:

$$
\left.\begin{aligned}
S_+ &= \sqrt{2s - a^\dagger a}\, a \\
S_- &= a^\dagger \sqrt{2s - a^\dagger a} \\
S_z &= s - a^\dagger a
\end{aligned}\right\}.
\tag{7.1.19}
$$

This is the *Holstein-Primakoff transformation*. The operators a and $a^\dagger$ are defined on sites while the excitations are defined on modes. We define the creation and annihilation operators of the *magnons of mode* k by the following relation[3]

$$
\left.\begin{aligned}
a_j^\dagger &= \frac{1}{\sqrt{N}} \sum_k e^{ik \cdot R_j} a_k^\dagger \\
a_j &= \frac{1}{\sqrt{N}} \sum_k e^{ik \cdot R_j} a_k
\end{aligned}\right\}.
\tag{7.1.20}
$$

The newly defined operators a_k and $a_k^\dagger$ satisfy the commutation relations

$$
\left.\begin{aligned}
[a_k, a_{k'}^\dagger] &= \delta_{k,k'} \\
[a_k, a_{k'}] &= [a_k^\dagger, a_{k'}^\dagger] = 0
\end{aligned}\right\}.
\tag{7.1.21}
$$

Elementary excitations are significant when they are not far away from the ground states, i.e., when the $n's$ are small. We then expand the square root in (7.1.19) in the number operators. When higher order quantities are neglected, S^+ contains a_k and $a_k a_{k'}^\dagger a_{k'}''$; S^- contains $a_k^\dagger$ and $a_k^\dagger a_{k'}^\dagger a_{k''}$. The Hamiltonian (7.1.6) can be expressed in a second quantized form.

$$
H = E_0 + \sum_k 2Jvs\,(1 - \gamma_k)\, a_k^\dagger a_k
$$

$$
+ \frac{vJ}{2N} \sum_{k,\kappa,k'} \left(\gamma_{k-\kappa} + \gamma_{k'} - 2\gamma_{k-\kappa-k'} \right) a_{k-\kappa}^\dagger a_{k'+\kappa}^\dagger a_{k'} a_k + \cdots .
\tag{7.1.22}
$$

---

[3]This is similar to the transformation from the Bloch wave function to the Wannier wave function. See Ref. [1] p. 120.

The second term in the right hand side is the magnon excitation, the energy of the magnon with wave vector $\boldsymbol{k}$ is

$$\hbar\omega_{\boldsymbol{k}} = 2J\nu s(1 - \gamma_{\boldsymbol{k}}). \qquad (7.1.23)$$

The third term in the right-hand side of (7.1.22) represents the interaction between a pair of magnons, in which a momentum exchange $\boldsymbol{\kappa}$ takes place. We have used $\gamma_{\boldsymbol{\kappa}} = \gamma_{-\boldsymbol{\kappa}}$ in obtaining (7.1.22).

If $J < 0$, then the energy of the system can be lowered by having nearest neighbors with antiparallel spins. We can expect that a state of lowest possible energy is one in which the intervening sublattices a and b with the same number of spins have $S^z = s$ and $S^z = -s$ respectively, and the total spin of the system is zero. This state is called the Néel state.[4] In comparison with the ferromagnet, we would suspect that the Néel state is the ground state of the Heisenberg Hamiltonian for $J < 0$. But in reality this state is not an eigenstate of the Hamiltonian. The point is that the nearest neighbors i and j must belong to different sublattices. Let us suppose that i has spin $+s$ and j has spin $-s$. The operator $S_i^+ S_j^-$ acting on $|s\rangle_i |-s\rangle_j$ gives 0, while $S_i^- S_j^+$ acting on it gives $\text{const}|s-1\rangle_i |-s+1\rangle_j$, both differ from $\text{const}|s\rangle_i |-s\rangle_j$, as it should be if the state $|s\rangle_i |-s\rangle_j$ is an eigenstate. We denote the Néel state as

$$|0, 0, \ldots, 0\rangle, \qquad (7.1.24)$$

and an arbitrary state by

$$|n_1, n_2, \ldots, n_N\rangle, \qquad (7.1.25)$$

where $n_i = s - s_i$ in sublattice a and $n_j = s_j + s$ in lattice b such that n denotes the *deviation* from the standard spin at a given site, and n is always non-negative. The creation and annihilation operator for sublattice a are denoted by $a_i^\dagger$ and a_i respectively, while those for

---

[4]Which sublattice chooses $S^z = s$? Why not the other way round? In reality the weak anisotropy field in the antiferromagnetic crystal helps the choice. We shall not go into details here.

sublattice b are denoted by $b_j^\dagger$ and b_j respectively. We see that

$$\left.\begin{aligned}
S_i^+ &\leftrightarrow a_i \\
S_i^- &\leftrightarrow a_i^\dagger \\
S_j^+ &\leftrightarrow b_j^\dagger \\
S_j^- &\leftrightarrow b_j
\end{aligned}\right\}. \qquad (7.1.26)$$

Here the symbol $\leftrightarrow$ means "is equivalent to". Consequently, $S_i^+ S_j^-$ and $S_i^- S_j^+$ are equivalent to $a_i b_j$ and $a_i^\dagger b_j^\dagger$ respectively, the former decreases the particle number by 2 and the latter increases the number by 2. This is a contrast to the ferromagnetic case where $S_i^+ S_j^-$ and $S_i^- S_j^+$ are equivalent to $a_i a_j^\dagger$ and $a_i^\dagger a_j$ respectively. They transfer particles but conserve the particle number.

Carrying out the Holstein-Primakoff transformation, we have for the sublattice a

$$\left.\begin{aligned}
S_i^+ &= (2s)^{\frac{1}{2}}(1 - \cdots)a_i \\
S_i^- &= (2s)^{\frac{1}{2}} a_i^\dagger (1 - \cdots) \\
S_i^z &= s - a_i^\dagger a_i
\end{aligned}\right\} \qquad (7.1.27)$$

and for sublattice b

$$\left.\begin{aligned}
S_j^+ &= (2s)^{\frac{1}{2}} b_j^\dagger (1 - \cdots) \\
S_j^- &= (2s)^{\frac{1}{2}}(1 - \cdots)b_j \\
S_j^z &= -s + b_j^\dagger b_j
\end{aligned}\right\}. \qquad (7.1.28)$$

In the above expressions we keep only the leading term in the parentheses, which is sufficient for subsequent deduction.

In order to get the mode operators, we use the transformation

$$\left.\begin{aligned}
c_k &= \frac{1}{\sqrt{N}} \sum_i e^{i\boldsymbol{k}\cdot\boldsymbol{R}_i} a_i \\
d_k &= \frac{1}{\sqrt{N}} \sum_j e^{i\boldsymbol{k}\cdot\boldsymbol{R}_j} b_j
\end{aligned}\right\}. \qquad (7.1.29)$$

This transformation results in the following spin operators of the sublattices:

$$
\left.
\begin{aligned}
S_i^+ &= \left(\frac{2s}{N}\right)^{\frac{1}{2}} \left(\sum_k e^{-i\boldsymbol{k}\cdot\boldsymbol{R}_i} c_k + \cdots\right) \\
S_i^- &= \left(\frac{2s}{N}\right)^{\frac{1}{2}} \left(\sum_k e^{i\boldsymbol{k}\cdot\boldsymbol{R}_i} c_k^\dagger + \cdots\right) \\
S_i^z &= s - \frac{1}{N} \sum_{k,k'} e^{-i(k-k')\cdot\boldsymbol{R}_i} c_k^\dagger c_k
\end{aligned}
\right\}
\tag{7.1.30}
$$

and

$$
\left.
\begin{aligned}
S_j^+ &= \left(\frac{2s}{N}\right)^{\frac{1}{2}} \left(\sum_k e^{-i\boldsymbol{k}\cdot\boldsymbol{R}_i} d_k^\dagger + \cdots\right) \\
S_j^- &= \left(\frac{2s}{N}\right)^{\frac{1}{2}} \left(\sum_k e^{i\boldsymbol{k}\cdot\boldsymbol{R}_i} d_k + \cdots\right) \\
S_j^z &= -s + \frac{1}{N} \sum_{k,k'} e^{-i(k-k')\cdot\boldsymbol{R}_i} d_k^\dagger d_k
\end{aligned}
\right\}
\tag{7.1.31}
$$

Up to terms bilinear in the operators H can then be written as

$$
H = -2NJ\nu s^2 + 2J\nu s \sum_k \left[\gamma_k \left(c_k^\dagger d_k^\dagger + c_k d_k\right) + \left(c_k^\dagger c_k + d_k^\dagger d_k\right)\right].
\tag{7.1.32}
$$

In contrast to the ferromagnetic case, the Hamiltonian is not diagonal in the particle number representation, since it involves operators $c^\dagger d^\dagger$ ($\Delta N = 2$) and cd ($\Delta N = -2$) which change the number of particles. To diagonalize the Hamiltonian we use the Bogoliubov transformation. Define the quasiparticle operators α_k, β_k and their conjugate operators:

$$
\left.
\begin{aligned}
\alpha_k &= u_k c_k - v_k d_k^\dagger, \quad \alpha_k^\dagger = u_k c_k^\dagger - v_k d_k \\
\beta_k^\dagger &= u_k d_k - v_k c_k^\dagger, \quad \beta_k = u_k d_k^\dagger - v_k c_k
\end{aligned}
\right\},
\tag{7.1.33}
$$

where u_k and v_k are real and satisfying

$$
u_k^2 - v_k^2 = 1.
\tag{7.1.34}
$$

With the help of (7.1.34) the commutation relations for operators α and β turn out to be

$$\left. \begin{aligned} [\alpha_{k}, \alpha_{k'}^{\dagger}] &= [\beta_{k}, \beta_{k'}^{\dagger}] = \delta_{kk'} \\ [\alpha_{k}, \alpha_{k'}] &= [\beta_{k}, \beta_{k'}] = 0 \\ [\alpha_{k}, \beta_{k'}] &= [\alpha_{k}, \beta_{k'}^{\dagger}] = \cdots = 0 \end{aligned} \right\}. \qquad (7.1.35)$$

The aim of the diagonalization is to achieve the following form for the Hamiltonian

$$H = \sum_{k} \lambda_{k} (\alpha_{k}^{\dagger} \alpha_{k} + \beta_{k}^{\dagger} \beta_{k}) + \text{const}, \qquad (7.1.36)$$

where λ_{k} is the energy eigenvalue of the quasiparticle. The condition (7.1.36) determines u_{k} and v_{k}. From (7.1.36) we obtain

$$[\alpha_{k}^{\dagger}, H] = -\lambda_{k} \alpha_{k}^{\dagger}, \quad [\alpha_{k}, H] = \lambda_{k} \alpha_{k}. \qquad (7.1.37)$$

The commutator $[\alpha_{k}^{\dagger}, H]$ can be calculated by using (7.1.32). Let

$$\omega_{0} = 2J\nu s, \quad \omega_{1} = 2J\nu s \gamma_{k}, \qquad (7.1.38)$$

then we have

$$[\alpha_{k}^{\dagger}, H] = -\mu_{k}(\omega_{0} c_{k}^{\dagger} + \omega_{1} d_{k}) - v_{k}(\omega_{0} d_{k} + \omega_{1} c_{k}^{\dagger}). \qquad (7.1.39)$$

Substituting the expression for $\alpha_{k}^{\dagger}$ in (7.1.33) into the right-hand side of the first equation of (7.1.37), we get

$$[\alpha_{k}^{\dagger}, H] = -\lambda_{k}(u_{k} c_{k}^{\dagger} - v_{k} d_{k}). \qquad (7.1.40)$$

Comparing (7.1.39) with (7.1.40), we find

$$\omega_{0} u_{k} + \omega_{1} v_{k} = \lambda_{k} u_{k}, \quad \omega_{1} u_{k} + \omega_{0} v_{k} = -\lambda_{k} u_{k}. \qquad (7.1.41)$$

The condition for the existence of non-zero solution to (7.1.41) is

$$\begin{vmatrix} \omega_{0} - \lambda_{k} & \omega_{1} \\ \omega_{1} & \omega_{0} + \lambda_{k} \end{vmatrix} = 0, \qquad (7.1.42)$$

i.e., $\lambda_{\boldsymbol{k}}^2 = \omega_0^2 - \omega_1^2$. Since the excitation energy of the quasiparticle is positive, we take the positive root

$$\lambda_{\boldsymbol{k}} = 2Jvs\left(1 - \gamma_{\boldsymbol{k}}^2\right)^{1/2}. \tag{7.1.43}$$

The same secular equation is obtained by calculating $[\beta_{\boldsymbol{k}}^\dagger, H]$ and using the condition for diagonalizing H. Substituting $\lambda_{\boldsymbol{k}}$ into (7.1.41) and solving for $u_{\boldsymbol{k}}$ and $v_{\boldsymbol{k}}$

$$u_{\boldsymbol{k}}^2 = \frac{1}{2}\left[\frac{1}{\left(1 - \gamma_{\boldsymbol{k}}^2\right)^{1/2}} + 1\right], \quad v_{\boldsymbol{k}}^2 = \frac{1}{2}\left[\frac{1}{\left(1 - \gamma_{\boldsymbol{k}}^2\right)^{1/2}} = 1\right]. \tag{7.1.44}$$

Expressing H (7.1.33) in terms of α and β, we have

$$H = -2NJvs(s+1) + \sum_{\boldsymbol{k}} 2Jvs\left(1 - \gamma_{\boldsymbol{k}}^2\right)^{1/2}\left(\alpha_{\boldsymbol{k}}^\dagger \alpha_{\boldsymbol{k}} + \beta_{\boldsymbol{k}}^\dagger \beta_{\boldsymbol{k}} + 1\right). \tag{7.1.45}$$

Denoting $\lambda_{\boldsymbol{k}}$ by $\hbar\omega_{\boldsymbol{k}}$, we have

$$\hbar\omega_{\boldsymbol{k}} = 2Jvs\left(1 - \gamma_{\boldsymbol{k}}^2\right)^{1/2} \quad \text{(antiferromagnetic case)}. \tag{7.1.46}$$

Comparing with (7.1.23) for the excitation energy of the ferromagnetic mode $\boldsymbol{k}$

$$\hbar\omega_{\boldsymbol{k}} = 2Jvs\left(1 - \gamma_{\boldsymbol{k}}\right) \quad \text{(ferromagnetic case)},$$

we find a striking difference. For simple cubic lattice with lattice constant a, we have for ka small

$$\left.\begin{array}{l} v\left(1 - \gamma_{\boldsymbol{k}}\right) \approx 2\left(ka\right)^2 \\[2mm] v\left(1 - \gamma_{\boldsymbol{k}}^2\right)^{1/2} \approx \dfrac{1}{\sqrt{3}}ka \end{array}\right\} \tag{7.1.47}$$

We find that for ferromagnetic spin wave the dispersion relation for low excitation is quadratic,

$$\hbar\omega_{\boldsymbol{k}} = 4Js\left(ka\right)^2 \quad \text{(ferromagnetic)}, \tag{7.1.48}$$

where the energy is proportional to the square of the momentum, and hence the quasiparticle is massive. The dispersion relation for anti-ferromagnetic excitation is linear,

$$\hbar\omega_{\boldsymbol{k}} = \frac{2}{\sqrt{3}} J s k a \quad \text{(antiferromagnetic)}, \tag{7.1.49}$$

where the energy is linearly dependent on the momentum, and hence the quasiparticle is massless.

After the Bogoliubov transformation the quasiparticle vacuum can be considered as the approximate ground state of the anti-ferromagnetic Heisenberg Hamiltonian. We denote this state by $|\rangle$, which by definition satisfies

$$\alpha_{\boldsymbol{k}} |\rangle = 0, \quad \beta_{\boldsymbol{k}} |\rangle = 0.$$

The spin wave theory gives an intuitive description of the physical process in spin systems in higher dimensions ($d = 2, 3$) , and is verified by experiments.

Next we concentrate on the 1-dimensional antiferromagnetic chain. Let us calculate the influence of the quantum fluctuation on the expectation value of the spin of a sublattice. The z-component of a spin of sublattice A is

$$\langle S_A^z \rangle = s - \left\langle a^\dagger a \right\rangle, \tag{7.1.50}$$

where the average is taken over the Bogoliubov ground state, i.e., the quasiparticle vacuum. The site operator a_i is the inverse of the mode operator $c_{\boldsymbol{k}}$ (7.1.29)

$$a_i = \frac{1}{\sqrt{N}} \sum_{\boldsymbol{k}} e^{-i\boldsymbol{k}\cdot\boldsymbol{R}_i} c_{\boldsymbol{k}}, \tag{7.1.51}$$

Therefore,

$$a_i^\dagger a_i = \frac{1}{N} \sum_{\boldsymbol{k},\boldsymbol{k}'} e^{-i(\boldsymbol{k}-\boldsymbol{k}')\cdot\boldsymbol{R}_i} c_{\boldsymbol{k}}^\dagger c_{\boldsymbol{k}'}. \tag{7.1.52}$$

The inverse of the Bogoliubov transformation is

$$\left. \begin{array}{l} c_{\boldsymbol{k}} = u_{\boldsymbol{k}} \alpha_{\boldsymbol{k}} - v_{\boldsymbol{k}} \beta_{\boldsymbol{k}}^\dagger \\ d_{\boldsymbol{k}}^\dagger = u_{\boldsymbol{k}} \beta_{\boldsymbol{k}}^\dagger - v_{\boldsymbol{k}} \alpha_{\boldsymbol{k}} \end{array} \right\}. \tag{7.1.53}$$

Substituting the expression for c_k into (7.1.52) and taking the expectation value of $a_i^\dagger a_i$ in the ground state, we get

$$\left\langle a_i^\dagger a_i \right\rangle = \frac{1}{N} \sum_{k,k'} e^{-i(k-k')\cdot R_i} \left\langle \left| \left(u_k \alpha_k^\dagger - v_k \beta_k \right) \left(u_{k'} \alpha_{k'} - v_{k'} \beta_{k'}^\dagger \right) \right| \right\rangle.$$

For the anti-ferromagnetic ground state, we have $\alpha_{k'}|\rangle = 0$ and $\langle | \alpha_k^\dagger = 0$. Therefore, the only non-vanishing expectation values in the above expression are

$$\left\langle \beta_k \beta_{k'}^\dagger \right\rangle = \delta_{kk'},$$

and we get

$$\left\langle a_i^\dagger a_i \right\rangle = \frac{1}{N} \sum_k v_k^2,$$

i.e.,

$$\langle S_A^z \rangle = s - \frac{1}{N} \sum_k \frac{1}{2} \left(\frac{1}{\left(1 - \gamma_k^2\right)^{1/2}} - 1 \right). \qquad (7.1.54)$$

For 2- and 3-dimensional lattices, the second term is finite. For 1-dimensional lattice, the second term is $-\int \frac{dk}{2\pi} \frac{1}{2k}$ for k small. The infrared divergence appears. This result is independent of the spin s. The conclusion is that the 1-dimensional antiferromagnetic chain ground state is *unstable* against the quantum fluctuations. The result obtained above in the classical spin wave theory for the 1-dimensional antiferromagnetic chain is subject to re-examination.

7.2 $O(3)$ Non-Linear Sigma Model, the Spontaneous Symmetry Breakdown and the Goldstone Theorem

The non-linear sigma model used to be a low-energy effective theory of particle physics. A scalar field in 1+1 dimensions n is a unit vector

$$n^2(x_\mu) = 1, \ \mu = 0, 1 \qquad (7.2.1)$$

in a N dimensional *internal* space. The most general form of the Lagrangian density involving the product of two partial derivatives satisfying the $O(N)$ symmetry[5] is

$$\mathscr{L} = \frac{1}{2g^2} \left| \partial_\mu \boldsymbol{n} \right|^2 . \qquad (7.2.2)$$

In the following, we concentrate on the case $N = 3$. We express the unit vector in terms of its components

$$\boldsymbol{n} \left(\pi_1, \pi_2, \sigma \right), \qquad (7.2.3)$$

where by (7.2.1)

$$\boldsymbol{\pi}^2 + \sigma^2 = 1,$$

or

$$\sigma = \left(1 - \boldsymbol{\pi}^2 \right)^{1/2} . \qquad (7.2.4)$$

The condition (7.2.1) limits the vector $\boldsymbol{n}$ on the 3-dimensional sphere with unit radius. Therefore, no matter how we select the vacuum state, the $O(3)$ symmetry cannot be respected by the vacuum. For instance, the choice

$$\boldsymbol{n}_0 = (0, 0, 1) \qquad (7.2.5)$$

is invariant only under a rotation about the 3-rd axis, but the rotational invariance about the 1-st and 2-nd axes is broken. This is a case that the dynamics of the field possesses a higher symmetry (in our case $O(3)$) and its vacuum state respects only a lower symmetry (in our case $O(1)$), which is called a spontaneous symmetry breaking. From (7.2.4) we have

$$\partial_\mu \sigma = \frac{\boldsymbol{\pi} \cdot \partial_\mu \boldsymbol{\pi}}{(1 - \boldsymbol{\pi}^2)^{1/2}},$$

---

[5]The $O(N)$ symmetry is an invariance with respect to the rotation about any axis in the N- dimensional (internal) space.

and the Lagrangian (7.2.2) becomes

$$\mathscr{L} = \frac{1}{2g^2}\left(\left|\partial_\mu\boldsymbol{\pi}\right|^2 + \frac{(\boldsymbol{\pi}\cdot\partial_\mu\boldsymbol{\pi})^2}{1-\boldsymbol{\pi}^2}\right). \tag{7.2.6}$$

This is the effective Lagrangian for the $\boldsymbol{\pi}$ field after the elimination of the σ field. The self-coupling of the $\boldsymbol{\pi}$ field appears, and there is no mass term for π_1 and π_2. This is a special case for which the *Goldstone theorem* applies. According to this theorem, the spontaneous breakdown of a continuous symmetry leads to massless particles. Consider a $O(3)$ symmetric theory. If the vacuum state is $(0,0,0)$, then it respects this symmetry. But in case when the constraint $\boldsymbol{n}^2 = 1$ leads to a choice of $(0,0,1)$ to be the vacuum, then *two* symmetry operations are spontaneously broken with one surviving. Correspondingly *two* massless particles appear. The Goldstone theorem [4, 5] states that "Corresponding to each spontaneously broken continuous symmetry transformation, there exists a massless particle in the theory." The massless particle is dubbed the Goldstone boson. In the following, we prove this theorem in the case of a scalar field (spin zero particle). The Lagrangian density of this field is

$$\mathscr{L} = \frac{1}{2}\partial_\mu\varphi^a\partial_\mu\varphi^a - V(\varphi), \tag{7.2.7}$$

where φ is the scalar field, $a = 1, 2, \ldots, N$ is internal symmetry index. Let φ_0^a be a constant field which minimizes $V(\varphi)$:

$$\left.\frac{\partial V}{\partial\varphi^a}\right|_{\varphi^a(x)=\varphi_0^a} = 0. \tag{7.2.8}$$

Since φ_0^a is a constant, its kinetic energy (the first term of $\mathscr{L}$) vanishes. Therefore, φ_0^a is the vacuum of the system. Expand $V(\varphi)$ around the vacuum, we have

$$V(\varphi) = V(\varphi_0) + \frac{1}{2}(\varphi-\varphi_0)^a(\varphi-\varphi_0)^b\left.\frac{\partial^2}{\partial\varphi^a\partial\varphi^b}V\right|_{\varphi=\varphi_0} + \cdots \tag{7.2.9}$$

There is no linear term, because V is minimum at φ_0. Compared with the harmonic oscillator case, where the potential energy is $\frac{1}{2}m\omega^2 x^2$, the mass squared of the quantum of the field (the spinless particle) can be obtained from the following $N \times N$ matrix:

$$\left(\frac{\partial^2}{\partial \varphi^a \partial \varphi^b} V \right) \Bigg|_{\varphi = \varphi_0} \equiv m_{ab}^2. \qquad (7.2.10)$$

Since φ_0 gives the minimum of $V(\varphi)$, the second derivative of V at φ_0 is positive definite. What we need to prove is that whenever a continuous symmetry is not satisfied by a ground state φ_0, it gives always a zero eigenvalue of m_{ab}^2.

In general a continuous symmetry transformation can be defined by

$$\varphi^a \rightarrow \varphi^a + \alpha \Delta^a(\varphi), \qquad (7.2.11)$$

where α is an infinitesimal parameter, and $\Delta^a(\varphi)$ is a function of all the components of φ. For a constant field the derivative term in $\mathscr{L}$ is zero. The invariance demands that V not to change under the transformation:

$$V(\varphi^a) = V(\varphi^a + \alpha \Delta^a),$$

i.e.,

$$\frac{\partial V(\varphi)}{\partial \varphi^a} \Delta^a(\varphi) = 0. \qquad (7.2.12)$$

Taking derivative with respect to φ^b and taking values at φ_0, we get

$$\left(\frac{\partial V}{\partial \varphi^a} \right)_{\varphi_0} \left(\frac{\partial \Delta^a}{\partial \varphi^b} \right)_{\varphi_0} + \Delta^a(\varphi_0) \left(\frac{\partial^2}{\partial \varphi^a \partial \varphi^b} V \right)_{\varphi_0} = 0.$$

The first term of the left-hand side vanishes, since V has a minimum at φ_0. Consequently the second term must vanish. If φ_0 does not vanish under (7.2.11), then $\Delta^a(\varphi_0) = 0$, the demand is automatically satisfied. If $\Delta^a(\varphi_0) \neq 0$, i.e., the continuous transformation is spontaneously broken, then it follows that

$$\left(\frac{\partial^2}{\partial\varphi^a\partial\varphi^b}V\right)_{\varphi_0} = 0, \tag{7.2.13}$$

i.e., the mass eigenvalue is zero. The theorem is proven.

Now we return to the spin wave theory of the antiferromagnets. The massless quasiparticles created by $\alpha^\dagger$ and $\beta^\dagger$ correspond to two Goldstone modes associated with the spontaneous breaking of $SO(3)$ down to $O(1)$, the rotation about the z-axis. It is illuminating to calculate the long range correlation $\langle \boldsymbol{S}(x)\cdot\boldsymbol{S}(0)\rangle$ for $x\to\infty$ in the spin wave theory. The result is

$$\langle \boldsymbol{S}(x)\cdot\boldsymbol{S}(0)\rangle \to \pm s^2\left[1 - \frac{1}{\pi s}\ln\frac{x}{a} + O\left(\frac{1}{s^2}\right)\right],$$

where a is the lattice constant. This indicates that the chain possesses only short range order. Disorder occurs at the length of order

$$\xi = ae^{\pi s}$$

at which the correlation tends to zero. The spontaneous symmetry breaking (long range order) is expected for which the correlation tends to $\pm s^2$ for $x\to\infty$. This means that the massless quasiparticles of the 1-dimensional antiferromagnet cannot be Goldstone bosons.

7.3   One-dimensional Quantum Antiferromagnetic Chain, Topological Phase Factor, Large-s Mapping Onto $O(3)$ Non-linear Sigma Model

The study of 1-dimensional antiferromagnetic spin chain began with *Bethe Ansatz* in 1931 [6, 7] for $s = \frac{1}{2}$. The *exact solution* provided by the Ansatz is special to one-dimension, and is not generalizable to higher dimensions. The conclusion for the $s = \frac{1}{2}$ ground state is that there is no long range order but there exists gapless excitation. We have seen that the spin wave theory gives the same conclusion. Beginning from 1950s, there have been more and more papers published. There were indications that $s = 1$ antiferromagnetic chain has massive excitations. F. D. M. Haldane [8, 9, 10] suggested that

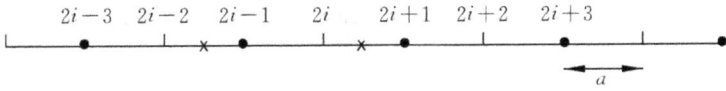

Fig. 7.2   One-dimensional antiferromagnetic chain.

integer spin antiferromagnetic Heisenberg chains have a finite gap, while only the half-odd-integer chains are gapless. This suggestion is dubbed the *Haldane conjecture* in the literature. In order to confirm the conjecture large number of papers have been published. There are solvable models, numerical studies and also proved theorems. The Lieb-Schultz-Mattis theorem [11] attempts at distinguishing the half-integral and integral spins. Before entering the discussion of the theorem, we follow Ref. [3] in establishing a low energy continuous limit of an antiferromagnetic chain of spin s. As shown in Fig. 7.2 each sublattice is populated with spins in approximately the same direction, and the spins of the two sublattices are approximately oppositely directed. Sites of sublattice A are denoted by dots, and those of sublattice B are denoted by short vertical lines. Vectors $\boldsymbol{\Omega}$ and $\boldsymbol{l}$ are defined at midpoints between sites of different sublattices denoted by crosses.

$$\left.\begin{aligned} \boldsymbol{\Omega}_{2i+\frac{1}{2}} &= \frac{\boldsymbol{S}_{2i+1} - \boldsymbol{S}_{2i}}{2s} \\ \boldsymbol{l}_{2i+\frac{1}{2}} &= \frac{\boldsymbol{S}_{2i+1} + \boldsymbol{S}_{2i}}{2a} \end{aligned}\right\} \tag{7.3.1}$$

Sites $2i$ and $2i + 1$ belong to two different sublattices, $\boldsymbol{\Omega}$ and $\boldsymbol{l}$ are defined at points between them. For a Néel state, $|\boldsymbol{\Omega}| = 1$ and $\boldsymbol{l} = 0$. $\boldsymbol{\Omega}$ is the Néel vector. We have from (7.3.1)

$$\left.\begin{aligned} a\boldsymbol{l}_{2i+\frac{1}{2}} + s\boldsymbol{\Omega}_{2i+\frac{1}{2}} &= \boldsymbol{S}_{2i+1} \\ a\boldsymbol{l}_{2i+\frac{1}{2}} - s\boldsymbol{\Omega}_{2i+\frac{1}{2}} &= \boldsymbol{S}_{2i} \end{aligned}\right\}. \tag{7.3.2}$$

This pair of expressions define the spin on sites in terms of the $\boldsymbol{\Omega}$ and $\boldsymbol{l}$ vectors to the left of the site. Equations (7.3.1) give

$$\left.\begin{aligned} \boldsymbol{\Omega} \cdot \boldsymbol{l} &= 0 \\ s^2 \boldsymbol{\Omega}^2 + a^2 \boldsymbol{l}^2 &= s(s+1) \end{aligned}\right\} \tag{7.3.3}$$

The vectors $\boldsymbol{\Omega}$ and $\boldsymbol{l}$ in these equations refer to the same site on the lattice. The second equation can also be written as

$$\Omega^2 = 1 + \frac{1}{s} - \frac{a^2 l^2}{s^2}, \tag{7.3.4}$$

so that under the large s limit $\Omega^2 \to 1$. From the commutation relations for the components of $\boldsymbol{S}$ it is possible to obtain the commutation relations for the components of $\boldsymbol{\Omega}$ and $\boldsymbol{l}$. Let the components be denoted by a, b and c, the commutation relations are

$$\left[\Omega^a_{2i+\frac{1}{2}}, \Omega^b_{2j+\frac{1}{2}} \right] = \frac{1}{4s^2} \left[S^a_{2i+1} - S^a_{2i}, S^b_{2j+1} - S^b_{2j} \right]$$

$$= \frac{i}{4s^2} \varepsilon^{abc} \delta_{ij} \left(S^c_{2i+1} + S^c_{2i} \right)$$

$$= \frac{ia}{2s^2} \varepsilon^{abc} \delta_{ij} l^c_{2i+\frac{1}{2}} \tag{7.3.5}$$

$$\left[l^a_{2i+\frac{1}{2}}, l^b_{2j+\frac{1}{2}} \right] = \frac{1}{4a^2} \left[S^a_{2i} + S^a_{2i+1}, S^b_{2j} + S^b_{2j+1} \right]$$

$$= \frac{1}{4a^2} \varepsilon^{abc} \delta_{ij} \left(S^c_{2i} + S^c_{2i+1} \right)$$

$$= \frac{i}{2a} \varepsilon^{abc} \delta_{ij} l^c_{2i+\frac{1}{2}} \tag{7.3.6}$$

$$\left[l^a_{2i+\frac{1}{2}}, \Omega^b_{2j+\frac{1}{2}} \right] = \frac{1}{4as} \left[S^a_{2i} + S^a_{2i+1}, S^b_{2j} - S^b_{2j+1} \right]$$

$$= \frac{i}{4as} \varepsilon^{abc} \delta_{ij} \left(S^c_{2i+1} - S^c_{2i} \right)$$

$$= \frac{i}{2a} \varepsilon^{abc} \delta_{ij} \Omega^c_{2j+\frac{1}{2}} \tag{7.3.7}$$

where ε^{abc} is the unit antisymmetric tensor. We take the continuity limit $a \to 0$, i.e.,

$$\frac{\delta_{ij}}{2a} \to \delta(x - y), \tag{7.3.8}$$

the commutations become

$$\left. \begin{array}{l} \left[l^a(x), l^b(y) \right] = i\varepsilon^{abc} l^c(x) \delta(x - y) \\ \left[l^a(x), \Omega^b(y) \right] = i\varepsilon^{abc} \Omega^c(x) \delta(x - y) \\ \left[\Omega^a(x), \Omega^b(y) \right] = 0 \end{array} \right\} \tag{7.3.9}$$

The first formula is the commutation relation for the components of the angular momentum (the generators of the rotation), and the second formula demonstrates that $\boldsymbol{\Omega}$ is a 3-dimensional vector.

In the following, we express the Hamiltonian of the Heisenberg model in terms of vectors $\boldsymbol{\Omega}$ and l, and take the continuity limit to get the Hamiltonian for the field theory. In the Heisenberg Hamiltonian, interactions between spins occur between nearest neighbors. Therefore, it is sufficient to sum only over *one* sublattice if it includes the interaction between a spin on site with both its left and right neighbors. We use $\boldsymbol{\Omega}$ and l to replace $\boldsymbol{S}$, then we get

$$H = J \sum_{2i} (\boldsymbol{S}_{2i} \cdot \boldsymbol{S}_{2i+1} + \boldsymbol{S}_{2i} \cdot \boldsymbol{S}_{2i-1})$$

$$= J \sum_{2i} (2a^2 l_{2i+\frac{1}{2}}^2 - s^2 \boldsymbol{\Omega}_{2i+\frac{1}{2}}^2 + a^2 l_{2i+\frac{1}{2}} \cdot l_{2i-\frac{3}{2}}$$

$$- s^2 \boldsymbol{\Omega}_{2i+\frac{1}{2}} \cdot \boldsymbol{\Omega}_{2i-\frac{3}{2}} - as l_{2i+\frac{1}{2}} \cdot \boldsymbol{\Omega}_{2i-\frac{3}{2}} + as \boldsymbol{\Omega}_{2i+\frac{1}{2}} \cdot l_{2i-\frac{3}{2}}).$$

$$(7.3.10)$$

The two terms involving $\boldsymbol{\Omega} \cdot l$ in the above equation can be replaced by

$$-2a^2 s \left\{ l_{2i+\frac{1}{2}} \cdot \left(\frac{\boldsymbol{\Omega}_{2i+\frac{3}{2}} - \boldsymbol{\Omega}_{2i-\frac{3}{2}}}{2a} \right) - \boldsymbol{\Omega}_{2i+\frac{1}{2}} \cdot \left(\frac{l_{2i+\frac{3}{2}} - l_{2i-\frac{3}{2}}}{2a} \right) \right\}.$$

Still adding two terms which both vanish because of the second equation of (7.3.3)

$$\frac{3}{2} \left[s^2 \boldsymbol{\Omega}_{2i+\frac{1}{2}}^2 + a^2 l_{2i+\frac{1}{2}}^2 - s(s+1) \right]$$

$$+ \frac{1}{2} \left[s^2 \boldsymbol{\Omega}_{2i-\frac{3}{2}}^2 + a^2 l_{2i-\frac{3}{2}}^2 - s(s+1) \right],$$

we can rewrite the Hamiltonian as

$$H = J \sum_{2i} \left\{ \left(\frac{5}{2} a^2 l_{2i+\frac{1}{2}}^2 + \frac{1}{2} a^2 l_{2i-\frac{3}{2}}^2 + a^2 l_{2i+\frac{1}{2}} \cdot l_{2i-\frac{3}{2}} \right) \right.$$

$$\left. + \frac{s^2}{2} \left(\boldsymbol{\Omega}_{2i+\frac{1}{2}}^2 + \boldsymbol{\Omega}_{2i-\frac{3}{2}}^2 - 2\boldsymbol{\Omega}_{2i+\frac{1}{2}} \cdot \boldsymbol{\Omega}_{2i-\frac{3}{2}} \right) \right.$$

$$-2a^2 s l_{2i+\frac{1}{2}} \cdot \left(\frac{\Omega_{2i+\frac{1}{2}} - \Omega_{2i-\frac{3}{2}}}{2a} \right)$$

$$+2a^2 s \Omega_{2i+\frac{1}{2}} \left(\frac{l_{2i+\frac{1}{2}} - l_{2i-\frac{3}{2}}}{2a} \right) \Bigg\}$$

$$-J2s(s+1).$$

In the above expression, the last term can be omitted because it is a constant. In the first circular parenthesis small quantity a^2 appear in all three terms. Therefore, we can neglect the difference between $l_{2i+\frac{1}{2}}$ and $l_{2i-\frac{3}{2}}$ which is proportional to a, when small quantities of order $O(a^3)$ are neglected. This leads to the parenthesis equal to $4a^2 l_{2i+\frac{1}{2}}^2$. The second circular parenthesis can be written as $4a^2 \left(\frac{\Omega_{2i+\frac{1}{2}} - \Omega_{2i-\frac{3}{2}}}{2a} \right)^2$. When we define the coordinate of the site $2i+\frac{1}{2}$ as x, this quantity is just $4a^2 \Omega'^2(x)$, where the prime $'$ represents taking derivative with respect to x. The last two terms of the curly bracket are summed to

$$-2a^2 s(l \cdot \Omega' - \Omega \cdot l') = -2a^2 s(l \cdot \Omega' + \Omega' \cdot l)$$

since from $\Omega \cdot l = 0$ (the first equation of 7.3.3) we have $\Omega \cdot l' + \Omega' \cdot l = 0$. The Hamiltonian H becomes under the continuity limit $2a \to dx$

$$H = \int dx \, (Ja) \left\{ 2l^2 + s^2 \Omega'^2 - s \left(l \cdot \Omega' + \Omega' \cdot l \right) \right\}.$$

The Hamiltonian density $\mathcal{H}$ can be written as

$$\mathcal{H} = \frac{v}{2} \left\{ g^2 \left(l - \frac{\Theta}{4\pi} \Omega' \right)^2 + \frac{1}{g^2} \Omega'^2 \right\}, \tag{7.3.11}$$

where

$$v = 2Jas, \quad g^2 = \frac{2}{s}, \quad \Theta = 2\pi s. \tag{7.3.12}$$

Under the large-s limit, $\Omega^2 = 1$. Therefore, the motion of Ω can be described by the angles (θ, ϕ) of the spherical coordinates of a point on the sphere with unit radius. The corresponding generalized momenta are

$$\left. \begin{aligned} \pi_\theta &= -i\hbar \frac{\partial}{\partial \theta} \\ \pi_\phi &= -i\hbar \frac{\partial}{\partial \phi} \end{aligned} \right\}. \qquad (7.3.13)$$

The next step is to express the Hamiltonian in terms of the generalized coordinates and their conjugate momenta. Since l components are the generators of rotations (components of angular momentum), they are

$$\left. \begin{aligned} l_x &= -\sin\phi \left(-i\hbar \frac{\partial}{\partial \theta} \right) - \cos\phi \cot\theta \left(-i\hbar \frac{\partial}{\partial \phi} \right) \\ &= -\sin\phi\, \pi_\theta - \cos\phi \cot\theta\, \pi_\phi \\ l_y &= \cos\phi \left(-i\hbar \frac{\partial}{\partial \theta} \right) - \sin\phi \cot\theta \left(-i\hbar \frac{\partial}{\partial \phi} \right) \\ &= \cos\phi\, \pi_\theta - \sin\phi \cot\theta\, \pi_\phi \\ l_z &= -i\hbar \frac{\partial}{\partial \phi} = \pi_\phi \end{aligned} \right\} \qquad (7.3.14)$$

The components of Ω and its derivatives are

$$\Omega = (\sin\theta \cos\phi, \ \sin\theta \sin\phi, \ \cos\theta),$$

$$\frac{\partial \Omega}{\partial \theta} = (\cos\theta \cos\phi, \ \cos\theta \sin\phi, \ -\sin\theta),$$

$$\frac{\partial \Omega}{\partial \phi} = (-\sin\theta \sin\phi, \ \sin\theta \cos\phi, \ 0).$$

Therefore, the components of Ω' are

$$\Omega' = (\cos\theta \cos\phi\, \theta' - \sin\theta \sin\phi\, \phi', \ \cos\theta \sin\phi\, \theta'$$
$$+ \sin\theta \cos\phi\, \phi', \ -\sin\theta\, \theta'). \qquad (7.3.15)$$

Substituting (7.3.14) and (7.3.15) back into (7.3.11), we obtain

$$\mathscr{H} = \frac{v}{2}g^2 \left\{ \left(\pi_\theta - \frac{\Theta}{4\pi}\phi' \sin\theta \right)^2 + \frac{1}{\sin^2\theta}\left(\pi_\phi + \frac{\Theta}{4\pi}\theta' \sin\theta \right)^2 \right\}$$

$$+ \frac{v^2}{2g^2}\left(\theta'^2 + \phi'^2 \sin^2\theta \right). \tag{7.3.16}$$

The corresponding Lagrangian density is

$$\mathscr{L} = \frac{v}{2g^2}\left[\left(\partial_\mu\theta \right)^2 + \sin^2\theta \left(\partial_\mu\phi \right)^2 \right] + \frac{\Theta}{8\pi}\sin\theta \varepsilon^{\mu\nu}\partial_\mu\theta\,\partial_\nu\phi, \tag{7.3.17}$$

where $\varepsilon^{\mu\nu}$ is the unit antisymmetric tensor, and $\mu, \nu = 0, 1$. The relation between $\mathscr{H}$ and $\mathscr{L}$ can be directly verified. In the following, we set $v = 1$. We find the conjugate momenta from (7.3.17) as

$$\left. \begin{aligned} \pi_\theta &= \frac{\partial\mathscr{L}}{\partial\dot\theta} = \frac{1}{g^2}\dot\theta + \frac{\Theta}{4\pi}\sin\theta\,\phi' \\ \pi_\phi &= \frac{\partial\mathscr{L}}{\partial\dot\phi} = \frac{1}{g^2}\dot\phi\sin^2\theta - \frac{\Theta}{4\pi}\sin\theta\,\theta' \end{aligned} \right\}. \tag{7.3.18}$$

After quantization θ and π_θ, ϕ and π_ϕ become operators satisfying the canonical commutation relations. Substituting (7.3.17) and (7.3.18) into the following expression

$$\mathscr{H} = \pi_\theta\dot\theta + \pi_\phi\dot\phi - \mathscr{L},$$

we obtain Eq. (7.3.16). The Lagrangian density $\mathscr{L}$ (7.3.17) can be expressed in terms of $\boldsymbol{\Omega}$. We have

$$d\boldsymbol{\Omega} = \hat{\boldsymbol\phi}d\theta + \hat{\boldsymbol\theta}\sin\theta d\phi,$$

where $\hat{\boldsymbol\theta}$ and $\hat{\boldsymbol\phi}$ are unit vectors in the direction of increasing θ and ϕ, and furthermore we have

$$\partial_\mu\boldsymbol{\Omega}\cdot\partial^\mu\boldsymbol{\Omega} = \left(\partial_\mu\theta \right)^2 + \sin^2\theta\left(\partial_\mu\phi \right)^2,$$

$$\boldsymbol{\Omega}\cdot\left(\partial_\mu\boldsymbol{\Omega}\times\partial_\nu\boldsymbol{\Omega} \right) = \sin\theta\,\partial_\mu\theta\,\partial_\nu\phi.$$

Using the above relations we obtain

$$\mathscr{L} = \frac{1}{2g^2} \partial_\mu \boldsymbol{\Omega} \cdot \partial^\mu \boldsymbol{\Omega} + \frac{\Theta}{8\pi} \varepsilon^{\mu\nu} \boldsymbol{\Omega} \cdot (\partial_\mu \boldsymbol{\Omega} \times \partial_\nu \boldsymbol{\Omega}). \qquad (7.3.19)$$

The first term in the right-hand side is the standard non-linear sigma model Lagrangian density (7.2.2). The second term is the Wess-Zumino term. From its form in (7.3.17) it is easy to see that it is a two-dimensional divergence. For this purpose we define the current

$$K^\mu = -\varepsilon^{\mu\nu} \cos\theta \, \partial_\nu \phi,$$

and it can be verified immediately that

$$\partial_\mu K^\mu = \varepsilon^{\mu\nu} \sin\theta \, \partial_\mu \theta \, \partial_\nu \phi.$$

A term proportional to a divergence added to the Lagrangian density does not influence the equation of motion. The reason is that its spatial volume integral can be transformed into a surface integral through the Gauss theorem, and finally reduced to a constant. In deriving the equation of motion from the variation of the action, a constant plays no role. It does not contribute when perturbation calculation is performed in the field theory. It has important non-perturbative effects. The Hilbert space for the 1-dimensional antiferromagnetic chain can be divided into different subspaces for different Θ values, which are connected by canonical transformations. This was first pointed by I. Affleck [3]. Consider the transformation

$$U = \exp\left(i\frac{\Theta}{4\pi} \int \mathrm{d}x \phi' \cos\theta \right) \equiv e^A. \qquad (7.3.20)$$

Through the Lagrangian density (7.3.19) we see that the Wess-Zumino term *redefines* the generalized momenta. Under the transformation U the generalized momentum changes in the following way

$$\pi \to U\pi U^{-1}.$$

Using the Baker-Campbell-Hausdorff formula

$$e^A B e^{-A} = B + [A, B] + \frac{1}{2!}[A, [A, B]] + \cdots , \qquad (7.3.21)$$

we obtain $[A, \pi_\theta(x)]$:

$$[A, \pi_\theta(x)] = i\frac{\Theta}{4\pi} \int dx' \frac{d\phi(x')}{dx'}[\cos\theta(x'), \pi_\theta(x)],$$

where the commutator is calculated to be

$$\begin{aligned}\left[\cos\theta\left(x'\right), \pi_\theta\left(x\right)\right] &= -\sin\theta\left(x'\right)\left[\theta\left(x'\right), \pi_\theta\left(x\right)\right] \\ &= -\sin\theta\left(x'\right)i\delta\left(x - x'\right).\end{aligned}$$

Consequently we obtain

$$[A, \pi_\theta\left(x\right)] = -\frac{\Theta}{4\pi} \int dx' \frac{d\phi\left(x'\right)}{dx'}\sin\theta\left(x'\right)\delta\left(x' - x\right) = -\frac{\Theta}{4\pi}\phi'\sin\theta.$$

Furthermore, we have

$$[A, [A, \pi_\theta\left(x\right)]] = 0,$$

and we have finally

$$\pi_\theta \to \pi_\theta - \frac{\Theta}{4\pi}\phi'\sin\theta. \qquad (7.3.22)$$

Similarly we have

$$\pi_\phi \to \pi_\phi + \frac{\Theta}{4\pi}\theta'\sin\theta. \qquad (7.3.23)$$

The conclusion is that the canonical transformation (7.3.20) redefines the generalized momenta, making them Θ-dependent.

In formulating the theory in terms of the path integral Euclidean space is usually used, where the Wess-Zumino term is

$$i\Theta \cdot \frac{1}{8\pi} \int \varepsilon^{ij} \boldsymbol{\Omega} \cdot (\partial_i \boldsymbol{\Omega} \times \partial_j \boldsymbol{\Omega}) \, \mathrm{d}^2 x \equiv i\Theta Q, \qquad (7.3.24)$$

where Q is called the topological charge, and it is an integer.[6] Since in the path integral of e^{-S_E} the factor $e^{-i\Theta Q}$ is involved, Θ appears as a phase angle, which is called the topological phase. From (7.3.12) $\Theta = 2\pi s$, we see that it depends on the spin on the lattice. As a phase angle the integer spin corresponds to $\Theta = 0$, and the half-integer spin corresponds to $\Theta = \pi$. Their physical properties are very different. It seems that the topological term involves the space derivative $\partial_1 \boldsymbol{\Omega}$, and is parity violating. But actually it does not appear in the ferromagnetic case, and in the anti-ferromagnetic case we have $e^{i\pi Q} = e^{-i\pi Q}$, and hence parity violation does not arise.

We now concentrate on the Lagrangian density for integer spin $\Theta = 0$ for the Néel vector $\boldsymbol{\Omega}$

$$\mathscr{L} = \frac{1}{2g^2} \partial_\mu \boldsymbol{\Omega} \cdot \partial_\mu \boldsymbol{\Omega}.$$

Assuming spontaneous breaking of symmetry, we have the vacuum state $(0, 0, 1)$. The field configuration close to the vacuum state is $(\Omega^1, \Omega^2, 1 - \Omega^1 \Omega^1 - \Omega^2 \Omega^2) \approx (\Omega^1, \Omega^2, 1)$. Therefore $\mathscr{L} \approx \frac{1}{2g^2} \sum_{i=1}^{2} (\partial_\mu \Omega^i)^2$, and the system is approximated by two Goldstone bosons. The sublattice magnetization $s\langle \Omega^3 \rangle$ is infra-red divergent [3] as in the spin wave theory, telling that the symmetry is in fact not spontaneously broken. To get a deeper understanding, the renormalization group method is used to give the scale dependence of the coupling constant g as[7]

$$\frac{\mathrm{d}}{\mathrm{d} \ln L} g^2 = \frac{g^4}{2\pi},$$

---

[6]See discussions accompanying (7.5.2).

[7]The renormalization group method provides an effective theory for low frequencies by integrating out the high frequency modes. During this process the coupling "constant" changes its effective value and becomes scale dependent [12].

and the result is an increase of the effective coupling with increasing scale:

$$g^2(L) \approx \frac{g_0^2}{1 - \frac{g_0^2}{2\pi}\ln L}.$$

We begin with a coupling constant $g_0^2 = \frac{2}{s}$ (7.3.12), which is very small in the large s limit. When L reaches a scale

$$\xi \approx e^{2\pi/g_0^2} = e^{\pi s},$$

the value of g^2 reaches $O(1)$. We normally expect that a field theory with this type of renormalization group behavior develops a mass gap due to the infrared fluctuations, with a mass $m \approx \xi^{-1}$. This is a special case to which Coleman's theorem [13] applies, which states that spontaneous symmetry breaking is impossible in a 1+1 dimensional theory when infrared divergence occurs. This applies to any value (integer or half-odd-integer) of the spin, or in our present context $\Theta = 0$ or π. The above discussion is based on the perturbation theory used in the renormalization group arguments. It is only the *non-perturbative effects*[8] of the topological term $\Theta = \pi$ which suppress the mass generation, with the massless boson not associated with the spontaneous symmetry breaking [Section 6, Ref 3].

7.4 Lieb-Schultz-Mattis Theorem

We have seen that in the large-s limit the theory of 1-dimensional antiferromagnetic chain maps onto different field theories with $\Theta = 0$ or π respectively. A clue about the behavior of half-integer spin chain comes from a theorem first proved by Lieb, Schultz and Mattis [11] for $s = \frac{1}{2}$ and later generalized by Affleck and Lieb to arbitrary half-integer s [14]. This theorem states that a half-integer spin chain respecting translational symmetry and rotational symmetry in the

---

[8]The Wess-Zumino term does not contribute to the perturbation theory, and the renormalization group method used in Coleman theorem involves perturbation calculations.

spin space either has a zero gap or has degenerate ground states with opposite parity.

Consider a finite antiferromagnetic chain of length L consisting of an *even* number of lattice distances, such that periodic conditions can be applied. The ground state $|\psi_0\rangle$ with energy E_0 respects parity (reflection about a site) and rotational invariance in the spin space. Construct a state $|\psi_1\rangle$ such that the following condition holds:

$$\langle \psi_1 | H - E_0 | \psi_1 \rangle = O\left(\frac{1}{L}\right). \tag{7.4.1}$$

Explicitly, ψ_1 is constructed by

$$|\psi_1\rangle = U |\psi_0\rangle, \tag{7.4.2}$$

where

$$U = \exp\left\{ \frac{i\pi}{l} \sum_{j=-l}^{l} (j+l) S_j^z \right\}. \tag{7.4.3}$$

It represents rotation of the spins between sites $j = -l$ and $j = l$ about z-axis by different angles: the spin at site $j = -l$ rotates by 0, the spin at site $j = -l+1$ rotates by $\frac{\pi}{l}, \ldots$, the spin at $j = 0$ rotates by π, $\cdots$, and the spin at $j = l$ rotates by 2π. Spins outside the interval $j = -l$ to $j = l$ do not rotate. The value of l is chosen to be $O(L)$. In the following, it will be proved that $|\psi_1\rangle$ has a very low energy, which approaches zero continuously when L becomes indefinitely large, and that $|\psi_1\rangle$ is orthogonal to $|\psi_0\rangle$. The second point is crucial, otherwise $|\psi_1\rangle$ can approach $|\psi_0\rangle$ as $L \to \infty$, which may not imply anything about the spectrum of the chain. The orthogonality implies that $|\psi_1\rangle$ can be a real excited state of $|\psi_0\rangle$ without a gap, if the first point is true.

The difference between the angles by which two neighboring spins rotate is very small, $\frac{\pi}{l} \ll 1$. Since the energy of the system is determined by the orientations of the spins on nearest neighbor sites, such construction costs very little energy. Specifically we have

$$
\left.
\begin{aligned}
U^\dagger S_i^+ S_{i+1}^- U &= \exp\left(-i\frac{\pi}{l}\right) S_i^+ S_{i+1}^- \\
U^\dagger S_i^- S_{i+1}^+ U &= \exp\left(i\frac{\pi}{l}\right) S_i^- S_{i+1}^+ \\
U^\dagger S_i^z S_{i+1}^z U &= S_i^z S_{i+1}^z
\end{aligned}
\right\}
\tag{7.4.4}
$$

Consequently,

$$
\begin{aligned}
\delta E &\equiv \langle \psi_1 | H - E_0 | \psi_1 \rangle = \left\langle \psi_0 \left| U^\dagger \left(H - E_0\right) U \right| \psi_0 \right\rangle \\
&= \frac{J}{2} \sum_{i=-l}^{l} \left\langle \psi_0 \left| \left[\exp\left(-i\frac{\pi}{l}\right) - 1\right] S_i^+ S_{i+1}^- \right. \right. \\
&\qquad + \left. \left. \left[\exp\left(i\frac{\pi}{l}\right) - 1\right] S_i^- S_{i+1}^+ \right| \psi_0 \right\rangle \\
&= \frac{J}{2} \left(\cos\frac{\pi}{l} - 1\right) \sum_{i=-l}^{l} \left\langle \psi_0 \left| S_i^+ S_{i+1}^- + S_i^- S_{i+1}^+ \right| \psi_0 \right\rangle \\
&\qquad - \frac{J}{2} i \sin\frac{\pi}{l} \sum_{i=-l}^{l} \left\langle \psi_0 \left| S_i^+ S_{i+1}^- - S_i^- S_{i+1}^+ \right| \psi_0 \right\rangle .
\end{aligned}
$$

Since $|\psi_0\rangle$ respects parity, the expectation values of $S_i^+ S_{i+1}^-$ and $S_i^- S_{i+1}^+$ are equal, such that

$$
\delta E = J \left(\cos\frac{\pi}{l} - 1\right) \sum_i \left\langle \psi_0 \left| S_i^+ S_{i+1}^- \right| \psi_0 \right\rangle .
$$

Furthermore, $|\psi_0\rangle$ respects the rotational invariance in the spin space, the expectation value of $S_i^+ S_{i+1}^-$ should be $\frac{2}{3}$ of the expectation value of $\boldsymbol{S}_i \cdot \boldsymbol{S}_{i+1}$. There are $2l$ pairs of nearest neighbors between the sites $-l$ to l. Let the coupling energy between a pair of nearest neighbors be ε_0, then it follows that

$$
J \sum_{i=-l}^{l} \left\langle \psi_0 \left| S_i^+ S_{i+1}^- \right| \psi_0 \right\rangle = \frac{2}{3} \varepsilon_0 \left(2l\right) ,
$$

and therefore

$$\delta E = \frac{2}{3}\varepsilon_0 \left(\cos\frac{\pi}{l} - 1\right)(2l) = O\left(\frac{1}{l}\right).$$

Since l is chosen to be $O(L)$, then we have $\delta E = O(L^{-1})$, and (7.4.1) has been proved. Next we prove that $|\psi_1\rangle$ is orthogonal to $|\psi_0\rangle$. Perform the parity transformation and a rotation about the y-axis in the spin space, taking $S_i^z \rightarrow -S_{-i}^z$. The transformation (7.4.3) becomes now

$$U = \exp\frac{\mathrm{i}\pi}{l}\sum_{j=-l}^{l}(j+l)(-S_{-j}^z).$$

Changing the dummy index for the summation j to $-j$, the right-hand side of the above expression becomes

$$\exp\left[\frac{\mathrm{i}\pi}{l}\sum_{j=-l}^{l}(j+l)S_j^z\right]\exp\left[-2\mathrm{i}\pi\sum_{j=-l}^{l}S_j^z\right] = U\exp\left[-2\mathrm{i}\pi\sum_{j=-l}^{l}S_j^z\right].$$

Since there are $2l+1$ spins between the sites $-l$ to l, we have

$$\exp\left(-2\mathrm{i}\pi\sum_{i=-l}^{l}S_j^z\right) = \begin{cases} = 1, & \text{for integer } s \\ = -1, & \text{for half-integer } s \end{cases}$$

Consequently for half-integer s we have $\langle\psi_1 \mid \psi_0\rangle = 0$. There are two possibilities left open by this result. Either there is a gapless excitation with parity opposite to the ground state, or there are two ground states of opposite parity in the case of spontaneous parity breakdown, with a gap above each. Both possibilities are realized in specific models [3]. I. Affleck characterized the foregoing arguments as a physicist's proof and that it is non-trivial to make the argument rigorous. At least a physicist's picture is that the 1-dimensional anti-ferromagnetic chain does not undergo spontaneous symmetry breaking because of the infrared divergence in the quantum fluctuations. For integer spin, there is a unique ground state with a finite gap.

For half-integer spin, either the ground state is degenerate due to spontaneous breaking of parity or the gap vanishes due to the non-perturbative effect of the topological term.

In experiments, quasi-one-dimensional spin -1 antiferromagnetic chain CsNiCl$_3$ was observed to possess a gap, while spin-$\frac{1}{2}$ antiferromagnetic chain CuCl$_2 \cdot$ 2(C$_5$H$_5$) possesses gapless excitation.

7.5   The Significance of the Topological Term

In the Euclidean space the Lagrangian density of the 1-dimensional antiferromagnetic Heisenberg chain is

$$\mathscr{L}_E = \frac{1}{2g^2}\left[(\partial_1 \boldsymbol{\Omega})^2 + (\partial_2 \boldsymbol{\Omega})^2\right] + i\frac{\Theta}{8\pi}\varepsilon_{ij}\left(\partial_i \boldsymbol{\Omega} \times \partial_j \boldsymbol{\Omega}\right), \qquad (7.5.1)$$

where $i, j = 1, 2$, and the Minkowski time coordinate $x_0 = -ix_2$. The field $\boldsymbol{\Omega}(x)$ represents a mapping from a point in Euclidean space onto a point on the 2-dimensional spherical surface of unit radius S_2 of the internal 3-space. We demand that the action is finite, i.e.,

$$\int d^2x \mathscr{L}_E\left[\boldsymbol{\Omega}\left(x\right)\right] = \text{finite.}$$

This condition requires that at the infinity of the Euclidean space the following condition holds,

$$\lim_{|x|\to\infty} \boldsymbol{\Omega}\left(x\right) = \boldsymbol{\Omega}_0,$$

where $\boldsymbol{\Omega}_0$ is a constant vector of the internal 3-space. This condition is necessary because non-vanishing derivatives of the field cannot guarantee the finiteness of the action in the integral over the infinite space. The constancy of $\boldsymbol{\Omega}$ at infinity renders possible the compactification of the space. Consider the 2-dimensional infinite plane E_2, to which a 2-sphere S_2[9] is tangent at a point (Fig. 7.3). The normal at this point intersects the sphere at O. Draw a straight line from O which intersects S_2 and E_2 at points P and P' respectively. Then the point P' on E_2 is projected onto the point

---

[9]This is a 2-sphere in the Euclidean 3-space, not the S_2 in the internal space.

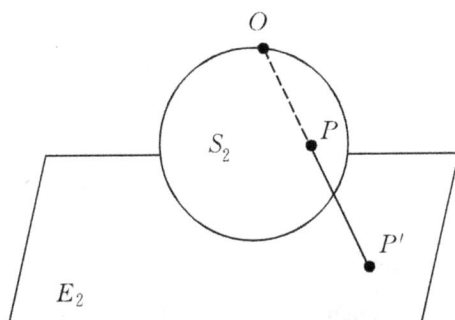

Fig. 7.3 The compactification of the infinite plane E_2.

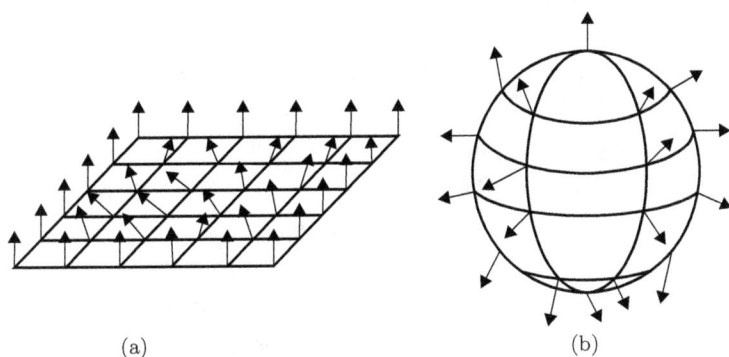

(a)   (b)

Fig. 7.4 (a) A configuration of spins on a Euclidean 2-space and (b) its mapping onto a compactified space S_2. Redrawn from Ref. [12].

P on S_2. In this way each point on the finite part of E_2 is projected onto a point on S_2 in a one-to-one correspondence. All points at infinity of E_2 are projected onto *one* unique point O of S_2. Therefore if the field $\boldsymbol{\Omega}$ has the same value at the infinity of E_2, the latter can be *compactified* to S_2.

Now consider the internal space of $\boldsymbol{\Omega}$, which is a unit vector with its orientation determined by angles (θ, ϕ). Its representative point lies also on S_2. A configuration of the field $\boldsymbol{\Omega}(x)$ with finite action is a smooth (differentiable) mapping from the compactified Euclidean space S_2 to the manifold S_2 of the order parameter. In Fig. 7.4 a configuration of spins and its projection onto the compactified S_2 are shown. In Fig. 7.5 the mapping from the compactified S_2 to

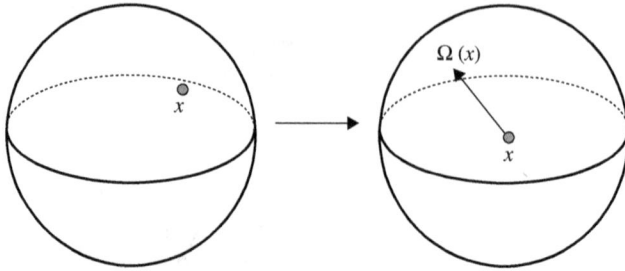

Fig. 7.5   The mapping $S_2 \xrightarrow{\Omega(x)} S_2$. Redrawn from Ref. [12].

the internal space S_2 is shown. The Pontryagin index (or winding number) Q of the spin configuration $\Omega(x)$ is defined by[10]

$$Q = \frac{1}{8\pi} \int d^2x \varepsilon^{ij} \, \Omega \cdot (\partial_i \Omega \times \partial_j \Omega) . \qquad (7.5.2)$$

It represents the number of times the configuration Ω winds the order of parameter space S_2 when the point x runs over all points on the compactified space S_2 in the mapping shown in Fig. 7.5. Consider the configuration represented by the instanton shown in Fig. 7.6. At the infinity the field points to the direction of the north pole indicated by Ω_0, while the field at the origin points to the south pole. When this configuration is projected onto the manifold of the field, it looks like the field of a monopole, as shown in Fig. 7.7. We see that for the instanton $Q = 1$.

---

[10]Let x_1, x_2 be the coordinates on the Euclidean 2-space S_2, e.g., $x_1 = \cos\theta$, $x_2 = \phi$. The surface element of S_2 is then $dx_1 dx_2$. The points lying within this surface element map onto the manifold space of Ω through the mapping $\Omega(x_1, x_2)$. The corresponding surface element in S_2 is $dx_1 dx_2 \Omega \cdot (\partial_1 \Omega \times \partial_2 \Omega)$. Since $\Omega^2 = 1$, both $\partial_1 \Omega$ and $\partial_2 \Omega$ are tangent to S_2 (manifold). When (x_1, x_2) runs over the entire surface S_2 of area 4π, the area swept by Ω on the manifold S_2 is

$$\int dx_1 dx_2 \, \Omega \cdot (\partial_1 \Omega \times \partial_2 \Omega) = \frac{1}{2} \int d^2x \varepsilon^{ij} \, \Omega \cdot (\partial_i \Omega \times \partial_j \Omega)$$

and therefore the winding number is

$$\frac{1}{4\pi} - \frac{1}{2} \int d^2x \varepsilon^{ij} \, \Omega \cdot (\partial_i \Omega \times \partial_j \Omega).$$

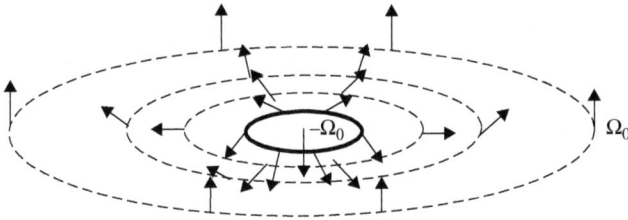

Fig. 7.6   A configuration of instanton in the Euclidean space-time. Redrawn from Ref. [12].

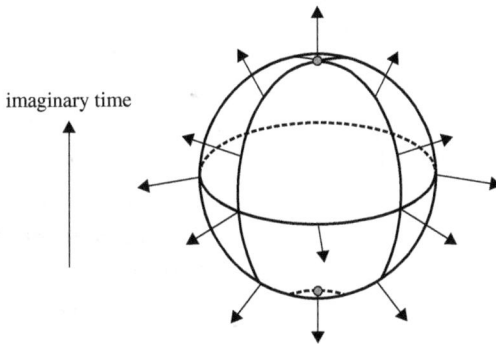

Fig. 7.7   The topology of this configuration in the manifold space of $\boldsymbol{\Omega}$ is the same as that for a monopole. Redrawn from Ref. [12].

For an anti-instanton, the "flux lines" in the manifold space of $\boldsymbol{\Omega}$ point inward, therefore $Q = -1$. The configurations $\boldsymbol{\Omega}(x)$ can be classified by their winding numbers, the mappings $S_2 \to S_2$ induced by different configurations $\boldsymbol{\Omega}(x)$ belong to different *homotopy class* with different topological properties characterized by their winding numbers.

We return to the Lagrangian density (7.5.1). The topological term contributes a phase to the action, and hence contributes a phase factor to the path integral

$$\mathrm{e}^{\mathrm{i}\Theta Q} = \mathrm{e}^{\mathrm{i}2\pi sQ} = (-1)^{2sQ}.$$

This factor is 1 for integer s, and for half integer s, its sign depends on the winding number, i.e., it is 1 for even winding numbers, and

−1 for odd winding numbers. The physics for integer and half-integer s is very different. This is the topological origin of Haldane's conjecture.

The presentation in this section follows Ref. [12].

7.6 The Θ Vacuum of the non-Abelian Gauge Fields

7.6.1 The non-Abelian gauge fields

The theory of non-Abelian gauge fields was created by C. N. Yang and R. L. Mills in 1954 [15]. In the theory of electromagnetic field the interaction of a charged fermion with the electromagnetic field is determined completely by the requirement of local gauge invariance. This paradigm was generalized by Yang and Mills to the $SU(2)$ gauge field. The gauge potential $A_\mu^a(x)$ has internal degrees of freedom $a = 1, 2, 3$ corresponding to these 3 generators $\tau^a/2$ of the $SU(2)$ group, where τ^a are the Pauli matrices. The fermion interacting with this gauge field is described by a 2-component spinor in the $SU(2)$ space, where each component is itself a Dirac spinor in the Minkowski space. The Lagrangian density of the free fermion field is

$$\mathscr{L}_0 = \bar{\psi}(x)\left(i\gamma^\mu\partial_\mu - m\right)\psi(x), \qquad (7.6.1)$$

where $\bar{\psi} = \psi^\dagger\gamma_0$ is the Dirac adjoint of ψ. $\mathscr{L}_0$ is invariant under the *global $SU(2)$* transformation.

$$\psi(x) \to \psi'(x) = \exp\left(-i\frac{\tau^a\theta^a}{2}\right)\psi(x), \qquad (7.6.2)$$

where $\theta^a(a = 1, 2, 3)$ are the parameters of the transformation. They are independent of the space-time coordinates, and hence the transformation is referred to as a global one. In order that the theory may be invariant under a *local $SU(2)$* transformation with parameters $\theta^a(x)$ being dependent on the space-time coordinates x, it is necessary to introduce the $SU(2)$ gauge field $A_\mu^a(x)$, the interaction between which and the fermion is determined completely by the requirement of the *local* gauge invariance. The fermion part of the Lagrangian density is to be generalized as

$$\bar{\psi}(x)\left\{i\gamma^{\mu}\left(\partial_{\mu}-ig\frac{T^{a}A_{\mu}^{a}}{2}\right)-m\right\}\psi, \qquad (7.6.3)$$

where g is the coupling constant. Under the local gauge transformation

$$U(x)=\exp\left(-i\frac{T^{a}\theta^{a}(x)}{2}\right) \qquad (7.6.4)$$

the fields $\psi(x)$ and $A_{\mu}^{a}(x)$ transform at the same time according to

$$\psi(x)\rightarrow\psi'(x)=U\psi(x) \qquad (7.6.5)$$

$$A_{\mu}^{a}(x)\rightarrow A_{\mu}^{'a}(x)=A_{\mu}^{a}(x)+\varepsilon^{abc}A_{\mu}^{c}(x)-\frac{1}{g}\partial_{\mu}\theta^{a}. \qquad (7.6.6)$$

The gauge field strengths $F_{\mu\nu}^{a}$ are related to the gauge potentials $A_{\mu}^{a}(x)$ by the following relations

$$F_{\mu\nu}^{a}=\partial_{\mu}A_{\nu}^{a}-\partial_{\nu}A_{\mu}^{a}+g\varepsilon^{abc}A_{\mu}^{b}A_{\nu}^{c}. \qquad (7.6.7)$$

In the Abelian gauge field theory (electromagnetism) the field strength is invariant, but in the non-Abelian gauge field theory the field strength transforms as

$$F_{\mu\nu}^{a}\rightarrow F_{\mu\nu}^{a}+\varepsilon^{abc}\theta^{b}F_{\mu\nu}^{c}. \qquad (7.6.8)$$

Comparing (7.6.6) with the transformation of Abelian gauge potential $A_{\mu}\rightarrow A_{\mu}+\partial_{\mu}\Lambda$ we see that the additional term $\varepsilon^{abc}A_{\mu}^{c}(x)$ appears because the non-Abelian potential is a vector in the $SU(2)$ space and this term is the effect of a rotation of the vector. At the same time the non-Abelian field strength is also a vector in the $SU(2)$ space and (7.6.8) also exhibits the effect of rotation of the vector. The inclusion of the last term of (7.6.7) is crucial in completing the framework of the theory. A strong requirement of symmetry, namely that $\theta^{a}(x)$ are arbitrary functions of x, imposes a strong limit on the form of interaction: it uniquely determines it. This is a very attractive feature of the theory. The standard model of fundamental interactions is built on the basis of non-Abelian gauge fields. The Yang-Mills field theory has far-reaching influence in modern theoretical physics.

The Lagrangian density of the system of $SU(2)$ doublet fermion field and the $SU(2)$ gauge field is

$$\mathcal{L} = -\frac{1}{4}F^a_{\mu\nu}F^{a\mu\nu} + \bar{\psi}\left\{i\gamma^\mu\left(\partial_\mu - ig\frac{\tau^a A^a_\mu}{2}\right) - m\right\}\psi. \qquad (7.6.9)$$

This Lagrangian density is invariant under $SU(2)$ local gauge transformations. We introduce the shorthand notations

$$A_\mu(x) \equiv A^a_\mu(x)\frac{\tau^a}{2}, \quad F_{\mu\nu}(x) = F^a_{\mu\nu}(x)\frac{\tau^a}{2}, \qquad (7.6.10)$$

where $A_\mu(x)$ and $F_{\mu\nu}(x)$ are now 2×2 matrices. The transformation relations (7.6.6) and (7.6.8) can now be rewritten as

$$\left. \begin{array}{l} A_\mu(x) \to A'_\mu(x) = UA_\mu U^{-1} + \frac{i}{g}U\partial_\mu U^{-1} \\ F_{\mu\nu}(x) \to F'_{\mu\nu}(x) = UF_{\mu\nu}U^{-1} \end{array} \right\} \qquad (7.6.11)$$

7.6.2 The equivalent classes of gauge transformations

In the classical electromagnetism the vacuum is described by $F_{\mu\nu} = 0$. The corresponding 4-potential A_μ is not necessarily zero, because

$$A_\mu = \frac{i}{g}U\partial_\mu U^{-1} \qquad (7.6.12)$$

is obtained by carrying out a gauge transformation $A_\mu \to A_\mu + \partial_\mu\theta$ on $A_\mu = 0$, where $\theta(x)$ is arbitrary. The 4-potential (7.6.12) describes the same $F_{\mu\nu} = 0$ state. It is called a *pure gauge*.

Since a gauge transformation admits significant arbitrariness, it is a usual practice to impose some restrictions. The gauge with vanishing zeroth component of the gauge potential is called the *temporal gauge*.

$$A_0(x) = 0. \qquad (7.6.13)$$

It is possible to carry out further gauge transformations on (7.6.13). If U is time independent, the transformed potential remains a temporal gauge:

$$A_0' = U A_0 U^{-1} + \frac{\mathrm{i}}{g} U \partial_0 U^{-1} = 0.$$

The first term on the right hand side vanishes because of (7.6.13), and the second term vanishes because U is time independent. A pure gauge potential in the temporal gauge has space components:

$$A_i\left(\boldsymbol{x}\right) = \frac{\mathrm{i}}{g} U \partial_i U^{-1}, \tag{7.6.14}$$

which depends on space coordinates only because U is time independent.

Consider $SU(2)$ gauge fields. The group element of $SU(2)$ group is

$$U = \mathrm{e}^{\mathrm{i}\theta\hat{\boldsymbol{n}}\cdot\frac{\boldsymbol{\tau}}{2}}, \tag{7.6.15}$$

where $\hat{\boldsymbol{n}}$ is a unit vector. U can also be written as

$$U = b_0 + \mathrm{i}\boldsymbol{b}\cdot\boldsymbol{\tau}, \tag{7.6.16a}$$

where

$$b_0 = \cos\frac{\theta}{2}, \quad \boldsymbol{b} = \hat{\boldsymbol{n}}\sin\frac{\theta}{2}. \tag{7.6.16b}$$

A group element can be characterized by 4 parameters $(b_0, \boldsymbol{b})$, which are connected by a relation

$$b_0^2 + \boldsymbol{b}^2 = 1. \tag{7.6.17}$$

Therefore, the manifold space of $SU(2)$ is a unit sphere S_3 in a 4-dimensional Euclidean space. An example of a gauge transformation independent of time is

$$U_1\left(\boldsymbol{x}\right) = \frac{\boldsymbol{x}^2 - \lambda^2}{\boldsymbol{x}^2 + \lambda^2} + \mathrm{i}\frac{2\lambda\boldsymbol{\tau}\cdot\boldsymbol{x}}{\boldsymbol{x}^2 + \lambda^2}, \quad \lambda > 0. \tag{7.6.18}$$

It is apparent that

$$U_n\left(\boldsymbol{x}\right) = \left[U_1\left(\boldsymbol{x}\right)\right]^n, \quad n = 1, 2, 3, \ldots \tag{7.6.19}$$

are all gauge transformations. The pure gauge potentials constructed by using (7.6.19) according to (7.6.14) are

$$A_i^{(n)}\left(\boldsymbol{x}\right) = \frac{i}{g} U_n\left(\boldsymbol{x}\right) \partial_i U_n^{-1}\left(\boldsymbol{x}\right). \tag{7.6.20}$$

They possess interesting topological properties. Define the winding number

$$n = \frac{ig^3}{24\pi^2} \int d^3x \; \mathrm{tr} \varepsilon^{ijk} A_i^{(n)}\left(\boldsymbol{x}\right) A_j^{(n)}\left(\boldsymbol{x}\right) A_k^{(n)}\left(\boldsymbol{x}\right), \tag{7.6.21}$$

where the A's are $SU(2)$ 2×2 matrices, tr is taking trace of the matrix. It can be shown that n is an integer. If the 3-dimensional space can be compactified into a 3-sphere S_3 in a 4-dimensional space[11] the gauge transformation (7.6.19) is a mapping from a point on S_3 onto a point on the $SU(2)$ manifold S_3; with n designating the times the manifold space is wound by the mapping. The gauge transformations U_n and the corresponding pure gauge potential $A_i^{(n)}$ are classified into equivalent classes, each class is characterized by a topological index (the winding number n). The potential $A_i^{(n)}(\boldsymbol{x})$ is transformed into $A_i^{(n+1)}(\boldsymbol{x})$ by U_1 :

$$A_i^{(n+1)}\left(\boldsymbol{x}\right) = \frac{i}{g} U_{n+1} \partial_i U_{n+1} = \frac{i}{g}\left(U_1 U_n\right) \partial_i \left(U_n^{-1} U_1^{-1}\right)$$

$$= \frac{i}{g} U_1 \partial_i U_1^{-1} + U_1 A_i^{(n)} U_1^{-1}. \tag{7.6.22}$$

Denoting by $|n\rangle$ the vacuum of class n, the above relation can be formally expressed as

$$U_1 \left| n \right\rangle = \left| n + 1 \right\rangle. \tag{7.6.23}$$

The foregoing presentation is based on Ref. [16].

---

[11] $\lim_{|\boldsymbol{x}| \to 0} A_i(x) = 0$ is therefore required.

7.6.3   The Θ vacuum

Which class does the real physical vacuum belong to ? The real vacuum must be gauge invariant. This demands a linear superposition

$$|\Theta\rangle = \sum_{n=-\infty}^{\infty} e^{-in\Theta} |n\rangle . \qquad (7.6.24)$$

It is gauge invariant under the transformation U_1, since an over all phase factor does not change the state.

$$U_1 |\Theta\rangle = \sum_{n=-\infty}^{\infty} e^{-in\Theta} |n+1\rangle = e^{i\Theta} |\Theta\rangle . \qquad (7.6.25)$$

The vacuum $|\Theta\rangle$ is also invariant under U_n, it is called the Θ vacuum. In order to understand the significance of Θ, we construct a gauge invariant time-ordered product of operators $T(O_1 \cdots O_p)$ and calculate its matrix element between two vacua with different Θ values. The result [16] turns out that it is proportional to $\delta(\Theta - \Theta')$, i.e., two worlds based on vacua with different values of Θ do not communicate with any physical means. In the path integral formalism of field theory the most basic quantity is the vacuum-to-vacuum matrix element in the presence of an external source J.[12] The transition amplitude between a vacuum state at $t = -\infty$ to a vacuum at $t = \infty$ under the presence of an external source J is

$$\langle \Theta \mid \Theta \rangle^J = \sum_{m,n} e^{im\Theta} e^{-in\Theta} \langle m \mid n \rangle^J$$

$$= \sum_{\nu} e^{i\nu\Theta} \sum_{n} \langle n+\nu \mid n \rangle^J , \qquad (7.6.26)$$

where a replacement of $\nu = m - n$ has been made. In the above expression, $\langle n + \nu \mid n \rangle$ is the transition between a vacuum with winding number n at $t = -\infty$ and a vacuum with winding number

---

[12]The presence of an external source is due to a purely technical reason that taking the derivative of the matrix element with respect to the external source gives the Green's function.

$n + \nu$ at $t = \infty$. From the instanton solution of the non-Abelian gauge field the vacua with different winding number are connected by an instanton. The kink solution discussed in Chapter 5 begins from a stable position at $\tau = -\infty$ and arrive at another stable position at $\tau = \infty$ through a potential barrier. Vacua with different winding numbers have different topological properties, and this is equivalent to a situation that there is a potential barrier between these states, and a tunneling takes place via the instanton solution. The constituent parts of the vacuum-to-vacuum transition amplitude (7.6.26) $\langle m \mid n \rangle$ are the tunneling between vacua of different winding numbers via the instanton. The transition amplitude (7.6.26) contains the physics of the system. The properties of the system is expressed through the "integrand" e^{-iS} or e^{-S_E} of the path integral. The phase factor $e^{i\nu\Theta}$ in (7.6.26) actually contributes a term added to the action. Remember the Lagrangian density of the 1-dimensional antiferromagnetic chain (7.3.19), the topological term of which is connected with the divergence of a current K^μ. According to the Bardeen identity [16], the scalar product of the dual of the gauge field strength $\tilde{F}_{a\mu\nu} = \frac{1}{2}\varepsilon_{\mu\nu\lambda\rho}F_a^{\lambda\rho}$ and the gauge potential $F_a^{\mu\nu}$ is a 4-divergence of a current.

$$F_a^{\mu\nu}\tilde{F}_{a\mu\nu} = \frac{1}{2}\varepsilon_{\alpha\beta\mu\nu}F_a^{\mu\nu}F_a^{\alpha\beta} = \partial^\mu K_\mu. \qquad (7.6.27)$$

For the case of $SU(2)$ we have

$$K_\mu = \varepsilon_{\mu\alpha\beta\gamma}A_a^\alpha \left[F_a^{\beta\gamma} - \frac{g}{3}\varepsilon_{abc}A_b^\beta A_c^\gamma \right], \qquad (7.6.28)$$

where a, b, c are the $SU(2)$ indices $1, 2, 3$, and $\mu, \alpha, \beta, \gamma$ are space-time indices. Performing the 4-dimensional volume integral of (7.6.27) and using Gauss theorem, we have

$$\int d^4x F_a^{\mu\nu}\tilde{F}_{a\mu\nu} = \int d^4x \partial^\mu K_\mu = \int_s d\sigma^\mu K_\mu. \qquad (7.6.29)$$

Here S can be considered as a 3-dimensional cylindrical surface enclosing the 4-dimensional space, the ends of which are the 3-dimensional space at $t = \pm\infty$, and the curved surface of the cylinder is situated at the spatial infinity. The field at infinity is the vacuum, $F_a^{\beta\gamma} = 0$, and in the temporal gauge $A_a^0 = 0$. Therefore, only the zeroth component of K_μ (6.7.28) is non-vanishing at infinity.

$$K_0 = -\frac{g}{3}\varepsilon_{ijk}\varepsilon_{abc}A_a^i A_b^j A_c^k = \frac{4}{3}ig\varepsilon_{ijk}\mathrm{tr}A^i A^j A^k. \qquad (7.6.30)$$

In the surface integral of (7.6.29) the integral over the cylindrical surface vanishes because K_μ has no spatial components. The only contribution comes from the integral over 3-space at $t = \pm\infty$:

$$\int d^4x F_a^{\mu\nu}\tilde{F}_{a\mu\nu} = \int d^3x K_0 \,|_{t=-\infty}^{t=+\infty} .$$

Comparing with the definition of n (7.6.21) and using (7.6.30), we have

$$\frac{g^2}{32\pi^2}\int d^4x F_a^{\mu\nu}\tilde{F}_{a\mu\nu} = n\,(+\infty) - n\,(-\infty) = \nu \qquad (7.6.31)$$

Therefore, the factor $e^{i\nu\Theta}$ in (7.6.26) is equivalent to contributing the following term to the action S in the path integral of e^{iS} :

$$S_\Theta = \Theta\frac{g^2}{32\pi^2}\int d^4x F_a^{\mu\nu}\tilde{F}_{a\mu\nu},$$

or contributing the following term to the Lagrangian density

$$\mathscr{L}_\Theta = \Theta\frac{g^2}{32\pi^2}F_a^{\mu\nu}\tilde{F}_{a\mu\nu}. \qquad (7.6.32)$$

As pointed out earlier, it is proportional to the 4-divergence $\partial^\mu K_\mu$, and therefore has no contribution to the equation of motion, nor does it contribute to the perturbation theory. In particle physics the quantum chromodynamics describing the strong interaction is constructed by using the $SU(3)$ gauge field. Its vacuum has a Θ value. It violates P (space inversion), T (time reflection), and also CP (joint charge conjugation-space inversion). This is called the strong CP problem [17]. In strong interaction processes the conservation of CP has been checked with very high precision, giving a very small upper limit of Θ. Can a $\Theta = 0$ result be obtained in the theory?

There have been attempts, but a satisfactory argument has not yet been achieved.

7.7 The Topological Term and the Anomalies

In quantum electrodynamics the electron mass sets scales for the theory, $\frac{\hbar}{mc}$ defines the length scale, $\frac{\hbar}{mc^2}$ defines the time scale, mc and mc^2 establish the scales for the momentum and energy. In quantum chromodynamics (QCD) quarks are massless.[13] In the Lagrangian density there is no other parameter with a dimension either. Such a theory is called a *scale invariant* theory. In carrying out the calculation of radiative corrections it is necessary to do renormalization. The renormalization theory introduces a mass (denoted by M), though the theory begins without a mass, nor is there any parameter with the dimension of mass. This is a characteristic feature of the renormalizable theories. The value of M cannot be given by the theory itself. For instance the quark-gluon coupling constant in QCD is given by the following formula:

$$\alpha_s(Q^2) = \frac{4\pi}{\left(11 - \frac{2}{3}n_f\right)\ln\frac{Q^2}{\Lambda^2}}, \tag{7.7.1}$$

where α_s is the equivalent of the fine structure constant α in QED, Q^2 is the square of the 4-momentum transfer of the physical process, n_f is the number of flavors of the quarks taking part in the process. There is no arbitrary parameter in the formula, Λ is equivalent to the parameter introduced by the renormalization condition. It is to be determined by experiment, the result being 150 MeV $\sim$ 200 MeV. Here the classical Lagrangian is scale invariant, and the scale invariance is only violated by the mass scale introduced via renormalization in calculating the quantum corrections. Such an occurrence is termed a *dimensional transmutation*. When a classical symmetry is violated by the quantum treatment (through

---

[13]The quark mass results from the spontaneous symmetry breaking in the electroweak interaction. Quarks are massless in the quantum chromodynamics describing the strong interaction.

path integral or perturbation theory), the symmetry is called *anomalous*.

The existence of anomalies has important physical implications. The example given above (7.7.1) is the *running coupling constant*, i.e., the coupling constant depends on the scale of Q^2 of the process. The QCD running coupling constant becomes small when the value of Q^2 becomes large enough. This is called the asymptotic freedom. Therefore, for processes with high momentum transfer at high energies perturbation theory can be applicable. The perturbative QCD has been receiving support from the experiments. Another example is the chiral anomaly, which leads to a Wess-Zumino term. Consider a Dirac field interacting with the electromagnetic field with the Lagrangian density

$$\mathscr{L} = \bar{\psi}\left[\gamma^\mu(\partial_\mu + \mathrm{i}A_\mu) + \mathrm{i}m\right]\psi. \tag{7.7.2}$$

It is invariant with respect to a global phase transformation

$$\psi \to \psi' = \mathrm{e}^{\mathrm{i}\lambda}\psi, \tag{7.7.3}$$

where λ is a constant. In the field theory, a symmetry corresponds to a conserved current, the Noether current. The current corresponding to the symmetry transformation (7.7.3) is

$$J_\mu = \mathrm{i}\bar{\psi}\gamma_\mu\psi, \tag{7.7.4}$$

which satisfies the following condition

$$\partial^\mu J_\mu = 0. \tag{7.7.5}$$

The conservation condition (7.7.5) is an equation of continuity:

$$\frac{\partial}{\partial t}J_0 - \nabla \cdot \boldsymbol{J} = 0. \tag{7.7.6}$$

By integrating over space and using Gauss theorem and using the fact that the field vanishes at infinity, we get

$$\frac{\partial}{\partial t}\int \mathrm{d}^3x\, J_0 = 0, \tag{7.7.7}$$

which is the conservation of the number of particles. We consider next the global chiral transformation

$$\psi \rightarrow \psi' = e^{i\nu\gamma_5}\psi, \tag{7.7.8}$$

where ν is a constant, and the Dirac matrix $\gamma_5 = i\gamma^0\gamma^1\gamma^2\gamma^3$. The corresponding current

$$J_\mu^5 = i\bar{\psi}\gamma_\mu\gamma_5\psi \tag{7.7.9}$$

satisfies

$$\partial^\mu J_\mu^5 = -2m\bar{\psi}\gamma_5\psi. \tag{7.7.10}$$

Therefore, J_μ^5 is a conserved current only in the $m = 0$ limit, and only then $\mathscr{L}$ is invariant with respect to the chiral transformation. This is the property of the classical Lagrangian, the field ψ has not yet been quantized. It is proved by S. Adler, J. Bell and R. Jackiw [17,18] that when quantum effects are considered, the following relations hold:

$$\partial^\mu J_\mu = 0$$
$$\partial^\mu J_5 = -2m\bar{\psi}\gamma_5\psi + \frac{i}{8\pi^2}F_{\mu\nu}\tilde{F}^{\mu\nu}. \tag{7.7.11}$$

It turns out that even for $m = 0$ the classical chiral phase symmetry is violated by the quantum effect. K. Fujikawa [19, 20] pointed out that the Jacobian of the integration measure of the path integral under the chiral phase transformation depends on the electromagnetic potential, and is therefore not gauge invariant. He derived directly the anomaly relation (7.7.11) without using perturbation theory. The foregoing result can be generalized to QCD. The field is a 3-component Dirac spinor ψ^a ($a = 1, 2, 3$), the gauge potential and gauge field strength A_μ and $F_{\mu\nu}$ are 3×3 matrices. The anomaly relation is now

$$\partial^\mu J_\mu^5 = \partial^\mu i\bar{\psi}^a\gamma_\mu\gamma_5\psi^a = \frac{i}{8\pi^2}\,\text{tr}\,F_{\mu\nu}\tilde{F}^{\mu\nu}. \tag{7.7.12}$$

Expressing the field strength in terms of the potential and reorganizing the above equation, we obtain

$$\partial^\mu \left(J^5_\mu - \frac{i}{4\pi^2} W_\mu \right) = 0, \tag{7.7.13}$$

where

$$W_\mu = \varepsilon_{\mu\nu\sigma\tau} \, \text{tr} \left(A^\nu \partial^\sigma A^\tau + \frac{2}{3} i A^\nu A^\sigma A^\tau \right). \tag{7.7.14}$$

Formally one can define a conserved charge

$$Q'_5 = \int d^3x \left(J^5_0 - \frac{i}{4\pi^2} W_0 \right), \tag{7.7.15}$$

but actually Q'_5 is not gauge invariant, such that there does not exist such a conserved quantity in physics. Under the gauge transformation

$$A_i \to A'_i = -iU\partial_i U^\dagger + U A_i U^\dagger \tag{7.7.16}$$

the chiral charge changes by

$$\Delta Q'_5 = -\frac{i}{4\pi^2} \int d^3x \varepsilon_{ijk} \left[\frac{1}{3} \text{tr} \left(U\partial_i U^\dagger \right) \left(U\partial_j U^\dagger \right) \left(U\partial_k U^\dagger \right) \right.$$

$$\left. + i\partial_i \, \text{tr} \left(U^\dagger \partial_j U A_k \right) \right]. \tag{7.7.17}$$

The second term in the bracket is a 3-divergence and can be transformed by Gauss theorem into a surface integral at infinity, and does not contribute to $\Delta Q'_5$. The first term is independent of the gauge potential, and it is the Pontryagin index or winding number (7.6.20)(7.6.21). The charge Q'_5 is invariant under a gauge transformation with winding number 0, but changes by an integer $2n$ under topological non-trivial gauge transformations with a finite winding number.

The foregoing discussion concerns the anomaly of a global symmetry. The existence of anomaly has physical implications. For example the Abelian chiral anomaly gives the decay rate of $\pi^0 \to 2\gamma$. From (7.7.11) we see that two photons are coupled to the pseudoscalar $\partial_\mu J_5^\mu$ which describes a pion. The ψ field describes charged quarks, the result can be checked by experiments. It turns out that not only the quark flavors (u and d) but also quark colors ($N_c = 3$) have to be included to give a good agreement with experiment. This is one of the evidences supporting the assumption of the color $SU(3)$ of quark dynamics. We have just discussed the anomaly of the scale invariance leading to the running coupling constant.

The existence of an anomaly of a local invariance, for instance the anomaly of local chiral invariance, can be dangerous to the theory. It can lead to the breakdown of local gauge invariance, thus destroying the renormalizability of the theory. Under such circumstances it is necessary to study the mechanism of the anomaly cancellation. This leads to constraints in choosing an underlying group for the theory, or a constraint on the number of fermions. In this context the anomaly of local invariance has become a useful element in formulating a theory, instead of a danger. Not any group can be chosen in formulating a theory; thus the choice can be rather restricted. This has been used in the grand unified theories and string theories.

References

[1] O. Madelung, *Introduction to Solid State Theory,* translated by B. C. Taylor, Berlin, Springer, 1978.

[2] W. J. Caspers, *Spin Systems*, Singapore, World Scientific, 1989.

[3] I. J. Affleck, *Phys. Condens. Matter* 1, 3047, 1989.

[4] J. Goldstone, *Nuovo Cimento* 19, 154, 1961.

[5] J. Goldstone, A. Salam and S. Weinberg, *Phys. Rev.* 127, 965, 1962.

[6] H A. Bethe, *Z. Phys.* 71,205, 1931.

[7] J. Lowenstein, in: L. B. Zuber and R. Stora , eds. *Recent Advances in Field Theory and Statistical Mechanics*, Amsterdam, North Holland, 1984.

[8] F. D. M. Haldane, *Phys. Lett. A* 93, 464, 1983.

[9] F. D. M. Haldane, *Phys. Rev. Lett.* 50, 1153, 1983.

[10] F. D. M. Haldane, *J. App. Phys.* 57, 3359, 1985.

[11] E. H. Lieb, T. Schultz and D. J. Mattis, *Ann. Phys. (N.Y.)* 16, 407, 1961.

[12] E. Fradkin, *Field Theories of Condensed Matter Systems*, Redwood City, Addison-Wesley, 1991.

[13] S. Coleman, *Comm. Math. Phys.* 31, 259, 1973.

[14] I. Affleck and E. H. Lieb, *Lett. Math. Phys.* 12, 57, 1986.

[15] C. N. Yang and R. L. Mills, *Phys. Rev.* 96, 191, 1954.

[16] R. Peccei, The Strong CP Problem, in: C. Jarlskog, ed. *CP Violations*, Singapore, World Scientific, 1989.

[17] S. L. Adler, *Phys. Rev.* 177, 2426, 1969.

[18] J. S. Bell and R. Jackiw, *Nuovo Cimento* 60A, 47, 1969.

[19] K. Fujikawa, *Phys. Rev. D* 21, 2848, 1980.

[20] K. Fujikawa, *Phys. Rev. Lett.* 42, 1195, 1979.

Chapter 8

Cavity Quantum Electrodynamics, van der Waals Forces and Casimir Effect

Cavity quantum electrodynamics (abbreviated cavity QED [1, 2, 4]) is a branch of quantum optics, dealing with the interaction of a single atom and a number of photons in an electromagnetic resonant cavity. In more than a decade genuine properties of both the atom and the field have been observed. The cavity in which the atom is placed can change the rate of spontaneous radiation of the atom, suppressing or enhancing it. The coupling between the atom and the cavity can shift the atomic energy levels. In correspondence, the electromagnetic field in the cavity (photons) also exhibit non-classical properties. Owing to the development of manufacturing high Q cavities and the atomic beam technique, the quantum non-demolition measurement of the number of photons in the cavity becomes possible. In a high Q cavity macroscopic coherence of fields, i.e., the Schrödinger cat can be observed. In Sections 8.1–8.6 we discuss the basic features of cavity QED.

The principle of causality is revealed in many branches in physics. When the electric field strength $E(t)$ in a dielectric changes with time, the displacement vector $D(t)$ can only depend on the values of E before the time t. Before the incident wave arrives at a scattering center there cannot be scattered waves. Such requirements must be expressed in the theory. Let $E(\omega)$ and $D(\omega)$ be the Fourier components of the vectors E and D. They are related by $\boldsymbol{D}(\omega) = \varepsilon(\omega) \, \boldsymbol{E}(\omega)$, where ε is the dielectric constant. The causality principle is then expressed in the analytic properties of the generalized susceptibility $\alpha(\omega) = \varepsilon(\omega) - 1$. Here $\alpha(\omega)$ is a function of a complex variable ω. The relation expressing the analytic properties of $\alpha(\omega)$ is called the dispersion relation. It was found by Kramers and Krönig in 1927 when they studied the causality requirement in the dispersion of light. The generalized susceptibility $\alpha(\omega)$ characterizes the response of a system to external disturbances. The study of the response leads to the well-known fluctuation-dissipation theorem, which is extensively used in dealing with various physical systems. This is discussed in Section 8.7.

The force between two neutral atoms is called the van der Waals force. The stationary state of an atom does not have a non-vanishing expectation value of the electric dipole moment, because such state has a definite parity. But an atom does have a spontaneous dipole moment different from zero as a result of fluctuation. This dipole moment then produces at the situation of another atom an electric field, which in its turn polarizes the second atom to produce an electric dipole moment. The dipole-dipole interaction is the origin of the van der Waals force. The fluctuation-dissipation theorem is then applied to the fluctuation of the electromagnetic field. The result is that the potential of such a force is inversely proportional to the 6th power of the interatomic distance R, if R is not too large. Retarded interaction must be taken into account when R is large and the potential turns out to be inversely proportional to the 7th power of R. These laws of proportionality is independent of the structure of atoms. Casimir had been thinking about the origin of the generality for some time. When he turned to Bohr for advice, Bohr answered that this was probably due to the zero-point energy. What Bohr had in mind was the vacuum fluctuation energy of the electromagnetic field. Having made use of this idea, Casimir succeeded to derive the inverse 7th power law. Encouraged by the success Casimir began to study the force between two uncharged metallic plate put parallel to each other. The forces arises because the vacuum fluctuation of electromagnetic field between the plates is different from that in free space. This is the Casimir effect.

Zero point fluctuations play an important role in developing the theory of Lamb Shift. Whether zero-point fluctuation is indispensable in the interpretation of Casimir effect, the Lamb shift and the van der Waals force is an interesting issue. We discuss these points in Sections 8.8–8.13.

8.1 Interaction between the Radiation Field and an Atom

There are systematic presentations on the interaction of an atom with the radiation field in textbooks on quantum mechanics. Usually the state of the radiation field is expanded in terms of the plane wave eigenstates. In the cavity QED, however, the electromagnetic field is subjected to the boundary conditions of the cavity. The formalism needs to be adapted to this situation. In the following we give only the main steps of derivation to facilitate the transition to the cavity QED [1, 2].

8.1.1 Quantization of the field of electron in an atom

The orthonormal eigenstates ψ_j of the electron in an atom are the solutions to the Schrödinger equation corresponding to the eigenvalues E_j:

$$\mathscr{H}_0\psi_j\left(\boldsymbol{r}\right) = \left(-\frac{\hbar^2}{2m}\nabla^2 + eV\right)\psi_j\left(\boldsymbol{r}\right) = E_j\psi_j\left(\boldsymbol{r}\right). \qquad (8.1.1)$$

In the second quantized formulation the electron field operator $\hat{\psi}\left(\boldsymbol{r}\right)$ is expanded in terms of the eigenstates ψ_j, the coefficient of ψ_j being the annihilation operator c_j of an electron in the state j:

$$\hat{\psi}\left(\boldsymbol{r}\right) = \sum_j c_j\psi_j\left(\boldsymbol{r}\right). \qquad (8.1.2)$$

Its hermitian conjugate is

$$\hat{\psi}^\dagger\left(\boldsymbol{r}\right) = \sum_j c_j^\dagger\psi_j^*\left(\boldsymbol{r}\right), \qquad (8.1.3)$$

where $c_j^\dagger$ is the operator of creation of an electron in the state j. The Hamiltonian operator of the electron field is

$$H_0 = \int \hat{\psi}^\dagger\left(r\right) \mathscr{H}_0 \hat{\psi}\left(r\right) d^3 r. \tag{8.1.4}$$

Substituting (8.1.2) and (8.1.3) into (8.1.4) and using the orthonormality of the eigenfunctions, we obtain

$$H_0 = \sum_j E_j c_j^\dagger c_j, \tag{8.1.5}$$

where $c_j^\dagger c_j$ is the operator of the number of electrons on state j. The vacuum state of the field $|0\rangle$ is defined by its property

$$c_j |0\rangle = 0. \tag{8.1.6}$$

The state of the field with one electron on the state j is $c_j^\dagger |0\rangle$. The fermion operator c_j and $c_j^\dagger$ satisfy the anti-commutation relations

$$\left.\begin{array}{l} \{c_i, c_j^\dagger\} = c_i c_j^\dagger + c_j^\dagger c_i = \delta_{ij} \\ \{c_i, c_j\} = \{c_i^\dagger, c_j^\dagger\} = 0 \end{array}\right\}. \tag{8.1.7}$$

These relations lead to the anti-commutation relations for the field operators.

$$\left.\begin{array}{l} \{\hat{\psi}^\dagger\left(r\right), \hat{\psi}\left(r'\right)\} = \delta^3\left(r - r'\right) \\ \{\hat{\psi}\left(r\right), \hat{\psi}\left(r'\right)\} = 0 \end{array}\right\}. \tag{8.1.8}$$

8.1.2   Quantization of the radiation field

The radiation field — the electromagnetic field in vacuum — can be described by the vector potential $A(r, t)$. Using the Coulomb gauge where the scalar potential of the radiation field can be set equal to zero,[1] therefore the field strengths are

$$
\left.\begin{array}{l}
\boldsymbol{B} = \nabla \times \boldsymbol{A} \\[2mm]
\boldsymbol{E} = -\dfrac{1}{c}\dfrac{\partial \boldsymbol{A}}{\partial t}
\end{array}\right\}
\tag{8.1.9}
$$

The Lorentz condition in the absence of scalar potential is

$$
\nabla \cdot \boldsymbol{A} = 0.
\tag{8.1.10}
$$

The vector potential of the quantized radiation field is

$$
\boldsymbol{A}(\boldsymbol{r},t) = \sum_k \left(\frac{2\pi\hbar c^2}{\omega_k}\right)^{1/2} \left[a_k \boldsymbol{u}_k(\boldsymbol{r})\,\mathrm{e}^{-i\omega_k t} + a_k^\dagger \boldsymbol{u}_k^*(\boldsymbol{r})\,\mathrm{e}^{i\omega_k t}\right],
\tag{8.1.11}
$$

where a_k, $a_k^\dagger$ are the annihilation and creation operator of a photon in state k respectively. The complex vector mode function $\boldsymbol{u}_k(\boldsymbol{r})$ satisfies

$$
\left(\nabla^2 + \frac{\omega_k^2}{c^2}\right)\boldsymbol{u}_k(\boldsymbol{r}) = 0,
\tag{8.1.12}
$$

and the transverse wave condition

$$
\nabla \cdot \boldsymbol{u}_k(\boldsymbol{r}) = 0.
\tag{8.1.13}
$$

The choice of the normalization factor in (8.1.11) makes the operators a_k and $a_k^\dagger$ dimensionless, and will lead to the photon picture of the radiation field (8.1.20). The mode functions $\boldsymbol{u}_k(\boldsymbol{r})$ form an

---

[1]The electrostatic field experienced by an electron within the atom is described by V, which is considered as an "external" field, and is not related to the radiation field. V is not quantized, and $\boldsymbol{A}$ is quantized. See Eq. (8.1.11)

orthonormal complete set.

$$\int u_k^* (r) \cdot u_{k'} (r) \, d^3 r = \delta_{kk'}. \qquad (8.1.14)$$

The mode functions satisfy the physical boundary conditions. For a cube of side L, the plane wave mode functions are

$$u_k (r) = L^{-3/2} \, \hat{e}^{(\lambda)} e^{ik \cdot r}, \qquad (8.1.15)$$

where $\hat{e}^{(\lambda)}$ is a unit polarization vector, $\lambda = 1, 2$ is the polarization index, $\hat{e}^{(1)}$, $\hat{e}^{(2)}$ and k are orthogonal to one another, as required by the transversality condition (8.1.13), and furthermore $k^2 = \frac{\omega^2}{c^2}$. The boundary condition that the mode functions vanish on the cube surfaces gives

$$k_x = \frac{2\pi n_x}{L}, \; k_y = \frac{2\pi n_y}{L}, \; k_z = \frac{2\pi n_z}{L}$$

$$n_x, n_y, n_z = 0, \pm 1, \pm 2, \dots. \qquad (8.1.16)$$

A mode is characterized by the polarization index and the wave vector k. Boson operators a_k and $a_k^\dagger$ satisfy the commutation relations

$$[a_k, a_{k'}^\dagger] = \delta_{kk'}, \; [a_k, a_{k'}] = [a_k^\dagger, a_{k'}^\dagger] = 0. \qquad (8.1.17)$$

The vacuum of radiation field $|0\rangle$ is defined via its property $a_k |0\rangle = 0$, and the state with one photon of state k is $a_k^\dagger |0\rangle$. The vector potential (8.1.11) gives the electric field strength

$$E (r, t) = i \sum_k (2\pi \hbar \omega_k)^{1/2} \left[a_k u_k (r) \, e^{-i\omega_k t} - a_k^\dagger u_k^* (r) \, e^{i\omega_k t} \right].$$

$$(8.1.18)$$

Substituting (8.1.18) and the corresponding expression for B into the energy for radiation field

$$H_{\text{rad}} = \frac{1}{8\pi} \int (E^2 + B^2) \, d^3 r, \qquad (8.1.19)$$

we obtain

$$H_{\text{rad}} = \sum_k \hbar \omega_k \left(a_k^\dagger a_k + \frac{1}{2} \right). \qquad (8.1.20)$$

8.1.3 Interacting electron field and radiation field, the rate of spontaneous emission

The interaction between electron field and radiation field can be obtained by generalizing the Hamiltonian (8.1.1):

$$\mathscr{H} = \frac{1}{2m}\left(\boldsymbol{p} - \frac{e}{c}\boldsymbol{A}\right)^2 + eV\left(\boldsymbol{r}\right)$$

$$= \mathscr{H}_0 - \frac{e}{m}\boldsymbol{A}\cdot\boldsymbol{p}, \tag{8.1.21}$$

where $\boldsymbol{A}$ is the vector potential of the radiation field. In the above we used $\nabla\boldsymbol{A} = 0$, and neglected $\boldsymbol{A}^2$, since we do not consider the two-photon process. The quantized Hamiltonian for the interaction between the radiation field and the atomic electron field is

$$H_{\text{int}} = \int \hat{\psi}^\dagger\left(\boldsymbol{r}\right)\left(-\frac{e}{m}\boldsymbol{A}\cdot\boldsymbol{p}\right)\hat{\psi}\left(\boldsymbol{r}\right)\mathrm{d}^3r, \tag{8.1.22}$$

and the Hamiltonian for the system is

$$H = H_0 + H_{\text{rad}} + H_{\text{int}}, \tag{8.1.23}$$

where H_0 and H_{rad} are given by (8.1.4) and (8.1.19) respectively. Substituting (8.1.2) and (8.1.11) into (8.1.22), we get

$$H_{\text{int}} = \sum_{i,j,k} c_i^\dagger c_j \left(g_{ij}^k a_k + g_{ij}^{k*} a_k^\dagger\right), \tag{8.1.24}$$

where i, j are the indices for electron states, k is the mode index for the radiation field, and g_{ij}^k represents the coupling between the electron with the radiation field,

$$g_{ij}^k = -\frac{e}{m}\left(\frac{2\pi\hbar}{\omega_k V}\right)^{1/2}\int \psi_i^* \boldsymbol{u}_k \cdot \boldsymbol{p}\psi_j \mathrm{d}^3r, \tag{8.1.25}$$

where $u_k = \hat{e}^\lambda(k)e^{i\vec{k}\cdot\vec{r}}$ is the plane wave mode of the photon and the sum over k in (8.1.24) means the sum over polarization index and the wave vector $\boldsymbol{k}$. The volume of the system is $V = L^3$. The interaction Hamiltonian H_{int} induces the transition of the atom from state j to state i while it absorbs or emits a photon k. For $\omega_k \approx E_i - E_j > 0$, the a_k term in (8.1.24) is relevant (photon absorption and atom excitation); and for $\omega_k \approx E_j - E_i > 0$, the $a_k^\dagger$ term is relevant (photon emission and atom de-excitation). We could have included the time evolution factors in the expansion of $\hat{\psi}$ and $\hat{A}$, and we would have instead of (8.1.24)

$$H_{\text{int}} = \sum_{i,j,k} c_i^\dagger c_j \left(g_{ij}^k a_k e^{-i\omega_k t} + g_{ij}^{k*} a_k^\dagger e^{i\omega_k t} \right) e^{-i(E_j - E_i)t/\hbar}.$$

In the discussion above, the "relevant" term (also called the resonance term) has an almost constant time dependence, for instance $e^{-i(\omega_k - E_i + E_j)t/\hbar}$ for the photon absorption, for which $\omega_k \approx E_i - E_j$, while the "irrelevant" term (also called the anti-resonant term) has a violently fluctuating time dependence, for instance $e^{i(\omega_k + E_i - E_j)t/\hbar}$ for the photon emission under the since condition $\omega_k \approx E_i - E_j$. Such fluctuating term averages to 0 over a very short time scale of $2\pi/\omega_k$, and can be dropped. This treatment is called "rotating wave approximation".

Since the wave length of an optical photon is very much larger than the dimension of the atom (the dimension of an atom ≈ 1, and the wave length of an optical photon $\approx 10^3$) it is legitimate to replace the factor $e^{ik \cdot r}$ of the plane wave mode function in (8.1.25) by $e^{ik \cdot R}$, where R is the position vector of the center of mass of the atom. Putting the center of mass of the atom at the origin, this factor becomes unity. The integral in (8.1.15) becomes

$$\int \psi_i^* \boldsymbol{p} \psi_j \mathrm{d}^3 r = \frac{im}{\hbar} \int \psi_i^* \left[\mathcal{H}_0, \boldsymbol{r} \right] \psi_j \mathrm{d}^3 r$$

$$= \frac{im}{\hbar} (E_i - E_j) \int \psi_i^* \boldsymbol{r} \psi_j \mathrm{d}^3 r,$$

and also

$$g_{ij}^k = i \left(\frac{2\pi\hbar}{\omega_k V} \right)^{1/2} \omega_{ji} \hat{e}^{(\lambda)} \cdot \boldsymbol{d}_{ij}, \qquad (8.1.26)$$

where

$$\omega_{ji} = \frac{E_j - E_i}{\hbar} \qquad (8.1.27)$$

and

$$d_{ij} = \int \psi_i^* e r \psi_j \mathrm{d}^3 r \qquad (8.1.28)$$

are respectively the angular frequency of the atomic transition and the transition matrix element of the dipole moment. In (8.1.26) we did not put $\omega_k = \omega_{ji}$, since in cavity QED "detuning" is usually used, i.e., the resonance frequency of the cavity ω_k is deliberately tuned to a value deviating from the transition frequency ω_{ji}.

For the radiation in free space, we have $\omega_k = \omega_{ji}$, and we have

$$g_{ij}^k = \mathrm{i} \left(\frac{2\pi\hbar\omega_k}{V} \right)^{1/2} \hat{e}_k^{(\lambda)} \cdot d_{ij}. \qquad (8.1.29)$$

The factor $\left(\frac{2\pi\hbar\omega_k}{V} \right)^{1/2}$ is the root-mean-square value of the vacuum electric field strength $\mathscr{E}_0$, as can be shown in the following. The operator of the electric field strength (8.1.18) does not commute with the photon number operator $a_k^\dagger a_k$, therefore the vacuum state (an eigenstate of photon number operator with eigenvalue 0) cannot be an eigenstate of $\boldsymbol{E}$. The expectation value of $\boldsymbol{E}$ in the vacuum state is zero, but that of $\boldsymbol{E}^2$ is different from zero. This is referred to as the *vacuum fluctuation*. The components of $\boldsymbol{E}$ do not commute with one another. From (8.1.20) we see that each mode of the radiation possesses zero point energy $\frac{1}{2}\hbar\omega_k$, and $\frac{1}{8\pi} \boldsymbol{E}_0^2$ contributes to one-half of the density of vacuum zero point energy, therefore

$$\frac{1}{8\pi} \left\langle \boldsymbol{E}_0^2 \right\rangle = \frac{1}{2} \frac{\frac{1}{2}\hbar\omega}{V},$$

i.e.,

$$\mathscr{E}_0 = \left\langle \boldsymbol{E}_0^2 \right\rangle^{1/2} = \left(\frac{2\pi\hbar\omega_k}{V} \right)^{1/2}. \qquad (8.1.30)$$

The physical process of the spontaneous emission of radiation becomes clear in the second quantized theory of the radiation field. The electron undergoes a transition due to its interaction with the vacuum fluctuation of the radiation field. The wave vector of the photon emitted in the spontaneous radiation in free space is not in any way restricted. The components of wave vectors given by (8.1.16) become continuous in the limit $L \to \infty$. Then the density of modes becomes

$$\rho_0(\omega) = \frac{\omega^2 V}{\pi^2 c^3}. \qquad (8.1.31)$$

According to Fermi's golden rule, the probability of spontaneous emission per unit time Γ_0 is[2]

$$\Gamma_0 = 2\pi \left(\frac{2\pi \hbar \omega_k}{V} \right) |d_{ij}|^2 \frac{\rho_0}{3}$$

$$= \frac{4\omega^3}{3\hbar c^3} |d_{ij}|^2. \qquad (8.1.32)$$

This is the well-known formula for dipole radiation.

In the following, we are going to discuss the cavity QED. The mode function is denoted by $\boldsymbol{f}$, which satisfies the boundary condition of the cavity. Since the dimension of an atom is much smaller than the wave length of the photon, the function $\boldsymbol{f}(\boldsymbol{R})$ can be taken out of the integral in (8.1.25). The coupling corresponding to (8.1.29) is

$$g_{ij}^k = i \left(\frac{2\pi \hbar \omega_k}{\mathcal{V}} \right)^{1/2} \boldsymbol{f}_k(\boldsymbol{R}) \cdot \boldsymbol{d}_{ij}, \qquad (8.1.33)$$

where $\mathcal{V}$ is the effective volume of the cavity,

$$\mathcal{V} = \int |\boldsymbol{f}_k(\boldsymbol{r})|^2 \, d^3 r. \qquad (8.1.34)$$

The interaction Hamiltonian (8.1.24) can be applied to the interaction of the atomic electron with the radiation field in free space and the field in the cavity as well. For the former g_{ij}^k is given by (8.1.29) and for the latter by (8.1.33).

---

[2] In ρ_0 two polarization states are taken into account. The factor $|\hat{\boldsymbol{e}} \cdot \boldsymbol{d}|^2$ in $|g_{ij}^k|^2$ gives $\frac{1}{3}$ when averaged over the direction of polarization vector.

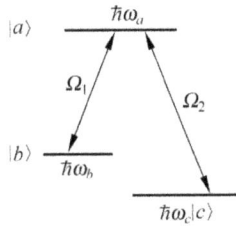

Fig. 8.1   Energy levels of Λ-configuration with resonant coupling fields.

8.1.4   Dark states, electromagnetically induced transparency

Consider a 3-level atom with a Λ-configuration of energy levels $|a\rangle$, $|b\rangle$ and $|c\rangle$ coupled by 2 resonant laser fields with Rabi frequencies Ω_1 and Ω_2 shown in Fig. 8.1. Dipole transition between $|b\rangle$ and $|c\rangle$ is not allowed. The Hamiltonian of the system is

$$H = H_0 + H_{\text{int}},$$
$$H_0 = \hbar\omega_a\,|a\rangle\langle a| + \hbar\omega_b\,|b\rangle\langle b| + \hbar\omega_c\,|c\rangle\langle c|,$$
$$H_{\text{int}} = -\frac{\hbar}{2}\left[\Omega_1 e^{-i(\omega_a-\omega_b)t}\,|a\rangle\langle b| + \Omega_2 e^{-i(\omega_a-\omega_c)t}\,|a\rangle\langle c|\right] + h.c.$$

$$(8.1.35)$$

The general form of a state of the atom is

$$|\psi(t)\rangle = c_a(t)e^{-i\omega_a t}\,|a\rangle + c_b(t)e^{-i\omega_b t}\,|b\rangle + c_c(t)e^{-i\omega_c t}\,|c\rangle.\qquad(8.1.36)$$

Substituting (8.1.36) into (8.1.35), we obtain the evolution equations for the coefficients

$$\dot{c}_a(t) = \frac{i}{2}\left(\Omega_1 c_a(t) + \Omega_2 c_b(t)\right),\ \ \dot{c}_b(t) = \frac{i}{2}\Omega_1 c_a(t),\ \ \dot{c}_c(t) = \frac{i}{2}\Omega_2 c_a(t).$$

$$(8.1.37)$$

Consider the initial conditions for a state $|\psi_{\text{dark}}\rangle$

$$c_a(0) = 0,\ \ c_b(0) = \Omega_2/\Omega,\ \ c_c(0) = \Omega_1/\Omega,\qquad(8.1.38)$$

where

$$\Omega = \left(\Omega_1^2 + \Omega_2^2\right)^{1/2}.\qquad(8.1.39)$$

Then by (8.1.37) all coefficients will be time independent. Then the state

$$|\psi_{\text{dark}}\rangle = \frac{\Omega_2 |b\rangle - \Omega_1 |c\rangle}{\Omega} \qquad (8.1.40)$$

will be a stationary state of the system, i.e., it remains in itself without transition to state $|a\rangle$ notwithstanding the resonant laser beams. It remains in the darkness. This phenomenon is called the coherent population trapping, arising from the destructive quantum interference between the transitions $|b\rangle \rightarrow |a\rangle$ and $|c\rangle \rightarrow |a\rangle$. If all the atoms are in the state $|b\rangle$ in the beginning with Ω_2 present and Ω_1 absent, we can adiabatically turn on Ω_1 and turn out Ω_2, the state remains dark with atoms transferred gradually to $|c\rangle$, while keeping number of atoms on $|b\rangle$ and $|c\rangle$ in proportion to Ω_2^2 and Ω_1^2. Such kind of population transfer is preferable because real transition to $|a\rangle$ does not happen, otherwise spontaneous emission from $|a\rangle$ could destroy the coherence.

This phenomenon is applied in the electromagnetically induced transparency [5, 6]. S.E. Harris' group [5] used Sr gas and L. Hau's group [6] used Na BEC with level diagram given in Fig. 8.2. The coupling beam Ω_c is resonant with the $|2\rangle \rightarrow |3\rangle$ transition, while a probe pulse Ω_p is tuned to the transition $|1\rangle \rightarrow |3\rangle$. The quantum interference occurs in a narrow interval of probe frequencies, with its width determined by the coupling laser power. In the absence of dephasing of the $|1\rangle \rightarrow |2\rangle$ (virtual) transition, the quantum

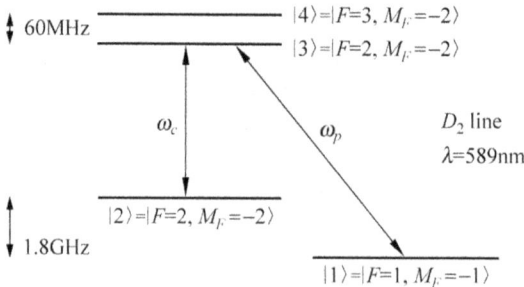

Fig. 8.2   Part of relevant Na atomic energy levels with laser beams coupling the states. Reprinted with permission from Macmillan Publishers Ltd: Nature (397, 594), copyright (1999).

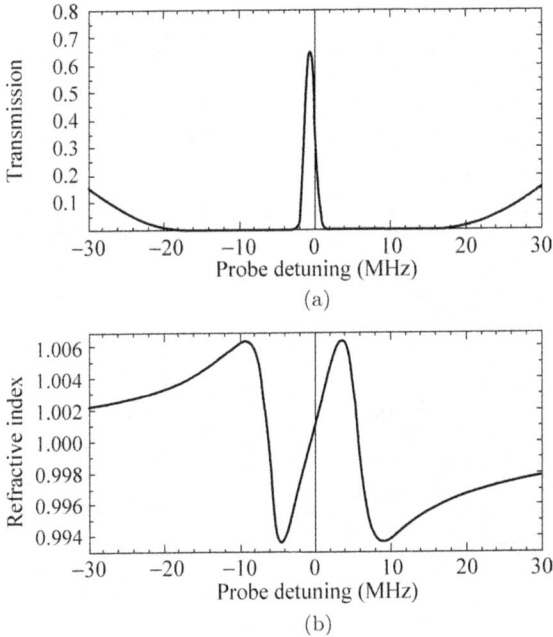

Fig. 8.3 Transmission and refractive index of the medium. (a) transmission as a function of the frequency detuning of probe pulse; (b) refractive index profile. The shift of curve center originated from the Stark shift of $|2\rangle \rightarrow |3\rangle$ transition. Reprinted with permission from Macmillan Publishers Ltd: Nature (397, 594), copyright (1999).

interference would be perfect, and the transmission at the resonance would be unity. A realistic calculation result of the transmission is given in Fig. 8.3(a). Figure 8.3(b) shows the refractive index for the probe beam as a function of the detuning. We see that not only was the medium consisting of the atoms with laser field Ω_2 coupling $|2\rangle$ and $|3\rangle$ is made transparent for the probe beam (the electromagnetically induced transparency EIT) because of the destructive quantum interference, but also it was made to possess a remarkable property of resulting in a very low group velocity for part of the probe beam transmitted, orders of magnitude lower than the previous experiments carried out with thermal atomic gases. A schematic illustration of the pulse delay is shown in Fig. 8.4. The probe pulse takes about 7 μs to traverse the 0.2 nm cloud of the cold atoms–1/10 millionth of the speed of light in vacuum. The group

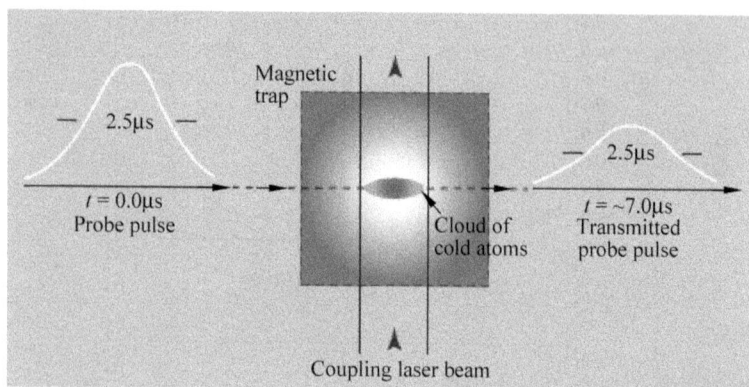

Fig. 8.4   Schematic illustration of the light pulse delay. Reprinted with permission from Macmillan Publishers Ltd: Nature (397, 559), copyright (1999).

velocity is determined by the variation with frequencies of the real part of the refractive index according to the relation

$$v_g = \frac{c}{n(\omega) + \omega \frac{dn}{d\omega}}, \qquad (8.1.41)$$

where $n(\omega)$ is the real part of refractive index, while the absorption coefficient α of the medium is determined by the imaginary part of the refractive index $\mathrm{Im}\, n$:

$$\alpha = 2\omega \mathrm{Im}\, n/c. \qquad (8.1.42)$$

The pulse delay measurement results are given in Fig. 8.5. The open circles represent the reference pulse, i.e., that without the medium, while the full circles represent the real pulse delayed by 7.05 μs in a 229 μm long atomic cloud. Compared with the delay in thermal gas, the maximum delay in the latter case was much less than the total duration of the pulse itself.

The underlying physics is that when a light signal propagates through the medium, its electric field strength polarizes the atoms. The polarization divided by the electric field strength amplitude gives the electric susceptibility χ_e, and the refractive index is given by $n = \sqrt{1 + 4\pi\chi_e}$. Our present problem is by far more subtle. The medium is not simply made of gas atoms with a Λ-configuration of energy levels, but rather of *light dressed atoms*. So the problem is not

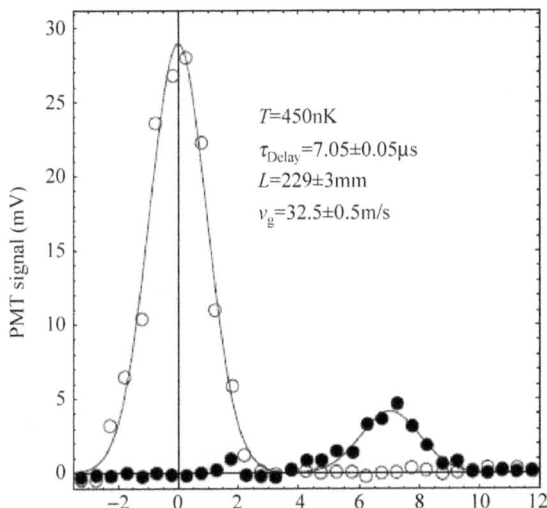

Fig. 8.5 Pulse delay measurement. Reprinted with permission from Macmillan Publishers Ltd: Nature (397, 594), copyright (1999).

one of medium of atoms under a given external light field, but one of a comprehensive system consisting of atoms with 3 quantum states coupled to 2 fields. The theory will be discussed briefly later.

The next progress was also achieved by L. Hau's group [8], where the physicists found a way to stop the pulse altogether, revive it, or make "multiple read-outs". Figure 8.6 shows the relevant energy levels and the fields, and the experimental set-up. Just as the pulse is contained completely in the BEC and before it re-emerges, the coupling beam is turned off. As a result, the speed of the probe beam becomes zero, and the pulse comes to a complete halt in the middle of the BEC. Light exists because it propagates. The key fact is that, as the pulse penetrates into the BEC, it turns to a quantum coherence pattern printed on the Na atoms, because the atoms are in a state of coherent superposition determined by the amplitude and phase of the probe field. Upon the sudden switching off of the coupling beam, the compressed probe pulse is halted, and the coherent information initially contained in the probe fields is now frozen in the atomic medium. Later the coupling field is turned back on, the probe pulse is regenerated: the stored coherence is read out and converted back into the probe radiation field again. The measurements of delayed

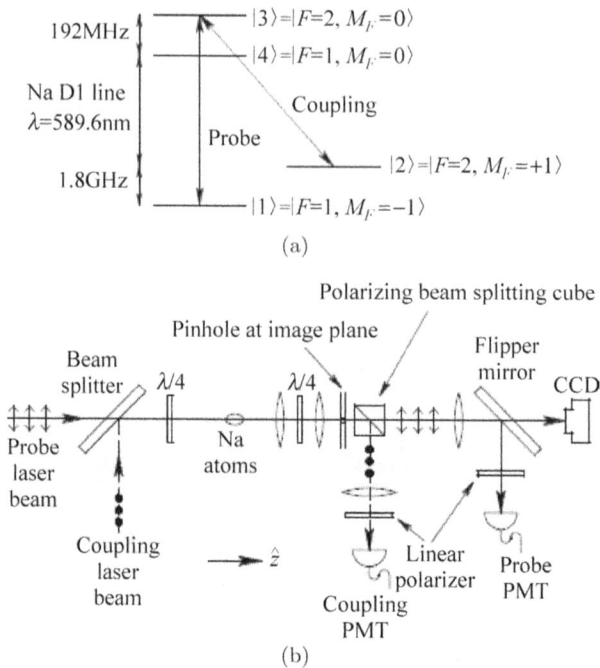

Fig. 8.6   (a) Relevant energy levels and the fields; (b) the experimental set-up. Reprinted with permission from Macmillan Publishers Ltd: [Nature] (409, 490), copyright (2001).

and revived probe pulses are shown in Fig. 8.7. Open circles represent the reference pulses, dashed curves and filled circles show measured intensities of the coupling and probe pulses. The cold atom cloud is 339 μm long and cooled to 0.9 μK. The pulse delay amounts to 11.8 μs.

The manipulation of the probe pulse goes beyond simple revivals. The coupling beam can be turned back on with an intensity higher (lower) than the original value. The amplitude of the revived pulse increases (decreases) and its temporal width decreases (increases), as shown in Fig. 8.8. The stored information of coherence in the atomic medium involves the ratio between the Rabi frequencies of the coupling and the probe fields. The intensity of the revived probe pulse is proportional to the intensity of the coupling field when it is turned back on. Furthermore, the *spatial width* of the revived pulse is determined by the distribution of atomic coherence which remains the same as the spatial width of the original probe pulse. The group

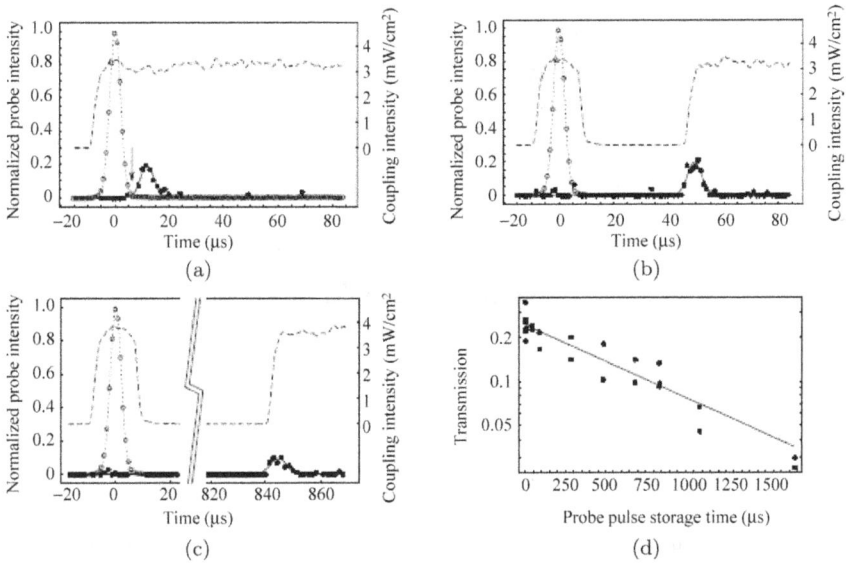

Fig. 8.7   Measurements of delayed and revived probe pulses. (a) pulse delay; (b) and (c) pulse halted and revived, coupling field turned off at $t = 6.3$ μs and turned back on at $t = 44.3$ μs and $t = 839.3$ μs respectively; (d) transmission of the probe pulse energy versus storage time. Reprinted with permission from Macmillan Publishers Ltd: [Nature] (409, 490), copyright (2001).

velocity of the pulse given by (8.1.41) is evaluated [9] to be

$$v_g \approx \frac{\hbar c \varepsilon_0}{2\omega_p} \frac{|\Omega_c|^2}{|\mu_{13}|^2 N}, \tag{8.1.43}$$

where μ_{13} is the electric dipole moment matrix element between states $|1\rangle$ and $|3\rangle$, ε_0 is the vacuum permittivity and N is the atomic density. The group velocity is proportional to the intensity of the coupling field. With a higher (lower) intensity the revived pulse propagates with a larger (smaller) group velocity, and consequently the pulse temporal width becomes smaller (larger), inversely proportional to the intensity. Dissipationless pulse storage and revival processes are only possible if the ratio between the rates of dissipative and coherence-preserving events is small. The ratio is found to be [10]

$$\frac{2\Gamma}{\Omega_c^2 + \Omega_p^2} \left(\frac{\dot{\Omega}_p}{\Omega_p} - \frac{\dot{\Omega}_c}{\Omega_c} \right), \tag{8.1.44}$$

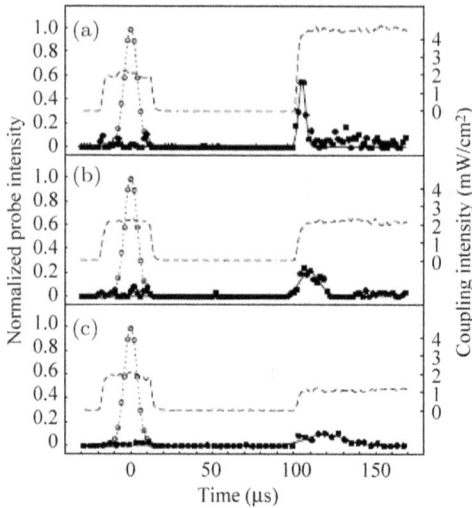

Fig. 8.8   Measurements of revived probe pulses for varying intensities of the second coupling pulse I_{c2} for constant intensity of the first coupling pulse I_{c1}. (a) $I_{c2} > I_{c1}$; (b) $I_{c2} = I_{c1}$; (c) $I_{c2} < I_{c1}$. Reprinted with permission from Macmillan Publishers Ltd: [Nature] (409, 490), copyright (2001).

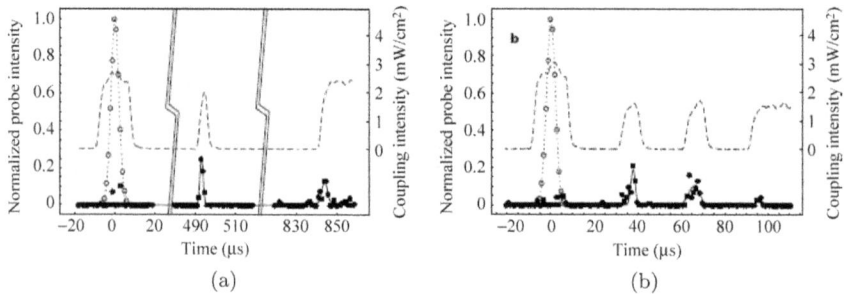

Fig. 8.9   Multiple read-outs. (a) two read-outs; (b) three read-outs. Reprinted with permission from Macmillan Publishers Ltd: [Nature] (409, 490), copyright (2001).

where Γ is the spontaneous decay rate from state $|3\rangle$. Numerical simulation [10] show that the probe field is adjusting to match the changes in the coupling field such that the two terms in the parenthesis almost cancel each other. By using a series of short coupling pulses multiple read-outs are possible. This is shown in Fig. 8.9.

In conclusion we stop briefly on the formulation of a theoretical paper [10] just to illustrate the nature of the problem. The two fields are described by $\Omega_p = -d_{13} \cdot E_p/\hbar$ and $\Omega_c = -d_{23} \cdot E_c/\hbar$ where E_p and E_c are slowly varying envelopes of the electric fields, both of which can be time and space varying, and d_{13} and d_{23} are the dipole matrix elements. The evolution of the wave functions of the states ψ_1, ψ_2, ψ_3 are given by the Gross-Pitaevskii equations

$$i\hbar\frac{\partial\psi_1}{\partial t} = \left[-\frac{\hbar^2}{2m}\frac{\partial^2}{\partial z^2} + V_1(z) + U_{11}\left|\psi_1\right|^2 + U_{12}\left|\psi_2\right|^2\right]\psi_1 + \frac{1}{2}\hbar\Omega_p^*\psi_3,$$

$$i\hbar\frac{\partial\psi_2}{\partial t} = \left[-\frac{\hbar^2}{2m}\frac{\partial^2}{\partial z^2} + V_2(z) + U_{22}\left|\psi_2\right|^2 + U_{12}\left|\psi_1\right|^2\right]\psi_2 + \frac{1}{2}\hbar\Omega_p^*\psi_3,$$

$$i\hbar\frac{\partial\psi_3}{\partial t} = \frac{1}{2}\hbar\Omega_p\psi_1 + \frac{1}{2}\hbar\Omega_c\psi_2 - i\frac{\Gamma}{2}\psi_3. \tag{8.1.45}$$

Here the external dynamics of ψ_3 (that under the kinetic, trap and interaction energies) is neglected in comparison with Ω_p, Ω_c and Γ, since the latter belong to the category of "fast motion" (3 orders of magnitude faster). The inclusion of Γ makes the evolution of ψ_3 non-hermitian. The evolution of the fields follows from the Maxwell's equations, in which the polarization of the medium is the source of the electric field. The evolution equations are

$$\left(\frac{\partial}{\partial z} + \frac{1}{c}\frac{\partial}{\partial t}\right)\Omega_p = -i\frac{f_{13}\sigma_0}{A}\frac{\Gamma}{2}N_c\psi_3\psi_1^*,$$

$$\left(\frac{\partial}{\partial z} + \frac{1}{c}\frac{\partial}{\partial t}\right)\Omega_c = -i\frac{f_{23}\sigma_0}{A}\frac{\Gamma}{2}N_c\psi_3\psi_2^*, \tag{8.1.46}$$

where σ_0 is the resonance absorption cross-section, N_c is the total number of atoms in the BEC, and A is the cross-sectional area of the BEC. ψ_3 is considered to be slowly varying, and by setting $\partial\psi_3/\partial t \to 0$ we can express ψ_3 in terms of ψ_1 and ψ_2:

$$\psi_3 \approx -\frac{i}{\Gamma}\left(\Omega_p\psi_1 + \Omega_c\psi_2\right). \tag{8.1.47}$$

By substituting (8.1.47) into (8.1.45) and (8.1.46) we obtain a system of coupled evolution equations for ψ_1, ψ_2, Ω_p and Ω_c. Numerical simulation gives detailed developments described in Ref. [8].

8.2 Jaynes-Cummings Model

The starting point of the Jaynes-Cummings model [11] is the coupling between a two-level atom and a single mode in the cavity, the coupling can be resonant or close to resonant. This model is highly idealized, but can be closely approximated by the coupling between a highly excited atom with a high Q cavity such that the mode in the cavity matches or nearly matches the transition between Rydberg states $|e\rangle \rightarrow |g\rangle$. The resonant coupling excludes the non-resonant atomic transitions and the non-resonant field modes. If a "circular" Rydberg state, i.e., a state with $l = n - 1$, is prepared the approximation can be very good, because the selection rule of the electric dipole radiation $|\Delta l| = 1$ allows a transition between the states $(n, l = n - 1)$ and $(n - 1, n - 2)$ only. Furthermore the Rydberg states have large transition matrix elements of the dipole moment to ensure a stronger coupling, for instance a state with $n = 50$ the "radius" of the atom is $2\,500$ times that of the ground state.

For a two-level atom, let $\omega_0 = \frac{1}{\hbar}(E_e - E_g)$, and let $E_e = \frac{1}{2}\hbar\omega_0$, $E_g = -\frac{1}{2}\hbar\omega_0$. Define the operators

$$\left.\begin{array}{l} b = |g\rangle \langle e| \\ b^\dagger = |e\rangle \langle g| \end{array}\right\} \tag{8.2.1}$$

which are the operators leading to the transitions $|e\rangle \rightarrow |g\rangle$ and $|g\rangle \rightarrow |e\rangle$ respectively. The definition (8.2.1) gives

$$bb^\dagger = |g\rangle \langle g|, \quad b^\dagger b = |e\rangle \langle e|.$$

Since the states $|e\rangle$ and $|g\rangle$ form a complete set, we have

$$bb^\dagger + b^\dagger b = 1. \tag{8.2.2}$$

The Hamiltonian of the atom-field system is

$$H = \frac{\hbar\omega_0}{2}\left(b^\dagger b - bb^\dagger\right) + \hbar\omega\left(a^\dagger a + \frac{1}{2}\right) + \hbar\Omega\left(\boldsymbol{R}\right)\left(ab^\dagger + a^\dagger b\right),$$

$$\tag{8.2.3}$$

where

$$\Omega\left(\boldsymbol{R}\right) = \frac{1}{\hbar}\left(\frac{2\pi\hbar\omega_k}{\mathscr{V}}\right)^{1/2}\boldsymbol{f}\left(\boldsymbol{R}\right)\cdot\boldsymbol{d}.$$

The phases of $|e\rangle$ and $|g\rangle$ are chosen such that $\Omega\left(\boldsymbol{R}\right)$ is real. Denoting the sum of the first two terms of (8.2.3) by H_0 (the Hamiltonian of the free fields) and the third term by H_{int} (Hamiltonian of the interaction between fields), we get after a simple calculation

$$[H_0, H_{\text{int}}] = 0. \tag{8.2.4}$$

provided $\omega_0 = \omega$. This implies that under the resonance condition the eigenstates of H can be constructed using the eigenstates of H_0, namely $|e, n\rangle$ and $|g, n+1\rangle$, where the index n represents the number of photons.

8.2.1 The eigenstates of the coupled atom-cavity system

Consider the resonance case, the level spacing $\hbar\omega_0$ coincides with the cavity mode photon energy $\hbar\omega$. In this case the eigenstates $|e, n\rangle$ and $|g, n+1\rangle$ of H_0 are degenerate. The eigenstates of H is

$$|\pm\rangle = \frac{1}{\sqrt{2}}[|e, n\rangle \pm |g, n+1\rangle]. \tag{8.2.5}$$

The result of operating H_{int} on the eigenstates of H_0 is

$$\left.\begin{array}{l} H_{\text{int}}\,|e, n\rangle = \hbar\Omega\sqrt{n+1}\,|g, n+1\rangle \\ H_{\text{int}}\,|g, n+1\rangle = \hbar\Omega\sqrt{n+1}\,|e, n\rangle \end{array}\right\} \tag{8.2.6}$$

from which we obtain the eigenvalues of H

$$E_\pm = \hbar\omega\left(n+1\right) \pm \hbar\Omega_0, \tag{8.2.7}$$

where

$$\Omega_0 = \Omega\sqrt{n+1} \tag{8.2.8}$$

is the vacuum Rabi frequency. This is the quantum mechanical two-level resonance phenomenon. If the atom is in the state $|e\rangle$ at time

$t = 0$, then the probability that it will remain in this state at time t is

$$P_e(t) = \cos^2 \Omega_0 t, \tag{8.2.9}$$

This is called the Rabi oscillation.

If there is a detuning, $w \neq w_0$, then the eigenstates $|e, n\rangle$ and $|g, n + 1\rangle$ of H_0 are no more degenerate. Since

$$\left.\begin{aligned} ab^\dagger \,|e, n\rangle &= 0, & ab^\dagger \,|g, n + 1\rangle &= \sqrt{n + 1}\,|e, n\rangle \\ a^\dagger b\,|e, n\rangle &= \sqrt{n + 1}\,|g, n + 1\rangle, & a^\dagger b\,|g, n + 1\rangle &= 0 \end{aligned}\right\} \tag{8.2.10}$$

we try to satisfy the Schrödinger equation using the following linear superpositions:

$$\left.\begin{aligned} |+, n\rangle &= \cos\theta\,|e, n\rangle - \sin\theta\,|g, n + 1\rangle \\ |-, n\rangle &= \sin\theta\,|e, n\rangle + \cos\theta\,|g, n + 1\rangle \end{aligned}\right\}. \tag{8.2.11}$$

We have then

$$H\,|+, n\rangle = \left[\frac{\hbar w_0}{2} + \hbar w\left(n + \frac{1}{2}\right) - \hbar\Omega\sqrt{n + 1}\tan\theta\right]\cos|e, n\rangle$$

$$- \left[-\frac{\hbar w_0}{2} + \hbar w\left(n + \frac{3}{2}\right) - \frac{\hbar\Omega\sqrt{n + 1}}{\tan\theta}\right]\sin|g, n + 1\rangle.$$

We demand $|+, n\rangle$ to be an eigenstate of H corresponding to the eigenvalue E_+.

$$H|+, n\rangle = E_+|+, n\rangle.$$

Comparing with the previous equation we can fix θ, which will depend on n.

$$E_+ = \frac{\hbar w_0}{2} + \hbar w\left(n + \frac{1}{2}\right) - \hbar\Omega\sqrt{n + 1}\tan\theta$$

$$= -\frac{\hbar w_0}{2} + \hbar w\left(n + \frac{3}{2}\right) - \frac{\hbar\Omega\sqrt{n + 1}}{\tan\theta}.$$

The detuning is defined by $\delta = \omega - \omega_0$, the above equation gives

$$\delta = -\Omega\sqrt{n+1}\left(\frac{1}{\tan\theta} - \tan\theta\right) = -2\Omega\sqrt{n+1}\frac{1}{\tan 2\theta},$$

and we have, after adding a subscript n to θ,

$$\left.\begin{array}{c} \tan 2\theta_n = -\dfrac{2\Omega\sqrt{n+1}}{\delta} \\[2mm] 0 \leq \theta_n \leq \frac{\pi}{2} \end{array}\right\} \tag{8.2.12}$$

and

$$E_+ = (n+1)\,\hbar\omega + \frac{\hbar}{2}\sqrt{4\Omega^2\,(n+1) + \delta^2}. \tag{8.2.13}$$

Similarly we have

$$H\,|-,n\rangle = E_-\,|-,n\rangle,$$

$$E_- = (n+1)\,\hbar\omega - \frac{\hbar}{2}\sqrt{4\Omega^2\,(n+1) + \delta^2}. \tag{8.2.14}$$

As a special case, $|g,0\rangle$ is an eigenstate of H,

$$E_{|g,0\rangle} = -\frac{\hbar\omega_0}{2} + \frac{\hbar\omega}{2} = \frac{\hbar}{2}\delta. \tag{8.2.15}$$

A special feature of cavity QED is that the root-mean-square value of the vacuum electric field is a function of position. It is zero on the cavity wall, and is maximum at the center of the cavity. Therefore, the coupling strength of the atom with the cavity is a function of the position of the center of mass of the atom.[3] Outside the cavity the coupling vanishes, $|e,n\rangle$ and $|g,n+1\rangle$ are the eigenstates of H. If the detuning is negative, the energy of $|e,n\rangle$ is larger than that of $|g,n+1\rangle$. Entering the cavity, they become $|+,n\rangle$ and $|-,n\rangle$ (cf. (8.2.11)), their energy difference grows gradually from the cavity wall to the cavity center. Figure 8.10(a)

---

[3]In Eq. (8.1.24) for the Hamiltonian of the interaction of the atom with the field in free space the coupling strength g_{ij}^k is independent of space coordinates. Here we denote the position dependent coupling strength by $\mathscr{G}$.

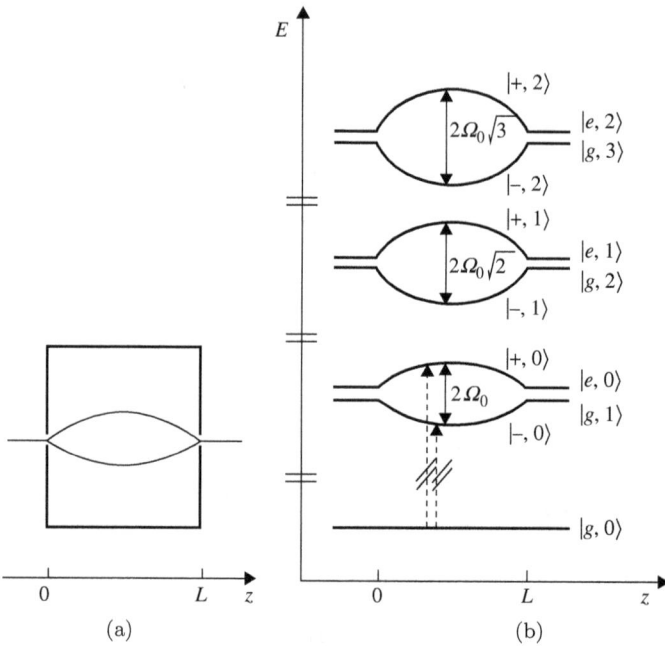

Fig. 8.10   The atom-cavity system (a) The cross-section of the cavity and the field mode; (b) Energy levels of the atom-cavity system, $|\delta| \ll \Omega_0$. Redrawn from Ref. [2].

shows the cross-section of a cylindrical cavity and a sinusoidal mode, for which the field vanishes at $z = 0$ and $z = L$. Figure 8.10(b) shows the variation of $E_{\pm}$ along the cavity axis (z-axis) for the case $|\delta| \ll \Omega_0$. The difference between E_+ and E_- reaches the maximum $2\hbar\Omega_0\sqrt{n+1}$ at the center of the cavity, where Ω_0 is the value of Ω at the cavity center.

8.2.2   The light shift of the atomic energy levels in the non-resonant case

How does the energy of an atom in the state $|e\rangle$ or $|g\rangle$ change after entering the cavity? When $|\delta| \gg \Omega_0$, we treat $\Omega\sqrt{n+1}/|\delta|$ as a small quantity to get

$$E_{\pm,n} = (n+1)\hbar\omega \pm \frac{\hbar}{2}|\delta| \pm \hbar\frac{\Omega^2(\boldsymbol{R})(n+1)}{|\delta|}. \qquad (8.2.16)$$

From (8.2.11) and (8.2.12) we see that in the case $\delta < 0$, the state $|+, n\rangle$ is closer to $|e, n\rangle$ and $|-, n\rangle$ is closer to $|g, n+1\rangle$, and the correspondence is opposite for $\delta > 0$. Therefore, the energy change of the system after the atom in a state $|e\rangle$ has entered the cavity is

$$\Delta E_e(n) = -\hbar \frac{\Omega^2(\boldsymbol{R})}{\delta}(n+1), \qquad (8.2.17)$$

which is valid irrespective of the sign of δ. This can be seen by inspecting (8.2.17) for the cases $\delta > 0$ and $\delta < 0$ separately. Similarly the energy change of the system after the atom in state $|g\rangle$ has entered the cavity is

$$\Delta E_g(n) = \frac{\hbar \Omega^2(\boldsymbol{R})}{\delta} n, \qquad (8.2.18)$$

which is also valid irrespective of the sign of δ. The energy shift of an atomic energy level due to the interaction with photons in the cavity is called the light shift. Figure 8.2 shows the energy level shift for $n = 1$, with the cavity situated between 0 and L, the z-axis being the cavity axis. Figure 8.11(a) and (b) give the level shifts for $\delta > 0$ and $\delta < 0$ respectively, ω_0 denotes the energy difference between $|e\rangle$ and $|g\rangle$ divided by $\hbar$ when the atom is not coupled to the cavity. For the case $\delta > 0$, the levels approach each other while for $\delta < 0$ the levels depart. The change of energy difference between $|e, n\rangle$ and $|g, n\rangle$ when the atom is coupled to the cavity can be obtained by

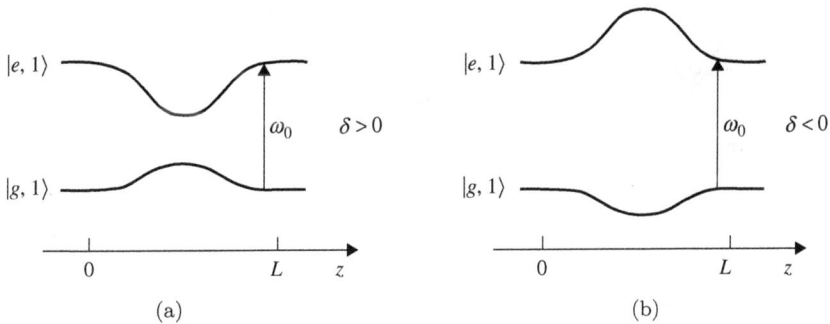

Fig. 8.11   The atomic energy level shift of $|e\rangle$ and $|g\rangle$ in a cavity with $n = 1$. (a) $\delta > 0$; (b) $\delta < 0$. Redrawn from Ref. [2].

taking the difference of (8.2.17) and (8.2.18) divided by $\hbar$:

$$\Delta\omega_0(n) = -\frac{\hbar\Omega^2(\boldsymbol{R})}{\delta}(2n+1). \qquad (8.2.19)$$

Let the field be in a Glauber coherent state

$$|\alpha\rangle = e^{-|\alpha|^2/2}\sum_n \frac{\alpha^n}{(n!)^{1/2}}|n\rangle, \qquad (8.2.20)$$

where n is the number of photons. The distribution of the number of photons in a coherent state is

$$P(n) = |\langle n \mid \alpha\rangle|^2 = \frac{|\alpha|^{2n}\,e^{-|\alpha|^2}}{n!}, \qquad (8.2.21)$$

which is the Poisson distribution with

$$\bar{n} = \langle\alpha|a^\dagger a|\alpha\rangle = |\alpha|^2. \qquad (8.2.22)$$

For the coherent state of the field, a complicated situation arises in the Rabi oscillation of the two level atom. Since Ω depends on n, the frequency of oscillation is different for different components n in the coherent state, and there will be interference between them. For a single frequency (8.2.9) gives

$$P_e(t) = \cos^2\Omega_0 t = \frac{1+\cos 2\Omega\sqrt{n+1}t}{2}.$$

For the coherent state we have by using the Poisson distribution

$$P_e(t) = \frac{1}{2}\left(1+\sum_n \frac{e^{-\bar{n}}\bar{n}^n}{n!}\cos 2\Omega\sqrt{n+1}t\right). \qquad (8.2.23)$$

Approximate calculation gives [12]

$$P_e(t) = \frac{1}{2}\left\{1+\cos 2\Omega(\bar{n}+1)^{1/2}t\,\exp\left[-\frac{\Omega^2 t^2\bar{n}}{2(\bar{n}+1)}\right]\right\}, \qquad (8.2.24)$$

i.e., the Rabi oscillation decays under an envelope with a lifetime

$$t_c \approx \frac{1}{\Omega}, \qquad (8.2.25)$$

where we have set $\bar{n} \approx \bar{n} + 1$. Similar to the Kepler wave packet constructed by using coherent states[4] the oscillation revives, with a time of revival

$$t_{\text{rev}} \approx \frac{2\pi}{\Omega} \bar{n}^{1/2}. \tag{8.2.26}$$

Since the frequencies are irrational numbers and are incommensurable, the revival cannot be complete. The decay and revival is the result of coherent superposition of the components of the coherent state, and is hence purely quantum mechanical. Such phenomenon has already been observed in micromasers [13].

Let the distribution in the photon number in the cavity be $p(n)$, then the probability that an excited atom will remain in the excited state at time t after entering the cavity at time $t = 0$ is

$$P_e(t) = \sum_n p(n) \cos^2\left(\Omega\sqrt{n+1}t\right).$$

The spontaneous radiation corresponds to the case of no photon present in the cavity initially, $n = 0$. Usually the coupling g is very small,[5] it is difficult to observe the Rabi oscillation. But for Rydberg states the coupling can be rather large, it can reach $10^3 \text{s}^{-1} \sim 10^6 \text{s}^{-1}$ for a principal quantum number 40. For such states the oscillation can be observed even in the case $n = 0$. In the experiment of Ref. [13] a small thermal field (2.5K) is set up in the cavity. The cavity is tuned to $21.6\,\text{GHz}$, corresponding to the $63P_{3/2} \rightleftharpoons 61D_{5/2}$ transition of the ^{85}Rb atom. Figure 8.12 shows the measured $P_e(t)$, the curve being the theoretical prediction. The distribution in number of photons in the cavity is given by the Boltzmann factor

$$p(n) = Ce^{-n\hbar\omega/kT},$$

---

[4]See Section 4.3.
[5]We do not consider the coordinate dependence of the coupling, and use the notation g for the coupling between the atom and the cavity field.

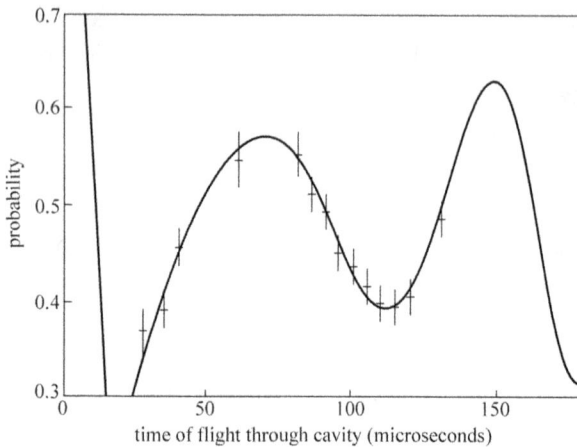

Fig. 8.12   The Rabi oscillation in a cavity. Reprinted with permission from [Gerhard Rempe, Herbert Walther, and Norbert Klein, *Phys. Rev. Lett.* 58, 353, 1987.] Copyright (2017) by the American Physical Society.

where the normalization constant can be obtained by performing the summation:

$$1 = \sum_{n=0}^{\infty} p(n) = C \sum_{n=0}^{\infty} e^{-n\hbar\omega/kT} = C\frac{1}{1 - e^{-\hbar\omega/kT}},$$

i.e.,

$$p(n) = (1 - e^{-\hbar\omega/kT})e^{-n\hbar\omega/kT}.$$

Rabi oscillation is clearly shown in Fig. 8.12, and the decay and revival are also observed.

8.2.3   The evolution of state in time

It is most convenient to treat the evolution problem in the interaction picture, where the evolution is determined by the interaction Hamiltonian only. If we represent $|e, n\rangle$ and $|g, n+1\rangle$ by a two-component wave function, then (8.2.6) can be written as

$$H_{\text{int}} \begin{pmatrix} |e, n\rangle \\ |g, n+1\rangle \end{pmatrix} = \begin{pmatrix} 0 & \hbar\Omega\sqrt{n+1} \\ \hbar\Omega\sqrt{n+1} & 0 \end{pmatrix} \begin{pmatrix} |e, n\rangle \\ |g, n+1\rangle \end{pmatrix}.$$

$$(8.2.27)$$

The time evolution operator is

$$U = \exp\left(-\frac{i}{\hbar}H_{\text{int}}t\right) = \exp\left[-i\begin{pmatrix} 0 & \Omega\sqrt{n+1} \\ \Omega\sqrt{n+1} & 0 \end{pmatrix}t\right]$$

$$= \cos\left(\Omega\sqrt{n+1}\,t\right) - i\begin{pmatrix} 0 & 1 \\ 1 & 0 \end{pmatrix}\sin\left(\Omega\sqrt{n+1}\,t\right)$$

$$= \begin{pmatrix} \cos\Omega\sqrt{n+1}\,t & -i\sin\Omega\sqrt{n+1}\,t \\ -i\sin\Omega\sqrt{n+1}\,t & \cos\Omega\sqrt{n+1}\,t \end{pmatrix}. \qquad (8.2.28)$$

If the atom is in the state $|e, n\rangle$ in entering the cavity, then the state of the atom at time t is

$$U\begin{pmatrix} 1 \\ 0 \end{pmatrix} = \begin{pmatrix} \cos\Omega\sqrt{n+1}\,t \\ -i\sin\Omega\sqrt{n+1}\,t \end{pmatrix}$$

provided the time t is smaller than the time τ for the atom to traverse through the cavity. This state can be written as

$$\psi(t) = \cos\Omega\sqrt{n+1}\,t\,|e, n\rangle - i\sin\Omega\sqrt{n+1}\,t\,|g, n+1\rangle. \qquad (8.2.29)$$

For $t \geq \tau$, it is sufficient to replace t by τ. If a measurement on the atom is carried out after it has passed through the cavity, then ψ will be reduced to $|e, n\rangle$ or $|g, n+1\rangle$ with probabilities $\cos^2\Omega\sqrt{n+1}\,\tau$ and $\sin^2\Omega\sqrt{n+1}\,\tau$ respectively.

8.3 The Suppression and Enhancement of Spontaneous Emission

Does the rate of spontaneous emission of an excited atom vary with its environment? If the spontaneous emission is the property of the excited atom *exclusively*, then the rate of emission will not vary with the environment. But the spontaneous emission is the result of interaction between the excited atom with the vacuum fluctuation of the radiation field, and the latter does depend on the environment. Consider an excited atom is situated between two parallel conducting plates or in a cavity, the vacuum electric field E has to satisfy the boundary condition. This will influence the rate of emission. The influence of the "zero point energy" of the electromagnetic field has been studied in the late 1940s, such as the Casimir-Polder force and Casimir effect.[6] The Casimir effect was only observed in 1958. Although the principle of the suppression and enhancement of the rate of spontaneous emission has become clear, but experimental demonstration was realized only in the mid-1980s. The wave length of a typical optical photon is of the order of μm, actually a fraction of it. Its energy is about 1 eV, and its frequency several times 10^{14}Hz. In order to influence the emission rate of an atom, the distance between two mirrors has to be of the order of μm. If a larger distance between the mirror is used, then the corresponding wave length of the emission has to be chosen. This can be realized by preparing Rydberg states of the atom. The first observation on the suppression of emission rate was due to Kleppner's group at MIT in 1985 [14]. As shown in Fig. 8.13 the distance between the mirrors is 0.2 mm, Cs atoms are excited to a circular Rydberg state before they enter the space between the mirrors. The radiation emitted by dipoles *parallel* to the surface of the mirrors forms a stationary wave of maximum wave length 0.4 mm. The distance between the mirrors d is adjusted. Ionization detector is used to measure the atoms leaving the space between mirrors, with the voltage applied adjusted just enough to ionize the atom in its initial Rydberg state. In case the detector gives a signal, the atom is still in its initial state without emitting a

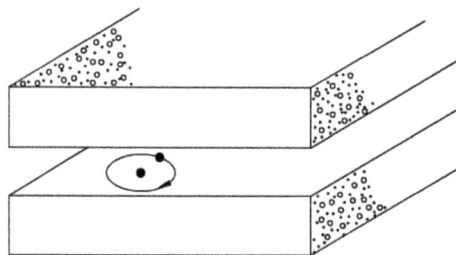

Fig. 8.13   An atom in the space between the mirrors. From Ref. [3].

---

[6]See Sections 8.9 and 8.11.

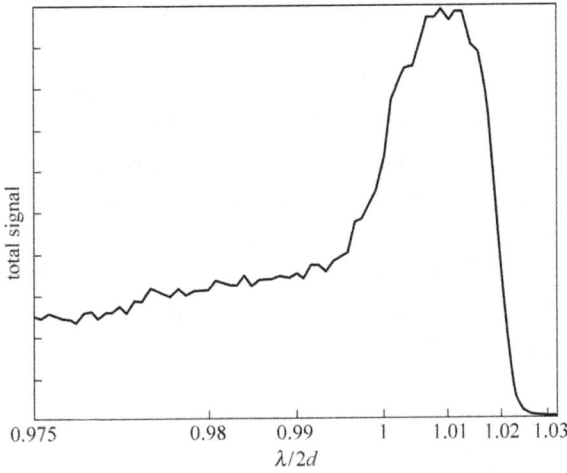

Fig. 8.14   The suppression of emission rate as a function of the distance between mirrors. Reprinted with permission from Randall G. Hulet, Eric S. Hilfer, and Daniel Kleppner, *Phys. Rev. Lett.* 55, 2137, 1985. Copyright (2017) by the American Physical Society.

photon. Figure 8.14 shows that when $\lambda/2d$ reaches the critical value 1 the emission rate is suddenly suppressed. From the time required for an atom to traverse the space between the mirrors, the emission rate is calculated to be reduced to $1/20$ its value in free space.

The suppression of radiation rate can be calculated by estimating the density of states per unit volume in a one-dimensional cavity. The expression given by Haroche and Raimond[7] is

$$D_c = \frac{\Delta\omega_c/2}{\pi V} \frac{1}{(\Delta\omega_c/2)^2 + (\omega - \omega_0)^2}. \qquad (8.3.1)$$

Where $\Delta\omega_c$ is the bandwidth of the cavity, related to the quality factor Q by $Q = \omega/\Delta\omega_c$. For $d \ll \lambda$, i.e., $\omega \gg \omega_0$ and a large detuning, (8.3.1) gives $D_c \approx \frac{1}{VQ\omega}$, and consequently

$$\Gamma_c = \frac{D_c}{D_0}\Gamma_0 \approx \frac{\lambda^3}{VQ}\Gamma_0, \qquad (8.3.2)$$

where we have used $D_0 = \rho_0/V = \omega^2/\pi^2 c^3$ given by (8.1.31).

---

[7]See Adv. in At. & Mol. Phys., vol. 20, 1985.

Meschede's group at the University of Munich [14] used a much smaller distance 1.1 μ between mirrors with a cut-off wave length 2.2μ. Cs atom excited to state 5d was used. The wave length of transition $5d \rightarrow 6p$ is 3.5 μm in the near infrared. The atoms passing through the space between mirrors were found to remain in the excited state. In addition a magnetic field having a component in the direction parallel to the surface of the mirrors was introduced. This component will induce precession of the dipole moment around it such that the dipole moment will acquire a component *perpendicular* to the mirror surface. Radiations from such a fluctuating dipole will propagate parallel to the mirror surface without being limited by the boundary condition, and its emission rate is no more suppressed. Figure 8.15 shows the relation between the survival of excited state and the direction of the magnetic field, the angle being that between the field and the normal to the mirror surface. The direction of the

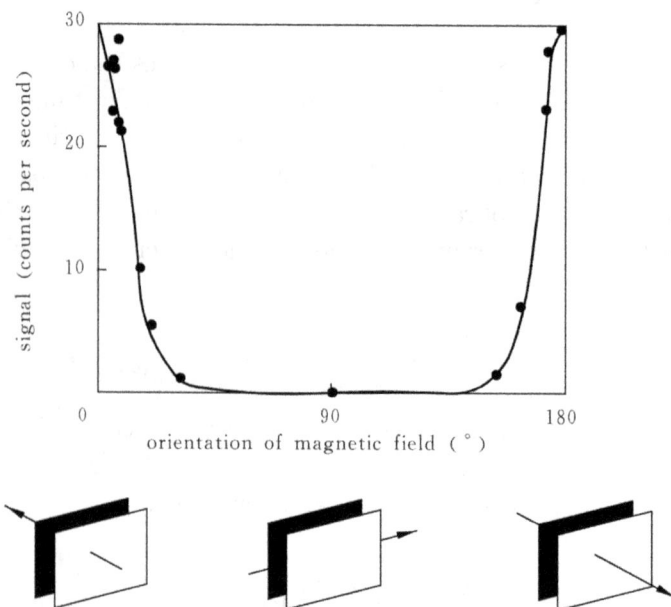

Fig. 8.15   Relation between the survival of the excited state and the direction of the magnetic field. Reprinted with permission from W. Jhe, A. Anderson, E. A. Hinds, D. Meschede, L. Moi, and S. Haroche, *Phys. Rev. Lett.* 58, 666, 1987. Copyright (2017) by the American Physical Society.

field relative to the mirror surface is shown in the pictures under the curve. From the figure we see that when the angle increases from 0 or decreases from π the survival of the excited states declines fast.

A resonant cavity can also enhance the vacuum fluctuation of the radiation field just as a cavity with a dimension smaller than the cut-off wave length suppresses it. The property of a cavity is determined by its Q value. The density of modes $(\Delta\omega_c)^{-1}$ represent the lifetime of a photon in the cavity. the structure of parallel mirrors is open at the ends and therefore it cannot have a high Q value. A superconducting cavity can have a Q value as high as 10^{11}, and the corresponding lifetime of a photon can be a few tenth of a second. An atom in the excited state in a high Q cavity undergoes Rabi oscillation. While an atom in magnetic resonance undergoes Rabi oscillation due to its interaction with an external field, the atom in a cavity undergoes Rabi oscillation due to its interaction with the photon emitted by the atom itself. The spontaneous emission in the free space is irreversible, while the Rabi oscillation in a cavity is reversible. In order to observe the enhancement of spontaneous emission in a cavity, a low Q cavity is to be used, such that the photon emitted can be promptly absorbed by the cavity wall and not to be reabsorbed by the atom while it is still in the cavity. The emission rate is enhanced by resonance. Let Γ_0 be the radiation rate in free space, Γ_c be the rate in the cavity, λ be the wave length of the photon and V be the cavity volume, then we have from the formula given by Haroche and Raimond (8.3.1) for the resonance case

$$\Gamma_c \approx \Gamma_0 \frac{Q\lambda^3}{V}. \tag{8.3.3}$$

Haroche's group at the l'Ecole Normale Supérieure [15] observed a 500 fold increase of the emission rate of the spontaneous emission from a Rydberg state of Na atom in a cavity of dimension of the *mm* order.

8.4 The Micromaser

When the lifetime of a photon $(\Delta\omega_c)^{-1} = Q/\omega$ in the cavity is larger than the time interval between the successive arrivals of atoms in the cavity, the photon emitted by an atom can interact with a late-arrived atom. When the field in a cavity accumulates, the coupling between the cavity and the atom characterized by $\Omega_0 = \Omega\sqrt{n+1}$ (8.2.8) increases accordingly and finally reaches a steady condition. Such a system is a new type of maser, called a micromaser [16]. Even when the stream of atoms is fairly weak, e.g., 100 atoms per second, the effect of the maser happens. For such a stream of atoms there can be no more than one atom in the cavity. Let the time of interaction of the atom with the cavity be τ, the temperature of the cavity be zero, the Q value be infinite, and the ionization detectors be ideal, the probability $P_n(m)$ that there are n photons in the cavity after m atoms passed was calculated by J. Krause, M. O. Scully and H. Walther [17]. From (8.2.9) we see that the probability that the atom-cavity system is in the state $|e, n\rangle$ after an atom in the state $|e\rangle$ has passed through the cavity with n photons in it is

$$P_{e,n}(\tau) = \cos^2 \Omega\sqrt{n+1}\,\tau \equiv c(n) \qquad (8.4.1)$$

and the probability that the system is in the state $|g, n+1\rangle$ is

$$P_{g,n+1}(\tau) = \sin^2 \Omega\sqrt{n+1}\,\tau \equiv s(n). \qquad (8.4.2)$$

When $m - 1$ atoms have passed through the cavity, the probability that there are n photons in the cavity is denoted by $P_n(m-1)$. When next atom (the m-th) has passed the probability that there are n photons in it is

$$P_n(m) = c(n)P_n(m-1) + s(n-1)P_{n-1}(m-1). \qquad (8.4.3)$$

This is a recurrence relation. Let there be no photon in the cavity initially, $P_0(0) = 1$. When one atom has passed, we have $P_0(1) = c(0)$, $P_1(1) = s(0)$. After two atoms have passed, we have

$$P_0(2) = c(0)P_0(1) = [c(0)]^2,$$
$$P_1(2) = c(1)P_1(1) + s(0)P_0(1) = s(0)[c(0) + c(1)],$$
$$P_2(2) = s(1)P_1(1) = s(0)s(1),$$

$$\cdots$$

Using the method of mathematical induction a general result was given in Ref. [11]:

$$P_n(m) = \prod_{i=0}^{n-1} s\,(i) \sum_{\substack{i_j=0 \\ (i_{m-1}\le\cdots\le i_n)}}^{n} \prod_{j=n}^{m-1} c\,(i_j), \qquad (8.4.4)$$

where

$$\sum_{\substack{i_j=0 \\ (i_{m-1}\le\cdots\le i_n)}}^{n} \prod_{j=n}^{m-1} c(i_j) = \begin{cases} \sum_{i_n=0}^{n}\sum_{i_{n+1}=0}^{i_n}\cdots\sum_{i_{m-1}=0}^{i_{m-2}}\prod_{j=n}^{m-1} c(i_j), & m>n \\ 1, & m=n \\ 0, & m<n, \quad \text{or } n<0 \end{cases}$$

Figure 8.16 shows $P_n(m)$ as a function of n for m taking values between 5 and 1000 for $g\tau = \Omega\sqrt{n+1}\,\tau = 0.4$. When the number of photons in the cavity reaches a value n_0 satisfying the condition that $\Omega\sqrt{n_0+1}\,\tau$ is an integral multiple of π, a steady state is reached,

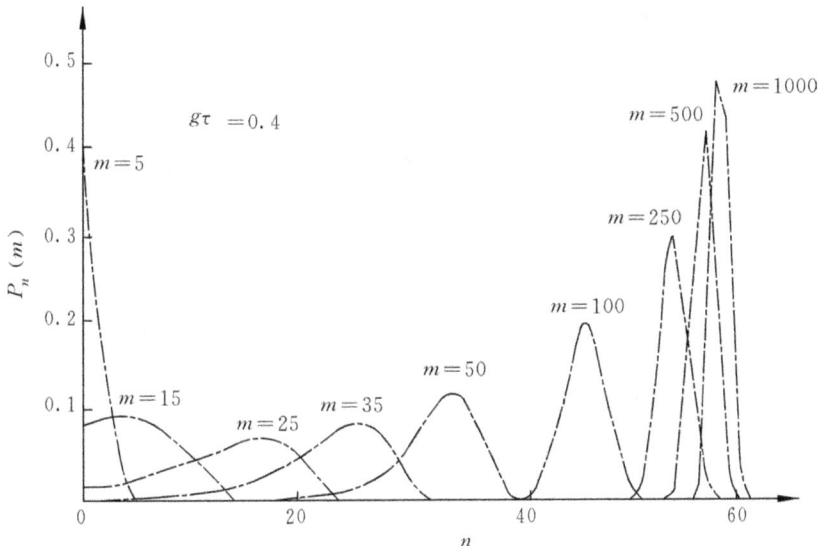

Fig. 8.16 The distribution in the number of photons in the cavity after m atoms have passed. Reprinted with permission from Joachim Krause, Marlan O. Scully, and Herbert Walther, *Phys. Rev. A* **36**, 4547, 1987. Copyright (2017) by the American Physical Society.

since by (8.2.29) the probability that an atom in state $|e\rangle$ entering and leaving the cavity remains in the excited state is unity, $P_{e,n_0}(\tau) = 1$. Such a state with completely definite number of photons (and hence with a completely indefinite phase) is highly non-classical, with a zero fluctuation of its amplitude, while for a classical field the fluctuation of amplitude is $\sqrt{\bar{n}}$.

Haroche's group [18] also created a maser with a two-photon process. To account for a two-photon process we have to restore the term $\frac{e^2}{2mc^2} \boldsymbol{A}^2$ in the Hamiltonian of interaction of the atom with the radiation field (8.1.21). Consider the energy level diagram of Rb atom shown in Fig. 8.17(a). The Rydberg state $40\,S_{\frac{1}{2}}$ decays predominantly to $39\,P_{\frac{3}{2}}$ in free space, but in a cavity tuned to

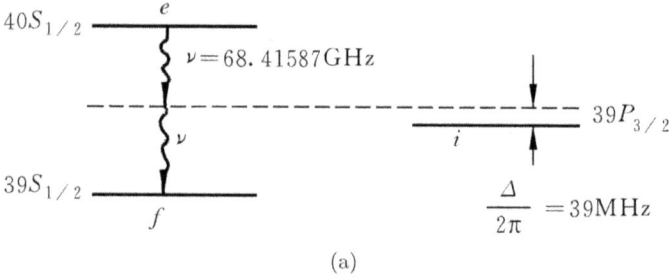

(a)

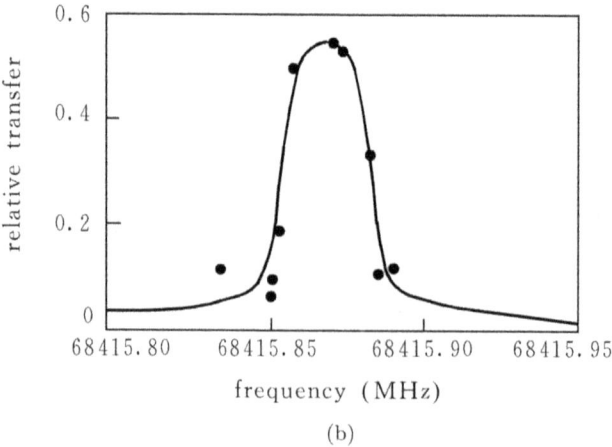

frequency (MHz)

(b)

Fig. 8.17 (a) Rydberg levels of Rb atom; (b) Two-photon maser. Reprinted with permission from M. Brune, J. M. Raimond, P. Goy, L. Davidovich, and S. Haroche, *Phys. Rev. Lett.* **59**, 1899, 1987. Copyright (2017) by the American Physical Society.

$\nu = 68.41587\,\text{GHz}$ this transition is suppressed, and the two-photon transition to $39\ P_{\frac{3}{2}}$ becomes possible, since the difference in energy of these two levels is equal to twice the resonant frequency of the cavity. Figure 8.17(b) shows the ratio of the number of atoms in the state $39\ P_{\frac{3}{2}}$ to the number in the state $40\ S_{\frac{1}{2}}$ for the atoms leaving the cavity as the function of the resonant frequency of the cavity. It is seen that the two-photon maser is realized around the resonance.

8.5   The Inverse Stern-Gerlach Effect

We discussed the energy shift of an atom in the state $|e\rangle$ or $|g\rangle$ upon entering the cavity. This energy is given by the Jaynes-Cummings Hamiltonian (8.2.3), including the energies of the cavity field, the internal state of the atom and the interaction between the atom and the field. The center-of-mass energy of the atom is not included. If this energy is included, then the total energy of the atom-cavity system must be conserved. Therefore, the energy shift given in Section 8.2.2 leads to a corresponding change of the center-of-mass energy, i.e., the atom experiences a force upon entering into the cavity. Let us assume that the motion of the center of mass (with position vector $\boldsymbol{R}$) is slow such that the internal motion can follow the change of $\boldsymbol{R}$, then the adiabatic approximation can be used. In the Born-Oppenheimer approximation the potential energy of the center-of-mass motion is just the energy of the internal motion (the fast motion) of the electron. The minus gradient of this potential energy is the force on the atom. C. Cohen-Tannoudji pointed out in 1992 that an atom entering into a light field with intensity gradient experiences a dipole force, here the light field can be that due to a small number of photons. As we can see from (8.2.17), (8.2.18) and Fig. 8.2 that the atom in state $|g\rangle$ does not experience any force on entering into an empty cavity, and an atom in state $|e\rangle$ experiences an attractive force for $\delta > 0$, and a repulsive force for $\delta < 0$. In a system of Rydberg atom-microwave cavity the Rabi coupling $\mathscr{G}$ is typically of the order 10^{-10} eV, the vacuum force of the cavity can be significant only when the velocity of the atom is fairly low (e.g., 1m/s, equivalent to about a temperature of 100 mK). The cavity vacuum can also produce a shift of energy level, called the "cavity Lamb shift".

Suppose that the cavity has a large positive detuning δ_0 initially. The force an atom in state $|e\rangle$ feels is small when it enters the cavity with a sufficiently small velocity. When the atom approaches the center of the cavity, the detuning is suddenly changed to a small negative value. Then the potential well is deepened such that the atom can be captured there. Haroche succeeded in 1991 to capture atoms with kinetic energy of order μK.

For the case of complete resonance ($\delta = 0, \theta = \frac{\pi}{4}$), Eq. (8.2.11) gives

$$\left.\begin{aligned}
|+,n\rangle &= \frac{1}{\sqrt{2}}\left(|e,n\rangle - |g,n+1\rangle\right), \\
|-,n\rangle &= \frac{1}{\sqrt{2}}\left(|e,n\rangle + |g,n+1\rangle\right),
\end{aligned}\right\}
\tag{8.5.1}$$

The states $|e,n\rangle$ and $|g,n+1\rangle$ are degenerate before the atom enters the cavity. Consider an atom in the state $|e\rangle$ entering the cavity with a definite number n of photons, the state of the system and its energy is respectively

$$|e,n\rangle = \frac{1}{\sqrt{2}}(|+,n\rangle + |-,n\rangle),
\tag{8.5.2}$$

and

$$E_{\pm,n} = (n+1)\hbar\omega \pm \hbar\Omega\sqrt{n+1}.
\tag{8.5.3}$$

During the motion of the atom the component $|+,n\rangle$ is repelled. If the center-of-mass kinetic energy is sufficiently small (smaller than the barrier height), this component can be reflected, while the component $|-,n\rangle$ is attracted and proceeds to move forward. In this way the coherent linear superposition of the wave function of the atom is "separated" into a reflected and a transmitted component (Englert, 1991).

The center-of-mass motion of the atom is governed by the Schrödinger equation

$$i\hbar\frac{\partial}{\partial t}\Psi_{e,n}\left(\boldsymbol{r},t\right) = -\frac{\hbar^2}{2m}\nabla^2\Psi_{e,n}\left(\boldsymbol{r},t\right) + E_{\pm,n}\Psi_{e,n}\left(\boldsymbol{r},t\right).
\tag{8.5.4}$$

Suppose that the width of the wave packet is much smaller than the dimension of the cavity initially, then the evolution of the wave packet is guided by the classical trajectory determined by the potential and at the same time the wave packet broadens. In case the initial kinetic energy of the center-of-mass is larger than the barrier height, the probability that the wave packet passes through the cavity is close to unity. The potential determining the motion of the center-of-mass depends on the number of photons in the cavity. The evolution of the cavity-atom system is adiabatic, and therefore the atom remains in the state $|e\rangle$ and the number of photons in the cavity does not change.

In the Stern-Gerlach experiment the trajectories of particles passing through the field with a gradient are determined by their spin. The gradient of the field serves as an instrument for analyzing the internal quantum state of the particles. In our example the trajectory of the atom depends on the gradient of the cavity field depending on the number of photons. The latter is the quantum state to be analyzed, and the atom serves as the analyzer. This example can be called the *inverse Stern-Gerlach experiment*. If the field is in a Glauber coherent state, i.e.,

$$|\alpha\rangle = \sum_n e^{-|\alpha|^2/2} \frac{\alpha^n}{n!} |n\rangle \equiv \sum_n c_n |n\rangle, \qquad (8.5.5)$$

where the average photon number $\bar{n} = |\alpha|^2$, the fluctuation of the photon number is $\Delta(n) = \sqrt{\bar{n}}$, the phase fluctuation of α is $\Delta(\phi) = 1/\Delta(n)$. Such a field in the cavity can be produced by connecting the cavity to a microwave source, and disconnecting it just before the atom arrives at the cavity. The wave packet of the atom is correspondingly

$$\Psi_e(\mathbf{r}, t) = \sum_n c_n \Psi_{e,n}(\mathbf{r}, t), \qquad (8.5.6)$$

where c_n is defined by (8.5.5), $\Psi_{e,n}(\mathbf{r}, t)$ is the solution of (8.5.4). The components with different n will eventually separate. Detection of the atom in some position causes the reduction of the wave packet,

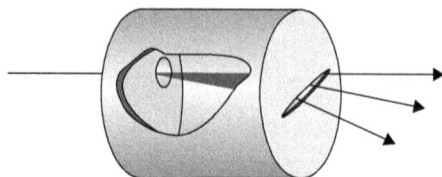

Fig. 8.18   The inverse Stern-Gerlach experiment. Redrawn from Ref. [2].

and the number of photons in the cavity completely determined. Figure 8.18 is the scheme of the experiment.

8.6   The Atom-Cavity Dispersive Phase Shift Effects

When the detuning is large enough, the atom passes through the cavity without energy exchange. The atom experiences an energy shift depending on the detuning δ according to (8.2.17) and (8.2.18). The interaction between the atom and the cavity leads to a phase shift for the cavity field, and this is called the "index effect" of the atom.[8] At the same time the wave packet of the atom also experiences a phase shift depending on the number of photons in the cavity, and the cavity is an "index medium" for the atom. Figure 8.9 shows the phase shift of the atom wave packet due to its interaction with the cavity containing a certain number of photons. The wave packet entering and leaving the cavity is represented by a and b respectively. The dotted line at b shows the would-be position of the packet in the absence of the cavity. The eigenstate $\Psi_{e,n}(\boldsymbol{r})$ of the atom in the internal state $|e\rangle$ coupled to the cavity field with n photons satisfies the energy eigenequation corresponding to the Schrödinger equation (8.5.4)[9]

$$\nabla^2 \Psi_{e,n}(\boldsymbol{r}) + \frac{2M}{\hbar^2}\left[E_{n,K} - E_{\pm,n}(\boldsymbol{r})\right]\Psi_{e,n}(\boldsymbol{r}) = 0, \qquad (8.6.1)$$

---

[8]The index refers to the refractive index. The atom serves as a medium, which gives the field a phase shift as the latter interacts with it. The inverse is also true. See (8.6.5) and (8.6.7).

[9]In this section the state of the atom is denoted by ψ, that of the field by Φ, and the state of the atom-cavity system by Ψ.

where $E_{\pm,n}(\boldsymbol{r})$ is the potential and $E_{n,K}$ is the eigenvalue

$$E_{n,K} = \left(n + \frac{1}{2}\right)\hbar\omega + \frac{\hbar\omega_0}{2} + \frac{\hbar^2 K^2}{2M}, \tag{8.6.2}$$

the three terms of it represent the energy of the field, the internal energy of the atom and the kinetic energy of the center of mass motion of the atom with momentum $\hbar K$ respectively. The state of the atom entering the cavity becomes $|+, n\rangle$ for $\delta < 0$, and its energy is given by (8.2.16):

$$E_{+,n} = (n+1)\hbar\omega - \frac{\hbar}{2}\delta - \hbar\frac{\Omega^2(\boldsymbol{r})(n+1)}{\delta}, \tag{8.6.3}$$

where we have changed $|\delta|$ to $-\delta$ for the present case. For $\delta > 0$ the atom is in the state $|-, n\rangle$ with the energy

$$E_{-,n} = (n+1)\hbar\omega - \frac{\hbar}{2}\delta - \hbar\frac{\Omega^2(\boldsymbol{r})(n+1)}{\delta}. \tag{8.6.4}$$

The form of (8.6.3) is now identical to that of (8.6.4) with the understanding that δ is negative in (8.6.3) while it is positive in (8.6.4). Therefore, we have

$$\frac{2M}{\hbar^2}\left(E_{n,K} - E_{\pm,n}(\boldsymbol{r})\right) = K^2 + \frac{2M\Omega^2(\boldsymbol{r})(n+1)}{\hbar\delta},$$

and the energy eigenequation becomes

$$\nabla^2 \Psi_{e,n}(\boldsymbol{r}) + N_{e,K}^2 K^2 \Psi_{e,n}(\boldsymbol{r}) = 0, \tag{8.6.5}$$

where

$$N_{e,K} = \left[1 + \frac{2M}{\hbar K^2}\frac{\Omega^2(\boldsymbol{r})(n+1)}{\delta}\right]^{1/2} \approx 1 + \frac{M}{\hbar K^2}\frac{\Omega^2(\boldsymbol{r})(n+1)}{\delta}. \tag{8.6.6}$$

Comparing (8.6.5) with the Maxwell wave equation for light propagation in a medium with refractive index $N(\boldsymbol{r})$,

$$\nabla^2 \psi_{e,n}(\boldsymbol{r}) + N^2(\boldsymbol{r})K^2\psi(\boldsymbol{r}) = 0, \tag{8.6.7}$$

we see that their form is identical. If the atom entering the cavity is initially in the state $|g\rangle$, then we have

$$N_{g,K}(\boldsymbol{r}) \approx 1 - \frac{M}{\hbar K^2} \frac{\Omega^2(\boldsymbol{r})n}{\delta}. \tag{8.6.8}$$

Comparing with the optical case we see that the phase shift for the atomic wave function caused by the refractive index $N_j(\boldsymbol{r})$ is

$$\Delta\phi_j(n) = \int [N_{j,K}(\boldsymbol{r}) - 1] K \mathrm{d}z, \tag{8.6.9}$$

where $j = (e,g)$, and the integral is taken along the atomic trajectory in the cavity. Substituting (8.6.6) and (8.6.8) into (8.6.9) and denoting $\frac{M}{\hbar K}$ by $\frac{1}{v}$ where v is the velocity of the atom, we obtain

$$\left.\begin{aligned} \Delta\phi_e &= (n+1)\,\varepsilon \\ \Delta\phi_g &= -n\varepsilon \end{aligned}\right\} \tag{8.6.10}$$

where

$$\varepsilon = \frac{1}{v\delta} \int \Omega^2(\boldsymbol{r})\,\mathrm{d}z. \tag{8.6.11}$$

Consider a quasi-monoenergetic atomic wave packet $\psi_{g,\bar{K},n}(\boldsymbol{r},t)$ formed by superimposing solutions $\Psi_{g,n}$ to the energy eigenequation, their K values being taken to be close to the average value $\bar{K}$. In this case the index dispersive effect of the cavity field can be neglected and the phase shift of the different components in the superposition is basically the same. We have then

$$\psi_{g,\bar{K},n}(\boldsymbol{r},t) = \psi_{g,\bar{K},0}(\boldsymbol{r},t)\,\mathrm{e}^{-in\varepsilon}. \tag{8.6.12}$$

In the right-hand side, $\psi_{g,\bar{K},0}$ is the wave packet in the absence of the cavity field. The first order effect of the cavity field is providing a phase shift to the atomic wave packet without changing appreciably its center position. In Fig. 8.19 the solid line at b corresponds to $\psi_{g,\bar{K},n}$ and the dashed line corresponds to $\psi_{g,\bar{K},0}$. In the inverse Stern-Gerlach effect where the centers of wave packets with different values of n follow different trajectories exhibiting a higher order effect, obtained when taking into account the dispersion (variation with K) of the matter wave index. The measurement of the phase shift

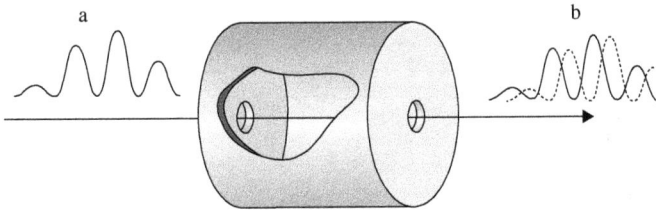

Fig. 8.19 Phase shift of the wave packet due to the cavity field. Redrawn from Ref. [2].

$\Delta\phi_j(n)$ is a more sensitive way of probing the atom-cavity interaction than the inverse Stern-Gerlach effect.

For the cavity field, the atom serves as an index medium. If the field is in a Glauber coherent state $|\alpha\rangle$ initially and an atom in the state $|g\rangle$ passes through the cavity, the state of the system is

$$|\Psi\rangle = \sum_n c_n \psi_{g,\bar{K},0}\left(\boldsymbol{r},t\right) e^{-in\varepsilon} |n\rangle, \qquad (8.6.13)$$

where the coefficients c_n is defined in (8.5.5). This state can be considered as the direct product of the unperturbed atom wave packet and the final state of the cavity field described by

$$|\Phi_g\rangle = \sum_n c_n e^{-in\varepsilon} |n\rangle = |\alpha e^{-i\varepsilon}\rangle, \qquad (8.6.14)$$

which is a phase shifted Glauber state. For an atom in the state $|e\rangle$ passing through the cavity, the field final state is

$$|\Phi_e\rangle = \sum_n c_n e^{i(n+1)\varepsilon} |n\rangle = e^{i\varepsilon} |\alpha e^{i\varepsilon}\rangle. \qquad (8.6.15)$$

The group led by Haroche carried out a measurement of the cavity induced atomic phase shift by Ramsey interferometry [19]. The experimental set up is shown in Fig. 8.20. The circular box CB prepares Rb atom in circular Rydberg state $n = 51$, R^+ and R^- are low Q cavities (called the Ramsey zones) fed by the microwave source S_2 with frequency ω_r including the frequency of transition $51c \rightarrow 50c$. The atom passes through the cavity R^- in a time interval of $\frac{1}{4}$ Rabi period such that an atom in state $|e\rangle$ becomes $\frac{1}{\sqrt{2}}|e\rangle - i\frac{1}{\sqrt{2}}|g\rangle$ on

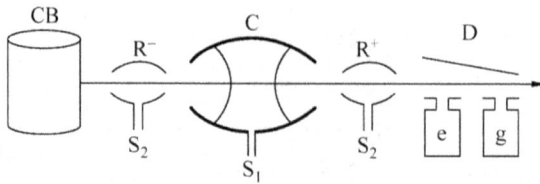

Fig. 8.20   Experimental setup for the detection of cavity induced atomic phase shift by Ramsey interferometry.

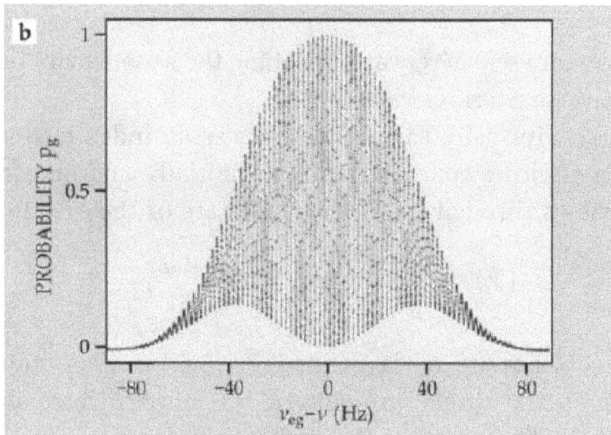

Fig. 8.21   Sweeping the microwave frequency through the transition frequency modulates of detecting ground state, laser cooled. Reprinted with permission from S. Haroche, M. Brune and J.-M. Raimond, *Physics Today* 66(1), 27, 2013. Copyright (2018) AIP Publishing. All rights reserved.

leaving the cavity (see Eq. (8.2.29)). The cavity C is fed by the source S_1 tuned off resonance such that the atom experiences a phase shift without undergoing a transition in passing through the cavity. When the atom passes through R^+ the components $|e\rangle$ and $|g\rangle$ mix again, and the atom is detected by the detector D. The signal received by the low energy detector g gives the transition rate $|e\rangle \rightarrow |g\rangle$. When the frequency ω_r is varied continuously a periodic change of the transition rate is observed, exhibiting the interference effect between atomic "paths". The transition $|e\rangle \rightarrow |g\rangle$ can take place either in the cavity R^- or in the cavity R^+. The different paths finally converge in the detector g giving rise to the Ramsey fringes. Figure 8.21 shows the Ramsey signal as a function of $(\omega_r - \omega_0)/2\pi$, the cavity C is

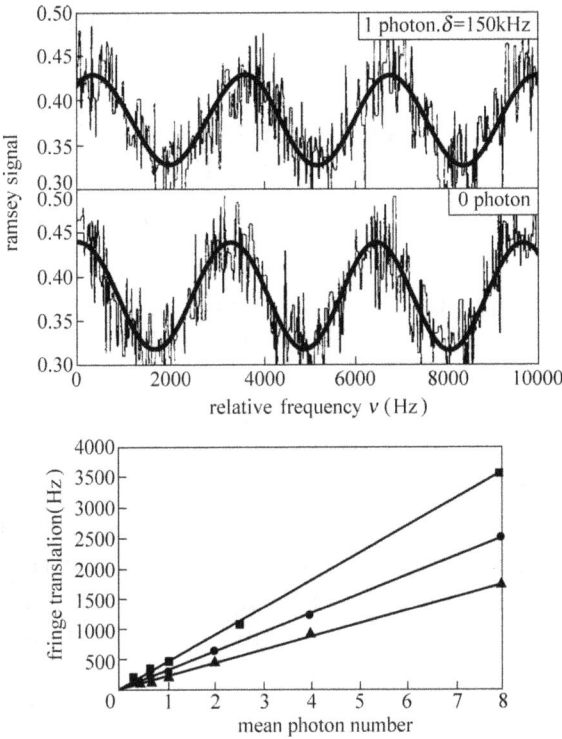

Fig. 8.22 The shift of Ramsey fringes due to presence of photons in the cavity. Reprinted with permission from [M. Brune, P. Nussenzveig, F. Schmidt-Kaler, F. Bernardot, A. Maali, J. M. Raimond, and S. Haroche, *Phys. Rev. Lett.* 72, 3339, 1994.] Copyright (2017) by the American Physical Society.

detuned by $\delta/2\pi = 150$ kHz from the atomic transition. When there are n photons existing in the cavity C, the components $|e,n\rangle$ and $|g,n\rangle$ of the superposition of the atomic wave function experience different phase shifts, namely $(n+1)\varepsilon$ and $-n\varepsilon$ respectively. This causes a shift in the Ramsey fringes. The shift is shown in Fig. 8.22, where the upper curve shows the Ramsey fringes when there is on the average one photon, and the lower curve shows the fringes with no photon in the cavity. Measuring the shift of Ramsey fringes gives the number of photons in the cavity. This is in fact a non-demolition measurement, since there is no change in the photon number during the experiment due to the dispersive coupling between the atom and the cavity field. Furthermore Haroche showed that m atoms are sufficient to measure a photon number between 0 and $2^m - 1$ in the cavity [20].

8.6.1   Quantum non-demolition detection (QND) of a single photon, QND of the birth and death of single photons

The ENS group has been developing the experimental setup and studying more sophisticated phenomena. In 1999 the ENS researchers detected single photons in a high finesse cavity in the quantum non-demolition regime, i.e., repeated measurements on the photon can be made during its lifetime, of the order of 1 ms [21]. The detection is realized using a Ramsey interferometer, the scheme of which is shown in Fig. 8.23(a) with the inset showing the three circular Rydberg states e, g, i of the probing atom (the meter atom). The atoms cross the cavity C one by one, the cavity mode stores the field to be measured (the signal field). The cavity is resonant at the transition frequency ν_{eg} of $e \Rightarrow g$. Atoms are subjected in the Ramsey zones R_1

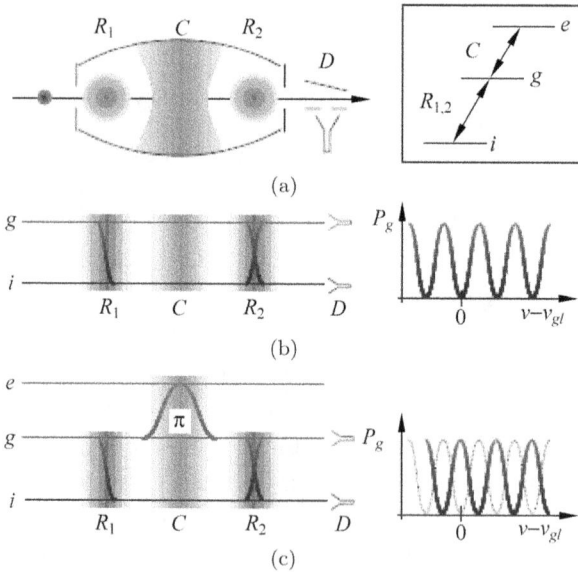

Fig. 8.23   Scheme of the interferometer with high-finesse cavity and Ramsey zones, and the resulting Ramsey fringes. Reprinted with permission from Macmillan Publishers Ltd: Nature (400, 239), copyright (1999).

and R_2 to pulses of frequency ν which is tuned in the vicinity of ν_{gi}, the transition frequency of $g \Rightarrow i$. Detector D downstream measures the state of the exit meter atom.

Let us assume that there is 1 photon in C and the meter atom enters C in the state g at the time $t = 0$. As the atom crosses C, the atom-cavity system undergoes a Rabi oscillation between $|g, 1\rangle$ and $|e, 0\rangle$ with frequency Ω. At time t the system is in the coherent superposition state

$$\cos \frac{\Omega}{2} t \; |g, 1\rangle + \sin \frac{\Omega}{2} t \; |e, 0\rangle . \tag{8.6.16}$$

If the total interaction time is $2\pi/\Omega$ (the atom experiences a 2π pulse), the atom ends up back to state g, after a full cycle of absorbing and re-emitting the photon. The photon remains unchanged, and the atom is left with a phase shift, becoming $e^{i\pi}|g, 1\rangle$. If the cavity is initially empty, then the system remains in the state $|g, 0\rangle$. Actually the atom has to cross the Ramsey zone R_1 first. The pulse of frequency ν (close to ν_{gi}) transforms $|g\rangle$ into $c_g|g\rangle + c_i|i\rangle$. In crossing the cavity C with 1 photon the system is in the state

$$c_g e^{i\pi} |g, 1\rangle + c_i |i, 1\rangle . \tag{8.6.17}$$

For an empty cavity the system is in the state

$$c_g |g, 0\rangle + c_i |i, 0\rangle . \tag{8.6.18}$$

The difference in the state is to be analyzed by the Ramsey zone R_2, in which the pulse mixes the state g and i once more. If the state of the meter atom exit from C is given by (8.6.18) (the cavity being empty) the mixing is shown in Fig. 8.23(b). After two mixings in R_1 and R_2 the exit meter atom can be either in state g or i. Let the probability of detecting the atom in g be P_g. The detected state g results from a quantum interference between two paths in which the atom crosses C in either g or i. When ν is scanned, the Ramsey fringes result. If there is one photon in C, the state of the exit atom is given by (8.6.17). The component $|g\rangle$ has now a phase shift π as shown in Fig. 8.23(c). The fringes are therefore shifted correspondingly.

The experiment was performed by using circular Rydberg states of Rb atoms, with $n = 49, 50, 51$ for i, g, e respectively. The cavity C

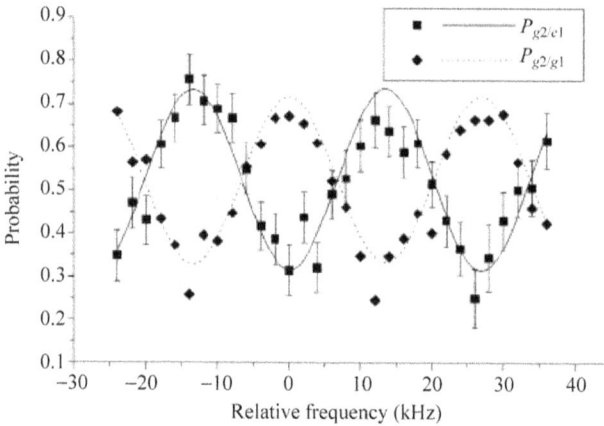

Fig. 8.24 Preparing and detecting one photon. Reprinted with permission from Macmillan Publishers Ltd: Nature (400, 239), copyright (1999).

is resonant at $e \Rightarrow g$ transition frequency of 51.1 GHz. The photon lifetime is 1 ms, long enough to permit repeated field measurements. The Ramsey zone cavities are resonant at the $g \Rightarrow i$ transition frequency of 54.3 GHz. The setup is cooled to 0.6 K or 1.2 K (average numbers of thermal photons is 0.02 and 0.15 respectively). The results of preparing and detecting one photon is represented in Fig. 8.24. A first atom prepares 0 or 1 photon and a second atom performs the interferometer recordings of conditional probabilities $P_{g2/e1}(\nu)$ (squares) and $P_{g2/g1}(\nu)$ (diamonds) of detecting the second atom in g provided the first has been found in e or g (leaving 0 or 1 photon in C respectively). These probabilities are reconstructed as a function of ν by averaging 250 correlated atom counts for each ν value.

QND measurements for one or zero photons can be used to lead to quantum logic gates based on cavity QED as well as to multi-atom entanglement.

While the photon lifetime of 1 ms is sufficient for it to be detected in QND regime, the record of birth, life and death of a single photon is much more demanding. Such a life story must be recorded in QND regime (i.e., many repeated measurements), and the lifetime of the photon in the cavity has to be extended to the order of a few seconds. To address this challenge, the ENS researchers had to improve the Q

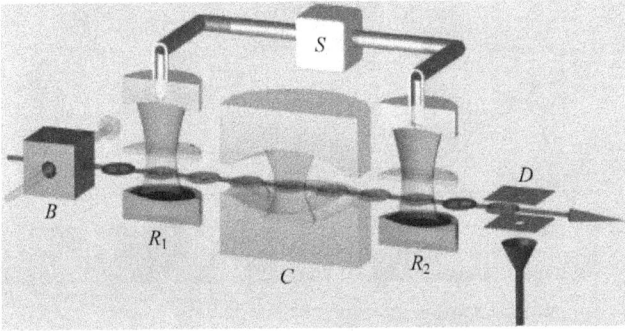

Fig. 8.25   Experimental Set-up. From Ref. [22].

factor of the cavity 100 fold up to 5.6×10^9. The sensor is again Rb atom in the $n = 50$ circular Rydberg state. The experimental setup is shown in Fig. 8.25. The cavity is cooled to $0.8\,\mathrm{K}$ and is enclosed in a box in order to shield it from thermal and static magnetic field. The cavity has a detuning from the $e \Rightarrow g$ transition, used to define the dispersive phase shift, which serves as the response of the sensor atom, depending on the number of photons in the cavity. The information of the phase shift, and eventually the photon number n, is to be read by the Ramsey interferometer. The Ramsey zone R_1 prepares the atom in the state $\frac{1}{\sqrt{2}}(|g\rangle + |e\rangle)$, which evolves at the exit of C into $\frac{1}{\sqrt{2}}(|g\rangle + e^{i\Phi(n,\delta)}|e\rangle)$, where Φ is the dispersive phase shift and δ is the detuning. When the detuning is set at a value such that $\Phi(1,\delta) - \Phi(0,\delta) = \pi$, the Ramsey pulse in R_2 ideally brings the atom into g if $n = 0$, and into e if $n = 1$. Therefore when e is recorded in the downstream detector, the implication is that there is a photon in the cavity, and when g is recorded, the cavity is empty. The cavity field is a superposition of $|0\rangle$ and $|1\rangle$, and the measurements on successive atoms projects the state onto $|0\rangle$ or $|1\rangle$. Since the measurements are carried out in QND regime, they can be repeated many times during the lifetime of the cavity field. The results on the birth, life and death of the a photon is shown in Fig. 8.26. The inset zooms into the region where the statistics of the detection events suddenly change, revealing quantum jumps of the cavity field due to thermal fluctuation. As seen from the figure there are misleading detection events which are not correlated to real photon number jumps due to

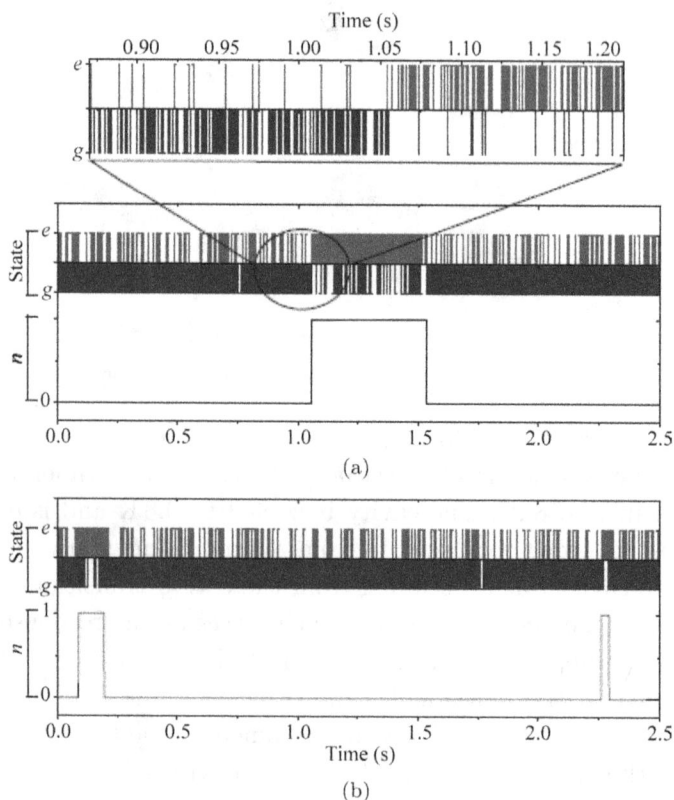

Fig. 8.26   Birth, life and death of a photon. (a) A long lived photon; (b) Two successive photons separated by a long time interval. Reprinted with permission from Macmillan Publishers Ltd: Nature (446, 297), copyright (2007).

imperfections in the setup. To reduce their influence on the inferred n value, a majority vote over 8 consecutive atoms is shown in the lower trace of Fig. 8.26(a), revealing the birth, life and death of an exceptionally long lived photon. In Fig. 8.26(b) similar signals showing two successive single photons, separated by a long time interval with cavity in vacuum.

In another experiment the researchers monitor the decay of a single-photon Fock state prepared at the beginning of each sequence. The QND probe atoms are sent across the cavity, leading to a single photon trajectory shown in Fig. 8.27(a) with the help of majority vote. Averages of 5, 15, and 904 similar quantum trajectories are

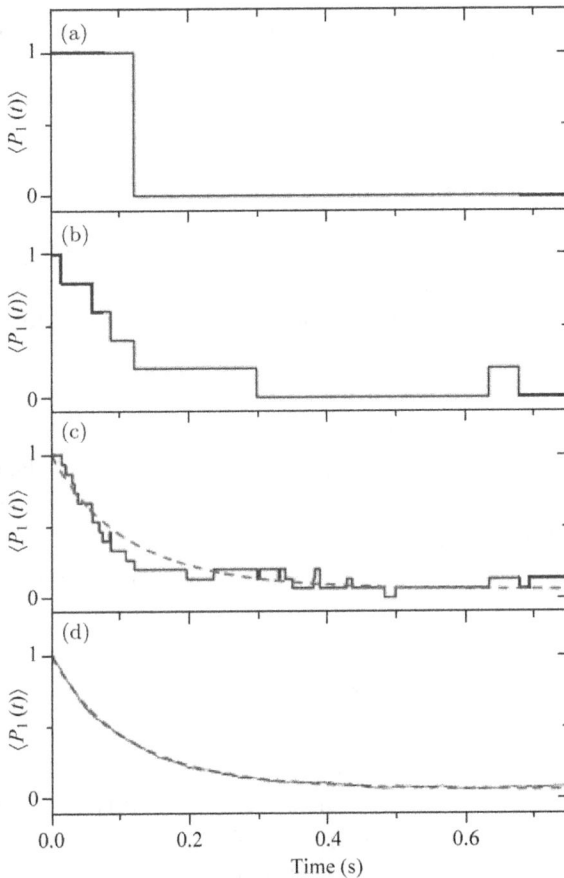

Fig. 8.27 Decay of the one-photon state. Reprinted with permission from Macmillan Publishers Ltd: Nature (446, 297), copyright (2007).

shown in Fig. 8.27(b-d), demonstrating a gradual transition from quantum randomness into a smooth exponential decay. Dotted curves represent solutions of the field master equation governing the decay of the photon in the cavity.

8.7 Entangled States in Cavity QED

Various entangled states can be constructed in cavity QED experiments with precision and versatility, and the weak interaction between the atom-cavity system and the environment is controllable.

Therefore, cavity QED has become a central paradigm for the study of open quantum systems, addressing the issue of decoherence on the boundary between quantum and classical mechanics, and has provided an arena for exploring quantum information manipulations [27].

The l'Ecole Normale Supérieure (ENS) group experiment discussed in Section 4.9.2 makes use of the fact that the phase shift of the cavity field depends strongly on the state (e or g) of the atom. This effect enables an initial superposition of atomic internal states to be transformed into an entanglement of the atomic internal states with the cavity field phase states (the phase cat). In a strong coupling regime, in which the vacuum Rabi frequency is much larger than both the dipole decay rate of the atom and the cavity field decay rate, we can expect the use of higher Q cavities enables to increase the average number of photons in the cavity, and the decoherence process as a function of the cat size can be explored in detail in the context of transition from the quantum regime to the classical one.

The predominant component of decoherence in cavity QED systems corresponds either to the absorption of photons by cavity walls or mirrors, or to the scattering into modes outside the cavity, called the "output channel". The system-environment entanglement is interrupted by a detection monitoring the output channel, and a collapse of wave function occurs. So far as the atom-cavity system is concerned, such output channel leads to a non-unitary evolution. The contribution of such channel to decoherence can be modeled by a stochastic dynamical evolution of a master equation for the density matrix, resulting in a "conditional quantum evolution" [24]. Detailed study of such processes may lead to a better understanding of how measurement and decoherence relate to the mesoscopic interface between quantum and classical mechanics.

The potential application of cavity QED to quantum information processing is based on the formation of entangled states. Turchette *et al.* [25] achieved an entangled state of fields in the cavity, when the phase shift of the "probe" depends on the mean photon number created by the "pump". Consider the case that the excited state $|e\rangle$

and the ground state $|g\rangle$ of the atom are coupled by the field with σ_+ polarization. The probe beam with σ_+ polarization and detuning Ω_a passing through the cavity experiences an additional phase shift due to atom-cavity interaction, while a beam with polarization σ_- receives a phase shift corresponding to an empty cavity. The difference of the phase shifts ϕ_a depends on the detuning Ω_a. When the latter is fixed, the phase shift has a non-linear dependence on the number of photons in the cavity controlled by a pump beam with detuning Ω_b and σ_+ polarization. It was found that the pump-probe coupling is manifest for mean photon number $n_b \approx 0.1$ with a 30% reduction of $|\phi_a|$ as n_b goes from 0.1 to 0.3. Let the probe-pump input be described by a direct product of coherent states

$$|\psi_{\text{in}}\rangle = |\alpha\rangle_a|\beta\rangle_b, \qquad (8.7.1)$$

with $|\alpha|$ and $|\beta| \ll 1$. The output state is

$$|\psi_{\text{out}}\rangle = |\alpha\rangle_a|\beta\rangle_b + |\alpha||\beta|[e^{i\Delta} - 1]|1\rangle_a|1\rangle_b, \qquad (8.7.2)$$

where $|1\rangle_a$ and $|1\rangle_b$ are one-photon state for the probe and pump respectively, and Δ is a parameter characterizing the entanglement, established through the intermediary atom.

A quantum phase gate can be implemented on the basis of the discussion above. The qubits are carried by polarization states $\sigma_\pm$ of the photons. When σ_+ polarized photon is coupled strongly with the atom while σ_- photon coupling can be neglected, the following table can be constructed:

$$
\begin{array}{ll}
\text{input} & \text{output} \\
|1^-\rangle_a|1^-\rangle_b \rightarrow |1^-\rangle_a|1^-\rangle_b \\
|1^+\rangle_a|1^-\rangle_b \rightarrow e^{i\phi_a}|1^+\rangle_a|1^-\rangle_b \\
|1^-\rangle_a|1^+\rangle_b \rightarrow e^{i\phi_b}|1^-\rangle_a|1^+\rangle_b \\
|1^+\rangle_a|1^+\rangle_b \rightarrow e^{i(\phi_a+\phi_b+\Delta)}|1^+\rangle_a|1^+\rangle_b.
\end{array}
$$

The parameters in the experiment [29] are: for Cs atom, $\bar{N} = 0.9$, $\Omega_a = +30\text{MHz}$, $\Omega_b = +20\text{MHz}$, $\phi_a \approx (17.5 \pm 1)°$, $\phi_b \approx (12.5 \pm 1)°$ and $\Delta = (16 \pm 3)°$.

The ENS group [26] achieved entanglement of two atoms in circular Rydberg states passing through a high Q microwave cavity successively. The first one is prepared in state $|e\rangle$ and the second one in $|g\rangle$. The initial state of the system is

$$|\Psi\rangle = |e_1, g_2, 0\rangle, \qquad (8.7.3)$$

where 0 denotes zero photon in the cavity. Let t_1 be the time for which the first atom interacts with the cavity such that $\Omega t_1 = \pi/2$, where Ω is the vacuum Rabi frequency. Between the passage of the two atoms across the cavity, the state of the system is described by

$$|\Psi'\rangle = \frac{1}{\sqrt{2}}(|e_1, g_2, 0\rangle - |g_1, g_2, 1\rangle), \qquad (8.7.4)$$

corresponding to maximum entanglement of the first atom and the cavity field. The second atom (prepared in state $|g\rangle$) enters the cavity after a time T, and the time of interaction with the cavity is set to be $t_2 = 2t_1$, so that $\Omega t_2 = \pi$. If the first atom left the cavity empty, the second atom passes through it without changing the field. If the first atom has emitted a photon, then the second atom absorbs it with certainty. The net result is that the final state of the system is described by

$$|\Psi_{\mathrm{EPR}}\rangle|0\rangle = \frac{1}{\sqrt{2}}(|e_1, g_2\rangle - |g_1, e_2\rangle)|0\rangle, \qquad (8.7.5)$$

where the atoms are entangled as an EPR state through the intermediary field, which is now decoupled from the atom. The foregoing discussion is confirmed by the ENS group experiment.

Similar procedures were used to create a tripartite entangled states of two atoms and the cavity field [27]. Rydberg Rb atoms are prepared in circular states, with principal quantum numbers 51, 50 and 49 for states $|e\rangle$, $|g\rangle$ and $|i\rangle$ respectively. The cavity is resonant on the $e \rightarrow g$ transition frequency of 51.1 GHz. Pulses from microwave source S with frequency 54.3 GHz corresponding to the $g \rightarrow i$ transition can be applied to the atom before it interacts with the cavity and after to produce programmed transitions. The full effective atom-cavity interaction time corresponds to 2π Rabi pulse.

Let the cavity be empty initially. An atom A_1 prepared in state $|e\rangle$ passes through the cavity in a time corresponding to a $\pi/2$ pulse, then the system becomes $\frac{1}{\sqrt{2}}(|e,0\rangle + |g,1\rangle)$. An atom A_2 in state $|g\rangle$ then enters, and is transformed into a state $\frac{1}{\sqrt{2}}(|g\rangle + |i\rangle)$ before entering the cavity by a pulse from S. It interacts with a 2π Rabi pulse in the cavity. If A_1 left one photon in the cavity, the A_2 coherence becomes $\frac{1}{\sqrt{2}}(-|g\rangle + |i\rangle)$. If A_1 left the cavity empty the A_2 coherence is unchanged. The resulting A_1-A_2-cavity quantum state is then

$$|\psi_{\text{triplet}}\rangle = \frac{1}{\sqrt{2}}[|e_1\rangle(|i_2\rangle + |g_2\rangle)|0\rangle + |g_1\rangle(|i_2\rangle - |g_2\rangle)|1\rangle] \quad (8.7.6)$$

which can be re-written as

$$|\psi_{\text{triplet}}\rangle = \frac{1}{2}[|i_2\rangle(|e_1,0\rangle + |g_1,1\rangle) + |g_2\rangle(|e_1,0\rangle - |g_1,1\rangle)] \quad (8.7.7)$$

describing an A_1-cavity EPR pair with phase conditioned by the state of A_2. Likewise, entangled states of two cavity modes were obtained [28]. These non-classical correlations serve to test theoretical models of entanglement and decoherence, and can have applications in quantum information processing.

In the foregoing we presented a brief introduction to the cavity QED with the aim of showing the basic physics. The readers can refer to [2] for further study.

8.8 Response of a System to External Disturbances, the Fluctuation-Dissipation Theorem

In the following, we are going to discuss the van der Waals force and the Casimir effect. We will need more understanding of the continuous media. For this purpose we present a brief introduction to the linear response theory and the fluctuation-dissipation theorem following the treatises by Landau and Lifshitz [29, 30].

8.8.1   The frequency dependence of the permittivity, the Kramers-Krönig dispersion relation

The permittivity ε and the permeability μ of a medium are constants in static problems. When the frequency of a time varying electromagnetic field is comparable to the frequency of oscillations of atoms or electrons, the polarization or magnetization of the medium undergoes fluctuations, leading to the change of ε and μ with the frequency (dispersion). From Maxwell equations we know that a time varying electromagnetic field necessarily varies with spatial position. The period of spatial variation is characterized by the wave length $\lambda = c/\omega$. If the frequency is sufficiently high such that the wave length $\lambda \lesssim a$ (the dimension of the atom), the macroscopic continuum description of the field breaks down. In the following, we restrict ourselves to the case that the frequency is high such that the dispersive effect is significant, but the continuum description of the field is still valid. The change in polarization of the medium involves only non-relativistic velocities $v \ll c$, therefore the relaxation time is a/v, and the corresponding wave length is $\lambda - ac/v \gg a$. This condition tells us that the continuum description is valid for relevant discussions. We assume that the field is not too strong such that the linear relation between $\boldsymbol{D}$ and $\boldsymbol{E}$ is valid. A time varying field $\boldsymbol{E}(t)$ leads to a time varying displacement vector $\boldsymbol{D}(t)$, and their most general relation can be written as

$$\boldsymbol{D}(t) = \boldsymbol{E}(t) + \int_0^\infty f(\tau)\, \boldsymbol{E}(t - \tau)\, \mathrm{d}\tau, \qquad (8.8.1)$$

where $f(\tau)$ is determined by the properties of the medium. The causality principle is already incorporated in (8.8.1): the displacement $\boldsymbol{D}(t)$ at any time can only depend on the value of $\boldsymbol{E}$ at earlier times, i.e., the lower limit of the integral is set equal to zero or the argument of $\boldsymbol{E}$ under the integral cannot be negative. We define the frequency dependent components of $\boldsymbol{D}$ and $\boldsymbol{E}$ by the Fourier transform

$$
\left.
\begin{aligned}
\boldsymbol{E}(t) &= \int_{-\infty}^{\infty} \boldsymbol{E}(\omega) \mathrm{e}^{-\mathrm{i}\omega t} \frac{\mathrm{d}\omega}{2\pi} \\
\boldsymbol{D}(t) &= \int_{-\infty}^{\infty} \boldsymbol{D}(\omega) \mathrm{e}^{-\mathrm{i}\omega t} \frac{\mathrm{d}\omega}{2\pi}
\end{aligned}
\right\}.
\tag{8.8.2}
$$

The relation between components $\boldsymbol{D}(\omega)$ and $\boldsymbol{E}(\omega)$ is obtained by using (8.8.1) and (8.8.2):

$$
\boldsymbol{D}(\omega)\mathrm{e}^{-\mathrm{i}\omega t} = \boldsymbol{E}(\omega)\mathrm{e}^{-\mathrm{i}\omega t} + \int_{0}^{\infty} \mathrm{d}\tau \boldsymbol{E}(\omega)\mathrm{e}^{-\mathrm{i}\omega(t-\tau)} f(\tau)
$$

$$
= \boldsymbol{E}(\omega)\mathrm{e}^{-\mathrm{i}\omega t} \left(1 + \int_{0}^{\infty} f(\tau)\,\mathrm{e}^{\mathrm{i}\omega\tau} \mathrm{d}\tau \right).
$$

Define $\varepsilon(\omega)$ by

$$
\varepsilon(\omega) = 1 + \int_{0}^{\infty} f(\tau)\,\mathrm{e}^{\mathrm{i}\omega\tau} \mathrm{d}\tau,
\tag{8.8.3}
$$

then we have

$$
\boldsymbol{D}(\omega) = \varepsilon(\omega)\boldsymbol{E}(\omega).
\tag{8.8.4}
$$

It is to be noticed that the time varying quantities $\boldsymbol{E}(t)$ and $\boldsymbol{D}(t)$ are real, but their Fourier components $\boldsymbol{E}(\omega)$ and $\boldsymbol{D}(\omega)$ as well as $\varepsilon(\omega)$ are complex. From (8.8.2) we know that the following conditions

$$
\boldsymbol{E}(-\omega) = \boldsymbol{E}^{*}(\omega), \quad \boldsymbol{D}(-\omega) = \boldsymbol{D}^{*}(\omega)
\tag{8.8.5}
$$

are necessary for $\boldsymbol{E}(t)$ and $\boldsymbol{D}(t)$ being real. Separating the real and imaginary parts of $\varepsilon(\omega)$

$$
\varepsilon(\omega) = \varepsilon'(\omega) + \mathrm{i}\varepsilon''(\omega),
\tag{8.8.6}
$$

we obtain from (8.8.3)

$$
\varepsilon(-\omega) = \varepsilon^{*}(\omega).
\tag{8.8.7}
$$

Expressing in terms of real and imaginary parts, this relation gives

$$
\varepsilon'(-\omega) = \varepsilon'(\omega), \quad \varepsilon''(-\omega) = -\varepsilon''(\omega),
\tag{8.8.8}
$$

i.e., the real part is an even function while the imaginary part is an odd function. When $\omega \to 0$, $\varepsilon''(0) = 0$ and $\varepsilon'(0) = \varepsilon(0) \equiv \varepsilon_0$, the static permittivity.

The lower limit of the integral in the right-hand side of (8.8.1) being zero guarantees the causality between $\boldsymbol{E}(t)$ and $\boldsymbol{D}(t)$, this will lead to the analytic properties of $\varepsilon(\omega)$ as a function of complex variable ω known as the "dispersion relation". The following discussion is restricted to a dielectric medium. The function $f(\tau)$ is finite for all values of τ. From (8.8.1) we see that τ is the difference between the arguments of $\boldsymbol{D}(t)$ and $\boldsymbol{E}(t - \tau)$. The value of $\boldsymbol{D}(t)$ depends on the value of $\boldsymbol{E}(t - \tau)$, but the degree of the dependence becomes weaker with increasing τ. This means that the dependence of $\boldsymbol{D}(t)$ on the value of $\boldsymbol{E}$ in the remote past becomes weaker and weaker, and $f(\tau)$ should be zero for τ sufficiently large (for instance, much larger than the relaxation time of the changing polarization). Consider the complex variable ω with real and imaginary parts separated:

$$\omega = \omega' + i\omega''. \tag{8.8.9}$$

The integral of (8.8.3) involves now a factor $e^{-\omega''\tau}$, which is an exponentially decreasing function of $\omega'' > 0$. Therefore, $\varepsilon(\omega)$ is single-valued and finite on the upper half plane of complex ω. Because of the properties of $f(\tau)$ stated above, Eq. (8.8.3) implies that $\varepsilon(\omega)$ is regular on the real axis, and approaches 1 for $\omega \to \infty$ because of the oscillating behavior of the factor $e^{-i\omega\tau}$ in the limit. We reach finally the conclusion that $\varepsilon(\omega)$ is analytic on the upper half plane including the real axis. It is the restriction $\tau \geq 0$ imposed by causality which makes $e^{-\omega''\tau}$ a decaying factor which leads to the analyticity of $\varepsilon(\omega)$. For a complex ω, (8.8.3) gives

$$\varepsilon^*(\omega) = 1 + \int_0^\infty f(\tau) e^{-i\omega^*\tau} d\tau = \varepsilon(-\omega^*). \tag{8.8.10}$$

This is a generalization of (8.8.7) for real ω. For $\omega = i\omega''$ on the imaginary axis, we have

$$\varepsilon(i\omega'') = \varepsilon^*(i\omega''), \tag{8.8.11}$$

which means that $\varepsilon(\omega)$ is real on the positive imaginary axis, but for ω on the negative imaginary axis no conclusion can be made because $\varepsilon(\omega)$ is not analytic there and a cut appears. From the fact that $\varepsilon(\omega')$ is an even function we infer that $\varepsilon(\omega)$ approaches 1 for $\omega \to -\infty$. When ω approaches infinity in *all* directions on the upper half plane, we find that the integral of (8.8.3) vanishes because of the behavior of $e^{-\omega''\tau}$ and $e^{-i\omega'\tau}$, and therefore $\varepsilon(\omega) \to 1$ in the limit. Consider the contour integral

$$\int_C \frac{\alpha(\omega)}{\omega - \omega_0} d\omega, \qquad (8.8.12)$$

where $\alpha(\omega) = \varepsilon(\omega) - 1$, and the contour C is shown in Fig. 8.28, where ω_0 is any fixed point on the real axis, and the large semicircle is to approach infinity. The integrand is analytic on the contour and in the area enclosed, and hence the integral vanishes according to Cauchy's theorem. The integral along the large (infinite) semicircle vanishes because $\varepsilon(\omega) \to 1$, the integral along the small semicircle equals to $-i\pi\alpha(\omega_0)$ by the residue theorem. Consequently the sum of the integral along the real axis (with an infinitesimal segment around ω_0 removed) and that along the small semicircle vanishes, i.e.,

$$P \int_{-\infty}^{\infty} \frac{\alpha(\omega)}{\omega - \omega_0} d\omega = i\pi\alpha(\omega_0),$$

where P means the principal value of the integral. Changing the notation ω to ξ and ω_0 to ω, we have

$$i\pi\alpha(\omega) = P \int_{-\infty}^{\infty} \frac{\alpha(\xi)}{\xi - \omega} d\xi.$$

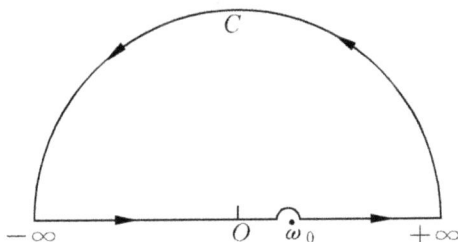

Fig. 8.28   The contour of integration of (8.8.12).

Separating the real and imaginary parts of α, we have

$$\left.\begin{array}{l} \alpha'(\omega) = \dfrac{1}{\pi} P \displaystyle\int_{-\infty}^{\infty} \dfrac{\alpha''(\xi)}{\xi - \omega} d\xi \\[4mm] \alpha''(\omega) = -\dfrac{1}{\pi} P \displaystyle\int_{-\infty}^{\infty} \dfrac{\alpha'(\xi)}{\xi - \omega} d\xi \end{array}\right\}. \tag{8.8.13}$$

This is a relation between the real and imaginary parts of α, called the dispersion relation (Kramers and Krönig, 1927). Expressing the dispersion relation in terms of $\varepsilon(\omega)$ we have

$$\left.\begin{array}{l} \varepsilon'(\omega) - 1 = \dfrac{1}{\pi} P \displaystyle\int_{-\infty}^{\infty} \dfrac{\varepsilon''(\xi)}{\xi - \omega} d\xi \\[4mm] \varepsilon''(\omega) - 1 = -\dfrac{1}{\pi} P \displaystyle\int_{-\infty}^{\infty} \dfrac{\varepsilon'(\xi) - 1}{\xi - \omega} d\xi. \end{array}\right\}. \tag{8.8.14}$$

Any function of a complex variable $\alpha(\omega)$ which is analytic on the upper half plane (including the real axis) and approaches zero for $\omega \to \infty$ in all directions satisfies the relations (8.8.13). $\alpha(\omega)$ is called the generalized susceptibility. The electric polarizability $\chi(\omega)$ and the magnetic susceptibility $\chi_m(\omega)$ belong to this category.

8.8.2   The correlation between fluctuations and the generalized polarizability

Consider a physical quantity describing a system under thermodynamic equilibrium or a subsystem of it. This quantity fluctuates around its average value. Let $x(t)$ be the difference between this quantity and its average value, then we have $\bar{x} = 0$. Generally speaking a correlation exists between $x(t)$ at different times, that means the value of x at t has an influence on the probability that x assumes a certain value at a later time t'. The statistical average defined by

$$\phi(t' - t) = \langle x(t)x(t') \rangle \tag{8.8.15}$$

is a measure of this correlation. $\phi(t'-t)$ depends only on the *difference* of t and t', and tends to zero for $t' - t \to \infty$. By definition, $\phi(t' - t)$

is an even function relative to an interchange of t and t'. For a quantum mechanical observable $\hat{x}(t)$ the definition of the correlation function is

$$\phi\left(t'-t\right) = \frac{1}{2}\left\langle \hat{x}(t)\hat{x}\left(t'\right) + \hat{x}\left(t'\right)\hat{x}(t)\right\rangle, \tag{8.8.16}$$

since in general $\hat{x}(t)$ and $\hat{x}(t')$ do not commute. The average above is taken in the quantum mechanical sense.

The spectral analysis of the correlation begins with the Fourier transform of $\hat{x}(t)$:

$$x(\omega) = \int_{-\infty}^{\infty} x(t)e^{i\omega t}dt, \tag{8.8.17}$$

$$x(t) = \int_{-\infty}^{\infty} x(\omega)e^{-i\omega t}\frac{d\omega}{2\pi}. \tag{8.8.18}$$

Substituting (8.8.18) into (8.8.16), we obtain

$$\phi\left(t'-t\right) = \int_{-\infty}^{\infty}\int_{-\infty}^{\infty}\left\langle x(\omega)x(\omega')\right\rangle e^{-i(\omega t + \omega' t')}\frac{d\omega d\omega'}{(2\pi)^2}. \tag{8.8.19}$$

Demanding the right-hand side being dependent on $t'-t$ only, we see that $\langle x(\omega)x(\omega')\rangle$ must contain a δ-function:

$$\langle x(\omega)x(\omega')\rangle = 2\pi\chi^2(\omega)\delta(\omega+\omega'), \tag{8.8.20}$$

where χ^2 is defined by (8.8.20) as the coefficient of the delta function. As the Fourier transform of a real $x(t)$, $x(\omega)$ is complex and satisfies

$$x(-\omega) = x^*(\omega).$$

From (8.8.20) we see that $\langle x(\omega)x(\omega')\rangle$ becomes $\langle x(\omega)x^*(\omega)\rangle$ for $\omega' = -\omega$, hence it is real and χ^2 is a real function of ω. By definition (8.8.19) $\langle x(\omega)x(\omega')\rangle$ is symmetrical with respect to the interchange of ω and ω', therefore

$$\chi^2(-\omega) = \chi^2(\omega). \tag{8.8.21}$$

Substituting (8.8.20) back into (8.8.19), we get

$$\phi(t) = \int_{-\infty}^{\infty} \chi^2(\omega) e^{-i\omega t} \frac{d\omega}{2\pi} \qquad (8.8.22)$$

and its inverse

$$\chi^2(\omega) = \int_{-\infty}^{\infty} \phi(t) e^{i\omega t} dt. \qquad (8.8.23)$$

From (8.8.15) we see that $\phi(0)$ is the mean square value of the fluctuating variable x. By using (8.8.21) and (8.8.22) we get

$$\langle x^2 \rangle = \int_{-\infty}^{\infty} \chi^2(\omega) \frac{d\omega}{2\pi} = \int_0^{\infty} \chi^2(\omega) \frac{d\omega}{\pi}, \qquad (8.8.24)$$

i.e., $\chi^2(\omega)$ is the spectral density of the mean square value of x. In the above discussion $x(t)$ is a classical variable. For a quantum mechanical observable we have

$$\frac{1}{2} \langle \hat{x}(\omega)\hat{x}(\omega') + \hat{x}(\omega')\hat{x}(\omega) \rangle = 2\pi \chi^2(\omega)\delta\left(\omega + \omega'\right). \qquad (8.8.25)$$

For simple systems, for instance quasi-stationary fluctuations, $\phi(t)$ and $\chi^2(\omega)$ can be explicitly given under certain simplifying assumptions.[10] In a general situation, we can relate the fluctuation (classical or quantum mechanical) to the response of the system to external disturbances — a quantity called the generalized polarizability. Consider the polarization of a medium under an external time varying electric field. We rewrite (8.8.3) as

$$\varepsilon(\omega) - 1 = \int_0^{\infty} f\left(\tau\right) e^{i\omega\tau} d\tau, \qquad (8.8.26)$$

---

[10]See Section 122 of Ref. [20]. In that book, $\chi^2(\omega)$ is simply written as $x^2(\omega)$. In order to avoid $x^2(\omega)$ being mistaken as $[x(\omega)]^2$ or the Fourier transform of $x^2(t)$, we adopt the notation $\chi^2(\omega)$. In fact, $\chi(\omega)$ never appears.

then $\varepsilon(\omega) - 1$ is an example of the generalized polarizability. For a physical quantity x coupled to an external generalized force $f(t)$, we add to the Hamiltonian a term

$$V = -xf(t). \tag{8.8.27}$$

We assume $\bar{x} = 0$ before the coupling. When the coupling is introduced, we have

$$\bar{x}(t) = \int_0^\infty \alpha(\tau)f(t-\tau)d\tau, \tag{8.8.28}$$

which depends only on the value of the force before time t (causality), as implemented by setting the lower limit of the integral to be zero. $\bar{x}(t)$ is called the response of the system to the disturbance of external force, $\alpha(\tau)$ is a function of time characterizing the properties of the system. The Fourier transform of $\alpha(t)$ is

$$\alpha(\omega) = \int_0^\infty \alpha(t)e^{i\omega t}dt, \tag{8.8.29}$$

which is related to $\bar{x}(\omega)$ and $f(\omega)$ by

$$\bar{x}(\omega) = \alpha(\omega)f(\omega). \tag{8.8.30}$$

$\alpha(\omega)$ is called the generalized polarizability. The linear relation between $\bar{x}(t)$ and $f(t-\tau)$ as well as between $\bar{x}(\omega)$ and $f(\omega)$ is valid only when the perturbation V is not very strong. The corresponding theory is called the linear response theory. Separating the real and imaginary parts of $\alpha(\omega)$, we have

$$\alpha(\omega) = \alpha'(\omega) + i\alpha''(\omega), \tag{8.8.31}$$

and

$$\alpha(-\omega) = \alpha^*(\omega). \tag{8.8.32}$$

The latter relation is necessary to keep $\alpha(t)$ real. It implies

$$\alpha'(-\omega) = \alpha'(\omega), \quad \alpha''(-\omega) = -\alpha''(\omega). \tag{8.8.33}$$

Because of the coupling with external force, there appears the correlation for x. The work done by the external force on the system causes its energy change. For a dissipative system this work done will be transformed to heat. The time rate of change of the energy is

$$\frac{dE}{dt} = \frac{\overline{\partial H}}{\partial t} = -\bar{x}\frac{df}{dt}. \tag{8.8.34}$$

For a single-frequency force

$$f(t) = \frac{1}{2}\left(f_0 e^{-i\omega t} + f_0^* e^{i\omega t}\right), \tag{8.8.35}$$

we have the corresponding $\bar{x}(t)$:

$$\bar{x}(t) = \frac{1}{2}\left[\alpha(\omega)f_0 e^{-i\omega t} + \alpha(-\omega)f_0^* e^{i\omega t}\right], \tag{8.8.36}$$

where we have used (8.8.30) and (8.8.32). The energy dissipated per unit time Q of the system under the action of the external force can be obtained by substituting (8.8.35) and (8.8.36) into (8.8.34):

$$Q = \frac{1}{4}i\omega\left(\alpha^* - \alpha\right)|f_0|^2 = \frac{1}{2}\omega\alpha''(\omega)\left|f_0^2\right|. \tag{8.8.37}$$

This important relation tells us that the dissipation is determined by the imaginary part of $\alpha(\omega)$. For realistic processes $Q > 0$, therefore the imaginary part of $\alpha(\omega)$ is positive for ω real and positive.

8.8.3   The fluctuation-dissipation theorem

An external force caused the fluctuation of the quantity x to which the force is coupled, their relation is expressed in terms of the generalized polarizability $\alpha(\omega)$. It also causes the dissipation Q in the system which is related to the imaginary part $\alpha''(\omega)$ of $\alpha(\omega)$. A question is immediately raised: Does the fluctuation possess intrinsic connection with the dissipation? This relation is characterized by the fluctuation-dissipation theorem. Let the system be situated in the quantum mechanical stationary state n. We calculate the correlation function of the fluctuation (8.8.25) averaged for the state n, i.e.,

$$\frac{1}{2}(\hat{x}(\omega)\hat{x}(\omega') + \hat{x}(\omega')\hat{x}(\omega))_{nn}$$

$$= \frac{1}{2}\sum_m [x(\omega)_{nm}x(\omega')_{mn} + x(\omega')_{nm}x(\omega)_{mn}]. \tag{8.8.38}$$

The right-hand side is summed over all energy states of the system. Since $\hat{x}(t)$ is time dependent, its matrix element has to be calculated by using time dependent wave functions, i.e.,

$$(\hat{x}(t))_{nm} = \int (\psi_n e^{-i\omega_n t})^* x(\psi_m e^{-i\omega_m t})\mathrm{d}^3 x$$

$$= x_{nm}e^{i(\omega_n - \omega_m)t}.$$

Consequently

$$x(\omega)_{nm} = \int_{-\infty}^{\infty} (\hat{x}(t))_{nm} e^{i\omega t}\mathrm{d}t = \int_{-\infty}^{\infty} x_{nm}e^{i(\omega_n - \omega_m + \omega)t}\mathrm{d}t$$

$$= 2\pi x_{nm}\delta(\omega_{nm} + \omega),$$

where x_{nm} and ω_{nm} are

$$x_{nm} = \int \psi_n^* x \psi_m \mathrm{d}^3 x,$$

$$\omega_{nm} = \omega_n - \omega_m.$$

Equation (8.8.38) becomes now

$$\frac{1}{2}\left(\hat{x}(\omega)\hat{x}(\omega') + \hat{x}(\omega')\hat{x}(\omega)\right)_{nn}$$

$$= 2\pi^2 \sum_m |x_{nm}|^2 \times [\,\delta(\omega_{nm} + \omega)\,\delta(\omega_{mn} + \omega')$$

$$+ \delta(\omega_{nm} + \omega')\,\delta(\omega_{mn} + \omega)], \tag{8.8.39}$$

where we have used the hermiticity of x, namely $x_{nm} = x_{mn}^*$. Because $\omega_{nm} = -\omega_{mn}$, the bracket of the right-hand side can be written as

$$\delta(\omega_{nm} + \omega)\,\delta(\omega + \omega') + \delta(\omega_{mn} + \omega)\,\delta(\omega + \omega'),$$

and we obtain on comparing the result with the right-hand side of (8.8.25)

$$\chi^2(\omega) = \pi \sum_m |x_{nm}|^2 \left[\delta\left(\omega + \omega_{nm}\right) + \delta\left(\omega + \omega_{mn}\right)\right]. \qquad (8.8.40)$$

The two delta functions on the right-hand side correspond to transitions to states higher and lower than n respectively.

The fluctuation of the quantum mechanical observable x is induced by its coupling to an external force, with the coupling term in the Hamiltonian given by

$$\hat{V} = -f\hat{x} = -\frac{1}{2}\left(f_0 e^{-i\omega t} + f_0^* e^{i\omega t}\right)\hat{x}. \qquad (8.8.41)$$

The system undergoes a transition $n \to m$ driven by the coupling with a probability per unit time

$$w_{mn} = \frac{\pi |f_0|^2}{2\hbar^2} |x_{mn}|^2 \left[\delta\left(\omega + \omega_{mn}\right) + \delta\left(\omega + \omega_{nm}\right)\right]. \qquad (8.8.42)$$

The rate of energy absorption by the system is

$$Q = \sum_m \hbar\omega_{mn} w_{mn}$$

$$= \frac{\pi}{2\hbar} |f_0|^2 \sum_m |x_{mn}|^2 \left[\delta\left(\omega + \omega_{mn}\right) + \delta\left(\omega + \omega_{nm}\right)\right] w_{mn}$$

$$= \frac{\pi}{2\hbar} |f_0|^2 \omega \sum_m |x_{mn}|^2 \left[\delta\left(\omega + \omega_{nm}\right) - \delta\left(\omega + \omega_{mn}\right)\right]. \qquad (8.8.43)$$

Comparing (8.8.43) with (8.8.37) we obtain

$$\alpha''(\omega) = \frac{\pi}{\hbar} \sum_m |x_{mn}|^2 \left[\delta\left(\omega + \omega_{nm}\right) - \delta\left(\omega + \omega_{mn}\right)\right]. \qquad (8.8.44)$$

Equations (8.8.40) and (8.8.44) characterize the correlation of fluctuation and dissipation respectively. They only differ by a sign of a delta function in the bracket. Their intimate relation can be found by averaging statistically using the Gibbs distribution

$$\rho_n = \exp\left(\frac{F - E_n}{kT}\right), \qquad (8.8.45)$$

where F is the free energy. Averaging over the state n, (8.8.40) gives

$$\langle \chi^2(\omega) \rangle = \pi \sum_{n,m} \rho_n \, |x_{nm}|^2 \left[\delta \left(\omega + \omega_{nm} \right) + \delta \left(\omega + \omega_{mn} \right) \right].$$

We can interchange the indices m and n in the second term because $|x_{nm}|^2 = |x_{mn}|^2$:

$$\langle \chi^2(\omega) \rangle = \pi \sum_{n,m} \left(\rho_n + \rho_m \right) |x_{nm}|^2 \, \delta \left(\omega + \omega_{nm} \right)$$

$$= \pi \sum_{n,m} \rho_n \left(1 + e^{\hbar \omega_{nm}/kT} \right) |x_{nm}|^2 \, \delta \left(\omega + \omega_{nm} \right)$$

$$= \pi \left(1 + e^{-\hbar \omega/kT} \right) \sum_{n,m} \rho_n \, |x_{nm}|^2 \, \delta(\omega + \omega_{nm}), \qquad (8.8.46)$$

where we have used the existence of the delta function in the last step. Treating (8.8.44) in the same way, we get

$$\langle \alpha''(\omega) \rangle = \frac{\pi}{\hbar} \left(1 - e^{-\hbar \omega/kT} \right) \sum_{n,m} \rho_n \, |x_{nm}|^2 \, \delta \left(\omega + \omega_{nm} \right). \qquad (8.8.47)$$

The negative sign before the second delta function in (8.8.44) was moved firstly in front of ρ_m, and subsequently it appears in front of the second term before the summation sign. Deleting the sign for statistical average, we have[11]

$$\chi^2(\omega) = \hbar \alpha''(\omega) \coth \frac{\hbar \omega}{2kT} = 2\hbar \alpha''(\omega) \left[\frac{1}{2} + \frac{1}{e^{\hbar \omega/kT} - 1} \right]. \qquad (8.8.48)$$

---

[11]In obtaining (8.8.46) and (8.8.47) we have used the Gibbs distribution for ρ_n and ρ_m. Although we delete the sign for taking the average, the distribution has done its job, the factor $e^{-\hbar \omega/kT}$ enters the two relations and finally appears in the theorem (8.8.49).

We carry out an operation $\int_0^\infty \frac{d\omega}{\pi}$ on both sides of the first equality and use (8.8.24), we get the *fluctuation-dissipation theorem* (Callan and Welton, 1951):

$$\langle x^2 \rangle = \frac{\hbar}{\pi} \int_0^\infty \alpha''(\omega) \coth \frac{\hbar\omega}{2kT} d\omega. \tag{8.8.49}$$

When multiplied by $\hbar\omega$, the first term in the bracket of the right-hand side of (8.8.48) gives the zero-point energy of the oscillator, and the second term gives the average energy of the oscillator at temperature T. The low temperature and high temperature limit of (8.8.48) is respectively

$$\chi^2(\omega) \xrightarrow{T \to 0} \hbar\alpha''(\omega),$$

$$\chi^2(\omega) \xrightarrow{T \to \infty} \frac{2kT}{\omega} \alpha''(\omega),$$

which shows that the fluctuation is quantum in nature at low temperature limit and classical in nature at high temperature limit. This theorem is very widely used. We will use it in the discussions of van der Waals force and the Casimir effect below. It has applications to quantum fluids, Bose fluid and Fermi fluid alike. The theorem can be easily generalized to the case when the system has several degrees of freedom, each coupled to a corresponding external force. Consider the coupling term for the degree of freedom i,

$$\hat{V} = -\bar{x}_i f_i(t), \tag{8.8.50}$$

where repeated indices mean a summation. The generalized polarizability is defined by the following linear relation:

$$\bar{x}_i(\omega) = \alpha_{ik}(\omega) f_k(\omega). \tag{8.8.51}$$

The rate of energy change of the system is

$$\dot{E} = -\dot{f}_i \bar{x}_i. \tag{8.8.52}$$

The spectral density of the correlation is

$$\frac{1}{2} \langle \hat{x}_i(\omega) \hat{x}_k(\omega') + \hat{x}_k(\omega') \hat{x}_i(\omega) \rangle = 2\pi (\chi_i \chi_k)(\omega) \delta(\omega + \omega'). \tag{8.8.53}$$

The introduction of $\chi_i\chi_k$ as a function of ω is similar to (8.8.20).[12] Average for the quantum state n gives

$$[(\chi_i\chi_k)(\omega)]_{nn} = \pi \sum_m \left[(x_i)_{nm}(x_k)_{mn}\delta(\omega + \omega_{nm})\right.$$

$$\left. + (x_k)_{nm}(x_i)_{mn}\delta(\omega + \omega_{mn})\right]. \tag{8.8.54}$$

To single-frequency perturbation

$$f_i(t) = \frac{1}{2}\left(f_{0i}e^{-i\omega t} + f_{0i}^* e^{i\omega t}\right) \tag{8.8.55}$$

the response of the system is

$$\bar{x}_i(t) = \frac{1}{2}\left[\alpha_{ik}(\omega)f_{ik}e^{-i\omega t} + \alpha_{ik}^*(\omega)e^{i\omega t}\right]. \tag{8.8.56}$$

Substituting (8.8.55) and (8.8.56) into (8.8.52), and taking average over a period $\frac{2\pi}{\omega}$, we obtain

$$Q = \frac{1}{4}i\omega\left(\alpha_{ik}^* - \alpha_{ki}\right)f_{0i}f_{0k}^*. \tag{8.8.57}$$

On the other hand, we calculate the energy absorbed by the system from the probability for the transition $n \to m$ per unit time:

$$Q = \frac{\pi}{2\hbar}\omega \sum_m f_{0i}f_{0k}^*\left[(x_i)_{nm}(x_k)_{mn}\delta(\omega + \omega_{nm})\right.$$

$$\left. - (x_i)_{nm}(x_k)_{mn}\delta(\omega + \omega_{mn})\right]. \tag{8.8.58}$$

On comparing the two above equations, we get

$$\alpha_{ik}^* - \alpha_{ki} = -\frac{2\pi i}{\hbar}\sum_m\left[(x_i)_{mn}(x_k)_{nm}\delta(\omega + \omega_{nm})\right.$$

$$\left. - (x_i)_{nm}(x_k)_{mn}\delta(\omega + \omega_{mn})\right]. \tag{8.8.59}$$

---

[12]To avoid the misunderstanding that $(x_ix_k)(\omega)$ is the Fourier transform of $x_i(t)x_k(t)$, we use here the notation $(\chi_i\chi_k)(\omega)$ different from Ref. [20]. It is related to $\langle x_ix_k\rangle$ by $\phi_{ik}(0) = \langle x_ix_k\rangle = \int_{-\infty}^{\infty}(\chi_i\chi_k)(\omega)\frac{d\omega}{2\pi}$.

Taking average over the Gibbs distribution, we get the generalized fluctuation-dissipation theorem:

$$(\chi_i \chi_k)(\omega) = \frac{1}{2} i\hbar \left(\alpha_{ki}^* - \alpha_{ik}\right) \coth \frac{\hbar \omega}{kT}. \qquad (8.8.60)$$

The left-hand side is the spectral density of $x_i x_k$.

8.9   The van der Waals Interaction

Consider two neutral atoms separated at a distance $R \gg a$, where a is the dimension of an atom. Assume that they do not possess magnetic moment, and the expected value of the electric dipole moment vanishes because of the conservation of parity. How does the interaction between them arise? Atom A has a spontaneous electric dipole moment due to fluctuation, the electric field of which at the position of atom B polarizes the latter, producing an induced electric dipole moment. The dipole-dipole interaction between the atoms has the nature of a fluctuation correlation. The first quantum mechanical derivation was given by F. London (1930), resulting in an interaction potential varying with the inverse sixth power of the distance $V \propto R^{-6}$. This is a long range force, in distinction from the short range force which decreases with distance exponentially. The R^{-6} dependence is universal in the sense that it is independent of the structure of the atoms. The experimental measurement of van der Waals force was carried out in 1940 by Overbeek using a suspension of quartz powder. He found that the R^{-6} dependence was verified for relatively small distance, and the potential V decreases faster than the inverse sixth power. Overbeek interpreted this finding as the effect of retardation. Casimir and Polder considered the retardation effect in 1948 and obtained $V \propto R^{-7}$, and the result is also universal. The interaction originates from the polarization of the atom, but the result turns out to be independent of the atomic structure, no matter retardation is taken into account or not. Is this accidental, or does it have a deeper reason? Casimir consulted Bohr, and got an advice that this is probably connected with the zero-point energy. Casimir then obtained the previous result of Casimir-Polder force starting from the change of zero-point energy.

Let us consider two spherically symmetrical atoms situated at r_1 and r_2 respectively with their relative distance $R = |r_1 - r_2| \gg a$ (dimension of an atom). Because of parity conservation the average value of the electric dipole moment of an atom vanishes. Due to fluctuation the atom 1 has a *spontaneous dipole moment* d_1^{sp}.[13] This dipole moment produces an electric field at the position of atom 2:

$$E_1(r_2) = \frac{3n(d_1^{\text{sp}} \cdot n) - d_1^{\text{sp}}}{R^3}, \tag{8.9.1}$$

where $n = \frac{r_2 - r_1}{|r_2 - r_1|}$. This field produces an induced electric dipole moment of atom 2:

$$d_2^{\text{ind}} = \alpha_2 E_1(r_2).$$

The energy of this induced moment in the field E_1 is

$$-\frac{1}{2} d_2^{\text{ind}} \cdot E_1(r_2) = -\frac{1}{2}\alpha_2 E_1^2(r_2). \tag{8.9.2}$$

The factor $\frac{1}{2}$ appears because d_2 is induced by E_1. Similarly the spontaneous electric dipole moment of atom 2 produces an induced electric dipole moment of atom 1, leading to the energy $-\frac{1}{2}\alpha_1 E_2^2(r_1)$. The total interaction energy is

$$U = -\frac{1}{2}\alpha_1 E_2^2(r_1) - \frac{1}{2}\alpha_2 E_1^2(r_2). \tag{8.9.3}$$

Taking average of the square of the fluctuating field, we have from (8.9.1)

$$\langle E_1^2(r_2) \rangle = \frac{3\langle (d_1^{\text{sp}} \cdot n)^2 \rangle + \langle (d_1^{\text{sp}})^2 \rangle}{R^6}. \tag{8.9.4}$$

The spontaneous dipole moment of a spherically symmetrical atom is isotropic, therefore

$$\langle (d_i^{\text{sp}} \cdot n)^2 \rangle = \langle d_{1x}^2 \rangle = \langle d_{1y}^2 \rangle = \langle d_{1z}^2 \rangle = \frac{1}{3}\langle d_1^2 \rangle,$$

---

[13] The spontaneous dipole moment is induced by the field fluctuation, and hence $\langle d^{\text{sp}} \rangle = 0$. But correlation of fluctuating quantities brings about average values of the interaction different from zero.

which leads to

$$\langle \boldsymbol{E}_1^2 (\boldsymbol{r}_2) \rangle = \frac{6}{R^6} \langle (d_{1z}^{sp})^2 \rangle,$$

and similarly

$$\langle \boldsymbol{E}_2^2 (\boldsymbol{r}_1) \rangle = \frac{6}{R^6} \langle (d_{2z}^{sp})^2 \rangle.$$

The interaction energy is then

$$U(R) = -\frac{3}{R^6} \left(\alpha_2 \langle (d_{1z}^{sp})^2 \rangle + \alpha_1 \langle (d_{2z}^{sp})^2 \rangle \right). \tag{8.9.5}$$

Expressing it in terms of the Fourier components, we have

$$U(R) = -\frac{3}{R^6} \int_{-\infty}^{\infty} \frac{d\omega}{2\pi} \left(\alpha_2(\omega) \langle d_{1z}^2(\omega) \rangle + \alpha_1(\omega) \langle d_{2z}^2(\omega) \rangle \right). \tag{8.9.6}$$

In the linear response theory the response $\boldsymbol{d}$ is induced by the fluctuating field, they are connected by the generalized polarizability α. According to the fluctuation-dissipation theorem

$$\left. \begin{aligned} \langle d_{1z}^2(\omega) \rangle &= \hbar \alpha_1''(\omega) \coth \frac{\hbar \omega}{2kT} \\[2mm] \langle d_{2z}^2(\omega) \rangle &= \hbar \alpha_2''(\omega) \coth \frac{\hbar \omega}{2kT} \end{aligned} \right\} \tag{8.9.7}$$

therefore (8.9.6) becomes

$$U(R) = -\frac{3\hbar}{R^6} \int_{-\infty}^{\infty} \frac{d\omega}{2\pi} \left(\alpha_2(\omega)\alpha_1''(\omega) + \alpha_1(\omega)\alpha_2''(\omega) \right) \coth \frac{\hbar \omega}{2kT}. \tag{8.9.8}$$

The imaginary part of α is an odd function of ω, and the hyperbolic cotangent function is also odd in its argument; therefore only the real part of α (an even function) in (8.9.8) contributes to the integral. The quantities in the parenthesis under the integral can be replaced by

$$\text{Re}\alpha_2(\omega)\text{Im}\alpha_1(\omega) + \text{Re}\alpha_1(\omega)\text{Im}\alpha_2(\omega) = \text{Im}\left[\alpha_1(\omega)\alpha_2(\omega)\right],$$

and we have

$$U(R) = -\frac{3\hbar}{2\pi R^6} \int_{-\infty}^{\infty} d\omega \text{Im}\left[\alpha_1(\omega)\alpha_2(\omega)\right] \coth\frac{\hbar\omega}{2kT}. \tag{8.9.9}$$

The polarizability of an atom in the ground state is

$$\alpha(\omega) = \sum_n \frac{e^2 f_{0n}}{m\left[(\omega_{n0}^2 - \omega^2) - i\omega\delta\right]}, \quad \delta \to 0_+, \tag{8.9.10}$$

where f_{0n} is the oscillator strength of the transition $0 \to n$,

$$\left. \begin{aligned} f_{0n} &= \tfrac{2m}{\hbar}\omega_{n0}\left|\sum_i (z_i)_{0n}\right|^2 \\ \omega_{n0} &= (E_n - E_0)/\hbar \end{aligned} \right\} \tag{8.9.11}$$

here z_i is the z coordinate of the electron i. Substituting (8.9.10) and (8.9.11) into (8.9.9), taking $T \to 0$ and carrying out a contour integral, we obtain finally

$$U(R) = -\frac{6e^4}{R^6} \sum_{n,n'} \frac{\left|\sum_i(z_i^{(1)})_{0n}\right|^2 \left|\sum_i(z_i^{(2)})_{0n}\right|^2}{E_n^{(1)} + E_n^{(2)} - E_0^{(1)} - E_0^{(2)}}. \tag{8.9.12}$$

The R^{-6} dependence is universal, and the sum depends on the shell structure of the atoms. This result coincides with that of F. London, who calculated the second order perturbation due to the dipole-dipole interaction

$$\frac{\boldsymbol{d}_1 \cdot \boldsymbol{d}_2 - 3\left(\boldsymbol{d}_1 \cdot \boldsymbol{n}\right)\left(\boldsymbol{d}_2 \cdot \boldsymbol{n}\right)}{R^3}.$$

Although $\boldsymbol{d}_1, \boldsymbol{d}_2$ are fluctuating quantities, i.e., $\langle \boldsymbol{d}_1 \rangle = \langle \boldsymbol{d}_2 \rangle = 0$, but $\langle \boldsymbol{d}_1 \cdot \boldsymbol{d}_2 \rangle$ and $\langle d_{1i}d_{2j} \rangle$ are different from 0.[14]

---

[14]The readers can refer to Ref. [30] for details.

8.10   The Effect of Retardation on the van der Waals Interaction

8.10.1   The field of an oscillating dipole[15]

When the distance R between atoms becomes comparable with the wave length of the radiation λ the retardation can no more be neglected. The fields of a charged particle moving with velocity v are derived from the Lienard-Wiechort potentials

$$\phi = \frac{e}{R - \frac{v \cdot R}{c}}, \quad A = \frac{ev}{c\left(R - \frac{v \cdot R}{c}\right)}, \tag{8.10.1}$$

where R is the radius vector from the charge to the point P where the potentials are to be measured, while the quantities on the right-hand sides are to be taken at a retarded time t' determined by the following relation:

$$t' + \frac{R(t')}{c} = t. \tag{8.10.2}$$

The time interval $t' - t$ is required for the propagation of the signal to cover the distance R. When the point of observation P is at a large distance from the system consisting of a number of point charges, we can fix the origin of coordinates at any point within the system shown in Fig. 8.29, where R_0 is the radius vector of P, r is the radius

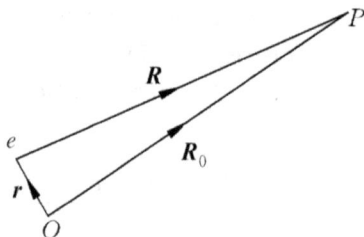

Fig. 8.29   Relation between the radius vectors of the charge and the point of observation.

---

[15]Please refer to Sections 63 and 72 of Ref. [31].

vector of a charge. Under the condition $R_0 \gg r$, we have

$$R = |\boldsymbol{R}_0 - \boldsymbol{r}| \approx R_0 - \boldsymbol{n} \cdot \boldsymbol{r}, \qquad (8.10.3)$$

where

$$\boldsymbol{n} = \frac{\boldsymbol{R}_0}{R_0}. \qquad (8.10.4)$$

The approximate relation (8.10.3) leads to the dipole radiation while the higher order terms of $\boldsymbol{n} \cdot \boldsymbol{r}$ and their derivatives lead to multipole radiations of higher order. Among the higher order multipole radiations the most important are the electric quadrupole radiation and the magnetic dipole radiation. Substituting (8.10.3) into the vector potential expression in (8.10.1) we obtain

$$A = \frac{e\boldsymbol{v}\left(t'\right)}{cR_0\left[1 - \frac{\boldsymbol{n} \cdot \boldsymbol{v}(t')}{c}\right]}. \qquad (8.10.5)$$

When the velocity of the charge $v \ll c$, the term $\frac{\boldsymbol{n} \cdot \boldsymbol{v}}{c}$ in the denominator can be neglected. Since $v \sim \frac{a}{2\pi/\omega}$, it follows that $a \sim \frac{2\pi v}{\omega} \ll \frac{2\pi c}{\omega} = \lambda$, and the approximation made above is equivalent to $a \ll \lambda$. Consequently

$$A = \frac{e\boldsymbol{v}\left(t'\right)}{cR_0} = \frac{1}{cR_0}\dot{\boldsymbol{d}}\left(t'\right), \qquad (8.10.6)$$

where $\boldsymbol{d}$ is the dipole moment of the charge

$$\boldsymbol{d} = e\boldsymbol{r}. \qquad (8.10.7)$$

Under the approximation $v \ll c$ (8.10.2) becomes

$$t' = t - \frac{R_0}{c}. \qquad (8.10.8)$$

The relation between the scalar potential ϕ and the dipole moment $\boldsymbol{d}$ can be obtained from the Lorentz condition

$$\nabla \cdot \boldsymbol{A} + \frac{1}{c}\frac{\partial \phi}{\partial t} = 0.$$

Substituting (8.10.6) into the condition and integrating with respect to t we get

$$\phi = -\nabla \cdot \frac{\boldsymbol{d}}{R_0}. \tag{8.10.9}$$

For the van der Waals interaction between atoms at a large distance the conditions are $R \gg a$, $\lambda \gg a$, but $R \sim \lambda$. The last condition brings complications that only for $R \gg \lambda$ the radiation from the charge can be considered as a plane wave. In that case, one can find $\boldsymbol{H}$ from $\boldsymbol{H} = \nabla \times \boldsymbol{A}$ and $\boldsymbol{E} = \boldsymbol{H} \times \boldsymbol{n}$ will follow. The last relation is valid only for plane waves. For $R \sim \lambda$ the radiation cannot be considered as a plane wave. Furthermore, in the Coulomb gauge $\nabla \cdot \boldsymbol{A} = 0$ it is sufficient to use the vector potential only and the fields are given by $\boldsymbol{H} = \nabla \times \boldsymbol{A}$ and $\boldsymbol{E} = -\frac{1}{c}\dot{\boldsymbol{A}}$, while in the present case the vector potential does not satisfy the relation $\nabla \cdot \boldsymbol{A} = 0$, and therefore both ϕ and $\boldsymbol{A}$ are needed. Now we derive the fields from (8.10.7) and (8.10.9)

$$\left. \begin{array}{c} \boldsymbol{H} = \dfrac{1}{c}\nabla \times \dfrac{\dot{\boldsymbol{d}}}{R_0} \\[3mm] \boldsymbol{E} = \nabla\nabla \cdot \dfrac{\boldsymbol{d}}{R_0} - \dfrac{1}{c^2}\dfrac{\ddot{\boldsymbol{d}}}{R_0} \end{array} \right\}. \tag{8.10.10}$$

The single-frequency component of $\boldsymbol{d}$ taken at a time t' is $\boldsymbol{d}(\omega)e^{-i\omega\left(t-\frac{R_0}{c}\right)} = \boldsymbol{d}(\omega) \cdot e^{-i\omega t + ikR_0}$. Substitution into (8.10.10) leads to

$$\left. \begin{array}{c} \boldsymbol{H}(\omega) = ik\boldsymbol{d}(\omega) \times \nabla\dfrac{e^{ikR_0}}{R_0} \\[3mm] \boldsymbol{E}(\omega) = k^2\boldsymbol{d}(\omega)\dfrac{e^{ikR_0}}{R_0} + (\boldsymbol{d}(\omega) \cdot \nabla)\nabla\dfrac{e^{ikR_0}}{R_0} \end{array} \right\}. \tag{8.10.11}$$

Carrying out the operation by ∇ we obtain

$$\boldsymbol{H}(\omega) = ik\boldsymbol{d}(\omega) \times \hat{\boldsymbol{n}}\left(\frac{ik}{R_0} - \frac{1}{R_0}\right)e^{ikR_0},$$

$$\boldsymbol{E}(\omega) = \boldsymbol{d}(\omega) \left(\frac{k^2}{R_0} + \frac{ik}{R_0^2} - \frac{1}{R_0^3} \right) e^{ikR_0}$$

$$+ \, \boldsymbol{n} \, (\boldsymbol{n} \cdot \boldsymbol{d}(\omega)) \left(-\frac{k^2}{R_0} - \frac{3ik}{R_0^2} + \frac{3}{R_0^3} \right) e^{ikR_0}. \qquad (8.10.12)$$

We express the above relation using components to demonstrate the relation between $E_i(\omega)$ and $d_j(\omega)$, and change R_0 to R:

$$E_i(\omega) = \mathscr{D}_{ij}^R (\omega, \boldsymbol{R}) \, d_j(\omega), \qquad (8.10.13)$$

where the retarded propagating function of the electromagnetic field is

$$\mathscr{D}_{ij}^R (\omega, \boldsymbol{R})$$

$$\equiv e^{ikR} \left[\left(\frac{k^2}{R} + \frac{ik}{R^2} - \frac{1}{R^3} \right) (\delta_{ij} - n_i n_j) + 2 \left(\frac{1}{R^3} - \frac{ik}{R^2} \right) n_i n_j \right],$$
$$(8.10.14)$$

Compared with (8.8.1), the relation (8.10.13) gives the electric field strength of a dipole when the effect of retardation is taken into account.

8.10.2 The fluctuation of electromagnetic field in a homogeneous medium[16]

To obtain the van der Waals force with retardation the correlation function $\langle E_i(\boldsymbol{r}_i) E_j(\boldsymbol{r}_2) \rangle$ of the components of electric field is of importance. It is connected with the correlation function $\langle A_i(\boldsymbol{r}_1) A_j(\boldsymbol{r}_2) \rangle$ of components of the vector potential. In the Hamiltonian the vector potential is coupled with the current density $\boldsymbol{j}$. In a homogeneous medium or in the vacuum, the perturbing $j_k(t)$ is connected to the response A_i via the generalized polarizability α_{ik}, which can be determined by the Maxwell equation. We choose the Coulomb gauge $\phi = 0$, in which the electric field strength is derived

---

[16] Please refer to Sections 7.5 and 7.6 of Ref. [32] for details.

from the derivative of the vector potential $\boldsymbol{E} = -\frac{1}{c}\dot{\boldsymbol{A}}$. Hence

$$\langle E_i\left(\boldsymbol{r}_i\right) E_j\left(\boldsymbol{r}_2\right)\rangle (\omega) = -\frac{\omega^2}{c^2} \langle A_i\left(\boldsymbol{r}_1\right) A_j\left(\boldsymbol{r}_2\right)\rangle (\omega). \qquad (8.10.15)$$

The relation between $\boldsymbol{A}$ and $\boldsymbol{j}$ is given by the Maxwell equation where time is Fourier transformed into frequency

$$\nabla \times \boldsymbol{H}\left(\boldsymbol{r},\omega\right) = \frac{4\pi}{c}\boldsymbol{j}\left(\boldsymbol{r},\omega\right) - \frac{i\omega}{c}\boldsymbol{D}\left(\boldsymbol{r},\omega\right). \qquad (8.10.16)$$

The second term on the right-hand side is the displacement current. In the medium we have

$$\left.\begin{array}{l} D_i(\omega) = \varepsilon_{ik}(\omega)E_k(\omega) \\ B_i(\omega) = \mu_{ik}(\omega)H_k(\omega) \end{array}\right\} \qquad (8.10.17)$$

In a homogeneous, isotropic and non-magnetic medium we have $\varepsilon_{ik}(\omega) = \varepsilon(\omega)\delta_{ik}$ and $\mu_{ik}(\omega) = \delta_{ik}$, and the relations between the field strengths and the vector potential are

$$\left.\begin{array}{l} \boldsymbol{B}\left(\boldsymbol{r},\omega\right) = \nabla \times \boldsymbol{A}\left(\boldsymbol{r},\omega\right) \\ \boldsymbol{E}\left(\boldsymbol{r},\omega\right) = \frac{i\omega}{c}\boldsymbol{A}\left(\boldsymbol{r},\omega\right) \end{array}\right\} \qquad (8.10.18)$$

Using the conditions for ε and μ in the medium and substituting (8.10.17) and (8.10.18) into (8.10.16), we get

$$\left[\left(\partial_i\partial_k - \delta_{ik}\nabla^2\right) - \frac{\omega^2}{c^2}\varepsilon(\omega)\delta_{ik}\right] A_k\left(\boldsymbol{r},\omega\right) = \frac{4\pi}{c}j_i\left(\boldsymbol{r},\omega\right). \qquad (8.10.19)$$

This equation can be conveniently solved by using the Green's function, which satisfies the following equation:

$$\left[\left(\partial_i\partial_j - \delta_{ij}\nabla^2\right) - \frac{\omega^2}{c^2}\varepsilon(\omega)\delta_{ij}\right] D_{jl}^{\mathrm{R}}\left(\boldsymbol{r}_1 - \boldsymbol{r}_2;\omega\right) = -4\pi\hbar\delta_{il}\delta^3\left(\boldsymbol{r}_1 - \boldsymbol{r}_2\right),$$
$$(8.10.20)$$

where the superscript R of the Green's function means retardation. The retardation effect is inherently incorporated in the Maxwell equation. Carrying out the Fourier transform of the Green's function, we have

$$D_{jl}^{\mathrm{R}}\left(\boldsymbol{r}_1 - \boldsymbol{r}_2, \omega\right) = \frac{1}{(2\pi)^3} \int e^{i\boldsymbol{k}\cdot(\boldsymbol{r}_1-\boldsymbol{r}_2)} D_{jl}^{\mathrm{R}}\left(\omega, \boldsymbol{k}\right) \mathrm{d}^3 k, \qquad (8.10.21)$$

and the transform of the Green's function satisfies

$$\left[k_i k_j - \delta_{ij} k^2 + \delta_{ij} \frac{\omega^2}{c^2} \varepsilon(\omega)\right] D_{jl}^{\mathrm{R}}\left(\omega, \boldsymbol{k}\right) = 4\pi\hbar\delta_{il}. \qquad (8.10.22)$$

It can be directly verified that the solution of the above equation is

$$D_{jl}^{\mathrm{R}}\left(\omega, \boldsymbol{k}\right) = 4\pi\hbar\frac{1}{\frac{\omega^2}{c^2}\varepsilon(\omega) - k^2}\left[\delta_{jl} - \frac{c^2}{\varepsilon(\omega)}\frac{k_j k_l}{\omega^2}\right]. \qquad (8.10.23)$$

This is the retarded Green's function in a homogeneous, isotropic and non-magnetic medium. The solution of the Equation (8.10.19) is

$$A_i\left(\boldsymbol{r}, \omega\right) = -\frac{1}{\hbar c}\int D_{ik}^{\mathrm{R}}\left(\boldsymbol{r}_1 - \boldsymbol{r}_2, \omega\right) j_k\left(\omega, \boldsymbol{r}_2\right) \mathrm{d}^3 r_2. \qquad (8.10.24)$$

The coupling between $\boldsymbol{A}$ and $\boldsymbol{j}$ in the Hamiltonian is

$$-\frac{1}{c}\int \boldsymbol{j}\left(\boldsymbol{r}\right) \cdot \boldsymbol{A}\left(\boldsymbol{r}\right) \mathrm{d}^3 r. \qquad (8.10.25)$$

From (8.10.24) and (8.10.25) we can read off that the generalized polarizability α_{ik} connecting the response A_i/c and the disturbance j_k is

$$\alpha_{ik}\left(\omega, \boldsymbol{r}_1 - \boldsymbol{r}_2\right) = -\frac{1}{\hbar c^2} D_{ik}^{\mathrm{R}}\left(\boldsymbol{r}_1 - \boldsymbol{r}_2, \omega\right). \qquad (8.10.26)$$

The correlation function of $\boldsymbol{A}$ is, according to the fluctuation-dissipation theorem

$$\langle A_i\left(\boldsymbol{r}_1\right) A_j\left(\boldsymbol{r}_2\right)\rangle\left(\omega\right) = -\mathrm{Im}D_{ik}^{\mathrm{R}}\left(\omega, \boldsymbol{r}_1 - \boldsymbol{r}_2\right) \coth\frac{\hbar\omega}{2kT}. \qquad (8.10.27)$$

The retarded Green's function $D_{ik}^{\mathrm{R}}(\omega, \boldsymbol{r}_1 - \boldsymbol{r}_2)$ is obtained by taking the Fourier transform of (8.10.23). Using the formula

$$\frac{1}{(2\pi)^3}\int \mathrm{d}^3 k \frac{4\pi e^{i\boldsymbol{k}\cdot\boldsymbol{R}}}{k^2 + \chi^2} = \frac{e^{-\chi R}}{R}$$

and making the substitution $\chi = \frac{\omega}{c}\sqrt{-\varepsilon(\omega)}$, we get

$$D^{\mathrm{R}}_{ik}(\omega, \boldsymbol{R}) = -\hbar \left(\delta_{ik} + \frac{c^2}{\varepsilon(\omega)\omega^2} \partial_i \partial_k \right) \frac{e^{-\frac{\omega}{c}\sqrt{-\varepsilon(\omega)}R}}{R}.$$

Substituting the above formula into Eq. (8.10.27) and Eq. (8.10.15), we get

$$\langle E_i(\boldsymbol{r}_i) E_j(\boldsymbol{r}_2) \rangle (\omega) = -\hbar \coth \frac{\hbar\omega}{2kT} \mathrm{Im} \left\{ \frac{1}{\varepsilon(\omega)} \left[\frac{\omega^2}{c^2} \delta_{ij} \varepsilon(\omega) + \partial_i \partial_j \right] \right.$$

$$\left. \times \frac{e^{-\left(\frac{\omega}{c}\right)\sqrt{-\varepsilon(\omega)}|r_1 - r_2|}}{|\boldsymbol{r}_1 - \boldsymbol{r}_2|} \right\} \qquad (8.10.28)$$

In vacuum, $\varepsilon(\omega) = 1$, $\sqrt{-\varepsilon(\omega)} = -i$,[17] and then

$$D^{\mathrm{R}}_{ik}(\omega, \boldsymbol{R}) = -\hbar \left(\delta_{ik} + \frac{c^2}{\omega^2} \partial_i \partial_k \right) \frac{\exp\left(i\frac{\omega}{c}R\right)}{R}.$$

Carrying out the partial derivative in the above formula, we get

$$D^{\mathrm{R}}_{ik}(\omega, \boldsymbol{R}) = -\hbar \frac{c^2}{\omega^2} \exp\left(i\frac{\omega}{c}R\right)$$

$$\times \left[\left(\frac{\omega^2}{c^2 R} + \frac{i\omega}{cR^2} - \frac{1}{R^3} \right) (\delta_{ik} - n_i n_k) \right.$$

$$\left. +2\left(\frac{1}{R^3} - \frac{i\omega}{cR^2} \right) n_i n_k \right]. \qquad (8.10.29)$$

Comparing (8.10.29) with (8.10.14), we have

$$\mathscr{D}^{\mathrm{R}}_{ij}(\omega, \boldsymbol{R}) = -\frac{\omega^2}{\hbar c^2} D^{\mathrm{R}}_{ij}(\omega, \boldsymbol{R}). \qquad (8.10.30)$$

The correlation $\langle E_i(\boldsymbol{r}_i) E_j(\boldsymbol{r}_2) \rangle$ can also be expressed as

$$\langle E_i(\boldsymbol{r}_i) E_j(\boldsymbol{r}_2) \rangle(\omega) = -\hbar \mathrm{Im} \mathscr{D}^{\mathrm{R}}_{ij}(\omega, \boldsymbol{R}) \coth \frac{\hbar\omega}{2kT}. \qquad (8.10.31)$$

---

[17]The negative sign before i originates from the fact that for a medium $\mathrm{Im}\,\varepsilon > 0$ to correspond to absorption. Please refer to pp. 321–322 of Ref. [33].

8.10.3 The Casimir-Polder interaction

Consider two atoms with polarizabilities $\alpha_1(\omega)$ and $\alpha_2(\omega)$ situated at $\boldsymbol{r}_1$ and $\boldsymbol{r}_2$ respectively. When the retardation is taken into account, the situation becomes more involved. Besides the correlation of fluctuations of the dipoles, there are also the correlation between the field fluctuations. Consider the component i of the dipole moment at $\boldsymbol{r}_1$:

$$
d_{1i}(\omega) = d_{1i}^{\mathrm{sp}} + \alpha_1(\omega)E_i\left(\boldsymbol{r}_1, \omega\right) + \alpha_1(\omega)\mathscr{D}_{ij}^{\mathrm{R}}\left(\omega, \boldsymbol{R}\right)d_{2j}^{\mathrm{sp}}
$$
$$
+ \alpha_1(\omega)\mathscr{D}_{ij}^{\mathrm{R}}\left(\omega, \boldsymbol{R}\right)\alpha_2(\omega)E_j\left(\boldsymbol{r}, \omega\right) + \cdots \qquad (8.10.32)
$$

The first term on the right-hand side is the spontaneous dipole moment, the second term is the dipole moment induced by the fluctuation field, the third term is the dipole induced by the electric field propagating from $\boldsymbol{r}_2$ to $\boldsymbol{r}_1$ (with retardation), and the fourth term is the dipole induced by the electric field of the dipole at $\boldsymbol{r}_2$ produced by the field fluctuation propagating to $\boldsymbol{r}_1$ (with retardation), etc. The electric field at $\boldsymbol{r}_1$ is

$$
\mathscr{E}_i\left(\boldsymbol{r}_1, \omega\right) = E_i\left(\boldsymbol{r}_1, \omega\right) + \mathscr{D}_{ij}^{\mathrm{R}}\left(\omega, \boldsymbol{R}\right)d_{2j}^{\mathrm{sp}} + \mathscr{D}_{ij}^{\mathrm{R}}\left(\omega, \boldsymbol{R}\right)\alpha_2(\omega)E_j\left(\boldsymbol{r}_2, \omega\right)
$$
$$
+ \mathscr{D}_{ij}^{\mathrm{R}}\left(\omega, \boldsymbol{R}\right)\alpha_2(\omega)\mathscr{D}_{ij}^{\mathrm{R}}\left(\omega, -\boldsymbol{R}\right)d_{1k}^{\mathrm{sp}} + \cdots . \qquad (8.10.33)
$$

The first term on the right-hand side is the fluctuating field at $\boldsymbol{r}_1$, the second term is the electric field of the spontaneous dipole moment at $\boldsymbol{r}_2$ propagating to $\boldsymbol{r}_1$ with retardation, the third term is the electric field of the dipole moment induced by the fluctuating field at $\boldsymbol{r}_2$ propagating to $\boldsymbol{r}_1$ with retardation, the fourth term is the field of a dipole propagating to $\boldsymbol{r}_1$ from $\boldsymbol{r}_2$ with retardation, while the dipole is induced by the electric field propagating to $\boldsymbol{r}_2$ from $\boldsymbol{r}_1$ (notice the argument $-\boldsymbol{R}$ in the argument of $\mathscr{D}_{ij}^{\mathrm{R}}$) of the spontaneous dipole moment at $\boldsymbol{r}_1$, etc. The interaction potential between the dipole moment and the electric field is

$$
U(R) = -\frac{1}{2}\int_{-\infty}^{\infty}\frac{d\omega}{2\pi}\langle\boldsymbol{d}_1\cdot\mathscr{E}(\boldsymbol{r}_1)\rangle(\omega), \qquad (8.10.34)
$$

where the average of the scalar product of two fluctuating quantities is to be taken. In order to emphasize the physics involved, we skip the complicated algebra and give only the steps and the results. Neglecting terms independent of R, and also higher order terms, and making use of the fluctuation-dissipation theorem, finally we obtain

$$U(R) = -\hbar \text{Im} \int_{-\infty}^{\infty} \frac{d\omega}{2\pi} \coth \frac{\hbar\omega}{2kT} (\alpha_1(\omega)\alpha_2(\omega) \mathscr{D}_{ij}^{\text{R}}(\omega, \boldsymbol{R}) \mathscr{D}_{ij}^{\text{R}}(\omega, \boldsymbol{R})).$$

$$(8.10.35)$$

In the static limit

$$E_i(r) = -\frac{3n_i n_j - \delta_{ij}}{R^3} d_{0j}$$

(8.10.35) gives

$$U(R) = -\frac{3\hbar}{\pi R^6} \int_0^{\infty} d\omega \coth \frac{\hbar\omega}{2kT} \text{Im}\,(\alpha_1(\omega)\alpha_2(\omega)),$$

which is the same as (8.9.9). In the low temperature limit, (8.10.35) gives

$$U(R) = -\frac{\hbar \alpha_1(0)\alpha_2(0)}{\pi R^2} \int_0^{\infty} d\omega \exp\left(-\frac{2\omega}{c}R\right) \left[\left(\frac{\omega}{c}\right)^4 + \frac{2}{R}\left(\frac{\omega}{c}\right)^3\right.$$

$$\left. + \frac{5}{R^2}\left(\frac{\omega}{c}\right)^2 + \frac{6}{R^3}\left(\frac{\omega}{c}\right) + \frac{3}{R^4}\right]$$

$$= -\frac{23\hbar c}{4\pi R^7} \alpha_1(0)\alpha_2(0). \qquad (8.10.36)$$

The force on an isolated atom is referred to as the Casimir-Polder force. This includes the force between two neutral atoms, and the force on an atom due to an infinite conducting plane or due to a semi-infinite dielectric medium, etc.

8.11   Zero-point Energy, Vacuum Fluctuation of the Electromagnetic Field and the van der Waals Interaction

Ten years before Heisenberg found the energy eigenvalues of a 1-dimensional harmonic oscillator, M. Planck gave an expression for the average energy of a harmonic oscillator with frequency ω in equilibrium with thermal radiations at temperature T[18]:

$$U = \frac{\hbar\omega}{e^{\hbar\omega/kT} - 1} + \frac{1}{2}\hbar\omega. \tag{8.11.1}$$

When $t \to 0, U \to \frac{1}{2}\hbar\omega$. Planck considered this "zero-point energy" without physical significance. Einstein pointed out that this was not so, because when $kT \gg \hbar\omega$ the average energy becomes up to terms to first order of $\hbar\omega/kT$,

$$U \approx \frac{\hbar\omega}{\frac{\hbar\omega}{kT} + \frac{1}{2}\left(\frac{\hbar\omega}{kT}\right)^2} + \frac{1}{2}\hbar\omega \approx kT - \frac{1}{2}\hbar\omega + \frac{1}{2}\hbar\omega = kT,$$

i.e., the zero-point energy is required to keep the relation $U \approx kT$ at high temperature. Einstein considered the existence of zero-point energy possible [34]. Zero-point energy is not only necessary for conceptual reasons as a consequence of the uncertainty relation in quantum mechanics, but is a measurable quantity in solid state physics. When the atoms on a crystal lattice oscillate randomly with a mean square displacement $\langle u^2 \rangle$, the intensity of X-rays scattered from the crystal is

$$I = I_0 \exp\left(-\frac{1}{3}\langle u^2 \rangle \mathbf{G}^2\right), \tag{8.11.2}$$

where $\mathbf{G}$ is the momentum transfer of the scattering, I_0 is the intensity of scattered X-rays from a rigid lattice. The exponential factor is called the Debye-Waller factor. The average potential energy of a classical oscillator of mass M at temperature T is $\frac{3}{2}kT$, hence by the relation

$$\langle U \rangle = \frac{3}{2}kT = \frac{1}{2}M\omega^2 \langle u^2 \rangle,$$

the intensity of scattered X-rays (8.11.2) can be written as

$$I = I_0 \exp\left[-\frac{2\langle U \rangle}{3M\omega^2}\mathbf{G}^2\right] = I_0 \exp\left[-\frac{kT}{M\omega^2}\mathbf{G}^2\right]. \tag{8.11.3}$$

---

[18]Compare it with (8.8.45).

If the zero-point energy is included, then in the limit $T \to 0$, we have

$$\langle U \rangle_0 = \frac{3}{2}\hbar\omega,$$

and then in the limit

$$I = I_0 \exp\left(-\frac{\hbar G^2}{M\omega}\right), \tag{8.11.4}$$

the exponent remains finite instead of zero as given by (8.11.3). Measurement confirms the existence of zero-point energy.

In field theory it is more complicated to handle the zero-point energy (or the vacuum fluctuation energy). Since for each mode of the field there is a zero-point energy $\frac{1}{2}\hbar\omega$, and the number of modes between ω and $\omega + d\omega$ in unit volume of free space is

$$dN_\omega = \frac{\omega^2}{\pi^2 c^3}d\omega,$$

we have the zero-point energy per unit volume in this frequency interval

$$\rho_0(\omega) = \frac{\omega^2}{\pi^2 c^3}\frac{1}{2}\hbar\omega d\omega, \tag{8.11.5}$$

and the total zero-point energy per unit volume is

$$E_0 = \int_0^\infty \rho_0(\omega)d\omega, \tag{8.11.6}$$

which is a divergent quantity. A method of handling it is simply to throw it away for "lack of physical significance", another method is to re-define the energy of the mode k

$$E_k = \frac{1}{2}\hbar\omega_k\left(a_k^\dagger a_k + a_k a_k^\dagger\right) \tag{8.11.7}$$

as the normal product of the operators[19]

$$E_k = \frac{1}{2}\hbar\omega_k : a_k^\dagger a_k + a_k a_k^\dagger := \hbar\omega_k a_k^\dagger a_k, \tag{8.11.8}$$

---

[19]The normal product is a re-ordering of operators in a product, such that creation operators stand to the left of annihilation operators. During each interchange of fermion operators a negative sign appears. The interchange of boson operators does not change the sign.

thus neglecting the zero-point energy equivalently. Realizing that the Dirac equation is unable to predict the Lamb shift in the hydrogen atomic spectra (a small splitting of the levels $2S_{\frac{1}{2}}$ and $2P_{\frac{1}{2}}$), H. A. Bethe pointed out that this might be a consequence of the interaction of the atomic electron with the vacuum fluctuation of the radiation field. In the calculation of this level shift Bethe subtracted the interaction energy between the vacuum fluctuation and a free electron (which is infinite) from the interaction energy with the bound electron (which is also infinite). The difference still diverges, albeit with a much milder behavior. It is only logarithmically divergent, i.e., it diverges with $\ln \omega$, which means a rather weak dependence on ω when the latter becomes large. Bethe took a cut-off frequency $\omega_c = mc^2/\hbar$, and his result for the Lamb shift came fairly close to the experimental data. Feynman showed that a relativistic treatment rendered the above mentioned difference finite, and agreed very well with the experimental value. This success is the starting point of the renormalization theory of quantum electrodynamics pioneered by R. Feynman, J. Schwinger and S. Tomonaga.

Another field of applying the vacuum fluctuation is initiated by H. B. G. Casimir. After obtaining the $V \propto R^{-7}$ result (8.10.36), he was wondering about the universality of the R^{-7}, where the effect of the atomic structure appears only as a coefficient multiplying the R^{-7} factor. In a walk with N. Bohr in 1947, he told Bohr about his result and also his puzzle about the universality. Bohr replied that this might be associated with the zero-point energy. Casimir then pursued the study of the change of zero-point energy and indeed reproduced the Casimir-Polder result [35, 36]. This encouraged Casimir to study other observable effect of the vacuum fluctuation, and one result was the well-known Casimir effect. We first discuss briefly the derivation of Casimir-Polder force [37]. Consider two neutral systems with internal degrees of freedom. Assume that the wave length of radiation is much larger than the dimension of the systems such that the dipole approximation is valid. The wave length is comparable with the distance between the systems such that the effect of retardation is important. System 1 is situated at r_1. The vacuum fluctuation field $E_0(\omega, r_1)$ at r_1 induces a spontaneous dipole moment $d_1^{sp}(\omega) = \alpha_1(\omega)E_0(\omega, r_1)$, which produces an electric field propagating to r_2

with retardation. Let us denote this field by $E_1(\omega, r_2)$, which is given by (8.9.13) and (8.9.14). As an order-of-magnitude estimate, we take only

$$E_1(\omega, r_2) \approx \frac{\alpha_1(\omega) E_0(\omega, r_1)}{R^3} g\left(\frac{\omega R}{c}\right), \qquad (8.11.9)$$

where g is a function of the dimensionless quantity $\omega R/c$, expressing the average over angular dependence. Within the range of $0 \leq \omega R/c \leq 1$, it is $O(1)$, and $g(x)/x$ is finite for $x \to 0$. The field at r_2 is the sum of the vacuum field and the field (8.11.9),

$$E_2(\omega, r_2) = E_0(\omega, r_2) + E_1(\omega, r_2),$$

and the dipole moment induced by this field is

$$d_2(\omega) = \alpha_2(\omega) [E_0(\omega, r_2) + E_1(\omega, r_2)],$$

with an interaction energy

$$U(\omega, R) \approx \frac{1}{2}\alpha_2(\omega) [E_0(\omega, r_2) + E_1(\omega, r_2)]^2.$$

We neglect E_0^2 (because it does not depend on R) and also E_1^2 (because it is a higher order term). Then

$$U(\omega, R) \approx \alpha_2(\omega) E_0(\omega, r_2) \cdot E_1(\omega, r_2)$$

$$= \alpha_2(\omega)\alpha_1(\omega) E_0(\omega, r_1) \cdot E_0(\omega, r_2) \frac{g\left(\frac{\omega R}{c}\right)}{R^3}. \qquad (8.11.10)$$

Assume that the field is enclosed in a volume $\mathscr{V}$, the distribution of the modes of the field is

$$N(\omega) = \mathscr{V}\frac{\omega^2}{c^2}d\omega$$

we get the interaction energy

$$U(R) = \frac{1}{R^3}\int \alpha_1(\omega)\alpha_2(\omega) E_0(\omega, r_1) \cdot E_0(\omega, r_2) g\left(\frac{\omega R}{c}\right) N(\omega)d\omega.$$

When $\omega < \frac{c}{R}$, i.e., $\lambda > R$, $\boldsymbol{E}_0(\omega, \boldsymbol{r}_1)$ can be replaced by $\boldsymbol{E}_0(\omega, \boldsymbol{r}_2)$, and furthermore we have

$$\left\langle \boldsymbol{E}_0^2 \left(\omega, \boldsymbol{r}_2\right) \right\rangle \cdot \mathscr{V} = \hbar\omega,$$

Then it follows that

$$U(R) \approx \frac{\hbar}{(cR)^3} \int_0^{c/R} g\left(\frac{\omega R}{c}\right) \alpha_1(\omega)\alpha_2(\omega)\omega^3 \mathrm{d}\omega. \qquad (8.11.11)$$

For R large, the integral gets the main contribution from the region $\omega < \frac{c}{R}$, and $\alpha(\omega)$ can be replaced by $\alpha(0)$. As a rough estimate, we have

$$U(R) \approx \frac{\hbar c}{R^7} \alpha_1(0)\alpha_2(0) \int_0^1 g\left(x\right) x^3 \mathrm{d}x$$

$$\approx \frac{\hbar c}{R^7} \alpha_1(0)\alpha_2(0), \qquad (8.11.12)$$

where the integral is of $O(1)$. So the R^{-7} dependence is reproduced. We now turn to the measurement of Casimir-Polder force as reported by the Yale University group [37]. The deflection of ground state Na atoms passing through a parallel-plate cavity was measured by the intensity of atoms transmitted through the cavity as a function of cavity plate separation. The existence of Casimir-Polder force is revealed by the modification of the ground state Lamb shift in the confined space because the vacuum of electromagnetic field is different from that of the free space. Since the vacuum field varies with position, this modified vacuum gives rise to a spatially varying Lamb shift whose gradient corresponds to an attractive Casimir-Polder force towards the cavity walls. For a spherical ground state atom between ideal parallel mirrors, the position dependent atom-cavity interaction potential is given by G. Barton [38] based on rigorous QED calculations:

$$U(z) = -\sum_e \frac{\pi |d_{eg}|^2}{6\varepsilon_0 L^3} \int_0^\infty \mathrm{d}\rho \frac{\rho^2 \cosh(2\pi\rho z/L)}{\sinh(\pi\rho)} \tan^{-1}\left(\frac{\rho\lambda_{eg}}{2L}\right),$$

$$(8.11.13)$$

where z is the distance of the atom from the center of cavity. The sum is over excited states, L is the plate separation, d_{eg} is the matrix element of the electric dipole operator between states e and g, and ρ is a dummy variable. This expression can be simplified in two extreme cases:

(1) The van der Waals limit where $L \ll \lambda$, with λ being the wave length of the dominant $3s \rightarrow 3p$ transition. In this limit retardation of the field can be neglected. Equation (8.11.13) can be rewritten as

$$U_{vdW} = -\frac{1}{4\pi\varepsilon_0} \frac{2 \langle g \,|\, d^2 \,|\, g \rangle}{3L^3} \sum_{\text{odd } n} \left[\frac{1}{\left(n - \frac{2z}{L}\right)^3} + \frac{1}{\left(n + \frac{2z}{L}\right)^3} \right],$$

$$(8.11.14)$$

which is the spherically averaged interaction energy of a static electric dipole coupled with its multiple electric images in the cavity walls. The condition of validity can be estimated as follows. Since the atom is in the ground state, the transition $3s \rightarrow 3p$ can only be a virtual one, and by the uncertainty principle the time of existence of the virtual photon is $\Delta t \le \frac{1}{\Delta\omega} = \frac{1}{2\pi c/\lambda} = \frac{\lambda}{2\pi c}$. The neglect of retardation is legitimate for $L < \frac{\lambda}{2\pi}$.

(2) For $L \gg \frac{\lambda}{2\pi}$, the retardation is important when the atom is not too close to one of the cavity walls. (8.11.13) is approximated by the Casimir-Polder potential

$$U_{CP} = -\frac{1}{4\pi\varepsilon_0} \frac{\pi^3 \hbar c \alpha(0)}{L^4} \left[\frac{3 - 2\cos^2 \frac{\pi z}{L}}{8 \cos^4 \frac{\pi z}{L}} \right].$$

$$(8.11.15)$$

The appearance of $\alpha(0)$ emphasizes the role of low frequency part of the vacuum spectrum in the modification of the Lamb shift. Comparing (8.11.14) with (8.11.15), we see the evolution of a L^{-3} dependence of the instantaneous van der Waals potential to the L^{-4} dependence of the Casimir-Polder potential with retardation. The gradient of this potential gives rise at large L to the Casimir-polder force between the atom and the cavity walls, which the experiment intends to measure. Here a comment is in order. The atom in an excited state can emit *real* photons which can exert strong influence on the atom due to the extensive lifetime. The

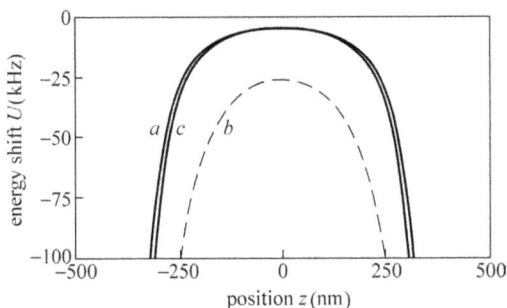

Fig. 8.30   Ground energy level shifts for a Na atom in a parallel plate cavity $1\,\mu m$ wide. Reprinted with permission from C. I. Sukenik, M. G. Boshier, D. Cho, V. Sandoghdar, and E. A. Hinds, *Phys. Rev. Lett.* 70, 560, 1993. Copyright (2017) by the American Physical Society.

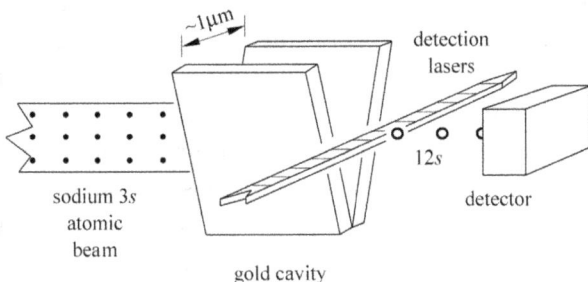

Fig. 8.31   Scheme of the experimental setup. Reprinted with permission from C. I. Sukenik, M. G. Boshier, D. Cho, V. Sandoghdar, and E. A. Hinds, *Phys. Rev. Lett.* 70, 560, 1993. Copyright (2017) by the American Physical Society.

cavity pulling effect in excited states is so large that it obscures the modification of the Lamb shift. Hence the exact measurement of the Casimir-Polder force can only be performed with the ground state atom. Figure 8.30 shows the ground state energy level shifts calculated for the Na atom in a cavity $1\,\mu m$ wide versus the distance away from the center of the parallel plate cavity. The exact potential (a), the instantaneous van der Waals potential (b) and the Casimir-Polder potential (c) are compared. This level shift is just the potential which the atom experiences between the parallel plate cavity. Figure 8.31 shows the scheme of the experimental setup. Na atoms effuse from a vertical oven slit 1 cm high into a vacuum $\sim 10^{-7}$ torr. They enter a wedge shaped gold cavity 3 cm high and 8 mm long and adjustable in width $0.5\,\mu m$ to $8\,\mu m$. They are deflected by the potential shown in Fig. 8.30. An atom that hits the cavity wall stick there with high probability. Those passing closer to the cavity center line will exit from the cavity. They are resonantly excited to the 12 s state by two superimposed laser beams and are field ionized and detected using a channel electron multiplier. The laser beams are focused so that the detection occurs over a small height ~200 µm, corresponding to a well defined cavity width. The intensity of the transmitted beam $I(L)$ at various cavity width L are measured. They are normalized to $I(6\,\text{µm})$ with the relative transmission $T(L) = \frac{I(L)}{I(6\,\text{µm})}$. The opacity is defined by $1/T(L)$. The result is plotted in Fig. 8.32, with curve (a) corresponding to the exact QED potential, curve (b) corresponding to the instantaneous van der Waals potential, and curve (c) corresponding to absence of interaction. These are the results of a Monte Carlo calculation in which atoms having a Maxwell-Boltzmann velocity distribution within the oven fly randomly into the cavity and propagate under whichever potential is assumed. In order to discriminate more effectively between the potential an extension down to $L = 0.7\,\text{µm}$ is shown in Fig. 8.33, where we see that the instantaneous van der Waals potential can be ruled out decisively. The Casimir-Polder potential with retardation is a good approximation to the QED interaction.

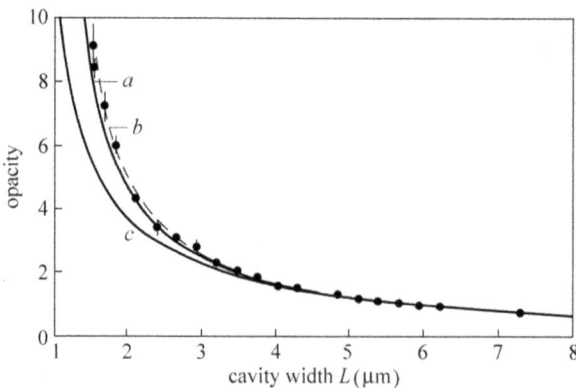

Fig. 8.32   Measured cavity opacity versus cavity widths. Reprinted with permission from C. I. Sukenik, M. G. Boshier, D. Cho, V. Sandoghdar, and E. A. Hinds, *Phys. Rev. Lett.* **70**, 560, 1993. Copyright (2017) by the American Physical Society.

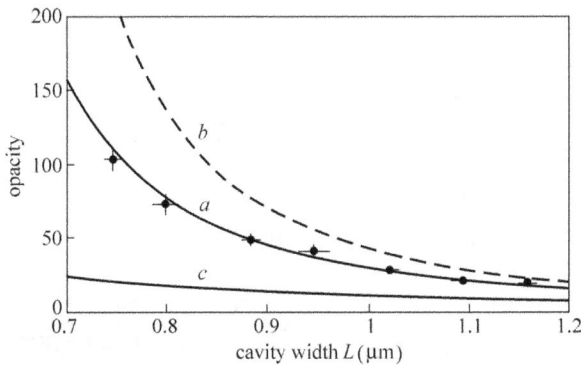

Fig. 8.33 Extension of the curves in Fig. 8.18 to L ranging from 0.7 μm to 1.2 μm. Reprinted with permission from C. I. Sukenik, M. G. Boshier, D. Cho, V. Sandoghdar, and E. A. Hinds, *Phys. Rev. Lett.* **70**, 560, 1993. Copyright (2017) by the American Physical Society.

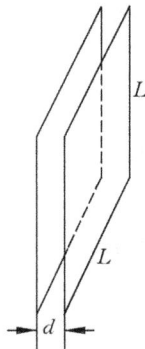

Fig. 8.34 Two perfect conducting plate of size $L \times L$.

8.12 The Casimir Effect

Consider two perfect conducting plate of size $L \times L$ separated by a distance d as shown in Fig. 8.34. Let the directions of the sides be the x and y axes, and direction of their separation be the z axis. In the volume between the plates the density of modes of the electromagnetic wave is

$$\omega_{lmn} = \pi c \left(\frac{l^2}{L^2} + \frac{m^2}{L^2} + \frac{n^2}{d^2} \right)^{1/2}, \qquad (8.12.1)$$

where l, m, n are integers or 0. The zero-point energy in the box is

$$E_0\left(d\right) = \sum_{l,m,n}{}' 2 \times \frac{1}{2}\hbar\omega_{lmn}$$

$$= \sum_{l,m,n}{}' \pi\hbar c \left(\frac{l^2}{L^2} + \frac{m^2}{L^2} + \frac{n^2}{d^2}\right)^{1/2}. \tag{8.12.2}$$

The factor 2 arises because of the two independent polarization states for each mode $l, m, n \neq 0$. The prime over the summation sign means that the factor 2 does not arise whenever one or more of the indices are equal to zero, because there is only one state of polarization for such modes.[20] Assumed that $d \ll L$, the sum over l and m can be replaced by integrals, i.e., there are no restrictions for k_x and k_y:

$$E_0 = \frac{\hbar c L^2}{\pi^2} \sum_n{}' \int_0^\infty \mathrm{d}k_x \int_0^\infty \mathrm{d}k_y \left(k_x^2 + k_y^2 + \frac{\pi^2 n^2}{d^2}\right)^{1/2}. \tag{8.12.3}$$

In a volume $L \times L \times d$, the zero-point energy in the free space is

$$E_0(\text{free}) = \frac{\hbar c L^2 d}{\pi^3} \int_0^\infty \mathrm{d}k_x \int_0^\infty \mathrm{d}k_y \int_0^\infty \mathrm{d}k_z \left(k_x^2 + k_y^2 + k_z^2\right)^{1/2}, \tag{8.12.4}$$

where all modes are unrestricted. $E_0 - E_0(\text{free})$ then represents the change in zero-point energy due to restriction of modes along the z-direction, namely,

$$U\left(d\right) = E_0 - E_0(\text{free})$$

$$= \frac{L^2 \hbar c}{\pi^3} \left[\sum_n{}' \int_0^\infty \mathrm{d}k_x \int_0^\infty \mathrm{d}k_y \left(k_x^2 + k_y^2 + \frac{\pi^2 n^2}{d^2}\right)^{1/2}\right.$$

$$\left. - \frac{d}{\pi} \int_0^\infty \mathrm{d}k_x \int_0^\infty \mathrm{d}k_y \int_0^\infty \mathrm{d}k_z \left(k_x^2 + k_y^2 + k_z^2\right)^{1/2}\right].$$

---

[20]If $n = 0$, then $k_z = \frac{\pi n}{d} = 0$, the wave propagates in the xy-plane. The polarization vector can only be parallel to the xy-plane and perpendicular to $\boldsymbol{k}$.

Changing to the polar coordinates $k_x^2 + k_y^2 = r^2$, $\int_0^\infty dk_x \int_0^\infty dk_y = \frac{\pi}{2} \int_0^\infty drr$, the above expression becomes

$$U(d) = \frac{L^2 \hbar c}{\pi^2} \frac{\pi}{2} \left[{\sum_n}' \int_0^\infty drr \left(r^2 + \frac{n^2 \pi^2}{d^2} \right)^{1/2} \right.$$

$$\left. - \frac{d}{\pi} \int_0^\infty dz \int_0^\infty dr\, r \left(r^2 + z^2 \right)^{1/2} \right].$$

Let $\xi = \frac{r^2 d^2}{\pi^2}$, $\eta = \frac{zd}{\pi}$, the above expression is rewritten as

$$U(d) = \frac{L^2 \hbar c \pi^2}{4d^3} \left[{\sum_n}' \int_0^\infty d\xi (\xi + n^2)^{1/2} \right.$$

$$\left. - \int_0^\infty d\eta \int_0^\infty d\xi (\xi + \eta^2)^{1/2} \right]. \tag{8.12.5}$$

This is the difference of two divergent integrals. Define

$$F(u) \equiv \int_0^\infty d\xi \, (\xi + u^2)^{1/2}, \tag{8.12.6}$$

then (8.12.5) can be written as

$$U(d) = \frac{\pi^2 \hbar c L^2}{4d^3} \left[\frac{1}{2} F(0) + \sum_{n=1}^\infty F(n) - \int_0^\infty d\eta F(\eta) \right]. \tag{8.12.7}$$

The difference between the series and the integral is given by Euler-Maclaurin formula:

$$\sum_{n=1}^\infty F(n) - \int_0^\infty d\eta F(\eta) = -\frac{1}{2} F(0) - \frac{1}{2!} B_2 F'(0) - \frac{1}{4!} B_4 F'''(0) - \cdots \tag{8.12.8}$$

where the Bernoulli number B_ν is defined as

$$\frac{y}{e^y - 1} = \sum_{\nu=0}^\infty B_\nu \frac{y^\nu}{\nu!}, \tag{8.12.9}$$

and $B_2 = \frac{1}{6}$, $B_4 = -\frac{1}{30}$, etc. The integral is still divergent, because for modes with wave length smaller than the atomic dimension the perfect conductor approximation is no more valid. It is necessary to

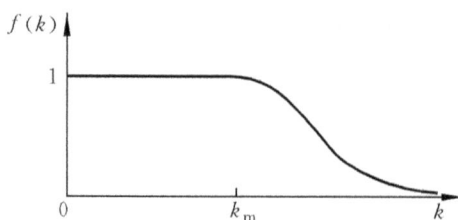

Fig. 8.35   The cut-off function.

introduce a cut-off function $f(k)$, which decreases rapidly to zero when the magnitude of the wave vector exceeds k_m, the reciprocal of the atomic dimension, and for $k \lesssim k_m$ the function $f(k) = 1$ (Fig. 8.35). Then the integral can be written as

$$F(u) = \int_0^\infty d\xi \, (\xi + u^2)^{1/2} f\left(\frac{\pi}{d}\sqrt{\xi + u^2}\right),$$

where the argument of f is the magnitude of the wave vector $\mathbf{k}$. Changing the variable of integration to $\zeta = \xi + u^2$, we have

$$F(u) = \int_{u^2}^\infty d\zeta \zeta^{1/2} f\left(\frac{\pi\sqrt{\zeta}}{d}\right), \qquad (8.12.10)$$

The dependence of F on u is via the lower limit exclusively, therefore

$$F'(u) = -2u^2 f\left(\frac{\pi u}{d}\right). \qquad (8.12.11)$$

In finding the higher derivatives of F with respect to u, we notice that $f(0) = 1$, and the higher derivatives of $f(k)$ at $k = 0$ are all zero. Therefore, $F(0) = 0$, $F'(0) = 0$, $F'''(0) = -4$, and all higher derivatives at $k = 0$ vanish. The Euler-Maclaurin series is found to be

$$\frac{U(d)}{L^2} = \frac{\hbar c \pi^2}{d^3} \frac{B_4}{4!} = -\frac{\pi^2}{720} \frac{\hbar c}{d^3}. \qquad (8.12.12)$$

This result is independent of the cut-off frequency, and is a "clean" result. The force on the parallel plates is

$$\mathscr{F} = -\frac{\pi^2}{240} \frac{\hbar c}{d^4}. \qquad (8.12.13)$$

This is a very weak force. If d is given in units of μm, the force is $-\frac{0.013}{[d(\mu m)]^4}$ dyne/cm^2. Sparnaay [39] succeeded in measuring the magnitude and the dependence on the separation of this force in 1958.

The presentation of this section is based on Ref. [40].

8.13 Cavity QED in Strong Coupling Regime

The essence of cavity QED is to enhance the effect of coupling between the atom and the photons. In free space, a photon interacts briefly when it passes by an atom. When both atom and photon are trapped inside a cavity, the coupling becomes much stronger. An excited atom can emit a photon and reabsorbs it repeatedly in a cavity. Physicists have been making effort to study cavity QED in strong coupling regime in recent years, and remarkable advance has been made. In such a regime, many Rabi oscillations occur before the atom or the photon escape from the cavity. If the "atom" can be stationary at a point of antinode of a standing wave in the cavity, the coupling can be optimized. this idea has been realized in a solid-state system by the Delft group (1. Chiorescu, J. E. Mooij *et al.*) [41] and the Yale group (A. Wallraff *et al.*) [42].

The Delft group studied the entanglement of a superconducting flux qubit (the "atom") and a SQUID, which serves as the harmonic oscillator mode of the cavity, as well as the measurement system for detecting the quantum state of the qubit.

Generation and control of the entangled system were achieved by performing microwave spectroscopy and detecting the Rabi oscillation of the coupled system. The device shown in Fig. 8.36(a) was made by electron beam lithography and metal evaporation. A large loop interrupted by two Josephson junctions (the SQUID) is merged with the smaller loop on the right hand side comprising three in-ling Josephson junctions (the flux qubit). By applying a perpendicular external magnetic field, the qubit is biased around half a flux quantum $\Phi_0/2$. The flux qubit is then a controllable two-level system with "spin-up" $|\uparrow\rangle$ and "spin-down" $|\downarrow\rangle$ states corresponding to persistent current flowing in "clockwise" and "anti-clockwise"

(a)

(b)

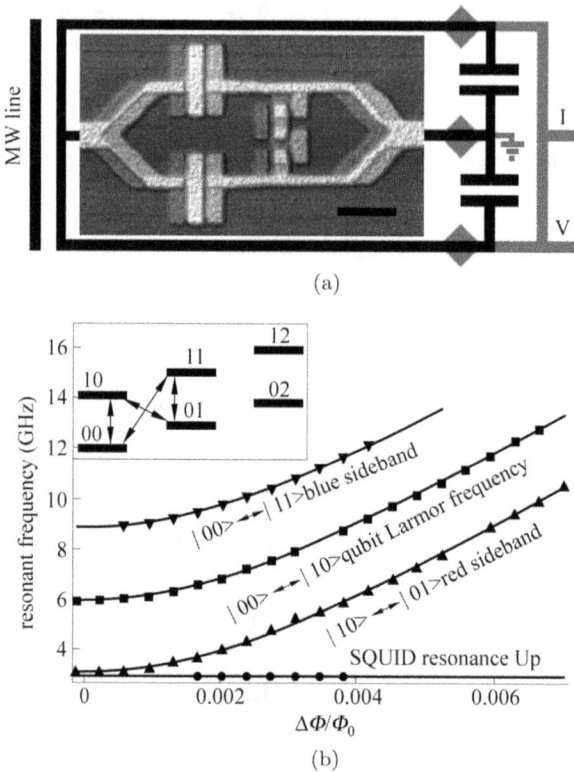

Fig. 8.36 Qubit-SQUID device and spectroscopy (a) Atomic force micrograph of the system; (b) Resonant frequencies versus $\Delta\Phi = \Phi - \Phi_0/2$. inset: levels of the system at a given bias. Reprinted with permission from Macmillan Publishers Ltd: Nature (431, 159), copyright (2004).

directions coupled by tunnelling.[21] The microwave field is provided by the coplanar waveguide (MW line) and couples inductively to the qubit. The current line (I) delivers the readout pulses, and the switching event[22] is detected on the voltage lin (V). The strong qubit-SQUID coupling enables to investigate the coupled dynamics of the qubit-harmonic oscillator system.

---

[21] See Section 6.3 and Fig. 6.8. More recent developments can be found in Ref. [43].
[22] The probability for the SQUID switching to a finite voltage regime, which depends on the excited state occupancy, rises or drops when Rabi oscillation occurs.

The dynamics of the qubit loop is determined by the coupling energy ε and the tunnelling splitting Δ. When the superconductor phase across the 3 junctions γ_q is equal to π, the loop is a symmetric double well, and the states $|\uparrow\rangle$ and $|\downarrow\rangle$ are degenerate, and $\varepsilon = 0$ at this symmetric point. The tunnelling breaks the degeneracy and the linear combinations $|\pm\rangle$ are now the energy eigenstates and have an energy difference Δ. A change of SQUID bias current (and correspondingly deviation of the flux from $\Phi_0/2$) leads to an energy difference between the energy eigenstates equal to $\sqrt{\varepsilon^2 + \Delta^2}$. The Hamiltonian of the qubit system is given by

$$H_q = -\varepsilon \frac{\sigma_z}{2} - \Delta \frac{\sigma_x}{2}, \tag{8.13.1}$$

The SQUID is described by the Hamiltonian

$$H_r = \hbar\omega_r \left(a^\dagger a + \frac{1}{2} \right), \tag{8.13.2}$$

where $\omega_r = 1/\sqrt{LC}$ with L and C the effective self-inductance and capacitance of the SQUID circuit, and $a^\dagger a$ the photon number operator. The qubit-SQUID coupling originates from the current distribution in the shared branches, and gives rise to the interaction Hamiltonian

$$H_{qr} = \lambda \sigma_x (a + a^\dagger), \tag{8.13.3}$$

where λ is the coupling strength. The state of the coupled system is denoted by $|\beta n\rangle$, where β represents the state of the qubit ($q = 0$ for the ground state and $q = 1$ for the excited state), and n represents state of the oscillator, $n = 0, 1, 2, \ldots$. The level diagram is sketched in the inset of Fig. 8.36(b). The system is initialized in the ground state $|00\rangle$. With successive resonant pulses controlled superposition of various $|\beta n\rangle$ states can be achieved. By applying a long microwave pulse with various frequencies one can measure the SQUID switching probability to the finite voltage regime, which is proportional to the excited state occupancy. The readout is performed by applying a short current pulse (I) and by monitoring whether the SQUID switches to the finite-voltage regime. Results are represented in Fig. 8.37. In the upper scan (after a π pulse) the system is first

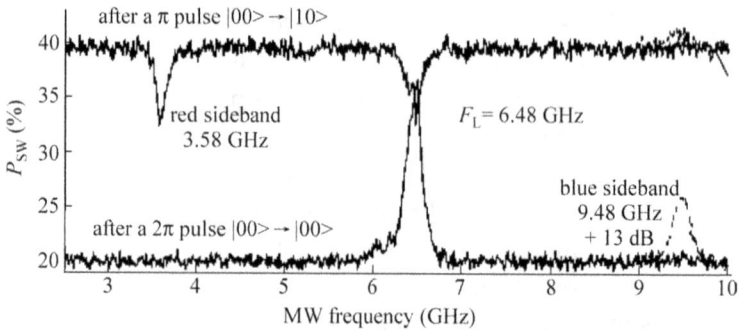

Fig. 8.37   Spectroscopic characterization of the energy levels. Reprinted with permission from Macmillan Publishers Ltd: Nature (431, 159), copyright (2004).

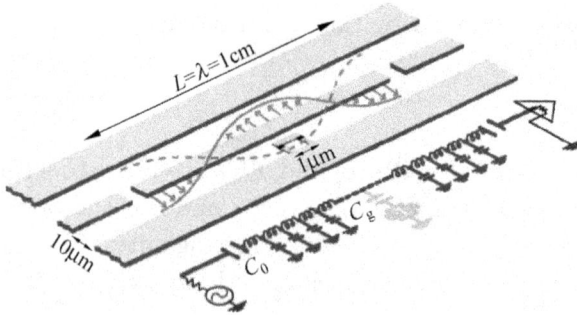

Fig. 8.38   Schematic layout of the coupled qubit-resonator system. Reprinted with permission from Alexandre Blais, Ren-Shou Huang, Andreas Wallraff, S. M. Girvin, and R. J. Schoelkopf, *Phys. Rev. A* 69, 062320, 2004. Copyright (2017) by the American Physical Society.

excited to $|10\rangle$, from which it decays to $|01\rangle$ (red sideband) or to $|00\rangle$ (the Larmor frequency). In the lower scan (after a 2π pulse) the system is rotated back to the ground $|00\rangle$ state, from which it is excited to the $|10\rangle$ or the $|11\rangle$ (blue sideband) states. A manifold of 3 resonance frequencies are observed in Fig. 8.36(b). Besides the Rabi oscillations between qubit states $|00\rangle \leftrightarrow |10\rangle$, the red sideband corresponding to Rabi oscillations $|10\rangle \leftrightarrow |01\rangle$ and the blue sideband corresponding to $|00\rangle \leftrightarrow |11\rangle$ are also observed. This gives convincing evidence that the qubit and the oscillator states are strongly coupled.

The Yale group used on-chip cavity and a Cooper pair box, shown in Fig. 8.38. The one-dimensional transmission line resonator consists of a lithographically fabricated full-wave section of superconducting coplanar waveguide. A Cooper pair box qubit is placed between the superconducting lines and is capacitively coupled to the center trace at an antinode of the voltage standing wave, yielding a strong electric dipole interaction between the qubit and a photon in the cavity. The Cooper pair box consists of two small Josephson junctions configured in a loop enabling tuning of the effective Josephson energy by an external flux [44]. Input and output signals are coupled to the resonator via the capacitive gaps in the center line from transmission lines which allow measurements of the amplitude and phase of the cavity state and the introduction of dc and rf pulses to manipulate the qubit states. The Hamiltonian of the cavity is again given by (8.12.2). At the operating temperature $T < 100\,\mathrm{nK}$ the resonator is nearly in its ground state with a thermal occupation $\langle a^\dagger a \rangle < 0.06$. The average photon lifetime in the cavity exceeds $100\,\mathrm{ns}$. The vacuum fluctuation of the resonator gives rise to an electric field between the central conductor and the ground plane $E_{\mathrm{rms}} \simeq 0.2\mathrm{Vm}^{-1}$, some hundred times larger than usual 3-dimensional cavities. The equivalent circuit of the coupled qubit-resonator system is sketched in Fig. 8.39.

A gate voltage V_{g} applied to the input port induces a gate charge n_{g} that controls the electrostatic energy of the Cooper pair box $E_{\mathrm{el}} = 4E_C(1-n_{\mathrm{g}})$, where E_C is the capacitive energy. The Josephson energy is $E_{\mathrm{J}} = E_{\mathrm{J\,max}} \cos \pi \frac{\Phi}{\Phi_0}$. As pointed out in Ref. [44], the relevant

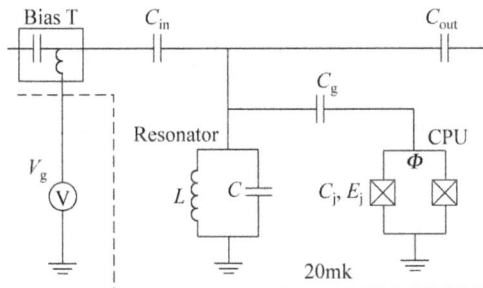

Fig. 8.39 Equivalent circuit of the coupled qubit-resonator system, CPB denotes the Cooper pair box. Reprinted with permission from Macmillan Publishers Ltd: Nature (431, 162), copyright (2004).

degree of freedom is the number of Cooper pairs on the island. In the charge regime $4E_C \gg E_J$ and restricting the gate charge to the range $n_g \in [0,1]$, only a pair of adjacent charge states on the island are relevant and the Hamiltonian is reduced to that of a two-level system:

$$H_a = -E_{el}\frac{\sigma_x}{2} - E_J\frac{\sigma_z}{2}, \qquad (8.13.4)$$

where the ground and excited states are represented by Pauli spinors $|\downarrow\rangle$ and $|\uparrow\rangle$. The energy separation $\hbar\omega_a$ between these states can be tuned by varying the external flux bias and the gate voltage. The Cooper pair box couples to photons stored in the resonator by electric dipole interaction via the coupling capacitance C_g. The vacuum voltage fluctuation V_{rms} of the resonator change the energy of a Cooper pair by an amount $\hbar g = dE_{rms}$, where d is the dipole moment. This coupled system is described by a Jaynes-Cummings Hamiltonian [44].

$$H_{JC} = H_a + H_r + \hbar g(a^\dagger \sigma^- + a\sigma^+). \qquad (8.13.5)$$

Let ω_r be the bare resonance frequency of the resonator. When the detuning $\Delta = \omega_a - \omega_r$ is equal to zero, the eigenstates of the coupled system are symmetric and antisymmetric superpositions $|\pm\rangle = (|0,\uparrow\rangle \pm |1,\downarrow\rangle)/\sqrt{2}$ with energies $E_\pm = \hbar(\omega_r \pm g)$, where the state is characterized by $|$resonator, qubit$\rangle$. In the case of large detuning, the Hamiltonian is diagonalized to be [44]

$$H \approx \hbar\left(\omega_r + \frac{g^2}{\Delta}\sigma_z\right)a^\dagger a + \frac{1}{2}\hbar\left(\omega_a + \frac{g^2}{\Delta}\right)\sigma_z. \qquad (8.13.6)$$

The transition frequency of the resonator is now conditioned by the qubit state, and therefore by measuring the transition frequency the qubit state can be determined. Similarly, the level separation in the qubit depends on the number of photons in the resonator.

The transmission of a microwave probe beam is measured as a function of its frequency ν_{RF}. In Fig. 8.40(a) the normalized transmission spectrum for large detuning is shown. The transition frequency $\omega_r - \frac{g^2}{\Delta}$ for the transition $|0,\downarrow\rangle \rightarrow |1,\downarrow\rangle$ is observed. For

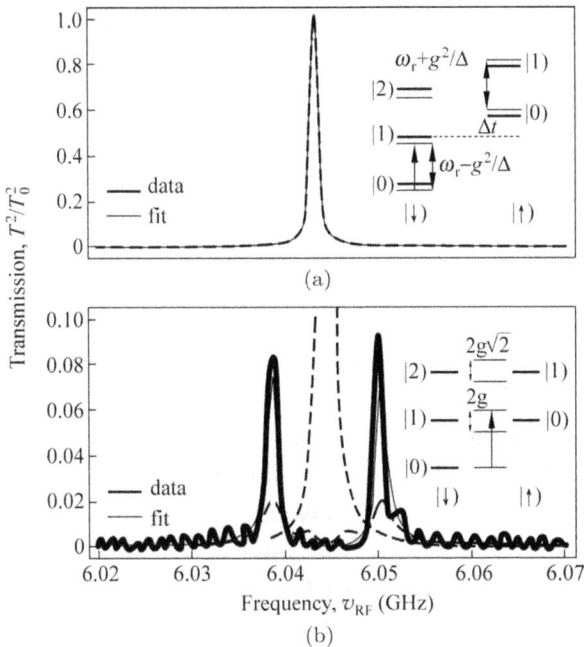

Fig. 8.40 (a) Transmission spectrum for large detuning; (b) Transmission for resonant case, showing Rabi mode splitting. The dashed line is the calculated transmission for $g = 0$ ($|0, \uparrow\rangle$ and $|1, \downarrow\rangle$ degenerate). Reprinted with permission from Macmillan Publishers Ltd: Nature (431, 162), copyright (2004).

the resonance case $\Delta = 0$ at $n_g = 1$, Fig. 8.40(b) shows the splitting of the Rabi frequencies, corresponding to the oscillation $|0, \downarrow\rangle \leftrightarrow |\pm\rangle$.

These experiments show that a high level of coherence and control is possible in Josephson circuits, a situation very significant for quantum communication and quantum computing.

8.14 Radiative Corrections

In photon absorption and emission processes, the states of radiation field (say number of photons) and of electrons (say their energies and momenta) undergo changes during the process. There are processes for which the initial and final states of the radiation field and the electron remain unchanged. The Lamb shift belongs to this category. These are called the radiative correction processes.

8.14.1   Spontaneous radiative correction: Free electron

We begin with the simple case of a free electron with momentum $\boldsymbol{p}$ and the radiation field vacuum $|0\rangle$. The state of the system is denoted by $|\boldsymbol{p}; 0\rangle$. Due to the interaction between the radiation field and the electron

$$H_{\text{int}}^{(1)} = -\frac{e}{mc} \boldsymbol{A} \cdot \boldsymbol{p}, \qquad (8.14.1)$$

the electron may emit a photon $\boldsymbol{k}, \hat{\boldsymbol{\varepsilon}}$, and its momentum becomes $\boldsymbol{p} - \hbar\boldsymbol{k}$ due to the conservation of momentum. The system is now denoted by $|\boldsymbol{p} - \hbar\boldsymbol{k}; \boldsymbol{k}, \hat{\boldsymbol{\varepsilon}}\rangle$. The photon is subsequently reabsorbed by the electron and the final state of the system bocomes again $|\boldsymbol{p}; 0\rangle$. Actually in the non-relativistic formalism of quantum electrodynamics, the energy of the intermediate state $\frac{1}{2m}(\boldsymbol{p} - \hbar\boldsymbol{k})^2 + \hbar\omega\,(\omega = kc)$ is not equal to the energy of the initial (and the final) state $\frac{1}{2m}\boldsymbol{p}^2$. The intermediate state can only exist for a very short period of time compatible with the deviation from energy conservation permitted by the uncertainty principle. The process is therefore virtual. Another interaction term

$$H_{\text{int}}^{(2)} = \frac{e^2}{2mc^2} \boldsymbol{A}^2 \qquad (8.14.2)$$

also contributes, where one operator $\boldsymbol{A}$ emits a photon, and the other operator $\boldsymbol{A}$ reabsorbs it. The processes are illustrated in Fig. 8.41.

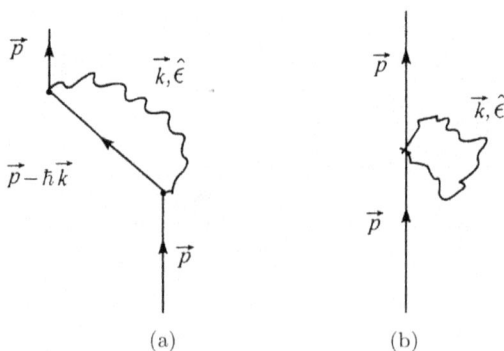

(a)   (b)

Fig. 8.41   Second order processes in coupling e for a free electron and the vacuum of the field.

The theoretical description of such processes can be carried out most conveniently in the interaction picture of quantum mechanics. The Hamiltonian of the system is

$$H = H_0 + H_{\text{int}}. \tag{8.14.3}$$

In this picture, eigenstates of H_0 are time independent. By the operation of H_{int} an eigenstate is transferred to another with a transition amplitude determined by the operator

$$\tilde{U}(t) = e^{iH_0 t/\hbar} e^{-iH_{\text{int}} t/\hbar}. \tag{8.14.4}$$

Since the initial and final states are identical, the relevant physical quantity of interest is the "self-persistence amplitude" $\langle \boldsymbol{p}; 0|\tilde{U}(t)|\boldsymbol{p}; 0\rangle$, with $H_{\text{int}} = H_{\text{int}}^{(1)} + H_{\text{int}}^{(2)}$ given by (8.14.1) and (8.14.2). As we can expect, this matrix element can only be a phase factor

$$\langle \boldsymbol{p}; 0|\tilde{U}(t)|\boldsymbol{p}; 0\rangle = e^{-i\frac{\delta E^{(1)} + \delta E^{(2)}}{\hbar} t} \tag{8.14.5}$$

$$\delta E^{(1)} = \sum_{\boldsymbol{k}, \varepsilon} \frac{\left| \langle \boldsymbol{p} - \hbar \boldsymbol{k}; \boldsymbol{k}, \hat{\varepsilon} | H_{\text{int}}^{(1)} | \boldsymbol{p}; 0\rangle \right|^2}{\frac{1}{2m} \boldsymbol{p}^2 - \frac{1}{2m} (\boldsymbol{p} - \hbar \boldsymbol{k})^2 - \hbar \omega}, \tag{8.14.6}$$

$$\delta E^{(2)} = \langle \boldsymbol{p}; 0| H_{\text{int}}^{(2)} | \boldsymbol{p}; 0\rangle, \tag{8.14.7}$$

where $\delta E^{(1)}$ and $\delta E^{(2)}$ are the energy shifts associated with the processes in Fig. 8.41(a) and 8.41(b) respectively. The energy shift of a free electron can be interpreted as the electron "dressed" by a cloud of virtual photons describing the state of the transverse field coupled to the electron and thus acquiring a correction in its mass δm. $\delta E^{(2)}$ is evaluated [45] in terms of the mode expansion of the transverse field strength $\mathcal{E}_\omega$ defined by

$$\boldsymbol{E}_\perp(\boldsymbol{r}) = \int d^3 k \sum_{\hat{\varepsilon}} i\mathcal{E}_\omega \left[\hat{\varepsilon} a_{\hat{\varepsilon}}(\boldsymbol{k}) e^{i\boldsymbol{k}\cdot\boldsymbol{r}} - \hat{\varepsilon} a_{\hat{\varepsilon}}^\dagger(\boldsymbol{k}) e^{-i\boldsymbol{k}\cdot\boldsymbol{r}} \right], \tag{8.14.8}$$

$$\delta E^{(2)} = \sum_{\boldsymbol{k}, \varepsilon} \frac{e^2 \mathcal{E}_\omega^2}{2m\omega^2}. \tag{8.14.9}$$

As to $\delta E^{(1)}$, the integral over $\boldsymbol{k}$ leads to divergence, which we shall address briefly in the case of bound electrons,

8.14.2 Spontaneous radiative correction: The bound electron

Another example of such processes is the reaction of the radiating field on the radiating atom. When an atom decays spontaneously from an excited state, it emits a photon into the radiating field, and subsequently this photon is reabsorbed by the atom. Correspondingly the atom returns to the excited state. The diagrams for this process look identical to Fig. 8.41, only the bound state of the electron is now denoted by b (specific) in the initial and final state, and by a (all possible bound states) in the intermediate state as shown in Fig. 8.42.

The essential difference between Fig. 8.42(a) and Fig. 8.41(a) is that here the intermediate state $|a; \boldsymbol{k}, \hat{\varepsilon}\rangle$ with the same energy as the initial state $|b, 0\rangle$ can appear when b is an excited state, and thus the process can be real. Correspondingly the phase similar to that defined in (8.14.5) can be complex:

$$\langle b; 0| \, \tilde{U}(t) \, |b; 0\rangle = e^{-i\left(\Delta - i\frac{\Gamma}{2}\right)t}, \qquad (8.14.10)$$

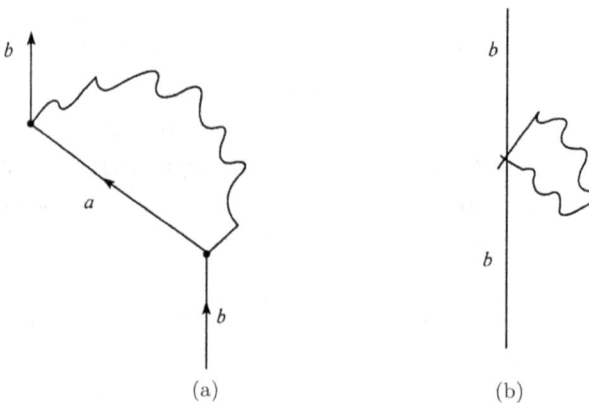

(a)   (b)

Fig. 8.42 Emission and reabsorption of a photon by an atom initially in the bound state b.

where $\hbar\Delta$ (the parallel of δE_1) represents the energy shift. Γ is evaluated [45] to be the probability per unit time of the transition $b \to$ all lower levels by real emission of a photon. The energy shift is

$$\hbar\Delta = \hbar\Delta_1 + \hbar\Delta_2, \tag{8.14.11}$$

$$\hbar\Delta_1 = \mathcal{P} \sum_a \sum_{\mathbf{k},\hat{\varepsilon}} \frac{\left| \langle a; \mathbf{k}, \hat{\varepsilon} | H_{\text{int}}^{(1)} | b; 0 \rangle \right|^2}{E_b - E_a - \hbar\omega}, \tag{8.14.12}$$

$$\hbar\Delta_2 = \langle b; 0 | H_{\text{int}}^{(2)} | b; 0 \rangle . \tag{8.14.13}$$

In (8.14.12) $\mathcal{P}$ denotes the principal part, used when the real transition takes place. As for $\hbar\Delta_1$, it diverges linearly, i.e., the integral is linearly proportional to the upper limit of integration over ω, which is to approach infinity. This problem has plagued quantum electrodynamics since 1930s. After the discovery of Lamb shift that the $2p_{1/2}$ and $2s_{1/2}$ levels of hydrogen atom have an energy difference of 1057 Mc/s while Dirac's theory predicts that they have the same energy, physicists had to face directly this problem. The removal of the divergence is accomplished by invoking the idea of mass renormalization. The first successful step was taken by H. A. Bethe. Actually we cannot have a "bare" electron in reality, it carries electromagnetic field (dressed by photons) as in the case of mass correction δm for the free electron. The observed kinetic energy of the electron should be [46]

$$T_{\text{obs}} = \frac{1}{2m_0} \left[1 - \frac{\delta m}{m_0} \right] p^2 = \frac{1}{2m_{\text{obs}}} p^2. \tag{8.14.14}$$

Using this kinetic energy in the Hamiltonian, the energy shift turns out to be logarithmically divergent. This represents a significant improvement, since the physics depends much less on the cutoff. By the argument that low energy physics must not depend sensitively on high energy virtual intermediate states, the cutoff of integration over ω can be chosen as mc^2. Such a choice gives a level shift 1040 Mc/s in excellent agreement with the measured value of 1057 Mc/s. Relativistic calculations of Feynman and Bethe give exact agreement without divergence when Dirac wave function is used.

Renormalization theory of quantum electrodynamics is one of the most prominent achievements in modern theoretical physics. R. P. Feynman, J Schwinger and S. Tomonaga were awarded the Nobel prize in 1965 for this achievement.

8.14.3 Stimulated radiative corrections

Suppose that in the initial and final states there are several identical photons $k, \hat{\varepsilon}$ and the atomic electron absorbs one of the photons and then reemits it in the same state, as shown in Fig. 8.43(a), or emits in a stimulated fashion a photon identical to the incident photons and then reabsorbs it, as shown in Fig. 8.43(b). The initial and final states of both the atom and the field are again identical. The only difference between these processes and the spontaneous radiation correction is that now the radiation field is not in the vacuum state. The Hamiltonian $H_{\text{int}}^{(2)}$ can also contribute in the first order, as in 8.14.1. The probability amplitude for the system to remain in the same state is very similar to (8.14.10)

$$\langle a; k, \hat{\varepsilon}, \text{ other photons}| \tilde{U}(t) \,|a; k, \hat{\varepsilon}, \text{ other photons}\rangle = e^{-i\left(\Delta' - i\frac{\Gamma'}{2}\right)t}.$$

$$(8.14.15)$$

The damping rate Γ represents the probability per unit time for the system to leave its initial state. Equivalently it is the rate of

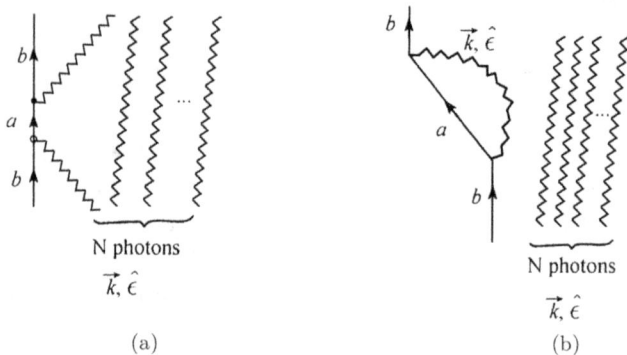

Fig. 8.43   (a) absorption and reemission of an incident photon; (b) induced emission of a photon and reabsorption of it with the photon in the same state of an incident photon.

forward scattering of one of the incident photons by the atom, a real process. It should be proportional to the total scattering cross-section. Actually it is the case due to the optical theorem which states that the total cross-section for the collision corresponding to initial state a is proportional to the imaginary part of the forward scattering amplitude $f_a(0)$:

$$\sigma_a^{tot} = \frac{4\pi}{k} Im f_a(0). \tag{8.14.16}$$

From the viewpoint of the atom Γ' is a radiative broadening of the level a caused by the N incident photons. In contrast to Γ of the spontaneous radiative correction which is an "intrinsic" quantity, Γ' depends on the characteristics of the incident beam (intensity, polarization, spectral distribution, etc.).

8.15 Interaction with Monochromatic Radiation, Bloch Sphere, and Light Shift

8.15.1 Population change and Bloch sphere

Consider a two-level atom interacting with a monochromatic electromagnetic field of frequency ω. The Hamiltonian of the system is

$$H = H_0 + H_1, \tag{8.15.1}$$

with the two levels being eigenstates of H_0

$$\begin{aligned} H_0 \left|1\right\rangle &= \hbar\omega_1, \\ H_0 \left|2\right\rangle &= \hbar\omega_2. \end{aligned} \tag{8.15.2}$$

The interaction Hamiltonian is

$$H_1 = e\boldsymbol{r} \cdot \boldsymbol{E}_0 \cos \omega t, \tag{8.15.3}$$

where $\boldsymbol{r}$ is the position of the electron with respect to the center of mass of the atom. The interaction mixes $\left|1\right\rangle$ and $\left|2\right\rangle$, and the system evolves according to the Schrödinger equation

$$H\psi = i\hbar \frac{\partial \psi}{\partial t}. \tag{8.15.4}$$

The solution has the form

$$\psi(\boldsymbol{r},t) = c_1 \, |1\rangle \, e^{-i\omega_1 t} + c_2 \, |2\rangle \, e^{-i\omega_2 t} \qquad (8.15.5)$$

with $|c_1|^2 + |c_2|^2 = 1$. Substituting (8.15.5) into (8.15.4) leads to the evolution equations for c_1 and c_2.

$$i \, \dot{c}_1 = \Omega \cos \omega t \cdot e^{-i\omega_0 t} c_2, \qquad (8.15.6)$$

$$i \, \dot{c}_2 = \Omega^* \cos \omega t \cdot e^{i\omega_0 t} c_1, \qquad (8.15.7)$$

where

$$\omega_0 = \omega_2 - \omega_1, \qquad (8.15.8)$$

and the Rabi frequency

$$\Omega = \frac{\langle 1| \, e \, \boldsymbol{r} \cdot \boldsymbol{E}_0 \, |2\rangle}{\hbar}. \qquad (8.15.9)$$

Assume that the electric field is along $\hat{e}_x$, the dipole moment induced is

$$-eD_x(t) = - \int \psi^*(t) e x \psi(t) \mathrm{d}^3 r. \qquad (8.15.10)$$

Using (8.15.5) gives

$$D_x = c_1 c_2^* X_{21} \, e^{i\omega_0 t} + c_1^* c_2 X_{12} \, e^{-i\omega_0 t}, \qquad (8.15.11)$$

where

$$X_{12} = \langle 1| \, x \, |2\rangle. \qquad (8.15.12)$$

The bilinear quantities $c_1 c_2^*$ and $c_1^* c_2$ are actually the non-diagonal elements of the density matrix of the 2-level system

$$\rho = |\psi\rangle \, \langle \psi| = \begin{pmatrix} |c_1|^2 & c_1 c_2^* \\ c_1^* c_2 & |c_2|^2 \end{pmatrix}. \qquad (8.15.13)$$

Define the detuning of the radiation from the atomic transition,

$$\delta = \omega - \omega_0 \qquad (8.15.14)$$

Then by defining new variables $\tilde{c}_1, \tilde{c}_2$

$$\begin{aligned} \tilde{c}_1 &= c_1 e^{-i\delta t/2}, \\ \tilde{c}_2 &= c_2 e^{i\delta t/2}, \end{aligned} \tag{8.15.15}$$

and correspondingly

$$\begin{aligned} \tilde{\rho}_{12} &= \tilde{c}_1 \tilde{c}_2^* = \rho_{12} e^{-i\delta t}, \\ \tilde{\rho}_{21} &= \tilde{c}_1^* \tilde{c}_2 = \rho_{21} e^{i\delta t}, \end{aligned} \tag{8.15.16}$$

we can express the induced dipole moment in terms of ω or ω_0 at our convenience.

$$\begin{aligned} -eD_x(t) &= -eX_{12}\{\rho_{12}\, e^{i\omega_0 t} + \rho_{21}\, e^{-i\omega_0 t}\} \\ &= -eX_{12}\{\tilde{\rho}_{12}\, e^{i\omega t} + \tilde{\rho}_{21}\, e^{-i\omega t}\}. \end{aligned} \tag{8.15.17}$$

The evolution equations for c_1 and c_2 (8.15.6) and (8.15.7) in the rotating wave approximation[23] become

$$i\,\dot{c}_1 = c_2 \frac{\Omega}{2} e^{i\delta t},$$

$$i\,\dot{c}_2 = c_1 \frac{\Omega}{2} e^{-i\delta t}. \tag{8.15.18}$$

Equations for $\tilde{c}_1$ and $\tilde{c}_2$ are obtained by using (8.15.15),

$$i\tilde{c}_1 = \frac{1}{2}\left(\delta\tilde{c}_1 + \Omega\tilde{c}_2\right),$$

$$i\tilde{c}_2 = \frac{1}{2}\left(\Omega\tilde{c}_1 - \delta\tilde{c}_2\right). \tag{8.15.19}$$

Defining twice the real and imaginary parts of the non-diagonal elements (called the coherences) of the density matrix $\tilde{\rho}$,

$$\begin{aligned} u &= \tilde{\rho}_{12} + \tilde{\rho}_{21}, \\ v &= -i\left(\tilde{\rho}_{12} - \tilde{\rho}_{21}\right), \end{aligned} \tag{8.15.20}$$

and the population imbalance

$$w = \rho_{11} - \rho_{22}, \tag{8.15.21}$$

---

[23] Please refer to Section 8.2.

we obtain the evolution equations for the system in compact form

$$\dot{u} = \delta v,$$
$$\dot{v} = -\delta u + \Omega w,$$
$$\dot{w} = -\Omega v. \tag{8.15.22}$$

Being directly related to the elements of the density matrix, (u, v, w) are the characteristics of the system. Define them as the components of the Bloch vector

$$\boldsymbol{R} = u\hat{e}_1 + v\hat{e}_2 + w\hat{e}_3. \tag{8.15.23}$$

Parameters Ω and δ characterize the fields, and are constant in time. They combine to define $\boldsymbol{W}$

$$\boldsymbol{W} = \Omega\hat{e}_1 + \delta\hat{e}_3. \tag{8.15.24}$$

Equations (8.15.22) can be written in the vector form

$$\dot{\boldsymbol{R}} = \boldsymbol{R} \times \boldsymbol{W}. \tag{8.15.25}$$

From (8.15.25) we see that $\dot{\boldsymbol{R}}$ is always orthogonal to $\boldsymbol{R}$, i.e., the length of $\boldsymbol{R}$ is a constant, and actually this constant is unity, as can be seen from a special case $u = v = 0$ and $w = 1$, for which $|\boldsymbol{R}| = 1$ and remains constant. The Bloch vector moves on a unit sphere, the Bloch sphere. The position vectors of points on Bloch sphere represent the states of the system in Hilbert space. At the poles are $\boldsymbol{R} = w\hat{e}_3$ with $w = \pm 1$ corresponding to $|1\rangle$ and $|2\rangle$ respectively, as shown in Fig. 8.44(a). States on the equator $\boldsymbol{R} = u\hat{e}_1 + v\hat{e}_2$ are shown in Figs. 8.44(b). From (8.15.25) we get $\dot{\boldsymbol{R}} \cdot \boldsymbol{W} = 0$, i.e., $\boldsymbol{R} \cdot \boldsymbol{W} =$ const. $\boldsymbol{W}$ is a constant vector and $\boldsymbol{R}$ is a vector of unit length, and consequently $\boldsymbol{R}$ moves around a cone about $\boldsymbol{W}$, as shown in Fig. 8.44(d). A special case for evolution $|1\rangle \rightarrow |2\rangle \rightarrow |1\rangle$ driven by a resonant field $\delta = 0$ i.e., $\boldsymbol{W} = \Omega\hat{e}_1$ is shown in Fig. 8.44(c) [47].

8.15.2 Light shift

Besides the population change on levels $|1\rangle$ and $|2\rangle$, the radiation also changes the energy of the levels. This is called the light shift.

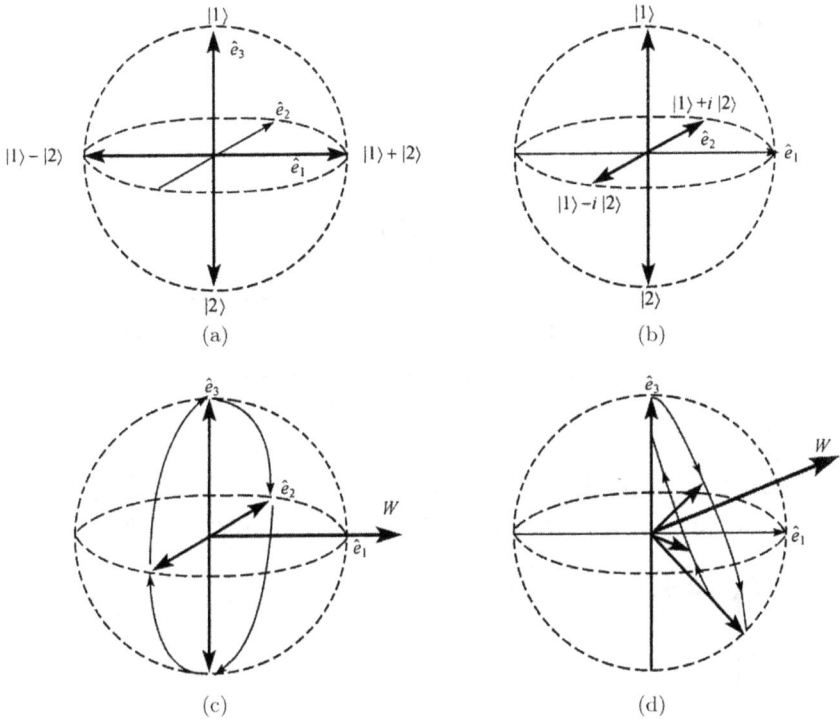

Fig. 8.44 The Bloch sphere. (a) and (b) show states $|1\rangle$, $|2\rangle$ and their linear superpositions, (c) shows the evolution $|1\rangle \rightarrow |2\rangle \rightarrow |1\rangle$ driven by the resonant field, (d) shows the general evolution for $\delta \neq 0$. From Ref. [47].

We rewrite (8.4.19) in matrix form

$$i\frac{\mathrm{d}}{\mathrm{d}t}\begin{pmatrix} \tilde{c}_1 \\ \tilde{c}_2 \end{pmatrix} = \begin{pmatrix} \delta/2 & \Omega/2 \\ \Omega/2 & -\delta/2 \end{pmatrix}\begin{pmatrix} c_1 \\ c_2 \end{pmatrix}. \qquad (8.15.26)$$

We look for solutions in the form

$$\begin{pmatrix} \tilde{c}_1 \\ \tilde{c}_2 \end{pmatrix} = \begin{pmatrix} a \\ b \end{pmatrix} e^{-i\lambda t}. \qquad (8.15.27)$$

Substituting (8.15.27) into (8.15.26) leads to the condition for the eigenvalue λ:

$$\begin{pmatrix} \delta/2 & \Omega/2 \\ \Omega/2 & -\delta/2 \end{pmatrix}\begin{pmatrix} a \\ b \end{pmatrix} = \lambda\begin{pmatrix} a \\ b \end{pmatrix}. \qquad (8.15.28)$$

i.e., the vanishing of the secular determinant

$$\begin{vmatrix} \delta/2 - \lambda & \Omega/2 \\ \Omega/2 & -\delta/2 - \lambda \end{vmatrix} = \lambda^2 - \left(\frac{\delta}{2}\right)^2 - \left(\frac{\Omega}{2}\right)^2 = 0. \quad (8.15.29)$$

The eigenvalues are

$$\lambda = \pm \frac{1}{2} \left(\delta^2 + \Omega^2\right)^{1/2}. \quad (8.15.30)$$

For $\Omega = 0$, the unperturbed eigenvalues are $\lambda = \pm\delta/2$, corresponding to two levels δ apart. These two states are the excited state at E_2 and the atom-photon state at $E_1 + \hbar w_1$, which are to be mixed by the radiation. At $\Omega = 0$, their separation is $E_2 - E_1 - \hbar w = \hbar (w_0 - w) = |\delta|$. With Ω finite the level separation becomes a function of Ω shown in Fig. 8.45(a) and (b) for $\delta < 0$ and $\delta > 0$ respectively. The effect is called the light shift, also called the a.c. Stark effect. Understanding

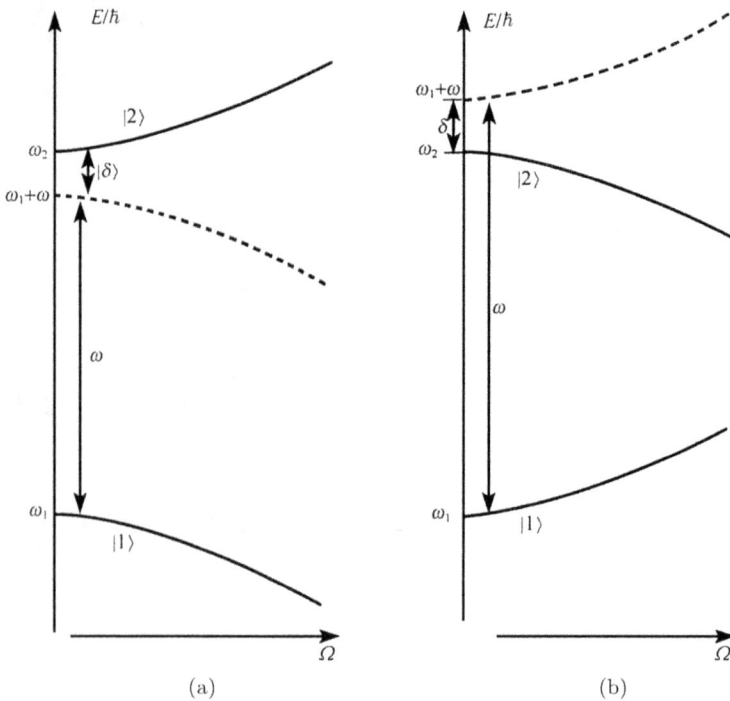

Fig. 8.45   a.c. Stark effect for negative and positive frequency detuning as functions of Rabi frequency. From Ref. [47].

this phenomenon is crucial for precision measurement of time, the atomic clock.

8.16 Renormalization in Quantum Field Theory

In discussing the radiative corrections in 8.14 we mentioned the renormalization theory of QED. In the following we elaborate on the main ideas of renormalization with the aim of introducing a more involved development of the renormalization group approach in the next section. We begin with the scalar boson field theory with Lagrangian density

$$\mathcal{L} = \frac{1}{2}(\partial\phi)^2 - m^2\phi^2 - \frac{\lambda}{4!}\phi^4. \qquad (8.16.1)$$

It is usually referred to as the ϕ^4 theory. It displays all relevant features of field theories, and can avoid complicated details of the gauge invariance of, for example, quantum electrodynamics. It also appears in some practical applications. For instance the d-dimensional Ising model in the low energy regime is described by the ϕ^4 theory. Without the interaction term, wave like fluctuations of the magnetization can be described. But phase transitions can only be represented when the ϕ^4 term is added.

Consider meson-meson scattering to the second order of λ, described by the following Feynman diagrams (Fig. 8.46).

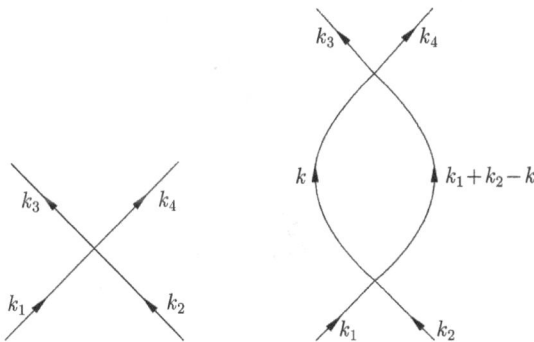

Fig. 8.46   Feynman diagrams for meson-meson scattering to second order of λ in ϕ^4 theory.

The amplitude of this process is given by

$$\mathcal{M} = -i\lambda + \frac{1}{2}(-i\lambda)^2 i^2 \int \frac{d^4k}{(2\pi)^4} \frac{1}{k^2 - m^2 + i\varepsilon} \frac{1}{(K-k)^2 - m^2 + i\varepsilon},$$
$$(8.16.2)$$

where

$$K = k_1 + k_2.$$

Here the k's are all 4-momenta. The integral is logarithmically divergent, as can be seen using power counting method. In a general picture, the perturbation theory works well in the first non-vanishing order, but is divergent in next order. In the 1930s and 1940s physicists struggled with the infinities in quantum field theory without success. The situation became even more tense when Lamb shift was discovered, that the fine structure of hydrogen atom was found to deviate from Dirac theory. The "next order" problem must be faced directly. H. A. Bethe introduced a cut-off of order mc^2 in the momentum integral, arguing that the electron-photon interaction at high energy can be neglected for low energy phenomena like hydrogen atomic spectrum. He obtained a result very close to the experimental value obtained by Lamb. As we mentioned in Section 8.14 the effort culminated in the triumph of renormalization theory of quantum electrodynamics in the late 1940s and early 1950s. Here we demonstrate the procedure using the ϕ^4 theory. In the language of A. Zee [48][24] the procedure is actually the "cutting off of our ignorance".

Is the introduction of a cut-off something unnatural, or "putting in by hands" Is it somehow "illegal"? It would be utterly unreasonable to insist that QED would hold to arbitrarily high energies. At the very least, with increasingly higher energies particles other than electrons and photons come in, and eventually QED becomes merely part of a larger electroweak theory. The modern view is that quantum field theory should be regarded as an effective low energy theory, valid up to some energy (or momentum in a Lorentz invariant theory) scale Λ. Then the calculated amplitude would depend on Λ. The amplitude

---

[24]We follow this reference closely in this subsection.

$\mathcal{M}$ expressed in terms of Λ is called the *regularized* one, which is now free from divergence. The problem is, which value of Λ should we choose in order to get the correct result? The solution lies in the assertion that the parameter λ, an unknown input of the theory, is actually dependent on Λ. This is called the scale dependence of the coupling constant. We hope that the dependence is just appropriate as to cancel the Λ in the final physical amplitude. Actually this can be realized for our case. The regularized expression of $\mathcal{M}$ is

$$\mathcal{M} = -i\lambda + i\lambda^2 C \left[\log \frac{\Lambda^2}{s} + \log \frac{\Lambda^2}{t} + \log \frac{\Lambda^2}{u} \right] + O(\lambda^3), \quad (8.16.3)$$

where $s = (k_1 + k_2)^2$ is the center of mass energy squared, $t = (k_1 - k_3)^2$ is one of the 4-momentum transfer squared, and $u = (k_1 - k_4)^2$ is the other alternative. Assume that we carry out measurement at the kinetic variables (s_0, t_0, u_o), the quantity in the square bracket is collectively denoted by L_0. We denote the resulting amplitude by $\mathcal{M}(s_0, t_0, u_o) = -i\lambda_R$ as a definition. λ_R and (s_0, t_0, u_o) are called the *renormalized coupling constant* and the *renormalization point* respectively. Then λ and λ_R are connected by

$$-i\lambda_R = -i\lambda + i\lambda^2 C \left[\log \frac{\Lambda^2}{s_0} + \log \frac{\Lambda^2}{t_0} + \log \frac{\Lambda^2}{u_0} \right] + O(\lambda^3)$$

$$\equiv -i\lambda + i\lambda^2 C L_0 + O(\lambda^3). \quad (8.16.4)$$

We observe that λ and λ_R differ by a quantity second order in λ, such that we can write (8.16.3) in the same approximation by eliminating λ in favor of λ_R as

$$\mathcal{M} = -i\lambda_R - i\lambda_R^2 C(L - L_0) + O(\lambda_R^3). \quad (8.16.5)$$

It is encouraging to note that the quantity in the parenthesis

$$L - L_0 = \log \frac{s_0}{s} + \log \frac{t_0}{t} + \log \frac{u_0}{u}$$

is now independent of Λ! Consequently we have

$$\mathcal{M} = -i\lambda_R + i\lambda_R^2 C \left[\log \frac{s_0}{s} + \log \frac{t_0}{t} + \log \frac{u_0}{u} \right] + O(\lambda_R^3). \quad (8.16.6)$$

This is a perfect looking expression, in which λ_R is a measured coupling constant at the kinetic variables L_0, and the C is a constant calculable in the regularization procedure. The expression contains well-defined quantities and is free from any divergence. This second step is called renormalization.

In quantum electrodynamics renormalization has brought spectacular success. Successive orders of radiative corrections can be calculated leading to predicted results of very high precision. Take the example of anomalous magnetic moment of the electron. It was a great triumph for Dirac when he predicted directly from his equation that the magnetic moment of the electron is one Bohr magneton. With improvements in experimental techniques, it became clear by the late 1940's that the magnetic moment of the electron was larger than the value calculated by Dirac by a factor of 1.00118 ± 0.00003. The challenge to quantum electrodynamics was to calculate this so-called anomalous magnetic moment. We begin with the matrix element of the electromagnetic current between two electron states $|p, s\rangle$ and $|p', s'\rangle$. Lorentz invariance and energy conservation lead to the following expression [48]:

$$\langle p', s' | J^\mu(0) | p, s \rangle = \bar{u}(p', s') \left[\gamma^\mu F_1(q^2) + \frac{i\sigma^{\mu\nu} q_\nu}{2m} F_2(q^2) \right] u(p, s),$$

(8.16.7)

where $q \equiv (p' - p)$. The functions $F_1(q^2)$ and $F_2(q^2)$, are known as form factors. Using the Gordon decomposition[25] we obtain to leading order of q:

$$\langle p', s' | J^\mu(0) | p, s \rangle$$

$$= \bar{u}(p', s') \left[\frac{(p + p')^\mu}{2m} F_1(0) + \frac{i\sigma^{\mu\nu} q_\nu}{2m} [F_1(0) + F_2(0)] \right] u(p, s).$$

(8.16.8)

The coefficient of the first term should correspond to the physical charge of the electron and by definition is equal to 1. Thus $F_1(0) = 1$.

---

[25] $\bar{u}(p')\gamma^\mu u(p) = \bar{u}(p') \left[\frac{(p'+p)^\mu}{2m} + \frac{i\sigma^{\mu\nu}(p'-p)_\nu}{2m} \right] u(p)$

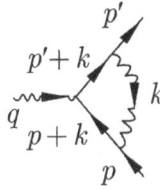

Fig. 8.47   Feynman diagram for the magnetic moment term.

The second term now tells us that the magnetic moment of the electron must be $1 + F_2(0)$ times the Bohr magneton. To evaluate the matrix element (8.16.9) to second order of the coupling constant e there are 5 Feynman diagrams, but for the magnetic moment term only one (Fig. 8.47) is relevant, all others are associated to the charge term. Schwinger met this challenge by using brilliant tricks [48] and in 1948 obtained the result $F_2(0) = \frac{\alpha}{2\pi}$, where α is the fine structure constant $\alpha = e^2/4\pi$. This had an electrifying impact on the physics community at that time. Kinoshita calculated radiative corrections to third order $(\alpha/\pi)^3$ and obtained the anomalous coefficient $a_e^{\text{th}} = 1159652359(282) \times 10^{-12}$ to be compared with the experimental value of 2012 $a_e^{\text{exp}} = 1259652180(76) \times 10^{-12}$. Numbers in parenthesis indicate uncertainties.

In such circumstances we have to keep in minds that QED is an effective field theory for low energies and that the perturbation series has the nature of an asymptotic expansion. In general an expansion in orders of a small parameter of a given expression may diverge. But a partial summation up to a limited number of terms can give a very good approximation to the given expression. An example illustrating this point was given in Ref. [49].

The process of renormalization is enlightening. But the next question is: are other field theories as lucky as the ϕ^4 theory? The answer is that some theories are renormalizable, as quantum electrodynamics (R. Feynman, J. Schwinger and S. Tomonaga, Nobel prize 1959) and the electroweak theory (G. 't Hooft and M. Veltman, Nobel prize 1999). But some others are not renormalizable, for example, the Fermi theory of weak interaction. A simple heuristic way of dimensional analysis helps in the understanding. In natural units in which $\hbar = 1$ and $c = 1$, length and time have the same dimension, the inverse of the dimension of mass (and of energy and momentum). We count dimension in terms of mass as they are used to thinking of energy scales. Since the action S appears in the path integral as e^{iS}, it is clearly dimensionless, thus the relation $S = \int d^4x \, \mathcal{L}$ implies that the Lagrangian density $\mathcal{L}$ has the same dimension as the 4th power of a mass. We will use the notation $[\mathcal{L}] = 4$ to indicate this. In this notation $[x] = -1$ and $[\partial] = 1$. Consider the scalar field theory $\mathcal{L} = \frac{1}{2}(\partial\phi)^2 - m^2\phi^2 - \frac{\lambda}{4!}\phi^4$. For the term $(\partial\phi)^2$ to have dimension 4, we see that $[\phi] = 1$. This then implies that $[\lambda] = 0$, the coupling λ is dimensionless. Applying this rule to the Lagrangian $\mathcal{L} = \bar{\psi}i\gamma_\mu\partial^\mu\psi + \ldots$ we see that $[\psi] = 3/2$. Looking at the coupling $f\phi\bar{\psi}\psi$ we see that the Yukawa coupling f is dimensionless. In contrast, in the Fermi theory of the weak interaction with $\mathcal{L} = G\bar{\psi}\gamma^\mu\psi\bar{\psi}\gamma_\mu\psi$ we see that the Fermi coupling G has dimension -2. From the Maxwell Lagrangian $-\frac{1}{4}F_{\mu\nu}F_{\mu\nu}$ we see that $[A_\mu] = 1$. The electromagnetic coupling $eA_\mu\bar{\psi}\gamma_\mu\psi$ tells us that e is dimensionless.

We are now ready for a heuristic argument regarding the nonrenormalizability of a theory. Consider Fermi's theory of the weak interaction. Imagine calculating the amplitude $\mathcal{M}$ for a four-fermion interaction, say neutrino-neutrino scattering at an energy much smaller than Λ. In lowest order, $\mathcal{M} \sim G$. The next order should be proportional to G^2, the factor to be multiplied by G^2 must have a dimension 2 in order to obtain for the second term the same dimension as the leading term, i.e., -2. The dominant scale of the theory is Λ, hence the second term must be $\sim G^2\Lambda^2$. Evaluation of $\mathcal{M}$ using Feynman rules leads to the same result. Successive higher orders leads to still stronger divergences. Such theories seem pathological. We know that the Fermi theory of weak interactions works very well when applied to low energy phenomena. Just like QED, it is the effective theory of weak interactions at low energies of a unified electroweak theory. Using the example given by Zee [48] we demonstrate how Fermi theory is obtained from a more comprehensive theory of vector boson of mass M coupled to a Dirac field with a dimensionless coupling constant g. The Lagrangian

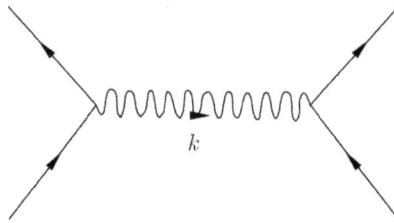

Fig. 8.48   Feynman diagram.

density is

$$\mathcal{L} = \bar{\psi}(i\gamma^\mu \partial_\mu - m)\psi - \frac{1}{4}F_{\mu\nu}F^{\mu\nu} + M^2 A_\mu A^\mu + gA_\mu \,\bar{\psi}\gamma^\mu\psi.$$

$$(8.16.9)$$

We calculate the fermion-fermion scattering amplitude generated by the Feynman diagram shown in Fig. 8.48 and obtain the result $(-ig)^2(\bar{u}\gamma^\mu u)[i/(k^2 - M^2 + i\varepsilon)](\bar{u}\gamma_\mu u)$. When $k \ll M$ the amplitude becomes $i(g^2/M^2)(\bar{u}\gamma^\mu u)(\bar{u}\gamma_\mu u)$. Interestingly enough, this is just what the Fermi theory with $\mathcal{L} = G\bar{\psi}\gamma^\mu\psi\bar{\psi}\gamma_\mu\psi$ would give, with a coupling constant $G = g^2/M^2$ of dimension -2! Being an effective theory with $k \ll M$ Fermi theory acquires a coupling constant of dimension 2 and becomes non-renormalizable.

In preparation for a discussion on the renormalization group, we formulate the problem in terms of path integral. The vertex we have just calculated originates from a path integral

$$\int D\phi\; \phi(x_1)\phi(x_2)\phi(x_3)\phi(x_4)\; e^{i\int d^dx[\frac{1}{2}(\partial\phi)^2 - m^2\phi^2 - \frac{\lambda}{4!}\phi^4]} \quad (8.16.10)$$

integrated over only the slow modes of field ϕ, where the functional integral $\int D\phi$ is restricted to the fields $\phi(x)$ whose Fourier transform $\varphi(k)$ vanishes for $k \geq \Lambda$.

8.17   Renormalization Group Approach

8.17.1   Renormalization group flow

The method of renormalization group provides powerful and efficient tools to explore interacting theories. It is most valuable in regimes where perturbation theory fails. It is dedicated to the study of phase transition and critical phenomena, where the mean field theory fails because of violent thermal and quantum fluctuations. The central effort of the renormalization group (RG) approach is to achieve a better understanding of the long range behavior of the theory, and to keep the short range fluctuations in the potential background (i.e., to integrate them over). We come back to the ϕ^4 theory and re-write (8.14.6), setting the renormalization point at $s_0 = t_0 = u_0 \equiv \mu^2$:

$$\mathcal{M} = -i\lambda_R(\mu) + i\lambda_R^2 C \left[\log \frac{\mu^2}{s} + \log \frac{\mu^2}{t} + \log \frac{\mu^2}{u} \right] + O(\lambda_R^3).$$

Why a particular choice of μ? The renormalized coupling constant $\lambda_R(\mu)$ is most convenient to choose when we discuss processes at the energy scale $s \sim \mu^2$, because then the quantity in the bracket would be a small correction. Then the interesting question arise: how does $\lambda_R(\mu)$ vary with the energy scale μ? Consider another renormalization point μ'. Then similar to (8.16.6) we have

$$\mathcal{M} = -i\lambda_R(\mu') + i\lambda_R^2 C \left[\log \frac{\mu'^2}{s} + \log \frac{\mu'^2}{t} + \log \frac{\mu'^2}{u} \right] + O(\lambda_R^3).$$

The relation between the coupling constants is

$$\lambda_R(\mu') = \lambda_R(\mu) + 3C\lambda_R^2(\mu) \log \frac{\mu'^2}{\mu^2} + O(\lambda_R^3).$$

For $\mu' \sim \mu$ we obtain a differential equation for λ_R:

$$\mu \frac{d\lambda_R}{d\mu} = 6C\lambda_R^2 + O(\lambda_R^3). \tag{8.17.1}$$

This is the flow equation for the ϕ^4 theory. It turns out that $C > 0$, so that the coupling constant increases with energy scale, it is "running". The renormalized charge for QED is calculated from the polarization of vacuum graph leading to [48]

$$e_R^2(\mu) = e^2 \frac{1}{1 + e^2\Pi(\mu^2)} \simeq e^2 \left[1 - e^2\Pi(\mu^2) + O(e^4) \right], \tag{8.17.2}$$

or

$$\mu \frac{d}{d\mu} e_R(\mu) = -\frac{1}{2} e^3 \mu \frac{d}{d\mu} \Pi(\mu^2) + O(e_R^5) = \frac{1}{12\pi^2} e_R^3 + O(e_R^5),$$

$$(8.17.3)$$

under the condition that $m \ll \mu \ll \Lambda$, where Λ is the regularization cut-off. We see that electromagnetism becomes very strong as we go to high energies or small distances. On the other hand $e_R(\mu) \to 0$ when μ becomes very small, corresponding to very long spatial range. The electron charge is screened by the virtual electron-positron pairs in the vacuum.

These results lead to the idea of "renormalization group flow". In general, in a quantum field theory with a coupling constant g, we have the renormalization group flow equation

$$\mu \frac{dg}{d\mu} = \beta(g), \qquad (8.17.4)$$

where we have neglected the subscript "R" of g. This equation is sometimes written as

$$\frac{dg}{dt} = \beta(g),$$

upon defining $t \equiv log(\mu/\mu_0)$. It is called the Gell-Mann and Low equation. If the theory happens to have several coupling constants $\{g_i\}, i = 1, \ldots, N$, then we have

$$\frac{dg_i}{dt} = \beta_i(g_1, \ldots g_N), \qquad (8.17.5)$$

We can think of $(g_1, \ldots, g_N)$ as the coordinate of a particle in N-dimensional space, t as time, and $\beta_i(g_1, \ldots, g_N)$ a position dependent velocity field. As we increase t, we can watch the flow of the particle. If there exists a point g^* at which $\beta(g^*)$ vanishes, then when the particle arrives at this point will stay there. If the velocity field around a fixed point g^* is such that the particle moves toward that point (and stays there) the fixed point is known as attractive or stable. Thus, to study the asymptotic behavior of a quantum field theory at high energies we "merely" have to find all its attractive fixed points under the renormalization group flow. In a given theory, we can typically see that some couplings are flowing toward larger values while others are flowing toward zero. The problem is whether it is difficult to calculate the function β. The case $g^* = 0$ is the most favorable one, for which perturbation theory is applicable. In 1960s ideas of recursively generating flow of coupling constants arose, and it took the insight of Kenneth Wilson (1982 Nobel Prize) to realize the full potential of this approach and to develop it into a prosperous field of research.

8.17.2 Physics on different energy scales

The renormalization group is the formalism that allows us to relate the physics on different length scales or, equivalently, physics on different energy scales. In condensed matter physics, one tends to think of length scales. Sometimes for a field theory a lowest length scale cut-off is required, for which one can choose the size of an atom. At the cut-off a continuum (field) theory does not make sense any more. In particle physics, one talks about energy scales. In the previous discussions a cut-off Λ has been used for the ϕ^4 theory and for quantum electrodynamics. Let us go back to the ϕ^4 theory continued to the Euclidean space for simplicity:

$$Z(\Lambda) = \int_\Lambda D\phi \, e^{-\int d^d x \mathcal{L}[\phi]}, \qquad (8.17.6)$$

where the notation $\int_\Lambda$ means that the integral is to be taken over configurations $\phi(x)$ whose Fourier transforms $\varphi(k) = 0$ for $|k| > \Lambda$. We now reduce our "resolving power" by letting $\Lambda \to \Lambda - \delta\Lambda$ with $\delta\Lambda > 0$. Dividing the field ϕ into slowly oscillating part ϕ_s and fast oscillating part ϕ_f, namely $\phi = \phi_s + \phi_f$, such that their Fourier transforms $\varphi_s(k)$ and $\varphi_f(k)$ are nonzero only for $|k| \leq \Lambda - \delta\Lambda$ and $\Lambda - \delta\Lambda \leq |k| \leq \Lambda$, respectively. In other words, $\Lambda - \delta\Lambda$ is the dividing point between slow and fast oscillating part of $\varphi(k)$. Substituting $\phi = \phi_s + \phi_f$ into (8.17.6) we obtain

$$Z(\Lambda) = \int_{\Lambda - \delta\Lambda} D\phi_s \, e^{-\int d^d x \mathcal{L}[\phi_s]} \int D\phi_f \, e^{-\int d^d x \mathcal{L}_1[\phi_s, \phi_f]}, \qquad (8.17.7)$$

where all ϕ_f dependent terms are included in $\mathcal{L}_1$. Imaging carrying out the functional integral over ϕ_f and denoting the result by $e^{-\int d^d x\, \delta\mathcal{L}[\phi_s]}$,

$$e^{-\int d^d x\, \delta\mathcal{L}[\phi_s]} \equiv \int D\phi_f\, e^{-\int d^d x \mathcal{L}_1[\phi_s, \phi_f]},$$

we get

$$Z(\Lambda) = \int_{\Lambda - \delta\Lambda} D\phi_s\, e^{-\int d^d x (\mathcal{L}[\phi_s] + \delta\mathcal{L}[\phi_s])}. \qquad (8.17.8)$$

We have re-written the theory in terms of slow components only! In reality the integral over ϕ_f can only be carried out perturbatively. Take a more general example

$$\mathcal{L} = \frac{1}{2}(\partial\phi)^2 + \sum_n \lambda_n \phi^n + \cdots,$$

which reduces to the ϕ^4 theory for $\lambda_2 = \frac{1}{2}m^2$ and $\lambda_4 = \lambda$. Since terms such as $\partial\phi_s \partial\phi_f$ is integrated to 0, we get

$$\int d^d x \mathcal{L}_1[\phi_s, \phi_f] = \int d^d x \left[\frac{1}{2}(\partial\phi_f)^2 + \sum_n \lambda_n \phi_f^n + \ldots \right].$$

By symmetry considerations $\delta\mathcal{L}(\phi_s)$ has the same form as $\mathcal{L}(\phi_s)$ but with different coefficients. Adding $\delta\mathcal{L}(\phi_s)$ to $\mathcal{L}(\phi_s)$ thus shifts[26] the coefficients of $\mathcal{L}(\phi_s)$ only! These shifts generate the flow of coupling constants discussed in last subsection. With this result as a perspective, we are tempted to compare (8.17.8) with the original functional integral (8.17.6), brought to the same integration limits, in order to see the explicit change of the "integrand". For this purpose we have to change $\int_{\Lambda - \delta\Lambda}$ in the latter to $\int_\Lambda$. This procedure is called the re-scaling. For convenience, introduce the positive real number $b < 1$ by $\Lambda - \delta\Lambda = b\Lambda$. In the integral $\int_{\Lambda - \delta\Lambda}$ we integrate over fields with $|k| < b\Lambda$. So we have to make a change of variable: Let $k = bk'$ so

---

[26]Terms like $(\partial\phi)^4$ can also be generated, but these are "irrelevant" and can be neglected. See next sub-section.

that $|k'| \leq \Lambda$. But then correspondingly we have to change $x = x'/b$ so that $e^{ikx} = e^{ik'x'}$. Substituting in, we obtain

$$\int \mathrm{d}^d x \mathcal{L}\left[\phi_s\right] = \int \mathrm{d}^d x' b^{-d} \left[\frac{1}{2} b^2 \left(\partial' \phi_s\right)^2 + \sum_n \lambda_n \phi_s^n + \cdots \right],$$

where $\partial' = \partial/\partial x' = \frac{1}{b} \frac{\partial}{\partial x}$. Correspondingly, ϕ' is defined such that

$$\int \mathrm{d}^d x \left(\partial \phi_s\right)^2 = \int \mathrm{d}^d x' \left(\partial' \phi_s'\right)^2, \quad \text{or}$$

$$b^{2-d} \left(\partial \phi_s\right)^2 = \left(\partial' \phi'\right)^2, \quad \text{i.e., } \phi' = b^{\frac{1}{2}(2-d)} \phi_s.$$

Then the last integral becomes

$$\int \mathrm{d}^d x' \left[\frac{1}{2} \left(\partial' \phi'\right)^2 + \sum_n \lambda_n b^{-d + \frac{n}{2}(d-2)} \phi'^n + \cdots \right].$$

Thus, integration over fast components leads to a change of the coupling constants

$$\lambda_n' = b^{-d + \frac{n}{2}(d-2)} \lambda_n. \tag{8.17.9}$$

To summarize, we aim at integrating over the fast oscillating components and at assessing its influence on the remaining degrees of freedom.

We discuss the significance of the important relation (8.17.9). In d-dimensional space the field ϕ has a dimension $[\phi] = L^{\frac{1}{2}(2-d)} = M^{-\frac{1}{2}(2-d)}$, and the dimension of λ_n is determined by the relation $L^d L^{n \cdot \frac{1}{2}(2-d)} [\lambda_n] = L^0$, giving

$$[\lambda_n] = L^{-\frac{1}{2}n(2-d)-d} = M^{\frac{1}{2}n(2-d)+d}.$$

It turns out that the exponent of scaling of λ_n is just its dimension in length. For the time being, let us ignore $\delta L(\phi_s)$ for simplicity. As we lower the resolving power, or we become interested in physics over longer distance scales, we can once again write $Z(\Lambda)$ as in (8.17.4) except that the couplings λ_n have to be replaced by λ_n'. Since $b < 1$, we see from (8.17.7) that the λ_n's with $(n/2)(d-2) - d > 0$ get smaller and smaller and can eventually be neglected. The corresponding fields ϕ_n are called irrelevant. Conversely the fields for which $(n/2)(d-2) - d < 0$ are called relevant, and those for which $(n/2)(d-2) - d = 0$ are called critical.

8.17.3 Anderson localization

To show the power of renormalization group, we consider Anderson localization [48] in condensed matter physics. It involves the study of disordered systems. Electrons in real materials scatter off the impurities inevitably present and effectively move in a random potential. The important question is whether the wave functions at a particular energy E extend over the entire system or are localized within a characteristic length scale $\xi(E)$. Clearly, this issue determines whether the material is a conductor or an insulator. Anderson and his collaborators made the surprising discovery that localization properties depend on D, the dimension of space, but not on the detailed form of the distribution of random potential of the impurities. For $D = 1$ and 2 all wave functions are localized, regardless of how weak the impurity potential might be. This is a highly nontrivial statement since a priori we might think that whether the wave functions are localized or not depends on the strength of the potential. In contrast, for $D = 3$, the wave functions are extended for E in the range $(-E_c, E_c)$. As E approaches the energy Ec (known as the mobility edge) from above, the localization length $\xi(E)$ diverges as $\xi(E) \sim 1/(E - Ec)^\mu$ with some critical exponent μ. Such a conclusion can also be reached by a heuristic argument [50] using renormalization group language.

Conductivity σ is defined by $J = \sigma E$ where J is the the current density. Conductance G is the inverse of resistance $G = 1/R = I/V$, where the I is the current. To relate σ and G, consider a lump of material, taken to be a cube of size L, with a voltage drop V across it. Then $I = JL^2 = \sigma EL^2 = \sigma(V/L)L^2 = \sigma LV$ and thus $G(L) = 1/R = I/V = \sigma L$. Next, go to two dimensions. Consider a thin sheet of material of length and width L and thickness $a \ll L$. Again, apply a voltage drop V over the length $L : I = J(aL) = \sigma EaL = \sigma(V/L)aL = \sigma Va$ and so $G(L) = 1/R = I/V = \sigma a$, independent of L. Finally, go on to one dimension: Consider a wire of length L and width and thickness a, thus $G(l) = I/V = J^2a/EL = \sigma a^2/L$. To sum up, for conductors we have $G(L) \propto L^{D-2}$. For insulators, we have $G(L) \propto e^{-L/\xi}$. So If we use the dimensionless conductance g, $g(L) \sim c\, e^{-L/\xi}$, and $L(dg/dL) = -(L/\xi)g(L) \approx g(L)\log g(L)$. Here we neglected $\log c$. The β function is then

$$\beta(g) \equiv \frac{L}{g}\frac{dg}{dL} = \begin{cases} D-2, & \text{for large } g \\ \log g, & \text{for small } g \end{cases}. \qquad (8.17.10)$$

Given (8.17.10) we can now make a plot of $\beta(g)$ as shown in Fig. 8.49.

We see that for $D=2$ (and $D=1$) the conductance $g(L)$ always flows toward 0 as we go to long distances regardless of where we start. In contrast, for $D=3$, if the initial value of g is greater than a critical g_c then $g(L)$ flows to infinity and the material is a metal, while if $g_0 < g_c$, the material is an insulator.

8.17.4 Path to phase transition and critical phenomena

The purpose of this subsection is to provide a link between the Gell-Mann and Low equation to the important topic on phase transition and critical phenomena. We follow closely [49].

We turned back to the analysis of Gell-Mann and Low equation. Begin with the fixed point, at which the system no longer changes

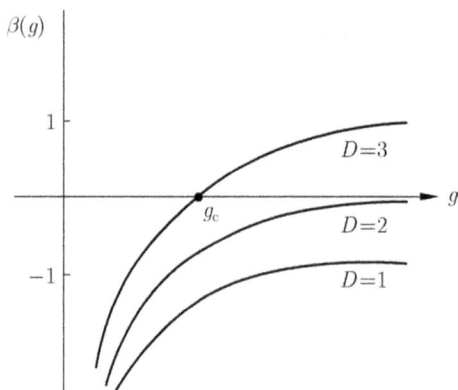

Fig. 8.49 β function for conductance versus dimensions. Reproduced with the kind permission of A. Altland and B. Simons, *Condensed Matter Field Theory*, Cambridge University Press, 2006.

under subsequent RG transformations. In particular, it remains unchanged under the change of space/time scale associated with the transformation. A system looks the same no matter how powerful a magnifying glass is used. Such a property of the system is referred to as *self-similar*. For such systems the magnification of any sub-region looks identical to the full system. An example of this is the fractal Julia set, one of which is shown in Fig. 8.50.

To each system one can attribute at least one intrinsic length scale, namely the length ξ determining the exponential decay of field correlation. However, the existence of a finite intrinsic scale cannot be consistent with the invariance under scale transformations. The only way out of this situation can only be: at the fixed point either $\xi = 0$ or $\xi = \infty$. The former possibility (complete disorder)

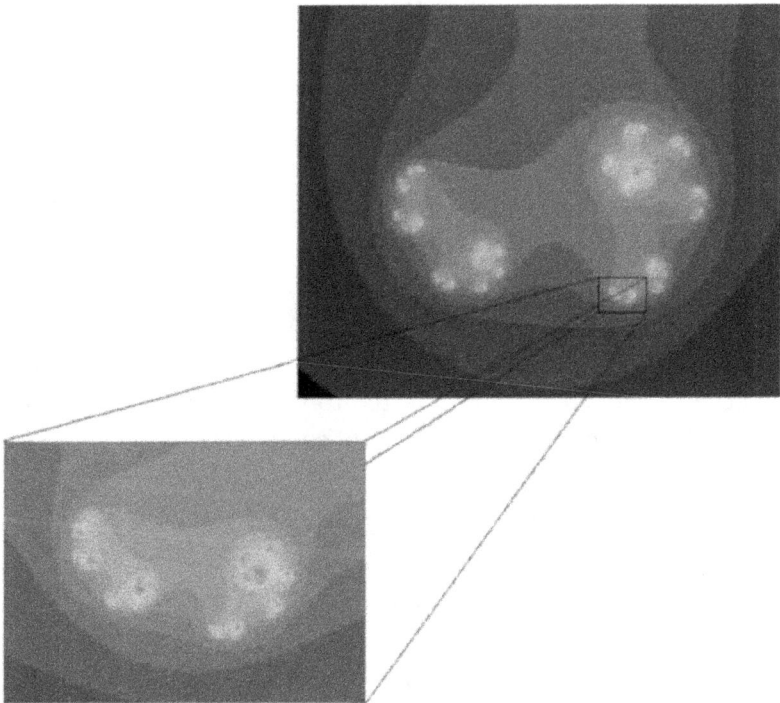

Fig. 8.50   A Julia set. Any sub-region of it contains the full information of the original set. Reproduced with the kind permission of A. Altland and B. Simons, *Condensed Matter Field Theory*, Cambridge University Press, 2006.

is not ineresting. The latter is just a hallmark of a second order phase transition. Thus we tentatively identify fixed points of the renormalization flow as candidates for transition points of the system. It is very important to study the behavior of the flow in the immediate vicinity of the fixed point in the manifold of coupling constants $\{g\} = (g_1, \ldots, g_a, \ldots g_N)$. We use the control variable $l = \ln b$, where b is defined in Section 8.17.2, the Gell-Mann Low equation is now written as

$$\frac{dg}{dl} = R(g). \qquad (8.17.11)$$

Let g^* be the fixed point. Close to it we expand R to linear order:

$$R(g) = R\left((g - g^*) + g^*\right) \approx W(g - g^*), \qquad (8.17.12)$$

since $R(g^*) = 0$, and we define the matrix W as

$$W_{ab} = \left(\frac{\partial R_a}{\partial g_b}\right)_{g=g^*}. \qquad (8.17.13)$$

This matrix is not symmetric, and therefore has left and right eigenvectors. Suppose that W has already been diagonalized, and we choose the left eigenstates[27]:

$$\phi_\alpha W = \lambda_\alpha \phi_\alpha, \qquad (8.17.14)$$

where the eigenvectors ϕ_α are row matrices and λ_α are the eigenvalues. The subcripts α are now referred to the basis of left eigenstates. Let v_α (a row matrix) be the α-th component of $g - g^*$ when represented in the basis $\{\phi_\alpha\}$:

$$v_\alpha = \phi_\alpha(g - g^*). \qquad (8.17.15)$$

The vector v_α possesses remarkable property:

$$\frac{dv_\alpha}{dl} = \phi_\alpha \frac{d}{dl}(g - g^*) = \phi_\alpha W(g - g^*) = \lambda_\alpha \phi_\alpha(g - g^*) = \lambda_\alpha v_\alpha. \qquad (8.17.16)$$

---

[27]This choice allows to conveniently express the flow of coupling constants.

The coefficients v_α change by a mere scaling factor λ_α under the RG procedure. The coefficients v_α are called scaling fields. Actually they are not fields, but simply a set of l-dependent coefficients. Integrating (8.17.16) we obtain

$$v_\alpha(l) \sim e^{l\lambda_\alpha}. \tag{8.17.17}$$

This result leads to a classification of three different types of scaling fields, which we have briefly mentioned at the end of Section 8.17.2.

1. For $\lambda_\alpha > 0$, the flow is directed away from the fixed point, i.e., the critical point. This is relevant scaling field, it drives the system away from the critical region. In Fig. 8.51 ϕ_2 is a relevant scaling field.
2. For $\lambda_\alpha < 0$, the flow is attracted by the fixed point. Scaling fields with this property are said to be irrelevant. In Fig. 8.51 ϕ_1 and ϕ_3 are irrelevant fields.
3. For $\lambda_\alpha = 0$, the scaling field is invariant under the RG flow. The fields are called marginal.

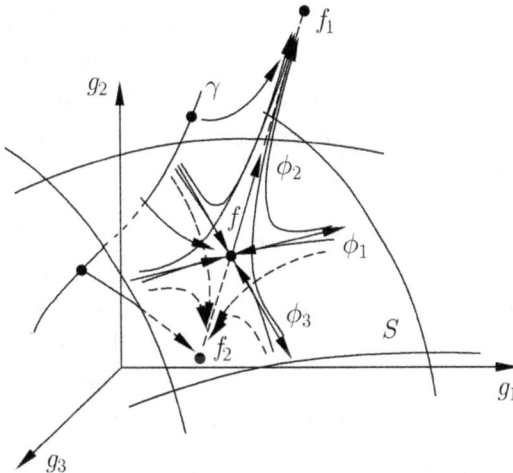

Fig. 8.51   RG flow in the vicinity of a fixed point. Reproduced with the kind permission of A. Altland and B. Simons, *Condensed Matter Field Theory*, Cambridge University Press, 2006.

Associated with the classification of scalint fields, there are three different types of fixed points.

1. Stable fixed points are those whose scaling fields are all irrelevant of at least marginal. They are called attractors: when a system in the vicinity of which is released, it will scale to the fixed point.

2. Complementary to the stable fixed points there are unstable fixed points: all scaling fields are relevant. The systems are driven away from this point, no matter how close the system managed to approach.

3. The generic class of fixed points are those with both relevant and irrelevant scaling fields. They are associated with phase transitions. In Fig. 8.51 the eigenvectors ϕ_1, ϕ_3 span the tangent space of a 2-dimensional manifold known as the critical critical surface denoted by S in the figure and defined through the vanishing of relevant field $\phi_2 = 0$. This critical manifold forms the *basin of attraction* of the fixed point, i.e., when a set of coupling constant g is fine-tuned so that $g \in S$, the expansion in terms of scaling fields *only contains the irrelevant components* and the system will be attracted to the fixed point. However, the smallest deviation from S may introduce a relevant component driving the system away exponentially from the fixed point. It can flow ending up with another fixed point. This picture actually suggests that systems with generic fixed points typically possess complementary stable fixed points, i.e., fixed points towards which the flow is directed after it has left the critical region.

The content of renormalization group approach to phase transition and critical phenomena is extremely rich, e.g., scaling and universality and the calculation of critical exponents. We give here a very brief introduction to the basic ideas. The most fundamental signature of a phase transition is its order parameter, $\mathcal{M}$, i.e., a quantity whose value unambiguously identifies the phase of the system, e.g., the magnetization for the ferromagnetic–paramagnetic transition. Transitions between different phases of matter fall into two large categories. In first-order phase transitions the order parameter exhibits a discontinuous jump across the transition line while, in the second-order transitions, the order parameter changes in a non-analytic but continuous manner. The order parameter $\mathcal{M}$ is coupled to a conjugate field, the magnetic field strength H, in the free energy F. $\mathcal{M}$ is obtained from the relation

$$\mathcal{M} = -\partial_H F,$$

where F is the free energy. At a second-order transition, $\mathcal{M}$ changes nonanalytically, which means that the second-order derivative, the magnetic susceptibility, $\chi = \partial_H \mathcal{M} = -\partial_H^2 F$ develops a singularity. This can be understood through the fluctuation-dissipation theorem. The susceptibility is intimately linked to the field fluctuation behavior of the system. For instance, in the d-dimensional Ising model we have the correlation function C defined by

$$C(\boldsymbol{r}_1 - \boldsymbol{r}_2) = \langle S(\boldsymbol{r}_1)S(\boldsymbol{r_2})\rangle - \langle S(\boldsymbol{r}_1)\rangle \langle S(\boldsymbol{r}_2)\rangle$$

$$\sim \exp\left(-\frac{|\boldsymbol{r}_1 - \boldsymbol{r_2}|}{\xi}\right),$$

$$\text{for } |\boldsymbol{r}_1 - \boldsymbol{r_2}| \gg \infty \qquad (8.17.18)$$

and χ is related to it through the partition function:

$$\chi = -\partial_H F \,|_{H=0} = T\partial_H^2 \ln Z \,|_{H=0}$$

$$= \beta \int \mathrm{d}^d r \mathrm{d}^d r' \left(\langle S(\boldsymbol{r})S(\boldsymbol{r}')\rangle - \langle S(\boldsymbol{r})\rangle \langle S(\boldsymbol{r}')\rangle\right).$$

The integrand, i.e., the expression in the bracket is just the correlation function, a measure of the fluctuation of the spin. A divergence of the susceptibility implies the proliferation of infinitely long-range field fluctuations.

Many other physical quantities share with the susceptibility the divergence property at the critical point. This has far-going implications. We have seen that right at the transition/fixed point, the system is self-similar. This implies that the behavior of its various characteristics must be described by *power laws*. We give here a heuristic argument. Consider a function $f(t)$, where f is an observable and t is a control parameter determining the distance to the transition point. In the immediate vicinity of the transition point, f is expected to "scale," i.e., under a change of the length scale $x \to x/b, t \to tb^{-D_t}$, where D_t is the dimension of t, the function f

must, at most, change by a factor reflecting its own scaling dimension D_f. The self-similarity demands

$$f(t) = b^{D_f} \, f(tb^{-D_t}). \tag{8.17.19}$$

This equation is satified by

$$f \sim t^{D_f / D_t},$$

as can be easily verified. The set of different exponents characterizing the relevant power laws occurring in the vicinity of the transition are known as critical exponents. They carry the same information of the flow of scaling fields in the vicinity of critical points. Critical exponents are fully universal; they are numbers depending, at most, on dimensionless characteristics such as the space-time dimensionality or number of components of the order parameter.

The most important critical exponents are:

1. α: In the vicinity of the critical temperature, the specific heat C scales as $C \sim |t|^{-\alpha}$, where $t = (T - Tc)/Tc$ measures the distance to the critical point. Note that a non-trivial statement has been made: although the phases above and below the transition are essentially different, the scaling exponents controlling the behavior of C are identical.

2. β: Approaching the transition temperature from below, the magnetization vanishes as $M \sim (-t)^{\beta}$.

3. γ: The magnetic susceptibility behaves as $\chi \sim |-t|^{\gamma}$.

4. δ: At the critical temperature, $t = 0$, the magnetization scales with the magnetic field as $M \sim |h|^{1/\delta}$.

5. ν: Upon approaching the transition point, the correlation length diverges as $\xi \sim |t|^{-\nu}$.

6. η: It characterizes the correlation function (8.17.18) at small distances $r \ll \xi$: $C(\boldsymbol{r}) \sim 1/|\boldsymbol{r}|^{d-2+\eta}$.

These exponents are not all independent. They are related by the following "scaling laws":

$$\nu(2 - \eta) = \gamma, \quad \text{(Fisher)}$$
$$\alpha + 2\beta + \gamma = 2, \quad \text{(Rushbrooke)}$$

$$\beta(\delta - 1) = \gamma, \quad \text{(Widom)}$$
$$2 - \alpha = \nu d. \quad \text{(Josephson)}.$$

Universality is another aspect of conceptual importance of the renormalization group approach. In fact, the majority of critical systems can be classified into a relatively small number of universality classes. Crudely speaking, for vast number of systems under study there are but a few types of RG flow. Remarkably, the origin of this universality lies in the critical surface in Fig. 8.51. On the curve denoted by γ in the coupling constant space are points representing different initial values of coupling constants. This curve intersects the critical surface at a cross. As we have discussed, points above or below the surface evolve and eventually direct away from the fixed point, because the scaling field at those points contain the relevant components. In the figure two trajectories approaching fixed points f_1 and f_2 are examples. The initial point starting from the cross on the critical point scales toward the critical point when the control parameters X_i vary and finally arrives at the point. For this particular set of coupling constants, the system is critical. As we look at it on larger and larger length scales, it will be attracted by the fixed point at S, i.e., it will display the universal behavior characteristic of this particular point. This is the origin of universality. Consider another system (with different material constants) which will generate a different trajectory $g_\alpha(\{X_i\}) = \gamma'$. However, as long as this trajectory intersects with S, it is guaranteed that the critical behavior will exhibit the same universal characteristics (controlled by the unique fixed point).

In fact a more far-reaching statement can be made. Given that there is an infinity of systems exhibiting transition behavior while there is only a very limited set of universality classes, many systems of very different microscopic morphology must have the same universal behavior. More formally, different microscopic systems must map onto the same critical low-energy theory. Examples of these coincidences include the equivalence of the Luttinger liquid to a Josephson junction, the equivalence of models of planar magnets to two-dimensional classical Coulomb plasmas, and the equivalence of the liquid–gas transition to the ferromagnetic transition. In all cases, "equivalence" means that the systems exhibit identical scaling behavior and, therefore, fall into the same universality class.

8.17.5 $O(n)$ Non-linear sigma model in d dimensions, weak coupling approximation [51]

In Section 7.3 we introduced the continuous version of one-dimensional AFM chain in terms of the Néel field $\widehat{\Omega}$ and a canting field l with the condition $\widehat{\Omega} \cdot l = 0$. The result is a Lagrangian consisting of a $O(3)$ non-linear sigma model (NLSM) plus a Wess-Zumino topological term. This is a special case of Haldane's mapping [52] in which the effective long wave length action os the quantum Heisenberg antiferromagnet in d dimensions is mapped into the NLSM of $d+1$ dimensions.

We discussed in Section 7.2 the spontaneous symmetry breaking aspect of NLSM. Actually it possesses the scaling renormalization group property. In the following we briefly discuss the effect of fluctuations on the ordered phase and the related infrared divergence in 1 and 2 spacetime dimensions. Subsequently we follow A. M. Polyakov [52]'s treatment leading to a narrow escape through the inferred divergence in a special case of $d = 2$, culminating in the quasi-long range order.

In the standard notation of a d-dimensional $O(n)$ NLSM the partition function is

$$Z = \int D\widehat{n} \exp\left[-\frac{\Lambda^{d-2}}{2f} \int d^d x \sum_{\mu,\alpha} (\partial_\mu n^\alpha)^2\right],$$

$$D\widehat{n} = \prod_\alpha Dn^\alpha \delta\left[1 - \sum_{\alpha=0}^{n-1} (n^\alpha)^2\right]. \tag{8.17.20}$$

The n orthonormal basis vectors are:

$$\widehat{e}^\alpha, \quad \alpha = 0, 1, \ldots n - 1$$
$$n^\alpha = \widehat{n} \cdot \widehat{e}^\alpha. \tag{8.17.21}$$

Haldane mapping relates the NLSM with the original Heisenberg model on a d-dimensional cubic lattice with $\aleph$ sites and lattice constant a under the condition that the Brillouin zone is now spherical with radius Λ, the number of degrees of freedom in the spherical zone is the same as that in the original cubic zone:

$$\aleph^{-1} \overset{\text{cube}}{\underset{|k[\mu]|<\pi/a}{\sum}} = \aleph^{-1} \overset{\text{sph}}{\underset{|k|<\Lambda}{\sum}} = 1,$$

which leads to values of Λ for different dimensions d:

$$\Lambda = \begin{cases} \pi/a, & d=1 \\ 2\sqrt{\pi}/a, & d=2 \\ \left(6\pi^2\right)^{1/3}/a, & d=3. \end{cases}$$

At low temperatures we expect that the ground configuration is close to an ordered state

$$\overline{\overline{n}} = \hat{e}^0. \tag{8.17.22}$$

The small fluctuations $\phi^a (a=1...n-1)$ are transverse to $\hat{e}^0$, and the field $\hat{n}$ is parametrized as

$$\hat{n} = \hat{e}^0 \sqrt{1 - \overline{\phi^2}} + \sum_{a=1}^{n-1} \phi^a \hat{e}^a, \tag{8.17.23}$$

where

$$\overline{\phi^2} = \sum_{a=1}^{n-1} (\phi^a)^2. \tag{8.17.24}$$

Substituting (8.17.23) into (8.17.20) and expanding the integrand in ϕ keeping terms up to ϕ^2, we obtain

$$Z = \int D\phi \exp\left(-S^{(2)}[\phi]\right),$$

$$S^{(2)}[\phi] = \frac{\Lambda^{d-2}}{2f} \int d^d x \exp\left[\sum_{\mu,a}(\partial_\mu \phi^a)^2\right] \tag{8.17.25}$$

Fourier transform ϕ

$$\phi^a_{\boldsymbol{k}} = \Lambda^d \int d^d x e^{-i\boldsymbol{k}\cdot\boldsymbol{x}} \phi^a(x)$$

gives

$$S^{(2)}[\phi] = \frac{1}{2f\Lambda^2 \aleph} \sum_{\boldsymbol{k},a} k^2 \phi^a_{\boldsymbol{k}} \phi^a_{-\boldsymbol{k}}. \qquad (8.17.26)$$

Calculation of the local flucuations using the partition function (8.17.25) gives

$$\left\langle |\delta\widehat{n}(0)|^2 \right\rangle \simeq \frac{1}{Z\aleph} \sum_{\boldsymbol{k},a} \int D\widehat{n} \phi^a_{\boldsymbol{k}} \phi^a_{-\boldsymbol{k}} \exp\left(-S^{(2)}[\phi]\right)$$

$$= \frac{(n-1)\Lambda^2 f}{(2\pi)^d} \lim_{\tilde{\Lambda}\to 0} \int_{\tilde{\Lambda}}^{\Lambda} \frac{d^d k}{|\boldsymbol{k}|^2}$$

$$\sim \begin{cases} (n-1)f \lim\limits_{\tilde{\Lambda}\to 0} \tilde{\Lambda}^{-1} \to \infty, & d=1 \\[2ex] -\dfrac{(n-1)f}{2\pi} \lim\limits_{\tilde{\Lambda}\to 0} \ln\tilde{\Lambda} \to \infty, & d=2 \qquad (8.17.27) \\[2ex] \dfrac{(n-1)f\Lambda}{2\pi^2}, & d=3 \end{cases}$$

The fluctuations diverge for $n \geq 2$ in 1 and 2 dimensions for any finite f. The infrared divergence invalidates the spontaneous symmetry breaking for 1 and 2 dimensions. This is in accordance with the Mermin and Wagner's theorem.

It is necessary then to abandon the assumption of broken symmetry and to work from the beginning using O(n) symmetry. In Polyakov's approach [38, 37] the parametrization (8.17.23) is replaced by

$$\widehat{n}(\boldsymbol{x}) = \widehat{n}^0(\boldsymbol{x})\sqrt{1 - \overline{\phi^2}} + \sum_{a=1}^{n-1} \phi^a \widehat{e}^a(\boldsymbol{x}),$$

$$\overline{\phi^2} = \sum_{a=1}^{n-1} (\phi^a)^2. \qquad (8.17.28)$$

Now both the field $\hat{n}$ and the fluctuations ϕ are x dependent, and keeping the 0th (longitudinal) component of $\hat{n}$ special, i.e., almost unimodular, the basis vectors $\hat{n}$ and $\hat{e}^a$ have to be also x dependent. We define the basis vectors to be

$$\hat{e}^0(\boldsymbol{x}) \equiv \hat{n}^0(\boldsymbol{x})$$
$$\hat{e}^\alpha(\boldsymbol{x}) \cdot \hat{e}^\beta(\boldsymbol{x}) = \delta_{\alpha\beta}, \quad \alpha = 0, 1, \dots n-1. \tag{8.17.29}$$

Greek indices include both longitudinal and the transverse components, while Latin indices refer to the transverse components only, $a, b = 1, \dots n-1$. Basis vectors at nearby point are related by the gauge potentials (affine connexions)

$$\tilde{A}_\mu^{\alpha\beta} \equiv \hat{e}^\alpha \cdot \partial_\mu \hat{e}^\beta. \tag{8.17.30}$$

These are the dyads (tensors of rank 2). The following useful relations follow directly from the above definition:

$$\partial_\mu \hat{n}^0 = \sum_a \tilde{A}_\mu^{a0} \hat{e}^a, \quad \partial_\mu \hat{e}^a = \sum_b \tilde{A}_\mu^{ab} \hat{e}^b - \tilde{A}_\mu^{a0} \hat{n}^0. \tag{8.17.31}$$

The ortho-normality condition (8.17.29) leads to

$$\sum_a \left(\tilde{A}_\mu^{a0} \right)^2 = \left(\partial_\mu \hat{n}^0 \right)^2. \tag{8.17.32}$$

Polyakov distinguishes the slow field $\hat{n}(\boldsymbol{x})$ and fast fields $\phi^a(\boldsymbol{x})$, the basis vectors $\hat{e}^\alpha(\boldsymbol{x})$ and the connexion potentials $\tilde{A}_\mu^{\alpha\beta}(\boldsymbol{x})$ are also slow fields. An intermediate momentum scale $\tilde{\Lambda}$ is chosen,

$$0 < \tilde{\Lambda} \ll \Lambda, \tag{8.17.33}$$

which separates between the fast fields, i.e.,

$$\phi^a(\boldsymbol{x}) = \sum_{\tilde{\Lambda} \le |k| \le \Lambda} e^{i\boldsymbol{k}\cdot\boldsymbol{x}} \phi_{\boldsymbol{k}}^a, \tag{8.17.34}$$

and the slow fields

$$X(\boldsymbol{x}) = \sum_{0 \le |k| \le \tilde{\Lambda}} e^{i\boldsymbol{k}\cdot\boldsymbol{x}} X_{\boldsymbol{k}}. \tag{8.17.35}$$

Substituting (8.17.28) into the NLSM Lagrangian, we obtain

$$
\begin{aligned}
\mathscr{L} &= \frac{\Lambda^{d-2}}{2f} \int d^d x \sum_{\mu=1}^{d} \partial_\mu \widehat{n} \cdot \partial_\mu \widehat{n} \\
&= \frac{\Lambda^{d-2}}{2f} \int d^d x \sum_{\mu,a,b} \left[\left(\partial_\mu \sqrt{1 - \phi^2} - \widetilde{A}_\mu^{a0} \phi^a \right)^2 \right. \\
&\quad \left. + \left(\partial_\mu \phi^a + \widetilde{A}_\mu^{ab} \phi^b + \widetilde{A}_\mu^{a0} \sqrt{1 - \phi^2} \right)^2 \right].
\end{aligned}
\tag{8.17.36}
$$

Expanding the Lagrangian to quadratic order in ϕ we obtain

$$
Z = \int_{\widetilde{\Lambda}} D\widehat{n}^0 \exp \left(- \int_{\widetilde{\Lambda}} d^d x \, \mathscr{L} \left[\widehat{n}^0 \right] \right) Z^{(2)} \left[\widehat{n}^0 \right], \tag{8.17.37}
$$

where (8.17.32) is used for yielding $\mathscr{L}[\widehat{n}^0]$, and $Z^{(2)}$ is the correction due to fast fields

$$
\begin{aligned}
Z^{(2)} &= \int_\Lambda D\phi \exp \left(- \int_\Lambda d^d x \, \mathscr{L}^{(2)} \left[\widehat{n}^0, \phi \right] \right), \\
\mathscr{L}^{(2)} &= \frac{\Lambda^{d-2}}{2f} \int d^d x \sum_{\mu,a,b} \left[\left(\partial_\mu \phi^a + \widetilde{A}_\mu^{ab} \phi^b \right)^2 \right. \\
&\quad \left. + \widetilde{A}_\mu^{a0} \widetilde{A}_\mu^{b0} \left(\phi^a \phi^b - \delta_{ab} \phi^2 \right) \right],
\end{aligned}
\tag{8.17.38}
$$

in which powers of ϕ higher than quadratic are neglected. The dependence of $\mathscr{L}^{(2)}$ on $\widetilde{A}_\mu^{ab}$ only through their derivatives, and will be neglected. Performing the Gaussian path integral in $Z^{(2)}$ we obtain

$$
Z^{(2)} \left[\widehat{n}^0 \right] \propto \exp \left[-\frac{1}{2} Tr \ln \left(\Pi_0 - \Pi_1 \right) \right], \tag{8.17.39}
$$

where the Π's are matrices in the fast mode space $|\boldsymbol{k}| > \widetilde{\Lambda}$

$$
(\Pi_0)_{\boldsymbol{k},\boldsymbol{k}'} = k^2 \delta_{\boldsymbol{k},\boldsymbol{k}'} \delta_{ab},
$$

$$
(\Pi_1)_{\boldsymbol{k},\boldsymbol{k}'} = \delta_{\boldsymbol{k},\boldsymbol{k}'} \left[\widetilde{A}_\mu^{a0} \widetilde{A}_\mu^{b0} - \delta_{ab} \sum_c \left(\widetilde{A}_\mu^{c0} \right)^2 \right]. \tag{8.17.40}
$$

Approximating $\ln(\Pi_0 - \Pi_1)$ by $\ln \Pi_0 + \ln(1 - \Pi_0^{-1}\Pi_1) \simeq \ln \Pi_0 - (\Pi_0^{-1}\Pi_1)$, we proceed to obtain

$$\frac{1}{2}Tr\left(\Pi_0^{-1}\Pi_1\right) = -\frac{1}{2\aleph}\sum_{\tilde{\Lambda}\leq|k|\leq\Lambda}\frac{1}{k^2}\sum_{\mu,a}\left[\left(\tilde{A}_\mu^{a0}\right)^2 - (n-1)\left(\tilde{A}_\mu^{a0}\right)^2\right]$$

$$\approx \frac{\Lambda^{d-2}\Delta_d}{2}\sum_{\mu}\left(\partial_\mu\hat{n}^0\right)^2, \tag{8.17.41}$$

where the dimensionless constant Δ_d is defined as

$$\Delta_d(\tilde{\Lambda}/\Lambda) = (n-2)\Lambda^{2-d}\int_{\tilde{\Lambda}\leq|k|\leq\Lambda}\frac{d^d k}{(2\pi)^d}\frac{1}{k^2} \tag{8.17.42}$$

$$=\begin{cases}\dfrac{n-2}{\zeta_1}\left[(\tilde{\Lambda}/\Lambda)^{d-2} - 1\right], & d=1 \\[2mm] -\dfrac{n-2}{\zeta_2}\ln(\tilde{\Lambda}/\Lambda), & d=2. \quad (8.17.43) \\[2mm] \dfrac{n-2}{\zeta_3}\left[1 - (\tilde{\Lambda}/\Lambda)^{d-2}\right], & d=3 \end{cases}$$

In (8.17.43) $\zeta_1 = \pi$, $\zeta_2 = 2\pi$, $\zeta_3 = 2\pi^2$. By incorporating the corrections into (8.17.39) with the right handed side given by (8.17.42), and replacing Λ by $\tilde{\Lambda}$ we get the new partition function

$$Z \approx \int_{\tilde{\Lambda}}D\hat{n}^0\exp\left[-\frac{\tilde{\Lambda}^{d-2}}{2\tilde{f}}\int_{\tilde{\Lambda}}d^d x\left(\partial_\mu\hat{n}^0\right)^2\right], \tag{8.17.44}$$

with the renormalized coupling constant $\tilde{f}$

$$\tilde{f} = (\tilde{\Lambda}/\Lambda)^{d-2}\frac{f}{1 - f\Delta_d(\tilde{\Lambda}/\Lambda)}. \tag{8.17.45}$$

We find out that by integrating out the fast fields we come back to the original form of Lagrangian with only the parameters changed! We observe that $d = 1$ and $d = 2$ still suffer from the infrared divergences. But just compare (8.17.43) with (8.17.27)! For $d = 1$ and $d = 2$ the ground state fluctuations diverge as $n - 1$, but here the infrared divergences of Δ_d occur only for $n \geq 3$. Now $n = 2$

is special. In $O(n)$ NLSM n is the degree of freedom of the order parameter, for instance for superfluidity or superconductivity $n = 2$ (amplitude and phase). It turns out that for $d = 2$ the continuum theory of $n = 2$ has a weak coupling phase with quasi-long range order, i.e., power law decaying correlations. $O(2)$ models and the Kosterlitz-Thouless transitions are important topics in superfluidity. The renormalization group treatment is physically enlightening. A more rigorous treatment involving the proof that higher order terms are either irrelevant or marginal, and can be found in specialized texts.

References

[1] D. F. Walls and G. J. Milburn, *Quantum Optics*, Berlin, Springer, 1994.

[2] S. Haroche and J. M. Raimond, In *Advances in Atomic, Molecular an Optical Physics*, Supplement 2 (Cavity Quantum Electrodynamics, edited by P. R. Berman), Academic Press, 1994.

[3] J. M. Raimond, M. Brune and S. Haroche, *Rev. Mod. Phys.* 73, 565, 2001.
Gerhard Rempe, Herbert Walther, and Norbert Klein, *Phys. Rev. Lett.* 58, 353, 1987.
W. Jhe, A. Anderson, E. A. Hinds, D. Meschede, L. Moi, and S. Haroche, *Phys. Rev. Lett.* 58, 666, 1987.

[4] S. Haroche and D. Kleppner, *Physics Today* 42(1), 24, 1989.

[5] K.-T. Boller, A. Imamoglu and S. E. Harris, *Phys. Rev. Lett.* 66, 2593, 1991.

[6] L. V. Hau, S. E. Harris, Z. Dutton and C. H. Behroozl, *Nature* 397, 594, 1997.

[7] Jon Marangos, *Nature* 397, 559, 1999.

[8] C. Liu, Z. Dutton, C. H. Behroozl and L. V. Hau, *Nature* 409, 490, 2001.

[9] S. E. Harris, J. E. Field and A. Kasapi, *Phys. Rev. A* 46, R29, 1992.

[10] Z. Dutton and L. V. Hau, *Phys. Rev. A* 70, 053831, 2004.

[11] M. Brune, J. M. Raimond, P. Goy, L. Davidovich and S. Haroche, *Rev. Mod. Phys.* 59, 1899, 1987.

[12] S. Haroche, In *New Trends in Atomic Physics*, Les Houches Summer School Session XXXVIII, edited by G. Grynberg and R. Stora, North Holland, Amsterdam, 1989.

[13] G. Rempe, H. Walther and N. Klein, *Phys. Rev. Lett.* 58, 353, 1987.

[14] R. G. Hulet, E. S. Hilfer and D. Kleppner, *Phys. Rev. Lett.* 55, 2137, 1985.

[15] W. Jhe, A. Anderson, E. A. Hinds, D. Meshede, L. Moi and S. Haroche, *Phys. Rev. Lett.* 58, 666, 1987.

[16] D. Meshede, H. Walther and G. Müller, *Phys. Rev. Lett.* 54, 551, 1985.

[17] J. Krause, M. O. Scully and H. Walther, *Phys. Rev. A* 36, 4547, 1987.

[18] M. Brune, J. M. Raimond, P. Goi, L. Davidovich and S. Haroche, *Phys. Rev. Lett.* 59, 1899, 1987.

[19] M. Brune, P. Nussenzveig, F. Schmide-Kaler, F. Bernardot, A. Maali, J. M. Raimond and S. Haroche, *Phys. Rev. Lett.* 72, 3339, 1994.

[20] S. Haroche, M. Brune and J. M. Raimond, *J. Phys. II (France)* 2, 659, 1992.

[21] S. Haroche, M. Brune and J.-M. Raimond, *Physics Today* 66(1), 27, 2013.

[22] S. Gleyes, S. Kuhr, C. Guelin, J. Bernu, S. Deliglise, U. B. Hoff, M. Brune, J. M. Raimond and S. Haroche, *Nature* 446, 297, 2007.

[23] H. Mabuchi and A. C. Doherty, *Science* 298, 1372, 2002.

[24] H. M. Wiseman and G. J. Milburn, *Phys. Rev. A* 47, 642, 1993.

[25] Q. A. turchette *et al.*, *Appl. Physics B* 60, S1, 1955.

[26] E. Hagley *et al.*, *Phys. Rev. Lett.* 79, 1, 1997.

[27] A. Rauschenbeutel *et al.*, *Science* 288, 2024, 2000.

[28] A. Rauschenbeutel *et al.*, *Phys. Rev. A* 64, 050301, 2001.

[29] L. D. Landau and E. M. Lifshitz, *Statistical Physics*, 3rd edition, Part I. Oxford, Pergamon Press, 1980.

[30] L. D. Landau and E. M. Lifshitz, *Electrodynamics of Continuous Media*, 2nd edition. Oxford, Pergamon Press, 1984.

[31] Yu. S. Barash and V. L. Ginzburg, *Sov. Phys. Uspekhi* 18, 305, 1975.

[32] L. D. Landau and E. M. Lifshitz, *The Classical Theory of Fields*, 4th edition. Oxford, Pergamon Press, 1975.

[33] E. M. Lifshitz and L. P. Pitaevskii, *Statistical Physics*, Part II. Oxford, Pergamon Press, 1980.

[34] P. W. Milonni and M.-L. Shih, *Am. J. Phys.* 59, 684, 1991.

[35] P. W. Milonni and M.-L. Shih, *Contemp. Phys.* 33, 313, 1992.

[36] H. B. G. Casimir and D. Polder, *Phys. Rev.* 73, 360, 1948.

[37] C. I. Sukenik, M. G. Boshier, D. Cho, V. Sandoghdar and E. A. Hinds, *Phys. Rev. Lett.* 70, 560, 1993.

[38] G. Barton, *Proc. Roy. Soc. London A* 410, 175, 1987.

[39] S. J. Sparnaay, *Physica* 24, 751, 1958.

[40] C. Itzykson and J.-B. Zuber, *Quantum Field Theory*, New York, McGraw Hill, 1980.

[41] I. Chiorescu *et al.*, *Nature* 431, 159, 2004.

[42] A. Wallraff *et al.*, *Nature* 431, 162, 2004.

[43] J. E. Mooij *et al.*, *Science* 285, 1036, 1999; C. H. van der Wal *et al.*, *Science* 290, 773, 2000.

[44] A Blais *et al.*, *Phys. Rev. A* 69, 062320, 2004.

[45] W. H. Louisell, *Quantum Statistical Properties of Radiation*, John Wiley, London, 1973.

[46] C. Cohen-Tannoudji, J. Dupont-Roc and G. Grynberg, *Atom-Photon Interactions*, John Wiley, New York 1992.

[47] C. J. Foot, *Atomic Physics*, Oxford University Press, 2005.

[48] Anthony Zee, *Quantum Field Theory in a Nutshell*, 2nd edition, Princeton University Press, 2010.

[49] A. Altland and B. Simons, *Condensed Matter Field Theory*, Cambridge University Press, 2006.

[50] E. Abrahams, P. W. Anderson, D. C. Licciardello and T. V. Ramakrishnan, *Phys. Rev. Lett.* 42, 673, 1979.

[51] A. Auerbach, *Interacting Electrons and Quantum Magnetism*, Springer Verlag, 1994.

[52] A. M. Polyakov, *Phys. Lett. B* 59, 79, 1975.

Chapter 9

The Quantum Hall Effect

In a period of 20 years the study of quantum Hall effect led to two Nobel awards in physics. It was found that the Hall resistance had quantized values expressed solely in terms of universal constants. This was rather exceptional in a research in condensed matter physics, where a measured quantity is usually dependent on the material. This phenomenon expresses some very basic regularities in physics. In Sections 9.2 and 9.3 we present some theoretical prerequisites for understanding the quantum Hall effect, namely the Landau levels and flux quantization. The integral quantum Hall effect (IQHE) and its interpretation is given in Section 9.4, including Laughlin's gauge invariance argument and Thouless's topological considerations on why this phenomenon is so basic.

The discovery of fractional quantum Hall effect (FQHE) was very demanding on the expertise of experimental physicists, and its theoretical interpretation was equally challenging. While the quantized value of the Hall resistance in the IQHE corresponds to the number of fully occupied Landau levels, why it is so special when a Landau level is only filled to 1/3, 2/5, etc. in the FQHE? To "write down" an approximate expression for the wave function of such a state demands not only the physics insight but also a persistence in guessing and justifying the trial wave function. Laughlin comprehended that such a state should be a strongly correlated state due to the interaction of electrons, and his research led to a wave function much more accurate than one could expect. Besides the ground state, the wave functions for quasiparticle excitations were also found. These states were predicted to have fractional charges. Large number of high quality results put together led to the understanding that the FQH states are those of a new type of quantum liquid.

The remaining sections deal with further developments of theoretical ideas with emphasis on the picture of composite fermions and composite bosons. The global phase diagram of quantum Hall matter is also presented.

There are many new and exciting developments of ideas in this new field. Our emphasis here is on the developments most intimately connected with the basic principles of quantum mechanics, and we have to refer the readers to the literature for many advanced topics.

9.1   The Classical Hall Effect

Consider a rectangular slab of conductor and apply an electric field E_x along its length (x-axis), the current density j_x produced is

$$j_x = nev, \qquad (9.1.1)$$

where n is the density of the current carrier, e and v its charge and velocity respectively. Applying a magnetic field B in the z-direction exerts a Lorentz force in the y-direction on the charge. When the flow becomes steady, a transverse potential difference U_H in the y-direction is developed, maintaining a transverse field $E_y = \frac{1}{c}vB$, which just balance the force $\frac{e}{c}v \times B$ due to the perpendicular magnetic field. When a steady state is reached, j_x, E_y and B are measured. They are related by

$$E_y = j_x \frac{B}{nec}.$$

Defining the Hall resistivity ρ_H by

$$\rho_H = \frac{E_y}{j_x}, \qquad (9.1.2)$$

then we have

$$\rho_H = \frac{B}{nec}, \qquad (9.1.3)$$

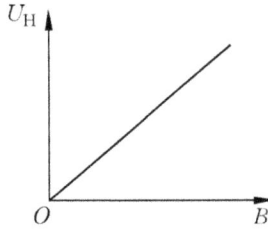

Fig. 9.1   The Hall voltage versus the magnetic field strength.

justifying the experimental result. From the quantities measured, the density of current carriers can be measured. If the current carrier is the electron, then E_y is along the $-y$ direction. If it is the hole, then E_y is in the $+y$ direction. This effect was discovered by Edwin Hall in 1879. The effect was named after him and the transverse voltage is called the Hall voltage. If the transverse width of the sample is L_y, then the Hall voltage $U_H = E_y L_y$. From (9.1.2) we see that U_H is proportional to B. The linear relation is shown in Fig. 9.1.

A century later, K. von Klitzing and collaborators [1, 2] discovered the integral quantum Hall effect (IQHE) in 1980, and D. Tsui, H. L. Störmer and A. Gossard [3] discovered the fractional quantum Hall effect (FQHE) in 1982. K. von Klitzing was awarded the Nobel prize in physics in 1985, D. Tsui, H. L. Störmer and R. Laughlin were awarded the Nobel prize in physics in 1998.

9.2   Electrons in Uniform Magnetic Field, the Landau Level

The quantum mechanical problem of the motion of electrons in uniform magnetic field was solved by L. D. Landau in 1930. The uniform magnetic field B along the z-axis is described by the vector potential[1]:

$$A_x = -By, \quad A_y = A_z = 0. \tag{9.2.1}$$

---

[1]It can be equivalently described by $A_y = Bx$, $A_x = A_z = 0$. These belong to the Landau gauge. A different choice $A_x = -\frac{y}{2}B$, $A_y = \frac{x}{2}B$, $A_z = 0$ is called the symmetric gauge.

The Schrödinger energy eigenequation is

$$\hat{H}\psi = \frac{1}{2m}\left[\left(\hat{p}_x + \frac{eB}{c}y\right)^2 + \hat{p}_y^2 + \hat{p}_z^2\right]\psi - \mu s_z B\psi = E\psi, \quad (9.2.2)$$

where s_z is the z-component of the electron spin, and $\mu s_z B$ is the electron Zeeman energy in the magnetic field. The Hamiltonian does not involve the x and z coordinates; hence p_x and p_z are constants of motion. The wave function is therefore of the form

$$\psi(x, y, z) = \exp\left[\frac{i}{\hbar}(p_x x + p_z z)\right]\chi(y). \quad (9.2.3)$$

Substituting (9.2.3) into (9.2.2) we obtain the equation for $\chi(y)$:

$$\chi''(y) + \frac{2m}{\hbar^2}\left[\left(E + \mu s_z B - \frac{p_z^2}{2m}\right) - \frac{1}{2}m\omega_c^2(y - y_0)^2\right]\chi(y) = 0, \quad (9.2.4)$$

where

$$y_0 = -\frac{cp_x}{eB}, \quad (9.2.5)$$

$$\omega_c = \frac{eB}{mc}, \quad (9.2.6)$$

ω_c is the cyclotron frequency. Equation (9.2.4) is the Schrödinger equation for a 1-dimensional harmonic oscillator, with the energy eigenvalue $E + \mu s_z B - \frac{p_z^2}{2m} = \hbar\omega_c(n + \frac{1}{2})$. Therefore

$$E_n = \left(n + \frac{1}{2}\right)\hbar\omega_c + \frac{p_z^2}{2m} - \mu s_z B, \quad (9.2.7)$$

The corresponding eigenfunction is

$$\chi_n(y) = \frac{1}{\pi^{1/4}a^{1/2}\sqrt{2^n n!}}\exp\left[-\frac{(y - y_0)^2}{2a^2}\right]H_n\left(\frac{y - y_0}{a}\right), \quad (9.2.8)$$

where H_n is the Hermite polynomial, and $a = \sqrt{\frac{\hbar}{m\omega_c}}$. It is interesting that p_x does not enter the expression for energy, but determines the position of equilibrium y_0 of the harmonic oscillator through (9.2.5). If the motion in the x-direction is not constrained, then p_x can vary continuously. If the motion in the x-direction is constrained by a length L_x, then p_x is discrete:

$$p_x = \frac{2\pi\hbar}{L_x}l, \quad l = 0, \pm 1, 2, \dots . \tag{9.2.9}$$

The corresponding positions of equilibrium for the oscillator in the y-direction is also discrete, with interval Δy_0 given by

$$\Delta y_0 = \frac{c}{eB}\frac{2\pi\hbar}{L_x}. \tag{9.2.10}$$

The number of equilibrium positions in a length L_y is

$$\frac{L_y}{\Delta y_0} = \frac{eBL_xL_y}{hc},$$

therefore the number of equilibrium positions per unit area in the $xy-$plane, i.e., the *density of states*, is

$$n_B = \frac{eB}{hc}. \tag{9.2.11}$$

To have an idea about the degeneracy of a Landau level, we consider the number of energy states per unit area of the sample for a completely filled Landau level $n_B = \frac{eB}{hc} = (2\pi a_0^2)^{-1}$, where $a_0 = (\frac{\hbar c}{eB})^{\frac{1}{2}}$ is the magnetic length. In typical QHE experiments a_0 is of the order of 50–100Å, n_B is therefore of the order of 10^{11} cm^{-2}. The sample area is typically 10^{-3} cm^2, and the degeneracy of a Landau level is 10^8.

If the electrons are constrained to move in the xy-plane, i.e., the degree of freedom in the z-direction is suppressed and the magnetic field is strong enough to polarize the electrons completely ($\mu s_z B$ is then a constant), the energy eigenvalue is simply

$$E_n = \left(n + \frac{1}{2}\right)\hbar\omega_c, \quad n = 0, 1, 2, \dots . \tag{9.2.12}$$

The energy levels are discrete, but highly degenerate with the degree of degeneracy given by n_B (9.2.11).

The problem can also be solved in symmetric gauge.

$$A_x = \frac{B}{2}y, \quad A_y = -\frac{B}{2}x, \quad A_z = 0. \tag{9.2.13}$$

The Hamiltonian is

$$\hat{H} = \frac{1}{2m}\left[\left(\hat{p}_x - \frac{eB}{2c}y\right)^2 + \left(\hat{p}_y + \frac{eB}{2c}x\right)^2\right], \tag{9.2.14}$$

The Zeeman energy of the completely polarized electrons (a constant) is not expli- citly written here. The scale of the harmonic oscillator with ω equal to the cyclotron frequency is called the magnetic length

$$a_0 = \left(\frac{\hbar}{m\omega_c}\right)^{1/2} = \left(\frac{\hbar c}{eB}\right)^{1/2}, \tag{9.2.15}$$

We use z and z^* as independent variables replacing x and y:

$$z = \frac{x + iy}{2a_0}, \quad z^* = \frac{x - iy}{2a_0}. \tag{9.2.16}$$

From (9.2.16) we obtain

$$\frac{\partial}{\partial z} = a_0\left(\frac{\partial}{\partial x} - i\frac{\partial}{\partial y}\right), \quad \frac{\partial}{\partial z^*} = a_0\left(\frac{\partial}{\partial x} + i\frac{\partial}{\partial y}\right). \tag{9.2.17}$$

The following useful relations can be obtained from (9.2.16) and (9.2.17):

$$\left.\begin{aligned} z^*z &= \frac{1}{4a_0^2}\left(x^2 + y^2\right) \\[2mm] z\frac{\partial}{\partial z} - z^*\frac{\partial}{\partial z^*} &= i\left(y\frac{\partial}{\partial x} - x\frac{\partial}{\partial y}\right) \\[2mm] \frac{\partial}{\partial z}\frac{\partial}{\partial z^*} &= a_0^2\left(\frac{\partial^2}{\partial x^2} + \frac{\partial^2}{\partial y^2}\right) \end{aligned}\right\} \tag{9.2.18}$$

Expressing the Hamiltonian (Eq. (9.2.14)) in terms of z and z^*, we have

$$\hat{H} = \frac{1}{2}\hbar\omega_c \left[-a_0^2 \left(\frac{\partial^2}{\partial x^2} + \frac{\partial^2}{\partial y^2} \right) + \frac{1}{4a_0^2}(x^2 + y^2) - i \left(y\frac{\partial}{\partial x} - x\frac{\partial}{\partial y} \right) \right]$$

$$= \frac{1}{2}\hbar\omega_c \left[-\frac{\partial}{\partial z}\frac{\partial}{\partial z^*} + z^*z - z\frac{\partial}{\partial z} + z^*\frac{\partial}{\partial z^*} \right]$$

$$= \frac{1}{2}\hbar\omega_c \left[\left(z^* - \frac{\partial}{\partial z} \right) \left(z + \frac{\partial}{\partial z^*} \right) + 1 \right].$$

Define operators a and $a^\dagger$ as

$$a = \frac{1}{\sqrt{2}} \left(z + \frac{\partial}{\partial z^*} \right), \quad a^\dagger = \frac{1}{\sqrt{2}} \left(z^* - \frac{\partial}{\partial z} \right), \tag{9.2.19}$$

Then the Hamiltonian can be written as

$$\hat{H} = \hbar\omega_c \left(a^\dagger a + \frac{1}{2} \right). \tag{9.2.20}$$

Operators a and $a^\dagger$ satisfy the commutation relations

$$[a, a^\dagger] = 1. \tag{9.2.21}$$

In the representation in which $n = a^\dagger a$ is diagonal, the energy eigenvalue is

$$E = \hbar\omega_c \left(n + \frac{1}{2} \right). \tag{9.2.22}$$

This is just the harmonic oscillator problem in second quantization form. For the Hamiltonian (9.2.14) there should be another conserved observable (quantum number) parallel to p_x in the Hamiltonian (9.2.2) in the Landau gauge. By inspecting the difference of vector potentials in the two gauges, we expect that the conserved observable is the angular momentum. Defining operators b and $b^\dagger$ by

$$b = \frac{1}{\sqrt{2}} \left(z^* + \frac{\partial}{\partial z} \right), \quad b^\dagger = \frac{1}{\sqrt{2}} \left(z - \frac{\partial}{\partial z^*} \right). \tag{9.2.23}$$

From the definition we find out that

$$\left.\begin{array}{l} [b, b^\dagger] = 1 \\ [b, a^\dagger] = [b, a] = 0 \\ [b^\dagger, a^\dagger] = [b^\dagger, a] = 0 \end{array}\right\}. \qquad (9.2.24)$$

Consequently, the eigenstates of $\hat{H}$ are simultaneous eigenstates of the operator $b^\dagger b$, viz.

$$\left.\begin{array}{l} \hat{H}\psi_{nm} = \hbar\omega_c\left(n + \dfrac{1}{2}\right)\psi_{nm} \\ b^\dagger b\psi_{nm} = m\psi_{nm} \end{array}\right\}. \qquad (9.2.25)$$

The eigenstates ψ_{0m} of $\hat{H}$ satisfies

$$a\psi_{0m} = 0,$$

i.e.,

$$\left(z + \frac{\partial}{\partial z^*}\right)\psi_{0m} = 0.$$

Solutions of this equation can be generated from the lowest state ψ_{00}, where

$$\psi_{00} = \text{const } e^{-zz^*}. \qquad (9.2.26)$$

Acting on (9.2.26) by $b^\dagger$ m times, we obtain

$$\psi_{0m} = (b^\dagger)^m \psi_{00} = \text{const } z^m e^{-zz^*}, \quad m = 0, 1, 2, \ldots, M. \qquad (9.2.27)$$

For the lowest Landau level states, m is the angular momentum quantum number

$$L_z\psi_{0m} = \hbar m\psi_{0m}.$$

This can be seen by acting the operator

$$L_z = -i\hbar\left(x\frac{\partial}{\partial y} - y\frac{\partial}{\partial x}\right) = \hbar\left(z\frac{\partial}{\partial z} - z^*\frac{\partial}{\partial z^*}\right) \qquad (9.2.28)$$

on ψ_{0m} and using the relations

$$L_z z^m = \hbar m z^m, \quad L_z e^{-zz^*} = 0.$$

The upper limit M of m can be determined in the following way.

We evaluate the expectation value of $x^2 + y^2$:

$$\frac{1}{4a_0^2} \left\langle x^2 + y^2 \right\rangle = \frac{\int_0^\infty (z^*z)^M e^{-2z^*z} z^* z \, dz dz^*}{\int_0^\infty (z^*z)^M e^{-2z^*z} dz dz^*} = \frac{M+1}{2},$$

where for the integral measure $2\pi r dr$ is used, $r^2 = zz^*$. Let the sample area be A,

$$A = \pi \left\langle x^2 + y^2 \right\rangle = 2\pi a_0^2 (M+1).$$

When M is large, the number of states per unit area n_B is

$$n_B = \frac{M}{A} = \frac{eB}{hc},$$

i.e.,

$$M = \frac{BA}{\Phi_0} = \frac{\Phi}{\Phi_0}, \tag{9.2.29}$$

where Φ is the flux through the sample, the result coincides with (9.2.11).

From Eqs. (9.2.19) and (9.2.23) it follows that

$$b^\dagger b = a^\dagger a + \frac{1}{\hbar} L_z.$$

Using this relation we obtain

$$b^\dagger b \psi_{nm} = m\psi_{nm} = n\psi_{nm} + \frac{1}{\hbar} L_z \psi_{nm},$$

i.e.,

$$L_z \psi_{nm} = \hbar(m-n)\psi_{nm}. \tag{9.2.30}$$

ψ_{nm} are eigenstates of L_z, the eigenvalue being $\hbar(m-n)$. From (9.2.28) we have

$$\left. \begin{array}{l} [L_z, b] = -\hbar b \\ [L_z, b^\dagger] = \hbar b^\dagger \end{array} \right\} \tag{9.2.31}$$

Therefore, b and $b^\dagger$ are respectively the lowering and raising operators of the eigenvalues of L_z by $\hbar$. Since the operators b and $b^\dagger$ do not change the quantum number n, the quantum number they lower and raise is m.

The simultaneous eigenfunctions of $\hat{H}$ and $b^\dagger b$ are

$$\psi_{nm} = \text{const}(b^\dagger)^m(a^\dagger)^n\psi_{00}$$

$$= \text{const}\left(z - \frac{\partial}{\partial z^*}\right)^m\left(z^* - \frac{\partial}{\partial z}\right)^n e^{-zz^*}. \qquad (9.2.32)$$

The states of the lowest Landau level are

$$\psi_{0m} = \text{const } z^m e^{-z^*z}, \quad m = 1, 2, \ldots, M. \qquad (9.2.33)$$

9.3 Quantization of Magnetic Flux

In Section 3.1 we discussed the flux quantization. The magnetic flux through a superconducting ring is quantized to be an integral multiple of $hc/2e$. C. N. Yang and N. Byers pointed out that this is not a new principle of physics, but a consequence of the gauge invariance and the single valuedness of the electron wave function. In Aharonov-Bohm effect an electron in a magnetic field free region can sense the existence of magnetic flux outside the region. When the flux changes by an integer number of the flux quantum $\Phi_0 = hc/e$ the interference fringes of electron waves shift by one fringe distance, i.e., the fringes look the same as those before the change. The flux quantization is also important in the physics of quantum Hall effect.

Let Φ be the flux passing through the origin of xy-plane along the z-axis. Then $\boldsymbol{B} = 0$ everywhere on the plane except at the origin. We can choose a vector potential

$$\boldsymbol{A} = \nabla\chi, \qquad (9.3.1)$$

which is valid everywhere except at the origin. The vector potential $\boldsymbol{A}$ satisfies the relation

$$\oint \boldsymbol{A} \cdot d\boldsymbol{l} = \Phi, \qquad (9.3.2)$$

where the closed contour encloses the origin. To obtain the explicit expression for χ, we assume that the contour is a circle of radius R

centered at the origin. Then

$$\oint \boldsymbol{A} \cdot \mathrm{d}\boldsymbol{l} = \frac{1}{R}\frac{\partial \chi}{\partial \phi}\bigg|_{r=R} \cdot 2\pi R = \Phi. \qquad (9.3.3)$$

Therefore,

$$\chi = \frac{\Phi}{2\pi}\phi.$$

where ϕ is the azimuthal angle. Consequently $\boldsymbol{A}$ can be written as

$$\boldsymbol{A} = \hat{\varphi}\frac{\Phi}{2\pi}\frac{1}{r},$$

where $\hat{\varphi}$ is a unit vector in the direction of increasing φ. In a gauge transformation the wave function ψ and the vector potential transform in the following way:

$$\left.\begin{array}{l} \boldsymbol{A} \to \boldsymbol{A}' = \boldsymbol{A} + \nabla\alpha \\[2mm] \psi \to \psi' = \exp\left(\dfrac{\mathrm{i}e}{\hbar c}\alpha\right)\psi \end{array}\right\}, \qquad (9.3.4)$$

where α is an arbitrary function of space coordinates. Since the scalar potential is not involved we can assume α to be time independent. If we choose $\alpha = -\chi$, we have

$$\left.\begin{array}{l} \boldsymbol{A}' = 0 \\[2mm] \psi' = \exp\left(-\dfrac{\mathrm{i}e}{\hbar c}\chi\right)\psi = \exp\left(-\mathrm{i}\dfrac{\Phi}{\Phi_0}\phi\right)\psi \end{array}\right\} \qquad (9.3.5)$$

The flux Φ previously involved with $\boldsymbol{A}$ has now slipped into ψ'. If the wave function describes an extended state, the coordinate ϕ can assume any value. If the angle varies continuously from ϕ to $\phi + 2\pi$ then the singlevaluedness of the wave function demands

$$\frac{\Phi}{\Phi_0} = m, \quad m = 0, \pm 1, \pm 2, \dots.$$

A flux equal to an integral multiple of the flux quantum can be *removed* by a gauge transformation, i.e., $m\Phi_0$ does not change the physics. It does not change the energy spectrum. Its effect is only to multiply the wave function by a phase factor.

Consider the state (9.2.33) in the lowest Landau level. The maximum value M of m is given by (9.2.29). Let the flux change adiabatically from Φ to $\Phi \pm \Phi_0$. The value m changes to $m \pm 1$. During the adiabatic process of flux change the Hamiltonian $\hat{H}(t)$ is a function of time. According to the adiabatic theorem the state (9.2.33) remains an "instantaneous stationary state" of $\hat{H}(t)$ during the evolution. When the flux change is completed the Hamiltonian does not differ at all from its form before the change, since a flux quantum can be *removed* by a gauge transformation. Consequently the state after the flux change is also an eigenstate of $\hat{H}$. The wave function ψ_{0m} now becomes $\psi_{0\,m\pm1}$. The change of angular momentum is due to the e.m.f. induced by the flux change. It is in the direction of decreasing (increasing) azimuthal angle, and therefore accelerates (decelerates) the electron and increases (decreases) the angular momentum. In the general case $m \to m \pm 1$ leads to a state still in the lowest Landau level.

9.4   The Integral Quantum Hall Effect

To observe the quantum Hall effect it is necessary to prepare a 2-dimensional electron gas (2DEG). This is necessary for the energy of the electrons to exhibit discrete Landau level (9.2.12). Electric field can be applied to restrict the electrons on a plane in a semiconductor. This can be realized in the Si metal oxide field effect tube (MOSFET) or heterostructure GaAs-Al$_x$Ga$_{1-x}$As [5]. The scheme of the experimental arrangement is shown in Figure 9.2.

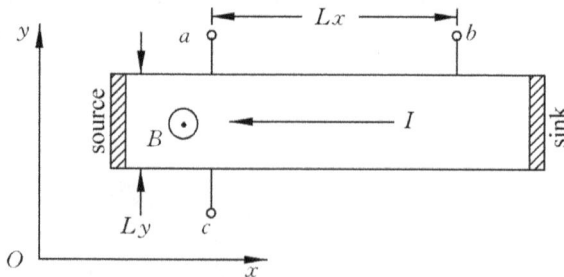

Fig. 9.2   The scheme of a quantum Hall effect measurement.

The longitudinal voltage U_L is applied between terminals a and b , the Hall voltage is measured between a and c. The magnetic field B is perpendicular to the page. Current I flows down the two-dimensional conductor of length L_x and width L_y. Measurements are carried out under low temperatures (few K) and strong magnetic field ($\lesssim$ 10T). The electrons are completely polarized to give a constant Zeeman energy. Pure samples are required. From the measured quantities the longitudinal resistance $R_{xx} = \frac{U_L}{I}$ and the Hall resistance $R_H = \frac{U_H}{I}$ are determined. Figure 9.3 shows the variation of ρ_{xx} (the longitudinal resistivity) and ρ_{xy} (the Hall resistivity) versus B. In the 2-dimensional case the current density is defined as the current per unit length

$$R_H = \frac{U_H}{I} = \frac{E_y L_y}{j_x L_y} = \frac{E_y}{j_x} = \rho_H. \tag{9.4.1}$$

i.e., the Hall *resistivity* is identical to the Hall *resistance*. Similarly, we see that the longitudinal resistivity is *proportional* to the longitudinal resistance.

$$R_x = \frac{U_x}{I} = \frac{E_x L_x}{j_x L_y} = \rho_x \frac{L_x}{L_y}.$$

The distinct feature of IQHE is that ρ_H does not bear a linear relation to B any more. von Klitzing and collaborators found that the Hall resistance R_H of a 2DEG exhibits a series of plateaus superimposed on a general linear trend. These plateaus appear on quantized values

$$R_H = \frac{h}{ie^2}, \quad i = 1, 2, 3, \dots. \tag{9.4.2}$$

The longitudinal resistance vanishes at values of B corresponding to the quantized Hall resistance. The quite unexpected feature is that the quantized resistance depends only on *universal constants* e and h, independent of materials. The values of the quantized Hall resistance can be very accurately measured (part per million

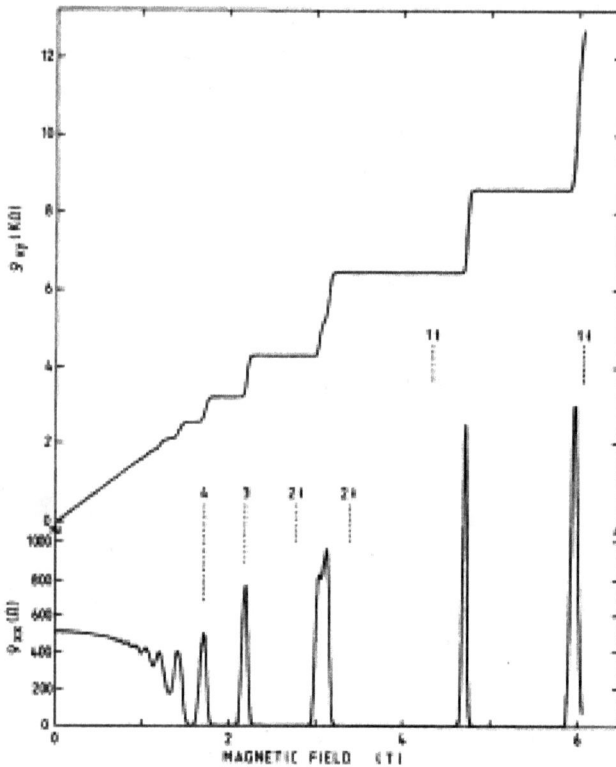

Fig. 9.3 ρ_{xx} and ρ_{xy} versus B in the IQHE. Reprinted with permission from K. von Klitzing, *Rev. Mod. Phys.* 58, 519, 1986. Copyright (2018) by the American Physical Society.

accuracy) under ideal conditions:

$$R_{\mathrm{H}} = \frac{25812.807557}{i}\Omega. \tag{9.4.3}$$

Since i is an integer, this effect is called the *integral* quantum Hall effect. It has already metrological applications for providing standard of resistance.

The IQHE can be understood on the basis of single electron behavior in solids. Equations (9.2.11) and (9.2.12) give the degeneracy of the Landau level in the sample:

$$N_B = n_B L_x L_y = \frac{e}{hc}BL_x L_y = \frac{\Phi}{hc/e}, \tag{9.4.4}$$

where Φ is the flux through the sample, the quotient hc/e in the denominator is the flux quantum

$$\Phi_0 = \frac{hc}{e}, \tag{9.4.5}$$

therefore

$$N_B = \frac{\Phi}{\Phi_0}. \tag{9.4.6}$$

This means that if one Landau level is filled completely, then each electron gets a share of one flux quantum. If i Landau levels are filled, the total number of electrons $N = iN_B$ is related to the flux by $N = i\Phi/\Phi_0$.

When one Landau level is filled completely the surface number density of electron is

$$n_s = \frac{eB}{hc}, \tag{9.4.7}$$

and the Hall resistance is

$$R_H = \frac{B}{n_s ec} = \frac{h}{e^2}. \tag{9.4.8}$$

If i Landau levels are completely filled, then

$$R_H = \frac{h}{ie^2}. \tag{9.4.9}$$

We have established the relation between the Hall resistance with the filling number of the Landau levels. Define the *filling fraction* of Landau level ν as

$$\nu = \frac{n_s}{n_B} = \frac{n_s}{eB/hc}, \tag{9.4.10}$$

which is the ratio of the number density of electrons to the degeneracy of the Landau level, characterizing the degree of occupation of the Landau level. The quantized value of the Hall resistance is obtained when $\nu = i$ (integers). The existence of plateau is a signature of the stability of the filled level. We express the relation between current, electric field strength and resistivity in case of general orientation of the axes as

$$\left. \begin{array}{l} E_x = \rho_{xx} j_x + \rho_{xy} j_y \\ E_y = \rho_{yx} j_x + \rho_{yy} j_y \end{array} \right\}. \tag{9.4.11}$$

Experiments establish the fact that $R_H = \frac{h}{ie^2}$ and $R_x = 0$, therefore ρ in the matrix form is

$$\rho = \begin{bmatrix} 0 & \frac{h}{ie^2} \\ -\frac{h}{ie^2} & 0 \end{bmatrix}. \tag{9.4.12}$$

This shows that the direction of current is always perpendicular to that of the field strength. The corresponding matrix of the conductivity $\sigma = \rho^{-1}$ is

$$\sigma = \begin{bmatrix} 0 & -\frac{ie^2}{h} \\ \frac{ie^2}{h} & 0 \end{bmatrix}. \tag{9.4.13}$$

The IQHE state $\nu = i$ is characterized by $\rho_{xx} = 0$ and $\sigma_{xx} = 0$.

We have seen that the quantized Hall resistance corresponds to the filling of energy levels. The question is, why is it independent of materials? Large numbers of theoretical investigations were carried out, considering the influence on the quantized value of Hall resistance of the periodic potentials, dimension of samples, finite temperature, interaction between electrons, impurities, etc. The conclusion is, as long as $\sigma_{xx} = 0$, these effects do not change the value of quantized Hall resistance [5]. Before discussing the formation of plateaus, we turn first to the removal of degeneracies of the Landau level by Hall voltage and by impurities.

In the foregoing discussions we did not take into account the periodic potential of the crystal. In a simple treatment the periodic potential can be considered as to lead to an effective mass and an effective dielectric constant. In this case the solutions to the Landau problem are the envelope functions of the Wannier wave functions.[2]

---

[2]See P. Y. Yu and M. Cardona, Fundamentals of Semiconductors, Berlin, Springer 1996, p. 152.

A more refined treatment leads to the magnetic translation and topological considerations (Section 9.4.1).

The degeneracy of the Landau level is lifted by the external field. In the case of quantum Hall effect, by the Hall voltage, and also by impurities. The Hall voltage is ELy, where E is the field strength. There should be an additional term for this electrostatic potential $-eEy$ in (9.2.2). This term can be incorporated into the shift of the origin y_0, now becoming

$$y_0 = -\frac{cp_x}{eB} + \frac{Emc^2}{eB^2}. \tag{9.4.14}$$

The additional energy $\delta\mathscr{E}$ corresponding to this voltage is

$$\delta\mathscr{E} = -eEy_0 = \frac{cEp_x}{B} - \frac{E^2mc^2}{B^2}, \tag{9.4.15}$$

which depends on p_x. Comparing with (9.2.12), where p_x does not enter the expression for the energy eigenvalue, we see that now it does. Since $0 \le y_0 \le L_y$ implies $0 \le |p_x| \le \frac{eBL_y}{c}$, we see that the degeneracy of Landau level is lifted. The p_x-dependence leads to a group velocity $v_g = \frac{dE}{dp_x} = \frac{cE}{B}$, which is the drift velocity of the electron in a cross E and B field.[3] The problem of impurities was investigated systematically [4]. It is sufficient to consider a δ-function impurity potential

$$V_I = \lambda\delta(x - x_0)\delta(y - y_0). \tag{9.4.16}$$

By expanding the eigenstates of $\hat{H} + V_I$ in terms of the eigenfunctions of $\hat{H}$,

$$\psi_{nk}(x, y) = L^{-\frac{1}{2}}e^{ikx}\varphi_{nk}(y - y_0), \tag{9.4.17}$$

---

[3]The motion of an electron in mutually perpendicular E and B fields is treated in standard text of electrodynamics. Let B be perpendicular to the plane of motion and E lie on the plane. The motion consists of a cyclotron motion with $\omega_c = \frac{eB}{mc}$ and a drift motion of the guiding center with drift velocity $v_d = \frac{cE \times B}{B^2}$.

where we have replaced $p_x/\hbar$ by k, perturbation theory gives the energy eigenvalues E

$$1 = \lambda \sum_{n,k} \frac{|\psi_{nk}(x_0, y_0)|^2}{E - E_{nk}}. \tag{9.4.18}$$

Prange [5] predicted that each E solving (9.4.18) lies between two unperturbed levels

$$E_{nk} = \hbar\omega\left(n + \frac{1}{2}\right) + \frac{cE\hbar k}{B}, \tag{9.4.19}$$

except possibly for one level lying below all unperturbed level when $\lambda < 0$ (attractive impurity potential). There is one state completely localized for each Landau level. For v_g small its energy lies well away ($\approx \lambda$) from the original Landau level position $n\hbar\omega_c$. An approximation to the lowest such state for $x_0 = y_0 = 0$ is

$$\psi_{\text{loc}} \propto e^{-(x^2+y^2)/4a_0^2} e^{ixy/2a_0^2}. \tag{9.4.20}$$

The energies of such states are shifted upward or downward from $n\hbar\omega$ for $\lambda > 0$ or $\lambda < 0$ respectively. All other levels ψ_{nk} solving (9.4.18) lie close to the unperturbed energy and are extended; there are $N-1$ of the extended states for each Landau level ($N = n_B L_x L_y$). Prange also gave the expression for the current carried by the extended states:

$$I = -\frac{e}{\hbar} \sum_n (E_{nk_{\max}} - E_{nk_{\min}}). \tag{9.4.21}$$

There is a tricky point that the impurities reduce the number of states carrying the current but the current remains the same as long as all extended states are occupied. The point is that electrons passing close to the impurity speed up and carry just enough more current as to compensate for the current subtracted due to the localized state. The density of states is shown in Fig. 9.4. In the figure each Landau level of extended states is split into a band with localized states populated between the Landau bands. When a Landau level is filled, the electrons in extended states contribute a Hall resistance h/e^2. Let the magnetic field B be decreased. According to (9.2.11) the number of electrons in the Landau level decreases. Due to Pauli principle,

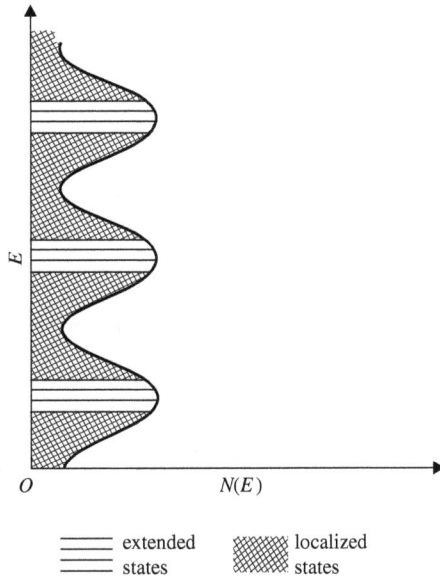

Fig. 9.4   Density of states in the presence of Hall voltage and impurities. Redrawn from Ref [5].

part of the electrons have to be arranged in the next Landau level. This would have led to a change of Hall resistance. But because of the existence of localized states between the Landau levels, the electrons are arranged first on the localized states. There will be no change in the Hall resistance because electrons on localized states do not participate in the conduction process. No variation of R_H while B varies — this is the plateau. As B decreases further, when the Fermi energy E_F has reached the extended states of the next Landau level such that electrons begin to populate on the extended states, the plateau ends and R_H begins to change.

To understand the universality of the quantized Hall resistance is a challenging problem. R. Laughlin attempted to interpret this point from gauge invariance in 1981 [6]. Consider a metallic loop of circumference L_x shown in Fig. 9.5. Magnetic field B is applied everywhere perpendicular to the surface of the loop (z-direction). Between the two lateral sides (y-direction) a voltage U_H is applied, the current along the loop I (x-direction) produces flux Φ threading the loop in the y-direction. Since $\rho_{xx} = 0$, there is no dissipation and the energy

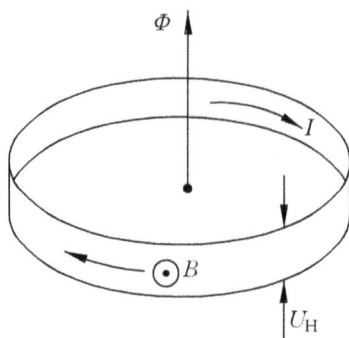

Fig. 9.5 Metallic loop of length L_x for illustrating the universality of Hall resistance.

is conserved. The relation between the current I, the flux Φ and the energy E is

$$I = c\frac{\partial E}{\partial \Phi} \qquad (9.4.22)$$

This relation can be simply derived. The current along the loop makes the loop a magnetic dipole with magnetic moment $\mu = \frac{I}{c}S$, where S is the area of the loop. Let Φ change by a small amount $\delta\Phi$, the corresponding change of the magnetic field is $\delta\Phi/S$, and the change of the energy of the magnetic dipole in this field is $\delta E = \mu\frac{\delta\Phi}{S} = \frac{I}{c}\delta\Phi$, which leads to (9.4.22). The wave function of the electron is (refer to (9.2.3))

$$\psi(x,y) = \exp\left(\frac{i}{\hbar}p_x x\right)\chi(y), \qquad (9.4.23)$$

where the single-valuedness of ψ demands $p_x = \frac{2\pi\hbar}{L_x}n$, where n is an integer. Different values of p_x correspond to different positions of equilibrium of the state along the y-direction. Let the flux Φ change by a flux quantum. From (9.3.5) we know that such a change is equivalent to a phase factor $e^{i\phi}$ multiplying the wave function, where from Fig. 9.5 we see that $\phi = 2\pi\frac{x}{L_x}$. The change of wave function (9.4.23) induced is $p_x \to p_x + 2\pi\frac{\hbar}{L_x}$. This corresponds to the situation that each state changes its value of p_x by $2\pi\frac{\hbar}{L_x}$, i.e., each electron replaces its neighbor of larger p_x. Effectively the electron with the least p_x value moves to the state with highest p_x value.

Correspondingly the position of equilibrium performs also the one-step shift. Effectively an electron moves from one end of the loop to the other end along the y-direction. The change of Φ induces an e.m.f. along the x-direction. Since the electric conductivity has only off-diagonal elements (9.4.13), this e.m.f. causes the electron motion in the y-direction. The energy change of the electron is eU_H. If there are i Landau levels completely filled, the energy change is

$$\Delta E = ieU_H. \tag{9.4.24}$$

Using (9.4.22) we have

$$I = c\frac{\Delta E}{\Phi_0} = c\frac{ie}{\Phi_0}U_H = \frac{ie^2}{h}U_H, \tag{9.4.25}$$

This gives

$$R_H = \frac{h}{ie^2}. \tag{9.4.26}$$

This demonstrates that the universality of the quantized Hall resistance is based on the principle of gauge invariance. Laughlin considers that the plateaus in IQHE is a measure of the electron charge . When FQHE was discovered, Laughlin made a daring guess that the quasiparticle excitation in FQHE should bear a fractional charge.

B. Halperin [6] used a similar approach as Laughlin's. In this paper Halperin has shown that there is a continuum of edge states at a physical boundary. In Fig. 9.6 an annular shaped film in region $r_1 < r < r_2$ is placed in a uniform magnetic field $\boldsymbol{B}$ perpendicular to the film and additional flux Φ is confined to the region $r < r_1$ through the hole. The wave function $\psi_{\nu,m}$ near the outer edge of the annulus satisfies a Schrödinger equation for a 1-dimensional harmonic oscillator with equilibrium positions r_m given by

$$B\pi r_m^2 = m\Phi_0 - \Phi, \tag{9.4.27}$$

where Φ_0 is the flux quantum, m is an integer corresponding to the angular momentum of the state, and ν is the number of nodes of the radial wave function. The vanishing of the wave function on the edge turns out to be a very strong demand. The energy $E_{\nu,m}$

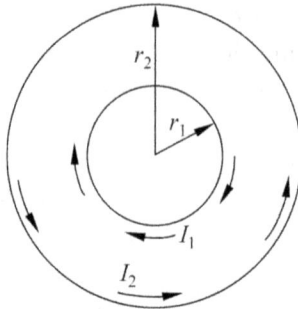

Fig. 9.6   Geometry of the sample. B is pointing out of the page. Reprinted with permission from B. I. Halperin, *Phys. Rev. B* 25, 2185, 1982. Copyright (2017) by the American Physical Society.

will monotonically increase as r_m increases towards r_2 for all values of ν. Halperin also pointed out that the current carried by the outer and inner edge states are in opposite directions. This can be seen from a semi-classical picture. B is pointed out of the page, the cyclotron motion of and electron is counter-clockwise. Close to the outer edge, an electron executing a cyclotron motion hits the edge, and is reflected back by the potential barrier. To continue its cyclotron motion, its guiding center moves along the edge in a clockwise sense. Hence the current is in the counter-clockwise sense as shown by I_2 in the figure. Similar analysis gives the current I_1 along the inner edge in clockwise sense. The edge current is therefore *chiral*.

Prange pointed out[4] that an arbitrary potential depending only on y can be added to $\hat{H}$ (9.2.2). At a very large potential rise representing a boundary, the solutions represent the edge states. These states possess a quasi-continuous spectrum which connects the bulk Landau levels with the energy at the top of the potential well, in agreement with Halperin's result [7].

Prange also pointed out that the emphasis on edge states may have led some to believe that QHE cannot exist without edge states or that all or most of the current is carried in edge states. Prange's point is that the bulk QHE guarantees the existence of the edge

---

[4]See Ref. [5] Chapter 1 note 8, and also problem 4.

states rather than the contrary. In fact, in some geometries, e.g., on the torus, the QHE presumably exist without edge states.

9.4.1 Magnetic translation, topological significance of the Hall conductance

Laughlin's theory of the IQHE seems to use a topological argument and the result is obtained without depending on specific details of the Hamiltonian. Another series of research was pioneered by D. J. Thouless [8, 9]. The interesting point is that the research was not undertaken to provide a topological basis for the Hall conductance, but it does provide one. The problem that Thouless and collaborators considered is an infinite 2-dimensional electron gas in a periodic potential under the action of a perpendicular magnetic field. This problem was studied by many authors previously [10–14]. One of the main results is the magnetic translation. The magnetic field is given by the vector potential in Landau gauge

$$A_x = 0, \quad A_y = Bx, \tag{9.4.28a}$$

and the one-particle Hamiltonian is[5]

$$H = \left[\frac{1}{2m} \left(-i\hbar \frac{\partial}{\partial x} \right)^2 + \frac{1}{2m} \left(-i\hbar \frac{\partial}{\partial y} - eBx \right)^2 \right] + U(x, y), \tag{9.4.28b}$$

where $U(x, y)$ is the periodic potential with lattice constants a and b,

$$U(x + a, y) = U(x, y + b) = U(x, y). \tag{8.4.29}$$

Without the magnetic field, the single particle wave function is given by the Bloch theorem

$$\psi_{k_1,k_2}(x, y) = e^{ik_1 x + ik_2 y} u_{k_1 k_2}(x, y), \tag{9.4.30}$$

---

[5]The electronic charge is $-e$.

where (k_1, k_2) is a vector in the reciprocal lattice, and $u_{k_1 k_2}$ satisfies the symmetry of lattice translations:

$$u_{k_1 k_2}(x + a, y) = u_{k_1 k_2}(x, y + b) = u_{k_1 k_2}(x, y). \qquad (9.4.31)$$

We have for the wave function ψ

$$T_a \psi = e^{i k_1 a} \psi, \qquad (9.4.32)$$

$$T_b \psi = e^{i k_2 b} \psi, \qquad (9.4.33)$$

where $T_a = e^{\frac{i}{\hbar} a p_x}$, $T_b = e^{\frac{i}{\hbar} b p_y}$ are the operator of lattice translations. When the magnetic field is included, T_a and T_b are no more symmetry operations because of the vector potential. It is necessary to find new symmetry operations suitable for this case. They are the magnetic translation operators. Many authors contributed to this, and in the following we follow Ref. [13].

We use the symmetric gauge

$$A_x = -\frac{1}{2} B y, \qquad A_y = \frac{1}{2} B x$$

and define the generators of magnetic translation

$$\Pi_i = p_i - \frac{e}{c} A_i - \frac{eB}{c} \varepsilon_{ij} x_j = p_i + \frac{e}{c} A_i. \qquad (9.4.34)$$

We find that

$$\left[\Pi_i, \quad p_j - \frac{e}{c} A_j \right] = 0.$$

Because of the chosen gauge this relation is valid for the same i, j components or different ones. Therefore

$$[\boldsymbol{\Pi}, \quad H_0] = 0, \qquad (9.4.35)$$

where

$$H_0 = \frac{1}{2m} \left(-i\hbar \nabla - \frac{e}{c} \boldsymbol{A} \right)^2.$$

The magnetic translation operators are

$$
\left.
\begin{array}{l}
\mathscr{T}_a = \exp\left(\dfrac{\mathrm{i}}{\hbar}a\Pi_x\right) \\[2mm]
\mathscr{T}_b = \exp\left(\dfrac{\mathrm{i}}{\hbar}b\Pi_y\right)
\end{array}
\right\}
\tag{9.4.36}
$$

The components of $\boldsymbol{\Pi}$ do not commute:

$$
[\Pi_i, \Pi_j] = -\frac{\mathrm{i}e\hbar}{c}\,\varepsilon_{ij}B.
\tag{9.4.37}
$$

The translation operators $\mathscr{T}_a$ and $\mathscr{T}_b$ do not commute either:

$$
\mathscr{T}_a\mathscr{T}_b = \exp\left(-\mathrm{i}2\pi\frac{\Phi}{\Phi_0}\right)\mathscr{T}_b\mathscr{T}_a,
\tag{9.4.38}
$$

where Φ is the flux through the area ab. When Φ is an integral multiple of the flux quantum Φ_0, $\mathscr{T}_a$ and $\mathscr{T}_b$ commute. The magnetic translation operators possess the remarkable property of commuting with the single particle Hamiltonian:

$$
H = \frac{1}{2m}\left[\left(-\mathrm{i}\hbar\frac{\partial}{\partial x} - \frac{e}{c}A_x\right)^2 + \left(-\mathrm{i}\hbar\frac{\partial}{\partial y} - \frac{e}{c}A_y\right)^2\right] + U(x,y),
\tag{9.4.39}
$$

$$
[\mathscr{T}_a,\ H] = [\mathscr{T}_b, H] = 0.
\tag{9.4.40}
$$

How does the vector potential become consistent with the lattice translation invariance of the periodic potential? The answer is that it has been included in the exponential of the magnetic translation operator, i.e., *gauge transformations* have been incorporated such that a linear increase of vector potential in all space can be replaced by a linear increase within a unit cell such that the periodic condition can be satisfied. We shall come back to this later. In the following we use the specific vector potential

$$A_x = 0, \quad A_y = Bx, \tag{9.4.41}$$

then the magnetic translation generators are:

$$\Pi_x = -i\hbar \frac{\partial}{\partial x},$$

$$\Pi_y = -i\hbar \frac{\partial}{\partial y} + \frac{e}{c} Bx.$$

We assume that

$$\Phi = \frac{cB}{hc} ab = \frac{p}{q}, \tag{9.4.42}$$

where p and q are integers prime to each other. We observe that although $\mathcal{T}_a$ and $\mathcal{T}_b$ do not commute in this case, but we have

$$[(\mathcal{T}_a)^q, \mathcal{T}_b] = 0, \tag{9.4.43}$$

such that equivalently a larger unit cell (qa, b) results, and it is possible to diagonalize simultaneously $H, (\mathcal{T}_a)^q$ and $\mathcal{T}_b$. Comparing with the magnetic translation invariance, we see that the Bloch condition is also generalized. Instead of (9.4.31) we have now

$$u_{k_1 k_2}(x + qa, y)e^{-2\pi i p y/b} = u_{k_1 k_2}(x, y + b) = u_{k_1 k_2}(x, y), \tag{9.4.44}$$

and instead of (9.4.32) we have

$$\mathcal{T}_a^q \psi = e^{ik_1 qa} \psi, \quad \mathcal{T}_b \psi = e^{ik_2 b} \psi. \tag{9.4.45}$$

From the periodic boundary condition we see that the magnetic Brillouin zone is

$$\left. \begin{aligned} 0 \le k_1 \le \frac{2\pi}{qa} \\ 0 \le k_2 \le \frac{2\pi}{b} \end{aligned} \right\} \tag{9.4.46}$$

$u_{k_1 k_2}(x, y)$ are eigenfunctions of a k-dependent Hamiltonian:

$$H'(k_1, k_2) = \frac{1}{2m}\left(-i\hbar\frac{\partial}{\partial x} + \hbar k_1\right)^2$$

$$+ \frac{1}{2m}\left(-i\hbar\frac{\partial}{\partial y} + \hbar k_2 - eBx\right)^2 + U(x, y). \quad (9.4.47)$$

In the following, we give only the result of Refs. [8, 9]. The original papers used Kubo's formula for the calculation of Hall conductance $\langle I_y \rangle / V_H$, which is effectively a velocity-velocity correlation, while the velocity operators $\frac{1}{m}(p_x + \frac{e}{c}A_x)$ and $\frac{1}{m}(p_y + \frac{e}{c}A_y)$ can be obtained from (9.4.47) by taking derivatives $\frac{\partial H'}{\partial k_1}$ and $\frac{\partial H'}{\partial k_2}$ respectively. The result turns out to be

$$\sigma_{xy} = \frac{ie^2}{2\pi\hbar}\int\left[\left\langle\frac{\partial u_\alpha}{\partial k_1}\Big|\frac{\partial u_\alpha}{\partial k_2}\right\rangle - \left\langle\frac{\partial u_\alpha}{\partial k_2}\Big|\frac{\partial u_\alpha}{\partial k_1}\right\rangle\right]\mathrm{d}k_1\mathrm{d}k_2$$

$$= \frac{e^2}{2\pi\hbar}\oint\left\langle u_\alpha\Big|\frac{\partial}{\partial \mathbf{k}}\Big|u_\alpha\right\rangle\cdot\mathrm{d}\mathbf{k}, \quad (9.4.48)$$

where α denotes (k_1, k_2) and the line integral is to be taken winding the boundary of the magnetic Brillouin zone. It represents the net phase change after a winding. Because of the single valuedness of the wave function, it should be 2π times an integer which leads to $\sigma_{xy} = n\frac{e^2}{\hbar}$. To be a topological invariant, it has to be independent of specific structure of the periodic potential. To see more clearly how periodic conditions can lead to the topology, we begin with a sample of dimensions (L_x, L_y).[6] Without the magnetic field the conditions are

$$\psi(L_x, y) = \psi(0, y), \quad \psi(x, L_y) = \psi(x, 0). \quad (9.4.49)$$

With the vector potential $A_x = 0$, $A_y = Bx$, we have to incorporate the gauge transformation mentioned in connection with the magnetic translation. We can glue the two ends in the x-direction to obtain a closed ribbon as in Laughlin's gauge invariance discussion, and then glue the two open y boundaries to form a torus. As the particle moves

---

[6]We follow here M. Stone in Ref. [15] p. 9.

out of the right-hand edge of the rectangular sample, it reenters from the left-hand edge, but with an additional phase

$$\psi(L_x, y) = e^{i\frac{eB}{\hbar c}L_x y}\psi(0, y), \quad \psi(x, L_y) = \psi(x, 0). \qquad (9.4.50)$$

The y-dependent phase factor resets the gauge potential A_y at the right-hand boundary back to its value at the left-hand boundary. There is a consistency requirement though. The relation between $\psi(L_x, L_y)$ and $\psi(0, 0)$ can be obtained by two different ways, viz. by using the first relation to get $\psi(L_x, 0)$ from $\psi(0, 0)$ and then by using the second relation to obtain $\psi(L_x, L_y)$ from $\psi(L_x, 0)$; or by reversing the order to obtain $\psi(0, 0) \rightarrow \psi(0, L_y) \rightarrow \psi(L_x, L_y)$. The two choices give different results unless

$$\exp\left(i\frac{eB}{\hbar c}L_x L_y\right) = 1,$$

i.e.,

$$\Phi = BL_x L_y = \frac{hc}{e}N = N\Phi_0.$$

Actually one is forced from beginning with a rectangular sample to incorporate the vector potential to satisfy the boundary condition by gluing the sample into a torus. Once a torus is obtained, the inevitable topology appears. It is the topological invariant that defines the first Chern class of the mapping of the Brillouin zone (a torus) onto the complex projective space of wave functions $u_\alpha(x, y)$. In the language of fiber bundle theory , the Hall conductance is the first Chern class of a $U(1)$ principal fiber bundle of the ground state wave function on the base manifold of a torus. The base manifold is the magnetic Brillouin zone, and the fibers are the single particle Bloch waves.

9.5   Fractional Quantum Hall Effect, Laughlin Wave Function

When 2DEG is subjected to a still lower temperature ($\sim$0.1K), and stronger magnetic field ($\sim$20T), and furthermore the sample is of very high purity, new regularities appear. Tsui, Störmer and

Gossard [3] discovered that quantized plateaus of Hall resistance appear at fractional filling fractions $\nu = \frac{1}{3}, \frac{2}{3}, \frac{2}{5}, \frac{3}{5}, \ldots$, etc. This is shown in Fig. 9.7. Corresponding to quantized plateaus of ρ_{xy} minimum values of ρ_{xx} appear. The FQHE brought surprise and puzzle. For IQHE stability is provided by the filling being integers, since adding one more electron necessitates a promotion to the next Landau level with an energy gap $\hbar \omega_c$. Now a Landau level is not yet completely filled, say filled only to 1/3. What makes such a state especially stable? There are still lot of space for the electrons in the same Landau level. The interpretation of IQHE is based on a picture of single particle behavior, very much similar to the case of many-electron problem in the atomic structure.

R. Laughlin realized that the fractionally filled state is a con-densate of the many-particle collective state, a strong correlation is

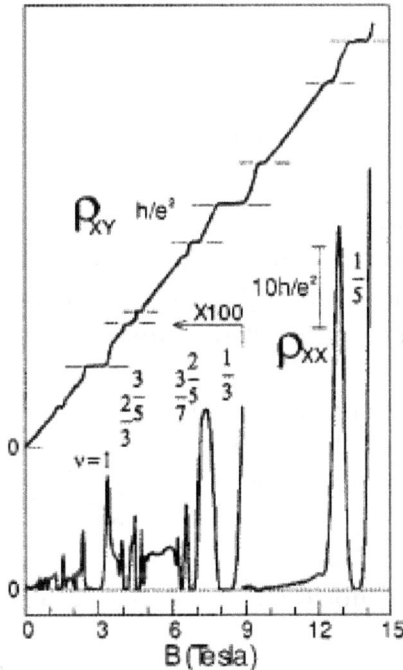

Fig. 9.7   The fractional quantum Hall effect: plateaus of ρ_{xy} appears at fractional filling fractions. Reprinted with permission from D. C. Tsui, *Rev. Mod. Phys.* 71, 891, 1999. Copyright (2018) by the American Physical Society.

produced by the electron-electron interaction. He proposed the wave functions of the ground state and the quasiparticle excitation. The theory is given in Ref. [16], but systematic research endeavors had been carried out before the publication, which are described in a chapter in Ref. [5]. Since this is an example how quantum mechanics can contribute to the progress of a new field of study, we present in the following a brief introduction of this development.

The Hamiltonian of the many particle system is

$$H = \sum_j \left\{ \frac{1}{2m} \left(\frac{\hbar}{i}\nabla + \frac{e}{c}\boldsymbol{A} \right)^2 + V(z_j) \right\} + \sum_{j<k} \frac{e^2}{|z_j - z_k|}, \qquad (9.5.1)$$

where $V(z_j)$ is the potential produced by the uniform background charge of the positive ions with the surface density σ.

$$V(z) = -\sigma e^2 \int \frac{\mathrm{d}^2 z'}{|z - z'|}, \qquad (9.5.2)$$

The vector potential is in the symmetric gauge

$$\boldsymbol{A} = \frac{B}{2}(y\hat{\boldsymbol{x}} - x\hat{\boldsymbol{y}}). \qquad (9.5.3)$$

Laughlin's attempt was not to "derive" the ground state wave function of N electrons for the fillilng fraction $\nu = 1/m$ directly from the Hamiltonian following the "reductionist" approach [3]. The FQHE is very much different from the IQRE for being the consequence of the collective effect of many electrons. Before writing down the wave functions of N electrons in the $\nu = 1/m$ state Laughlin began with the study of the quantized motion of a small number of electrons, from which he acquired the physics intuition necessary in making generalization to the many electron case.

9.5.1 Quantized motion of a small number of electrons

We begin with the two-electron problem. Suppose the Coulomb potential between the electrons is small compared to the cyclotron energy $\hbar\omega_c$. It is then possible to require the two-body wave functions to be comprised solely of single particle wave functions of the lowest Landau level

$$|m\rangle = \frac{1}{(2^{m+1}\pi n!)^{1/2}} z^m \exp\left(-\frac{1}{4}|z|^2\right),$$

where the unit of length is $a_0 = (\frac{\hbar c}{eB})^{\frac{1}{2}}$, the magnetic length, and $z_i = x_i + iy_i$ is the complex coordinates of the i-th electron.[7] Since the Hamiltonian is azimuthally symmetric, the state must be an eigenstate of the *relative* angular momentum. The wave function has to be antisymmetric under the interchange of z_1 and z_2. The wave function is then

$$\psi_k = (z_1 - z_2)^{2k+1} e^{-\frac{1}{4}(|z_1|^2 + |z_2|^2)}. \tag{9.5.4}$$

The power of the factor $(z_1 - z_2)$ has to be odd to satisfy Pauli principle; hence the relative angular momentum of the cluster is $2k+1$. To check whether this wave function is a good approximation, the case $k = 0$ and $\hbar\omega_c = e^2/a_0$ (a_0 is the magnetic length) is compared to the exact solution shown in Fig. 9.8. The condition $\hbar\omega_c = e^2/a_0$ corresponds to a magnetic field strength 6T, while the experiment for GeAs used 15T, for which the approximation should be much better.

Consider now the 3-electron problem. This is the case from which Laughlin drew his ideas. The Hamiltonian is now

$$H = \frac{1}{2m} \sum_{j=1}^{3} \left[-i\hbar\nabla_j + \frac{e}{c}A_j\right]^2 + \frac{e^2}{r_{12}} + \frac{e^2}{r_{23}} + \frac{e^2}{r_{31}}. \tag{9.5.5}$$

Define the center-of-mass and relative coordinates of the system as

$$\left.\begin{aligned}
\bar{z} &= \frac{z_1 + z_2 + z_3}{3} \\
z_a &= \sqrt{\frac{2}{3}}\left[\frac{z_1 + z_2}{2} - z_3\right] \\
z_b &= \frac{1}{\sqrt{2}}(z_1 - z_2)
\end{aligned}\right\} \tag{9.5.6}$$

---

[7]The definition of z_i is different from Section 9.2 by a factor 2.

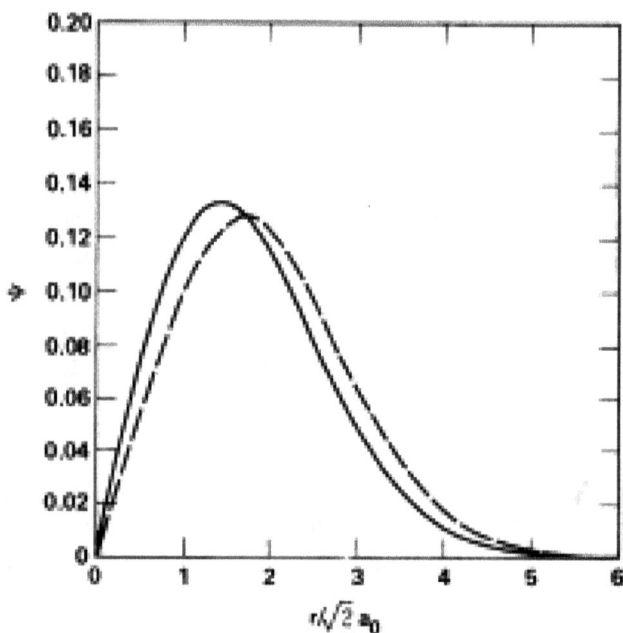

Fig. 9.8 Comparison of the approximate wave function (solid) with the exact wave function (dashed). Reprinted with permission from R. B. Laughlin, *Phys. Rev. B* 27, 3383, 1983. Copyright (2018) by the American Physical Society.

Excluding the center-of-mass degree of freedom, the Hamiltonian of the internal motion is given by

$$H_0 = \frac{1}{2m}\left[-i\hbar\nabla_a + \frac{e}{c}A_a\right]^2 + \frac{1}{2m}\left[-i\hbar\nabla_b + \frac{e}{c}A_b\right]^2$$

$$+ \frac{e^2}{\sqrt{2}}\left(\frac{1}{|z_b|} + \frac{1}{\left|\frac{1}{2}z_b + \frac{\sqrt{3}}{2}z_a\right|} + \frac{1}{\left|\frac{1}{2}z_b - \frac{\sqrt{3}}{2}z_a\right|}\right). \qquad (9.5.7)$$

Pauli's principle demands the wave function to be antisymmetric under the interchange of any two electrons. The interchange $z_1 \leftrightarrow z_2$ leads to

$$\left.\begin{array}{c} z_b \rightarrow -z_b \\ z_a \pm iz_b \rightarrow z_a \mp iz_b \end{array}\right\} \qquad (9.5.8)$$

The interchange $z_1 \leftrightarrow z_3$ gives[8]

$$\left.\begin{aligned}
z_a &\to -\frac{1}{2}z_a - \frac{\sqrt{3}}{2}z_b \\[4pt]
z_b &\to -\frac{\sqrt{3}}{2}z_a + \frac{1}{2}z_b \\[4pt]
z_a + iz_b &\to e^{-i\frac{2\pi}{3}}(z_a - iz_b) \\[4pt]
z_a - iz_b &\to e^{i\frac{2\pi}{3}}(z_a + iz_b)
\end{aligned}\right\} \tag{9.5.9}$$

and the interchange $z_2 \to z_3$ leads to

$$\left.\begin{aligned}
z_a &\to -\frac{1}{2}z_a + \frac{\sqrt{3}}{2}z_b \\[4pt]
z_b &\to \frac{\sqrt{3}}{2}z_a + \frac{1}{2}z_b \\[4pt]
z_a + iz_b &\to e^{i\frac{2\pi}{3}}(z_a - iz_b) \\[4pt]
z_a - iz_b &\to e^{-i\frac{2\pi}{3}}(z_a + iz_b)
\end{aligned}\right\}. \tag{9.5.10}$$

We also observe that $z_a^2 + z_b^2$ is invariant under any interchange. The Pauli principle is satisfied if the wave function is *odd* in z_b, and *symmetric* under rotation by $\pm 2\pi/3$ in the *ab* plane. The rotation is already explicit in (9.5.9) and (9.5.10). To get the wave function explicitly, Laughlin begins with the demands that it should be constructed with the single particle wave functions of the lowest

---

[8]The inverse of (9.5.6) is

$$z_3 = \bar{z} - \sqrt{\frac{2}{3}}z_a,$$

$$z_1 = \bar{z} + \sqrt{\frac{1}{6}}z_a + \frac{1}{\sqrt{2}}z_b,$$

$$z_2 = \bar{z} + \sqrt{\frac{1}{6}}z_a - \frac{1}{\sqrt{2}}z_b.$$

From these relations (9.5.9) and (9.5.10) follow.

Landau level, it satisfies the Pauli principle and is an eigenstate of the angular momentum. Then Laughlin chooses the following orthonormal basis functions:

$$|m,n\rangle = \frac{1}{\sqrt{2^{6m+4n+1}(3m+n)!n!\pi^2}} \left[\frac{(z_a+iz_b)^{3m} - (z_a-iz_b)^{3m}}{2i} \right].$$
$$\times [z_a^2 + z_b^2]^n e^{-\frac{1}{4}(|z_a|^2+|z_b|^2)}$$

$$(9.5.11)$$

These wave functions span the lowest Landau level, and satisfy the Pauli principle explicitly, and are eigenstates of angular momentum with eigenvalues $M = 2n + 3m$. The question is now how good are these functions for being *energy* eigenfunctions? Two states $|m = 3, n = 0\rangle$ and $|m = 1, n = 3\rangle$ have the same angular momentum $M = 9$. For these the expectation value of H_0 are respectively $0.722e^2/a_0$ and $0.867e^2/a_0$. Since these states mix, the off-diagonal matrix element is calculated to be

$$|\langle 3,0|H_0|1,3\rangle| = 0.0277\frac{e^2}{a_0},$$

much smaller than the diagonal elements. The exact eigenvalues of the corresponding energy eigenstates are $0.717\frac{e^2}{a_0}$ and $0.872\frac{e^2}{a_0}$ respectively. These values can be obtained also from the diagonalization of the energy matrix trivially.

Can one proceed in this way up to more and more electrons to find the required wave function of FQH states? Apparently this is not the way to go. The most important information which Laughlin acquired is the unusual stability of the 3-electron states $|m, 0\rangle$, which is required by the $\nu = \frac{1}{3}$ FQH state. To see this, an external potential is now added

$$V = \frac{\alpha}{2}(|z_1|^2 + |z_2|^2 + |z_3|^2)$$

$$= \frac{3\alpha}{2}|\bar{z}|^2 + \frac{\alpha}{2}(|z_a|^2 + |z_b|^2), \qquad (9.5.12)$$

which is equivalent to a pressure exerted on the 3-electron cluster to test its stability. The functional form is chosen for convenience, and

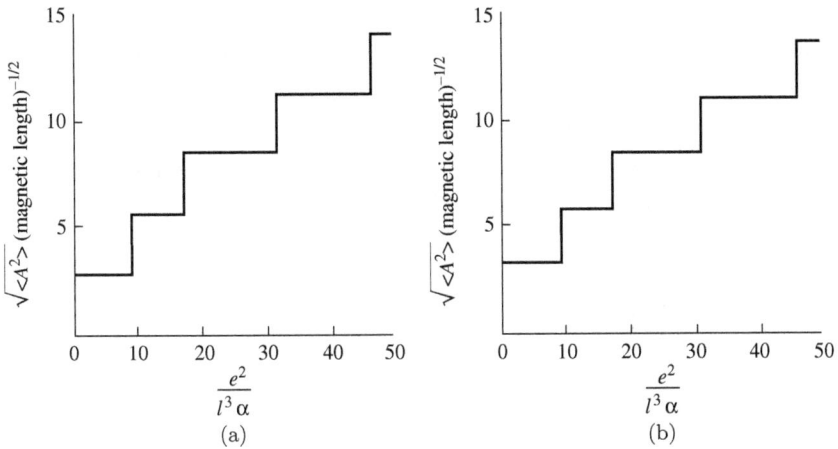

Fig. 9.9 (a) Total angular momentum M; (b) Root-mean-square of area of the lowest energy state of a 3-electron cluster under external pressure as a function of the parameter α. Redrawn from Ref. [17].

is not crucial. It gives

$$\left\langle m, n \left| \frac{\alpha}{2}(|z_a|^2 + |z_b|^2) \right| m', n' \right\rangle = \delta_{mm'}\delta_{nn'}(M+2)\alpha. \qquad (9.5.13)$$

The relation between M of the cluster with the lowest energy against $\frac{1}{\alpha}$ is plotted in Fig. 9.9(a). For the cluster in a state of angular momentum M under external pressure the total energy is

$$E_{\text{tot}}(M) = E_M + (M+2)\alpha.$$

Compare with a state of smaller M',

$$E_{\text{tot}}(M') = E_{M'} + (M'+2)\alpha.$$

Since in the state M the electrons are further apart, the internal energy is smaller, while the slope of the second term is larger. When we increase α in the state M, the pressure energy increases faster. When a point is reached for which $E_{\text{tot}}(M) = E_{\text{tot}}(M')$, further increase in α will find the cluster in a state M'. This explains the step behavior. The important point is that for an interval of $\frac{1}{\alpha}$ the value of M is stabilized at a multiple of 3 until a value of $\frac{1}{\alpha}$ is reached for stepping up/down.

This means that only $|m, 0\rangle$ states are stable. The external pressure can only change from a $|m, 0\rangle$ state to another with a different m. States $|m, n\rangle$ with $n \neq 0$ do not have a chance to participate. For any value of α the lowest energy state has $n = 0$. A companion graph shows the relation between the area of the 3-electron cluster and the pressure parameter in Fig. 9.9(b). The area operator is

$$A = \frac{1}{2}\text{Im}\left[\left(\frac{z_1 + z_2}{2} - z_3\right)(z_1 - z_2)^*\right]$$

$$= \frac{\sqrt{3}}{4i}(z_a z_b^* - z_b z_a^*), \tag{9.5.14}$$

and the expectation value is

$$\langle m, n| A^2 |m', n'\rangle = \delta_{mm'}\delta_{nn'}\frac{3}{4}[(3m)^2 + (M + 2)] \tag{9.5.15}$$

The same step-like behavior results. This can be seen from Fig. 9.10 in which the probability of finding either of the electrons 1 or 2 with electron 3 fixed at the position marked $\times$ and the center-of mass fixed at the position marked Δ. The states with $n = 0$ has a compact structure and the three electrons avoid one another in a way mimicking a crystal. These states are the most stable ones against the external pressure.

9.5.2   Laughlin wave function for the $\nu = \frac{1}{m}$ states

From the above discussions we see that the unique feature of the state $|m, 0\rangle$ is its tendency to exclude the electron density probability from the region $r_{ij} = 0$. A good approximate wave function should have *deep nodes*. The approximate wave functions in liquid ^{3}He make use of a Slater determinant multiplied by the factor $\prod_{j<k}^{N} f(z_j - z_k)$, where f goes to zero when its argument goes to zero. The Slater determinant is used for reducing the kinetic energy of the system, and is not required in the present context since the strong magnetic field has already done the job. Therefore, the wave function can be written

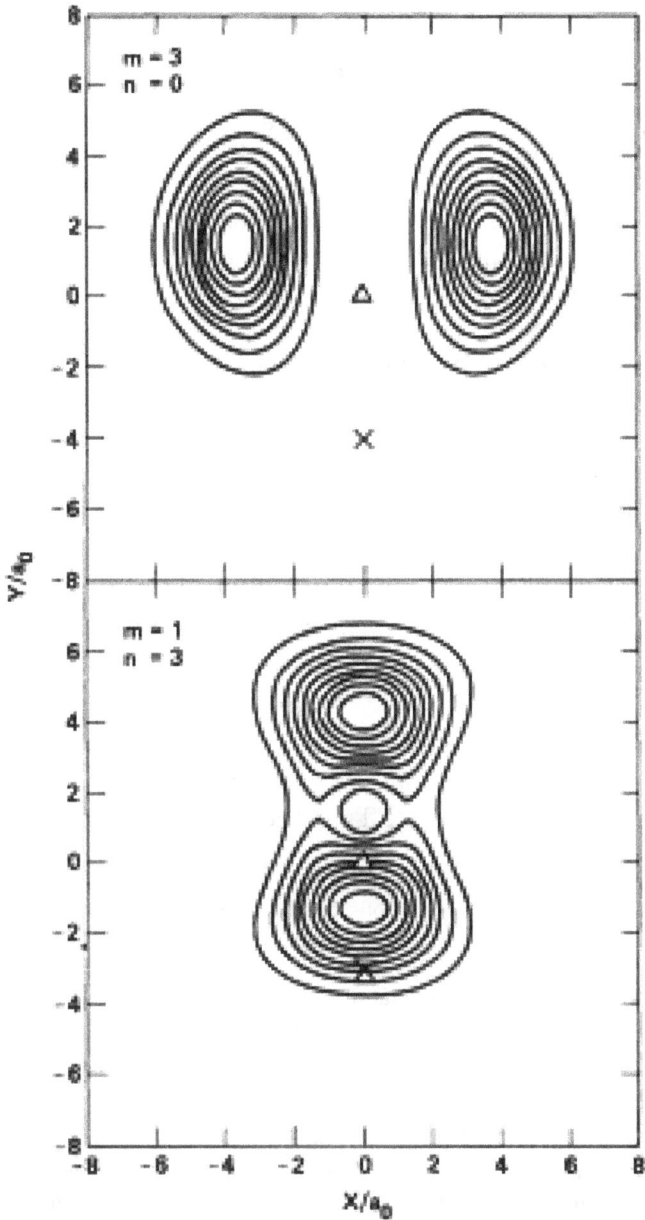

Fig. 9.10 Charge density of several $|m,n\rangle$ states with the center-of-mass and one of the electrons fixed at positions marked by Δ and $\times$ respectively. Reprinted with permission from R. B. Laughlin, *Phys. Rev. B* 27, 3383, 1983. Copyright (2018) by the American Physical Society.

in the form of the Jastrow product of *antisymmetric* functions f^9:

$$\psi(z_1, z_2, \ldots, z_N) = \prod_{j<k}^{N} f(z_j - z_k) \exp\left(-\frac{1}{4}\sum_{l}^{N} |z_l|^2\right). \quad (9.5.16)$$

Strong arguments are invoked which eventually fix the form of the Jastrow product. The constraints are the same as imposed on the 3-electron problem.

(1) The many-body wave function is comprised solely of the single-particle wave functions of the lowest Landau level. This requires $f(z)$ to be *analytic*.[10]
(2) The wave function is totally antisymmetric, i.e., $f(z)$ is *odd*.
(3) The wave function is an *eigenstate* of total angular momentum. This implies that $\prod_{j<k}^{N} f(z_j - z_k)$ must be a polynomial in the particle coordinates $z_1, z_2, \ldots, z_N$ of degree M,

$$M = \frac{N(N-1)}{2}m \quad (9.5.17)$$

where M is the total angular momentum. Mixtures of states $|m\rangle$ are not allowed. The only function satisfying all the constraints is $f(z) = z^m$, where m is odd. Thus the wave function is completely

---

[9]The Jastrow form of wave functions has been used in liquid helium theory to account for correlations between the atoms. The exponential factor is included without a loss of generality, because it can be factored into an exponential of $|z_j - z_k|^2$ and another exponential of the center-of-mass coordinates.

[10]Girvin and Jach [24] showed that the space of lowest Landau level with the Hamiltonian

$$H = -\frac{2}{m}\nabla_z\nabla_{\bar{z}} + \frac{1}{2}\omega_c$$

is equivalent to a space of holomorphic functions. Actually since our wave function depends only on z but not on z', we have $\nabla_{\bar{z}}\psi(z, \bar{z}) = 0$, which implies that $H\psi = \frac{1}{2}\omega_c\psi$, and the equation $\nabla_{\bar{z}}\psi(z, \bar{z}) = 0$ is the Cauchy-Riemann equation for analytic functions. The above equations are solved by

$$\psi(z, \bar{z}) = f(z)\exp\left(-\frac{eB}{4}|z|^2\right).$$

determined to be of the Jastrow form:

$$|m\rangle \equiv \psi_m(z_1, z_2, \ldots, z_N) = \prod_{j<k}^{N}(z_j - z_k)^m \exp\left(-\frac{1}{4}\sum_{l}^{N}|z_l|^2\right).$$

$$(9.5.18)$$

To justify this form of wave function, Laughlin calculated the overlap of the approximate wave functions of 3 electrons with the exact wave functions for three kinds of interactions, viz. $\frac{1}{r}$, $-\ln r$, and $e^{-r^2/2}$ for values of m from 1 to 13, and found overlaps close to unity. The physical significance of the parameter m can be found by a proper interpretation of the wave function. The modulus squared of the wave function can be interpreted as a *classical probability distribution* function

$$|\psi_m(z_1, z_2, \ldots, z_N)|^2 = e^{-\beta\Phi(z_1, z_2, \ldots, z_N)}, \qquad (9.5.19)$$

where Φ is the classical potential energy, and $\frac{1}{\beta}$ is the fictitious temperature. The left-hand side of (9.5.19) is

$$\prod_{j<k}|z_j - z_k|^{2m}\, e^{-\frac{1}{2}\sum|z_l|^2} = e^{2m\sum_{j<k}\ln|z_j-z_k|-\frac{1}{2}\sum|z_l|^2}.$$

Putting $\beta = \frac{1}{m}$, we see that

$$\Phi(z_1, z_2, \ldots, z_N) = -2m^2\sum_{j<k}\ln|z_j - z_k| + \frac{m}{2}\sum_{l}^{N}|z_l|^2. \quad (9.5.20)$$

This is the potential energy of a 2-dimensional one-component plasma: particles of charge m repelling one another via logarithmic interactions, the natural *"Coulomb"* interaction in two dimensions, and being attracted by a uniform neutralizing background of "charge" density $\sigma_n = \frac{1}{2\pi a_0^2}$. In a plasma electrical neutrality implies that the electron density of the state $|m\rangle$ is

$$\rho(z_1) = \frac{N \int |\psi_m(z_1, z_2, \ldots, z_N)|^2 d^2 z_2 \cdots d^2 z_N}{\langle m \mid m \rangle}, \qquad (9.5.21)$$

which must be equal to $\frac{1}{m}$ times the "charge" density of the equivalent plasma

$$\rho_m = \frac{1}{2\pi m a_0^2}.$$

Remember that the surface charge density of a filled Landau level is $\frac{1}{2\pi a_0^2}$ (9.4.7), then m^{-1} is just the filling fraction of Landau level:

$$\nu = \frac{1}{m}.$$

The plasma analogy is physically appealing, and will be used in the discussion on quasiparticle excitation in Section 9.5.3. There is, however, a simpler way of determining m. Select a coordinate z_1. The product $\prod_{j<k}^{N}(z_j - z_k)^m$ gives the highest power $m(N-1)$ of z_1. Comparing with the single particle wave function (9.5.4) of the lowest Landau level and the maximum value $M = \frac{\Phi}{\Phi_0} = n_B A$ of m, we have for large N

$$mN = M = n_B A,$$

i.e., $m = \frac{n_B}{N/A} = \frac{1}{\nu}$.

The FQH states are characterized by an incompressible quantum fluid. Is this compatible with the one-component plasma analogy? Actually the electrical neutrality guarantees the uniformity down to distance of the order of interparticle spacing. For even smaller length scales (e.g., several times a_0), we have to make sure that the plasma is really a liquid. The dimensionless characteristic parameter Γ for the plasma with charge Q and temperature $1/\beta$ is $\Gamma = 2\beta Q^2$. For our present case $\beta = 1/m$ and $Q = m$, then $\Gamma = 2m$. Extensive Monte-Carlo studies of two-dimensional plasma have shown that for $\Gamma \geqslant 140$ the plasma is a solid, and otherwise it is a liquid. For FQH states $\nu = 1, \frac{1}{3}, \frac{1}{5}, \ldots$, $\Gamma = 2, 6, 10, \ldots$, the equivalent plasma is unquestionably a liquid.

The theory is also tested with competing approximate wave functions. It is generally argued that due to the mutual repulsion the electrons tend to keep away from one another at some definite distance. Figure 9.11 gives an idea about this. Are the FQH states a Wigner crystal? It is almost apparent that they are not, because Hall current exists. Figure 9.11 gives the radial distribution $g(|z_1 - z_2|)$ defined by

$$g(|z_1 - z_2|) = \frac{N(N-1)}{\rho_m^2 \langle m \mid m \rangle} \int |\psi_m(z_1, z_2, \ldots, z_N)|^2 \, \mathrm{d}^2 z_3 \cdots \mathrm{d}^2 z_N$$

(9.5.22)

for $m = 1, 3, 5$. They exhibit liquid behavior: a hole region of size $\sim a_0$ around the origin and rapid healing and converging to 1 beyond this region. The wave function ψ_m is compared with the Hartree-Fock Wigner crystal wave function at the same density in Fig. 9.12. The difference between their behavior at small x makes ψ_m much more favored energetically than the crystal wave function [17].

The Laughlin wave function was "written" down starting from physical arguments. But actually it is much more accurate than expected. For Coulomb interaction it is a very good approximation,

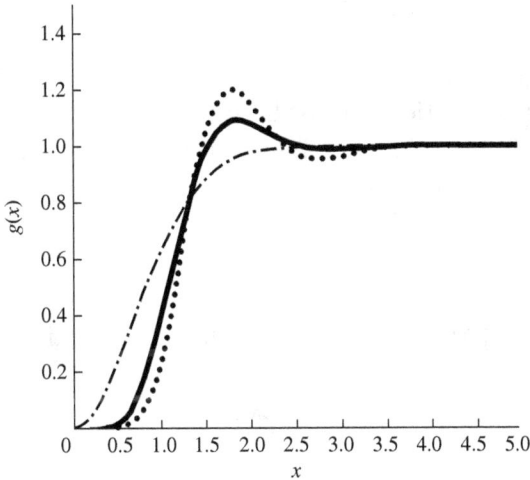

Fig. 9.11   Radial distribution function for ψ_m for $m = 1$ (dotted), $m = 3$ (solid) and $m = 5$ (dashed) versus the reduced variable $x = |z_1 - z_2|/\sqrt{2m}$ see (9.5.22). Redrawn from Ref. [17].

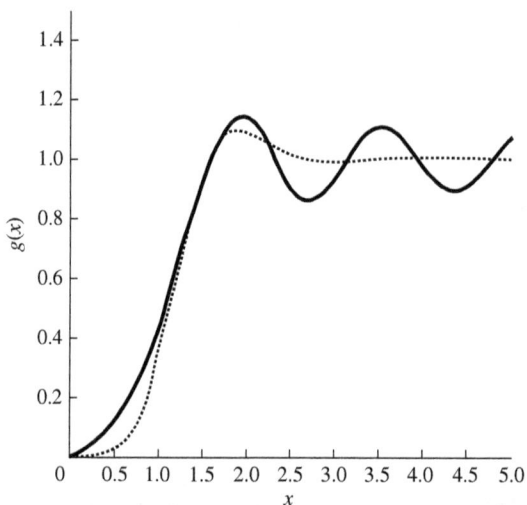

Fig. 9.12   Comparison of radial distribution function of ψ_m (dashed) with that of Wigner crystal wave function (solid) at the same density, plotted against x. Redrawn from Ref. [17].

and for a class of short range repulsive pseudopotentials the Laughlin states are *exact* ground states [18].

9.5.3   Quasiparticle excitations

Consider an infinitely thin solenoid which carries one flux quantum is introduced adiabatically into the system and it pierces the sample plane at $z = 0$. How does the Laughlin wave function change under this flux change? We expand the $(z_j - z_k)^m$ factor in ψ_m to get

$$\psi_m(z_1, \ldots, z_N) = \sum_{\{k_1 \cdots k_n\}} C_{k_1 \cdots k_N} z_1^{mk_1} \cdots z_N^{mk_N} e^{-\frac{1}{4}|z_j|^2}, \quad (9.5.23)$$

the integers $\{k_j\}$ run from 0 to N with the constraint

$$\sum_{j=1}^{N} k_j = \frac{1}{2}N(N-1). \quad (9.5.24)$$

During the flux change the single particle state changes from $z^n e^{-|z|^2/4}$ to $z^{n+1} e^{-|z|^2/4}$. If we ignore the change in the coefficients C, the flux change leads to a factor $\prod_{j=1}^{N} z_j$ multiplying the Laughlin wave function. This observation motivates Laughlin's choice for the wave function $\psi_m^{(+)}(z_0, \{z_j\})$ of the quasihole state by

$$\psi_m^{(+)}(z_0;\, z_1, \ldots, z_N) = \exp\left(-\frac{1}{4}\sum_i z_i^* z_i\right) \prod_i (z_i - z_0) \prod_{j<k}(z_j - z_k)^m,$$

$$(9.5.25)$$

which has the angular momentum $M_m^{(+)} = M_m + N$. The amplitude vanishes whenever the coordinates z_j of any of the electron approaches z_0, the effect being a depletion of the charge density. Therefore, the state is a quasihole.

The quasiparticle can be constructed by introducing a solenoid carrying a flux quantum which tends to weaken the local magnetic field. The angular momentum of the quasiparticle excited state is $M_n^{(-)} = M_n - N$. The corresponding change of Laughlin wave function is achieved by a derivative operator, and Laughlin gave

$$\psi_m^{(-)}(z_0;\, z_1, \ldots, z_N)$$

$$= \exp\left(-\frac{1}{4}\sum_i z_i^* z_i\right) \prod_i \left(2\frac{\partial}{\partial z_j} - z_0^*\right) \prod_{j<k}(z_j - z_k)^m.$$

$$(9.5.26)$$

Elementary excitations (quasiparticles) above the ground state ψ_m possess very peculiar properties, namely they have fractional charges. This can be illustrated in the diagram shown in Fig. 9.13. Suppose that a solenoid with flux perpendicular to the plane of the sample pass through a point, this is meant to be the position of the quasiparticle. Let the flux in the solenoid change adiabatically by a flux quantum. According to the adiabatic theorem the wave function remains to be an eigenstate of $\hat{H}(t)$. When the change is completed this eigenstates becomes quite different from its original form ψ_m. Since a flux quantum can be "removed" by a gauge transformation, the Hamiltonian $\hat{H}$ remains the same as that before change. The eigenstate which we have just talked about is an elementary excitation. In order to judge the properties of this excitation, draw a cylinder with the solenoid as its axis. The state ψ_{0m} becomes $\psi_{0\ m\pm 1}$ after the flux change. This change leads to a

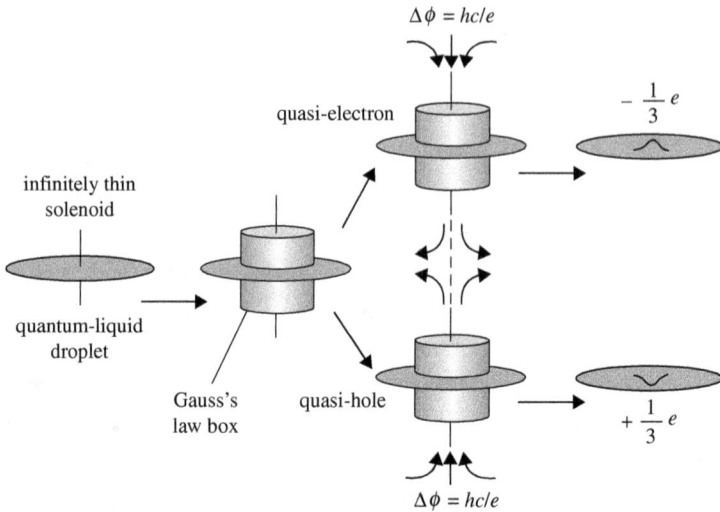

Fig. 9.13   Illustration of the fractional charge of quasiparticle excitations. Redrawn from Ref. [17].

changed expectation value of r^2 for a single particle state:

$$\langle r^2 \rangle = 2a_0^2(m \pm 1).$$

This means that one single particle state has entered or left the cylinder. The filling $\nu = \frac{1}{3}$ means that each state has a share of $e/3$. Therefore, the charge of a quasiparticle or quasihole is $\pm e/3$. D. Haldane [19] showed that the charge of a quasiparticle excitation above a ground state with filling fraction $\nu = \frac{p}{q}$ is $\pm e/q$.

The charge of the quasiparticle can also be inferred by calculating the classical distribution in the 2-dimensional one-component plasma. Let there be a quasihole at z_0. From the wave function (9.5.25) we obtain

$$e^{-\beta \Phi'} = \left| \psi_m^{(+)}(z_0; \, z_1, z_2, \ldots, z_N) \right|^2$$

where

$$\Phi' = -2m^2 \sum_{j<k} \ln |z_j - z_k| + \frac{m}{2} \sum_l^N |z_l|^2 - 2m \sum_i^N \ln |z_i - z_0|.$$

There is a phantom charge $\frac{1}{m}$ at z_0. Since the equivalent plasma screens perfectly, the charge surrounding z_0 is therefore $-\frac{1}{m}$, i.e., $-\frac{e}{m}$, where e is the charge of the electron.

Remember that in the illustration of Fig. 9.13 a solenoid of flux is put in the position of the would-be-quasiparticle excitation. An alternative description of the quasiparticles is in terms of the vortex, to which we come back in the next subsection.

9.5.4 Collective modes in the incompressible quantum Hall liquid

Many condensed matter systems possess collective modes of excitations. Sound waves are coherent oscillations in the media existing in liquids and gases, where the pressure change induced by density fluctuation serves as the restoring force. The wave lengths of sound waves are much larger than the inter-particle spacing and within one period of oscillation there are many collisions between the particles for maintaining local equilibrium. Such modes of excitations are called hydrodynamical.

Phonons in solids are of a different nature. They are the quantized lattice vibrations, true elementary excitations in the system in the sense that they are eigenstates of the system Hamiltonian. Coherent states of phonons can be formed which behave very much like density oscillations, e.g., sound waves in media, just as the Schrödinger coherent state of quantum harmonic oscillator behaves like a classical oscillator. Since a solid can support shearing stress, there are longitudinal as well as transverse phonons.

Before discussing the low energy excitation of the FQH liquid, we have to stay briefly on liquid helium. The superfluid helium (HeII) is a quantum liquid exhibiting many remarkable properties. There are also phonon excitations in HeII. L. D. Landau was the first who pointed out the difference between phonon as an elementary excitation and the sound wave, though their speeds of propagation are fairly close. In liquid argon, there are sound waves, but no phonon exists. According to Feynman's theory phonons in superfluid helium remain well defined for their wave length approaching the interparticle spacings. Due to the absence of low lying *single particle excitations*, phonons are elementary excitations in the *collisionless* regime, in contrast to the collision controlled sound waves.

In the FQH liquid, the particles are charged and are under a strong magnetic field. The Lorentz force perpendicular to the electron velocity can provide *transverse* restoring force. Plasma physicists refer to the longitudinal mode as the "upper hybrid mode" and to the transverse mode as the "lower hybrid mode". Actually the magnetic field mixes the longitudinal and transverse motion.

The importance of Feynman's contribution to the liquid helium theory can be understood from the background of the study prior to 1953 which is characterized in a paper by D. Pines [20].[11] Feynman decided to begin a complete study of liquid helium from first principles with attempts to answer the most important questions. The question which concern our problem here is the low lying excited states of HeII. Electrons in solids can give rise to low energy single particle excitations, for instance a hole under the Fermi sea and a particle above it. Landau developed a theory of quantum hydrodynamics and obtained the energy E versus momentum q dispersion relation of low energy excitations (phonons and rotons) by analyzing early experiments on the specific heat and second sound of HeII. Landau viewed the dispersion relation as something to be determined from experiment. What Feynman did was an "intuitive derivation of wave functions".[12] He used a series of closely reasoned arguments that led to the conclusion that liquid ^{4}He does not have low-lying excitations other than phonons because of *Bose statistics*. He also argued that [21] the wave function representing an excitation in HeII must be of the form

$$\psi_{\text{ext}} = \sum_j f(\boldsymbol{r}_j) \cdot \psi_0, \qquad (9.5.27)$$

where ψ_0 is the ground state wave function. The form of $f(\boldsymbol{r})$ and the excited state energy are then determined by variational principle.

---

[11]In this paper readers can find comments by Feynman on the previous studies (London, Tisza, Landau).

[12]Such is the name of a paragraph in his book "Statistical Mechanics" (W. A. Benjamin 1972), where a chapter is devoted to superfluidity.

He found that $f(r) = e^{i q \cdot r}$ so that an excitation with momentum q has a wave function

$$\psi_q = \sum_j e^{i q \cdot r_j} \psi_0 \equiv \rho_q \psi_0, \qquad (9.5.28)$$

and the corresponding energy is

$$E_q = \frac{q^2}{2m S_q}. \qquad (9.5.29)$$

This is the Bijl-Feynman formula. Here ρ_q is the Fourier transform of the density $\rho(r)$,[13] where S_q is the Fourier transform of the two-particle correlation function

$$S_q = \frac{1}{n} \langle \psi_0 | \rho_q^\dagger \rho_q | \psi_0 \rangle, \qquad (9.5.30)$$

where n is the number density of the He atoms. S_q is called the static structure factor of the liquid that determines the elastic scattering of neutrons or X-rays and can be determined experimentally. For large q it approaches unity, and for small q it has the form $S_q = \frac{q}{2ms}$, where s is the sound velocity. The dispersion relation is now $E_q = sq$. A remarkable feature is that the excitation energy of collective excitations is determined by a static property of the *ground state:* S_q is the expectation value of $\rho_q^\dagger \rho_q$ for the ground state ψ_0. We shall see that this feature remains for the collective excitation in the FQH states and also for the Bose-Einstein condensates of the gaseous atoms.

Feynman went on to consider the motion of the fluid as a whole. Consider a wave function in the following form:

$$\Psi = \psi_0 e^{i \sum_j^N S(r_j)}, \qquad (9.5.31)$$

where ψ_0 is the wave function of the ground state (real) and S is a real function of space coordinates. Ψ can be rewritten as

$$\Psi = e^{i \int d^2 r S(r) \rho(r)} \psi_0,$$

---

[13] For $\rho(r) = \sum_j \delta(r - r_j)$, the Fourier transform is $\rho_q = \sum_j e^{i q \cdot r_j}$. In second quantized form $\rho_q^\dagger = \sum_k a_{q+k}^\dagger a_k$.

where $\rho(r)$ is the density distribution of the liquid.[14] Notice that the phase factor of this form involves now the function S of a single radius vector r. The ground state is normalized by $\psi_0^\dagger \psi_0 = n$, the number density of the fluid. The current density j of the wave function can be calculated, and the superfluid velocity $v_s = \dfrac{j}{n}$ is then given by

$$v_s(r) = \frac{\hbar}{m} \nabla S(r). \qquad (9.5.32)$$

The steady flow satisfies the equation of continuity $\nabla \cdot v_s(r) = 0$ leading to

$$\nabla^2 S(r) = 0, \qquad (9.5.33)$$

and also (9.5.32) gives the irrotational flow condition

$$\nabla \times v_s = 0. \qquad (9.5.34)$$

Consider the circulation

$$\Gamma = \oint v_s \cdot dr = \int_S \nabla \times v_s \cdot dS, \qquad (9.5.35)$$

which vanishes if the region is simply connected because of (9.5.34). Using what we discussed above, Feynman resolved a paradox in the superfluid flow. Assume that the liquid helium under the superfluid transition temperature[15] is in a bucket which rotates with an angular velocity ω. Let H and H' be the Hamiltonian in the laboratory frame and the frame rotating together with the bucket respectively. Then

$$H' = H - \omega \cdot L,$$

where

$$L = \sum_i r_i \times p_i.$$

The equilibrium situation is obtained by minimizing H'. Notice that it is only in the rotating frame that the walls cannot provide energy to the system. In this frame the liquid is subjected to the Coriolis

---

[14]When $\rho(r) = \sum_j \delta(r - r_j)$, Ψ recovers its previous form (9.5.31).
[15]Under this temperature normal fluid and superfluid coexist.

force which acts as a transverse probe applied to the system. A normal liquid would respond to this probe and would be dragged by the walls and rotate at the same angular velocity. One would expect that the superfluid would not respond to the transverse probe and would stay at rest, because it has no viscosity. But actually the average flow pattern of the fluid resembles closely that of an ordinary rotating liquid! The answer is that the superfluid develops point defects, the *vortices,* where the *wave function vanishes,* and the surrounding region becomes multiply connected. The curl of v_s vanishes everywhere except at such point defects. The circulation becomes

$$\Gamma = \oint v_\mathrm{s} \cdot \mathrm{d}r = \frac{\hbar}{m} \oint \nabla^S \cdot \mathrm{d}r = \frac{\hbar}{m} 2\pi n, \qquad (9.5.36)$$

since the wave function is single valued, and $\tilde{S}$ can only change by an integral multiple of 2π. Therefore, we can set $S(r) = \lambda\varphi$, where φ is the azimuthal angle. Consequently,

$$v_\mathrm{s} = \frac{\hbar}{mr}\lambda\hat{\varphi}, \qquad (9.5.37)$$

where $\hat{\varphi}$ is a unit vector in the direction of increasing φ. The vortex carries quantized value of angular momentum. Because of the existence of the angular momentum carrying vortices H' is minimized to yield a steady state of the rotating liquid helium which appears as a rotating normal liquid. The point defect is seen to be a singular point, at which the superfluid velocity is singular and the wave function vanishes.

We return finally to the collective excitation in the FQH liquid [22]. The system bears a strong resemblance to superfluid helium in that it is a quantum fluid and exhibits dissipationless current flow. Although electrons are fermions, no low energy single particle excitation exists because the strong magnetic field splits the Landau levels by energy gaps, inter-Landau-level transitions correspond to the upper-hybrid mode in plasma physics. Intra-Landau-level excitations have much smaller energies of the order of Coulomb energy $\frac{e^2}{\kappa a_0}$, while the kinetic energy of the lowest Landau level is 0, disregarding the zero point energy. We follow Feynman's theory in considering the two modes

$$\Psi_k = P_1 \rho_k \psi \qquad (9.5.38)$$

and

$$\Phi_k = P_0 \rho_k \psi, \qquad (9.5.39)$$

where P_n is the projection operator for the n-th Landau level and ψ is the exact ground state of the system. Owing to the interaction between electrons, ψ is no more exactly in the lowest Landau level, but is still largely in it. From the construction Ψ_k and Φ_k are the eigenstates of the kinetic energy operator T :

$$\begin{aligned} T\Phi_k &= 0\Phi_k, \\ T\Psi_k &= \hbar\omega_c\,\Psi_k \end{aligned} \qquad (9.5.40)$$

Ψ_k corresponds to the upper-hybrid magnetoplasmon mode and Φ_k to the lower-hybrid magnetoplasmon mode. To ensure that k can be used to characterize a quantum state, the Landau gauge is used and a gauge transformation is incorporated in the boundary condition.

For the magnetophonons (magnetorotons) we have the excitation mode energy $\Delta(k)$,

$$\Delta(k) = \frac{\bar{f}(k)}{\bar{S}(k)}, \qquad (9.5.42)$$

where

$$\bar{f}(k) = \frac{1}{n} \langle \psi | \Lambda_k^\dagger (\mathscr{H} - E_0) \Lambda_k | \psi \rangle \qquad (9.5.43)$$

is the projected oscillator strength,

$$\bar{S}(k) = \frac{1}{n} \langle \psi | \Lambda_k^\dagger \Lambda_k | \psi \rangle \qquad (9.5.44)$$

is the projected static structure factor, and

$$\Lambda_k = P_0 \rho_k P_0 \qquad (9.5.45)$$

is the projected density operator. Using the single particle wave functions in the lowest Landau level the projected density operator

A_k as well as $\bar{f}(k)$, $\bar{S}(k)$ are calculated [23, 24]:

$$\bar{f}(k) = \frac{\nu}{2\pi} \int \frac{\mathrm{d}^2 q}{(2\pi)^2} v(q) \int \mathrm{d}^2 r [g(r) - 1] \mathrm{e}^{-|k|^2/2}.$$
$$\times \left[\mathrm{e}^{\mathrm{i}\boldsymbol{q}\cdot\boldsymbol{r}} \left(\mathrm{e}^{(kq^* - k^* q)/2} - 1 \right) + \mathrm{e}^{\mathrm{i}(\boldsymbol{k}+\boldsymbol{q})\cdot\boldsymbol{r}} \left(\mathrm{e}^{k\cdot q} - \mathrm{e}^{k^* q} \right) \right],$$

$$(9.5.46)$$

$$\bar{S}(k) = S(k) - (1 - \mathrm{e}^{-|k|^2/2}),$$

$$(9.5.47)$$

where q and k are complex number representations of $\boldsymbol{q}$ and $\boldsymbol{k}$; $g(r)$ is the 2-point correlation function related to the Fourier transform of $S(k)$, the (unprojected) static structure factor of the *ground* state. The scale of $\bar{f}(k)$, and therefore the scale of the collective excitation energy, is set solely by the scale of the Coulomb interaction $v(q)$, since the kinetic energy is completely quenched. Just as in the case of HeII, the excitation energy is expressed in terms of the static properties of the *ground* state. Unlike the case of HeII, the $S(k)$ here cannot be measured. Laughlin's wave function is used to evaluated $S(k)$. The result of the collective excitation energy for $\nu = \frac{1}{3}, \frac{1}{5}, \frac{1}{7}$ (solid lines) is given in Fig. 9.14. The striking features are: The collective mode is not a massless mode, i.e., $E(k)$ does not vanish linearly for small k. This is due to the incompressibility of the system at rational fractional filling factors. The deep minima of $\Delta(k)$ occurs analogous to the roton minimum in HeII, corresponding to a peak in $\bar{S}(k)$ at the wave vector associated with the mean particle spacing. The mass gap decreases with decreasing ν, this is the precursor to the Wigner crystal instability predicted to occur for $\nu = \frac{1}{7} - \frac{1}{10}$.

The upper hybrid mode is also discussed briefly in Ref. [22].

Finally we consider the vortex excitation in the QH liquid. In HeII and in gaseous Bose-Einstein condensates a vortex is a kind of collective superfluid flow pattern. In the QH liquid the state of electrons is intimately connected with the magnetic field. These quantized vortices are in fact Laughlin's fractionally charged particles. We can rewrite the wave function for the excited

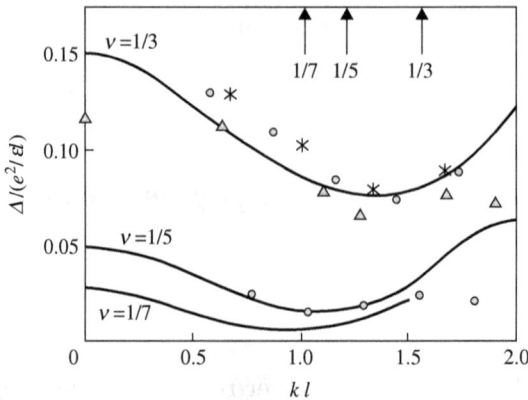

Fig. 9.14   Collective excitation energy for $\nu = \frac{1}{3}, \frac{1}{5}, \frac{1}{7}$ (solid lines). Small system numerical results are compared: crosses $\times (N = 7, \nu = \frac{1}{3}$, spherical system), triangles $\triangle$ $(N = 6, \nu = \frac{1}{3}$, hexagonal unit cell), solid dots $\cdot (N = 9, \nu = \frac{1}{3})$, $(N = 7, \nu = \frac{1}{5})$. Reprinted with permission from D. C. Tsui, *Rev. Mod. Phys.* 71, 891, 1999 and H. L. Störmer, *Rev. Mod. Phys.* 71, 875, 1999. Copyright (2017) by the American Physical Society.

state as[16]

$$\Phi = \exp\left(\sum_{j=1}^{N} i\varphi(\boldsymbol{r}_j) + \ln|\boldsymbol{r}_j - \boldsymbol{r}_0|\right)\psi_m, \qquad (9.5.48)$$

where ψ_m is the Laughlin wave function (9.5.18). In complex number representation we have

$$z_j - z_0 = |\boldsymbol{r}_j - \boldsymbol{r}_0|\,e^{i\varphi(\boldsymbol{r}_j)},$$

the wave function is actually

$$\Phi = \prod(z_j - z_0)\psi_m,$$

which is just the wave function for a quasihole (9.5.25).

----

[16]Here $\boldsymbol{r}_0$ is the position of the singularity (vortex core), where the wave function vanishes. In order to account for this the term $\ln|\boldsymbol{r}_j - \boldsymbol{r}_0|$ on the exponential is included, which is equivalent to a factor $|\boldsymbol{r}_j - \boldsymbol{r}_0|$ multiplying ψ_m. Its effect is to keep the wave function vanishing at $\boldsymbol{r}_0$ and rising linearly to the core boundary. Further away this factor is suppressed.

There have been experimental demonstrations of the fractional charges of quasiparticles [25–27]. Reference [25] gave the longitudinal conductivity in the neighborhood of Hall states $\nu = 2/3, 4/3, 5/3$. The results are shown in Fig. 9.15. The longitudinal resistance is proportional to the number of quasiparticles, and depend therefore on temperature,

$$\rho_{xx}\left(T\right) = \rho_{xx}\left(\infty\right)e^{-\Delta/kT},$$

where Δ is the energy necessary to excite a quasiparticle, and $\sigma_{xx} = \frac{(e/q)^2}{h}$ is the point where the straight lines merge. This is not yet a direct measurement. In the mid 1980s, D. C. Tsui suggested to determine the charge of quasiparticles by measuring the shot noise of its current. Provided that the current carriers are uncorrelated in their flow, the shot noise at $T = 0$ is given by Schottky formula $S = 2qI$, where q is the charge of the carriers

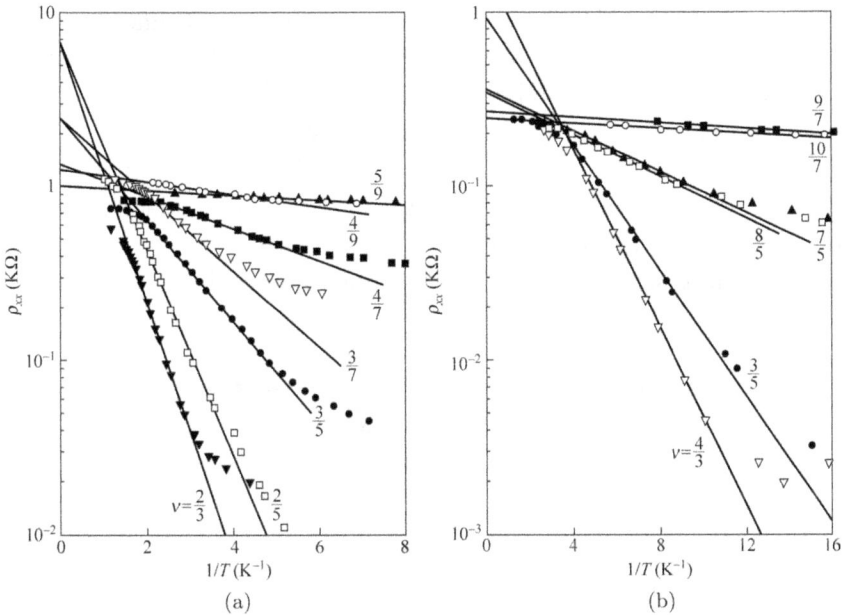

Fig. 9.15   The longitudinal Hall conductivity of FQH states at different temperatures. Reprinted with permission from R. G. Clark, J. R. Mallett, S. R. Haynes, J. J. Harris, and C. T. Foxon, *Phys. Rev. Lett.* 60, 1747, 1988. Copyright (2017) by the American Physical Society.

and I is the "backscattered" current. The difficulty in obtaining a quasiparticle current is that they are usually trapped by impurities and do not carry a current in the bulk. X. G. Wen [28] pointed out that at the boundaries of the system the quasiparticles can form edge states and flow freely in one direction along the edge, thus providing a 1-dimensional channel. C. L. Kane and M. P. A. Fisher [29] suggested to realize a quasiparticle flow by using a quantum point contact shown in Fig. 9.16. The quantum point contact is a narrow channel in a FQH liquid. Quasiparticles flow along the edges as indicated by the arrows. At the narrow constriction the quasiparticles can tunnel across the strait in the Hall liquid between the 1-dimensional channels. In a continuous flow of a current how can one distinguish between a given number of particles of charge e and three times as much particles with charge $e/3$? The best way is to measure current fluctuations — the noise. The temporal fluctuation can distinguish them as Kane and Fisher describe: small hailstones falling on the tin roof of a house can be distinguished from hailstones 3 times larger in size but 3 times less frequent. Since the number of quasiparticles is not large and the probability of tunneling is fairly small, the flow of quasiparticles can be considered uncorrelated. A Weizmann Institute group [30] and a French group [31] measured the minute fluctuations in the current (the shot noise) and obtained direct determination of the quasiparticle charge $e/3$ for the quantum Hall state $\nu = \frac{1}{3}$. Experimental sophistication includes fine lithography technology for

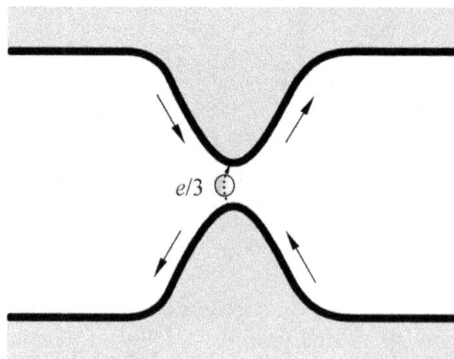

Fig. 9.16   A quantum point contact of a 2-dimensional electron gas.

preparing the quantum point contact and the filtering of the noise in external circuitry. Reference [31] reported an arrangement of 2DEG to ensure the tunneling occurs for the quasiparticles and not by other accidental events. Reference [32] reported a direct measurement using the shot noise of a charge $e/5$ of the quasiparticle of the $\nu = 2/5$ of FQH state.

9.5.5 Gapless edge states of fractional quantum Hall liquids

In Section 9.4 we discussed the edge excitation of the IQH states. Halperin's theory can be characterized by a chiral 1-dimensional Fermi liquid theory. Actually as X. G. Wen [29] pointed out that the FQH states also support gapless excitations, and it turns out that [33] the transport in FQH states is also governed by the edge excitations. The very rich internal structures of the topological orders reveal themselves through the edge states [34]. Since FQH states are intrinsically many-body states, the edge excitations in FQH states cannot be constructed from single particle states, i.e., the approach of Fermi liquid theory cannot be applied at the outset,[17] and most importantly, the final result shows that the theory differs from the Fermi liquid theory. In the following we consider the simplest case $\nu = \frac{1}{m}$ FQH state and use the hydrodynamical approach following [35]. The starting point is the understanding that these states are incompressible quantum liquid which does not have low energy bulk excitations. The only mode of low-lying excitations is the surface wave on the droplet. Consider a droplet of Hall liquid with filling fraction ν confined by a smooth potential well shown in Fig. 9.17. The electric field of the potential (in radial direction) generates a persistent current flowing along the edge

$$\boldsymbol{j} = \sigma_{xy}\hat{z} \times \boldsymbol{E}, \qquad (9.5.49)$$

---

[17]Roughly speaking the Fermi liquid theory treats the quasiparticle in one-to-one correspondance to the free fermion, thus taking into account the interactions between the fermions while keeping the single particle notion. It is a very successive theory within its range of applications.

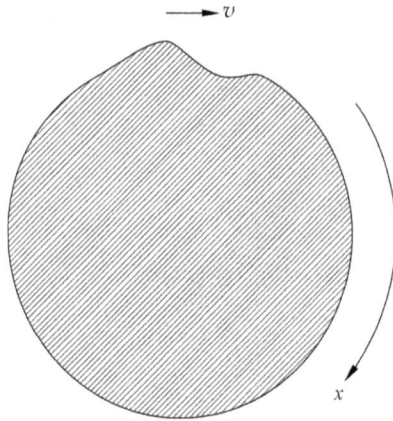

Fig. 9.17   A droplet of quantum Hall liquid.

where σ_{xy} is the Hall conductance

$$\sigma_{xy} = \nu \frac{e^2}{h}.$$ (9.5.50)

The drift velocity of the electron under the crossed $\boldsymbol{E}$ and $\boldsymbol{B}$ fields is

$$v = \frac{cE}{B}.$$ (9.5.51)

The corresponding current density is $j = nev$. With the filling fraction $\nu = n \cdot 2\pi a_0^2 = n\frac{hc}{eB}$ we obtain exactly (9.5.50). Let $h(x)$ be the displacement of the edge (deviation from the constant radius), x is the coordinate along the edge. The one-dimensional density $\rho(x) = nh(x)$ describes the edge wave, propagating with velocity v, where the two-dimensional electron density in the bulk $n = \frac{\nu}{2\pi a_0^2}$. The wave propagation is described by the wave equation

$$\partial_t \rho - v \partial_x \rho = 0.$$ (9.5.52)

The energy of the edge wave is

$$E = \int dx \rho(x) \frac{1}{2} eEh(x) = \int dx \rho^2(x) \frac{\pi v}{\nu}.$$ (9.5.53)

The edge wave in the momentum space ρ_k is defined as

$$\rho(x) = \frac{1}{\sqrt{L}} \int dk e^{-ikx} \rho_k, \qquad (9.5.54)$$

where L is the circumference of the droplet. Then (9.5.52) and (9.5.53) become

$$\dot{\rho}_k = ivk\rho_k, \qquad (9.5.55)$$

$$H = 2\pi \frac{v}{\nu} \sum_{k>0} \rho_k \rho_{-k}, \qquad (9.5.56)$$

where we have denoted the energy by H. Consider ρ_k as the generalized coordinate, and denote its conjugate momentum by p_k, we have

$$\frac{\partial H}{\partial p_k} = \dot{\rho}_k = ivk\rho_k. \qquad (9.5.57)$$

Comparing with (9.5.56) we get

$$p_k = -i\frac{2\pi}{k\nu}\rho_{-k}, \qquad (9.5.58)$$

and we see that

$$H = iv \sum_{k>0} k\rho_k p_k. \qquad (9.5.59)$$

Using the other canonical equation, we have

$$-\frac{\partial H}{\partial \rho_k} = \dot{p}_k = -\frac{2\pi}{\nu} v\rho_{-k}, \qquad (9.5.60)$$

which coincides with the result of taking directly the time derivative of (9.5.58). We can now quantize the theory by assuming

$$[\rho_k, p_{k'}] = i\delta_{kk'}, \qquad (9.5.61)$$

where $\hbar = 1$ is assumed. Combining (9.5.58) and (9.5.61), we obtain

$$[\rho_k, \rho_{k'}] = \frac{\nu}{2\pi} k\delta_{k+k'}. \qquad (9.5.62)$$

The values of k are determined by the circumference of the droplet L,

$$k = \frac{2\pi}{L}\kappa, \tag{9.5.63}$$

where κ is an integer. The quantization condition (9.5.61) implies that

$$[H, \rho_k] = kv\rho_k. \tag{9.5.64}$$

The algebra (9.5.62) is called the $U(1)$ *Kac-Moody algebra*. We also have

$$[H, p_k] = -kvp_k. \tag{9.5.65}$$

Relations (9.5.64) and (9.5.65) imply for any eigenstate ψ of H corresponding to the eigenvalue E that

$$H(\rho_k\psi) = (E + kv)(\rho_k\psi),$$
$$H(p_k\psi) = (E - kv)(p_k\psi).$$

We see that the energy spectrum of the edge wave is, after restoring $\hbar$,

$$E_k = \hbar kv. \tag{9.5.66}$$

This is the dispersion relation for low momenta, and we see that the edge excitation is gapless, and that the wave propagates with velocity v. The Kac-Moody algebra Eqs. (9.5.62),(9.5.63) and (9.5.64) characterize a one-dimensional single branch free phonon theory. These relations provide a complete description of the low lying edge excitations of the simple FQH states $\nu = \frac{1}{m}$.

The foregoing discussions concern neutral (uncharged) excitations. The charged excitations are created by electron operators $\psi^\dagger$. It creates on the edge a localized charge and should satisfy

$$[\rho(x),\ \psi^\dagger(x')] = \delta(x - x')\psi^\dagger(x'). \tag{9.5.67}$$

This suggests that ψ can be expressed as an exponential:

$$\psi(x) \propto e^{iC\phi(x)}, \tag{9.5.68}$$

where C is a constant and $\phi(x)$ is a scalar field. From (9.5.67) we obtain

$$[\rho(x),\ \mathrm{i}C\phi(x')] = \delta(x - x'), \qquad (9.5.69)$$

or in momentum representation

$$[\rho_k,\ \mathrm{i}C\phi_{k'}] = \delta_{k+k'}. \qquad (9.5.70)$$

Comparing with the Kac-Moody algebra (9.5.62) we find

$$k\phi_k = 2\pi\mathrm{i}\rho_k, \qquad (9.5.71)$$

if we choose $C\nu = 1$. Finally we have

$$\psi(x) \propto \exp\left[\mathrm{i}\frac{1}{\nu}\phi(x)\right] \qquad (9.5.72)$$

$$\rho(x) = \frac{1}{2\pi}\partial_x\phi(x). \qquad (9.5.73)$$

Operators of both neutral and charged excitations are now expressed in terms of a single scalar field, and it turns out that this field can be treated with relative ease. To identify that $\psi(x)$ is an electron operator, we have to show that $\psi(x)\psi(x')$ and $\psi(x')\psi(x)$ *anti-commute* for $x \neq x'$. By Baker-Campbell-Hausdorff formula

$$e^A e^B = e^B e^A e^{[A,B]}, \qquad (9.5.74)$$

where $[A, B]$ is a c-number. It follows that

$$\psi(x)\psi(y) = \psi(y)\psi(x) \exp\left(-\frac{1}{\nu^2}[\phi(x), \phi(y)]\right). \qquad (9.5.75)$$

The Kac-Moody commutator is fairly complicated to evaluate. We give here only the result [32]

$$[\phi(x), \phi(y)] = i\pi\nu\mathrm{sgn}(x - y). \tag{9.5.76}$$

Therefore

$$\psi(x)\psi(y) = (-1)^{\frac{1}{\nu}}\psi(y)\psi(x), \quad x \neq y. \tag{9.5.77}$$

The conclusion is: ψ is a fermion operator if $\frac{1}{\nu}$ is an odd integer. For Laughlin states $\nu = \frac{1}{m}$ we have a single branch of edge excitations described by the hydrodynamical. approach. This seemingly simple outcome is really rewarding. The free chiral boson field $\phi(x)$ is easily handled, because the problem is exactly soluble [37]. Putting $L = 2\pi$ and $e = \nu = 1$, the Lagrangian for the scalar field $\phi(x)$ is

$$\mathcal{L} = \frac{1}{8\pi}\left[(\partial_0\phi)^2 - (\partial_\sigma\phi)^2\right], \tag{9.5.78}$$

where the scalar field satisfies the chiral constraint

$$(\partial_0 - \partial_\sigma)\phi = 0. \tag{9.5.79}$$

The equation of motion for ϕ is

$$(\partial_0 - \partial_\sigma)(\partial_0 + \partial_\sigma)\phi = 0. \tag{9.5.80}$$

The free chiral field has a space-time dependence of $x - vt$, e.g., the phonon propagator is

$$\langle\phi(x, t)\phi(0, 0)\rangle = -\nu\ln(x - vt). \tag{9.5.81}$$

The electron propagator is [37]

$$G(x, t) = \langle T(\psi^\dagger(x, t)\psi(0, 0))\rangle$$
$$= \exp\left(\frac{1}{\nu^2}\langle\phi(x, t)\phi(0, 0)\rangle\right) \propto \frac{1}{(x - vt)^m}. \tag{9.5.82}$$

This is a really surprising result. For the Fermi liquid theory, the electron propagator is proportional to $\frac{1}{(x-vt)}$. This implies that the electrons on the edge of the FQH liquid are strongly correlated such that they cannot be described by the Fermi liquid theory. Such a new type of electron state is called a chiral Luttinger liquid. The exponent $m = \frac{1}{\nu}$ of the edge electron propagator is determined by the *bulk state* (the filling fraction). Such an exponent is a *topological* number that is independent of electron interactions, edge potentials, etc. It can be regarded as a quantum number that characterizes the topological orders in the bulk FQH states [34]. In the momentum space the electron propagator has the form

$$G(k,\omega) \propto \frac{(vk+\omega)^{m-1}}{vk-\omega}. \tag{9.5.83}$$

The anomalous exponent can be measured in tunneling experiments. The tunneling density of states is given by

$$N(\omega) \propto |\omega|^{m-1},$$

implying a differential conductance $\frac{dI}{dV} \propto V^{m-1}$ for a metal insulator FQH junction.

The foregoing discussion is only the beginning part of Wen's approach to the edge states, serving the purpose of illustrating how a theory is started, and many subsequent developments followed [36].

The physics of the edge states in FQH liquid is a good example of a way of solving problems. Textbook examples teach us that a Hamiltonian is written down first, and then eigenproblem is solved, exactly or by using approximations. There are problems for which we do not know the Hamiltonian, like in the edge states problem. X. G. Wen adopted another approach. Since the FQH liquid is incompressible, massless edge excitations should be possible. Physics intuition dictates to set up a plausible equation of motion. Then a tentative Lagrangian is found which leads to the equation of motion. Then the canonical dynamical variables can be identified and quantization rules set up. Once the Kac-Moody algebra is obtained, large numbers of problems can be formulated and theoretical deductions can be made. In the case of Laughlin wave function, the Hamiltonian is known, but to handle the strong correlation between electrons by a direct attack is formidable or even impossible. Try to enter the problem from simple systems and extract the crucial feature of the solution wanted, then through ingenious guess based on physics intuition the real solution is at hand. We have again Feynman's "intuitive derivation of the wave function". Physics intuition helps in many important occasions.

As for the edge states of FQH liquid, there is an alternative approach. Y. Yu and collaborators formulated a simplified microscopic theory based on composite fermions [37].

9.5.6 Hierarchy states of the fractional quantum Hall effect

Fractional quantum Hall effect occurs in multiple series, involving fractions of odd denominator only. For $\nu < 1$ we have the states $1/3$, $2/3$, $2/5$, $3/5$, $3/7$, $4/7$, $4/9$, $5/9,\ldots$. To understand this multiplicity, D. H. M. Haldane [38] proposed a hierarchy model in which a sequence of successive ground states is derived from a parent state. Let us consider the Laughlin state $\nu = \frac{1}{m}$. It is the result of condensation of the strongly correlated electrons to form an incompressible quantum liquid. If we have enough number of quasiparticle or quasihole excitations of the parent $\nu = \frac{1}{m}$ state, these elementary excitations can re-condense to form Laughlin type correlated states— daughters of the hierarchy. The daughter state derived in this manner has also fractionally charged quasiparticle or quasihole excitations, which can again condense to form granddaughter states, etc.

The degree of the *homogeneous* polynomial for electron coordinates $\{z_j\}$ in the Laughlin wave function ψ_m is $mN(N-1)$. Let us call this quantity divided by N, i.e., the degree of any one of the coordinates, which is $m(N-1)$ the *polynomial degree* of the Laughlin wave function. In thermodynamic limit this polynomial degree is mN which is equal to the number of total flux in units of flux quanta. The polynomial degree of the quasihole/quasiparticle wave function is

$$N_\phi = m(N-1) + \alpha N_{qp}, \tag{9.5.84}$$

where N_{qp} is the number of quasiparticles/quasiholes, and $\alpha = +1$ for quasiholes, $\alpha = -1$ for quasiparticles. On the other hand, when the quasiparticle/quasihole condense to form correlated states, the wave function is

$$\Psi_m^p(Z_1,\ldots,Z_{N_{qp}},\; z_1,\ldots,z_N) = \sum_{i<j}^{N_{qp}}(Z_i - Z_j)^p\psi_m,$$

where $\{Z_j\}$ are the complex coordinates of the condensed quasi-particles/quasiholes. The quasiparticle excitations has nominal Bose statistics, therefore p must be an *even* integer. The degree of the homogeneous polynomial of Ψ_m^p is

$$\alpha N_{qp}(N_{qp} - 1)p + mN(N - 1),$$

hence the polynomial degree of this state is

$$N_\phi = \alpha\frac{N_{qp}(N_{qp} - 1)}{N}p + m(N - 1). \qquad (9.5.85)$$

Comparing (9.5.85) with (9.5.84) we obtain

$$N = (n_{qp} - 1)p. \qquad (9.5.86)$$

The Eqs. (9.5.84) and (9.5.86) form the first level hierarchy equations. The filling fraction of the first level daughter is, in the thermodynamic limit,

$$\nu = \frac{N}{N_\phi} = \frac{1}{m + \frac{\alpha}{p}}. \qquad (9.5.87)$$

The sequence is given by the continued fraction

$$\nu = \cfrac{1}{m + \cfrac{\alpha_1}{p_1 + \cfrac{\alpha_2}{p_2 + \cdots}}}, \qquad (9.5.88)$$

where m is odd, $\alpha_i = \pm1$, p_i are even integers. All FQHE hierarchy states can find their places in this sequence.

9.5.7 The composite Fermion, Jain's construction

The construction of the quasihole state wave function (9.5.25) suggests a different interpretation of Laughlin's wave function proposed by J. K. Jain [39], leading to a composite fermion picture. The basic qualitative features are:

(1) Electrons capture $2p$ flux quanta to become composite fermions. Since a thin solenoid of flux leads to a vanishing wave function at the point of flux, this is equivalent to a screening of the electrons from one another.

(2) Due to the screening, composite fermions are *weakly interacting*, thus enabling a standard single particle description of the FQH liquid.

In the case of Laughlin wave function $m = 3$, each electron has a share of 3 flux quanta. If each electron nucleates 2 flux quanta to become a composite fermion, then the composite fermions experience an effective magnetic field which is the remaining field after the flux nucleation by the electrons, i.e., each composite fermion has a share of one flux quantum. The effective filling fraction is $\nu_{\text{eff}} = \frac{n}{n_{\text{eff}}} = 1$. The FQH state $\nu = \frac{1}{3}$ for electrons becomes an IQH state $\nu = 1$ for composite fermions. Expressed in terms of wave functions, the foregoing argument starts from rewriting ψ_m in the form

$$\psi_m(z_1 \cdots z_N) = \sum_{i<j} (z_i - z_j)^{m-1} \chi_1(z_1 \cdots z_N), \qquad (9.5.89)$$

where

$$\chi_1(z_1 \cdots z_N) = \sum_{i<j} (z_i - z_j) e^{-\sum_l \frac{1}{4}|z_l|^2}. \qquad (9.5.90)$$

The phase of $(z_i - z_j)^{m-1}$, i.e., $(m-1)\varphi_{ij}$ represents an even number $(m-1)$ of flux quanta attached to each coordinate z_i where an electron is present. It is a crucial feature of this picture that the electrons bind to an even number of flux quanta and they retain their fermion character. This approach allows for a simple description of the hierarchy states. The hierarchy states can be classified as:

$$\nu = \frac{n}{2n+1} = \frac{1}{3}, \frac{2}{5}, \frac{3}{7}, \dots \quad \text{equivalent to } n_\phi = 2n + 1$$

$$\nu = \frac{n}{2n-1} = \frac{2}{3}, \frac{3}{5}, \frac{4}{7}, \dots \quad \text{equivalent to } n_\phi = 2n - 1$$

$$\nu = \frac{n}{4n+1} = \frac{1}{5}, \frac{2}{7}, \frac{3}{13}, \dots \qquad \text{equivalent to } n_\phi = 4n+1$$

$$\nu = \frac{n}{4n-1} = \frac{2}{7}, \frac{3}{11}, \dots \qquad \text{equivalent to } n_\phi = 4n-1$$

$n_\phi = 2n+1$ can be interpreted that each electron is coupled to 2 flux quanta such that the remaining effective magnetic field is one flux quantum per composite fermion. $n_\phi = 2n-1$ is interpreted as an effective field which is in the opposite direction of the external field after each electron has nucleated two flux quanta. The result is also an IQHE for the composite fermion. Other states can be interpreted similarly. The particle-hole symmetry gives additional states

$$\nu = 1 - \frac{n}{2pn \pm 1}.$$

The Jain's construction establishes a connection between the FQHE with the IQHE, providing a unified view. More than that, it provides the possibility of studying a strong correlated state through methods adequate for weak interacting systems. An interesting example is the $\nu = \frac{1}{2}$ state [40]. This is *not* a proper Quantum Hall state. It can be transformed to a state of composite fermions by coupling each electron with two flux quanta. Then the remaining field is *zero*. In the absence of a magnetic field one could expect that, for a system without impurity scattering, there should be a well-defined *Fermi surface* for the composite fermions. This is a *Hall metal* state. In the theoretical representation the nucleation of magnetic flux is realized by a gauge field. Before any prediction about this state can be made, the gauge fluctuation must be taken into account. This is possible because the composite fermions are now weakly interacting, and can be dealt with by a mean field theory and the random phase approximation, etc. Interaction dependent behavior of such a system was discussed in detail in Ref. [40].

9.6 The Landau-Ginzburg Theory of the Fractional Quantum Hall Effect

S. Girvin and A. MacDonald [41] discovered that by carrying out a gauge transformation on Laughlin wave function it is possible to obtain the wave function of a system of *bosons*, and this wave function has a quasi-long range order. This suggests that a Landau-Ginzburg theory might be formulated for the quantum Hall system. If it is the case, then the low energy stationary and transport properties of the quantum Hall system can be discussed conveniently with this theory.

9.6.1 Zhang-Hansson-Kivelson mapping, the Chern-Simons-Landau-Ginzburg action

S. C. Zhang, H. Hansson and S. Kivelson [42] proposed a gauge transformation which maps a quantum Hall state into a state of a system of bosons. The Hamiltonian of a system of electrons in the fields of 4-potential $(A_0, \mathbf{A})$ and interacting via the scalar potential $V(x_i - x_j)$ is

$$H = \frac{1}{2m} \sum_i \left[\mathbf{p}_i - \frac{e}{c} \mathbf{A}(x_i) \right]^2 + \sum_i eA_0(x_i) + \sum_{i<j} V(x_i - x_j).$$
(9.6.1)

The eigenfunction $\Psi(x_1, \dots, x)$ satisfies the Schrödinger equation

$$H\Psi(x_1, \dots, x_n) = E\Psi(x_1, \dots, x_n),$$
(9.6.2)

and $\Psi(x_1, \dots, x_n)$ is anti-symmetric with respect to the interchange of any pair of coordinates. We carry out a gauge transformation

$$\Psi(x_1, \dots, x_n) = U\Phi(x_1, \dots, x_n)$$
(9.6.3)

where

$$U = \exp\left[-i \sum_{i<j} \frac{\theta}{\pi} \alpha_{ij} \right].$$
(9.6.4)

Here θ is a parameter and α_{ij} is the angle between the x-axis and the line from the electron j to the electron i. The Hamiltonian after the transformation is

$$H' = \frac{1}{2m} \sum_i \left[\mathbf{p}_i - \frac{e}{c} \mathbf{A}(x_i) - \frac{e}{c} \mathbf{a}(x_i) \right]^2$$

$$+ \sum_i eA_0(x_i) + \sum_{i<j} V(x_i - x_j),$$
(9.6.5)

where

$$a(x_i) = \frac{\Phi_0}{2\pi} \frac{\theta}{\pi} \sum_{j \neq i} \nabla \alpha_{ij}. \tag{9.6.6}$$

When the parameter θ takes the value

$$\theta = (2k+1)\pi, \tag{9.6.7}$$

where k is an integer, the new eigenfunction Φ satisfies

$$H'\Phi(x_1, \ldots, x_n) = E\Phi(x_1, \ldots, x_n). \tag{9.6.8}$$

The eigenvalues in Eqs. (9.6.3) and (9.6.8) are identical. The important new feature of the eigenfunction Φ is that it is *symmetrical* under the interchange of any pair of coordinates. This change of symmetry originates from the angle α in the transformation U (9.6.4). The interchange $i \leftrightarrow j$ leads to

$$\alpha_{ji} \rightarrow \alpha_{ij} + \pi. \tag{9.6.9}$$

Therefore, the interchange under the condition $\theta = (2k+1)\pi$ leads to a factor $e^{-i(2k+1)\pi} = -1$. The transformation result (9.6.5) is proved as follows. We substitute (9.6.3) into (9.6.2), and then multiply from the left by U^{-1}. Since U commutes with any function of the coordinates x_i, and furthermore direct calculation using (9.6.4) gives

$$U p_i U^{-1} = p_i + \hbar \frac{\theta}{\pi} \sum_{i<j} \nabla \alpha_{ij} = p_i + \frac{e}{c} \frac{\Phi_0}{2\pi} \sum_{i<j} \nabla \alpha_{ij}$$

$$= p_i + \frac{e}{c} a(x_i)$$

then we have

$$U \left[p_i - \frac{e}{c} A(x_i) - \frac{e}{c} a(x_i) \right] U^{-1} = p_i - \frac{e}{c} A(x_i).$$

This proves (9.6.5). The change of symmetry property of a wave function, i.e., the change of statistics, is called a *statistical transmutation*. The gauge field $a(x_i)$ is called the statistical gauge field. It does not change the eigenvalue, and only changes an antisymmetric wave function Ψ to a symmetric Φ. Having established the identity of eigenvalues of Ψ and Φ, it is now possible to continue further discussions with the *boson* representation. It is convenient to use the second quantized form of the theory and the path integral method. We introduce the boson field operators $\varphi, \varphi^\dagger$ satisfying the commutation relation

$$[\varphi(x), \varphi^\dagger(y)] = \delta^2(x - y), \tag{9.6.10}$$

where x and y are 2-dimensional vectors. The second quantized form of the Hamiltonian is

$$H = \int d^2x \varphi^\dagger(x) \left[\frac{1}{2m} \left(-i\hbar\nabla - \frac{e}{c}\boldsymbol{A}(x_i) - \frac{e}{c}\boldsymbol{a}(x_i) \right)^2 + eA_0(x) \right] \varphi(x)$$

$$+ \frac{1}{2} \int d^2x d^2y \delta\rho(x) V(x - y) \delta\rho(y). \tag{9.6.11}$$

The density operator $\rho(x) = \varphi^\dagger(x)\varphi(x)$, and $\delta\rho = \rho(x) - \bar{\rho}$ is the deviation of the local density from the average density $\bar{\rho}$. The second quantized form of the statistical gauge field is

$$a^\alpha(x) = -\frac{\Phi_0}{2\pi} \frac{\theta}{\pi} \varepsilon^{\alpha\beta} \int d^2y \frac{x^\beta - y^\beta}{|x - y|^2} \rho(y). \tag{9.6.12}$$

To get this expression, we can put $x_i \equiv x$, $y_i \equiv y$, and then calculate $\nabla_x \alpha_{xy}$ in the following way. Calculating $d\alpha_1$ due to an increment dx_1, we get $\frac{dx_1 \sin\alpha}{|x-y|}$, therefore $(\nabla_x \alpha_{xy})_1 = \frac{y_2 - x_2}{|x-y|^2}$. Similarly, $(\nabla_x \alpha_{xy})_2 = -\frac{y_1 - x_1}{|x-y|^2}$. In (9.6.12) $\varepsilon^{\alpha\beta}$ is the antisymmetric tensor of second rank. From (9.6.12) we see that $\boldsymbol{a}$ is completely determined by $\rho = \varphi^\dagger \varphi$, and therefore it is not an independent dynamical quantity. But in the theoretical framework it is necessary to know its equation of motion for deriving the Lagrangian and the action of the system. We obtain the equation of motion directly from (9.6.12). We use the Landau-Lorentz gauge[18] $\partial_\alpha a^\alpha = 0$, then (9.6.12) is the solution of

---

[18]In (3+1)-dimensional theory this is the Lorentz gauge, and in (2+1)-dimensional theory it is called the Landau-Lorentz gauge.

the following differential equation

$$\varepsilon^{\alpha\beta}\partial_\alpha a_\beta(x) = \Phi_0 \frac{\theta}{\pi}\rho(x). \tag{9.6.13}$$

The equation does not involve the evolution in time. We take the time derivative

$$\varepsilon^{\alpha\beta}\partial_\alpha \dot{a}_\beta(x,t) = \Phi_0 \frac{\theta}{\pi}\dot{\rho}(x,t) = -\Phi_0 \frac{\theta}{\pi}\partial_\alpha j^\alpha. \tag{9.6.14}$$

The time derivative of the density ρ can be expressed in terms of the divergence of the conserved current j^α by the help of the equation of continuity $\dot{\rho} + \partial_\alpha j^\alpha = 0$. We get

$$\varepsilon^{\alpha\beta}\dot{a}_\beta(x,t) = -\Phi_0 \frac{\theta}{\pi}j^\alpha. \tag{9.6.15}$$

Equations (9.6.13) and (9.6.15) together give the equation of motion for a . The spatial and temporal derivatives of a are given by the density and current density of the field φ, they can be derived from the *Chern-Simons* Lagrangian:

$$\mathscr{L} = \frac{e}{2}\left(\frac{\pi}{\theta}\right)\frac{1}{\Phi_0}\varepsilon^{\mu\nu\rho}a_\mu\partial_\nu a_\rho - ea_\mu j^\mu. \tag{9.6.16}$$

The 3-dimensional vector $a_\mu(a_0, a)$ contains the zero component (time component) a_0, which is introduced as a Lagrange multiplier. The variation of $\mathscr{L}$ with respect to a_0 put equal to zero gives (9.6.13), and its variation with respect to a put equal to zero gives

$$\varepsilon^{\alpha\beta}(\partial_\beta a_0 - \partial_t a_\beta) = \Phi_0 \frac{\theta}{\pi}j^\alpha.$$

In a gauge in which $a_0 = 0$ it is reduced to (9.6.15). Combining (9.6.16) with the Lagrangian of the φ field and including the interaction in the Lagrangian for φ, we obtain the Chern-Simons-Landau-Ginzburg (CSLG) Lagrangian [43][19]:

$$S = S_a + S_\varphi = \int d^3x \mathscr{L}_a + \int d^3x \mathscr{L}_\varphi, \tag{9.6.17}$$

---

[19] A Bose liquid is described by the Landau-Ginzburg theory. Here the Chern-Simons statistical gauge field has been included such that quantum Hall liquid can be described by the Landau-Ginzburg theory. Hence the name CSLG.

where

$$\mathcal{L}_a = \frac{e\pi}{2\theta\,\Phi_0}\varepsilon^{\mu\nu\rho}a_\mu\partial_\nu a_\rho, \tag{9.6.18}$$

$$\mathcal{L}_\varphi = \varphi^\dagger\left(i\hbar\partial_t - e\left(A_0 + a_0\right)\right)\varphi - \frac{1}{2m}\left|\left(-i\hbar\nabla - \frac{e}{c}\boldsymbol{A} - \frac{e}{c}\boldsymbol{a}\right)\varphi\right|^2$$

$$- \frac{1}{2}\int d^3y\,\delta\rho(x)V(x-y)\delta\rho(y). \tag{9.6.19}$$

The thermodynamical properties and electromagnetic response are all included in the path integral (partition function)

$$Z\left[A_\mu\right] = \int \left[da_\mu\right]\left[d\varphi\right]\exp\left(iS_a\left[A_\mu\right] + iS_\varphi\left[A_\mu + a_\mu, \varphi\right]\right). \tag{9.6.20}$$

Under the condition that $\theta = (2k+1)\pi$, the boson description discussed above is completely equivalent to the original many-electron quantum Hall problem. Since the boson quantum superfluid theory is well developed, we can use many of the ready methods, e.g., the mean field theory and the random phase approximation in discussing the quantum Hall problem. In the ordinary vector notation (9.6.13) can be written as

$$\nabla \times \boldsymbol{a}(x) = (2k+1)\,\Phi_0 n_s(x), \tag{9.6.21}$$

Here θ is explicitly set equal to $(2k+1)\pi$, and the ρ has been rewritten as n_s (the surface density).

Going from the anti-symmetric wave function to the symmetric wave function, from the fermion to boson description does not imply any change of the physics. The system described by $\Phi(x_1, \ldots, x_n)$ interacts with the statistical gauge field $\boldsymbol{a}(x)$, besides $(\varphi, \boldsymbol{A})$. From (9.6.21) it follows that

$$\oint \boldsymbol{a}(x) \cdot d\boldsymbol{l} = (2k+1)\,\Phi_0 n. \tag{9.6.22}$$

The closed integral encloses an area containing n electrons, i.e., the role that the statistical gauge field plays is to impose $2k+1$ flux quanta on each electron. Imposing an odd number of flux quanta on an electron will change its statistical property, i.e., change it from a fermion to a boson, while imposing an even number of flux quanta does not change the statistical property. We have discussed the Jain's construction and composite fermions which correspond to the latter case.

9.6.2 The mean field solution, phenomenology of the fractional quantum Hall effect

We look for a mean field solution of the Lagrangian (9.6.19). We shall find that there is a close correspondence between the QHE and the superfluid phenomenology. Let the potential A_μ describe a uniform magnetic field pointing to the $-z$ direction perpendicular to the plane of the system

$$\varepsilon^{\alpha\beta}\partial_\alpha A_\beta = -B. \tag{9.6.23}$$

A mean field solution can be guessed to be

$$\varphi(x) = (\bar{\rho})^{\frac{1}{2}}, \quad \boldsymbol{a}(x) = -\boldsymbol{A}(x), \quad a_0(x) = 0. \tag{9.6.24}$$

It is straightforward to check that this solution satisfies all equations derived from the CSLG Lagrangian provided a condition is satisfied. Eq. (9.6.13) has the form under the mean field

$$\varepsilon^{\alpha\beta}\partial_\alpha a_\beta = \Phi_0 \frac{\theta}{\pi}\bar{\rho}.$$

Substituting in (9.6.24) and using (9.6.23), we obtain

$$B = \Phi_0 \frac{\theta}{\pi}\bar{\rho}.$$

Denoting the flux density B/Φ_0 by ρ_A, we find that the filling fraction is

$$\nu = \frac{\bar{\rho}}{\rho_A} = \frac{\pi}{\theta} = \frac{1}{2k+1}. \tag{9.6.25}$$

The condition for the mean field solution to hold is just $\theta = (2k+1)\pi$, the same condition for the validity of fermion-boson mapping, corresponding to the filling fraction of the FQHE state.

If we distribute uniformly the flux carried by the bosons, then the flux just cancels the original magnetic field, and the bosons will

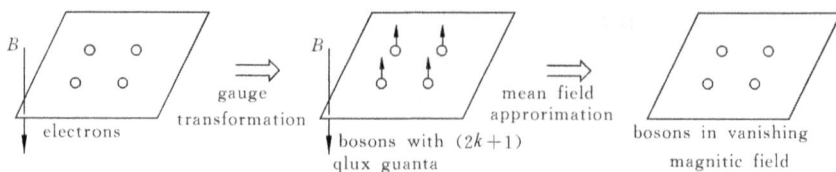

Fig. 9.18   Alternative description of the FQH state $\nu = \frac{1}{2k+1}$.

not experience any magnetic field. The result of carrying out the gauge transformation is shown in Fig. 9.18. Under sufficiently low temperature and in the absence of magnetic field, the system of *composite bosons*[20] undergoes the Bose-Einstein condensation. Now the quantum Hall system of composite bosons can be compared with a superconducting or superfluid system, which does not only provide a new understanding of the Quantum Hall phenomena, but also gives new theoretical predictions.

We begin with the interpretation why the FQH state $\nu = \frac{1}{2k+1}$ is an incompressible quantum liquid. A superconductor forbids the existence of magnetic flux in the bulk (the Meissner effect). According to (9.6.21), any change in the electron density will be accompanied by the change in the gauge field a, i.e., by the appearance of net local flux. This is forbidden by the Meissner effect. Consequently under the condition $\nu = \frac{1}{2k+1}$ there can be no change in the density of the liquid, and the state is incompressible.

A question frequently raised is that now the composite bosons do not experience any magnetic field. How can the Hall effect occur without a magnetic field? The point is that a composite boson does not only carry charge, but also magnetic flux $(2k + 1)\Phi_0$. When the bosons move along the x-axis the electric current is $I = e\frac{dN}{dt}$, where dN is the number passing through a line in the y-direction in time dt. The flux current carried by the bosons is

$$\frac{d\Phi}{dt} = (2k + 1)\, \Phi_0 \frac{dN}{dt}.$$

---

[20]The bosons carrying magnetic flux with them are called composite bosons, similar to the flux carrying composite fermions.

According to Faraday's law such a rate of flux change gives rise to a transverse potential difference

$$U_{\mathrm{H}} = \frac{1}{c}\frac{d\Phi}{dt} = \frac{1}{c}(2k+1)\,\Phi_0\frac{I}{e} = (2k+1)\,\frac{h}{e^2}I,$$

and consequently the transverse resistance is $\rho_{xy} = \frac{U_{\mathrm{H}}}{I} = (2k+1)\frac{h}{e^2}$.

In the composite boson description the $\nu = \frac{1}{3}$ FQHE is equivalent to composite bosons each carrying $3\Phi_0$ in the absence of magnetic field, and the IQHE with $\nu = 1$ is equivalent to composite bosons each carrying Φ_0 in the absence of magnetic field. In the absence of magnetic field bosons condense at low temperatures. Both cases are identical.

Using the parallelism of the composite boson description with superfluidity we can study the quasiparticle excitation of the FQH states. The result is the states with fractional charge and that these states obey the "fractional statistics" [44]. Using the analogy with the superconductors of the second class we can interpret the plateaus in terms of the vortices pinned by impurities. Using the analogy with the superfluids collective excitations of the FQH states can be obtained [43].

For a series of special filling states $\nu = 1$, $\nu = \frac{1}{3},\ldots$, the longitudinal resistance vanishes. This is a state of complete conductivity. This is not easy to understand in the fermion description. In the equivalent boson description this is natural, because we have now a Bose-Einstein condensate of charged bosons, which forms a complete conductor.

The response to external electromagnetic field can be obtained step by step. First by performing the path integral with respect to the φ field, we obtain the effective action for the electromagnetic potentials $A_\mu + a_\mu$. Then by integrating with respect to a_μ we obtain the effective action S_A for the external potentials A_μ. The general electromagnetic response of the system $D_{\mu\nu}(q)$ is obtained from the correlation function

$$D_{\mu\nu}(q) = \frac{\delta^2 S_A}{\delta A_\mu(-q)\delta A_\nu(q)}.$$

9.6.3   The algebraic off-diagonal long-range order

The CSLG action not only leads to the entire FQHE phenomenology, but also leads to the derivation of the Laughlin wave function and the off-diagonal long-range order of Girvin and MacDonald. We observe that the equation obtained by the variation of $\mathcal{L}_a$ with respect to a_0 is[21]

$$i\varepsilon^{\alpha\beta}[q_\alpha a_\beta(\boldsymbol{q},\omega) - q_\beta a_\alpha(\boldsymbol{q},\omega)] = \frac{2\theta}{e}\rho(\boldsymbol{q},\omega).$$

Rearranging this equation by using the gauge condition $q_\alpha a^\alpha = 0$, we obtain

$$a_\alpha(\boldsymbol{q},\omega) = \frac{2\theta}{e}\varepsilon^{\alpha\beta}\frac{iq_\beta}{q^2}\rho(\boldsymbol{q},\omega). \tag{9.6.26}$$

The functional integration with respect to a is equivalent to inserting in the integral the δ-functional $\delta[a_\alpha(\boldsymbol{q},\omega) - \frac{2\theta}{e}\varepsilon^{\alpha\beta}\frac{iq_\beta}{q^2}\rho(\boldsymbol{q},\omega)]$; hence $\mathcal{L}_\phi$ in (9.6.19) becomes in the $(\boldsymbol{q},\omega)$ space

$$\mathcal{L}_\varphi^{\text{eff}} = \varphi^\dagger(\boldsymbol{q},\omega)\left[\omega - \frac{1}{2m}q^2\right]\varphi(\boldsymbol{q},\omega) - \frac{1}{2}\delta\rho(-\boldsymbol{q},-\omega)V(q)\delta\rho(\boldsymbol{q},\omega)$$

$$- 2\theta\delta\rho(-\boldsymbol{q},-\omega)\varepsilon^{\alpha\beta}\frac{iq_\alpha}{q^2}j_\beta(\boldsymbol{q},\omega) - \frac{2\bar{\rho}\theta^2}{m}\delta\rho(-\boldsymbol{q},-\omega)\frac{1}{q^2}\delta\rho(\boldsymbol{q},\omega), \tag{9.6.27}$$

where

$$j_\alpha(x) = \frac{ie}{2m}(\varphi^\dagger\partial_\alpha\varphi - \varphi\partial_\alpha\varphi^\dagger). \tag{9.6.28}$$

The third term of (9.6.27) comes from the $\partial_\alpha a^\alpha$ term in developing the kinetic energy term in (9.6.19), and the fourth term comes from the $a_\alpha a^\alpha$ term. In the further discussion we need to introduce a crucial approximation that we neglect the existence of vortices in the boson field when we discuss the properties of the ground state. Then j^α is pure longitudinal, i.e., $j^\alpha \propto \partial^\alpha\Theta$, where Θ is the phase of the boson field. Therefore, $\varepsilon^{\alpha\beta}q_\alpha q_\beta = 0$, and the third term

---

[21]In the $(\boldsymbol{q},\omega)$ representation we take $\hbar = c = 1$, hence $\Phi_0 = \frac{2\pi}{e}$.

vanishes. Now $\mathscr{L}^{\text{eff}}$ becomes fairly simple: the second term is the interaction term, and the fourth term becomes by the substitution $\delta\rho(x) = \sum_i \delta(x - x_i) - \bar{\rho}$ and $a_0^{-2} = 2\pi\,\bar{\rho}/\nu$[22]

$$\frac{\bar{\rho}}{2m}\frac{2\pi}{\nu^2}\int \mathrm{d}x\mathrm{d}y\delta\rho\,(x)\ln|x - y|\,\delta\rho(y)$$

$$= \frac{1}{m}\frac{1}{a_0^2}\left[\sum_{i<j}\ln|x_i - x_j|^{1/\nu} - \frac{1}{4a_0^2}\sum_i |x_i|^2\right]. \qquad (9.6.29)$$

Here the two-dimensional Fourier transform

$$\int \mathrm{d}^2 q\frac{e^{iq\cdot x}}{q^2} = -2\pi\ln|x|$$

has been used. Now it is possible to derive the algebraic off-diagonal long range order. Since $\varphi(x) = \sqrt{\rho(x)}e^{i\Theta}$, the correlation fluctuation of $\varphi(x)$ depends on the correlation fluctuation of $\Theta(x)$ when we neglect the fluctuation of the amplitude part. Substituting $\varphi(x) = \sqrt{\rho(x)}e^{i\Theta}$ into (9.6.27) we have

$$\mathscr{L}^{\text{eff}} = \delta\rho(-q, -\omega)i\omega\,\Theta(q,\omega) - \frac{\bar{\rho}}{2m}\Theta(-q,-\omega)q^2\,\Theta(q,\omega)$$

$$- \frac{1}{2}\delta\rho(-q,-\omega)\left[V(q) + \frac{4\pi^2\bar{\rho}}{m\nu^2}\frac{1}{q^2}\right]\delta\rho(q,\omega). \qquad (9.6.30)$$

Integrating with respect to $\delta\rho$ we obtain the effective Lagrangian for the Θ field

$$\mathscr{L}_\Theta^{\text{eff}} = \frac{1}{2}\Theta(-q,-\omega)\left[\frac{\omega^2}{V(q) + \frac{4\pi\bar{\rho}}{m\nu^2}\frac{1}{q^2}} - \frac{\bar{\rho}}{m}q^2\right]\Theta(q,\omega). \qquad (9.6.31)$$

We can read directly from $\mathscr{L}_\Theta^{\text{eff}}$ the propagator, which is the inverse of the square bracket in the above expression. Introducing

---

[22]$a_0^{-2} = eB = \frac{2\pi B}{\Phi_0} = 2\pi\rho_A = 2\pi\bar{\rho}/\nu$, a_0 is the magnetic radius of gyration.

the cyclotron frequency $\omega_c = \frac{eB}{m}$ we have[23]

$$\langle \Theta(-\boldsymbol{q}, -\omega) \Theta(\boldsymbol{q}, \omega) \rangle = \frac{\frac{2\pi}{\nu}\omega_c \frac{1}{q^2} + V(q)}{\omega^2 - \omega_q^2 + i\delta}, \qquad (9.6.32)$$

where

$$\omega_q^2 = \omega_c^2 + \frac{\bar{\rho}}{m}q^2 V(q). \qquad (9.6.33)$$

Integrating (9.6.32) with respect to ω, we obtain the static correlation function

$$\langle \Theta(-\boldsymbol{q}) \Theta(\boldsymbol{q}) \rangle$$

$$= -i \int \frac{d\omega}{2\pi} \langle \Theta(-\boldsymbol{q}, -\omega) \Theta(\boldsymbol{q}, \omega) \rangle$$

$$= -i \oint \frac{d\omega}{2\pi} \left(\frac{2\pi\omega_c}{\nu q^2} + V(q) \right) \left[\frac{1}{\omega - \omega_q + i\delta} - \frac{1}{\omega + \omega_q - i\delta} \right]$$

$$= -\frac{1}{2\nu}\frac{2\pi}{q^2} + O\left(\frac{1}{q} \right) \qquad (9.6.34)$$

From (9.6.34) we get the correlation function of the order parameter[24]:

$$\langle \varphi^\dagger(x)\varphi(y) \rangle = \bar{\rho}\langle e^{i\Theta(x) - i\Theta(y)} \rangle$$

$$= \bar{\rho}\,(\, 1 + i\langle \Theta(x) \rangle + i\langle \Theta(y) \rangle$$

$$+ \frac{1}{2}\langle \Theta(x)\Theta(y) \rangle + \frac{1}{2}\langle \Theta(y)\Theta(x) \rangle + \cdots)$$

---

[23] At the pole of the propagator the square bracket in (9.6.31) vanishes. Consequently, we put the square bracket equal to zero, solve for ω, and denote the root by ω_q, we obtain the pole of the propagator. The result is (9.6.33).
[24] Use $\langle \Theta \rangle = 0$ and (9.6.34). In the derivation we neglect $V(q)$, which is not important for long range correlations, and does not influence our subsequent discussions.

$$\approx \bar{\rho} e^{\langle \Theta(x)\Theta(y)\rangle} = \bar{\rho} \exp\left(-\frac{1}{2\nu}\int \frac{2\pi}{q^2} e^{iq(x-y)}\right) \mathrm{d}^2 q$$

$$= \bar{\rho} \exp\left[-\frac{1}{2\nu}\ln|x-y|\right] = \bar{\rho}\,|x-y|^{-1/\nu}\,.$$

$$(9.6.35)$$

This is Girvin and Mac Donald's result using the Laughlin wave function. The result (9.6.35) is obtained directly from the CSLG action and the mean field theory *without* reference to Laughlin wave function. Therefore, the Laughlin wave function should also be derivable from the theory. In fact Eq. (9.6.29) did appear in the plasma analogy in the interpretation of Laughlin wave function (9.5.20). Reference [42] gives the derivation of Laughlin wave function as well as the wave functions of the quasihole and quasiparticle.

Under a gauge transformation the phase of $\langle \varphi^\dagger(x)\varphi(y)\rangle$ changes, it is not gauge invariant because $\varphi^\dagger(x)\varphi(y)$ is not a physical observable. Actually a gauge invariant order parameter $\langle O^\dagger(x)O(y)\rangle$ can be defined [45], which reduces to $\langle \varphi^\dagger(x)\varphi(y)\rangle$ in the Landau-Lorentz gauge, but in other gauges $O(x)$ is in general not a local function of the field operator $\varphi(x)$.

9.6.4 Topological order of the fractional quantum Hall fluid

The quantum Hall fluid is an example of strongly correlated electron systems in two spatial dimensions, and a fascinating feature of such systems is that they possess topological order. In many cases in condensed matter physics the order is associated with a symmetry breaking described by Landau's theory. In the quantum Hall fluid there is no identifiable symmetry breaking. While the Zhang-Hansson-Kivelson effective theory (Section 9.6) is formulated in the form of a Ginzburg-Landau theory with a Chern-Simons term, another effective theory was formulated by X.G. Wen and A. Zee [46] containing only pure Chern-Simons term. This formulation is more convenient for studying hierarchical FQH states. These two forms of effective theory are related by a duality transformation [47]. The Chern-Simons action is expressed in terms of the Chern-Simons potential a_μ as

$$S = \int d^3x \frac{k}{4\pi} \epsilon^{\mu\nu\lambda} a_\mu \partial_\nu a_\lambda, \qquad (9.6.36)$$

where k is an integer. Compare this action, for example, with the action of a scalar field

$$S = \int d^3x \sqrt{g} (g^{\mu\nu} \partial_\mu \phi \partial_\nu \phi - m^2 \phi^2), \qquad (9.6.37)$$

where $g^{\mu\nu}$ is the metric tensor. For the Chern-Simons action there is no reference to the space-time metric whatsoever, i.e., no ruler and clock scale. The pure Chern-Simons action depends only on the topology of space time. This independence of space-time metric has deep consequences: any perturbation that depend on the metric cannot affect the Chern-Simons term. For instance disorder and imputities do depend on the metric, they cannot affect the Chern-Simons term, and hence the long distance physics of the Hall liquid, such as the charge and statistics of the quasi-particles in the fluid.

In the effective theory we are only interested in the physics at large space and time intervals, that is, at small wave numbers and low frequencies. Terms in the action are organized according to powers of derivatives and powers of the field variables. In 2+1 dimensional space-time, the current has mass dimension 2, the gauge potential has dimension 1, because the gauge potential is coupled to matter fields according to the gauge principle. The Chern-Simons term $\epsilon^{\mu\nu\lambda} a_\mu \partial_\nu a_\lambda$ has therefore dimension 3, and is the term having lowest possible mass dimension. The Mawell term has dimension 4 because it has two derivatives and two powers of the gauge potential. Therefore it scales as L^{-4} or M^4, and is less important at long distance. The long distance physics is determined purely by the Chern-Simons term, provided $k \neq 0$. For quantum fluid with $\nu = 1$ we have $k = 1$, and mathematical argument shows that k must be an integer.

The Laughlin $\nu = 1/m$ states represent the simplest quantum Hall states containing only one component of the incompressible fluid. From the wave function

$$\Psi_m = \left[\prod (z_i - z_j)^m \right] e^{-\frac{1}{4} \sum |z_i|^2},$$

we know that when the j-th electron is moved around the i-th electron the wave function acquires a phase factor $e^{i2\pi m}$. More general QH states with filling fraction such as $\nu = 2/5, 3/7, \ldots$ contain more than one component of the imcompressible fluids and have more complicated topological orders. They are characterized by a symmetrical matrix K with integer elements and a vector Q with integer entries. When the particle of the i-th component is moved around a particle of the j-th component, the wave function acquires a phase factor $e^{i2\pi k_{ij}}$, where k_{ij} is an element of the matrix K. An entry Q_i of the vector Q represents the charge (in units of e) carried by the particles of the i-th component of the fluid. All physical properties associated with the topological orders can be determined in terms of K and Q, e.g., $\nu = Q^T K^{-1} Q$. The Laughlin states $1/m$ has $K = m$ and $Q = 1$, while the $\nu = 2/5$ states is described by $K = \left(\begin{smallmatrix} 3 & 2 \\ 2 & 3 \end{smallmatrix} \right)$ and $Q = \left(\begin{smallmatrix} 1 \\ 1 \end{smallmatrix} \right)$.

Physically the topological order manifests itself in various aspects of the QH fluid, for instance the ground state degeneracy and the anomalous exponent in the edge states (Section 6.5.5). Actually the ground state degeneracy of QH fluid on a closed 2-dimensional surface of genus G is given by k^G, where k is the parameter in (9.6.36). The genus measures the number of holes or handles, e.g., the sphere has genus 0, the torus has genus 1. The topology dependence is here explicit. Some hierarchical states with the same filling fraction can have different ground state degeneracies, indicating different topological orders in these states. Hierarchy states can be constructed by different schemes giving different wave functions for the same filling fraction. Whether these wave functions belong to the same universality class can be judged from their K and Q characterization. If two different constructions lead to the same K and Q, then these two constructions generate the same FQH liquid.

9.7   The Global Phase Diagram of the Quantum Hall Effect

We have seen that the filling fraction is the key parameter of the quantum Hall effect. For a definite value of ν there correspond definite values of the longitudinal resistance ρ_{xx} and Hall resistance ρ_{xy}. S. Kivelson, D. H. Lee and C. S. Zhang [48, 49] established the global phase diagram of the quantum Hall effect. A "law of corresponding states" is derived. This law is a set of approximate relations between the properties of the system at one value of the magnetic field and the properties of the same system at another value of the magnetic field. Qualitative considerations are given below:

(1) The Landau-level addition transformation:

$$\nu \leftrightarrow \nu + 1. \tag{9.7.1}$$

The symbol $\leftrightarrow$ implies that there exists a correspondence between the properties of the system at two different values of ν. This is based on the observation that a filled Landau level acts as an inert background which does not much affect the physics in the higher partially filled Landau levels. This transformation relates the different integral quantum Hall states.

(2) The particle-hole transformation for $\nu < 1$:

$$\nu \leftrightarrow 1 - \nu \tag{9.7.2}$$

This means that to create a state ν on the basis of the vacuum is equivalent to removing from a filled Landau level (by (9.7.1) it corresponds to the vacuum) an amount of electrons of filling fraction ν.

(3) The flux attachment transformation:

$$\frac{1}{\nu} \leftrightarrow \frac{1}{\nu} + 2 \tag{9.7.3}$$

The meaning of this transformation can be understood from a simple example $3 \leftrightarrow 1 + 2$, which means that the FQHE $\nu = \frac{1}{3}$ corresponds to the IQHE $\nu = 1$. By attaching a flux of $2\Phi_0$ to each electron one gets a corresponding state.

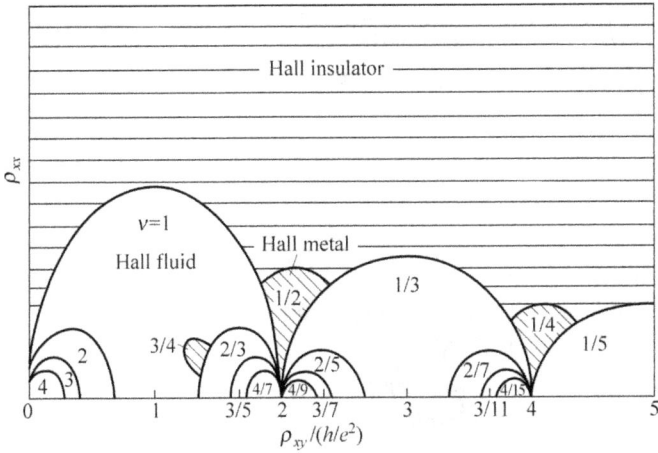

Fig. 9.19   The global phase diagram of the quantum Hall effect. Reprinted with permission from Steven Kivelson, Dung-Hai Lee, and Shou-Cheng Zhang, *Phys. Rev. B* 46, 2223, 1992. Copyright (2017) by the American Physical Society.

Another ingredient of this theory is the selection rule, which states that the insulator-quantum Hall state transition happens only through the $\nu = 1$ IQH state. All higher ν states have to undergo transitions to lower ν states and then eventually the $\nu = 1$ state goes over to the insulator phase under increased impurities or decreased magnetic field. The derivation is based on the "floating up" scenario.[25] The phase diagram is shown in Fig. 9.19. In the phase diagram ρ_{xx} represents the disorder in the crystal (impurities, defects, phonon excitations) and changes with the temperature. The general trend of ρ_{xy} is proportional to B. When impurities exist, there are two different phases in the limit $T \to 0$. One is the Hall

---

[25]Shortly following the discovery of the IQHE it became evident that there exist extended states below the Fermi energy, because the longitudinal conductivity is in general finite. The Anderson localization has been well-known in solid state physics which states that all 2-dimensional electron system are insulating when the disorder strength becomes larger than a critical value. To explain the transition from the QH phase to insulating phase, Laughlin and Khmel'nizkii proposed the "floating up" picture that when the magnetic field decreases, the energies of the extended states float up from the center of the Landau level. As B approaches 0 the extended states energies rise above the Fermi energy, rendering all states below the Fermi energy localized.

liquid phase with $\rho_{xx} \to 0$ and $\sigma_{xx} \to 0$ as we discussed in connection with Eq. (9.4.13), and the other is the insulator phase $\rho_{xx} \to \infty$, $\sigma_{xx} \to 0$. This phase is different from ordinary insulators. When the defect in the sample is increased to pass over a certain value, a very high value of the voltage is required to keep the longitudinal current. When T is lowered, the voltage required *increases*. This is typical for an insulator. At the same time the Hall voltage is independent of temperature and rises linearly with the magnetic field, this is a characteristic of Hall liquid. This phase is called the Hall insulator. It has been discovered by D. Tsui's group [50].

The state $\nu = \frac{1}{2}$ was mentioned in Section 9.5 in connection with the composite fermion picture. From the flux attachment transformation (9.7.3) we know that $\nu = \frac{1}{4}$ is a corresponding state, and the particle-hole transformation (9.7.2) gives the corresponding state $\nu = \frac{3}{4}$ of the latter.

While the principle of corresponding states is supported by most experiments, the selection rule seems to be at odds with many experiments. The group of D. Tsui [51] carried out systematic study on the insulator to IQHE transition. They used a 2-dimensional system confined in a Ge/SiGe quantum well. The experimental signature of an insulator-IQHE transition is that the longitudinal resistivity ρ_{xx} is increasing with temperature in the QHE phase while it is decreasing with increasing temperature in the insulating phase, as shown in Fig. 9.20. At the transition point (B_C^L in Fig. 9.20(a)), ρ_{xx} is independent of temperature. With increasing B, ρ_{xy} first develops a very narrow plateau at $\nu = 3$, then at $\nu = 2$ and finally at $\nu = 1$. This is a direct transition between the insulating phase and $\nu = 3$ IHQ state at B_C^L. At higher fields the transition between QH $\nu = 1$ state to the insulator is shown in Fig. 9.20(b). Here the transition at B_C occurs at a point where the ρ_{xx} curves merge. All the transition points between QH liquid and insulators (solid dots) and transitions between QH states of different ν values (open dots) are shown in Fig. 9.21(a) in a $n - \nu^{-1}$ plot, where the sizes of dots represent expected inaccuracy. To demonstrate the transition of a Landau level as a function of B, the same data is plotted in Fig. 9.21(b) in a $n - B$ plot. At higher B, the data points follow precisely the straight lines

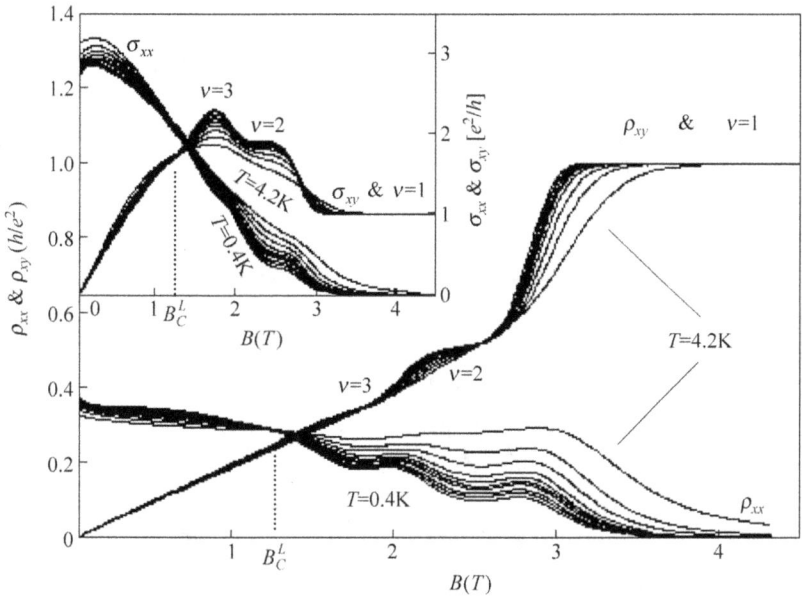

(a)

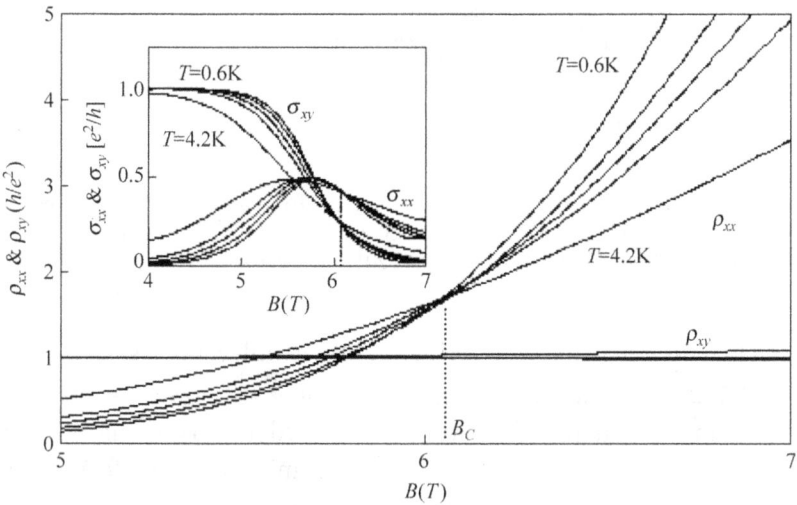

(b)

Fig. 9.20   Magnetic field dependence of ρ_{xx}, ρ_{xy} for temperature between 0.6 K and 4.2 K. In the insets σ_{xx}, σ_{xy} vs. B are plotted. (a) $n = 1.2 \times 10^{11} \text{cm}^{-2}$; (b) $n = 0.8 \times 10^{11} \text{cm}^{-2}$. Reprinted with permission from M. Hilke, D. Shahar, S. H. Song, D. C. Tsui, and Y. H. Xie, *Phys. Rev. B* 62, 6940, 2000. Copyright (2017) by the American Physical Society.

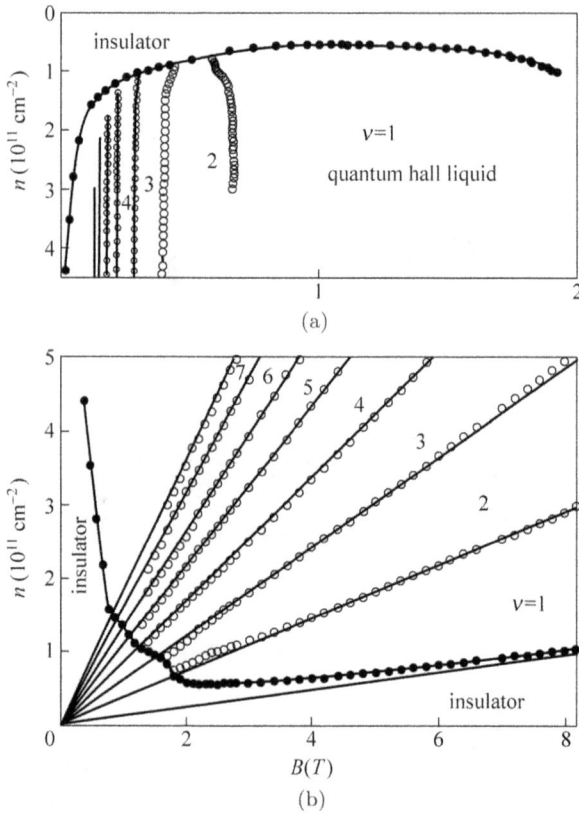

Fig. 9.21   Phase diagrams (a) $n - \nu^{-1}$ plot; (b) $n - B$ plot. Reprinted with permission from M. Hilke, D. Shahar, S. H. Song, D. C. Tsui, and Y. H. Xie, *Phys. Rev. B* 62, 6940, 2000. Copyright (2017) by the American Physical Society.

$n = \left(i + \frac{1}{2}\right)\frac{eB}{h}$ (the boundaries between different ν states) where $i = 0, 1, 2, \ldots$. The insulating phase below the lowest energy level $i = 0$ can be understood in terms of disorder broadening of energy levels. As the energy levels are broadened, minibands are formed. The states at the center are extended, and those away from the center are localized. At $T = 0$ and the Fermi energy is below the lowest energy level, E_F is pinned to the localized states and the system becomes insulating. As B is lowered, the gaps between the energy level are decreased and levels overlap. Some states from higher energy levels increase the density of states below the lowest energy level

($i = 0$) which lead to an elevation of thee insulator phase boundary. All higher ν states undergo transition to the insulating phase *directly*.

The theoretical study of the insulator to IQHE transitions was carried out by D. N. Sheng and Z. Y. Weng [52, 53]. A tight-binding lattice model is used. With the hopping constant set equal to 1 as the unit of energy, there are two parameters in the model. One is the magnetic flux per plaquette of the lattice $\phi = \frac{2\pi}{M}$. The integer M corresponds to the filling fraction. For instance $M = 8$ gives a maximum filling fraction $n_\nu = 8$. The other parameter is the disorder strength W. At each lattice site a disorder potential is chosen randomly from $-W/2$ to $W/2$. The Hall conductance is calculated by using the Kubo formula and the result for $M = 8$ is shown in Fig. 9.22. At weaker disorder ($W = 1$), 3 IQHE plateaus at $\sigma_H = \nu \frac{e^2}{h}$ ($\nu = 1, 2, 3$) are seen corresponding to 4 Landau levels centered at the jumps of σ_H for $E \leq 0$ (half the entire band). At increasing disorder the plateaus are deteriorated in a one-by-one order from high to low energies. The $\nu = 3$ plateau is destroyed at $W = 2$ and

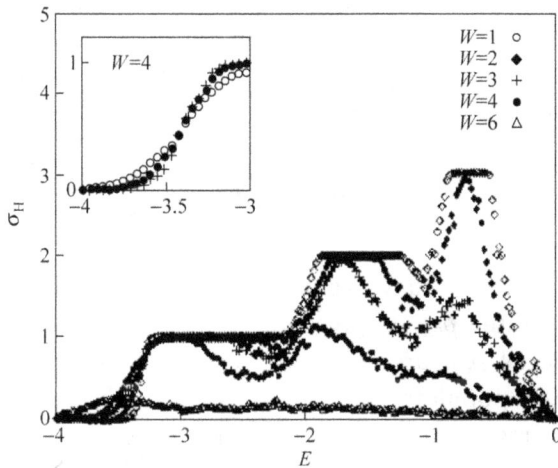

Fig. 9.22 Calculated σ_H versus energy E for disorder strength W from 1 to 6. Inset shows $\nu = 1$ IQHE plateau at $W = 4$ with different lattice size from $8 \times 8(\diamond), 16 \times 16(\bullet), 24 \times 24(*)$. Reprinted with permission from D. N. Sheng and Z. Y. Weng, *Phys. Rev. Lett.* **78**, 318, 1997. Copyright (2017) by the American Physical Society.

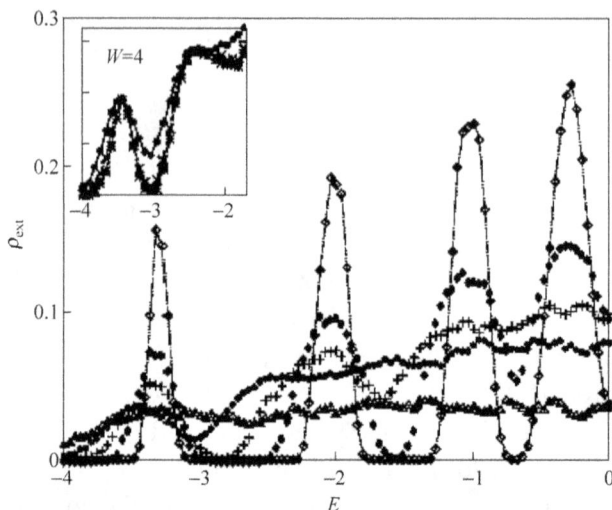

Fig. 9.23   Calculated density of extended states for different disorder strengths. Reprinted with permission from D. N. Sheng and Z. Y. Weng, *Phys. Rev. Lett.* 78, 318, 1997. Copyright (2017) by the American Physical Society.

the $\nu = 1$ plateau breaks down at $W = 6$. The key to understand this evolution of IQHE plateaus is the topological considerations of D. J. Thouless discussed in Section 9.4. The quantized Hall conductance is a sum of Chern numbers carried by extended states below the Fermi energy:

$$\sigma_H(E_F) = \frac{e^2}{h} \sum_{\epsilon(m)<E_F} C^{(m)},$$

while localized states have vanishing Chern numbers. The density of extended states ρ_{ext} is presented in Fig. 9.23. At $W = 1$ well defined peaks of ρ_{ext} are located at centers of Landau bands with localized states populated in between. Each peak corresponds to the rapid growth region of σ_H. When the Fermi energy reaches a point when all extended states in a peak are included, a quantized plateau is obtained. Total Chern number for each of the three lower energy peaks is +1, and this is the reason that plateaus with $\nu = 1, 2, 3$ are reached when the Fermi energy moves from low energy end towards the high energy. The Hall conductance at $E = 0$ (band half filled) should be zero. Therefore, the last peak closest to the band center carries a Chern number -3, leading to a fall of σ_H from $\nu = 3$ to zero. The conservation of topological number is clearly demonstrated. During the process of destruction of σ_H plateau under increasing W the extended levels separating different plateaus merge and then disappear without "floating up" in energy, and the lowest extended level band is the last to vanish at a critical value W_c of the disorder strength. The critical strength W_c decreases with decreasing magnetic field and can be extended to 0 at vanishing field limit. The picture of destruction of IQHE persists into the weak field limit.

The phase diagram given by the tight binding model is presented in Fig. 9.24. There are some interesting features. Starting from strong magnetic field region of the insulating phase (denoted by II in the figure) and reducing B continuously at a fixed electron density, the curve A is obtained which cuts through different IQH phases. As B is reduced $\frac{W}{W_c(B)}$ increases because W_c decreases with a decreasing B. The insulating phase II is characterized by a Hall resistance ρ_{xy} persisting in a quantized value $\frac{h}{e^2}$ even when the longitudinal resistance ρ_{xx} has increased many times, while the insulating phase I is characterized by a classical value of $\rho_{xy} = \frac{1}{n_\nu}\frac{h}{e^2} = \frac{B}{nec}$.

The regime I is believed to approach an Anderson insulator at vanishing magnetic field, while in the regime II σ_{xx} and σ_{xy} are found to satisfy a one-parameter scaling law believed to be a consequence related to the quantum phase transition [51].

9.8 Quantum Spin Hall Effect

M. Dyakonov and V. Perel [54] predicted that when a voltage is applied to a semiconductor strip, electrons of given spin would concentrate on one side of the strip and electrons of opposite spin would concentrate on the other side, while no transverse Hall voltage is developed. This phenomenon is based on the spin dependent scattering of electrons on impurities. When it was realized that the spin degree of freedom can be used to carry, store, and manipulate information, research in this direction resurges. S. Murakami, N. Nagaosa and Shou-Cheng Zhang [55], and independently J. Sinova

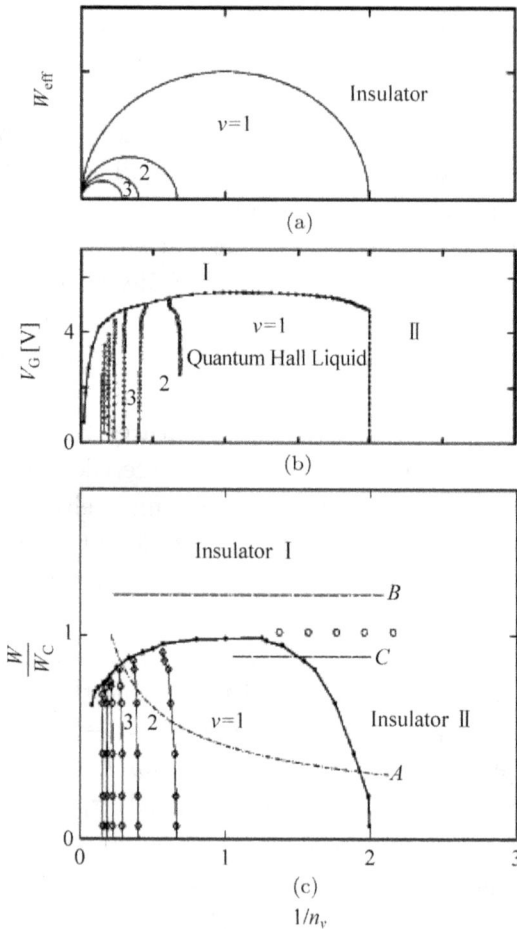

Fig. 9.24 Phase diagram given by the tight binding lattice model. Reprinted with permission from D. N. Sheng and Z. Y. Weng, *Phys. Rev. B* 62, 15363, 2000. Copyright (2017) by the American Physical Society.

and A. MacDonald and collaborators [56] predicted that a strong spin Hall effect occurs even without impurities. This impurities-independent version was named *"intrinsic"* spin Hall effect, and has been pursued much more vigorously both theoretically and experimentally. As a consequence of an earlier work [57] the authors of Ref. [55] pointed out that the spin transport is dissipationless. Just like the charge Hall effect, the spin Hall effect also has a quantum counterpart [58–60]. It turns out that the spin quantum Hall system is a new state of matter that exists in the absence of magnetic field with edge states developing where carriers with opposite spins move in opposite directions on a given edge of the sample as shown in Fig. 9.25. The system is realized in HgTe/(Hg,Cd)Te quantum wells with a well width larger than a critical value. In such kind of quantum wells there are 4 relevant bands close to the Fermi level, viz. the E1 band with 2 spin states of the s orbital $|s, \uparrow\rangle, |s, \downarrow\rangle$ and HH1 band consisting of the p orbitals $|p_x + ip_y, \uparrow\rangle, |p_x - ip_y, \downarrow\rangle$ with $m_J = 3/2$ and $m_J = -3/2$ respectively. The wave function is a 4-component spinor

$$
\begin{pmatrix}
\left| s, m_J = \dfrac{1}{2} \right\rangle \\[2mm]
\left| p, m_J = \dfrac{3}{2} \right\rangle \\[2mm]
\left| s, m_J = -\dfrac{1}{2} \right\rangle \\[2mm]
\left| p, m_J = -\dfrac{3}{2} \right\rangle
\end{pmatrix}. \tag{9.8.1}
$$

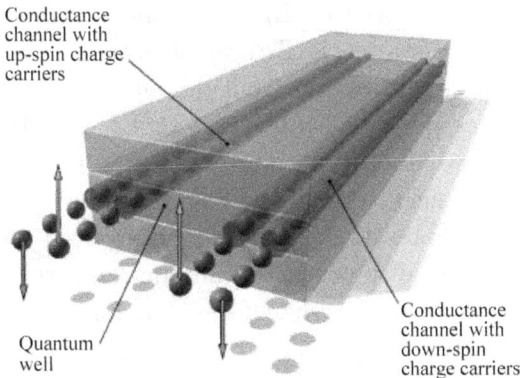

Fig. 9.25   Schematic of the spin polarized edge channels in a quantum spin Hall insulator. Reprinted with permission from Markus König, Steffen Wiedmann, Christoph Brüne, Andreas Roth, Hartmut Buhmann, Laurens W. Molenkamp, Xiao-Liang Qi, and Shou-Cheng Zhang, *Science*, 318, 766, 2007. Copyright (2017) by the AAAS.

The effective Hamiltonian near the Γ point, the center of the Brillouin zone is given by Ref. [60]

$$H_{\text{eff}}(k_x, k_y) = \begin{pmatrix} H(\vec{k}) & 0 \\ 0 & H^*(-\vec{k}) \end{pmatrix}, \qquad (9.8.2)$$

$$H(\vec{k}) = \epsilon(\vec{k}) + d_i(\vec{k})\sigma_i, \qquad (9.8.3)$$

where σ_i are the Pauli matrices, and

$$d_1 + id_2 = A\,(k_x + ik_y) \equiv Ak_+,$$
$$d_3 = M - B(k_x^2 + k_y^2),$$
$$\epsilon_k = C - D(k_x^2 + k_y^2). \qquad (9.8.4)$$

Here k_x and k_y are components in the plane of the 2-dimensional electron gas, A, B, C, D are constants specific of the system. Spin-orbit coupling is built in through the $\vec{d}(\vec{k})\cdot\vec{\sigma}$ term of the Hamiltonian. In (9.8.2), the upper block refers to "spin up" and the lower block refers to "spin down". $d_1 + id_2$ couples s and p orbitals, and is therefore odd in k, while d_3 does not couple them, and is even in k. The diagonal blocks of H_{eff} are time reversal transform of each other, and the system is T reversal invariant. This invariance is crucial for the quantum spin Hall system. The parameter M in $d_3\sigma_3$ discriminates the s and p orbitals, or E1 and HH1 bands. For conventional semiconductors such as CdTe $M > 0$, i.e., E1 lies above HH1. Semi-metals such as graphene has $M = 0$, although the bands have a diferent interpretation. In the "band inverted" semiconductor such as HgTe with the thickness of the quantum well d larger than a critical value $d > d_c$, HH1 lies above E1, for which $M < 0$. For $d < d_c$, M is positive: a topological quantum phase transition occurs at the critical thickness d_c.

The quantum spin Hall phase occurs in the band inverted regime $M < 0$ $(d > d_c)$, characterized by the existence of helical edge states. To see this we return to the Hamiltonian (9.8.2) and concentrate on the upper block at first. It can be diagonalized to give two separated bands under appropriate condition (9.8.8):

$$E_\pm(\boldsymbol{k}) = \epsilon(\boldsymbol{k}) \pm d(\boldsymbol{k}), \tag{9.8.5}$$

where $d(\boldsymbol{k})$ is the norm of the 3-vector $\boldsymbol{d}(\boldsymbol{k})$:

$$d(\boldsymbol{k}) = \sqrt{d_\alpha(\boldsymbol{k})d^\alpha(\boldsymbol{k})}. \tag{9.8.6}$$

For

$$E_+(\boldsymbol{k}) - E_-(\boldsymbol{k}) = 2d(\boldsymbol{k}) > 0 \tag{9.8.7}$$

for all $\boldsymbol{k}$, the system is an insulator. The gap opening condition is for $\boldsymbol{k} \in$ BZ (Brillouin zone)

$$\min E_+(\boldsymbol{k}) > \max E_-(\boldsymbol{k}). \tag{9.8.8}$$

The authors calculated the Hall conductivity σ_{xy} by using Kubo's formula and found that in the insulator phase it is a topological invariant defined on the first Brillouin zone [59]. Although the single electron states in the present problem are very different from the Landau levels, the quantization of transverse conductivity shares the same topological origin. To show the behavior of edge states, we define the Hamiltonian on a strip with the periodic boundary condition in y-direction and open boundary condition in x-direction, with wave function vanishing at $x = 0$ and $x = L$, where L is the width of the sample. In this case, k_y is a good quantum number and the single particle energy is a function $E_m(k_y)$, $m = 1, 2, \ldots, 2L$. The energy spectrum obtained in a 2D Luttinger model with tight-binding regularized version of the vector $d_\alpha(\boldsymbol{k})$ [59] is shown in Fig. 9.26. For a given k_y there are $2L$ states, two of these are localized, while the others are extended. When the Fermi level lies in the bulk energy gap (represented in Fig. 9.26 by a horizontal dot-dashed line), each crossing of it with the edge state spectrum (solid and dashed curves) denoted by solid and empty circles defines two edge states on the left and right boundary of the sample and with opposite spin directions.

To study the behavior of edge states under an infinitesimal electric field, consider the Laughlin-Halperin gauge argument [61, 62]. Our sample with adopted boundary conditions can be considered as a cylinder with its axis in the x-direction as shown in Fig. 9.27. When

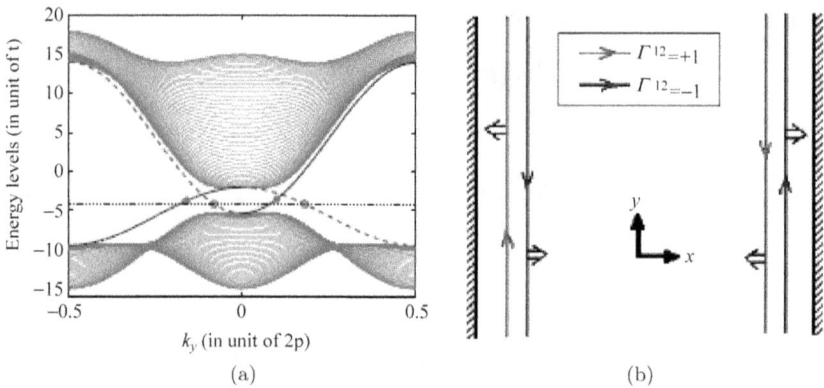

(a)   (b)

Fig. 9.26   (a) Energy spectrum; (b) the schematic diagram of the edge states. Reprinted with permission from Xiao-Liang Qi, Yong-Shi Wu, and Shou-Cheng Zhang, *Phys. Rev. B* 74, 085308, 2006. Copyright (2017) by the American Physical Society.

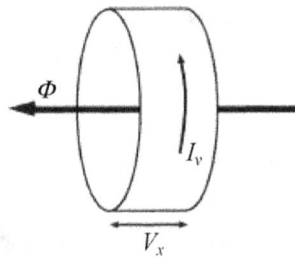

Fig. 9.27   Sample with electric field in y-direction (along the strip) induced by a change in flux Φ.

the flux $\Phi(t)$ threading the cylinder is increased adiabatically from $\Phi(0) = 0$ to $\Phi(T) = 2\pi$, the electric field induced in the y-direction at time t is

$$E_y(t) = -\frac{\partial A_y(t)}{\partial t} = \frac{1}{L}\frac{\partial \Phi(t)}{\partial t}, \qquad (9.8.9)$$

where $\boldsymbol{A}$ is the vector potential associated with the flux. The convention is chosen such that $E_y > 0$ when Φ increases. The effect of the flux change is represented by the replacement $k_y \to k_y - A_y$ in the Hamiltonian. Correspondingly each single particle wave function

$$\phi_{m,k_y}(x, y) = u_{m,k_y}(x, y)e^{ik_y y}$$

is to be replaced by

$$\phi_{m,k_y}(x,y;t) = u_{m,k_y - A_y(t)}(x,y)e^{i(k_y - A_y)y}.$$

At time T the adiabatic evolution results in

$$|m, k_y\rangle \rightarrow \left|m, k_y + \frac{2\pi}{L}\right\rangle, \qquad (9.8.10)$$

where L is the length of the sample in x-direction. The ground state of the system

$$|G\rangle = \prod_{E_m(k) \leq \mu} |m, k_y\rangle \qquad (9.8.11)$$

evolves as the consequence of the flux change to the final state

$$|G'\rangle = \prod_{E_m(k) \leq \mu} \left|m, k_y + \frac{2\pi}{L}\right\rangle. \qquad (9.8.12)$$

Since the bulk part of $|G\rangle$ is a product of states with all k's, it does not change under a such a translation, so that the only difference between $|G\rangle$ and $|G'\rangle$ occurs to the edge states near the Fermi Level represented by the solid and hollow circles for particle and hole excitations respectively in Fig. 9.26(a). The single particle energy change due to a change in k_y is

$$\delta E \simeq v_y \delta k = 2\pi \hbar v_y / L. \qquad (9.8.13)$$

Therefore each edge state on the Fermi surface with $v_y > 0$ will move out of the Fermi sea becoming a particle excitation, while each one with $v_y < 0$ will move into the Fermi sea becoming a hole excitation. The density distribution of these edge states $|\phi(i_x)|^2$ is calculated to determine whether they lie on the left (L) or right (R) edge. The state $|G'\rangle$ is expressed as a particle-hole state

$$|G'\rangle = \prod_{i=1}^{n} c_{iL+}^\dagger c_{iR-}^\dagger c_{iL-} c_{iR+} |G\rangle, \qquad (9.8.14)$$

where the label $\pm$ refers to the quantum number Γ^{12}. In the presence of spin-orbit coupling the spin is not conserved. The spin operator is decomposed into a "conserved" part [59] Γ^{12} which commutes with the Hamiltonian, and a "non-conserved part" Γ^{34} which represents the precession effect: $S^z = -\frac{1}{2}\Gamma^{12} - \Gamma^{34}$. From (9.8.14) it can be seen that the net effect of adiabatically turning on a flux is to transfer the edge states with $\langle S^z \rangle = -1/2$ from the right edge to the left one and the edge states with $\langle S^z \rangle = 1/2$ in the opposite sense. This leads to an accumulation of different spin density on the boundary. Such an accumulation can be considered as a spin Hall current induced by E_y (9.8.9)

$$j_x = \sigma_{xy}^{(s)} E_y. \tag{9.8.15}$$

The spin Hall conductivity is calculated in Ref. [59] to be

$$\sigma_{xy}^{(s)} = \frac{2e^2}{h}. \tag{9.8.16}$$

L.W. Molenkamp and H. Buhman [60] developed sophisticated technology for fabricating HgTe quantum wells. The actual existence of the edge channels is revealed in small samples by measuring the longitudinal resistance as a function of the gate voltage by which the Fermi level is scanned through the gap. The charge transport is in the Landauer-Büttiker ballistic regime, and the mean free path is about 1μm. The arrangement of measuring terminals is given in Fig. 9.28, the longitudinal resistance measured is $R_{12,34} = V_{23}/I_{14}$. The longitudinal conductance is also quantized [58] to be $G = 2e^2/h$. As pointed out in Ref. [60] a Landauer-Büttiker type of calculation gives the same value, $R_{12,34} = h/2e^2$. In the experiment samples with

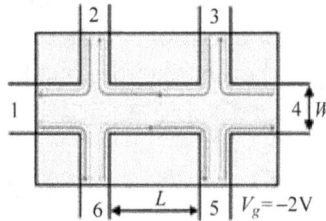

Fig. 9.28   Sample and measurement terminals. Reprinted with permission from Markus König, Steffen Wiedmann, Christoph Brüne, Andreas Roth, Hartmut Buhmann, Laurens W. Molenkamp, Xiao-Liang Qi, and Shou-Cheng Zhang, *Science*, 318, 766, 2007. Copyright (2017) by the AAAS.

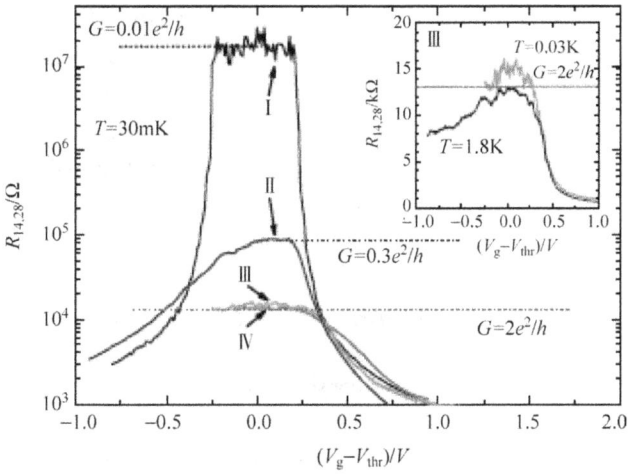

Fig. 9.29 Longitudinal resistance of quantum well samples. Reprinted with permission from M. König, S. Wiedmann, C. Brüne, A. Roth, H. Buhmann, L. W. Molenkamp , X-L Qi, and Shou-Cheng Zhang, Science, 318, 766, 2007. Copyright (2017) by the AAAS.

width smaller than the critical value (the normal electronic structure, $d = 5.5$ nm $< d_c = 6.3$ nm) show basically zero longitudinal conductance (curve I in Fig. 9.29). For all devices with inverted electronic structure ($d = 7.3$ nm $> d_c$) the longitudinal conductance becomes much higher. For a large sample of 20.0 $\mu rmm \times 13.3$ μm (curve II in Fig. 9.29) where the dimension is larger than the mean free path, the conductance has not yet reached the quantized value. For mush shorter samples ($L = 1.0$ μm, curves III and IV in Fig. 9.29) the longitudinal conductance actually reached the predicted value of $G = 2e^2/h$, demonstrating the existence of the spin edge channels in the inverted electronic structure of the quantum well.

9.9 Topology in the Physics of $2 + 1$ Dimensions, Anomalous Quantum Hall Effect

The study of gauge field theory in $2 + 1$ dimensions initiated by Deser, Jackiw and Templeton [64] was motivated by its connection with gauge theories in $3 + 1$ dimension at high temperatures. The study led to the remarkable result that topologically non-trivial, gauge invariant terms provide masses for the gauge fields. Moreover, it led to exotic properties of the relativistic fermions coupled to the gauge field: fermion zero mode, fractional quantum numbers, and abnormal vacuum current. The relevance of these aspects induced a flourishment of subsequent research on 2-dimensional spatial systems. In the following, we begin with the $3+1$ dimensional chiral anomaly before introducing the $2+1$ dimensional "anomaly".

9.9.1   The $3+1$ dimensional chiral anomaly

The following discussion is based on a textbook by Itzykson and Zuber [65].

Consider the relativistic fermion field described the Lagrangian density

$$\mathcal{L} = \overline{\psi}(i\gamma_\mu \partial^\mu - m)\psi, \quad \mu = 0, 1, 2, 3 \tag{9.9.1}$$

where the $\gamma's$ are 4 by 4 matrices

$$\gamma^0 = \begin{pmatrix} 1 & 0 \\ 0 & -1 \end{pmatrix}, \quad \gamma^i = \begin{pmatrix} 0 & \sigma_i \\ -\sigma_i & 0 \end{pmatrix}, \quad i = 1, 2, 3 \tag{9.9.2}$$

satisfying anti-commutation relations

$$\{\gamma^\mu, \gamma^\nu\} = 2\eta^{\mu\nu}, \quad diag(\eta) = (1, -1, -1, -1), \tag{9.9.3}$$

where $\sigma_i's$ are the Pauli matrices. The Dirac adjoint $\overline{\psi}$ is defined as $\psi^\dagger \gamma_0$. The squares of the $\gamma's$ are $\gamma^{02} = -\gamma^{i2} = 1$. We further define a cousin of the $\gamma's$

$$\gamma_5 = i\gamma^0 \gamma^1 \gamma^2 \gamma^3 = \begin{pmatrix} 0 & 1 \\ 1 & 0 \end{pmatrix}, \tag{9.9.4}$$

which anticommutes with γ^μ:

$$\{\gamma_5, \gamma^\mu\} = 0. \tag{9.9.5}$$

Now we can decompose ψ into the left- and right-hand parts:

$$\psi_L = \frac{1}{2}(1 - \gamma_5)\psi, \quad \psi_R = \frac{1}{2}(1 + \gamma_5)\psi, \quad \psi = \psi_R + \psi_L \tag{9.9.6}$$

They are eigenstates of γ_5 with eigenvalues -1 and $+1$ respectively. We find that the mass term of $\mathcal{L}$ mixes the left- and right-hand parts:

$$m\overline{\psi}\,\psi = m\overline{\psi}_L\psi_R + m\overline{\psi}_R\psi_L, \qquad (9.9.7)$$

while the kinetic energy term does not:

$$\overline{\psi}(i\gamma_\mu\partial^\mu)\psi = \overline{\psi}_L(i\gamma_\mu\partial^\mu)\psi_L + \overline{\psi}_R(i\gamma_\mu\partial^\mu)\psi_R \qquad (9.9.8)$$

For the fermions being a spinor representation of $SU(2)$, for example

$$\psi = \begin{pmatrix} u \\ d \end{pmatrix} \qquad (9.9.9)$$

we see that the kinetic energy part of the Lagrangian has a higher chiral symmetry of $SU(2)_L \otimes SU(2)_R$, in which left- and right-hand parts can transform independently, while the mass term breaks this higher symmetry down to the original isospin $SU(2)$.

An interesting question arises: *Is a symmetry of classical theory necessarily a symmetry of the quantum theory?* Historically field theory theorists took as almost self-evident the answer *yes*. On the classical level, a symmetry exists if a transformation keeps the action I invariant. On the quantum level the symmetry exists if it keeps the path integral $\int [D\phi] \exp (iI\,[\phi]))$ invariant. Does the former imply the latter? The answer is yes only when the measure is invariant under the transformation. In our case "classical" theory means the equation of motion $(i\gamma_\mu\partial^\mu - m)\,\psi = 0$ level. From the Dirac equation it follows that the vector current $j^\mu = \overline{\psi}\gamma^\mu\psi$ is conserved: The symmetry is associated with the gauge invariance, i.e. the conservation of fermion number (the charge). In contrast the axial current $j_5^\mu = \overline{\psi}\gamma^\mu\gamma_5\psi$ is not conserved: $\partial_\mu\,j_5^\mu = 2m\gamma_5\psi \neq 0$. In classical theory only *massless* fermions abide the chiral $SU(2)_L \otimes SU(2)_R$ symmetry. Can this classical symmetry for massless fermions continue to survive in quantum theory? The question can only be answered by practice. The story begins with the the calculation of triangle Feynman diagram with 2 vector vertices which can be coupled to photons (k_1, k_2) and an axial vertex which can be coupled to a pion as shown in the Fig 9.30. It is denoted by $\Delta^{\lambda\mu\nu}(k_1, k_2)$, describing the quantum

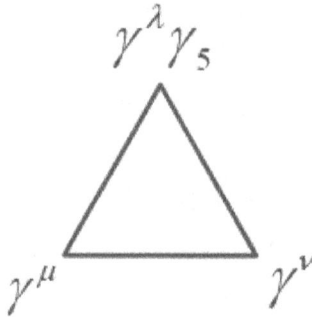

Fig. 9.30   The triangle diagram. The axial vertex is coupled to a pion of momentum q, and the vector vertices are coupled to photons of momenta k_1 and k_2 respectively.

process $\langle 0 | T J_5^\lambda (0) J^\mu (x_1) J^\nu (x_2) | 0 \rangle$, i.e. $\pi^0 \to 2\gamma$. The vector current is conserved in the process if $k_{1\mu} \Delta^{\lambda\mu\nu} (k_1, k_2) = 0$.

In the calculation a 4D Minkowski integral can be evaluated after a Wick rotation to make the integral 4D Euclidean, then integrated and rotated back. The result is

$$k_{1\mu} \Delta^{\lambda\mu\nu} (k_1, k_2) = \frac{i}{8\pi^2} \varepsilon^{\lambda\nu\sigma\tau} k_{1\sigma} k_{2\tau} \neq 0. \qquad (9.9.10)$$

This would be simply a disaster! The conservation of *vector* current is based on the gauge symmetry implying the conservation of fermion number. What is wrong? Let us go back to the integral $\Delta^{\lambda\mu\nu} (k_1, k_2)$ itself. In calculating amplitudes given by Feynman rules it is a common practice to shift the variable of integration. But when is this legitimate? Let us consider

$$\int_{-\infty}^{\infty} dp f (p + a) - \int_{-\infty}^{\infty} dp f (p) = \int_{-\infty}^{\infty} dp a \frac{df}{dp} + \cdots$$

$$= a (f (\infty) - f (-\infty)).$$

The shift is legitimate when

$$f (\infty) = f (-\infty)$$

i.e. when the integral is finite, or at most logarithmically divergent. Applying this shift technique to evaluate $\Delta^{\lambda\mu\nu} (k_1, k_2)$, we find that it depends on the shift a! The result does not make any sense! In other words, the integral is ill-defined, and by power counting it is actually linearly divergent. It turns out that the Feynman rules do not suffice to determine the process. The situation can be saved by hitting two birds with one stone: by invoking the physics demand of vector current conservation

$$k_{1\mu}\Delta^{\lambda\mu\nu}(a, k_1, k_2) = 0, \qquad (9.9.11)$$

which happens when

$$a = -(k_1 - k_2)/2,$$

Then $\langle 0 | T J_5^\lambda(0) J^\mu(x_1) J^\nu(x_2) | 0 \rangle$ is given by $\Delta^{\lambda\mu\nu}(a, k_1, k_2)$ with a given above, and the vector current is conserved, while the axial current is not:

$$q_\lambda \Delta^{\lambda\mu\nu}(a, k_1, k_2) = \frac{i}{2\pi^2} \varepsilon^{\lambda\nu\sigma\tau} k_{1\sigma} k_{2\tau} \neq 0. \qquad (9.9.12)$$

Therefore the classical symmetry associated with the conservation of axial current for massless fermions is broken in the quantum theory. The mystery was uncovered by K. Fujikawa who pointed out [65] that the Jacobian of the integration measure of the path integral under the chiral phase transformation depends on the electromagnetic potential and is therefore not gauge invariant. He derived the anomaly without using perturbation theory.

In case of the fermions coupled to an electromagnetic gauge field, the Lagrangian density

$$\mathcal{L} = \overline{\psi} i \gamma^\mu (\partial_\mu - ieA_\mu) \psi \qquad (9.9.13)$$

gives rise to the axial current divergence

$$\partial_\mu J_5^\mu = \frac{e^2}{(4\pi)^2} \varepsilon^{\mu\nu\lambda\sigma} F_{\mu\nu} F_{\lambda\sigma}, \qquad (9.9.14)$$

which does not vanish, and is given the name of the chiral anomaly. The left- and right-hand currents have therefore non-vanishing divergences:

$$\partial_\mu J_L^\mu = -\frac{1}{2}\frac{e^2}{(4\pi)^2}\varepsilon^{\mu\nu\lambda\sigma}F_{\mu\nu}F_{\lambda\sigma}, \quad \partial_\mu J_R^\mu = \frac{1}{2}\frac{e^2}{(4\pi)^2}\varepsilon^{\mu\nu\lambda\sigma}F_{\mu\nu}F_{\lambda\sigma},$$

$$(9.9.15)$$

The vector current remains conserved. If the fermion is massive, then the Lagrangian density $\mathcal{L} = \overline{\psi}\ i\gamma^\mu[(\partial_\mu - ieA_\mu) - m]\psi$ gives

$$\partial_\mu J_5^\mu = 2m\overline{\psi}i\gamma_5\psi + \frac{e^2}{(4\pi)^2}\varepsilon^{\mu\nu\lambda\sigma}F_{\mu\nu}F_{\lambda\sigma}, \qquad (9.9.16)$$

the Adler-Bell-Jackiw equation. In conclusion, the quantum theory breaks the classical conservation of currents, unless one *defines* the shift of integration variable to make the vector current conserved, while the axial current acquires an "anomalous" divergence.

To convince our readers, it is helpful to remind about the story of calculating the lifetime of neutral pion. The calculation based on the conserved axial current in current algebra gave the result that the decay amplitude $\mathcal{T}$ at zero momentum transfer vanishes: $\mathcal{T}(0) = 0$. The pion decay amplitude $\mathcal{T}\ (m_\pi^2)$ should then be very much suppressed, contradicting the experimental results ($\Gamma = 7.37\,eV$). The triangular Feynman diagram with axial vector anomalous current coupled to light flavors of quarks gives the correct result, if a factor 3 is included to account for the color degrees of freedom of the quarks! But one should be cautious when the axial vector current is coupled to a gauge field. The existence of an anomaly of a local invariance can be dangerous to the theory, because the local anomaly can lead to breakdown of gauge invariance, making the theory non-renormalizable. Study of anomaly cancellation became an important part of related theories. This turns out to become a virtue, because it significantly limit the possible choice of gauge symmetries!

9.9.2   2 + 1 dimensional Dirac Field interacting with Gauge Field

In 2+1 dimensional space time the matrices involved in the theory of a relativistic fermion are three 2×2 ones, and there is no γ_5 possible.

It was a surprise that the study of $2 + 1$ dimensional systems of fermions and gauge fields brought about new anomalous phenomena, receiving much attention in the early 1980s. They are:

1. the asssociated fermion zero modes,
2. induced currents of anormal parity,
3. unusual quantum numbers.

They are manifestations of the role of topological structures in quantum physics, i.e. not all virtues of the classical theory are inherited in the quantum theory, and the emergence of anomalies is connected with topology.

In the following we introduce several papers revealing these results, with the emphasis on clarifying concepts and the steps leading to these concepts in the theoretical derivations, while avoiding some of the detailed calculations. We begin with a paper by Deser, Jackiw and Templeton [66].

Consider a fermion field coupled to a $U(1)$ gauge field in $(2+1)$-dimensions with Lagrangian density

$$\mathcal{L}(x) = \overline{\psi}(x) \left(i\gamma^\mu D_\mu - m \right) \psi(x), \quad \mu = 0, 1, 2, \tag{9.9.17}$$

where 2×2 matrices $\gamma^\mu = (\sigma^3, i\sigma^1, i\sigma^2)$ and the σ^i are the Pauli matrices, and the covariant derivatives $D_\mu = \partial_\mu - ieA_\mu$.

The parity transformation is somewhat unusual:

$$\boldsymbol{r} = (x, y), \quad \boldsymbol{r}' = (-x, y), \tag{9.9.18}$$

Here only one of the two spatial coordinates changes sign. The purpose of this transformation is to interchange the left-hand to the right-hand. The fields transform under the parity as

$$\mathcal{P}A^0 \left(\boldsymbol{r} \right) \mathcal{P}^{-1} = A^0 \left(\boldsymbol{r}' \right),$$
$$\mathcal{P}A^1 \left(\boldsymbol{r} \right) \mathcal{P}^{-1} = -A^1 \left(\boldsymbol{r}' \right), \tag{9.9.19}$$
$$\mathcal{P}A^2 \left(\boldsymbol{r} \right) \mathcal{P}^{-1} = A^2 \left(\boldsymbol{r}' \right),$$
$$\mathcal{P}\psi \left(\boldsymbol{r} \right) \mathcal{P}^{-1} = \sigma^1 \psi \left(\boldsymbol{r}' \right). \tag{9.9.20}$$

They tranform under time reversal as

$$\mathcal{T}A^0(t)\,\mathcal{T}^{-1} = A^0(-t),$$
$$\mathcal{T}\boldsymbol{A}(t)\,\mathcal{T}^{-1} = -\boldsymbol{A}(-t), \tag{9.9.21}$$
$$\mathcal{T}\psi(t)\,\mathcal{T}^{-1} = \sigma^2\psi(-t). \tag{9.9.22}$$

Under charge conjugation the fields transform as

$$\mathcal{C}A_\mu\mathcal{C}^{-1} = -A_\mu,$$
$$\mathcal{C}\psi\mathcal{C}^{-1} = \sigma^1\psi^\dagger. \tag{9.9.23}$$

It is important to notice that the mass term in the Lagrangian *changes sign* under both parity and time reversal:

$$\mathcal{P}: \overline{\psi}\psi = \psi^\dagger\sigma^3\psi \to \psi^\dagger\sigma^1\sigma^3\sigma^1\psi = -\overline{\psi}\psi,$$
$$\mathcal{T}: \overline{\psi}\psi \to \psi^\dagger\sigma^2\sigma^3\sigma^2\psi = -\overline{\psi}\psi. \tag{9.9.24}$$

The theory is $\mathcal{PT}$ and also $\mathcal{CPT}$ invariant.

9.9.3 Topologically non-trivial vacuum current with abnormal parity

In the following we follow Redlich [67, 68], where the author considers a fermion field coupled to a non-Abelian SU(2) gauge field in $2+1$ dimensions. A non-Abelian gauge field can be subjected to a gauge transformation with a winding number n (see Section 7.6.2). The purpose is to demonstrate the possible conflict between gauge invariance and the space-time reversal invariance. The result of a compromise will lead to anomaly. The functional integral of the system is

$$Z = \int d\overline{\psi}d\psi dA \exp iI\,[A, \psi], \tag{9.9.25}$$

where the action I is given by

$$I\,[A, \psi] = \int d^3x \left[tr F^2/2 + i\overline{\psi}\gamma^\mu\left(\partial_\mu - A_\mu\right)\psi\right], \quad \mu = 0, 1, 2. \tag{9.9.26}$$

The action is invariant under space-time reversal and $SU(2)$ local gauge transformations. The effective action can be obtained by integrating out the fermion degrees of freedom [67]:

$$Z = \int dA \exp\left\{ i \left(\int d^3x \, tr F^2 / 2 + I_{eff}\left[A\right] \right) \right\},$$

$$I_{eff}\left[A\right] = -i \ln \det \left(\gamma^\mu \left(\partial_\mu - A_\mu \right) \right). \tag{9.9.27}$$

Under a homotopically non-trivial gauge transformation of winding number n,

$$\det \left(\gamma^\mu \left(\partial_\mu - A_\mu \right) \right) \rightarrow (-1)^{|n|} \det \left(\gamma^\mu \left(\partial_\mu - A_\mu \right) \right),$$

the effective action becomes

$$I_{eff}\left[A\right] \rightarrow I_{eff}\left[A\right] \pm \pi \left|n\right|. \tag{9.9.28}$$

In the functional integral formulation of the theory, the gauge invariance demands that $\exp[iI_{eff}]$ can only change by an integral multiple of 2π. For even n in (9.9.28) this brings no change to $\exp[iI_{eff}]$, but for odd n this changes its sign. This is unacceptable. One attempts to use gauge invariant regularization (e.g. Pauli-Villars [63]

$$I_{eff}^R\left[A\right] = \lim_{M \to \infty} \left(I_{eff}\left[A\right] - I_{eff}\left[A, M\right] \right) \tag{9.9.29}$$

leading eventually to

$$I_{eff}^R\left[A\right] = I_{eff}\left[A\right] \pm W\left[A\right], \tag{9.9.30}$$

where $W\left[A\right]$ is the Chern-Simons second characteristic class

$$W\left[A\right] = \frac{e^2}{8\pi^2} \int d^3x \varepsilon^{\mu\alpha\beta} \left[A_\mu^a F_{\alpha\beta}^a - \frac{e}{3} \varepsilon^{abc} A_\mu^a A_\alpha^b A_\beta^c \right], \tag{9.9.31}$$

which changes by $\pm \pi \left|n\right|$ under gauge transformation and therefore can cancel the change of I_{eff}. But the price to pay is that W gives rise to an *additional term* of the ground state current

$$\langle j_a^\mu \rangle = \frac{\delta W[A]}{\delta A_{\mu a}} = \pm c \frac{e}{8\pi} \varepsilon^{\mu\alpha\beta} F_{\alpha\beta}^a;$$

$$c = 1(QED), \quad c = \frac{1}{2}(SU(2)) \tag{9.9.32}$$

This current is parity abnormal, because it it a *pseudovector*. The $\pm$ sign originates from the sign of the Pauli-Villar regulator mass M in (9.9.29), when the limit of infinity is taken, the mass disappears, but the sign survives. For QED and a static magnetic field B,

$$\langle j^0 \rangle = \pm \frac{e}{4\pi} B, \quad Q = \int d^2x \langle j^0 \rangle = \pm \frac{e}{4\pi} \Phi. \tag{9.9.33}$$

For a flux quantum, $\Phi = 2\pi$, the induced charge is equal to $e/2$, a fraction of charge appears. The anomaly in $2+1$D appears as a parity non-conserving topological vacuum current, which can carry fractional charge. In $3+1$D the anomaly appears in the divergence of axial current. In both cases the anomaly causes a physical ground state current to violate a symmetry of the action, a spontaneously broken symmetry.

It seems that the $2+1$D anomaly arises because of the gauge transformation with winding numbers. Actually begin with QED of massless fermions in 2+1D the calculation of vacuum current involves divergence, and the gauge invariant regularization also gives rise to the topological term. Using non-Abelian gauge theory, the author aimed at showing the odd-even effect of the gauge transformation.

9.9.4   Relation between fractional charge, vacuum current and zero mode

The following discussion is based on a paper by R. Jackiw [70]. For static background fields in the temporal gauge the Dirac Hamiltonian possesses a conjugate-symmetric spectrum with zero mode if the field satisfies certain requirements. The massless Dirac Hamiltonian H is

$$H = \boldsymbol{\alpha} \cdot (\boldsymbol{p} - e\boldsymbol{A}), \quad \alpha_1 = -\sigma_2, \quad \alpha_2 = \sigma_1. \tag{9.9.34}$$

The matrix $\beta = \sigma_3$ is used as the conjugation matrix. The eigenequation is

$$\boldsymbol{\alpha} \cdot (\boldsymbol{p} - e\boldsymbol{A})\psi_E = E\psi_E \qquad (9.9.35)$$

Since $\{H, \sigma_3\} = 0$, we have

$$H\sigma_3\psi_E = -E\sigma_3\psi_E,$$

i.e. $\sigma_3\psi_E = \psi_{-E}$ is the charge conjugate state. Charge conjugation symmetry is the necessary condition for the existence of zero modes, and it holds when the mass term is absent. Assume that zero mode ψ_0 exists:

$$\boldsymbol{\alpha} \cdot (\boldsymbol{p} - e\boldsymbol{A})\psi_0 = 0, \quad \psi_0 = \begin{pmatrix} u \\ v \end{pmatrix} \qquad (9.9.36)$$

The vector potential in the Coulomb gauge is assumed to be single-valued and well behaved at the origin. We can thus introduce a so that

$$A^i = \varepsilon^{ij}\partial_j a, \quad \boldsymbol{B} = \hat{z}B, \quad B = -\nabla^2 a, \qquad (9.9.37)$$

Dirac equation becomes decoupled:

$$(\partial_x + i\partial_y)u - e(\partial_x + i\partial_y)au = 0,$$
$$(\partial_x - i\partial_y)v + e(\partial_x - i\partial_y)av = 0. \qquad (9.9.38)$$

We can then form self-conjugate solutions $\begin{pmatrix} u \\ 0 \end{pmatrix}$ and $\begin{pmatrix} 0 \\ v \end{pmatrix}$. Since $(\partial_x \pm i\partial_y)(x \pm iy) = 0$, it is obvious that the decoupled equations (9.9.38) have solutions

$$u = \exp(ea) f(x + iy),$$
$$v = \exp(-ea) g(x - iy), \qquad (9.9.39)$$

where f and g are arbitrary entire functions. Whether these are acceptable solutions depends on whether they are normalizable, which in turn depends on the large r behavior of a. If a grows sufficiently rapidly at large distance, then *either* u or v will be normalizable, and there exist one or more isolated zero-energy bound states, the multiplicity depending on how many different forms for

f or g may be taken. It is useful to classify the various possibilities in terms of the total flux, which is also proportional to the total induced charge:

$$\langle j^0 \rangle = \pm \frac{e}{4\pi} B,$$

$$Q = \int d^2 r \, \langle j^0 \rangle = \pm \frac{e}{4\pi} \int d^2 r \, B = \pm \frac{e}{2} \Phi,$$

$$\Phi = \frac{1}{2\pi} \int d^2 r \, B = -\frac{1}{2\pi} \int d\mathbf{r} \times \mathbf{\nabla} a. \qquad (9.9.40)$$

The last integral is over the circle at infinity. Since the potential is single valued, the flux may also be presented as

$$\Phi = -\frac{1}{2} \int_0^{2\pi} d\theta \, r \frac{\partial}{\partial r} a \bigg|_{r=\infty} \qquad (9.9.41)$$

When Φ vanishes because a goes to zero at large distances, the modes u and v (9.9.39) are not normalizable. On the other hand, nonvanishing flux, arising from the persistence of the vector potential at infinity will allow normalization of (9.9.39), and isolated zero-energy modes are present. Consider the case of constant magnetic field in symmetric gauge and Landau gauge respectively:

$$a^{\mathrm{I}} = -\frac{1}{4} r^2 B,$$

$$a^{\mathrm{II}} = \frac{1}{2} x^2 B, \qquad (9.9.42)$$

In the former case $exp(ea)$ is square integrable, and f may be any integer power. Thus the zero-energy states are infinitely degenerate:

$$\psi_{0(\pi)}^{\mathrm{I}} = e^{-eBr^2/4} r^n e^{in\theta}, \qquad n = 0, 1, \ldots . \qquad (9.9.43)$$

These are the Landau levels with quantized angular momentum. In the latter case, we find $exp(ea)$ to be square integrable in x, and f must be chosen to be continuum normalizable in y. Thus again we find infinite degeneracy, parametrized by a continuous variable k:

$$\psi_{0(k)}^{II} = e^{-eBx^2/2_e k(x+iy)}, \qquad (9.9.44)$$

These are the Landau levels with quantized linear momentum in the y-direction. Finite flux is obtained in a vortex magnetic field

$$a \underset{r\to\infty}{\sim} - \Phi \ln r. \qquad (9.9.45)$$

Assuming $\Phi > 0$, $u \sim_{r\to\infty} r^{-e\Phi}$ is normalizable for $e\Phi > 1$. Also f may be any integer power less than $e\Phi - 1$. Thus there are $[e\Phi - 1]$ normalized zero-energy states:

$$\psi_{0(n)} = \exp(ea)r^n e^{in\theta}, \quad n = 0, 1, \ldots, [e\Phi - 1], \qquad (9.9.46)$$

where $[\nu]$ represents the largest integer less than ν. When $e\Phi = N$, there are $N - 1$ zero modes. The mode $e\Phi = 1$ is not normalizable because of the measure. By compactifying the manifold R_2 to S_2 this can be avoided, and the number of zero modes for $e\Phi = N$ becomes N. This satisfies the Atiyah-Singer index theorem that the number of zero modes is determined by the topological properties of the background fields.

   Jackiw [70] concluded: "Thus we have closed the circle between zero modes, unexpected quantum numbers, and vacuum currents in $2 + 1$ dimensions. The signal for topologically interesting effects is nonvanishing flux, and its magnitude measures the degeneracy of the zero modes." However, he warned that the electron is massive, so that one should be cautious in applications. In the next subsection we study a model due to G. Semenoffwhere the Dirac particle mass is "generated".

9.9.5   A condensed matter simulation of the $2 + 1$ D anomaly

Semenoff [71] considered the tight-binding description of massive electrons on a planar honeycomb lattice. Haldane did the same [72], but with a further purpose. The band structure exhibits two degeneracy points per Brillouin zone and the low-energy behavior is obtained in the continuum limit around these two points, where two species of relativistic $(2+1)$-dimensional fermions emerge. Then, in accordance with general considerations, external magnetic fields induce zero modes in the electron energy spectrum, and the resulting degenerate ground states are fractionally charged. The Bravais lattice is triangular and contains two sites, A and B, described in Fig. 9.31. We consider a slightly more general diatomic system by assuming that sites A and B are occupied by distinct types of atoms. We parametrize the difference of energies of electrons localized on A and B by $\pm\Delta$. The purpose is to generate a mass for the initially massless electrons. By setting Δ to zero the graphene model is recovered.

Basis vectors connect the site B with the neighboring sites A are

$$\boldsymbol{a}_1 = \left(\frac{1}{2}, \frac{1}{2\sqrt{3}}\right) a$$

$$\boldsymbol{a}_2 = \left(-\frac{1}{2}, \frac{1}{2\sqrt{3}}\right) a$$

$$\boldsymbol{a}_3 = \left(0, \frac{-1}{\sqrt{3}}\right) a$$

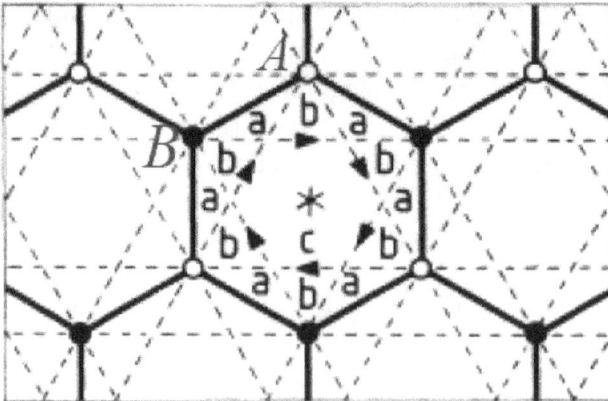

Fig. 9.31   The honeycomb lattice as a superposition of two inter-penetrating triangular sublattices. Open (solid) points represent A and B sites. Reprinted with permission from F. D. M. Haldane, *Phys. Rev. Lett.* 61, 2015, 1988. Copyright (2017) by the American Physical Society.

where a is the lattice spacing. The three vertices of a sublattice are connected by

$$b_1 = \left(-\frac{1}{2}, \frac{\sqrt{3}}{2}\right) a = a_2 - a_3$$

$$b_2 = \left(-\frac{1}{2}, -\frac{\sqrt{3}}{2}\right) a = a_3 - a_1$$

$$b_3 = (1,0)\, a = a_1 - a_2.$$

The notation for the vectors is taken from [72]. The Brilluoin zone is a hexagon in momentum space shown in Figure 9.32. It has two sets of equivalent corners marked by red and blue points. Two of the representatives are $K_1\left(\frac{4\pi}{3\sqrt{3}a}, 0\right)$ and $K_2\left(-\frac{4\pi}{3\sqrt{3}a}, 0\right)$.

In the tight-binding approximation and where only nearest-neighbor interactions are retained, the Hamiltonian is

$$H = t \sum_{j\in B, m=1,2,3} \left[c_{j+a_m}^\dagger\, c_j + h.c.\right] + \Delta \sum_{i\in A, j\in B} \left[c_i^\dagger c_i - c_j^\dagger\, c_j\right].$$

$$(9.9.47)$$

Here $c_j^\dagger$ and c_j are the creation and destruction operators for electrons localized on sites j, t is the hopping parameter between neighboring

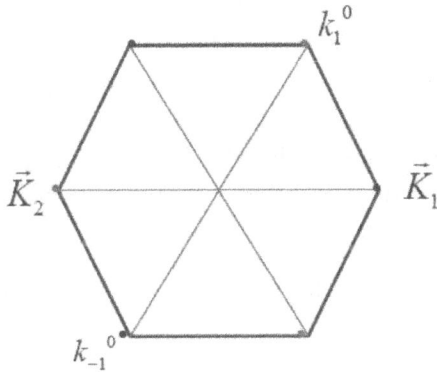

Fig. 9.32   The Brillouin zone with two sets of equivalent corners marked by red and blue points. Two of the representatives are K_1 and K_2.

sites. For the moment, we ignore the electron spin. Using the Fourier transform

$$c_i = \sum_k e^{i\boldsymbol{k}\cdot\boldsymbol{r}_i} \psi_A(\boldsymbol{k}), \quad c_j = \sum_k e^{i\boldsymbol{k}\cdot\boldsymbol{r}_j} \psi_B(\boldsymbol{k}) \qquad (9.9.48)$$

in (9.9.47) results in

$$H = \sum_k \left(\psi_A^\dagger(\boldsymbol{k}), \psi_B^\dagger(\boldsymbol{k}) \right) \begin{pmatrix} \Delta & t\sum_j e^{i\boldsymbol{k}\cdot\boldsymbol{a}_j} \\ t\sum_j e^{-i\boldsymbol{k}\cdot\boldsymbol{a}_j} & -\Delta \end{pmatrix} \begin{pmatrix} \psi_A(\boldsymbol{k}) \\ \psi_B(\boldsymbol{k}) \end{pmatrix}$$

$$= \sum_k \left(\psi_A^\dagger(\boldsymbol{k}), \psi_B^\dagger(\boldsymbol{k}) \right) h_0\left(\boldsymbol{k}\right) \begin{pmatrix} \psi_A(\boldsymbol{k}) \\ \psi_B(\boldsymbol{k}) \end{pmatrix}. \qquad (9.9.49)$$

where $h_0\left(\boldsymbol{k}\right)$ can be rewritten as

$$h_0\left(\boldsymbol{k}\right) = i\sum_i \left[\cos\left(\boldsymbol{k}\cdot\boldsymbol{a}_i\right)\sigma_1 - \sin\left(\boldsymbol{k}\cdot\boldsymbol{a}_i\right)\sigma_2 \right] + \Delta\sigma_3. \qquad (9.9.50)$$

The Hamiltonian can easily be diagonalized to give the energy eigenvalues

$$E(\boldsymbol{k}) = \pm \left(\Delta^2 + t^2 \sum_i \left| e^{i\boldsymbol{k}\cdot\boldsymbol{a}_i} \right|^2 \right)^{\frac{1}{2}},$$

which is manifestly charge conjugation symmetric. With one electron per site (one electron per unit cell per spin degree of freedom), the negative-energy states (valence band) are filled and the positive-energy states (conduction band) are empty. The separation of the conduction and valence bands is minimal at the zeros of $\sum_i \exp\left(e^{i\boldsymbol{k}\cdot\boldsymbol{a}_i} \right)$. These occur for instance at $\boldsymbol{K}_1 = \left(\frac{4\pi}{3\sqrt{3}a}, 0 \right)$ and $\boldsymbol{K}_2 = \left(-\frac{4\pi}{3\sqrt{3}a}, 0 \right)$ as can be immediately verified, and all other equivalent points, which are the corners of the Brillouin zone. In the continuum (low energy) limit ($a \to 0$), only electron states near the Brilluoin zone corners, e.g. K_1 and K_2 participate in the dynamics. We obtain two species of fermions as linear superpositions of ψ_A and

ψ_B. Close to K_1 we set $\boldsymbol{q}_1 = \boldsymbol{k} - \boldsymbol{k}_1$ and for $|\boldsymbol{q}|$ small we have

$$\sum_i \cos \boldsymbol{k} \cdot \boldsymbol{a}_i \simeq \frac{3}{2} q_x, \qquad \sum_i \sin \boldsymbol{k} \cdot \boldsymbol{a}_i \simeq \frac{3}{2} q_y,$$

Consequently, close to K_1 we have

$$h_0^{(1)}(\boldsymbol{q}) \approx \begin{pmatrix} \Delta & -\frac{3}{2}tq_+ \\ -\frac{3}{2}tq_- & -\Delta \end{pmatrix}. \tag{9.9.51a}$$

Similarly, close to K_2 we have

$$h_0^{(2)}(\boldsymbol{q}) \approx \begin{pmatrix} \Delta & -\frac{3}{2}tq_- \\ -\frac{3}{2}tq_+ & -\Delta \end{pmatrix}. \tag{9.9.51b}$$

This is the species doubling found by G. Semenoff [71]. They have identical energy spectrum:

$$E_\pm = \pm \left[\Delta^2 + \frac{9}{4}t^2 q^2 \right]^{\frac{1}{2}} \tag{9.9.52}$$

Under the space inversion (9.9.18), (9.9.20) we have

$$q_+ = q_x + iq_y \rightarrow -q_-, \tag{9.9.53}$$

and consequently

$$h_0^{(1)}(\boldsymbol{q}) \rightarrow h_0^{(2)}(\boldsymbol{q}). \tag{9.9.54}$$

They interchange under parity. Under time reversal

$$h_0^{(1,2)}(\boldsymbol{q}) \rightarrow h_0'^{(1,2)}(-\boldsymbol{q}) = h_0^{(1,2)}(\boldsymbol{q}), \tag{9.9.55}$$

both possess time reversal invariance. Under charge conjugation symmetry,

$$\sigma_3 h_0^{(1,2)} \sigma_3 = h_0^{(1,2)}, \tag{9.9.56}$$

both are charge conjugation symmetric. This symmetry leads to the existence of zero modes at the corners of the Brillouin zone when

$\Delta \to 0$. The separately conserved induced currents are

$$j_1^\mu = \left\langle \frac{1}{2} \left[\overline{\psi}_1, \gamma^\mu \psi_1 \right] \right\rangle = \frac{e}{4\pi} \varepsilon^{\mu\nu\lambda} F_{\nu\lambda}(x) \ \text{sgn} \ (m) + \cdots,$$

$$j_2^\mu = \left\langle \frac{1}{2} \left[\overline{\psi}_2, \gamma^\mu \psi_2 \right] \right\rangle = -\frac{e}{4\pi} \varepsilon^{\mu\nu\lambda} F_{\nu\lambda}(x) \ \text{sgn} \ (m) + \cdots.$$

where we have included a factor of 2 for the spin degeneracy. Therefore, because of the fermion doubling, the electric current $j_+^\mu = j_1^\mu + j_2^\mu$ vanishes. The odd combination is nonzero:

$$j_-^\mu = j_1^\mu - j_2^\mu = \frac{e}{2\pi} \varepsilon^{\mu\nu\lambda} F_{\nu\lambda}(x) sgn \ (m) + \cdots \qquad (9.9.57)$$

Thus the model gives successfully the induced current with abnormal parity (9.9.32). There the general considerations (the Atiyah-Singer index theorem) imply that in an external magnetic field with flux $\Phi = n$, there are $2n$ zero-energy modes. If we take the electron spin into account, each of these bound states is a spin doublet. The ground state is then $(4n + 1)$-fold degenerate: the charges of these states are $Q = -2e\,|\Phi|,\ -2e\,|\Phi| + e, \ldots 2e\,|\Phi|$. The spinless state (all spins paired) with charge $Q = \pm 2e\,|\Phi|$ is easily obtained by the introduction of a small positive (negative) chemical potential which guarantees that all zero-energy levels are occupied (unoccupied). Alternatively, if the electrons are subject to the Zeeman interaction which has the effect of raising the energy of those whose spins are aligned parallel with a uniform external magnetic field and of lowering the energy of those antialigned, the system would be uncharged but would have $2n$ unpaired spins. The latter scenario would be observable as an anisotropic electron spin-resonance amplitude [71]. In the massless limit $\beta \to 0$ or $m \to 0$ we recover the graphene model. The Hamiltonian then describes two species of massless Dirac particle, represented in the energy band diagram as Dirac cones at the corners of the Brilluoin zone, shown in Figure 9.33.

Haldane discussed in his model the quantum Hall effect. We discussed in 9.4.1 the topological significance of the integral quantum Hall conducance (Thouless, Kohmoto, Nightingale, den Nijs, TKNN for short). We have a set of N occupied Bloch wave functions $\{u_m(\boldsymbol{k})\}$

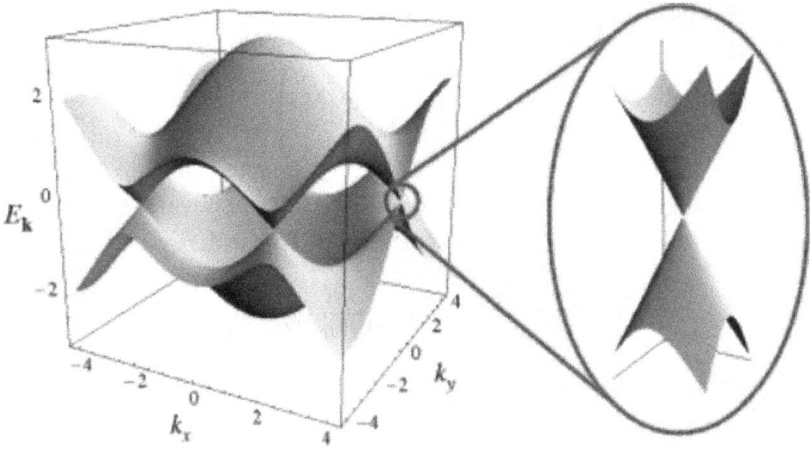

Fig. 9.33   Dirac cones at the corners of the Brilluoin zone.

of the valence band which define the many body ground state. The number n in the quantum Hall conductance is the first Chern number of the Brilluoin zone torus

$$n = \frac{1}{2\pi} \int_{BZ} d\boldsymbol{S} \cdot \mathcal{F},$$

$$\mathcal{F} = \nabla \times \mathcal{A}, \qquad (9.9.58)$$

$$\mathcal{A} = \sum_{m=1}^{N} u_m(\boldsymbol{k}) |i\nabla_k| u_m(\boldsymbol{k}),$$

where $\mathcal{F}$ and $\mathcal{A}$ are Berry curvature and Berry connexion respectively. Now we attempt to calculate the contribution from the low energy levels at the corners of the Brilluoin zone. The Hamiltonian close to the corner $\boldsymbol{K}_1$ is given by (9.9.51a),

$$h_0^{(1)}(q) \approx \begin{pmatrix} \Delta & -\frac{3}{2}tq_+ \\ -\frac{3}{2}tq_- & -\Delta \end{pmatrix} \simeq -\frac{3}{2}t\boldsymbol{q} \cdot \boldsymbol{\sigma} + \Delta \cdot \sigma_3, \qquad (9.9.59)$$

the energy of the valence band levels is

$$E_{valence} = -\sqrt{9t^2q^2/4 + \Delta^2} \equiv -\lambda, \qquad (9.9.60)$$

The corresponding eigenstates are

$$\psi_-^{(1)} = \frac{1}{\sqrt{N}} \begin{pmatrix} -\frac{3}{2}tq_+ \\ \Delta + \lambda \end{pmatrix},$$

with N a normalization factor. The Berry curvatures acquires contributions from the distinct corners at K_1 and K_2 as

$$\Omega_-^{(1)} = -\Omega_-^{(2)} = -\frac{9t^2q^2/4}{2\left(9t^2q^2/4 + \Delta^2\right)}. \tag{9.9.61}$$

The corresponding contributions to the first Chern number are

$$\frac{1}{2\pi} \int d^2q \, \Omega^{(1,2)}(q) = \mp \frac{1}{2}, \tag{9.9.62}$$

with a zero sum,

$$n = \frac{1}{2\pi} \sum_{i=1,2} \int d^2q \, \Omega^{(1,2)}(q) = 0.$$

The Hall conductance vanishes as it should be, because it is time reversal odd, and the model is time reversal invariant.

9.9.6   Anomalous quantum Hall effect

It was believed that the occurence of the quantum Hall effect is associated with the Landau levels under presence of a uniform external field. Haldane showed [72] that a quantum Hall effect may also result from breaking of time-reversal symmetry (i.e., magnetic ordering) without any net magnetic fiux through the unit cell of a periodic 2D system. In this case, the electron states retain their usual Bloch state character. Haldane introduced a second real hopping term t_2 between second-neighbor sites (i.e., between nearest-neighbor sites on the same sublattice). Hereafter we redefine the hopping between the nearest neighbor as t_1. To break time-reversal invariance, Haldane added a periodic local magnetic-flux density $B(r)$ in the z direction normal to the 2D plane, with the full symmetry of the lattice, and with zero total flux through the unit cell (Fig. 9.33). Since the net flux per unit cell vanishes, the vector potential $A(r)$ can be chosen to be periodic. The effect of this local field is to multiply the matrix element for hopping between next nearest neighbouring sites by the $U(1)$ phase factor $\exp[i(e/\hbar)\int \boldsymbol{A} \cdot d\boldsymbol{r}]$ where the integral is along the hopping path, so that forward and backward hopping differ by a phase factor $\exp(\pm i\phi)$. In the case when the local fluxes are arranged as in Fig 9.34, we have $\phi = -2\pi\Phi/3\Phi_0$. The forward direction is defined by the arrow in the Fig 9.31. Introducing t_2 leads to new terms in the Hamiltonian

$$2t_2 \cos\phi \left[\sum_i \cos \boldsymbol{k} \cdot \boldsymbol{b}_i\right] I - 2t_2 \sin\phi \left[\sum_i \sin \boldsymbol{k} \cdot \boldsymbol{b}_i\right] \sigma_3.$$

The last terms combines with the mass term Δ in the original Hamiltonian, and consequently

$$\begin{aligned} H\left(\boldsymbol{k}\right) = {} & 2t_2 \cos\phi \left[\sum_i \cos \boldsymbol{k} \cdot \boldsymbol{b}_i\right] I \\ & + t_1 \sum_i \left[\cos\left(\boldsymbol{k} \cdot \boldsymbol{a}_i\right)\sigma_1 + \sin\left(\boldsymbol{k} \cdot \boldsymbol{a}_i\right)\sigma_2\right] \\ & + \left[\Delta - 2t_2 \sin\phi \left[\sum_i \sin \boldsymbol{k} \cdot \boldsymbol{b}_i\right]\right] \sigma_3. \end{aligned} \qquad (9.9.63)$$

We expand the Hamiltonian in the neighborhood of the band extrema at the zone corners $\boldsymbol{k}_\alpha^0$ to linear order in $\delta\boldsymbol{k} = \boldsymbol{k} - \boldsymbol{k}_\alpha^0$, and make the Landau-Peierls substitution $\hbar\delta\boldsymbol{k} \rightarrow \boldsymbol{\Pi}$ in order to include electromagnetic fields. Here $\boldsymbol{\Pi} = (\Pi^x, \Pi^y)$ is the dynamical

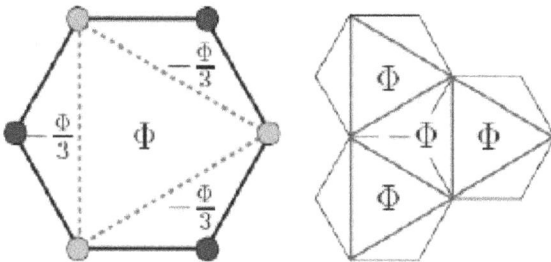

Fig. 9.34   Local fluxes with vanishing total flux through a unit cell.

momentum with components

$$\Pi^{x,y} = P_{x,y} - eA_{x,y} \tag{9.9.64}$$

satisfying the commutation relation

$$[\Pi^x, \Pi^y] = i\hbar e B_0. \tag{9.9.65}$$

where B_0 is the weak magnetic field in the z-direction, For week B_0, coupling between the two distinct zone corners can be neglected. The two independent Hamiltonians H_α are

$$H_\alpha = c\left(\Pi_\alpha^x \sigma_2 - \Pi_\alpha^y \sigma_1\right) + m_\alpha c^2 \sigma_3, \tag{9.9.66}$$

where $c = \frac{3}{2} t_1 |a_i| / \hbar$ and $m_\alpha c^2 = \Delta - 3\sqrt{3}\, \alpha t_2 \sin\phi$; Π_α^x and Π_α^y are hermitian operators pertaining to the distinct corners of the Brilluoin zone defined through the factor $e^{-i\boldsymbol{k}_\alpha^0 \cdot \boldsymbol{a}_i}$ by

$$\Pi_\alpha^x + i\Pi_\alpha^y = \frac{2}{3} \sum_i e^{-i\boldsymbol{k}_\alpha^0 \cdot \boldsymbol{a}_i} \left(\boldsymbol{a}_i \cdot \boldsymbol{\Pi}\right) \frac{1}{|\boldsymbol{a}_i|}, \tag{9.9.67}$$

and they satisfy the commutation relation

$$[\Pi_\alpha^x, \Pi_\alpha^y] = i\alpha\hbar e B_0. \tag{9.9.68}$$

Comparing with equation (9.9.34), we see that (9.9.66) is the Hamiltonian for a massive $(2+1)$-D "Dirac" particle interacting with a magnetic field, for which the "light" velocity c and the mass m_α are defined in terms of the lattice. The energy spectrum of (9.9.66) for $B_0 = 0$ is that of a "relativistic" particle:

$$\epsilon_{\alpha\pm}\left(\boldsymbol{k}\right) = \pm\left[(\hbar c k)^2 + \left(m_\alpha c^2\right)^2\right]^{\frac{1}{2}}. \tag{9.9.69}$$

This is a parallel of (9.9.52). For non-vanishing B_0, *relativistic* Landau levels are obtained as follows:

$$\epsilon_{\alpha n\pm} = \pm\left[\left(m_\alpha c^2\right)^2 + n\hbar |eB_0| c^2\right]^{\frac{1}{2}}, \quad n = 1, 2, \ldots$$

$$\epsilon_{\alpha 0} = \alpha m_\alpha c^2 \mathrm{sgn}\left(eB_0\right). \tag{9.9.70}$$

Every $n \geq 1$ level that evolves out of the upper band as B_0 is turned on is balanced by a level that evolves from the lower band. However, the $n = 0$ "zero-mode" energy (actually it is not 0) is not symmetric under $B_0 \to -B_0$. It evolves from the *upper* band if $\alpha m_\alpha e B_o$ is *positive*, and from the *lower* band if it is *negative*.

In the time-reversal symmetric case, $t_2 \sin \phi = 0$, the two masses m_+ and m_- are equal $m_+ = m_-$, and the sum of the Landau-level spectra derived from the two distinct zone corners is particle-hole symmetric, and invariant under $B_0 \to -B_0$. In this case, $\sigma^{xy} = 0$ by time-reversal invariance, as we obtained in the last subsection. When $\Delta = 0$ we have $m_\alpha c^2 = -3\sqrt{3}\alpha t_2 \sin \phi$. and $m_+ = -m_-$. We can already expect that for this case we would have $\Omega_-^{(1)} = \Omega_-^{(2)}$ and $n = 1$.

Haldane proceeded [72] to calculate the charge density induced by the magnetic field B_0, by turning on continuously the external field, at the same time varying the Fermi level so at all times it lies in a gap. Comparison of the occupation numbers of the Landau levels obtained this way with those obtained by continuously applying the field to the time-reversal invariant system $(m_+ = m_-)$ shows that they differ by the complete filling of one Landau level. The application of a weak external magnetic field to a system where $m_+ = -m_-$ induces an extra field dependent ground state charge density $\pm e^2 B_0/h$ relative to the field independent charge density when these parameters have the same sign. By taking its derivative with respect to B_0 the Hall conductance is obtained. The result is $\nu e^2/h$, and $\nu = \frac{1}{2}[\text{sgn}(m_-) - \text{sgn}(m_+)]$. We summarize: in the time reversal invariant but parity violating case, m_+ and m_- have the same sign, we have normal semiconductor/insulator phase, while in the opposite (parity invariant but time reversal violating) case, m_+ and m_- have opposite sign, we have a topologically non-trivial quantum Hall phase. The occurence of Hall conductance without Landau level and in absence of an extrernal magnetic field is termed the anomalous Hall effect.

It seems extremely difficult to physically realize the local flux arrangement in Haldane model. The group led by Q.K. Xue [73] reported the observation of the quantum anomalous Hall effect in thin films of Cr-doped $(Bi,Sb)_2Te_3$, a magnetic topological insulator. At zero magnetic field, the anomalous Hall resistance reaches the

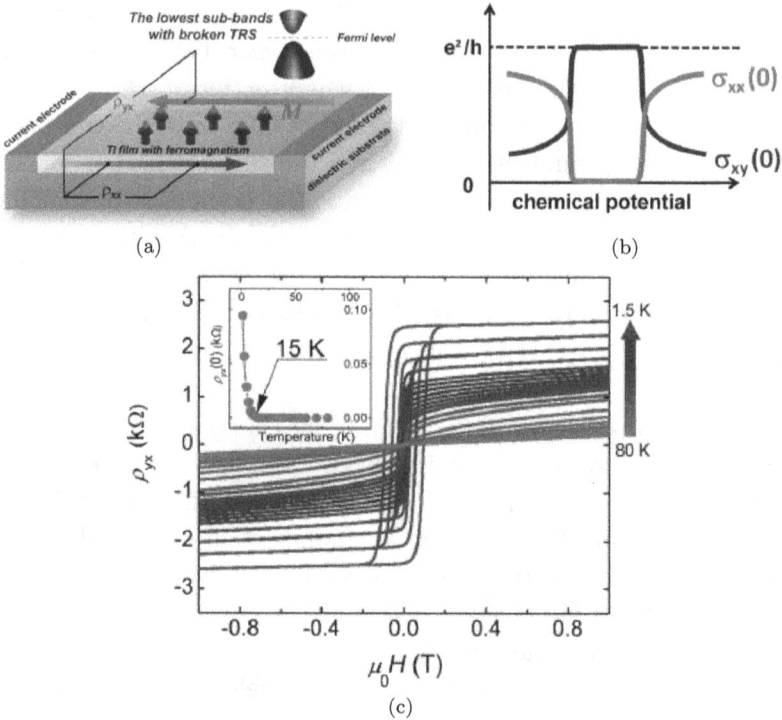

(a)

(b)

(c)

Fig. 9.35   Observation of the quantum anomalous quantum Hall effect. (a) A schematic drawing depicting the principle of the quantum anomalous Hall effect in a TI thin film with ferromagnetism. (b) A schematic drawing of the expected chemical potential dependence of zero field σ_{xx} (in red) and σ_{xy} (in blue). (c) Magnetic field dependence of ρ_{yx} curves of the $Cr_{0.15}(Bi_{0.1}Sb_{0.9})_{1.85}Te_3$ film measured at different temperatures (from $80\,K$ to $1.5\,K$). The inset shows the temperature dependence of zero field ρ_{yx}, which indicates a Curie temperature of $\sim 15\,K$, From [73].

predicted quantized value of h/e^2, accompanied by a considerable drop of the longitudinal resistance. Here the time reversal invariance is broken by doping magnetic particles in topological insulators.

9.10   Topology in Condensed Matter Physics, the Topological Insulators and Superconductors

In physics phases can be understood using Landau's approach, which characterizes states in terms of underlying symmetries that are spontaneously broken. For example, a crystalline solid breaks translation symmetry, a magnet breaks rotation symmetry. A superconductor breaks the more subtle gauge symmetry. The pattern of symmetry breaking leads to a unique order parameter, which assumes a nonvanishing expectation value only in the ordered state. The study of the quantum Hall effect has led to a different classification paradigm based on the notion of topological order. The state responsible for the quantum Hall effect does not break any symmetries, but it defines a topological phase in the sense that certain fundamental properties such as the quantized value of the Hall conductance and the number of gapless boundary modes are insensitive to smooth changes in material parameters and cannot change unless the system passes through a quantum phase transition. The key concept of topology in physics is that of a "smooth deformation". In mathematics, one considers smooth deformations of shapes without the violent action of creating a hole. In physics, one can consider general Hamiltonians of many-particle systems with an energy gap separating the ground state from the excited states. In this case, one can define a smooth deformation as a change in the Hamiltonian which does not close the bulk gap. If we put in contact two pieces of materials belonging to the same topological class, the interface between them does not need to support gapless states. On the other hand, if we put in contact two quantum states belonging to different topological classes, or put a topologically nontrivial state in contact with the vacuum, the interface must support gapless states. Under certain driving forces edge currents emerges. In Section 9.4 we discussed the edge current flowing along the edge of a circular ring, i.e. the interface between the quantum Hall material and the vacuum. Importantly, the edge currents are chiral in the sense that they propagate in one direction only along the edge. We explained this in terms of the semi-classical cyclotron orbits in Section 9.4 (Fig. 9.36a). The chiral current in the energy band diagram is shown in Fig. 9.36b. The edge currents are non-dissipative: they cannot be scattered by a non-magnetic impurity, beause there are no states available for backscattering (see Fig. 9.37a).

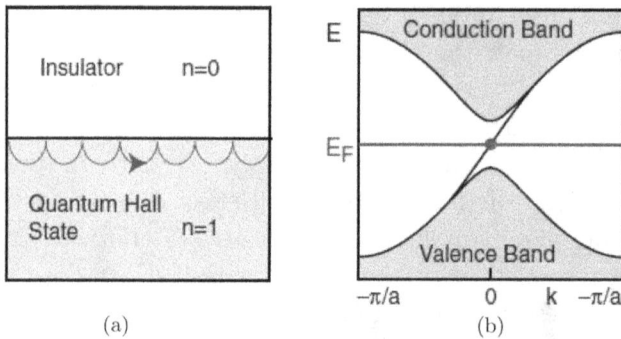

Fig. 9.36   (a) Skippping cyclotron orbits on the edge, (b) a single edge state connects the conduction band with the valence band. Reprinted with permission from M. Z. Hasan and C. L. Kane, *Rev. Mod. Phys.* 82, 3045, 2010. Copyright (2017) by the American Physical Society.

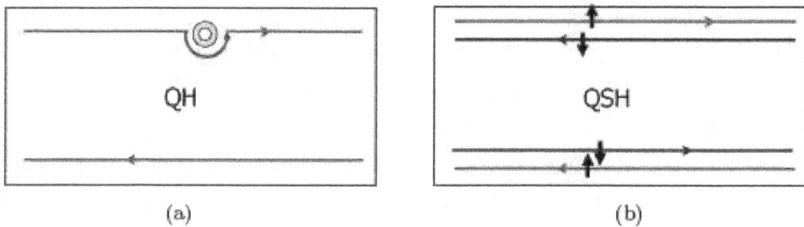

Fig. 9.37   Edge currents in a quantum Hall system (a) and a quantum spin Hall system. Reprinted with permission from Xiao-Liang Qi and Shou-Cheng Zhang, *Rev. Mod. Phys.* 83, 1057, 2001. Copyright (2017) by the American Physical Society.

By changing the Hamiltonian near the surface the dispersion of the edge states can be modified. For instance, $E(k)$ could develop a kink (as shown in the following figure)

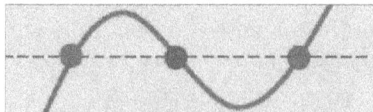

so that the edge states intersect E_F three times — twice with a positive group velocity and once with a negative group velocity. The difference, $N_R - N_L$, between the number of right and left moving modes, however, cannot change and is determined by the topological structure of the bulk states. This is summarized by the bulk-boundary correspondence, the topological structure of the bulk states. This is summarized by the bulk-boundary correspondence

$$N_R - N_L = \Delta n$$

where Δn is the difference in the Chern number across the interface.

9.10.1 The quantum spin Hall insulator, topological classifications

In section 9.8 we introduced the developments of the quantum spin Hall effect from the band theoretical point of view. These new quantum states belong to a class which is invariant under time reversal, with spin-orbit coupling playing an essential role. All time reversal invariant insulators in nature (without ground state degeneracy) fall into two distinct classes, classified by a $\mathcal{Z}_2$ topological order parameter. The topologically nontrivial state (the topological insulator) has a full insulating gap in the bulk, but has gapless edge or surface states consisting of an odd number of Dirac fermions. The 2D topological insulator is synonymously called the quantumspin Hall insulator. There exist excellent review articles on this subject, for instance that by M.Z. Hasan and C.L. Kane [74] and by Xiao-Liang Qi and Shou-Cheng Zhang [75]. In the following we are concerned with some of the most basic concepts related to quantum mechanics.

Suppose in the quantum Hall system the electrons are spin-polarized, the edge currents do not involve the spin. The forward and backward movers exist on the upper and lower edge of a Hall bar separately (Fig. 9.37a). In the quantum spin Hall system both spins of the electrons take part such that the edge currents consist of electrons of both spin directions. The upper edge supports a forward mover with spin up and a backward mover with spin down, and conversely for the lower edge (Fig. 9.37b).

Since the Hall conductance is odd under time reversal, the topologically nontrivial Hall states described can only occur when time reversal symmetry is broken. However, the spin-orbit interaction allows a different topological class of insulating band structures when time reversal symmetry is unbroken while the spin-orbit interaction plays the critical role [78]. Let us consider the time reversal symmetry for spin 1/2 particles, for which the time reversal transformation is represented by the antiunitary operator

$$\Theta = e^{i\pi S_y/\hbar}K, \tag{9.10.1}$$

where S_y is the spin component operator and K is the complex conjugation. An antiunitary operator is defined by the following relation

$$(\Theta\Psi, \Theta\Phi) = (\Psi, \Phi)^* \tag{9.10.2}$$

We observe that the exponential factor in (9.10.1) is actually real such that for spin 1/2 electrons

$$\Theta^2 = e^{i2\pi S_y/\hbar} = -1. \tag{9.10.3}$$

For a time reversal invariant system the following relation holds for an antiunitary operator Θ [77]

$$[\mathcal{H}, \Theta] = 0. \tag{9.10.4}$$

Consequently for a stationary state χ of a time reversal invariant Hamiltonian with energy ε_χ we have

$$\mathcal{H}\Theta\chi = \Theta\mathcal{H}\chi = \varepsilon_\chi\Theta\chi,$$

i.e. the state $\Theta\chi$ has the same energy eigenvalue as the state χ. This leads to an important constraint, known as Kramers' theorem, that all eigenstates of a time reversal invariant Hamiltonian are at least twofold degenerate. This follows because if a nondegenerate state existed then $\Theta\chi = c\chi$ for some constant c. This would mean $\Theta^2\chi = |c|^2\chi$, which is not allowed because $|c|^2 \neq -1$ as (9.10.3) demands. In the absence of spin-orbit interactions, Kramers' degeneracy is simply the degeneracy between up and down spins. In the presence of spin-orbit interactions, however, it has nontrivial consequences. A time reversal invariant Bloch Hamiltonian must satisfy

$$\Theta\mathcal{H}(\boldsymbol{k})\Theta^{-1} = \mathcal{H}(-\boldsymbol{k}). \tag{9.10.5}$$

One can classify the equivalence classes of Hamiltonians satisfying this constraint that can be smoothly deformed without closing the

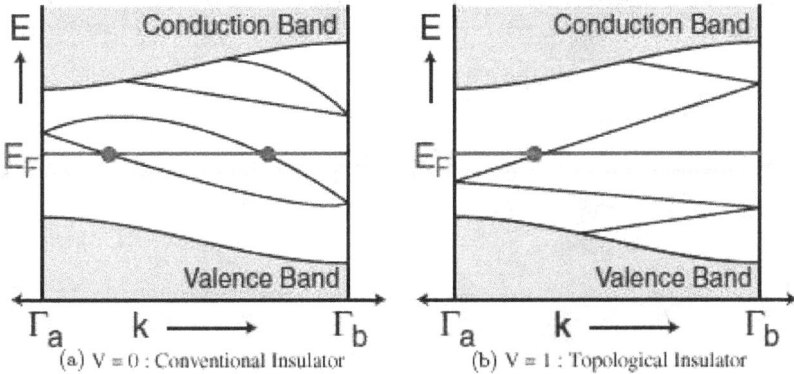

Fig. 9.38   Electronic dispersion between two boundary Kramers degenerate points Γ_a and Γ_b. In (a) the number of surface states crossing the Fermi energy E_F is even, whereas in (b) it is odd. An odd number of crossings leads to topologically protected boundary states.

energy gap. There is an invariant with two possible values, $\nu = 0$ or 1 [76]. The fact that there are two topological classes can be understood by appealing to the bulk-boundary correspondence. In Fig. 9.38 we plot analogous to Fig. 9.36 the electronic states associated with the edge of a time reversal invariant 2D insulator as a function of the crystal momentum along the edge. Only half of the Brillouin zone $0 < k_x < \pi/a$ is shown because time reversal symmetry requires that the other half $-\pi/a < k_x < 0$ is a mirror image. Γ_a and Γ_b are the "time reversal invariant momenta" 0 and π/a respectively.

If the edge states are present then Kramers' theorem requires that they are twofold degenerate at the time reversal invariant momenta $k_x = 0$ and π/a which is the same as $-\pi/a$. Away from these special points, labeled $\Gamma_{a,b}$ in Fig. 9.38, a spin-orbit interaction will split the degeneracy. There are two ways the states at $k_x = 0$ and π/a can connect. In Fig. 9.38(a) they connect pairwise. In this case the edge states can be eliminated by pushing all of the bound states out of the gap. Between $k_x = 0$ and π/a, the bands intersect E_F an even number of times. In contrast, in Fig. 9.38(b) the edge states cannot be eliminated. The bands intersect E_F an odd number of times. Which of these alternatives occurred depends again on the

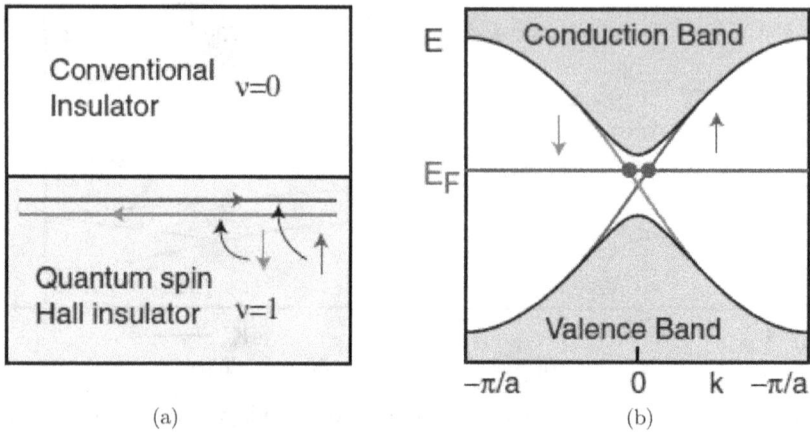

Fig. 9.39 Edge states in quantum spin Hall insulator (a) interface between topologically different materials (b) the edge state dispersion

edge-bulk correspondance. Since each band intersecting E_F at k_x has a Kramers partner at $-k_x$, the correspondance relates the pairs of edge modes intersecting E_F to the change in the Z_2 invariants across the interface,

$$N_K = \Delta\nu \text{ mod } 2.$$

In the case of one pair of edge currents the up and down spins under an applied electric field flow in opposite directions perpendicular to the electric field $J_x^\uparrow - J_x^\downarrow = \sigma_{xy}^s E_y$ with the quantum spin Hall conductivity $\sigma_{xy}^s = e/2\pi$. The edge states is shown in Fig. 9.39, in parallel to Fig. 9.36 for the quantum Hall effect.

But here remains a subtle point about the dissipationless edge currents in the topological insulators. We can still argue that no back scattering on a non-magnetic impurity is possible, because the back scattering route is on the other edge, but the spin-orbit interaction can flip the spin, such that the back route is on the same edge. Here we need a pure quantum mechanical argument: the different paths of back scattering interferes destructively. A forward-moving electron with spin up on the edge can make either a clockwise or a counterclockwise turn around the impurity (Fig. 9.40). Since only spin down electrons can propagate backwards, the electron

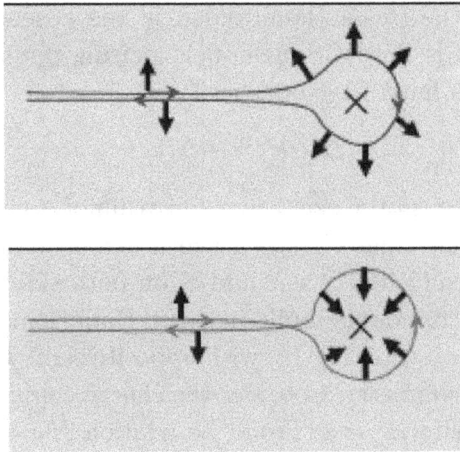

Fig. 9.40 Destructive interference of two paths of back scattering leads to perfect transmission.

spin has to rotate adiabatically, either by an angle of π or $-\pi$, i.e. into the opposite direction. Consequently, the two paths differ by a full $\pi - (-\pi) = 2\pi$ rotation of the electron spin. However, the wave function of a spin-1/2 particle picks up a negative sign under a full 2π rotation. Therefore, two backscattering paths related by time reversal always interfere destructively, leading to perfect transmission.

9.10.2 Topological Superconductors, Majorana fermions

In the BCS mean-field theory of a superconductor the Hamiltonian for a system of spinless electrons may be written in the form

$$H - \mu N = \frac{1}{2} \sum_k \left(c_k^\dagger \; c_{-k} \right) \mathcal{H}_{BdG}(k) \begin{pmatrix} c_k \\ c_{-k}^\dagger \end{pmatrix}, \qquad (9.10.6)$$

where $c_k\dagger$ is an electron creation operator and the Bogoliubov-de Gennes Hamiltonian $\mathcal{H}_{BdG}$ is a 2×2 block matrix,

$$\mathcal{H}_{BdG} = \begin{pmatrix} \mathcal{H}_0(k) - \mu & \Delta \\ \Delta^* & -(\mathcal{H}_0(k) - \mu) \end{pmatrix}. \qquad (9.10.7)$$

Here, $\mathcal{H}_0(\boldsymbol{k})$ is the Bloch Hamiltonian in the absence of superconductivity and Δ is the BCS mean-field pairing potential, which for spinless particles must have odd parity

$$\Delta(-\boldsymbol{k}) = -\Delta(\boldsymbol{k}). \qquad (9.10.8)$$

For a uniform system the excitation spectrum of a superconductor is given by the eigenvalues of $\mathcal{H}_{BdG}$, which exhibit a superconducting energy gap. Since (9.10.6) has c and $c^\dagger$ on both sides of $\mathcal{H}_{BdG}$ there is an inherent redundancy built into the Hamiltonian. For $\Delta = 0$, $\mathcal{H}_{BdG}$ includes two copies of $\mathcal{H}_0$ with opposite sign. There is a signal of particle-hole symmetry. Consider the charge conjugation operator $\Xi = \sigma_1 K$, satisfying $\Xi^2 = 1$. From the relation $\mathcal{H}_0(-\boldsymbol{k}) = -\mathcal{H}_0(\boldsymbol{k})^*$ and the odd parity of the real $\Delta(\boldsymbol{k})$ one can actually prove that the Bogoliubov-de Gennes Hamiltonian $\mathcal{H}_{BdG}$ has an intrinsic particle-hole symmetry expressed by

$$\Xi \mathcal{H}_{BdG}(\boldsymbol{k}) \Xi^{-1} = -\mathcal{H}_{BdG}(-\boldsymbol{k}). \qquad (9.10.9)$$

It follows that every eigenstate of $\mathcal{H}_{BdG}$ with energy E has a partner at $-E$. These two states are redundant because the Bogoliubov quasiparticle operators associated with them satisfy $\Gamma_E^\dagger = \Gamma_{-E}$, i.e. creating a quasiparticle with energy E has the same effect as removing one with energy $-E$. The particle-hole symmetry constraint (9.10.9) has a similar structure to the time-reversal constraint in (9.10.5), so it is natural to consider the classes of $\mathcal{H}_{BdG}$ Hamiltonians that can be continuously deformed into one another without closing the energy gap. In the simplest case, the classification spinless fermions can be shown to be $\mathcal{Z}_2$ in one dimension and $\mathcal{Z}$ in two dimensions. This can be most easily understood by appealing to the bulk-boundary correspondence. We consider the one-dimensional superconductor [77]. At its end there may or may not be discrete bound states within the energy gap (Fig. 9.41a).

If they are present, then every state at $+E$ has a partner at $-E$ (Fig. 41b). Such finite-energy pairs are not topologically protected because they can simply be pushed out of the energy gap. However, a single unpaired bound state at $E = 0$ (Fig. 41c) is protected because it cannot move away from $E = 0$. The presence or absence of such

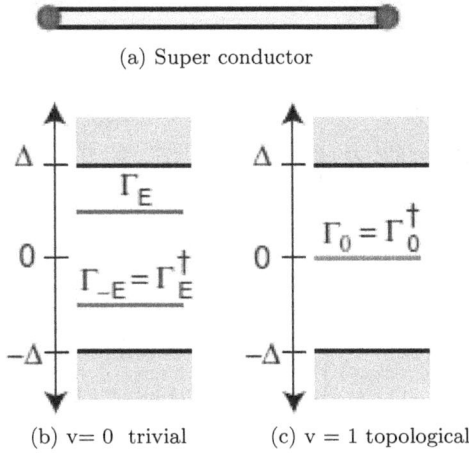

(a) Super conductor

(b) v= 0 trivial   (c) v = 1 topological

Fig. 9.41 Trivial and topological superconductor energy spectrum.

a zero mode is determined by the $\mathcal{Z}_2$ topological class of the bulk 1D superconductor. The Bogoliubov quasiparticle states associated with the zero modes are fascinating objects: the Majorana fermions due to Kitaev [78]. In the following we give a brief introduction to Majorana fermions. Dirac equation for a free fermion is

$$(i\partial^\mu \gamma_\mu - m)\,\psi = 0. \tag{9.10.10}$$

We use Bjorken-Drell convention

$$\gamma_0 = \beta, \quad \gamma_k = \beta\alpha_k$$

γ_0 is hermitian, $(\gamma_0)^2 = 1$, and γ_k are anti-hermitian, charged particles are described by complex wave functions, as in the Schrödinger equation, this can be seen from the expression for current

$$j^\mu = \overline{\psi}\gamma^\mu\psi,$$

where $\overline{\psi} = \psi^\dagger\gamma_0$. Majorana attempted to describe a neutral spin $1/2$ fermion which is the anti-particle of itself, hence should be charge neutral. As a particle of spin $1/2$, the equation remains to be Dirac's. For the Dirac spinor to be real, γ_μ must be pure imaginary. Majorana found the representation

$$\widetilde{\gamma}^0 = -\sigma_2 \otimes \sigma_1, \quad \widetilde{\gamma}^1 = i\sigma_1 \otimes 1, \quad \widetilde{\gamma}^2 = i\sigma_3 \otimes 1, \quad \widetilde{\gamma}^3 = i\sigma_2 \otimes \sigma_2,$$

i.e.

$$\tilde{\gamma}^0 = \begin{pmatrix} 0 & 0 & 0 & -i \\ 0 & 0 & -i & 0 \\ 0 & i & 0 & 0 \\ i & 0 & 0 & 0 \end{pmatrix}, \quad \tilde{\gamma}^1 = \begin{pmatrix} 0 & 0 & i & 0 \\ 0 & 0 & 0 & i \\ i & 0 & 0 & 0 \\ 0 & i & 0 & 0 \end{pmatrix},$$

$$\tilde{\gamma}^2 = \begin{pmatrix} i & 0 & 0 & 0 \\ 0 & i & 0 & 0 \\ 0 & 0 & -i & 0 \\ 0 & 0 & 0 & -i \end{pmatrix}, \quad \tilde{\gamma}^3 = \begin{pmatrix} 0 & 0 & 0 & -i \\ 0 & 0 & i & 0 \\ 0 & -i & 0 & 0 \\ i & 0 & 0 & 0 \end{pmatrix}. \qquad (9.10.11)$$

With this representation of γ's the real Majorana spinor ψ_M satisfying Dirac equation describes a neutral spin $1/2$ particle whose anti-particle is its own identity. Majorana speculated that neutrinos might be such fermions. The reasons appear very compelling even judged from the present day physics. In weak interactions which is the only kind that neutrinos take part in, all neutrinos are left-handed and all anti-neutrinos are right-handed. Therefore ν and $\bar{\nu}$ can be considered as the left-handed and right-handed components of the Majorana neutrino. In particle physics neutrino is an "elementary particle". In nuclear physics search for the neutrinoless double beta decay e.g. $Ge^{76} \rightarrow Se^{76} + 2e^-$ has been a hot topic, but firm experimental evidence is still lacking.

In condensed matter physics the Majorana fermions can only appear as emergent degrees of freedom in certain system of interacting electrons. Due to the particle-hole redundancy in the 1D superconductor discussed above the quasiparticle operators of the zero modes satisfy $\Gamma_0^\dagger = \Gamma_0$. Thus, a quasiparticle is its own antiparticle — the defining feature of a Majorana fermion. Majorana zero modes must always come in pairs. For instance, a 1D superconductor has two ends, and a well separated pair defines a degenerate two-level system, whose quantum state is stored nonlocally. This has profound implications. Consider a 1D chain of N sites [78], with electron operators $a_j, a_j^\dagger$. One can define Majorana operators

$$c_{2j-1} = a_j + a_j^\dagger, \quad c_{2j} = (a_j - a_j^\dagger)/i, \quad j = 1, 2, \ldots, N, \quad (9.10.12)$$

satisfying

$$c_m^\dagger = c_m, \quad \{c_l \cdot c_m\} = 2\delta_{lm}, \quad l, m = 1, 2, \ldots, 2N. \quad (9.10.13)$$

Notice the Majorana operators here are fermion ones satisfying anti-commutation relations, but any operator squared gives 1 instead of 0. Roughly speaking, c_{2j-1} and c_{2j} correspond to the real and imaginary parts of the Dirac operator a_j. Note that c_{2j-1} and c_{2j} belong to same site on chain. Majorana fermion is essentially half of an ordinary Dirac fermion. In superconductors charge is not conserved, but one can define *conserved fermionic parity* as charge model 2, since 2 free electrons can pair to form a Cooper pair whose charge is no more counted. Since a single Majorana operator cannot appear as a term in a normal Hamiltonian conserving the fermionic parity, so most Majorana fermions appear in pairs.

Kitaev model begins with a quantum wire of N sites on the surface of a superconductor. Such proximity effect is necessary because a one dimensional system cannot be a superconductor by itself. In order to simplify the model, it is assumed that only one component of spin is present. The superconductor must then be the p-wave triplet pairing type. The Hamiltonian is

$$\mathcal{H} - \mu N = \sum_j \left[-t \left(a_j^\dagger a_{j+1} + a_{j+1}^\dagger a_j \right) - \mu \left(a_j^\dagger a_j - \frac{1}{2} \right) \right.$$
$$\left. + \Delta a_j a_{j+1} + \Delta^* a_j^\dagger a_{j+1}^\dagger \right]. \quad (9.10.14)$$

Here t is the hopping amplitude, μ the chemical potential and $\Delta = |\Delta| e^{i\theta}$ the induced superconducting pairing. The phase θ of the pairing parameter can be hidden in the definition of Majorana operators, so that instead of (9.10.12) we have now

$$c_{2j-1} = e^{i\frac{\theta}{2}}a_j + e^{-i\frac{\theta}{2}}a_j^\dagger, \quad c_{2j} = (e^{i\frac{\theta}{2}}a_j - e^{-i\frac{\theta}{2}}a_j^\dagger)/i, \quad j = 1, 2, \ldots, N.$$

$$(9.10.15)$$

In terms of the Majorana operators, $\mathcal{H}$ becomes

$$\mathcal{H} = \frac{i}{2}\sum_j[-\mu c_{2j-1}c_{2j} + (t + |\Delta|)c_{2j}c_{2j+1} + (-t + |\Delta|)c_{2j+1}c_{2j+2}].$$

$$(9.10.16)$$

We discuss only two special cases.

(1) $|\Delta| = t = 0$, $\mu < 0$.
    Then

$$\mathcal{H} = -\mu\sum_j\left(a_j^\dagger a_j - \frac{1}{2}\right) = -\frac{i}{2}\mu\sum_j c_{2j-1}c_{2j}.$$

The Majorana operators c_{2j-1}, c_{2j} from the same site j are paired together (Fig. 9.42a). The ground state has the occupation number 0.

(2) $|\Delta| = t > 0$, $\mu = 0$. Then

$$\mathcal{H} = it\sum_j c_{2j}c_{2j+1}.$$

Now the Majorana operators c_{2j}, c_{2j+1} from different sites are paired together (see Fig. 9.42b). One can define new annihilation and creation operators spaning neighboring sites $2j$ and $2j + 1$

$$\gamma_j = \frac{1}{2}(c_{2j} + ic_{2j+1}), \quad \gamma_j^\dagger = \frac{1}{2}(c_{2j} - ic_{2j+1})$$

The Hamiltonian becomes

$$\mathcal{H} = 2t\sum_{j=1}^{N-1}\left(\gamma_j^\dagger\gamma_j - \frac{1}{2}\right).$$

Ground states satisfy the condition $\gamma_j|\psi\rangle = 0$ for $j = 1, \ldots, N - 1$. There are two orthogonal states $|\psi_0\rangle$ and $|\psi_1\rangle$ with this property.

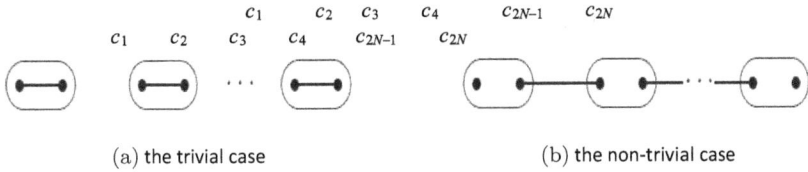

(a) the trivial case   (b) the non-trivial case

Fig. 9.42   Two types of paring

Indeed, the Majorana operators c_1 and c_{2N} remain unpaired (i. e. do not enter the Hamiltonian), so we can write

$$-ic_1c_{2N}|\psi_0\rangle = |\psi_0\rangle, \quad -ic_1c_{2N}|\psi_1\rangle = -|\psi_1\rangle. \quad (9.10.17)$$

We define the operator for fermion parity by

$$P = \prod_j (-ic_{2j-1}c_{2j}). \quad (9.10.18)$$

We find that $|\psi_0\rangle$ has an even fermionic parity (i.e. it is a superposition of states with even number of electrons) while $|\psi_1\rangle$ has an odd parity. In the ground state the Majorana fermions at the two ends are coupled in a non-local way.

The general case is discussed in [78].

References

[1] K. von Klitzing, G. Dorda and M. Pepper, *Phys. Rev. Lett.* 45, 494, 1980.

[2] K. von Klitzing, *Rev. Mod. Phys.* 58, 519, 1986.

[3] D. C. Tsui, H. L. Störmer and A. C. Gossard, *Phys. Rev. Lett.* 48, 1559, 1982; 50, 1953, 1983.

[4] D. C. Tsui, *Rev. Mod. Phys.* 71, 891, 1999; H. L. Störmer, *Rev. Mod. Phys.* 71, 875, 1999.

[5] R. E. Prange and S. Girvin (eds.), *The Quantum Hall Effect*, 2nd edition, New York, Springer, 1990.

[6] R. Laughlin, *Phys. Rev. B* 23, 5632, 1981.

[7] B. I. Halperin, *Phys. Rev. B* 25, 2185, 1982.

[8] D. J. Thouless, Topological Considerations, in Ref. [5]

[9] D. J. Thouless, M. Kohmoto, M. P. Nightingale and M. den Nijs, *Phys. Rev. Lett.* 49, 405, 1982.

[10] Q. Niu, D. J. Thouless and Y. S. Wu, *Phys. Rev. B* 31, 3372, 1985.

[11] P. G. Harper, *Proc. Phys. Soc. (London) A* 265, 317, 1955.

[12] G. E. Zil'berman, *Sov. Phys. JETP* 5, 208, 1957; 6, 299, 1957.
[13] J. Zak, *Phys. Rev.* 134, A, 1602, 1964.
[14] D. G. Hofstadter, *Phys. Rev. B* 14, 2239, 1976.
[15] M. Stone (ed.), *Quantum Hall Effect*, Singapore, World Scientific, 1992.
[16] R. Laughlin, *Phys. Rev. Lett.* 50, 141,1983.
[17] R. Laughlin, Incompressible Quantum Fluid, in Ref. [5].
[18] S. A. Trugman and S. Kivelson, *Phys. Rev. B* 31, 5280, 1985.
[19] F. D. M. Haldane, *Phys. Rev. Lett.* 51, 605, 1983.
[20] D. Pines, *Phys. Today* 42(2), 61, 1989.
[21] R. P. Feynman, *Phys. Rev.* 94, 262, 1954.
[22] S. M. Girvin, Collective Excitations in Ref. [4].
[23] S. M. Girvin and T. Jach, *Phys. Rev. B* 29, 5617, 1984.
[24] S. M. Girvin, A. H. MacDonald and P. M. Platzman, *Phys. Rev. Lett.* 54, 581, 1985; *Phys. Rev. B* 33, 2481, 1986.
[25] R. G. Clark, J. R. Mallet, S. R. Haynes, J. J. Harris and C. T. Foxon, *Phys. Rev. Lett.* 60, 1747, 1988.
[26] J. R. Simmons, H. P. Wei, L. W. Engel, D. C. Tsui and M. Shagan, *Phys. Rev. Lett.* 63, 1731, 1989.
[27] A. M. Chang and J. E. Cunningham, *Solid State Comm.* 72, 651, 1989.
[28] X. G. Wen, *Phys. Rev. Lett.* 64, 2206, 1990.
[29] C. L. Kane and M. P. A. Fisher, *Phys. Rev. Lett.* 72, 724, 1994.
[30] L. Saminadayar, D. C. Glaffli, Y. Jin and B. Eflienne, *Phys. Rev. Lett.* 79, 2526, 1997.
[31] R. dePicciotto, M. Reznikov, M. Heiblum, V. Umansky, G. Bunin and D. Mahalu, *Nature* 389, 162, 1997.
[32] M. Reznikov, R. diPicciotto, T. G. Griffth, M. Heiblum and V. Umansky, *Nature* 399, 238, 1999.
[33] C. L. Kane and M. P. A. Fisher, in *Perspective in Quantum Hall Effects*, ed. S. Das Sarma, A. Pinczuk, New York, Wiley-Interscience, 1997.
[34] X. G. Wen, *Int. J. Mod. Phys.* B5, 1641, 1991.
[35] X. G. Wen, *Int. J. Mod. Phys.* B6, 1711, 1992.
[36] X. G. Wen, *Phys. Rev. B* 41, 12838, 1990.
[37] Y. Yu, W. J. Zhang and Z. Y. Zhu, *Phys. Rev. B* 56, 13279, 1997; W. J. Zhang and Y. Yu, *Phys. Rev. Lett.* 79, 3242, 1997.
[38] F. D. M. Haldane, The Hierarchy, Numerical Results, in Ref. [5].
[39] J. K. Jains, *Phys. Rev. Lett.* 63, 199, 1989.
[40] B. I. Halperin, P. A. Lee and N. Read, *Phys. Rev. B* 47, 7312, 1993.
[41] S. Girvin and A. MacDonald, *Phys. Rev. Lett.* 58, 1, 1987.
[42] S. C. Zhang, H. Hansson and S. Kivelson, *Phys. Rev. Lett.* 62, 82, 1989; 62, 980, 1989.

[43] S. C. Zhang, *Int. J. Mod. Phys. B* 6, 25, 1992.

[44] F. Wilczek, Anyons, *Scientific American* 264 (5), 58, 1991.

[45] E. Fradkin, Field Theories of Condensed Matter Systems, Redwood City, Addison-Wesley, 1991, Section 10.9.

[46] X. G. Wen and A. Zee, *Nucl. Phys.* B15, 135, 1990; A. Zee, arXir: cond-mat/9501002.

[47] M. P. A. Fisher and D. H. Lee, *Phys. Rev. Lett.* 63, 903, 1989; X. G. Wen and A. Zee, *Int. J. Mod. Phys.* B4, 437, 1990.

[48] S. Kivelson, D. H. Lee and S. C. Zhang, *Phys. Rev. B* 46, 2223, 1992.

[49] S. Kivelson, D. H. Lee and S. C. Zhang, *Scientific American* 269 (3), 86, 1996.

[50] M. Hilke, D. Shahar, S.-H. Song, D. C. Tsui, Y. H. Xie and D. Monroe, *Nature* 395, 675, 1998.

[51] S.-H. Song, D. Shahar, D. C. Tsui, Y. H. Xie and D. Monroe, *Phys. Rev. Lett.* 78, 2200, 1997;
M. Hilke, D. Shahar, S.-H. Song, D. C. Tsui and Y. H. Xie, *Phys. Rev. B* 62, 6940, 2000.

[52] D. N. Sheng and Z. Y. Weng, *Phys. Rev. Lett.* 78, 318, 1997; 80, 580, 1998.

[53] D. N. Sheng and Z. Y. Weng, *Phys. Rev. B* 62, 15363, 2000.

[54] M. Dyakonov and V. Perel, *JETP Letters* 13, 467, 1971; *Phys. Lett. A* 35, 459, 1971.

[55] S. Murakami, N. Nagaosa and Shou-Cheng Zhang, *Science* 301, 1348, 2003.

[56] J. Sinova D. Culcer, Q. Niu, N.A. Sinitsyn, T. Jungwirth and A. MacDonald, *Phys. Rev. Lett.* 92, 126603, 2004.

[57] Shou-Cheng Zhang and Jiangping Hu, *Science* 294, 823, 2001.

[58] B. Andrei Bernevig, Taylor L. Hughes and Shou-Cheng Zhang, *Science* 314, 1757, 2006.

[59] Xiao-Liang Qi, Yong Shi Wu and Shou-Cheng Zhang, *Phys. Rev. B* 74, 085308, 2006.

[60] Markus König, S. Wiedmann, C. Brüne, A. Roth, H. Buhmann, L.W. Molenkamp, X.-L. Qi and S.-C. Zhang, *Science* 318, 766, 2007.

[61] R. B. Laughlin, *Phys. Rev. B* 23, 5632, 1981.

[62] B. I. Halperin, *Phys. Rev. B* 25, 2185, 1982.

[63] S. Murakami, N. Nagaosa and Shou-Cheng Zhang, *Phys. Rev. B* 69, 235206, 2004.

[64] S. Deser, R. Jackiw and S. Templeton, *Ann. Phys.* 140, 372, 1982, reprinted in [66].

[65] C. Itzykson and J.-B. Zuber, *Quantum Field Theory*, McGraw and Hill, New York, 1980.

[66] S. Deser, R. Jackiw, S. Templeton, reprinted in Ann Phys, 281, 409, 2000.

[67] A. N. Redlich, *Phys. Rev. Lett.* 52, 18, 1984.

[68] A. N. Redlich, *Phys. Rev. D* 29, 2366, 1984.

[69] T. P. Cheng and L. F. Li, Gauge Field Theory, Clarendon Press, Oxford, 1884.

[70] R. Jackiw, *Phys. Rev. D* 29, 2375, 1984.

[71] G. Semenoff, *Phys. Rev. Lett.* 53, 2449, 1984.

[72] F. D. M. Haldane, *Phys. Rev. Lett.* 61, 2015.

[73] Cui-Zu Chang *et al.*, Science Express 1234414, 14 March 2013.

[74] M. Z. Hasan and C. L. Kane, *Rev. Mod. Phys.* 82, 3045, 2010.

[75] Xiao-Liang Qi and Shou-Cheng Zhang, *Rev. Mod. Phys.* 83, 1057, 2011.

[76] C. L. Kane and E. J. Mele, *Phys. Rev. Lett.* 95, 226801, 2005.

[77] E. Merzbacher, Quantum Mechanics 2nd edition, John Wiley and Sons Inc., New York 1970.

[78] A. Kitaev, arXiv cond-mat/0010440, 2000.

[79] K. von Klitzing, *Rev. Mod. Phys.* 58, 519, 1986.

[80] D. C. Tsui, *Rev. Mod. Phys.* 71, 891, 1999.

[81] R. B. Laughlin, *Phys. Rev. B* 27, 3383, 1983.

Chapter 10

Bose-Einstein Condensation

In 1924 S.N. Bose succeeded to derive the Planck's law of black body radiation, treating photons as identical particles with their numbers not conserved. Subsequently Einstein generalized the problem to an ideal gas of identical particles with their numbers conserved, and obtained far-reaching conclusions. When the temperature of the gas is below a critical value, the gas atoms accumulate on the lowest energy level and the number on this single level can become macroscopic. This phenomenon is called the *Bose-Einstein condensation*. This was the beginning of Bose-Einstein statistics. Einstein himself was somewhat skeptical about the condensation phenomenon. He wrote to P. Ehrenfest: "It is pretty, but is it correct?" [1] He then turned his back on the problem forever. In 1938 the superfluidity of liquid helium at low temperature (below 2.2 K) was discovered. F. London pointed out that superfluidity by its nature could be a quantum statistical phenomenon and calculated the transition temperature for liquid helium to be 3.2 K. Since then the Bose-Einstein condensation began to attract the interest of physics community. Examples of Bose-Einstein condensates were: the superfluid phase of liquid helium, Cooper pairs in superconductivity and the excitons in semiconductors. Owing to the development of experimental technique, an effort to prepare Bose-Einstein condensate of gas atoms began in the early 1980s. Physicists characterize this effort as the struggle for the Holy

Grail. Eventually Bose-Einstein condensates of alkali gas atoms
^{87}Rb,^{23}Na and ^{7}Li were found in 1995 [2–4]. This success caused
an excitement among physicists. The sophistication of experimental
methods, the exactness of the results did impress them. Anyway the
Bose-Einstein condensation has been a well-established phenomenon
and results of many research works have been accumulated. Why
are the Bose-Einstein condensates of gas atoms so special? The
reason was that it has been difficult to treat the existing by then
systems of condensates more or less exactly in the theory, either
because the interaction is too strong or because the environment too
complicated. Among the exactly observed phenomena it was difficult
to distinguish between the consequence of the Bose-Einstein statistics
and that of the interaction of the particles in the system. The gas
atom condensates belong to the category of weakly interacting Bose
gas, and is relatively easily handled in the theory. The detailed
comparison between theory and experiment can then be made to
facilitate a deeper understanding of the nature of this important
phenomenon. This is also the reason why physics journals have
been publishing many papers on Bose-Einstein condensation, both
theoretical and experimental since the second half of 1995. The
1997 Nobel Prize in Physics was awarded to S. Chu, C. Cohen-
Tannoudji and W.D. Phillips for the development of key experimental methods of laser cooling and the trapping of gases, and
the 2001 Nobel Prize in Physics was awarded to E.A. Cornell,
W. Ketterle and C.E. Wieman for achievement of Bose-Einstein
condensation in dilute gases and the fundamental studies of the
condensates [5, 6].

From a system of Bose gas to a Bose-Einstein condensate a phase
transition takes place. The question immediately arising is: what is
the order parameter? The answer to this question is the condensate
wave function appearing in the off-diagonal long range order, which
concerns also the superconductivity phase transition. This concept
is discussed in Section 10.2.

The fact that a system containing macroscopic number of atoms
can be described by a single particle wave function is firmly based on
the coherence of all the particles in the phase of the wave function

which implies the spontaneous breaking of gauge symmetry. This point is discussed in Section 10.3.

The interaction between atoms in a Bose-Einstein condensate is a central issue. The problem for a homogeneous condensate was solved in the mid 1950s, and the result is outlined in Section 10.4. The condensates under experimental study is trapped by a potential, usually anisotropic. The theoretical interpretation of experimental results has to be adapted to this situation. We discuss this adaptation in Sections 10.5 and 10.6. The Gross-Pitaevskii equation and the quantum phase dynamics are included in Section 10.5. Finally we turn to the vortices and their role in the stability of Bose-Einstein condensates with attractive interactions in Section 10.7. In Section 10.8 we discuss the spinor condensate, where the spin degree of freedom of the condensate is liberated when optical trap is used. The use of optical lattices produced by laser beams propagating in opposite directions opens up a new phase of cold quantum gas research, of which a remarkable example is the quantum phase transition. We discuss this aspect in Section 10.9. By using Feshbach resonance the interaction between atoms can be tuned at will by applying an external magnetic field. This leads to a new direction in the study of ultracold Fermi gas, including the BEC-BCS crossover. We discuss this progress in Section 10.10.

10.1 Some Fundamental Relations Pertaining to the Bose-Einstein Condensation

The Bose-Einstein condensation (abbreviated as BEC) occurs when atoms populate on the lowest energy level (zero momentum state for a system of free bosons) with a finite density n_0:

$$n_0 \equiv \frac{1}{V} \langle a_0^\dagger a_0 \rangle > 0, \tag{10.1.1}$$

where a_k is the operator of annihilation of a particle with momentum $\hbar\mathbf{k}$, and a_0 corresponds to the operator for the zero momentum state; V is the volume of the gas, and a statistical average is taken. The number of atoms in the Bose-Einstein condensate is macroscopic. For a system of interacting bosons, BEC is also defined by (10.1.1), while the interaction is to be included in the Hamiltonian when the average is taken.

10.1.1   The BEC as a quantum statistical phenomenon

A question frequently raised is the following. By classical statistics, the ratio of particles occupying energy levels E_1 (the ground state) and E_2 (any excited state) is given by the Boltzmann factor:

$$N_2 : N_1 = e^{-(E_2 - E_1)/kT},$$

where k is the Boltzmann constant. When $kT \ll E_2 - E_1$ we can have $N_1 \gg N_2$. Can this be considered as the BEC? We answer the question by considering a specific example. Consider the superfluid helium in a volume of a cube of sides 1cm, with the boundary represented by potential barrier of infinite height. We make a rough estimate.[1] The energy difference between the ground state and the first excited state is $\Delta E = 2.48 \times 10^{-30}$ erg, the corresponding temperature $\frac{\Delta E}{k}$ is 10^{-14} K. In order to achieve $N_1 \gg N_2$ it is necessary that the temperature of the gas should be $T \ll 10^{-14}$ K. The temperature of the λ-transition of liquid helium is 2.18 K. It is clear that the classical statistics cannot predict the BEC. As F. London pointed out, the BEC occurring in liquid helium is the result of the quantum statistics, which becomes necessary when the density of particles is larger than the quantum concentration

$$n_Q = \left(\frac{mkT}{2\pi\hbar^2} \right)^{3/2}. \qquad (10.1.2)$$

The thermal de Broglie wave length of a particle with mass m is

$$\lambda = \left(\frac{2\pi\hbar^2}{mkT} \right)^{1/2}, \qquad (10.1.3)$$

and therefore the density of a system with one atom in a volume λ^3 is just n_Q. This is the condition for the indistinguishability of identical

---

[1]See Ref. [7], p. 201.

particles. For liquid helium at low temperatures this condition is satisfied. The average number of particles $N(\varepsilon, T)$ with energy ε at temperature T is given by the Bose-Einstein distribution

$$N\left(\varepsilon, T\right) = \frac{1}{e^{(\varepsilon-\mu)/kT} - 1}, \qquad (10.1.4)$$

where μ is the chemical potential. For ideal boson gas, $\mu = kT \ln \frac{n}{n_Q}$ for $n < n_Q$.[2] When $T \to 0$, $N(\varepsilon = 0) = N$, the total number of particles. We have $N = -kT/\mu$ by (10.1.4), i.e., $\mu = -kT/N$ which is negative. For $N = 10^{22}, T = 1\text{K}, \mu \approx -1.4 \times 10^{-38}$ erg. In our example of liquid helium in a cube of 1 cm^3, the ground state has $\varepsilon_1 = 0$, $\varepsilon_1 - \mu = 1.4 \times 10^{-38}$ erg, the first excited has $\varepsilon_2 - \mu = 2.48 \times 10^{-30}$ erg. At $T \sim 1\text{K}$, both states have $\varepsilon - \mu \ll kT$, since $k = 1.38 \times 10^{-16}$ erg $\cdot$ K^{-1}. Then we have

$$N(\varepsilon, T) \approx \frac{kT}{\varepsilon - \mu},$$

i.e., $N_2 : N_1 = (\varepsilon_1 - \mu) : (\varepsilon_2 - \mu) \sim 10^8$. It is seen that the population on the ground state being macroscopic is caused by the extreme proximity of the chemical potential to the ground state energy. This is the result of Bose-Einstein statistics.

10.1.2 The Bose-Einstein temperature

Let $N_e(T)$ be the total number of particles on all of the excited states, which is a function of the temperature. It is determined by the Bose-Einstein distribution and the density of states $D(\varepsilon)$ of free particles:

$$D\left(\varepsilon\right) = \frac{V}{4\pi^2}\left(\frac{2m}{\hbar^2}\right)^{3/2} \varepsilon^{1/2}, \qquad (10.1.5)$$

from which it follows that

$$N_e(T) = \int N\left(\varepsilon, T\right) D\left(\varepsilon\right) d\varepsilon.$$

---

[2]See Ref. [7], p. 121.

The lower limit of the integral is the energy of the first excited state. The chemical potential μ can be neglected in (10.1.4) for $|\mu| \ll kT$. Let $x = \frac{\varepsilon}{kT}$, we have from (10.1.4) and (10.1.5)

$$N_e(T) = \frac{V}{4\pi^2}\left(\frac{2m}{\hbar^2}\right)^{3/2}(kT)^{3/2}\int_0^\infty \frac{x^{1/2}\mathrm{d}x}{e^x - 1}, \tag{10.1.6}$$

where the lower limit has been replaced by 0, since $D(\varepsilon = 0) = 0$ rules out contribution from the ground state. The integral is evaluated to be $1.036\sqrt{\pi},^3$ hence

$$N_e(T) = V2.612n_Q(T),$$

i.e.,

$$\frac{N_e(T)}{N} = 2.612\frac{n_Q(T)}{n}, \tag{10.1.7}$$

where $n = \frac{N}{V}$ is the particle number density of the system. Substituting (10.1.2) for n_Q in the right-hand side, we have

$$\frac{N_e(T)}{N} = \left(\frac{T}{T_E}\right)^{3/2}, \tag{10.1.8}$$

where

$$T_E = \frac{2\pi\hbar^2}{km}\left(\frac{N}{2.612V}\right)^{3/2} \tag{10.1.9}$$

is the Bose-Einstein temperature. From (10.1.7) we see that BEC occurs only when $n \geq 2.612n_Q$, and it is necessary that $T \leq T_E$ to meet this demand. Figure 10.1 shows the relation between $N_0(T)$, $N_e(T)$ and $\frac{T}{T_E}$, where N_0 is the number of particles of the condensate.

The discussion above is based on the thermal equilibrium, and for the case of homogeneous gas without interaction. A general argument given by A. Leggett [8], is that the tendency of bosons to cluster is very generic, not restricted by thermal equilibrium or the absence of interactions. The statistical argument is direct: consider the distribution of N atoms among p states. For distinguishable particles,

---

^{3}See Ref. [7], p. 204.

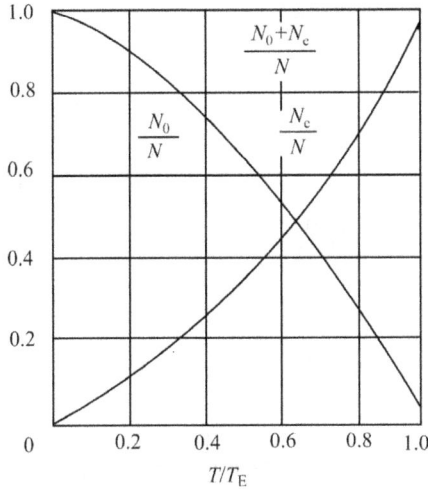

Fig. 10.1   The number of particles in the condensate, the number in the excited states versus the temperature.

the number of different ways of distributing them into a partition $\{n_i\}$, where n_i is the number of atoms in state i, is $N!/\prod_{i=1}^{p} n_i!$. For bosons, there is one and only one way of distributing them for a given $\{n_i\}$. The percentage of "clustered" occupancy is higher for the boson case than for the distinguishable particles. The difference becomes important when $N \gtrsim p$ (the degeneracy condition). For $N = p = 2$, states involving double occupancy are 50% of the whole for distinguishable particles, while for bosons they are 66.7%.

10.1.3   Thermodynamic properties of Bose gas

In the following, we list the relevant formulas, some of which are given without derivation. Details can be found in textbooks, for instance Refs. [7, 8].

When the system is in equilibrium with a heat reservoir at temperature T, the probability $\mathscr{P}(\varepsilon)$ that the system is in the energy state ε is proportional to the Boltzmann factor $e^{-\beta\varepsilon}$, where $\beta = 1/kT$. Therefore, the probability is

$$\mathscr{P}(\varepsilon) = \frac{e^{-\beta\varepsilon}}{Q}, \qquad (10.1.10)$$

where Q is the canonical partition function

$$Q(T) = \sum_s e^{-\beta \varepsilon_s}, \qquad (10.1.11)$$

the sum is carried out over all the energy levels ε_s of the system. All thermodynamic quantities of the system can be obtained by taking thermal average. For instance, the internal energy U is

$$U = \frac{\sum_s \varepsilon_s e^{-\beta \varepsilon_s}}{Q} = kT^2 \frac{\partial}{\partial T} \ln Q. \qquad (10.1.12)$$

The Helmholtz free energy of the system is related to the partition function by

$$F(V,T) = -\frac{1}{\beta} \ln Q(V,T), \qquad (10.1.13)$$

All thermodynamic functions can be derived from $F(V,T)$:

$$\text{Pressure } P = -\frac{\partial F}{\partial V}, \qquad (10.1.14)$$

$$\text{Entropy } S = -\frac{\partial F}{\partial T}, \qquad (10.1.15)$$

$$\text{Internal energy } U = F + TS, \qquad (10.1.16)$$

$$\text{Heat capacity } C_V = \frac{\partial U}{\partial T}. \qquad (10.1.17)$$

When the system can exchange particles as well as energy with the reservoir, the probability $\mathscr{P}(N, \varepsilon)$ that the system has a particle number N and energy ε is determined by the Gibbs factor $e^{\beta(\mu N - \varepsilon)}$, where μ is the chemical potential,

$$\mu(T, V, N) = \left(\frac{\partial F}{\partial N}\right)_{T,V}. \qquad (10.1.18)$$

It can also be expressed by

$$\mu = \left(\frac{\partial U}{\partial N}\right)_{S,V} = -T\left(\frac{\partial S}{\partial N}\right)_{U,V}. \qquad (10.1.19)$$

When two systems in thermal and diffusive (exchange of particles) contact has reached equilibrium, their chemical potentials are equal.

The Gibbs factor gives

$$\mathscr{P}(N,\varepsilon) = \frac{e^{\beta(\mu N - \varepsilon)}}{\mathscr{Q}}, \qquad (10.1.20)$$

where the grand partition function $\mathscr{Q}$ is

$$\mathscr{Q}(\mu, V, T) = \sum_{N=0}^{\infty} \sum_{s(N)} e^{\beta(N\mu - \varepsilon_s(N))}, \qquad (10.1.21)$$

An important quantity in statistical physics is the *fugacity*, or absolute activity as it is sometimes called, denoted by z :

$$z = e^{\beta\mu}. \qquad (10.1.22)$$

The grand partition function $\mathscr{Q}$ can be expressed in terms of the partition function Q_N where the number of particles is explicitly indicated:

$$\mathscr{Q}(z, V, T) = \sum_{N=0}^{\infty} z^n Q_N(V, T), \qquad (10.1.23)$$

Thermodynamic quantities can also be derived from the grand partition function. Taking into account that $\frac{\partial}{\partial\mu} = \frac{\partial z}{\partial\mu} = \beta z \frac{\partial}{\partial z}$, the average number of particles is

$$\langle N \rangle = \frac{\sum_N \sum_{s(N)} N e^{\beta(N\mu - \varepsilon_s(N))}}{\mathscr{Q}} = \frac{kT}{\mathscr{Q}} \frac{\partial \mathscr{Q}}{\partial\mu} = z \frac{\partial}{\partial z} \ln \mathscr{Q}. \qquad (10.1.24)$$

To obtain the equation of state of the system we define the specific volume v by $v = V/\langle N \rangle$, i.e., the volume a particle occupies on the average, then

$$\frac{1}{v} = \frac{\langle N \rangle}{V} = z \frac{\partial}{\partial z} \left(\frac{1}{V} \ln \mathscr{Q} \right). \qquad (10.1.25)$$

From the grand canonical ensemble description we have[4]

$$\frac{PV}{kT} = \ln \mathscr{Q}(z, V, T). \qquad (10.1.26)$$

---

[4]See the Section 7.3 of Ref. [9].

Equations (10.1.25) and (10.1.26) are the equations of state involving the fugacity as a parameter. The grand partition function of an ideal gas is

$$\mathscr{Q} = \frac{1}{1 - ze^{-\beta\varepsilon}}, \tag{10.1.27}$$

and we obtain after substituting (10.1.27) into (10.1.24)

$$\langle N \rangle = \frac{1}{\frac{1}{z}e^{\beta\varepsilon} - 1} = \frac{1}{e^{\beta(\varepsilon-\mu)} - 1},$$

which is just the Bose-Einstein distribution function. By using the saddle-point integration we get an important relation[5]:

$$\frac{1}{V}\ln Q_N = \frac{1}{V}\ln \mathscr{Q}(z) - \frac{1}{v}\ln z. \tag{10.1.28}$$

For free bosons the energy-momentum relation is

$$\varepsilon_p = \frac{1}{2m}p^2,$$

and the average number of particles in the ground state is by the Bose-Einstein distribution

$$\langle N_0 \rangle = \frac{z}{1 - z}. \tag{10.1.29}$$

Since $N_0 \geq 0$, we have $0 \leq z \leq 1$. For an ideal boson gas $\beta\mu = -\frac{1}{N}$, therefore, for $T \to 0$ we have $z \to 1$ under the thermodynamic limit. Treating p as a continuous variable, and using $\sum_p \to \frac{1}{h^3}V\int \mathrm{d}^3p$, we have from (10.1.26),

$$\frac{P}{kT} = \frac{1}{V}\ln \mathscr{Q} = \frac{4\pi}{h^3}\int_0^\infty \mathrm{d}p\,p^2 \ln\left(1 - ze^{-\beta p^2/2m}\right) - \frac{1}{V}\ln(1 - z), \tag{10.1.30}$$

where the contribution to $\mathscr{Q}$ from $p = 0$ has been separated from the integral, since the volume element of the momentum space vanishes

---

[5] See Ref. [10], pp. 6~7.

at $p = 0$. The contribution is $\mathcal{2} \to \frac{1}{1-z}$ which leads to the second term of (10.1.30). The integral of the first term is reduced to

$$g_{5/2}(z) \equiv -\frac{4}{\sqrt{\pi}} \int_0^\infty dx x^2 \ln \left(1 - z e^{-x^2} \right) = \sum_{l=1}^\infty \frac{z^l}{l^{5/2}}, \qquad (10.1.31)$$

i.e.,

$$\frac{P}{kT} = \frac{1}{\lambda^3} g_{5/2}(z) - \frac{1}{V} \ln (1 - z), \qquad (10.1.32)$$

where λ is the thermal de Broglie wave length $\lambda = \left(\frac{2\pi\hbar^2}{mkT} \right)^{\frac{1}{2}}$. Define

$$g_{3/2}(z) \equiv z \frac{\partial}{\partial z} g_{5/2}(z) \equiv \sum_{l=1}^\infty \frac{z^l}{l^{3/2}}, \qquad (10.1.33)$$

Equation (10.1.25) then gives

$$\frac{1}{v} = \frac{1}{\lambda^3} g_{3/2}(z) + \frac{1}{V} \frac{z}{1-z}. \qquad (10.1.34)$$

The parametric equations of state (10.1.32) and (10.1.34) have a specific feature that the first terms of their right-hand side are contributions from all states with $p \neq 0$ while the second terms are contributions from the condensate state $p = 0$. From (10.1.34) the critical condition for the occurrence of BEC can be obtained. The function $g_{3/2}(z)$ is shown in Fig. 10.2. Its derivative diverges at $z = 1$, but the function itself is finite, i.e.,

$$g_{3/2}(1) = \sum_{l=1}^\infty \frac{1}{l^{3/2}} = 2.612. \qquad (10.1.35)$$

Equation (10.1.34) can be rewritten with the help of (10.1.29) as

$$\lambda^3 \frac{\langle N_0 \rangle}{V} = \frac{\lambda^3}{v} - g_{3/2}(z). \qquad (10.1.36)$$

This means that $\frac{\langle N_0 \rangle}{V}$ is different from zero only when the temperature and the specific volume satisfies the relation

$$\frac{\lambda^3}{v} > g_{3/2}(1) \qquad (10.1.37)$$

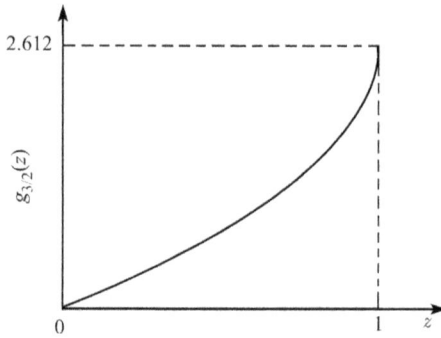

Fig. 10.2   The function $g_{\frac{3}{2}}(z)$.

When this condition is fulfilled, macroscopic number of particles condense in the state $p = 0$, i.e., Bose-Einstein condensation occurs. For a given specific volume v the critical condition (10.1.37) is

$$\frac{\lambda_c^3}{v} = g_{3/2}(1),\qquad(10.1.38)$$

For the critical value of λ, the corresponding critical temperature is

$$T_{\mathrm{E}} = \frac{2\pi\hbar^2}{mk\left[vg_{3/2}(1)\right]^{2/3}},\qquad(10.1.39)$$

and this is the Bose-Einstein temperature (10.1.9). For a given temperature T, (10.1.38) gives the critical specific volume

$$v_{\mathrm{c}} = \frac{\lambda^3}{g_{3/2}(1)} = \frac{1}{g_{3/2}(1)}\left(\frac{2\pi\hbar^2}{mkT}\right)^{3/2}.\qquad(10.1.40)$$

In order to discuss further the properties of the equation of state, we need to solve for $z(v, T)$ as a function of v and T. Equation (10.1.34) is solved by graphic method in Fig. 10.3(a). For large and finite V, z is given as function of v/λ^3 in Fig. 10.3(b). In the limit $V \to \infty$ we have

$$z = \begin{cases} 1, & \dfrac{\lambda^3}{v} \geq g_{3/2}(1) \\[3mm] \text{the root of eqn. } g_{3/2}(z) = \dfrac{\lambda^3}{v}, & \dfrac{\lambda^3}{v} \leq g_{3/2}(1). \end{cases}\qquad(10.1.41)$$

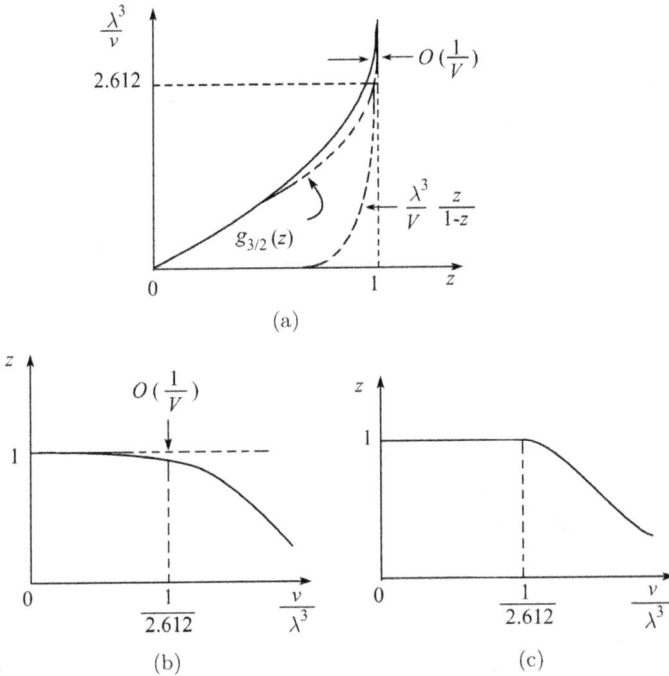

Fig. 10.3 Fugacity of free Bose gas. (a) graphic solution of (10.1.34); (b) fugacity of the Bose gas; (c) fugacity in the limit $V \to \infty$.

The result is shown in Fig. 10.3(c). Equation (10.1.36) is now rewritten as

$$\frac{\langle N_0 \rangle}{N} = 1 - \frac{v}{\lambda^3} g_{3/2}(z).$$

From (10.1.41) we see that the ratio of the number of atoms in the ground state to the total number is

$$\frac{\langle N_0 \rangle}{N} = \begin{cases} 0, & \frac{\lambda^3}{v} \le g_{3/2}(1) \\ 1 - \left(\frac{T}{T_E}\right)^{3/2} = 1 - \frac{v}{v_c}, & \frac{\lambda^3}{v} \ge g_{3/2}(1). \end{cases} \tag{10.1.42}$$

The ratio of atoms populated in the ground state to the total number is zero when $T > T_E$ or $v < v_c$, the ratio is finite when $T < T_E$ or $v > v_c$, and the ratio approaches 1 when $T \to 0$.

The second term of the right-hand side of (10.1.32) is the contribution from the ground state. When $T > T_E$, $z < 1$, this term approaches zero for $V \to \infty$. When $T < T_E$, z differs from 1 by $O(\frac{1}{V})$, this term is $\frac{1}{V}(\ln \frac{1}{V} +$ constant$)$ and approaches zero for $V \to \infty$. Therefore we have

$$\frac{P}{kT} = \begin{cases} \frac{1}{\lambda^3} g_{5/2}(z), & v > v_c \\[2mm] \frac{1}{\lambda^3} g_{5/2}(1), & v < v_c. \end{cases} \tag{10.1.43}$$

The atoms populated on the ground state do not contribute to the pressure, no matter the condensation takes place or not. When the specific volume v decreases from above while remaining larger than v_c, z increases and hence P increases also with $g_{5/2}(z)$. As v further decreases from v_c, P ceases to vary with v and remains constant. Figure 10.4 shows two isothermals of the ideal Bose gas. When v decreases and passes the critical point, the constancy of the pressure is the revelation of the coexistence of the normal gas phase and the condensate phase. This is very similar to the liquid-gas phase transition. From (10.1.42) we see that $\frac{\langle N_0 \rangle}{N} \to 1$ and $v \to 0$ as $T \to 0$. The normal gas phase has a finite specific volume, while the condensate phase has a vanishing specific volume. At the critical

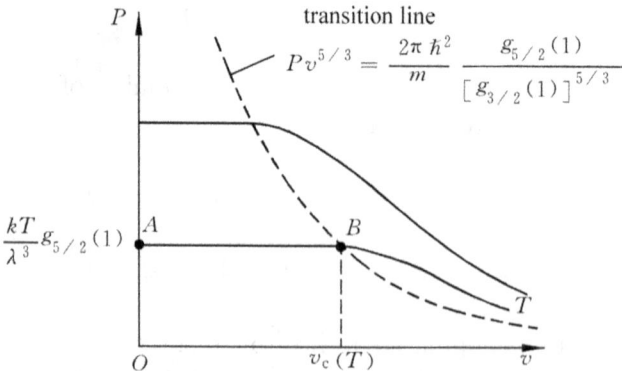

Fig. 10.4   Isothermals of ideal Bose gas.

point, we have

$$\frac{\mathrm{d}P}{\mathrm{d}T} = \frac{5}{2}\frac{kg_{5/2}(1)}{\lambda^3} = \frac{L}{T\Delta v}. \tag{10.1.44}$$

In (10.1.44) the second equality is based on the Clapeyron equation, $\Delta v = v_{\mathrm{c}} - 0 = v_{\mathrm{c}}$ is the difference of specific volume of the two phases, and L is the latent heat per particle:

$$L = \frac{g_{5/2}(1)}{g_{3/2}(1)}\frac{5}{2}kT. \tag{10.1.45}$$

Therefore, the BEC of *ideal* gas is a first order phase transition. When the interaction between atoms becomes strong enough the conclusion may be invalidated, for example the λ transition of helium has no latent heat. In the literature BEC is often characterized as a condensation in the momentum space. This is correct in emphasizing the cause of the condensation as the Bose-Einstein statistics rather than the force between the atoms. If a Bose gas is in the gravitational field, then the spatial separation of the two phases occurs when BEC takes place. This is pointed by W. Lamb and A. Nordsieck in 1941 [11].

The internal energy of the system is

$$\frac{U}{N} = \frac{3}{2}Pv = \begin{cases} \dfrac{3}{2}\dfrac{kTv}{\lambda^3}g_{5/2}(z), & v > v_{\mathrm{c}} \\[2mm] \dfrac{3}{2}\dfrac{kTv}{\lambda^3}g_{5/2}(1), & v < v_{\mathrm{c}}. \end{cases} \tag{10.1.46}$$

In the following, we calculate the entropy of the system. The relations between the thermodynamic potential $G(T,P,\mu)$ with P and S are

$$p = -\left.\frac{\partial G}{\partial V}\right|_T, \quad S = -\left.\frac{\partial G}{\partial T}\right|_V.$$

Consider the thermodynamic limit in the case $T > T_{\mathrm{E}}$,

$$P = \lim\left(-\frac{G(T,V,\mu)}{V}\right) = \frac{kT}{\lambda^3}g_{5/2}(z).$$

The entropy per unit volume of the system is

$$\lim\frac{\langle S\rangle}{T} = \lim\left(-\frac{1}{V}\left.\frac{\partial G}{\partial T}\right|_{V,\mu}\right) = \frac{\partial}{\partial T}\frac{kT}{\lambda^3}g_{5/2}(z).$$

From

$$\frac{T}{\lambda^3} = T\left(\frac{2\pi\hbar^2}{mkT}\right)^{-3/2} = T^{5/2}\left(\frac{2\pi\hbar^2}{mk}\right)^{-3/2}$$

and

$$z\frac{\partial}{\partial z}g_{5/2}(z) = g_{3/2}(z) = \frac{\lambda^3\langle N\rangle}{V},$$

we get

$$\lim \frac{\langle S\rangle}{V} = \frac{5}{2}\frac{k}{\lambda^3}g_{5/2}(z) - \frac{k\langle N\rangle}{V}\ln z.$$

Therefore,

$$\frac{S}{Nk} = \begin{cases} \dfrac{5}{2}\dfrac{v}{\lambda^3}g_{5/2}(z) - \ln z, & v > v_c \\[2mm] \dfrac{5}{2}\dfrac{v}{\lambda^3}g_{5/2}(1), & v < v_c. \end{cases} \qquad (10.1.47)$$

When $T \to 0$, $z \to 1$ and $v \to 0$, and therefore $S = 0$, in accordance with the third law of thermodynamics. We see that the entropy of the condensate atoms is zero. The entropy per particle of the system below the critical temperature is

$$\frac{S}{N} = \frac{v}{v_c}s = \left(\frac{T}{T_E}\right)^{3/2}s,$$

where s is the entropy per particle of the *normal phase*. At the critical point

$$\frac{S}{N} = \frac{5}{2}k\frac{v_c}{\lambda^3}g_{5/2}(1),$$

i.e.,

$$s = \frac{g_{5/2}(1)}{g_{3/2}(1)}\frac{5}{2}k. \qquad (10.1.48)$$

The latent heat of phase transition is therefore

$$L = T\Delta s = Ts = \frac{g_{5/2}(1)}{g_{3/2}(1)}\frac{5}{2}kT,$$

a result identical to (10.1.45).

10.2 Order Parameter of Bose-Einstein Condensation, the Off-Diagonal Long Range Order

O. Penrose and L. Onsager [12, 13] first proposed an order parameter for an interacting boson system. The occurrence of BEC is characterized by the existence of off-diagonal long range order. C. N. Yang [14] made a thorough study and generalized the concept to include the case of interacting fermion systems. Consider a many-particle system with translational invariance, the single-particle density matrix can be defined by

$$\rho_1(\boldsymbol{x}, \boldsymbol{y}) = \langle \psi^\dagger(\boldsymbol{x})\psi(\boldsymbol{y})\rangle = \frac{1}{V}\sum_{\boldsymbol{p},\boldsymbol{q}} e^{-i(\boldsymbol{p}\cdot\boldsymbol{x} - \boldsymbol{q}\cdot\boldsymbol{y})}\left\langle a_{\boldsymbol{p}}^\dagger a_{\boldsymbol{q}}\right\rangle, \qquad (10.2.1)$$

where the field operators are expanded in terms of the plane wave states with $a_{\boldsymbol{q}}$ as the operator of annihilation of a particle with wave vector $\boldsymbol{q}$, and the average means an ensemble average. The single-particle density matrix is also a correlation function, that between the annihilation of a particle at $\boldsymbol{y}$ and the creation of a particle at $\boldsymbol{x}$. The operator of the total momentum of the system is

$$\boldsymbol{P} = \sum \boldsymbol{p}\, a_{\boldsymbol{p}}^\dagger a_{\boldsymbol{p}}. \qquad (10.2.2)$$

From the translational invariance of the system follows the relation

$$[\boldsymbol{P}, H] = 0. \qquad (10.2.3)$$

We calculate the following average

$$\left\langle \left[\boldsymbol{P}, a_{\boldsymbol{p}}^\dagger a_{\boldsymbol{q}}\right]\right\rangle = \frac{\mathrm{tr}\left(\mathrm{e}^{-\beta H}\boldsymbol{P} a_{\boldsymbol{p}}^\dagger a_{\boldsymbol{q}} - \mathrm{e}^{-\beta H} a_{\boldsymbol{p}}^\dagger a_{\boldsymbol{q}}\boldsymbol{P}\right)}{\mathrm{tr}\,\mathrm{e}^{-\beta H}}$$

$$= \frac{\mathrm{tr}\left(\left[\mathrm{e}^{-\beta H}, \boldsymbol{P}\right] a_{\boldsymbol{p}}^\dagger a_{\boldsymbol{q}}\right)}{\mathrm{tr}\,\mathrm{e}^{-\beta H}}$$

$$= 0,$$

where the second equality is obtained by rewriting the second term on the numerator by using $\mathrm{tr}ABC = \mathrm{tr}CAB$ and the third equality is obtained by using (10.2.3). Besides, we can get directly

$$\left[P, a_p^\dagger a_q \right] = (P - q) a_p^\dagger a_q,$$

therefore we have

$$\left\langle a_p^\dagger a_q \right\rangle = \delta_{pq} \left\langle N_p \right\rangle, \qquad (10.2.4)$$

where

$$N_p = a_p^\dagger a_p.$$

Substituting back into (10.2.1) we obtain

$$\begin{aligned} \rho_1(\boldsymbol{x}, \boldsymbol{y}) &= \frac{1}{V} \sum_p e^{-i\boldsymbol{p}\cdot(\boldsymbol{x}-\boldsymbol{y})} \left\langle N_p \right\rangle \\ &= \frac{\langle N_0 \rangle}{V} + \int \frac{d^3 p}{(2\pi)^3} e^{-i\boldsymbol{p}\cdot(\boldsymbol{x}-\boldsymbol{y})} \left\langle N_p \right\rangle. \end{aligned} \qquad (10.2.5)$$

In the above we have separated the term for $\boldsymbol{p} = 0$ from the sum and treated $\boldsymbol{p}$ as a continuous variable. Formally the integral includes the contribution from $\boldsymbol{p} = 0$, but it is automatically removed by the p^2 factor in the volume element $d^3 p$. The second term approaches zero when $|\boldsymbol{x}-\boldsymbol{y}| \to \infty$. This can be seen from the heuristic argument that the exponential factor oscillates violently in the limit, but a rigorous proof does exist.[6] Therefore,

$$\rho_1(\boldsymbol{x}, \boldsymbol{y}) \xrightarrow{|\boldsymbol{x}-\boldsymbol{y}| \to \infty} \frac{\langle N_0 \rangle}{V} \qquad (10.2.6)$$

In deriving (10.2.6) the statistics which the system abides has not been used. Therefore, for both bosons at temperature $T > T_E$ and fermions in general we have $\rho_1(\boldsymbol{x}, \boldsymbol{y}) \xrightarrow{|\boldsymbol{x}-\boldsymbol{y}| \to \infty} 0$ in the thermodynamic limit. The limit does not vanish only for bosons at temperature $T < T_E$. The condition (10.2.6) does not mean that there is a long range order in the usual sense for the boson system below the critical temperature. As the criterion for the existence of BEC, the importance of the condition (10.2.6) lies in the fact

---

[6]See Ref. [14], p. 304.

that for an interacting system the momentum of a particle ceases to be a constant of motion and N_0 does not commute with the Hamiltonian, but as an ensemble average a non-vanishing $\frac{\langle N_0 \rangle}{V}$ can still be considered as characterizing the existence of BEC. Penrose and Onsager proposed earlier that the characteristics of the existence of BEC is

$$\langle \psi^\dagger(\boldsymbol{x})\psi(\boldsymbol{y}) \rangle \to f^*(\boldsymbol{x})f(\boldsymbol{y}),$$

in the limit $|\boldsymbol{x} - \boldsymbol{y}| \to \infty$, i.e., the correlation is given by a single function of one position vector. They adopted f as the order parameter of BEC. The choice adopted now is

$$\langle \psi^\dagger(\boldsymbol{x})\psi(\boldsymbol{y}) \rangle \xrightarrow{|\boldsymbol{x}-\boldsymbol{y}|\to\infty} \langle \psi^\dagger(\boldsymbol{x}) \rangle \langle \psi(\boldsymbol{y}) \rangle. \tag{10.2.7}$$

The complex function $\langle \psi(\boldsymbol{x}) \rangle$ has an amplitude and a phase, it is called the "wave function of the condensate". For an interacting fermion system it is shown in [14] that the criterion for the existence of BEC is that the two-particle density matrix $\langle \psi^\dagger(\boldsymbol{x}'_1)\psi^\dagger(\boldsymbol{x}'_2)\psi(\boldsymbol{x}_1)\psi(\boldsymbol{x}_2) \rangle$ has off-diagonal long range order:

$$\langle \psi^\dagger(\boldsymbol{x}'_1)\psi^\dagger(\boldsymbol{x}'_2)\psi(\boldsymbol{x}_1)\psi(\boldsymbol{x}_2) \rangle \xrightarrow[\boldsymbol{x}_1 \approx \boldsymbol{x}_2, \boldsymbol{x}'_1 \approx \boldsymbol{x}'_2]{|\boldsymbol{x}_1 - \boldsymbol{x}'_1| \to \infty} \varphi^*(\boldsymbol{x}'_1, \boldsymbol{x}'_2)\varphi(\boldsymbol{x}_1, \boldsymbol{x}_2) \neq 0. \tag{10.2.8}$$

In the above, the sign " $\approx$ " means that the distance between the two positions is microscopic. Similar to (10.2.6), the finite limit of (10.2.8) does not mean a long range order for $|\boldsymbol{x}_1 - \boldsymbol{x}'_1| \to \infty$ of the system. It means only that the pair at $\boldsymbol{x}_1$ and the pair at $\boldsymbol{x}'_1$ are described by the same wave function φ. The order is called "off-diagonal" because (10.2.6) concerns the non-diagonal matrix element $\langle \boldsymbol{x} | \rho_1 | \boldsymbol{y} \rangle$ and (10.2.7) involves the non-diagonal matrix element $\langle \boldsymbol{x}'_1\boldsymbol{x}'_2 | \rho_2 | \boldsymbol{x}_1\boldsymbol{x}_2 \rangle$. The long-range order in a solid can be described in terms of the language of classical physics. It corresponds to the diagonal matrix element $\langle \boldsymbol{x}_1\boldsymbol{x}_2 | \rho_2 | \boldsymbol{x}_1\boldsymbol{x}_2 \rangle$ in quantum mechanics. The off-diagonal long-range order has no classical correspondence, and is a purely quantum mechanical concept. The single-particle density matrix ρ_1 and the 2-particle density matrix

ρ_2 of a system of free fermions do not possess long range order. Only when macroscopic number of pairs of fermions are coupled to pairs and populate a single energy level there appears the off-diagonal long-range order.

Being the criterion of the very existence of BEC, the ODLRO must lead to fundamental properties of the condensates. It is actually the case. For superconductivity the ODLRO in ρ_2 leads to the Meissner effect and flux quantization. This was shown by G.L. Sewell [15], and H.T. Nieh, G. Su and B.H. Zhao [16]. For superfluidity, the ODLRO in ρ_1 leads to the irrotational flow in singly connected regions and the quantization of circulation in multiply connected regions.[7] This has been shown by G. Su and M. Suzuki [17]. The proof given in Refs. [16, 17] ascribes a single origin to all the relevant phenomena: the phase coherence of the condensate wave function.

10.3   The Nature of Bose-Einstein Condensation: The Spontaneous Symmetry-Breaking and the Phase Coherence

There are two closely connected concepts relevant to the BEC, namely the spontaneous symmetry breaking and the phase coherence. The order parameter $\langle \psi(\boldsymbol{x}) \rangle$ of BEC is the ensemble average of the field operator:

$$\langle \psi(\boldsymbol{x}) \rangle = \frac{\mathrm{tr}\left(e^{-\beta(\hat{H}-\mu\hat{N})}\psi(\boldsymbol{x})\right)}{\mathrm{tr}\,e^{-\beta(\hat{H}-\mu\hat{N})}}. \qquad (10.3.1)$$

The numerator can be formally written as $\sum_s \langle s \mid \psi(\boldsymbol{x}) \mid s \rangle$, where s characterizes the eigenstate of the grand canonical Hamiltonian $\hat{H} - \mu\hat{N}$. A subtle point is that, for the annihilation operator $\psi(\boldsymbol{x})$, its expectation value in any eigenstate of the operator $\hat{N}$ is zero. We can argue that an eigenstate of $\hat{H} - \mu\hat{N}$ is not necessarily an eigenstate of $\hat{N}$, because in a large system the spectrum of $\hat{H} - \mu\hat{N}$ is highly degenerate, and we can construct the eigenstates of $\hat{H} - \mu\hat{N}$

---

[7]See (10.3.8) and (10.3.9).

using states of different particle number N, such that $\langle \psi(\boldsymbol{x}) \rangle$ need not be zero. But this does not help actually, since the trace in (10.3.1) is independent of the choice of basis. We can choose the eigenstates with definite N and get the zero order parameter, and no change of basis can save the situation. This is a revelation of a fundamental principle: the direct connection between the global phase invariance under the transformation

$$\psi(\boldsymbol{x}) \to e^{i\alpha} \psi(\boldsymbol{x}) \tag{10.3.2}$$

and the conservation of the number of particles. In taking the ensemble average each $\psi = re^{i\theta}$ is cancelled by a $\psi = re^{i(\theta+\pi)}$. This is similar to the van Leuven theorem in magnetism. To calculate the spontaneous magnetization of a ferromagnet we take the ensemble average

$$\langle \boldsymbol{M} \rangle = \frac{\operatorname{tr}(e^{-\beta \hat{H}} \boldsymbol{M})}{\operatorname{tr} e^{-\beta \hat{H}}}. \tag{10.3.3}$$

In the absence of external magnetic field the Hamiltonian $\hat{H}$ is invariant with respect to the interchange $\boldsymbol{M} \to -\boldsymbol{M}$. Therefore, to each $\boldsymbol{M}$ there exists a $-\boldsymbol{M}$ occurring with the same probability which cancels $\boldsymbol{M}$. We then get always a vanishing spontaneous magnetization. This result contradicts the fact because we do have finite spontaneous magnetization under the Curie temperature. This apparent paradox is solved in the following argument. The ground state of a ferromagnet is characterized by a definite spontaneous magnetization $\boldsymbol{M}$. The existence of a non-vanishing $\boldsymbol{M}$ does not abide the rotational symmetry of the system. Of course the ground state can choose this $\boldsymbol{M}$ or another one of the same magnitude but a different direction, because they give rise to the same energy of the state. But once a *definite* direction in the space is chosen, the ground state is no more rotational invariant. Can all the elementary magnetic moments in a ferromagnet change their direction momentarily to a newly chosen one? This costs no energy after all. Watching this to happen we need to wait for a time of the order of the Poincaré cycle of the system. The ensemble average in statistical mechanics has a meaning for a finite relaxation time. Therefore, in carrying out the ensemble average the occurrence of $\boldsymbol{M}$ and $-\boldsymbol{M}$ cannot be equally probable. The mathematical expression of this argument is that we introduce an external magnetic field $\boldsymbol{B}$, whose direction is the same as the *prescribed* final magnetization $\boldsymbol{M}$, then carry out the ensemble average and then take the external field away after the average. Then the magnetization per unit volume would be

$$\frac{\langle \boldsymbol{M} \rangle}{V} \equiv \lim_{|\boldsymbol{B}| \to 0} \lim_{V \to \infty} \frac{1}{V} \frac{\mathrm{tr} \, e^{-\beta(\hat{H} - \mu \hat{N} - \boldsymbol{M} \cdot \boldsymbol{B})} \boldsymbol{M}}{\mathrm{tr} \, e^{-\beta(\hat{H} - \mu \hat{N} - \boldsymbol{M} \cdot \boldsymbol{B})}}. \qquad (10.3.4)$$

This argument can be borrowed to the BEC. Introduce a c-number external source $\eta(\boldsymbol{x})$, and the grand canonical Hamiltonian becomes a functional of ψ and η :

$$\overset{\wedge}{\mathcal{E}} [\psi, \eta] = \hat{H} - \mu \hat{N} - \int \mathrm{d}^3 x [\psi(\boldsymbol{x}) \eta(\boldsymbol{x}) + \psi^\dagger(\boldsymbol{x}) \eta^*(\boldsymbol{x})], \qquad (10.3.5)$$

and we take the average first and then let the source to become 0 :

$$\langle \psi(\boldsymbol{x}) \rangle \equiv \lim_{\eta \to 0} \lim_{V \to \infty} \frac{1}{V} \frac{\mathrm{tr} \left[e^{-\beta \overset{\wedge}{\mathcal{E}} [\psi, \eta]} \psi(\boldsymbol{x}) \right]}{\mathrm{tr} \, e^{-\beta \overset{\wedge}{\mathcal{E}} [\psi, \eta]}}. \qquad (10.3.6)$$

If we reverse the order of averaging and taking the limit, the result would be zero. For a system possessing a certain symmetry, its ground state may not abide the symmetry of the system. This is dubbed the spontaneous breaking of the symmetry. We have discussed it in Chapter 7. A non-vanishing order parameter (condensate wave function) has a phase. Once a phase is chosen, the invariance with respect to the global phase transformation is lost, and a spontaneous breaking of the global phase invariance occurs.

With the use of spontaneous breakdown of global phase invariance, A. Leggett disagrees Refs. [18–20]. He considers that the superposition $\Psi = \sum_N a_N \psi_N$[8] is not the physical correct description of the system in any circumstance, and the definition of the order

---

[8]This is a coherent superposition. The coherent state (a state with definite phase) is an example of such a superposition. The grand canonical ensemble states with different particle numbers constitute a statistical mixture, they have no phase coherence with one another.

parameter as $\langle \psi(\boldsymbol{r},t) \rangle$ is liable to generate pseudoproblems and is best avoided. Leggett's definition of order parameter is $\psi(\boldsymbol{r},t) = \sqrt{N_0(t)}\chi_0(\boldsymbol{r},t)$, where χ_0 is the single particle state into which the condensation occurs.

Another concept is the phase coherence. The order parameter can be written as

$$\langle \psi(\boldsymbol{x}) \rangle = f(\boldsymbol{x})e^{i\theta(\boldsymbol{x})}, \qquad (10.3.7)$$

where the amplitude $f(\boldsymbol{x})$ is the square root of the density of the condensate. As the wave function of the condensate, it is *macroscopic*. Not only the number of particles in the condensate is macroscopic, but also the dimension that the wave function extends is macroscopic. It is different from the wave function of a many particle system in that it is not a function of the radius vectors of all the particles as $\Psi(\boldsymbol{x}_1, \boldsymbol{x}_2, \ldots, \boldsymbol{x}_N)$, but a function of a single radius vector. The phase has a certain freedom that a *constant* phase α can be added to $\theta(\boldsymbol{x})$ without changing the physics. But the phases $\theta(\boldsymbol{x}_1)$ and $\theta(\boldsymbol{x}_2)$ have definite relative values as long as the points $\boldsymbol{x}_1$ and $\boldsymbol{x}_2$ are in the condensate, irrespective of their distance. P.W. Anderson explains this in an example [20]. Let the system be divided into K identical boxes, and the number of particles on the ground state $\boldsymbol{k} = 0$ in each be N_0/K. Let A_i be the operator of annihilation of a particle on the ground state in the box i. Then we have

$$a_0 = \frac{1}{\sqrt{K}} \sum_i A_i,$$

$$\left\langle a_0^\dagger a_0 \right\rangle = N_0,$$

$$\left\langle A_i^\dagger A_i \right\rangle = \frac{N_0}{K}.$$

From the first two of these relations we get

$$\left\langle a_0^\dagger a_0 \right\rangle = \frac{1}{K} \sum_{i,j} \left\langle A_i^\dagger A_j \right\rangle = \frac{1}{K} \sum_i \left\langle A_i^\dagger A_i \right\rangle + \frac{1}{K} \sum_{i \neq j} \left\langle A_i^\dagger A_j \right\rangle$$

$$= \frac{N_0}{K} + \frac{1}{K} \sum_{i \neq j} \left\langle A_i^\dagger A_j \right\rangle.$$

For very large K the second term is dominant, therefore the phases in different boxes must be *coherent*. We have discussed wave function of the form (10.3.7) in Chapter 6, where we found the relation between the superfluid velocity and the gradient of the phase

$$v_s = \frac{\hbar}{m} \nabla \theta. \tag{10.3.8}$$

Taking a closed integral along a line in the superfluid, we get

$$\oint v_s \cdot ds = \frac{\hbar}{m} \oint \nabla \theta \cdot ds$$

Because of the single valuedness of the wave function, the phase can only change by an integral multiple of 2π after traversing a closed contour. Therefore, we have

$$\oint v_s \cdot ds = \frac{2\pi\hbar}{m}\kappa, \quad \kappa = 0, \pm 1, \pm 2, \ldots \tag{10.3.9}$$

This is the Onsager-Feynman quantization of circulation, a fundamental attribute of the BEC. In the expression (10.3.7) the phase $\theta(x)$ is a definite function. The state is called a "phase locked state". Similarly, a "number locked state" $\psi(N)$ can be defined as:

$$\psi(\theta) = \sum_N c_N e^{i\theta N} \psi(N), \tag{10.3.10}$$

where c_N is peaked at a certain value of N_0 and has a certain fluctuation.[9]

---

[9]The coherent state is often used to describe the state of a condensate:

$$|z\rangle = e^{-\frac{1}{2}|z|^2} \sum_n \frac{1}{n!} z^n |n\rangle,$$

where $|z\rangle = |z|e^{i\theta}$ and $|n\rangle$ is a state of definite particle umber n. It is a special case of (10.3.10).

Let us calculate the expectation value of the field $\hat{\psi}$ in the state $\psi(\theta)$:

$$\langle \hat{\psi} \rangle = (\psi(\theta), \hat{\psi}\psi(\theta)).$$

Using (10.3.10) we find

$$\langle \hat{\psi} \rangle = f e^{i\theta},$$

because the operator $\hat{\psi}$ connects the term $e^{i\theta N}$ to its right to the term $e^{i\theta(N+1)}$ to its left. This goes back to (10.3.7), justifying the construction (10.3.10). The operators N and θ are conjugate to each other, i.e.,

$$N = i\frac{\partial}{\partial \theta} \quad \text{or} \quad \theta = -i\frac{\partial}{\partial N}. \qquad (10.3.11)$$

In quantum mechanics the question on the quantum phase operator has been a controversial one. The problem is that as an operator $e^{i\theta}$ is not unitary, or θ is not Hermitian. Besides, the uncertainty relation $\Delta N \cdot \Delta \theta \geq 1$ is also problematic. Since the maximum value of $|\Delta \theta|$ is π, we would have violated the uncertainty relation if we demand a very small ΔN [21]. We will elaborate on this problem in chapter 12. The problem can be serious in a microscopic system. In the present context N is a macroscopic quantity, and its fluctuation is also macroscopic. Hence the relations (10.3.11) and the uncertainty relation can be used. The equations of motion for N and θ give the Ginzburg-Landau supercurrent equation and London equation [20, 22]. In a review on the theory of superfluidity P.W. Anderson[10] pointed out that "Coherence phenomena can be treated perfectly satisfactorily using the Penrose-Onsager ODLRO scheme. Unfortunately, all of those who until recently have chosen to work with this theory have made an additional assumption which is not necessary, and which makes the amount of work one has to go through in order to derive results considerably greater." The assumption is that one should always work with states with fixed, definite total particle number. The phase coherence of a Bose-Einstein condensate

---

[10]See Ref. [22], p. 231.

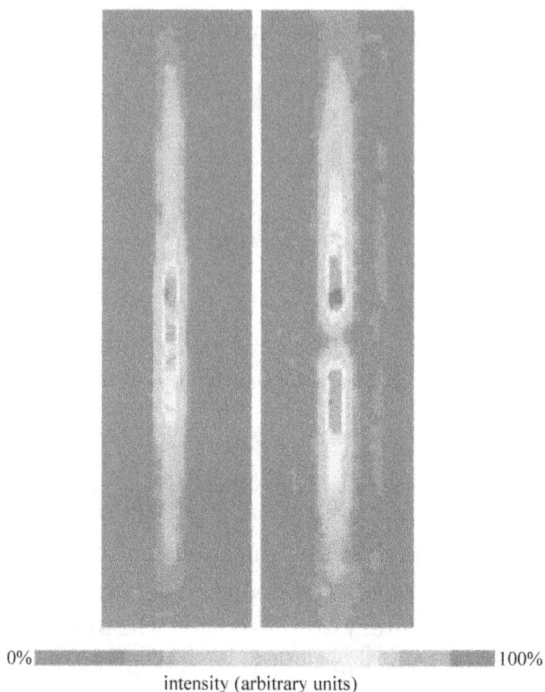

was experimentally demonstrated by W. Ketterle's group [23]. It can be considered as Young's experiment for matter waves. A cigar-shaped condensate of Na atoms is formed in a magneto-optical trap. The condensate is split-up into two by a laser focused in a plane as shown in Fig. 10.5. The magneto-optical trap is then removed, and the two condensate clouds fall freely and expand. Absorption probe measures the density of atoms in a slice in the overlapping region of the two clouds, and the interference fringes are seen as shown in Fig. 10.6. The fringe spacing is larger and the fringes more curved in the case when the initial separation of the two clouds is smaller Fig. 10.6(a). This demonstrates that phase coherence is well preserved during the formation, separation and free falling

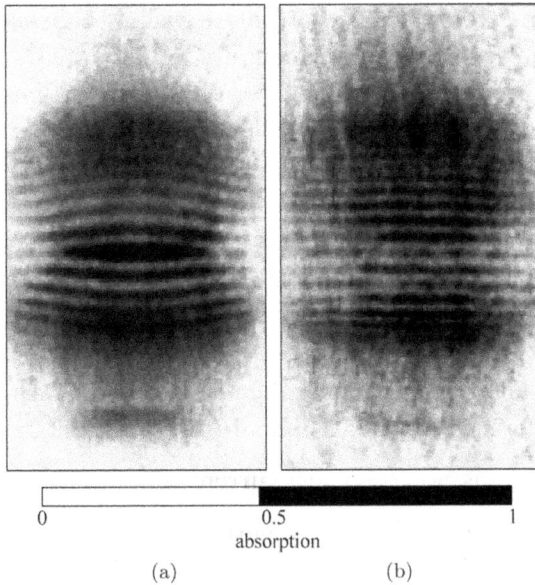

Fig. 10.6   Interference pattern of two expanding condensates, original distance between the condensates was (a) 32 μm, (b) 35 μm. Reprinted with permission from M. R. Andrews, C. G. Townsend, H.-J. Miesner, D. S. Durfee, D. M. Kurn, W. Ketterle, *Science*, 275, 637, 1997. Copyright (2017) by the AAAS.

of the condensate. The fringe spacing and curvature can be fully accounted for by the non-linear interaction between the atoms of the condensate [24].

The application of quantum optical methods in the study of BEC brought interesting results. W.D. Phillips, L. Deng and collaborators [25] used stationary wave of light as a grating on which the matter wave (the condensate) undergoes Bragg scattering, and a Mach-Zehnder Bragg interferometer for matter waves was constructed. Is it possible to prepare a condensate with a prescribed distribution of the phase? A phase imprinting method was created by W.D. Phillips' group [26]. The energy level shift of an atom in the light field is called the *light shift*. A light shift δE produces a phase shift $\delta\phi = -\frac{i}{\hbar}t\delta E$, where t is duration of the light pulse. Hence, it is possible to adjust the intensity distribution of light $I(x, y)$ and the light pulse duration t to obtain the phase distribution of the xy-plane, which can be measured by the Mach-Zehnder interferometer. Phase of the condensate (matter wave) is not only a well-established theoretical concept, but also can be experimentally manipulated.

The physics of the phase has been the subject of intensive study. The quantum phase dynamics will be discussed in Section 10.5.4. Concerning the relative phase of two condensates, P.W. Anderson raised a question [27]: Let two sets of liquid He be prepared at opposite ends of the earth and brought together, being connected through a Josephson junction. Will there be a Josephson current? The Josephson current is proportional to $\sin \delta$, where δ is the phase difference between the two condensates. They are prepared far away from each other, and did not have a chance of getting to know each other. What the phase difference can be? There are two possible answers:

Answer 1. There is no Josephson current on any trial.

Answer 2. In general a Josephson current flows, but the corresponding relative phase is random from one trial to another.

Anderson himself favors the second answer. A. Leggett and F. Sols [18] give the following answer. Consider two condensates of equal number of atoms $N_1 = N_2 = \frac{N_0}{2}$ initially. When they are brought together, the particle numbers change because of possible Josephson tunneling. Let $N_2 - N_1 = 2N$, where N is the number of transferred particles from condensate 1 to 2, and the relative phase be $\varphi = \varphi_1 - \varphi_2$. The Hamiltonian of the system is[11]

$$H = \frac{1}{2}E_J(1 - \cos \varphi) + \frac{1}{2}E_C N^2. \qquad (10.3.12)$$

Here the first term is the coupling energy, the second term is the capacitive energy. For a large homogeneous condensate E_C is very small, such that the coupling energy dominates. This means that φ can be well-defined, leaving a less well-defined particle number N. The subsequent evolution depends on the dephasing process:

(1) For long enough time of evolution, dephasing takes place and as a result two condensates with well-defined number of particles are formed. When Josephson contact is established between them, the Hamiltonian (10.3.12) applies. Then the system relaxes to the

---

[11] See Section 6.2.

ground state $\varphi = 0$. Hence there will be no Josephson current. This is answer 1.

(2) If the quantum dephasing is very small, and the two sub-systems had been subjected to some unknown potential difference, then according to Josephson relation the phase changes. They acquire a relative phase but it is random from trial to trial. This is answer 2.

Leggett and Sols conclude: "Under certain conditions the relative phase of two condensates can be meaningful, even when they are physically separated but the conditions are stringent."

Interesting studies were carried out by Y. Castin and J. Dalibard [28] and J. Javanainen and S.M. Yoo [29]. The following is a brief sketch of the paper [28]. Two independent Bose-Einstein condensates each initially contain a well defined number of atoms. It will be shown that they will appear coherent in an experiment which measures the interference between these condensates, i.e., the phase coherence is developed dynamically in the process of measurement. Figure 10.7 depicts a thought experiment. Atoms leaking from two trapped condensates A and B are detected in the output channels $+$ and $-$ of a 50%-50% beam splitter. First assume that the condensates have definite phase and the beams are described by $|\psi_0| e^{i\phi_A}$ and $|\psi_0| e^{i\phi_B}$ respectively. The intensities in the outputs of the beam splitter are

$$I_+ = 2 |\psi_0|^2 \cos^2 \phi, \quad I_- = 2 |\psi_0|^2 \sin^2 \phi, \qquad (10.3.13)$$

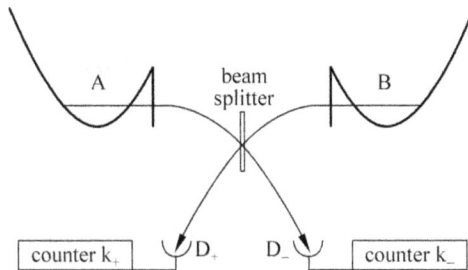

Fig. 10.7   A thought experiment: atoms leaking from two trapped condensates A and B are detected downstream the beam splitter. Reprinted with permission from Yvan Castin and Jean Dalibard, *Phys. Rev.* A55, 4330, 1997. Copyright (2017) by the American Physical Society.

where

$$\phi = \frac{1}{2}(\phi_A - \phi_B). \tag{10.3.14}$$

The probability that the counts in detectors are k_+ and k_- respectively (with $k_+ + k_- = k$) for a phase difference 2ϕ between the condensates is

$$P(k_+, k_-, \phi) = \frac{k!}{k_+!k_-!}(\cos\phi)^{2k_+}(\sin\phi)^{2k_-}. \tag{10.3.15}$$

Now let the two condensates be a number definite Fock state $|\frac{N}{2}, \frac{N}{2}\rangle$ (Anderson's original proposed condition). How is the outcome of the interference experiment? Define the phase state [18] after Leggett and Sols,

$$|\phi\rangle_N = \frac{1}{\sqrt{2^N N!}}(\hat{a}^\dagger e^{i\phi} + \hat{b}^\dagger e^{-i\phi})^N |0\rangle, \tag{10.3.16}$$

which describes an EPR state with definite $N = N_A + N_B$ and $\phi = \phi_A - \phi_B$. Any state $|\psi\rangle$ with N particles can be expressed in terms of the set of phase states

$$|\psi\rangle = \int_{-\pi/2}^{\pi/2} \frac{d\phi}{\pi} c(\phi) |\phi\rangle_N. \tag{10.3.17}$$

The Fock state $|\frac{N}{2}, \frac{N}{2}\rangle$ can be expressed as

$$\left|\frac{N}{2}, \frac{N}{2}\right\rangle = c_0 \int_{-\pi/2}^{\pi/2} \frac{d\phi}{\pi} |\phi\rangle_N, \tag{10.3.18}$$

where

$$c_0 = 2^{N/2} \frac{(N/2)!}{\sqrt{N!}}. \tag{10.3.19}$$

This shows that the phase of the Fock state is completely random. Let us begin to register the counts measuring the interference of the Fock state. After k particles have been registered ($k_+ + k_- = k$, all large), the Fock state becomes

$$|\Psi(k_+,k_-)\rangle \propto (\hat{a}+\hat{b})^{k_+}(\hat{a}-\hat{b})^{k_-}\left|\frac{N}{2},\frac{N}{2}\right\rangle$$

$$= c_0\int\frac{\mathrm{d}\phi}{\pi}(\cos\phi)^{k_+}(\sin\phi)^{k_-}|\phi\rangle_{N-K}. \qquad (10.3.20)$$

For $k_\pm \gg 1$, the integrand can be approximated by a Gaussian function around the maxima ϕ_0 and $-\phi_0$, where

$$k_+ = k\cos^2\phi_0, \quad k_- = k\sin^2\phi_0. \qquad (10.3.21)$$

The integral is evaluated to be

$$|\Psi(k_+,k_-)\rangle \propto \int_{-\pi/2}^{\pi/2}\mathrm{d}\phi\left[e^{-k(\phi-\phi_0)^2}+(-1)^{k_-}\,e^{-k(\phi+\phi_0)^2}\right]|\phi\rangle_{N-K}.$$

$$(10.3.22)$$

Thus we conclude that the initially flat distribution $c(\phi)$ =const has become a double Gaussian with maxima at $\pm\phi_0$ determined by (10.3.21). A phase difference has developed dynamically in the process of measurement. Actually the detection of one atom influences the distributions for the detection of subsequent atoms, such that the overall distribution collapses to the fringe pattern. It is the very detection process which selects a phase for the interference. It remains to be emphasized for each trial with large enough k, k_+, k_- the resulting phase ϕ_0 varies randomly.

In the first experiment on the phase difference between condensates [23], a trapped condensate was first split into two condensates by a laser beam. The two condensates were released from the trap, freely expanded and were finally overlaid. Their interference confirms the phase coherence between the condensates. After the experiment the condensates disperse. M. Saba *et al.* [30] used Bragg laser beams to extract small samples from two condensates and allowed the samples to interfere. Therefore the relative phase of the condensates can be monitored repeatedly in real time. This is the first step toward realizing an atomic interferometer, which is supposed to have a very high sensitivity and a low noise. Even if the condensates are in states with poorly defined relative phase, they can interfere with each other, and a relative phase is "created" in the measurement process projecting the system on a coherent state with a well defined relative phase, as we discussed earlier in this section. The basis of the experiment is that the structure factor of two neighboring BECs in inelastic scattering of light shows interference fringes in momentum space [31]. During an inelastic scattering the target atom receives a momentum transfer q and an energy transfer E from the photon. The rate of the inelastic scattering is proportional to the structure factor $S(q, E)$. A measurement of the scattered atom as a function of q and E gives the structure factor. When a system of two BECs with identical order parameters and a phase difference ϕ between them undergoes inelastic scattering, the structure factor is expressed as [31]

$$S(q, E) = 2 \left[1 + \cos \left(\frac{md}{qh} (E - \frac{q^2}{2m}) + \phi \right) \right] S_0(q, E), \quad (10.3.23)$$

where $S_0(q, E)$ is the structure factor of a single BEC. Therefore at high momentum transfer with E fixed, the structure factor, or the number of scattered photons exhibits oscillating behavior with period $2h/d$ when q is scanned. If q is kept constant and the relative phase ϕ is time dependent, for instance due to a difference in the depth of the wells trapping the two condensates such that $\phi = t\Delta V/h$, the number of scattered atoms will vary in time at the same rate of the relative phase. Saba *et al.* used inelastic scattering with selected recoil momentum of the atoms to obtain atom wave packets from the two condensates, i.e., of the same coherent ensemble, affecting only a small fraction of the atoms. The two components interfere, giving a stream of scattered atoms with intensities oscillating in time. This process is equivalent to the beating of two atom lasers. The experimental tool used to impart a precise momentum to atoms in a BEC is Bragg scattering. Two counterpropagating laser beams parallel to the displacement of the two condensates with wavevectors k_1 and k_2 hit the atoms so that atoms absorbing the photon k_1 and emitting k_2 under stimulation acquire a recoil momentum $\hbar(k_1 - k_2)$, provided that the energy transfer matches the atom recoil energy. When the atoms from the first condensate reach the second one, the two streams of atoms moving with the same momentum q will overlap and interfere. In the experiment two cigar-shaped BECs of Na atoms were prepared in a double well optical trap and were illuminated by the Bragg laser beams. The Bragg scattered atoms escape from the trap, because the recoil energy is much greater than the trap barrier. In the stream of outcoupled atoms the spatial modulations in the absorption images reflect temporal oscillation in the number of outcoupled atoms, as shown, in Fig. 10.8.

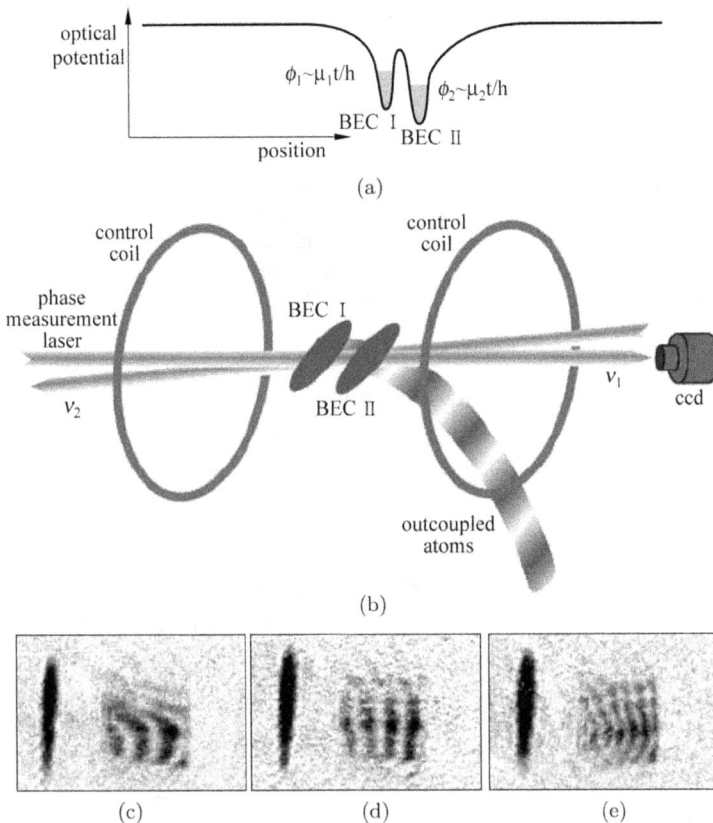

Fig. 10.8 (a) BECs confined in a double well trap. (b) Experimental scheme. (c)~(e) Absorption images showing the optical density of the atomic clouds, with magnetic field gradients 11.5G/cm (c), −0.77 G/cm (e). The difference in magnetic field between the two wells modifies the energy offset and therefore changes the beat frequency. Reprinted with permission from M. Saba, T. A. Pasquini, C. Sanner, Y. Shin, W. Ketterle, D. E. Pritchard, *Science*, 307, 1945, 2005. Copyright (2017) by the AAAS.

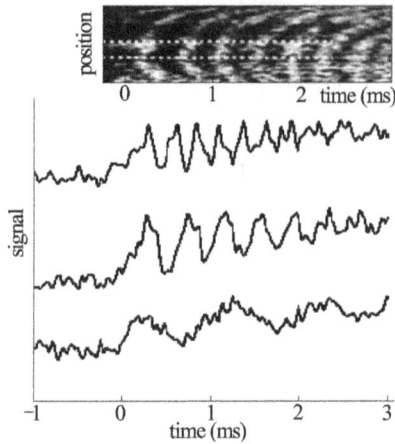

Fig. 10.9   Continuous optical readout of the relative phase of the condensates. The upper panel shows the optical signal image detected by the CCD camera. In the lower panel the signals with different magnetic field gradients are plotted versus time. The offset of traces of optical signals with different magnetic field gradients is made for clarity. Reprinted with permission from M. Saba, T. A. Pasquini, C. Sanner, Y. Shin, W. Ketterle, D. E. Pritchard, *Science*, 307, 1945, 2005. Copyright (2017) by the AAAS.

For each atom outcoupled from the condensate, a photon was transferred from one beam to the other one. Therefore information contained in the stream of outcoupled atoms is also contained in the scattered photons and can be gathered in real time by monitoring the intensity of one of the beams by a charge-coupled device camera in streaking mode. This is shown in Fig. 10.9.

The experimental result represents a real progress toward practical applications of atom interferometry. J. Javanainen commented [92] that the situation reminds lasers in the 1960s that they are a solution looking for problems.

10.4   Weakly Interacting Bose Gas: Homogeneous Condensate

BEC can occur when there is no interaction between atoms, it is the consequence of the symmetry of wave function of a system of bosons. In this sense the Bose-Einstein phase transition can be considered as a transition of purely quantum mechanical origin. But the physics of BEC depends decisively on the interaction between atoms. We have explained that for ideal Bose gas the BEC is a first order phase transition, but for the liquid He we cannot make this conclusion because of the relatively strong interaction. The superfluidity of the condensate of gaseous atoms depends crucially on the interaction. In superfluidity there is a critical velocity v_{cr}. When the fluid velocity becomes larger than this value, excitations can occur which lead to energy transfer, and viscosity appears. The state of superflow is thus violated. The critical velocity of an ideal Bose gas is zero, which means a non-viscous flow is unstable against the motion of the fluid, no matter how small the velocity is.[12] The stability of a condensate of gaseous atoms depends on the nature of interaction, attractive or repulsive.

We consider the Bose gas without external field, its condensate is homogeneous. The case $T = 0$ is a quantum mechanical many-body problem, and the finite temperature case needs in addition statistical mechanics.

10.4.1 Bogoliubov Theory of Weakly Interacting Bose Gas

N.N. Bogoliubov created his theory of weakly interacting Bose gas in 1947 in which quasiparticles are introduced by carrying out a transformation which is called subsequently Bogoliubov transformation. The theory is widely used in the theory of superfluidity and superconductivity. At $T = 0$ the ground state of the ideal Bose gas is $N_0 = N$, $N_i = 0$ $(i \neq 0)$. In the presence of interaction, some pairs can appear on states $(\mathbf{k}_1, -\mathbf{k}_1), (\mathbf{k}_2, -\mathbf{k}_2), \ldots$ as demanded by the conservation of momentum during the 2-body interaction. Weakly interacting Bose gas in its ground state can have any number of such pairs, the coherent superposition of which leads to a minimization of the energy. The second quantized form of the interaction is

$$\frac{1}{2} \int \psi^\dagger(\mathbf{x})\psi^\dagger\left(\mathbf{x}'\right) V\left(\mathbf{x} - \mathbf{x}'\right) \psi\left(\mathbf{x}'\right) \psi(\mathbf{x}) \mathrm{d}^3x \mathrm{d}^3x'. \qquad (10.4.1)$$

---

[12]See Ref. [32], p. 75 and Ref. [11], p. 35.

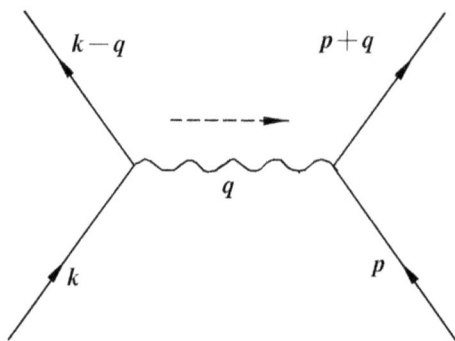

Fig. 10.10   Transition induced by the two-body interaction.

It becomes in the momentum space

$$\sum_{k,\,p,\,q} \frac{V_q}{2} a^\dagger_{p+q} a^\dagger_{k-q} a_k a_p, \qquad (10.4.2)$$

where V_q is the Fourier component of $V(x)$ for momentum transfer q, and $V_q = V_{-q}$. Figure 10.10 shows the transition induced by the interaction. The unperturbed ground state is $|\Phi_0(N)\rangle$ (the vacuum), and the result of operating a_0 and $a^\dagger_0$ is respectively

$$\left.\begin{aligned} a_0|\Phi_0(N)\rangle &= \sqrt{N}\,|\Phi_0(N-1)\rangle \\ a^\dagger_0|\Phi_0(N)\rangle &= \sqrt{N+1}\,|\Phi_0(N+1)\rangle \end{aligned}\right\} \qquad (10.4.3)$$

Due to the existence of the condensate, the operator a_0 does not annihilate the vacuum as in the usual field theory where we have $a|\text{vac}\rangle = 0$, $\langle\text{vac}|a^\dagger = 0$, and the vacuum expectation value of a normal product of operators vanishes. Here because of (10.4.3) the vacuum expectation value of the normal product of a_0 and $a^\dagger_0$ does not vanish such that the many-particle problem cannot be founded on the basis of usual field theory. Consider that N_0 is a macroscopic quantity, $N_0 - 1$, N_0, and $N_0 + 1$ do not differ very much, and $|N_0 - 1\rangle$, $|N_0\rangle$, and $|N_0 + 1\rangle$ do not differ much either. Consider the relations

$$a^\dagger_0 a_0 |\Phi_0(N_0)\rangle = N_0|\Phi_0(N_0)\rangle$$

and

$$a_0 a_0^\dagger \left| \Phi_0 \left(N_0 \right) \right\rangle = (N_0 + 1) \left| \Phi_0 \left(N_0 \right) \right\rangle \approx N_0 \left| \Phi_0 \left(N_0 \right) \right\rangle,$$

we observe that it is possible to make the following replacement:

$$a_0 \to \sqrt{N_0}, \ a_0^\dagger \to \sqrt{N_0}, \qquad (10.4.4)$$

i.e., we treat them as c-numbers. All other annihilation operators a_k ($k \neq 0$) give, when operating on the ground state $\left| \Phi_0(N) \right\rangle = \left| N, 0, \ldots, 0 \right\rangle$, the conventional result

$$a_k \left| N_0, 0, \ldots, 0 \right\rangle = 0, \quad k \neq 0. \qquad (10.4.5)$$

The boson field operator is now

$$\hat{\psi}(\boldsymbol{x}) = \sqrt{n_0(\boldsymbol{x})} + \hat{\varphi}(\boldsymbol{x}), \qquad (10.4.6)$$

where

$$\hat{\varphi}(\boldsymbol{x}) = \frac{1}{\sqrt{V}} \sum_{\boldsymbol{k}}{}' \mathrm{e}^{\mathrm{i} k \cdot \boldsymbol{x}} a_{\boldsymbol{k}}. \qquad (10.4.7)$$

Here $n_0(\boldsymbol{x})$ is the particle number density on the state $\boldsymbol{k} = 0$, the prime on the summation sign means that the state $\boldsymbol{k} = 0$ is not included, and operator $\hat{\varphi}$ contains annihilation operators for all states $\boldsymbol{k} \neq 0$. Then the particle number on states $\boldsymbol{k} \neq 0$

$$\hat{N}' = \sum_{\boldsymbol{p}}{}' a_{\boldsymbol{p}}^\dagger a_{\boldsymbol{p}} \qquad (10.4.8)$$

is no more conserved. Instead, an extra condition must be introduced

$$\langle \hat{N}' \rangle = N - N_0. \qquad (10.4.9)$$

In writing the Hamiltonian we can classify the interaction terms according to whether one or both of $\boldsymbol{p}$ and $\boldsymbol{q}$ belongs to the category of $\boldsymbol{k} = 0$, and the result consists of 8 terms named (a) to (h) according to their order of appearance:

$$\hat{H} = \underbrace{\sum_{p}{}'(\varepsilon_p^0 + N_0 V_0}_{(a)} + \underbrace{N_0 V_p)a_p^{\dagger}a_p}_{(b)} + + \sum_{p}{}'\underbrace{\frac{N_0 V_p}{2}(a_p^{\dagger}a_{-p}^{\dagger}}_{(c)} + \underbrace{a_p a_{-p})}_{(d)}$$

$$+ \underbrace{\frac{1}{2}N_0^2 V_0}_{(e)} + \sum_{pq}{}'\sqrt{N_0}\frac{V_q}{2}(\underbrace{a_{p+q}^{\dagger}a_q a_p}_{(f)} + \underbrace{a_{p+q}^{\dagger} + a_{-q}^{\dagger}a_p}_{(g)})$$

$$+ \sum_{kpq}{}'\underbrace{\frac{V_q}{2}a_{p+q}^{\dagger} + a_{k-q}^{\dagger}a_k a_p,}_{(h)} \tag{10.4.10}$$

where ε_p^0 is the free particle energy

$$\varepsilon_p^0 = \frac{1}{2m}p^2. \tag{10.4.11}$$

Figure 10.11 depicts contributions from (a) to (h), where solid lines represent particles with non-vanishing momenta, dashed lines represent particles with zero momentum, wavy lines represent the interaction. Physical significance of the related contributions can be read from the diagrams. For example, Fig. 10.11 (c) represents the process in which two particles from the condensate interact with a momentum transfer p, and as the result change into particles with momenta $-p$ and p outside the condensate. Diagrams containing one dashed line have a factor $\sqrt{N_0}$ originating from the replacement of a_0 or $a_0^{\dagger}$ by $\sqrt{N_0}$. Diagrams with two dashed lines have a factor

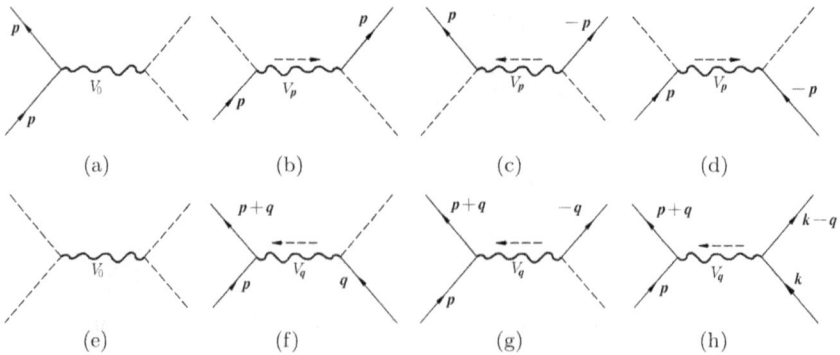

Fig. 10.11   Transitions induced by the two-body interaction.

N_0. Diagrams with 4 dashed lines have a factor N_0^2. Because of momentum conservation there cannot be a diagram with 3 dashed lines and one solid line. From (10.4.10) we see that $\sum_p {}' a_p^\dagger a_p$ does not commute with $\hat{H}$. To find the energy spectrum of the interacting system it is necessary to diagonalize the Hamiltonian under the supplementary condition (10.4.9), i.e., to diagonalize the grand canonical Hamiltonian

$$\hat{K} = \hat{H}(N_0) - \mu \hat{N}' \qquad (10.4.12)$$

where the chemical potential μ serves as the Lagrange multiplier. This method was suggested by N.M. Hugenholtz and D. Pines [33]. Let eigenvalues of operators $\hat{K}$, $\hat{H}(N_0)$ and $\hat{N}'$ be denoted by $E_0'(N_0, \mu)$, $E_0(N_0, \mu)$ and $N'(N_0, \mu)$ respectively. We can use

$$N'(N_0, \mu) + N_0 = N \qquad (10.4.13)$$

to find $\mu(N, N_0)$. Substituting μ into $E_0(N_0, \mu)$ and using

$$\frac{\mathrm{d}}{\mathrm{d}N_0} E_0 = 0 \qquad (10.4.14)$$

we can find N_0. In solving this problem the condition $N - N_0 \ll N_0$ is not assumed, so that the result is applicable for the case of large depletion. The case $N - N_0 \ll N_0$ is simpler to handle. There are two useful relations [33] which can be used to give μ directly. The quantum mechanical ground state has zero entropy $S = 0$, therefore, Eq. (10.1.19) gives $\mu = \frac{\partial E_0(V,N)}{\partial N}$. Since $E_0' = E_0 - \mu N' = E_0 + \mu N_0 - \mu N$, it follows that $\frac{\partial E_0'}{\partial N_0} = \frac{\partial E_0}{\partial N_0} + \mu$. The first term of the right hand side vanishes for the ground state, and we obtain finally

$$\mu = \frac{\partial E_0}{\partial N} = \frac{\partial E_0'}{\partial N_0}. \qquad (10.4.15)$$

Only terms proportional to N_0 and N_0^2 are kept in $\hat{K}$ in the *weak coupling* approximation of Bogoliubov:

$$\hat{K} = \sum_q {}' \tilde{\varepsilon}_q a_q^\dagger a_q + \sum_q {}' N_0 \frac{V_q}{2} \left(a_q^\dagger a_{-q}^\dagger + a_q a_{-q} \right) + \frac{1}{2} N_0^2 V_0,$$

$$(10.4.16)$$

where

$$\tilde{\varepsilon}_q = \varepsilon_q^0 + N_0 V_0 + N_0 V_q - \mu. \tag{10.4.17}$$

The second term in (10.4.16) is not diagonal to the occupation number space and needs diagonalization. To this end, we consider first the evolution of the operator a_p. Direct calculation gives

$$i\hbar \frac{da_p}{dt} = [a_p, \hat{K}]$$

$$= \tilde{\varepsilon}_p a_p + N_0 V_p a_{-p}^\dagger.$$

The first term on the right-hand side after the second equality sign can be expected. If the second term does not exist, the equation would be the time-dependent Schrödinger equation. Begin with a_p, it gets mixed with $a_{-p}^\dagger$ during evolution because of the interaction as the second term indicates. From this hint we define the quasiparticle operators α_p and $\alpha_p^\dagger$ to implement the mixing:

$$\left. \begin{array}{l} \alpha_p = u_p a_p - v_p a_{-p}^\dagger \\ \alpha_p^\dagger = u_p a_p^\dagger - v_p a_{-p} \end{array} \right\}, \tag{10.4.18}$$

where u_p and v_p are real functions of p. The aim of this transformation is to make $\hat{K}$ diagonal in the quasiparticle number space. In order that the operators α_p and $\alpha_p^\dagger$ satisfy the commutation relations

$$\left[\alpha_p, \alpha_q^\dagger \right] = \delta_{pq}, \tag{10.4.19}$$

it is necessary that

$$u_p^2 - v_p^2 = 1. \tag{10.4.20}$$

Diagonalization demands that

$$\hat{K} = \text{const} + \sum_q{}' \omega_q \alpha_q^\dagger \alpha_q$$

and this leads to

$$[\alpha_p, \hat{K}] = \omega_p \alpha_p, \quad \left[\alpha_p^\dagger, \hat{K} \right] = -\omega_p \alpha_p^\dagger. \tag{10.4.21}$$

Substituting (10.4.18) into (10.4.16) and using (10.4.21), we obtain

$$\left[a_p^\dagger, \hat{K}\right] = -u_p\left(\tilde{\varepsilon}_p a_p^\dagger + N_0 V_p a_{-p}\right) - v_p\left(\tilde{\varepsilon}_p a_{-p} + N_0 V_p a_p^\dagger\right)$$
$$= -\omega_p\left(u_p a_p^\dagger - v_p a_{-p}\right),$$

consequently

$$\tilde{\varepsilon}_p u_p + N_0 V_p v_p = \omega_p u_p,$$
$$N_0 V_p u_p + \tilde{\varepsilon}_p v_p = -\omega_p v_p.$$

The condition for the existence of a non-zero solution to this system of equation is

$$\omega_p^2 = \tilde{\varepsilon}_p^2 - N_0^2 V_p^2, \tag{10.4.22}$$

$$\left.\begin{array}{l} u_p^2 = \dfrac{1}{2}\left[1 + \dfrac{\tilde{\varepsilon}_p}{\omega_p}\right] \\[3mm] v_p^2 = \dfrac{1}{2}\left[\dfrac{\tilde{\varepsilon}_p}{\omega_p} - 1\right] \end{array}\right\}. \tag{10.4.23}$$

After diagonalization $\hat{K}$ becomes

$$\hat{K} = \sum_q{}'\left\{\omega_q \alpha_q^\dagger \alpha_q + \frac{1}{2}\left(\omega_q - \tilde{\varepsilon}_q\right)\right\} + \frac{1}{2}N_0^2 V_0. \tag{10.4.24}$$

The transformation (10.4.18) is called the Bogoliubov transformation, and its inverse is

$$\left.\begin{array}{l} a_p = u_p \alpha_p - v_p \alpha_{-p}^\dagger \\[2mm] a_p^\dagger = u_p \alpha_p^\dagger - v_p \alpha_{-p}. \end{array}\right\} \tag{10.4.25}$$

Bogoliubov transformation leads to the quasiparticle occupation space. The vacuum state is called the *quasiparticle vacuum*, denoted by $|0\rangle$, satisfying the relation

$$\alpha_p |0\rangle = 0. \tag{10.4.26}$$

The significance of the transformation can be understood in the following. The ground state of the system at $T = 0$ consists of macroscopic number of particles populated in the $p = 0$ state and also a number of pairs $(q, -q)$ on states $q \neq 0$ because of the interaction. For a definite q, a pair may exist, and may not exist. Therefore, in order to produce a quasiparticle with momentum q, we can use $a_q^\dagger$ if the pair state q is unoccupied, and we have to use a_{-q} if the pair state is occupied such that the particle with momentum q survives when its partner with momentum $-q$ is annihilated.. In order to determine w_p we need to determine μ. The relation (10.4.15) can be used for this purpose. The leading term of E'_0 is $\frac{1}{2}N_0^2 V_0$, and we get

$$\mu = \frac{\partial E'_0}{\partial N_0} = N_0 V_0. \tag{10.4.27}$$

Using (10.4.17) and (10.4.22) we obtain

$$\tilde{\varepsilon}_q = \frac{1}{2m}q^2 + N_0 V_q, \tag{10.4.28}$$

$$w_q = \left(\frac{q^2 N_0 V_q}{m} + \frac{q^4}{4m^2}\right)^{1/2}. \tag{10.4.29}$$

From (10.4.22) we see that $w_p < \tilde{\varepsilon}_p$, so that the term $\frac{1}{2}(w_q - \tilde{\varepsilon}_q)$ in the vacuum state energy is negative. We discussed in Section 10.2 the number locked state and the phase locked state that have the same energy because they are related by a Fourier transform. When different number of pair states of different momenta are coherently superimposed with the particle vacuum to form the physical vacuum of the interacting system, the energy of the system is lowered at the price of tolerating the particle number fluctuation around the average value. The interaction chooses such a state as the ground state of the condensate.

Taking the low momentum limit of (10.4.29), we obtain the dispersion relation

$$w_q = q\sqrt{\frac{N_0 V_0}{m}}, \tag{10.4.30}$$

which is a collective phonon state with the sound velocity

$$s_B = \left(\frac{N_0 V_0}{m}\right)^{1/2}.$$

The high momentum limit gives

$$\omega_q = \tilde{\varepsilon}_q = \frac{q^2}{2m} + N_0 V_q, \qquad (10.4.31)$$

where the first term is the particle kinetic energy, the second term is the average potential energy of a non-condensate particle. Therefore $\tilde{\varepsilon}_q$ is the single particle Hartree-Fock energy. Bogoliubov theory can be used in discussing the response of the condensate to external perturbations, the motion of superfluid and so on [32].

10.4.2 Non-ideal Bose gas

The study of $T \neq 0$ non-ideal Bose gas begins with the calculation of the partition function. One of the methods is the cluster integral in statistical mechanics [9]. We can also determine the energy spectrum of the system, and then obtain the partition function directly by $Q_N(V,T) = \sum_n e^{-\beta \varepsilon_n}$. When the interaction potential is simple or made simple by approximations so that the energy spectrum is obtained by a perturbation calculation, such an approach is feasible. Actually the interaction potential between atoms is rather complicated. It has a strongly repulsive core, and is called the hard core potential. Because the core is rather small, the pseudopotential approximation can be used. It leads to an *effective potential* expressed in terms of a low energy phase shift parameter. This makes the perturbation calculation possible. The theory of dilute Bose gas of hard spheres is given by C.N. Yang, T.D. Lee and K. Huang in the late 1950s [34, 35].[13] In the quantum theory of scattering the phase shift in low energy scattering is characterized by a parameter, the scattering length, and is independent of the detailed shape of the potential. For $r \to \infty$ the scattering wave function is

$$r\psi_\infty = \text{const} \left(\sin kr + \tan \eta_0 \cos kr \right),$$

---

[13]See Chapters 10 and 13 of Ref. [9].

where k is the magnitude of the wave vector, η_0 is the phase shift. For the low energies $k \to 0$ we have

$$\psi_\infty \xrightarrow{k \to 0} \text{const}\left(1 + \frac{\tan \eta_0}{kr}\right),$$

where

$$\tan \eta_0 = -ka.$$

Here a is the scattering length, $a > 0$ for a repulsive potential and $a < 0$ for an attractive potential. The hard sphere potential with core radius a has a scattering length equal to the core radius. There are 3 parameters for non-ideal gas at low temperatures, namely the thermal de Broglie wave length λ, average distance between atoms $v^{\frac{1}{3}}$ and the scattering length a. In the pseudopotential method a/λ and $a/v^{\frac{1}{3}}$ are both small parameters, the former being the condition for low temperature and the latter for diluteness. The equation for hard sphere potential scattering and the boundary conditions are given below:

$$\left.\begin{array}{ll}(\nabla^2 + k^2)\psi = 0, & r > a \\ \psi(r) = 0, & r < a\end{array}\right\}. \qquad (10.4.32)$$

E. Fermi proved in 1936 that for the wave function in the region $r > a$ the scattering problem is equivalent to that due to an effective potential

$$V = \frac{4\pi a\hbar^2}{m}\sum_{i<j}\delta^3(\mathbf{r}_i - \mathbf{r}_j)\frac{\partial}{\partial r_{ij}}r_{ij}. \qquad (10.4.33)$$

Then we have achieved the following. First the interaction between particles is reduced to that of hard sphere potential, and then the boundary condition in (10.4.32) is incorporated in the effective potential. It is called the pseudopotential method because there is an operator of taking derivative besides the dependence on relative distance. The operator $\frac{\partial}{\partial r}r$ gives the result $\frac{\partial}{\partial r}r\psi = \psi + r\frac{\partial\psi}{\partial r}$ when operating on a wave function ψ. Under the condition that there is a delta function $\delta(r)$ in front, we have $\frac{\partial}{\partial r}r\psi \to \psi$ if ψ is regular at the origin, which means $\frac{\partial}{\partial r_{ij}} r_{ij}$ can be replaced by 1. In calculations involving first order perturbation the pseudopotential operates on the unperturbed wave function only, and $\frac{\partial}{\partial r_{ij}} r_{ij}$ can actually be replaced by 1. In higher order results its role is non-trivial and leads to important results [34, 35]. We restrict ourselves to first order perturbation, then the Hamiltonian is

$$H = \frac{\hbar^2}{2m} \left(\nabla_1^2 + \cdots + \nabla_n^2 \right) + \frac{4\pi a \hbar^2}{m} \sum_{i<j} \delta^3(\boldsymbol{r}_i - \boldsymbol{r}_j), \qquad (10.4.34)$$

The unperturbed state is characterized by $\{n_{\boldsymbol{p}}\}$, where $n_{\boldsymbol{p}}$ is the number of particles on the state with momentum $\boldsymbol{p}$:

$$\Phi_n = \{ \cdots n_{\boldsymbol{p}} \cdots \}. \qquad (10.4.35)$$

The expectation value of energy in the state Φ_n is[14]

$$E_n = (\Phi_n, H \Phi_n) = \sum_{\boldsymbol{p}} \frac{p^2}{2m} n_{\boldsymbol{p}} + \frac{4\pi a \hbar^2}{mV} \left(N^2 - \frac{1}{2} \sum_{\boldsymbol{p}} n_{\boldsymbol{p}}^2 \right).$$

$$(10.4.36)$$

The occupation number of states $\boldsymbol{p} \neq 0$ is very low at low temperatures, and $\sum n_{\boldsymbol{p}}^2$ for $\boldsymbol{p} \neq 0$ states in the parenthesis can be neglected. Hence

$$E_n = \sum_{\boldsymbol{p}} \frac{p^2}{2m} n_{\boldsymbol{p}} + \frac{4\pi a \hbar^2}{mV} \left(N^2 - \frac{1}{2} n_0^2 \right). \qquad (10.4.37)$$

The partition function is then

$$Q_N = \sum_{\{n_{\boldsymbol{p}}\}} \exp \left\{ -\beta \left[\sum_{\boldsymbol{p}} \frac{p^2}{2m} n_{\boldsymbol{p}} + \frac{4\pi a \hbar^2}{mV} \left(N^2 - \frac{1}{2} n_0^2 \right) \right] \right\}.$$

$$(10.4.38)$$

---

[14]See Appendix A of Ref. [9].

Further calculation gives[15]

$$
\frac{1}{N}\ln Q_N =
\begin{cases}
\dfrac{v}{\lambda^3}g_{5/2}(z) - \ln z - \dfrac{2a\lambda^2}{v}, & v > v_c, T > T_E \\[2ex]
\dfrac{v}{\lambda^3}g_{5/2}(1) - 2\dfrac{a\lambda^2}{v}\left[1 - \dfrac{1}{2}\left(1 - \dfrac{v}{v_c}\right)^2\right], & v < v_c, T < T_E
\end{cases}
$$

$$(10.4.39)$$

From this partition function the macroscopic properties of the system follow. For instance, for $T > T_E$ we have

$$
\lambda^3\frac{P}{kT} = g_{5/2}(z) - \frac{2a}{\lambda}\left(g_{3/2}(z)\right)^2 + \cdots
$$

$$
\rho = z\frac{\partial}{\partial z}\left(\frac{P}{kT}\right) = \lambda^{-3}\left(g_{3/2}(z) - \frac{4a}{\lambda}g_{3/2}(z)g_{1/2}(z) + \cdots\right).
$$

10.5   Weakly Interacting Bose Gas: Inhomogeneous Condensate

10.5.1   Dependence of the properties of the condensation on the external fields

In the experimental study of BEC of atomic gases, magnetic field and magneto-optical trap or optical trap are used. When external field exists the density of condensates is a function of position. Properties of condensates in an external field, such as the critical temperature, fraction of particles in the condensate, heat capacity of the Bose gas and so on, all depend on the form of the potential [36]. Consider first N ideal Bose gas atoms in the external field $V(\boldsymbol{r})$ at temperature T. The gas atoms are distributed by their energies ε according to

$$
n_\varepsilon = \frac{g_\varepsilon}{e^{(\varepsilon-\mu)/kT} - 1}, \tag{10.5.1}
$$

where g_ε is the degree of degeneracy of the energy level, μ is the chemical potential. The energy of the ground state is taken to be zero. Assume that the interval between energy levels is much less

---

[15]See Chapter 12 of Ref. [9].

than kT, i.e., $\varepsilon_{i+1} - \varepsilon_i \ll kT$. The density of states depends on the potential of external field $V(\boldsymbol{r})$, which can be estimated in the semi-classical approximation. There is one energy state in a volume h^3 of the phase space. A particle of energy ε is enclosed by the potential in a volume $\mathscr{V}(\varepsilon)$, and the volume element of the momentum space is $4\pi p^2 dp = 4\pi m p d\varepsilon = 4\pi m (2m)^{\frac{1}{2}} \sqrt{\varepsilon - V(\mathrm{r})} d\varepsilon$. Hence the density of states is

$$\rho(\varepsilon) = \frac{(2\pi)(2m)^{3/2}}{h^3} \int_{\mathscr{V}(\varepsilon)} \sqrt{\varepsilon - V(\boldsymbol{r})} d^3 r. \tag{10.5.2}$$

The relation between the chemical potential and the total number of particles can be determined using

$$N = N_0 + \int_0^\infty n_\varepsilon \rho(\varepsilon) \, d\varepsilon. \tag{10.5.3}$$

Since $\rho(0) = 0$, the integral does not contain the number of particles in the ground state.

$$N = N_0 + \sum_{j=1}^\infty \exp\left(\frac{j\mu}{kT}\right) \int_0^\infty \rho(\varepsilon) \exp\left(-\frac{j\varepsilon}{kT}\right) d\varepsilon, \tag{10.5.4}$$

Taking the leading term and performing the integral, we obtain $\mu(T)$.

The critical temperature T_c can be found by putting $N_0 = 0$ and $\mu = 0$ in (10.5.3) and solving for T, similar to the case of free particles. The total energy of the system is

$$E(T) = \int_0^\infty \varepsilon n_\varepsilon \rho(\varepsilon) \, d\varepsilon. \tag{10.5.5}$$

The heat capacity can be found from (10.5.5):

$$C(T) = \frac{\partial E(T)}{\partial T}$$

$$= \frac{1}{kT} \int_0^\infty \frac{\varepsilon \rho(\varepsilon)}{g_\varepsilon} (n_\varepsilon)^2 \left[\frac{\partial \mu}{\partial T} + \frac{\varepsilon - \mu}{T}\right] \exp\left(\frac{\varepsilon - \mu}{kT}\right) d\varepsilon. \tag{10.5.6}$$

Table 10.5.1   Influence of the potential on the properties of the condensates.

potential	T_c	$\dfrac{N_0}{N}$	$C(T_{c-})/Nk$
3-dimensional box	$\dfrac{2\pi\hbar^2}{km}\left(\dfrac{N}{2.612V}\right)^{\frac{2}{3}}$	$1 - \left(\dfrac{T}{T_c}\right)^{\frac{3}{2}}$	1.92
isotropic harmonic oscillator $V(r) = \varepsilon_1\left(\frac{r}{a}\right)^2$	$\sqrt{\dfrac{2}{m}}\dfrac{\hbar}{k}\left(\dfrac{N}{1.202}\right)^{\frac{1}{3}}\left(\dfrac{\varepsilon_1}{a^2}\right)^{\frac{1}{2}}$	$1 - \left(\dfrac{T}{T_c}\right)^{3}$	10.82

Anisotropic power potentials are studied in Ref. [36]. We give in Table 10.5.1 an example to illustrate the influence of the potential on various parameters.

We generalize the above results to weakly interacting Bose gas. From the two-body interaction potential (10.4.34) we obtain the average interaction energy per particle $E_{\text{int}} = \frac{2\pi a\hbar^2}{m}\left(\frac{dN}{d\mathcal{V}}\right)$, where $\left(\frac{dN}{d\mathcal{V}}\right)$ is the particle number density.[16] The Hamiltonian in the semi-classical approximation can be written as

$$H(\boldsymbol{p}, \boldsymbol{r}) = \frac{p^2}{2m} + V(\boldsymbol{r}) + \frac{2\pi a\hbar^2}{m} n(\boldsymbol{r}). \qquad (10.5.7)$$

The particle density in the trap $n(\boldsymbol{r})$ is

$$\frac{dN}{d\mathcal{V}} = \sum n_\varepsilon |\psi_\varepsilon|^2,$$

which is difficult to calculate precisely. Using the semi-classical approximation we have

$$dN = \frac{1}{h^3} n(\boldsymbol{p}, \boldsymbol{r}) d^3 p \, d^3 r, \qquad (10.5.8)$$

where

$$n(\boldsymbol{p}, \boldsymbol{r}) = \frac{1}{e^{[(\boldsymbol{p}^2/2m)+V(\boldsymbol{r})-\mu]/kT} - 1}. \qquad (10.5.9)$$

---

[16]From (10.4.38) we find the total interaction energy $\frac{2\pi a\hbar^2}{m\mathcal{V}}N^2$ for $n_0 = N$. Therefore, the interaction energy per particle is $\frac{2\pi a\hbar^2}{m}\frac{N}{\mathcal{V}}$.

The particle number density is

$$\left(\frac{\mathrm{d}N}{\mathrm{d}\mathscr{V}}\right)_{\boldsymbol{p},\boldsymbol{r}} = \frac{1}{h^3}n(\boldsymbol{p},\boldsymbol{r})\mathrm{d}^3p.$$

Integration over momentum gives

$$\left(\frac{\mathrm{d}N}{\mathrm{d}\mathscr{V}}\right)_{\boldsymbol{r}} = \frac{4\pi}{h^3}\int p^2 n(\boldsymbol{p},\boldsymbol{r})\mathrm{d}p. \qquad (10.5.10)$$

Let $x = \frac{p^2}{2mkT}$, the above expression becomes

$$\left(\frac{\mathrm{d}N}{\mathrm{d}\mathscr{V}}\right)_{\boldsymbol{r}} = \frac{2}{\sqrt{\pi}}\frac{1}{\lambda^3}\int_0^\infty x^{1/2}\frac{1}{e^{x+(V(\boldsymbol{r})/kT)-(\mu/kT)}-1}$$

$$= \frac{2}{\sqrt{\pi}}\frac{1}{\lambda^3}\sum_{t=0}^\infty \exp\left[-t\left(\frac{V}{kT}-\frac{\mu}{kT}\right)\right]\int_0^\infty e^{-xt}x^{1/2}\mathrm{d}x,$$

$$(10.5.11)$$

where λ is the thermal de Broglie wave length, and $\frac{1}{1-y} = \sum_{t=0}^\infty y^t$ has been used. Carrying out the integration over x we obtain

$$\left(\frac{\mathrm{d}N}{\mathrm{d}\mathscr{V}}\right)_{\boldsymbol{r}} = \frac{1}{\lambda^3}\sum_{t=1}^\infty e^{t[\mu-V(\boldsymbol{r})]/kT}\frac{1}{t^{3/2}}. \qquad (10.5.12)$$

When $\mu \lesssim 0$, the sum can be replaced by its leading term $t=1$, and (10.5.7) becomes

$$H(\boldsymbol{r},\boldsymbol{p}) = \frac{p^2}{2m} + V_{\text{eff}}(\boldsymbol{r}),$$

where

$$V_{\text{eff}}(\boldsymbol{r}) = V(\boldsymbol{r}) + \frac{2\pi\hbar^2 a}{m\lambda^3}e^{-V(\boldsymbol{r})/kT}. \qquad (10.5.13)$$

Results for a cylindrical potential can be found in Ref. [36].

It is interesting to consider BEC in lower dimensions. We begin from (10.5.4). The second term on the right-hand side gives N_{ex}, the number of particles on the excited states. It depends critically on the density of states $\rho(\varepsilon)$. For free particles in 3 dimensions $\rho(\varepsilon) \propto \sqrt{\varepsilon}$. The integral gives $N_{\text{ex}} \propto (kT)^{\frac{3}{2}}g_{\frac{3}{2}}(z)$. This is the first term of

(10.1.34). For a two-dimensional Bose gas, the volume element of the momentum space is $2\pi p dp \propto d\varepsilon$, and the level density $\rho(\varepsilon)$ is independent of ε, hence

$$N_{\mathrm{ex}} \propto kT g_1(z).$$

The Bose function is defined by $g_p(z) = \sum_{l=1}^{\infty} \frac{z^l}{p^l}$, and we see that $g_1(1)$ diverges. This means that under finite temperature the number of particles on the excited states is not limited, and BEC of ideal Bose gas cannot happen in 2 dimensions. The crucial role is played by the density of states. When an external field exists, particles of a given energy is restricted in a corresponding volume, which makes T_{c} finite [37]. Specifically, for an isotropic harmonic oscillator potential of frequency ω in 3 dimensions, the volume $\mathscr{V}(\varepsilon)$ containing the particles depends on the energy ε. Semiclassical calculation gives $\mathscr{V}(\varepsilon) \propto \varepsilon^{3/2}$, and the volume element of the momentum space $\propto \varepsilon^{1/2}$, which together leads to the phase space volume $\propto \varepsilon^2$, and hence $N_{\mathrm{ex}} \propto (kT)^3 g_3(z e^{-\frac{3}{2}\frac{\hbar\omega}{kT}})$. Owing to the property of the function g_3, N_{ex} is suppressed, and this eases the condensation. For two-dimensional case, $\rho(\varepsilon) \propto \varepsilon$, $N_{\mathrm{ex}} \propto (kT)^2 g_2(z e^{-\frac{\hbar\omega}{kT}})$, and N_{ex} remains finite under finite temperature, and BEC can take place in 2 dimensions with a trap.

10.5.2   Bose-Einstein condensation of weakly interacting Bose gas in a trap

The results of Section 10.4 on the statistical mechanics of homogeneous interacting Bose gas have to be adapted to the existence of an external field. The Thomas-Fermi model in atomic physics [38] is an example of treating similar problems using the *local density approximation*. It is called the *statistical theory of the atom*. J. Oliva [39] used this method to study BEC in a trap. In the Thomas-Fermi model the potential produced by the Coulomb field of the nucleus jointly with the equilibrium distribution of the electrons $V(\boldsymbol{r})$ is related to the local Fermi momentum $p_{\mathrm{F}}(\boldsymbol{r})$ by the condition

$$\frac{1}{2m} p_{\mathrm{F}}^2(\boldsymbol{r}) + V(\boldsymbol{r}) = \text{const}, \qquad (10.5.14)$$

where the dependence of the Fermi momentum on the position is conditioned by the variation of electron density over space. Oliva suggested that in a trap the inhomogeneous condensate in a diffusive equilibrium, the sum of the *intrinsic* chemical potential $\bar{\mu}$ which is a functional of the condensate density distribution, and the potential V has to be a constant μ, the total chemical potential:

$$\bar{\mu}\left[\rho\left(\boldsymbol{r}\right)\right] + V\left(\boldsymbol{r}\right) = \mu. \qquad (10.5.15)$$

If the density distribution varies relatively smoothly, the non-local functional $\bar{\mu}[\rho(\boldsymbol{r})]$ can be replaced by a *local* function of the radius vector $\bar{\mu}(\boldsymbol{r})$. This is the local density approximation. We have then

$$\bar{\mu}\left(\boldsymbol{r}\right) + V\left(\boldsymbol{r}\right) = \mu. \qquad (10.5.16)$$

Using the free energy derived from (10.4.39), Oliva calculated the density distribution of hydrogen atoms with definite spin orientation in a trap.

The condition of validity of the local density approximation is given by T.T. Chou, C.N. Yang and L.H. Yu [40, 41]. A relation which can be verified experimentally for the repulsive interaction, and the momentum distribution of the Bose gas in the trap are also given. Consider the Bose gas in a harmonic oscillator potential $V = \frac{1}{2}m\omega^2 r^2$. There are 4 parameters with the dimension of length. Besides the thermal de Broglie wave length and the scattering length there are the classical amplitude of a harmonic oscillator of frequency ω with energy kT denoted by L_1 and the size of the ground state of the oscillator in the trap denoted by L_2:

$$\left.\begin{aligned} L_1 &= \left(2\pi m\omega^2 \beta\right)^{-1/2} \\ L_2 &= \left(\hbar/m\omega\right)^{1/2} \end{aligned}\right\}, \qquad (10.5.17)$$

where $\beta = \frac{1}{kT}$. In the local density approximation the space is divided into cells of volume between L_1^3 and L_2^3 :

$$L_1^3 \gg \text{volume of the cell} \gg L_2^3.$$

Within a cell the potential is considered as a constant. The fugacity $z = e^{\beta\mu}$ in a homogeneous system is now replaced by

$$\zeta(\boldsymbol{r}) = e^{\beta\bar{\mu}(\boldsymbol{r})} = e^{\beta\mu}e^{-\beta V(\boldsymbol{r})} = ze^{-\beta V(\boldsymbol{r})}, \qquad (10.5.18)$$

where z is a constant in the system. The grand partition function is

$$\mathscr{D} = \prod_{p} \frac{1}{1 - \zeta e^{-\beta\varepsilon_p}}. \qquad (10.5.19)$$

The density distribution of the homogeneous Bose gas (without the condensate) is derived from Eq. (10.4.39)

$$\rho(\boldsymbol{r}) = \lambda^{-3} g_{3/2}(z) \left(1 - \frac{4a}{\lambda} g_{1/2}(z)\right),$$

is now replaced by

$$\rho(\boldsymbol{r}) = \lambda^{-3} g_{3/2}(\zeta) \left(1 - \frac{4a}{\lambda} g_{1/2}(\zeta)\right), \qquad (10.5.20)$$

where the $\boldsymbol{r}$ dependence of ρ is determined by $\zeta(\boldsymbol{r})$. For the case $a = 0$ we can solve ρ exactly, providing a check for the local density approximation. The spherically symmetrical total density distribution of the system $\rho_t(r)$ is

$$\rho_t(r) = \langle r \,|D|\, r \rangle, \qquad (10.5.21)$$

where the density operator D is

$$D = \frac{ze^{-\beta H}}{1 - ze^{-\beta H}} = \sum_{l=1}^{\infty} z^l e^{-\beta Hl}. \qquad (10.5.22)$$

The matrix element of $e^{-\beta Hl}$ is known.[17] Substitution of the matrix element into (10.5.22) gives

$$\rho_t(r) = \varepsilon^{3/2}\lambda^{-3} \sum_{l=1}^{\infty} (\sinh l\varepsilon)^{-3/2} z^l e^{-\sigma^2 \tanh l\varepsilon/2}, \qquad (10.5.23)$$

---

[17]See R.P. Feynman: Statistical Mechanics, Reading, W.A. Benjamin Inc. 1972, p. 49.

where

$$\varepsilon = \frac{\hbar\omega}{kT}, \qquad \sigma = \frac{r}{L_2}. \tag{10.5.24}$$

Denoting the series in Eq. (10.5.23) by $\sum_l a_l$, we get the condition for convergence of the series:

$$\left(\frac{a_{l+1}}{a_l}\right) = z \left[\frac{e^{(l+1)\varepsilon}}{e^{l\varepsilon}}\right]^{-\frac{3}{2}} = ze^{-(3/2)\varepsilon} \equiv z_1 < 1. \tag{10.5.25}$$

The series diverges for any value of r for $z_1 = 1$. Comparing with the case of homogeneous Bose gas we know that this divergent part corresponds to the condensate. When we separate this part out, the remaining part is the density distribution of the thermal gas. Taking the logarithm of the general term of (10.5.23), we get

$$\ln a_l = l(\ln z) - \frac{3}{2}\ln\sinh l\varepsilon - \sigma^2\tanh\frac{l\varepsilon}{2}$$

$$= l\left(\ln z - \frac{3}{2}\varepsilon\right) + \left(\frac{3}{2}\ln 2 - \sigma^2\right) + c_l,$$

where $c_l \to 0$ for $l \to \infty$. Therefore, for l sufficiently large,

$$a_l \to 2^{3/2}e^{-\sigma^2}z_1^l e^{c_l}.$$

Rewriting a_l as

$$a_l = 2^{3/2}e^{-\sigma^2}z_1^l + 2^{3/2}e^{-\sigma^2}z_1^l\left(e^{c_l} - 1\right),$$

we get

$$\sum_l a_l = 2^{3/2}e^{-\sigma^2}\frac{z_1}{1-z_1} + 2^{3/2}e^{-\sigma^2}\sum_l z_1^l\left(e^{c_l} - 1\right).$$

Substituting back into (10.5.23) we have

$$\rho_t(r) = \frac{z_1}{1-z_1}|\psi_0(r)|^2 + \rho(r). \tag{10.5.26}$$

The first term on the right-hand side diverges at $z_1 = 1$, representing the density distribution of the condensate, and

$$\psi_0(r) = 2^{3/4}e^{-r^2/2L_2^2} \tag{10.5.27}$$

is the ground state wave function of the harmonic oscillator. The second term

$$\rho(r) = \lambda^{-3} (2\varepsilon)^{3/2} \sum_{l=1}^{\infty} z_1^l \left\{ \frac{1}{(1 - e^{-2l\varepsilon})^{3/2}} e^{-\sigma^2 \tanh \frac{\varepsilon l}{2}} - e^{-\sigma^2} \right\}$$

(10.5.28)

is the normal gas part. This implies that the condensate is distributed according to the single particle ground state wave function in the harmonic oscillator potential. The sum in (10.5.28) is convergent for $z_1 = 1$. The expressions (10.5.26)–(10.5.28) are exact, and can be used in justifying the degree of exactness of the local density approximation. We consider the case $\varepsilon \ll 1$. In the sum of (10.5.28), terms for $\varepsilon l > 1$ can be neglected due to the cancellation of the two terms in the curly bracket. For terms with $\varepsilon l < 1$ the following replacement can be done:

$$1 - e^{-2l\varepsilon} \to 2l\varepsilon, \quad \tanh \frac{\varepsilon l}{2} \to \frac{\varepsilon l}{2},$$

and the term $e^{-\sigma^2}$ can be neglected. The sum then becomes

$$\sum_{l}^{\infty} \frac{\left[z_1 \exp\left(-\frac{1}{2}\beta m\omega^2 r^2\right)\right]^l}{l^{3/2}} = g_{3/2}\left[z_1 \exp\left(-\frac{1}{2}\beta m\omega^2 r^2\right)\right]$$

and consequently

$$\rho_t(r) \approx \frac{z_1}{1 - z_1} |\psi_0(r)|^2 + \lambda^{-3} g_{3/2}\left[z_1 \exp\left(-\frac{1}{2}\beta m\omega^2 r^2\right)\right].$$

(10.5.29)

The second term on the right-hand side is just the normal gas density given by (10.5.20) with z replaced by z_1. For $\varepsilon \ll 1$ the error caused by this replacement can be neglected. For the gas phase, $1 - z_1 = O(1)$, the second term in (10.5.29) dominates, and we see that the local density approximation works well. In the condensate phase, $1 - z_1 \approx O(N^{-1})$, therefore we can set $z_1 = 1$ in the second term, which makes it identical with the exact expression. The condensate phase needs more elaborations. For $a = 0$, (10.5.28) gives $\rho(r) = \lambda^{-3}g_{3/2}(ze^{-\beta\frac{1}{2}m\omega^2 r^2})$. Adding more particles to the system increases the value of z. Note that the largest value of the argument of $g_{\frac{3}{2}}$ is 1. In the process of adding particles to the system the argument at the origin reaches 1 earlier than elsewhere. Adding more particles induces condensation at the origin. In the local density approximation the condensate distribution is of the delta function type, while the exact distribution is $|\psi_0(r)|^2$. This difference demonstrates the nature of the approximation. When the space is divided into cells, the volume of the cell is much larger than L_2^3, and the size L_2 of $|\psi_0(r)|$ shrinks to a point relative to the cell.

The Hamiltonian of system for $a = 0$ is

$$H = \frac{1}{2m}p^2 + \frac{1}{2}m\omega^2 x^2,$$

possessing an interchange symmetry

$$p \leftrightarrow x, \quad \frac{1}{m} \leftrightarrow m\omega^2.$$

The partition function $Q_N(V,T) = \mathrm{tr}\ e^{-\beta H}$ is invariant with respect to the interchange. Therefore, the momentum distribution must be $|\psi_0(p)|^2$. For $a \neq 0$, the momentum distribution is given in Ref. [41].

For $a > 0$, (10.5.28) can be written as

$$\rho(r) = \lambda^{-3}g_{3/2}(ze^{-\beta V(r)-4a\lambda^2\rho(r)}), \tag{10.5.30}$$

up to first order of $\frac{a}{\lambda}$. Condensation begins at $r = 0$ when more particles are added to the system. The local density approximation gives the local chemical potential

$$\bar{\mu}(r) = kT\ln(ze^{-\beta V}) = kT\ln z - V(r), \tag{10.5.31}$$

in contrast to the homogeneous case $\mu = kT\ln z$. The density of the saturated gas given by (10.5.30) is $\rho_0 = \lambda^{-3}g_{3/2}(1)$, hence the condensate density is $\rho_c = \rho_t - \rho_0$. The Helmholtz free energy of the system is given by $A = -\frac{1}{\beta}\ln Q_N$ with Q_N given by (10.4.39):

$$\frac{A}{V} = -kT\lambda^{-3}g_{5/2}(1) + 2a\lambda^2 kT\rho_t^2 - a\lambda^2 kT\rho_c^2, \tag{10.5.32}$$

from which we get the chemical potential $\bar{\mu} = \frac{\partial}{\partial \rho_t} \left(\frac{A}{V} \right) |_{V,T}$:

$$\bar{\mu} = \left[4a\lambda^2 \rho_t - 2a\lambda^2 \left(\rho_t - \rho_0 \right) \right] kT$$
$$= \frac{a4\pi\hbar^2}{m} \left(\rho_t + \rho_0 \right).$$

Equating this expression to (10.5.31) we obtain

$$kT \ln z - V(r) = 4\pi a \left(\rho_t + \rho_0 \right) \hbar^2 / m.$$

At the boundary of the condensate $r = r_0$, $\rho_c = 0$, $\rho_t = \rho_0$, hence

$$V(r) + 4\pi a \rho_t(r)\hbar^2/m = V(r_0) + 4\pi a \rho_0 \hbar^2/m. \qquad (10.5.33)$$

This relation is valid for any potential, and can be checked experimentally.

10.5.3 Gross-Pitaevskii equation

When external field exists, the field operator $\hat{\psi}$ is the sum of the condensate wave function Ψ and the field operator $\hat{\varphi}$ of the non-condensate states:

$$\hat{\psi}(\boldsymbol{x}) = \Psi(\boldsymbol{x}) + \hat{\varphi}(\boldsymbol{x}), \qquad (10.5.34)$$

which is the parallel of (10.4.6). We have

$$\Psi(\boldsymbol{x}) = \langle \hat{\psi}(\boldsymbol{x}) \rangle, \qquad (10.5.35)$$

and

$$\langle \hat{\psi}(\boldsymbol{x}) \rangle = 0. \qquad (10.5.36)$$

In the following, we derive the equations of evolution and the eigenequations for Ψ and $\hat{\varphi}$ using the mean field approximation starting from the equation of motion of ψ. The second quantized

Hamiltonian of the system is

$$\hat{H} = \int d^3x \hat{\psi}^\dagger \left(-\frac{\hbar^2}{2m} \nabla^2 + V \right) \hat{\psi}$$

$$+ \frac{1}{2} \int d^3x d^3x' U\left(\boldsymbol{x} - \boldsymbol{x}'\right) \hat{\psi}^\dagger(\boldsymbol{x})\hat{\psi}(\boldsymbol{x})\hat{\psi}^\dagger(\boldsymbol{x}')\hat{\psi}(\boldsymbol{x}').$$

$$\text{(10.5.37)}$$

Under the approximation of weakly interacting Bose gas we have

$$U\left(\boldsymbol{r}\right) = \frac{4\pi\hbar^2 a}{m} \delta^3\left(\boldsymbol{r}\right). \tag{10.5.38}$$

Using (10.5.38) the second term on the right-hand side of (10.5.37) can be written as $\int d^3x \frac{2\pi\hbar^2 a}{m} \hat{\psi}^\dagger(\boldsymbol{x})\hat{\psi}(\boldsymbol{x})\hat{\psi}^\dagger(\boldsymbol{x}')\hat{\psi}(\boldsymbol{x}')$. The equation of motion for $\hat{\psi}$ is then

$$i\hbar \frac{\partial \hat{\psi}}{\partial t} = -\frac{\hbar^2}{2m} \nabla^2 \hat{\psi} + V(\boldsymbol{x})\hat{\psi} + \frac{4\pi\hbar^2 a}{m} \hat{\psi}^\dagger \hat{\psi}\hat{\psi}. \tag{10.5.39}$$

Substituting (10.5.34) into (10.5.39) and taking average through out the equation considering the relations $\langle\varphi\rangle = \langle\varphi^\dagger\rangle = 0$, $\langle\varphi^\dagger\varphi\rangle = n'(\boldsymbol{r},t)$, we obtain

$$i\hbar \frac{\partial \Psi}{\partial t} = -\frac{\hbar^2}{2m} \nabla^2 \Psi + V(\boldsymbol{x})\Psi + \frac{4\pi\hbar^2 a}{m}(2n' + |\Psi|^2)\Psi, \tag{10.5.40}$$

where n' is the density of the non-condensate. This is the equation of evolution for the condensate. It is non-linear because of the term $|\Psi|^2\Psi$.

To find the equation for φ we need the mean field approximation. Consider the product of field operators

$$\hat{\psi}^\dagger\hat{\psi}\hat{\psi} = (\Psi + \hat{\varphi}^\dagger)(\Psi + \hat{\varphi})(\Psi + \hat{\varphi})$$

$$= |\Psi|^2 \Psi + 2|\Psi|^2 \hat{\varphi} + \Psi\hat{\varphi}\hat{\varphi} + \Psi^2\hat{\varphi}^\dagger + 2\Psi\hat{\varphi}^\dagger\hat{\varphi} + \hat{\varphi}^\dagger\hat{\varphi}\hat{\varphi}.$$

We replace the bilinear form $\hat{\varphi}^\dagger\hat{\varphi}$ by its expectation value n' in the linearization, and replace $\hat{\varphi}\hat{\varphi}$ by 0. In linearizing $\hat{\varphi}^\dagger\hat{\varphi}\hat{\varphi}$ we notice that the operator $\hat{\varphi}^\dagger$ is to be "contracted" with both operators $\hat{\varphi}$, hence $\hat{\varphi}^\dagger\hat{\varphi}\hat{\varphi} \to 2n'\hat{\varphi}$. This kind of approximation neglects fluctuations of some operator product, and the result is to be handled with caution. The operator $\hat{\psi}^\dagger \hat{\psi}\hat{\psi}$ is reduced in the following way

$$\hat{\psi}^\dagger \hat{\psi}\hat{\psi} \rightarrow 2n'\hat{\varphi} + 2n'\Psi + \Psi^2\hat{\varphi}^\dagger + 2|\Psi|^2\,\hat{\varphi} + |\Psi|^2\,\Psi, \qquad (10.5.41)$$

and becomes linear. Subtracting (10.5.40) from (10.5.39) we obtain

$$i\hbar\frac{\partial\hat{\varphi}}{\partial t} = -\frac{\hbar^2}{2m}\nabla^2\hat{\varphi} + V\left(\boldsymbol{r}\right)\hat{\varphi} + \frac{4\pi\hbar^2 a}{m}[2(|\Psi|^2 + n')\hat{\varphi} + \Psi^2\hat{\varphi}^\dagger].$$

$$(10.5.42)$$

Equations (10.5.40) and (10.5.42) are coupled. Usually n' is neglected in comparison with $|\Psi|^2$ under very low temperatures, and (10.5.40) becomes

$$i\hbar\frac{\partial\Psi}{\partial t} = -\frac{\hbar^2}{2m}\nabla^2\Psi + V\left(\boldsymbol{r}\right)\Psi + \frac{4\pi\hbar^2 a}{m}|\Psi|^2\Psi, \qquad (10.5.43a)$$

the well-known Gross-Pitaevskii equation [42–44] (abbreviated as G.P. equation in the following). In the applications of this equation we have to remember its nature of the mean field approximation.

To find the equation for stationary states we begin with the time dependence of the order parameter

$$\Psi(\boldsymbol{x}, t) = \langle\hat{\psi}(\boldsymbol{x})e^{-iHt/\hbar}\rangle = \Psi(\boldsymbol{x})e^{-i\mu t/\hbar},$$

where $\mu = E(N) - E(N-1) \sim \partial E/\partial N$ is the chemical potential. We obtain finally with the help of (10.5.43) the Gross-Pitaevskii equation for stationary states

$$-\frac{\hbar^2}{2m}\nabla^2\Psi + V\left(\boldsymbol{r}\right)\Psi + \frac{4\pi\hbar^2 a}{m}|\Psi|^2\Psi = \mu\Psi. \qquad (10.5.43b)$$

Gross-Pitaevskii equation has been applied to various problems with great success. E.H. Lieb *et al.* [45] contributed to the rigorous proof of this equation. The authors proved rigorously that $2\pi\hbar^2 a\rho/m$ a is a **lower bound** of the interaction energy per particle, while earlier lower bound derived was actually much larger than this.

Furthermore E.H. Lieb *et al.* [46] went a step further, giving a rigorous proof of the Gross-Pitaevskii energy functional for $a > 0$.

10.5.4 The quantum phase dynamics

The ground state of a condensate of interacting atoms is usually assumed to be a coherent state with a fixed phase. Actually such a state is not an eigenstate even of the simple Hamiltonian for a homogeneous interacting condensate. The phase diffusion becomes inevitable. For a condensate at $T = 0$ the phase has to undergo quantum diffusion because a constant phase is inconsistent with the atom number conservation. M. Lewenstein and L. You [47] studied the quantum phase diffusion of a Bose-Einstein condensate. The second quantized Hamiltonian is

$$\mathscr{H} = \int d^3r \, \hat{\Psi}^\dagger(r) \left[-\frac{\hbar^2}{2M}\nabla^2 + V_t(r) - \mu \right] \hat{\Psi}(r)$$

$$+ \frac{u_0}{2} \int d^3r \, \hat{\Psi}^\dagger(r) \, \hat{\Psi}^\dagger(r) \, \hat{\Psi}(r) \, \hat{\Psi}(r), \qquad (10.5.44)$$

where $u_0 = \frac{4\pi\hbar^2 a}{M}$, and a is the scattering length. The conservation of the average particle number is incorporated through the term $\mu \hat{N} = \mu \int d^3r \, \hat{\Psi}^\dagger(r) \, \hat{\Psi}(r)$, where the Lagrange multiplier μ is the chemical potential. In the Bogoliubov approximation we set

$$\hat{\Psi}(r) = \sqrt{N}\psi_0(r) + \delta\hat{\Psi}(r), \qquad (10.5.45)$$

where ψ_0 is the ground state condensate wave function (assumed real) normalized to $\int d^3r |\psi_0(r)|^2 = 1$, $\delta\hat{\Psi}(r)$ is the quantum fluctuation. We substitute (10.5.45) into (10.5.44), and neglect terms of 3rd and 4th order in $\delta\hat{\Psi}$. The first order term vanishes provided $\psi_0(r)$ is the solution of the Gross-Pitaevskii equation

$$[\mathscr{L} + u_0\rho(r)]\psi_0(r) = 0, \qquad (10.5.46)$$

where

$$\left. \begin{aligned} \mathscr{L} &= -\frac{\hbar^2}{2M}\nabla^2 + V_t(r) - \mu \\ \rho(r) &= N\psi_0^2(r) \end{aligned} \right\} \qquad (10.5.47)$$

The Hamiltonian becomes then a bilinear form of $\delta\hat{\Psi}$ and $\delta\hat{\Psi}^\dagger$ besides a constant term

$$\mathscr{H} = \int d^3r\, \delta\hat{\Psi}^\dagger \mathscr{L}\delta\hat{\Psi} + \frac{u_0}{2}\int d^3r \rho(\boldsymbol{r})\left(\delta\hat{\Psi}^\dagger\delta\hat{\Psi}^\dagger + \delta\hat{\Psi}\delta\hat{\Psi} + 4\delta\hat{\Psi}^\dagger\delta\hat{\Psi}\right).$$

$$(10.5.48)$$

Introducing the quasiparticle annihilation and creation operators

$$\left.\begin{aligned} g_k &= \int d^3r\left[U_k(\boldsymbol{r})\delta\hat{\Psi}(\boldsymbol{r}) + V_k(\boldsymbol{r})\delta\hat{\Psi}^\dagger(\boldsymbol{r})\right] \\ g_k^\dagger &= \int d^3r\left[U_k^*(\boldsymbol{r})\delta\hat{\Psi}^\dagger(\boldsymbol{r}) + V_k^*(\boldsymbol{r})\delta\hat{\Psi}(\boldsymbol{r})\right] \end{aligned}\right\}$$

$$(10.5.49)$$

U and V are functions of $\boldsymbol{r}$ and in general complex for a trapped condensate. They correspond to the u_p and v_p for the homogeneous case (Section 10.4.1) and are sometimes called the trap modes. Operators g_k and $g_k^\dagger$ satisfy the bosonic commutation relations

$$\left[g_k, g_{k'}^\dagger\right] = \delta_{kk'} \quad \text{and} \quad [g_k, g_{k'}] = 0, \qquad (10.5.50)$$

which lead to the biorthonormality condition following from (10.5.49):

$$\int d^3r\,(U_k(\boldsymbol{r})U_{k'}^*(\boldsymbol{r}) - V_k(\boldsymbol{r})V_{k'}^*(\boldsymbol{r})) = \delta_{kk'}. \qquad (10.5.51)$$

We expect that after diagonalization $\mathscr{H}$ becomes

$$\mathscr{H} \sim \sum_k \hbar\omega_k g_k^\dagger g_k. \qquad (10.5.52)$$

This form is tentative, and will be made more exact in the subsequent development. We have

$$[\mathscr{H}, g_k] = -\hbar\omega_k, \quad \left[\mathscr{H}, g_k^\dagger\right] = \hbar\omega_k. \qquad (10.5.53)$$

This is what we expect, because g_k operating on an eigenstate of $\mathscr{H}$ decreases its eigenvalue by $\hbar\omega_k$. Calculating $[\mathscr{H}, g_k]$ using (10.5.48) and (10.5.49) and comparing the result with (10.5.53), we obtain

$$
\left.
\begin{aligned}
[\mathscr{L}+2u_0\rho(\boldsymbol{r})]\,U_k(\boldsymbol{r}) - u_0\rho(\boldsymbol{r})V_k(\boldsymbol{r}) &= \hbar\omega_k U_k(\boldsymbol{r}) \\
[\mathscr{L}+2u_0\rho(\boldsymbol{r})]\,V_k(\boldsymbol{r}) - u_0\rho(\boldsymbol{r})U_k(\boldsymbol{r}) &= -\hbar\omega_k V_k(\boldsymbol{r})
\end{aligned}
\right\}. \tag{10.5.54}
$$

These are the Bogoliubov-de Gennes equation for the quasiparticle wave functions. Under time reversal, we have

$$
\left.
\begin{aligned}
U_k &\leftrightarrow U_k^*, \ V_k \leftrightarrow V_k^* \\
\delta\hat{\Psi} &\leftrightarrow \delta\hat{\Psi}^\dagger, \ g_k \leftrightarrow g_k^\dagger
\end{aligned}
\right\}. \tag{10.5.55}
$$

Therefore, the time reversal invariance demands that corresponding to the solution (U_k, V_k) for ω_k, there exists a solution (V_k^*, U_k^*) for $-\omega_k$. Equation (10.5.54) have a unique solution $U_0(\boldsymbol{r}) = V_0^*(\boldsymbol{r}) \propto \psi_0(\boldsymbol{r})$ with $\omega_0 = 0$. This is the Goldstone mode resulting from the global $U(1)$ symmetry breaking.

It is tempting to define the annihilation operator of the zero mode according to (10.5.49), namely

$$
\hat{P} = \int \mathrm{d}^3 r \psi_0(\boldsymbol{r})[\delta\hat{\Psi}(\boldsymbol{r}) + \delta\hat{\Psi}^\dagger(\boldsymbol{r})]. \tag{10.5.56}
$$

We find that this operator is *Hermitian*, and as such cannot be associated with either annihilation or creation operator of the zero mode. We also find that this operator commutes with the Hamiltonian;

$$
[\hat{P}, \mathscr{H}] = \int \mathrm{d}^3 r \{\psi_0(\mathscr{L}+u_0\rho)\delta\hat{\Psi} - \psi_0(\mathscr{L}+u_0\rho)\delta\hat{\Psi}^\dagger\} = 0 \tag{10.5.57}
$$

because of the Gross-Pitaevskii equation (10.5.46). It also commutes with g_k and $g_k^\dagger$ for $k \neq 0$, e.g.

$$
[g_k, \hat{P}] = \int \mathrm{d}^3 r \psi_0 [U_k - V_k] = 0, \tag{10.5.58}
$$

because of the biorthonormality condition (10.5.51). We can now make the expression (10.5.52) exact. $\mathscr{H}$ must be a bilinear form, and must commute with $\hat{P}$ such that (10.5.53) and (10.5.54) remain valid. The sum does not include $k = 0$, and the zero mode is taken care of by a term $\frac{\alpha}{2}\hat{P}^2$ with α to be determined. The correct form is

$$\mathscr{H} = \frac{\alpha}{2}\hat{P}^2 + \sum_{k \neq 0} \hbar\omega_k g_k^\dagger g_k. \tag{10.5.59}$$

The zero mode represents a collective motion associated with the spontaneous symmetry breaking of the global phase invariance.[18] $\hat{P}$ can be designated the momentum operator of this mode, and its canonical conjugate position operator is

$$\hat{Q} = i\int d^3r\,\Phi_0(r)[\delta\hat{\Psi}(r) - \delta\hat{\Psi}^\dagger(r)], \tag{10.5.60}$$

which has to satisfy

$$[\hat{Q},\hat{P}] = i, \quad [\hat{Q},g_k] = 0, \quad k \neq 0, \tag{10.5.61}$$

and consequently from (10.5.59)

$$[\hat{Q},\mathscr{H}] = i\alpha\hat{P}. \tag{10.5.62}$$

From this demand it follows that

$$2\int d^3r\,\Phi_0(r)\psi_0(r) = 1, \tag{10.5.63}$$

$$\int d^3r\,\Phi_0(r)\,[U_k + V_k] = 0, \quad \text{for } k \neq 0, \tag{10.5.64}$$

and

$$[\mathscr{L} + 3u_0\rho(r)]\,\Phi_0(r) = \alpha\psi_0(r). \tag{10.5.65}$$

The solution to (10.5.65) with parameters from the JILA trap is given in Fig. 10.12; the solution to (10.5.46) and (10.5.65) gives $\mu = 1.769\hbar\omega_x$ and $\alpha = 1.129\hbar\omega_x$.

The creation and annihilation operators of the zero mode are given by

$$g_0 = \frac{\hat{P} - i\hat{Q}}{\sqrt{2}}, \quad g_0^\dagger = \frac{\hat{P} + i\hat{Q}}{\sqrt{2}}, \tag{10.5.66}$$

---

[18]This is discussed systematically in Ref. [48], especially Section 10.6.

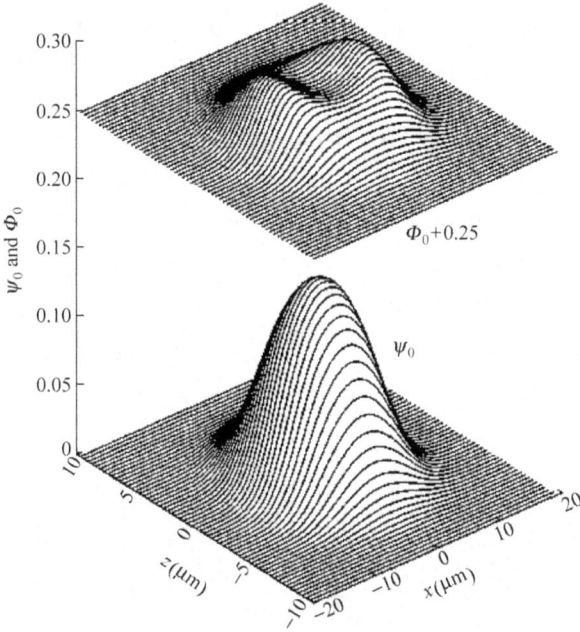

Fig. 10.12 The wave functions $\Phi_0(\boldsymbol{r})$ and $\psi_0(\boldsymbol{r})$ for JILA trap. $N = 2000$, $a = 5.2$nm, $(\omega_x : \omega_y : \omega_z) = (1 : 1 : 8^{\frac{1}{2}})(10\text{Hz})$. Reprinted with permission from M. Lewenstein and L. You, *Phys. Rev. Lett.* 77, 3489, 1996. Copyright (2017) by the American Physical Society.

where the commutation relations provide the guide. The zero mode functions are

$$U_0(\boldsymbol{r}) = \frac{1}{\sqrt{2}}\left(\psi_0(\boldsymbol{r}) + \Phi_0(\boldsymbol{r})\right), \quad V_0(\boldsymbol{r}) = \frac{1}{\sqrt{2}}\left(\psi_0(\boldsymbol{r}) - \Phi_0(\boldsymbol{r})\right).$$

(10.5.67)

Only when these functions are included, the set of (U_k, V_k) does become *complete*:

$$\left.\begin{array}{l} \displaystyle\sum_{k=0}^{\infty} \left[U_k(\boldsymbol{r})U_k^*(\boldsymbol{r}') - V_k(\boldsymbol{r})V_k^*(\boldsymbol{r}')\right] = \delta^3(\boldsymbol{r} - \boldsymbol{r}'), \\[4mm] \displaystyle\sum_{k=0}^{\infty} \left[U_k(\boldsymbol{r})V_k^*(\boldsymbol{r}') - V_k(\boldsymbol{r})U_k^*(\boldsymbol{r}')\right] = 0. \end{array}\right\}$$

(10.5.68)

The biorthonormality condition (10.5.51) is now extended to include $k = 0$. The quantum fluctuation $\delta \Psi$ (10.5.45) can be expanded as

$$\delta \Psi(\boldsymbol{r}) = \sum_{k=0}^{\infty} \left[U_k(\boldsymbol{r}) g_k - V_k^*(\boldsymbol{r}) g_k^\dagger \right]. \tag{10.5.69}$$

The contribution from the zero-mode is

$$U_0 g_0 - V_0 g_0^\dagger = -i\psi_0 \hat{Q} + \Phi_0 \hat{P}, \tag{10.5.70}$$

and the expansion of the field operator is

$$\hat{\Psi}(\boldsymbol{r},t) = \sqrt{N}\psi_0(\boldsymbol{r}) - i\psi_0 \hat{Q} + \Phi_0 \hat{P} + \sum_{k \neq 0}^{\infty} \left[U_k(\boldsymbol{r}) g_k - V_k^*(\boldsymbol{r}) g_k^\dagger \right].$$
$$\tag{10.5.71}$$

The dynamics of the zero mode is obtained by the equations of motion. From (10.5.59) we have

$$[\hat{P}, \mathscr{H}] = 0,$$

and consequently

$$\hat{P}(t) = \hat{P}(0), \quad \text{a constant}, \tag{10.5.72}$$

Another equation of motion

$$i\frac{\partial \hat{Q}}{\partial t} = [\hat{Q}, \mathscr{H}] = i\alpha \hat{P} \tag{10.5.73}$$

is solved by

$$\hat{Q}(t) = \hat{Q}(0) + \alpha \hat{P}(0)t. \tag{10.5.74}$$

10.6 Bose-Einstein Condensation in Anisotropic Potential Traps

The stationary states and their evolution in time of a Bose-Einstein condensate are described by the solutions of the G.P. equation. Because of the non-linearity of the equation, one has to rely on numerical solutions [49, 50]. Simple approximate solutions are available, from which important conclusions can be drawn. G. Baym and C.J. Pethick [51] studied the ^{87}Rb condensate obtained in the original experiment of C. Wieman and E. Cornell's group. The potential trap is anisotropic with axial frequency ω_z^0 and transverse frequency $\omega_\perp^0 = \frac{\omega_z^0}{\sqrt{8}}$. The corresponding oscillator amplitudes are $a_z = (\frac{\hbar}{m\omega_z^0})^{\frac{1}{2}}$ and $a_\perp = (\frac{\hbar}{m\omega_\perp^0})^{\frac{1}{2}}$. In the absence of interactions the wave function of the single particle state in the trap is

$$\Psi_0\left(r\right) = \frac{1}{\pi^{3/4} a_\perp a_z^{1/2}} \exp\left[-\frac{1}{2\hbar} m\left(\omega_\perp^0 r_\perp^2 + \omega_z^0 z^2\right)\right], \qquad (10.6.1)$$

where $r_\perp$ is the projection of r on the xy-plane. The density distribution $\rho_0(r) = N\Psi_0^2(r)$ of this wave function is Gaussian. The scattering length of ^{87}Rb atoms is $a \approx 100 a_0$, where a_0 is the Bohr radius of hydrogen atom. The positive value of a means a repulsive interaction. The interaction energy per unit volume of the gas is $\frac{2\pi\hbar^2 a}{m} \frac{N}{R^3}$. The repulsive interaction leads to a decrease of the density relative to that of the ideal gas. Since the transverse potential is weaker, the gas extends on the x–y-plane first, and then in the z direction. The size of the gas cloud depends on an equilibrium between the interaction energy and the potential energy of the trap. We disregard the anisotropy for the time being and denote the radius of the gas cloud by R, then $\rho \sim N/R^3$. The potential energy of a particle in the trap is $\frac{1}{2} m\omega_\perp^2 R^2$, the interaction energy of the particle with the rest of the atoms is $\frac{4\pi\hbar^2 a}{m} \frac{N}{R^3}$. An equilibrium value of R results

$$R^5 = 8\pi\frac{\hbar^2}{m^2\omega_\perp^2} aN = aa_\perp^4 8\pi N \equiv a_\perp^5 \zeta^5, \qquad (10.6.2)$$

where

$$\zeta = \left(\frac{8\pi N a}{a_\perp}\right)^{1/5} \qquad (10.6.3)$$

is the dimensionless scale of the condensates in a trap, and the dimension of the condensate is $a_\perp \zeta$. In the experiment of NIST group $\zeta \gg 1$. The average kinetic energy of the particle in the trap is $T = \hbar^2/2mR^2$; hence the ratio between the kinetic energy to the interaction energy (or the trap potential energy) is

$$\frac{\bar{T}}{\bar{V}} \approx \frac{\hbar^2}{mR^2 m\omega_\perp^2 R^2} = \left(\frac{a_\perp}{R}\right)^4 = \zeta^{-4},$$

which is very small. In order to further explore the influence of the interaction, we use the wave function of the condensate Ψ normalized to $\int |\Psi(r)|^2 d^3r = N$. The G.P. energy functional is

$$E[\Psi(r)] = \int d^3r \left\{ \frac{\hbar^2}{2m}|\nabla\Psi(r)|^2 + \frac{m}{2}[(\omega_\perp^0)^2 r_\perp^2 + (\omega_z^0)^2 z^2]|\Psi(r)|^2 \right.$$

$$\left. + \frac{2\pi\hbar^2 a}{m}|\Psi(r)|^4 \right\}. \tag{10.6.4}$$

For the first approximation we adopt the form of the ground state wave function without interaction, with the effective trap frequencies $\omega_\perp$ and ω_z as variational parameters:

$$\Psi(r) = \pi^{-3/4}\omega_\perp^{1/2}\omega_\perp^{1/4}\left(\frac{m}{\pi\hbar}\right)^{3/4}\exp\left[-\frac{m}{2\hbar}\left(\omega_\perp r_\perp^2 + \omega_z z^2\right)\right]. \tag{10.6.5}$$

Substituting (10.6.5) into (10.6.4), we obtain $E(\omega_\perp, \omega_z)$. Minimizing E by varying $\omega_\perp$ we obtain $\omega_\perp = \omega_\perp^0/\Delta$, where $\Delta \propto N^{1/2}a^{1/4}$. Comparison with the experimental parameter of the Wieman-Cornell group gives for $N = 10^4$ the following results: $\omega_z/\omega_z^0 = 0.40 \sim 0.55$ and $\omega_\perp/\omega_\perp^0 = 0.16 \sim 0.26$, depending on the trap parameters. Because of the repulsive interaction the wave function extends considerably. The extension will be remarkable if $N = 10^6$ is attained.

Minimizing the energy functional (10.6.4) under fixed N leads to the G.P. equation:

$$\left[-\frac{\hbar^2}{2m}\nabla^2 + \frac{1}{2}m\left(\omega_\perp^{02}r_\perp^2 + \omega_z^{02}z^2\right) + \frac{4\pi\hbar^2 a}{m}|\Psi(r)|^2 \right]\Psi(r) = \mu\Psi(r), \tag{10.6.6}$$

where μ is the chemical potential. Defining the dimensionless radius vector r_1 by $r_1 = \frac{r}{a_\perp\zeta}$, the G.P. equation takes the following form,

$$\left[-\frac{1}{\zeta^4}\nabla_1^2 + r_{1\perp}^2 + \lambda^2 z_1^2 + |f(\boldsymbol{r}_1)|^2\right] f(\boldsymbol{r}_1) = \nu^2 f(\boldsymbol{r}_1), \qquad (10.6.7)$$

where $\lambda = \omega_z^0/\omega_\perp^0$ and

$$\nu^2 = \frac{2\mu}{\zeta^2 \hbar \omega_\perp^0}. \qquad (10.6.8)$$

For sufficiently large N the kinetic energy term can be neglected, and the dimensionless wave function can be directly obtained:

$$f(r_1^2) = \nu^2 - r_{1\perp}^2 - (\lambda z_1)^2, \qquad (10.6.9)$$

where the equality holds when the right-hand side is non-negative, and otherwise $f = 0$.

The trial wave function (10.6.5) gives a Gaussian momentum distribution

$$\mathscr{P}(\boldsymbol{p}) = \left|\int d^3r \exp\left(-i\frac{\boldsymbol{p}\cdot\boldsymbol{r}}{\hbar}\right)\Psi(\boldsymbol{r})\right|^2$$

$$\propto \exp\left[-\frac{1}{m\hbar}\left(\frac{p_\perp^2}{\omega_\perp} + \frac{p_z^2}{\omega_z}\right)\right], \qquad (10.6.10)$$

The mean square value of the momentum is

$$\langle p^2 \rangle = \frac{m\hbar}{2}(\omega_z + 2\omega_\perp). \qquad (10.6.11)$$

In the approximation of neglecting the kinetic energy in comparison with the interaction energy and the trap potential energy, we would have directly from the Gross-Pitaevskii equation (10.6.6)

$$|\Psi(\boldsymbol{r})|^2 = \begin{cases} \frac{m}{4\pi\hbar a}(\mu - V(\boldsymbol{r})), & \mu > V(\boldsymbol{r}) \\ 0, & \text{otherwise} \end{cases} \qquad (10.6.12)$$

This is the Thomas-Fermi approximation for BEC valid for large N, $N\frac{a}{a_{HO}} \gg 1$, where $a_{HO} = \sqrt{\frac{\hbar}{m\omega}}$, and ω is the trap angular frequency. Using the time dependent G.P. equation (10.5.43), Stringari [61]found modes of collective excitation of a trapped BEC. The ground state density $\rho_0 = |\Psi_0(\boldsymbol{r})|^2$ of BEC is given by (10.6.12). Let there be a deviation $\delta\rho$ from the ground state density

$$\rho(\boldsymbol{r},t) = \rho_0(\boldsymbol{r}) + \delta\rho(\boldsymbol{r})e^{-i\omega t}, \tag{10.6.13}$$

where ω is the angular frequency of the excitation mode to be determined. After substituting the stationary state wave function of the condensate

$$\Psi(\boldsymbol{r},t) = \sqrt{\rho(\boldsymbol{r},t)}e^{-i\theta(\boldsymbol{r})}e^{-i\mu t/\hbar} \tag{10.6.14}$$

into the G.P. equation (10.5.43), we obtain a pair of equations corresponding to the real and imaginary parts of Ψ in accordance with the Thomas-Fermi approximation:

$$\frac{\partial \rho}{\partial t} = -\nabla \cdot (\rho \boldsymbol{v}), \tag{10.6.15}$$

$$m\frac{\partial \boldsymbol{v}}{\partial t} + \nabla \left(V + \frac{4\pi\hbar^2 a}{m}\rho - \mu\right) = 0, \tag{10.6.16}$$

where

$$\boldsymbol{v} = \frac{\hbar}{m}\nabla\theta. \tag{10.6.17}$$

Let V be the isotropic harmonic potential

$$V = \frac{1}{2}m\omega_0^2 r^2. \tag{10.6.18}$$

Substituting (10.6.14) into (10.6.16) and (10.6.17) and eliminating $\boldsymbol{v}$, considering $\delta\rho$ and $\boldsymbol{v}$ as small quantities, we obtain

$$\omega^2 \delta\rho(\boldsymbol{r}) = -\frac{1}{2}\omega_0^2 \nabla \cdot (R^2 - r^2)\nabla(\delta\rho), \tag{10.6.19}$$

where R is the boundary of the BEC in Thomas-Fermi approximation defined by

$$V(R) = \frac{m}{2}\omega_0^2 R^2 = \mu. \tag{10.6.20}$$

Equation (10.6.19) can be solved by an expansion in spherical harmonics $Y_{lm}(\vartheta, \varphi)$, leading to the mode frequencies

$$\omega(n, l) = \omega_0(2n^2 + 2nl + 3n + l)^{\frac{1}{2}}, \tag{10.6.21}$$

where $n = 1, 2, 3, \ldots$ and $l = 0, 1, 2, \ldots, n - 1$. Stringari also gave modes of collective oscillations in the case of anisotropic trap and considered the influence of the kinetic energy by using sum rules.

It can be well understood that a liquid drop, and in the same sense a nucleus, can have collective oscillations of this kind, for instance a surface quadrupole oscillation. Here we see that a dilute Bose gas can also support such collective excitations when Bose-Einstein condensation takes place in it. This can be anticipated already from Eq. (10.6.16), which is a hydrodynamical equation of irrotational flow.

For anisotropic trap with time dependent parameters Yu. Kagan, E.L. Surkov and G.V. Shlyapnikov [52–54] found an approximate solution to G.P. equation with universal scaling using local density approximation and neglecting the kinetic energy. The condensate wave function is expressed in terms of scaling parameters $b_i(t)$ ($i = x, y, z$) which can be determined by solving *classical* equations of motion. Problems such as evolution of the condensate wave function under external disturbances and under variation of trap parameters can be dealt with by using such wave functions.

10.7 Vortices and the Stability of Bose-Einstein Condensates against Attractive Interactions

One of the experimental signatures of the formation of a superfluid HeII is the possibility of vortex formation in it. We have seen in Section 10.3 that for the condensate wave function $\Psi = \sqrt{n(\boldsymbol{r})}e^{i\theta(\boldsymbol{r})}$ the superfluid velocity is

$$\boldsymbol{v}_{\mathrm{s}} = \frac{\hbar}{m}\nabla\theta,$$

which is apparently irrotational:

$$\nabla \times \boldsymbol{v}_{\mathrm{s}} = 0.$$

The velocity field around a vortex line is in the azimuthal direction, and the magnitude of the velocity is inversely proportional to distance from the vortex line. This guarantees that the integral of

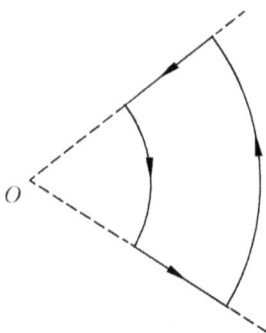

Fig. 10.13   Path of integration: the intersection of vortex line is denoted by O.

v_s around a closed contour shown in Fig. 10.13 vanishes such that the field is irrotational. If the closed integral includes the vortex line, then the quantization of the circulation (10.3.9)

$$\oint v_s \cdot ds = \frac{2\pi\hbar}{m}\kappa, \quad \kappa = 1, \pm 1, \pm 2, \ldots$$

gives the phase of the wave function characterized by the quantum number κ:

$$\theta = \kappa\phi, \tag{10.7.1}$$

where ϕ is the azimuthal angle. This can be seen by considering the component of the gradient in the direction of increasing ϕ,

$$\nabla_\phi = \hat{\phi}\frac{1}{r_\perp}\frac{\partial}{\partial\phi}, \tag{10.7.2}$$

where $\hat{\phi}$ is the unit vector in the direction of increasing ϕ. Therefore

$$v = \frac{\hbar}{m}\nabla\theta = \frac{\hbar\kappa}{mr_\perp}\hat{\phi}, \tag{10.7.3}$$

which leads to the quantization of circulation. In the state character-ized by the quantum number κ the component of angular momentum in the z direction is $N\kappa\hbar$. Vortices in superfluid He have been experimentally observed. Attempts have been made in laboratories to observe vortices in gaseous Bose-Einstein condensates. The existence of vortices would have been a direct proof that the condensate is a superfluid. First attempts were made in deforming the trap and making the trap to rotate during the stage of evaporation cooling, hoping that the condensate can be formed in the state of a vortex. The trial was unsuccessful. J.E. Williams and M.J. Holland suggested [55] using the two hyperfine states of ^{87}Rb which are coupled by a two-photon process. There are two traps for each of the species, with the centers of the traps separated by a small distance. In the beginning of the experiment there is a condensate of one of the species. When the coupling laser beams are operating, the traps are rotated about the midpoint between their centers with an adequate frequency. After a certain period of time the laser beams are shut down and rotation of traps stopped. A condensate of the second species with vortices is expected which coexists with the condensate of the first species (without vortex). This suggestion was realized by the group of E. Cornell and C. Wieman with some modifications [56]. K.W. Madison, F. Chevy, W. Wolleben and J. Dalibard [57] obtained vortices in a single species of the condensate in a slightly anisotropic trap uniformly rotated. The authors reported observation of no vortex first, then successively one, two, three and four vortices. W. Ketterle's group [58] observed vortex lattices in a rotating Bose condensed gas. In most favorable situations up to 140 vortices with lifetime of several seconds were observed. W. Ketterle's group [59] demonstrated the existence of the critical velocity of the super-fluid flow in the condensate. The superfluid nature of the gaseous Bose-Einstein condensates is thus experimentally established.

The existence of vortices plays an important role in the stability of condensates with attractive interaction between the atoms. Consider the wave function of a vortex state

$$\Psi(\boldsymbol{r}) = \psi(\boldsymbol{r})e^{i\kappa\phi},$$

where ψ is a real function. Its gradient is

$$\nabla \Psi(\boldsymbol{r}) = e^{i\kappa\phi}\nabla\psi + \hat{\phi}\frac{i\kappa}{r}e^{i\kappa\phi}\psi, \qquad (10.7.4)$$

in which the first term has only radial and axial components, because ψ is a function of r and z only. Hence

$$|\nabla \Psi|^2 = |\nabla\psi|^2 + \frac{\kappa^2}{r_\perp^2}|\psi|^2 .$$

Substituting the above expression into the energy functional we obtain

$$E\left[\psi\right] = \int d^3r \left[\frac{\hbar^2}{2m}|\nabla\psi|^2 + \frac{m}{2}\left(\omega_\perp^2 r_\perp^2 + \omega_z^2 z^2 + \frac{\hbar^2\kappa^2}{m^2 r_\perp^2}\right)|\psi(r)|^2\right.$$
$$\left. + \frac{2\pi\hbar^2 a}{m}|\psi(r)|^4\right].\tag{10.7.5}$$

The vortex brings the "centrifugal" term. The $r_\perp^2$ in the denominator does not introduce any trouble because the wave function is zero on the vortex line. F. Dalfovo and S. Stringari [54] carried out numerical computations on this problem, using dimensionless coordinates $r_1 = \frac{1}{a_\perp}r$, $z_1 = \frac{1}{a_\perp}z$, dimensionless energy $E_1 = (\hbar\omega_\perp)^{-1}E$ and normalized wave function $\psi_1(r_1)$. Starting from the trial wave function $\psi_1(r_1, t)$, where t is a parameter characterizing the computation steps, we compute the energy functional E_1/N. Advancing to the next step by setting Δt, we obtain

$$\frac{\partial}{\partial t}\psi_1\left(r_1, t\right) = -\frac{\bar{\delta}E_1/N}{\bar{\delta}\psi_1\left(r_1, t\right)},$$

where $\bar{\delta}$ represents the functional derivative constrained by the normalization of ψ_1. Hopefully a smaller E_1/N can be obtained. Repeat the process until E_1/N converges to a value which is considered the variational minimum. Figure 10.14 shows the dimensionless wave function for ^{87}Rb without interaction between the atoms (dashed line). The solid lines from upside down represent wave functions with interaction of a system of $N = 100, 200, 500, 1000, 2000, 5000$ and 10000 respectively. The influence of interaction on the extension of wave function is very noticeable, especially on the xy plane where the trap is weaker. Figure 10.15 depicts the wave function at $N = 5000$ with dashed line denoting the case of no interaction, the solid line denoting the exact wave function, and the dash-dotted line denoting the Thomas Fermi approximate wave function (10.6.9). It is seen that for $N = 5000$ the Thomas-Fermi approximation is already very good. Figure 10.16 shows the influence of a vortex, (a) gives the wave function contours of the $\kappa = 0$ state, the numbers labeling the

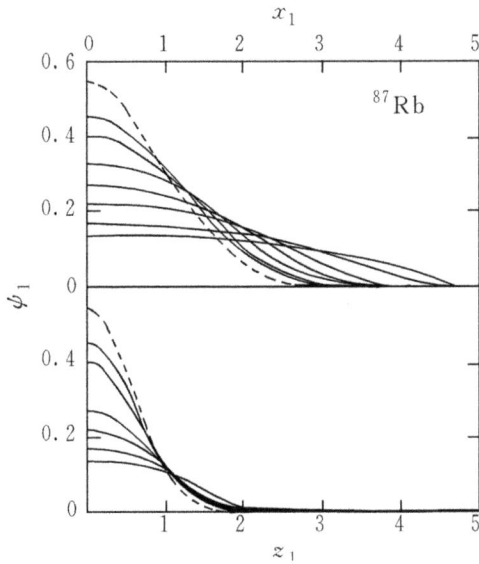

Fig. 10.14 Dimensionless wave function of ^{87}Rb condensate, demonstrating the influence of interaction. Reprinted with permission from F. Dalfovo and S. Stringari, *Phys. Rev.* A53, 2477, 1996. Copyright (2017) by the American Physical Society.

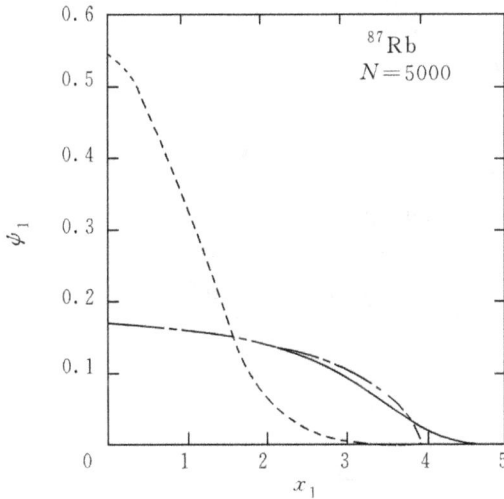

Fig. 10.15 Comparison between the numerical solution with the Thomas-Fermi approximation. Reprinted with permission from F. Dalfovo and S. Stringari, *Phys. Rev.* A53, 2477, 1996. Copyright (2017) by the American Physical Society.

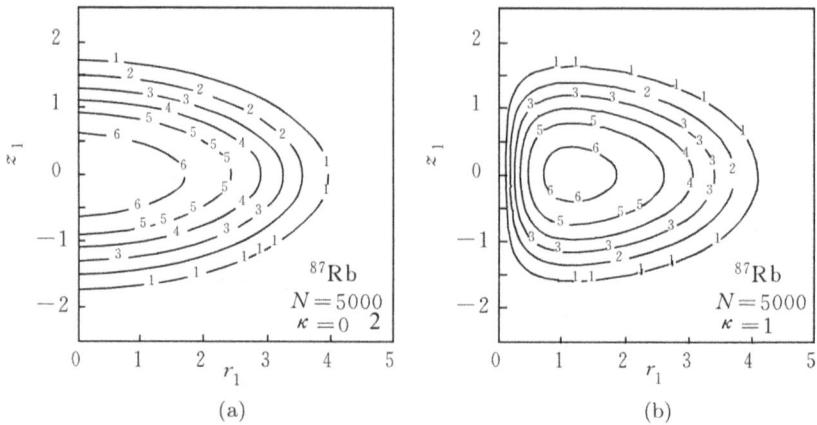

Fig. 10.16   Wave function contours (a) No vortex; (b) With a $\kappa = 1$ vortex. Reprinted with permission from F. Dalfovo and S. Stringari, *Phys. Rev.* A53, 2477, 1996. Copyright (2017) by the American Physical Society.

contours are in an arbitrary unit, and (b) gives the case for $\kappa = 1$, demonstrating that the vortex pushes away the wave function from its axis.

Condensate with attractive interaction between atoms provides special interest. From (10.4.37) we see that the interaction energy is proportional to a/v, where v is the specific volume. The total energy can be made lower and lower by decreasing the specific volume for a condensate with $a < 0$. Therefore a homogeneous condensate with attractive interaction tends to collapse, and cannot be stable. When a trapping potential exists the zero-point energy offers the possibility of having a stable condensate with attractive interaction if the number of particles is lower than a critical value. The result of calculation for ^{7}Li gaseous condensate is shown in Fig. 10.17, where dashed line represents wave function without interaction and solid lines from upside down represent wave functions with $n = 200, 500, 1000$. With increasing N the density at the center of the condensate increases, and the speed of convergence becomes slower and slower in the computation. For $N > 1400$ the computation for E_1/N does not converge any more. One influence of the vortex is again pushing the wave function away from its axis, as shown in Fig. 10.18(a). Another influence of the vortex is to enhance the stability of the condensate.

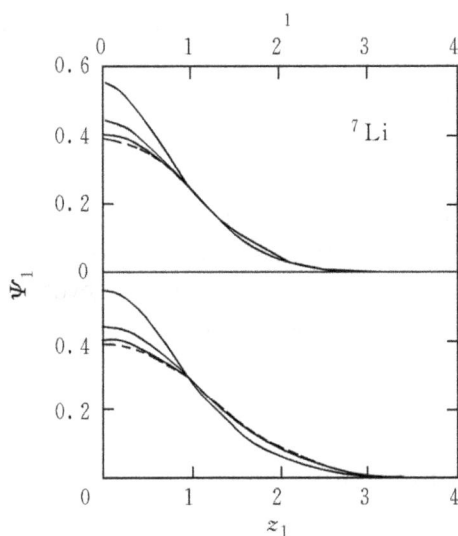

Fig. 10.17 Condensate wave function for ^{7}Li, exhibiting the influence of interaction between atoms. Reprinted with permission from F. Dalfovo and S. Stringari, *Phys. Rev.* A53, 2477, 1996. Copyright (2017) by the American Physical Society.

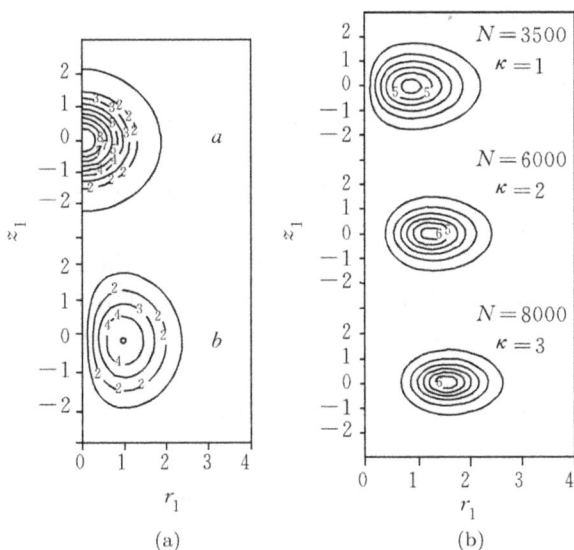

Fig. 10.18 Wave function contours (a) comparison between the wave functions without vortex ($\kappa = 0$) and with a vortex $\kappa = 1$ for $N = 1000$. (b) comparison between different combinations of κ and N. Reprinted with permission from F. Dalfovo and S. Stringari, *Phys. Rev.* A53, 2477, 1996. Copyright (2017) by the American Physical Society.

Figure 10.18(b) gives the wave function contours for different N and κ. For $\kappa = 1$, N can reach 4000, for $\kappa = 2$, N can reach 6500 and for $\kappa = 3$, N can reach 8300.

Many theoretical issues were addressed on the basis of the Gross-Pitaevskii equations, the time independent one and the time-dependent one. They are applicable when the condensate is already formed. The physics of creation and destruction of the condensate is a big challenge. There were many discussions devoted to this problem [47], but besides the evolution close to statistical equilibrium the problem is still far from being solved.

10.8 The Spinor Condensate

The use of optical trap [60] opened up a new direction in the study of confined gaseous condensates. In conventional magnetic trap the spin degree of freedom of the atom is frozen, since only atoms with a single value of the hyperfine spin component are confined. The atoms behave as if they are spinless. In an optical trap the spin nature of the atoms can be manifested. The theoretical study of the spinor condensate was pioneered by T.L. Ho [61, 62]. Alkali atoms ^{23}Na, ^{39}K and ^{87}Rb have nuclear spin 3/2, and therefore can have hyperfine spin 1 or 2. Energetically the $f = 2$ multiplets have excitation energies much higher than the trap potential energy. At low temperatures the two-body scattering between two $f = 1$ atoms can lead only to hyperfine states within this multiplet. We can therefore concentrate on the lower multiplet $f = 1$ with $m_f = -1, 0.1$. The scattering channel can have hyperfine spins $F = 0$ or 2, while hyperfine spin 1 is ruled out because of the demand of symmetry. The interaction potential between two atoms is given by

$$V(\mathbf{r}_1 - \mathbf{r}_2) = \delta^3(\mathbf{r}_1 - \mathbf{r}_2) \sum_{F=0,2} g_F \mathscr{P}_F, \qquad (10.8.1)$$

where $g_F = \frac{4\pi\hbar^2 a_F}{M}$, with the scattering length a_F and mass of the atom M, and $\mathscr{P}_F$ the projection operator onto the channel with hyperfine spin F. The projection operators have the following properties:

$$\mathscr{P}_0 + \mathscr{P}_2 = 1, \quad \mathscr{P}_0 \mathscr{P}_2 = \mathscr{P}_2 \mathscr{P}_0 = 0, \quad \mathscr{P}_0^2 = \mathscr{P}_0, \quad \mathscr{P}_2^2 = \mathscr{P}_2. \tag{10.8.2}$$

The projection operators can be explicitly expressed in terms of $\boldsymbol{F}_1 \cdot \boldsymbol{F}_2$ where $\boldsymbol{F}_1$ and $\boldsymbol{F}_2$ are the hyperfine spin operators of atom 1 and 2 respectively. Since $\boldsymbol{F}_1 \cdot \boldsymbol{F}_2 = \frac{1}{2}[F(F+1) - 2f(f+1)]$, we have

$$\lambda_0 \equiv \boldsymbol{F}_1 \cdot \boldsymbol{F}_2 \,|_{F=0} = -2,$$
$$\lambda_2 \equiv \boldsymbol{F}_1 \cdot \boldsymbol{F}_2 \,|_{F=2} = 1.$$

Therefore,

$$\boldsymbol{F}_1 \cdot \boldsymbol{F}_2 = -2\mathscr{P}_0 + \mathscr{P}_2.$$

Using (10.8.2) we get

$$\mathscr{P}_0 = \frac{1}{3}(1 - \boldsymbol{F}_1 \cdot \boldsymbol{F}_2), \quad \mathscr{P}_2 = \frac{1}{3}(2 + \boldsymbol{F}_1 \cdot \boldsymbol{F}_2). \tag{10.8.3}$$

Dropping the delta function in (10.8.1) we have

$$V = g_0 \frac{1}{3}(1 - \boldsymbol{F}_1 \cdot \boldsymbol{F}_2) + g_2 \frac{1}{3}(2 + \boldsymbol{F}_1 \cdot \boldsymbol{F}_2)$$

$$= c_0 + c_2 \boldsymbol{F}_1 \cdot \boldsymbol{F}_2, \tag{10.8.4}$$

where

$$c_0 = \frac{1}{3}(g_0 + 2g_2), \quad c_2 = \frac{1}{3}(g_2 - g_0). \tag{10.8.5}$$

The Hamiltonian of the spinor condensate in the second quantized form is then

$$\mathscr{H} = \int d^3r \left[\sum_a \left(\frac{\hbar^2}{2M} \right) \nabla \psi_a^\dagger \cdot \nabla \psi_a + \sum_a U \psi_a^\dagger \psi_a \right.$$

$$\left. + \frac{c_0}{2} \sum_{a,a'} \psi_a^\dagger \psi_{a'}^\dagger \psi_{a'} \psi_a + c_2 \sum_{aa'bb'} (\psi_a^\dagger \boldsymbol{F} \psi_b) \cdot (\psi_{a'}^\dagger \boldsymbol{F} \psi_{b'}) \right], \tag{10.8.6}$$

where ψ_a is the field operator for annihilation of an atom in the hyperfine component state a, $a = -1, 0, 1$, and U is the trap potential. For convenience we give explicitly the matrix form of the hyperfine spin operators for the multiplet $f = 1$ (in units $\hbar = 1$):

$$F_x = \begin{pmatrix} 0 & 1 & 0 \\ 1 & 0 & 1 \\ 0 & 1 & 0 \end{pmatrix}, \quad F_y = \frac{i}{\sqrt{2}} \begin{pmatrix} 0 & -1 & 0 \\ 1 & 0 & -1 \\ 0 & 1 & 0 \end{pmatrix}, \quad F_z = \begin{pmatrix} 1 & 0 & 0 \\ 0 & 0 & 0 \\ 0 & 0 & -1 \end{pmatrix}.$$
$$(10.8.7)$$

The spin rotation operator

$$\mathscr{U}(\alpha, \beta, \tau) = e^{-iF_z\alpha} e^{-iF_y\beta} e^{-iF_z\tau} \tag{10.8.8}$$

is expressed in terms of the Euler angles α, β and τ:

$$e^{-iF_z\alpha} = \begin{pmatrix} e^{-i\alpha} & 0 & 0 \\ 0 & 1 & 0 \\ 0 & 0 & e^{i\alpha} \end{pmatrix}, \quad e^{-iF_z\alpha} = \begin{pmatrix} e^{-i\tau} & 0 & 0 \\ 0 & 1 & 0 \\ 0 & 0 & e^{i\tau} \end{pmatrix},$$

$$e^{-iF_y\beta} = \begin{pmatrix} \cos^2\frac{\beta}{2} & -\frac{1}{\sqrt{2}}\sin\beta & -\cos^2\frac{\beta}{2} \\ \frac{1}{\sqrt{2}}\sin\beta & \cos\beta & -\frac{1}{\sqrt{2}}\sin\beta \\ -\cos^2\frac{\beta}{2} & \frac{1}{\sqrt{2}}\sin\beta & \cos^2\frac{\beta}{2} \end{pmatrix}. \tag{10.8.9}$$

We will see that for different values of on the parameters g_0 and g_2 the ground state structure, the collective excitation, and the vortex structure show distinctly different behaviors.

10.8.1 Ground state structure

We write the condensate wave function as

$$\Psi_a(r) \equiv \langle \hat{\psi}_a(r) \rangle = \sqrt{n(r)} \zeta_a(r), \tag{10.8.10}$$

where n is the density and ζ_a is a normalized spinor characterizing the hyperfine spin state a and also the phase, $\zeta^\dagger\zeta = 1$. The ground state is determined by minimizing the energy under the constraint of a fixed particle number, i.e., $\delta \langle H - \mu N \rangle = 0$,

$$H - \mu N = \int \mathrm{d}^3 r \left(\frac{\hbar^2}{2M} (\nabla \sqrt{n})^2 + \frac{\hbar^2}{2M} |\nabla \zeta_a|^2 n - [\mu - U(\boldsymbol{r})]n \right.$$
$$\left. + \frac{n^2}{2} [c_0 + c_2 \langle \boldsymbol{F} \rangle^2] \right), \tag{10.8.11}$$

where

$$\langle \boldsymbol{F} \rangle = \zeta_a^\dagger \boldsymbol{F} \zeta_a.$$

It is obvious that all spinors related to one another by a gauge transformation $e^{i\theta}$ and spin rotations $\mathscr{U}(\alpha, \beta, \tau)$ (10.8.8) are degenerate in energy. There are two distinct cases.

(1) Polar state[19]
For $g_2 > g_0$, we have $c_2 > 0$. From (10.8.11) we see that the energy is minimized by $\langle \boldsymbol{F} \rangle^2 = 0$. This can be achieved, for instance by choosing the hyperfine spin component $m_f = 0$, i.e.,

$$e^{i\theta} \begin{pmatrix} 0 \\ 1 \\ 0 \end{pmatrix},$$

for which $\langle F_x \rangle = \langle F_y \rangle = \langle F_z \rangle = 0$. Applying the spin rotation operator to it we obtain

$$\zeta = e^{i\theta} \mathscr{U} \begin{pmatrix} 0 \\ 1 \\ 0 \end{pmatrix} = e^{i\theta} \begin{pmatrix} -\frac{1}{\sqrt{2}} e^{-i\alpha} \sin\beta \\ \cos\beta \\ \frac{1}{\sqrt{2}} e^{i\alpha} \sin\beta \end{pmatrix}. \tag{10.8.12}$$

The minimization of $H - \mu N$ gives

$$\delta \langle H - \mu N \rangle = \int \mathrm{d}^3 r \left(-\frac{\hbar^2}{2M} \frac{\nabla^2 \sqrt{n}}{\sqrt{n}} - [\mu - U(\boldsymbol{r})] + c_0 n \right) \delta n = 0,$$

---

[19] The name is adapted from the analog of the polar state in liquid ^{3}He.

which leads to the ground state density

$$n^0(\boldsymbol{r}) = \frac{1}{c_0}[\mu - U(\boldsymbol{r}) - W(\boldsymbol{r})], \qquad (10.8.13)$$

where

$$W(\boldsymbol{r}) = \frac{\hbar^2}{2M} \frac{\nabla^2 \sqrt{n^0}}{\sqrt{n^0}}. \qquad (10.8.14)$$

Note that ζ (10.8.12) is independent of the Euler angle τ. The symmetry group of the polar state is therefore $U(1) \times S^2$, where $U(1)$ refers to the phase θ and S^2 refers to the unit sphere on the surface of which the spin quantization axis is given by the angles α, β.

(2) Ferromagnetic state

For $g_0 > g_2$, we have $c_2 < 0$. The energy is minimized by demanding $\langle \boldsymbol{F} \rangle^2 = 1$. This can be achieved by choosing the hyperfine spin component $m_f = 1$, or more generally

$$\zeta = e^{i\theta} \mathcal{U} \begin{pmatrix} 1 \\ 0 \\ 0 \end{pmatrix} = e^{i(\theta - \tau)} \begin{bmatrix} e^{-i\alpha} \cos^2 \dfrac{\beta}{2} \\ \sqrt{2} \cos \dfrac{\beta}{2} \sin \dfrac{\beta}{2} \\ e^{i\alpha} \sin^2 \dfrac{\beta}{2} \end{bmatrix}. \qquad (10.8.15)$$

In calculating $\delta \langle H - \mu N \rangle$ the term proportional to n^2 is $\frac{n^2}{2}(c_0 + c_2) = \frac{n^2}{2} g_2$ which leads to $g_2 n \delta n$. Finally we get

$$n^0(\boldsymbol{r}) = \frac{1}{g_2}[\mu - U(\boldsymbol{r}) + W(\boldsymbol{r})]. \qquad (10.8.16)$$

The direction of the spin is $\langle \boldsymbol{F} \rangle = \hat{\boldsymbol{z}} \cos \beta + \sin \beta (\cos \alpha \, \hat{\boldsymbol{x}} + \sin \alpha \, \hat{\boldsymbol{y}})$. The combination $(\theta - \tau)$ in (10.8.15) displays a local "spin-gauge" symmetry.[20] The symmetry group of the ferromagnetic state

---

[20]See Ref. [62]. Let the condensate wave function be $\Psi_m(\boldsymbol{r},t) = \zeta_m(\boldsymbol{r},t) \Phi(\boldsymbol{r},t)$, where m is the hyperfine spin component along the quantization axis $\hat{\boldsymbol{n}}$. A local gauge transformation $e^{i\chi(\boldsymbol{r},t)}$ on Ψ can be undone by a local spin rotation $e^{-i\frac{\chi(\boldsymbol{r},t)}{F} \hat{\boldsymbol{n}} \cdot \boldsymbol{F}}$. This is the local gauge-spin symmetry.

is $SO(3)$. The difference in symmetry will lead to a fundamental difference between the properties of their vortices. The condensates of ^{23}Na and ^{87}Rb are probably polar and ferromagnetic states respectively.

10.8.2 Collective modes of trapped spinor condensate

The Hamiltonian (10.8.6) leads to the equation of motion for the field operator

$$i\hbar\partial_t\hat{\psi}_m = -\frac{\hbar^2}{2M}\nabla^2\hat{\psi}_m + [U(r) - \mu]\hat{\psi}_m + c_0\sum_a(\hat{\psi}_a^\dagger\hat{\psi}_a)\hat{\psi}_m$$

$$+ c_2\sum_{a,b}(\hat{\psi}_a^\dagger \mathbf{F}\hat{\psi}_b) \cdot \mathbf{F}\hat{\psi}_m. \tag{10.8.17}$$

To study the elementary excitations, we write $\hat{\psi}_m = \Psi^0 + \hat{\phi}_m$ and linearize (10.8.17) about the ground state Ψ^0.

(1) Polar state

We write $\Psi^0 = \sqrt{n}\begin{pmatrix} 0 \\ 1 \\ 0 \end{pmatrix}$, use (10.8.13) and $\langle \mathbf{F}\rangle = 0$, (10.8.17) then gives

$$i\hbar\partial_t\begin{pmatrix} \hat{\phi}_0 \\ -\hat{\phi}_0^\dagger \end{pmatrix} = -\frac{\hbar^2}{2M}\nabla^2\begin{pmatrix} \hat{\phi}_0 \\ \hat{\phi}_0^\dagger \end{pmatrix} + W(\mathbf{r})\begin{pmatrix} \hat{\phi}_0 \\ \hat{\phi}_0^\dagger \end{pmatrix} + n^0c_0\begin{pmatrix} \hat{\phi}_0 + \hat{\phi}_0^\dagger \\ \hat{\phi}_0 + \hat{\phi}_0^\dagger \end{pmatrix},$$

$$\tag{10.8.18}$$

$$i\hbar\partial_t\begin{pmatrix} \hat{\phi}_1 \\ -\hat{\phi}_{-1}^\dagger \end{pmatrix} = -\frac{\hbar^2}{2M}\nabla^2\begin{pmatrix} \hat{\phi}_1 \\ \hat{\phi}_{-1}^\dagger \end{pmatrix} + W(\mathbf{r})\begin{pmatrix} \hat{\phi}_1 \\ \hat{\phi}_{-1}^\dagger \end{pmatrix} + n^0c_2\begin{pmatrix} \hat{\phi}_1 + \hat{\phi}_{-1}^\dagger \\ \hat{\phi}_1 + \hat{\phi}_{-1}^\dagger \end{pmatrix}.$$

$$\tag{10.8.19}$$

In the linearization, we substitute

$$\hat{\psi}^\dagger\hat{\psi} \rightarrow \Psi^{0*}\hat{\phi} + \hat{\phi}^\dagger\Psi^0$$

and take into account that $\Psi^0 \propto \zeta_0$ such that only $\hat{\phi}_0$ and $\hat{\phi}_0^\dagger$ contribute in the density fluctuations

$$\delta\hat{n}(\mathbf{r}) = \sqrt{n^0(\mathbf{r})}(\hat{\phi}_0 + \hat{\phi}_0^\dagger).$$

The spin fluctuations are

$$\delta \hat{M}_+ \equiv \delta(\hat{M}_x + i\hat{M}_y) = \sqrt{n^0(\boldsymbol{r})}(\hat{\phi}_1 + \hat{\phi}^\dagger_{-1}),$$
$$\delta \hat{M}_- = \delta \hat{M}^\dagger_+.$$

Denoting the frequencies of the excitation modes $\hat{\phi}_0$ and $\hat{\phi}_\pm$ as ω_0 and $\omega_\pm$, and substituting $\phi \propto e^{i(\boldsymbol{k}\cdot\boldsymbol{x} - \omega t)}$ into (10.8.18) and (10.8.19) we obtain a system of coupled linear homogeneous equations. The condition for the existence of non-zero solution leads to (for the homogeneous case $(W = 0)$

$$\hbar\omega_0 = \sqrt{\varepsilon_{\boldsymbol{k}}(\varepsilon_{\boldsymbol{k}} + 2c_0 n_0)}, \quad \hbar\omega_\pm = \sqrt{\varepsilon_{\boldsymbol{k}}(\varepsilon_{\boldsymbol{k}} + 2c_2 n_0)},$$

where $\varepsilon_{\boldsymbol{k}} = \frac{\hbar^2 k^2}{2M}$. In a harmonic trap, Eq. (10.8.18) is identical to the equation for collective excitation of a scalar boson condensate [62]. Using Thomas-Fermi approximation for the condensate density $n^0(\boldsymbol{r}) = [\mu - U(\boldsymbol{r})]/c_0$ and neglecting W, Eqs. (10.8.18) and (10.8.19) can be written as

$$\partial_t \delta\hat{n} = \nabla(c_0 n^0 \nabla \delta\hat{n}), \quad \partial_t \delta\hat{M}_\pm = \nabla(c_2 n^0 \nabla \delta\hat{M}_\pm). \qquad (10.8.20)$$

Notice that the spin wave modes obey the same equation as the density modes except that c_0 is replaced by c_2, and therefore the quantum numbers and wave functions of spin modes are identical with those of the density modes and their frequencies are related by

$$\omega_\pm^2 = \frac{c_2}{c_0}\omega_0^2 = \frac{a_2 - a_0}{2a_2 + a_0}\omega_0^2. \qquad (10.8.21)$$

(2) Ferromagnetic state

Taking $\Psi^0 = \sqrt{n}\begin{pmatrix} 1 \\ 0 \\ 0 \end{pmatrix}$, using $\langle \boldsymbol{F} \rangle = \hat{\boldsymbol{z}}$, and substituting $\hat{\Psi}_m = \Psi^0 + \hat{\phi}_m$ into (10.8.17), we obtain

$$i\hbar\partial_t \begin{pmatrix} \hat{\phi}_1 \\ \hat{\phi}_0 \\ \hat{\phi}_{-1} \end{pmatrix} = -\frac{\hbar^2}{2M}\nabla^2 \begin{pmatrix} \hat{\phi}_1 \\ \hat{\phi}_0 \\ \hat{\phi}_{-1} \end{pmatrix} + W(\boldsymbol{r}) \begin{pmatrix} \hat{\phi}_1 \\ \hat{\phi}_0 \\ \hat{\phi}_{-1} \end{pmatrix} + n^0 \begin{pmatrix} g_2(\hat{\phi}_1 + \hat{\phi}^\dagger_1) \\ 0 \\ 2|c_2|\hat{\phi}_{-1} \end{pmatrix}.$$

$$(10.8.22)$$

To linear order in $\hat{\phi}$, the density, spin and quadrupolar spin fluctuations are given by $\delta\hat{n} = \sqrt{n_0}(\hat{\phi}_1 + \hat{\phi}_1^\dagger)$, $\delta\hat{M}_- = \sqrt{n_0}\hat{\phi}_0^\dagger$, and $\delta\hat{M}_-^2 = 2\sqrt{n_0}\hat{\phi}_{-1}^\dagger$ respectively. In the homogeneous case ($W = 0$) the frequencies of these modes are respectively $\hbar\omega_1 = \sqrt{\varepsilon_k(\varepsilon_k + 2g_2 n_0)}$, (a Bogoliubov spectrum), $\hbar\omega_0 = \varepsilon_k$ (a free particle spectrum), and $\hbar\omega_{-1} = \varepsilon_k + 2c_2 n^0$ (a free particle like spectrum with a gap).

10.8.3 Intrinsic stability of vortices in the ferromagnetic state

The superfluid velocity is given by $\boldsymbol{v}_\mathrm{s} = \frac{\hbar}{M}\zeta^\dagger\nabla\zeta$. We allow the Euler angles to vary with position (spin texture). The superfluid velocity turns out to be

$$\left.\begin{array}{l} \boldsymbol{v}_\mathrm{s} = \dfrac{\hbar}{M}\nabla\theta, \text{ for polar state} \\[2ex] \boldsymbol{v}_\mathrm{s} = \dfrac{\hbar}{M}[\nabla(\theta - \tau) - \cos\beta\nabla\alpha], \text{ for ferromagnetic state} \end{array}\right\}$$

$$(10.8.23)$$

The superfluid velocity for ferromagnetic state differs from that for the polar state in that it depends on the spin rotations. This leads to an interesting situation for the vortices in the ferromagnetic state. If too much vortex energy is stored in one spin component, the system can get rid of it by spin rotation, i.e., such vortex is unstable. This can be demonstrated by choosing a state for which $\tau = 0, -\alpha = \theta = m\phi$ (m is a positive integer) and $\beta = \pi t$ (t is a parameter varying from 0 to 1). The state is now described by

$$\Psi(t) = \sqrt{n^0}\begin{pmatrix} e^{i2m\phi}\cos^2\left(\dfrac{\pi t}{2}\right) \\[2ex] e^{im\phi}\sqrt{2}\sin\dfrac{\pi t}{2}\cos\dfrac{\pi t}{2} \\[2ex] \sin^2\dfrac{\pi t}{2} \end{pmatrix}. \qquad (10.8.24)$$

We begin with

$$\zeta(t = 0) = \begin{pmatrix} e^{i2m\phi} \\ 0 \\ 0 \end{pmatrix}$$

characterizing a vortex with circulation $2m$ in the $\hat{z}$ direction and continuously deform the state by varying the Euler angle β locally through t and end up with

$$\zeta(t = 1) = \begin{pmatrix} 0 \\ 0 \\ 1 \end{pmatrix}$$

characterizing a condensate with spin -1 without vortex. The spin texture $\langle \boldsymbol{F} \rangle$ of the state (10.8.24) is

$$\langle \boldsymbol{F} \rangle = \cos(\pi t)\hat{z} + \sin(\pi t)[\cos(m\phi)\hat{\boldsymbol{x}} + \sin(m\phi)\hat{\boldsymbol{y}}].$$

The $2m$ vortices are unstable topologically. Multiplying (10.8.24) by $e^{i\phi}$, we get the "evolution" in t as

$$\zeta(0) = \begin{pmatrix} e^{i(2m+1)\phi} \\ 0 \\ 0 \end{pmatrix} \rightarrow \zeta(1) = \begin{pmatrix} 0 \\ 0 \\ e^{i\phi} \end{pmatrix},$$

characterizing an evolution from a $2m + 1$ vortex in the $\hat{z}$ direction to a vortex of unit circulation in the $-\hat{z}$ direction. The vortex of unit circulation is stable against the change of spin texture. The difference of the polar and ferromagnetic states has its origin in their first homotopy group, $\Pi_1(U(1) \times S^2) = Z$ and $\Pi_1(SO(3)) = Z_2$. The polar phase has infinitely many distinct defects of different winding, whereas the ferromagnetic phase has only one defect with nontrivial winding.

10.8.4 Coreless vortices

Consider a condensate with

$$
\zeta = \begin{pmatrix} \cos^2 \dfrac{\beta}{2} \\[2mm] \sqrt{2}\mathrm{e}^{\mathrm{i}\phi} \sin \dfrac{\beta}{2} \cos \dfrac{\beta}{2} \\[2mm] \mathrm{e}^{2\mathrm{i}\phi} \sin^2 \dfrac{\beta}{2} \end{pmatrix},
$$

where $\beta(r)$ is an increasing function of r starting from $\beta = 0$ at $r = 0$. The spin texture and superfluid velocity of the condensate are both cylindrically symmetrical, $\langle \boldsymbol{F} \rangle = \cos \beta \hat{\boldsymbol{z}} + \sin \beta [\cos \phi \hat{\boldsymbol{x}} + \sin \phi \hat{\boldsymbol{y}}]$ and $\boldsymbol{v}_\mathrm{s} = \frac{\hbar}{mr}(1 - \cos \beta)\hat{\phi}$. When $\beta(r)$ reaches $\frac{\pi}{2}$, $\boldsymbol{v}_\mathrm{s}$ is the superfluid velocity of a vortex, but as $r \rightarrow 0$, $\beta \rightarrow 0$ the superfluid velocity vanishes instead of becoming divergent. The spin texture, the topological instability and the superfluid velocity bear a close resemblance to the A-phase of superfluid ^{3}He.

10.8.5 Fragmented condensates

Previous discussions are based on the mean field theory. By going beyond the mean field theory we can explore many interesting properties of a Bose system with internal degrees of freedom. We follow [63] to discuss the fragmented and single condensates of spin-1 Bose gas. The Hamiltonian of a spin-1 Bose gas in an external magnetic field is

$$
\hat{H} = \int \hat{\psi}_\mu^\dagger \left(-\frac{\hbar^2}{2M}\delta_{\mu\nu}\nabla^2 - \gamma \boldsymbol{B} \cdot \boldsymbol{F}_{\mu\nu} \right) \hat{\psi}_\nu d^3r
$$

$$
+ \frac{1}{2} \int \hat{\psi}_\mu^\dagger \hat{\psi}_\alpha^\dagger \hat{\psi}_\beta \hat{\psi}_\nu \left(c_0 \delta_{\alpha\beta}\delta_{\mu\nu} + c_2 \boldsymbol{F}_{\mu\nu} \cdot \boldsymbol{F}_{\alpha\beta} \right) d^3r. \qquad (10.8.25)
$$

The field operator $\hat{\psi}_\mu$ is expanded as

$$
\hat{\psi}_\mu = \frac{1}{\Omega^{1/2}} \sum_{\boldsymbol{k}} \mathrm{e}^{\mathrm{i}\boldsymbol{k}\cdot\boldsymbol{r}} a_\mu(\boldsymbol{k}), \qquad (10.8.26)
$$

where Ω is the volume of the system. We denote $a_\mu(0)$ simply by a_μ, and the part of $\hat{H}$ containing a_μ only by $\hat{H}_0$, thus

$$\hat{H} = \hat{H}_0 + \hat{H}',$$
$$\hat{H}_0 = \frac{c_2}{2\Omega}\left(\boldsymbol{F}^2 - 2N\right) - \gamma \boldsymbol{B}\cdot\boldsymbol{F} + \frac{c_0}{2\Omega}N(N-1), \qquad (10.8.27)$$

where $\boldsymbol{F} = a_\mu^\dagger \boldsymbol{F}_{\mu\nu}a_\nu$, $N = a_\mu^\dagger a_\mu$. $\hat{H}_0$ is used to study the ground state (condensate), and $\hat{H}'$ is used to study the condensate depletion. We denote the ground state of $\hat{H}_0$ by $|F\rangle$. From (10.8.27) it follows that

$$|F\rangle = |F = S, F_z = S\rangle, \qquad (10.8.28)$$

where S is the integer that minimizes

$$\left\langle \hat{H}_0 \right\rangle_F = \frac{c_2}{2\Omega}S(S+1) - \gamma BS + \frac{c_0}{2\Omega}N(N-1) - \frac{c_2 N}{\Omega}. \qquad (10.8.29)$$

Before elaborating the properties of the eigenstates, we mention a paper [64] by Law, Pu and Bigelow, who first treated the ground state of S=1 spinor condensate beyond mean field theory, in absence of an external magnetic field. The Hamiltonian is the same as (10.8.27), without the $\boldsymbol{B}$ term. The spin dependent part is then

$$\hat{H}_a = \frac{c_2}{2\Omega}[S(S+1) - 2N]. \qquad (10.8.30)$$

For $c_2 > 0$, the ground state is $|F\rangle = |0, 0\rangle$ with eigenvalue $E_a = -c_2 N/\Omega.|F\rangle$ is expanded in terms of the Fock states $|N_1, N_0, N_{-1}\rangle$ as

$$|F\rangle = \sum_{k=0}^{[N/2]} A_k |k, N - 2k, k\rangle. \qquad (10.8.31)$$

A recursion formula for A_k is given in Ref. [62]. The average numbers of components in $|F\rangle$ are all equal: $\langle N_1 \rangle = \langle N_0 \rangle = \langle N_{-1} \rangle = N/3$. The distribution A_k is almost uniform versus k, therefore the number fluctuations are very large, although the total number is fixed. The authors of Ref. [64] pointed out that the super-Poisson distribution of particle numbers is a common feature for low energy eigenstates of $\hat{H}_a$ for $c_2 > 0$. Such a state is called a collective spin state by the authors because it cannot be expressed as product states of individual atoms.

For $c_2 < 0$, the system has $2N + 1$ degenerate ground states

$$|F\rangle_{m_l} = |N, m_l\rangle, \quad m_l = 0, \pm 1, \ldots, \pm N, \tag{10.8.32}$$

with eigenvalue $E_a = \frac{c_2}{2\Omega} N(N - 1)$. In terms of Fock states the expansion is

$$|N, m_l\rangle = \sum_k B_k^{(m_l)} |k + m_l, N - 2k - m_l, k\rangle. \tag{10.8.33}$$

The distribution $B_k^{(m_l)}$ is sub-Poissonian, its width decreases for increasing m_l. Both A_k and B_k are shown in figures in Ref. [62] for specific examples.

Now we return to the ground state (10.8.28) [63], concentrating on the case $c_2 > 0$, in the presence of a magnetic field. To understand the structure and properties of this state, we notice that because of the Bose statistics, the N-body singlet of a spin-1 Bose gas is unique; because exchange of any two bosons leads to a state identical to the original one. The operator creating a singlet pair is

$$\Theta^\dagger = -2a_1^\dagger a_{-1}^\dagger + a_0^{\dagger 2}. \tag{10.8.34}$$

The ground state $|F\rangle = |S, S\rangle$ is composed of S bosons with $f_z = 1$ and $(N - S)/2$ singlet pairs, viz.

$$|S, S\rangle = \frac{1}{\sqrt{f(Q, S)}} a_1^{\dagger S} \Theta^{\dagger \frac{N-S}{2}} |vac\rangle. \tag{10.8.35}$$

The normalization factor $f(Q, S)$ is [63]

$$f(Q, S) = S! Q! 2^Q \frac{(2Q + 2S + 1)!!}{(2S + 1)!!} \tag{10.8.36}$$

The single particle density matrix of $|F\rangle$ is diagonal

$$(\hat{\rho}^F)_{\alpha\beta} = \langle a_\beta^\dagger a_\alpha \rangle_F = N_\alpha \delta_{\alpha\beta}, \tag{10.8.37}$$

with

$$N_1 = \frac{N(S+1) + S(S+2)}{2S+3}, \quad N_0 = \frac{N-S}{2S+3},$$

$$N_{-1} = \frac{(N-S)(S+1)}{2S+3}. \tag{10.8.38}$$

To characterize the system, single particle density matrix is insufficient, because systems with identical single particle density matrix can have very different two-particle correlations. We notice that because of the identities $\hat{N}_{-1} = \hat{N}_1 - S$ (because in singlet pairs $+1$ and -1 components are equally represented) and $\hat{N}_0 = N + S - 2\hat{N}_1$, all two-body correlations can be expressed in terms of $(\Delta\hat{N}_1)^2 \equiv \langle(a_1^\dagger a_1 - \langle a_1^\dagger a_1\rangle^2)\rangle = \langle(a_1^\dagger a_1)^2\rangle - \langle a_1^\dagger a_1\rangle^2$. For example, $(\Delta\hat{N}_1)^2 = (\Delta\hat{N}_0)^2/4 = (\Delta\hat{N}_{-1})^2$, $\langle\hat{N}_0\hat{N}_{-1}\rangle = \langle\hat{N}_1\rangle(N+3S) - 2\langle\hat{N}_1^2\rangle SN - S^2$, etc. Using (10.8.36) it can be shown that

$$\left(\Delta\hat{N}_1\right)^2 = \left(\frac{N}{2S+3}\right)^2 \left(\frac{S+1}{2S+5}\right) + \left(\frac{3N}{(2S+3)^2}\right)\left(\frac{S+1}{2S+5}\right)$$

$$+ \left(\frac{S+1}{2S+5}\right)\left(\frac{S^2-3S}{(2S+3)^2}\right). \tag{10.8.39}$$

The properties of the state $|F\rangle = |S, S\rangle$ depends on the magnetization S/N. For $S = 0$, we have the state $|0,0\rangle$ first studied by Law, Pu and Bigelow [62]. It has equal average particle number of components $N_1 = N_0 = N_{-1} = N/3$ and enormous fluctuations $\Delta N_\alpha \sim N$. As S increases both N_0 and ΔN_α decrease rapidly.

When S becomes macroscopic N_0 and ΔN_α become $O(1)$, i.e., zero in the thermodynamic limit, whereas $N_{\pm 1} = (N \pm S)/2$ remain macroscopic. The state $|F\rangle$ is a fragmented condensate, because the single particle density matrix has more than one macroscopic eigenvalue for all magnetization. Setting $\Delta N \sim \sqrt{N}$ (the Poisson distribution) as the crossover from "super-fragmented state" ($\Delta N_\alpha \sim N$) to "coherent fragmented state" ($\Delta N_\alpha \sim 1$), Eq. (10.8.39) implies that the crossover is given by $S/N < 1/\sqrt{8N}$ for $N, S \gg 1$. Thus, super-fragmentation can be achieved only for S/N very small, and coherent fragmentation is more generic and easier to realize.

Consider the coherent fragmentaion $|C\rangle$ for large S for which

$$N_{\pm 1} = \frac{N \pm S}{2}, \quad N_0 = 0.$$

$$|C\rangle = \frac{1}{\sqrt{N!}} \left(\sqrt{\frac{N_1}{N}} a_1^\dagger + \sqrt{\frac{N_{-1}}{N}} a_{-1}^\dagger \right)^N |vac\rangle. \qquad (10.8.40)$$

It can be approximated by the Fock state

$$|S, S\rangle \to |N_1, 0, N_{-1}\rangle = \frac{a_1^{\dagger N_1} a_{-1}^{\dagger N_{-1}}}{\sqrt{N_1! N_{-1}!}} |vac\rangle. \qquad (10.8.41)$$

Within the Fock space (i.e., the total N is fixed) $\hat{H}_0$ can be written as

$$\hat{H}_0 = \frac{c_2}{2\Omega} (N_1 - N_{-1})^2 - \gamma B (N_1 - N_{-1}). \qquad (10.8.42)$$

State (10.8.41) amounts to dropping a_0 operator in the operator Θ. This can be considered as a consequence of the bosonic enhacement for large N: adding a boson to a state consisting of N bosons is associated with a factor $\sqrt{N+1}$. It is enlightening to compare the Fock state (10.8.40) with the single condensate (10.8.40).

The single particle density matrix of $|C\rangle$ is

$$\hat{\rho}^C = \begin{pmatrix} N_1 & \sqrt{N_1 N_{-1}} \\ \sqrt{N_1 N_{-1}} & N_{-1} \end{pmatrix}, \qquad (10.8.43)$$

to be contrasted with that of $|F\rangle$:

$$\hat{\rho}^F = \begin{pmatrix} N_1 & 0 \\ 0 & N_{-1} \end{pmatrix}. \qquad (10.8.44)$$

We notice that the operator $a_\mu^\dagger a_\nu$, whose expectation value is to appear on the non-diagonal elements of the single particle matrix, changes the spin component when acting on a given spin state. Therefore the absence of non-diagonal elements in the fragmented state originates from the conservation of spin. It is suggestive that a spin changing gradient of magnetic field might restore the phase coherence of a condensate by recovering non-diagonal elements of the density matrix.

Consider the external magnetic field

$$\boldsymbol{B}(\boldsymbol{r}) = B_0 \left(\hat{z} + G' \left[x\hat{x} - z\hat{z} \right] \right), \tag{10.8.45}$$

where G' characterizes the strength of the field gradient. By performing a unitary transformation leading to a choice of local field direction as the quantization axis, the resulting Hamiltonian becomes [63]

$$\tilde{H}_0 = -\frac{\varepsilon}{2} \left(a_1^\dagger a_{-1} + H.c. \right) + \frac{c_2}{2\Omega} \left(N_1 - N_{-1} \right)^2 - \gamma B \left(N_1 - N_{-1} \right), \tag{10.8.46}$$

in which the spin non-conserving operators $a_1^\dagger a_{-1} + H.c.$ appear and where $\varepsilon = \hbar^2 G'^2 / 2M$. The ground state of $\tilde{H}_0$ is [63]

$$|\psi\rangle = \left(\pi \eta^2 \right)^{-1/4} \sum_{l=-N_1}^{N_{-1}} e^{-l^2 / 2\eta^2} e^{\xi l} \, |l\rangle \,, \tag{10.8.47}$$

where $|l\rangle = |N_1 + l, N_0, N_{-1} - l\rangle$, $\eta^4 = \varepsilon \sqrt{N_1 N_{-1}} \Omega / 4c_2$, $\xi = (N_{-1}^{-1} - N_1^{-1})/4$. The density matrix of this state is

$$\hat{\rho}_\psi = \begin{pmatrix} N_1 & \sqrt{N_1 N_{-1}} e^{-1/4\eta^2} \\ \sqrt{N_1 N_{-1}} e^{-1/4\eta^2} & N_{-1} \end{pmatrix}. \tag{10.8.48}$$

The eigenvalues of $\hat{\rho}_\psi$ are

$$\lambda_\pm = \frac{1}{2} \left[N \pm \sqrt{N^2 e^{-1/2\eta^2} + S^2(1 - e^{-1/2\eta^2})} \right]. \tag{10.8.49}$$

For zero field gradient, $\eta \to 0$, $\lambda_\pm \to \frac{1}{2}(N \pm S)$, and the condensate is fragmented. For larger field gradients, say $\eta \sim 5$, the system is essentially a single condensate.

In the presence of a trap elaborate analysis is given in Ref. [63], where possibility of observing fragmented condensate in trapped gases is discussed.

10.9 Cold Bosonic Atoms in Optical Lattices

Two oppositely propagating laser beams form a stationary wave, providing a periodic potential for an atom experiencing the a.c. Stark effect. Wells of the periodic potential form arrays of "optical lattices", and depending on the number of beams the lattices can be 1, 2, or 3-dimensional. An atom in a potential well is in a vibrational motion, and tunneling between neighboring wells leads to a spectrum of band structure. Let the potential acting on an atom be

$$V_0(\boldsymbol{x}) = \sum_{j=1}^{3} V_{j0}\sin^2 kx_j, \qquad (10.9.1)$$

where k is the wave vector of the laser, connected with the wave length λ by $k = 2\pi/\lambda$. The lattice constant is $a = \lambda/2$. The parameter V_0 is proportional to the product of the atomic polarizability if the atom and the laser intensity. D. Jaksch *et al.* [65] showed that the dynamics of cold bosonic atoms loaded in an optical lattice realizes a Bose-Hubbard model, put forward first by M.P.A. Fisher *et al.* [66], with parameters of the model controlled by the laser intensity and configurations. The Hamiltonian for the interacting bosonic atoms in the periodic potential $V_0(\boldsymbol{x})$ and a slowly varying trap potential $V_T\boldsymbol{x}$ is

$$H = \int d^3x \psi^\dagger(\boldsymbol{x}) \left(-\frac{\hbar^2}{2m}\nabla^2 + V_0(\boldsymbol{x}) + V_T(\boldsymbol{x}) \right) \psi(\boldsymbol{x})$$
$$+ \frac{1}{2}\frac{4\pi a_s \hbar^2}{m} \int d^3x \psi^\dagger(\boldsymbol{x})\psi^\dagger(\boldsymbol{x})\psi(\boldsymbol{x})\psi(\boldsymbol{x}), \qquad (10.9.2)$$

where $\psi(\boldsymbol{x})$ is the boson field operator and a_s is the s-wave scattering length. Expanding $\psi(\boldsymbol{x})$ in terms of the Wannier basis, restricting to the lowest vibrational state at each site, we have

$$\psi(\boldsymbol{x}) = \sum_i b_i w(\boldsymbol{x} - \boldsymbol{x}_i), \qquad (10.9.3)$$

where $w(\boldsymbol{x} - \boldsymbol{x}_i)$ is the Wannier function an atom localized at the site i, and b_i the annihilation operator of an atom at site i. The operator

b_i and its adjoint $b_i^\dagger$ satisfy the commutation relation

$$\left[b_i, b_j^\dagger\right] = \delta_{ij}. \tag{10.9.4}$$

Substituting (10.9.3) into (10.9.2) and using (10.9.4), we reduce the Hamiltonian (10.9.2) to that of a Bose-Hubbard model

$$H = -J \sum_{\langle i,j \rangle} b_i^\dagger b_j + \sum_i \varepsilon_i \hat{n}_i + \frac{1}{2} U \sum_i \hat{n}_i(\hat{n}_i - 1). \tag{10.9.5}$$

The first term describes the hopping of an atom from a site to a neighboring site, with the sum performed over a pair of nearest neighbors the "hopping constant" J is given by

$$J = \int d^3x\, w^*(\boldsymbol{x} - \boldsymbol{x}_i) \left[-\frac{\hbar^2}{2m} \nabla^2 + V_0(\boldsymbol{x}) \right] w(\boldsymbol{x} - \boldsymbol{x}_j). \tag{10.9.6}$$

The second term is the energy at the sites, with the number operator $\hat{n}_i = b_i^\dagger b_i$ and ε_i given by

$$\varepsilon_i = \int d^3x\, V_T(\boldsymbol{x}) |w(\boldsymbol{x} - \boldsymbol{x}_i)|^2 \approx V_T(\boldsymbol{x}_i), \tag{10.9.7}$$

which is the energy offset of each site. The kinetic energy operator and the optical potential are not included, because they give rise to a constant independent of the sites. The third term is the interaction energy, with U given by

$$U = \frac{4\pi a_s \hbar^2}{m} \int d^3x\, |w(\boldsymbol{x})|^4. \tag{10.9.8}$$

For a given optical potential the parameters J and U can be readily evaluated. The Wannier function $w(\boldsymbol{x}) = w(x)w(y)w(z)$ can be determined from a 1-dimensional band structure calculation. The optical potential and the parameters J and U versus V_0 are shown in Figs. 10.19(a) and (b) respectively. The natural unit of energy in this problem is the "recoil energy" $E_R = \hbar^2 k^2 / 2m$, which is the energy of recoil when an atom at rest absorbs a photon of the laser beam.[21] In terms of the recoil energy the oscillation frequency of an atom in the

---

[21] Actually the laser beams are detuned from atomic resonances in order to diminish loss of atoms.

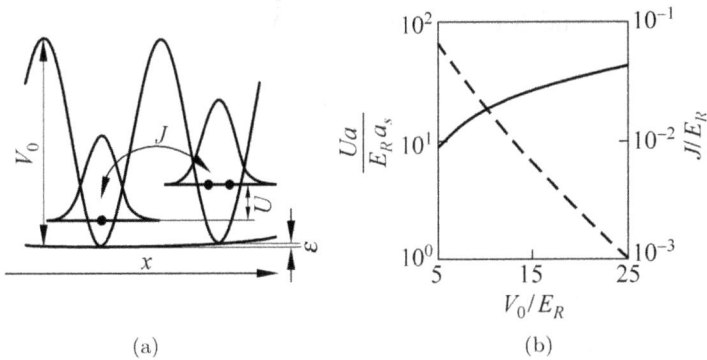

Fig. 10.19 (a) The optical potential; (b) $Ua/E_R a_s$ (solid line, axis on the left) and J/E_R (dashed line, axis on the right) versus V_0/E_R. Reprinted with permission from D. Jaksch, C. Bruder, J. I. Cirac, C. W. Gardiner, and P. Zoller, *Phys. Rev. Lett.* 81, 3108, 1998. Copyright (2017) by the American Physical Society.

potential well is $w = 2\sqrt{E_R V_0}/\hbar$. The energy difference between the lowest band and the higher one is $\hbar w$. The oscillator wave function size is $a_0 = \sqrt{\hbar/mw}$.

The physics of the Bose-Hubbard model is the interplay between J and U. Let us suppose that there is one atom per site, shown as a black dot in the left well, Fig. 10.19(a). If it hops to the right well and pairs with the atom originally sitting there, the interaction energy between them raises the energy of the system proportional to U. The hopping itself diminishes the energy proportional to J. The ground state of the system is determined by the competition between J and U. If V_0 is large enough such that the effect of U overwhelms that of J, hopping becomes difficult. The ground state of the system is then obtained by distributing all atoms evenly over the lattice sites to minimize the interaction energy. Let the total number of atoms be N and the number of sites be M, and the average number of atoms per site $n = N/M$ be an integer. In the limit of very large U there is no hopping, and n is definite, without number fluctuation. Consequently there is no phase coherence among the sites. Such a state is called the Mott insulator, described by

$$|\Psi_{MI}\rangle \propto \prod_{i=1}^{M} (b_i^\dagger)^n |0\rangle. \qquad (10.9.9)$$

On the other hand, decrease in V_0 facilitates hopping, the particle number fluctuation at all sites increases, and so does the phase coherence among sites. A superfluid phase appears. In the limit of very large J the ground state of the system is

$$|\Psi_{SF}\rangle \propto \left(\sum_{i=1}^{M} b_i^\dagger\right)^N |0\rangle. \qquad (10.9.10)$$

This state is a single condensate. The Bose-Hubbard model predicts phase transition from a superfluid phase to a Mott insulator phase at low temperatures with increasing ratio of U/J. The critical value of the transition for the phase $n = 1$ is at $U/zJ \approx 5.8$, where z is the number of nearest neighbors for a homogenous system [66]. Such phase transition occurs due to the variation of parameters in the quantum Hamiltonian. It can happen at very low temperatures, and is called quantum phase transition, thus distinguishing it from the thermodynamic phase transitions.

A 2-dimensional mean field calculation was performed in Ref. [65]. The ground state wave function to begin with is

$$|\Psi_{MF}\rangle = \prod_i |\varphi_i\rangle. \qquad (10.9.11)$$

where the wave function at site i is expanded in terms of Fock state $|n\rangle$ basis:

$$|\varphi_i\rangle = \sum_i f_n^{(i)} |n\rangle_i. \qquad (10.9.12)$$

The grand canonical Hamiltonian $H - \mu \sum_i \hat{n}_i$ is minimized with respect to $f_n^{(i)}$, where the chemical potential μ is the Lagrange multiplier to ensure fixed total number of atoms:

$$\langle \Psi_{MF}|H|\Psi_{MF}\rangle - \mu\langle \Psi_{MF}|\sum_i \hat{n}_i|\Psi_{MF}\rangle \to \text{min.} \qquad (10.9.13)$$

In case the solution to the problem turns out to be such that $|\varphi_i\rangle$ is a single Fock state $|n\rangle_i$ and consequently the number fluctuation at this site is 0, this is the signature of a Mott insulator. A superfluid phase appears in case that the solution is a superposition of many Fock

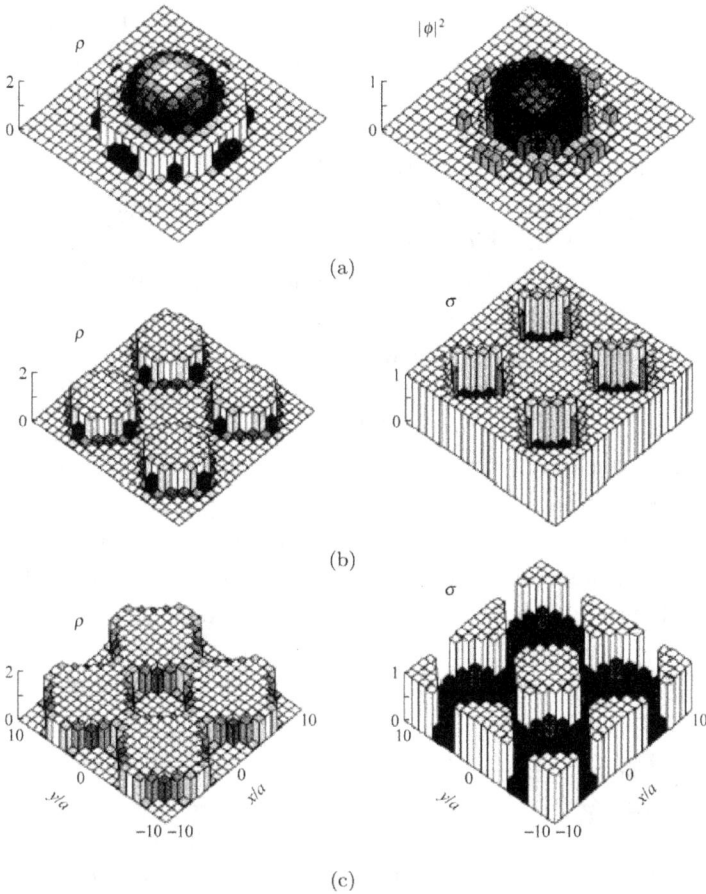

Fig. 10.20 (a) Mott insulator and superfluid phases in a 2-dimensional optical potential and a harmonic trap; (b) Superlattice in 2D, $\mu = 25J$; (c) Same, with $\mu = 35J$. Reprinted with permission from D. Jaksch, C. Bruder, J. I. Cirac, C. W. Gardiner, and P. Zoller, *Phys. Rev. Lett.* 81, 3108, 1998. Copyright (2017) by the American Physical Society.

states with non-vanishing mean field $\phi_i \equiv \langle b_i \rangle \neq 0$. The calculated results are shown in Fig. 10.20. The density $\rho(x,y) = \langle \hat{n}(x,y) \rangle$ and the superfluid component $|\phi(x,y)|^2$ in an optical lattice with a slowly varying trap potential are plotted. The lattice points are $\left(\frac{x}{a}, \frac{y}{a}\right) = (i,j), i,j = 0, \pm 1, \ldots$. Figure 10.20(a) shows a Mott insulator phase with $\rho = 2$ at the central part of the trap surrounded by another insulator phase with $\rho = 1$ and superfluid rings in between the insulating phases. This can be understood in terms of the trap potential $\varepsilon_i = V_T(\boldsymbol{x})$ which leads to an effective local chemical potential $\mu(\boldsymbol{x}) = \mu - \varepsilon_i$. In between the insulator phases the number of particles deviates from integers 1 or 2. These "decimal" parts of the atoms are free to hop between sites.

By using interfering laser beams at different angles one can produce superlattice, in which the offset of the optical potential is modulated periodically in space on a scale larger than the lattice period. Figures 10.20(b) and (c) show the density $\rho(x, y)$ and the number fluctuation $\sigma(x, y)$ of Mott structures formed in a superlattice. With μ increasing atoms form bridges between superlattice wells.

In Fig. 10.19(b) we noticed that the parameters J and U are determined once V_0 is fixed. They cannot vary independently. One can choose atoms in different internal states $|g_1\rangle$ and $|g_2\rangle$ such that they have atomic polarizability of different signs, and they experience optical potentials shifted relatively by $\lambda/4$, as shown in Fig. 10.21(a). The result is a 2-species Bose-Hubbard Hamiltonian

$$H = -\left(J \sum_{\langle i,j \rangle} a_i^\dagger b_j + H.c.\right) + \sum_i \varepsilon_{ii} a_i^\dagger a_i + \sum_j (\varepsilon_j - \delta) b_j^\dagger b_j$$

$$+ \frac{U_{aa}}{2} \sum_j a_i^{\dagger 2} a_i^2 + \frac{U_{bb}}{2} \sum_j b_j^{\dagger 2} b_j^2 + U_{ab} \sum_{\langle i,j \rangle} a_i^\dagger a_i b_j^\dagger b_j, \quad (10.9.14)$$

where operators a and b refer to internal states $|g_1\rangle$ and $|g_2\rangle$ respectively. When atoms of different internal states are coupled by Raman beams, hopping between adjacent cells is induced with the coupling

$$J = \frac{1}{2} \int \mathrm{d}^3 x\, w_a^*(x) \Omega_{\mathrm{eff}}(x) w_b \left(x - \frac{\lambda}{4}\right),$$

where Ω_{eff} is the effective two-photon Rabi frequency. The Raman detuning $-\delta$ is introduced for atoms in the state $|g_2\rangle$, shifting the chemical potential of species b relative to a. This is used to generate checkerboard patterns, e.g., a Mott insulator phase for species a and a superfluid phase for species b, as shown in Fig. 10.21(b).

The superfluid to Mott insulator transition of cold Bose gas in optical lattice was observed by the University of Munich group led

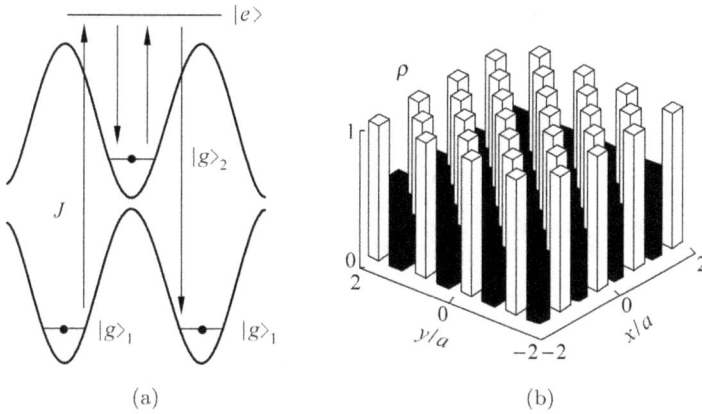

(a)   (b)

Fig. 10.21   (a) Atomic level scheme with shifted optical potentials; (b) Coexisting Mott insulator phase and superfluid phase, with parameters $\mu = 25J, U_{aa} = U_{bb} = 45J, U_{ab} = 0, \delta = -25J, \varepsilon_i = 0$. Reprinted with permission from D. Jaksch, C. Bruder, J. I. Cirac, C. W. Gardiner, and P. Zoller, *Phys. Rev. Lett.* 81, 3108, 1998. Copyright (2017) by the American Physical Society.

by T. Hänsch [67]. ^{87}Rb atoms in hyperfine state $F = 2, m_F = 2$ are used. Spherical BEC with a Thomas-Fermi radius of 26 μm is formed in a magnetic trap, and is subsequently transferred into the optical lattice by slowly increasing the intensity of the laser beams. The wave length of the laser beams is 852 nm, with optical potential amplitude up to $22E_R$. The oscillation frequency at the well of the optical potential is 30 kHz. The Gaussian cross-section of the laser beams produces the slowly varying potential V_T, with the harmonic oscillator frequency 65 Hz. The number of lattice sites is 150 000, holding 2×10^5 atoms of the BEC.

After some final value of the laser intensity is reached, the combined trapping potential is suddenly turned off, the atomic cloud is allowed to expand freely and absorption image is taken to check for phase coherence. When the system is in the superfluid regime (small lattice potential depth), and all atoms are delocalized, a high-contrast 3-dimensional interference pattern is obtained (shown in Fig. 10.22). When the lattice potential depth is increased, the interference pattern changes markedly, as shown in Fig. 10.23. Initially the strength of higher-order interference maxima increases as the potential depth is increased, because more and more atoms

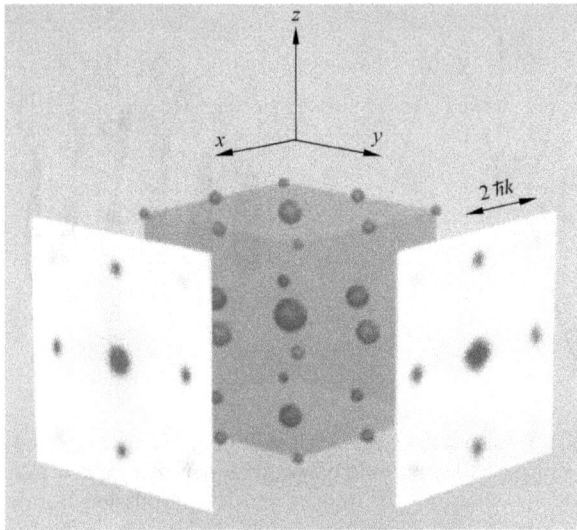

Fig. 10.22   Schematic 3-dimensional interference pattern with measured absorption images taken along two directions, $V_0 = 10E_R$. Reprinted with permission from Macmillan Publishers Ltd: Nature 415, 39. Copyright (2002) by Macmillan Publishers Ltd.

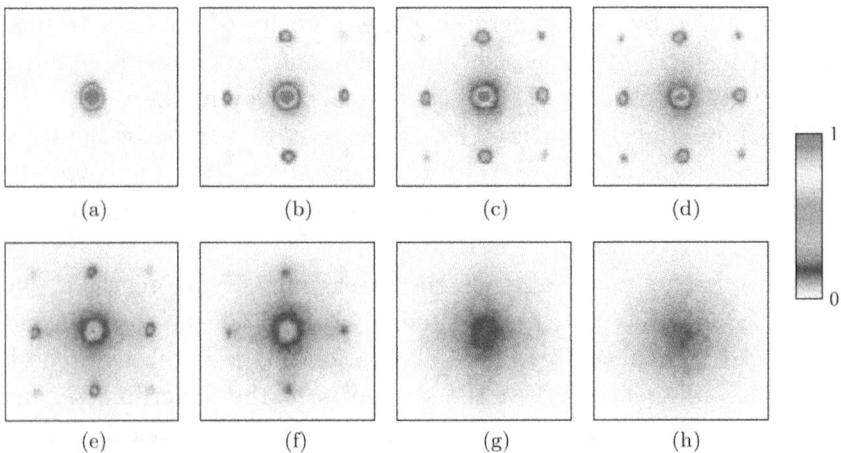

Fig. 10.23   Absorption images of interference patterns. Values of V_0 (in units of E_R) are: a, 0; b, 3; c, 7; d, 10; e, 13; f, 14; g, 16; h, 20. Reprinted with permission from Macmillan Publishers Ltd: Nature 415, 39. Copyright (2002) by Macmillan Publishers Ltd.

are localized at the lattice sites. At $V_0 = 13E_R$ the higher order maxima cease to increase (Fig. 10.23(e)), and at the same time an incoherent background of atoms gains more and more strength, until at a potential depth of $22E_R$ the interference pattern disappears altogether. A notable feature during the evolution process is that when the interference pattern is still visible no broadening of the interference peaks can be detected until they completely vanish in the incoherent background.

A very interesting feature of the Mott insulator is that phase coherence can be restored very rapidly when the optical potential depth is lowered again to a value when the system is superfluid. After ramp-up to $22E_R$ in 80 ms, V_0 is kept constant for 20 ms. It is ramped down to $9E_R$ in a time t (Fig. 10.24(a)). After only 4 ms of ramp-down time the interference pattern is visible again, and after 14 ms of the ramp-down time the interference peaks have narrowed to their steady state value. (Fig. 10.24(b), black dots). In contrast, a phase incoherent state obtained by applying a magnetic field gradient during the ramp-up period when the system is still superfluid does not show any restoration of phase coherence although the same procedure of ramp-sown of potential depth is repeated exactly (open circles in Fig. 10.24(b)).

The studies of BEC on optical lattices provide possibilities to probe strongly coupled many-body systems, Bose or Fermi alike. In such systems the experimental parameters can be precisely adjusted. Combined with the application of Feshbach resonance to regulate the interaction between atoms, clean and precise experimental results covering large parameter range can be obtained. The Zürich ETH group (M. Köhl *et al.* [68]) filled the optical lattice with ^{40}K atoms of two hyperfine spin states (Feshbach resonance at 224G) and studied the energy level population under different interaction strengths, mapping out the Fermi surfaces. For ideal Fermi gas, the first band (corresponding to momentum within the first Brillouin zone, a square of length $2\hbar k, k = \frac{1}{\pi}$ (lattice constant), for 2D square lattice) is to be completely filled before the second band begins to be filled. With interaction between the atoms the situation is different. Depending on the strength of interaction, the second band begins

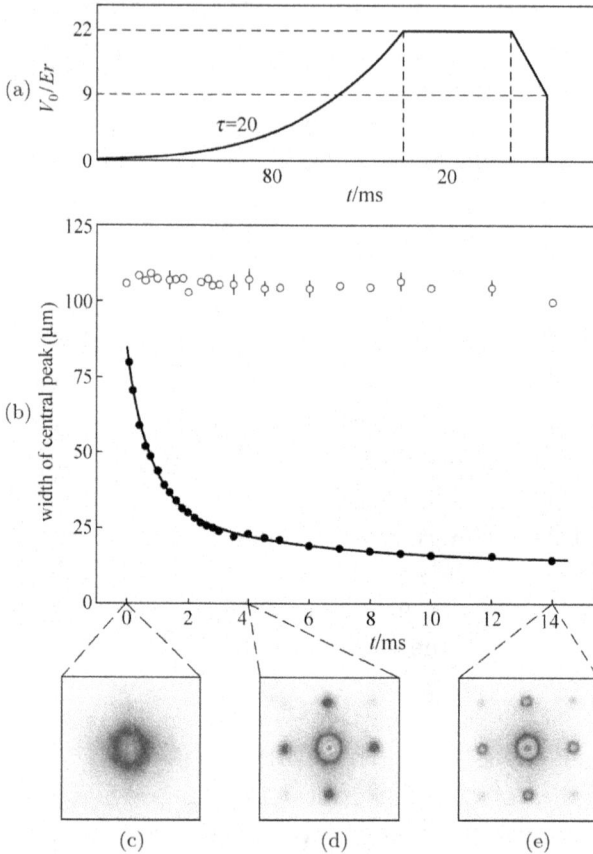

Fig. 10.24   (a) Experimental sequence of varying V_0; (b) Widths of the central interference peak for different ramp-down time t; (c)~(e) Absorption images for different values of the ramp-down time. Reprinted with permission from Macmillan Publishers Ltd: Nature 415, 39. Copyright (2002) by Macmillan Publishers Ltd.

to be filled sooner or later when the first band still has vacancies. In the experiment the magnetic field is set at a certain value. Then trapping potentials is adiabatically turned off and the magnetic field is removed, the gas cloud is allowed to expand freely for imaging. The image reveals the momentum distribution of atoms. In the case of weak interaction, the distribution is a small circle for low density of atoms. With increasing density the circle develops extensions toward the Bragg planes and finally transforms to a square (the first Brillouin

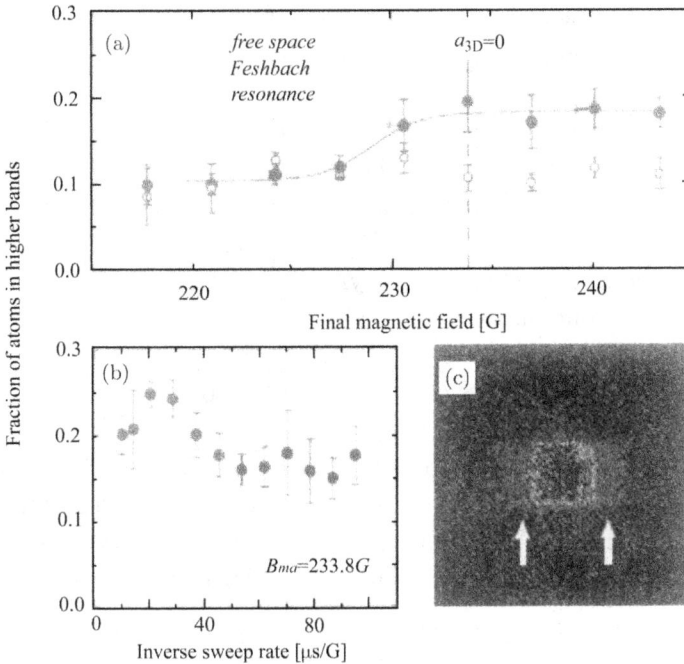

Fig. 10.25 Interaction-induced transition between Bloch bands. Arrows indicate momenta in the weak lattice direction beyond the first Brillouin zone corresponding to the second energy band. Reprinted with permission from Michael Köhl, Henning Moritz, Thilo Stöferle, Kenneth Günter, and Tilman Esslinger, *Phys. Rev. Lett.* 94, 080403, 2005. Copyright (2017) by the American Physical Society.

zone) filled uniformly. In order to study the effect of interaction the optical lattice is modified such that $\omega_x = 2\pi \times 50\,\text{kHz}$ and $\omega_x = \omega_y = 2\pi \times 62$ kHz. The sweep across the Feshbach resonance goes from the side of $a > 0$ to the side of $a < 0$, and the fraction of atoms transferred to the higher band is determined. For final magnetic field values well above the Feshbach resonance a significant increase of the number of atoms in the higher bands along the weak axis of the lattice, demonstrating the interaction induced coupling between the lowest bands (Fig. 10.25).

The experimental results are useful for theoretical study of the change of Fermi surface in strong interaction regime.

Finally, we turn to the problem of interference of an array of independent Bose-Einstein condensates. The experimental study was made possible by loading the condensates in an optical lattice [69]. This is a generalization of the case discussed in Section 10.3 [28]. The optical lattice can be formed by laser beams intersecting at an angle θ (Fig. 10.25(a)) with the effect of increasing the period of the lattice form $\lambda/2$ to $d = \frac{\lambda}{2\sin\theta/2}$. This enables to achieve $d = 2.7$ μm by using laser beams of wave length $\lambda = 532$ nm intersecting at an angle $\theta = 0.20$ rad. In the experiment ^{87}Rb atoms in the state $F = m_F = 2$ are contained in a cylindrically symmetric magnetic trap forming a cigar shaped condensate with a Thomas-Fermi length of 84 μm and a radius of 6 μm. The potential of the optical lattice is ramped up and about 30 condensates each containing $\sim 10^4$ atoms are produced. Due to the long lattice period and the large height $V_0 \approx 600 E_R$ of the potential barrier between the sites the Rabi frequency for tunneling is less than 10^{-4} Hz and the tunneling is negligible on the time scale of the experiment. At full lattice depth the axial motion of the gas at each site is frozen, the harmonic oscillator ground state length scale is

$$l = \sqrt{\frac{\hbar}{2m\omega_z}} = 120\,\text{nm} \ll d.$$

In a typical experiment the lattice potential is ramped to $600 E_R$ in $\tau_{\text{ramp}} = 200$ ms. Initially before the formation of the optical lattice the chemical potential μ is constant across the system. Due to the Thomas-Fermi distribution the number of atoms is different in different positions. This difference is just balanced by the interaction energy such that μ is kept constant. Once the potential of the optical lattice is raised above say $100 E_R$ the tunneling between sites is negligible and the occupation on sites becomes fixed. Further increase of the potential leads to compression of the atomic clouds on sites and therefore to the increase of interaction energy which can no more be balanced by the change of particle number. The chemical potential becomes a function of lattice sites. The evolution in time of the condensate wave functions on different sites leads to different phases, and after a sufficiently long time the phases on different sites differ significantly.

After holding atoms on the lattice for another $\tau_{\text{hold}} = 500$ ms, the optical and the magnetic trap are switched off, and the density

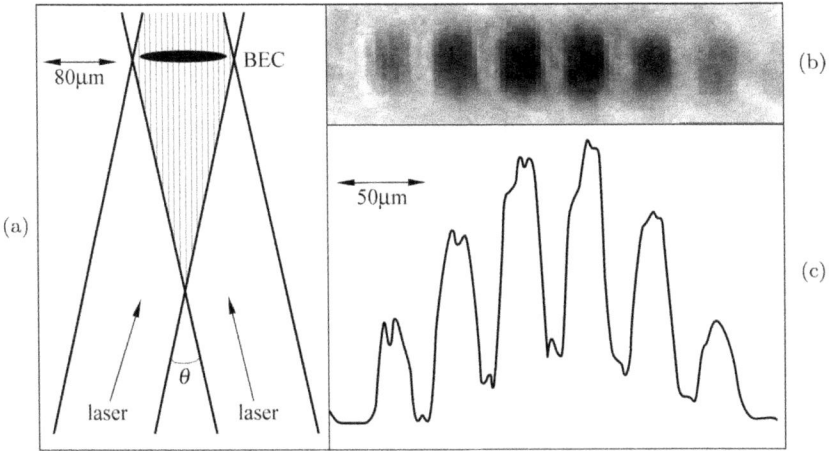

Fig. 10.26 (a) Optical lattice and the independent BECs formed (b) Absorption image of the cloud after expansion (c) Axial density profile of the cloud, radially averaged over the central $25\,\mu m$. The fit gives $A_1 = 0.64$. Reprinted with permission from Zoran Hadzibabic, Sabine Stock, Baptiste Battelier,Vincent Bretin, and Jean Dalibard, *Phys. Rev. Lett.* 93, 180403, 2004.] Copyright (2017) by the American Physical Society.

distribution of the cloud is recorded by absorption imaging after $t = 22$ ms time-of -flight expansion. Despite the fact that the phases of different BECs are completely uncorrelated, the images show interference fringes in general. A specially spectacular example is give in Figs. 10.26(b) and (c) where the contrast reaches 60%.

The result can be analyzed in the following way. Consider a 1D array of N condensates at position $z_d = nd(n = 1, \ldots, N)$, and assume that each condensate is in a coherent state described by an amplitude α_n and phase ϕ_n. After expansion for a time t the atom density at position z is given by

$$I(z) \propto \left| \sum_{n=1}^{N} \alpha_n e^{i\phi_n} e^{im(z-z_n)^2/2\hbar t} e^{-(z-z_n)^2/Z_0^2} \right|^2, \qquad (10.9.15)$$

where $Z_0 = \hbar t/ml$. The periodic exponential is the time development factor: to reach point z from site z_n in time t, the atom has an energy $\frac{1}{2}m\frac{(z-z_n)^2}{t^2}$. The Gaussian factor is the result of expansion for time t of the harmonic oscillator ground state wave function. The momentum scale of a condensate on site is $\hbar/l$ and for time t its length scale becomes $\hbar t/ml \equiv Z_0$. The Gaussian distribution at this time is therefore $e^{-(z-z_0)^2/Z_0^2}$. We assume that all sites contain the same number of atoms $\alpha_n = \alpha$, and the expansion time is large such that $l, z_n \ll Z_0, \sqrt{\hbar t/m}$. In this case, (10.9.15) becomes

$$I(z) \propto N\alpha^2 \exp(-2z^2/Z_0^2)F(z), \qquad (10.9.16)$$

where

$$F(z) = 1 + \sum_{n=1}^{N-1} A_n \cos(B_n + 2\pi nz/D), \qquad (10.9.17)$$

$$D = ht/md.$$

A_n and B_n are respectively the amplitude and phase of the n-th harmonic of $F(z)$, given by the modulus and argument of $\frac{2}{N}\sum_{j=n+1}^{N} e^{i(\phi_j - \phi_{j-n})}$. To the first harmonic,

$$F(z) = 1 + A_1 \cos(B_1 + 2\pi z/D).^{22} \qquad (10.9.18)$$

In Fig. 10.27(a) results for 200 consecutively taken images are summarized. In most cases the fringe contrast A_1 is significant with a mean value $\langle A_1 \rangle = 0.34$. The phase B_1 is randomly distributed between 0 and 2π. Consequently, no periodic modulation survives if average over the 200 profiles is taken. More phase correlated condensates can be obtained by ramping up the lattice potential more rapidly $\tau_{\mathrm{ramp}} = 3$ ms and the condensates are released immediately before dephasing advances significantly. In this case (Fig. 10.27(b)), the fringe phases B_1 are not randomly distributed and the sum over 200 images still shows interference pattern of significant contrast.

The foregoing arguments can be extended to the 2D and 3D cases. In conclusion the single-shot result of the interference of an array of

---

[22]This can be seen by considering $\frac{1}{N}|\sum_{n=1}^{N} e^{i\phi_n} e^{im(z-z_n)^2/2\hbar t}|^2 = \frac{1}{N}(\sum)^*(\sum)$. The squared terms sum up to give 1. The first harmonic is given by the sum of cross-products for which k and l differ by 1. The ϕ dependent part gives $\frac{2}{N}\sum_{j=2}^{N} e^{i(\phi_j - \phi_{j-1})}$ and the z-dependent part gives $e^{imdz\hbar t} = e^{i2\pi z/D}$, where a common phase factor independent of z is omitted. The final result gives $A_1 \cos(B_1 + 2\pi z/D)$.

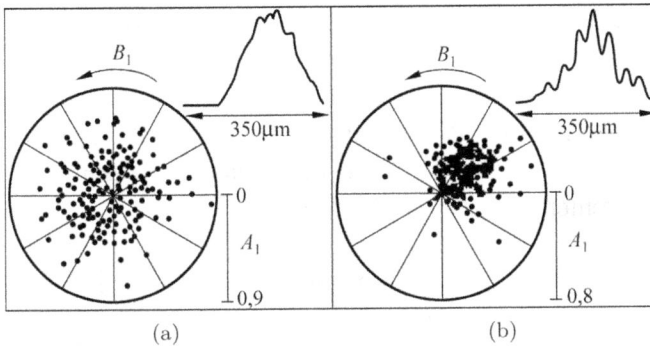

Fig. 10.27 Polar plots of the fringe amplitudes and phases (A_1, B_1) for 200 images obtained under same experimental conditions. (a) Phase uncorrelated condensates. (b) Phase correlated condensates. Insets: axial density profiles averaged over 200 images. Reprinted with permission from Zoran Hadzibabic, Sabine Stock, Baptiste Battelier, Vincent Bretin, and Jean Dalibard, *Phys. Rev. Lett.* 93, 180403, 2004.] Copyright (2017) by the American Physical Society.

phase-uncorrelated BE condensates is random in the sense that the contrast (determined by A_1) can be any value within a certain range, most of them show contrasts of considerable magnitude. The phase B_1 determines a shift along the z-axis of the pattern. Its randomness implies that when an average over many images will result in a structureless profile. The lesson learned from this experiment is that one should be careful not to take the presence of interference pattern only as a definite sign of phase coherence.

10.10 Feshbach Resonance and Resonance Superfluidity

The low energy atomic scattering is characterized by the scattering length, which can be tuned by adjusting an applied magnetic field under the condition of the existence of a Feshbach resonance. Not only the strength of the interaction can be tuned, but also the repulsive and attractive nature of the interaction can be changed. This possibility opens up a large field of experimental study of the physics of cold quantum gas, both Bose gas and Fermi gas. Fermi atoms can form molecules, they can also form pairs under many-body interactions, similar to the formation of BCS pairs. When the temperature is low enough, molecules can form Bose-Einstein condensate, because a molecule formed by two Fermi atoms is a Bose particle. Fermi atom pairs can also condense when the temperature is lower than the critical value and possess superfluid behavior. The experimental realization of these phenomena represents one of the main achievements in physics in 2004.

10.10.1 The Feshbach resonance

In a quantum mechanical scattering problem, the low energy scattering amplitude f of a particle with wave vector k is given by

$$f = \frac{1}{2ik}(e^{2i\delta_0} - 1). \qquad (10.10.1)$$

The low energy scattering is dominated by the s-wave, and δ_0 is the s-wave phase shift. The s-wave scattering length a is defined by

$$\tan\delta_0 = -ka. \qquad (10.10.2)$$

When the phase shift is small, we can use the approximate expression

$$f \approx \frac{\delta_0}{k} = -a. \qquad (10.10.3)$$

The scattering cross-section σ is given by

$$\sigma = 4\pi|f|^2 = 4\pi a^2. \qquad (10.10.4)$$

Low energy scattering between two atoms is characterized by only one parameter, viz. the s-wave scattering length. When the relative coordinates of the two particles $r = r_1 - r_2$ are used, the scattering problem can be reduced to a single-particle problem, with the scattering potential given by $V(r)$. Typical interaction between atoms is shown in Fig. 10.28, where R is the relative distance between two atoms. When two atoms approach each other and enter the interaction range, the interaction begins. When the distance becomes very small, a strong repulsion makes the atoms to separate, and the scattering process is completed. The time duration in which the atoms stay in the interaction range is usually very brief, of the order of the ratio between the range of interaction and the relative velocity of the atoms. This is typical in potential scattering. Sometimes the

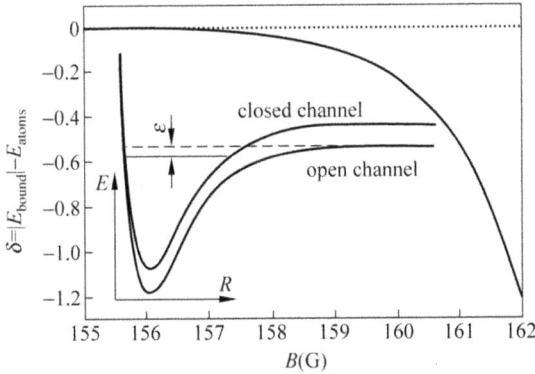

Fig. 10.28 Interaction potential between atoms. The potential depends on the spin of atoms. For low energy scattering there is an open channel and also a closed channel.

cross-section shows a broad resonance behavior, and this is referred to as a shape resonance. The scattering length depends on the nature of atoms. The interaction is assumed to be weak, $a > 0$ when the interaction is repulsive, and $a < 0$ when the interaction is attractive.

The interaction between atoms depends on the orientation of the spin of valence electrons. There is one valence electron in the atom of alkali metals. So far as the interaction between two Na atoms is concerned, the potential well for parallel spin of the two valence electrons is lower than that for antiparallel spin of the electrons. The hyperfine interactions between electrons is given by

$$V_{\text{hf}} = \sum_{i=1,2} \frac{a_{\text{hf}}}{\hbar}^2 \, s_i \cdot i_i = \frac{a_{\text{hf}}}{2\hbar^2} [\boldsymbol{S} \cdot \boldsymbol{I} + \boldsymbol{s} \cdot \boldsymbol{i}], \qquad (10.10.5)$$

where $\boldsymbol{s}$ and $\boldsymbol{i}$ denote the electron and nuclear spin respectively, and capital letters denote the sum and small letters denote the difference of the corresponding angular momenta of two electrons or two nuclei. The $\boldsymbol{s} \cdot \boldsymbol{i}$ term can cause the spin flip of an electron, so the spin state of the electrons can change from parallel to antiparallel or vice versa. Let us assume that two atoms with parallel electron spin collide with low energy. The spin state becomes antiparallel when the atoms are in the interaction region. Now the atoms observe a deeper potential well and get trapped in it. If there exists a level of bound state in the potential well for antiparallel spin with energy close to that of

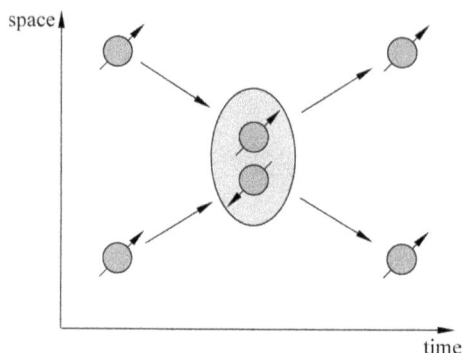

Fig. 10.29   Illustration of a Feshbach resonant atomic collision.

the scattering state (in Fig. 10.28 a bound state level with a small energy difference ε is shown), the atoms can live temporarily in the potential well for antiparallel spin as a bound state, until the hyperfine interaction flip the spin again. Then the atoms becomes free again and the scattering process is now complete. For a pair of atoms with very small kinetic energy, the channel of parallel spin is an open channel, while the channel for antiparallel spin is a closed channel. This process is illustrated by Fig. 10.29. When $\varepsilon = 0$, the energy of the bound state (in the closed channel) is equal to that of the scattering state (in the open channel), and a resonance occurs. The scattering process with ε small is called a Feshbach resonance scattering. Since the atoms stay in the interaction range for a time much longer than that for a usual potential scattering, the width of the resonance is much narrower (by several order of magnitude) compared with that of the shape resonance. This is the consequence of the uncertainty relation for energy in quantum mechanics. When the resonance scattering dominates, the scattering length changes with ε given by

$$a = a_0 - \frac{\Gamma}{2k\varepsilon},$$ (10.10.6)

where a_0 is the scattering length far away from the resonance (or the background scattering length), Γ is the width of the resonance. From (10.10.6) we see that when the resonance is approached, $a \to \pm\infty$.

The argument given above aims at providing a rough picture of the resonance scattering. For a rigorous treatment of two-channel scattering, see the article by Duine and Stoof [93]. The open and closed channels are actually linear superpositions of the singlet and triplet channels.

A remarkable feature is that the position of the bound state in the closed channel, namely the parameter ε can be tuned by an applied magnetic field. Let the magnetic moment of the bound state be μ_0, that of the single atom be μ_i, (therefore the magnetic moment of the scattering state is $2\mu_i$). The difference of Zeeman energy of the scattering state and the bound state in the applied magnetic field B is consequently $(2\mu_i - \mu_0)B$. Therefore

$$\varepsilon(B) = (2\mu_i - \mu_0)(B - B_0), \tag{10.10.7}$$

where B_0 is the magnetic field at which the resonance occurs. By adjusting the magnetic field around B we can change the sign of ε. From (10.10.6) we see that the second term can be a positive or negative number of arbitrarily large magnitude. Figure 10.30 shows the variation of the scattering length of ^{40}K atoms with magnetic field.

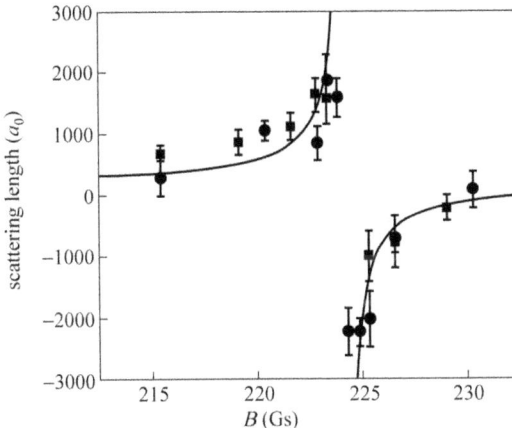

Fig. 10.30 Scattering length of ^{40}K atoms as a function of applied magnetic field. Reprinted with permission from C. A. Regal and D. S. Jin, *Phys. Rev. Lett.* 90, 230404, 2003. Copyright (2017) by the American Physical Society.

At the resonance $a \to \pm\infty$. Does it imply an interaction of infinite strength? Does it lead to an infinitely large scattering cross-section it does not. The reason is that (10.10.3) and (10.10.4) are both approximate, valid for δ_0 very small. The relation between scattering amplitude and the phase shift is given by (10.10.1):

$$f = \frac{1}{2ik}(-2\sin^2\delta_0 + 2i\sin\delta_0\cos\delta_0) = -\frac{1}{2ik}\left(\frac{2k^2a^2 + 2ika}{1 + k^2a^2}\right)$$

$$= -a\frac{1 - ika}{1 + k^2a^2}. \tag{10.10.8}$$

The real part of f determines the energy shift of the scattering state, and the imaginary part determines the reciprocal of the time of scattering. For ka large, $f \to -i/k, \sigma \to 4\pi/k^2$. The scattering cross-section does not depend on the scattering length, i.e., independent of interaction between atoms. it is called the unitarity limit of cross-section. Due to unitarity of scattering process, which implies the conservation of probability, the scattering cross-section can never overtake this limit. The cross-section has this unique value at resonance, quite independent of the atomic species. This is termed a *universal behavior*. The interaction energy at resonance is given by Ho and Mueller [72]

$$|\varepsilon_{int}| = \frac{3k_BTn}{2}\left(\frac{n\lambda^3}{2^{3/2}}\right), \tag{10.10.9}$$

where k_B is the Boltzmann constant, T is the gas temperature, n the number density of gas and λ is the thermal de Broglie wave length

$$\lambda = \left(\frac{2\pi\hbar^2}{mk_BT}\right)^{1/2}. \tag{10.10.10}$$

The interaction energy at resonance also shows the universal behavior, independent of the scattering length. The points in Fig. 10.30 represent results of the experiment by C. Regal and D. Jin [71] of NIST using radio-frequency spectroscopy, the quantity measured being the interaction energy. We see that the interaction energy is determined by the scattering length far away from the resonance, but becomes independent of the scattering length when

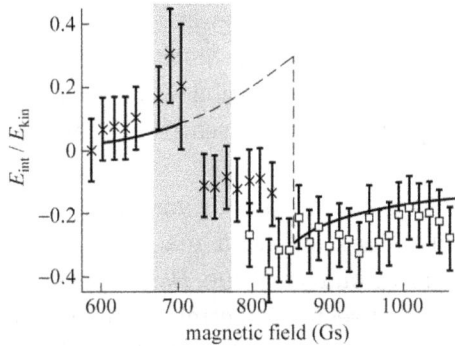

Fig. 10.31 Ratio of interaction energy to kinetic energy of a mixture of the two lowest hyperfine states of ^{6}Li gas. Reprinted with permission from C. A. Regal and D. S. Jin, *Phys. Rev. Lett.* 90, 230404, 2003. Copyright (2017) by the American Physical Society.

a resonance is approached. The ENS group led by C. Salomon [73] measured directly the ratio of the interaction energy to the kinetic energy as a function of the magnetic field. The universal behavior is clearly illustrated in the resonance region, as shown in Fig. 10.31. The experimental data agree well with theory [72].

10.10.2 The degenerate Fermi gas

Alkali atoms with even mass number are fermions, for instance ^{6}Li, ^{40}K, etc. A gas becomes degenerate when its density is larger than the quantum concentration n_Q:

$$n_Q = \left(\frac{mk_B T}{2\pi\hbar^2}\right)^{3/2}. \qquad (10.10.11)$$

Under this condition the average distance between two atoms is identical to the thermal de Broglie wave length (10.10.9): $n_Q = \lambda^3$. The gaseous atoms become indistinguishable identical particles. The degeneracy temperature T_{deg} expressed in terms of the gas density is then

$$T_{\text{deg}} = \frac{2\pi\hbar^2}{mk_B} n^{2/3}. \qquad (10.10.12)$$

To cool a Fermi gas is a difficult task. A Fermi gas confined in a magnetic trap is spin polarized. Low energy scattering of fermions is dominated by p-wave, because s-wave scattering does not comply with Pauli exclusion principle. The much lower probability of p-wave scattering makes the thermalization very inefficient. The use of Feshbach resonance enhances the scattering length, and makes the cooling more efficient.

The physics of degenerate Fermi gas on both sides of Feshbach resonance (by both sides we mean magnetic field strengths larger or smaller than the resonance value B_0) is very rich. Consider the Feshbach resonance between two Fermi atoms first. At the resonance the bound state (a molecule composed of two Fermi atoms) in the closed channel emerges. When the magnetic field strength continues to decrease, the binding energy of the molecule increases. The spatial extension of the molecule decreases correspondingly. Above the resonance the atoms are not bound.

The situation is different for a many-body system of Fermi atoms. The BCS-BEC crossover theory [74, 75] gives the following picture. When the Feshbach resonance is approached from the $B < B_0$ side, the spatial extension of the bound state increases, and reaches the average distance between atoms at the resonance. All kinds of gases of Fermi atoms of the same density have identical properties at the Feshbach resonance, and this is the universality. Above the resonance, the fluctuation between the open and close channels continues to bind the atoms into pairs. The remaining atoms reach equilibrium through collisions. In this situation atom pairs coexist with free atoms. Far away from the resonance in the $a < 0$ and $B > B_0$ side, weak attraction leads to the formation of BCS pairs. The pairs formed by two Fermi atoms are bosons themselves. Therefore under low enough temperatures large quantities of pairs can also condensate to from a Bose-Einstein condensate. Such condensate has superfluid properties. The BCS transition temperature depends sensitively on the interaction strength between atoms. The weak coupling theory gives

$$T_c \sim T_F e^{-\frac{\pi}{2|a|k_F}}, \qquad (10.10.13)$$

where k_F is the Fermi momentum, $T_F = \varepsilon_F/k_B$ is the Fermi temperature, which is very close to the degeneracy temperature. The transition temperature is rather low when the interaction is weak.

The BCS pair has a very loose spatial structure, its extension is much larger than the average distance between two atoms. When the magnetic field strength is decreased to approach the resonance value, the interaction becomes stronger, and the transition temperature becomes larger. The weak coupling theory leading to (10.10.13) is no more valid. Close to resonance the pair size becomes comparable to the average distance between atoms. On the $a > 0$ and $B < B_0$ side the energy of bound state (molecule) is lower than that of the scattering state, and the binding energy and the size of the bound state increases and decreases respectively with the decreasing magnetic field. Far away from the resonance the size of bound state is comparable to the atomic size. When the temperature is lower than the Bose-Einstein temperature, the molecules form Bose-Einstein condensate. When the temperature is raised above the critical temperature, the superfluid property is destroyed by the thermal fluctuation, and the Bose-Einstein condensate does not exist anymore. The atomic pairing can still exist however, but the phase coherence between the pairs is lost. This situation is referred to as the pseudogap. The phase diagram is shown in Fig. 10.32. The inset shows the situation for two free Fermi atoms, where ΔE represents

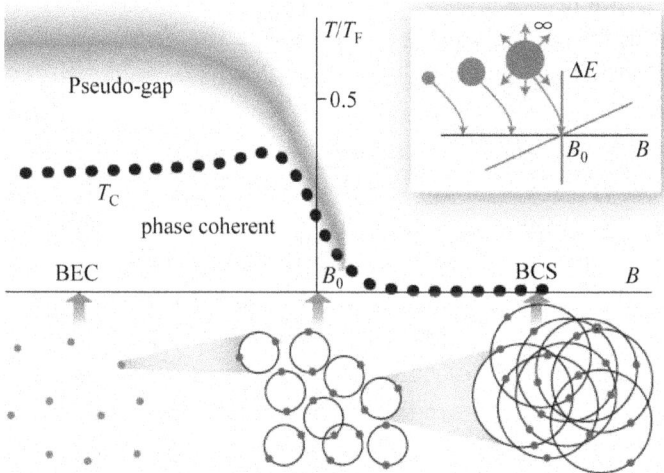

Fig. 10.32   Phase diagram of the BEC-BCS crossover. From Ref. [75].

the energy difference between the bound state and scattering state. The main diagram shows the situation for many-body Fermi atoms. T_C represents the BEC (or superfluid) transition temperature. On the lower side of the diagram is given the scale of the atom pairs. On the left of the diagram we find the molecules, BEC of molecules exists below T_c, while above T_c pseudogap exists. Above the grey line pairs can hardly exist. On the $a < 0$ side close to the resonance there are pairs of two Fermi atoms with size comparable with the average distance between atoms. When the temperature is low enough, these pairs can also form condensates. One expects that such "pair condensate" is very different from the condensate of BCS pairs. Physicists are very interested in studying the superfluidity of pair condensates formed owing to the Feshbach resonance, now called the "resonance superfluidity".

10.10.3   Molecules formed by Fermi atoms and the molecular condensates

The discovery of the condensate of molecules formed by Fermi atoms was considered a milestone of the study of cold quantum gas. Besides the significance of its own, the molecular condensate is also considered as the bridgehead for the march toward the superfluid of the pair condensates.

To obtain the molecular condensate, the first requisite is the availability of long lived molecules. In 2003 D. Jin's group at JILA [76] succeeded to transform 50% of ^{40}K gas into molecules of ^{40}K$_2$. Mixture of equal amount of atoms in internal states $|f = 9/2, mf = -5/2\rangle$ and $|f = 9/2, mf = -9/2\rangle$ was used. The magnetic field was decreased from the values on the $a < 0$ side of the Feshbach resonance to the values on the $a > 0$ side at $T = 150$ nK. Radio-frequency dissociation spectroscopy was used to detect the molecules formed. This method enables to obtain the binding energy and the wave function of the molecules. States with different hyperfine spin components were separated by using Stern-Gerlach method, and pictures were taken of the gas by laser absorption imaging. Their results were in good agreement with the theory. The left image of Fig. 10.33 shows both components of $m_f = -5/2, -9/2$ of the gas

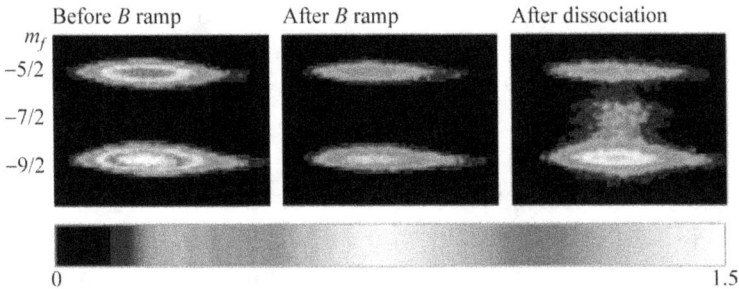

Fig. 10.33 The laser absorption image of different spin states of the atoms before the magnetic field ramp (left) and after the field ramp (middle) and after the radio-frequency dissociation (right). Reprinted with permission from Macmillan Publishers Ltd: Nature 424, 47, Copyright (2003) by Macmillan Publishers Ltd.

before the sweep of magnetic field. The middle image shows that both components are lost over half after the magnetic field has been ramped down to the $a > 0$ region. The molecules are invisible by the imaging laser. The atoms resulted in the radio-frequency dissociation are in the states $m_f = -7/2$ and $-9/2$. The right image consists of the faint new component $m_f = -7/2$ and a much stronger $m_f = -9/2$ component, part of which consists of the original free atoms and the remaining part contains the product of dissociation. An important part of the research is the reversibility of the atom molecule transformation during the magnetic field ramp.

The group of R. Grimm in Innsbruck, C. Salomon's group at ENS in Paris and R. Hulet's group at Rice University in Houston all obtained molecules composed of Fermi atoms. The atoms are long-lived, of the order of a second. Molecules formed by Bose atoms live much shorter, of the order of ms. The reason is that the spins of two valence electrons of a molecule are antiparallel. The main mechanism of dissociation is the collision of the molecule with an atom, resulting in 3 free atoms. The spin of the valence electron of the atom colliding with the molecule must be parallel to the spin of one of the valence electrons in the molecule. Therefore the 3-body collision cannot be realized in a s-wave scattering due to Pauli's principle, since two electrons with antiparallel spins cannot be in a state with even orbital angular momentum. The collision rate is suppressed. This phenomenon is called the Pauli blocking.

The dissociation of a molecule formed by two Bose atoms is not restricted by Pauli blocking, and has therefore a much higher rate of dissociation. A long lift-time of the molecules is very important for the formation of molecular condensates.

The condensates of molecules formed by two Fermi atoms were obtained in winter 2003 by the groups of R. Grimm (Innsbruck), of D. Jin (JILA, Boulder, Colorado) and of W. Ketterle (MIT). The result aroused great interest, not only because the molecule has rich internal structure which can be probed and studied, but slso the molecular condensate can serve as the source of pair condensate when the magnetic field is ramped up to the $a < 0$ side of the resonance. D. Jin's group [77] used a mixture of two hyperfine states $|f = 9/2, m_f = -7/2\rangle$ and $|f = 9/2, m_f = -9/2\rangle$ of the ^{40}K gas, with the Feshbach resonance at $B_0 = 202.10G$. The magnetic field was ramped down slowly in a period of 10ms from 278G to 201.54G. The gas expanded freely for 20 ms after the confining optical trap was removed. In a freely expanding gas, the spatial distribution expresses directly the momentum distribution. When the magnetic field was ramped down at $T = 1.54T_c$ (the upper left picture of Fig. 10.34) the momentum distribution did not show the presence of the condensate. The lower left curve was the intersection of the momentum distribution. The right pictures of Fig. 10.34 were obtained at $T = 0.49T_c$. The momentum distribution showed clearly the formation of the molecular condensate. The intersection of the distribution showed the momentum peak of the condensate on top of the Gaussian background. In the experiment it was found that 88% of the gas atoms could be transformed into molecules.

10.10.4   Condensates of the pairs of Fermi atoms

Report on the progress of obtaining the pair condensate began early in 2004, from the groups of D. Jin [78], R. Grimm [79] and W. Ketterle [80]. Since pairs of Fermi atoms in the $a < 0$ side of the resonance are formed owing to many-body interactions, they are no more observable when the confining trap is removed and the gas begins to expand. The method used by Grimm's group is "optical in-situ imaging", and the spatial distribution observed

(a)

(b)

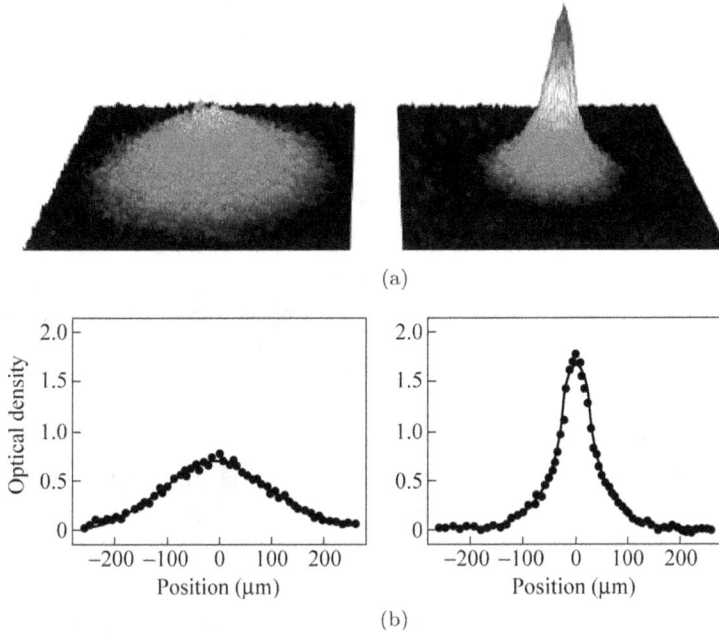

Fig. 10.34 The time-of-flight image of the molecular cloud, above the critical Bose-Einstein temperature (left) and below (right). Reprinted with permission from Macmillan Publishers Ltd: Nature 426, 537. Copyright (2003) by Macmillan Publishers Ltd.

cannot be transformed into momentum distribution. The existence of condensate can only be implied by indirect judgement, by comparing the result with theory. D. Jin's group invented a "pairwise projection technique", transforming the pairs into molecules by ramping down the magnetic field. The momentum distribution of the condensate can be inferred from the observed momentum distribution of the molecules. The Feshbach resonance of the hyperfine states $|f = 9/2,\ m_f = -7/2\rangle$ and $|f = 9/2,\ m_f = -9/2\rangle$ of the ^{40}K occurs at 202.1G, with a width of 7.8G. In the experiment of D. Jin, the magnetic field was set initially in the $a < 0$ side close to the resonance. The magnetic field was ramped down to the $a > 0$ side with a speed slow enough so that molecules can efficiently form, but still too brief for particles to collide or move significantly in the trap. The confining optical trap was then removed and time-of

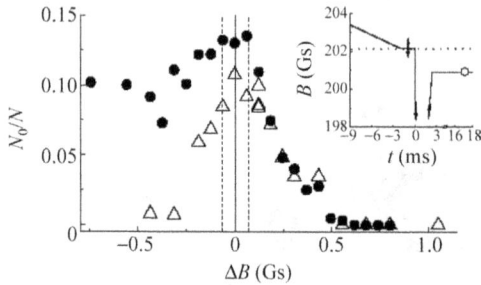

Fig. 10.35 Measured condensate fraction versus magnetic field detuning. Reprinted with permission from C. A. Regal, M. Greiner, and D. S. Jin, *Phys. Rev. Lett.* **92**, 040403, 2004. Copyright (2017) by the American Physical Society.

flight expansion images were taken. The projection always results in 60% to 80% of the atom sample appearing as molecules. When the initial temperature T was varied, a threshold curve relating T/T_F and $B - B_0$ was found, under which the momentum distribution of molecules was strongly peaked at zero momentum. The authors interpreted this as the evidence of pair condensate in the $a < 0$ side, which was projected to a molecular condensate. Figure 10.35 shows the relation between condensate fraction with the initial magnetic field detuning for $T/T_F = 0.08$, $T_F = 0.35$ μK. The inset shows the change of magnetic field with time. The field begins in the $a < 0$ side, it decreased toward the resonance (dotted line), and is hold for different times in the vicinity of resonance (two-sided arrow). After a hold time the optical trap is turned off and the magnetic field is quickly lowered to project the atom gas onto the molecular gas. After free expansion, the molecules are imaged on the $a > 0$ side (circle). The hold time is 2 ms for black dots in the figure, and 30 ms for triangles. Condensation of fermionic atom pairs is seen near and on either side of the resonance. Data taken for different hold times indicate that the pair condensed state has a significantly longer lifetime near the Feshbach resonance and on the $a < 0$ side. Figure 10.36 shows the time-of-flight images taken after the projection. From left to right the images correspond to $B - B_0$ values of 0.12, 0.25, and 0.55G, initial temperature $T/T_F = 0.07$. The condensate fractions are respectively 0.10, 0.05, and 0.01. Close to the resonance on the $a < 0$ side there are more pairs, and far away

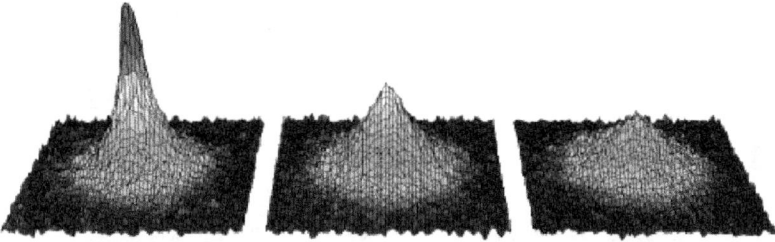

Fig. 10.36   Time of flight images of the gas cloud. Reprinted with permission from C. A. Regal, M. Greiner, and D. S. Jin, *Phys. Rev. Lett.* 92, 040403, 2004. Copyright (2017) by the American Physical Society.

from resonance the critical temperature is too low so that very few atom pairs can be observed.

W. Ketterle's group [80] also used the pairwise projection technique to study ^{6}Li. The maximum condensate fraction obtained was as high as 80%. Since atom pairs on the $a < 0$ side have rather loose structure, question can be raised that whether a molecule is formed by an original atom pair or by any two atoms of correct spin orientation that happen to be nearby? The latter possibility does not lead necessarily to a molecule with zero momentum, and hence a significant fraction of molecules do not condense. While experiment shows a high condensate fraction, Ketterle *et al.* then argue that there must be significant number of molecules even on the $a < 0$ (BCS) side. Although a molecule has an energy higher than that of the scattering state on the $a < 0$ side, it cannot be stable as judged by the two-body interaction. Since there are free Fermi atoms populated on the low momentum levels, the dissociation of molecules is suppressed by Pauli Principle. The dissociation is more probable only when the molecule has much higher energy. Such metastable molecules then appear as stable molecules when the magnetic field is ramped down. This argument casts doubt whether the pair condensate exists or not.

A different experimental approach was adopted by R. Grimm's group [81]. Direct measurement of the energy gap of the Fermi atom pairs on the $a < 0$ side was carried out using radio-frequency spectroscopy. Three hyperfine states of ^{6}Li (denoted by $|1\rangle$, $|2\rangle$ and $|3\rangle$ roughly corresponding to the nuclear spin components $m_I = 1, 0, -1$

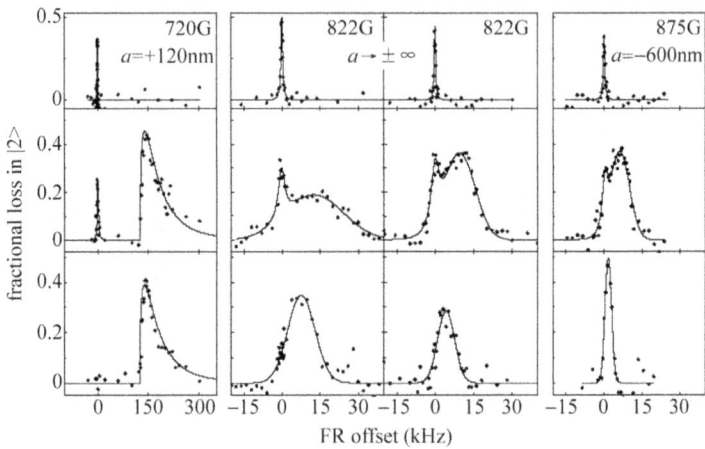

Fig. 10.37   Radio-frequency spectrum at different temperatures. Reprinted with permission from C. Chin, M. Bartenstein, A. Altmeyer, S. Riedl, S. Jochim, J. Hecker Denschlag, R. Grimm , *Science*, 305, 1128, 2004. Copyright (2017) by the AAAS.

respectively) were used. Consider equal mixture of $|1\rangle$ and $|2\rangle$. If there is no pairing between these states, then a frequency for resonance absorption $|2\rangle \rightarrow |3\rangle$ can be observed. If there is pairing between $|1\rangle$ and $|2\rangle$, then another frequency corresponding to the dissociation of $|1\rangle$ and $|2\rangle$ is observed. Figure 10.37 gives the radio-frequency spectrum on both sides of the resonance. $B = 720G$ is on the BEC side, corresponding to $a = 120$ nm. $B = 822G$ and $837G$ are the lower and upper limits of the Feshbach resonance. $B = 875G$ is on the BCS side, corresponging $a = -600$ nm. Sice the temperature of the gas changes as the magnetic field changes, a temperature T' is defined as that of the gas in the extreme BEC limit. When the field is ramped up to the BCS side, the temperature will decrease. The first row of the Fig. 10.38 represents data for $T' = 6T_F$, $T_F = 15$ µK. We see that there is only one absorption frequency corresponding to the transition $|2\rangle \rightarrow |30\rangle$ at all magnetic field values. This frequency is taken to be the origin of coordinates. The middle row represents data for $T' = 0.5T_F$, $T_F = 3.4$ µK. On the side, BEC an additional broad peak shows up, it is the peak corresponding to the binding energy of the molecule. In the resonance region and on the BCS side another absorption peak is found. This correspond to the energy

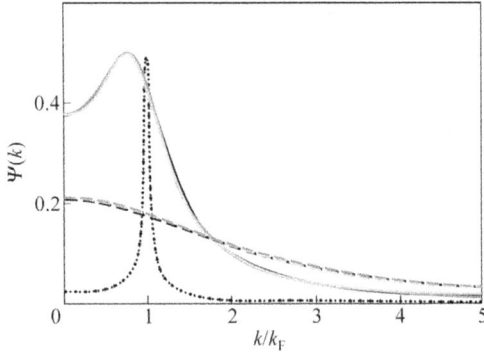

Fig. 10.38 Order parameter in different regions across the resonance. From Ref. [70].

gap of the Fermi atom pair. The lowest row represent data for $T' < 0.2T_F$, $T_F = 1.2 \ \mu K$. Free atoms cease to exist. On the BEC side the binding energy of the molecule is observed, and in the resonance region and on the BCS side, only the energy gap of the Fermi atom pair is observed. It is emphasized that the energy scale for the BEC side is different from that for the resonance and BCS side. The molecular binding energy is much greater than the energy gap of atom pairs. The line shape of the radio-frequency spectrum was worked out by Torma's group of Helsinki and the result lent strong support to Grimm's measurement.

Physicists now have confidence in the existence of the fermion superfluid in the strongly interacting regime made up of large fermion pairs. Tin-Lun Ho [75] reviewed the situation, and in two theoretical papers [82, 83] Ho and Diener offered strong support to the experimental studies [78, 80]. The authors carried out crossover theory calculations [70] using 3 kinds of potentials, namely V_I a square well, V_{II} a well with a high barrier, and V_{III} a delta function potential. Since V_{II} can accommodate a very long lived quasi-bound state, it has the major characteristics of two-channel resonance model.

To describe the fraction of condensate molecules, the operator creating a fermion pair of momentum q is defined

$$D_q^\dagger(x) = \sum\nolimits_{k,\alpha\beta} f_{k,\alpha\beta}(x) a_{k+q/2,\alpha}^\dagger , a_{-k+q/2,\beta}^\dagger /2, \qquad (10.10.14)$$

where $x = (k_F a_s)^{-1}$ describes the distance from the resonance, the variable characterizing the state of the system during crossover, $f_{k,\alpha\beta}(x)$ is the Fourier transform of the pair wave function:

$$f_{\alpha\beta}(r; x) = \Omega^{-1/2} \sum_k e^{ik \cdot r} f_{k,\alpha\beta}(x), \qquad (10.10.15)$$

where Ω is the volume of the system. The state of a condensate with N zero-momentum pairs is described by

$$|x\rangle = \mathcal{N} D_{q=0}^{\dagger N}(x)|vac\rangle \qquad (10.10.16)$$

and the pairing ground state is given by

$$|\Psi(x)\rangle = \mathcal{N} \prod_{k,\alpha\beta} \left(u_k(x) + v_{k,\alpha\beta}(x) a_{k,\alpha}^\dagger a_{-k,\beta}^\dagger \right) |vac\rangle, \qquad (10.10.17)$$

where u and v are related to $f_{k,\alpha\beta}(x)$ by

$$f_{k,\alpha\beta}(x) = \varsigma v_{k,\alpha\beta}(x)/u_k, \qquad (10.10.18)$$

where ς is a normalization constant. During the pairwise projection the system evolves from x_0 to x. The number of condensed and uncondensed molecules emergent at x are

$$N_0 = \langle D_0^\dagger(x) D_0(x)\rangle_{x_0}, \quad N_{ex} = \sum_{q\neq 0} \langle D^\dagger q(x) D_{qx}\rangle_{x_0}$$

respectively. We have

$$N_0 = |\langle D_0(x)\rangle|^2 = \left| \sum_k f_{k,\alpha\beta}(x) \Psi_{k,\alpha\beta}(x_0)/2 \right|^2, \qquad (10.10.19)$$

where $\Psi_{k,\alpha\beta}^*(x_0) = \langle a_{k,\alpha}^\dagger a_{-k,\beta}^\dagger\rangle_{x_0}$ is the order parameter of the condensate. We have also

$$N_{ex} = \sum_{k,q} |f_{k,\alpha\beta}(x)|^2 n_{q,\alpha} n_{q-2k,\beta}/2,$$

where $n_k = \langle a_{k,\alpha\alpha}^\dagger k, \alpha\rangle_{x_0}$. The expression (10.10.19) shows that N_0 is the overlap of the initial order parameter $\Psi_{\alpha\beta}$ with the final wave function. The condensate fraction is therefore

$$\frac{N_0}{N_m} = \frac{|\sum_k f_k(x) \Psi_k^*|^2}{|\sum_k f_k(x)_k^*|^2 + \sum_{k,q} |f_k(x)|^2 n_q n_{q-2k}}, \qquad (10.10.20)$$

where the total number of molecules N_m is $N_m = N_0 + N_{ex}$.

The crossover picture is characterized by the following Hamiltonian

$$H = \sum_{\pmb{k},\alpha} \varepsilon_{\pmb{k}} a_{\pmb{k},\alpha} a_{\pmb{k},\alpha}$$
$$+ \sum_{\pmb{k},\pmb{k}',\pmb{q}} V(\pmb{k} - \pmb{k}') a^{\dagger}_{\pmb{q}/2+\pmb{k},\uparrow} a^{\dagger}_{\pmb{q}/2-\pmb{l},\downarrow} a_{\pmb{q}/2-\pmb{k}',\downarrow} a_{\pmb{q}/2+\pmb{k}',\downarrow},$$

$$(10.10.21)$$

where V is given by the three types of potentials enumerated above. Expressing the relevant parameters u and v as usual in the BCS theory,

$$|u_{\pmb{k}}|^2 = (E_{\pmb{k}} + \xi_{\pmb{k}})/2E_{\pmb{k}}, \ |v_{\pmb{k}}|^2 + (E_{\pmb{k}} - \xi_{\pmb{k}})/2E_{\pmb{k}},$$

where

$$\xi_{\pmb{k}} = \varepsilon_{\pmb{k}} - \mu, \ E_{\pmb{k}} = \sqrt{\xi^2 + \Delta_{\pmb{k}}^2},$$

we obtain the gap equation

$$\Delta_{\pmb{k}} = -\sum_{\pmb{k}'} V(\pmb{k} - \pmb{k}') \Delta_{\pmb{k}'}/(2E_{\pmb{k}'}), \qquad (10.10.22)$$

the order parameter $\Psi_{\pmb{k}} = u_{\pmb{k}} v_{\pmb{k}} = \Delta_{\pmb{k}}/2E_{\pmb{k}}$ and the occupation number $n_{\pmb{k}} = v_{\pmb{k}}^2$. The chemical Potential μ is determined by the relation

$$n = \Omega^{-1} \sum_{\pmb{k}} n_{\pmb{k}} = n(T = 0, \mu). \qquad (10.10.23)$$

The gap equation is solved numerically for different potentials and the relation (10.10.23) is inverted so that quantities $\Delta_{\pmb{k}}, v_{\pmb{k}}, v_{\pmb{k}}, \Psi_{\pmb{k}}$ and $n_{\pmb{k}}$ are obtained as functions of the parameters of the potentials and the density n. All these potentials give rise to identical properties (such as chemical potential, energy gap, and coherence factors) near resonance, demonstrating universality in this regime. We quote from [82] the figure for order parameter $\Psi(k)$ in Fig. 10.38, where dashed, solid and dotted curves denote $x = 2$ (BEC region), $x = 0$ (resonance) and region, a strong signal for BCS pairing. In the BEC region we see the wave function of the molecule, and in the resonance region something in between. The pair wave function is obtained from (10.10.18). The important result is that the wave function on the BEC side has significant overlap with the wave function near the resonance, and therefore the molecular component in the BCS side close to the resonance is unimportant in contributing to the projected condensate fraction N_0/N_m.

The foregoing discussions on experimental studies are very suggestive of the superfluid behavior of the pairs of Fermi atoms in the strong coupling regime, and they agree well with theoretical predictions; but they do not provide unambiguous evidence. The smoking gun for the superfluid behavior is provided by the experimental demonstration of vortex lattice in a strongly interacting rotating Fermi gas, since it is the direct consequence of the macroscopic wave function describing the superfluid. The demonstration is given by W. Ketterle's group [84]. A 50%–50% mixture of the lowest hyperfine spin states of ^{6}Li were used. between which there is a broad Feshbach resonance at 834G. The BCS-BEC crossover region is defined by $1/k_F|a| < 1$, lying between 780G and 925G. The cigar shaped atomic cloud of Li is stirred by two beams of non-resonant laser parallel to its axis for an interval of 300 to 500 ms at 812G, the field was ramped down in 100 ms to 729G (the BEC side), 833G (resonance) and 853G (BCS side). After some time for equilibration the confining traps were withdrawn and the gas cloud was allowed to expand freely for 2 ms before the magnetic field was quickly ramped to 735G. The cloud was imaged after an additional 9 ms of expansion at this field. This measure is necessary for enhancing the contrast because pairs of Fermi atoms are very unstable during expansion of the cloud, and under 735G the pairs have become stable molecules such that the vortex lattice can be preserved. Question can be raised that whether the vortex lattice is already there before the ramp or formed under 735G? The time of the magnetic field ramp is much shorter than the time of formation of the vortex lattice (determined to be several hundreds ms. The images of vortex lattices formed across the resonance is shown in Fig. 10.39, and the lifetime of the vortex lattice versus the magnetic field/interaction strength is shown in Fig. 10.40.

R. Grimm [85] characterized this work as a "final and spectacular proof of superfluidity". The superfluidity of ultracold Fermi gas serves as a model for physics of quantum matter in general, for

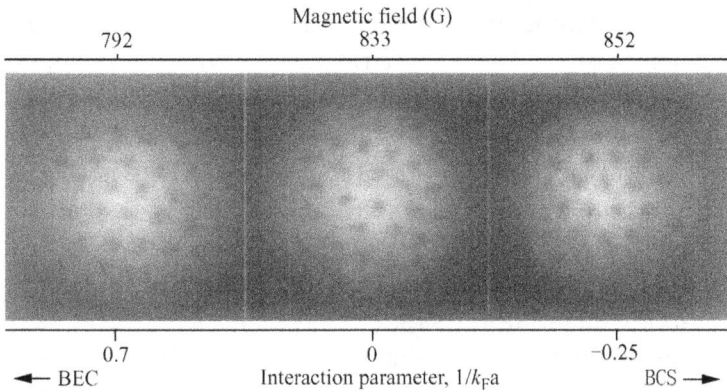

Fig. 10.39 Vortex lattices in the BEC-BCS crossover region. Reprinted with permission from Macmillan Publishers Ltd: Nature 435, 1047. Copyright (2005) by the Macmillan Publishers Ltd.

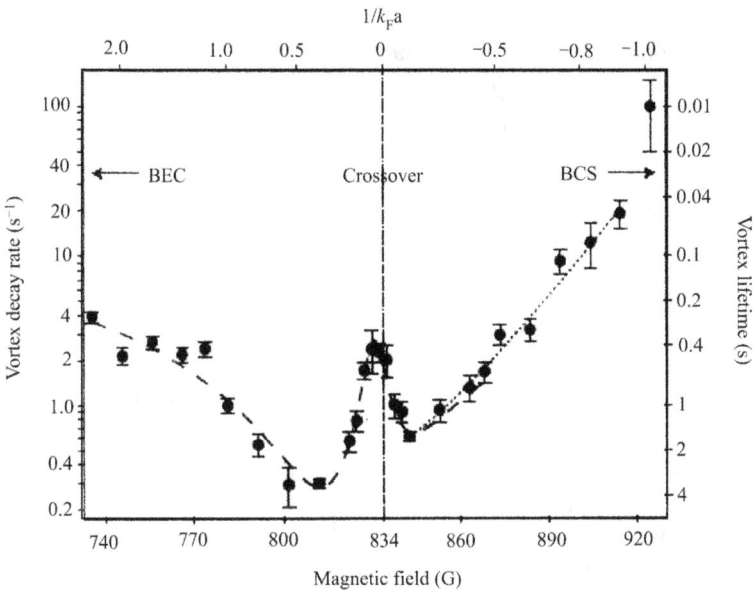

Fig. 10.40 Decay rate and lilfetime of the vortex lattice versus magnetic field and interaction strength. The dashed and dotted lines are Gaussian and exponential fitss and the position of Feshbach resonance is indicated by the vertical dash-dotted line. Reprinted with permission from Macmillan Publishers Ltd: Nature 435, 1047. Copyright (2005) by the Macmillan Publishers Ltd.

instance high temperature superconductivity, neutron stars and quark-gluon plasma. We compare the ratio of transition temperature to the Fermi temperature T_c/T_F for different systems: it is 10^{-4} for normal superconductors, 10^{-3} for superfluid ^{3}He, 10^{-2} for high T_c superconductors, and 0.3 for ^{6}Li superfluid. This is the highest temperature superfluid ever known. Owing to the purity of the system and versatile and precise manipulation possible, the study of ultracold quantum gas has opened new prospects for different fields of many-body quantum physics.

The theoretical and experimental studies of Bose-Einstein condensation and of physics of the quantum gas in general continue to flourish, a number of review articles [86–88], proceedings of summer schools [89, 90], and also monographs [91, 92] have been published.

References

[1] D. Kleppner, *Physics Today* 49(8), 11, 1996.

[2] M. H. Anderson, J. R. Ensher, M. R. Mathews, C. E. Wieman and E. A. Cornell, *Science* 269, 198, 1995.

[3] K. B. Davies, M.-O. Mewes, M. R. Anderson, N. J. van Druten, D. S. Durfee, D. M. Kurn and W. Ketterle, *Phys. Rev. Lett.* 75, 3969, 1995.

[4] C. C. Bradley, C. A. Sackett, J. J. Toulett and R. G. Hulet, *Phys. Rev. Lett.* 75, 1687, 1995.

[5] E. A. Cornell and C. E. Wieman, *Rev. Mod. Phys.* 74, 875, 2002.

[6] W. Ketterle, *Rev. Mod. Phys.* 74, 1131, 2002.

[7] C. Kittel and H. Kroemer, *Thermal Physics*, 2nd edition, San Fransisco, W. H. Freeman 1980.

[8] A. Leggett, in *Bose-Einstein Condensation*, edited by A. Griffin, D. W. Snoke and S. Stringari, New York, Cambridge University Press, 1995.

[9] K. Huang, *Statistical Mechanics*, 2nd edition, New York, John Wiley, 1987.

[10] K. Huang, in *Studies in Statistical Mechanics*, vol. II, edited by J. de Boer and G. E. Uhlenbeck, Amsterdam, North Holland 1964.

[11] K. Huang, in *Bose-Einstein Condensation*, edited by A. Griffin, D. W. Snoke and S. Stringari, New York, Cambridge University Press, 1995.

[12] O. Penrose, *Philos. Mag.* 42, 1373 1951.

[13] O. Penrose and L. Onsager, *Phys. Rev.* 104, 576, 1956.

[14] C. N. Yang, *Rev. Mod. Phys.* 34, 4, 1962.

[15] G. L. Sewell, *J. Statistical Physics*, 61, 415, 1990; *J. Math. Phys.* 38, 2053, 1997.

[16] H. T. Nieh, G. Su and B. H. Zhao, *Phys. Rev. B* 51, 3760, 1996.

[17] G. Su and M. Suzuki, *Phys. Rev. Lett.* 86, 2708, 2001.

[18] A. Leggett and F. Sols, *Found. of Phys.* 21, 353, 1993.

[19] A. Leggett, *Rev. Mod. Phys.* 73, 307, 2001.

[20] P. W. Anderson, *Rev. Mod. Phys.* 38, 298, 1966.

[21] M. Nieto, *Physica Scripta* T48, 5, 1993.

[22] P. W. Anderson, *Basic Notions of Condensed Matter Physics*, Menlo Park, Benjamin-Cummings, 1984.

[23] M. R. Andrews, C. G. Townsend, H.-J. Miesner, D. S. Durfee, D. M. Kurn and W. Ketterle, *Science* 275, 637, 1997.

[24] W. M. Liu, B. Wu and Q. Niu, *Phys. Rev. Lett.* 84, 2294, 2000.

[25] M. Kozuma, L. Deng, E. W. Hagley, J. Wen, R. Litwak, K. Helmerson, S. L. Rolston and W. D. Phillips, *Phys. Rev. Lett.* 82, 871, 1999.
Y. Torii, Y. Suzuki, M. Kozuma, T. Kuga, L. Deng and E. W. Hagley, cond-mat/9908160, 1999.

[26] J. Denschlag, J. E. Simsarian, D. L. Feder, C. W. Clark, L. A. Collins, J. Cubizolles, L. Deng, E. W. Hagley, K. Helmerson, W. P. Reinhardt, S. L. Rolston, B. I. Schneider and W. D. Phillips, *Science* 297, 97, 2000.

[27] P. W. Anderson, in *The Lessons of Quantum Theory*, edited by J. de Boer, E. Dal and O. Ulfbeck, Amsterdam, North-Holland 1986.

[28] Y. Castin and J. Dalibard, *Phys. Rev. A* 55, 4330, 1997.

[29] J. Javanainen and S. M. Yoo, *Phys. Rev. Lett.* 76, 161, 1996.

[30] M. Saba *et al.*, *Science* 307, 1945, 2005.

[31] L. Pitaevskii and S. Stringari, *Phys. Rev. Lett.* 83, 4237, 1999.

[32] P. Nozières and D. Pines, *The Theory of Quantum Liquids*, vol. II, Redwood City, Addison-Wesley, 1990.

[33] N. M. Hugenholtz and D. Pines, *Phys. Rev.* 116, 489, 1959.

[34] T. D. Lee and C. N. Yang, *Phys. Rev.* 105, 767, 1957.

[35] T. D. Lee, K. Huang and C. N. Yang, *Phys. Rev.* 106, 1135, 1957.

[36] V. Bagnato and D. E. Pritchard and D. Kleppner, *Phys. Rev. A* 35, 4354, 1987.

[37] V. Bagnato and D. Kleppner, *Phys. Rev. A* 44, 7439, 1991.

[38] L. D. Landau and E. M. Lifshitz, *Quantum Mechanics*, 3rd edition, Oxford, Pergamon, 1977.

[39] J. Oliva, *Phys. Rev. B* 39, 4197, 1989.

[40] T. T. Chou, C. N. Yang and L. H. Yu, *Phys. Rev. A* 53, 4257, 1996.

[41] T. T. Chou, C. N. Yang and L. H. Yu, *Phys. Rev. A* 55, 1179, 1997.

[42] V. L. Ginzburg and L. P. Pitaevskii, *Sov. Phys. — JETP*, 7, 858, 1958.

[43] E. P. Gross, *Nuovo Cimento* 20, 454, 1961.

[44] L. P. Pitaevskii, *Sov. Phys. — JETP*, 13, 451, 1961.

[45] E. H. Lieb and J. Yngvason, *Physical Review Letters* 80, 2504, 1998.

[46] E. H. Lieb, R. Seiringer and J. Yngvason, *Physical Review A* 61, 043602, 2000.

[47] M. Lewenstein and L. You, *Phys. Rev. Lett.* 77, 3489, 1996.

[48] J. P. Blaizot and G. Ripka, *Quantum Theory of Finite Systems*, Cambridge, MA, MIT Press 1986.

[49] M. Edwards and K. Burnett, *Phys. Rev. A* 51, 1382, 1995.

[50] P. A. Ruprecht, M. J. Holland, K. Burnett and M. Edwards, *Phys. Rev. A* 51, 4704, 1995.

[51] G. Baym and C. J. Pethick, *Phys. Rev. Lett.* 76, 6, 1996.

[52] Yu. Kagan, E. L. Surkov and G. V. Shlyapnikov, *Phys. Rev. A* 54, R1753, 1996.

[53] Yu. Kagan, E. L. Surkov and G. V. Shlyapnikov, *Phys. Rev. A* 55, R18, 1997.

[54] F. Dalfovo and S. Stringari, *Phys. Rev. A* 53, 2477, 1996.

[55] J. E. Williams and M. J. Holland, *Nature* 401, 568, 1999.

[56] M. R. Mattews, B. P. Anderson, P. C. Haljan, D. S. Hall, C. E. Wieman and E. A. Cornell, *Phys. Rev. Lett.* 83, 2498, 1999.

[57] K. W. Madison, F. Chevy, W. Wohlleben and J. Dalibard, *Phys. Rev. Lett.* 84, 806, 2000.

[58] J. R. Ado-Shaeer, C. Raman, J. M. Vogels and W. Ketterle, *Science* 292, 476, 2001.

[59] C. Raman, M. Köhl, R. Onofrio, D. S. Durfee, C. E. Kuklewicz, Z. Hadzibabic and W. Ketterle, *Phys. Rev. Lett.* 83, 2502, 1999.

[60] D. M. Stamper-Kurn, M. R. Andrews, A. P. Chikkatur, S. Inouye, H.-J. Miesner, J. Stenger and W. Ketterle, *Phys. Rev. Lett.* 80, 2027, 1998.

[61] T.-L. Ho, *Phys. Rev. Lett.* 81, 742, 1998.

[62] T.-L. Ho and V. B. Shenoy, *Phys. Rev. Lett.* 77, 2596, 1996 and 77, 3276, 1996.

[63] T.-L. Ho and S. K. Yip, *Phys. Rev. Lett.* 84, 4031, 2000.

[64] C. K. Law, H. Pu and N. P. Bigelow, *Phys. Rev. Lett.* 81, 5257, 1998.

[65] D.Jaksch, C. Bruder, J. I. Cirac, C. W. Gardiner and P. Zoller, *Phys. Rev. Lett.* 81, 3108, 1998.

[66] M. P. A. Fisher, P. B. Weichman, G. Grinstein and D. S. Fisher, *Phys. Rev. B* 40, 546, 1989.

[67] M. Greiner, O. Mandel, T. Esslinger, T. W. Hänsch and I. Bloch, *Nature* 415, 39, 2002.

[68] M. Köhl *et al.*, *Phys. Rev. Lett.* 94, 0804403, 2005.

[69] Z. Hadzibabic, S. Stock, B. Battelier, V. Bretin and J. Dalibard, *Phys. Rev. Lett.* 93, 180403, 2004.

[70] M. Randeria *et al.*, *Phys. Rev. B* 41, 327, 1990. arXiv cond-mat/0405174.
[71] C. Regal and D. Jin, *Phys. Rev. Lett.* 90, 230404, 2003. arXiv condmat/0303079.
[72] Tin-Lun Ho and E. Mueller, *Phys. Rev. Lett.* 92, 160404, 2004
[73] T. Bourdel *et al.*, cond -mat/0303079.
[74] P. Nozieres and S. Schmidt-Rink, *J. Low Temp. Phys.* 59, 195, 1985.
[75] Tin-Lun Ho, *Science* 305, 1114, 2004.
[76] C. Regal *et al.*, *Nature* 424, 47, 2003.
[77] M. Greiner *et al.*, *Nature* 426, 537, 2003.
[78] C. Regal *et al.*, *Phys. Rev. Lett.* 92, 040403, 2004.
[79] M. Bartenstein *et al.*, *Phys. Rev. Lett.* 92, 120401, 2004.
[80] M. Zwierlein *et al.*, *Phys. Rev. Lett.* 92, 120403, 2004.
[81] C. Chin *et al.*, *Science* 305, 1128, 2004.
[82] R. Diener and Tin-Lun Ho, Cond-mat/0404517.
[83] R. Diener and Tin-Lun Ho, cond-mat/0405174.
[84] M. Zwierlein *et al.*, *Nature* 435, 1047, 2005.
[85] R. Grimm, *Nature* 435, 1035, 2005.
[86] S. Stringari, *Phys. Rev. Lett.* 77, 2360, 1996.
[87] A. S. Parkins and D. F. Walls, *Phys. Reports* 303, 1, 1998.
[88] F. Dalfovo, S. Giorgini, L. P. Pitaevskii and S. Stringari, *Rev. Mod. Phys.* 71, 463, 1999. A. Leggett, *Rev. Mod. Phys.* 73, 307, 2001.
[89] M. Inguscio, S. Stringari and C. E. Wieman, editors, *Bose-Einstein Condensation in Atomic Gases*, International School of Physics "Enrico Fermi", course 140. Amsterdam, IOP Press, 1999.
[90] C. M. Savage and M. P. Das, editors, *Bose-Einstein Condensation*, Singapore, World Scientific, 2000.
[91] C. J. Pethick and H. Smith, *Bose-Einstein Condensation in Dilute Gases*, New York, Cambridge University Press, 2001.
[92] L. Pitaevskii and S. Stringsri, *Bose-Einstein Condensation*, Oxford University Pres, 2003.
[93] R. A. Duine and H. T. C. Stoof, *Phys. Reports*, 396, 115, 2004.

Chapter 11

The Yangian Commutation Relations in Quantum Mechanics

Since this chapter and henceforth, we concentrate on introducing the Yang-Baxter system.[1] It is the general theory of a large class of non-linear quantum integrable models dealing with many-body problems in low dimensions. In the past three decades the research in this direction has had far-reaching developments, and the direction has become a prosperous branch of mathematical physics. Many papers in the literatures appear rather mathematical in the form, since many mathematicians have been actively involved in the research and reached great achievements. But at the same time many problems in theoretical physics are intimately connected with it, including even the hydrogen atom. In Chapter 4 we introduced Pauli's $SO(4)$ symmetry of the hydrogen atom, emphasizing that the dynamical symmetry itself is sufficient to give the spectrum of bound states. In this chapter we will be able to see that the hydrogen atom possesses the Yangian symmetry, i.e., it possesses a kind of algebraic structure of infinite dimension, beyond the framework of Lie algebra. This may represent a deeper understanding of the hydrogen atom problem.

---

[1]Earlier articles on Yang-Baxter equation are assembled in Ref. [1]. The theories of Yangian and its representation can be found in Refs. [2–4].

We would like to point out that the terminology " Yang-Baxter equation" was firstly introduced by the Faddeev School and Yangian by Drinfeld.

11.1　Tensor Operators for the Hydrogen Atom and the Yangian

We discussed in Section 5.3 the $SO(4)$ dynamical symmetry of the hydrogen atom. The generators of the group are the angular momentum operators I_α ($\alpha = 1, 2, 3$) and the scaled Runge-Lenz operators B_α. $\boldsymbol{I}$ and $\boldsymbol{B}$ components belong to a special realization of the following $SO(4)$ commutation relations:

$$\left[I_\alpha, I_\beta\right] = i\varepsilon_{\alpha\beta\gamma} I_\gamma, \tag{11.1.1}$$

$$\left[B_\alpha, B_\beta\right] = i\varepsilon_{\alpha\beta\gamma} I_\gamma, \tag{11.1.2}$$

$$\left[I_\alpha, B_\beta\right] = i\varepsilon_{\alpha\beta\gamma} B_\gamma, \tag{11.1.3}$$

where $\varepsilon_{\alpha\beta\gamma}$ is the anti-symmetrical tensor of third rank, $\varepsilon_{123} = 1$, and an interchange of any pair of its indices gives rise to a sign change. Let $I_\alpha = L_\alpha$ and $B_\alpha = M'_\alpha$ then we go back to the relations given in Section 4.3 with $\hbar = 1$. In the following, we begin with the relations of $(11.1.1) \sim (11.1.3)$ to develop a more general theory and come back to the hydrogen atom.

Define the tensor operator

$$\boldsymbol{J} = \frac{1}{2i} \boldsymbol{I} \times \boldsymbol{B}. \tag{11.1.4}$$

The commutation relations between the operators I_α and J_β are simple:

$$\left[I_\alpha, J_\beta\right] = \varepsilon_{\beta\mu\nu} \left[I_\alpha, \frac{1}{2i} I_\mu B_\nu\right]$$

$$= \varepsilon_{\beta\mu\nu} \left(\left[I_\alpha, I_\mu\right] \frac{1}{2i} B_\nu + I_\mu \left[I_\alpha, \frac{1}{2i} B_\nu\right]\right)$$

$$= \frac{1}{2} \varepsilon_{\beta\mu\nu} \left(\varepsilon_{\alpha\mu\lambda} I_\lambda B_\nu + \varepsilon_{\alpha\nu\lambda} I_\mu B_\lambda\right)$$

$$= \frac{1}{2}(I_\alpha B_\beta - I_\beta B_\alpha)$$

$$= i\varepsilon_{\alpha\beta\gamma} J_\gamma. \tag{11.1.5}$$

In the above derivation we have used the properties of the tensor $\varepsilon_{\alpha\beta\gamma}$:

$$\varepsilon_{\alpha\beta\mu}\varepsilon_{\gamma\delta\mu} = \delta_{\alpha\gamma}\delta_{\beta\delta} - \delta_{\alpha\delta}\delta_{\beta\gamma}. \tag{11.1.6}$$

The commutation relations of (11.1.5) are similar to (11.1.3), but the commutators between J_α and J_β are more complicated:

$$[I_\alpha, J_\beta] = -\frac{1}{4}\varepsilon_{\alpha\mu\nu}\varepsilon_{\beta\sigma\tau}\left[I_\mu B_\nu, I_\sigma B_\tau\right]$$

$$= -\frac{1}{4}\varepsilon_{\alpha\mu\nu}\varepsilon_{\beta\sigma\tau}\left([I_\mu, I_\sigma] B_\nu B_\tau + I_\sigma [I_\mu, B_\tau] B_\nu \right.$$

$$\left. + I_\mu [B_\nu, I_\sigma] B_\tau + I_\mu I_\sigma [B_\nu, B_\tau]\right). \tag{11.1.7}$$

Using (11.1.1)~(11.1.3), (11.1.6), and also $\varepsilon_{\alpha\beta\mu}\varepsilon_{\alpha\beta\nu} = 2\delta_{\mu\nu}$, we obtain

$$\varepsilon_{\alpha\mu\nu} I_\mu I_\nu = \frac{1}{2}\varepsilon_{\alpha\mu\nu}\left[I_\mu, I_\nu\right] = \frac{1}{2}i\varepsilon_{\alpha\mu\nu}\varepsilon_{\mu\nu\lambda} I_\lambda$$

$$= i\delta_{\alpha\lambda} I_\lambda = iI_\lambda, \tag{11.1.8}$$

and (11.1.7) is simplified as

$$[J_\alpha, J_\beta] = -\frac{i}{4}\varepsilon_{\alpha\beta\gamma} I_\gamma \left[\boldsymbol{I}^2 - \boldsymbol{B}^2 - 1\right]. \tag{11.1.9}$$

For the hydrogen atom, by putting $\hbar = 1$ and $m = 1$ in (4.3.20) and (4.3.26) we have $\boldsymbol{B} = \boldsymbol{M}'$, and hence

$$\boldsymbol{B}^2 = -\left(\boldsymbol{I}^2 + 1\right) - \frac{\kappa^2}{2E}. \tag{11.1.10}$$

Substituting (11.1.10) into (11.1.9), we get

$$[J_\alpha, J_\beta] = -\frac{1}{4}i\varepsilon_{\alpha\beta\gamma} I_\gamma \left(2\boldsymbol{I}^2 + \frac{\kappa^2}{2E}\right). \tag{11.1.11}$$

In the above formula the commutator involves the energy eigen-value, and furthermore the right-hand side cannot be expressed as a linear combination of I_α and J_β while it is a non-linear combination instead. For this purpose we have to calculate the commutation relation between (11.1.11) and J_α. In convenience, we use the $(+, -, 3)$ form of the components of a vector instead of $(\alpha = 1, 2, 3)$, for instance

$$I_+ = I_1 + iI_2, \quad I_- = I_1 - iI_2.$$

The component form of (11.1.4) is

$$\left. \begin{aligned} J_\pm &= \mp \frac{1}{2} \left(L_\pm B_3 - L_3 B_\pm \right) \\ J_3 &= \frac{1}{4} \left(L_+ B_- - L_- B_+ \right) \end{aligned} \right\}. \tag{11.1.12}$$

The commutation relations between the components of $\boldsymbol{I}$ are:

$$[I_3, I_\pm] = \pm I_\pm, [I_+, I_-] = 2I_3. \tag{11.1.13}$$

The commutation relations between the components of $\boldsymbol{I}$ and $\boldsymbol{J}$ become

$$\left. \begin{aligned} [I_\pm, J_\pm] &= 0, \quad [I_3, J_3] = 0 \\ [I_3, J_\pm] &= [J_3, I_\pm] = \pm J_\pm \\ [I_+, J_-] &= [J_+, I_-] = 2J_3 \end{aligned} \right\}. \tag{11.1.14}$$

Besides, we have also

$$\left. \begin{aligned} \left[\boldsymbol{I}^2, J_\pm \right] &= \pm 2 \left(I_3 J_\pm - J_3 I_\pm \right) \\ \left[\boldsymbol{I}^2, J_3 \right] &= I_+ J_- - J_+ I_- \end{aligned} \right\}. \tag{11.1.15}$$

Equation (11.1.9) gives

$$\left. \begin{aligned} [J_3, J_\pm] &= \mp \frac{1}{4} \left(2\boldsymbol{I}^2 + \frac{k^2}{2E} \right) I_\pm \\ [J_+, J_-] &= -\frac{1}{2} \left(2\boldsymbol{I}^2 + \frac{k^2}{2E} \right) I_3 \end{aligned} \right\}. \tag{11.1.16}$$

A higher order commutation relation can be obtained by using (11.1.16) and (11.1.15):

$$[J_\pm, [J_3, J_\pm]] = \mp \frac{1}{4} \left[J_\pm, \left(2\mathbf{I}^2 + \frac{k^2}{2E} \right) I_\pm \right]$$

$$= \mp \frac{1}{2} \left[J_\pm, \mathbf{I}^2 \right] I_\pm$$

$$= (I_3 J_\pm - J_3 I_\pm) I_\pm.$$

Using (11.1.14) we obtain other component forms of this relation.

$$[J_\pm, [J_3, J_\pm]] = (I_3 J_\pm - J_3 I_\pm) I_\pm$$

$$= (J_\pm I_3 - I_\pm J_3) I_\pm$$

$$= I_\pm (I_3 J_\pm - J_3 I_\pm)$$

$$= I_\pm (J_\pm I_3 - I_\pm J_3). \tag{11.1.17}$$

Similarly we have

$$[J_3, [J_+, J_-]] = I_3 (I_+ J_- - J_+ I_-)$$

$$= I_3 (J_- I_+ - I_- J_+)$$

$$= (I_+ J_- - J_+ I_-) I_3$$

$$= (J_- I_+ - I_- J_+) I_3. \tag{11.1.18}$$

We can go further to calculate $[J_\pm, [J_+, J_-]]$ and $[J_3, [J_3, J_\pm]]$. For this purpose we introduce

$$K = [J_3, J_\pm],$$

then we have the Jacobi identities

$$[I_\mp, [J_\pm, K]] + [J_\pm, [K, I_\mp]] + [K, [I_\mp, J_\pm]] = 0. \tag{11.1.19}$$

The third term of the above formula can be rewritten as

$$[K, \mp 2J_3] = \mp 2 [[J_3, J_\pm], J_3]$$

$$= \mp 2 [J_3, [J_\pm, J_3]]$$

by using (11.1.14). In the second term of (11.1.19) we get

$$[K, I_\mp] = [J_3 J_\pm, I_\mp] - [J_\pm J_3, I_\mp]$$
$$= [J_\pm, I_\mp] J_\pm - J_\pm [J_3, I_\mp] = \pm [J_\pm, J_\mp],$$

and hence the second term of (11.1.19) becomes

$$\pm [J_\pm, [J_\pm, J_\mp]] = [J_\pm, [J_+, J_-]],$$

while the first term can be rewritten as

$$[I_\mp, I_\pm (J_\pm I_3 - I_\pm J_3)]$$
$$= [I_\mp, I_\pm] (J_\pm I_3 - I_\pm J_3) + I_\pm [I_\mp, (J_\pm I_3 - I_\pm J_3)]$$
$$= -2I_3 (J_\pm I_3 - I_\pm J_3) + I_\pm (J_\pm I_\mp - I_\pm J_\mp)$$

by using (11.1.15). Substituting these results into the identity (11.1.19), we obtain

$$[J_\pm, [J_+, J_-]] \pm 2 [J_3, [J_3, J_\pm]]$$
$$= 2I_3 (J_\pm I_3 - I_\pm J_3) + I_\pm (I_\pm J_\mp - J_\pm I_\mp). \qquad (11.1.20)$$

Taking into account (11.1.14) and the Jacobi identity, we see that there is only one independent relation among Eqs. (11.1.17), (11.1.18) and (11.1.20).

In the above derivation we see that the angular momentum operator I and the tensor operator $J = \frac{1}{2i} I \times B$ together form a set satisfying Eqs. (11.1.13), (11.1.14), (11.1.17), (11.1.18) and their implication (11.1.20). The relations satisfied by the components I form a Lie algebra, the commutation relations between I and J components (11.1.14) demonstrate the vector nature of J, while the commutation relations (11.1.17)–(11.1.20) for the components of J are more involved and work in a tensor space. The generators of the algebras obey the addition rule while there appears direct product of two generators in J. These algebraic relations have deep significance, which we shall elaborate step by step in the following.

The set of operators I, J originates from the Lie algebra $SL(2)$, and they satisfy commutation relations of Eq. (11.1.13), (11.1.14),

(11.1.17)–(11.1.20). The set formed from I, J together with the co-product (see later) is called $Y(SL(2))$. Notice that I is not limited to be the angular momentum operators and J is not limited to be $\frac{1}{2i} L \times B$. Such kind of algebraic structure is named as Yangian after the suggestion by V.G. Drinfeld in 1985 [2] to acknowledge the great contribution of Professor Chen Ning Yang in the research on the many-body integrable models (1967) [5]. The introduction of Yangian began with the algebraic structure of Yang-Baxter systems related to exact solutions in one-dimensional quantum many-body systems, and evolved into a high-lighted direction of mathematical physics. In many problems it is connected with abstract concepts including unknown physical effects in quantum tensor space, but it is also involved in the hydrogen atom problem as we have just seen.

11.2   The Yangian Algebras

We generalize the algebraic relations discussed in the last section to the general Yangian commutation relations:

(1)
$$\left[I_\lambda, I_\mu\right] = C_{\lambda\mu\nu} I_\nu, \quad \lambda, \mu, \nu = 1, 2, 3, \ldots \qquad (11.2.1)$$

where $C_{\lambda\mu\nu}$ is the structure constant. In the case $C_{\lambda\mu\nu} = i\varepsilon_{\lambda\mu\nu}$ and $\lambda, \mu, \nu = 1, 2, 3$, (11.2.1) is reduced to (11.1.13).

(2)
$$\left[I_\lambda, J_\mu\right] = C_{\lambda\mu\nu} J_\nu, \quad \lambda, \mu, \nu = 1, 2, 3, \ldots \qquad (11.2.2)$$

When $C_{\lambda\mu\nu} = i\varepsilon_{\lambda\mu\nu}$ and $\lambda, \mu, \nu = 1, 2, 3$, (11.2.2) is reduced to (11.1.14).

(3)
$$\left[J_\lambda, \left[J_\mu, I_\nu\right]\right] - \left[I_\lambda, \left[J_\mu, J_\nu\right]\right] = a_{\lambda\mu\nu\alpha\beta\gamma} \left\{I_\alpha, I_\beta, I_\gamma\right\}, \qquad (11.2.3)$$

where

$$a_{\lambda\mu\nu\alpha\beta\gamma} = \frac{1}{4!} C_{\lambda\alpha\sigma} C_{\mu\beta\tau} C_{\nu\gamma\rho} C_{\sigma\tau\rho}, \qquad (11.2.4)$$

$$\{x_1, x_2, x_3\} = \sum_{\substack{i,j,k=1,2,3 \\ i \neq j \neq k}} x_i x_j x_k. \qquad (11.2.5)$$

The expression of (11.2.5) is a symmetric combination of three quantities.

(4)
$$\left[\left[J_\lambda, J_\mu\right], \left[I_\sigma, J_\tau\right]\right] + \left[\left[J_\sigma, J_\tau\right], \left[I_\lambda, J_\mu\right]\right]$$
$$= \left(a_{\lambda\mu\nu\alpha\beta\gamma} C_{\sigma\tau\nu} + a_{\sigma\tau\nu\alpha\beta\gamma} C_{\lambda\mu\nu}\right)\left\{I_\alpha, I_\beta, J_\gamma\right\}. \qquad (11.2.6)$$

When $C_{\lambda\mu\nu} = i\varepsilon_{\lambda\mu\nu}$ and $\lambda, \mu, \nu = 1, 2, 3$, Eq. (11.2.3) is automatically satisfied. The proof goes as follows. Using Jacobi identity we rewrite the second term in the left-hand side of Eq. (11.2.3) in the form as

$$\left[J_\lambda, \left[J_\mu, I_\nu\right]\right] - \left[I_\lambda, \left[J_\mu, J_\nu\right]\right]$$
$$= \left[J_\lambda, \left[J_\mu, I_\nu\right]\right] + \left[J_\mu, \left[J_\nu, I_\lambda\right]\right] + \left[J_\nu, \left[I_\lambda, J_\mu\right]\right]$$
$$= i\varepsilon_{\mu\nu\sigma}\left[J_\lambda, J_\sigma\right] + i\varepsilon_{\nu\lambda\sigma}\left[J_\mu, J_\sigma\right] + i\varepsilon_{\lambda\mu\sigma}\left[J_\nu, J_\sigma\right]. \qquad (11.2.7)$$

When any two of the indices λ, μ, ν equal, one of the three terms on the right-hand side vanishes and the other two terms cancel each other. For instance the third term vanishes for $\lambda = \mu$ and the first two terms cancel each other. Let us consider the first term for λ, μ, ν all different. In order that $\varepsilon_{\mu\nu\sigma}$ do not vanish, the index σ must be equal to λ, but in this case the commutator vanishes. The same is true for the other two terms. We conclude therefore that for $\lambda, \mu, \nu = 1, 2, 3$ the left-hand side of (11.2.7) is identically zero. Consider the right-hand side of (11.2.3) in the case $C_{\lambda\mu\nu} = i\varepsilon_{\lambda\mu\nu}$ and $\lambda, \mu, \nu = 1, 2, 3$. Let the fixed indices λ, μ, ν be all different, the summation indices α, β, γ must be all different and equal to a respective index among λ, μ, ν, in order that $a_{\lambda\mu\nu\alpha\beta\gamma}$ (hereafter called a for short) is different from zero. The possible combinations of the indices for a nonzero a can only be $\alpha = \mu$, $\beta = \nu$, $\gamma = \lambda$ or $\alpha = \nu$, $\beta = \lambda$, $\gamma = \mu$. Therefore the right-hand side of Eq. (11.2.3) should be

$$a_{\lambda\mu\nu\alpha\beta\gamma}\left\{I_\alpha, I_\beta, I_\gamma\right\}$$
$$= \frac{i^3}{24}\varepsilon_{\lambda\mu\nu}\varepsilon_{\mu\nu\lambda}\varepsilon_{\nu\lambda\mu}\varepsilon_{\nu\lambda\mu}\left\{I_\mu, I_\nu, I_\lambda\right\}$$
$$+ \frac{i^3}{24}\varepsilon_{\lambda\nu\mu}\varepsilon_{\mu\lambda\nu}\varepsilon_{\nu\mu\lambda}\varepsilon_{\mu\nu\lambda}\left\{I_\nu, I_\lambda, I_\mu\right\}.$$

Let λ, μ, ν be in the standard order, then the coefficient of the first term is $\frac{i^3}{24}$, and that of the second term is $-\frac{i^3}{24}$. Since $\{I_\mu, I_\nu, I_\lambda\}$ is a symmetric combination; hence the two terms cancel. Let two of the indices among λ, μ, ν be equal, and the third be different, e.g., $\lambda = \mu \neq \nu$. Let σ be another index different from λ and ν. Among the four antisymmetric tensors in the right-hand side of the last equation let the first be $\varepsilon_{\lambda\sigma\nu}$ or $\varepsilon_{\lambda\nu\sigma}$, the demand that the indices in the fourth tensor be all different (otherwise it vanishes) requires that the last index in the first and the second tensors must be different. Therefore, corresponding to the two choices for the first tensor the second tensor is $\varepsilon_{\lambda\nu\sigma}$ or $\varepsilon_{\lambda\sigma\nu}$ (the first index being λ is due to the prescribed $\lambda = \mu$). The fourth tensor is $\varepsilon_{\nu\sigma\lambda}$ corresponding to both choices, and the third tensor is similarly $\varepsilon_{\nu\sigma\lambda}$. Therefore we have

$$a_{\lambda\lambda\nu\alpha\beta\gamma}\{I_\alpha, I_\beta, I_\gamma\}$$

$$= \frac{i^3}{24}(\varepsilon_{\lambda\sigma\nu}\varepsilon_{\lambda\nu\sigma}\varepsilon_{\nu\sigma\lambda}\varepsilon_{\nu\sigma\lambda}\{I_\sigma, I_\nu, I_\sigma\}$$

$$+ \varepsilon_{\lambda\nu\sigma}\varepsilon_{\lambda\sigma\nu}\varepsilon_{\nu\sigma\lambda}\varepsilon_{\sigma\nu\lambda}\{I_\nu, I_\sigma, I_\sigma\}).$$

Among the antisymmetric tensors in the two terms above, three pairs are identical, and the fourth pair appears with different signs. The curly brackets equal to each other because they are symmetrical combinations. The above expression vanishes. If $\lambda = \mu = \nu$, then the demand that σ, τ, ρ be all different, but none of them equal to λ cannot be realized. Then at least one of the tensors must be zero, i.e., $a_{\lambda\lambda\lambda\alpha\beta\gamma}\{I_\alpha, I_\beta, I_\gamma\} = 0$. In the foregoing we have proved that for $SL(2)$, Eq. (11.2.3) is reduced to the identity $0 = 0$.

Consider (11.2.6) for $SL(2)$. In the left-hand side there are four fixed indices in two pairs: (λ, μ) and (σ, τ). If one pair consists of equal indices, for instance $\lambda = \mu$, then the left-hand side vanishes. The first term of the right-hand side $a_{\lambda\lambda\nu\alpha\beta\gamma}C_{\sigma\tau\nu}\{I_\alpha, I_\beta, I_\gamma\} = 0$. The reason is the same as that given for the case $\lambda = \mu$ for (11.2.3). The second term contains $\varepsilon_{\lambda\lambda\nu} = 0$ for self-evident reason. The case $\sigma = \tau$ is similar. Then Eq. (11.2.6) is also an identity $0 = 0$.

If an index in the first pair is equal to an index in the second pair, e.g., $\lambda = \tau$, then the remaining two indices can be either equal

$\sigma = \mu$ or unequal $\sigma \neq \mu$. The left-hand side of Eq. (11.2.6) for the case $\lambda = \tau$ is

$$\left[[J_\lambda, J_\mu], [I_\sigma, J_\lambda]\right] + \left[[J_\sigma, J_\lambda], [I_\lambda, J_\mu]\right],$$

which becomes by using the commutation relations of $\boldsymbol{I}$ and $\boldsymbol{J}$ components

$$i\varepsilon_{\sigma\lambda\rho}\left[J_\rho, [J_\mu, J_\lambda]\right] + i\varepsilon_{\lambda\mu\rho}\left[J_\rho, [J_\lambda, J_\sigma]\right], \qquad (11.2.8)$$

where λ, μ, σ are fixed indices and ρ the summation index. The condition that the first term does not vanish is $\sigma \neq \lambda \neq \rho$, $\mu \neq \lambda$. Hence the summation index ρ must be μ. The condition that the second term does not vanish is $\lambda \neq \mu \neq \rho$, $\lambda \neq \sigma$, and hence ρ must be σ. The non-vanishing part of the sum (11.2.8) is, under the condition $\sigma \neq \mu$

$$i\varepsilon_{\sigma\lambda\mu}\left[J_\mu, [J_\mu, J_\lambda]\right] + i\varepsilon_{\lambda\mu\sigma}\left[J_\sigma, [J_\lambda, J_\sigma]\right]$$
$$= i\varepsilon_{\lambda\mu\sigma}\left(\left[J_\mu, [J_\mu, J_\lambda]\right] + \left[J_\sigma, [J_\lambda, J_\sigma]\right]\right), \qquad \sigma \neq \mu.$$

A point to be noted is, there is no summation index in the above equation, i.e., the repeating indices are *not* meant to be summed. When $\mu = \sigma$ (11.2.8) becomes

$$2i\varepsilon_{\lambda\mu\rho}\left[J_\rho, [J_\lambda, J_\mu]\right], \qquad \sigma = \mu.$$

Summarizing the above, we have that for $\lambda = \tau$ the left-hand side of Eq. (11.2.6) equal to

$$\begin{cases} i\varepsilon_{\sigma\lambda\mu}\left[J_\mu, [J_\mu, J_\lambda]\right] + [J_\sigma, [J_\lambda, J_\sigma]], & \sigma \neq \mu \\ 2i\varepsilon_{\lambda\mu\rho}\left[J_\rho, [J_\lambda, J_\mu]\right], & \sigma = \mu, \text{ summed over } \rho \end{cases}$$

Correspondingly, for $\lambda = \tau$ the right-hand side of (11.2.6) is equal to

$$i\left(a_{\lambda\mu\nu\alpha\beta\gamma}\varepsilon_{\sigma\lambda\nu} + a_{\sigma\lambda\nu\alpha\beta\gamma}\varepsilon_{\lambda\mu\nu}\right)\left\{I_\alpha, I_\beta, J_\gamma\right\}$$

We note that λ, μ, σ are the fixed indices and $\nu, \alpha, \beta, \gamma$ are the summation indices. Substituting the definition of a (11.2.4) into the above expression, carrying out the summation and using Eq. (11.1.6), we obtain

$$
\begin{cases}
\dfrac{i}{24}\varepsilon_{\lambda\mu\sigma}(\{I_\sigma, I_\lambda, J_\sigma\} - \{I_\mu, I_\lambda, J_\mu\} & \sigma \neq \mu \\[2mm]
\qquad + \{I_\mu, I_\mu, J_\lambda\} - \{I_\sigma, I_\sigma, J_\lambda\}), \\[2mm]
\dfrac{i}{24}2\varepsilon_{\lambda\mu\rho}\left(\{I_\rho, I_\lambda, J_\mu\} - \{I_\mu, I_\rho, J_\lambda\}\right), & \sigma = \mu
\end{cases}
$$

Since $\{I_\alpha, I_\beta, I_\gamma\}$ is a symmetric combination, Eq. (11.2.6) gives

$$
[J_\nu, [J_\lambda, J_\mu]] = \frac{1}{4}\left(J_\mu I_\lambda - I_\mu J_\lambda\right) I_\nu, \tag{11.2.9}
$$

$$
[J_\mu, [J_\lambda, J_\mu]] - [J_\nu, [J_\lambda, J_\nu]] = \frac{1}{4}(I_\nu I_\lambda J_\nu - I_\mu I_\lambda J_\mu
$$

$$
+ I_\mu I_\lambda J_\mu - I_\nu I_\lambda J_\nu). \tag{11.2.10}
$$

In the above equations λ, μ, ν are fixed indices $1, 2, 3$.

If we let $\lambda = \sigma$, or $\mu = \sigma$, or $\mu = \tau$ among the two pairs of indices (λ, μ) and (σ, τ) in Eq. (11.2.6), the result is the same as in the discussion for the case $\lambda = \tau$. If we use $(+, -, 3)$ to label the components, then (11.2.9), (11.2.10) can be written as five relations

$$
[J_3, [J_+, J_-]] = \frac{1}{4}\left(J_- I_+ - I_- J_+\right) I_3, \tag{11.2.11}
$$

$$
[J_+, [J_3, J_+]] = \frac{1}{4}I_+\left(J_+ I_3 - I_+ J_3\right), \tag{11.2.12}
$$

$$
[J_-, [J_3, J_-]] = \frac{1}{4}I_-\left(J_- I_3 - I_- J_3\right), \tag{11.2.13}
$$

$$
2\,[J_3, [J_3, J_+]] + [J_+, [J_+, J_-]] = \frac{1}{4}(2I_3(J_+ I_3 - I_+ J_3)
$$

$$
+ I_+(J_+ I_- - I_+ J_-)), \tag{11.2.14}
$$

$$
2\,[J_3, [J_3, J_+]] + [J_-, [J_-, J_+]] = \frac{1}{4}(2I_3(J_- I_3 - I_- J_3)
$$

$$
+ I_-(J_- I_+ - I_- J_+)). \tag{11.2.15}
$$

We rescale $J_\alpha \to \frac{1}{2}J_\alpha$ and take into account relations such as $J_3(J_-I_+ - I_-J_+) = (I_+J_- - J_+I_-)J_3$, then Eqs. (11.2.11)–(11.2.15) are just Eqs. (11.1.17)–(11.1.20). From (11.2.2) and the Jacobi identity we can prove that among (11.2.11)–(11.2.15) only one is independent.

The above discussion has demonstrated that (11.2.1), (11.2.2), (11.2.3) and (11.2.6) are the general Yangian commutation relations given by Drinfeld. They can be applied to any simple Lie algebra. In the case of $SL(2)$, $C_{\lambda\mu\nu} = i\varepsilon_{\lambda\mu\nu}$, $\lambda, \mu, \nu = 1, 2, 3$, they reduce to the algebraic relations satisfied by the angular momentum operator $\boldsymbol{L}$ and $\boldsymbol{J} = \frac{1}{2i}\boldsymbol{L} \times \boldsymbol{B}$ for the hydrogen atom, i.e., operators $\boldsymbol{I}, \boldsymbol{J}$ of the hydrogen atom is the simplest realization of the Yangian $Y(SL(2))$.

11.3   Other Realizations of $Y(SL(2))$ in Quantum Mechanics

The presentation of this section is based on Ref. [4].

From the fundamental relations of Yangian (Eqs. (11.1.13), (11.1.14), (11.1.18) or (11.1.17)) it is easy to reach the following results: If I_α and J_α ($\alpha = +, -, 3$) satisfy the commutation relations of $Y(SL(2))$, then I_α and $-J_\alpha$ must also satisfy the Yangian. At the same time I_α and $\mu I_\alpha + J_\alpha$ also satisfy the Yangian. It is obvious that I_α and $-J_\alpha$ satisfy the Yangian because Eqs. (11.1.14), (11.1.17) and (11.1.18) contain only odd numbers of J_α. Carrying out a translation on $J_\pm$ in (11.1.14) $J_\pm \to J_\pm + \mu I_\pm$, where μ is an arbitrary parameter or a Casimir operator of Yangian. In fact, we have

$$[I_3, J_\pm + \mu I_\pm] = [I_3, J_\pm] + \mu [I_3, I_\pm]$$
$$= \pm J_\pm \pm \mu I_\pm = \pm (J_\pm + \mu I_\pm).$$

Equation (11.1.14) is preserved after the translation. Inspecting (11.1.18), we see that $[J_+, J_-]$ becomes after translation

$$[J_+ + \mu I_+, J_- + \mu I_-] = [J_+, J_-] + 2\mu^2 I_3 + 4\mu J_3,$$

and that $[J_3 + \mu I_3, [J_+ + \mu I_+, J_- + \mu I_-]]$ is calculated to be $\mu[I_3, [J_+, J_-]] + [J_3, [J_+, J_-]]$. The increment of the commutator $[J_3, [J_+, J_-]]$ due to the translation is $\mu[I_3, [J_+, J_-]]$, which becomes after using the Jacobi identity

$$-\mu\left[J_+, [J_-, I_3]\right] - \mu\left[J_-, [I_3, J_+]\right] = -\mu\left([J_+, J_-] + [J_-, J_+]\right) = 0.$$

The increment of the right-hand side of (11.1.18) due to the translation is

$$\mu\left(I_+I_- - I_+I_-\right)I_3 = 0.$$

Similarly the invariance of (11.1.17) during the translation can be shown. In the following, we discuss examples of the realization of $Y(SL(2))$ in terms of simple quantum mechanical operators.

For system of two angular momenta we consider two angular momentum operators $\boldsymbol{j}_1$ and $\boldsymbol{j}_2$ coupled to be total angular momentum $\boldsymbol{I}$:

$$\boldsymbol{I} = \boldsymbol{j}_1 + \boldsymbol{j}_2, \tag{11.3.1}$$

and we define

$$\boldsymbol{J} = -i\boldsymbol{j}_1 \times \boldsymbol{j}_2. \tag{11.3.2}$$

Operators $\boldsymbol{j}_1$ and $\boldsymbol{j}_2$ satisfy the commutation relation

$$\left[j_{i\alpha}, j_{i\beta}\right] = i\varepsilon_{\alpha\beta\gamma}j_{i\gamma}\delta_{ij}, \quad i, j = 1, 2. \tag{11.3.3}$$

Since any component of $\boldsymbol{j}_1$ commutes with any component of $\boldsymbol{j}_2$, the components of the total angular momentum satisfy the $SL(2)$ commutation relations of Eq. (11.1.13). It is easy to verify (11.1.14):

$$\begin{aligned}
i\left[I_\alpha, J_\beta\right] &= \left[j_{1\alpha} + j_{2\alpha}, \varepsilon_{\beta\sigma\tau}j_{1\sigma}j_{2\tau}\right] \\
&= \varepsilon_{\beta\sigma\tau}\left([j_{1\alpha}, j_{1\sigma}]j_{2\tau} + j_{1\sigma}[j_{2\alpha}, j_{2\tau}]\right) \\
&= i\varepsilon_{\beta\sigma\tau}\left(\varepsilon_{\alpha\sigma\rho}j_{1\rho}j_{2\tau} + \varepsilon_{\alpha\tau\rho}j_{1\sigma}j_{2\rho}\right) \\
&= i\left(j_{1\alpha}j_{2\beta} - j_{1\beta}j_{2\alpha}\right),
\end{aligned}$$

i.e.,

$$\left[I_\alpha, I_\beta\right] = i\varepsilon_{\alpha\beta\gamma}J_\gamma. \tag{11.3.4}$$

Equations (11.1.17) and (11.1.18) are verified as follows:

$$\begin{aligned}
\left[J_\alpha, J_\beta\right] &= -\varepsilon_{\alpha\rho\sigma}\varepsilon_{\beta\mu\nu}\left[j_{1\rho}j_{2\sigma}, j_{1\mu}j_{2\nu}\right] \\
&= -\varepsilon_{\alpha\rho\sigma}\varepsilon_{\beta\mu\nu}\left([j_{1\rho}, j_{1\mu}]j_{2\sigma}j_{2\nu} + j_{1\rho}j_{1\mu}[j_{2\sigma}, j_{2\nu}]\right)
\end{aligned}$$

$$= -i\varepsilon_{\alpha\rho\sigma}\varepsilon_{\beta\mu\nu}\left(\varepsilon_{\rho\mu\tau}j_{1\tau}j_{2\sigma}j_{2\nu} + \varepsilon_{\sigma\nu\tau}j_{1\rho}j_{1\mu}j_{2\tau}\right)$$

$$= i\varepsilon_{\alpha\rho\sigma}(j_{1\beta}j_{2\sigma}j_{2\rho} - \delta_{\beta\rho}j_{2\sigma}(\boldsymbol{j}_1\cdot\boldsymbol{j}_2)$$

$$+ \delta_{\sigma\beta}j_{1\rho}\boldsymbol{j}_1\cdot\boldsymbol{j}_2 - j_{1\rho}j_{1\sigma}j_{2\beta}).$$

In the above we have used the expansion of the commutator of composite operators:

$$[AB, CD] = [A, C]\,BD + B\,[A, D]\,C + A\,[B, C]\,D + AC\,[B, D],$$

and also the commutators of the components of $\boldsymbol{j}_1, \boldsymbol{j}_2$. Further reduction needs the following relation:

$$\varepsilon_{\alpha\rho\sigma}j_\rho j_\sigma = \frac{1}{2}\varepsilon_{\alpha\rho\sigma}\left[j_\rho, j_\sigma\right] = \frac{i}{2}\varepsilon_{\alpha\rho\sigma}\varepsilon_{\rho\sigma\tau}j_\tau = ij_\alpha.$$

Denoting $(I_1)^2 + (I_2)^2 + (I_3)^2$ defined in (11.3.1) by $\boldsymbol{I}^2$, we get

$$[J_\alpha, J_\beta] = -i\varepsilon_{\alpha\beta\sigma}\,(\boldsymbol{j}_1\cdot\boldsymbol{j}_2)\,I^\sigma + i\left(j_{1\alpha}j_{2\beta} - j_{2\alpha}j_{1\beta}\right). \tag{11.3.5}$$

The definition (11.3.2) leads to $i\varepsilon_{\alpha\beta\gamma}J_\gamma = i(j_{1\alpha}j_{2\beta} - j_{2\alpha}j_{1\beta})$, and furthermore we have $\boldsymbol{j}_1\cdot\boldsymbol{j}_2 = \frac{1}{2}[\boldsymbol{I}^2 - j_1(j_1+1) - j_2(j_2+1)] \equiv \frac{1}{2}[\boldsymbol{I}^2 - \zeta]$, so we obtain

$$\left.\begin{array}{l} [J_\alpha, J_\beta] = -\varepsilon_{\alpha\beta\sigma}J_\sigma - \dfrac{i}{2}\varepsilon_{\alpha\beta\sigma}\left(\boldsymbol{I}^2 - \zeta\right)I_\sigma \\ \zeta = j_1(j_1 + 1) + j_2(j_2 + 1) \end{array}\right\}. \tag{11.3.6}$$

Expressing in terms of components $(+, -, 3)$, we have

$$\left.\begin{array}{l} [J_3, J_\pm] = \pm\left\{J_\pm - \dfrac{i}{2}\left(\boldsymbol{I}^2 - \zeta\right)I_\pm\right\} \\ [J_+, J_-] = 2J_3 - \left(\boldsymbol{I}^2 - \zeta\right)I_3 \end{array}\right\}. \tag{11.3.7}$$

Using (11.1.15) we get finally

$$\left.\begin{array}{l} [J_3\,[J_+, J_-]] = [J_3, \boldsymbol{I}^2]\,I_3 = (J_+I_- - I_+J_-)\,I_3 \\ [J_\pm, [J_3, J_\pm]] = \mp\dfrac{i}{2}\,[J_\pm, \boldsymbol{I}^2 I_\pm] = (I_3J_\pm - J_3I_\pm)\,I_\pm \end{array}\right\}. \tag{11.3.8}$$

These relations are identical with (11.1.17), (11.1.18).

Consider a system of two spins $\frac{1}{2}$, $\boldsymbol{I} = \boldsymbol{S}_1 + \boldsymbol{S}_2$, and the components of $\boldsymbol{J}$ are

$$
\left.
\begin{aligned}
J_3 &= \frac{1}{2}\left(S_1^+ S_2^- - S_1^- S_2^+\right) \\
J_\pm &= \pm\left(S_1^3 S_2^\pm - S_1^\pm S_2^3\right)
\end{aligned}
\right\}. \tag{11.3.9}
$$

If we adopt another way of normalization in defining I and $J(+,-)$ components:

$$
J_\pm = \frac{1}{\sqrt{2}}\left(J_1 \pm i J_2\right),
$$

its components (denoted now by Q_3 and $Q_\pm$) expressed in terms of S_1 and S_2 are

$$
\left.
\begin{aligned}
Q_3 &= S_1^+ S_2^- - S_1^- S_2^+ \\
Q_\pm &= \pm\frac{1}{\sqrt{2}}\left(S_1^3 S_2^\pm - S_1^\pm S_2^3\right)
\end{aligned}
\right\}. \tag{11.3.10}
$$

In the relations involving total angular momentum of the system, the operator $\boldsymbol{Q}$ has direct physical significance. Let $|\uparrow\downarrow\rangle$ be the state in which the components along the quantization axis z of spin 1 and spin 2 are $+\frac{1}{2}$ and $-\frac{1}{2}$ respectively, and so on for similar notations. The total spin of the system can be 1, the triplet (with components $1, 0, -1$), or 0, the singlet (with component 0). The wave functions of the triplet states are

$$
\left.
\begin{aligned}
\Psi_{1,1} &= |\uparrow\uparrow\rangle \\
\Psi_{1,0} &= \frac{1}{\sqrt{2}}\left(|\uparrow\downarrow\rangle + |\downarrow\uparrow\rangle\right) \\
\Psi_{1,-1} &= |\downarrow\downarrow\rangle
\end{aligned}
\right\} \tag{11.3.11}
$$

The subscripts of Ψ denote the total spin and its z-component. The wave function of the singlet state is

$$
\Psi_{0,0} = \frac{1}{\sqrt{2}}\left(|\uparrow\downarrow\rangle - |\downarrow\uparrow\rangle\right). \tag{11.3.12}
$$

S^+ and S^- are the raising and lowering operators of the z- component of the spin respectively:

$$\left.\begin{array}{ll} S^+\left|\downarrow\right\rangle = \left|\uparrow\right\rangle, & S^+\left|\uparrow\right\rangle = 0 \\ S^-\left|\uparrow\right\rangle = \left|\downarrow\right\rangle, & S^-\left|\downarrow\right\rangle = 0 \end{array}\right\}. \qquad (11.3.13)$$

Operating between the three triplet states $I^+ = S_1^+ + S_2^+$ and $I^- = S_1^- + S_2^-$ are just the raising and lowering operators of the z-component of the spin. The "transition operators" operating between the triplet and singlet states are just $Q^\pm$ and Q^3. Consider the operation of Q^+ on $\Psi_{1,-1}$:

$$Q^+ \Psi_{1,-1} = \frac{1}{\sqrt{2}} \left(S_1^3 S_2^+ - S_1^+ S_2^3 \right) \left|\downarrow\downarrow\right\rangle = \frac{1}{\sqrt{2}} \left(-\left|\downarrow\uparrow\right\rangle + \left|\uparrow\downarrow\right\rangle \right) = \Psi_{0,0}.$$

The transition between spin states induced by $I^\pm$ and $Q^\pm$ and Q^3 is shown in Fig. 11.1.

Another example is the coupling between spin $\frac{1}{2}\sigma$ and orbital angular momentum $\boldsymbol{L}$ in the form

$$\left.\begin{array}{l} \boldsymbol{j}_1 = \boldsymbol{L}, \ \ \boldsymbol{j}_2 = \dfrac{1}{2}\boldsymbol{\sigma} \\[2mm] \boldsymbol{I} = \boldsymbol{L} + \dfrac{1}{2}\boldsymbol{\sigma} \\[2mm] \boldsymbol{J} = -\mathrm{i}\boldsymbol{L} \times \dfrac{1}{2}\boldsymbol{\sigma} \end{array}\right\} \qquad (11.3.14)$$

with the commutation relations:

$$\left[\sigma_\alpha, \sigma_\beta\right] = 2\mathrm{i}\varepsilon_{\alpha\beta\gamma}\sigma_\gamma,$$
$$\left[L_\alpha, L_\beta\right] = \mathrm{i}\varepsilon_{\alpha\beta\gamma}L_\gamma,$$

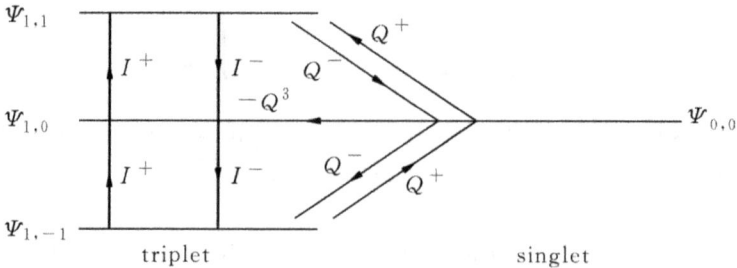

Fig. 11.1   Raising and lowering operators of the total spin of the system.

where $\alpha, \beta, \gamma = 1, 2, 3$ By using $[AB, C] = A\{B, C\} - \{A, C\}B$ ($\{B, C\} = BC + CB$) it can be proved directly that the set (11.3.14) satisfies the commutation relations for $Y(SL(2))$.

In the following we discuss the realization of Y(SL(2)) through the operator combining rotation (angular momentum) and translation (momentum), we denote $(L_1)^2 + (L_2)^2 + (L_3)^2$ by $\boldsymbol{L}^2$, then

$$\left.\begin{array}{l} \boldsymbol{I} = \boldsymbol{L}, \\ \boldsymbol{J} = \boldsymbol{L}^2 \boldsymbol{p} \end{array}\right\} \tag{11.3.15}$$

are also generators of $Y(SL(2))$. Because the commutation relations between the coordinate $\boldsymbol{r}$ and $\boldsymbol{L}$

$$\left[L_\alpha, r_\beta\right] = \mathrm{i}\varepsilon_{\alpha\beta\gamma} r_\gamma \tag{11.3.16}$$

and that between $\boldsymbol{p}$ and $\boldsymbol{L}$

$$\left[L_\alpha, p_\beta\right] = \mathrm{i}\varepsilon_{\alpha\beta\gamma} p_\gamma \tag{11.3.17}$$

are similar, therefore $\boldsymbol{I} = \boldsymbol{L}$, $\boldsymbol{J} = \boldsymbol{L}^2 \boldsymbol{r}$ also satisfy Yangian relations that can be proved in the following way.

The commutation relations between $\boldsymbol{I}$ and $\boldsymbol{J}$ components are simple:

$$\left[I_\alpha, J_\beta\right] = \left[L_\alpha, \boldsymbol{L}^2 p_\beta\right] = \mathrm{i}\boldsymbol{L}^2 \varepsilon_{\alpha\beta\gamma} p_\gamma = \mathrm{i}\varepsilon_{\alpha\beta\gamma} J_\gamma.$$

In calculating $[I_\alpha, J_\beta]$ we need $[\boldsymbol{L}^2, p_\beta]$, which is directly evaluated:

$$\begin{aligned} \left[\boldsymbol{L}^2, p_\beta\right] &= L_\alpha \left[L_\alpha, p_\beta\right] + \left[L_\alpha, p_\beta\right] L_\alpha \\ &= \mathrm{i}\varepsilon_{\alpha\beta\sigma} \left(L_\alpha p_\sigma + p_\sigma L_\alpha\right) \\ &= -2\mathrm{i}\varepsilon_{\beta\alpha\sigma} L_\alpha p_\sigma - 2p_\beta. \end{aligned} \tag{11.3.18}$$

The commutation relations between $\boldsymbol{J}$ components are calculated as follows:

$$\begin{aligned} \left[J_\alpha, J_\beta\right] &= \left[\boldsymbol{L}^2 p_\alpha, \boldsymbol{L}^2 p_\beta\right] = \boldsymbol{L}^2 \left[\boldsymbol{L}^2, p_\beta\right] p_\alpha + \boldsymbol{L}^2 \left[p_\alpha, \boldsymbol{L}^2\right] p_\beta \\ &= 2\mathrm{i}\boldsymbol{L}^2 \left(\varepsilon_{\alpha\mu\nu} L_\mu p_\nu p_\beta - 2\mathrm{i}p_\alpha p_\beta - \alpha \leftrightarrow \beta\right). \end{aligned}$$

Here $\alpha \leftrightarrow \beta$ means the anti-symmetrization of the indices α, β. Since p_α commutes with p_β, the anti-symmetrization of the indices of $p_\alpha p_\beta$

gives zero. Therefore

$$[J_\alpha, J_\beta] = 2\mathrm{i}L^2\left(\varepsilon_{\alpha\mu\nu}L_\mu p_\nu p_\beta - \alpha \leftrightarrow \beta\right).$$

In the above relation α and β are fixed indices, μ and ν are summation indices. We are concerned only with the case $\alpha \neq \beta$. For the indices of $\varepsilon_{\alpha\mu\nu}$ being all different (otherwise it is zero), we have either $\mu = \beta$ or $\nu = \beta$. Therefore

$$[J_\alpha, J_\beta] = 2\mathrm{i}L^2\left(\varepsilon_{\alpha\beta\nu}L_\beta p_\nu p_\beta + \varepsilon_{\alpha\mu\beta}L_\mu p_\beta p_\beta - \alpha \leftrightarrow \beta\right).$$

Using the antisymmetric property of ε, then we get finally

$$[J_\alpha, J_\beta] = 2\mathrm{i}L^2\varepsilon_{\alpha\beta\nu}\left[(L \cdot p)p_\nu - L_\nu p^2\right]. \qquad (11.3.19)$$

In the further calculations the following relations are useful.

$$\left.\begin{array}{l} [L \cdot p, L_\alpha] = 0, [L \cdot p, p_\alpha] = 0 \\ [p^2, L_\alpha] = 0 \\ [L \cdot p, J_\alpha] = [L \cdot p, L^2 p_\alpha] = 0 \end{array}\right\} \qquad (11.3.20)$$

We obtain from (11.3.19)

$$[J_+, J_-] = -2\mathrm{i}\,[J_1, J_2] = 4L^2\left[(L \cdot p)p_3 - L_3 p^2\right],$$

and the subsequent calculations are direct.

$$\begin{aligned} [J_3, [J_+, J_-]] &= -4\left[L^2 p_3, L^2 p^2 L_3\right] = -4L^2\left[p_3, L^2\right]p^2 L_3 \\ &= -8L^2 p^2 L_3 \left(\mathrm{i}\left(L_1 p_2 - L_2 p_1\right) + p_3\right). \end{aligned}$$

From the other hand we have

$$\begin{aligned} I_3\left(J_- I_+ - I_- J_+\right) &= I_3\left(I_+ J_- - I_- J_+ - 2J_3\right) \\ &= -2L_3 L^2 \left(\mathrm{i}\left(L_1 p_2 - L_2 p_1\right) + p_3\right). \end{aligned}$$

Therefore, we get

$$[J_3, [J_+, J_-]] = 4p^2 I_3\left(J_- I_+ - I_- J_+\right). \qquad (11.3.21)$$

Similarly we obtain

$$\begin{aligned} [J_\pm, [J_3, J_\pm]] &= 4p^2 L^2 L_\pm\left(-L_\pm p_3 + L_3 p_\pm \mp p_\pm\right) \\ &= 4p^2 I_\pm\left(J_\pm I_3 - I_\pm J_3\right). \end{aligned} \qquad (11.3.22)$$

Since p^2 is a Casimir operator of the algebra, the factor $4p^2$ can be eliminated by redefining J. Thus the algebraic relations of the Yangian $SL(2)$ have been verified.

In this example there are two Casimirs p^2 and $L \cdot p$ that commute with not only L, but also J defined in (11.3.15). The physical meaning of $L \cdot p$ is the monopole charge that can be understood in a simple way. Under electro-magnetic field in order that the same commutation relations of angular momentum still remain valid, we should replace $L = r \times p$ by

$$L' = r \times (p - A) - g\frac{r}{r}$$

where $g = \frac{1}{e}(c = \hbar = 1)$ and A is Wu-Yang monopole potential [6]:

$$(A_r)_a = (A_\theta)_a = 0, \quad (A_\varphi)_a = \frac{g}{r \sin \theta}(1 - \cos \theta)$$

$$(A_r)_b = (A_\theta)_b = 0, \quad (A_\varphi)_b = \frac{-g}{r \sin \theta}(1 + \cos \theta)$$

The regions a and b include the equator with the north pole and south pole respectively. It is easy to verify that the new angular momentum L' satisfies the same commutation relations as that for L. Obviously, we have

$$L' \cdot r = -g$$

If we understand the L in (11.3.15) to be L' then (11.3.20) indicates that monopole is also a Yangian invariant.

Before ending the section we would like to emphasize that for systems involving more spins we can define

$$I = \sum_{i=1}^{N} I_i, \quad \left[I_{i\alpha}, I_{j\beta}\right] = i\varepsilon_{\alpha\beta\gamma} I_{ir} \delta_{ij}$$

and

$$J = \sum_{i=1}^{N} \mu_i I_i - i \sum_{i<j} I_i \times I_j$$

where μ_i are N free parameters. Direct calculation can verify that the set I and J satisfies the $Y(SL(2))$ commutation relations [7].

11.4 The One-dimensional Chain Model of Long-range Interaction [8–10]

Consider N spin $\frac{1}{2}$ operators situated at the N sites of a one-dimensional lattice, with spin operators on different sites commute with each other:

$$[S_{i\alpha}, S_{j\beta}] = 0, \quad i \neq j. \tag{11.4.1}$$

The condition of spin $\frac{1}{2}$ gives

$$\left(S_i^+\right)^2 = \left(S_i^-\right)^2 = 0. \tag{11.4.2}$$

Operators $\boldsymbol{I}$ and $\boldsymbol{J}$ are constructed as follows:

$$\left.\begin{array}{l} I_\alpha = \displaystyle\sum_{i=1}^{N} S_{i\alpha}, \quad \alpha = +, -, 3 \\[2mm] J_\pm = \mp \displaystyle\sum_{i\neq j}^{N} w_{ij} S_i^\pm S_j^3 = \pm \displaystyle\sum_{i\neq j}^{N} w_{ij} S_i^3 S_j^\pm \\[2mm] J_3 = \dfrac{1}{2} \displaystyle\sum_{i\neq j}^{N} w_{ij} S_i^+ S_j^- \end{array}\right\} \tag{11.4.3}$$

Here w_{ij} satisfies the relations

$$\left.\begin{array}{l} w_{ij} = -w_{ji} \\ \Delta_{ijk} = w_{ij}w_{jk} + w_{jk}w_{ki} + w_{ki}w_{ij} = -1 \end{array}\right\} \tag{11.4.4}$$

This model was first introduced by Haldane and Shastry [8, 9]. The expression for $J_\pm$ is consistent with the antisymmetric property of w_{ij} and that operators on different sites commute. The long-ranged interaction is incorporated in w_{ij}. The simplest solution satisfying (11.4.4) is

$$w_{ij} = \begin{cases} 1, & i > j, \quad i, j = 1, 2, \dots, N \\ 0, & i = j, \quad i, j = 1, 2, \dots, N \\ -1, & i < j, \quad i, j = 1, 2, \dots, N \end{cases} \tag{11.4.5}$$

In order to verify that $\boldsymbol{I}$ and $\boldsymbol{J}$ defined in the way above satisfy Yangian relations, we need to calculate

$$[J_3, J_+] = \frac{1}{2} \sum_{i,j,k,l} w_{ij} w_{kl} \left[S_i^+ S_j^-, S_k^+ S_l^3 \right]$$

$$= -\frac{1}{2} \sum_{i,j,k,l} w_{ij} w_{kl} \left(S_i^+ S_k^+ S_j^- \delta_{lj} - 2 S_i^+ S_j^3 S_l^3 \delta_{jk} - S_k^+ S_i^+ S_j^- \delta_{il} \right)$$

$$= \frac{1}{2} \sum_{i,j,k,l} w_{ij} \left(2 w_{jk} S_i^+ S_j^3 S_k^3 + (w_{jk} + w_{ki}) S_i^+ S_k^+ S_j^- \right).$$

$$(11.4.6)$$

Further, we calculate $[J_+, [J_3, J_+]]$ using (11.4.6):

$$[J_+, [J_3, J_+]] = \frac{1}{2} \sum_{i,j,k,l,m} w_{ij} w_{lm}$$

$$\times \left[S_l^+ S_m^3, 2 w_{jk} S_i^+ S_j^3 S_k^3 + (w_{jk} + w_{ki}) S_i^+ S_k^+ S_j^- \right]$$

$$= -\frac{1}{2} \sum_{i,j,k,l,m} w_{ij} w_{lm} \left\{ S_l^+ \left[2 w_{jk} S_i^+ S_j^3 S_k^3 \delta_{mi} + (w_{jk} + w_{ki}) \right. \right.$$

$$\times (\delta_{im} + \delta_{km} - \delta_{jm}) S_i^+ S_k^+ S_j^- \Big] k$$

$$\times \left[-2 w_{jk} \left(\delta_{jl} S_i^+ S_j^+ S_k^3 + \delta_{kl} S_i^+ S_j^3 S_k^+ \right) \right.$$

$$\left. +2 (w_{jk} + w_{ji}) \delta_{jl} S_i^+ S_k^+ S_j^3 \right] S_m^3 \Big\}$$

$$= -\frac{1}{2} \sum_{i,j,k,l,m} w_{ij} \left\{ 2 w_{jk} w_{li} S_i^+ S_l^+ S_j^3 S_k^3 \right.$$

$$+ (w_{jk} + w_{ki}) (w_{li} + w_{lk} - w_{lj})$$

$$\times S_i^+ S_k^+ S_l^+ S_j^- - 2 w_{jk} w_{jl} S_i^+ S_j^+ S_k^3 S_l^3$$

$$- 2 w_{jk} w_{kl} S_i^+ S_j^3 S_k^+ S_l^3$$

$$\left. +2 (w_{jk} + w_{ki}) w_{jl} S_i^+ S_k^+ S_j^3 S_l^3 \right\}$$

$$= -\frac{1}{2} \sum_{i,j,k,l,m} \left\{ (2w_{il}w_{lk}w_{ji} - 2w_{ij}w_{jk}w_{jl} - 2w_{ik}w_{kj}w_{jl} \right.$$

$$+ 2w_{ik}(w_{kj} + w_{ji})w_{kl})S_i^+ S_j^+ S_k^3 S_l^3$$

$$\left. + w_{ij}(w_{jk} + w_{ki})(w_{li} + w_{lk} - w_{lj})S_i^+ S_k^+ S_l^+ S_j^- \right\}.$$

Noticing that under the summation sign as the coefficient of $S_j^3 S_k^3$ the indices j and k can be interchanged,

$$w_{ik}w_{ji}w_{kl} \overset{k \leftrightarrow l}{=} w_{il}w_{ji}w_{lk},$$

we obtain

$$[J_+, [J_3, J_+]] = -\frac{1}{2} \sum_{i,j,k,l} \left\{ 2(2w_{il}w_{ji}w_{lk} - w_{ij}w_{jk}w_{jl} \right.$$

$$- w_{ik}w_{jl}w_{kj} + w_{ik}w_{kj}w_{kl})S_i^+ S_j^+ S_k^3 S_l^3$$

$$\left. + w_{ij}(w_{jk} + w_{ki})(w_{li} + w_{lk} - w_{lj})S_i^+ S_k^+ S_j^- \right\}. \tag{11.4.7}$$

In the coefficient of the first term in the right-hand side we notice that as the coefficients of the symmetrical product of operators $S_i^+ S_j^+$ and $S_k^3 S_l^3$ the indices can be interchanged:

$$w_{ji}w_{kl}w_{li} \underset{\longrightarrow}{k \leftrightarrow l} w_{ji}w_{ik}w_{kl},$$

$$-w_{ij}w_{jk}w_{jl} \underset{\longrightarrow}{i \leftrightarrow j} w_{ji}w_{li}w_{ik}.$$

Consequently in (11.4.7) the following relation is valid:

$$2w_{il}w_{ji}w_{lk} - w_{ij}w_{jk}w_{jl} = w_{ji}(w_{ik}w_{kl} + w_{kl}w_{li} + w_{li}w_{ik}) = w_{ji}\Delta_{ikl}.$$

As the coefficient of $S_i^+ S_j^+ S_k^3 S_l^3$ the pairs of indices (i,j) and (k,l) can be separately interchanged, and we have

$$w_{ik}w_{lj}w_{kl} \underset{\longrightarrow}{i \leftrightarrow j, k \leftrightarrow l} -w_{ik}w_{lj}w_{kl},$$

consequently

$$\sum_{i,j,k,l} w_{ik}w_{lj}w_{kl}S_i^+ S_j^+ S_k^3 S_l^3 = 0.$$

This leads to

$$w_{ik}w_{jl}w_{kj} - w_{ik}w_{kj}w_{kl} = w_{ik}(w_{jk}w_{kl} + w_{kl}w_{lj} + w_{lj}w_{jk})$$
$$- w_{ik}w_{lj}w_{kl}$$
$$= w_{ik}\Delta_{jkl} - w_{ik}w_{lj}w_{kl},$$

where Δ is defined by (11.4.4).

In the coefficient of the second term in (11.4.7) we have

$$w_{ij}(w_{jk} + w_{ki})(w_{li} + w_{lk} - w_{lj})$$
$$= w_{lk}(w_{ij}w_{jk} + w_{jk}w_{ki} + w_{ki}w_{ij}) - w_{lk}w_{jk}w_{ki}$$
$$+ w_{ij}(w_{jk} + w_{ki})w_{li} - w_{lj}w_{ij}(w_{jk} + w_{ki})$$
$$= w_{lk}\Delta_{ijk} - w_{li}w_{ji}w_{jk} + w_{ij}w_{jk}w_{li} + w_{lj}w_{ij}w_{ki}$$
$$- w_{lj}w_{ij}w_{jk} - w_{lj}w_{ij}w_{ki}.$$

Carrying out cyclic permutation of the indices i, j, k, the above expression becomes

$$w_{lk}\Delta_{ijk} + w_{jk}w_{li}w_{ij} + w_{jk}w_{ij}w_{jl} + w_{jk}w_{jl}w_{li}$$
$$= w_{lk}\Delta_{ijk} + w_{jk}\Delta_{ijl}.$$

Substituting the two coefficients calculated above into (11.4.7), we obtain

$$[J_+, [J_3, J_+]] = -\frac{1}{2}\sum_{i,j,k,l}\left\{2(w_{ji}\Delta_{ikl} - w_{ik}\Delta_{jkl})S_i^+ S_j^+ S_k^3 S_l^3\right.$$
$$\left. + (w_{lk}\Delta_{ijk} + w_{jk}\Delta_{ijl})S_i^+ S_k^+ S_l^+ S_j^-\right\},$$

On the other hand, we calculate from (11.2.12)

$$I_+(I_3 J_+ - J_3 I_+) = I_+(J_+ I_3 - I_+ J_3)$$
$$= -\sum_{i,j,k,l}\left\{w_{jk}S_i^+ S_j^+ S_k^3 S_l^3 + \frac{1}{2}w_{kl}S_i^+ S_j^+ S_k^+ S_l^-\right\}$$
$$= -\sum_{i,j,k,l}w_{jk}\left\{S_i^+ S_j^+ S_k^3 S_l^3 - \frac{1}{2}S_i^+ S_k^+ S_l^+ S_j^-\right\}.$$

$$(11.4.8)$$

If we demand (11.1.17) be valid, i.e., we demand that $[J_+, [J_3, J_+]] = I_+(I_3 J_+ - J_3 I_+)$, the right-hand sides of Eqs. (11.4.7) and (11.4.8) should be identical, i.e.,

$$\frac{1}{2} \sum_{i,j,k,l} \left\{ 2(w_{ji} \Delta_{ikl} - w_{ik} \Delta_{jkl}) S_i^+ S_j^+ S_k^3 S_l^3 \right.$$

$$\left. + (w_{lk} \Delta_{ijk} + w_{jk} \Delta_{ijl}) S_i^+ S_k^+ S_l^+ S_j^- \right\}$$

$$= \sum_{i,j,k,l} w_{jk} \left\{ S_i^+ S_j^+ S_k^3 S_l^3 - \frac{1}{2} S_i^+ S_k^+ S_l^+ S_j^- \right\}, \qquad (11.4.9)$$

which we prove in the following.

When $i \neq j \neq k \neq l$, we take $\Delta_{ijk} = -1$, then we have $\Delta_{ikl} = \Delta_{jkl} = -1$, it follows that

$$\sum_{\substack{i,j,k,l \\ i \neq j \neq k \neq l}} (w_{ik} - w_{ji} - w_{jk}) S_i^+ S_j^+ S_k^3 S_l^3 = 0.$$

Hence (11.1.17) is valid under the condition $\Delta_{ijk} = -1$. Similarly, when $i \neq j \neq k$, the left-hand side of (11.4.8) gives

$$\sum_{\substack{i,j,k,l \\ i \neq j \neq k \neq l}} (w_{lk} - w_{jk} - w_{jk}) S_i^+ S_k^+ S_l^+ S_j^- = 0.$$

$$\frac{1}{2} \sum_{\substack{i,j,l \\ i \neq j \neq l}} \left\{ 2 \left(w_{ji} \Delta_{iil} S_i^+ S_i^3 S_j^+ S_l^3 + (w_{ji} \Delta_{ijl} - w_{ij} \Delta_{jjl}) S_i^+ S_j^+ S_j^3 S_l^3 \right) \right.$$

$$\left. + w_{lj} \Delta_{ijj} S_i^+ S_j^+ S_j^- S_l^+ \right\}$$

$$+ \sum_{\substack{i,j,l \\ i \neq j \neq k}} \left\{ 2 \left(w_{ji} \Delta_{iki} - w_{jk} \Delta_{ijk} \right) S_i^+ S_i^3 S_j^+ S_k^3 \right.$$

$$+ 2 \left(w_{ji} \Delta_{ikj} - w_{ik} \Delta_{jkj} \right) S_i^+ S_j^+ S_j^3 S_k^3$$

$$+ 2 \left(w_{ji} \Delta_{ikk} - w_{jk} \Delta_{jkk} \right) S_i^+ S_j^+ \left(S_k^3 \right)^2$$

$$\left. + (w_{jk} \Delta_{ijk} + w_{jk} \Delta_{ijj}) S_i^+ S_k^+ S_l^+ S_j^- \right\}$$

$$= \frac{1}{2} \sum_{\substack{i,j,l \\ i \neq j \neq k}} \left\{ 2(w_{ji} \Delta_{iik} + w_{ij} \Delta_{ijk} - w_{ji} \Delta_{iik} + w_{ji} \Delta_{iki} \right.$$

$$- \Delta_{jk}\Delta_{ijk} + w_{ij}\Delta_{ijk} - w_{jk}\Delta_{iik})S_i^+ S_i^3 S_j^+ S_k^3$$

$$+ 2\left(w_{ji}\Delta_{ikk} - w_{ik}\Delta_{ikk}\right)S_i^+ S_j^+ \left(S_k^3\right)^2$$

$$+ \left(w_{kj}\Delta_{ijj} + w_{jk}\Delta_{ijk} + w_{jk}\Delta_{ijj}\right)S_i^+ S_k^+ S_j^+ S_j^-\Big\}$$

$$= \frac{1}{2}\sum_{\substack{i,j,l \\ i\neq j\neq k}} \Big\{ 2(2w_{ij}\Delta_{ijk} - w_{jk}\Delta_{ijk} - w_{jk}(w_{ij}w_{jk} + w_{jk}w_{ki}$$

$$+ w_{ki}w_{ij}) - w_{ik}w_{ij}w_{jk})S_i^+ S_i^3 S_j^+ S_k^3$$

$$- 2w_{ik}\left(w_{ij}w_{jk} + w_{jk}w_{ki} + w_{ki}w_{ij}\right)S_i^+ S_j^+ \left(S_k^3\right)^2$$

$$+ w_{jk}\Delta_{ijk}S_i^+ S_k^+ S_j^+ S_j^-\Big\}$$

$$= \frac{1}{2}\sum_{\substack{i,j,l \\ i\neq j\neq k}} \Big\{ 2\left((2w_{ij} - w_{jk} - w_{ik})\Delta_{ijk} - w_{ij}w_{ik}w_{jk}\right)S_i^+ S_i^3 S_j^+ S_k^3$$

$$- w_{ij}\Delta_{ijk}S_i^+ S_j^+ \left(S_k^3\right)^2 + w_{ij}\Delta_{ijk}S_i^+ S_k^+ S_j^+ S_j^- \Big\}$$

$$= \frac{1}{2}\sum_{\substack{i,j,l \\ i\neq j\neq k}} \Big\{ (w_{jk} + w_{ik})\Delta_{ijk}S_i^+ S_j^+ S_k^3$$

$$- 2w_{ik}\Delta_{ijk}S_i^+ S_j^+ \left(S_k^3\right)^2 + w_{jk}\Delta_{ijk}S_i^+ S_k^+ S_j^+ S_j^- \Big\}. \quad (11.4.10)$$

We have used the following equalities for spin $\frac{1}{2}$:

$$S_i^+ S_i^3 = -\frac{1}{2}S_i^+,$$

$$\sum_{\substack{i,j,l \\ i\neq j\neq k}} w_{ij}w_{ik}w_{jk}S_i^+ S_j^+ S_k^3 = 0.$$

Now consider the right-hand side of (11.4.9), i.e.,

$$\sum_{\substack{i,j,l \\ i\neq j\neq k,l}} w_{jk}\left\{ S_i^+ S_j^+ S_k^3 S_l^3 - \frac{1}{2}S_i^+ S_k^+ S_l^+ S_j^- \right\}$$

$$= \sum_{\substack{i,j,l \\ i\neq j\neq k}} (w_{ji}S_i^+ S_j^+ S_i^3 S_k^3 + w_{jk}S_i^+ S_i^3 S_j^+ S_k^3 + w_{jk}S_i^+ S_j^+ S_j^3 S_k^3$$

$$+ w_{jk}S_i^+ S_j^+ \left(S_k^3\right)^2 - \frac{1}{2}w_{jk}S_i^+ S_k^+ S_j^+ S_j^-)$$

$$= \sum_{\substack{i,j,l \\ i \neq j \neq k}} \left\{ (w_{ji} + w_{jk} + w_{ik}) S_i^+ S_i^3 S_j^+ S_k^3 \right.$$

$$\left. + 2w_{jk} S_i^+ S_j^+ \left(S_k^3\right)^2 - \frac{1}{2} w_{jk} S_i^+ S_k^+ S_j^+ S_j^- \right\}$$

$$= -\frac{1}{2} \sum_{\substack{i,j,l \\ i \neq j \neq k}} \left\{ (w_{jk} + w_{ik}) S_i^+ S_j^+ S_k^3 - 2w_{ik} S_i^+ S_j^+ \left(S_k^3\right)^2 \right.$$

$$\left. + w_{jk} S_i^+ S_k^+ S_j^+ S_j^- \right\}. \tag{11.4.11}$$

Comparing (11.4.10) and (11.4.11), i.e., both sides of (11.4.9), we find that the condition of their equality under $i \neq j \neq k$ is

$$\Delta_{ijk} = 1.$$

Then we have proved that (11.4.3) satisfies the $Y(SL(2))$ commutation relations under all possible summations over the indices.

It is easy to verify that the parametrized solution for w_{jk} satisfying (11.4.4) is [10]

$$w_{jk} = \frac{z_j + z_k}{z_{jk}}, \quad z_{jk} = z_j - z_k. \tag{11.4.12}$$

Substituting (11.4.12) into (11.4.4) we have

$$\Delta_{ijk} = \frac{1}{z_{ij} z_{jk} z_{ki}} \{ (z_i + z_j)(z_j + z_k) z_{ki} + (z_i + z_k)(z_j + z_k) z_{ij}$$

$$+ (z_i + z_k)(z_i + z_j) z_{jk} \}$$

$$= \frac{1}{z_{ij} z_{jk} z_{ki}} \left(z_j^2 z_{ki} + z_k^2 z_{ij} + z_i^2 z_{jk} \right) = -1.$$

When we choose

$$z_k = \exp\left(ik \frac{2\pi}{N} \right) \tag{11.4.13}$$

then

$$w_{jk} = i \cot(j - k) \frac{\pi}{N}, \tag{11.4.14}$$

and the corresponding J_α is

$$J_\alpha = -\mathrm{i} \sum_{\substack{j,k \\ j \neq k}}^{N} \cot \frac{(j-k)\,\pi}{N} \varepsilon_{\alpha\beta\gamma} S_{j\beta} S_{k\gamma}. \tag{11.4.15}$$

It corresponds to a model of long-range interaction [8–10]. We see this in observing that in the construction of $\boldsymbol{J}$ the coefficient $\cot \frac{(j-k)\pi}{N}$ of a pair of different lattice sites j and k depends on the relative distance between these sites, and the interaction endures between distant sites. The physical meaning of $\boldsymbol{J}$ in H-S model has been explored in Refs. [8–13] we note that $\boldsymbol{J}$ is not only regarded as the symmetry, but also the persistent spin currents for the model as pointed out in Ref. [12].

11.5 The Hubbard Model [13]

Let a_j and b_j be the fermion operators annihilating the spin $\frac{1}{2}$ and spin $-\frac{1}{2}$ respectively on the site j, and $a_j^\dagger$ and $b_j^\dagger$ be the corresponding creation operators. They satisfy the following anti-commutation relations.

$$\left.\begin{aligned} \{a_j, a_k^\dagger\} &= \delta_{jk} \\ \{b_j, b_k^\dagger\} &= \delta_{jk} \\ \{a_j, b_k\} &= \{a_j, b_k^\dagger\} = 0 \end{aligned}\right\}. \tag{11.5.1}$$

The spin on lattice sites and the total spin of the system can be expressed in terms of the operators:

$$\left.\begin{aligned} S^+ &= \sum_{j=1}^{+} a_j^\dagger b_j \\ S^- &= \sum_{j=1}^{+} b_j^\dagger a_j \\ S^3 &= \frac{1}{2} \sum_{j=1}^{+} (a_j^\dagger a_j - b_j^\dagger b_j) \end{aligned}\right\}. \tag{11.5.2}$$

They satisfy the following commutation relations

$$\left[S^3, S^\pm\right] = \pm S^\pm, \quad \left[S^+, S^-\right] = 2S^3. \tag{11.5.3}$$

These relations can be easily proved by using the anti-commutation relations (11.5.1) for the operators a and b, and the construction (11.5.2) of the operator S. Again we define I_α to be S_α,

$$I_\alpha = S_\alpha, \tag{11.5.4}$$

and define the construction of J_α as

$$\left.\begin{aligned}
J_+ &= \sum_{i,j} \theta_{ij} a_i^\dagger b_j - U \sum_{\substack{i,j \\ i \neq j}} \varepsilon_{ij} I_i^+ I_j^3 \\
J_- &= \sum_{i,j} \theta_{ij} b_i^\dagger a_j + U \sum_{\substack{i,j \\ i \neq j}} \varepsilon_{ij} I_i^- I_j^3 \\
J_3 &= \frac{1}{2} \sum_{i,j} \theta_{ij} \left(a_i^\dagger a_j - b_i^\dagger b_j\right) + U \sum_{i,j} \varepsilon_{ij} I_i^+ I_j^-.
\end{aligned}\right\}, \tag{11.5.5}$$

where

$$\begin{aligned}
\theta_{ij} &= \delta_{i,j-1} - \delta_{i,j+1} \\
\varepsilon_{ij} &= \begin{cases} 1, & i > j \\ 0, & i = j \\ -1, & i < j \end{cases}
\end{aligned} \tag{11.5.6}$$

The terms involving coupling constant U on the right-hand side of (11.5.5) are the components of $\sum_{i \neq j} \boldsymbol{S}_i \times \boldsymbol{S}_j$, and terms involving θ_{ij} are the hopping terms moving the spin from a lattice site to a neighboring site. The hopping terms can only be expressed in terms of operators a and b, but not in terms of the total spin $\boldsymbol{S}$. The proof that $\boldsymbol{I}$ and $\boldsymbol{J}$ satisfy the Yangian is rather involved, and we present here only the main points. For details the original literature is recommended [13].

The commutation relations between I and J components can be proved by using the anti-commutation relations between a and b and the following formula

$$[AB, C] = A\,[B, C] + [A, C]\,B = A\,\{B, C\} - \{A, C\}\,B.$$

The commutator $[J_3, J_\pm]$ is calculated directly to give

$$[J_3, J_\pm] = A_\pm + U B_\pm + U^2 C_\pm, \qquad (11.5.7)$$

where

$$A_+ = \sum_{i,k,l} \theta_{ij}\theta_{jk} a_i^+ b_k,$$

$$B_+ = -\frac{1}{2} \sum_{i,k,l} \theta_{ij} \{ (\varepsilon_{jk} + \varepsilon_{ik})\, a_i^\dagger b_j I_k^3 + \frac{1}{2}\, (3\varepsilon_{kj} - \varepsilon_{ki})\, I_k^+ a_i^\dagger a_j$$

$$+ \frac{1}{2}\, (\varepsilon_{kj} - 3\varepsilon_{ki})\, I_k^+ b_i^\dagger b_j \}.$$

$$C_+ = \frac{1}{2} \sum_{i,j,k} \varepsilon_{ij} \left\{ 2\varepsilon_{jk} I_i^+ I_j^3 I_k^3 - (\varepsilon_{ki} - \varepsilon_{kj})\, I_i^+ I_k^+ I_j^- \right\},$$

$$A_- = -\,(A_+)^+,$$

$$B_- = -\,(B_+)^+,$$

$$C_- = -\frac{1}{2} \sum_{i,j,k} \varepsilon_{ij} \left\{ 2\varepsilon_{jk} I_i^- I_j^3 I_k^3 - (\varepsilon_{ki} - \varepsilon_{kj})\, I_i^- I_k^- I_j^+ \right\}.$$

Consider the Jacobi identity

$$[[J_3, J_-], I_+] + [[J_-, I_+], J_3] + [[I_+, J_3], J_-] = 0.$$

We find that the second term vanishes because $[J_-, I_+] = -2J_3$, and we transform the third term by using $[I_+, J_3] = -J_+$. The identity becomes

$$[J_+, J_-] = [[J_3, J_-], I_+].$$

Direct calculation of $[J_+, J_-]$ gives

$$[J_+, J_-] = A_3 + U B_3 + U^2 C_3, \qquad (11.5.8)$$

where

$$A_3 = \sum_{i,j,k} \theta_{ij}\theta_{jk}\left(a_i^\dagger a_k - b_i^\dagger b_k\right),$$

$$B_3 = -\frac{1}{2}\sum_{i,j,k}\theta_{ij}\left\{(\varepsilon_{jk}+\varepsilon_{ik})\left(b_i^+ a_j I_k^+ - a_i^\dagger b_j I_k^-\right)\right.$$

$$\left.+2(\varepsilon_{ij}-\varepsilon_{jk})\left(a_i^\dagger a_j - b_i^\dagger b_j\right)I_k^3\right\},$$

$$C_3 = -\frac{1}{2}\sum_{i,j,k}\varepsilon_{ij}\left\{2\varepsilon_{jk}I_i^3 I_j^3 I_k^3 + (\varepsilon_{ki}-\varepsilon_{kj})I_i^- I_j^+ I_k^3\right\}.$$

On the basis of (11.5.7) we can get

$$[J_+,[J_3,J_+]] = \mathscr{A} + U\mathscr{B} + U^2\mathscr{C} + U^3\mathscr{D}, \qquad (11.5.9)$$

where

$$\mathscr{A} = 0, \quad \mathscr{B} = 0,$$

$$\mathscr{C} = -\frac{1}{2}\sum_{i,j,k,l}\left\{(\bar{\Delta}_{ijk}+\bar{\Delta}_{ikl}+\varepsilon_{kl}\varepsilon_{li}+\varepsilon_{ij}\varepsilon_{jk}+\varepsilon_{ij}\varepsilon_{kl}-\varepsilon_{jk}\varepsilon_{il})\right.$$

$$\times I_i^+ a_j^\dagger b_l I_k^3 + I_i^+ I_k^+\left[\left(\varepsilon_{il}(\varepsilon_{ik}+\varepsilon_{kl})+\frac{1}{2}(\varepsilon_{jk}-3\varepsilon_{lk})\right)\right.$$

$$\times\left(\varepsilon_{ki}+\frac{1}{2}(\varepsilon_{il}-\varepsilon_{lj})\right)a_j^\dagger a_l + \left(\varepsilon_{ij}(\varepsilon_{ki}-\varepsilon_{kj})+\frac{1}{2}(3\varepsilon_{jk}-\varepsilon_{lk})\right)$$

$$\left.\left.\times\left(\varepsilon_{ki}-\frac{1}{2}(\varepsilon_{il}-\varepsilon_{ij})\right)b_j^\dagger b_l\right]\right\}$$

$$-\frac{1}{4}\sum_{i,j,k,l}\theta_{jk}(\varepsilon_{ij}\varepsilon_{jk}+\varepsilon_{jk}\varepsilon_{ki})I_i^+ a_j^\dagger b_k,$$

$$\mathscr{D} = \frac{1}{2}\sum_{i,j,k,l}\left\{2(\varepsilon_{jl}-\varepsilon_{kl})\bar{\Delta}_{ijk}I_i^+ I_j^+ I_k^3 I_l^3 + (\varepsilon_{jl}-\varepsilon_{jk})\bar{\Delta}_{ikl}I_i^+ I_j^+ I_k^+ I_l^+\right\}.$$

In the above the notation $\bar{\Delta}$ is defined as

$$\bar{\Delta}_{ijk} = \varepsilon_{ij}\varepsilon_{jk} + \varepsilon_{jk}\varepsilon_{ki} + \varepsilon_{ki}\varepsilon_{ij}.$$

The expression to be compared with (11.5.9) is

$$I_+(I_3 J_+ - J_3 I_+) = D_+ + U E_+, \qquad (11.5.10)$$

and the result of direct calculation is

$$D_+ = \sum_{i,j,k,l} \theta_{jl} \left\{ I_i^+ a_j^\dagger b_l I_k^3 + \frac{1}{2} I_i^+ I_k^+ \left(b_j^\dagger b_l - a_j^\dagger a_l \right) \right\},$$

$$E_+ = -\sum_{i,j,k,l} \varepsilon_{jl} \left\{ I_i^+ I_j^+ I_l^3 I_k^3 + \frac{1}{2} I_i^+ I_k^+ I_j^+ I_l^- \right\}.$$

The comparison gives

$$D_+ = \mathscr{C}, \quad E_+ = \mathscr{D},$$

which leads to

$$[J_+, [J_3, J_+]] = U^2 I_+ (I_3 J_- - J_3 I_+). \qquad (11.5.11)$$

Other relations can be obtained in the similar way.

The Hamiltonian of the Hubbard model is

$$H = -\sum_{i=1}^{N} \left(a_i^\dagger a_{i+1} + a_{i+1}^\dagger a_i + b_i^\dagger b_{i+1} + b_{i+1}^\dagger b_i \right)$$

$$- U \sum_{i=1}^{N} \left(a_i^\dagger a_i - \frac{1}{2} \right) \left(b_i^\dagger b_i - \frac{1}{2} \right). \qquad (11.5.12)$$

In the following, we prove that the Yangian generators (11.5.4) and (11.5.5) commute with H when N approaches infinity, which means that the one-dimensional infinite Hubbard chain possesses the

$Y(SL(2))$ symmetry [12]. We rewrite Eq. (11.5.12) as

$$H = -\sum_i \tau_{ij} \left(a_i^\dagger a_j + b_i^\dagger b_j \right) - U \left(\sum_i \left(a_i^\dagger a_i b_i^\dagger b_i - \frac{1}{4} \right) - \frac{N}{2} \right),$$

$$(11.5.13)$$

where

$$\tau_{ij} = \delta_{i+1,j} - \delta_{i-1,j}.$$

We obtain by the direct calculation

$$\left[\sum_i \tau_{ij} \left(a_i^\dagger a_j + b_i^\dagger b_j \right), I_+ \right] = \sum_i \tau_{ij} \left(a_i^\dagger b_j - a_i^\dagger b_j \right) = 0,$$

$$\left[\sum_i a_i^\dagger a_i b_i^\dagger b_i, I_+ \right] = \sum_i \left(a_i^\dagger b_i - a_i^\dagger b_i \right) = 0$$

and therefore we have $[H, I_+] = 0$. From $[H, I_-] = -[H, I_+]^\dagger$ and the Jacobi identity we can prove that $[H, I_-] = 0$ and $[H, I_3] = 0$.

Among the relation $[H, \boldsymbol{J}] = 0$ we prove $[H, J_+] = 0$ as an example. We have

$$[H, J_+] = - \left[\sum_{i,j}^N \tau_{ij} \left(a_i^\dagger a_j + b_i^\dagger b_j \right), \sum_{k,l}^N \theta_{kl} a_k^\dagger b_l \right]$$

$$+ U \left\{ \left[\sum_{i,j}^N \tau_{ij} \left(a_i^\dagger a_j + b_i^\dagger b_j \right), \sum_{k,l}^N \varepsilon_{kl} I_k^+ I_l^3 \right] \right.$$

$$\left. - \left[\sum_{i,j}^N a_i^\dagger a_i b_i^\dagger b_i, \sum_{i,j}^N \theta_{jk} a_j^\dagger b_k \right] \right\}$$

$$+ U^2 \left[\sum_i^N a_i^\dagger a_i b_i^\dagger b_i, \sum_{i,j}^N \varepsilon_{jk} I_j^+ I_k^3 \right]. \qquad (11.5.14)$$

We can prove by the direct calculation that the coefficients of the zeroth power and the second power terms in (11.5.14) vanish when the chain length $N \to \infty$, while the coefficient of the first power consists of two terms that are respectively

$$\left[\sum_{i,j}^{N} \tau_{ij} \left(a_i^\dagger a_j + b_i^\dagger b_j \right), \sum_{k,l}^{N} \varepsilon_{kl} I_k^+ I_l^3 \right]$$

$$= \sum_{i,j,k} \tau_{ij} \left\{ (\varepsilon_{jk} - \varepsilon_{ik}) a_i^\dagger b_j I_k^3 + \frac{1}{2} (\varepsilon_{jk} - \varepsilon_{ki}) I_k^+ \left(a_i^\dagger a_j - b_i^\dagger b_j \right) \right\}$$

$$= \sum_{i,k} \left\{ (\varepsilon_{i+1,k} - \varepsilon_{ik}) a_i^\dagger b_{i+1} I_k^3 + (\varepsilon_{i-1,k} - \varepsilon_{ki}) a_i^\dagger b_{i-1} I_k^3 \right.$$

$$+ \frac{1}{2} (\varepsilon_{k,i+1} - \varepsilon_{ki}) I_k^+ \left(a_i^\dagger a_{i+1} - b_i^\dagger b_{i+1} \right)$$

$$\left. + \frac{1}{2} (\varepsilon_{k,i-1} - \varepsilon_{ki}) I_k^+ \left(a_i^\dagger b_{i-1} - b_i^\dagger b_{i-1} \right) \right\}$$

$$= -\frac{1}{2} \sum_i \left\{ a_i^\dagger b_{i+1} \left(b_i b_i^\dagger - b_i^\dagger b_i \right) - a_i^\dagger b_{i+1} + a_i^\dagger b_{i-1} \left(b_i^\dagger b_i - b_i b_i^\dagger \right) \right.$$

$$\left. + a_i^\dagger b_{i-1} - a_i^\dagger b_{i+1} a_{i+1}^\dagger a_{i+1} + a_i^\dagger b_{i-1} a_{i-1}^\dagger a_{i-1} \right\}$$

$$= \sum_i a_i^\dagger \left\{ (b_{i+1} - b_{i-1}) b_i^\dagger b_i - b_{i+1} a_{i+1}^\dagger a_{i+1} + b_{i-1} a_{i-1}^\dagger a_{i-1} \right\},$$

and

$$- \left[\sum_i a_i^\dagger a_i b_i^\dagger b_i, \sum_{j,k} \theta_{jk} a_j^\dagger b_k \right]$$

$$= -\sum_{i,j} \theta_{ij} \left\{ a_i^\dagger b_j b_i^\dagger b_i - a_i^\dagger b_j a_i^\dagger a_j \right\}$$

$$= -\sum_i a_i^\dagger \left\{ b_{i+1} \left(b_i^\dagger b_i - a_{i+1}^\dagger a_{i+1} \right) - b_{i-1} \left(b_i^\dagger b_i - a_{i-1}^\dagger a_{i-1} \right) \right\}$$

$$= -\sum_i a_i^\dagger \left\{ (b_{i+1} - b_{i-1}) b_i^\dagger b_i - b_{i+1} a_{i+1}^\dagger a_{i+1} + b_{i-1} a_{i-1}^\dagger a_{i-1} \right\},$$

and they cancel each other. Other commutators $[H, J_-]$ and $[H, J_3]$ can be proved to be zero in the similar fashion. This example shows that while the Lie algebra describes the rotation, the Yangian describes the composition of rotation and the hopping.

In the above we gave several examples of the realization of Yangian in terms of the usual quantum mechanical operators. Some of them are single particle systems, and the others many-body systems. Some are in the first quantized form and the others in second quantized form. The Yangian prescribes the commutation relations only, and leaves a relatively large degree of freedom of making choices in finding realizations. Yangian is a new symmetry operation that includes the Lie algebra as a sub-algebra, and provides a mechanism of constructing an algebra of infinite dimensions. For example in $\boldsymbol{J} = \boldsymbol{L}^2\boldsymbol{p}$ or $\boldsymbol{J} = \boldsymbol{L}^2\boldsymbol{r}$, $\boldsymbol{L}^2$ is rotational invariant, and $\boldsymbol{p}$ is a translation operator, which goes beyond the region of rotational operations. For the Hubbard model, we see from Eq. (11.5.5) that in the $\boldsymbol{J}$ operators there are tensor operators $\boldsymbol{S}_i \times \boldsymbol{S}_j$ characterizing a rotation on a given site, and there are also terms characterizing hopping between the sites. It leads to an algebra of infinite dimensions. On the other hand, a translation over many sites can be realized by a number of consecutive hoppings between neighboring sites. This helps us to comprehend the nature of the Yangian associated with $SL(2)$. It is an algebra of infinite dimensions, but consists of finite number of generators. Two sets of fundamental operators $\boldsymbol{I}$ and $\boldsymbol{J}$ determine the behavior of all operators of higher orders. Such a phenomenon originates from a kind of strong intrinsic constraint in the system, not all elements of higher orders are independent. Such constraint is very cute in the sense that only $\boldsymbol{I}$ and $\boldsymbol{J}$ are the independent generators. It is clear that one cannot find Yangian in any arbitrary system, but a large class of models are connected with the Yangian. Examples will be given in the next chapter on the Yang-Baxter system.

We began with an example in quantum mechanics — the hydrogen atom — to lead to the Yangian algebra. In the subsequent chapters we will see that it is connected with many models and phenomena in physics.

Before ending the chapter we would like to introduce briefly Yangian associated with $SU(3)$.

11.6 Yangian Transition between the Singlet and Octet in $SU(3)$

In 11.3 we have proved that the $Y(SL(2))$ operators serve as the transition operators between the Spin-singlet and triplet. This idea is easy to be extended to $SU(3)$. From the basic course of particle physics we are familiar with the $SU(3)$ symmetry for elementary particles.

The $SU(3)$ algebra is defined by

$$[F_\lambda, F_\mu] = i f_{\lambda\mu\nu} F_\nu, \tag{11.6.1}$$

where $\lambda, \mu, \nu = 1, 2, \ldots, 8$ and the structure constants $f_{\lambda\mu\nu}$ are anti-symmetric with respect to any two indices of (11.6.2)

$$f_{123} = 1, \quad f_{458} = f_{678} = \frac{\sqrt{3}}{2},$$

$$f_{147} = f_{246} = f_{257} = f_{345} = -f_{156} = -f_{367} = \frac{1}{2}.$$

The 3-dimensional representation of $SU(3)$ is formed by the well-known Gell-mann matrices $\lambda_\nu = 2F_\nu$:

$$\lambda_1 = \begin{bmatrix} 0 & 1 & 0 \\ 1 & 0 & 0 \\ 0 & 0 & 0 \end{bmatrix}, \quad \lambda_2 = \begin{bmatrix} 0 & -i & 0 \\ i & 0 & 0 \\ 0 & 0 & 0 \end{bmatrix}, \quad \lambda_3 = \begin{bmatrix} 1 & 0 & 0 \\ 0 & -1 & 0 \\ 0 & 0 & 0 \end{bmatrix}$$

$$\lambda_4 = \begin{bmatrix} 0 & 0 & 1 \\ 0 & 0 & 0 \\ 1 & 0 & 0 \end{bmatrix}, \quad \lambda_5 = \begin{bmatrix} 0 & 0 & -i \\ 0 & 0 & 0 \\ i & 0 & 0 \end{bmatrix}, \quad \lambda_6 = \begin{bmatrix} 0 & 0 & 0 \\ 0 & 0 & 1 \\ 0 & 1 & 0 \end{bmatrix}$$

$$\lambda_7 = \begin{bmatrix} 0 & 0 & 0 \\ 0 & 0 & -i \\ 0 & i & 0 \end{bmatrix}, \quad \lambda_8 = \frac{1}{\sqrt{3}} \begin{bmatrix} 1 & 0 & 0 \\ 0 & 1 & 0 \\ 0 & 0 & -2 \end{bmatrix}. \tag{11.6.2}$$

Define

$$I_\pm = F_1 \pm iF_2, \quad U_\pm = F_6 \pm iF_7, \quad V_\pm = F_4 \mp iF_5 \\ I_3 = F_3, \quad I_8 = \frac{2}{\sqrt{3}}F_8 \left.\vphantom{\begin{matrix}a\\b\end{matrix}}\right\} \tag{11.6.3}$$

The commutation relation for $SU(3)$ can be written in the form:

$$\left. \begin{aligned} &[I_3, I_\pm] = \pm I_\pm, [I_+, I_-] = 2I_3, [I_8, I_\alpha] = 0(\alpha = \pm, 3) \\ &[I_3, U_\pm] = \mp\frac{1}{2}U_\pm, [I_8, U_\pm] = \pm U_\pm, [U_+, U_-] = -I_3 + \frac{3}{2}I_8 \\ &[I_3, V_\pm] = \mp\frac{1}{2}V_\pm, [I_8, V_\pm] = \mp V_\pm, [V_+, V_-] = -\left(I_3 + \frac{3}{2}I_8\right) \\ &[I_\pm, U_\mp] = [U_\pm, V_\mp] = [V_\pm, I_\mp] = 0 \\ &[I_\pm, U_\pm] = \pm V_\mp, [U_\pm, V_\pm] = \pm I_\mp, [V_\pm, I_\pm] = \pm U_\mp \end{aligned} \right\}. \tag{11.6.4}$$

With the notations

$$U_3 = -\frac{1}{2}I_3 + \frac{3}{4}I_8, \quad V_3 = -\frac{1}{2}I_3 - \frac{3}{4}I_8 \tag{11.6.5}$$

we have

$$\left. \begin{aligned} &[U_3, U_\pm] = \pm U_\pm, \quad [U_+, U_-] = 2U_3 \\ &[V_3, V_\pm] = \pm V_\pm, \quad [V_+, V_-] = 2V_3 \end{aligned} \right\} \tag{11.6.6}$$

and

$$Q = I_3 + \frac{1}{2}Y, \tag{11.6.7}$$

where Q is charge operator and Y stands for hypercharge. The quarks and anti-quarks form meson, for example, the η^0 (Octet) and $\eta^{0'}$ (singlet). Both of them possess $I = I_3 = Y = Q = 0$:

$$|\eta^0\rangle = \frac{1}{\sqrt{6}}(-|u\bar{u}\rangle - |d\bar{d}\rangle + 2|s\bar{s}\rangle), \tag{11.6.8}$$

$$|\eta^{0'}\rangle = \frac{1}{\sqrt{3}}(|u\bar{u}\rangle + |d\bar{d}\rangle + |s\bar{s}\rangle), \tag{11.6.9}$$

where $|u\bar{u}\rangle = |u\rangle_1 \cdot |\bar{u}\rangle_2$ and similar for others are product of states for quark and anti-quark.

$$\left.\begin{array}{l}
|u\rangle : \; I = \dfrac{1}{2}, \; I_3 = \dfrac{1}{2}, \; Y = \dfrac{1}{3}, \; Q = \dfrac{2}{3} \\[2mm]
|d\rangle : \; I = \dfrac{1}{2}, \; I_3 = -\dfrac{1}{2}, \; Y = \dfrac{1}{3}, \; Q = -\dfrac{1}{3} \\[2mm]
|s\rangle : \; I = 0, \; I_3 = 0, \; Y = -\dfrac{2}{3}, \; Q = -\dfrac{1}{3} \\[2mm]
|\bar{u}\rangle : \; I = \dfrac{1}{2}, \; I_3 = -\dfrac{1}{2}, \; Y = -\dfrac{1}{3}, \; Q = -\dfrac{2}{3} \\[2mm]
|\bar{d}\rangle : \; I = \dfrac{1}{2}, \; I_3 = \dfrac{1}{2}, \; Y = -\dfrac{1}{3}, \; Q = \dfrac{1}{3} \\[2mm]
|\bar{s}\rangle : \; I = 0, \; I_3 = 0, \; Y = \dfrac{2}{3}, \; Q = \dfrac{1}{3}
\end{array}\right\} \qquad (11.6.10)$$

The nonvanishing actions are

$$\left.\begin{array}{l}
I_+ |d\rangle = |u\rangle , \; I_- |u\rangle = |d\rangle , \; U_- |d\rangle = |s\rangle \\[1mm]
U_+ |s\rangle = |d\rangle , \; V_+ |u\rangle = |s\rangle , \; V_- |s\rangle = |u\rangle
\end{array}\right\}. \qquad (11.6.11)$$

Now we define the operators

$$I_\lambda^{(1)} = \sum_i F_i^\lambda , \qquad (11.6.12)$$

$$J_\lambda = \sum_{i=1}^N \mu_i F_i^\lambda + \beta f_{\lambda\mu\nu} \sum_{i \neq j}^N W_{ij} F_i^\mu F_j^\nu , \quad W_{ij} = -W_{ji}, \qquad (11.6.13)$$

where

$$\left[F_i^\lambda, F_j^\mu \right] = i f_{\lambda\mu\nu} F^\nu \delta_{ij} \qquad (11.6.14)$$

the index i here represents different sort of flavors. Substituting (11.6.12) and (11.6.13) into the Yangian commutation relations

(11.2.1), (11.2.2) and (11.2.6), we find that they satisfy $Y(SU(3))$, as exhibited below.

(1) $A_{ijk} = W_{ij}W_{jk} + W_{jk}W_{ki} + W_{ki}W_{ij} = -1$ (no summation over repeating indices $i \neq j \neq k$, $W_{ij} = -W_{ji}$)

(2) F_i^λ take the Gell-Mann matrices ($\lambda = 1, 2, \ldots, 8$ in fundamental representation).

(3) μ_i arbitrary parameters.

In terms of the notations

$$\bar{I}_\pm = J_1 \pm iJ_2, \quad \bar{v}_\pm = J_6 \pm iJ_7, \quad \bar{U}_\pm = J_4 \pm iJ_5,$$

$$\bar{I}_3 = J_3, \qquad \bar{I}_8 = \frac{2}{\sqrt{3}}J_8 \tag{11.6.15}$$

and setting $\beta = -i\rho$, the $Y(SU(3))$ can be sufficiently realized by

$$
\left.
\begin{aligned}
&I_\pm^{(1)} = \sum_{i=1} I_i^\pm, \quad U_\pm^{(1)} = \sum_i U_i^\pm, \quad U_\pm^{(1)} = \sum_i V_i^\pm, \\
&I_3^{(1)} = \sum_i I_i^3, \quad I_8^{(1)} = \sum_i I_i^8 \\
&\bar{I}_\pm = \sum_{i=1}^N \mu_i I_i^\pm \mp 2\rho \sum_{i\neq j}^N W_{ij}\left(I_i^\pm I_j^3 + \frac{1}{2}U_i^\mp V_j^\mp\right) \\
&\bar{U}_\pm = \sum_i \mu_i v_i^\pm \pm \rho \sum_{i\neq j} W_{ij}\left[U_j^\pm\left(I_j^3 - \frac{3}{2}Y_j\right) + I_i^\mp V_j^\mp\right] \\
&\bar{V}_\pm = \sum_i \mu_i V_i^\pm \pm \rho \sum_{i\neq j} W_{ij}\left[V_i^\pm\left(I_j^3 + \frac{3}{2}Y_j\right) - I_i^\mp U_j^\mp\right] \\
&\bar{I}_3 = \sum_i \mu_i I_i^3 + \rho \sum_{i\neq j} W_{ij}\left[I_i^+ I_j^- - \frac{1}{2}\left(U_i^+ U_j^- + V_i^+ V_j^-\right)\right] \\
&\bar{I}_8 = \sum_i \mu_i Y_i + \rho \sum_{i\neq j} W_{ij}\left[U_i^+ U_j^- - V_i^+ V_j^-\right]
\end{aligned}
\right\}
$$

$$\tag{11.6.16}$$

About the Yangian commutation relations we would like to point out that for $Y(SU(2))$, Eqs. (11.2.1)–(11.2.3) are those to be satisfied, while (11.2.6) gives an identity 0=0. For other Yangian algebras, for example $Y(SL(n))$, relations of (11.2.1), (11.2.2) and (11.2.3) are those to be satisfied, while (11.2.6) is not independent.

For $N = 2$, i.e., two quarks, we set $W_{12} = 1(W_{21} = -1)$. Acting $\bar{I}_8$ on $\left|\eta^0\right\rangle$ we obtain

$$I_8 \left|\eta^0\right\rangle = -\frac{1}{3}(\mu_1 - \mu_2)\left|\eta^0\right\rangle - \frac{\sqrt{2}}{3}(\mu_1 - \mu_2 + 3\rho)\left|\eta^{0'}\right\rangle$$

$$= \sqrt{r^2 - 4\rho r + 6\rho^2}\left(\cos\theta\left|\eta^0\right\rangle + \sin\theta\left|\eta^{0'}\right\rangle\right), \quad (11.6.17)$$

where $r = \mu_2 - \mu_1$ and $\cos\theta = \frac{1}{\sqrt{3}}\frac{r}{\sqrt{r^2-4\rho r+6\rho^2}}$.

The relation of (11.6.17) tells that the operator I_8 makes the transition from $\left|\eta^0\right\rangle$ to the mixture state of $\left|\eta^0\right\rangle$ and $\left|\eta^{0'}\right\rangle$. It is a counterpart of the transition from ψ_{10} to ψ_{00} for a two spin$-\frac{1}{2}$ system.

References

[1] Jimbo M. (Ed), *Yang-Baxter Equation in Integrable Systems*, Singapore, World Scientific, 1990.

[2] Drinfeld V. G., *Soviet Math. Dokl.*, **22**, 254, 1985 and **36**, 212, 1988.

[3] Drinfeld V. G., *Quantum Groups*, in proc. Intern. Congress of Mathematicians, Berkeley, USA 1986, edited by Cleason A.M., American Mathematical Society, 1987. pp. 269–291.

[4] Chari A. and Pressley A., *A Guide to Quantum Groups*, Cambridge University Press, Cambridge, 1994.

[5] Yang C. N., *Phys. Rev. Lett.* **19**, 1312, 1967; *Phys. Rev.* **168**, 1920, 1968.

[6] Wu T. T. and Yang C. N., *Nucl. Phys.* **B107**, 365, 1976.

[7] Ge. M.-L. , Xue K. and Cho Y. M., *Phys. Lett.* **A260**, 484, 1999.

[8] Haldane F. D. M., *Phys. Rev. Lett.* **60**. 639, 1988.

[9] Shastry B. S., *Phys. Rev. Lett.* **60**, 635, 1988.

[10] Haldane F. D. M., Physics of the ideal semion gas: Spinons and quantum symmetries of the integrable Haldane-Shastry spin-chain, *Proc. 16th Taniguch Symp. Condensed Matter*, Berlin, Springer, 1994. Haldane F. D. M., Ha Z. N. C., Talstra J. C., Bernard D., Pasquier V., *Phys. Rev. Lett.* **69**, 2021, 1992.

[11] Bernard D., *Commun. Math. Phys.* **137**, 191, 1991.

[12] Bernevig B. A., *Giuliano D. and Laughlin R.B., Coordinnte representation of the Two-spin wavefunction and spinon Interaction,* Cond-mat/0011270, **16**, 2000.

[13] Uglov d. B. and Korepn V. E., *Phys. Lett.* **A190**, 238, 1994.

Chapter 12

RTT Relation and the Yang-Baxter Equation, Physical Significance of the Quantum Algebras

In the last chapter we introduced the Yangian from the hydrogen atom problem, and discussed a number of examples of the realizations of Yangian. The existence of such algebraic structure is deeply rooted and is connected with a general principle. The hydrogen atom problem is simple, because it is linear. The importance of Yangian lies on the fact that it represents the nature of a large class of *non-linear* models. For non-linear problems it is not possible to obtain the solution by superimposing the perturbative solutions. For some models of low dimensions there exist exact solutions. It is desirable to have a general theory which on one hand gives commutation relations in the form of Eqs. (11.1.13), (11.1.14), (11.1.17) and (11.1.18) and on the other hand gives the principle of constructing Hamiltonians. In this way one can establish the link between the new algebraic relations and the conserved Hamilton system on a deeper level. To ensure the generality of the discussions, it is desirable to pursue model independent analysis without using any specific realization of the commutators. This chapter is an elementary introduction of the systematic works of Faddeev, Kulish, Reshetikhin, Sklyanin, Takhtajan, and their collaborators [1–3, 9, 23, 24] and also the applications to physics.

12.1 Commutation Relations in Direct Product Form of Matrices

The presentation of this section is based on Refs. [1, 2]. The general references on Quantum Groups and related papers and books can be found in Refs. [3–26].

The most fundamental commutation relation in quantum mechanics is that between the canonically conjugate pair of observables x, p:

$$[x_\alpha, p_\beta] = \mathrm{i}\delta_{\alpha\beta}, \tag{12.1.1}$$

where we set $\hbar = 1$, and from which commutation relations between other observables can be derived, for instance

$$\left[L_\alpha, L_\beta\right] = \mathrm{i}\varepsilon_{\alpha\beta\gamma}L_\gamma, \tag{12.1.2}$$

the commutation relation between components of angular momentum, and others like

$$[L_\alpha, x_\beta] = \mathrm{i}\varepsilon_{\alpha\beta\gamma}L_\gamma,$$

$$[L_\alpha, p_\beta] = \mathrm{i}\varepsilon_{\alpha\beta\gamma}p_\gamma,$$

and so on. In the quantum field theory the canonical commutation relations corresponding to (12.1.1) have different forms for different fields. For example for scalar field we have

$$[\varphi\left(\boldsymbol{x}, t\right), \dot{\varphi}\left(\boldsymbol{y}, t\right)] = \delta^3\left(\boldsymbol{x} - \boldsymbol{y}\right), \tag{12.1.3}$$

and for the spinor field we have

$$\left[\psi_\alpha\left(\boldsymbol{x}, t\right), \psi_\beta^+\left(\boldsymbol{y}, t\right)\right] = \delta^3\left(\boldsymbol{x} - \boldsymbol{y}\right)\delta_{\alpha\beta}, \tag{12.1.4}$$

and so on. Is it possible to construct a general framework of the commutation relation which can cover all possible forms as its realizations?

The central feature of a commutation relation is the order of the operators. In order to distinguish the order of operators we can introduce a 2×2 matrix L, the matrix elements of which are quantum mechanical operators. To distinguish the order of operators we can introduce a parameters u,

$$L(u) = \begin{bmatrix} L_{11}(u) & L_{12}(u) \\ L_{21}(u) & L_{22}(u) \end{bmatrix} \tag{12.1.5}$$

such that $L(u)L(v)$ is different from $L(v)L(u)$. The space spanned by the matrix is called the "auxiliary space", each element is a quantum mechanical operator in the Hilbert space. In order that each operator in front has a chance to form a product with any operator behind, we use the direct product of matrices.

$$A = \begin{bmatrix} a_{11} & a_{12} \\ a_{21} & a_{22} \end{bmatrix}, \quad B = \begin{bmatrix} b_{11} & b_{12} \\ b_{21} & b_{22} \end{bmatrix},$$

$$A \otimes B = \begin{bmatrix} a_{11}b_{11} & a_{11}b_{12} & a_{12}b_{11} & a_{12}b_{12} \\ a_{11}b_{21} & a_{11}b_{22} & a_{12}b_{21} & a_{12}b_{22} \\ a_{21}b_{11} & a_{21}b_{12} & a_{22}b_{11} & a_{22}b_{12} \\ a_{21}b_{21} & a_{21}b_{22} & a_{22}b_{21} & a_{22}b_{22} \end{bmatrix}. \tag{12.1.6}$$

The row and column of an matrix element of $A \otimes B$ are represented by two indices each:

$$(A \otimes B)_{ij;kl} = A_{ik}B_{jl}. \tag{12.1.7}$$

Each matrix element of $A \otimes B$ consists of an element of A followed by an element of B, such that the difference between $A \otimes B$ and $B \otimes A$ will lead to the commutation relation.

In the following we discuss a specific form of $L(u)$:

$$L(u) = \begin{bmatrix} 1 & 0 \\ 0 & 1 \end{bmatrix} + u^{-1} \begin{bmatrix} a & b \\ c & d \end{bmatrix}, \tag{12.1.8}$$

where u is called the spectral parameter. First we calculate $L(u) \otimes L(v)$ and $L(v) \otimes L(v)$:

$$L(u) \otimes L(v) = \left(\begin{bmatrix} 1 & 0 \\ 0 & 1 \end{bmatrix} + u^{-1} \begin{bmatrix} a & b \\ c & d \end{bmatrix} \right) \otimes \left(\begin{bmatrix} 1 & 0 \\ 0 & 1 \end{bmatrix} + v^{-1} \begin{bmatrix} a & b \\ c & d \end{bmatrix} \right)$$

$$
= \begin{bmatrix} 1 & & & \\ & 1 & & \\ & & 1 & \\ & & & 1 \end{bmatrix} + u^{-1} \begin{bmatrix} a & 0 & b & 0 \\ 0 & a & 0 & b \\ c & 0 & d & 0 \\ 0 & c & 0 & d \end{bmatrix}
$$

$$
+ v^{-1} \begin{bmatrix} a & b & 0 & 0 \\ c & d & 0 & 0 \\ 0 & 0 & a & b \\ 0 & 0 & c & d \end{bmatrix} + u^{-1}v^{-1} \begin{bmatrix} a^2 & ab & ba & b^2 \\ ac & ad & bc & bd \\ ca & cb & da & db \\ c^2 & cd & dc & d^2 \end{bmatrix},
$$

$$
\tag{12.1.9}
$$

$$
L(v) \otimes L(u) = \begin{bmatrix} 1 & & & \\ & 1 & & \\ & & 1 & \\ & & & 1 \end{bmatrix} + v^{-1} \begin{bmatrix} a & 0 & b & 0 \\ 0 & a & 0 & b \\ c & 0 & d & 0 \\ 0 & c & 0 & d \end{bmatrix}
$$

$$
+ u^{-1} \begin{bmatrix} a & b & 0 & 0 \\ c & d & 0 & 0 \\ 0 & 0 & a & b \\ 0 & 0 & c & d \end{bmatrix} + v^{-1}u^{-1} \begin{bmatrix} a^2 & ab & ba & b^2 \\ ac & ad & bc & bd \\ ca & cb & da & db \\ c^2 & cd & dc & d^2 \end{bmatrix}.
$$

$$
\tag{12.1.10}
$$

The existence of the spectral parameter reveals their difference. In the expression for L, the composition of operators is definite, such that (12.1.9) and (12.1.10) differ only in spectral parameters, while the order of operators in the corresponding matrix elements of their last terms is the same. In order that operators can appear in different order, we introduce the interchange matrix

$$
P = \begin{bmatrix} 1 & 0 & 0 & 0 \\ 0 & 0 & 1 & 0 \\ 0 & 1 & 0 & 0 \\ 0 & 0 & 0 & 1 \end{bmatrix}. \tag{12.1.11}
$$

It is easy to verify that operating from the left by P on any 4×4 matrix interchanges the second and third row, while operating from the right interchanges the second and third column. Define a c-number matrix $\check{R}$,

$$
\check{R}(u - v) = I + (u - v)P, \tag{12.1.12}
$$

where I is the unit matrix. Operating $\check{R}$ from the left and from the right on (12.1.9) and (12.1.10) can produce different orders of operators for the corresponding matrix elements. Equating the result will lead to commutation relations. We demonstrate this in the following:

$$\check{R}(u - v)\,(L(u) \otimes L(v))$$

$$= I + u^{-1}\begin{bmatrix} a & 0 & b & 0 \\ 0 & a & 0 & b \\ c & 0 & d & 0 \\ 0 & c & 0 & d \end{bmatrix} + v^{-1}\begin{bmatrix} a & b & 0 & 0 \\ c & d & 0 & 0 \\ 0 & 0 & a & b \\ 0 & 0 & c & d \end{bmatrix}$$

$$+\, u^{-1}v^{-1}\begin{bmatrix} a^2 & ab & ba & b^2 \\ ac & ad & bc & bd \\ ca & cb & da & db \\ c^2 & cd & dc & d^2 \end{bmatrix}$$

$$+\, (u - v)\left(\begin{bmatrix} 1 & 0 & 0 & 0 \\ 0 & 0 & 1 & 0 \\ 0 & 1 & 0 & 0 \\ 0 & 0 & 0 & 1 \end{bmatrix} + u^{-1}\begin{bmatrix} a & 0 & b & 0 \\ c & 0 & d & 0 \\ 0 & a & 0 & b \\ 0 & c & 0 & d \end{bmatrix}\right.$$

$$\left. +\, v^{-1}\begin{bmatrix} a & b & 0 & 0 \\ 0 & 0 & a & b \\ c & d & 0 & 0 \\ 0 & 0 & c & d \end{bmatrix} + u^{-1}v^{-1}\begin{bmatrix} a^2 & ab & ba & b^2 \\ ca & cb & da & db \\ ac & ad & bc & bd \\ c^2 & cd & dc & d^2 \end{bmatrix}\right),$$

$$(L(v) \otimes L(u))\check{R}(u - v)$$

$$= I + v^{-1}\begin{bmatrix} a & 0 & b & 0 \\ 0 & a & 0 & b \\ c & 0 & d & 0 \\ 0 & c & 0 & d \end{bmatrix} + u^{-1}\begin{bmatrix} a & b & 0 & 0 \\ c & d & 0 & 0 \\ 0 & 0 & a & b \\ 0 & 0 & c & d \end{bmatrix}$$

$$+\, u^{-1}v^{-1}\begin{bmatrix} a^2 & ab & ba & b^2 \\ ac & ad & bc & bd \\ ca & cb & da & db \\ c^2 & cd & dc & d^2 \end{bmatrix}$$

$$(+u - v) \left(\begin{bmatrix} 1 & 0 & 0 & 0 \\ 0 & 0 & 1 & 0 \\ 0 & 1 & 0 & 0 \\ 0 & 0 & 0 & 1 \end{bmatrix} + v^{-1} \begin{bmatrix} a & b & 0 & 0 \\ 0 & 0 & a & b \\ c & d & 0 & 0 \\ 0 & 0 & c & d \end{bmatrix} \right.$$

$$\left. + u^{-1} \begin{bmatrix} a & 0 & b & 0 \\ c & 0 & d & 0 \\ 0 & a & 0 & b \\ 0 & c & 0 & d \end{bmatrix} + u^{-1}v^{-1} \begin{bmatrix} a^2 & ba & ab & b^2 \\ ac & bc & ad & bd \\ ca & da & cb & db \\ c^2 & dc & cd & d^2 \end{bmatrix} \right).$$

Equating both expressions,

$$\check{R}(u - v)(L(u) \otimes L(v)) = (L(v) \otimes L(u))\check{R}(u - v), \qquad (12.1.13)$$

we can easily verify the following result:

$$(v^{-1} - u^{-1}) \begin{bmatrix} 0 & [a,b] + b & [b,a] - b & 0 \\ [c,a] + c & [c,b] + d - a & [d,a] & [d,b] - b \\ [a,c] - c & [a,d] & [b,c] - d + a & [b,d] + b \\ 0 & [c,d] - c & [d,c] + c & 0 \end{bmatrix} = 0.$$

$$(12.1.14)$$

We have set the spectral parameter of $\check{R}$ to be $u - v$. As a result of (12.1.13), among the eight matrices on both sides of the equality only the terms with the coefficients u^{-1} and v^{-1} make sense that gives (12.1.14). From (12.1.14) we get the following commutation relations:

$$\left. \begin{array}{ll} [a,b] = -b, & [a,d] = 0 \\ [a,c] = c, & [b,d] = -b \\ [c,b] = a - d, & [c,d] = c \end{array} \right\}. \qquad (12.1.15)$$

Equation (12.1.8) is one choice for $L(u)$. If we choose an alternative form

$$L(u) = \begin{bmatrix} 1 & 0 \\ 0 & \lambda \end{bmatrix} + u^{-1} \begin{bmatrix} a & b \\ c & d \end{bmatrix}, \qquad (12.1.16)$$

then we obtain the following commutation relations:

$$\left. \begin{array}{ll} [a,b] = -b, & [a,d] = 0 \\ [a,c] = c, & [b,d] = -\lambda b \\ [c,b] = \lambda a - d, & [c,d] = \lambda c \end{array} \right\}. \qquad (12.1.17)$$

Quantum mechanical operators can be chosen to realize these commutation relations. If we set in (12.1.16) $\lambda = 0, d = 0$, then relations in (12.1.17) become

$$[a, c] = c, \quad [a, b] = -b, \quad [c, b] = 0. \tag{12.1.18}$$

It is easy to verify that

$$a = \frac{\partial}{\partial q}, c = e^q, b = e^{-q} \tag{12.1.19}$$

or

$$a = i\frac{\partial}{\partial q}, c = e^{-iq}, b = e^{iq} \tag{12.1.20}$$

is realization of Eq. (12.1.18), if q and p satisfy the Heisenberg commutation relation $[q, p] = i$. If we set in (12.1.16) $\lambda = 1$, i.e., we return to (12.1.8), and choose

$$a = -L_3, \quad = L_+, \quad c = L_-, \quad d = L_3 \tag{12.1.21}$$

or

$$a = L_3, \quad b = L_-, \quad c = L_+, \quad d = -L_3, \tag{12.1.22}$$

we see that Eq. (12.1.15) becomes

$$\begin{aligned}[L_3, L_\pm] &= \pm L_\pm, \\ [L_+, L_-] &= 2L_3.\end{aligned} \tag{12.1.23}$$

Here we emphasize that the auxiliary space works for spin $\frac{1}{2}$ does not mean the representation of the quantum operator $\boldsymbol{L}$ to be 2-dimensional. In contrast, as quantum operator acting in Hilbert space, $\boldsymbol{L}$ can have representations of arbitrary dimensions.

The interchange matrix plays an important role in establishing the general framework of commutation relations. In the subsequent discussions we will find out that the matrix $\check{R}(u)$ plays an important role in handling problems of interacting identical particles. Next we elaborate the properties of the interchange matrix. The matrix of

(12.1.11) can be expressed in terms of the direct product of Pauli matrices,

$$P = \frac{1}{2} \left(\boldsymbol{I} + \boldsymbol{\sigma} \otimes \boldsymbol{\sigma} \right), \tag{12.1.24}$$

where

$$\boldsymbol{\sigma} \otimes \boldsymbol{\sigma} = \sigma^1 \otimes \sigma^1 + \sigma^2 \otimes \sigma^2 + \sigma^3 \otimes \sigma^3. \tag{12.1.25}$$

The direct products of the Pauli matrices are:

$$\sigma^1 \otimes \sigma^1 = \begin{bmatrix} & & & 1 \\ & & 1 & \\ & 1 & & \\ 1 & & & \end{bmatrix}, \quad \sigma^2 \otimes \sigma^2 = \begin{bmatrix} & & & -1 \\ & & 1 & \\ & 1 & & \\ -1 & & & \end{bmatrix},$$

$$\sigma^3 \otimes \sigma^3 = \begin{bmatrix} 1 & & & \\ & -1 & & \\ & & -1 & \\ & & & 1 \end{bmatrix},$$

where the matrix elements in the empty space are zero. Substituting into (12.1.24) we recover (12.1.11).

$$P = \begin{bmatrix} 1 & & & \\ & 0 & 1 & \\ & 1 & 0 & \\ & & & 1 \end{bmatrix}. \tag{12.1.26}$$

In (12.1.24) the direct product represents the product of operators in different spaces. For example, for a system of two spin $\frac{1}{2}$ particles, the two spin operators operate in different spaces. If we label the Pauli operators of the spin of two particles by σ_1 and σ_2, and denote the spin states of the system by the conventional notation $|\uparrow\downarrow\rangle$, then (12.1.24) is the interchange operator for the spin in quantum mechanics

$$P = \frac{1}{2} \left(\boldsymbol{I} + \boldsymbol{\sigma}_1 \cdot \boldsymbol{\sigma}_2 \right).$$

In order to demonstrate the operation of P on the spin state of the system, we rewrite P as

$$P = \frac{1}{2}\left[I + \frac{1}{2}\left(\sigma_1^+\sigma_2^- + \sigma_1^-\sigma_2^+\right) + \sigma_1^3\sigma_2^3\right]. \qquad (12.1.27)$$

It is easy to verify

$$P\left|\uparrow\uparrow\right\rangle = \left|\uparrow\uparrow\right\rangle,$$

$$P\frac{1}{\sqrt{2}}\left(\left|\uparrow\downarrow\right\rangle + \left|\downarrow\uparrow\right\rangle\right) = \frac{1}{\sqrt{2}}\left(\left|\downarrow\uparrow\right\rangle + \left|\uparrow\downarrow\right\rangle\right),$$

$$P\left|\downarrow\downarrow\right\rangle = \left|\downarrow\downarrow\right\rangle,$$

$$P\frac{1}{\sqrt{2}}\left(\left|\uparrow\downarrow\right\rangle - \left|\downarrow\uparrow\right\rangle\right) = \frac{1}{\sqrt{2}}\left(\left|\downarrow\uparrow\right\rangle - \left|\uparrow\downarrow\right\rangle\right).$$

The interchange operator interchanges the two particles; it gives an additional factor 1 when operating on triplet states (symmetrical), and a factor -1 when operating on the singlet state (antisymmetrical). An important property of the interchange operator (matrix) is

$$P^2 = I. \qquad (12.1.28)$$

Twice interchange is equivalent to an identity transformation.

Let us return to (12.1.13). The auxiliary space of $L(u)$ belongs to $SL(2)$, being associated with the matrix representation of spin$\frac{1}{2}$. $L(u) \otimes L(v)$ involves two spin spaces, the P matrix term of $\check{R}(u-v)$ matrix interchanges the two spaces. To discuss (12.1.13) further, we write it in terms of its matrix elements.

$$\check{R}(u-v)_{ij,mn}\left(L(u) \otimes L(v)\right)_{mn,kl} = \left(L(v) \otimes L(u)\right)_{ij,mn}\check{R}(u-v)_{mn,kl}, \qquad (12.1.29)$$

where we label the matrix elements by indices $+$ and $-$.

$$L = \begin{bmatrix} L_{++} & L_{+-} \\ L_{-+} & L_{--} \end{bmatrix}.$$

Using (12.1.7) we express the matrix element of the direct product in terms of the matrix element of L:

$$\check{R}(u-v)_{ij,mn}L(u)_{mk}L(v)_{nl} = L(v)_{im}L(u)_{jn}\check{R}(u-v)_{mn,kl}. \qquad (12.1.30)$$

We introduce a new matrix $R(u)$:

$$R(u)_{ij,mn} = \check{R}(u)_{ji,mn}. \tag{12.1.31}$$

Equation (12.1.30) can be written as the product of three 4×4 matrices:

$$R(u-v)\overset{1}{L}(u)\overset{2}{L}(v) = \overset{2}{L}(v)\overset{1}{L}(u)R(u-v), \tag{12.1.32}$$

where

$$\overset{1}{L}(u) = L(u) \otimes I,$$
$$\overset{2}{L}(u) = I \otimes L(u),$$

and I is a 2×2 unit matrix. Equation (12.1.32) is proved as follows. Expressing it in terms of matrix elements, we have

$$R(u-v)\overset{1}{L}(u)\overset{2}{L}(v) = R(u-v)_{ij,mn} \left(L(u) \otimes \boldsymbol{I} \right)_{mn,st} \left(\boldsymbol{I} \otimes L(v) \right)_{st,kl}$$

$$= R(u-v)_{ij,mn} L(u)_{ms}\delta_{nt}\delta_{sk} L(v)_{tl}$$

$$= R(u-v)_{ij,mn} L(u)_{mk} L(v)_{nl},$$

$$\overset{2}{L}(v)\overset{1}{L}(u)R(u-v) = \left(\boldsymbol{I} \otimes L(v) \right)_{ij,mn} \left(L(u) \otimes \boldsymbol{I} \right)_{mn,st} R(u-v)_{st,kl}$$

$$= \delta_{im}L(v)_{jn}L(u)_{ms}\delta_{nt}R(u-v)_{st,kl}$$

$$= L(v)_{jn}L(u)_{is}R(u-v)_{sn,kl}.$$

Interchanging the free indices i and j, and replacing the summation indices (n,s) on the right-hand side by (m,n), we have

$$R(u-v)_{ij,mn}L(u)_{mk}L(v)_{nl} = L(v)_{im}L(u)_{jn}R(u-v)_{nm,kl}. \tag{12.1.33}$$

Comparing with (12.1.30), we find the relation between $\check{R}$ and R in (12.1.31). Write the matrix form of $\check{R}$ in the following

$$\check{R} = I + uP = \begin{bmatrix} 1 & & & \\ & & 1 & \\ & 1 & & \\ & & & 1 \end{bmatrix} + u \begin{bmatrix} 1 & & & \\ & 1 & & \\ & & 1 & \\ & & & 1 \end{bmatrix}$$

and interchange the matrix element $(ij; kl)$ with $(ji; kl)$, we get R. Actually this involves only the interchange of the second row with the third row, the elements in the first row being $(++; kl)$, those in the second row being $(+-; kl)$, those in the third row being $(-+; kl)$ and those in the fourth row being $(--; kl)$. Consequently we have

$$R = \begin{bmatrix} 1 & & & \\ & & 1 & \\ & 1 & & \\ & & & 1 \end{bmatrix} + u \begin{bmatrix} 1 & & & \\ & 1 & & \\ & & 1 & \\ & & & 1 \end{bmatrix} = P + uI, \qquad (12.1.34)$$

i.e.,

$$R = P\check{R}, \quad \check{R} = PR. \qquad (12.1.35)$$

We concentrate on the non-vanishing matrix elements of R and $\check{R}$. For $\check{R}$, they are $\check{R}_{ij,kl} = \check{R}_{++,++}, \check{R}_{+-,+-}, \check{R}_{+-,-+}, \check{R}_{-+,+-}, \check{R}_{-+,-+}, \check{R}_{--,--}$. What they have in common is $i+j = k+l$, implying the spin conservation, which we will see later.

If the interchange P operates on two of the spins of many particles, we need to label them, for instance P_{12} interchanges particles 1 and 2. An important property of the interchange operator is

$$P_{12}P_{13}P_{23} = P_{23}P_{13}P_{12}. \qquad (12.1.36)$$

This is shown in Fig.12.1

Particles $(1,2,3)$ arrive in the state $(3,2,1)$ via interchanges P_{23}, P_{13}, P_{12} in the order indicated. They can arrive in the same state

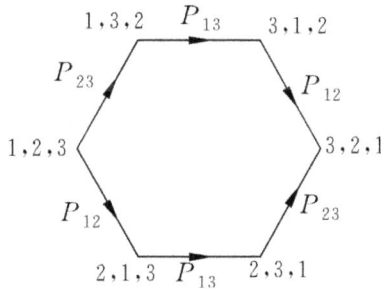

Fig. 12.1   Picture representation of Eq. (12.1.36).

via interchanges in the order P_{12}, P_{13}, P_{23}. Similarly we can prove

$$P_{23}P_{12} (1, 2, 3) = P_{23} (2, 1, 3) = (3, 1, 2),$$
$$P_{13}P_{23} (1, 2, 3) = P_{13} (1, 3, 2) = (3, 1, 2).$$

Therefore $P_{23}P_{12} = P_{13}P_{23}$, and so on. To sum up, we have

$$P_{23}P_{12} = P_{13}P_{23} = P_{12}P_{13}, \tag{12.1.37}$$

$$P_{23}P_{13} = P_{13}P_{12} = P_{12}P_{23}. \tag{12.1.38}$$

Since $R(u)$ is linearly related to P, it has to satisfy a similar relation:

$$R(u)_{12}R(u+v)_{13}R(v)_{23} = R(v)_{23}R(u+v)_{13}R(u)_{12}. \tag{12.1.39}$$

This can be proved by direct calculation:

$$R(u)_{12}R(u+v)_{13}R(v)_{23}$$
$$= (u + P_{12})\left[(u+v) + P_{13}\right](v + P_{23})$$
$$= uv(u+v) + v(u+v)P_{12} + uvP_{13} + u(u+v)P_{23}$$
$$+ (u+v)P_{12}P_{23} + vP_{12}P_{13} + uP_{13}P_{23} + P_{12}P_{13}P_{23}.$$

Using (12.1.38) we obtain the terms linear in u and v as $vP_{12}P_{13} + uP_{13}P_{23} = (u+v)P_{12}P_{13}$. Taking into account (12.1.36), we get the result identical with the right-hand side of (12.1.39). $R(u)$ is a generalization of the interchange operator. Operating in the spin space it satisfies a similar relation as P does, though it has a different spectral parameter in a different space, and also $[R(u)]^2 \neq 1$. Later, we will see that the spectral parameter has the meaning of the momentum. Therefore, the solution for $R(u)$ has a dynamical content. In the above we derived the equation (12.1.39) from a simple form of $R(u)$ (12.1.34). Equation (12.1.39) has solutions of many different forms, and (12.1.34) is the simplest one. Actually (12.1.39) has a very general meaning. It is the *Yang-Baxter equation* [2–5].

12.2   The RTT Relation

In the previous section we discussed the auxiliary space $L(u)$ as 2×2 matrices, which leads to 4×4 matrices for $\check{R}(u)$. Actually the auxiliary space can have any dimensions, e.g., $L(u)$ can be a $M \times M$ matrix and correspondingly $\check{R}(u)$ is a $M^2 \times M^2$ matrix. In the expression for R exhibiting its interchange property, $\overset{1}{L}(u)$ and $\overset{2}{L}(v)$ belong to different spaces, and $R(u-v)$ operates between these spaces.

In physical problems there are systems with many lattice sites. The observables (operators) on each site satisfy conventional commutation relation, but observables on different sites commute with one another. Matrix $L(u)$ can be defined on sites, for instance the matrix on site j is $L_j(u)$, and its matrix elements are operators on site j. Operators on different sites commute, and therefore we have

$$\left[L_j^{ab}(u), L_k^{cd}(v) \right] = 0, \quad j \neq k, \tag{12.2.1}$$

where a, b and c, d are indices for matrix elements of the auxiliary space. As matrices, $L_j(u)$ and $L_k(v)$ do not commute in general. For a system of N lattice sites $L(u)$ is generalized to $T(u)$.

$$T(u) = L_N(u) L_{N-1}(u) \cdots L_2(u) L_1(u) \equiv \overset{\overset{\curvearrowright}{N}}{\underset{j=1}{\prod}} L_j(u). \tag{12.2.2}$$

The arrow above the summation sign denotes the direction of increasing order of L matrices. $T(u)$ is called the global transition matrix. The relation given by (12.1.33) $R(u-v)_{ij,mn} L(u)_{mk} L(v)_{nl} = L(v)_{im} L(u)_{jn} R(u-v)_{nm,kl}$ can now be generalized to

$$R(u-v)_{ij,mn} T(u)_{mk} T(v)_{nl} = T(v)_{jm} T(u)_{in} R(u-v)_{nm,kl}, \tag{12.2.3}$$

the matrix form of which is

$$R(u-v) \overset{1}{T}(u) \overset{2}{T}(v) = \overset{2}{T}(v) \overset{1}{T}(u) R(u-v) \tag{12.2.4}$$

To prove (12.2.3), we begin with the case $N = 2 : T(u) = L_2(u) L_1(u)$. The left-hand side of (12.2.3) is

$$R(u-v)_{ij,mn} L_2(u)_{ms} L_1(u)_{sk} L_2(v)_{nr} L_1(v)_{rl}.$$

We change the order of matrix elements of L_2 and L_1 using the fact that operators on different sites commute, and then use (12.1.33) to change RLL into LLR, the left-hand side becomes

$$
\begin{aligned}
L_2(v)_{jm} & L_2(u)_{in} R(u-v)_{nm,sr} L_1(u)_{sk} L_1(v)_{rl} \\
&= L_2(v)_{jm} L_2(u)_{in} L_1(v)_{mr} L_1(u)_{ns} R(u-v)_{sr,kl} \\
&= L_2(v)_{jm} L_1(v)_{mr} L_2(u)_{in} L_1(u)_{ns} R(u-v)_{sr,kl} \\
&= T(v)_{jm} T(u)_{in} R(u-v)_{nm,kl},
\end{aligned}
$$

which is just the right-hand side of (12.2.3). The subscript i in $L_i(u)$ indicates the quantum space where i-th particle is situated. Having proved that (12.2.2) satisfies (12.2.4) for the case $N = 2$, we multiply $T(u)$ by the matrix $L(u)$ of another site and prove the above statement for the case $N = 3$, and so on. By iteration we can prove the statement for the general case. Equation (12.2.3) is called the RTT relation. Compared with (12.1.30), the change is only the replacement of L by T. L characterizes the local relation on a site while the RTT relation characterizes the global relation. RTT relation is the fundamental relation in the theory of Yang-Baxter systems, and is also the starting point of connecting different models in physics.

To solve the RTT relation for a given $R(u)$ means to find the matrix elements of T satisfying the RTT relation. The matrix elements can be expressed in terms of the fundamental operators in quantum mechanics. In the simple example discussed in the previous section, the matrix $L(u)$ in (12.1.22) realizing the Eq. (12.1.8) is

$$
L(u) = \begin{bmatrix} 1 & 0 \\ 0 & 1 \end{bmatrix} + u^{-1} \begin{bmatrix} L_3 & L_- \\ L_+ & -L_3 \end{bmatrix}, \tag{12.2.5}
$$

which satisfies the RTT relation determined by the simplest matrix R, $R(u) = uI + P$. We rewrite $\boldsymbol{L}$ as $\boldsymbol{S}$:

$$
S_{\pm} = \frac{1}{2} \left(L_1 \pm iL_2 \right),
$$

and then $L(u)$ can be rewritten as

$$
L(u) = I + u^{-1} \boldsymbol{S} \cdot \boldsymbol{\sigma}, \tag{12.2.6}
$$

by using Pauli matrices, where $\boldsymbol{S}$ are operators in Hilbert space and $\boldsymbol{\sigma}$ matrices in the auxiliary space. Defining the quantum mechanical operator $\boldsymbol{S}$ and the corresponding $L(u)$ on lattice sites, we have

$$L_j(u) = I + u^{-1}\boldsymbol{S}_j \cdot \boldsymbol{\sigma}, \qquad (12.2.7)$$

The commutation relations of $\boldsymbol{S}_j$ are

$$[S_{j\alpha}, S_{k\beta}] = \mathrm{i}\varepsilon_{\alpha\beta\gamma}S_{j\gamma}\delta_{jk}. \qquad (12.2.8)$$

$L_j(u)$ is called the local transition matrix. The global transition matrix is

$$T(u) = \left(I + u^{-1}\boldsymbol{S}_N \cdot \boldsymbol{\sigma}\right)\left(I + u^{-1}\boldsymbol{S}_{N-1} \cdot \boldsymbol{\sigma}\right) \cdots \left(I + u^{-1}\boldsymbol{S}_1 \cdot \boldsymbol{\sigma}\right). \qquad (12.2.9)$$

Using

$$\sigma_\alpha \sigma_\beta = \delta_{\alpha\beta} + \mathrm{i}\varepsilon_{\alpha\beta\gamma}\sigma_\gamma,$$

we can expand $T(u)$ to get

$$T(u) = I + u^{-1}\left(\sum_{j=1}^{N} \boldsymbol{S}_j\right) \cdot \boldsymbol{\sigma}$$

$$+ u^{-2}\left\{\mathrm{i} \sum_{j,k,j\neq k} \varepsilon_{\alpha\beta\gamma}S_{j\alpha}S_{k\beta}\sigma_\gamma + \sum_{j,k,j>k} \boldsymbol{S}_j \cdot \boldsymbol{S}_k\right\} + \cdots$$

$$= I + u^{-1}\boldsymbol{S} \cdot \boldsymbol{\sigma} + u^{-2}\mathrm{i} \sum_{j,k,j\neq k} \boldsymbol{S}_j \times \boldsymbol{S}_k \cdot \boldsymbol{\sigma}$$

$$+ u^{-2} \sum_{j,k,j>k} \boldsymbol{S}_j \cdot \boldsymbol{S}_k + \cdots, \qquad (12.2.10)$$

where $\boldsymbol{S} = \sum_j^N \boldsymbol{S}_j$ is the total spin.

In the foregoing discussions we constructed the global transition matrix $T(u)$ from the local transition matrix $L(u)$, and proved the validity of the RTT relation of (12.2.4). But the solution of a RTT relation in general does not necessarily factorize into a product of local transition matrices.

The global transition matrix in (12.2.9) can be expanded in an infinite series in u^{-1} in (12.2.10), it can also be a finite series. The key issue is that the RTT relation is solvable for a given R matrix.

The RTT relation is a constraint condition for the transition matrix T for a given $R(u)$. Further, matrix R is also conditional: it should be a solution of the Yang-Baxter equation because of the associativity.

12.3 Yang-Baxter Equation

Starting from a simple $R(u) = P + uI$ and operating it in certain order on different spaces, the self-consistency requirement leads to the Yang-Baxter equation (abbreviated YBE) of (12.1.39). The general validity of YBE should be independent of the choice of a specific $R(u)$. We change the form of (12.2.4) to accommodate to the multi-space discussion.

$$R_{12}(u_1 - u_2)\overset{1}{T}(u_1)\overset{2}{T}(u_2) = \overset{2}{T}(u_2)\overset{1}{T}(u_1)R_{12}(u_1 - u_2). \qquad (12.3.1)$$

Consider the product of matrices in three spaces $\overset{1}{T}(u_1)\overset{2}{T}(u_2)\overset{3}{T}(u_3)$. For the auxiliary spaces, the associative law of the matrix product has to be satisfied (as the general property of Quantum Field Theories).

$$\overset{1}{T}(u_1)\left(\overset{2}{T}(u_2)\overset{3}{T}(u_3)\right) = \left(\overset{1}{T}(u_1)\overset{2}{T}(u_2)\right)\overset{3}{T}(u_3), \qquad (12.3.2)$$

This general requirement leads to the YBE satisfied by $R(u)$.

Multiplying both sides of (12.3.1) from the left and the right by $R_{12}^{-1}(u_1 - u_2)$, we obtain

$$R_{12}^{-1}(u_1 - u_2)\overset{2}{T}(u_2)\overset{1}{T}(u_1) = \overset{1}{T}(u_1)\overset{2}{T}(u_2)R_{12}^{-1}(u_1 - u_2). \qquad (12.3.3)$$

We observe that R_{12} and R_{12}^{-1} operate on space 1 and 2 only, and hence they commute with $\overset{3}{T}$, and in a product like $\overset{1}{T}\overset{2}{T}\overset{3}{T}$ we will not use parenthesis because the associative law is valid. In the following proof we will not indicate explicitly the spectral parameters in R_{ij}

and $\overset{i}{T}$. We begin with $R_{23}R_{13}R_{12}\overset{1\ 2\ 3}{TTT}$. Using the RTT relation and the fact that R_{12} commutes with $\overset{3}{T}$, it becomes $R_{23}R_{13}\overset{2\ 1\ 3}{TTT}R_{12}$. Handling the first two factors in the similar way, we obtain

$$R_{23}R_{13}R_{12}\overset{1\ 2\ 3}{TTT} = \overset{3\ 2\ 1}{TTT}R_{23}R_{13}R_{12}. \qquad (12.3.4)$$

Using the relation

$$[R_{12}R_{13}R_{23}]^{-1} = R_{23}^{-1}R_{13}^{-1}R_{12}^{-1},$$

and operating the left-hand side of it on the left-hand side of (12.3.4), the right-hand side of it on the right-hand side of (12.3.4), we get

$$[R_{12}R_{13}R_{23}]^{-1}R_{23}R_{13}R_{12}\overset{1\ 2\ 3}{TTT} = \overset{1\ 2\ 3}{TTT}[R_{12}R_{13}R_{23}]^{-1}R_{23}R_{13}R_{12}.$$

This is identical with

$$\left[(R_{12}R_{13}R_{23})^{-1}R_{23}R_{13}R_{12}, \overset{1\ 2\ 3}{TTT}\right] = 0.$$

Since $\overset{1\ 2\ 3}{TTT}$ is a product of matrices in three spaces, the quantity commuting with it must be a c-number, i.e.,

$$R_{23}R_{13}R_{12} = cR_{12}R_{13}R_{23}.$$

Taking the determinant of the product of matrices in three spaces and using $\det ABC = \det A \det B \det C$, we find that $c = 1$. Therefore,

$$R_{12}\left(u_1 - u_2\right)R_{13}\left(u_1 - u_3\right)R_{23}\left(u_2 - u_3\right)$$

$$= R_{23}\left(u_2 - u_3\right)R_{13}\left(u_1 - u_3\right)R_{12}\left(u_1 - u_2\right). \qquad (12.3.5)$$

The proof is completed. Equation (12.3.5) is in a somewhat abstract form, and the labels of auxiliary spaces originate from the labels of matrix elements. Comparing with the RTT relation (12.2.3), Eq. (12.3.5) has to be rewritten as

$$R(u)_{ji}^{ml}R\left(u+v\right)_{kl}^{nt}R(v)_{nm}^{rs} = R(v)_{kj}^{lm}R\left(u+v\right)_{li}^{rn}R(u)_{mn}^{st}. \qquad (12.3.6)$$

We have replaced in the above relation $u_1 - u_2, u_1 - u_3, u_2 - u_3$ by $u, u+v, v$ respectively, noting that $(u_1 - u_2) + (u_2 - u_3) = (u_1 - u_3)$.

The notation for matrix element $R_{ji,ml}$ is replaced by R_{ji}^{ml}. The both sides of (12.3.6) are not products of three matrices. If it were so, there would be two free superscripts and two free subscripts. In both sides of (12.3.6) there are three free subscripts i, j, k and three free superscripts t, r, s. The others l, m, n are summation indices, and each superscript is to be summed with a subscript. The situation arises from the following fact. Both sides operate on the product of three T matrices, each of which has one superscript and one subscript. These indices "dispense" three superscripts and three subscript in the product of R matrices. The relation between (12.3.6) and (12.3.5) is the same as that between (12.2.3) and (12.3.4). Correspondingly $\check{R}(u)_{ij}^{kl} = R(u)_{ji}^{kl}$, satisfying

$$\check{R}(u)_{ij}^{ml} \check{R}(u+v)_{lk}^{nt} \check{R}(v)_{mn}^{rs} = \check{R}(v)_{jk}^{lm} \check{R}(u+v)_{il}^{rn} \check{R}(u)_{nm}^{st}. \quad (12.3.7)$$

The relation with space label is

$$\check{R}_{12}(u)\check{R}_{23}(u+v)\check{R}_{12}(v) = \check{R}_{23}(v)\check{R}_{12}(u+v)\check{R}_{23}(u). \quad (12.3.8)$$

To understand the equivalence between the space label notation and the matrix indices notation, we use the diagram method shown in Fig. 12.2. $\check{R}(u)_{ij}^{lk}$ represents a particle with internal degree of freedom i collides with a particle with internal degree of freedom j with a relative rapidity u, and after the collision the internal degrees of freedom become k and l respectively. If this internal degree of freedom is the spin, then by spin conservation $i + j = k + l$. Space

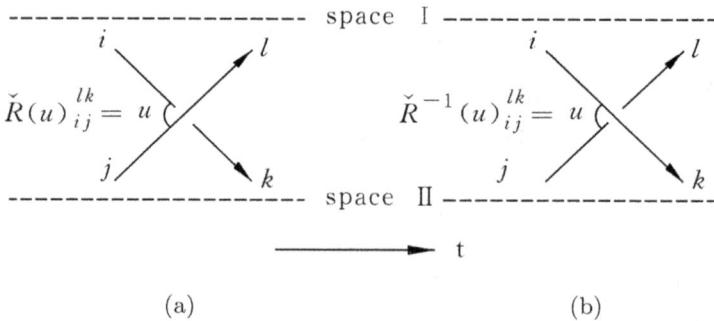

Fig. 12.2   Diagram for $\check{R}$ and $\check{R}^{-1}$.

is labeled by I and II, for instance after collision the particle in final state l is in space I and that with final state k is in space II, the crossing in the figure represents the collision. $\check{R}$ and $\check{R}^{-1}$ are represented differently by the different crossings of the lines representing the particles. The convention is: for $\check{R}$ the particle with final state in space I lies on top in the crossing while for $\check{R}^{-1}$ the opposite way round. This is determined by the relation $\check{R}\check{R}^{-1} = I$. Figure 12.3 depicts the case where collision happens twice due to $\check{R}$ and $\check{R}^{-1}$. The two lines representing particles are such that one always lies on top. These lines can be stretched and become two parallel lines without crossing at all, which is the identity I. Diagrams like this depends only on the number of crossings and property of the crossings, being "overcrossing" or "undercrossing". Continuous deformation of the lines without changing the crossing property is a topological transformation. We see that $\check{R}\check{R}^{-1}$ and $\check{R}^{-1}\check{R}$ are topologically equivalent to I. Notice that $\check{R} = PR$.

The RTT relation

$$R(u - v)_{ij}^{mn} T(u)_m^k T(v)_n^l = T(v)_j^m T(u)_i^n R(u - v)_{nm}^{kl} \qquad (12.3.9)$$

is shown diagrammatically in Fig. 12.4 Those two crossings represent $R(u - v)_{ij}^{mn}$ and $R(u - v)_{nm}^{kl}$ respectively. In the left diagram, R_{ij}^{mn} operates on T_n^l and T_m^k, transferring particle 1 (rapidity u) from the state m to state k, and particle 2 (rapidity v) from the state n to state l, the dashed line represents the labels of quantum spaces (not indicated). m and n are summation indices. The right diagram is interpreted similarly to the left diagram. Their equality means that

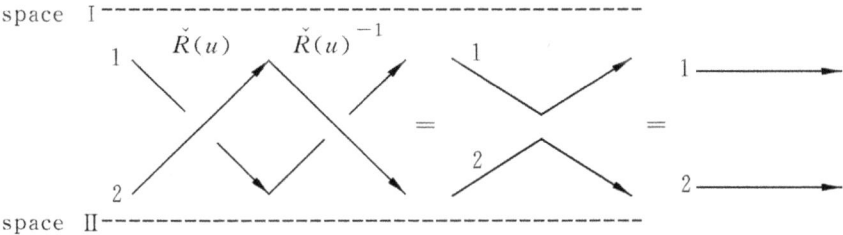

Fig. 12.3   Diagram showing $\check{R}\check{R}^{-1} = 1$.

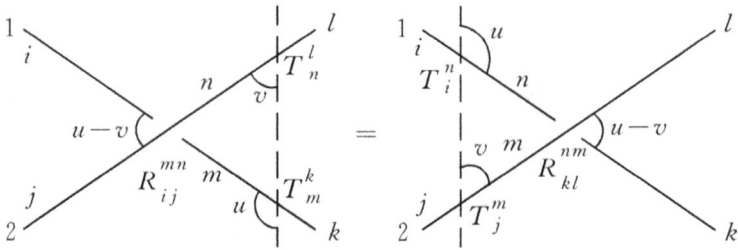

Fig. 12.4   The RTT relation.

moving continuously the dashed line in the left diagram changes the latter to the right diagram without changing the topological properties of the diagrams.

The process of finding a solution is the following. First solve the YBE, obtaining a solution $R(u)$. Then solve the RTT relation for the given $R(u)$, then one gets the matrix elements of the matrix $T(u)$ as a realization of quantum mechanical operators. The realization has to ensure $T(u)$ expanded to a certain order of u to satisfy the RTT relation. Eq. (12.2.10) gives an example of such a realization.

The importance of the RTT relation is that it does not only give the commutation relation, but also the conserved quantities of the quantum system, including the Hamiltonian, i.e., it fixes the dynamics of the system. The RTT relation is given by L.D. Faddeev and collaborators [1–3, 9] in establishing the second quantized inverse scattering method. It was developed further in mathematics by V.G. Drinfeld [6, 7] to a new type of algebraic theory.

Before going to the next paragraph let us obtain an interesting consequence of the RTT relation with given simple form of the permutation,

$$R_{12} = P_{12} = \frac{1}{2}I + 2\sum_{i=1}^{3} t_i(1)t_i(2), \quad t_i = \frac{1}{2}\sigma_i, \tag{12.3.10}$$

where $t_i(1)$ indicates the operator acts on the first auxiliary space and

$$[t_j(a), t_k(b)] = i\delta_{ab}\varepsilon_{jkl}t_l(a), \quad a, b = 1, 2. \tag{12.3.11}$$

The RTT relation (12.3.1) becomes

$$(u - v)\left(P_{12} \overset{1}{T}(u) \otimes \overset{2}{T}(v) - \overset{1}{T}(v) \otimes \overset{2}{T}(u) P_{12} \right)$$

$$= \overset{1}{T}(v) \overset{2}{T}(u) - \overset{1}{T}(u) \overset{2}{T}(v). \tag{12.3.12}$$

By multiplying P_{12} on the left-hand side of (12.3.12) and noting $(P_{12})^2 = I$ we obtain

$$(u - v)\left[\overset{1}{T}(u), \overset{2}{T}(v) \right] = P_{12}\left(\overset{1}{T}(v) \overset{2}{T}(u) - \overset{1}{T}(u) \overset{2}{T}(v) \right). \tag{12.3.13}$$

Since $t_i (i = 1, 2, 3)$ is the usual spin operator we introduce 4-dimensional vector.

$$T_i(u) = \mathrm{tr}(t_i T(u)), \qquad T_0(u) = \mathrm{tr} T(u). \tag{12.3.14}$$

Multiplyng $t_i(1) t_j(2)$ on the left-hand side of the (12.3.13) and taking trace to space 1 and 2 respectively, on account of (12.3.10) we obtain

$$(u - v)[T_i(u), T_j(v)] = \mathrm{tr}_{12} t_i(1) t_j(2) P_{12} \cdot \left(\overset{1}{T}(v) \overset{2}{T}(u) - \overset{1}{T}(u) \overset{2}{T}(v) \right). \tag{12.3.15}$$

Noting that $\mathrm{tr}(t_i) = 0$, $t_j t_k = i t_{jkl} t_l \ (j \neq k)$ we get

$$(u - v) \left[T_i(u), T_j(v) \right]$$

$$= \frac{1}{2} \left[T_i(v), T_j(u) \right] - \left[T_i(u), T_j(v) \right]$$

$$+ \frac{i}{4} \varepsilon_{ijk} [T_0(u), T_k(v) - T_0(v), T_k(u)$$

$$+ T_k(v) T_0(u) - T_k(u) T_0(v)]. \tag{12.3.16}$$

On the other hand by interchanging u and v in the (12.3.13) and making comparison with (12.3.13) itself we see

$$\left[\overset{1}{T}(u), \overset{2}{T}(v) \right] = \left[\overset{1}{T}(v), \overset{2}{T}(u) \right], \tag{12.3.17}$$

which is obvious, because $\overset{1}{T}$ and $\overset{2}{T}$ belong to the different auxiliary spaces. Multiplying $t_i(1) t_j(2)$ and $t_i(1)$, respectively, and taking trace with respect to two spaces, i.e., tr_{12}, we have

$$[T_i(u), T_j(v)] = [T_i(v), T_j(u)] , \qquad (12.3.18)$$

$$[T_i(u), T_0(v)] = [T_i(v), T_0(u)] .$$

Substituting (12.3.18) into (12.3.16) one obtains

$$[T_i(u), T_j(v)] = \frac{i}{2}\varepsilon_{ijk} \left(T_0(u)T_k(v) - T_0(v), T_k(u)\right) \diagup (u - v).$$
$$(12.3.19)$$

Furthermore, left-multiplying $t_i(1)$ to (12.3.13) and taking trace to spaces 1 and 2 leads to

$$(u - v)[T_i(u), T_0(v)] = \frac{1}{2}\left([T_i(v), T_0(u)] - [T_i(u), T_0(v)]\right)$$

$$+ i\varepsilon_{ikl}(T_i(v)T_k(u) - T_l(u)T_k(v)). \quad (12.3.20)$$

On account of (12.3.19) we derive

$$[T_i(u), T_0(v)] = i\varepsilon_{ijk}(T_k(v), T_j(u) - T_k(u), T_j(v))/(u - v).$$
$$(12.3.21)$$

It is easy to write (12.3.19) and (12.3.21) to a compact form with defining

$$j_0 = \tfrac{1}{2}T_0, \quad j_k = T_k, \quad k = 1, 2, 3$$
$$(12.3.22)$$
$$j_\mu = g_{\mu\nu}j^\nu, \, g_{\mu\nu} = \text{diag}(1, -1, -1, -1).$$

On account of tr (13.3.13) there holds

$$[T_0(u), T_0(v)] = 0. \qquad (12.3.23)$$

Summarizing (12.3.19), (12.3.21) and (12.3.23) we finally have

$$[j_\mu(u), j_\nu(v)] = [j_\mu(v), j_\nu(u)] = \frac{\frac{i}{2}\varepsilon_{\mu\nu\rho\sigma}\{j^\rho(u)j^\sigma(v) - j^\rho(v)j^\sigma(u)\}}{(u - v)}$$
$$(12.3.24)$$

that is the 4-current form of the global transition matrix. We emphasize that the currents j_μ take the Minkowski Lorentz metric (12.3.22) and contain the spectral parameters u and v. Therefore, the parametrized current is an extension of the loop algebra and the consequence of RTT relation. The 4-current form of RTT relation was first presented by Lipatov [27].

12.4   Sets of Conserved Quantities, the Hamiltonian

In mechanics when all the integrals of motion (the conserved quantities) are known, the system is called completely integrable. In quantum mechanics the corresponding system is called quantum integrable. The transition matrix introduced in Section 12.2 is usually closely connected with the family of conserved quantities. The RTT relation expressed in terms of $\check{R}$ is

$$\check{R}(u - v)\,(T(u) \otimes T(v)) = (T(v) \otimes T(u))\,\check{R}(u - v). \qquad (12.4.1)$$

$\check{R}$ is a non-singular matrix, so its inverse exists. Therefore, we have

$$T(u) \otimes T(v) = \check{R}^{-1}(u - v)\,(T(v) \otimes T(u))\,\check{R}(u - v).$$

Since $\mathrm{tr}A \otimes B = \mathrm{tr}A\,\mathrm{tr}B$, we have after taking trace of both sides of the above equation

$$\mathrm{tr}T(u)\mathrm{tr}T(v) = \mathrm{tr}T(v)\mathrm{tr}T(u),$$

i.e.,

$$[\mathrm{tr}T(u), \mathrm{tr}T(v)] = 0. \qquad (12.4.2)$$

Define

$$\tau(u) = \mathrm{tr}T(u) = \sum_{n} u^{-n}\tau^{(n)}, \qquad (12.4.3)$$

From the above two relations we get

$$\left[\tau^{(n)}, \tau^{(m)}\right] = 0. \qquad (12.4.4)$$

In (12.4.3) the limit of the summation was not given. They depend on the form of the solution $T(u)$ of the RTT relation. Sometimes we take $\infty > n \geq 0$; sometimes the sum is taken over positive integers. In general the sum is taken over the range from $-\infty$ to $+\infty$.

From the above discussion we see that when a physical realization of $T(u)$ is known, by taking the trace of the matrix and expanding in powers of the spectral parameter it is possible to obtain all mutually commuting conserved quantities $\tau^{(n)}$ of the system. We can choose one or a linear combination of some of them to be the Hamiltonian. Such a choice is usually guided by the physical significance. Sometimes the classical limit of the system can provide some reference. Besides, a system satisfying the RTT relation is definitely quantum integrable. Although a concrete realization often is not simple, the theoretical framework is very definite. The RTT relation does not only give the commutation relations of algebras, but also the complete set of conserved quantities, and serves to justify the completely quantum integrability of the system.

If $\tau(u)$ is expanded in an infinite power series of u^{-1}, there are infinite number of conserved quantities. The classical soliton solution has infinite conserved quantities. Roughly speaking here we have its quantum counterpart (even the statement has only verified at low level). When the expansion stops at a finite order, then we have a finite number of conserved quantities that usually occurs in more than one-dimensions. Its number depends on the degree of freedom of the system. The typical example is Chaplygin top [8].

We can use $\mathrm{tr}T(u)$, or $\ln \mathrm{tr}T(u)$ in the expansion in comparison to its classical correspondence. For one-dimensional lattice model with long-range interaction there is a specific form. In the following as an example we discuss the Toda lattice with nearest neighbor interaction. The local transition matrix on the site n is

$$L_n(u) = \begin{bmatrix} 1 & 0 \\ 0 & 0 \end{bmatrix} + u^{-1} \begin{bmatrix} -p_n & e^{iq_n} \\ e^{-iq_n} & 0 \end{bmatrix}. \qquad (12.4.5)$$

The global transition matrix is

$$T(u) = \prod_{n=1}^{N} L_n(u) = \begin{bmatrix} A(u) & B(u) \\ C(u) & D(u) \end{bmatrix}.$$

Because the constant matrix in $L_n(u)$ is $\begin{bmatrix} 1 & 0 \\ 0 & 0 \end{bmatrix}$ it leads to $D(u) = 0$. Actually, we have

$$
T(u) = \left\{ \begin{bmatrix} 1 & 0 \\ 0 & 0 \end{bmatrix} + u^{-1} \begin{bmatrix} -p_N & e^{iq_N} \\ e^{-iq_N} & 0 \end{bmatrix} \right\}
$$

$$
\left\{ \begin{bmatrix} 1 & 0 \\ 0 & 0 \end{bmatrix} + u^{-1} \begin{bmatrix} -p_{N-1} & e^{iq_{N-1}} \\ e^{-iq_{N-1}} & 0 \end{bmatrix} \right\}
$$

$$
\cdots \left\{ \begin{bmatrix} 1 & 0 \\ 0 & 0 \end{bmatrix} + u^{-1} \begin{bmatrix} -p_1 & e^{iq_1} \\ e^{-iq_1} & 0 \end{bmatrix} \right\}. \tag{12.4.6}
$$

In the following, we calculate the lower order terms in the expansion of $T(u)$ in the power series of u^{-1}. Taking trace, we have

$$
\mathrm{tr}T(u) = A_0 + u^{-1}A_1 + u^{-2}A_2 + \cdots \tag{12.4.7}
$$

where

$$
A_0 = 1,
$$

$$
A_1 = -\sum_{n=1}^{N} p_n,
$$

$$
A_2 = \sum_{n,m,n>m}^{N} p_n p_m + \sum_{n=1}^{N-1} e^{i(q_{n+1}-q_n)}.
$$

The calculation of A_2 is as follows. The two matrices contributing to order u^{-2} can be neighbors on the lattice, or not be. If they are neighbors, the contribution is $\sum_n p_{n+1}p_n + \sum_n \exp[i(q_{n+1} - q_n)]$; and if they are not, there is a constant matrix $\begin{pmatrix} 1 & 0 \\ 0 & 0 \end{pmatrix}$ between the matrices, resulting in $\sum_{n,m,n>m} p_n p_m$. In order to get a classical correspondence, we take $\ln \mathrm{tr}T(u)$ first and then expand. Using

$$
\ln(1+x) = x - \frac{1}{2}x^2 + \cdots,
$$

the corresponding x in (12.4.7) is

$$
x = u^{-1}A_1 + u^{-2}A_2 + \cdots,
$$

and therefore we have

$$\ln \mathrm{tr} T(u) = u A_1 + u^{-2} \left(A_2 - \frac{1}{2} A_1^2 \right) + \cdots$$

$$= -uP - u^{-2} H + \cdots .$$

Since $-\frac{1}{2} A_1^2 = -\frac{1}{2} \left(\sum_{n=1}^{N} p_n \right)^2 = -\sum_{n,m,n>m}^{N} p_n p_m - \frac{1}{2} \sum_{n=1}^{N} p_n^2$, the first term cancels with the first term of A_2, then the conserved quantity is

$$\left. \begin{array}{l} -P = -\displaystyle\sum_{n=1}^{N} p_n \\[2ex] +H = \displaystyle\sum_{n=1}^{N} \left(\frac{1}{2} p_m^2 + e^{i(q_n - q_{n-1})} \right) \\[1ex] \cdots \end{array} \right\}, \qquad (12.4.8)$$

where P is the total momentum, and H is the Hamiltonian of the Toda lattice. In H there only appears the interaction between the nearest neighbors. If we use $\begin{bmatrix} 1 & 0 \\ 0 & 1 \end{bmatrix}$ instead of $\begin{bmatrix} 1 & 0 \\ 0 & 0 \end{bmatrix}$ in choosing the local transition matrix, long-range interaction appears.

To summarize, there are three steps in the above process: (1) choose $R(u)$, and find the transition matrix $L(u)$; (2) ascribe subscripts of the lattice sites to L to form the global transition matrix $T(u) = \prod_{1}^{\overleftarrow{N}} L_n(u)$; (3) Expand $\mathrm{tr} T(u)$ or $\ln \mathrm{tr} T(u)$ in power series of u^{-1}, and the coefficients of various orders are the conserved quantities. The general method was created by Faddeev and collaborators in their quantum inverse scattering theory. Under special circumstances the Hamiltonian can be directly found from the R matrix, and the process is then greatly simplified. The approach was presented by Baxter [5]. Let us introduce briefly the basic idea.

Each element of the matrix $L(u)$ is an operator in the Hilbert space. If we take the matrix element between the quantum states a, b of Eq. (12.1.33), the result will be

$$R(u)_{ij}^{nm} \left(L_{ac}(u+v) \right)_n^t \left(L_{cb}(v) \right)_m^s = \left(L_{ac}(v) \right)_j^m \left(L_{cb}(u+v) \right)_i^n R(u)_{nm}^{ts}.$$

$$(12.4.9)$$

Define matrix $\hat{R}(u)$:

$$\hat{R}(u)_{ij}^{mn} = R(u)_{ij}^{mn}, \tag{12.4.10}$$

which interchanges the two following indices of R, while $\check{R}$ interchanges the foregoing indices of R. We define further

$$\tilde{R}(u)_{na}^{ct} = (L_{ac}(u))_n^t, \tag{12.4.11}$$

its indices come from the auxiliary space (n, t), and also from the Hilbert space (a, c). Equation (12.4.9) can now be written as

$$\hat{R}(u)_{ij}^{mn} \tilde{R}(u+v)_{na}^{ct} \tilde{R}(v)_{mc}^{bs} = \tilde{R}(v)_{ja}^{cm} \tilde{R}(u+v)_{ic}^{bn} \hat{R}(u)_{nm}^{st}. \tag{12.4.12}$$

On comparing it with the YBE equation (12.3.7), we see that $\tilde{R}$ satisfies the same equation as $\hat{R}$. Therefore, the simplest solution for $\tilde{R}$ is

$$\tilde{R}(u) = \hat{R}(u). \tag{12.4.13}$$

Under a special circumstance, namely that the range of values taken by the auxiliary space indices $i, j, k, \ldots$ is the same as that taken by the Hilbert space indices, we can use $\tilde{R}$ to define the global transition matrix:

$$\left\{ T^{(N)}(u)_{\boldsymbol{a}}^{\boldsymbol{b}} \right\}_{i_1}^{i_{N+1}} = (L_{a_1 b_1}(u))_{i_1}^{i_2} (L_{a_2 b_2}(u))_{i_2}^{i_3} \cdots (L_{a_N b_N}(u))_{i_N}^{i_{N+1}}, \tag{12.4.14}$$

where

$$\boldsymbol{a} = (a_1, a_2, \ldots, a_N), \quad \boldsymbol{b} = (b_1, b_2, \ldots, b_N).$$

In (12.4.14) the Hilbert space indices depend on lattice sites, there is no summation between sites, while there is the product of N matrices in the auxiliary space, and the relevant indices are summed over. Using the definition of (12.4.11), we have

$$\left\{ T^{(N)}(u)_{\boldsymbol{a}}^{\boldsymbol{b}} \right\}_{i_1}^{i_{N+1}} = \hat{R}(u)_{i_1 a_1}^{b_1 i_2} \hat{R}(u)_{i_2 a_2}^{b_2 i_3} \cdots \hat{R}(u)_{i_N a_N}^{b_N i_{N+1}}. \tag{12.4.15}$$

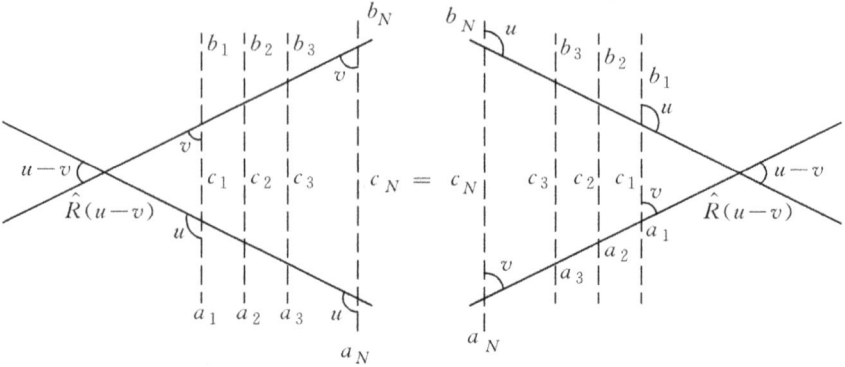

Fig. 12.5   Diagram of the RTT relation.

The RTT relation is

$$\hat{R}(u-v)\left(T^{(N)}(u)_{\boldsymbol{a}}^{\boldsymbol{c}} \otimes T^{(N)}(v)_{\boldsymbol{c}}^{\boldsymbol{b}}\right) = \left(T^{(N)}(v)_{\boldsymbol{a}}^{\boldsymbol{c}} \otimes T^{(N)}(u)_{\boldsymbol{c}}^{\boldsymbol{b}}\right)\hat{R}(u-v).$$
$$(12.4.16)$$

Since $\hat{R}$ operates in the auxiliary space, it does not influence the indices in the Hilbert space. The diagrams shown in Fig. 12.5 for demonstrating Eq. (12.4.16) is an extension of Fig. 12.4, In the figure a_i and b_j are free indices.

In any one of the diagrams we translate the dashed lines to cross over the intersection point while keeping their relative distances from the crossing, then we get the other diagram, those two diagrams representing the two sides of (12.4.16) are equal. Taking trace in the auxiliary space, we get [8]

$$\left(\tau^{(N)}(u)\right)_{\boldsymbol{a}}^{\boldsymbol{b}} \equiv \operatorname{tr}\left(T^{(N)}(u)\right)_{\boldsymbol{a}}^{\boldsymbol{b}} = \left\{T^{(N)}(u)_{\boldsymbol{a}}^{\boldsymbol{b}}\right\}_{i_1}^{i_1}$$
$$= \hat{R}(u)_{i_1 a_1}^{b_1 i_2}\hat{R}(u)_{i_2 a_2}^{b_2 i_3} \cdots \hat{R}(u)_{i_{N-1}a_{N-1}}^{b_{N-1}i_N}\hat{R}(u)_{i_N a_N}^{b_N i_1}.$$
$$(12.4.17)$$

Since $\check{R}$ and $\hat{R}$ have no difference in the sense of YBE, we now demand

$$\hat{R}\left(u = 0\right) = I,\qquad\qquad (12.4.18)$$

which is the equivalence of $\check{R}(u = 0) = I$ which we imposed on the solution of YBE. When the range of values taken by indices in the auxiliary space and the Hilbert space is the same, we use

$$\hat{R}(0)^{b_1 i_2}_{i_1 a_1} = (I \otimes I)^{b_1 i_2}_{i_1 a_1} = \delta_{i_1 b_1} \delta_{a_1 i_2},$$

where $\hat{R}$ is a $M^2 \times M^2$ matrix, I is the unit matrix, to obtain

$$\left\{ T^{(N)}(u)^{b}_{a} \right\}^{i_1}_{i_1} = \underbrace{\delta^{b_1}_{i_1} \delta^{i_2}_{a_1} \cdot \delta^{b_2}_{i_2} \delta^{i_3}_{a_2} \cdots \cdots \delta^{b_N}_{i_N} \delta^{i_1}_{a_N}}_{2N \text{ Kronecker } \delta\text{'s}} = \underbrace{\delta^{b_1}_{a_N} \delta^{b_2}_{a_1} \delta^{b_3}_{a_2} \cdots \delta^{b_N}_{a_{N-1}}}_{N \text{ Kronecker } \delta\text{'s}}.$$

(12.4.19)

$$\left(T^{(N)}(u)^{b}_{a} \right)^{i_1}_{i_1} \Big|_{u=0} = \delta^{b_1}_{a_2} \delta^{b_2}_{a_3} \cdots \delta^{b_{N-1}}_{a_N} \delta^{b_N}_{a_1} \qquad (12.4.20)$$

Then we have

$$\frac{d}{du} \ln \left\{ \tau^{(N)}(u) \right\}^{b}_{a} \Big|_{u=0}$$

$$= \left\{ \tau^{(N)}(u)^{-1} \right\}^{b}_{c} \frac{d}{du} \left\{ \tau^{(N)}(u) \right\}^{c}_{a} \Big|_{u=0}$$

$$= \sum_{i=1}^{N} \left(\delta^{b_1}_{a_1} \delta^{b_2}_{a_2} \cdots \delta^{b_{i-1}}_{a_{i-1}} \frac{d}{du} \hat{R}(u)^{b_i b_{i+1}}_{a_i a_{i+1}} \Big|_{u=0} \delta^{b_{i+2}}_{a_{i+2}} \cdots \delta^{b_N}_{a_N} \right).$$

(12.4.21)

Written in the form of direct product, it becomes

$$\frac{d}{du} \ln \tau^{(N)}(u) \Big|_{u=0} = \sum_{i=1}^{N} \overset{(1)}{I} \otimes \overset{(2)}{I} \otimes \cdots \otimes \overset{(i-1)}{I} \otimes \frac{d\hat{R}(u)^{(i,i+1)}}{du} \Big|_{u=0}$$

$$\otimes I^{(i+2)} \otimes \cdots \otimes I^{(N)}. \qquad (12.4.22)$$

Equation (12.4.22) gives the coefficient (operator) of the first order in u in the expansion of $\ln \tau^{(N)}(u)$. Some models adopt it as the Hamiltonian. By this choice the local Hamiltonian given by (12.4.22) [5, 8] is

$$H_{i,i+1} = \frac{d}{du} \hat{R}(u)^{(i,i+1)} \Big|_{u=0}, \qquad (12.4.23)$$

while the Hamiltonian of the system is

$$H = \sum_{i=1}^{N} H_{i,i+1}. \tag{12.4.24}$$

Since $\hat{R}(u)$ and $\check{R}(u)$ both are solutions of the YBE, we can also choose

$$H_{i,i+1} = \frac{\mathrm{d}}{\mathrm{d}u} \check{R}(u)^{(i,i+1)}\Big|_{u=0}. \tag{12.4.25}$$

The above discussion has proved that when the dimension of the quantum space is the same as the dimension of the auxiliary space for a class of integrable systems, for instance the space for the states and the auxiliary space of spin $\frac{1}{2}$ are both with dimension 2, and there is a shortcut in choosing the coefficient of the first order in the expansion of $\ln \tau(u)$ as the Hamiltonian instead of taking all the steps outlined by the quantum inverse scattering methods. It is sufficient to choose the solution of YBE, localize it, take the derivative with respect to u and then set $u = 0$, finally the Hamiltonian is obtained.

Since now $\check{R}(u)$ interchanges two neighboring spaces, the simplest solution $\check{R}(u) = I + uP$ in the case of a lattice is

$$\check{R}(u)^{(i,i+1)} = I + uP_{i,i+1}.$$

The coefficient of the first order in u is P, therefore

$$H = \sum_{i=1}^{N} H_{i,i+1} = \sum_{i=1}^{N} P_{i,i+1}. \tag{12.4.26}$$

The interchange operator in (12.4.26) can be expressed as

$$P_{i,i+1} = \frac{1}{2}(1 + \boldsymbol{\sigma}_i \cdot \boldsymbol{\sigma}_{i+1}),$$

in the case of spin $\frac{1}{2}$, so we obtain

$$H = \frac{1}{2}\sum_{i=1}^{N}(1 + \boldsymbol{\sigma}_i \cdot \boldsymbol{\sigma}_{i+1}). \tag{12.4.27}$$

This is the Hamiltonian of the isotropic Heisenberg chain model. For a spin different from $\frac{1}{2}$ it is more complicated to express $P_{i,i+1}$ in terms of the spin operators. The above prescription can be applied to the spin models with nearest neighbor interactions.

Finally we comment on the expansion in the powers of the spectral parameter. Earlier we considered an expansion in powers of u^{-1}, and in the above we used expansion in powers of u. Because of the RTT relation, we expand T in powers of u^{-1} if we choose positive powers of u in R, and vice versa. In many problems in physics the poles of certain analytic functions and the residue at the poles are of importance. This is the reason to get the Laurent series, and the two possible choices explained above. If the function is analytic everywhere, the physics becomes trivial.

The readers can refer to Refs. [1, 2, 5, 8] for the contents of this section.

12.5   The Quantum Determinant, and the Co-product

When the auxiliary space is $SL(2)$, the transition matrix $T(u)$ is a 2×2 matrix, but its matrix elements are operators. For a c-number matrix, its inverse depends on the determinant of the matrix. To find the inverse of a transition matrix we need to find the corresponding "determinant" of the transition matrix. It should commute with any matrix element (operator). Denote this determinant by $\det T(u)$, and call it the "quantum determinant".

$$\det T(u) = T_{11}(u)T_{22}(u-1) - T_{12}(u)T_{21}(u-1), \qquad (12.5.1)$$

It satisfies

$$[\det T(u), T_{ab}(v)] = 0. \qquad (12.5.2)$$

Before proving (12.5.2) we need to derive another relation. In the RTT relation

$$\check{R}(u-v)\,(T(u) \otimes T(v)) = (T(v) \otimes T(u))\,\check{R}(u-v),$$

when we take

$$\check{R}(u) = I + uP,$$

1002 *Frontier Problems in Quantum Mechanics*

and use the expression for the interchange matrix P:

$$P^{cd}_{ab} = \delta^d_a \delta^c_b, \quad a, b, c, d = 1, 2. \tag{12.5.3}$$

RTT relation gives

$$(u - v)\left[T_{bc}(u), T_{ad}(v)\right] + T_{ac}(u)T_{bd}(v) - T_{ac}(v)T_{bd}(u) = 0. \tag{12.5.4}$$

We set $a = b, c = d$ and consider that u and v are arbitrary, then we have

$$[T_{ac}(u), T_{ac}(v)] = 0. \tag{12.5.5}$$

To prove (12.5.2), we take the matrix element T_{11} as an example. Using (12.5.5), we have

$$\begin{aligned}
[\det T(u), T_{11}(v)] &= T_{11}(u)\left[T_{22}(u-1), T_{11}(v)\right] \\
&\quad - [T_{12}(u), T_{11}(v)]T_{21}(u-1) \\
&\quad - T_{12}(u)\left[T_{21}(u-1), T_{11}(v)\right].
\end{aligned}$$

Using the relation $[T_{bc}(u-1), T_{ad}(v)] = (u-v-1)^{-1}(T_{ac}(v)T_{bd}(u) - T_{ac}(u)T_{bd}(v))$ as implied by (12.5.4) substituted into the above equation, we get

$$\begin{aligned}
[\det &T(u), T_{11}(v)] \\
&= (u-v-1)^{-1}T_{11}(u)\left(T_{12}(v)T_{21}(u-1) - T_{12}(u-1)T_{21}(v)\right) \\
&\quad - (u-v)^{-1}(T_{11}(v)T_{12}(u) - T_{11}(v)T_{12}(u))T_{21}(u-1) \\
&\quad - (u-v-1)^{-1}T_{12}(u)(T_{11}(v)T_{21}(u-1) - T_{11}(u-1)T_{21}(v)) \\
&= (u-v-1)^{-1}(u-v)^{-1}\{(u-v)[T_{11}(v), T_{12}(u)] \\
&\quad + T_{11}(u)T_{12}(v) - T_{11}(v)T_{12}(u)\}T_{21}(u-1) \\
&= 0.
\end{aligned}$$

The last step was the result of putting $b = c = 1, a = 1, d = 2$ in (12.5.4). Similarly we can prove that $\det T(u)$ commutes with all other matrix elements of $T(v)$. Obviously $\det T(u)$ commutes with

$\mathrm{tr}T(v) = T_{11}(v) + T_{22}(v).$

$$[\det T(u), \mathrm{tr}T(v)] = 0. \tag{12.5.6}$$

We expand $T_{ab}(u)$ and $\det T(u)$:

$$T_{ab}(u) = I + \sum_{n=1}^{\infty} u^{-n} T_{ab}^{(n)}, \tag{12.5.7}$$

$$\det T(u) = C_0 + \sum_{n=1}^{\infty} u^{-n} C_n. \tag{12.5.8}$$

Equation (12.5.2) gives

$$\left[C_n, T_{ab}^{(m)} \right] = 0, \tag{12.5.9}$$

and consequently

$$\left[C_n, \tau^{(m)} \right] = 0. \tag{12.5.10}$$

$\{C_n\}$ comprises a family of conserved quantities

$$[C_n, C_m] = 0, \tag{12.5.11}$$

which commute with $\{\tau^{(m)}\}$. The difference between C_n and $\tau^{(m)}$ is that C_n commutes with $T_{ab}^{(m)}$, but in general τ_n does not commute with $T_{ab}^{(m)}$. Noting that the family of conserved quantities consists of mutually commuting quantities, there is no requirement that they should commute with $T_{ab}^{(m)}$. The Hamiltonian of some models is connected with C_n, and others with $\tau^{(m)}$.

Since $\det T(u)$ commutes with T_{ab}, we can introduce $(T(u))^{-1}$:

$$(T(u))^{-1} = (\det T(u))^{-1} \begin{bmatrix} T_{22}(u-1) & -T_{12}(u-1) \\ -T_{21}(u-1) & T_{11}(u-1) \end{bmatrix}. \tag{12.5.12}$$

We used $u - 1$ in defining det, actually $u - a$ can do as well, where a is an arbitrary constant. Another essential element of the Yangian is the *co-product*. It is an operation [6, 7], denoted by Δ. Operating on the product of two operators it gives

$$\Delta(AB) = \Delta(A)\Delta(B), \qquad (12.5.13)$$

Its operation on the elements (operators) of the transition matrix is defined as [2, 3]

$$\Delta(T_{ab}) = \sum_{c} T_{ac} \otimes T_{cb} \equiv T_{ac} \otimes T_{cb}. \qquad (12.5.14)$$

We use the direct product to represent operators belonging to independent quantum spaces. When two direct products of Bose operators are multiplied we have

$$(A \otimes B)(C \otimes D) = AC \otimes BD, \qquad (12.5.15)$$

where orders of operators are kept in both spaces. The definition of co-product is required to guarantee that when Δ operates on the product of operators/elements of $T(u)$ the RTT relation remains to be valid in the tensor space of the quantum spaces:

$$\Delta[T_{bc}(u), T_{ad}(v)]$$
$$= \Delta[T_{bc}(u)T_{ad}(v) - T_{ad}(v)T_{bc}(u)]$$
$$= \Delta(T_{bc}(u))\,\Delta(T_{ad}(v)) - \Delta(T_{ad}(v)T_{bc}(u))$$
$$= (T_{bn}(u) \otimes T_{nc}(u))(T_{am}(v) \otimes T_{md}(v))$$
$$\quad - (T_{an}(v) \otimes T_{nd}(v))(T_{bm}(u) \otimes T_{mc}(u)).$$

In the above we have used (12.5.13) and (12.5.14), and repeating indices n and m are to be summed. Using (12.5.15) we have

$$\Delta[T_{bc}(u), T_{ad}(v)]$$
$$= T_{bn}(u)T_{am}(v) \otimes T_{nc}(u)T_{md}(v)$$
$$\quad - T_{am}(v)T_{bn}(u) \otimes T_{nc}(u)T_{md}(v)$$
$$\quad + T_{am}(v)T_{bn}(u) \otimes T_{nc}(u)T_{md}(v)$$
$$\quad - T_{am}(v)T_{bn}(u) \otimes T_{md}(v)T_{nc}(u).$$

The second and third terms on the right-hand side of the above equation were deliberately added, their sum being zero. The dummy indices m and n in the fourth term have been interchanged. Therefore, we have

$$\Delta\left[T_{bc}(u), T_{ad}(v)\right]$$
$$= [T_{bn}(u), T_{am}(v)] \otimes T_{nc}(u)T_{md}(v)$$
$$+ T_{am}(v)T_{bn}(u) \otimes [T_{nc}(u), T_{md}(v)].$$

Multiplying the above equation by $(u - v)$, and applying (12.5.4) to the resultant right hand side, we get

$$(u - v)\Delta\left[T_{bc}(u), T_{ad}(v)\right]$$
$$= (T_{an}(v)T_{bm}(u) - T_{an}(u)T_{bm}(v)) \otimes T_{nc}(u)T_{md}(v)$$
$$+ T_{am}(v)T_{bn}(u) \otimes (T_{mc}(v)T_{nd}(u) - T_{mc}(u)T_{nd}(v))$$
$$= T_{an}(v)T_{bm}(u) \otimes T_{nc}(u)T_{md}(v)$$
$$- T_{an}(u)T_{bm}(v) \otimes T_{nc}(u)T_{md}(v)$$
$$+ T_{am}(v)T_{bn}(u) \otimes T_{mc}(v)T_{nd}(u)$$
$$- T_{am}(v)T_{bn}(u) \otimes T_{mc}(u)T_{nd}(v).$$

Interchanging the dummy indices n and m in the first term, we find that it cancels the fourth term. The result is

$$(u - v)\Delta\left[T_{bc}(u), T_{ad}(v)\right]$$
$$= (T_{am}(v) \otimes T_{mc}(v))(T_{bn}(u) \otimes T_{nd}(u))$$
$$- (T_{an}(u) \otimes T_{nc}(u))(T_{bm}(v) \otimes T_{md}(v))$$
$$= \Delta\left(T_{ac}(v)\right)\Delta\left(T_{bd}(u)\right) - \Delta\left(T_{ac}(u)\right)\Delta\left(T_{bd}(v)\right)$$
$$= \Delta(T_{ac}(v)T_{bd}(u) - T_{ac}(u)T_{bd}(v)).$$

This is just the result of operating Δ on (12.5.4), i.e., the co-product conserves the RTT relation. The definition of co-product (12.5.14) has a deeper significance. We know that the RTT relation gives the commutation relations between the operators/elements of $T_{ab}^{(n)}$, and they are realized in the same quantum space, the " vector" space. To be a complete operation, there should be a definition of an operation on the quantum *tensor* space which should satisfy a certain closure of the operations. Therefore, the introduction of the co-product is necessary, because it gives the rule for extension from the vector space to the tensor space. From the point of view of the RTT relation, the co-product Δ defined should guarantee that both sides of the RTT relation must be equal under the Δ operation, i.e., the relation should be valid in the quantum tensor space. This was the actual meaning in the foregoing proof. Strictly speaking, Yangian consists of two parts, one is the commutation relations, and the other is the co-product. Starting from the RTT relation there should be a consistency between the two parts: once we find the $T_{ab}(u)$ satisfying the RTT relation, the co-product is defined according to (12.5.14), everything is self-consistent.

When the auxiliary space is chosen to be 2×2, the general form of the expansion of $T(u)$ in powers of u^{-n} is

$$T(u) = I + \sum_{n=1}^{\infty} u^{-n} \begin{bmatrix} T_{11}^{(n)} & T_{12}^{(n)} \\ T_{21}^{(n)} & T_{22}^{(n)} \end{bmatrix}. \tag{12.5.16}$$

It is natural to use $(+, -, 3, 0)$ components of a 2×2 matrix:

$$\left.\begin{aligned} T_+^{(n)} &= T_{12}^{(n)}, & T_-^{(n)} &= T_{21}^{(n)} \\ T_3^{(n)} &= T_{22}^{(n)} - T_{11}^{(n)}, & T_0^{(n)} &= T_{22}^{(n)} + T_{11}^{(n)} \end{aligned}\right\} \tag{12.5.17}$$

$T(u)$ can be expressed in terms of them:

$$T(u) = I + \sum_{n=1}^{\infty} u^{-n} \begin{bmatrix} \frac{1}{2}\left(T_0^{(n)} + T_3^{(n)}\right) & T_+^{(n)} \\ T_-^{(n)} & \frac{1}{2}\left(T_0^{(n)} - T_3^{(n)}\right) \end{bmatrix}. \tag{12.5.18}$$

Introduce

$$\left.\begin{aligned} I_\pm &= T_\pm^{(1)}, & I_3 &= \frac{1}{2}T_3^{(1)} \\ J_\pm &= T_\pm^{(2)}, & J_3 &= \frac{1}{2}T_3^{(2)} \end{aligned}\right\} \tag{12.5.19}$$

then the transition matrix can be expressed as

$$T(u) = I + u^{-1} \begin{bmatrix} \frac{1}{2}T_0^{(1)} - I_3 & I_+ \\ \\ I_- & \frac{1}{2}T_0^{(1)} + I_3 \end{bmatrix}$$

$$+ u^{-2} \begin{bmatrix} \frac{1}{2}T_0^{(2)} - J_3 & J_+ \\ \\ J_- & \frac{1}{2}T_0^{(2)} + J_3 \end{bmatrix}$$

$$+ \sum_{n=3}^{\infty} u^{-n} \begin{bmatrix} \frac{1}{2}\left(T_0^{(n)} - T_3^{(n)}\right) & T_+^{(n)} \\ \\ T_-^{(n)} & \frac{1}{2}\left(T_0^{(n)} + T_3^{(n)}\right) \end{bmatrix}.$$

$$(12.5.20)$$

Using the notation of (12.5.20), it is easy to get the coefficients of various orders in the expansion of $\det T(u)$ (12.5.8).

$$\left. \begin{aligned} C_0 &= 1 \\ C_1 &= T_0^{(1)} = \mathrm{tr}T^{(1)} \\ C_2 &= T_0^{(2)} - \frac{1}{2}\left(2I_3^2 + I_+I_- + I_-I_+\right) + T_0^{(1)}\left(1 + \frac{1}{2}T_0^{(1)}\right) \\ &= T_0^{(2)} - I^2 + T_0^{(1)}\left(1 + \frac{1}{2}T_0^{(1)}\right) \\ &\cdots \end{aligned} \right\}$$

$$(12.5.21)$$

In Section 12.2, starting from $\check{R} = I + uP$, and applying the solution to the RTT relation $L_j(u) = I + u^{-1}S_j \cdot \sigma$ as local transition matrix, we constructed the global transition matrix $T(u)$, and its expansion in powers of u^{-1} (12.2.10) is

$$T(u) = I + u^{-1}S \cdot \sigma + u^{-2}\mathrm{i} \sum_{i,j,i\neq j} S_i \times S_j \cdot \sigma$$

$$+ u^{-2} \sum_{i,j,i>j} S_i \cdot S_j + \cdots.$$

Comparing with the general form (12.5.20), we have

$$T_0^{(1)} = 0, \quad T_0^{(2)} = \boldsymbol{I}^2, \tag{12.5.22}$$

$$I_\pm = S_\mp, \quad I_3 = -S_3, \tag{12.5.23}$$

$$J_\pm = i \sum_{i,j,i\neq j} (\boldsymbol{S}_i \times \boldsymbol{S}_j)_\mp, \quad J_3 = -i \sum_{i,j,i\neq j} (\boldsymbol{S}_i \times \boldsymbol{S}_j)_3. \tag{12.5.24}$$

Comparing (12.5.22) with (12.5.21), we get $C_0 = 1$, $C_1 = 0$, $C_2 = 0$. Therefore, to the order u^{-2} we have $\det T(u) = 1$. It can be rigorously proved that $\det T(u) = 1$. In the following, we study a specific case of the co-product. Introducing the quantum direct product symbol $\otimes$, which is to be distinguished from the direct product in the auxiliary space, we have

$$\Delta (T_{mn}(u)) = \sum_k (T_{mk}(u) \otimes T_{kn}(u)), \tag{12.5.25}$$

where T_{mn} is the matrix element of the transfer matrix $T(u)$. In the example we discussed above, $m, n = a, b = 1, 2$. Equation (12.5.25) is valid for any value of u, consequently we can consider its expansion in u^{-1}, u^{-2}, i.e., by substituting $T(u) = \sum_{n=0} u^{-n} T^{(n)}$ into the above equation. In the case of $\det T(u) = 1$ (i.e., for $Y(SL(2))$), we have

$$\Delta (T_{ab}(u)) = \sum_{c=1,2} (T_{ac}(u) \otimes T_{cb}(u)). \tag{12.5.26}$$

Owing to (12.5.23) and (12.5.24), $\boldsymbol{I}$ corresponds to the operator coefficient of u^{-1} and $\boldsymbol{J}$ corresponds to that of u^{-2}, and consequently it is sufficient to have an expansion up to u^{-1} and u^{-2}:

$$\left.\begin{aligned}
\Delta\left(T_{ab}^{(0)}\right) &= \sum_c \left(T_{ac}^{(0)} \otimes T_{cb}^{(0)}\right) \\
\Delta\left(T_{ab}^{(1)}\right) &= \sum_c \left(T_{ac}^{(1)} \otimes T_{cb}^{(0)} + T_{ac}^{(0)} \otimes T_{cb}^{(1)}\right) \\
\Delta\left(T_{ab}^{(2)}\right) &= \sum_c \left(T_{ac}^{(2)} \otimes T_{cb}^{(0)} + T_{ac}^{(0)} \otimes T_{cb}^{(2)} + T_{ac}^{(1)} \otimes T_{cb}^{(1)}\right)
\end{aligned}\right\} \tag{12.5.27}$$

The first equation of (12.5.27) gives $\Delta(1) = 1 \otimes 1$, as one expects. The other two equations of (12.5.27) become

$$\left. \begin{array}{l} \Delta\left(T_{ab}^{(1)}\right) = T_{ab}^{(1)} \otimes 1 + 1 \otimes T_{ab}^{(1)} \\[2mm] \Delta\left(T_{ab}^{(2)}\right) = T_{ac}^{(2)} \otimes 1 + 1 \otimes T_{ac}^{(2)} + \sum_c \left(T_{ab}^{(1)} \otimes T_{cb}^{(1)}\right) \end{array} \right\} \qquad (12.5.28)$$

Taking into account $T_0^{(n)} = T_{11}^{(n)} + T_{22}^{(n)}$, $T_3^{(n)} = T_{22}^{(n)} - T_{11}^{(n)}$, and substituting

$$T_{\pm}^{(1)} = I_{\pm},$$

$$T_{22}^{(1)} = -T_{11}^{(1)} = I_3,$$

$$T_{\pm}^{(2)} = (-2h)J_{\pm}, \quad T_3^{(2)} = \left(-\frac{4}{h}\right)J_3$$

into the above equations, we have

$$\Delta\left(I_\alpha\right) = I_\alpha \otimes 1 + 1 \otimes I_\alpha, (\alpha = \pm, 3), \qquad (12.5.29)$$

$$\Delta\left(J_\pm\right) = J_\pm \otimes 1 + 1 \otimes J_\pm \mp \frac{h}{2}(I_3 \otimes I_\pm - I_\pm \otimes I_3)$$

$$- \frac{h}{4}\left(I_\pm \otimes T_0^{(1)} + T_0^{(1)} \otimes I_\pm\right), \qquad (12.5.30)$$

$$\Delta\left(J_3\right) = J_3 \otimes 1 + 1 \otimes J_3 + \frac{h}{4}(I_+ \otimes I_- - I_- \otimes I_+)$$

$$- \frac{h}{4}\left(I_3 \otimes T_0^{(1)} + T_0^{(1)} \otimes I_3\right), \qquad (12.5.31)$$

$$\left. \begin{array}{l} \Delta T_0^{(1)} = T_0^{(1)} \otimes 1 + 1 \otimes T_0^{(1)} \\[2mm] \Delta(T_0^{(2)}) = T_0^{(2)} \otimes 1 + 1 \otimes T_0^{(2)} + 2I_3 \otimes I_3 \\[2mm] \qquad + I_+ \otimes I_- + I_- \otimes I_+ + \frac{1}{2}T_0^{(1)} \otimes T_0^{(1)} \end{array} \right\}.$$

$$(12.5.32)$$

If we choose $T_0^{(1)} = 0$, $T_0^{(2)} = \boldsymbol{I}^2$, then we get

$$\left.\begin{aligned}
\Delta\left(I_\lambda\right) &= I_\lambda \otimes 1 + 1 \otimes I_\lambda \\
\Delta\left(J_\lambda\right) &= J_\lambda \otimes 1 + 1 \otimes J_\lambda + \tfrac{h}{2}\mathrm{i}\varepsilon_{\lambda\mu\nu}I_\nu \otimes I_\mu \\
&\quad - \frac{h}{4}\left(I_\lambda \otimes T_0^{(1)} + T_0^{(1)} \otimes I_\lambda\right)
\end{aligned}\right\}, \qquad (12.5.33)$$

where $\lambda, \mu, \nu = 1, 2, 3$, and

$$\begin{aligned}
\Delta\left(\boldsymbol{I}^2\right) &= \Delta\left(I_3^2 + \frac{1}{2}I_+I_- + \frac{1}{2}I_-I_+\right) \\
&= (\Delta I_3)\,\Delta\left(I_3\right) + \frac{1}{2}\Delta\left(I_+\right)\Delta\left(I_-\right) + \frac{1}{2}\Delta\left(I_-\right)\Delta\left(I_+\right) \\
&= \boldsymbol{I}^2 \otimes 1 + 1 \otimes \boldsymbol{I}^2 + 2I_3 \otimes I_3 + I_+ \otimes I_- + I_- \otimes I_+.
\end{aligned}$$

Finally we obtain the definition of co-product originally given by Drinfeld:

$$\left.\begin{aligned}
\Delta\left(\boldsymbol{I}_\lambda\right) &= I_\lambda \otimes 1 + 1 \otimes I_\lambda \\
\Delta\left(\boldsymbol{J}_\lambda\right) &= J_\lambda \otimes 1 + 1 \otimes J_\lambda + \tfrac{h}{2}C_{\lambda\mu\nu}I_\nu \otimes I_\mu
\end{aligned}\right\}. \qquad (12.5.34)$$

In our example $C_{\lambda\mu\nu} = \mathrm{i}\varepsilon_{\lambda\mu\nu}$. In general $C_{\lambda\mu\nu}$ is the anti-symmetric tensor of the given semi-simple Lie algebra.

With the foregoing discussions, we can now give a concise definition of the Yangian associated with $SL(2)$. Let $R(u)$ satisfy the YBE and has the form of a polynomial in u, and $T(u)$ satisfies the RTT relation, then the coefficients/operators of powers u^{-1} and u^{-2} in the power series expansion of $T(u)$ comprise the Yangian generators; their commutation relations are determined by the RTT relation, and the RTT relation is consistent with the co-product. Commutation relations and the co-product are the two essential elements of the Yangian. If we obtain the commutation relations of $\{I_\alpha\}$ (associated with u^{-1}) and $\{J_\alpha\}$ (associated with u^{-2}) from the RTT relation, then the co-product is valid automatically. When $\det T(u) = 1$, the Yangian is $Y(SL(2))$, when $\det T(u) \neq 1$, the Yangian is $Y(GL(2))$ which is related to Haldane-Shastry model through the RTT relation [54].

Drinfeld discovered the following important property of the Yangian. If $R(u)$ is in a polynomial form of u (including $\check{R} = I + uP$ discussed above), it is called a rational solution of the YBE. For rational solutions, it is sufficient to know I_α and J_α (altogether 6 generators) in order to unambiguously determine *all* elements of higher orders $T_\alpha^{(n)}$ ($n \geq 3; \alpha = +, -, 3$) and their commutation relations. $T_\alpha^{(n)}$ comprise an algebra of *infinite* dimensions, but this algebra is determined by a *finite* number of generators (in the present case 6). This can be compared to the relation between Lie group and Lie algebra. A Lie group is determined by the Lie algebra, the first derivatives with respect to the group parameters of a group element at the origin. They are called the generators of the Lie algebra, and they determine the behavior of the Lie group. Now $T(u)$ is determined not only by $T_\alpha^{(1)}$ ($\alpha = +, -, 3$) but also by $T_\alpha^{(2)}$. Another type of solutions to YBE is the trigonometric solution which possesses periodicity in the parameter u. We will discuss it later.

The contents of this section are based on Refs. [6, 7]. The books regarding quantum groups are listed in Refs. [9–26].

As an extension of the $SL(2)$ algebra, the representation theory of $Y(SL(2))$ was worked out by V. Chari and A. Pressley [12]. We would like to emphasize that $Y(SL(2))$ is an infinite algebra, but it possesses finite dimensional representations. For details, Ref. [12] gives the complete discussions.

12.6 Expansion of the RTT Relation and the Commutation Relations

For $\check{R} = I + uP$, the RTT relation gives Eq. (12.5.4):

$$(u - v)[T_{bc}(u), T_{ad}(v)] + T_{ac}(u)T_{bd}(v) - T_{ac}(v)T_{bd}(u) = 0.$$

Substituting (12.5.16) in (12.5.4) and making comparison for the different powers in u and v, we see that the elements of $T^{(0)}$ must be c-numbers. Since the RTT relation admits transformations, we can carry out diagonalization of $T^{(0)}$ and in general we get

$$T^{(0)} = \begin{bmatrix} 1 & 0 \\ 0 & \lambda \end{bmatrix}, \tag{12.6.1}$$

where λ is an arbitrary constant. For non-vanishing λ we can choose

$$T^{(0)} = \begin{bmatrix} 1 & 0 \\ 0 & 1 \end{bmatrix}. \tag{12.6.2}$$

Operators of higher orders satisfy

$$\left[T_{bc}^{(n+1)}, T_{ad}^{(m)} \right] - \left[T_{bc}^{(n)}, T_{ad}^{(m+1)} \right] + T_{ac}^{(n)} T_{bd}^{(m)} - T_{ac}^{(m)} T_{bd}^{(n)} = 0. \tag{12.6.3}$$

We observe that setting $a = b, c = d$ and let m changes for a fixed n, then we get

$$\left[T_{ab}^{(n)}, T_{ab}^{(m)} \right] = 0. \tag{12.6.4}$$

Interchanging m and n, a and b, c and d in (12.6.3) and subtracting the resulting equation from (12.6.3), we get

$$\left[T_{ab}^{(n)}, T_{cd}^{(m)} \right] = \left[T_{ab}^{(m)}, T_{cd}^{(n)} \right]. \tag{12.6.5}$$

For different values of a and b (for $SL(2)$ they take 0, $+$, $-$, and 3), m and n Eq. (12.6.3) can be written explicitly as

$$\left. \begin{aligned} & \left[T_{\alpha}^{(1)}, T_{\alpha}^{(2)} \right] = 0, \quad \alpha = +, -, 3 \\ & \left[T_{3}^{(1)}, T_{\pm}^{(k)} \right] = \left[T_{3}^{(k)}, T_{\pm}^{(1)} \right] = \pm 2 T_{\pm}^{(k)}, \quad k = 1, 2 \\ & \left[T_{+}^{(1)}, T_{-}^{(k)} \right] = \left[T_{+}^{(k)}, T_{-}^{(1)} \right] = T_{3}^{(k)}, \quad k = 1, 2 \end{aligned} \right\}, \tag{12.6.6}$$

$$\left. \begin{aligned} & \left[T_{0}^{(1)}, T_{\alpha}^{(k)} \right] = \left[T_{0}^{(k)}, T_{\alpha}^{(1)} \right] = 0, \quad \alpha = \pm, 3, 0, \quad k = 1, 2 \\ & \left[T_{0}^{(2)}, T_{\pm}^{(2)} \right] = \pm \left(T_{3}^{(1)} T_{\pm}^{(2)} - T_{3}^{(2)} T_{\pm}^{(1)} \right) \\ & \left[T_{0}^{(2)}, T_{3}^{(2)} \right] = 2 \left(T_{+}^{(1)} T_{-}^{(2)} - T_{+}^{(2)} T_{-}^{(1)} \right) \end{aligned} \right\}, \tag{12.6.7}$$

$$\left. \begin{aligned} & T_{\pm}^{(n+1)} = \frac{1}{2} \left(\pm \left[T_{3}^{(2)}, T_{\pm}^{(n)} \right] + T_{0}^{(n)} T_{\pm}^{(1)} - T_{0}^{(1)} T_{\pm}^{(n)} \right), \quad n \geq 2 \\ & T_{3}^{(n+1)} = \left[T_{+}^{(n)}, T_{-}^{(2)} \right] + \frac{1}{2} \left(T_{0}^{(n)} T_{3}^{(1)} - T_{0}^{(1)} T_{3}^{(n)} \right), \quad n \geq 2 \end{aligned} \right\}, \tag{12.6.8}$$

$$\left[T_{\alpha}^{(n)}, T_{\alpha}^{(m)} \right] = 0, \alpha = \pm, 3, 0, m, n \geq 2. \tag{12.6.9}$$

Furthermore, we have

$$
\left.\begin{aligned}
&\left[T_\alpha^{(m)}, T_\beta^{(n)}\right] = \left[T_\alpha^{(n)}, T_\beta^{(m)}\right], \quad \alpha \neq \beta, \quad m < n, \quad n > 2 \\
&\left[T_0^{(n+1)}, T_\pm^{(m)}\right] - \left[T_0^{(n)}, T_\pm^{(m+1)}\right] \pm \left(T_3^{(m)} T_\pm^{(n)} - T_3^{(n)} T_\pm^{(m)}\right) = 0 \\
&\left[T_0^{(n+1)}, T_3^{(m)}\right] - \left[T_0^{(n)}, T_3^{(m+1)}\right] + 2\left(T_+^{(m)} T_-^{(n)} - T_+^{(n)} T_-^{(m)}\right) = 0 \\
&\left[T_3^{(n+1)}, T_\pm^{(m)}\right] - \left[T_3^{(n)}, T_\pm^{(m+1)}\right] \pm \left(T_0^{(m)} T_\pm^{(n)} - T_0^{(n)} T_\pm^{(m)}\right) = 0 \\
&\left[T_+^{(n+1)}, T_-^{(m)}\right] - \left[T_+^{(n)}, T_-^{(m+1)}\right] \\
&\quad + \frac{1}{2}\left(T_0^{(m)} T_3^{(n)} - T_0^{(n)} T_3^{(m)}\right) = 0, \quad m < n, \quad n > 2
\end{aligned}\right\}
$$

$$(12.6.10)$$

In the above relations only (12.6.6)–(12.6.9) are independent. Relations in (12.6.10) can be derived from them after laborious calculations.

Eqations (12.6.6) and (12.6.7) give the commutation relations between the matrix elements/operators of $T^{(1)}, T^{(2)}$ ($\alpha = +, -, 3$), Eq. (12.6.8) is an important recurrence relation. When $T^{(1)}, T^{(2)}$ are known, we can then use (12.6.8) to give $T^{(3)}$, and then $T^{(4)}$, etc. This means that $T^{(1)}$ and $T^{(2)}$ determine all $T^{(n)}$, $n \geq 3$. Equation (12.6.9) is a constraint to higher order $T^{(n)}$. Substituting (12.6.8) into (12.6.9), we get

$$
\left.\begin{aligned}
&\left[T_\pm^{(m)}, \left[T_3^{(2)}, T_\pm^{(n)}\right]\right] \pm \left[T_\pm^{(m)}, T_0^{(n)}\right] T_\pm^{(1)} = 0 \\
&2\left[T_3^{(m)}, \left[T_+^{(n)}, T_-^{(2)}\right]\right] + \left[T_3^{(m)}, T_0^{(n)}\right] T_3^{(1)} = 0, m, n \geq 2
\end{aligned}\right\}
$$

$$(12.6.11)$$

By using (12.6.7) for $m = n = 2$, these relations become

$$
\left[T_\pm^{(2)}, \left[T_3^{(2)}, T_\pm^{(2)}\right]\right] = \left(T_3^{(1)} T_\pm^{(2)} - T_3^{(2)} T_\pm^{(1)}\right) T_\pm^{(1)} \tag{12.6.12}
$$

$$
\left[T_3^{(2)}, \left[T_+^{(2)}, T_-^{(2)}\right]\right] = \left(T_+^{(1)} T_-^{(2)} - T_+^{(2)} T_-^{(1)}\right) T_3^{(1)} \tag{12.6.13}
$$

From the commutators of $T_{\pm}^{(1)}$ and (12.6.13) found by using (12.6.6) it follows that

$$2\left[T_{\pm}^{(2)},\left[T_{+}^{(2)},T_{-}^{(2)}\right]\right] \pm \left[T_3^{(2)},\left[T_3^{(2)},T_{\pm}^{(2)}\right]\right]$$

$$= 2\left(T_+^{(1)}T_-^{(2)} - T_+^{(2)}T_-^{(1)}\right)T_{\pm}^{(1)} \pm \left(T_3^{(1)}T_{\pm}^{(2)} - T_3^{(2)}T_{\pm}^{(1)}\right)T_3^{(1)}.$$

$$(12.6.14)$$

Equation (12.6.6) with $k = 2$, Eqs. (12.6.12), (12.6.13) are the Yangian commutation relations Eqs. (11.1.13), (11.1.14), (11.1.17), (11.1.18).

It is interesting to observe that $T_0^{(2)}$ is not arbitrary because of the RTT relation, being constrained by (12.6.7). When the physical realization of $\boldsymbol{I}$ and $\boldsymbol{J}$ is known, we can determine $T_0^{(2)}$ by $T_0^{(2)} = T_{22}^{(2)} + T_{11}^{(2)}$ (12.5.17) and then determine C_2 by using (12.5.21). It is the "center" of the algebra, since $\det \boldsymbol{T}(u)$ commutes with all operators $T_{ab}(u)$.

It is not easy to find a realization of $\boldsymbol{I}$ and $\boldsymbol{J}$ by solving directly Eqs. (11.1.13), (11.1.14), (11.1.17), (11.1.18) and then to find $T_0^{(2)}$. The one-dimensional chain with long range interaction is a successful example. In the next section we use again the hydrogen atom as an example [28, 29]. Being a single-body problem, it can be easily handled. We can have a concrete understanding of the RTT relation and operators of various orders $T(u)$, and the commutation relations.

For the content of this section the readers can referred to Ref. [30].

12.7   The Hydrogen Atom and the RTT Relation

We go back to the hydrogen atom to see how it is related to YBE and the RTT relation. First we re-define the operator $\boldsymbol{J}$, using the original Runge-Lenz vector $\boldsymbol{M}$ instead of its scaled counterpart $\boldsymbol{M}'$. Our purpose is to reveal the scaling role of the Hamiltonian. Define

$$\boldsymbol{J} = \boldsymbol{L} \times \boldsymbol{b} \equiv -\boldsymbol{L} \times \frac{\mathrm{i}}{\sqrt{2}}\boldsymbol{M}, \qquad (12.7.1)$$

where $\boldsymbol{M}$ is given in Section 4.3.

Obviously the commutation relations between $\boldsymbol{L}$ and $\boldsymbol{J}$ remain to be

$$\left[L_\alpha, J_\beta\right] = i\varepsilon_{\alpha\beta\gamma}J_\gamma. \tag{12.7.2}$$

Furthermore, we have

$$\left[J_\alpha, J_\beta\right] = i\varepsilon_{\alpha\beta\gamma}L_\gamma\left(HL^2 - b^2 - H\right).$$

We set $m = 1, \hbar = 1$, and take into account that (see Chapter 4)

$$b^2 = -\frac{1}{2}M^2 = -\frac{1}{2}\left(2HL^2 + 2H + k^2\right), \tag{12.7.3}$$

then we get

$$\left[J_\alpha, J_\beta\right] = i\varepsilon_{\alpha\beta\gamma}L_\gamma\left(HL^2 + \frac{1}{2}k^2\right). \tag{12.7.4}$$

Equation (12.7.1) in $(+, -, 3)$ components can be recast to

$$\left.\begin{aligned} J_\pm &= \pm\frac{1}{\sqrt{2}}\left(L_3M_\pm - L_\pm M_3\right) \\ J_3 &= \frac{1}{2\sqrt{2}}\left(L_+M_- - L_-M_+\right) \end{aligned}\right\} \tag{12.7.5}$$

Using the commutation relations between $\boldsymbol{I}^2$ and $\boldsymbol{J}$

$$\left.\begin{aligned} \left[\boldsymbol{I}^2, J_\pm\right] &= \pm 2\left[I_3 J_\pm - J_3 I_\pm\right] \\ \left[\boldsymbol{I}^2, J_3\right] &= I_+ J_- - J_+ I_- \end{aligned}\right\} \tag{12.7.6}$$

we obtain

$$\left.\begin{aligned} \left[J_\pm, \left[J_3, J_\pm\right]\right] &= (-4H) I_\pm \left(I_3 J_\pm - J_3 I_\pm\right) \\ \left[J_3, \left[J_+, J_-\right]\right] &= (-4H) I_3 \left(I_+ J_- - J_+ J_-\right) \end{aligned}\right\}. \tag{12.7.7}$$

Derivation of these relations is parallel to that in Section 11.1, with the factor $-4H$ explicitly appearing here. It can be removed by redefining $\boldsymbol{J}$ for bound states $E < 0$, therefore $-4H$ is positive. In Chapter 4 we have seen that rescaling is the key step in and determining the energy spectrum of the hydrogen atom.

Equation (12.7.1) can be written as

$$J_\alpha = -\frac{i}{\sqrt{2}}\varepsilon_{\alpha\beta\sigma}L_\beta M_\sigma$$

$$= -\frac{i}{\sqrt{2}}\left[\varepsilon_{\alpha\beta\sigma}\varepsilon_{\sigma\tau\rho}L_\rho L_\tau p_\rho - i\left(\boldsymbol{L}\times\boldsymbol{p}\right)_\alpha + \frac{k}{r}\left(\boldsymbol{L}\times\boldsymbol{r}\right)_\alpha\right].$$

Through the direct calculation we get

$$J_\alpha = -\frac{i}{\sqrt{2}}\left[\frac{k}{r}\left(\boldsymbol{L}\times\boldsymbol{r}\right)_\alpha - L^2 p_\alpha\right]. \tag{12.7.8}$$

In Section 11.3 (Eq. (11.3.15)), we have seen that L_α and $L^2 p_\alpha$ comprise the Yangian generators. Here we have $\frac{k}{r}\left(\boldsymbol{L}\times\boldsymbol{r}\right)_\alpha$ in addition. For the hydrogen atom, $L^2\boldsymbol{p}$ does not commute with H, while here J_α (12.7.8) is not only a Yangian generator, but also commutes with H.

In Section 11.3, we discussed about the fact that $\boldsymbol{J}\to\boldsymbol{J}+F\boldsymbol{I}$ behaves the same way as $\boldsymbol{J}$ in satisfying various commutation relations of the Yangian, where F is a Casimir operator of the Yangian. Now we carry out a translation of $\boldsymbol{J}$:

$$\boldsymbol{J}\to\mathscr{J}=\boldsymbol{J}+F\boldsymbol{L}=\boldsymbol{L}\times\boldsymbol{b}+F\boldsymbol{L}, \tag{12.7.9}$$
$$[F,\boldsymbol{L}]=[F,\boldsymbol{M}]=0. \tag{12.7.10}$$

Now the commutation relation for $\mathscr{J}$ is

$$[\mathscr{J}_\alpha,\mathscr{J}_\beta]=i\varepsilon_{\alpha\beta\sigma}\left\{\left(2HL^2+\frac{1}{2}k^2\right)L_\sigma+2F\mathscr{J}_\sigma\right\}. \tag{12.7.11}$$

All Yangian commutation relations (e.g., (12.7.7)) are valid for $\mathscr{J}$. Following (12.5.19), we connect I_α, $\mathscr{J}_\alpha$ and $T_\alpha^{(1)}, T_\alpha^{(2)}$,

$$\left.\begin{array}{ll}T_\pm^{(1)}=I_\pm=L_\pm, & T_3^{(1)}=I_3=L_3\\ T_\pm^{(2)}=G\mathscr{J}_\pm, & T_3^{(2)}=G\mathscr{J}_3\end{array}\right\} \tag{12.7.12}$$

where G is also a Casimir operator. Because of the relation

$$\left[T_3^{(2)},T_\pm^{(2)}\right]=2G^2\left[\mathscr{J}_3,\mathscr{J}_\pm\right]$$
$$=\pm 2G^2\left\{\left(2HI^2+F^2+\frac{1}{2}k^2\right)I_\pm+2F\mathscr{J}_\pm\right\} \tag{12.7.13}$$

we can get $T_\pm^{(3)}$ via the recurrence relation (12.6.8):

$$
\begin{aligned}
T_\pm^{(3)} &= \frac{1}{2}\left(\pm\left[T_3^{(2)}, T_\pm^{(2)}\right] + T_0^{(2)}T_\pm^{(1)} - T_0^{(1)}T_\pm^{(2)}\right) \\
&= \frac{1}{2}\left\{\left[2G^2\left(2HI^2 + F^2 + \frac{1}{2}k^2\right) + T_0^{(2)}\right]I_\pm \right.\\
&\quad\left. + \left(4G^2F - T_0^{(1)}G\right)\mathcal{J}_\pm\right\}.
\end{aligned}
\tag{12.7.14}
$$

Similarly we obtain

$$
\begin{aligned}
T_3^{(3)} &= \left[T_+^{(2)}, T_-^{(2)}\right] + \frac{1}{2}\left(T_0^{(2)}T_3^{(1)} - T_0^{(1)}T_3^{(2)}\right) \\
&= \left\{2G^2\left(2HI^2 + F^2 + \frac{1}{2}k^2\right) + T_0^{(2)}\right\}I_3 \\
&\quad + \left(4G^2F - T_0^{(1)}G\right)\mathcal{J}_3.
\end{aligned}
\tag{12.7.15}
$$

From (12.5.21) we see that $T_0^{(1)} = C_1$ which commutes with any $T_\alpha^{(n)}$ ($\alpha = \pm, 3$), so we can choose it as the combination of G and F:

$$
T_0^{(1)} = 4GF
\tag{12.7.16}
$$

$T_0^{(2)}$ is constrained by (12.6.7):

$$
\left.\begin{aligned}
\left[T_0^{(2)}, \mathcal{J}_\pm\right] &= \pm\left(I_3\mathcal{J}_\pm - \mathcal{J}_3 I_\pm\right) \\
\left[T_0^{(2)}, \mathcal{J}_3\right] &= 2\left(I_+\mathcal{J}_- - \mathcal{J}_+ I_-\right)
\end{aligned}\right\}
\tag{12.7.17}
$$

Since $\mathcal{J}$ and J differ by FI, and $[I^2, I_\alpha] = 0$, then (12.7.6) is also valid for $\mathcal{J}$. Therefore

$$
T_0^{(2)} = 2I^2 + Q,
\tag{12.7.18}
$$

where Q is a set of operators commuting with $(I, \mathcal{J})$. Since H commutes with $I, \mathcal{J}$, we can choose

$$
T_0^{(2)} = -2G^2\left(2HI^2 + F^2 + \frac{1}{2}k^2\right).
\tag{12.7.19}
$$

The aim of the choice is that together with (12.7.16) it can lead to $T_\alpha^{(3)} = 0$. This can be seen by substituting (12.7.12) and the choice for $T_0^{(1)}, T_0^{(2)}$ into (12.7.14) and (12.7.15), and consequently $T_\alpha^{(n)} = 0$ for all $n \geq 3$. This makes the expansion of the transition matrix stop at u^{-2}, and break up automatically, thus making the physical requirement that the hydrogen atom problem has only a finite number of conserved quantities realized. Comparing Eqs. (12.7.18) and (12.7.19), we get

$$G = (-2H)^{-1/2}. \tag{12.7.20}$$

$T_0^{(1)}$ commutes with all the generators, and F in (12.7.16) has not yet been fixed. We demand

$$T_0^{(1)} = H \tag{12.7.21}$$

and thus have fixed F:

$$F = \frac{1}{4G} T_0^{(1)} = \pm \frac{i}{2\sqrt{2}} H^{3/2}. \tag{12.7.22}$$

To sum up, all operators appearing in the expansion of the transition matrix are:

$$\left. \begin{array}{l} T_\pm^{(1)} = I_\pm = L_\pm, \quad T_3^{(1)} = 2L_3, \quad T_0^{(1)} = H \\ T_\pm^{(2)} = \pm i \, (2H)^{-1/2} \, \mathscr{J}_\pm, \quad T_3^{(2)} = \pm i \, (2H)^{-1/2} \, \mathscr{J}_3 \\ T_0^{(2)} = I^2 + H^{-1} \left(\frac{1}{2} k^2 - \frac{1}{8} H^3 \right) \end{array} \right\} . \tag{12.7.23}$$

All coefficients of higher orders vanish. The generators of the Yangian are:

$$\left. \begin{array}{l} \mathbf{I} = \mathbf{L} \\ \mathscr{J} = -\frac{i}{\sqrt{2}} \mathbf{L} \times \mathbf{M} \mp \frac{i}{2\sqrt{2}} H^{3/2} \mathbf{L} \end{array} \right\} . \tag{12.7.24}$$

The coefficients of expansion of the quantum determinant are

$$\left. \begin{array}{l} C_1 = T_0^{(1)} = H \\ C_2 = T_0^{(2)} - I^2 + \frac{1}{2} T_0^{(1)} \left(1 + \frac{1}{2} T_0^{(1)} \right) \\ \quad = \frac{1}{2} k^2 H^{-1} + \frac{1}{2} H - \frac{1}{8} H^2 \\ \quad \cdots \end{array} \right\} . \tag{12.7.25}$$

The remaining $C_n(n > 2)$ are all functions of H. Explicit representation of $T(u)$ is

$$
T(u) = I + u^{-1} \begin{bmatrix} \frac{H}{2} - I_3 & I_+ \\ I_- & \frac{H}{2} + I_3 \end{bmatrix}
$$

$$
+ u^{-2} \begin{bmatrix} \frac{1}{2}T_0^{(2)} \mp i\,(2H)^{-1/2}\,\mathscr{J}_3 & \pm i\,(2H)^{-1/2}\,\mathscr{J}_+ \\ \pm i\,(2H)^{-1/2}\,\mathscr{J}_- & \frac{1}{2}T_0^{(2)} \pm i\,(2H)^{-1/2}\,\mathscr{J}_3 \end{bmatrix}.
$$

$$(12.7.26)$$

We have seen that the hydrogen atom is a simplest example for model satisfying the RTT relation. Its Hamiltonian is not determined by the RTT relation (i.e., its solution $T(u)$); rather it is determined by the non-trivial term C_1 of lowest order of the quantum determinant, and all terms of higher orders $C_n(n \geq 2)$ are functions of H. According to the general form of the RTT relation, the expansion of $T(u)$ in powers of u^{-1} should have infinite order, but in special models it may be of finite order corresponding to finite number of conserved quantities. In this example we meet the revelation of Yangian in the hydrogen atom problem. For more details refer to Ref. [28].

In this chapter we limit ourselves to the rational solution of the YBE. The algebra related to the single periodic (trigonometric) solutions is related to the quantum algebras, with which many problems in low-dimensional quantum theory are connected. We will discuss the relevant questions in the next chapter.

12.8 Representations of Yangian and the Hydrogen Atomic Spectrum [29]

From Sections 11.1~11.3 we see that the orbital angular momentum operators $\boldsymbol{L} = \boldsymbol{I}$ and the scaled Runge-Lenz vector $\boldsymbol{B}$ constitute a $SO(4)$ algebra. Their linear combinations

$$
\boldsymbol{L}_1 = \frac{1}{2}\,(\boldsymbol{L} + \boldsymbol{B}), \quad \boldsymbol{L}_2 = \frac{1}{2}\,(\boldsymbol{L} - \boldsymbol{B}) \tag{12.8.1}
$$

1020   Frontier Problems in Quantum Mechanics

constitute two commuting $SO(2)$ algebras. Besides these linear combinations we can also introduce new tensor operators

$$\boldsymbol{J} = \frac{\mathrm{i}}{4}\boldsymbol{L} \times \boldsymbol{B}. \tag{12.8.2}$$

Obviously the set formed by $\boldsymbol{I}$ and $\boldsymbol{J}$ exceeds the category of Lie algebra. Operators of this set act on a tensor space, which contains the intersection of the spaces spanned by $\boldsymbol{I}$ and $\boldsymbol{B}$. Notice that (12.8.2) differs from (11.1.4) by a factor of 2. Compared with the standard Yangian relation (11.2.11), (11.1.4) has to be rescaled by a factor $\frac{1}{2}$. Yangian independently works in tensor space and has its own representation theory [12]. We now ask whether such representation determines the correct spectrum of the hydrogen atom. This is to test if the representation theory of $Y(SL(2))$ works for hydrogen atom.

We begin with the combination of orbital angular momentum and scaled Runge-Lenz vector with the absence of magnetic monopole:

$$\boldsymbol{L} \cdot \boldsymbol{B} = 0, \tag{12.8.3}$$

for the absence of monopole.

We now have

$$\boldsymbol{L}_1^2 = \boldsymbol{L}_2^2 = \frac{1}{4}\left(\boldsymbol{L}^2 + \boldsymbol{B}^2\right). \tag{12.8.4}$$

In vector space, this symmetry leads to the hydrogen atomic spectrum, as was shown in Section 4.3.2. We now start from $Y(SU(2))$ or $\{\boldsymbol{L}, \boldsymbol{J}\}$ to rederive the energy spectrum. We shall make use of

$$\boldsymbol{J}^2 = -\frac{1}{16}\left\{\boldsymbol{L}^2\left(\boldsymbol{B}^2 - 1\right) - (\boldsymbol{L} \cdot \boldsymbol{B})^2\right\} = -\frac{1}{16}\boldsymbol{L}^2\left(\boldsymbol{B}^2 - 1\right).$$
$$\tag{12.8.4a}$$

Let the quantum numbers of $\boldsymbol{L}^2 = (\boldsymbol{L}_1 + \boldsymbol{L}_2)^2$ be $l = 2k, 2k - 1, \ldots, 0$. In general, the weights can be

$$l = 2k - p, \quad p = 0, 1, \ldots, 2k, \tag{12.8.5}$$

where $\boldsymbol{L}_1^2 = \boldsymbol{L}_2^2 = k(k+1)$. It is easy to find the eigenvalues of $\boldsymbol{L}_1 \cdot \boldsymbol{L}_2$ to be $\frac{1}{2}[l(l+1) - 2k(k+1)]$. Because of

$$\boldsymbol{B}^2 = \boldsymbol{L}_1^2 + \boldsymbol{L}_2^2 - 2\boldsymbol{L}_1\boldsymbol{L}_2, \tag{12.8.6}$$

we find the eigenvalue of B^2 to be $4k(k+1) - l(l+1)$, so that the eigenvalue of J^2 is

$$J(J+1) = -\frac{1}{16}l(l+1)\left[4k(k+1) - l(l+1) - 1\right]. \qquad (12.8.7)$$

Also we have

$$B^2 = -\left(L^2 + \frac{\kappa^2}{2E} + 1\right) \qquad (12.8.8)$$

and from (12.8.8) and (12.8.4) we have

$$J^2 = \frac{1}{16}L^2\left(L^2 + \frac{\kappa^2}{2E} + 2\right) \qquad (12.8.9)$$

or

$$J^2 = \frac{1}{16}l(l+1)\left[l(l+1) + \frac{\kappa^2}{2E} + 2\right]. \qquad (12.8.10)$$

Comparing (12.8.10) and (12.8.7), leads to

$$\frac{\kappa^2}{2E} = -\left[4k(k+1) + 1\right] \qquad (12.8.11)$$

i.e.,

$$E = -\frac{\kappa^2}{2(2k+1)^2} = -\frac{\kappa^2}{2n^2}, \qquad (12.8.12)$$

where $n = 2k+1$, with k taking values $0, \frac{1}{2}, 1, \frac{3}{2}, \dots$. Substituting (12.8.5) into (12.8.7), we obtain

$$J^2 = -\frac{1}{16}(2k-p)(2k-p+1)[4k(k+1)$$

$$- (2k-p)(2k-p+1) - 1]. \qquad (12.8.13)$$

This implies that the eigenvalues of $Y(SU(2))$ for the hydrogen atom also gives the correct energy spectrum in the quantum tensor space.

Since the set $\{L, J\}$ satisfies $Y(SU(2))$, and $Y(SU(2))$ has its own independent representation theory, then how is (12.8.13) for the hydrogen atom connected with the representation theory of $Y(SU(2))$? Now J is composed from the direct products of the generators of two generators of Lie algebras. Let e_m be the basis of $SU(2)$ with the weight m, then we have to choose the basis of the tensor product as [12]:

$$\Omega_p = \sum_{i=0}^{p} (-1)^i \frac{(m-i)!\,(n-p+i)!}{m!\,(n-p)!} e_{m-i} \otimes e_{n-p+i},$$

$$0 \le p \le \min(m,n). \qquad (12.8.14)$$

Substituting $\boldsymbol{L} = \boldsymbol{L}_1 + \boldsymbol{L}_2$ and $\boldsymbol{B} = \boldsymbol{L}_1 - \boldsymbol{L}_2$ into (12.8.2) we get

$$\boldsymbol{J} = a\boldsymbol{L}_1 + b\boldsymbol{L}_2 - \frac{\mathrm{i}}{2}\boldsymbol{L}_1 \times \boldsymbol{L}_2, \qquad (12.8.15)$$

where $a = -\frac{1}{4}$, $b = \frac{1}{4}$ for the hydrogen atom. It is easy to prove that $\boldsymbol{J}$ defined in (12.8.15) satisfies the Yangian commutation relations for any parameters a and b. Notice that

$$\boldsymbol{J}^2 = \frac{1}{2}\left(J_+ J_- + J_- J_+\right) + J_3^2, \qquad (12.8.16)$$

Apply $\boldsymbol{J}^2$ on Ω_p, we obtain the eigenvalues

$$\boldsymbol{J}^2 = l_1\,(l_1+1)\,a^2 + l_2\,(l_2+1)\,b^2 + \left[2\,(l_1-p)\,(l_2-p) - p\,(p+1)\right]ab$$

$$- \frac{1}{4}\left[(l_1+l_2)l_1 l_2 + \frac{p}{4}(2l_1 + 2l_2 + 1 - p)\right.$$

$$\left. [4l_1 l_2 + 2 - p(2l_1 + 2l_2 + 1 - p)]\right]. \qquad (12.8.17)$$

This expression can be considered as a direct consequence of the $Y(SU(2))$ representation theory due to Chari and Pressley, and what we have been doing is to calculate the eigenvalue of $\boldsymbol{J}^2$.

Substituting $b = -a = \frac{1}{4}$ into $\boldsymbol{J}^2$, taking into account that $l_1 = l_2 = k$ because of $l_1^2 = l_2^2$, we observe that (12.8.17) becomes

$$\boldsymbol{J}^2 = \frac{1}{16}\left\{2k\,(k+1) - 2\,(k-p)^2 + p\,(p+1) - 8k^3 - p\,(4k+1-p)\right.$$

$$\cdot\left.[4k^2 + 2 - p(4k+1-p)]\right\}$$

$$= \frac{1}{16}\,(2k-p)\left\{1 - 2k\,(2k+p)\,4k^2 p + 2k^2 p - 2kp + p^2\right.$$

$$\left. - p\left[4k^2 + 2 - p\,(4k+1-p)\right]\right\}$$

$$= -\frac{1}{16} (2k - p) (2k - p + 1)$$

$$\times [4k (k + 1) - (2k - p) (2k - p + 1) - 1]$$

which is identical with (12.8.13). This demonstrates that the result we obtained from the specific model for the hydrogen atom coincides with what the $Y(SU(2))$ representation theory has produced, which implies that Yangian representation theory of Chari-Pressley does lead to the correct result of the energy spectrum of the hydrogen atom, viewed from the quantum tensor space.

In the following, we discuss the case including a magnetic monopole. The Hamiltonian of the system is given by

$$\mathscr{H} = \frac{\pi^2}{2\mu} + \frac{1}{2\mu} \frac{q^2}{r^2} - \frac{\kappa}{r}, \quad \pi = p - Ze\boldsymbol{A}. \tag{12.8.18}$$

Now $so(4)$ is composed of the angular momentum $\boldsymbol{L}$ and the rescaled Runge-Lenz vector $\boldsymbol{B}$:

$$\boldsymbol{L} = \boldsymbol{r} \times \boldsymbol{\pi} - q\frac{\boldsymbol{r}}{r}, \quad \boldsymbol{B} = \frac{i}{\sqrt{2\mu E}} \left(\frac{1}{2} (\boldsymbol{\pi} \times \boldsymbol{L} - \boldsymbol{L} \times \boldsymbol{\pi}) - \frac{\mu\kappa}{r} \boldsymbol{r} \right),$$

$$\tag{12.8.19}$$

where μ is the reduced mass; $q = Zeg$, g is the magnetic charge; $\kappa = Ze^2$; $\boldsymbol{A}$ is the monopole potential given by Wu and Yang [53]. It is easy to calculate

$$\boldsymbol{L} \cdot \boldsymbol{B} = \boldsymbol{B} \cdot \boldsymbol{L} = L_1^2 - L_2^2 = q\sqrt{-\frac{\mu\kappa^2}{2E}} \tag{12.8.20}$$

$$L_1^2 = \frac{1}{4} \left[\left(q + \sqrt{-\frac{\mu\kappa^2}{2E}} \right)^2 - 1 \right] = l_1 (l_1 + 1) \tag{12.8.21}$$

and determine

$$q = |l_1 - l_2| \tag{12.8.22}$$

and the spectrum is

$$\varepsilon_n = -\frac{\mu\kappa^2}{2} \frac{1}{n^2}, \quad n = l_1 + l_2 + 1 \tag{12.8.23}$$

This demonstrates that ε_n is degenerate with respect to q. Since l_1 and l_2 take values of $0, \frac{1}{2}, 1, \frac{3}{2}, \ldots$, so that

$$q = \begin{cases} 0, 1, \ldots, n-1, & n = \text{integer} \\ \dfrac{1}{2}, \dfrac{3}{2}, \ldots, n-1, & n = \text{halfinteger} \end{cases} \qquad (12.8.24)$$

At the same time, although $l = l_1 + l_2 - p = n - p + 1$ takes value in the same range as q does, this by no means implies necessarily that $q = l$. (Note that l is connected with the quantum number of $\boldsymbol{L}^2$, i.e., $l(l+1)$) For example, we can have the following combinations.

$$
\left.
\begin{array}{lll}
n = 1 & l_1 = l_2 = 0 & l = q = 0 \\[4pt]
n = \dfrac{3}{2} & l_1 = \dfrac{1}{2}, l_2 = 0 \ \ or \ \ l_1 = 0, l_2 = \dfrac{1}{2} & l = q = \dfrac{1}{2} \\[10pt]
n = 2 & \begin{cases} l_1 = 1, l_2 = 0 \ \ or \ \ l_1 = 0, l_2 = 1 & l = 1, 0, \quad q = 1 \\[8pt] l_1 = l_2 = \dfrac{1}{2} & l = 1, 0, \quad q = 0 \end{cases} \\[24pt]
n = \dfrac{5}{2} & \begin{cases} l_1 = \dfrac{3}{2}, l_2 = 0 \ \ or \ \ l_1 = 0, l_2 = \dfrac{3}{2} & l = \dfrac{3}{2}, \dfrac{1}{2}, \quad q = \dfrac{3}{2} \\[8pt] l_1 = 1, l_2 = \dfrac{1}{2} \ \ or \ \ l_1 = \dfrac{1}{2}, l_2 = 1 & l = \dfrac{3}{2}, \dfrac{1}{2}, \quad q = \dfrac{1}{2} \end{cases} \\[24pt]
n = 3 & \begin{cases} l_1 = 2, l_2 = 0 \ \ or \ \ l_1 = 0, l_2 = 2 & l = 2, 1, 0, \quad q = 2 \\[8pt] l_1 = \dfrac{3}{2}, l_2 = \dfrac{1}{2} \ \ or \ \ l_1 = \dfrac{1}{2}, l_2 = \dfrac{3}{2} & l = 2, 1, \quad q = 1 \\[8pt] l_1 = l_2 = 1 & l = 2, 1, 0, \quad q = 0 \end{cases}
\end{array}
\right\}
$$

$$(12.8.25)$$

We see that from $n = 3$ onward, when $q \neq 0$ there are two quintets ($l = 2$) and two triplets ($l = 1$). They cannot be distinguished by l and $l_3 = m$, but nevertheless can be distinguished by q. When $q \neq 0$ we have

$$\boldsymbol{J}^2 = -\frac{1}{16}\left\{\boldsymbol{L}^2\left(\boldsymbol{B}^2 - 1\right) - (\boldsymbol{L} \cdot \boldsymbol{B})^2\right\}, \qquad (12.8.26)$$

where L and B are given by (12.8.20). From (12.8.21) we get

$$B^2 = - \left(L^2 + \frac{\mu \kappa^2}{2E} + 1 \right) + q^2. \qquad (12.8.27)$$

Consequently

$$J^2 = \frac{1}{16} \left\{ L^4 + \frac{\mu \kappa^2}{2E} \left(L^2 - q^2 \right) + \left(2 - q^2 \right) L^2 \right\}. \qquad (12.8.28)$$

On substituting (12.8.28) in L^2 we obtain the eigenvalue ρ of J^2:

$$\rho = \frac{1}{16} \{ l (l+1)^2 - (l_1 + l_1 + 1)^2 \left(l (l+1) - q^2 \right)$$

$$+ \left(2 - q^2 \right) l(l+1) \}. \qquad (12.8.29)$$

By using $l_1 + l_2 = l + p$, we get

$$\rho = - \frac{1}{16} \{ [(l+1)(2p+1) + p^2][l(l+1) - q^2] - 2l(l+1) \}. \qquad (12.8.30)$$

It can be shown that (12.8.30) is consistent with (12.8.17). Actually, since

$$l (l+1) - q^2 = (l_1 + l_2 + p) (l_1 + l_2 - p + 1) - (l_1 - l_2)^2$$

$$= (2l_1 - p) (2l_2 - p) + l_1 + l_2 - p$$

$$= 4l_1 l_2 - 2 (l_1 + l_2) p + l_1 + l_2 - p + p^2$$

and

$$(l + 1) (2p + 1) + p^2 = 2 (l_1 + l_2 - p) p + l_1 + l_2 + p + p^2 + 1$$

$$= l_1 + l_2 + 1 + 2 (l_1 + l_2) p - p^2 + p.$$

Then (12.8.29) becomes

$$\rho = - \frac{1}{16} \{ - l_1 (l_1 + 1) - l_2 (l_2 + 1) + 2l_1 l_2 + 4 (l_1 + l_2) l_1 l_2$$

$$+ (4l_1 l_2 - 1) \left[2 (l_1 + l_2) p - p^2 + p \right] - \left[2 (l_1 + l_2) p - p^2 + p \right]^2$$

$$+ 2 (2l_1 + 2l_2 + 1) p - 2p^2 \}$$

$$= \frac{1}{16} \{ l_1 (l_1 + 1) + l_2 (l_2 + 1) - [2 (l_1 - p) (l_2 - p) - p (p + 1)]$$

$$- 4 (l_1 + l_2) l_1 l_2 - p (2l_1 + 2l_2 + 1 - p) [4l_1 l_2 + 2$$

$$- p(2l_1 + 2l_2 + 1 - p)] \} \tag{12.8.31}$$

This is exactly what (12.8.17) gives for $a^2 = b^2 = \frac{1}{4}$. Obviously, the eigenvalue of $\boldsymbol{J}^2$ depends on q when there exists a monopole. Suppressing a common factor $\frac{1}{16}$, we give in the following examples for the eigenvalues of $\boldsymbol{J}^2 = \rho$:

$n = 1$	$l = q = 0 \, (p = 0)$	$\rho = 0$
$n = \frac{3}{2}$	$l = q = \dfrac{1}{2} \, (p = 0)$	$\rho = (l + 1) [l (l + 1) - q^2] - 2l (l + 1)$
		$= \dfrac{3}{4}$
$n = 2$	$l = 1 \, (p = 0), 0 \, (p = 1), q = 1$	$\rho = 2 \, (l = 1), 4 \, (l = 0)$
	$l = 1, 0; q = 0$	$\rho = 0 \, (l = 1), 0 \, (l = 0)$
	$l = \dfrac{3}{2} \, (p = 0), \dfrac{1}{2} \, (p = 1) \, ; q = \dfrac{3}{2}$	$\rho = \dfrac{15}{4} \left(l = \dfrac{3}{2} \right), \dfrac{37}{4} \left(l = \dfrac{1}{2} \right)$
	$l = \dfrac{3}{2}, \dfrac{1}{2}; q = \dfrac{1}{2}$	$\rho = -\dfrac{5}{4} \left(l = \dfrac{3}{2}, \dfrac{1}{2} \right)$
$n = 3,$	$l = 2 \, (p = 0), 1 \, (p = 1), 0 \, (p = 2) \, ; q = 2$	$\rho = 6 \, (l = 2), 18 \, (l = 1), 36 \, (l = 0)$
	$l = 2, 1; l = 1$	$\rho = -3 \, (l = 2, l = 1)$
	$l = 2, 1, 0, q = 0$	$\rho = -6 \, (l = 2), -10 \, (l = 1), 0 \, (l = 0)$

In concluding this paragraph, we would like to point out an interesting result. As we pointed out earlier, when $\boldsymbol{J}$ satisfies the Yangian commutation relations, any translation $\boldsymbol{J} \to \lambda \boldsymbol{I} + \boldsymbol{J}$ will not alter the commutation relations. Applying this to the present situation and carrying out a translation in Eq. (12.8.2), then the eigenvalue ρ' of $\boldsymbol{J}'^2$ ($b = a = -\frac{1}{4}$ in Eq. (12.8.17)) is connected with the following eigenvalue

$$\boldsymbol{J}' = F\boldsymbol{I} + \boldsymbol{J}, \boldsymbol{I} = \boldsymbol{L}_1 + \boldsymbol{L}_2 \tag{12.8.32}$$

$$\rho' = \rho + \{ 2l_1 t(l_1 + 1)a + 2l_2(l_2 + 1)b$$

$$+ [2(l_1 - p)t(l_2 - p) - p(p + 1)(a + b)]F$$

$$+ [l_1(l_1 + 1) + l_2(l_2 + 1) + 2(l_1 - p)(l_2 - p) - p(p + 1)]F^2 \} . \tag{12.8.33}$$

On the other hand, we have

$$J'^2 = J^2 + 2F I \cdot J + F^2 I^2. \tag{12.8.34}$$

Substituting (12.8.34) into (12.8.2) and using

$$I \cdot J = -\frac{1}{4} L \cdot B \tag{12.8.35}$$

we obtain

$$J'^2 = J^2 - \frac{1}{2} (L \cdot B) F + F^2 L^2. \tag{12.8.36}$$

Taking notice of

$$2l_1 (l_1 + 1) a + 2l_2 (l_2 + 1) b + [2 (l_1 - p) (l_2 - p) - p (p + 1)] (a + b)$$
$$= -\frac{1}{4} [2l_1 (l_1 + 1) - 2l_2 (l_2 + 1)] = -\frac{1}{2} (l_1 - l_2) (l_1 + l_2 + 1).$$

The equation (12.8.33) can be rewritten as

$$\rho' = \rho - \frac{1}{2} (l_1 - l_2) (l_1 + l_2 + 1) F + (l_1 + l_2 - p) (l_1 + l_2 - p + 1) F^2. \tag{12.8.37}$$

Because of the magnetic monopole, (12.8.36) gives

$$\rho' = \rho - \frac{1}{2} q \sqrt{-\frac{\mu \kappa^2}{2\varepsilon}} F + l (l + 1) F^2, \tag{12.8.38}$$

which is identical to (12.8.37). With the help of $l = l_1 + l_2 - p$ and $q = |l_1 - l_2|$ we get again

$$\varepsilon = -\frac{\mu \kappa^2}{2} \frac{1}{(l_1 + l_2 + 1)^2}. \tag{12.8.39}$$

This demonstrates that the energy spectrum of the hydrogen atom does not change under the translation $J \to J + F I$.

12.9 Yangian and the Bell Basis

For a system of two spins A and B, we denote by $|\uparrow\downarrow\rangle$ the state $|\uparrow\rangle_A|\downarrow\rangle_B$, then the Bell basie are defined by

$$\left.\begin{aligned}
|\Psi^\pm\rangle &= \frac{1}{\sqrt{2}}\left(|\uparrow\downarrow\rangle \pm |\downarrow\uparrow\rangle\right) \\
|\Phi^\pm\rangle &= \frac{1}{\sqrt{2}}\left(|\uparrow\uparrow\rangle \pm |\downarrow\downarrow\rangle\right), \left(|\Psi^-\rangle = \Psi_{00}\right)
\end{aligned}\right\} \qquad (12.9.1)$$

They represent maximally entangled states. Among the four states, $|\psi^-\rangle$ is in a sense special. It is the spin singlet state of the system. We are going to show that the other three states (the spin triplet) can be generated by acting Yangian operators on $|\Psi^-\rangle$.

We introduce the Yangian operators

$$\boldsymbol{J} = a\overrightarrow{\boldsymbol{S}}_1 + b\overrightarrow{\boldsymbol{S}}_2 - i\overrightarrow{\boldsymbol{S}}_1 \times \overrightarrow{\boldsymbol{S}}_2, \qquad (12.9.2)$$

or specifically

$$\left.\begin{aligned}
J_+ &= aS_1^+ + bS_2^+ - \left(S_1^+ S_2^3 - S_1^3 S_2^+\right) \\
J_- &= aS_1^- + bS_2^- + \left(S_1^- S_2^3 - S_1^3 S_2^-\right) \\
J_3 &= aS_1^3 + bS_2^3 + \tfrac{1}{2}\left(S_1^+ S_2^- - S_1^- S_2^+\right)
\end{aligned}\right\}, \qquad (12.9.3)$$

where a and b are arbitrary parameters, and $\boldsymbol{S}_1$ and $\boldsymbol{S}_2$ operate on states of particles A and B, respectively. $J_\pm$ are connected with the Cartesian components J_1 and J_2 by

$$J_\pm = J_1 \pm iJ_2. \qquad (12.9.4)$$

Operating $J_\pm$ on $|\Psi^-\rangle$, we have

$$\left(aS_1^+ + bS_2^+\right)\frac{1}{\sqrt{2}}\left(|\uparrow\downarrow\rangle - |\downarrow\uparrow\rangle\right) = \frac{(b-a)}{\sqrt{2}}|\uparrow\uparrow\rangle,$$

$$\left(aS_1^- + bS_2^-\right)\frac{1}{\sqrt{2}}\left(|\uparrow\downarrow\rangle - |\downarrow\uparrow\rangle\right) = -\frac{(b-a)}{\sqrt{2}}|\downarrow\downarrow\rangle,$$

$$\left(S_1^+ S_2^3 - S_1^3 S_2^+\right) \frac{1}{\sqrt{2}} \left(|\uparrow\downarrow\rangle - |\downarrow\uparrow\rangle\right) = \frac{1}{\sqrt{2}} \left(\frac{-1}{2} |\uparrow\uparrow\rangle - \frac{1}{2} |\uparrow\uparrow\rangle\right)$$

$$= -\frac{1}{\sqrt{2}} |\uparrow\uparrow\rangle,$$

$$\left(S_1^- S_2^3 - S_1^3 S_2^-\right) \frac{1}{\sqrt{2}} \left(|\uparrow\downarrow\rangle - |\downarrow\uparrow\rangle\right) = \frac{1}{\sqrt{2}} \left(-\frac{1}{2} |\downarrow\downarrow\rangle - \frac{1}{2} |\downarrow\downarrow\rangle\right)$$

$$= -\frac{1}{\sqrt{2}} |\downarrow\downarrow\rangle,$$

$$J_+ \Psi_{00} = \frac{(b - a + 1)}{\sqrt{2}} |\uparrow\uparrow\rangle,$$

$$J_- \Psi_{00} = \frac{a - b - 1}{\sqrt{2}} |\downarrow\downarrow\rangle = -\frac{(b - a + 1)}{\sqrt{2}} |\downarrow\downarrow\rangle,$$

$$\left(a S_1^3 + b S_2^3\right) \Psi_{00} = \frac{1}{\sqrt{2}} \left\{ \frac{a}{2} |\uparrow\downarrow\rangle + \frac{a}{2} |\downarrow\uparrow\rangle - \frac{b}{2} |\uparrow\downarrow\rangle - \frac{b}{2} |\downarrow\uparrow\rangle \right\}$$

$$= \frac{1}{\sqrt{2}} \left(\frac{a - b}{2}\right) \left(|\uparrow\downarrow\rangle + |\downarrow\uparrow\rangle\right),$$

$$\frac{1}{2} \left(S_1^+ S_2^- - S_1^- S_2^+\right) \Psi_{00} = \frac{1}{2\sqrt{2}} \left(-|\uparrow\downarrow\rangle - |\downarrow\uparrow\rangle\right) = -\frac{1}{2\sqrt{2}} \left(|\uparrow\downarrow\rangle + |\downarrow\uparrow\rangle\right).$$

Consequently we obtain

$$\left.\begin{array}{l}
J_3 \Psi_{00} = \dfrac{(a - b - 1)}{2} \dfrac{1}{\sqrt{2}} \left(|\uparrow\downarrow\rangle + |\downarrow\uparrow\rangle\right) = -\dfrac{(b - a + 1)}{2} \Psi^+ \\[3mm]
J_1 \Psi_{00} = \dfrac{(b - a + 1)}{2} \dfrac{1}{\sqrt{2}} \left(|\uparrow\uparrow\rangle - |\downarrow\downarrow\rangle\right) = \dfrac{b - a + 1}{2} \Phi^- \\[3mm]
J_2 \Psi_{00} = \dfrac{(b - a + 1)}{2i} \dfrac{1}{\sqrt{2}} \left(|\uparrow\uparrow\rangle + |\downarrow\downarrow\rangle\right) = \dfrac{b - a + 1}{2i} \Phi^+
\end{array}\right\}.$$

$$(12.9.5)$$

If $a - b - 1 \neq 0$, i.e., $|\Psi^-\rangle$ is not a sub-representation of $Y(SU(2))$, we have the situation represented in the chart in Fig. 12.6.

$$\left.\begin{array}{l}
J_1 |\Psi^-\rangle \longrightarrow |\Phi^-\rangle \\
J_2 |\Psi^-\rangle \longrightarrow |\Phi^+\rangle \\
J_3 |\Psi^+\rangle \longrightarrow |\Psi^+\rangle
\end{array}\right\}$$

$$(12.9.6)$$

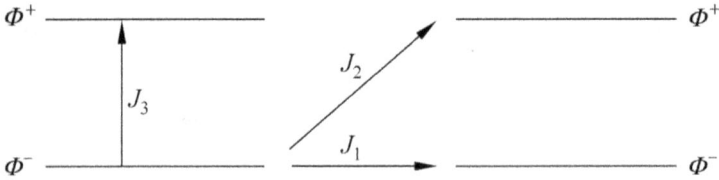

Fig. 12.6   Operating J_1, J_2, J_3 on Bell state.

The above discussion says that $|\Phi^\pm\rangle$ and $|\Psi^+\rangle$ can be generated by operating J on $|\Psi^-\rangle$, or in terms of the transition for Ψ^- and other states

$$
\left.
\begin{aligned}
J_3 &= -\lambda\,|\Psi^+\rangle\langle\Psi^-| \\
J_1 &= \lambda\,|\Phi^-\rangle\langle\Psi^-| \\
J_3 &= -\mathrm{i}\lambda\,|\Phi^+\rangle\langle\Psi^-|
\end{aligned}
\right\}. \tag{12.9.7}
$$

From (12.9.6) and (12.9.7), it is obvious that the operators $J_\pm, J_3$ given by (12.9.3) make the transitions between the spin-singlet and triplet. For long time, the transition is well-known in electrodynamics that can be given rise by the multipole expansion of vector potential $A(r,t)$. However, here we introduce the lowering and raising operators constructed in terms of the spin operators themselves. This example shows that the Bell basis can be generated naturally by J which, of course, may be caused by electromagnetic excitation in special way.

12.10   Transition from S- To P-wave Superconductivity

The usual superconductivity (SC) is described by Cooper pair theory, i.e., the S-wave pairing of two electrons. On the other hand, there is p-wave superconductivity theory, say, presented by Balian and Werthamer (BW) [32]. Although the p-wave theory has been applied to superfluid ^{3}He [33–35], rather than p-wave superconductivity. As an exercise we would like to discuss that in terms of what kind of operator we are able to make transitions from s-wave SC to p-wave SC.

The basis of two particles with spin$\frac{1}{2}$can be written in the form

$$|W\rangle = \sum_{\beta,\gamma=1}^{2} W_{\beta\gamma} |\beta\rangle_1 |\gamma\rangle_2 , \qquad (12.10.1)$$

where $|\beta\rangle_1$ and $|\gamma\rangle_2$ denote the spin states for 1-st and 2-nd particles respectively. The orthogonality of $|\beta\rangle$ and $|\gamma\rangle$ gives

$$W_{\beta\gamma} =_1 \langle\beta|_2\langle\nu|W\rangle = \langle\beta,\nu \mid W\rangle. \qquad (12.10.2)$$

Since β and γ take over the indices 1 and 2 in the two fixed spaces, the matrix form of $W_{\beta\gamma}$ can be expressed by a 2×2 matrix Φ.

$$\Phi(1,1) = \begin{bmatrix} 1 & 0 \\ 0 & 0 \end{bmatrix}, \quad \Phi(1,2) = \begin{bmatrix} 0 & 1 \\ 0 & 0 \end{bmatrix},$$

$$\Phi(2,1) = \begin{bmatrix} 0 & 0 \\ 1 & 0 \end{bmatrix}, \quad \Phi(2,2) = \begin{bmatrix} 0 & 0 \\ 0 & 1 \end{bmatrix}.$$

Acting the composed operator $A_1 B_2$ on the tensor state $|W\rangle$ we have

$$A_1 B_2 |W\rangle = |V\rangle = \sum_{\beta,\nu} W_{\beta\nu} (A_1 |\beta\rangle_1)(B_2 |\nu\rangle_2) = \sum_{\sigma,\tau} V_{\sigma\tau} |\sigma\rangle_1 |\tau\rangle_2$$

$$= \sum_{\beta,\nu,\sigma,\tau} W_{\beta\tau} A_{\sigma\beta} B_{\tau\nu} |\sigma\rangle_1 |\tau\rangle_2$$

$$= \sum_{\beta,\nu,\sigma,\tau} A_{\sigma\beta} W_{\beta\nu} \left(\tilde{B}_{\nu\tau}\right)_{\nu\tau}, \qquad (12.10.3)$$

i.e.,

$$V_{\sigma\tau} = \sum_{\beta,\nu} A_{\sigma\beta} W_{\beta\nu} \left(\tilde{B}_{\nu\tau}\right)_{\nu\tau}, \qquad (12.10.4)$$

where $\tilde{B}$ stands for the transverse of the matrix B.

Recalling

$$I_\alpha = S_1^\alpha \otimes 1 + 1 \otimes S_2^\alpha , \qquad (12.10.5)$$

$$J_\alpha = aS_1^\alpha \otimes 1 + b1 \otimes S_2^\alpha - i\nu\epsilon_{\alpha\mu\nu} S_1^\mu \otimes S_2^\nu , \qquad (12.10.6)$$

where $\alpha, \mu, \nu = 1, 2, 3$ and a, b, ν arbitrary parameters. From (12.10.4) it follows

$$A_1 B_2 \,|W\rangle = \sum_{\beta,\nu} \left(AW\widetilde{B} \right)_{\beta\nu} |\beta\rangle_1 \,|\nu\rangle_2 \,. \tag{12.10.7}$$

Formally, we denote

$$A_1 B_2 \,(W) = AW\widetilde{B}. \tag{12.10.8}$$

In terms of (12.10.5)~(12.10.8) we obtain

$$\left.\begin{aligned} I_\alpha\,(W) &= S^\alpha W + W\widetilde{S}^\alpha \\ J_\alpha\,(W) &= aS^\alpha W + bW\widetilde{S}^\alpha - \tfrac{i\nu}{2}\varepsilon_{\alpha\mu\nu}\left(S^\mu W\widetilde{S}_\nu - S^\nu W\widetilde{S}_\mu\right). \end{aligned}\right\} \tag{12.10.9}$$

With the notation of (12.10.2) the singlet takes the form

$$\varphi\,(0,0) = \frac{1}{\sqrt{2}}\left(\varPhi\,(1,2) - \varPhi\,(2,1)\right) = \frac{1}{\sqrt{2}}\begin{bmatrix} 0 & 1 \\ -1 & 0 \end{bmatrix}, \tag{12.10.10}$$

whereas the triplet reads

$$\left.\begin{aligned} \varphi\,(1,1) &= \varPhi\,(1,1) = \begin{bmatrix} 1 & 0 \\ 0 & 0 \end{bmatrix} \quad \varphi\,(1,-1) = \varPhi\,(2,2) = \begin{bmatrix} 0 & 0 \\ 0 & 1 \end{bmatrix} \\ \varphi\,(1,0) &= \frac{1}{\sqrt{2}}\left(\varPhi\,(1,2) + \varPhi\,(2,1)\right) = \frac{1}{\sqrt{2}}\begin{bmatrix} 0 & 1 \\ 1 & 0 \end{bmatrix}. \end{aligned}\right\} \tag{12.10.11}$$

Hence

$$\begin{aligned} J_+\,(\varphi\,(0,0)) = \frac{1}{\sqrt{2}}\Big\{ & a\begin{bmatrix} 0 & 1 \\ 0 & 0 \end{bmatrix}\begin{bmatrix} 0 & 1 \\ -1 & 0 \end{bmatrix} + b\begin{bmatrix} 0 & 1 \\ -1 & 0 \end{bmatrix}\begin{bmatrix} 0 & 0 \\ 1 & 0 \end{bmatrix} \\ & - \frac{\nu}{2}\left(\begin{bmatrix} 0 & 1 \\ 0 & 0 \end{bmatrix}\begin{bmatrix} 0 & 1 \\ -1 & 0 \end{bmatrix}\begin{bmatrix} 1 & 0 \\ 0 & -1 \end{bmatrix}\right. \\ & \left. - \begin{bmatrix} 1 & 0 \\ 0 & -1 \end{bmatrix}\begin{bmatrix} 0 & 1 \\ -1 & 0 \end{bmatrix}\begin{bmatrix} 0 & 0 \\ 1 & 0 \end{bmatrix}\right)\Big\} \end{aligned}$$

$$= -\frac{1}{\sqrt{2}} (a - b - \nu) \begin{bmatrix} 1 & 0 \\ 0 & 0 \end{bmatrix}$$

$$= -\frac{1}{\sqrt{2}} (a - b - \nu) \varphi(1, 1). \tag{12.10.12}$$

Similarly we have

$$J_- (\varphi(0, 0)) = \frac{1}{\sqrt{2}} (a - b - \nu) \varphi(1, -1), \tag{12.10.13}$$

$$J_3 (\varphi(0, 0)) = \frac{1}{2} (a - b - \nu) \varphi(1, 0). \tag{12.10.14}$$

Now we turn to discuss the imagined transition between S- and P-wave SC. For a two-electron system it should be totally antisymmetric by interchanging. The spin-angular part of the wave functions are as the following [32].

S-wave:

$$\phi_{J=0, m=0} = Y_{00} \varphi(0, 0), \tag{12.10.15}$$

P-wave (BW type [32, 33]):

$$\Psi_{J=0, m=0} = \frac{1}{\sqrt{3}} \left(Y_{1,1} \varphi(1, -1) - Y_{1,0} \varphi(1, 0) + Y_{1,-1} \varphi(1, 1) \right), \tag{12.10.16}$$

where $Y_{l,m}$ is the spherical harmonics. In the momentum space we have

$$\left. \begin{array}{l} Y_{00} = \dfrac{1}{\sqrt{4\pi}} Y_{1,0} = \sqrt{\dfrac{3}{4\pi}} \cos\theta_k = \sqrt{\dfrac{3}{4\pi}} \hat{k}_z \\[2mm] Y_{1,\pm1} = \mp\sqrt{\dfrac{3}{8\pi}} \sin\theta_k e^{\pm i\phi_k} = \mp\sqrt{\dfrac{3}{8\pi}} \hat{k}_\pm \\[2mm] \hat{k}_\pm = \hat{k}_x \pm i\hat{k}_y \end{array} \right\}, \tag{12.10.17}$$

where $\hat{\boldsymbol{k}}$ is univector in k-space. The (12.10.15) and (12.10.16) can be recast into

$$\phi_{J=0,m=0} = \frac{1}{\sqrt{2}} \begin{bmatrix} 0 & Y_{00} \\ -Y_{00} & 0 \end{bmatrix}, \qquad (12.10.18)$$

$$\Psi_{J=0,m=0} = \frac{1}{\sqrt{2}} \begin{bmatrix} Y_{1-1} & -\dfrac{1}{\sqrt{2}}Y_{10} \\ -\dfrac{1}{\sqrt{2}}Y_{00} & Y_{11} \end{bmatrix} \qquad (12.10.19)$$

$$= \frac{1}{\sqrt{8\pi}} \begin{bmatrix} \hat{k}_- & -\hat{k}_z \\ -\hat{k}_z & -\hat{k}_+ \end{bmatrix}.$$

Using (12.10.12) we get

$$\hat{k}_- J_+ (Y_{00}\varphi(0,0)) = \hat{k}_- Y_{00} J_+(\varphi(0,0)) = \sqrt{\frac{1}{8\pi}}(b-a+\nu) \begin{bmatrix} \hat{k}_- & 0 \\ 0 & 0 \end{bmatrix},$$

$$\hat{k}_+ J_- (Y_{00}\varphi(0,0)) = \sqrt{\frac{1}{8\pi}}(b-a+\nu) \begin{bmatrix} 0 & 0 \\ 0 & -\hat{k}_+ \end{bmatrix},$$

$$\hat{k}_3 J_3 (Y_{00}\varphi(0,0)) = \sqrt{\frac{1}{8\pi}}(b-a+\nu) \begin{bmatrix} 0 & -\hat{k}_+ \\ -\hat{k}_z & 0 \end{bmatrix}. \qquad (12.10.20)$$

Summarizing (12.10.20) we have

$$(\boldsymbol{k} \cdot \boldsymbol{J}) \, \phi_{J=0,m=0} = (b-a+\nu) \, \Psi_{J=0,m=0}. \qquad (12.10.21)$$

We would like to emphasize that the BW wave function $\Psi_{J=0,m=0}$ was originally used to describe the superfluidity of ^{3}He, now we borrow it to describe a hypothetical case of superconductivity. If someday the P-wave superconductivity is discovered, and it is observed that there is a weak transition from the S-wave to the P-wave superconductivity, then the transition, may be described by the action of a Yangian operator.

When $b - a = -1$, (by setting $\nu = 1$) the transition from S-wave to P-wave superconductivity is forbidden, while for $b - a = 1$, the transition is allowed. We observe that different values of parameters can induce different regimes of transitions. For special values of the parameters we have one-directional transition, and this is determined by the irreversibility of transitions induced by Yangian operators. We know that Lie algebra operates work in linear space and is reversible, while Yangian operates in quantum tensor space, it includes interactions between different linear spaces. How can the parameters a and b be controlled physically deserves further study.

12.11   Quantum Algebras

As was shown in Sections 12.2–12.3 the RTT relation in (12.2.3) plays the important role in determining quantum integrable systems. The associativity in (12.3.2) leads to the YBE, and we were only concerned with the simplest rational solution of YBE, i.e., $\check{R}(u) = I + uP$, where P is permutation. Actually there are three types of solutions to $\check{R}(u)$: rational, trigonometric and elliptic [35]. For a given solution of $\check{R}(u)$ there should be correspondingly algebraic structure given rise by RTT relation. In convenience we write the RTT relation in the form

$$\check{R}\left(xy^{-1}\right)\left[T\left(x\right)\otimes T\left(y\right)\right] = \left[T\left(y\right)\otimes T\left(x\right)\right]\check{R}\left(xy^{-1}\right), \qquad (12.11.1)$$

where instead of u and ν the spectral parameters $x = e^u$ and $y = e^\nu$ have been used. When $u,\nu \to \infty$ and $u - \nu \to \infty$, we have

$$\check{R}\left(x \to \infty\right) = S, \quad \check{R}\left(x = 1\right) = I(\check{R} = PR), \qquad (12.11.2)$$

$$\lim_{x \to \infty} T(x) = T = \|T_{ab}\|. \qquad (12.11.3)$$

The asymptotic behavior of (12.11.1) takes the form (summation over repeating indices)

$$S\left(T \otimes T\right) = \left(T \otimes T\right)S, \qquad (12.11.4)$$

or

$$S_{ij,mn}\left(T \otimes T\right)_{mn,kl} = \left(T \otimes T\right)_{ij,mn}S_{mn,kl}. \qquad (12.11.5)$$

Noting that

$$\left(A \otimes B\right)_{mn,kl} = A_{mk}B_{nl},$$

the component form of (12.11.5) is

$$S_{ij,mn}T_{mk}T_{nl} = T_{im}T_{jn}S_{mn,kl}, \tag{12.11.6}$$

where $i, j, \ldots = 1, 2, \ldots$, for $SU(2)$ auxiliary space, S satisfies

$$S_{ij,ml}S_{nt,lk}S_{vs,mn} = S_{lm,jk}S_{vn,il}S_{st,nm} \tag{12.11.7}$$

with three incoming indices i, j, k and outgoing ones v, s, t, the (12.11.7) is called braid relation [37–39].

A simple solution of (12.11.7) can be checked directly:

$$
S =
\begin{array}{c}

\begin{array}{cccc}
11 & 12 & 21 & 22
\end{array}
\\
\begin{array}{c}
11 \\ 12 \\ 21 \\ 22
\end{array}
\left[
\begin{array}{cccc}
q & 0 & 0 & 0 \\
0 & 0 & 1 & 0 \\
0 & 1 & q - q^{-1} & 0 \\
0 & 0 & 0 & q
\end{array}
\right],
\end{array}
\tag{12.11.8}
$$

where the matrix is labeled by the indices i, j for column and m, l for row. In solving (12.11.7) it allows a free parameter q in S. When $q = 1$ we have $S = P$, so S is nothing but a deformation of permutation P.

It is easy to obtain solution of YBE

$$\check{R}_{ij,ml}(x)\check{R}_{nt,lk}(xy)\check{R}_{vs,mn}(y) = \check{R}_{lm,jk}(y)\check{R}_{vn,il}(xy)\check{R}_{st,nm}(x) \tag{12.11.9}$$

with given solution of Braid Group (12.11.8). The direct check verifies that

$$
\check{R}(x) =
\begin{bmatrix}
\text{sh}(u + v) & 0 & 0 & 0 \\
0 & \text{sh}u & \text{sh}v & 0 \\
0 & \text{sh}v & \text{sh}u & 0 \\
0 & 0 & 0 & \text{sh}(u + v)
\end{bmatrix}
= xs - x^{-1}s^{-1}
\tag{12.11.10}
$$

satisfies (12.11.9), where $x = e^u$ and $q = e^v$. We see that there is one deformation parameter q in S, whereas there appears dynamic parameter x besides q in $\check{R}(x)$. As $u, v \to 0$ from (12.11.10) it follows

$$\check{R}(x)\,|_{u \to 0, v \to 0} = uI + vP = u\left(I + \frac{v}{u}P\right), \tag{12.11.11}$$

i.e., $\frac{v}{u}$ plays the role in comparison to the simplest rational form of $\check{R}(u)$ in (12.1.12), since (12.11.9) allows an arbitrary overall factor. Equation (12.11.10) is called trigonometric solution of YBE that is still true for $x = e^{iu}$ and $q = e^{iv}$.

Equipped with S (4×4 matrix) given by (12.11.8) we shall look for the allowed T (2×2 matrix) due to (12.11.6). Exhausting all the free indices $i, j, \ldots = 1, 2$ in (12.11.6), for instance, for $i = 1$, $j = 2$, $k = l = 1$, there is only non-vanishing element $S_{12,21} = 1$ on the left hand side and $S_{11,11} = q$ on the right hand side, hence the dummy indices should be $m = 2, n = 1$. This leads to

$$T_{21}T_{11} = qT_{11}T_{21}. \tag{12.11.12}$$

Recalling

$$T = \begin{bmatrix} a & b \\ c & d \end{bmatrix} = \begin{bmatrix} T_{11} & T_{12} \\ T_{21} & T_{22} \end{bmatrix}, \tag{12.11.13}$$

then (12.11.12) can be written as $ac = q^{-1}ca$. We obtain

$$\left. \begin{array}{l} ab = q^{-1}ba, \, ac = q^{-1}ca, \, bc = cb \\ bd = q^{-1}db, \, cd = q^{-1}dc \\ ad - da = \left(q^{-1} - q\right)bc \end{array} \right\}. \tag{12.11.14}$$

The latest relation can be recast to

$$ad - q^{-1}bc = da - qbc \equiv \det_q T, \tag{12.11.15a}$$

which is the q-deformed form of $\det_q T$.

The $\det_q T$ commutes with a, b, c and d, for example, by virtue of (12.11.15a) we get

$$[\det_q T, a] = (ad - q^{-1}bc)a - a(da - qbc)$$
$$= qabc - q^{-1}bca = 0, \tag{12.11.15b}$$

where $bca = q^2 abc$ has been used. Therefore, $\det_q T$ is the center of the algebra formed by a, b, c and d. The commutativity

$$[\det_q T, A] = 0, \quad A = a, b, c, d \tag{12.11.16}$$

allows to give the inverse of T:

$$T^{-1} = (\det_q T)^{-1} \begin{bmatrix} d & -qb \\ -q^{-1}c & a \end{bmatrix}. \qquad (12.11.17)$$

For the historical reason the set $T = \{a, b, c, d\}$ is called quantum group associated with $SL(2)$. In fact, it is not the group defined in the usual group theory: actually, it belongs to a q-deformed Hopf algebra.

The simplest choice of (12.11.13) is

$$T = \begin{bmatrix} a & b \\ 0 & a^{-1} \end{bmatrix}. \qquad (12.11.18)$$

In this case the only survived commutation relation is

$$ba = qab, \qquad (12.11.19)$$

for which the realization can be made through

$$b = \exp\{i\alpha\hat{p}\},$$
$$a = \exp\{i\beta x\}.$$

This leads to

$$q = e^{i\alpha\beta\hbar}$$

by virture of $[\hat{p}, x] = -i\hbar$. This example gives nothing new but is equivalent to the canonical commutation relation between $\hat{p}$ and x.

The quantum group of (12.11.14) associated with $SL(2)$ is given by the asymptotic behavior of the solution of YBE, i.e.,

$$\check{R}(x) = xS - x^{-1}S \qquad (12.11.20)$$

through (12.11.6) as $x \to \infty$. Now substituting (12.11.20) into the (12.11.1) and supposing $T(x)$ to take the form:

$$T(x) = xL_+ + xL_-, \quad L_\sigma = \|(L_\sigma)_{ij}\|, \quad \sigma = \pm, i, j = 1, 2$$

we obtain the known Reshetikhin-Faddeev-Takhtajan (RFT) relations [38]

$$S(L^{\pm} \otimes L^{\pm}) = (L^{\pm} \otimes L^{\pm})S, \qquad (12.11.21)$$

$$S(L^{+} \otimes L^{-}) = (L^{-} \otimes L^{+})S, \qquad (12.11.22)$$

where S is given by (12.11.8). Substituting Eq. (12.11.8) into Eqs. (12.11.21)–(12.11.22) leads to

$$L_{12}^{\sigma} L_{12}^{\sigma} = 0 \qquad (12.11.23)$$

$$\left[L_{ii}^{\sigma}, L_{jj}^{\sigma} \right] = 0, \quad i, j = 1, 2, \sigma = \pm \qquad (12.11.24)$$

$$L_{ii}^{\sigma} L_{12}^{+} = q^{+\sigma\varepsilon(i)} L_{12}^{+} L_{ii}^{\sigma} \qquad (12.11.25)$$

$$L_{ii}^{\sigma} L_{21}^{-} = q^{-\sigma\varepsilon(i)} L_{21}^{-} L_{ii}^{\sigma} \qquad (12.11.26)$$

$$\left[L_{12}^{+}, L_{21}^{-} \right] = \left(q - q^{-1} \right) \left(L_{22}^{-} L_{11}^{+} - L_{22}^{+} L_{11}^{-} \right) \qquad (12.11.27)$$

$$\varepsilon(1) = -1, \quad \varepsilon(2) = +1,$$

where no summation over the repeating indices.

From Eq. (12.11.23) it follows that

$$L_{+} = \begin{bmatrix} L_{11}^{+} & L_{12}^{+} \\ 0 & L_{22}^{+} \end{bmatrix}, \quad L_{-} = \begin{bmatrix} L_{11}^{-} & 0 \\ L_{21}^{-} & L_{22}^{-} \end{bmatrix}. \qquad (12.11.28)$$

Equation (12.11.24) shows that the diagonal elements of $L_{\pm}$ are commutative. The $L_{\pm}$ satisfying (12.11.23)–(12.11.28) should take the forms:

$$L_{-} = \begin{bmatrix} k & \left(Q - Q^{-1} \right) X_{+} \\ 0 & K^{-1} \end{bmatrix}, \quad L_{+} = \begin{bmatrix} K^{-1} & 0 \\ -\left(Q - Q^{-1} \right) X_{-} & K \end{bmatrix}, \qquad (12.11.29)$$

where $Q = q^{-1}$, K and $X_{\pm}$ satisfy

$$K X_{\pm} K^{-1} = Q^{\pm 1} X_{\pm}, \qquad (12.11.30)$$

$$[X^{+}, X^{-}] = \frac{K^{2} - K^{-2}}{Q - Q^{-1}}. \qquad (12.11.31)$$

Defining $K = Q^{X_3} = e^{\gamma X_3}$, then

$$[X_+, X_-] = \frac{\mathrm{sh}\,(2\gamma X_3)}{\mathrm{sh}2\gamma}. \qquad (12.11.32)$$

The algebra defined by (12.11.30) and (12.11.31) together with the coproduct

$$\Delta(K) = 1 \otimes K + K \otimes 1$$
$$\Delta(X_\pm) = X_\pm \otimes K + K^{-1} \otimes X_\pm \qquad (12.11.33)$$

and the antipode S

$$S(1) = 1, \quad S(K) = K^{-1}, \quad S(X_\pm) = -Q^{\pm 1} X_\pm \qquad (12.11.34)$$

as well as the co-unit:

$$\varepsilon(1) = 1, \varepsilon(K) = \varepsilon(X_\pm) = 0 \qquad (12.11.35)$$

forms the quantum algebra associated with $SL(2)$. The (12.11.30)–(12.11.35) were firstly presented by V.G. Drinfild [6, 7] and M. Jimbo [40] independently, meanwhile L.D. Faddeev also presented similar idea in an unpublished paper. Actually, there is more general form of $L_\pm$ satisfying (12.11.21) and (12.11.22). More developments can be found in a book [17] edited by M. Jimbo.

12.12   Bi-frequency Harmonic Oscillators and the Commutation Relations of Quantum Algebras

In the literatures of Yang-Baxter systems, Yangian and the quantum algebras belong to the structure of quantum groups. In the following, we study the physical significance of simple forms of the quantum algebras from the point of view of quantum mechanics. Some illustrative examples will be given below.

Considering a system of harmonic oscillators, the second-quantized form of its Hamiltonian is

$$H_0 = \omega_1 a^\dagger a + \omega_2 b^\dagger b, \qquad (12.12.1)$$

where the zero-point energies have been suppressed, and $a^\dagger$, a and $b^\dagger$, b represent creation and annihilation operators of oscillators with frequencies ω_1 and ω_2 respectively. These operators satisfy the commutation relations

$$[a, a^\dagger] = 1, \quad [b, b^\dagger] = 1, \tag{12.12.2}$$

while a and b are mutually independent, anyone of $a, a^\dagger$ commutes with anyone of $b, b^\dagger$. The second-quantized operator ψ and its adjoint $\psi^\dagger$ are represented as

$$\psi = \begin{pmatrix} a \\ b \end{pmatrix}, \quad \psi^\dagger = (a^\dagger, b^\dagger). \tag{12.12.3}$$

The Hamiltonian H can be expressed as

$$H_0 = \psi^\dagger \begin{bmatrix} \omega_1 & 0 \\ 0 & \omega_2 \end{bmatrix} \psi = (\omega_1 \omega_2)^{1/2} \psi^\dagger \begin{bmatrix} q^{1/2} & 0 \\ 0 & q^{-1/2} \end{bmatrix} \psi, \tag{12.12.4}$$

where

$$q = \omega_1/\omega_2.$$

If the interaction between the oscillators with some field is introduced, for example the oscillators are charged and are situated in a resonant light field, transitions can take place. Transition between the two states of oscillators is induced by the operators $S^+ = a^\dagger b$ and $S^- = b^\dagger a$. Expressed in terms of ψ and $\psi^\dagger$ they become

$$\left. \begin{aligned} S^+ &= \psi^\dagger X^+ \psi \\ S^- &= \psi^\dagger X^- \psi \end{aligned} \right\} \tag{12.12.5}$$

where

$$X^+ = \begin{bmatrix} 0 & 1 \\ 0 & 0 \end{bmatrix}, \quad X^- = \begin{bmatrix} 0 & 0 \\ 1 & 0 \end{bmatrix}. \tag{12.12.6}$$

i.e., the well-known Pauli matrices σ^+ and σ^-. The reason that we introduce new notations is that we are going to construct a set of generators of an algebra K, X^+, and X^-. The other member of the set is

$$K = \begin{bmatrix} q^{1/2} & 0 \\ 0 & q^{-1/2} \end{bmatrix}, \qquad (12.12.7)$$

and operators related to it are:

$$\left. \begin{array}{c} K^{-1} = \begin{bmatrix} q^{-1/2} & 0 \\ 0 & q^{1/2} \end{bmatrix}, \; K^2 = \begin{bmatrix} q & 0 \\ 0 & q^{-1} \end{bmatrix}, \; K^{-2} = \begin{bmatrix} q^{-1} & 0 \\ 0 & q \end{bmatrix} \\[2em] 2X_3 = \begin{bmatrix} 1 & 0 \\ 0 & -1 \end{bmatrix} = \dfrac{K^2 - K^{-2}}{q - q^{-1}} \end{array} \right\}$$

$$(12.12.8)$$

Then H_0 can be rewritten as

$$H_0 = (\omega_1 \omega_2)^{1/2} \psi^\dagger K \psi. \qquad (12.12.9)$$

From the well-known commutation relations of σ^+, σ^- and σ_3 we obtain the formal commutation relations for X^+, X^- and X_3.

$$\left. \begin{array}{c} [X_3, X^\pm] = \pm X^\pm \\[1em] [X^+, X^-] = \dfrac{K^2 - K^{-2}}{q - q^{-1}} \end{array} \right\}. \qquad (12.12.10)$$

The first relation of these is not what we actually need, because we need K instead of X_3. For this end we write $q = e^\gamma$, then we have

$$K = \begin{bmatrix} e^{\gamma/2} & 0 \\ 0 & e^{-\gamma/2} \end{bmatrix} = e^{\gamma X_3} = q^{X_3}. \qquad (12.12.11)$$

In the following, we calculate $K X^\pm K^{-1}$.

$$KX^\pm K^{-1} = \left(1 + \gamma X_3 + \frac{1}{2!} \gamma^2 X_3^2 + \cdots \right)$$

$$\times X^\pm \left(1 - \gamma X_3 + \frac{1}{2!} \gamma^2 X_3^2 - \cdots \right)$$

$$= X^\pm + \gamma \left[X_3, X^\pm \right] - \frac{1}{2} \gamma^2 X_3 \left[X^\pm, X_3 \right]$$

$$- \frac{1}{2} \gamma^2 \left[X_3, X^\pm \right] X_3 + \cdots$$

$$= X^{\pm} \pm \gamma X^{\pm} + \frac{1}{2}\gamma^2 \left[X_3, \left[X_3, X^{\pm} \right] \right] \pm \cdots$$

$$= \left(1 \pm \gamma + \frac{1}{2}\gamma^2 \pm \cdots \right) X^{\pm}$$

$$= \mathrm{e}^{\pm\gamma} X^{\pm} = q^{\pm 1} X^{\pm}.$$

It is equivalent to the first relation of (12.12.10) in view of the definition (12.12.11). Together with (12.12.10) it formly constitutes the commutation relations for the quantum algebra:

$$\left.\begin{array}{l} KX^{\pm}K^{-1} = q^{\pm 1}X^{\pm} \\[2mm] \left[X^+, X^- \right] = \dfrac{K^2 - K^{-2}}{q - q^{-1}} = \dfrac{q^{2X_3} - q^{-2X_3}}{q - q^{-1}} \equiv [2X_3]_q \end{array}\right\} . \qquad (12.12.12)$$

This quantum algebra is connected with $SU(2)$. The set $\{X^+, X^-, K\}$ forms quantum algebra associated with $SU(2)$, and K is connected with X_3 through (12.12.11). It seems that the above approach prefers complications instead of simplicity, without promoting the $SU(2)$ algebra a bit. In fact, what we have done is to demonstrate through a simple example that if we choose X^+, X^- and K as basis, the original commutation relations of $SU(2)$ become "q-deformed" relations of (12.12.12). In the second relation of (12.12.12), $[2X_3]_q$ means the q-deformation of $2X_3$. In the above example we used the 2×2 matrix representation of $SU(2)$ for X^+, X^- and K, the new algebra is equivalent to $SU(2)$ whether deformation exists or not. But for a general representation of $SU(2)$, the algebra (12.12.12) differs essentially from $SU(2)$. It reduces to $SU(2)$ only when $q \to 1$, and this is because

$$\left[X^+, X^- \right] = \frac{q^{2X_3} - q^{-2X_3}}{q - q^{-1}} = \frac{\sinh 2\gamma X_3}{\sinh \gamma}, \qquad (12.12.13)$$

and we get the conventional relation

$$\left[X^+, X^- \right] = 2X_3,$$

only when $\gamma \to 0$, i.e., $q \to 1$. The above example looks trivial, but the relations obtained (12.12.12) are general. They reveal the essence of the q-deformation.

The second relation of (12.12.12) can be expressed in second quantized form, but we shall show that for single-fermion (12.12.10) reduces to the usual Lie algebra $SL(2)$, i.e., without q-deformation. This means that in order to generate q-deformed commutation relations we have to use current algebra form with multi-fermion [41]. By using $[\psi_\alpha, \psi_\beta^\dagger]_\pm = \delta_{\alpha\beta}$, we have

$$
\begin{aligned}
\left[\psi^\dagger X^+\psi, \psi^\dagger X^-\psi\right] &= (X^+)_{\alpha\beta}(X^-)_{\sigma\tau}\left[\psi_\alpha^\dagger\psi_\beta, \psi_\sigma^\dagger\psi_\tau\right] \\
&= (X^+)_{\alpha\beta}(X^-)_{\sigma\tau} \\
&\quad \times \left(\psi_\alpha^\dagger\left[\psi_\beta, \psi_\sigma^\dagger\right]\psi_\tau + \psi_\sigma^\dagger\left[\psi_\alpha^\dagger\psi_\tau\right]\psi_\beta\right) \\
&= \psi_\alpha^\dagger[X_+, X_-]_{\sigma\tau}\psi_\tau \\
&= \psi^\dagger\frac{K^2 - K^{-2}}{q - q^{-1}}\psi \\
&= \psi^\dagger[2X_3]_q\psi.
\end{aligned}
\tag{12.12.14}
$$

Supposing that ψ is a fermionic operator, then (12.12.13) can be reduced to the usual commutation relation of $SU(2)$ since

$$
\psi^+X_3\psi = \psi_\alpha^+(X_3)_{\alpha\beta}\psi_\beta = \frac{1}{2}\left(\psi_1^+\psi_1 - \psi_2^+\psi_2\right) = \frac{1}{2}(n_1 - n_2)
\tag{12.12.15}
$$

then

$$
K = e^{\frac{\gamma}{2}(n_1 - n_2)}
$$

$$
KX_+K^{-1} = e^{\frac{\gamma}{2}(n_1 - n_2)}\psi_1^+\psi_2 e^{-\frac{\gamma}{2}(n_1 - n_2)}
$$

$$
= \left(e^{\frac{\nu}{2}n_1}\psi_1^+ e^{-\frac{\nu}{2}n_1}\right)\left(e^{-\frac{\nu}{2}n_2}\psi_2 e^{\frac{\nu}{2}n_2}\right).
$$

For any operators A and B we have the Baker-Campbell-Hausdorff theorem,

$$
e^A B e^{-A} = B + [A, B] + \frac{1}{2!}[A, [A, B]] + \frac{1}{3!}[A, [A, [A, B]]] + \cdots,
$$

therefore,

$$e^{\frac{\gamma}{2}n_1}\psi_1^+ e^{-\frac{\gamma}{2}n_1} = e^{\frac{\gamma}{2}}\psi_1^+$$

$$e^{-\frac{\gamma}{2}n_2}\psi_2 e^{\frac{\gamma}{2}n_2} = e^{\frac{\gamma}{2}}\psi_2.$$

We thus obtain

$$KX_+K^{-1} = e^{\gamma}\psi_1^+\psi_2 = qX_+, \ q = e^{\gamma} \qquad (12.12.16)$$

whereas

$$K^2 - K^{-2} = e^{\gamma(n_1-n_2)} - e^{-\gamma(n_1-n_2)}$$
$$= \left(e^{\gamma n_1} - e^{-\gamma n_1}\right) - \left(e^{\gamma n_2} - e^{-\gamma n_2}\right).$$

On account of

$$e^{\gamma n} = 1 + n\left(e^{\gamma} - 1\right),$$

for single fermion, we get

$$K^2 - K^{-2} = (n_1 - n_2)\left(e^{\gamma} - e^{-\gamma}\right) = \left(q - q^{-1}\right)(n_1 - n_2),$$

i.e.,

$$[X_+, X_-] = 2X_3. \qquad (12.12.17)$$

Obviously the single fermion cannot give rise to the q-deformation.

In this paragraph we have obtained the q-commutation relations (12.12.10) through a trivial example. In this example the q-commutator is only a formal expression. However, it works in more general problem for arbitrary representations of $SU(2)$ operators. To complete the quantum algebras we should have the co-product as discussed before.

For (12.12.10), the co-product was also introduced for the quantum algebra. By definition we need to have

$$\Delta(AB) = \Delta(A)\Delta(B).$$

But how can the co-product be defined so that (12.12.12) is self-consistent? Let us show that the co-product

$$\Delta(K) = K \otimes K, \quad \Delta\left(K^{-1}\right) = K^{-1} \otimes K^{-1} \qquad (12.12.18)$$

$$\Delta\left(X_{\pm}\right) = X_{\pm} \otimes K + K^{-1} \otimes X_{\pm} \qquad (12.12.19)$$

are consistent with q-commutation relations.

Operating Δ on the first equation of (12.12.12), we get

$$\Delta\left(KX^{\pm}\right) = \Delta(K)\Delta\left(X^{\pm}\right) = (K \otimes K)\left(K \otimes X^{\pm} + X^{\pm} \otimes K^{-1}\right)$$
$$= K^2 \otimes KX^{\pm} + KX^{\pm} \otimes Z,$$

$$\Delta\left(X^{\pm}K\right) = \Delta\left(X^{\pm}\right)\Delta(K) = \left(K \otimes X^{\pm} + X^{\pm} \otimes K^{-1}\right)(K \otimes K)$$
$$= K^2 \otimes X^{\pm}K + X^{\pm}K \otimes Z.$$

Comparing the two equations and using $KX^{\pm} = q^{\pm}X^{\pm}K$, we obtain

$$\Delta\left(KX^{\pm}\right) = q^{\pm}\Delta\left(X^{\pm}K\right).$$

Similarly we can prove that

$$\Delta\left([X_+, X_-]\right) = \Delta\left(X_+\right)\Delta\left(X_-\right) - \Delta\left(X_-\right)\Delta\left(X_+\right)$$

$$= \frac{1}{q - q^{-1}}\left(\Delta\left(K^2\right) - \Delta\left(-K^{-2}\right)\right). \qquad (12.12.20)$$

therefore, we conclude that (12.12.12) is self-consistent for the co-product.

The quantum algebras belong to a category of the q-deformed Hopf algebra and can be realized through multifermions. Interested readers are referred to Refs. [6, 7, 23, 24, 40, 41].

12.13   The Translation Operators for the Coherent States and the Hofstadter model

Consider the motion of an electron in the xy-plane, with a stationary uniform magnetic field B_0 along the z-axis, i.e., the problem of Landau levels. With the electronic charge $-e$ and in units of

$\hbar = c = 1$, the Hamiltonian is

$$H = \frac{1}{2m}(\boldsymbol{p} + e\boldsymbol{A})^2 \equiv \frac{1}{2m}\pi^2, \qquad (12.13.1)$$

where

$$\boldsymbol{p} = (-\mathrm{i}\partial_x, -\mathrm{i}\partial_y), \quad \boldsymbol{A} = (A_x, A_y),$$

$$\partial_x A_y - \partial_y A_x = B_0.$$

We introduce

$$\pi^{\pm} = \frac{1}{\sqrt{2}}(\pi_x \pm \mathrm{i}\pi_y), \qquad (12.13.2)$$

with the commutation relations

$$\left[\pi^-, \pi^+\right] = eB_0. \qquad (12.13.3)$$

The unitary translation operator on the xy-plane is

$$U(\alpha) = \mathrm{e}^{\alpha\pi^+ - \alpha^*\pi^-}, \qquad (12.13.4)$$

where α is a complex parameter whose real and imaginary parts characterize the coordinates of translation on the xy-plane. The operator in (12.13.4) is also related to coherent states. We define the boson creation and annihilation operators in terms of $\pi^{\pm}$:

$$b^{\dagger} = \pi^+/\sqrt{eB_0}, \quad b = \pi^-/\sqrt{eB_0},$$

with the commutation relation

$$\left[b, b^{\dagger}\right] = 1.$$

Operating $U(\alpha)$ on the vacuum state $|0\rangle$ of the system of bosons, we get the coherent state $|\alpha\rangle$.

$$U(\alpha)|0\rangle = \mathrm{e}^{\alpha\pi^+}|0\rangle \equiv |\alpha\rangle.$$

We carry out two consecutive translations on the xy-plane in different orders, namely, $U(\alpha)U(\beta)$ and $U(\beta)U(\alpha)$. They lead to different results because of the presence of the magnetic field. Using a special case of the Baker-Campbell-Hausdorff formula

$$e^A e^B = \exp\left(A + B + \frac{1}{2}[A, B]\right),$$

which is valid for $[A, [A, B]] = [B, [A, B]] = 0$, we calculate the commutator

$$\frac{1}{2}\left[\alpha\pi^+ - \alpha\pi^*, \beta\pi^+ - \beta^*\pi^-\right] = ieB_0 \left(\mathrm{Re}\alpha\mathrm{Im}\beta - \mathrm{Im}\alpha\mathrm{Re}\beta\right)$$

$$= -ieB_0 \times \text{area enclosed by } \alpha \text{ and } \beta$$

$$= -ie\Phi = -i2\pi\frac{\Phi}{\Phi_0},$$

where Φ is the magnetic flux passing through the area, and $\Phi_0 = \frac{2\pi}{e}$ is the flux quantum. We then obtain the Heisenberg-Weyl relation

$$U(\alpha) U(\beta) = q^2 U(\beta) U(\alpha), \tag{12.13.5}$$

where

$$q = \exp\left(-i2\pi\frac{\Phi}{\Phi_0}\right). \tag{12.13.6}$$

If the ratio Φ/Φ_0 is P/Q, where P and Q are prime numbers, then

$$q^Q = 1.$$

From the definition of $U(\alpha)$ we know that

$$U(-\alpha) = (U(\alpha))^{-1}. \tag{12.13.7}$$

Introducing operators [42]

$$\left.\begin{aligned}
X^+ &= \frac{1}{q - q^{-1}}\left(U(-\alpha) + U(-\beta)\right) \\
X^- &= \frac{1}{q - q^{-1}}\left(U(\alpha) + U(\beta)\right) \\
q^{-1}K^2 &= U(-\alpha) U(\beta) \\
qK^{-2} &= U(-\beta) U(\alpha)
\end{aligned}\right\} \tag{12.13.8}$$

and using (12.13.5) and (12.13.7), we obtain

$$\left.\begin{aligned} [X^+, X^-] &= \frac{K^2 - K^{-2}}{q - q^{-1}} \\ KX^\pm K^{-1} &= q^{\pm 1} X^\pm \end{aligned}\right\}, \qquad (12.13.9)$$

which are the commutation relations for the quantum algebra of (12.12.12). When the unitary operator U satisfies the Heisenberg-Weyl relation (12.13.5), we can define by using operators $(X^\pm, K)$ in Eq. (12.13.8), which are the generators of the quantum algebra connected with $SU(2)$. Equation (12.13.5) can also have other realizations, for instance, by using the operator of magnetic translation

$$t(\boldsymbol{a}) = e^{i\boldsymbol{\Pi} \cdot \boldsymbol{a}} \qquad (12.13.10)$$

where $\prod$ is magnetic translation operator:

$$\prod = \boldsymbol{p} + e\boldsymbol{A} + e\boldsymbol{r} \times \boldsymbol{B}, \quad \boldsymbol{B} = B_0 \boldsymbol{e}_3 \qquad (12.13.11)$$

we have

$$t(\boldsymbol{a}) t(\boldsymbol{b}) = \exp\left(-i2\pi \frac{\Phi}{\Phi_0}\right) t(\boldsymbol{b}) t(\boldsymbol{a}). \qquad (12.13.12)$$

Direct calculation gives

$$[H, \prod] = 0, \qquad (12.13.13)$$

i.e.,

$$[H, t(\boldsymbol{a})] = 0. \qquad (12.13.14)$$

where H is given by (12.13.1) that leads to Landau levels.

Similarly we can define generators of the quantum algebra $X^\pm$ and K by using (12.13.8) with U replaced by $t(\boldsymbol{a})$ in (12.13.10) and we have [43]

$$[H, X^\pm] = [H, K] = 0, \qquad (12.13.15)$$

which implies that the Hamiltonian commutes with the generators of the quantum algebra, or in other words H has the symmetry of the quantum algebra. The equivalence of (12.12.8) and (12.13.9) is based on the area enclosed by $\boldsymbol{\alpha}$ and $\boldsymbol{\beta}$, when the plane is periodic in the fundamental translations, q is determined by the unit cell, and the periodic network (12.13.9) has clear-cut definition.

With this preparation let us turn to the quantum algebraic interpretation of the Hofstadter model based on the work of Wiegmann and Zabrodin [42]. Consider electrons moving on a lattice in the xy-plane, with constant uniform magnetic field $\boldsymbol{B}$ along the z direction. The electron can jump from a lattice site to one of the nearest neighbors. The Hamiltonian is

$$H = \sum_{\langle n,m \rangle} \exp \left[i \frac{1}{2} \boldsymbol{B} \cdot (\boldsymbol{n} \times \boldsymbol{m}) \right] c_n^\dagger c_m, \qquad (12.13.16)$$

where $\langle n,\ m \rangle$ implies that n,m are nearest neighbors. The hopping amplitude $t_{n,m}$ is set to be 1 and $\frac{1}{2}\boldsymbol{n} \times \boldsymbol{m}$ is the area of the triangle with sides $\boldsymbol{n}$ and $\boldsymbol{m}$, $\frac{1}{2}\boldsymbol{B} \cdot (\boldsymbol{n} \times \boldsymbol{m})$ is the magnetic flux through the triangle. Summing the flux over all lattice sites, we obtain the total flux $\varPhi$.

$$\sum_{\langle n,m \rangle} \frac{1}{2} \boldsymbol{B} \cdot (\boldsymbol{n} \times \boldsymbol{m}) = \varPhi. \qquad (12.13.17)$$

The state with an electron at site $\boldsymbol{j}$ is

$$|\boldsymbol{j}\rangle = c_{\boldsymbol{j}}^\dagger |0\rangle. \qquad (12.13.18)$$

Let

$$A_{n,m} = \frac{1}{2} \boldsymbol{B} \cdot (\boldsymbol{n} \times \boldsymbol{m}), \qquad (12.13.19)$$

and define the translation operator

$$T_\mu (\boldsymbol{j}) = \exp \left(i A_{j,j+\mu} \right) |\boldsymbol{j}\rangle \langle \boldsymbol{j} + \boldsymbol{\mu}|, \qquad (12.13.20)$$

and

$$T_\mu = \sum_j T_\mu (\boldsymbol{j}) = \sum_j \exp \left(i A_{j,j+\mu} \right) |\boldsymbol{j}\rangle \langle \boldsymbol{j} + \boldsymbol{\mu}|. \qquad (12.13.21)$$

Studying the results of consecutive translations in different orders, we have

$$T_n T_m = \sum_{j,k} \exp\left[i\left(A_{j,j+n} + A_{k,k+m}\right)\right] |j\rangle \langle j+n| k \rangle \langle k+m|$$

$$= \sum_j \exp\left(iA_{j,j+n}\right) \exp\left(iA_{j+n,j+m+n}\right) |j\rangle \langle j+m+n|,$$

where the ortho-normal property of the state vectors

$$\langle j+n \mid k\rangle = \delta_{k,\,j+n},$$

has been used. Similarly we have in the reverse order

$$T_m T_n = \sum_j \exp\left(iA_{j,j+m}\right) \exp\left(iA_{j+m,j+m+n}\right) |j\rangle \langle j+m+n|.$$

Using (12.13.19) and taking into account

$$j \times (j+n) + (j+n) \times (j+m+n) = j \times (n+m) + n \times m,$$

we obtain

$$T_n T_m = \exp\left[\frac{i}{2} B \cdot (n \times m)\right] \sum_j$$

$$\times \exp\left[\frac{i}{2} B \cdot j \times (n \times m)\right] |j\rangle \langle j+m+n|$$

$$= \exp\left[\frac{i}{2} B \cdot (n \times m)\right] T_{n+m} = \exp\left(\frac{i}{2}\Phi\right) T_{n+m},$$

and

$$T_m T_n = \exp\left(-\frac{i}{2}\Phi\right) T_{m+n}.$$

If n is along the x direction and m along the y direction, the preceding relation gives

$$T_x T_y = q^2 T_y T_x, \tag{12.13.22}$$

where

$$q^2 = e^{i2\pi\Phi/\Phi_0}. \tag{12.13.23}$$

The Hamiltonian (12.13.16) can now be written as [42]:

$$H = -i \sum_{\mu=\pm x, \pm y} \sum_{j} \exp\left(i A_{j,j+\mu}\right) |j\rangle \langle j + \mu| \tag{12.13.24}$$

$$= T_x + T_{-x} + T_y + T_{-y}.$$

Following (12.13.8) we define

$$X^+ = \frac{1}{q - q^{-1}} \left(T_{-x} + T_{-y}\right),$$

$$X^- = -\frac{1}{q - q^{-1}} \left(T_x + T_y\right), \tag{12.13.25}$$

$$T_{-x}T_y = q^2 K^{-2}, \quad T_{-y}T_x = q^{-2}K^2.$$

Then $X^{\pm}$ and K satisfy the quantum algebra relations of (12.12.12), and H can be expressed in terms of generators of the quantum algebra:

$$H = i\left(q - q^{-1}\right)\left(X^- - X^+\right). \tag{12.13.26}$$

The physics of the model discussed above is well-known in condensed matter physics. What we have shown is that it is a realization of the quantum algebra. This understanding has induced further studies in this direction [44].

12.14 The Cyclic Representation of the Quantum Algebras

In quantum mechanics the process of quantization is realized by replacing the classical Poisson bracket $\{\,,\}$ by $\frac{1}{i\hbar}[\,,]$, where $[\,,]$ is the quantum commutator. But this recipe fails to work with the quantization of the phase. In analytical mechanics for an integrable system after a canonical transformation there appear the action variable J and its conjugate angle variable ϕ, with their Poisson bracket $\{J, \phi\} = 1$. In the quantization $J \to \hat{N}$ (the particle number), and $\phi \to \hat{\phi}$, the phase operator. In the representation $\hat{N} = a^{\dagger}a$, it is natural to write $a = e^{i\hat{\phi}}\sqrt{\hat{N}}$ and $a^\dagger = \sqrt{\hat{N}}e^{-i\hat{\phi}}$. But we will find that $e^{i\hat{\phi}}$ is not unitary. Let us calculate the average of the commutator $[\hat{N}, \hat{\phi}] = i$ over the state $|n\rangle$. We find the inconsistent result

$$\langle n|\, [\hat{N}, \hat{\phi}]\, |n\rangle = n\langle n|\hat{\phi}|n\rangle - n\langle n|\hat{\phi}|n\rangle = 0.$$

There have been many discussions and debates over the problem of quantum phase operator. The most well-known are the Susskind-Glogower (abbreviated as S-G) theory and the Pegg-Barnett (P-B) theory [45]. The S-G theory uses the non-unitary exponential operators while keeping the requirement that the number of bosons are unrestricted. The P-B theory, on the other hand guarantees the unitarity of exponential operators by imposing a restriction on the maximum number of particles. (Actually the particles are no longer bosons.) In the following, we use P-B theory as an example to explain the cyclic representation of the quantum algebra for $q^P = 1$. We emphasize that the phase quantization may not be reliable in Physics and no intention to solve the phase quantization in this book. We only use an example to explain the cyclic representation of quantum algebras that is useful in Mathematical Physics.

We expand the phase eigenstate $|\phi_m\rangle$ in terms of the number eigenstate $|n\rangle$, with the maximum number of particles set to be S:

$$|\phi_m\rangle = \frac{1}{\sqrt{S+1}} \sum_{n=0}^{S} e^{in\phi_m} |n\rangle. \tag{12.14.1}$$

In order that the phase eigenstates are orthogonal and normalized, i.e.,

$$\langle \phi_m \mid \phi_n \rangle = \delta_{mn}, \tag{12.14.2}$$

the phase eigenstates must satisfy

$$\phi_m = \phi_0 + \frac{2m\pi}{S+1}, \tag{12.14.3}$$

where ϕ_0 is an arbitrary real number. We demonstrate in the following that (12.14.2) can be obtained from (12.14.3),

$$\langle \phi_n \mid \phi_m \rangle = (S+1)^{-1} \sum_{l,k=0}^{S} e^{i(l\phi_m - k\phi_n)} \langle l \mid k \rangle$$

$$= (S+1)^{-1} \sum_{k=0}^{S} e^{ik(\phi_m - \phi_n)} = (S+1)^{-1} \sum_{k=0}^{S} q^{k(m-n)},$$

where

$$q = \exp\left(i\frac{2\pi}{S+1}\right), \quad q^P = 1, \quad P = S+1. \tag{12.14.4}$$

Obviously, q satisfies

$$\sum_{n=0}^{S} q^n = 0.$$

Therefore, (12.14.2) holds. The phase operator is

$$\hat{\phi} = \sum_{m=0}^{S} \phi_m \mid \phi_m \rangle \langle \phi_m \mid . \tag{12.14.5}$$

Operating $e^{i\hat{\phi}}$ on $|n\rangle$, we obtain

$$e^{i\hat{\phi}} \mid n \rangle = (S+1)^{1/2} \sum_{m=0}^{S} e^{-in\phi_m} e^{i\hat{\phi}} \mid \phi_m \rangle$$

$$= (S+1)^{1/2} \sum_{m=0}^{S} e^{-i(n-1)\phi_m} \mid \phi_m \rangle .$$

By comparing the above equation with Eq. (12.14.1), we get

$$e^{i\hat{\phi}} \mid n \rangle = \mid n-1 \rangle , \quad n \geq 1. \tag{12.14.6}$$

But for $n = 0$, we have instead

$$e^{i\hat{\phi}} |0\rangle = (S+1)^{-1/2} \sum_{m=0}^{S} e^{i\hat{\phi}} |\phi_m\rangle$$

$$= (S+1)^{-1/2} \sum_{m=0}^{S} e^{i(S+1)\phi_0} e^{-iS\phi_m} |\phi_m\rangle$$

$$= e^{i(S+1)\phi_0} |S\rangle. \tag{12.14.7}$$

The operator $e^{i\hat{\phi}}$ leads to a decrease of the particle number, but on operating on $|0\rangle$ it gives the state with maximum particle number $|S\rangle$ instead of 0. Summarizing the above results, we have:

$$\left. \begin{array}{l} e^{i\hat{\phi}} |n\rangle = |n-1\rangle, \quad n \neq 0 \\ e^{i\hat{\phi}} |0\rangle = e^{i(S+1)\phi_0} |S\rangle \\ e^{-i\hat{\phi}} |n\rangle = |n+1\rangle, \quad n \neq s \\ e^{-i\hat{\phi}} |S\rangle = e^{-i(S+1)\phi_0} |0\rangle \end{array} \right\} \tag{12.14.8}$$

$$e^{i\hat{\phi}} = |0\rangle\langle 1| + |1\rangle\langle 2| + \cdots + |S-1\rangle\langle S| + \exp(i(S+1)\phi_0)|S\rangle\langle 0|$$

$$e^{i\hat{\phi}} e^{-i\hat{\phi}} = e^{-i\hat{\phi}} e^{i\hat{\phi}} = 1. \tag{12.14.9}$$

It should be emphasized that (12.14.9) is not obvious. Its validity is based on the restriction of the maximum number of particles. This is connected with the existence of the minimum number of particles, namely S. If there exists a minimum number of particles without a maximum number, then if $e^{i\hat{\phi}}|0\rangle = 0$, the result of operating $e^{-i\hat{\phi}} e^{i\hat{\phi}}$ on $|0\rangle$ is different from $e^{i\hat{\phi}} e^{-i\hat{\phi}}|0\rangle$. Given a minimum number as well as a maximum number, operations of (12.14.8) and Eq. (12.14.9) are consistent with one another. If we set the number n in the expansion of (12.14.1) in the range $-\infty$ to $+\infty$, no problem arises, but it remains to interpret the meaning of negative values of n. For example they correspond to number of particles with negative charge and we have an effective theory. The operation of $e^{i\hat{\phi}}$ and $e^{-i\hat{\phi}}$ is shown in Fig. 12.7.

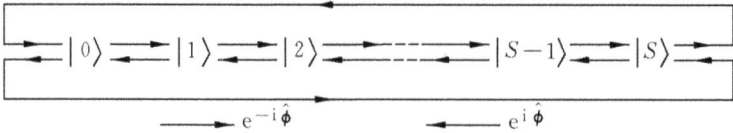

Fig. 12.7   Operation of $e^{i\hat\phi}$ and $e^{-i\hat\phi}$ on number states.

In the above we introduced the P-B theory. In the following, we re-iterate in another language, demonstrating that P-B theory is related to the cyclic representation of the quantum algebra [46, 47]. We define the operator

$$q^{\hat N}|n\rangle = q^{n+\eta}|n\rangle,\qquad (12.14.10)$$

where η is an arbitrary complex number and q satisfies $q^{S+1}=1$. According to Eqs. (12.14.10) and (12.14.8) it follows that

$$e^{i\hat\phi}q^{\hat N}|n\rangle = q^{n+\eta}e^{i\hat\phi}|n\rangle = \begin{cases} q^\eta e^{i(S+1)\phi_0}|S\rangle, & n=0 \\ q^{n+\eta}|n-1\rangle, & n\neq 0 \end{cases}$$

$$q^{\hat N}e^{i\hat\phi}|n\rangle = \begin{cases} q^\eta e^{i(S+1)\phi_0}q^{-1}|S\rangle, & n=0 \\ q^{n+\eta}q^{-1}|n-1\rangle, & n\neq 0 \end{cases},$$

where $q^S=q^{-1}$ has been used because of $q^{S+1}=1$, i.e., q is root of unit. Comparison of the above gives

$$e^{i\hat\phi}q^{\hat N} = qq^{\hat N}e^{i\hat\phi}.\qquad (12.14.11)$$

Similarly we can prove that

$$q^{\hat N}e^{-i\hat\phi} = qe^{-i\hat\phi}q^{\hat N}.\qquad (12.14.12)$$

Defining

$$U = e^{i\hat\phi},\ V = q^{\hat N},\ q = Q^2,\qquad (12.14.13)$$

the relation of (12.14.11) becomes

$$UV = Q^2VU,\quad Q = \exp\left(\frac{\pi i}{S+1}\right).\qquad (12.14.14)$$

Motivated by (12.13.8) we introduce the generators of quantum algebra through

$$U^{-1}V = Q^{-1}K^{-2} \quad \text{or} \quad V^{-1}U = QK^2, \tag{12.14.15}$$

$$\left. \begin{aligned} X_+ &= \frac{1}{Q - Q^{-1}}(U^{-1} + V^{-1}) \\ X_- &= \frac{1}{Q - Q^{-1}}(U + V) \end{aligned} \right\}. \tag{12.14.16}$$

Recalling (12.14.9) and $V^{-1}U = Q^2 U V^{-1}$ which is the consequence of (12.14.14), it is easy to check that $X_\pm$ and K satisfy the commutation relations of quantum algebra associated with $SL(2)$.

$$\left. \begin{aligned} K X_\pm K^{-1} &= Q^{\pm 1} X_\pm \\ [X_+, X_-] &= \frac{K^2 - K^{-2}}{Q - Q^{-1}}, \quad Q = \exp\left(i\frac{\pi}{S+1}\right) \end{aligned} \right\}. \tag{12.14.17}$$

Equipped with (12.14.17) we are able to set up the Fock states such that for arbitrary η,

$$\left. \begin{aligned} K|Z\rangle &= Q^{Z+\eta+\eta_0+\frac{1}{2}}|Z\rangle \\ X_+|Z\rangle &= [Z + \eta + \eta_0 + 1]_Q|Z+1\rangle \\ X_-|Z\rangle &= [Z + \eta + \eta_0]_Q|Z-1\rangle \end{aligned} \right\} \tag{12.14.18}$$

and verify

$$\left. \begin{aligned} (K X_\pm K^{-1})|Z\rangle &= Q^{\pm 1} X_\pm |Z\rangle \\ [X_+, X_-]|Z\rangle &= \frac{1}{(Q - Q^{-1})}\left(Q^{2\beta} - Q^{-2\beta}\right)|Z\rangle, \quad \beta = \alpha - \frac{1}{2} \end{aligned} \right\} \tag{12.14.19}$$

with $\eta_0 = \frac{1}{2}(S+1)$, and $\alpha = Z + \eta + \eta_0 + 1$. Thus we obtain

$$[X_+, X_-]|Z\rangle = \frac{1}{(Q - Q^{-1})}\left(K^2 - K^{-2}\right)|Z\rangle. \tag{12.14.20}$$

Acting $(X_+)^m$ on $|Z\rangle$ we have

$$(X_+)^m |Z\rangle = \frac{[\alpha + m]_Q!}{[\alpha]_Q!}|Z\rangle. \tag{12.14.21}$$

From $[\alpha + m]_Q = \frac{Q^{\alpha+m} - Q^{-(\alpha+m)}}{Q - Q^{-1}} = \frac{1}{\sin\frac{\pi}{S+1}} \sin[(\frac{\pi}{S+1})(\alpha + m)]$ it

follows $[\alpha+m]_Q = 0$ for $\alpha+m = S+1$, i.e., when $m+Z+\eta = \frac{1}{2}(S-1)$
we have

$$(X_+)^m \, |Z\rangle = 0.$$

The (12.14.18) is related to the cyclic representation of quantum algebra for Q being root of unity. There elegant discussions on this question, see Kac and De Concini [46, 47].

Before ending this paragraph let us consider another possibility to introduce the q-deformation of boson operators. Define operators

$$\left. \begin{aligned} a^\dagger &= [\hat{N}]^{1/2} e^{-i\hat{\phi}} \\ a &= e^{i\hat{\phi}} [\hat{N}]^{1/2} \end{aligned} \right\}, \qquad (12.14.22)$$

where

$$[\hat{N}]^{1/2} = \frac{q^{\hat{N}} - q^{-\hat{N}}}{q - q^{-1}}. \qquad (12.14.23)$$

From (12.14.13) we have

$$e^{i\hat{\phi}} = a[\hat{N}]^{-1/2}, \quad e^{-i\hat{\phi}} = [\hat{N}]^{-1/2} a^\dagger. \qquad (12.14.24)$$

From (12.14.13) and (12.14.15) we have

$$a^\dagger a = [\hat{N}], \quad aa^\dagger = [\hat{N} + 1]. \qquad (12.14.25)$$

The calculation of $aa^\dagger$ above needs $q^{-\hat{N}} e^{-i\hat{\phi}} = q^{-1} e^{-i\hat{\phi}} q^{-\hat{N}}$, which in turn is obtained from (12.14.11) and $e^{i\hat{\phi}} = (e^{-i\hat{\phi}})^{-1}$. Calculating directly the commutation relation of $a, a^\dagger$, we get [48–50]

$$aa^\dagger - qa^\dagger a = [\hat{N} + 1] - q[\hat{N}] = \frac{q^{-\hat{N}+1} - q^{-\hat{N}-1}}{q - q^{-1}} = q^{-\hat{N}},$$

i.e.,

$$\left. \begin{aligned} aa^\dagger - qa^\dagger a &= q^{-\hat{N}} \\ aa^\dagger - q^{-1}a^\dagger a &= q^{\hat{N}} \end{aligned} \right\} \qquad (12.14.26)$$

From Eqs. (12.14.15) and (12.14.8) we obtain

$$
\left.\begin{aligned}
a\,|n\rangle &= ([n+\eta])^{1/2}\,|n-1\rangle, \quad n \neq 0 \\
a\,|0\rangle &= [\eta]^{1/2}\,e^{i(S+1)\phi_0}\,|S\rangle \\
a^\dagger\,|n\rangle &= ([n+\eta+1])^{1/2}\,|n+1\rangle, \quad n \neq S \\
a^\dagger\,|S\rangle &= [\eta]^{1/2}\,e^{-i(S+1)\phi_0}\,|0\rangle \\
q^{S+1} &= 1
\end{aligned}\right\}
\tag{12.14.27}
$$

The eigenvalue of $[\hat{N}]$ is given by

$$
[\hat{N}]\,|n\rangle = \frac{q^n - q^{-n}}{q - q^{-1}}\,|n\rangle \equiv [n]\,|n\rangle.
\tag{12.14.28}
$$

The operation of a and $a^\dagger$ on the number states is shown in Fig. 12.8. Because of $q^{S+1} = 1$, P-B theory defines q-deformed boson operators, states $|n\rangle$ provide a cyclic representation with the total number of states set to be $S+1$.

The q-deformed boson representation (12.14.26) is a realization of the quantum algebra. C.P. Sun and H.C. Fu [48], L.C. Biedenharn [49] and A.J. Macfarlane [50] and other authors independently proved the result concerning two kinds of q-deformed bosons. Let i ($i = 1, 2$) be the index distinguishing the two kinds of bosons. Operators with different indices i commute with each other, while those with the same index satisfy commutation relations (12.14.26).

$$
\left.\begin{aligned}
a_i a_i^\dagger - q a_i^\dagger a_i &= q^{-\hat{N}_i} \\
a_i a_i^\dagger - q^{-1} a_i^\dagger a_i &= q^{\hat{N}_i}
\end{aligned}\right\}
\tag{12.14.29}
$$

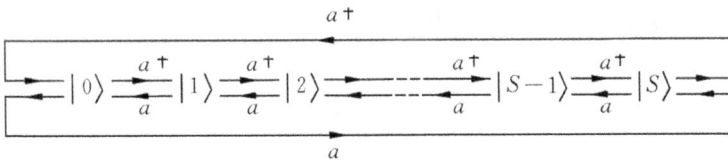

Fig. 12.8   Operations of a, $a^\dagger$ on number states.

We can construct the q-deformed $SL(2)$ operators by using a_i and $a_i^\dagger$:

$$X^+ = a_1^\dagger a_2, X^- = a_2^\dagger a_1, J_3 = \frac{1}{2}(\hat{N}_1 - \hat{N}_2), \quad K = q^{J_3}. \quad (12.14.30)$$

We obtain the commutation relations for X^+ and X^- by using (12.14.29):

$$\begin{aligned}
\left[X^+, X^-\right] &= a_1^\dagger a_2 a_2^\dagger a_1 - a_2^\dagger a_1 a_1^\dagger a_2 \\
&= a_1^\dagger a_1 \left(q a_2^\dagger a_2 + q^{-\hat{N}_2}\right) - a_2^\dagger a_2 \left(q a_1^\dagger a_1 + q^{-\hat{N}_1}\right) \\
&= [\hat{N}_1] q^{-\hat{N}_2} - [\hat{N}_2] q^{-\hat{N}_1} \\
&= \frac{q^{\left(\hat{N}_1 - \hat{N}_2\right)} - q^{-\left(\hat{N}_1 - \hat{N}_2\right)}}{q - q^{-1}} = \frac{q^{2J_3} - q^{-2J_3}}{q - q^{-1}} \\
&= \frac{K^2 - K^{-2}}{q - q^{-1}}. \quad (12.12.31)
\end{aligned}$$

This is exactly the commutation relation of the quantum algebra $U_q(SL(2))$. The introduction of quantum algebra was connected with the XXZ model, where the q-deformation was considered by the interaction $S_i^3 S_{i+1}^3$. In the examples of Sections 12.13, q was introduced by the external magnetic field. From the preceding example we see that the quantum algebra is handled more intuitively and efficiently by the q-boson representation. An interesting question arises: If we let $S \to \infty$ in the P-B theory, shall we obtain the result of the S-G theory? K. Fujikawa's answer [51] is: If we set $S \to \infty$ in the P-B theory, the topological nature of the resulting theory is still different from that of the S-G theory. These theories are different in their different indices(dim ker). It is tempting to consider a similar question. One of the regularization schemes in quantum field theory is to fix a parameter Λ and do calculation with it. We then set $\Lambda \to \infty$ in the final result. We know that the nature of the cyclic representation is such that the theory in the limit $S \to \infty$ is different from a theory without an upper limit from the outset. How should we consider the above mentioned regularization scheme?

The phase quantization discussed above serves only as an exercise to illustrate the cyclic representation. For the quantization of phase (and phase difference) there are many different points of view. Debates among different views have been going on. Some authors even argue that phase quantization does not exist. It needs more extensive studies in order to clarify the issue. Readers interested are advised to read the review article [52].

References

[1] L. D. Faddeev, in *Integrable Models in* $1 + 1$ *Dimensional Quantum Field Theories*, Les Houches Lectures 1982. Amsterdam, Elsvier, 1984.

[2] L. D. Faddeev, Lectures on Quantum Inverse Scattering Method, in *Nankai Lectures on Mathematical Physics* 1987, edited by X.C. Song, Singapore, World Scientific, 1992.

[3] P. P. Kulish and E. K. Sklyanin, *Lecture Notes in Phys.*, Springer, 151, 61, 1982.

[4] C. N. Yang, *Phys. Rev. Lett.* 19, 1312, 1967.

[5] C. N. Yang, *Phys. Rev.* 168, 1920, 1968. A. B. Zamoldchikov and A. B. Zamolodchikov, *Ann. Phys.* 120, 253, 1979.

[6] V. G. Drinfeld, *Soviet Math. Dokl.* 22, 254, 1985 and 36, 212, 1988.

[7] V. G. Drinfeld, Quantum Groups, in *Proc. Intern. Congress of Mathematicians, Berkeley, USA 1986*, edited by A.M. Gleason, *American Mathematical Society*, 1987. pp. 269–291.

[8] R. J. Baxter, *Exactly Solved Methods in Statistical Mechanics*, London, Academic Press, 1982.

[9] E. K. Sklyanin, in *Quantum Groups and Quantum Integrable Systems*, edited by M. L. Ge and B. H. Zhao, Nankai Lectures on Mathematical Physics, World Scientific, Singapore, 1991.

[10] S. Shnider and S. Sternberg, "Quantum Groups" *Graduate Texts in Mathematical Physics*, Ed. E. Lieb. P International Press, 1991.

[11] S. Majid, *Foundations of Quantum Group Theory*, Cambridge University press, Cambridge, 1995.

[12] V. Chari and A. Pressley, *A Guide to Quantum Groups*, Cambridge University Press, Cambridge, 1994.

[13] M. Chaichian and A. Demichev, *Introduction to Quantum Groups*, World Scientific, Singapore, 1998.

[14] A. Klimyk and K. Schmudgen, "Quantum Groups and Their Representations" Texts and Monographs in Physics, Series Editors, R. Balian, Lieb E., Reshetichin N., Thirring W., Springer, 1997.

[15] L. C. Biedenharn and M. A. Lohe, *Quantum Group Symmetry and q-Tensor Algebras*, World Scientific, Singapore, 1995.

[16] L. H. Kauffman, *Knots and Physics*, World Scientific, Singapore, 1991.

[17] Jimbo M. *Quantum Groups and Yang-Baxter Equations* (in Japanese), Springer, Tokyo, 1991.

[18] P. P. Kulish (Ed.), "Quantum Groups", *Lecture Notes in Mathematics*, Springer-Verlag, 1992.

[19] C. N. Yang and M. L. Ge (Eds.), *Braid Group, Knot Theory and Statistical Mechanics*, World Scientific, Singapore, 1989.

[20] C. N. Yang and M. L. Ge (Eds.), *Braid Group, Knot Theory and Statistical Mechanics II*, World Scientific, Singapore, 1994.

[21] Z. Q. Ma, *Yang-Baxter Equation and Quantum Enveloping Algebras*, World Scientific, Singapore.

[22] M. L. Ge and K. Xue, "Yang-Baxter equation", (in Chinese) Shanghai Scientific & Technical Pub., Shanghai, 1999. "Yang-Baxter equation in quantum Mechanics" (in Chinese) Shanghai Scientific and Technological Education Pub. House, Shanghai, 1998.

[23] Nankai Symposium. *Introduction to Quantum Groups and Integrable Massive Models of Quantum Field Theory*, (Eds.) M. L. Ge and B. H. Zhao, World Scientific, Singapore, 1990.

[24] Nankai Symposium. *Quantum Groups and Quantum Integrable Systems*, (Ed.) M. L. Ge, World Scientific, Singapore, 1992.

[25] Nankai Symposium. *New Developments of Integrable Systems and Long-Ranged Interaction Modles*, Eds. M. L. Ge and Y. S. Wu, World Scientific, Singapore, 1995.

[26] Nankai Symposium. *Quantum Groups, Integrable Statisfical Models and Knot Theory*, Eds. M. L. Ge and H. J. De Vega, World Scientific, Singapore, 1993.

[27] L. N. Lipatov, *Talkat Nankai Symposium*, Nankai University, Tianjin, 1994, China.

[28] M. L. Ge, K. Xue and Y. M. Cho, *Phys. Lett.* A249, 358, 1998, A260, 484, 1999.

[29] C. M. Bai, M. L. Ge and K. Xue, *J. Stat. Phys.* 102, 545, 2001.

[30] Z. F. Wang, M. L. Ge and K. Xue, *J. Phys.* A30, 5023, 1997.

[31] T. T. Wu and C. N. Yang, *Phys. Rev.* D12, 3845, 1975; *Nucl. Phys.* B107, 365, 1976.

[32] R. Balian and N. R. Werthamer, *Phys. Rev.* 131, 1553, 1963.

[33] D. M. Lee, *Rev. Mod. Phys.* 69, 645, 1997.

[34] A. J. Leggett, *Rev. Mod. Phys.* 47, 331, 1975.

[35] P. W. Anderson and P. Morel, *Phys. Rev.* 123, 1911, 1961.

[36] A. A. Belavin and V. G. Drinfeld, Triangle Equations and Simple Lie Algebras, S. P. Novikov (Ed.), *Mathematical Physics Review* (Section C, Vol. 4). Amsterdam, OPA, 1984.

[37] V. F. R. Jones, *Commun. Math. Phys.* 125, 459, 1987; L. H. Kauffman, see [16]; Wadati M. see [19]; Kohno T. see [19].

[38] L. Takhtajan, *Letures on Quantum Groups*, in [23].

[39] V. V. Bazhanov and Yu. G. Stroganov, see [17].

[40] M. Jimbo, *Lett. Math. Phys.* 10, 63, 1985; 11, 247, 1986; *Commun Math. Phys.* 102, 537, 1986.

[41] H. J. de Vega, *Int. J. Mod. Phys.* A4, 237, 1989; A5, 1611, 1989.

[42] P. B. Wiegmann and A. Z. Zabrodin, *Phys. Rev. Lett.* 73, 1134, 1994.

[43] G. H. Chen, L. M. Kung and M. L. Ge, *Phys. Rev.* B53, 9540, 1996; Chen G. H. and M. L. Ge, *Phys. Rev.* B54, 7654, 1996.

[44] Y. Hatsugai, M. Kohmoto and Y. S. Wu, *Phys. Rev. Lett.* 73, 1134, 1994.

[45] D. T. Pegg and S. M. Barnett, *Phys. Rev.* A39, 1665, 1989; S. M. Barnett and D. T. Pegg, *J. Mod. Opt.* 36, 7, 1989.

[46] De Concini, V. G. Kac and Colloque Dixmier, *Progress in Math*, 92, 471, 1990; Jimbo M., Topics from Representations of Uq(g)-An Introduction Guide to Physicists, in Ref. [24].

[47] C. P. Sun and M. L. Ge, q-Boson Realization theory of quantum algebras and Its applications to YBE, in Ref. [24].

[48] C. P. Sun and H. C. Fu, *J. Phys.* A22, L983, 1989.

[49] L. C. Biedenharn, *J. Phys.* A22, L873, 1989.

[50] A. J. Macfarlane, *J. Phys.* A22, 4581, 1989.

[51] K. Fujikawa, *Phys. Rev.* A52, 3299, 1995.

[52] Yu. I. Vorontsov, *Physics-Uspekhi* 45(8), 907, 2002.

[53] T. T. Wu and C. N. Yang, *Nucl. Phys.* B107, 1976, 365–380; *Phys. Rev.* D14, 1976, 437–445.

[54] F. D. M. Haldane, *Phys. Rev. Lett.* 6, 635, 1988; B. S. Stastry, *Phys. Rev. Lett.* 60, 639, 1988.